REFERENCE HANDBOOK ON THE DESERTS OF NORTH AMERICA

REFERENCE HANDBOOK ON THE DESERTS OF NORTH AMERICA

Edited by GORDON L. BENDER

GP

GREENWOOD PRESS

WESTPORT, CONNECTICUT ● LONDON, ENGLAND

Library of Congress Cataloging in Publication Data
Main Entry under title:

Reference handbook on the deserts of North America.

   Bibliography: p.
   Includes index.
   1. Deserts—North America—Handbooks, manuals, etc.
I. Bender, Gordon Lawrence, 1918–
GB612.R43   917.3'0954   80-24791
ISBN 0-313-21307-0 (lib. bdg.)

Library of Congress Catalog Card Number: 80-24791
ISBN: 0-313-21307-0

First published in 1982

Greenwood Press
A division of Congressional Information Service, Inc.
88 Post Road West, Westport, Connecticut 06881

Printed in the United States of America

10 9 8 7 6 5 4 3 2 1

CONTENTS

Figures   vii

Tables   xiii

1. Introduction: *Gordon L. Bender*   1

2. The Great Basin: *Don Fowler and David Koch*   7
   CONTRIBUTORS: CADY JOHNSON, THOMAS
   LUGASKI, HAMILTON VREELAND, AND PATRICIA
   VREELAND

   Appendix 2. Recent Native Fishes of the Great
            Basin   64

3. The Northern Great Basin Region:
   Don McKenzie   67

   Appendixes:
   3-A.   A Checklist of the Plants of the
          Northern Great Basin Region of Oregon   83
   3-B.   A Checklist of the Vertebrate Animals of
          the Northern Great Basin Region of
          Oregon   96

4. The Mojave Desert: *Peter Rowlands,*
   Hyrum Johnson, Eric Ritter, and Albert Endo   103

   Appendixes:
   4-A.   Quantitative Summary of the Vascular Plants
          of the Mojave Desert in California   146
   4-B.   A Checklist of Birds of Death Valley
          National Monument   148
   4-C.   A Checklist of Fishes, Amphibians,
          Reptiles, and Mammals of Death Valley
          National Monument   152
   4-D.   A Checklist of Amphibians and Reptiles
          of the Joshua Tree National Monument   155
   4-E.   A Checklist of the Mammals of the
          Joshua Tree National Monument   157
   4-F.   A Checklist of Birds of the Joshua Tree
          National Monument   159

5. The Sonoran Desert: *Frank S. Crosswhite and*
   Carol D. Crosswhite   163

   Appendixes
   5-A.   Mammals of the Sonoran Desert   296
   5-B.   Birds of the Sonoran Desert   303
   5-C.   Reptiles and Amphibians of the Sonoran
          Desert   310
   5-D.   Freshwater Fishes of the Sonoran Desert   317

6. The Chihuahuan Desert: *Fernando*
   Medellín-Leal   321

   Appendixes
   6-A.   A Checklist of the Amphibians and
          Reptiles of the Big Bend National Park   373
   6-B.   A Checklist of the Birds of Big Bend
          National Park   376
   6-C.   A Mammals Field List of Big Bend
          National Park   379

7. The Arctic Desert: *David M. Hickok,*
   Joseph C. LaBelle, and Robert G. Adler   383

   Appendix 7. Mammals of the Arctic National
            Wildlife Refuge   404

8. Animal Adaptations: *Neil F. Hadley*   405

9. Plant Adaptations: *Otto T. Solbrig*   419

10. North American Desert Riparian Ecosystems:
    Robert D. Ohmart and Bertin W. Anderson   433

    Appendixes
    10-A. Habitat Relationships of Native Reptiles
           and Amphibians of North American
           Deserts   467
    10-B. Habitat Relationships of Native Birds in
           North American Deserts   470

11. Sand Dunes in the North American Deserts:
    Roger S. U. Smith   481

    Appendixes
    11-A. Sources of Maps and Aerial Photographs   525
    11-B. Sources of Wind and Climatic Data   526

12. Desert Varnish: *Carleton B. Moore
    and Christopher Elvidge*   527

13. Research Areas and Facilities: *Gordon L. Bender*   537

    Appendixes
    13-A. Publications of the Committee on Desert
           and Arid Zone Research   541
    13-B. Publications of the Committee on Arid
           Lands (COAL)   542
    13-C. Desert Research Institutes   543

    13-D. Federal Experiment Stations   545
    13-E. Natural Areas and Landmarks   546
    13-F. State Natural Area Programs   549
    13-G. IBP Validation Sites   550
    13-H. Biosphere Reserves   551
    13-I. National Parks and Monuments   553

14. Desertification: *Gordon L. Bender*   555

Author Index   561

Index to Common Names   572

Index to Scientific Names   577

Subject Index   586

About the Contributors   593

FIGURES

1-1. Map of the deserts of North America — 2

1-2. Photograph of the Mojave Desert, California — 4

1-3. Photograph of the Sonoran Desert, Arizona — 4

1-4. Photograph of the Great Basin Desert, Utah — 5

1-5. Photograph of the Chihuahuan Desert, Texas — 5

2-1. Outline map of the Great Basin Desert — 7

2-2. Mountain ranges of the Great Basin — 7

2-3. Pluvial lakes and glaciers during post-Sanganon times — 8

2-4. Major Paleozoic tectonic elements of the western United States — 10

2-5. Thickness and paleogeographic map of the Cambrian — 12

2-6. Thickness and paleogeographic map of the Ordovician — 12

2-7. Thickness and paleogeographic map of the Silurian — 13

2-8. Thickness and paleogeographic map of the Devonian — 13

2-9. Diagram of inferred sequence of events during the Antler Orogeny in north-central Nevada — 14

2-10. Thickness and paleogeographic map of the Mississippian — 15

2-11. Thickness and paleogeographic map of the Pennsylvanian — 15

2-12. Diagram of inferred sequence of events following the Antler Orogeny in north-central Nevada — 16

2-13. Detail of Mississippian, Pennsylvanian, and Permian formations in the Antler Orogeny in north-central Nevada — 17

2-14. Sketch map of the Late Mesozoic showing truncation of Paleozoic geosynclinal and deformational trends — 18

2-15. Present configuration of plate boundaries in the Pacific Northeast and western North America — 19

2-16. Implication of plate tectonics for the Cenozoic tectonic evolution of western North America — 20

2-17. Average January sea-level pressure patterns over North America — 23

2-18. Average July sea-level pressure patterns over North America — 23

2-19. Photograph of northern desert shrub ecosystem — 37

2-20. Photograph of Playa and Dry Lake Zone, Black Rock Desert — 43

2-21. Photograph of sand dunes — 44

2-22. Map of Great Basin principal archeological sites — 47

2-23. Map of Great Basin tribal distribution — 49

2-24. Map of Great Basin linguistic distribution — 51

3-1. Map of the Northern Great Basin Region — 68

3-2. Topographic map of Steens Mountain   71
3-3. Photograph of Catlow Rim   72
3-4. Photograph of the High Steens from the south   73
3-5. Photograph of Pike Creek flow   74
3-6. Photograph of Kiger Gorge   75
3-7. Photograph of small cirques   76
3-8. Schematic profile of south-facing slope of Steens Mountain   77

4-1. A digital mosaic of ten Landsat scenes   104
4-2. Physiographic map of the Mojave Desert   105
4-3. Outline map of the Mojave Desert   106
4-4. Profile diagram of the Mojave Desert at about 36° 30′ N   107
4-5. Profile diagram of the Mojave Desert at about 34° 47′N   107
4-6. Mean annual precipitation in the Mojave Desert   108
4-7. Winter rainfall in millimeters in the Mojave Desert   109
4-8. Snowfall in millimeters in the Mojave Desert   109
4-9. Percent annual precipitation as winter rain in the Mojave Desert   110
4-10. Percent annual precipitation as summer rain in the Mojave Desert   110
4-11. Scatter diagram showing seasonality of precipitation and days when the temperature falls below freezing   111
4-12. Mean January minimum/mean July maximum temperatures   113
4-13. Potential evapotranspiration/mean annual precipitation   113
4-14. Aerial photograph showing doughnut-shaped clones of creosote bush   128
4-15. Aerial photograph showing doughnut-shaped clones of Mojave yucca   128
4-16. Photograph of abstract-geometric petroglyphs   132
4-17. Photograph of Von Trigger Rock shelter   134
4-18. Photograph of rock ring in the El Paso Mountains   135

4-19. Photograph of rock alignment, Eureka Valley   135

5-1. Map of the geology of the Sonoran Desert by rock type   165
5-2. Map of the geology of the Sonoran Desert by age of rocks   169
5-3. Satellite imagery of the Salton Trough   174
5-4. Satellite imagery of the Colorado River delta   175
5-5. Satellite imagery of the Pinacate lava field and northwest trending mountains   177
5-6. Satellite imagery showing agricultural and urban land in the Southern Basin and Range Province   179
5-7. Satellite imagery of the Gila River   182
5-8. Map of annual precipitation in the Sonoran Desert   184
5-9. Map of divisions of the Sonoran Desert   213
5-10. Map of subdivisions of the Sonoran Desert   214
5-11. Photograph of vegetation in the Arizona Upland Division   219
5-12. Photograph of south-facing slope in the Arizona Upland Division   221
5-13. Photograph of the Sonoran Plains Division   222
5-14. Photograph of the sarcocaulescent ''swollen trunk'' syndrome   225
5-15. Photograph of the Vizcaíno Division   227
5-16. Map of historic geographic patterns of the O'odham   251
5-17. Subsistence patterns of the O'odham   257

6-1. Map of the boundaries of the Chihuahuan Desert   324
6-2. Map of cities and towns of the Chihuahuan Desert   326
6-3. Map of proposed limits of the Chihuahuan Desert   328
6-4. Map of the topography of the Chihuahuan Desert   329
6-5. Vertical sections of the Chihuahuan Desert   330

6-6.   Map of the physiographic provinces of the Chihuahuan Desert   332
6-7.   Map of mountain systems and geomorphological features of the Chihuahuan Desert   335
6-8.   Map of hydrography of the Chihuahuan Desert   336
6-9.   Map of the Bolsones   340
6-10.   Mean annual temperature in the Chihuahuan Desert   343
6-11.   Mean annual precipitation in the Chihuahuan Desert   344
6-12.   Map of climatic regions of the Chihuahuan Desert according to the Meigs system   346
6-13.   Map of climatic regions of the Chihuahuan Desert according to a modified Köppen system   348
6-14.   Climate aridity index of the Chihuahuan Desert   351
6-15.   Distribution of Emberger's aridity index in the Chihuahuan Desert   352
6-16.   Superficial lithology of the Chihuahuan Desert   353
6-17.   Map of the soils of the Chihuahuan Desert   356
6-18.   Map of the vegetation of the Chihuahuan Desert   358
6-19.   Map of communications in the Chihuahuan Desert   364
6-20.   Map of tourism, special phenomena, and events in the Chihuahuan Desert   367
6-21.   Map of the desertification risk in the Chihuahuan Desert   368

7-1.   Map of the Arctic Desert   384
7-2.   Photograph of the Arctic slope near the mouth of the Kellik River   384
7-3.   Photograph of the Arctic coastal plain   385
7-4.   Photograph of low center polygons in the Arctic   389
7-5.   Photograph of rolligon tracks on the Arctic tundra   390
7-6.   Photograph of a beaded stream in the Arctic   391

7-7.   Photograph of the Pingo south of the Colville River Delta   392
7-8.   Photograph of windblown snow formations in the Arctic   395
7-9.   Photograph of a whiteout in the Arctic   396
7-10.   Photograph of an Arctic musk ox   399
7-11.   Photograph of moist tundra in the Arctic   400
7-12.   Photograph of the petroleum camp at Prudhoe Bay, Alaska   402

8-1.   Circulatory system of a sphinx moth and body temperatures during flight   407
8-2.   Projected thermal behavior of experimental plumages under extreme conditions   409
8-3.   Relation of oxygen to ambient temperature of summer and winter peccaries   411
8-4.   Effect of hydration state of the scorpion on ileal sodium and potassium concentrations   413

9-1.   Flow of water from soil to atmosphere   421
9-2.   Schematic representation of plant branch and root structure   423
9-3.   Hypothetical rate of photosynthesis versus leaf nitrogen content of a desert plant   426
9-4.   Hypothetical light-saturated rates of photosynthesis of plants versus leaf nitrogen in water limited and unlimited habitats   427
9-5.   Diagramatic representation showing the development of superficial and deep systems of roots   430

10-1.   Three types of riparian ecosystems in the Sonoran Desert   434
10-2.   Foliage volume characteristics of six vegetation structural types of riparian plant communities along the lower Colorado River   441
10-3.   Photograph of a broad riparian ecosystem near the Arizona-California border   443

10-4   Photograph of Washington palms at Alkali Spring, Arizona   444

10-5.   Photograph of entrenchment along Sonoita Creek   457

10-6.   Photograph of vertical entrenchment on Sonoita Creek   458

10-7.   Photograph of entrenchment in the San Simon Valley   459

10-8.   Aerial photograph of entrenchment conditions in the San Simon Valley   459

10-9.   Photograph of year-old cottonwood and willow trees, lower Colorado River   462

10-10. Photograph of two-year-old cottonwood trees along the lower Colorado River   462

11-1.   Index map of dune fields in U.S. and Mexican deserts   482

11-2.   Diagram of the relations between wind velocity, height, and shear velocity   484

11-3.   Diagram of relations between wind, vegetation, and sand supply that generate different dune forms   485

11-4.   Vertical aerial photograph of longitudinal dunes on the Moenkopi Plateau   486

11-5a. Oblique aerial view across Eureka Valley   487

11-5b. Ground view northward along the crest of Eureka Valley   487

11-6.   Telephoto view of star-shaped dunes in Panamint Valley   488

11-7.   Photograph of Saratoga Springs dunes from the west   488

11-8.   Oblique aerial photograph in Saline Valley   489

11-9.   Stereogram of longitudinal dunes in Devil's Playground   490

11-10a. Oblique aerial photograph of Kelso Dunes from the north   491

11-10b. Ground view toward the west along the Kelso Dunes   491

11-11a. Ground view of Cat Mountain from the south   492

11-11b. View down Cat Mountain   492

11-12. Oblique aerial photograph across stabilized U-shaped dunes in the western Mojave Desert   493

11-13. Oblique aerial photograph across low dunes in the Coachella Valley   494

11-14. Vertical aerial photograph of longitudinal dunes on Superstition Mountain   494

11-15. Vertical U-2 aerial photograph of the Algodones dune chain   495

11-16. Oblique aerial photograph along longitudinal dunes in the Algodones chain   496

11-17. Ground view along longitudinal dunes in the Algodones dune chain   497

11-18. Oblique aerial photograph across three "megabarchans"   497

11-19. View across a barchan in the Algodones dune chain   498

11-20. Vertical aerial photograph of barchan swarm on an intradune flat at the U.S.-Mexican border   499

11-21. Topographic map of Bruneau Dunes   500

11-22. Photograph of Bruneau Dunes from the northeast   500

11-23a. Oblique aerial photograph across closely-spaced barchans in Silver State Valley   501

11-23b. Ground view across dunes in Silver State Valley   501

11-24a. Vertical aerial photograph of U-shaped and barchan dunes in the Little Humbolt River valley   502

11-24b. Oblique air photo across the Little Humbolt River valley   502

11-25. Oblique aerial photograph across depressions and fixed dunes in the Estancia Valley   503

11-26. Orbital view of dunes in northwestern Sonora, Mexico   513

11-27. Stereogram and topographic map of star-shaped dunes in western Gran Desierto   514

11-28. Ground view toward the north across large star-shaped dunes in the western Gran Desierto   515

11-29. Ground view along transverse

dunes and low star-shaped dunes
at Sierra del Rosario 515

11-30. Ground view across sand-mantled
hills of El Capitan 516

11-31. Ground view across sand-mantled
hills of Sierra del Rosario 516

11-32. Oblique aerial photograph across
barchans east of the Columbia
River 518

11-33. Ground view along U-shaped
dunes near Moses Lake 518

12-1. Transmitted light photomicrograph
of desert varnish on rock at
Phoenix, Arizona 529

12-2. Electron microprobe x-ray maps of
desert varnish on schist from
Mummy Mountain 529

12-3. En-pH diagram showing stability
field relations of iron and
manganese 531

12-4. Photograph of defaced petroglyph
in a manganese-stained cliff near
Montezuma's Castle National
Monument 534

14-1. Map of extent of desertification in
arid and semiarid lands of North
America 556

TABLES

2-1. Area extent of major Great Basin Pleistocene lakes — 9

2-2. Temperature means and extremes at selected Great Basin stations, 1933-1974 — 24

2-3. Precipitation and snowfall means and extremes at selected Great Basin stations, 1935-1974 — 24

2-4. Great Basin terrestrial ecosystems — 31

2-5. Great Basin Indian languages — 50

3-1. Climatological data for selected collecting stations, northern Great Basin region, Oregon — 70

3-2. Average snow depths and water equivalence, Steens Mountain — 70

4-1. Coverage of each landform type, Mojave Desert — 114

4-2. Soil subgroups for major landforms, Mojave Desert regions — 115

4-3. Vascular plant species, Mojave Desert regions — 116

4-4. Ten most important plant families, California Desert — 116

4-5. Selected classifications of California Desert vegetation — 118

4-6. Vegetation types within the California Desert, with a summary of the ranges of climatological variables associated with each — 120

4-7. Perennial plant species presence data for fourteen California Desert dune systems — 126

4-8. A summary of the distribution of biota in the zonal biotic communities of southern Nevada — 129

4-9. Total vascular plants and vertebrates found in biotic communities — 129

4-10. Example of Mojave Desert (California) site types: Small-scale systematic sample, 1976-1978 — 136

4-11. Example of Mojave Desert (California) site features: Small-scale systematic sample, 1976-1978 — 136

4-12. Example of Mojave Desert (California) site artifacts: Small-scale systematic sample, 1976-1978 — 136

4-13. Important Mojave Desert archaeological site/environmental associations — 137

5-1. Sonoran Desert vegetation — 212

7-1. Pleistocene Epoch in North America — 387

8-1. Comparative summary of lipid-hydrocarbon/body weight ratios, hydrocarbon composition, and cuticular permeability of desert tenebrionid beetles — 411

10-1. Summary of the dependency of the avifauna on riparian ecosystems in each North American desert — 447

10-2. Summary of the mean number of species and densities of birds in ten types of wetland habitats along the lower Colorado River — 449

10-3. Breeding bird densities, species richness, and bird species diversity in various riparian communities in North American deserts — 451

11-1. Dune fields in U.S. and Mexican deserts — 483

INTRODUCTION

GORDON L. BENDER

THE DESERTS OF southwestern United States and Mexico have been recognized for many years by both scientists and nonscientists. Journals of early pioneers tell of the dangers and dread of attempting to cross the feared Forty Mile or Black Rock deserts. Others relate the sufferings encountered in traveling the infamous Devils Highway in Mexico. There was no question in their minds but that they were in a desert.

Recognition of the polar regions as deserts came much later. Fistrup (1953) published an article on wind erosion in the Arctic desert, and Smiley and Zumberge (1974) brought much of the available information together in their book, *Polar Deserts and Modern Man*. Since that time the polar regions have been generally accepted as deserts (figure 1-1).

DEFINITION OF A DESERT

Although deserts have been recognized for some time, there is little agreement as to what actually constitutes one. Arguments over an acceptable definition have gone on endlessly, with no clear agreement being attained. Deserts have been defined in terms of the types of vegetation, types of soil, and variability in climatic factors.

Some approach the question solely from the standpoint of how much precipitation is received in a year's time, suggesting that fewer than ten inches or fewer than six inches per year qualifies an area as a desert. Others have suggested that what is important is the relationship between precipitation and evaporation. They consider an area to be a desert if the evaporation in a year's time exceeds the precipitation received in the same time.

A complicating factor is seasonality of the precipitation received. Some areas receive adequate amounts of precipitation in a period of a few weeks, which is followed by an extended dry period of several months. By the end of this period, conditions are definitely desert-like.

It seems that any acceptable definition must include consideration of the following factors: the amount of precipitation received, the distribution of this precipitation over a calendar year, the amount of evaporation, the mean temperature during the designated period, and the amount and the utilization of the radiation received.

INDEXES OF DRYNESS

A number of persons have devised formulas or equations for determining the degree of dryness in specific areas. Köppen (1931) suggested a relationship among precipitation, annual mean temperature, and seasonality of rainfall. He recognized that the values would be different in regions of summer rain, winter rain, or regions with no clearly defined rainfall period.

DeMartone (1926) suggested an Aridity Index,

$$I = \frac{n \cdot p}{t + 10},$$

where n is the number of rainy days, p is the mean precipitation per day, and t is the mean temperature for the period selected.

Emberger (1955) suggested a Quotient of Dryness,

$$Q = 100P/(M + m)(M - m),$$

where P is precipitation, M is the mean maximum temperature in the hottest month, and m is the mean minimum temperature in the coldest month.

Budyko (1956) suggested the Radiational Index of Dryness,

$$RID = R/LP$$

where R is the radiative balance at the earth's surface, P is the amount of rainfall, and L is the latent heat of vaporization.

Thornthwaite (1948) suggested a Moisture Index,

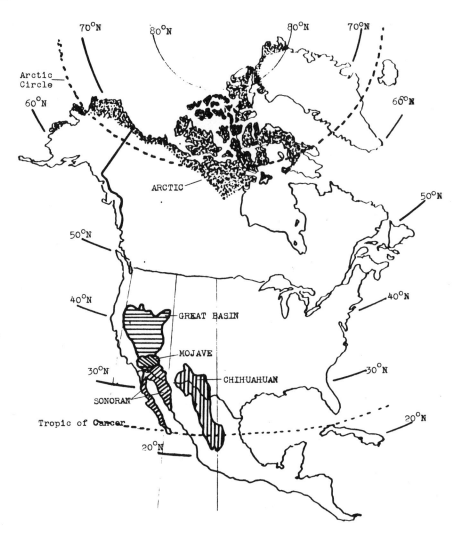

Figure 1-1. Map of the Deserts of North America.
Prepared by Gordon L. Bender.

$$I_m = 100\,[(S - d)/PE],$$

where S is the excess of monthly rainfall over potential evapotranspiration and soil water storage during the wet months; d is the deficit of rainfall plus available soil moisture below potential evapotranspiration during the dry months; and PE is the potential evapotranspiration based on temperature and hours of daylight.

Meigs (1953) adopted the Thornthwaite Index of Aridity, chose different values of the index to define different degrees of aridity, and added seasonal distributions of rainfall and temperature as important factors. Meigs's system has been widely used in preparing maps of desert areas.

Many of these indexes were developed primarily for hot, subtropical deserts. They can be used for polar deserts but the numerical index usually must be changed. Bovis and Barry (1974) point out that the Budyko Radiational Index of Dryness of 3.3 for a subtropical desert corresponds very closely

to one of <0.5 for a polar desert, due to the fact that the radiative regimes are quite different even though the moisture conditions are very similar.

Each of the proposed indexes has certain advantages and disadvantages, and each will result in slightly different geographic limits of desert areas. Sellers (1965) and Wallen (1962) have reviewed the various indexes and their applicability. As you read the following chapters, you will find that the authors have utilized several of them.

LOCATION AND AGE OF NORTH AMERICAN DESERTS

The major deserts of the world, including those of North America, are located in two rather well-defined bands 20°–30° north and south of the equator and at the poles. The subtropical deserts are explained by the worldwide circulation of air, which creates dry high pressure areas in these bands, resulting in reduced precipitation. Local factors such as cold ocean currents, rain shadows on the lee side of mountain ranges, and the influence of large surrounding land masses amplify the aridity. Polar deserts occur because they are remote from sources of moisture.

The age of North American deserts is in dispute. Johnson (1968) and Rzedowski (1962) considered them to be old. Blair (1977) suggested that the aridity of the Sonoran Desert originated in late Pliocene and Quaternary periods of geological history following the elevation of the Sierra Nevada mountains. Climatic events of the Pleistocene were considered to have resulted principally in localized movements of vegetation.

Van Devender (1977) studied radiocarbon dated pack-rat middens in the Chihuahuan, Sonoran, and Mojave deserts and reported that the change from woodland to desert or grassland occurred about 8000 years ago. Wells (1977) considers that the Chihuahuan Desert is the product of the Holocene interglacial climate and is less than 11,500 years old. Van Devender and Worthington (1977), studying the herpetofauna of Howell's Ridge Cave, reported a pronounced drying of climate 4000 to 5000 years ago, with minor fluctuations since that time. Axelrod (1979) has summarized the existing literature.

MOUNTAIN ISLANDS

Throughout the deserts are scattered numerous mountain ranges that are more or less mesic islands surrounded by a sea of aridity. Should these ranges be considered as part of the desert? In this book they will be discussed as part of the desert environment on the basis of their close interrelationship with the desert, moisture relations, and the movements of organisms up and down their slopes in response to changes in climate.

ADAPTATIONS

Organisms that live in a desert are subjected to severe environmental conditions, among them extremes of temperature, high solar radiation, and lack of water. Adaptations have evolved that enable desert organisms to survive. Interestingly these adaptations are very similar in both plants and animals. The adaptations to temperature regimes illustrate this.

The thermal relationships of animals include tolerating high temperatures, modifying the heat load by surface irregularities such as scales of spiny lizards, tubercles on beetle elytra, or the development of boundary layers of hair, feathers, fur, or setae that serve as a buffer to heat flow. Behavioral adaptations such as burrowing, seeking shade, nocturnality, and body orientation also modify the heat load. If all else fails, the animal may aestivate until environmental conditions improve.

Plants, too, can tolerate high temperatures. Heat load is modified by trichomes, spines, fluted stems, rough, textured, waxy layers, and boundary layers of spines that buffer heat flow. The orientation of stems, leaves, or pads modifies the heat load and seasonality of growth avoids the extreme desert conditions. Chapters 8 and 9 discuss these and other adaptations in detail.

RIPARIAN SITUATIONS

Every desert mentioned in this book has ribbons of life-giving water that flow through the aridity and sustain life. These are the desert streams with their associated riparian vegetation along their borders.

These rivers served early settlers as routes of travel, sources of water, sources of game and firewood and as shelter from the extremes of the desert. Today, man still occupies and depends on them, as do many plants and animals. Yet despite their importance, they have been little studied and are usually barely mentioned in books on desert areas. Chapter 10 examines these riparian ecosystems in detail and emphasizes their importance to desert life.

SAND DUNES

When most people think of deserts, they envision mile after mile of barren, wind-swept dunes extending from horizon to horizon. Some deserts of the world do have enormous areas of dunes. The North American deserts, however, have dunes occupying only about 6 percent of the desert area. For that reason they have been overlooked in the desert biome programs and virtually disregarded. They are, however, very interesting and unique areas of the North American deserts. Chapter 11 synthesizes the available information on these dune areas.

DESERT VARNISH

In many deserts of the world, rocks that have been exposed on the ground surface for many years are covered with a thin, dark layer known as desert varnish. Chapter 12 discusses the chemical nature of this layer and the conditions under which it is formed. A method of producing artificial desert varnish is also described.

DESERTIFICATION

Many of the major desert areas of the world are increasing in size. In addition, once-productive areas are developing desert-like characteristics. This phenomenon has been labeled desertification. It is so important worldwide that international conferences have been held to consider ways to combat it. National plans of action to combat desertification have been prepared, and international agreements have been signed to bring the resources of the world to bear on the problem. Chapter 14 considers specific examples of desertification, its indicators, and the efforts being employed to combat it. Chapters 4, 6, and 10 discuss desertification as it relates to the particular geographic areas examined in those chapters.

Figure 1-2. The Mojave Desert, California.
Photograph by Gordon L. Bender.

Figure 1-3. The Sonoran Desert, Arizona Upland.
Photograph by Gordon L. Bender.

Figure 1-4. The Great Basin Desert, Desert Experimental Range, Utah.
Photograph by Gordon L. Bender.

Figure 1-5. The Chihuahuan Desert, Big Bend National Park, Texas.
Photograph by Gordon L. Bender.

THE DESERTS OF NORTH AMERICA

Although each is considered a desert, there is tremendous diversity in the geology, biology, and degree of understanding and use by man. The Mojave Desert, the smallest and the closest to large centers of population, is undoubtedly the most used and abused by man. Rowlands in chapter 4 details the biology of the Mojave Desert and discusses its abuses by man. The California Desert Plan is a concerted effort to protect the values of the desert while allowing approved use of it at the same time (figure 1-2).

The Sonoran Desert undoubtedly is the most beautiful and the most studied of the North American deserts. There is a tremendous literature in existence, attested to by the extensive bibliography that Crosswhite and Crosswhite have prepared for chapter 5. The University of Arizona Office of Arid Lands Studies has published a number of bibliographies of the Sonoran Desert, as well as of individual plants such as jojoba and guayule. The popularity of the peninsula of Baja California, the Anza-Borrego Desert, the Salton Sea, and the Organ Pipe National Monument as tourist attractions and of the Southwest in general as a haven for retirees documents the attraction of the Sonoran Desert for people (figure 1-3).

The Great Basin Desert differs from the other deserts of the Southwest in that it is considered a "cold" desert, although it can become very hot in the summer. A large desert, it is interesting physiographically because of the large number of small mountain ranges (nearly two hundred) found within its borders. Fowler and Koch in chapter 2 discuss the geology, biology, climate, and human history in the entire Great Basin, while McKenzie in chapter 3 concentrates on the Steens Mountain area, one of the truly outstanding areas in the northern Great Basin (figure 1-4).

The Chihuahuan Desert is the least studied and the least understood of the southwestern deserts. It is a very large desert located principally in Mexico, with only a small extension into the United States. Medellín-Leal in chapter 6 pulls together the available information and makes available the Mexican literature, which has often been overlooked in publications on the Chihuahuan Desert. The series of detailed maps is particularly outstanding (figure 1-5).

The Arctic only recently has been considered a cold polar desert. A very extensive literature exists on the Arctic because of the many years of study at the Arctic Research Laboratory at Point Barrow, the University of Alaska, and a number of arctic institutes. However, until recently, the literature did not refer to the Arctic as a polar desert. Hickok, LaBelle, and Adler in chapter 7 outline its physiography and present a general overview of our most recent desert in terms of general recognition.

REFERENCES

Axelrod, D. I. 1979. Age and origin of Sonoran Desert vegetation. *Occasional Papers of the California Academy of Sciences, San Francisco. Publ.* 132:1-74.

Blair, W. F., and Hulse, A. 1977. The biota: The dependent variable. In *Convergent evolution in warm deserts*, ed. G. H. Orians and O. T. Solbrig. Stroudsburg, Pa.: Dowden, Hutchinson & Ross.

Bovis, M. J., and Barry, R. G. 1974. A climatological analysis of north polar desert areas. In *Polar deserts and modern man*, ed. T. L. Smiley and J. H. Zumberge. Tucson: University of Arizona Press.

Budyko, M. I. 1956. Teplovoi balans zemnoi poverkhnosti. *Gidrometeorologicherkoe izdatel'stov.* Leningrad.

de Martonne, E. 1926. Areisme et indices d'aridité. *Academie des Sciences, Paris. Comptes Rendus* 182:1395-1398.

Emberger, L. 1955. Afrique du nord-ouest. In *Plant ecology, review of research*, pp. 219-249. Arid Zone Research 6. Paris: Unesco.

Fistrup, G. 1953. Wind erosion within the Arctic deserts. *Geografisk Tidsskrift* 52:51-65.

Johnson, A. W. 1968. The evolution of desert vegetation in western North America. In *Desert Biology*, ed. G. W. Brown Jr. Vol. 1. New York: Academic Press.

Köppen, W. 1931. *Gundriss der klimakunde*, 2d ed. Berlin: Walter D. Gruyter Co.

Meigs, P. 1953. World distribution of arid and semi-arid homoclimates. In *Reviews of research on arid zone hydrology*, pp. 203-210. Arid Zone Programme 1. Paris: Unesco.

Rzedowski, J. 1962. Contribuciones a la fitogeografica floristica e historica de Mexico I. Algunas consideraciones acerca del elemento endemico en la flora Mexicana. *Bol. Soc. Bot. Mexico* 27:52-65.

Sellers, W. 1965. *Physical climatology.* Chicago: University of Chicago Press.

Smiley, T. L., and Zumberge, J. H. 1974. *Polar deserts and modern man.* Tucson: University of Arizona Press.

Thornthwaite, C. W. 1948. An approach towatd a rational classification of climate. *Geographical Review* 38:55-94.

Van Devender, T. R. 1977. Holocene woodlands in the southwestern deserts. *Science,* 14 October, pp. 189-192.

Van Devender, T. R., and Worthington, R. D. 1977. The herpetofauna of Howell's Ridge cave and the paleoecology of the northwestern Chihuahuan desert. In *Transactions of the symposium of biological resources of the Chihuahuan Desert region*, ed. R. Warner. Transactions and Proceedings, no. 3. Washington, D.C.: National Park Service, U.S. Department of the Interior.

Wallen, C. C. 1962. Climatology and hydrometeorology with special regard to the arid lands. In *Problems of the arid zone*, pp. 53-81. Arid Zone Research 18. Paris: Unesco.

Wells, P. V. 1977. Post-glacial origin of the present Chihuahuan desert less than 11,500 years ago. In *Transactions of the symposium on the biological resources of the Chihuahuan Desert region*, ed. R. Warner. Transactions and Proceedings, no. 3. Washington, D.C.: National Park Service, U.S. Department of the Interior.

2

THE GREAT BASIN

DON FOWLER AND DAVID KOCH

CONTRIBUTORS: CADY JOHNSON, THOMAS LUGASKI, HAMILTON VREELAND, AND PATRICIA VREELAND

THE GREAT BASIN is a vast inland region of the western United States covering an area of about 51,800,000 hectares (200,000 square miles). It is bounded on the north by the Columbia Plateau, on the west by the Sierra Nevada, on the south by the Sonoran Desert section of the Basin and Range Province, and on the east by the Wasatch Range and the Colorado Plateau. The Great Basin comprises the largest section of the Basin and Range Physiographic Province and covers small portions of eastern California, portions of southern Oregon, most of Nevada and western Utah, and a portion of Idaho (figure 2-1). The Great Basin was named by John Frémont, who in 1844 recognized that the region has no outlet to the sea, ending some twenty years of speculation by early explorers as to the courses and outlets of various river systems within the region (Cline 1963).

As Morrison (1965) points out, the Great Basin in fact contains more than 150 desert basins separated from each other by over 160 north-south trending mountain ranges (figure 2-2). In the nineteenth century Clarence Dutton (1880) characterized these ranges as an "army of caterpillars marching to Mexico." Elevations range from the 4340 meter high White Mountain in western Nevada to 85 m below sea level in Death Valley, California. The region generally slopes downward from north to south. Elevations of basin floors range from 1615 to 1840 m in eastern Nevada and 1160 to 1525 m in western Nevada and western Utah to less than 760 m in the southern Great Basin and 85 m below sea level in Death Valley (Morrison 1965). Most of the basins are closed and contain playa remnants of pluvial lakes, which existed during Pleistocene times (figure 2-3) (Morrison 1965;

Figure 2-1. *Outline map showing the extent of the Great Basin Desert.*
Prepared by Gordon L. Bender.

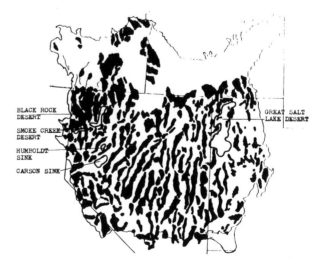

Figure 2-2. *Mountain ranges of the Great Basin.*
Major sinks and named desert areas are indicated. Adapted from *Intermountain flora-vascular plants of the Intermountain West USA,* Cronquist, Holmgren, Holmgren and Reveal, New York: Hafner Publishing Co., 1972. Figure 56 reprinted with the permission of The New York Botanical Garden.

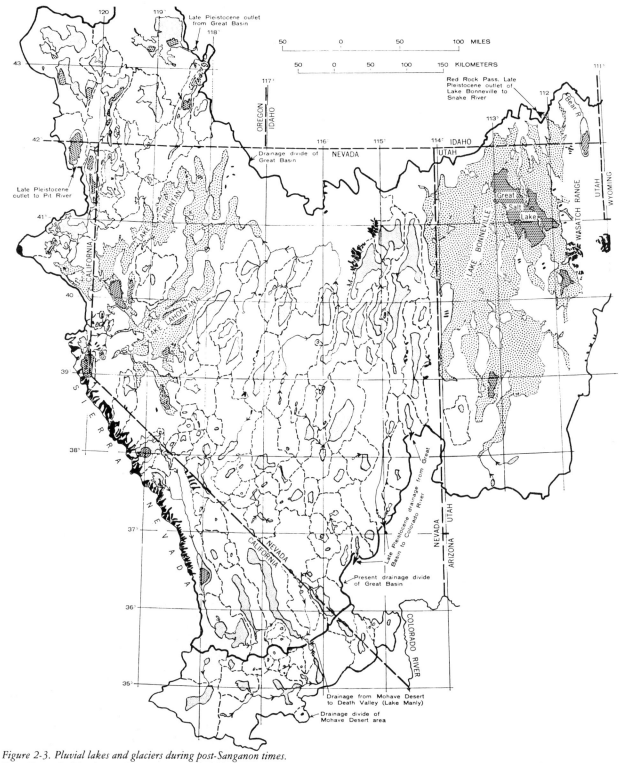

Figure 2-3. Pluvial lakes and glaciers during post-Sanganon times.
From Roger B. Morrison, "Quaternary Geology of the Great Basin," in *The Quaternary of the United States*, ed. by H. E. Wright and David G. Frey (copyright© 1965 by Princeton University Press), Fig. 1, p. 266. Reprinted by permission of Princeton University Press.

Note: Map showing maximum expansion of the pluvial lakes and glaciers within the Great Basin during post-Sangamon time. Glaciers are in solid black, existing lakes in darkest stipple, and pluvial lakes in lighter stipple. Dashed lines show boundaries of drainage basins; thin solid lines (with arrows) indicate overflow connections between basins. Data on glaciers are from Atwood (1909), Blackwelder (1931, 1934), Blackwelder and Matthes (1938), Sharp (1938), and Morrison (unpublished field notes); on pluvial lakes from Miller (1946), Hubbs and Miller (1948), Crittenden (1963a), Snyder *et al.* (1964), and Morrison (unpublished compilations).

Mifflin and Wheat 1979). Table 2-1 lists the largest of the Pleistocene lakes, including lakes Lahontan and Bonneville.

The climate of the Great Basin is semiarid to arid, due principally to the rain-shadow effect of the Sierra Nevada and, to a lesser extent, of other high mountain ranges, such as the Toiyabe Range, within the basin proper. The region has had a long and complex pattern of paleoclimatic change during, and since, the Pleistocene (Antevs 1952, 1955; Mehringer 1977; Weide and Weide 1977). Concomitant with climatic changes were changes in biogeography, principally elevational shifts of biological communities. Humans first entered the Great Basin around 12,000 B.P. (and possibly earlier) and established patterns of Archaic foraging cultures, which persisted into historic times (Aikens 1978).

STRUCTURAL GEOLOGIC HISTORY

The structural geologic history of the Great Basin is a complex and controversial subject. Interpretations of the style, magnitude, and chronology of structural events in the Great Basin region have evolved slowly over the past hundred years and are constantly being reevaluated. Our primary objective here is to present the field evidence upon which structural interpretations are based so that as more data become available for the Great Basin region, it will not be necessary to abandon one overgeneralized conceptual model in favor of another. Regional stratigraphic studies form the basis for the more profound structural interpretations in favor today.

Cordilleran Geosyncline

The geologists of the Fortieth Parallel Survey (King 1878:247, 342) were the first to record the major lithologic types and their distribution along a transect of the Great Basin. Turner (1909:255–256) was probably the first to recognize facies changes in the Paleozoic rocks of Nevada (Roberts 1964). Nolan (1928) first pointed out that carbonate rocks predominate in eastern Nevada, and volcanic and clastic rocks predominate in western Nevada. The fossil record allows us to correlate rocks of differing lithology and thus obtain a picture of the paleogeography of a region. Limestones and clean sandstones are characteristic of shallow-water or shelf environments, traditionally known as the miogeosyncline; bedded cherts, greywackes, and volcanic rocks are thought to be deposited offshore in deeper water, an environment known as the eugeosyncline. Although environments of deposition are more satisfactorily explained in terms of marginal trenches, island arcs, interarc basins, and so on, the older terminology is so firmly entrenched in the literature that we retain it here. In the Great Basin, the boundary between paleozoic eugeosynclinal and miogeosynclinal rocks trends north-northeast across Nevada, roughly east of an irregular line between meridians 116° and 117° (Roberts 1964) (figure 2-4). Hotz and Willden (1955) recognized a transitional assemblage of rocks characterized by clastic, volcanic, and carbonate elements. Transitional rocks have been recognized in the northern Shoshone Range, at Antler Peak, in parts of Eureka and Nye counties (Nevada), in the Sonoma Range, Edna Mountain, and the Osgood Mountains (Roberts 1964).

The fundamental assumption upon which the structural interpretations of Paleozoic events in the Great Basin are based is that the eugeosynclinal suite of rocks—graywacke, bedded chert, basaltic and andesitic volcanics, and ultramafic rocks—represent oceanic crust and associated sediments (Maxwell 1974). The Cordilleran geosyncline consists of the entire suite of miogeosynclinal, transitional, and eugeosynclinal rocks, which presumably were deposited along the ancient margin of North America.

Pretectonic Stratigraphy

Evidence for the precise time of initiation of the Cordilleran geosyncline for the most part lies outside the Great Basin and will not be considered here. The Cordilleran geosyncline

TABLE 2-1
Area extent of major Great Basin Pleistocene lakes

LAKE	TOTAL BASIN AREA (KM²)	PLUVIAL LAKE AREA (KM²)
Spring Valley, Nev.	4222	868
Ruby Lake, Nev.	3368	1232
Toyabee Lake, Nev.	3220	584
Deep Springs Lake, Calif.	519	107
Lake Lahontan, Nev.-Calif.	107,000	21,390
Lake Bonneville, Utah-Nev.	140,000	51,900
Warner Valley, Ore.	4900	1310

SOURCE: Time, space and intensity in Great Basin paleo-ecological models, M. L. Weide and D. L. Weide, *Desert Research Institute Publications in the Social Sciences* 12: 79-112.

clearly originated before earliest Cambrian time. (Monger et al., 1972 and Stewart, 1972 look at some problems with the Precambrian geology of the Cordilleran geosyncline.) Within the Great Basin, Precambrian sedimentary rocks are not commonly exposed, and the structural events discussed here all involve Paleozoic rocks.

Early Cambrian time in the Great Basin region is represented by generally light-colored, well-rounded, silica-cemented quartz arenites (sandstones). Local formation names for these early Cordilleran sediments include the Prospect Mountain Quartzite (in central and eastern Nevada), the Tepeats Sandstone (in southern Nevada), and the Osgood Mountain Quartzite (in north-central Nevada). These quartzites represent the basal lithologies in the basin and range province upon which there was almost uninterrupted sedimentation until the close of the Ordovician period (figures 2-5 and 2-6). In eastern Nevada and Utah, post-Early Cambrian

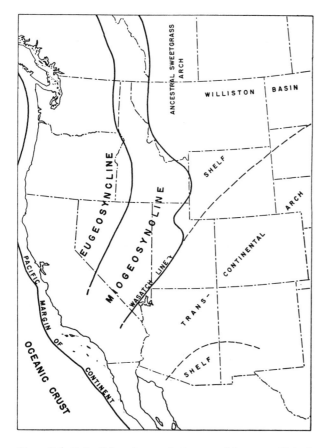

Figure 2-4. Major Paleozoic tectonic elements of the western United States.
The eugeosynclinal boundary is farther east than shown in Permian time. The Wasatch line though southern Nevada has been called the Las Vegas line (Welsh 1959). From *Structural Geology of North America* by A. J. Eardley. 2nd Edition. Copyright 1951 by Harper & Row, Publishers, Inc., and copyright © 1962 by A. J. Eardley. Figure 6.1 (after Welsh, 1959). Reprinted by permission of the publisher.

miogeosynclinal sedimentation is recorded by such units as the Cambrian Eldorado dolomite and Hamburg dolomite and the Ordovician Pogonip Group. The shale and carbonate units that comprise the miogeosynclinal assemblage are generally fine grained; the absence of any coarse clastic component suggests that the craton lay far to the east. Coeval deposition of a deep water, clastic, and volcanic (eugeosynclinal) facies in western Nevada may represent an island arc system generated in response to a subduction zone somewhere to the west. The true thickness of these western assemblage rocks is not known, since they outcrop only as fault-bounded slivers in the zones of great tectonic disturbance.

One of the better-known rock units in the Great Basin is the mid-Ordovician Eureka Quartzite, a most ubiquitous quartz arenite. Based on grain size distributions (Webb 1958), we can conclude that the Eureka Quartzite was transported by longshore currents southward from the Precambrian shield in central Alberta to much of Nevada and Utah. The Eureka Quartzite is limited in its western extent by the uplift of the Manhattan geanticline (Nolan 1928) along the western edge of the carbonate bank. It appears therefore that a north-northeasterly trending structural hinge of sorts has existed in central Nevada since the early Paleozoic and roughly coincides with the boundary between miogeosynclinal and eugeosynclinal Cordilleran rock facies. Evidence for Early Middle Ordovician uplift along an axis transverse to the general trend of the Cordilleran geosyncline is found in the Cortez Window, where the Eureka Quartzite rests quite unconformably on the Cambrian Hamburg dolomite. Some 1525 m of Upper Cambrian and Lower Ordovician beds are absent at this locality. Regional studies (Webb 1958) have led to the recognition of the Uinta-Cortez axis of uplift, where erosion removed pre-Middle Ordovician sediments prior to the deposition of the Eureka Quartzite.

Mid-Ordovician eugeosynclinal rocks are represented by the Valmy formation and its more easterly equivalent, the Vinini formation. The Vinini formation contains some carbonates that are characteristic of the eastern miogeosynclinal sequence; however, both the Valmy and Vinini contain Steinman Trinity type (ophiolite) rocks, which may represent oceanic crust. Correlations of Lower Paleozoic eugeosynclinal rocks are based largely on graptolite fauna.

Above the Eureka Quartzite lies a thick conformable sequence of carbonate rocks of Upper Ordovician through Lower Devonian age. Representative silurian rocks include the widespread laketown dolomite and Roberts Mountain formation, both of which are almost universally bounded above and below by unconformities (Langenheim and Larson 1973). Figure 2-7 is a highly generalized thickness map of silurian rocks. The general absence of lower and upper silurian rocks suggests either deformation at the close of the Ordovician or a lowering of sea level (Lintz 1978). Devonian carbonate rocks accumulated to depths in excess of 1219 m in the Eureka area (figure 2-8). Upper Ordovician through Lower Devonian rocks from the eugeosyncline to the west are lithologically similar to the Valmy and Vinini formations and are difficult to distinguish from them in the field. In a general way, the pattern of westward additions of successively younger belts of thick sediments and associated volcanics continued in the western United States throughout Paleozoic and Mesozoic time (Maxwell 1974), although sedimentation was interrupted periodically by tectonic events.

Paleozoic Structural History

ANTLER OROGENY

The broad geosyncline in which the three rock assemblages were laid down persisted with only local disturbances until Late Devonian time (Roberts 1964). Figure 2-9 shows diagrammatically the sequence of events that led to the formation of the Antler orogenic belt between the 116° and 118° meridians. Intense folding and faulting in this belt culminated in the Roberts Mountain thrust fault in early Mississippian time. Merriam and Anderson (1942) were the first to recognize the Roberts Mountain thrust, their evidence being juxtaposed facies of Lower Paleozoic rocks in the Roberts Mountains. The eastern margin of Paleozoic eugeosynclinal rocks today is the present extent of Klippen and Early Paleozoic cherts, shales, and volcanics, lying above a thick carbonate sequence of equivalent age (Roberts et al. 1958; King 1969a, 1969b). The miogeosynclinal carbonate rocks appear in windows 100 km west of the easternmost Klippen (Maxwell 1974). The Klippen are remnants of the Roberts Mountain thrust plate. Rocks of the transitional assemblage occur in windows beneath the Roberts Mountain thrust plate and as fault slivers within the thrust plate (Roberts et al. 1958; Kay and Crawford 1964). If the root zone of the Roberts Mountain thrust lies west of transitional rocks in the Sonoma Range, Edna Mountain, and the Osgood Mountains, then eastward-directed horizontal displacement on that great fault would amount to 150 km. Gravity sliding off the uplifted western eugeosyncline aided by a fluid pressure mechanism (Hubbert and Rubey 1959) is a logical mechanism for the production of faults of such awesome magnitude.

Thrust sheets of the Antler orogeny are stacked in an imbricate fashion. In general, these slices increase in age westward across the belt of thrusting (Grant 1974; Bortz 1959). These relationships suggest that the uplift occurred in spasms that advanced progressively eastward.

Coarse clastic rocks were shed eastward and westward from the Antler orogenic belt during its development, and these rocks have come to be known as the overlap assemblage (Roberts and Lehner 1955). The locus of eugeosynclinal deposition shifted concomitantly westward to western Nevada and eastern California (Eardley 1962) (figures 2-10, 2-11, 2-12). Within the Antler orogenic belt, the clastic rocks and associated lenticular limestones overlap folded and faulted carbonate, transitional, and silicious and volcanic assemblage strata that were involved in the Antler orogeny (Roberts 1964). The early Pennsylvanian Ely Basin (Coogan 1964) developed on the east flank of the Antler axis between intervals of mild uplift, as evidenced by bracketing intervals of nondeposition. In southern Nevada, the Permian system lies unconformably on rocks that range in age from Ordovician to Middle Pennsylvanian and is thought to record the encroachment of Permian seas on the remaining ridges of the Antler orogenic belt. The base of the Permian system in north-central Nevada is placed within the Havallah formation (Ketner 1967).

During latest Permian time, eugeosynclinal rocks containing small, tectonically emplaced slices of serpentinite (Davis 1973) were thrust eastward over rocks of a continental shelf environment. The eugeosynclinal rocks piled up on the west margin of the Antler orogenic belt (Roberts 1964) (figure 2-13). This second great deformational episode, which is recorded in rocks of the Great Basin, is known as the Sonoma orogeny, and its most salient structural feature is the Golconda thrust. The easternmost extent of Golconda thrusting is marked by a Klippe of battle conglomerate (a member of the overlap assemblage), which caps Havingdon Peak in the northern Shoshone Range and is in fault contact with the Valmy formation.

POST-SONOMA, PRE-TERTIARY STRATIGRAPHY AND TECTONICS

Following the Sonoma orogeny, the Antler orogenic belt remained a positive area and furnished sediments to the flanking seas during latest Permian and early Mesozoic time (Silberling and Roberts 1962). Conflicting paleogeographic interpretations for Middle and Late Triassic time have been made by Muller (1949) and by Silberling and Roberts (1962). Muller (1949) implied the persistence of a marine basin in northeastern Nevada throughout the Triassic, based on his interpretations of facies relationships between the Winnemucca and Augusta sequence of post-Sonoma rocks. In northwestern Nevada, he considered the Augusta sequence a relatively offshore facies of the Winnemucca sequence. However, Silberling and Roberts (1962) disagree. Silberling and Roberts (1962) note that there is no record of marine Middle and Upper Triassic rocks in the Great Basin east of central Nevada and that continental Upper Triassic rocks have been reported in southeastern Elko County, Nevada (Wheeler et al. 1949). Silberling and Roberts (1962) indicate that the major portion of the Triassic fine-grained terrigenous sediment supplied to western Nevada was derived from the craton, which lay far to the east of the beveled Antler-Sonoma orogenic belt. Subsequent work (Silberling and Wallace 1969) in the Humboldt

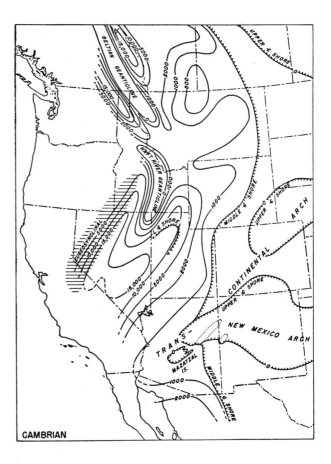

Figure 2-5. Thickness and paleogeographic map of the Cambrian.
From *Structural Geology of North America,* by A. J. Eardley, 2nd Edition. Copyright 1951 by Harper & Row, Publishers, Inc. and copyright in 1962© by A. J. Eardley. Figure 6.2. Reprinted by permission of the publisher.

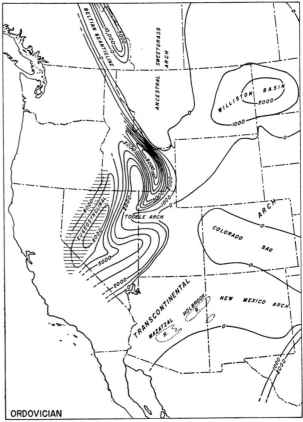

Figure 2-6. Thickness and paleogeographic map of the Ordovician.
From *Structural Geology of North America,* by A. J. Eardley, 2nd Edition. Copyright 1951 by Harper & Row, Publishers, Inc. and copyright in 1962© by A. J. Eardley. Figure 6.3. Reprinted by permission of the publisher.

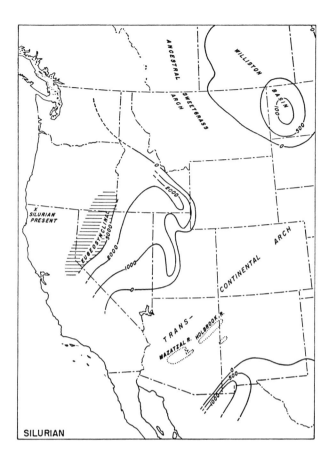

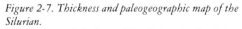

Figure 2-7. Thickness and paleogeographic map of the Silurian.

From *Structural Geology of North America*, by A. J. Eardley, 2nd Edition. Copyright 1951 by Harper & Row, Publishers, Inc. and copyright in 1962© by A. J. Eardley. Figure 6.4. Reprinted by permission of the publisher.

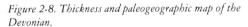

Figure 2-8. Thickness and paleogeographic map of the Devonian.

The Antler Orogenic belt, Stansbury anticline, and Beaverhead dome made their appearance in Late Devonian. Most of the sediments are Middle Devonian. From *Structural Geology of North America*, by A. J. Eardley, 2nd Edition. Copyright 1951 by Harper & Row, Publishers, Inc. and copyright © 1962 by A. J. Eardley. Figure 6.5. Reprinted by permission of the publisher.

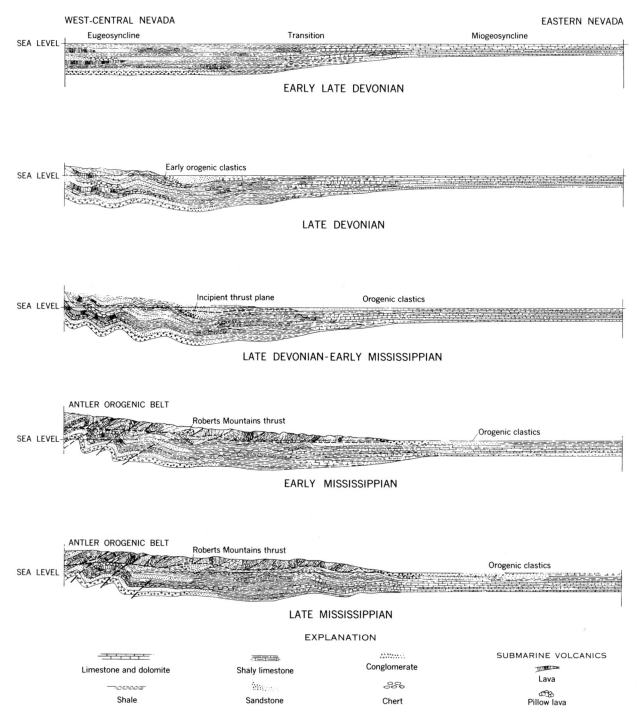

WEST-CENTRAL NEVADA   EASTERN NEVADA

SEA LEVEL
Eugeosyncline   Transition   Miogeosyncline

EARLY LATE DEVONIAN

SEA LEVEL
Early orogenic clastics

LATE DEVONIAN

SEA LEVEL
Incipient thrust plane   Orogenic clastics

LATE DEVONIAN-EARLY MISSISSIPPIAN

ANTLER OROGENIC BELT
SEA LEVEL
Roberts Mountains thrust
Orogenic clastics

EARLY MISSISSIPPIAN

ANTLER OROGENIC BELT
SEA LEVEL
Roberts Mountains thrust
Orogenic clastics

LATE MISSISSIPPIAN

EXPLANATION

Limestone and dolomite   Shaly limestone   Conglomerate   SUBMARINE VOLCANICS
                                                                                 Lava
Shale   Sandstone   Chert   Pillow lava

Figure 2-9. Diagram showing inferred sequence of events during the Antler Orogeny in north-central Nevada.
From *Geology of the Antler Peak Quadrangle,* R. J. Roberts, U.S. Geological Survey, Professional Paper 459-A, 1964, figure 2, p. 10.

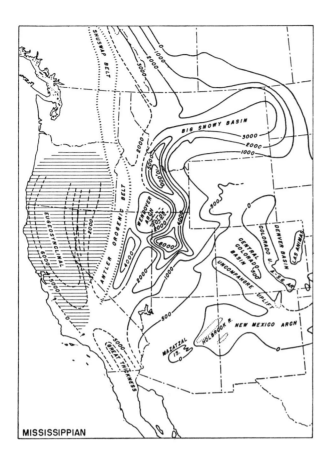

MISSISSIPPIAN

Figure 2-10. Thickness and paleogeographic map of the Mississippian.
A.-S.G. AR. is Apishapa-Sierra Grande arch. Uncompahgre and Colorado uplifts first became emergent in latest Mississippian and developed into major ranges in Early Pennsylvanian. From *Structural Geology of North America,* by A. J. Eardley, 2nd Edition. Copyright 1951 by Harper & Row, Publishers, Inc. and copyright© 1962 by A. J. Eardley. Figure 6.6. Reprinted by permission of the publisher.

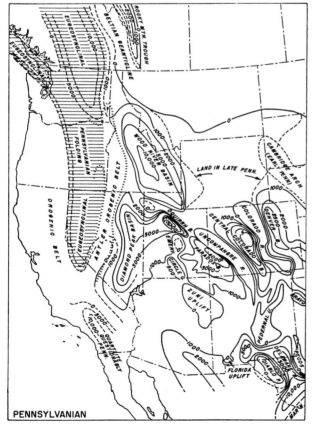

PENNSYLVANIAN

Figure 2-11. Thickness and paleogeographic map of the Pennsylvanian.
From *Structural Geology of North America,* by A. J. Eardley, 2nd Edition. Copyright 1951 by Harper & Row, Publishers, Inc. and copyright© 1962 by A. J. Eardley. Figure 6.7. Reprinted by permission of the publisher.

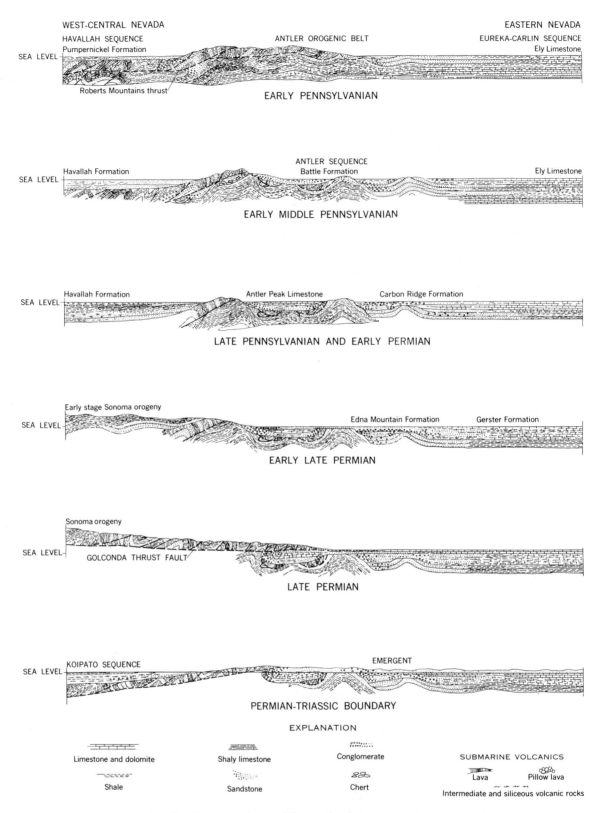

WEST-CENTRAL NEVADA
HAVALLAH SEQUENCE
Pumpernickel Formation

ANTLER OROGENIC BELT

EASTERN NEVADA
EUREKA-CARLIN SEQUENCE
Ely Limestone

SEA LEVEL

Roberts Mountains thrust

EARLY PENNSYLVANIAN

Havallah Formation

ANTLER SEQUENCE
Battle Formation

Ely Limestone

SEA LEVEL

EARLY MIDDLE PENNSYLVANIAN

Havallah Formation

Antler Peak Limestone

Carbon Ridge Formation

SEA LEVEL

LATE PENNSYLVANIAN AND EARLY PERMIAN

Early stage Sonoma orogeny

Edna Mountain Formation

Gerster Formation

SEA LEVEL

EARLY LATE PERMIAN

Sonoma orogeny

SEA LEVEL

GOLCONDA THRUST FAULT

LATE PERMIAN

KOIPATO SEQUENCE

EMERGENT

SEA LEVEL

PERMIAN-TRIASSIC BOUNDARY

EXPLANATION

Limestone and dolomite   Shaly limestone   Conglomerate   SUBMARINE VOLCANICS

Shale   Sandstone   Chert   Lava   Pillow lava

Intermediate and siliceous volcanic rocks

Figure 2-12. Diagram showing inferred sequence of events following the Antler Orogeny in north-central Nevada.
From *Geology of the Antler Peak Quadrangle,* R. J. Roberts, U.S. Geological Survey, Professional Paper 459-A., 1964, figure 4, p. 12.

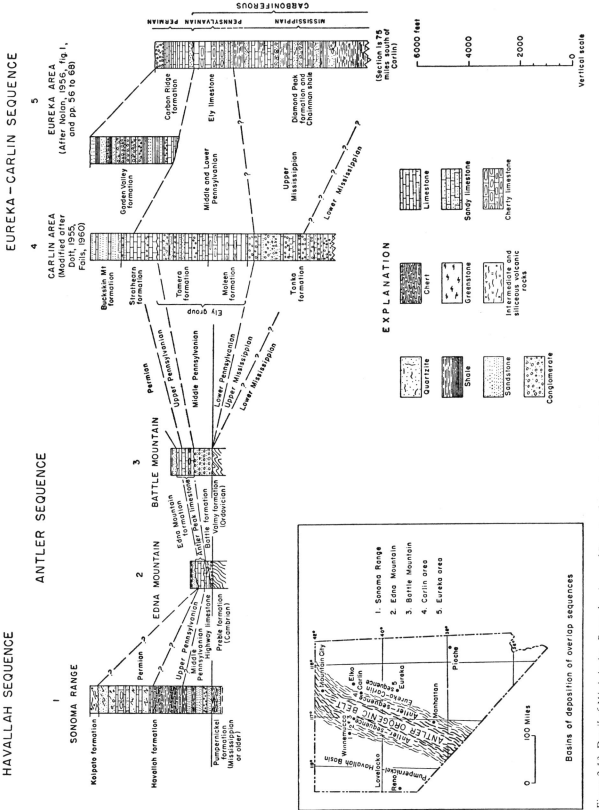

Figure 2-13. Detail of Mississippian, Pennsylvanian, and Permian formations involved in the Antler Orogeny of north-central Nevada.
From *Geology of the Antler Peak Quadrangle,* R. J. Roberts, U.S. Geological Survey, Professional Paper 459-A, 1964, figure 10.

Range, Nevada, indicates that in Middle-Late Triassic time, the drainage from some large portion of the eastern Great Basin, the Colorado Plateau, and Rocky Mountain provinces found access through northern Nevada to the sea. The Luning embayment of Ferguson and Muller (1949) was the eastern extension of a major Upper Triassic depositional basin that included a large part of the area of the present Sierra Nevada. The Luning formation represents the initiation of sedimentation in the Luning embayment; the mid-Jurassic Dunlap formation (Silberling 1959) represents the initiation of folding of the older Mesozoic rocks and the eradication of the Luning embayment as a basin of marine deposition.

SEVIER OROGENY

The Jurassic and Cretaceous periods in the Great Basin are characterized by regional uplift and a transition from marine to continental conditions, although an inland sea invaded Utah in Middle Jurassic time. Major shifts in sediment source areas marked the beginning of orogenic activity in the eastern Great Basin (Stokes 1963) and the end of the Cordilleran geosyncline. Subsequently sediments were shed eastward from a highly deformed fold and thrust belt that developed during Cretaceous time along the eastern margin of the Great Basin.

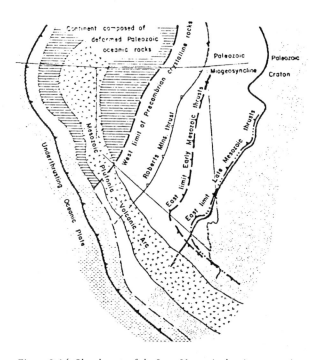

Figure 2-14. Sketch map of the Late Mesozoic showing truncation of Paleozoic geosynclinal and deformational trends by a Mesozoic Plutonic-volcanic arc of Andean type.
From J. C. Maxwell, "Early Western Margin of the United States," in *The Geology of Continental Margins,* edited by C. A. Burk and C. L. Drake, Springer-Verlag, New York, 1974. Figure 11 (after Burchfie and Davis, 1972, figure 7). Reprinted by permission of the publisher.

This zone of deformation has come to be known as the Sevier orogenic belt (Armstrong 1963) and is illustrated in figure 2-14.

CENOZOIC STRUCTURAL HISTORY

The Cenozoic geologic history of the Great Basin is most conveniently cast in the light of modern theories of plate tectonics, clearly elucidated by Atwater (1970). The sea floor retains a record of periodic reversals of the earth's magnetic field, preserved as polarized bands that are symmetrically disturbed and increase in age outward from linear spreading centers such as the mid-Atlantic ridge and the east Pacific rise (figure 2-15). Just as new oceanic crust is created at the spreading centers, old oceanic crust is consumed in subduction zones, which usually lie along continental margins. The interaction of oceanic crust with continental crust is thought to be the mechanism responsible for the major geologic features of continental margins.

Three main explanations have been advanced to account for the present topographic relief in the Great Basin (Nolan 1943):

1. The ranges and valleys are limited by normal faults that are due to tensional forces.
2. They are limited by reverse faults or by superficial normal faults caused by regional compression.
3. The valleys have been formed by erosion.

King (1870) put forth the earliest published theory for the origin of the individual mountain ranges in the Great Basin when he speculated that they are a series of eroded folds, similar to the Appalachians. Shortly thereafter, Gilbert (1872, 1874) proposed a block-faulting hypothesis, based almost entirely on physiographic evidence. He emphasized that if the front of a mountain range is linear and straight and cuts indiscriminately across the rock structure, it must be limited by a fault. Spurr (1901) took issue with King and Gilbert in favor of an erosional origin, noting that numerous faults within the mountain blocks were not reflected in the topography.

Nolan (1943) provides a critical review of the evolution of thought on the origin of basin and range structure. He concludes his discussion with a statement that most, if not all, of the major ranges in the province are bounded by faults on either one or both sides but that simple tensional forces alone are not adequate to explain the observed block faulting.

Paleocene and Eocene rocks are not widely distributed in the Great Basin; where they are present, they consist primarily of nonvolcanic fluvatile and lacustrine sediments. Commencing in the earliest Oligocene were widespread eruptions of chiefly ignimbrites in central Nevada, which formed a blanket 1000 to 2000 feet thick, predating most major normal faulting in

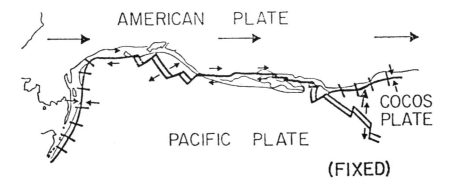

Figure 2-15. Present configuration of plate boundaries in the Pacific Northeast and western North America.
The map is a mercator projection about the pole of relative motion between the Pacific and American Plates, 53°N, and 53°W (Morgan 1968). Transform faults between the two plates lie on small circles about the pole of relative motion; thus, in this projection they form hoizontal lines. From Tanya Atwater, *Geological Society of America Bulletin* (1970) 81:3513-3536. Figure 4, p. 3513. Reprinted by permission of the publisher.

the area (Armstrong 1963). Similar volcanic rocks are as young as Pliocene in southern Nevada.

Overlying the widespread ignimbrite sections is a heterogeneous collection of clastic units, volcanic-rich sediments, basalts, other volcanics, and lacustrine sediments that were deposited during the development of basin and range structure.

Earlier faults appear to have followed a northeasterly or easterly trend, later faults are generally north-south, and it is along them that the greatest displacements have occurred (Armstrong 1963). In central Nevada, major normal faulting began in Late Miocene time (Deffeyes 1959); similar timing of normal faulting has been cited for other areas in the Great Basin (Armstrong 1963). It is tempting to infer a cause-effect relationship between the voluminous ignimbrite eruptions and normal faulting, but the apparent lag of 10 million years between the two events casts some doubt on the relationship.

Slemmons (1967) has summarized Pliocene and Quaternary crustal movements of the basin and range province; active and recently active faults appear to follow and thus rejuvenate older structural and geomorphic features. Observed displacements on historic faults in Dixie Valley, Nevada, indicate that a right-slip component is present, suggesting a genetic relationship between faulting in the Great Basin and the well-known San Andreas fault system of California. According to Atwater (1970), the San Andreas fault system was born in the Oligocene epoch, some 32 million years ago. Prior to that time, a subduction zone had existed off the California coast, but since the Oligocene, the mode of interaction at the plate boundary has changed from subduction to right-lateral movement in an ever-widening zone (figure 2-16). The possibility that the southern limb of the east Pacific rise was subducted and thus caused distension of the overlying crust is an attractive hypothesis for the origin of basin and range structure and is still being

tested (1978). (Albers 1967 has estimated the amount of right-lateral displacement across the soft western margin of the United States by mapping the oroclinal flexures shown by equal-thickness lines of numerous rock units, notably Ordovician sediments. Of particular significance is the fact that the major part of the inferred 100-mile horizontal displacement is taken up by bending, rather than by movement along any of the major faults of the region. Again, regional stratigraphic studies form the basis for many of the more profound structural interpretations in the Great Basin.)

PLEISTOCENE LAKES AND MODERN DRAINAGE SYSTEMS

Lake Bonneville

Pleistocene Lake Bonneville was by far the largest of the pluvial lakes in the Great Basin (figure 2-3). At its maximum it covered some 50,000 km² of Utah, Nevada, and Idaho and had a maximum depth of 330 m (Gilbert 1890; Hubbs and Miller 1948; Morrison 1965; Flint 1971). Lake Bonneville's ancient lake beds were first recognized by members of the Stansbury expedition (1852) and a monograph was prepared and published by Gilbert (1890). Present-day lakes are the highly saline Great Salt and Sevier lakes and the freshwater Utah and Bear lakes (figure 2-1).

The Great Salt Lake is a shallow, highly saline lake covering 4533 km² and is the largest remnant of Lake Bonneville. Great Salt Lake's total dissolved salts varies from 137 to 277 grams per liter. Its depth varies from a low mean depth of 9 m in 1940 to about 11 m mean depth today (Houghton 1976; Edmondson 1963; Flint 1971). There are no living fish

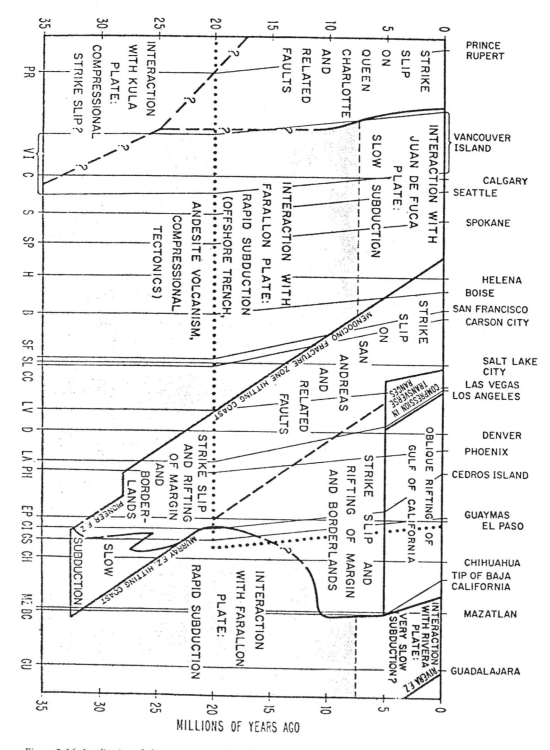

Figure 2-16. Implication of plate tectonics for the Cenozoic tectonic evolution of western North America.

From Tanya Atwater, *Geological Society of America Bulletin* (1970) 81: 3513-3536. Figure 17, p. 3530. Reprinted by permission of the publisher.

found in the lake, but other vertebrates utilize the algae and invertebrates found there today. Great Salt Lake is fed by Bear, Weber, and Jordan rivers.

Bear Lake, in the northeastern portion of the Bonneville Basin on the border between Utah and Idaho, is about 60 m deep near the eastern shore, with an average depth of 9 to 12 m. The lake is drained by the Bear River, which flows southwest into the Great Salt Lake.

Utah Lake is a shallow, turbid, eutrophic freshwater lake with a maximum surface area of approximately 237,500 hectares. The average depth is 2.7 m. Water inflow comes from the Provo River, Spanish Fork River, small streams, and underwater springs. Outflow is controlled by a dam on the Jordan River. The depth of the lake varies widely during the course of the year. Total dissolved solids vary during the year from 600 ppm to 2000 ppm with an average value of 900 ppm.

Sevier Lake is frequently listed as a dry lake. When water is present, it is highly saline.

Lake Lahontan and Adjacent Areas

Pluvial Pleistocene Lake Lahontan covered portions of northwestern Nevada, southern Oregon, and northeastern California (figure 2-3). At its greatest extent, Lake Lahontan covered an area of about 22,000 km^2 and was fed from a hydrographic basin of about 116,000 km^2. Lake Lahontan had a maximum depth of 270 m in the area of present-day Pyramid Lake, Nevada. Although numerous early explorers, hunters, and trappers (Cline 1963; Elliot 1973) passed through northern Nevada, little of the lake bed geology was documented. John C. Frémont (1845, 1849) carried out the first mapping of the region and first noted its geology. In the early 1850s the Beckwith and Ingalls parties traversed portions of the Lake Lahontan basin but paid little attention to the lake beds. In 1859, J. H. Simpson's party, exploring a wagon route across the Great Basin, included a geologist, Henry Engelmann, who first noted the presence of ancient waterlines and calcareous tufa deposits along the borders of the Carson Desert (Beckwith 1855; Simpson 1876; Russell 1885). The Williamson report of the Wheeler survey of 1879 gives a brief account of explorations around Pyramid Lake in 1865–1866. In 1867 Clarence King organized and led the exploration of the fortieth parallel, which included detailed study of most of the area covered by Lake Lahontan. A descriptive geology of the area was published by Arnold Hague in 1877, which for the first time mentions Lake Lahontan. In 1878 King published a systematic geology of the region in which he described the shorelines and deposits of Lake Lahontan. In 1881 Israel Cook Russell began surveying the Lake Lahontan area (Russell 1883, 1885). Russell's works have been the standard sources on Lake Lahontan ever since, but his findings have been modified somewhat by the researches of Jones (1925),

Broecker and Orr (1958), and Morrison (1964). There are pluvial Pleistocene Lakes in the other valleys adjacent to Lake Lahontan. These have been reviewed by Hubbs and Miller (1948a), La Rivers (1962), Snyder, Hardman, and Zdenek (1964), Feth (1964), Hubbs, Miller, and Hubbs (1974), and Houghton (1976). The remnants of Lake Lahontan include Pyramid Lake, Walker Lake, Soda Lakes, Carson Lake, and Winnemucca Lake. Winnemucca, located just northeast of Pyramid Lake, was a viable lake until 1938, when losses of water from the Truckee River by a diversion canal at Derby Dam, Nevada, over a period of thirty-two years caused a drop in the level of both Pyramid and Winnemucca lakes to a point where Winnemucca Lake became extinct (La Rivers 1962).

The main rivers occurring in the Lahontan drainage basin are the Walker River, which empties into Walker Lake; the Carson River, which empties into the Carson Desert near Fallon, Nevada; the Truckee River, which empties into Pyramid Lake; and the Humboldt River, which ends in the Humboldt sink at the northwestern end of the Carson Desert.

HUMBOLDT RIVER SYSTEM

The Humboldt River, named by Frémont in 1849, is the longest (320 km) river in the Lahontan basin. It ends in the Humboldt Lake and sink. Often the water, which originates in the Ruby Mountains to the east in Elko County, never reaches the sink, being used or lost to evaporation along the way. The Humboldt River has numerous tributaries that feed into it. The longest is Reese River, which originates in the Toiyabe Mountains of northwestern Nye County, some 160 miles south of the Humboldt. During Pleistocene times the Humboldt system may have included drainages from pluvial Lakes Dixie, Buffalo, Gilbert, Diamond, Newark, Hubbs, Gale, Steptoe, Franklin, Clover, and Waring.

TRUCKEE RIVER SYSTEM

The Truckee River has its origin in the Sierra Nevada at Lake Tahoe and then winds its way 175 km down the mountains and through the valleys to a desert lake some 760 m below, Pyramid Lake. Lake Tahoe is an oligotrophic lake, pH 6.8, low productive rate, with low concentrations of ions and 75 parts per million (ppm) total dissolved solids (La Rivers 1962; Lugaski 1977). However, on its way to Pyramid Lake the Truckee River picks up large amounts of nutrients and dissolved solids. Pyramid Lake, on the other hand, is extremely entrophic, pH 9.15, high productive rate and total dissolved solids about 5500 ppm (La Rivers 1962; Koch 1972; Lugaski 1977). The limnological characteristics of this system have been studied by Russell (1885), Jones (1925), Hutchinson (1937), La Rivers (1962), Koch (1973), and others. There are numerous tributaries to the Truckee River; some originate in the Sierra Nevada, but others, such as Steamboat Creek, which has its origin at Washoe Lake, Nevada, originate within the mountains and valleys of the basin.

CARSON RIVER SYSTEM

The Carson River has its origins in the Sierra Nevada in California and ends in the Carson sink some 150 miles away. Along the way much of the water is diverted for irrigation, especially at Lahontan Dam near Fallon, Nevada, and only a small amount of water currently reaches the sink. Big Soda Lake occurs along the route of the Carson River; it has been investigated by Russell (1885), Hutchinson (1933), La Rivers (1962), Breese (1968), Koenig et al. (1971), Axler et al. (1978) and Kimmel et al. (1978). Big Soda Lake has a pH of 9–10 and about 110,000 ppm dissolved solids (Hutchinson 1933; La Rivers 1962).

WALKER RIVER SYSTEM

The Walker River originates in the Sierra Nevada and terminates 135 miles below at Walker Lake, Nevada. For much of its course the Walker River is made up of two forks that join in Mason Valley, Nevada. Most of the water of the Walker River is diverted for irrigation and residential use before it reaches Walker Lake, and as a result the lake has been diminishing in the last hundred years. The depth of the lake is currently less than half of what it was when surveyed by Russell in the early 1880s (Russell 1885; La Rivers 1962). The chemical characteristics of Walker Lake are similar to those of Pyramid Lake. Walker Lake's pH is in the 9 range, while its total dissolved solids are often over 8500 ppm.

Adjacent Areas of the Great Basin

North of Pyramid Lake there is little water except in numerous small streams, springs, ponds, and a few small lakes. South and southeast of the Lake Lahontan system, little permanent water is found with the exception of numerous springs and creeks in Big Smoky Valley, which contained at one time pluvial Lakes Toiyabe and Tonopah. Numerous other pluvial lakes occurred in the region, such as Lake Columbus, Lake Hot Creek, Lake Railroad, Lake Reveille, and others (Russell 1885; Hubbs and Miller, 1948; La Rivers, 1962; Feth 1964; Snyder, Hardman, and Zdenek 1964; Hubbs, Miller, and Hubbs 1974).

Amargosa River and Death Valley System

The Amargosa River of southwestern Nevada and southeastern California originates in the Timber Mountains of southern Nye County, Nevada, and terminates some 280 km away in the southern end of Death Valley. This river, which more often than not is only a series of puddles, begins at the 2260 m mark in the mountains and ends technically at Bad Water, Death Valley, some 85 m below sea level. Some 80 km south of Beatty, Nevada, the Amargosa River receives water from Ash Meadows to the east.

This area also contained a large number of pluvial lakes; the largest was Lake Manley in Death Valley itself (Miller 1946, 1948; Snyder, Hardman, Zdenek 1964). At one time this system was connected to the Owens Valley system, pluvial Lake Owens.

White River and Colorado River System

The White River in southeastern Nevada during pluvial times drained into the Moapa River and then into the Colorado River. The White River today is discontinuous over most of its length. At one time this river was nearly 320 km long and consisted of two forks. The western fork originates southwest of Ely, Nevada, runs into Pahranagat Valley, and then into Warm Springs Valley and the Moapa River. The eastern fork is some 80 km shorter and runs by Meadow Valley Wash into the Moapa River. Today much of this drainage is perennially dry.

CLIMATE

Great Basin climates are structured by geographical factors and atmospheric circulation patterns. Climates are characterized as arid to semiarid, but such definitions are, in part, functions of both latitude and elevation.

The principal geographical factors are latitude, elevation, and continentality. The hydrographic Great Basin spans nearly 8 degrees of latitude, from about 36° to about 44° N, a lineal distance of over 880 km. Hence the northern and central Great Basin has temperatures characteristic of middle latitudes and the southern portion is subtropical. The Great Basin is characterized by high mountain ranges on the west perimeter, the Sierra Nevada, and on the east, the Wasatch Ranges. In between are over two hundred separate ranges with elevations ranging up to 3965 m. There is also a general downslope from north to south, with basin floors between 1220 and 1830 m in the north and less than 915 m in the south. The basin and range physiography affects climate in two ways: a sharp drop in temperature with increasing elevation and an increase of precipitation with increasing elevation. This climatic phenomenon, of course, greatly influences the vertical succession of ecological zones.

The third influential geographical influence on climate is continentality. This is expressed in dryness and large temperature variations (Houghton et al. 1975). The Sierra Nevada and the Southern Cascade ranges form a barrier to the maritime winds off the Pacific Ocean and create a rain-shadow effect on the east of those ranges. Most storms crossing the Great Basin derive from the Pacific Ocean. As moisture-bearing winds cross the Pacific mountain ranges, they are forced to rise, cool, and condense, losing much moisture as precipitation on the westward side of the mountains. This rain-shadow effect is the principal factor influencing the aridity of the Great Basin (Houghton et al. 1975:6).

The temperature regime within the basin is also influenced by the blockage of maritime air. The Great Basin has a continental climate with well-defined seasons because the area responds quickly to changes in solar radiation (Thomas et al. 1970). The often considerable diurnal temperature fluctuation (in the summer as much as 60°F) is a related phenomenon (see table 2-2).

Atmospheric Circulation

The Great Basin lies in the belt of prevailing westerly winds. In the winter the upper air winds (above 10,000 feet) tend to move nearly due west to east; in the summer, the pattern tends to be from southwest to northeast. In winter the jetstream may often be directly over the Great Basin.

The winter surface level atmospheric pressure pattern is indicated in figure 2-17. The summer pressure is very different (figure 2-18). The result of this seasonally variant pattern is that most precipitation in the Great Basin falls during the winter season, and most of it as snowfall at higher elevations. Summer precipitation generally derives from moisture-bearing winds from the Gulf of Mexico, Gulf of California, and the tropical Pacific Ocean rather than from the Pacific areas west of the Great Basin (Houghton et al. 1975:9–10).

Storms passing over or occurring in the Great Basin are of three types, occurring at different seasons: Pacific fronts, Great Basin lows, and summer thundershowers (Houghton 1969). Pacific fronts occur principally in the winter; often the heaviest precipitation is on the west slopes of the mountain barriers. Great Basin lows develop along cold fronts moving inland; usually northerly winds bring polar air, and often moisture,

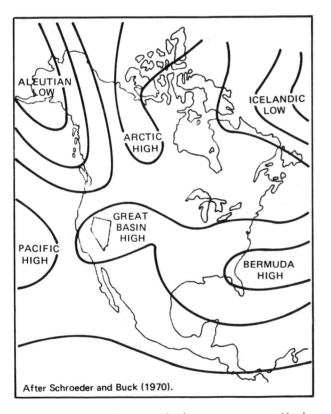

After Schroeder and Buck (1970).

Figure 2-17. Average January sea-level pressure pattern over North America.
The Aleutian and Icelandic lows are well developed, and the Pacific and Bermuda highs are weaker and farther south than in summer. Pressure is generally high over the cold continent, causing seasonal highs such as the Great Basin and Arctic highs to form. From "Fine Weather," M. J. Schroeder and C. C. Buck, *U.S. Forest Service Agricultural Handbook* No. 360, 1970, figure 5.

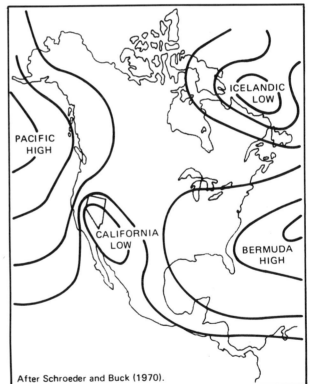

After Schroeder and Buck (1970).

Figure 2-18. Average July sea-level pressure patterns over North America.
The Pacific and Bermuda highs are strong and rather far north. The Icelandic low is weak, pressure is generally low over the continent, and intense heat in the Southwest causes formation of the California low. From "Fine Weather," M. J. Schroeder and C. C. Buck, *U.S. Forest Service Agricultural Handbook* No. 360, 1970, figure 7.

southward. The result is that heavier amounts of precipitation fall within the Great Basin than during the occurrence of Pacific fronts (Houghton et al. 1975:16).

The summer thunderstorm pattern over the Great Basin is very similar to that over much of the arid West—an influence of tropical air, with hot days and surface heating. This causes scattered afternoon and evening thunderstorms to occur if there is sufficient moisture in the air.

Houghton et al. (1975:19) summarize the general atmospheric circulation over the Great Basin as follows:

Winter: periods of fair weather associated with Great Basin highs, but Pacific fronts also move inland in the prevailing westerlies aloft, bringing precipitation and frequent weather changes.

Spring: precipitation and changeable weather are common largely because of Great Basin lows, which are frequent in the spring. In late

TABLE 2-2
Temperature means and extremes at Selected Great Basin stations, 1933-1974

ANNUAL AVERAGE: (°F.)	ELKO, NEV.	ELY, NEV.	RENO, NEV.	LAS VEGAS, NEV.	WENDOVER, UTAH	MILFORD, UTAH	SALT LAKE CITY, UTAH
Record Mean[a]	45.8	44.2	50.0	65.9	52.6	49.2	51.7
Maximum	62.6	60.4	65.3	80.1	63.9	65.2	62.7
Minimum	29.0	27.9	34.6	51.6	41.3	33.2	40.7
Extremes							
High	102 (8/1967)	99 (6/1954)	103 (8/1972)	116 (7/1963)	112 (7/1939)	105 (6/1970)	107 (7/1960)
Low	−28 (12/1972)	−28 (12/1972)	−16 (12/1972)	8 (1/1963)	−19 (2/1933)	−32 (12/1972)	−18 (1/1963)

SOURCE: National Oceanic and Atmospheric Administration (1975)

[a]Periods for record mean values for temperature and precipitation are: Elko (1910–1974); Ely (1939–1974); Reno (1888–1974); Las Vegas (1937–1974); Wendover (1912–1974); Milford (1909–1974): Salt Lake City (1874–1974); period for extremes is 1935–1974.

TABLE 2-3
Precipitation and snowfall means and extremes at selected Great Basin stations, 1935-1974 (in inches)

PRECIPITATION AND SNOWFALL	ELKO, NEV.	ELY, NEV.	RENO, NEV.	LAS VEGAS, NEV.	WENDOVER, UTAH	MILFORD, UTAH	SALT LAKE CITY, UTAH
Precipitation:							
Record mean[a]	8.90	8.67	7.66	3.94	4.82	8.56	15.62
Record annual high	16.24 (1941)	13.52 (1941)	11.75 (1940)	10.72 (1941)	10.13 (1941)	13.20 (1941)	21.11 (1968)
Record annual low	5.51 (1959)	4.22 (1974)	1.55 (1947)	1.11 (1968)	2.47 (1960)	4.48 (1950)	9.36 (1939)
Snowfall:							
Record mean	39.30	47.00	26.10	1.60	7.50	44.00	57.10
Annual high	69.10 (1970-1971)	65.00 (1970-1971)	59.30 (1951-1952)	13.40 (1974-1975)	22.40 (1961-1962)	60.10 (1951-1952)	117.30 (1951-1952)
Annual low	12.50 (1937-1938)	12.10 (1950-1951)	5.40 (1940-1941)			16.00 (1940-1941)	18.50 (1939-1940)

SOURCE: National Oceanic and Atmospheric Administration (1975).

[a]Periods for record mean values for temperature and precipitation are: Elko (1910-1974); Ely (1939-1974); Reno (1888-1974); Las Vegas (1937-1974); Wendover (1912-1974); Milford (1909-1974); Salt Lake City (1874-1974); period for extremes is 1935-1974.

spring, the subtropical high brings drought to the southern Great Basin.

Summer: clear dry weather is dominant due to the subtropical high and the lack of available moisture from the Pacific Ocean. Precipitation occurs as occasional showers in moist air from the tropical oceans. Thunderstorm frequency increases over the northeastern section of the Great Basin.

Fall: fair weather is dominant, with frequent Great Basin highs, but Pacific fronts and Great Basin lows cause precipitation and changeable weather, especially in the late fall.

Precipitation

The variable nature of Great Basin climate is reflected in the variability of annual precipitation at different locales in different years. For example, in the western Great Basin in Reno, Nevada, the mean annual precipitation is 7.20 inches per year, but recorded amounts have varied from 1.55 inches in 1947 to 13.73 inches in 1890. The relation of latitude and elevation to mean annual precipitation can be indicated by reference to table 2-3, which lists stations in the western, central, and southern portions of the Great Basin.

FISHES

Lake Lahontan and Adjacent Areas

FOSSIL FISHES

A number of shark and sharklike fossils have been described from the western portion of the Great Basin. The oldest is an edestid shark (*Helicoprion nevadensis*) from the Permian described by Wheeler (1939), and several others are known from the Mesozoic (Wemple 1906; Davidson 1919). David (1941) described *Leptolepis nevadensis* from the Cretaceous of central Nevada. Cenozoic fossil fishes from the western Great Basin are somewhat better known but still sparse. Cope (1872) described *Amyzon mentalis* (family Catostomidae), the Nevada Amazon sucker, and *Trichophanes hians* (family Aphredoderidae), the Nevada pirate perch from the Miocene coal beds of Osino, Nevada. La Rivers (1964) described the oldest known salmonid in North America belonging to the genus *Salmo (Salmo cyniclope)* from the Miocene deposits of Rabbit Hole, near the southern edge of the Black Rock Desert, Nevada. He also described a Miocene cyprinid (*Gila esmeralda*) from the Coaldale area of central Nevada (La Rivers 1966). Lugaski (1978) has recently described a Miocene cyprinodont fish (*Fundulus lariversi*) from central Nevada. Lucas (1900) described a Pliocene cyprinid (*"Leuciscus" turneri*) from the "Esmeralda" formation of Nevada (Lugaski, 1977). Two other fossil fish, the Nevada stickleback (*Gasterosteus doryssus*) and the Nevada killifish

(*Fundulus nevadensis*), have been described from diatomite beds near Hazen, Nevada (Jordan 1907, 1908, 1925; Hay 1907; Eastman 1917; Miller 1958, 1965; La Rivers 1962). La Rivers (1966) described a salmonid fossil (*Salmo esmeralda*) from central Nevada of Pliocene age. Cavender (1969) has reported a *Salvelinus* Pliocene fossil from near Fallon, Nevada. Lugaski (1979) has described a Pliocene fossil cyprinid (*Gila traini*) from Jersey Valley, Nevada. Jordan (1924) described the remains of *Cottus beldingi*, a species currently found in the Lahontan Basin, from Lake Lahontan sediments near Fallon, Nevada. La Rivers (1962) reviewed the fossil fish bones found in Lake Lahontan sediments. They include *Salmo clarki*, *Gila bicolor*, and *Chasmistes* sp., which are found in the area today.

RECENT FISHES

The recent native fishes of the western portion of the Great Basin, which includes the Lake Lahontan drainage and adjacent areas, have been reviewed by Snyder (1917), Hubbs and Miller (1948), Miller and Hubbs (1960), Miller (1943, 1946, 1948, 1949, 1950, 1952, 1958, 1965), La Rivers (1952, 1962), La Rivers and Trelease (1952), Hubbs, Miller, and Hubbs (1974), and Smith (1978). Hubbs and Miller (1948) report five families, ten genera, thirteen species, and about thirty kinds or forms of native fishes known in the Lahontan system. These numbers remain fairly accurate despite lumping and splitting of taxa that have taken place subsequently.

Appendix 2 includes only native fishes found in the area today or in the past and excludes introduced species (Hubbs and Miller, 1948a, 1948b, 1972; Miller 1948; Miller and Hubbs 1960; La Rivers 1962; Lugaski 1972; Cavender 1978).

Lake Bonneville and Adjacent Areas

The ichthyofauna of the Lake Bonneville comprised five families, twelve genera, and twenty-two native species as tabulated by Hubbs and Miller (1948a) but due to taxonomic reordering the number of species has been reduced to seventeen native species by Smith (1978). Of these native species, only two genera and fourteen species were listed as endemic by Hubbs and Miller (1948a), which indicates that the Bonneville system was recently connected with other outside systems and there was a faunal transfer between them and Lake Bonneville.

FOSSIL FISHES

There have been only a few fossil fish reported from the Lake Bonneville area and the eastern Great Basin in general. Paleoichthyofauna from the Lake Bonneville area have been reported by Stokes et al. (1964) and Smith et al. (1968). These include seven species that currently or once were found in Utah and Bear lakes of the Bonneville Basin. The seven fossil fishes include a salmonid (*Salmo clarkii*), two whitefish (*Prosopium gemmiferum* and *Prosopium spilonotus*), a cyprimid (*Gila atraria*), a sucker (*Catostomus ardens*), and two sculpins (*Cottus bairdii* and *Cottus extensus*).

RECENT FISHES

The recent native fishes of the eastern Great Basin have been reviewed by Cope and Yarrow (1875), Snyder (1919, 1921, 1924), Tanner (1936), Hubbs and Miller (1948a), McConnell, Clark, and Sigler (1957), and Sigler and Miller (1963) and are listed in Appendix 2.

AMPHIBIANS AND REPTILES

Paleontology

There are few fossil records of pre-Cenozoic amphibians and reptiles in the Great Basin, with the exception of Dinosaurian records for the eastern Great Basin (Colbert 1968) and the Ichthyosaurians found in the western Great Basin (Leidy 1868; Merriam 1902, 1905, 1906, 1908, 1910a, 1911). The Cenozoic fossil amphibian and reptile records for the Great Basin are also sparse. Hecht (1960) describes new Leptodactylid, *Eorubeta nevadensis,* from the Eocene deposits of the Sheep Pass formation of White Pine County. Taylor (1941) describes a new palodytid, *Miopelodytes gilmorei,* from the Miocene Elko Shales near Elko, Nevada. La Rivers (1953, 1966) has described two fossil frogs from the Lower Pliocene of the western Great Basin: *Rana johnsoni* from near Virginia City, Nevada, and *Rana plax* from near Lovelock, Nevada. Zweifel (1956) describes a new pelobatid, *Scaphopus alexanderi* from the Lower Pliocene in Fishlake Valley, Nevada. Ruben (1971) has described a fossil snake, *Coluber constrictor,* from the Pliocene Truckee formation of western Nevada, although this now appears to be an undescribed species of *Thamnophis* (Lugaski, unpublished data). Brattstrom (1954, 1958) has described numerous fossil amphibians and reptiles from Gypsum Cave, Nevada, and Smith Creek Cave, Nevada. These include *Rana pipiens, Bufo punctatus, Heloderma suspectum, Crotaphytus collaris, Phrynosoma platyrhinos, Sauromalus obesus, Cnemodophorus tigris, Lampropeltis getulus, Lampropeltis pyromelana, Masticophis flagellum, Masticophis taeniata, Pituophis catenifer, Crotalus atrox, Crotalus mitchelli* and *Crotalus viridis* from Gypsum Cave; *M. flagellum* and *P. catenifer* from Smith Creek Cave. Heizer (1942) and Hattori (in press) have described remains of the Pacific pond turtle, *Clemmys marmorata,* from archaeological sites in the western Great Basin.

Recent Amphibians and Reptiles

The recent amphibians and reptiles of the Great Basin are relatively well known. Baird and Girard (1852a, 1852b), Girard (1852, 1858), Baird (1859), Cope (1868, 1875, 1883a, 1883b, 1889, 1892, 1893, 1895, 1896, 1900), Yarrow (1875), Yarrow and Henshaw (1878), Simpson (1876), Boulenger (1884, 1885, 1896), Garman (1884), Steineger (1893, 1895), Van Denbergh (1895, 1897, 1922), Van Denbergh and Slevin (1918, 1919, 1921), Taylor (1912), Richardson (1915), Slevin (1928, 1934), Pack (1930), Klauber (1930, 1956), Woodbury (1931, 1952), Woodbury and Smart (1950), Burt (1933), Linsdale (1938, 1940), La Rivers (1942), Stebbins (1951, 1954), Banta (1961, 1963, 1965a, 1965b, 1967), Banta and Tanner (1964), Tanner and Jorgensen (1963), and Lugaski, Vreeland and Vreeland (1978), to name a few, have all made major contributions to the understanding of amphibian and reptile distribution in the Great Basin. Banta (1965c) has published a bibliography of the herpetology of the state of Nevada that includes most of the citations mentioned, plus many more. Tanner (1978) has published a discussion of the zoogeography of Great Basin herpetology in which he states there are few, if any, endemic amphibians and reptiles in Utah. This is true for the Great Basin as a whole, with the exception of some subspecies due to geographical and semireproductive isolation. Most Great Basin amphibian and reptile species had their origins in areas south and southeast of the Great Basin. Those amphibians and reptiles endemic to the Great Basin include the Black toad (*Bufo boreas exsul*) from Deep Springs, Inyo County, California; the Amargosa toad (*Bufo boreas nelsoni*) from the Amargosa River region of southern Nevada; the extinct Vegas Valley leopard frog (*Rana pipiens fisheri*) from Vegas Valley, Nevada; the western red-tailed skink (which has a blue tail in Great Basin desert mountains), *Eumeces gilberti rubricaudatus,* from the mountains of the southwestern Great Basin; the Great Basin skink (*Eumeces skiltonianus utahensis*), which occurs over the entire Great Basin and into southern Idaho; the Panamint alligator lizard (*Gerrhonotus panamintinus*) from the Panamint Mountains of the southwestern Great Basin; the Utah Mountain kingsnake (*Lampropeltis pyromelana infralabialis*) from the eastern Great Basin; and the Great Basin rattlesnake (*Crotalus viridis lutosus*) from throughout the Great Basin. Recently Tanner and Banta (1977) described the Lahontan leopard lizard (*Crotaphytus wislizeni maculosus*), from the Lahontan Basin of the western Great Basin. There are probably numerous other undescribed subspecies of amphibians and reptiles in the Great Basin, which will be described in the future when more is known about the morphology, evolution, and distribution of those amphibians and reptiles currently known from the Great Basin, many of which have been poorly analyzed due to insufficient specimens.

Probably there are not more than one hundred different

species and subspecies of amphibians and reptiles found in the Great Basin today. Most of these have migrated and evolved naturally into their present habitats. There are two notable exceptions: the red-legged frog (*Rana aurora*), which has been transplanted into several areas of the Great Basin, and the bullfrog (*Rana catesbeiana*), which was transplanted into numerous desert springs for commercial production in the past. These introductions compete with native species and other endemic wildlife, especially the native fishes, which they often eliminate.

BIRDS

Paleontology

Selander (1965) concludes that the consensus among avian paleontologists is that most modern bird species in the Great Basin arose during the Pleistocene. There are only a few fossil records for birds from the Great Basin during the Late Cenozoic. Merriam (1911) described a fossil form of a goose (*Branta*) from the Pliocene deposits of Thousand Creeks, Nevada. Burt (1929) described a Pliocene goose (*Branta esmeralda*) from Fish Lake Valley, Nevada, and Brodkorb (1958) describes some fossil birds from the northern Great Basin in Oregon (*Nettion bunkeri, Lophortyx shotwelli*, and *Bartramia umatilla*). Harrington (1933) mentions the occurrence of the California condor (*Gymnogyps californianus*) and a vulture (*Teratornis* sp.) from Gypsum Cave, Nevada. Howard (1935) described an eagle (*Spizaëtus willetti*), an extinct condor (*Breagyps clarki*), and a vulture (*Coragyps occidentalis*) from Smith Creek Cave, White Pine County, Nevada. Howard (1952) completed the description of the Smith Creek Cave avifauna. This avifauna included the three birds already mentioned, a giant condor (*Teratornis incredibilis*), the California condor (*G. californianus*), a spotted owl (*Strix occidentalis*), sage grouse (*Centrocercus urophasianus*), and prairie falcon (*Falco mexicanus*). Over 50 percent of the avifauna remains studied were raptors, 22 percent were water birds (mainly surface feeding ducks), 19 percent were grouse, and 8 percent were passerines with remains of one swift, one nighthawk, and one dove found. Twelve percent of the birds found in Smith Creek Cave are considered to be extinct. Numerous bones of the turkey vulture (*Cathartes aura*) were found on the floor of the cave. *Cathartes* bones were found only in the upper levels of the cave. This form appears to replace the *Coragyps*, much the same as was found in the Rancho La Brea pits (Howard and Miller 1939, Howard 1952).

From Fossil Lake, Oregon, on the extreme northwestern edge of the Great Basin, Howard (1946) describes a number of Pleistocene fossil birds. These avifauna consist of some sixty-eight different species of birds, 91 percent of which are waterbirds. These include the extinct cormarant (*Phalacrocorax macropus*) and the extinct flamingo (*Phoenicopterus copei*). Of the total number of species found, only about 25 percent appear to be extinct. Howard (1958) has also described a Pleistocene cormorant (*Phalacrocorax auritus*) from Crypt Cave, Nevada, which is identical to forms found in the Great Basin today. Howard (1964) also describes a new species of pygmy goose (*Ana bernicula oregonesis*) from Fossil Lake, Oregon, and Jehl (1967) describes fourteen species from two areas near Fossil Lake, Oregon. Hall (1940) describes three species from Lake Lahontan Tufa beds at Rattlesnake Hill near Fallon, Nevada. These include the white pelican (*Pelecanus erythrorhynchos*), the double-crested cormorant (*Phalacrocorax auritus*), and a Canada goose (*Branta canadensis*). Jennings (1957) reported bird remains from Danger Cave near Wendover, Utah, and Aikens (1970) reported bird remains from Hogup Cave near the northwestern corner of Great Salt Lake; all appear to be represented by living species.

Recent Birds

Baird (1853) published a list of the recent birds collected on the Stansbury expedition to the Great Salt Lake in 1849–1850; Remy and Brenchley (1861) published a list of 28 birds from Utah; Merriam (1873) noted 176 Utah birds collected by the F. V. Hayden survey; Henshaw (1874a, 1874b, 1875) published a list of 214 birds collected by the Wheeler survey; Ridgeway (1877) described Great Basin birds collected by the King survey; Bendire (1877) noted some southeastern Oregon birds; Hoffman (1881) of the Wheeler survey described birds he collected from the Great Basin in 1871 and summarized previous work of Ridgeway, Henshaw, and Yarrow. Fisher (1893) reported on the ornithology of the Death Valley expedition of 1891 under the leadership of C. Hart Merriam and fellow scientists of the U.S. Biological Survey; Hanford (1903) studied the birds of Washoe Lake, Nevada; Hanna (1904) and Taylor (1912) published notes on birds of Humboldt County, Nevada; Van Rossem (1931) describes avifauna of the Spring Mountains of southern Nevada. Linsdale (1936, 1951) compiled the first general monographs on Nevada birds.

Behle (1941, 1943, 1944, 1948, 1955, 1960, 1976, 1978) has produced a number of major works on selected areas of Utah and the intermountain avifauna in general, several of these in conjunction with others, Behle and Ghiselin (1958) and Behle et al. (1958).

Woodbury, Cottam, and Sugden (1949) produced a checklist of 436 birds found in Utah, and Woodbury and Cottam in conjunction with Hayward and Frost (1976) produced a monograph on Utah birds. Johnson (1965, 1970, 1973, 1974, 1975, 1978) has been instrumental in updating and advancing our knowledge of Nevada avifauna.

Hayward, Kilpack, and Richards (1963) list 192 species of birds from the Nevada Test Site. Larrison, Tucker, and Jollie (1967) describe the birds of Idaho and include numerous records of Great Basin species. In recent years Alcorn (1940a, 1940b, 1941, 1943, 1946), Richards (1962), Ryser (1963), Austin and Bradley (1966, 1968), Lugaski, Worley, and Kleiner (1972), Lawson (1973a, 1973b, 1974, 1975, 1977), Austin and Rea (1976), and others have added to the list of birds occurring in the Great Basin.

Behle (1978) indicates that there are about 154 species of resident land birds in the Great Basin, with perhaps twice that number being visitors. Johnson (1975, 1978), Behle (1978), and Kelleher (1970) have analyzed the patterns of avian distribution in montane regions of the Great Basin. Behle (1978) has found that the number of permanent Boreal species within the areas of Utah studied averages 40 percent of the total Boreal species observed. Johnson (1975) reports that within the Great Basin the number of permanent Boreal species averages 17 percent of the total Boreal species observed; in the Cascade-Sierra Mountain series he finds that an average of 24 percent of the Boreal species are permanent residents, as is the case in the Rocky Mountain series. Johnson (1975, 1978), Behle (1978), and Kelleher (1970) point out that the Boreal avifauna in the Cascade-Sierra Mountain series is closely related to the avifauna found in California; the rest of the Great Basin avifauna are more closely related to the Rocky Mountain and Great Plains avifauna, with elements of the Mojave and Sonoran Desert avifauna important in the southern Great Basin.

Nappe and Klebenow (1973) have recently reviewed the rare and endangered birds of Nevada. This list includes the white pelican (*Pelecanus erythrorhynchos*), the white-faced ibis (*Plegadis chihi*), goshawk (*Accipiter gentilis atricapillus*), ferruginous hawk (*Buteo regalis*), southern bald eagle (*Haliaeetus leucocephalus leucocephalus*), osprey (*Pandion haliaetus carolinensis*), prairie falcon (*Falco mexicanus*), American peregrine falcon (*Falco peregrinus anatum*), pigeon hawk (*Falco columbarius*), sharp-tailed grouse (*Pedioecetes phasianellus columbianus*), greater sandhill crane (*Grus canadensis tabida*), and the western yellow-billed cuckoo (*Coceyzus americanus occidentalus*). All are endangered due to habitat destruction, killing, and overuse of pesticides. The plight of the white-faced ibis has become more severe in recent years. They have three main nesting sites in the United States, two of them in the Great Basin (at Great Salt Lake in Utah and in Lahontan Valley, Nevada), in addition to smaller areas both inside and outside the Great Basin. In recent years as many as 10,000 ibis produced 3300 nests in the Lahontan Valley. However, recent droughts and court decisions sharply reducing water allocations to the Lahontan Valley have reduced this number to about 400 nesting pairs, with only 100 nests being successful. This situation has become the norm in recent years. As the human population of the Great Basin continues to grow, greater and greater demand is placed on the limited water supplies, and many marsh and avian habitats available only a few years ago are quickly drying up as that water is used elsewhere.

MAMMALS

Paleontology

The Great Basin had a variety of mammalian species during the Cenozoic. There are few if any Early Tertiary records. The best paleontological material comes from Miocene and Pliocene deposits in the northern portion of the Great Basin. Pleistocene mammalian fossils are found mainly in cave deposits, once inhabited by man; the rest are found in sedimentary deposits left behind by the numerous Pleistocene pluvial lakes. Merriam (1907, 1910b, 1911, 1913, 1914a, 1914b, 1915a, 1915b, 1916a, 1916b, 1917, 1918), Furlong (1910, 1932, 1935a, 1935b, 1943), Gidley (1908), Kellog (1910), Stock (1921, 1926, 1951), Stirton (1929, 1931, 1932a, 1932b, 1934, 1935, 1936), Hall (1929, 1930a, 1930b), Wilson (1936, 1938), Curry (1939), Henshaw (1942), MacDonald 1948, 1956a, 1956b, 1959, 1966), Reed (1958), Shotwell (1961, 1967a, 1967b, 1970), Stokes and Condie (1961), Clark, Dawson, and Wood (1964), Mawby (1967, 1968a, 1968b), and Orr (1969) describe numerous species derived from the latter half of the Cenozoic era. Extinct species include elephants and their various relatives, camels, horses, rhinoceroses, saber-toothed cats, and many other small and large mammals. A number of genera evolved in or near the Great Basin, including the horse (Simpson 1961). Changes in climate and habitats, and possibly hunting by man (Martin 1963), led to the extinction of numerous species, genera, and even families of mammals. Late Pleistocene mammal remains were earlier reported in association with human artifacts in Gypsum Cave and at Tule Springs, Nevada, in the southern Great Basin, including ground-sloth (*Northrotherium shastense*), mammoth (*Mammuthus columbi*), camel (*Camelops* sp.), and horse (*Equus* sp.) (Harrington 1933; Mawby 1967). However, these associations subsequently have been disproven (Heizer and Berger 1970; Wormington and Ellis 1967). Other Late Pleistocene mammal localities include a small fauna described by MacDonald (1956a) from Wichman, Nevada, which included elephant, horse, and camel but no evidence of human artifacts (Hibbard et al. 1965). Additional discussion of Late Pleistocene mammals in the Great Basin can be found in Hay (1927).

Recent Mammals

The distribution of Great Basin mammals is generally well known. Burt (1934) published a list of fifty-nine mammals from southern Nevada; Bailey (1936) describes Oregon mammals and life zones; Hall (1946) describes the mammals of Nevada, which included all of the known mammal species at that time. He reported five faunal areas (Sierra Nevadan, northern Great Basin, central Rocky Mountain, lower Sonoran-Lahontan Lake Basin, and Bonneville Basin), with numerous centers of differentiation. Durrant (1952) describes

247 species of mammals found in Utah. Jorgensen and Hayward (1965) list the mammal species occurring on the Nevada Test Site, with data on distribution, activity, and behavior. In his guide to Idaho mammals, Larrison (1967) provides information on life history and activities of the mammals found in the Great Basin section of Idaho. Brown (1971, 1978) discusses the distribution of Boreal mammal species on the mountain tops of the various Great Basin ranges. Armstrong (1977), reviewing the work of Durrant (1952), used numerical methods to analyze the distribution of 126 mammal species native to Utah. An overall distribution of Great Basin mammals can be found in Hall and Kelson (1959). There are numerous smaller works on distribution of mammals in the Great Basin, such as Deacon et al. (1964).

VEGETATION

Paleobotany

The little we know about most of pre-Tertiary floras of the Great Basin comes from Paleozoic and Mesozoic localities in the eastern Great Basin. Tidwell (1967) reports a lowland, swampy vegetation from the lowermost Pennsylvanian Manning Canyon shale of north-central Utah. This flora includes *Lepidodendran, Mariopteris, Sphenopteris, Rhodea, Neuropteris, Crossopteris, Calamites,* and *Asterophyllites.*

A humid, lowland, warm-temperate to subtropical climate is indicated in the eastern Great Basin by one of the few Cretaceous fossil floras found in the Blackhawk formation of central Utah (Parker 1968, Tidwell et al. 1972).

Parker (1968) describes two genera of palms, *Sabalites* and *Geonomites*, three fern genera, a variety of gymnosperms, including *Sequoia, Ginkgo, Protophyllocladus,* and *Araucaria,* and various angiosperms, including *Dryophyllum, Juglans, Cinnamomum, Magnolia, Menispermum, Ficus, Myrtophyllum, Salix, Trapa,* and *Nymphaeites.*

Tertiary Period

The Tertiary period witnessed a dramatic change in the climatic regime of the Great Basin, with its associated changes in vegetation. Pre-Tertiary climate was warm and moist, with subtropical to warm temperate. The withdrawal of the Cretaceous epicontinental seas, increasing the uplift of the Sierra Nevada, Cascade ranges, and the Great Basin in general, led to three trends in Great Basin climate that extend through the Tertiary and into the Quaternary: a general cooling of temperature (often interrupted by warmer periods); progressive drying due principally to the increasing rain shadow effect of the Sierra Nevada and Cascade ranges; and fluctuation between glaciopluvial and warm-dry periods, especially in the Pleistocene (Antevs 1948, 1952).

The members of the Eocene flora of the Great Basin in many cases are the same genera present in the Great Basin flora today (Tidwell et al. 1972). Axelrod (1950, 1958) believes that the evidence from fossil floras of the central Rocky Mountains suggests that sufficient geomorphic and climatic heterogeneity existed in that region for the evolution of plants adapted to drier condition by mid-Eocene, and this Madro-Tertiary geoflora became established over the semiarid southwestern interior region by mid-Oligocene. Chaney (1944a) also thought the Eocene to be dry, based on the report of a fossil cactus from the Eocene of Utah, but the form has since been shown to be a sedge (Brown 1959; Becker 1962).

From the north-cental Great Basin near Jarbidge, Nevada, Axelrod (1966) describes an upper Eocene Copper Basin flora of 42 genera, mainly (60 percent) montane conifers, but also numerous rosaceous shrubs and other smaller angiosperms. The data suggest the Copper Basin flora represented a conifer-deciduous hardwood forest with a close relationship to the coast redwood and spruce-hemlock forests of northwestern California, all indicating a cool-temperate climate. From nearby, Axelrod (1966) describes an Upper Eocene Bull Run flora consisting of 22 genera, mainly montane conifers (99.5 percent of fossils found) and few deciduous hardwoods. Axelrod (1966) estimates that the Copper Basin flora existed at 1067 to 1220 m elevation and the Bull Run flora at 1220 m elevation. In his summary, Axelrod (1966) emphasizes increasing altitude within the Great Basin and Far West with altitudinal zonation of climate exerting control of these Early Tertiary forests.

Potbury (1935) and McGinitie (1941) indicate that subtropical to warm-temperate Eocene forests existed on the western slopes of the Sierra Nevada. Axelrod (1949, 1950) reports a warm-temperate to subtropical flora once thought to be Oligocene but now placed in Early Miocene (Everndan and James 1964), the Sutro flora of western Nevada. Axelrod (1949, 1950, 1966) suggests that during the Eocene-Oligocene-Early Miocene at altitudes above 1220 m elevation, floras were characterized by montane conifer forests; areas around 900 m elevation were characterized by conifer-deciduous hardwood forests, and areas near 300 m elevation were characterized by deciduous hardwood forests.

Numerous floras of Miocene age (Everndan and James 1964) have been reported from the Great Basin and adjacent areas: Upper Cedarville flora (LaMotte 1936), Esmeralda flora (Berry 1927; Axelrod 1940), Alvord Creek flora, Sutro flora, Pyramid flora, Trapper Creek, Chlorophagus flora, Middle-gate flora (Axelrod 1944, 1956, 1957, 1964, 1976), Succor Creek flora (Chaney and Axelrod 1959; Graham 1965), Trout Creek flora (McGinitie 1933), and Fingerrock Wash flora (Wolfe 1964).

Chaney (1940, 1947) thought that the floras of the Early Miocene showed a shift of subtropical forests southward and a temperate forest occupying Oregon and the Great Basin. Axelrod (1950) has postulated two completely different geofloras in the Great Basin during the Miocene. A hardwood-deciduous and conifer forest of the Arcto-Tertiary geoflora was present in the lowlands of the Great Basin, while the Madro-Tertiary geoflora, made up of semiarid oak woodland, chaparral, thorn forest, and semidesert vegetation, was found in the southern Great Basin. The Arcto-Tertiary geoflora was a temperate flora with a suggested rainfall of 890 to 1270 millimeters (mm) distributed over the year. The Madro-Tertiary geoflora was similar to that flora now existing in the semiarid and arid parts of the southwestern United States. They existed in a climate of hot summers and mild winters, with annual rainfall of 380 to 630 mm (Axelrod 1950).

By Late Miocene and Early Pliocene, Madro-Tertiary plants numerically increased in the central and northern Great Basin, eliminating the Arcto-Tertiary forests from the slopes and leaving dominant oak, conifer woodland, and chaparral over the lowlands (Axelrod 1950). The increased elevation of the Sierra Nevada and Cascade ranges by the end of the Miocene caused a decreased amount of moisture in the northern portions of the Great Basin, which led to the development of a xeric woodland association in western Nevada and eastern Oregon (Chaney 1944b; Wolfe 1964).

Numerous Pliocene floras (Everndan and James 1964) have been reported from the Great Basin and adjacent areas: the sequoia-dendron forest, Alturus flora, Verdi flora, Aldrich Station flora, Fallon flora (Axelrod 1944b, 1956, 1958, 1962, 1976), Late Pliocene floras east of the Sierra Nevada (Axelrod and Ting 1960), and Cache Valley flora (Brown 1949). During the Pliocene the Sierra Nevada underwent its greatest uplift (Louderback 1907, 1924; but see Lovejoy 1969 for contrasting points of view). Now with the Sierra Nevada block uplifted some 1500 to 1800 m in the north and 2300 to 2500 m in the south, the rain shadow that was created exerted a greater and greater influence. The Great Basin as a whole was also disrupted by a series of extensive block faults (Louderback 1923, 1924, 1926; Nolan 1943; King 1959; Morrison 1965), which created interior mountain ranges. In the Early Pliocene, temperatures were distinctly higher and more uniform over the year (Antevs 1952), but by Middle Pliocene time, precipitation was greatly reduced to that found currently within the Great Basin, with a large contrast between summer and winter temperatures. The Late Pliocene saw temperatures similar to those of the present, but then a cooling trend began, leading to the ice ages of the Pleistocene.

Axelrod (1940, 1948, 1950, 1976) indicates that the flora of the Great Basin was one of a savanna and grassland assemblage, with riparian communities composed of *Acer, Populus, Prunus,* and *Salix.* The central portion of the Great Basin was a semiarid woodland containing *Juglans, Purshia, Rhus,* and live oak communities, along with chaparral communities that included *Arctostaphylos, Cercocarpus, Ceanothus, Mahonia, and Prunus.*

Quaternary Period

The flora of the Great Basin at the beginning of the Pleistocene was essentially the same as it is today. The Pleistocene epoch was a time of four major glacial advances, three of which can be recognized in the Great Basin (Flint 1971; Morrison 1964, 1965).

Well-preserved geomorphic features, glacial moraines, cirques, and other glacial deposits show that glaciation was extensive on the eastern flank of the Sierra Nevada, in the Ruby Mountains, on the Snake Range and several other ranges of the central Great Basin, and in the Unita and Wasatch mountains of central and northeastern Utah. During this time, a series of pluvial lakes formed within the various valleys, with Lake Bonneville in western Utah being the largest and Lake Lahontan in northwestern Nevada being somewhat smaller (Antevs 1948, 1952; Birkeland 1962; Birman 1964; Blackwelder 1931, 1934; Flint 1971; Gilbert 1890; Houghton 1976; Morrison 1964, 1965; Putnam 1950; Richmond 1960, 1961, 1962, 1964; Russell 1883, 1885, 1897; Sharp 1938, 1940; Sharp and Birman 1963).

Toward the end of the Pleistocene, the climate became drier, the glaciers melted, streams and rivers ceased or decreased their flow, and the evaporation of the lakes exceeded the inflow. The high point in the pluvial period ended 8000 to 12,000 years ago (Russell 1885; Morrison 1964, 1965; Broecker and Orr 1958).

Modern Ecosystems

Within the Great Basin there are a number of plant ecosystems. Their composition is a function of temperature, latitude, elevation, rainfall, and geology. A general classification system is shown in table 2-4. However, there are numerous local exceptions (Vreeland et al. 1979). Since the advent of Euro-Americans in the nineteenth century, overgrazing and other land-altering activities have caused various changes in several plant communities (Young et al. 1974). The impact of overgrazing is graphically illustrated by contrasting the ecosystems on the U.S. Department of Energy's Nevada Test Site in southern Nevada (O'Farrell and Emery 1976) with those in adjacent areas. The test site has been closed to grazing for over forty years. The most notable ecological difference is the ratio of sagebrush (*Artemisia* spp.) and shadscale (*Artiplex* sp.) to grasses and other plant forms, the former being much lower on the test site than outside it.

As population and land-altering activities increase, plant communities continue to change. The current status of flora and fauna are documented by Lugaski et al. (1979), H. Vreeland et al. (1978), and P. Vreeland et al. (1979).

The classifications shown in the table generally are consistent with and follow the guidelines of Billings (1951). Modi-

TABLE 2-4
Great Basin terrestrial ecosystems

ALPINE ECOSYSTEM
    Sierra Alpine Zone
    Basin Range Alpine Zone
        Limber Pine-Bristlecone Pine Community
        Alpine Tundra Community
    Wasatch Alpine Zone

WESTERN PINE-SPRUCE-FIR-DECIDUOUS-FOREST ECOSYSTEM
    Wasatch Zone
        Pine-Deciduous Community
        White Fir-Douglas Fir-Blue Spruce Community
        Englemann Spruce-Subalpine Fir Community
    Basin Range Zone
        Mountain Mahogany Community
        Upper Sagebrush Community
    Sierran Zone
        Pine-Fir Community
        Red Fir Community
        Lodgepole-Pine-Mountain Hemlock Community
        White Bark Pine Community

PINYON-JUNIPER ECOSYSTEM
    1. Pinyon-Juniper Community

DESERT SHRUB ECOSYSTEM
    Northern Desert Shrub Zone
        Sagebrush Community
        Rabbit Brush Community
        Shadscale Community
        Winter Fat Community
        Hop Sage-Coleogyne Community
        Mat Saltbush Community
        Gray Molly Community

    Salt Desert Shrub Zone
        Greasewood Community
        Greasewood-Shadscale Community
        Saline-Alkaline Community
        Rabbitbrush Community

    Southern Desert Shrub Zone
        Desert Saltbush Community
        Creosote Bush Community
        Joshua Tree Community
        Mesquite Community
    Transition Desert Shrub Zone
        Creosote Bush-Boxthorn Community
        Hop Sage-Boxthorn Community
        Boxthorn-Atriplex Community
        Artemisia Community

fications discussed are further detailed in Lugaski, Vreeland, and Vreeland (in preparation). Except in a few rare instances, there are no sharp dividing lines between or among the various categories. Most taxonomic references are in accordance with Cronquist et al. (1972). Where there is substantial disagreement in the literature as to the proper scientific name, or when there has been a recent change in nomenclature, the alternate or old name is shown in parentheses. During the past century, many botanists and ecologists have studied the Great Basin (Reifschneider 1964). General ecological overviews of the Great Basin are contained in Wheeler (1967, 1971, 1978).

Alpine Ecosystem

Great Basin ecosystems are conditioned by the topographical features, air and soil temperatures, wind movement, precipitation, evaporation, and length of growing season. In the alpine ecosystem, the frost-free growing season seldom exceeds ninety days (Sampson 1925).

Although called the "timberless zone" by Sampson (1925), Billings (1978) stresses the fallacy of this, noting that frequently alpine flora can be found in pockets at lower elevations. Generally, however, alpine ecosystems within the Great Basin are confined to the highest mountain peaks. Shallow soil and few frost-free nights make these zones unsuitable for all but the hardiest tree species. Those that do occur are stunted and prostrate.

SIERRA ALPINE ZONE

This zone is found above the timberline of the Sierra Nevada, often around permanent snowbanks. Temperatures within this zone are generally warmer than on many of the basin ranges at a comparable altitude. Precipitation, primarily in the form of snow, is much greater. Soil is poor and shallow, covered mostly by summit screes and alpine fell-fields. Shrub associations are well developed, occupying open areas above red fir stands (Smith 1973); however, perennial herbaceous vegetation is generally dominant.

The Sierra Alpine Community consists mostly of herbaceous vegetation occurring in rocky fell-fields above the timber line. Commonly occurring species include *Oxyria digyna, Rhodiola integrifolia, Ranunculus oxynotus, Primula suffrutescens, Polemonium pulcherrimum, Penstemon davidacnii, Hulsca algida,* and *Erigeron trifidus* (Billings 1951). Kennedy (1911, 1912), lists *Phlox dejecta, Cassiope mertensiana, Polemonium montrosense, Salix caespitosa, Elephantella attolens, Mimulus implexus, Hulsea caespitosa, Ribes inebrians,* and *Rhodiola integrifolia.* These plants share a low, matlike appearance and deep roots that penetrate the rocky crevices on which they occur. Smith (1973) reports *Muhlenbergia richardsonis, Carex*

helleri, Juncus parryi, Draba densifolia, Antennaria alpina, and *Phacelia frigida* in alpine fell-fields in the Lake Tahoe Basin. Shrubs include *Arctostaphylos patula, Ceanothus velutinus, Chrysolepis sempervirens,* and *Quercus vaccinifolia.*

BASIN-RANGE ALPINE ZONE

On the some 160 mountain ranges in the Great Basin, there are a variety of "alpine islands" (Billings 1978), unlike floristically as well as ecologically. Many species occurring in the islands are endemic to a particular peak, making it difficult to classify individual synusia. However, for simplification, two communities will be considered representative of this zone: the limber pine-bristlecone pine community and the alpine tundra community.

Moisture and temperature are two key environmental factors affecting the Basin Range Alpine Zone. Annual precipitation in this zone is generally less than in the Sierra Alpine Zone. Pacific storms drop most of their moisture when crossing the Sierra Nevada, leaving the interior basin ranges largely dependent upon convectional summer rains, although some peaks of the basin ranges are sufficiently high to catch moisture from the Pacific storm tracts. Winter temperatures characteristic of many basin ranges are colder than the Sierra Nevada.

LIMBER PINE-BRISTLECONE PINE COMMUNITY

This community is generally located below 3050 m in the Basin Range Alpine Zone. An open subalpine forest species, *Pinus flexilis* (limber pine), is dominant in the northern ranges and at lower elevations, although Critchfield and Allenbaugh (1969) state that this species is rare or absent from several ranges in northern Nevada and southern Idaho. *Pinus logaeva* (aristata) (bristlecone pine) becomes more prevalent at higher elevations in southern ranges (Billings 1951). Stands in the central Great Basin may contain either or both species and may be extensive or patchy.

Distribution and slow rate of growth of the long-lived bristlecone pine may be limited in part by the short growing season at high elevations. Schulze et al. (1967) found that net photosyntheses declined sharply in winter while respiration continued at a high rate. The resulting negative CO_2 balance for this species was not corrected until considerable summertime photosynthesis had taken place. Only after this deficit had been made up was the plant able to produce new growth. As elevation increases and temperatures decline, the growing season becomes shorter; thus Schulze et al. (1967) believe that the upper distributional limit of this species may be controlled by the plant's inability to compensate for the large CO_2 deficit.

Clokey (1951) reports a pure stand of bristlecone pine marking the timberline in the Charleston Mountains, Clark County, Nevada. At lower elevations of this range, *Pinus flexilis* is an associated tree species. Bradley and Deacon (1965) have found *Abies concolor* (white fir) and *Juniperus communis* (juniper) to occur together with limber pine and bristlecone pine at lower elevations of the Sheep and Spring ranges of southern Nevada. Limber pine also forms pure stands at lower elevations in the Charleston Mountains (Clokey 1951).

In the White Mountains, large stands of the limber pine-bristlecone pine community occur. Bristlecone pine forest appears to be best developed on soils that are dolomitic in origin (Lloyd and Mitchell 1973). On soils of this type, sagebrush and other understory shrubs are generally absent. If shrubs are present, the bristlecone forest is less dense. Associated shrubs and forbs include *Erigeron clokeyi, Arenaria kingii, Eriogonum gracilipes, Haplopappus acaulis, H. suffruticosus,* and *Chrysothamus viscidiflorus.* Limber pine in the White Mountains occurs most abundantly on granitic soils where bristlecones sparsely appear. Limber pine is generally confined to elevations below 3350 m, while bristlecone forests may extend to 3500 m.

In central Nevada, limber pine stands in the Toiyabe Range occur between 2900 and 3350 m on northwest facing slopes. Much of this range had been logged, and these reported stands consist mostly of second-growth timber (Linsdale et al. 1952).

Limber pine is reported as "the prevalent pine in the East Humboldt (Ruby) Mountains, Nevada, and frequent in the Wahsatch [sic] and Uintas; [1980 m] to [3350 m] in altitude" (Heller 1910). No mention is made of bristlecone pine or other associated species from this early sighting. A reported occurrence of bristlecone pine in 1879 is found in Billings (1954). Stands on the slopes of Prospect Mountain near Eureka, Nevada, were logged extensively for timber to be used in nearby mining operations.

ALPINE TUNDRA COMMUNITY

At elevations above 3350 m, this community consists of low shrubs and perennial herbaceous species. Examples within the Great Basin may be found in the Toiyabe, White, and Ruby mountains, the Deep Creek Mountains, and the Snake Range (Billings 1951). Commonly found species in this community type include *Phacelia alpina, Ivesia gordonii, Draba oligosperma, Eriogonum neglectum* and *Hulsea nana* (Sampson 1925) and *Astragalus platytropis, Castilleja lapidicola, Erigeron* sp., *Primula parryi, P. nevadensis,* and *Saxifraga caespitosa.* Grasses and sedges include *Agropyron scribneri, Carex* sp., *Deschampsia cespitosa, Festuca ovina,* and *Poa rupicola.* According to Cronquist (1972), community composition within the Alpine Tundra varies more than in any other community in the Great Basin. Many species of *Primula* and *Eriogonum* are endemic to their particular ranges.

In the White Mountains, alpine tundra occurs both above and below the timberline (Lloyd and Mitchell 1973). Sparse cover on dolomitic soils consists mostly of *Eriogonum gracilipes, Erigeron pygmaeus,* and *Phlox covillei.* On granitic substrates, tundra communities are more dense, dominated by

Trifolium monoense and *Eriogonum ovalifolium,* with associated grass and sedge species.

Went (1964) lists twenty-six species occurring above the timberline in the Charleston Mountains of southern Nevada. Forty-five percent of these plants are endemic. Germination studies conducted on selected species by Went (1964) show that the restriction of plant occurrence to the alpine zone depends upon the required low temperatures and long chilling of seeds resulting from the short growing season.

Although considered "pseudoalpine" by Bradley and Deacon (1965), because of lack of distinctive alpine vegetation on Charleston, Hayford, and Sheep peaks of southern Nevada, they are here categorized in the Alpine Tundra Community. Clokey (1951) terms these areas "above timber line." The low-growing shrubs, forbs, and grasses scattered over these exposed areas above 3050 m are alpine in an ecological sense, if not in true floristic content. At the same elevation in more protected sites on these peaks, bristlecone pines can be found (Bradley and Deacon 1965).

WASATCH ALPINE ZONE

Separated from each other by low-lying deserts, islands of alpine tundra can be found on high mountain peaks of the Wasatch region. According to Cronquist et al. (1972), this zone contains few endemic species, in contrast to the many endemics found in the Basin Range Alpine Zone. Alpine tundra can be found at elevations above 3200 m in the Wasatch Mountains, above 3660 m in the Utah Plateaus, and over 3960 m in the LaSal Range in southeastern Utah. A short growing season, freezing night temperatures, and poor, shallow soils render this an unfavorable environment for trees. Low-growing shrubs, forbs, and grasses are the principal components of the Wasatch Alpine Zone. In general, Wasatch Alpine Zone flora is similar to the alpine tundra of the central Rocky Mountains (Billings 1951).

WASATCH ALPINE COMMUNITY

This community type, consisting of low-growing woody and herbaceous species, is located above the timberline of high peaks in the Wasatch and LaSal ranges and the Utah plateaus. Rocky areas are often characterized by *Aquilegia scopulorum, Castilleja applegatei, Cirsium* sp., *Erigeron leiomerum, Fransera speciosa, Potentilla crinita,* and *Ribes cereum.* Dry meadow areas contain grasses, sedges, rushes, and forbs, which include *Agropyron scribneri, Carex* sp., *Festuca ovina, Juncus drummondii, Leucopoa kingii, Aquilegia scopulorum, Arenaria* sp., *Clematis pseudoalpina, Erigeron* sp., (*Lewisia pygmaea, Penstemon* sp., *Phlox* sp., and *Selaginella densa.* Moist areas around melting snowbanks support different species such as *Dodecatheon alpinum, Lupinus argenteus, Polygonum bistortoides, Primula parryi,* and *Viola nephrophylla* (Cronquist et al. 1972).

Western Pine-Spruce-Fir-Deciduous Forest Ecosystem

This ecosystem can be found on the Wasatch and Sierra Nevada mountain ranges marking, respectively, eastern and western boundaries of the Great Basin and also on the many basin ranges lying between. These highlands receive considerably more precipitation and colder temperatures than do the communities of the Desert Shrub Ecosystem. Although three separate zones are considered in this ecosystem, there is considerable species overlap between and among them. Occasionally two zones may occur on the same range, especially toward the extreme eastern and western limits of the Great Basin. For example, both Basin Range and Sierran Zones can be found on the Carson Range, located just east of the Sierra Nevada.

WASATCH ZONE

Floristically, this zone is similar to synusia of the Rocky Mountains. Three community types are representative of this zone: pine-deciduous; white fir-douglas fir-blue spruce; and Englemann spruce-subalpine fir communities. General altitudinal ranges will be given with each community description; however, these vary depending on slope direction. Geographically the Wasatch Zone can be found along the eastern boundary of the Great Basin, on the Wasatch Range and the Wasatch Plateau, although partial community extensions occur on neighboring mountains.

PINE-DECIDUOUS FOREST COMMUNITY

This community can be found in the Wasatch Range at elevations of about 1525 to 2290 m where the desert shrub communities are replaced by deciduous and semideciduous chaparral woodland. Dominant woody species include *Quercus gambelii* (gambel oak), *Acer grandidentatum* (big-tooth maple), and *A. glabrum*. *Pinus ponderosa* (ponderosa pine) is an important community component in southernmost areas where it gradually replaces the chaparral. Ponderosa pine can also be found in areas of hydrothermally altered rock together with gambel oak (Salisbury 1964). Because of its small size, gambel oak lacks value as an ornamental, but Indians found the scrubby tree useful for medicinal reasons. They made a lotion from boiled bark and used it to treat skin sores or took it internally for gastrointestinal disorders and malaria (Krochmal and Krochmal 1973). *Cercocarpus ledifolium* (mountain mahogany) is frequent throughout this community. Common shrubs include *Artemisia arbuscula*, *A. tridentata*, *Ceonothus velutinus*, *Prunus virginiana*, *Purshia tridentata*, *Ribes cereum*, *Rosa woodsii*, *Sambucus cerulea*, and *Symphoricurpos oreophilus*. Frequently occurring herbaceous species include *Agropyron* sp., *Delphinium nelsonii*, *Elymus cinereus*, *Geranium fremontii*, *Lithophyagma parviflora*, *Poa pratensis*, and *Wyethia amplexicaulis* (Cronquist et al. 1972).

WHITE FIR-DOUGLAS FIR-BLUE SPRUCE COMMUNITY

This community, found in elevations between 2290 and 2900 m, receives more precipitation than any other in the Wasatch Zone (Cronquist 1972). As a result, a dense coniferous forest consisting primarily of *Abies concolor* (white fir), *Pseudotsuga menziesii* (Douglas fir), and *Picea pungens* (blue spruce) is the dominant vegetation. Douglas fir, the principal dominant, is present in drier sites, while white fir and blue spruce generally can be found in moister locations. Along streams, *Populus tremuloides* (trembling aspen) often forms pure stands together with *Acer glabrum*, *Betula occidentalis*, and *Salix* sp. *Pinus ponderosa* (ponderosa pine) often forms an important community component at lower elevations, while *Pinus flexilis* can be found on dry, exposed slopes at the upper community limit. *Pinus contorta* var. *latifolia* (lodgepole pine) is present in the northern Wasatch range.

Because of forest density and consequent limited light penetration, understory vegetation is generally sparse. Common shrubs in this community include *Amelanchier alnifolia*, *Berberis repens*, *Ceanothus fendleri*, *Holodiscus dumosus*, *Pachistima myrsinites*, *Physocarpus malvaceus*, *Rubus parviflorus*, *Sorbus scopulina*, and *Symphoricarpos oreophilus*. Forbs and grasses include *Agropyron spicatum*, *Bromus marginatus*, *Poa pratensis*, *Antennaria* sp., *Castilleja miniata*, *Delphinium occidentale*, *Gilia aggregata*, *Monardella odoratissima*, *Potentilla* sp., *Rudbeckia occidentalis*, *Scrophularia lanceolata*, and *Smilacina* sp.

ENGLEMANN SPRUCE-SUBALPINE FIR COMMUNITY

Extending from 2900 to 3200 m, this subalpine community may be either dense woodland or alternate with grassy meadows and shrubby vegetation (Billings 1951). *Picea engelmannii* (Englemann spruce) and *Abies lasiocarpa* (subalpine fir) are the dominant trees; other coniferous components include *Pinus flexilis* on dry, rocky sites and *Pinus longaeva* at higher elevations. Englemann spruce is the common timberline tree, often becoming stunted krummholz. *Populus tremuloides* extends into this community from the White Fir-Douglas Fir-Blue Spruce Community below. Common shrubs include *Ribes montigenum*, *Artemisia tridentata*, *Chrysothamnus viscidiflorus*, *Juniperus communis*, *Sambucus racemosa*, and *Symphoricarpos oreophilus*. Examples of herbaceous vegetation that occur in this community are *Agropyron* sp., *Calamagrostic canadensis*, *Deschampsia cespitosa*, *Festuca* sp., *Poa* sp., Muhlenbergia sp., *Stipa columbiana*, *Achillea millefolium*, *Agoseris* sp., *Artemisia frigida*, *Aster foliaceus*, *Castilleja* sp., *Circium* sp., *Erigeron* sp., *Frasera speciosa*, *Lupinus* sp., *Penstemon rydbergii*, *Trifolium* sp., and *Viola* sp.

The open meadowland of this community contains several

species of shrub common to the Northern Desert Shrub Communities. These plus the many palatable grasses have resulted in heavy overgrazing of these areas by cattle and sheep and the consequent erosion and flash-flooding problems in the canyons below (Billings 1951). Overgrazing of the native flora has also brought about the intrusion of undesirable species of little or no forage value. The density of *Artemisia tridentata* and *Chrysothamnus viscidiflorus* in overgrazed areas has increased, while more palatable shrubs have decreased according to studies by Cottman and Evans (1945).

BASIN-RANGE ZONE

Situated eastward from the Sierran Zone, a series of mountains ranging in elevation from 1525 to over 3350 m extend in a north-south direction across the Great Basin to west-central Utah. These basin ranges receive less precipitation than the Sierra Nevada and consequently support different vegetation communities. However, certain coniferous species that are also found in the Sierran Zone do occur: *Pinus jeffreyi*, *P. Ponderosa*, *P. contorta*, *P. monticola*, and *Abies concolor* (Billings 1954; Little 1956). Two community types are associated with the Basin Range Zone: Mountain Mahogany and Upper Sagebrush communities.

MOUNTAIN MAHOGANY COMMUNITY

Cercocarpus ledifolius (mountain mahogany) may occur in dense stands between 2440 and 3050 m. Although this scrubby tree is frequently found in conjunction with shrubs of the genera *Symphoricarpos* and *Holodiscus*, it can be often found in pure stands on the drier ridges, slopes, and gravelly areas. When more moisture is present, mountain mahogany is replaced by *Pinus ponderosa* as the dominant tree (Clokey 1951). Clokey (1951) and Medin (1960) consider mountain mahogany to be a good indicator of considerable soil depth.

According to Linsdale et al. (1952) mountain mahogany is the principal tree on the eastern slopes of the Toiyabe Range in central Nevada from 2290 to 2900 m. Stands also occur in the Charleston Mountains in southern Nevada (Clokey 1951), the White Mountains of the west-central part of Nevada (Lloyd and Mitchell 1973), and the Ruby Mountains of eastern Nevada (Heller 1909), in addition to most mountain ranges within the Great Basin. At lower elevations, *Pinus monophylla* is an associated tree species. Trees on the Toiyabe Range average 5.5 m in height (Linsdale et al. 1952), although they can attain heights of up to 10.5 m (Billings 1954). Generally mountain mahogany is found where there is some wind protection. Thick litter under dense stands often prevents the growth of most herbaceous species. The very hard heartwood of this species is used for carving and fuel (Billings 1954). Certain species provide excellent browse for livestock and winter forage for big game animals (Medin 1960).

UPPER SAGEBRUSH COMMUNITY

This community is dominated by shrubs and herbs, giving many mountain peaks a bald appearance throughout the Basin Range Zone. The absence of trees in these areas may be due to climatic shifts (Billings and Mark 1957) or drying trends (Vasek 1966). *Artemisia tridentata* subsp. *vaseyana* (sagebrush) and other shrubs extend to elevations of over 3050 m, exhibiting a greater range of elevational tolerances than many species of trees. The climate at these elevations is much colder than that of the low-lying desert sagebrush communities. Precipitation is generally two to three times the amount received in lower elevations (Billings 1951). Vegetation in the Upper Sagebrush communities is generally denser than sagebrush communities found at lower elevations. Associated woody species include *Holodiscus dumosus*, *Symphoricarpos* sp., and *Cerocarpus ledifolius* (Billings 1951). Occasionally *Populus tremuloides* may be found in wet areas of this community along with many species of perennial forbs and grasses.

SIERRAN ZONE

The Sierran Zone of the Western Pine-Spruce-Fir-Deciduous Forest Ecosystem marks the western boundary of the Great Basin. It is located along the eastern slopes of the Sierra Nevada and the nearby Carson Range.

Four major community types can be found within this zone. Species overlap is not uncommon, not only with trees but also with shrubs and forbs. Woods and Brock (1964) consider a forest as a functional ecological unit, due largely to the interrelationships of its many mycorrhizal components. One such species, *Sarcodes sanguinea*, occurs in three of the four communities within this zone (P. Vreeland et al. 1976), resulting perhaps from the overlap of its associated autotrophic species.

YELLOW PINE-WHITE FIR COMMUNITY

This community extends from about 1525 to 2290 m and is characterized by *Pinus ponderosa*, *P. jeffreyi* (both considered yellow pine), and *Abies concolor* (white fir) as principal dominants, with *P. lambertiana* and *Calocedrus decurrens* commonly associated tree species. Understory shrubs found in this community include the genera *Ceanothus*, *Arctostaphylos*, and *Quercus*. Ponderosa pine, considered by Billings (1954) to be the most important western timber tree, is widely distributed from Canada to Mexico. It can be found in eastern, southern, and western Nevada at elevations ranging from 2240 to 2750 m. Areas of hydrothermally altered rock in the Carson Range are often occupied solely by *Pinus ponderosa*, with no understory present (Billings 1950).

Once considered to be a variety of the ponderosa pine, the jeffrey pine is a separate and distinct species (Heller 1912). *Pinus jeffreyi* is the predominant yellow pine in western

Nevada. The range of the jeffrey pine is not as extensive as ponderosa pine; it extends only from southern Oregon to lower California (Billings 1954). In Nevada, it is confined to the western portion of the state from southern Washoe to Mineral counties at elevations from 1525 to 2380 m.

The *Pinus washoensis* (Washoe pine) bears a close resemblance to both the ponderosa and the jeffrey pines and is considered to be a possible hybrid of the two (Billings 1954). The type specimen was located on Mount Rose in the Sierra Nevada in 1938 by Mason and Stockwell (1945) in a small grove, elevation approximately 2440 m.

The range of the white fir, one of the most common trees in mixed coniferous forests of the Sierra Nevada, is from southern Oregon to Mexico. *Abies concolor* occurs at 1830 to 2590 m elevation in eastern and western Nevada. The eastern trees are characterized by shorter needles than are those found in the Sierra Nevada (Billings 1954).

RED FIR COMMUNITY

Extending above the Yellow Pine-White Fir Community, this community type is normally found at elevations between 2285 and 2590 m (Billings 1954), although Smith (1973) reports red fir at the 1950 m elevation in the Tahoe Basin. The dominant tree, *Abies magnifica* (red fir), is the largest true fir in America (Billings 1954). This species, along with frequently occurring subdominants *Pinus contorta* (*murrayana*) and *Pinus monticola*, can be found in areas of deep soil and gentle slope. *Populus tremuloides* stands can be found in moist areas within this community (Critchfield and Allenbaugh 1969). Associated understory plants include *Amelanchier pallida*, *Spiraea densiflora*, and *Ceanothus cordulatus*.

LODGEPOLE PINE-MOUNTAIN HEMLOCK COMMUNITY

The dominant species, *Pinus contorta* (*murrayana*) (lodgepole pine) and *Tsuga mertensiana* (mountain hemlock), can be found from 1830 to 2830 m, often forming dense stands around subalpine meadows. This community is replaced by the Yellow Pine-White Fir Community as slope and elevation increase. *Salis, Prunus,* and *Ribes* are common genera found in the understory. *Pinus monticola* (western white pine) and *Pinus albicaulis* (whitebark pine) are frequently associated tree species (Billings 1954). In the Great Basin, lodgepole pine is confined to the Sierra Nevada, the neighboring Carson Range and Virginia mountains, and the Brawley Peaks area of Mineral County, Nevada. Mountain hemlock is a common tree at 8000 to 9000 feet in the Sierra Nevada, occurring in abundance on Slide Mountain and Mount Rose in southern Washoe County, Nevada (Billings 1954).

WHITE BARK PINE COMMUNITY

This community occurs between 2600 and 3200 m and is characterized by open stands of *Pinus albicaulis* (white bark pine) with a dense understory of herbaceous and shrubby vegetation, including *Penstemon, Ribes,* and *Epilobium.* Associated tree species include *Pinus flexilis, P. contorta,* and *Tsuga mertensiana.* As elevation increases, these species often become stunted and prostrate.

The presence of *P. albicaulis* has been often overlooked in areas of the Great Basin because of its similarity with *P. flexilis.* Although it had been reported in several mountain ranges of northern Nevada for seventy years or so, recent collections have confirmed its presence (Critchfield and Allenbaugh 1969). The species also occurs in the Ruby Mountains of eastern Nevada, Cedarhill Canyon near Virginia City, Mount Grant in Mineral County, and Pine Mountain in Humboldt County, Nevada (Billings 1954). White bark pine is the timberline tree of the northern Sierra Nevada.

Pinyon-Juniper Ecosystem

This ecosystem is considered the dominant forest type with respect to area in the intermountain region (Cronquist et al. 1972; West et al. 1978). Elevation ranges from 1525 to 2440 m for the Pinyon-Juniper Ecosystem, although these limits vary greatly. Precipitation is generally in excess of 300 mm per year. Only one zone is representative of this ecosystem type in the Great Basin; however, species diversity is considerable (Harner and Harper 1976). Various external factors, such as fire and grazing patterns, may have had a substantial effect as to the distribution of this ecosystem in Nevada (Blackburn and Tueller 1970). The topography in which this community occurs is varied, ranging from gentle, rolling hills to steep mountain slopes, rocky canyons, and narrow ridges. Soil is usually well-drained sandy loam.

PINYON-JUNIPER ZONE

This zone occupies major portions of the Great Basin area and is found in Utah, Arizona, and Nevada (Odum 1971). In general, the elevation of this zone is highest in the west-central portion of the Great Basin and declines both westward toward the Sierra Nevada and eastward to the Wasatch front-high plateaus (West et al. 1978). The northern limit of this zone is generally considered south of the Humboldt River in Nevada, although east of the Nevada-Utah border it extends north to the Raft River Mountains and southern Idaho (Critchfield and Allenbaugh 1969). West et al. (1978) attribute the absence of this zone from northwestern Nevada to be largely a factor of unstable temperature inversions over this area due to Pacific frontal systems. Approximately 20 percent of the state of Utah and northern Arizona foothills are covered by the Pinyon-Juniper Zone (Woodbury 1947).

One phytocoenosis, the Pinyon-Juniper Community, comprises this zone. Considered "pygmy conifers" by Woodbury (1947), this community consists of open stands of low, scrubby conifers dominated by *Pinus monophylla* (pinyon) and several species of *Juniperus* (juniper). While both pinyon and juniper are commonly present in this association, juniper is usually more abundant at lower elevations, and pinyon becomes more numerous as elevation increases. *Cercocarpus ledifolius* can be frequently associated in higher elevations while *Betula occidentalis, Salix* sp., and *Populus* sp. occur in wetter areas of this community type. *Artemesia tridentata* is the dominant shrub in this community throughout much of the Great Basin, although other important shrubs include the genera *Chrysothamnus, Ephedra, Ribes, Purshia, Ceanothus Tetradymia, Sambucus,* and *Symphoricarpos.* In southern Nevada *Yucca baccata* and *Agave utahensis* are commonly associated woody species (Bradley and Deacon 1965). *Quercus* is an important woody component in Utah. Many species of forb and grass are also present. Although the lower limit of this community is generally around 1525 m, this community may contain *Pseudotsuga menziesii* var. *glauca* in western Utah, along with *Abies concolor* and *Pinus ponderosa.* Disturbance by fire can cause the occurrence of *Cowania mexicana, Falugia paradoxia, Quercus gambelli, Q. turbinella, Ameranchier utahensis, Prunus fasciculata,* and *Arctostaphylos pungens* (Bradley and Deacon 1965). *Juniperus* has been said to invade sagebrush communities in lower elevations as a result of overgrazing (Cottam and Stewart 1940). Pinyon-juniper vegetation is considered unimportant economically but provides erosion protection, wood for fencing and fuel, and pine nuts, a major component of the diet of some western Indian tribes.

Desert Shrub Ecosystem

According to Whittaker (1970) deserts can be divided into categories based on a number of environmental factors, such as temperature, precipitation, latitude, and altitude and the resulting different vegetational groupings. The Desert Shrub Ecosystem is the category describing the overall physiognomy of much of the Great Basin Desert. Based largely upon geographic location, four zones are considered within this ecosystem: Northern Desert Shrub Zone, Salt Desert Shrub Zone, Southern Desert Shrub Zone, and Transition Desert Shrub Zone.

NORTHERN DESERT SHRUB ZONE

This zone consists of seven distinct community types based upon the dominant species: sagebrush, rabbitbrush, shadscale, winter fat, hop sage-coloegyne, mat saltbrush, and gray molly (figure 2-19). According to Clements (1920) most of the

Figure 2-19. Northern desert shrub ecosystem near Winnemucca, Nevada.
Photograph by H. Vreeland and P. Vreeland.

dominant shrubs in this zone belong to the families Asteraceae and Chenopodiaceae. This northern desert zone can be considered cool-temperate desert scrub (Whittaker 1970) or semi-desert. Precipitation is generally greater than the other zones within the Desert Shrub Ecosystem. Temperature ranges between 46° C in summer to −24° C in winter, with diurnal fluctuations frequently between 15 and 18 degrees (Billings 1951). The topography of this zone is dominated by many relatively small basins separated from each other by faultblock mountain ranges in the northern portion of the Great Basin. The seven community types found in this zone frequently overlap and dovetail as environmental factors vary, resulting in a mosaic of synusia.

Overgrazing has been a major factor affecting species composition in many communities within this zone. Bunch grass, once a major nonwoody component throughout much of this area, is replaced by less palatable grasses and shrubs. Robertson (1947), Tueller (1973), and others offer solutions for restitution of this valuable range land.

SAGEBRUSH COMMUNITY

Dominated by *Artemisia tridentata* (big sagebrush), this community is considered the climatic climax of desert areas with annual precipitation in excess of 175 mm (Cronquist et al. 1972). This community replaces the shadscale community as elevation increases above 1370 m in the north and 1675 m in the southern areas. *A. tridentata* is less drought resistant and salt tolerant than shadscale community vegetation (Billings 1951). As a result, the sagebrush community stretches over vast expanses of the northern Great Basin, covering both valleys and mountain ranges. According to Beetle (1960), *A. tridentata* communities cover approximately 60.7 million hectares in the western United States. This community occupies more area within the Great Basin than does any other vegetation type (Billings 1951).

The typical sagebrush community consists of nonspiny shrubs from 0.3 to 2.0 m high, perennial and annual grasses, and forbs. In addition to *Artemisia*, woody genera present include *Chrysothamnus, Ephedra, Purshia, Ribes, Symphoricarpos,* and *Tetradymia.* Shrubs account for approximately 20 percent of the ground cover in this community (Billings 1951).

Perennial grasses, an important community component, include *Agropyron, Oryzopsis, Poa, Sitanion,* and *Stipa.* Numerous studies have shown that overgrazing has greatly reduced the percentage cover of bunchgrass and increased that of *A. tridentata* in portions of this community (Robertson 1947; Cottam 1961; Christensen and Johnson 1964; Driscoll 1964). An alien annual, *Bromus tectorum,* has invaded overgrazed areas also. This less palatable grass has prevented the establishment of more desirable species (Cronquist et al. 1972).

A. tridentata was a useful plant for primitive Indians tribes in the Great Basin. Bark was shredded, twisted, and fashioned into moccasins and other items of clothing. A tea from the leaves was used to treat headache and gastrointestinal disorders. Rheumatism discomforts were eased by applying a poultice of *Artemisia* leaves, and newborn infants were bathed in a wash made from boiling the leaves (Krochmal and Krochmal 1973).

Sagebrush is important winter forage for cattle, sheep, big game animals, and game birds. Grazing is known to increase the density of sagebrush, often at the expense of perennial grasses. The use of fire and herbicides to control the intrusion of sagebrush onto desert range lands is cautioned against by Beetle (1960), who points out that both grasses and sagebrush are important dietary components of many livestock and game animals who browse either or both on winter ranges depending upon winter climatic conditions.

RABBITBRUSH COMMUNITY

Chrysothamnus puberulus (little rabbitbrush) is frequently dominant in large areas above sagebrush communities, often to the exclusion of other vegetation (Shantz 1925). *Chrysothamnus* communities in the Northern Desert Shrub Zone frequently arise as a result of overgrazing or fire in a Sagebrush Community (Young and Evans 1974) because of this shrub's method of propagation by root sprouting. Once established, rabbitbrush communities continue to dominate, preventing the recovery of more palatable bunch grasses and shrubs (Young and Evans 1974). Studies by McKell and Chilcote (1957) indicate that plant vigor, seed production, and shoot elongation of rabbitbrush are greatly increased following removal of competing vegetation; thus the Rabbitbrush Community, once established, is self-perpetuating. According to Billings (1945), rabbitbrush near Pyramid Lake is associated with *Distichlis stricta, Echinopsilon hyssopifolius, Salsola kali, Heliotropium curassivacum,* and other species considered to be of a transitory nature in community succession.

SHADSCALE COMMUNITY

Atriplex confertifolia (shadscale) is the dominant shrub in this community, which occupies many valley bottoms in western Nevada and Utah, perhaps because of this species' high salt tolerance and low moisture requirements (Billings 1949). According to Billings (1949) the largest shadscale community extends from the Carson Desert region of west-central Nevada south and east to the mountains of Death Valley and southern Nevada. Another large shadscale area can be found in western Utah, with many smaller communities extending into eastern Oregon and southern Idaho. *Atriplex* is dominant in the Red Desert (elevation 2130 m), considered by Jaeger (1957) to be outside the Great Basin since it is not an area of interior drainage but drains partially into the Pacific and also into the Atlantic oceans. Other shrubs which occur in conjunction with *Atriplex confertifolia* are *A. spinescens, Chrysothamnus viscidiflorus, Ephedra nevadensis, Ceratoides (Eurotia) lanata, Sar-*

cobatus baileyi, Tetradymia glabrata, Gutierrezia sarothrae, Grayia spinosa, and several other species of *Atriplex.* These low, gray-green shrubs are generally spiny and cover less than 10 percent of the ground area (Cronquist 1972). Forbs and grasses, both annual and perennial, are few and appear only after sufficient precipitation, which averages 114 mm annually in Nevada and almost 203 mm annually in Utah for this community (Billings 1951). *Halogeton glomeratus,* an introduced weed common in this community, is poisonous to sheep and spreads rapidly in disturbed areas. Many of the palatable grasses have been replaced by shrubs as a result of overgrazing.

Although of questionable fertility according to Shantz (1925), soils in the shadscale community can be reclaimed for agriculture. These efforts have gone on for many years. What was once described as "one of the most desolate and arid spots on this continent" (Blanchard 1907) is now a productive agricultural community surrounding Fallon, Nevada. Water for this reclamation was diverted approximately 64 km from the Truckee River to provide irrigation. Although complicated by high salt concentrations in the soil, the area now produces many hay and food crops and supports a sizable community.

Billings (1949) points out the difficulty in classifying the shadscale communities because of the discontinuity of associated species in various areas due to environmental variations. However, he does consider shadscale to be a recognizable community type in that it is a natural area with marked biological characteristics.

WINTER FAT COMMUNITY

Ceratoides (Eurotia) lanata often becomes dominant in areas where shadscale has been killed or removed (Shantz 1925). Frequently confused with gray molly because of morphological similarity, this shrub is found on soils with relatively high subsurface salt content and consequently develops shallow (0.3 to 0.6 m) root systems compared with most woody desert vegetation. North and east of Pyramid Lake, pure stands of winter fat can be found. In such cases, there is no associated vegetation present (Billings 1945). Although this plant provides excellent winter forage for cattle and sheep, it is far less abundant than greasewood or shadscale communities.

HOPSAGE-COLEOGYNE COMMUNITY

This community, dominated by *Grayia spinosa* (hopsage) and *Coleogyne ramossissima,* forms a transition zone between northern and southern desert areas on soils relatively alkali free (Shantz 1925).

MAT SALT BUSH COMMUNITY

On soils with a relatively uniform distribution of high alkali concentrations, *Atriplex corrugata* (mat salt bush) dominates where sagebrush and shadscale are unable to grow (Shantz 1925). This shrub forms a low, matlike cover with great expanses of bare ground between plants. Because of the high alkali concentrations, there are rarely other species present.

GRAY MOLLY COMMUNITY

Areas dominated by *Kochia* sp. (gray molly) are not extensive. *K. americana* codominates with shadscale in the central Bonneville Basin (Cronquist 1972). The shrub tends to occupy soils that are light in color but with a very heavy texture. High alkali concentrations at a depth of about 0.3 m prevent the development of plant roots (Shantz 1925).

SALT DESERT SHRUB ZONE

This zone differs from the other geographically distinguished zones of the Desert Shrub Ecosystem in that it is differentiated based on the high salt concentration in the soil. Many interior drainage areas within the Great Basin collect runoff from the surrounding hills, where it evaporates leaving behind deposits of soluble salts (Shantz 1925). According to Stälfelt (1960), soils in large areas of Utah contain excessive amounts of chlorides, sulfates, carbonates, and bicarbonates of sodium, potassium, calcium, and magnesium with a high concentration of sodium chloride. Crystalline deposits of chlorides and sulfates of sodium, calcium, and magnesium comprise "white alkali," while those of sodium carbonate cause a dark soil coloration, thus the name *black alkali* (Shantz 1925). The latter is considered more harmful to vegetation. The pH of these soils varies depending on the proportion of salts, ranging from weakly to strongly alkaline. In areas of extreme alkalinity, no vegetation occurs on the mud-encrusted soil.

The Salt Desert Shrub Zone consists of four community types: Greasewood, Greasewood-Shadscale, Saline-Alkaline, and Rabbitbrush Communities. These plant associations are relatively bright green in color when compared with the sage, gray-green characteristic of the northern desert shrub zone. Generally vegetation in the Salt Desert Shrub Zone receives adequate moisture since the water table is usually only 30 to 60 cm below the soil surface (Shantz 1925).

GREASEWOOD COMMUNITY

Sarcobatus vermiculatus (greasewood) occurs on soil containing high concentrations of salts and where the water table is high at least part of the year. These shrubs are evenly spaced from 1.2 to 2 m apart, with expanses of bare ground between. *Sarcobatus* roots often extend to depths of 3 to 4.5 m, and this shrub is considered to be a good indicator of subsurface ground water (Flowers 1934). Chenopodiaceae is the dominant plant family in the Greasewood Community (Flowers 1934).

GREASEWOOD-SHADSCALE COMMUNITY

Sarcobatus vermiculatus (greasewood) forms communities with *Atriplex confertifolia* (shadscale) at higher elevations from the Greasewood Community. This vegetation association often marks a transition between the Northern Desert and Salt Desert Shrub Zones. The two species are quite different morphologically. *Sarcobatus vermiculatus* has a bright green color and is somewhat taller than the gray-green, low, hemispheric *Atriplex confertifolia*. Both species also have different moisture requirements. Shantz (1925) explains this unusual association by differences in root structure. *Sarcobatus vermiculatus* is deeper rooted and is able to utilize groundwater found below the dry, alkaline soil inhabited by more shallow root systems of *Atriplex confertifolia*. Some associated shrubs in the Carson Desert Greasewood-Shadscale Community are *Artemisia spinescens, Ceratoides (Eurotia) lanata*, and *Lycium cooperi*. Although often lacking in this community, herbaceous flora may include *Oryzopsis hymenoides, Sphaeralcea ambigua*, and *Hermidium alipes*, along with annuals of the genera *Cryptantha, Coldenia, Gilia, Eriogonum*, and *Glyptopleura* (Billings 1945).

According to Stutz (1978), *Atriplex* is able to invade playas formed by the recession of inland seas such as Lake Bonneville and Lake Lahontan because of this species' ability to hybridize rapidly. These hybrids are able to tolerate extremely high salt concentrations and the resulting physiological drought in such habitats. Not only is *Atriplex* hybridizing with other species of the same genus, but Stutz (1978) reports interspecific hybridization with *Artemisia tridentata* in western Utah and northern Nevada. The resulting seeds produce viable offspring with characteristics of both parents. Thus it can be concluded that this hostile environment seems conducive to the rapid evolution of species.

SALINE-ALKALINE COMMUNITY

According to Stälfelt (1960) the draining of saline soils, either naturally or by man, results in leaching of mainly the neutral salts, leaving the alkali salts. The resulting soil is strongly alkaline, with excessive accumulations of sodium carbonate. If humus is present, this will be dispersed by the sodium carbonate, resulting in black alkali soils. In areas of poor soil permeability, dissolved salts remain on the surface where they are precipitated. In such cases the soil is white, resulting from a high surface concentration of sodium carbonate.

Saline-alkaline communities can be found surrounding the receding shores of large, prehistoric inland seas or alkali playas that mark the locations of dry lake beds. The vegetation of such areas is characterized by high tolerance for salts. *Atriplex*, for example, can be found in areas with concentrations of between 0.5 and 6.5 percent, perhaps because of the rapid evolution Stutz (1978) found to be occurring within this genus. Very few species exhibit such a wide range of salt tolerance; therefore little vegetation is found in such extreme saline-alkaline areas. *Allenrolfea occidentalis, Salicornia utahensis*, and *S. rubra* are found in damp, extremely saline-alkaline areas in Nevada and Utah around the Great Salt Lake, while *Distichlis spincata* and *Sporobolus airoides* are grasses that occur on less extreme soil conditions. The grasses are excellent livestock forage (Shantz 1925).

RABBITBRUSH COMMUNITY

Chrysothamnus graveolens (rabbitbrush) is often scattered over areas occupied by the grasses *Distichlis spicata* and *Sporobolus airoides* and is considered an indicator of groundwater and soils with a relatively low alkali content compared with other communities within the Salt Desert Shrub Zone (Shantz 1925). Stands of *Chrysothamnus graveolens* often become dense thickets.

According to Flowers (1934), *Chrysothamnus* is one of the most prominent genera in the Salt Desert Shrub Zone. Extensive communities are located in Utah in the Toole and Blue Springs valleys. On Stansbury Island in the Great Salt Lake, *Chrysothamnus* is found in association with greasewood and sagebrush. The plant is also frequently dominant on sand dunes and often occurs along the junction of alkaline and nonalkaline soils. Grasses and forbs are important occupants of the areas between the evenly spaced rabbitbrush. Other species found in this community include *Bromus tectorum, Sporobolus airoides, Distichlis spicata, Atriplex* sp., *Ceratoides (Eurotia) lanata, Sarcobatus vermiculatus, Opuntia* sp., and *Artemisia* sp. (Flowers 1934).

SOUTHERN DESERT SHRUB ZONE

This zone occurs in southwestern Utah and low, warm valleys south of the thirty-seventh parallel in Nevada (Shantz 1925). Cacti, yucca, and green shrubs characterize this zone in contrast to the gray-green shrubs of the Northern Desert Shrub Zone. Some species such as *Atriplex*, however, do overlap the two zones. Maximum temperatures of 49° C have been reported with frost-free periods exceeding 120 days (Shantz 1925). Annual precipitation ranges from as little as 50 mm in Utah to as much as 130 to 260 mm in parts of Nevada.

DESERT SALTBRUSH COMMUNITY

Atriplex polycarpa (desert saltbush), often referred to as desert sage because of its resemblance to *Artemisia tridentata*, forms uniform gray-green stands that sometimes reach a height of 1 to 1.2 m when sufficient moisture is available. This species, often clumping into thickets, has been used as an indicator of tillable land since it grows best on fine, loamy soil. Where water and soil conditions are poor, the plant will appear low and widely spaced (Shantz 1925). The Desert Saltbush Community is dominated by members of Chenopodiaceae, with *Grayia spinosa, Atriplex canescens, A. confertifolia*, and *Ceratoides (Eurotia) lanata* as important shrubs (Odum 1971). *Artemisia spinescens, Gutierrezia sarothrae*, and *Tetradymia axillaris* are other shrubs present.

This community is widespread throughout the southern

portion of the Great Basin, especially in the valley floors between 150 and 1280 m (Bradley and Deacon 1965) but can occasionally be found as high as 1830 m elevation. Soils generally contain a high percentage of calcium carbonate, and soil surface may be layered with "desert pavement." Temperature extremes over a ten-year period range from −13° C to 47° C, and precipitation tends to be sporadic. *Larrea divaricata* (creosote bush) and *Franseria dumosa* (bur-sage) are the dominant shrubs, with *Yucca schidegera, Atriplex* sp., *Dalea* sp., *Salvia* sp., *Ephedra* sp., and several cacti, namely *Opuntia* sp. and *Ferocactus acanthodes,* also present. *Larrea divaricata* provides an overstory from 0.6 to 1.5 m high, while *Franseria dumosa,* a low-growing shrub, forms the understory dominant (Billings 1951). Following periods of precipitation, numerous small desert annuals appear that can complete their life cycle in a very short period and whose seeds may lie dormant several years until another wet period (Went 1948, 1949). Winter annuals that rely on sufficient fall rains include *Oenothera* sp., *Gilia* sp., and *Phacelia fremontii. Euphorbia* sp. and *Amaranthus fimbiratus* and others germinate following summer rains (Cronquist et al. 1972). Precipitation in this community ranges from an annual mean of 38 mm in Death Valley to 200 mm at Searchlight, Nevada, elevation 1067 m, considered by Billings (1951) to be the upper elevation limit of the Creosote Bush Community. Because of the scarcity of water and forage throughout most of this community, grazing is almost nonexistent. According to Beatley (1976) this community type is prevalent over much of the Nevada Test Site.

JOSHUA TREE COMMUNITY

Bradley and Deacon (1965) consider this community to be part of the blackbrush community since both *Yucca brevifolia* (Joshua tree) and *Coleogyne ramosissima* (blackbrush) are frequently found as codominants. The Joshua Tree Community forms forests south of the thirty-seventh parallel in southern Nevada above the *Coleogyne* community of Bradley and Deacon (1965) (Shantz 1925). In this community, *Yucca brevifolia* dominates, often reaching heights of 4.5 m, with *Coleogyne* and *Grayia spinosa, Ephedra* sp., *Ceratoides (Eurotia) lanata, Artemisia spinescens, Dalea* sp., *Tetradymia* sp., *Agave utahensis,* and *Haplopapus* sp. important shrubs present. Although cacti are not as abundant as in the Creosote Bush Community (Bradley and Deacon 1965), grasses are more important.

MESQUITE COMMUNITY

Prosopis pubescens and *P. juliflora* (mesquite) occur in occasionally thick stands, along with *Atriplex confertifolia* in this community type, indicating an availability of subsurface moisture (Bradley and Deacon 1965). According to Jaeger (1940) roots of *Prosopis* may extend 15 to 18 m through the soil into the water table. Pods that ripen in early fall are a favorite forage for livestock and small mammals. This versatile plant also provides a traditional staple in the diet of southern Indians and Mexicans. Pods are ground, shaped into a loaf, and baked by burying in hot sand for several hours. Mesquite bark was pounded into fabric by primitive tribes, wood was used as fuel, and a beverage made from the beans was brewed by the Mexicans (Krochmal and Krochmal 1973).

Mesquite communities occur from the Colorado and Mojave deserts to Death Valley and extend into Mexico. *Prosopis* is either a shrub or small tree up to 6 m tall, which may become all but buried in sand hummocks that form at its base. The resulting hummock provides refuge for many small mammals who live among the roots. Commonly associated species in this community type include *Solidago spectabilis, Cleome sparsifolia, Suaeda torreyana, Oligomeris linifolia,* and *Elymus cinereus* (Beatley 1976).

Transition Desert Shrub Zone

According to Meyer (1978), a floristic transition zone is one where "a high proportion of indigenous plant species reach a distributional limit." Species found in the Mojave Desert do not normally occupy the Transition Desert Shrub Zone, largely due to cold air accumulations in the lowlands at night. However, it is possible for a species with a wide range of tolerances and sufficient genetic variability to migrate into a favorable microenvironment and consequently form an island in an otherwise hostile element (Meyer 1978). The higher mountains within this zone are considered by West et al. (1978) to be areas of greatest environmental and floristic diversity.

According to Beatley (1976), transition communities in the southern Great Basin characterize the change in vegetational composition between the southern, lower Mojave Desert and the more northern Great Basin Desert, which lies at a higher elevation. This transition is further influenced by the close proximity of the Sonoran Desert.

The Nevada Test Site is perhaps the most studied area with respect to floral composition within the Transition Desert Shrub Zone. Beatley has compiled several checklists on plant distribution, biotic studies have been conducted by Allred et al. (1963), ecological aspects have been considered (O'Farrell and Emery 1976), and effects of underground detonation have been studied by Tueller et al. (1974) and Tueller and Clark (1976), to mention just a few. However, many areas remain scientifically unexplored within this zone, which contains numerous endemic, endangered, and relic species (Clokey 1951; Bradley 1967; Beatley 1976).

Classification of communities within the Transition Desert Shrub Zone is based upon topographical position, species composition, and the open or closed drainage pattern of the basin. Four communities are considered characteristic of this zone: Creosote bush-boxthorn, Hop sage-boxthorn, Boxthorn-shadscale, and Sagebrush.

CREOSOTE BUSH-BOXTHORN COMMUNITY

Beatley (1976) has found associations of *Larrea* sp. (creosote bush) and *Lycium* (boxthorn) sporadically within the 1270 to 1370 m elevations of the transition zone. Rainfall and temperature minimums seem to control the distribution of these and associated species that are usually found only in the Mojave Desert. The transition zone marks the northernmost extension of *Ambrosia dumosa, Psorothamnus fremontii,* and *Krameria parvifolia,* which are associated species in this community type. Other species found here, as well as in communities of the Northern Desert Shrub Zone, include *Ceratoides (Eurotia) lanata, Tetradymia axillaris, T. glabrata,* and *Artemisia spinescens.*

HOPSAGE-BOXTHORN COMMUNITY

This community, considered by Beatley (1976) most characteristic of the transition region, occupies the floors of closed drainage basins and lower mountain slopes within the 1220 to 1525 m elevations of the Desert Shrub Zone. Major shrub components of this community, accounting for 40 to 60 percent of the total shrub cover, are *Grayia spinosa* (hopsage) and *Lycium andersonii* (boxthorn). Other associated woody species include *Ceratoides (Eurotia) lanata, Tetradymia glabrata, Artemisia spinescens, Atriplex canescens, Chrysothamnus viscidiflorus, Ephedra nevadensis, Yucca brevifolia,* and *Opuntia* sp. Among herbaceous perennials present are *Mirabilis pudica, Sphaeralcea ambigua,* and *Stephanomeria parryi.* Grasses include *Oryzopsis hymenoides* and *Sitanion hystrix,* but their cover value is small. Total coverage for this community type, considering shrubs, forbs, and grasses, is between 32 and 37 percent (Beatley 1976).

Lycium pallidum-Grayia spinosa associations may be found at lower elevations (945 to 990 m) than *Grayia spinosa-Lycium andersonii.* In one such area, Beatley (1976) found that *L. pallidum* comprised 75 percent of the shrub cover, with 20 percent attributable to *G. spinosa.* Thus this association is almost entirely composed of these two species, in contrast with the greater diversity found among woody species in the *G. spinosa-L. andersonii* community.

BOXTHORN-SHADSCALE COMMUNITY

Lycium shockleyi (boxthorn), an endemic often overlooked in southern Nevada, almost always occurs with *Atriplex confertifolia* (shadscale) in the Transition Desert Shrub Zone of the Great Basin. Other associated woody species include *Sarcobatus vermiculatus* and *Larrea* sp. Beatley (1976) found this species sporadically dispersed on the Nevada Test Site, its restricted distribution largely attributable to climatic factors. However, environmental measurements taken by Beatley (1976) show considerable variability and no pattern development correlating distribution with climatic variables.

SAGEBRUSH COMMUNITY

Artemisia tridentata (big sagebrush) is the dominant woody species of this community, comprising 95 percent of the total shrub cover in areas of the Nevada Test Site (Beatley 1976). This species, together with other important shrub components, *Ceratoides (Eurotia) lanata, Chrysothamnus viscidiflorus, C. nauseosus, Ephedra viridis, Tetradymia glabrata,* and *Cowania mexicana,* account for a total shrub cover of 32.6 percent (Beatley 1976). *Prunus andersonii, Symphoricarpos longiflorus,* and *Purshia glandulosa* may also be found occasionally in this community type. A great diversity in herbaceous perennials in this community is attributed by Beatley (1976) to the various soil origins in the region; however, their cover is sparse. *Stipa comata* and *Sitanion hystrix,* both perennial grasses, contributed greatly to the nonwoody species cover. Frequently associated herbaceous genera include *Erigeron, Caulanthus, Stanleya, Chenopodium, Astrogalus, Lupinus, Mentzelia, Sphaeralcea, Camissonia, Gilia,* and *Erigonum.*

Occasionally *Artemisia nova* (black sagebrush) is found to be dominant over *A. tridentata.* Characteristic associated species are similar to the *A. tridentata*-dominated areas along with total percentage cover by woody species (Beatley 1976).

At higher elevations, usually above 1830 m, *Pinus monophylla* and *Juniperus osteosperma* are associated with the Sagebrush Community, giving the area an open woodland-shrub appearance. Woody species occurring in this association include many found at lower elevations, plus *Arctostaphylos pungens, Quercus gambelii, Ceanothus* sp., *Amelanchier utahensis, Cercocarpus ledifolium,* and *Leptodactylon pungens,* which are normally associated with higher elevations.

In general, the Sagebrush Community composition described here as a representative of the Transition Desert Shrub Zone is perhaps more typical of the northern desert. It is mentioned here largely because of the importance Beatley (1976) ascribes to it on the Nevada Test Site and partly because a few associated species occur in the transition desert that are not found in the northern zone.

Playa and Dry Lake Zone

In this land of interior drainage, valleys between the mountain ranges of the Great Basin often form collecting basins for ephemeral streams formed by runoff from the surrounding high lands. Dry lake beds or playas mark the retreating shores of prehistoric lakes that once covered large portions of the Great Basin. As they evaporated, these bodies of water left behind large alkali deposits that support virtually nothing living, plant or animal. One of these, the Great American Desert, is an area approximately 80 by 160 km vacated by the decline of the Great Salt Lake. Wind and alkali "render this area one of the most forbidding deserts on this continent" (Flowers 1934). Farther westward, the Smoke

Creek and Black Rock deserts mark the retreat of prehistoric Lake Lahontan, which once covered more than 20,000 km², mostly within Nevada (Wheeler 1978) (figure 2-20). These deserts became known to pioneers of the 1850s (who chose to traverse them) as the death route. The alternate trail westward crossed the dreaded Forty Mile Desert, a sink formed by the terminus of the Carson River. Because of the intense heat, lack of shade, lack of forage for the livestock, and absence of water, these alkali-encrusted playas resulted in untold hardships for early pioneers journeying westward. An early traveler, Israel Russell, wrote in 1881 of his experience: "The scenery on the larger playas is peculiar, and usually desolate in the extreme, but yet is not without its charms. In crossing these wastes the traveler may ride for miles over a perfectly level floor, with an unbroken skyline before him and not an object in sight to cast a shadow on the oceanlike expanse" (Wheeler 1978).

Botanically speaking, the playa and dry lake community is a wasteland. As early as 1872, Haskill stated in his account of the agricultural resources of Nevada to the U.S. Geological Survey,

Black Rock Valley, forty miles west of Humboldt City, contains 350,000 acres of sagebrush and alkali flats, and volcanic matter lines the outskirts. This valley is almost entirely destitute of vegetation"

(Haskill 1872). Shreve (1942) notes the absence of plants on the playa formed by Lake Bonneville, the largest dry lake bed in the Great Basin. William Wallace, a correspondent for a San Francisco newspaper, wrote of his journey across the Forty Mile Desert in 1858: "The road there is hard and smooth for the greater portion of the way, but the plain around for thousands of acres together is leafless and lifeless, white arid plains without water, upon which the sun glares. I never saw [such] a desert before and I do not wish to see another [Wheeler 1971].

Soils are saline to great depths; frequently this salt is relatively pure and is mined as table salt. Most of the Great American Desert is barren, but *Allenrolfea occidentalis* (pickleweed) occupies widely scattered hummocks along the borders and occasionally appears further interior (Flowers 1934). *Salcornia* sp., *Distichlis spicata*, and *Suaeda* sp. are frequently occurring herbaceous species found around the edges, along with *Sarcobatus* (greasewood) and *Atriplex* (shadscale) shrubs as distance increases.

Mosses may occur on soils with a high salt content. Bordering the Great Salt Lake where sodium chloride exceeds 90 percent of the soluble salts, *Tortula* sp., *Pteryagoneurum* sp., *Pottia* sp., *Funaria* sp., and *Bryum* sp. were found (Flowers 1973).

Figure 2-20. Playa and Dry Lake Zone, Black Rock Desert, northwestern Nevada.
Photograph by H. Vreeland and P. Vreeland.

OTHER CATEGORIES

Other plant categories present within the Great Basin have not been mentioned with respect to a particular community type. These plants are either widespread in occurrence or have been little studied so that their phytosociology is poorly understood. One such plant group is the desert lichens. An early study near Reno, Nevada, found numerous lichens on soil and rocks, but the trees and shrubs were "as bare as a new and freshly painted hitching post" (Herre 1911). However, certain lichens were abundant on trees in other areas. Herre concluded that his collection of fifty-seven species comprised the lichen flora around Reno. Currently Hugh Mozingo from the University of Nevada Biology Department is studying lichen in the Great Basin, and it is hoped that his work will fill this void in understanding the distribution of this plant group.

Mosses are another plant group of widespread distribution but somewhat sporadic occurrence within the Great Basin. According to Flowers (1973) mosses may be found in a variety of desert habitats, including soils with high salt content. *Tortula* is a frequently found genus; *T. ruralis* has a wide distribution ranging from the hot, dry lowlands to the mountain peaks. Other *Tortula* species occupy areas around the base of shrubs, among beds of cacti, and occasionally in dense communities on open soil. *Pterygoneurum, Crossidium,* and *Barbula* are other frequently found genera, along with six species of *Bryum,* three of *Grimmia,* and two species of *Funaria. Aloina pilfera* and *Pottia nevadensis* are rare or infrequent species, confined largely to deserts. Around springs and in moist habitats, mosses flourish. A greater variety of species can be found in large populations in mountainous regions with greater precipitation (Flowers 1973). In addition to his work with lichens, Hugh Mozingo is contributing to our limited knowledge of the taxonomy and distribution of desert mosses.

Sand dunes are an interesting desert phenomenon. Occurring in isolated areas throughout the Great Basin, dunes as high as several hundred feet can be found, arising like mirages from the desert floor (figure 2-21). Around the Great Salt Lake, dunes are comprised of either calcareous oolitic sand or fine-grained siliceous sand (Flowers 1934). Both types are dotted with shrubs and tufts of grass, with a variety of herbs between. Around the Great Salt Lake, *Atriplex* sp., *Sarcobatus* sp., *Gutierrezia microcephala,* and *Chrysothamnus* sp. are frequently found dune shrubs. Grasses and forbs include *Sporobolus airoides, Distichlis spicata, Bromus tectorum, Poa* sp., *Atriplex hastata,* and *Salsola pestifer* (Flowers 1934). *Juniperus utahensis* can be found on the northern end of the Stansbury Island dunes. The moss, *Tortula ruralis,* is also of importance here.

Figure 2-21. Sand dunes.
Photograph by H. Vreeland and P. Vreeland.

In the Carson Desert, sand dunes support large stands of *Parosela (Dalea) polyadenia* (smoke brush), while on those near Washoe Lake, *Prunus andersonii* (desert peach) occurs. North of Reno, *Chrysothamnus* sp. is found on dunes, often together with *Grayia spinosa, Ceratoides (Eurotia) lanata, Atriplex canescens,* and *Artemisia tridentata* (Ting 1961).

On dunes in Churchill County, Nevada, the Greasewood-Shadscale Community of the surrounding Salt Desert Shrub Zone is replaced by a *Parosela (Dalen) polyadenia, Atriplex canescens, Tetradymia comosa* association. A variety of annual forbs that are absent in the Greasewood-Shadscale Community are found on the dunes, including *Tiquila (Coldenia) nuttallii, Cryptantha circumscissa,* and *Gilia leptomera* (Billings 1941). Billings (1941) attributes this change in community composition largely to edaphic factors such as soil composition, pH, and organic content variations.

The sand dunes north of Winnemucca, Nevada, are quite extensive. They were visited by members of the Northern Nevada Plant Society in June 1978 who reported the following species present: *Phacelia bicolor, Abronia* sp., *Eriogonum ovalifolium, Oxytheca dendroidea, Psorothamnus kingii, Stanleya elata, Rumex venosus,* and the rare *Oryctes nevadensis.* All of these plants were in bloom at the time, covering the dunes with splashes of color.

Smith, in chapter 11 of this volume, discusses the dune fields in the Great Basin.

Fungi make an important contribution to the desert ecology. Nutrient cycling is greatly enhanced by fungal decomposition of plant materials. The role of mycorrhija in interplant relationships is poorly understood. Soil fungi contribute to the stabilization of sand dunes and prevention of erosion. Desert fungi and decomposition currently are being studied by Don Prusso of the University of Nevada, Reno. Patricia Vreeland and Hamilton Vreeland from the same institution are working on the relationship of mycorrhija and achlorophyllous plants in the Sierra Nevada.

This classification scheme of desert vegetation groups synusia by zones and zones within ecosystems is an attempt to present an overview of plant community relationships as they exist in the Great Basin. It should be remembered that this system is based on generalities and on existing literature. Most certainly exceptions in species composition can be found for each plant community. Although overview vegetation mapping does exist for most of the Great Basin (Cronquist et al. 1972; Billings 1951; and others) and certain portions have been thoroughly studied (Beatley 1976), most of this area remains ecologically unexplored. Much work must still be done before an analytical vegetation composition can be presented at the community level of organization. Even then there will be exceptions, for the desert is a land of extremes where even an individual hot spring, for example, may be a separate ecosystem.

PREHISTORY

Early and Paleo-Indian Cultures

The advent of human occupation in the New World and the Great Basin region of North America is a subject of intense debate and reevaluation (MacNeish 1978). For several decades, available archeological evidence suggested that the peopling of the New World occurred toward the end of the Pleistocene epoch 12,000-14,000 B.G. (Haynes 1969). However, recent radiometric determinations in the Arctic, Pennsylvania, Texas, California, Mexico, and South America have extended the time range backward by ten and possibly twenty millennia (available carbon 14 dates are summarized by MacNeish 1978).

Within the physiographic Great Basin, various Early but undated lithic complexes have been reported by several researchers. These usually occur as surface sites on beach terraces or strand lines around several of the Pleistocene lakes in the region (Aikens 1978). To date (1979), none of these lithic complexes have been found in stratigraphic contexts or with datable organic materials. Reports of lithic artifacts from the Calico Mountains of southeastern California (Simpson 1976) with an estimated age of more than 50,000 B.G. are not accepted by many authorities. Earlier reports of associations of Pleistocene megafauna remain with artifacts at Tule Springs, Gypsum Cave, and Smith Creek Cave (Harrington 1933; Harrington and Simpson 1961), all in southern Nevada, have subsequently been disproven (Heizer and Berger 1970; Wormington and Ellis 1967; Bryan 1972). However, a recently completed intensive study of Ranchlabrean megafauna remains, and artifactual material in and around China Lake in the Mojave Desert has suggested that human occupation in that region may have occurred as early as 45,000 B.G. (Davis 1978). Neither the artifacts nor the faunal remains have yet been dated radiometrically or by stratigraphic means (Davis 1978).

A major hypothesis growing out of the Davis study is that a pre-Clovis lithic tradition developed very early in the Mojave Lakes area, subsequently developed into the Clovis tradition, and then spread out of the Great Basin into the Great Plains and Southwest regions. In these latter regions (Plains and Southwest) Clovis tradition sites, primarily "kill sites"—artifacts associated with extinct mammoth remains—are well dated at between 12,000 and 10,000 B.G. (Wedel 1978). Isolated finds of Clovis projectile points occur widely in the Great Basin but not, as yet, in directly datable or stratigraphic situations (Tuohy 1974). It has generally been assumed that the Clovis tradition was brought into, or diffused into, the Great Basin from the Great Plains (Watters 1979 reviews alternative hypotheses). The Davis (1978) hypothesis, if verified by subsequent research, would radically alter this view.

The Clovis lithic tradition is generally included within the

Paleo-Indian stage of cultural development in the New World (Jennings 1978), generally dated at between about 12,000 and 8,000 B.G,

Western Archaic Tradition

Post-Pleistocene prehistoric cultures in the Great Basin, dated from about 10,000 B.G. to, in some areas, historic times, are generally subsumed within the western Archaic tradition (Jennings 1978). Within the general framework of New World culture history, Archaic refers to prehorticultural or nonhorticultural (depending upon the region under discussion) technoeconomic patterns of adaptations to local and regional environments, in which subsistence is based on foraging—hunting and gathering of available resources. The long time depth of such patterns in the Great Basin had been suspected since the 1920s (C. Fowler 1977; D. Fowler 1977) but was only firmly established by Jennings (1957) and others after the advent of carbon 14 dating.

Within the overall western Archaic tradition, various subregional adaptive patterns are recognized. In the northeastern Great Basin, a pattern of nomadic foraging has been demonstrated at numerous rock shelter and open sites, especially at Danger Cave (Jennings 1957), Hogup Cave (Aikens 1970), Deadman, Promontory, and Black Rock caves (Steward 1937; Smith 1941) and Swallow Shelter and related sites (Dalley 1976). Dates from these several sites range from 11,000 B.P. to at least 500 B.G., although not all sites exhibit the full range of time, and there are hiatuses in many deposits, probably reflecting climatic-ecological shifts and cultural responses to those shifts (Simms 1977; Aikens 1978). Artifacts, including a variety of projectile point types, milling stones, basketry, nets, bone tools, and floral and faunal remains, suggest a pattern of people moving through an annual round. Major resources included seeds, roots, small animals, and some larger ungulates, such as bison, antelope, deer, and mountain sheep. At both Hogup and Danger caves, a major food resource was pickleweed (*Allenrolfea occidentalis*). It is clear from the evidence that rock-shelter sites were not permanent occupation loci but rather stopping places during the seasonal round.

The subsistence patterns reflected in these sites are very similar to those practiced by historic Indian bands in the area, and it has generally been assumed that the patterns are of great antiquity (Jennings 1957; see C. Fowler 1977 for a review of this model and its implications).

Farther west across the northern Great Basin at various rock-shelter sites, a generally similar pattern has been observed (figure 2-22). Among these sites are Deer Creek Cave (Shutler and Shutler 1963), South Fork Shelter (Heizer et al. 1968), Newark Cave (Fowler 1968), Dirty Shame rock shelter (Aikens et al. 1977), Fort Rock Caves (Bedwell 1973), Eastgate and Wagon Jack shelters (Heizer and Baumhoff 1961),

Gatecliff Shelter and Hidden Cave (currently under excavation), Last Supper Cave (Layton 1979), and the lower levels of Leonard rock shelter (Heizer 1951). Again, no single site reflects the full time range from about 11,000 B.G. to historic times, and artifactual remains vary somewhat, as do floral and faunal remains, but the general pattern is similar to that of the northeastern area.

Recent studies of the distributions of prehistoric sites in whole valleys, rather than studies of single sites, has further demonstrated the seasonal transhuman patterns of subsistence. These sites include the Reese River Valley (Thomas 1973), Suprise Valley (O'Connell 1975), Owens Valley (Bettinger 1977, 1979), and Warner Valley (Weide 1974). In Surprise Valley, the existence of semisubterranean pithouses, dated between 6,000 and 5,000 B.P., suggests a pattern of permanent or semipermanent valley-floor small villages from which the people ranged, occupying temporary camps in other ecozones within the valley when resources became available (O'Connell 1975). The pattern does not, however, continue into later times.

A further variant of the overall western Archaic pattern existed around the shores of remnants of Pleistocene Lake Lahontan in the western Great Basin—the Humboldt sink (Heizer and Napton 1970), Pyramid Lake (Tuohy 1980), Winnemucca Lake (Hattori 1980), and probably Honey Lake, although the last is not well known archeologically. Although not a part of Lake Lahontan, Eagle Lake (Pippin et al. 1979) in northeastern California exhibits a similar pattern. At these various locales, the emphasis was on lacustrine resources: fish, waterfowl, and marsh plants. The Lovelock culture, as seen in sites around the Humboldt sink, includes artifacts reflecting this Lacustrine pattern, among them, bone and wood fishhooks, setlines, duck decoys of tule covered with duck skins, milling stones for processing seeds, and roots.

In the south-central Great Basin, Archaic cultural patterns were similar to those farther north. Recent exacavations and reexcavations in Council Hall, Smith Creek, and Kachina caves and Amy's Shelter (Tuohy and Rendell 1979) in the Snake Valley area of southeastern Nevada have demonstrated an 11,000-year sequence, although no single site yielded the full sequence. An earlier report of *Equus* sp. bones in association with artifacts (Harrington 1934) has been disproven. At the O'Malley Shelter site, east of Caliente, Nevada, occupation began about 7000 v.P., although there were hiatuses between 6500 and 4600 B.G. and between 3000 and 1000 B.G. (Fowler, Madsen, and Hattori 1973). Etna Cave, south of Caliente, was earlier reported to contain Pleistocene horse dung and artifacts of early Holocene age (Wheeler 1942). This report has also been subsequently disproven: Etna Cave has a basal date of 5750 v.P. (Fowler 1973). The reputed association of artifacts with the dung of Pleistocene sloth (Harrington 1933) in Gypsum Cave near Las Vegas, Nevada, has also been disproven (Heizer and Berger 1970).

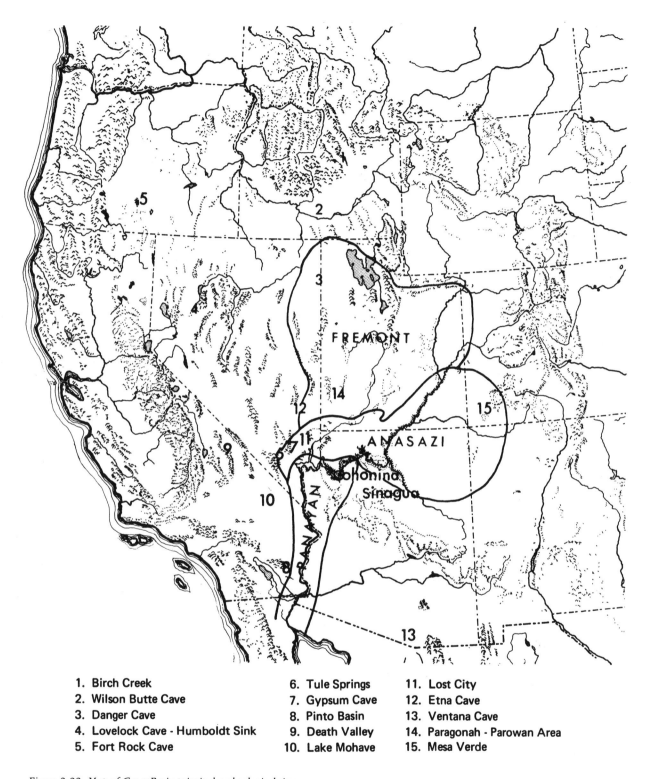

1. Birch Creek	6. Tule Springs	11. Lost City
2. Wilson Butte Cave	7. Gypsum Cave	12. Etna Cave
3. Danger Cave	8. Pinto Basin	13. Ventana Cave
4. Lovelock Cave - Humboldt Sink	9. Death Valley	14. Paragonah - Parowan Area
5. Fort Rock Cave	10. Lake Mohave	15. Mesa Verde

Figure 2-22. Map of Great Basin principal archeological sites.
From D. D. Fowler, *The Archeology of Newark Cave, White Pine County,* Nevada, Desert
Research Institute, University of Nevada, *Social Science Publication* (1968) no. 3, map 2.
Reprinted by permission of the publisher.

Post-Pleistocene Climate and Great Basin Prehistory

A central concern in Great Basin prehistoric studies for several decades has been the relationship(s) between postulated Holocene climatic change and prehistoric cultures. The initial framework for Holocene climatic change was established by Ernst Antevs (1948, 1952, 1955), who postulated an anithermal, altithermal, and medithermal sequence. For a number of years, archeological debate centered on whether the climatic regime during Antevs's altithermal period was severe enough, due to drought, to force out-migration of prehistoric peoples from all, or parts, of the Great Basin (Heizer and Baumhoff 1965). Paleoenvironmental and archeological research during the past two decades has now clearly demonstrated that Antevs's scheme is far too coarse grained to account for Holocene climatic change; rather, climatic and associated ecological changes must be considered on a basin-by-basin basis (Mehringer 1977; Mifflin and Wheat 1979; Weide and Weide 1977). Also although drying trends are ascertainable for some areas, the impacts were insufficient to force abandonment of large areas (Fagan 1974; Aikens 1978).

Puebloid Cultures

While the western Archaic cultural patterns continued into historic times in the central and western Great Basin, in the eastern and southern areas a new pattern developed between about 400 to 800 A.L. and continued until about 1300 A.D. The pattern was based in part on domesticated plants, principally the cultigens corn, beans, and cucurbits, originally developed in Mexico and diffused northward over several millennia. Reliance, at least in part, on cultigens permitted the development of small, pueblo settled villages, particularly along the Wasatch front in Utah and as far west as Snake Valley, which is bisected by the Utah-Nevada border. Farther south this overall pattern is also discernible along the lower Virgin River Valley and its tributary, the Muddy River.

Two variants of the pattern developed. One, the Virgin Branch Anasazi culture, developed along the Virgin River from its confluence with the Colorado River upstream to present-day Zion National Park in southern Utah (Shutler 1961; Aikens 1965). The second variant, originally designated as the Fremont culture (Morss 1931), occurred over much of present-day Utah north of the Glen Canyon, extending into the Uintah Basin, western Colorado, along the Wasatch front, and as far north as the delta of the Bear River at the north end of the Great Salt Lake (Aikens 1978). Recently the Fremont culture west of the Wasatch front has been redesignated as the Sevier culture (Madsen and Lindsey 1977). According to Madsen and Lindsey (1977; cf. Madsen et al. 1979) the Sevier culture is characterized by a settlement pattern of small villages located on alluvial fans adjacent to marsh or riverine ecosystems and by temporary encampments in other ecozones. Subsistence was based on foraging, supplemented by maize horticulture. Architecture consists of deep, clay-lined semisubterranean pit houses and rectangular surface storage structures. Characteristic artifacts include distinctive forms of basketry and other textiles, decorated, unfired small anthropomorphic clay figurines, various forms of well-made pottery, troughed grinding slabs or metates, and a variety of bone implements and ornaments. The Sevier culture began about 1300 to 1500 B.P. and ended about 650 V.P,

The Virgin Branch Anasazi and the Sevier and Fremont cultures ended during the same period that other southwestern Pueblo cultures ended or were severely disrupted, possibly due to climatic shifts and erosion. The fate of the Puebloid cultures in the Great Basin has been discussed for two decades (see Fowler, Madsen, and Hattori 1973, and Madsen et al. 1979 for summaries and weighing of the arguments), but the matter remains unclear. It is clear that the Puebloid cultures were succeeded by the cultures carried by the historic Numic-speaking peoples, especially the southern Paiute. These cultures artifactually are very different in textiles, ceramics, and lithic technology from the Puebloid cultures. In the O'Malley and Conaway sites in southeastern Nevada, Sevier and Virgin Branch ceramics occur in levels dated at about 900 A,D. succeeded by a mixture of Sevier, Virgin Branch, and Paiute ceramics dated about 1150 I.D., with Paiute ceramics occurring only in upper levels above, dated as late as about 1700 A. L. (Fowler, Madsen, and Hattori 1973). This suggests that carriers of the Puebloid cultures were in the area, then Puebloid and Paiute together, then only Paiute. Whether the Puebloid folk migrated elsewhere or were amalgamated into the Paiute is not known at present. What is clear is that Numic-speaking peoples with a distinctive, basically western Archaic culture pattern were in the area by at least 1000 A.D. and remained into historic times.

ETHNOGRAPHY

The historic native American peoples of the Great Basin must be placed within an anthropological context (figure 2-23). Anthropologists traditionally discuss North American Indian ethnographic cultures in terms of culture areas. Culture areas are usually defined as delimited geographic regions within which material culture, subsistence practices, and sometimes social organization and ideological systems are generally similar, when compared with similar phenomena in adjacent areas or regions. On this basis, in historic times (post-1776-1820 I.D,) the Great Basin culture area was not coextensive with the hydrographic or physiographic Great Basin but was much larger. The area included much of the Snake River Plain in Idaho, the Bridger Basin and adjacent mountains in Wyoming, all of present-day Utah, and portions of western Colo-

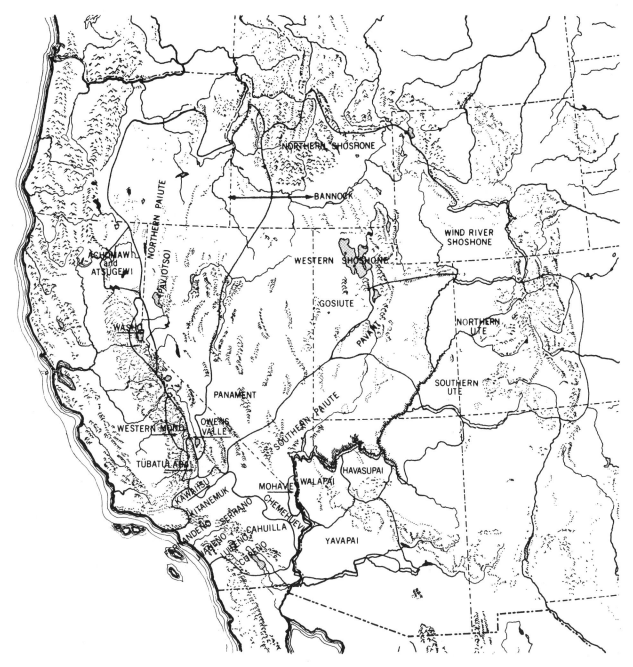

Figure 2-23. Map of Great Basin tribal distribution.
From D. D. Fowler, Great Basin Social Organization, Desert Research Institute, University of
Nevada, *Publication in Social Science and Humanities* (1966) 1:56-76, map 4. Reprinted with
permission of the publisher.

rado, as well as the hydrographic Great Basin proper. On the west, cultures intergraded into the California culture area; on the northwest into the Plateau culture area, and on the east and northeast into the Plains culture area. For present purposes, we will discuss the distributions and cultures of Great Basin Indian groups as they existed about 1820-1850 A,D., the period of intensive white exploration and initial settlement.

Languages spoken by Great Basin native peoples derive from two very different language families, Hokan and Uto-Aztekan. The Washo, whose territory centered on Lake Tahoe on the Nevada-California border, speak a Hokan language related to other Hokan languages farther west in California. Other Great Basin peoples speak Uto-Aztekan languages as shown in table 2-5.

TABLE 2-5
Great Basin Indian languages

I. Hokan family: Washo
II. Uto-Aztekan family
    1. Southern Numic: Ute-Kawaaisu
    2. Central Numic: Panamint-Shoshoni
    3. Western Numic: Mono-Bannock

Numic Peoples

Numic-speaking people were widespread over the Great Basin culture area (figure 2-24). One group of central Numic speakers, the Comanche, had, by 1750 A.L.! migrated from the Wyoming-Uintah Basin area to the southern Great Plains and had become, culturally, Plains Indians.

The Numic languages (the term *Numa* is derived from natives' terms for themselves) are separated by linguists into three language pairs. Southern Numic includes Kawaiisu, spoken by a small group in southeastern California, and Ute, dialects of which are spoken by the Chemehuevi, several southern Paiute bands, and the various bands of northern and southern Ute. The distinction between Ute and southern Paiute is basically a cultural one; Ute are those groups principally east of the Wasatch Mountains who adopted the use of horses as early as about 1730 A.L. and became, culturally, Plains Indians; southern Paiute are those Ute-dialect-speaking groups who remained without horses until well into the nineteenth century. Central Numic includes Panamint Shoshoni, spoken by a group near Death Valley, and Shoshone proper, spoken by various western Shoshoni groups (including the Gosiute), and eastern and northern Shoshoni groups. Western Numic includes Mono, spoken by the Mono of eastern California, and Bannock, which includes various northern Paiute bands and the Bannock proper, who in early historic times moved from eastern Oregon to Idaho and joined forces with horse-using Shoshone groups in that region.

Culturally the Ute, eastern and northern Shoshoni, and the Bannock were Great Basin people who after the advent of horses took on Plains culture traits: bison hunting, skin tepees, the travois, military patterns, and ceremonials, such as the Sun Dance, a Plains phenomenon. They will be considered here only peripherally as they affected other groups within the Great Basin proper.

Prehorse and historic nonhorse Great Basin cultures, as described by Steward (1938) and others, were basically continuations of those cultures that archeologically are identified as western Archaic. In the more arid reaches of the Great Basin, such as the Mojave Desert and the Great Salt Lake Desert, aboriginal population estimates range from one person per 250 km² to one person per 100 km². In ecologically richer areas, population estimates range from one person per 20 km² in the Uintah Basin to one person per 10 km² in the Reese River Valley of central Nevada. In highly favorable areas, such as the Washo territory centering on Lake Tahoe and the Truckee Meadows (site of present-day Reno, Nevada), the aboriginal population may have been as high as 1.5 persons per one km². A similar figure is estimated for Owens Valley in eastern California.

Numic and Washo Culture

The basic Great Basin aboriginal social unit was the "kin clique" (Fowler 1966), one to three or four nuclear families related through bilateral kinship ties. Numic kin cliques who centered their activities in a specific valley, or around a lake, were food-named. That is, they were known for a specific food resource found in an area, such as "cui-ui eaters" at Pyramid Lake and "trout eaters" at Walker Lake. Washo kin cliques were usually known by their informal leaders' names, "so-and-so's bunch."

Subsistence activities followed an annual round. Beginning in the spring, seed and root resources were collected as they became available in the various ecozones of the valleys and adjacent uplands. Small and large game (rodents, deer, antelope, mountain sheep) were taken when available throughout the year. Rabbits and antelope were taken in communal drives. Where available, fish were taken with spears, dip nets, and hooks. Waterfowl were taken by a variety of techniques, including the use of tule duck decoys. Pinyon nuts were a staple, collected in the fall at large communal gatherings and stored for winter use. The Washo subsistence pattern differed somewhat from the rest of the Great Basin. Extensive fishing took place at Lake Tahoe in the spring and early summer. In the late summer some Washo groups traveled to the west slope of the Sierra Nevada to harvest acorns.

Summer dwellings were simple brush wickiups. Winter village houses, located usually in the pinyon-juniper zone, were of more substantial brush and bough contruction. Material culture was simple and portable: baskets, nets, milling stones,

Figure 2-24. Map of Great Basin linguistic distribution.
From D. D. Fowler, Great Basin Social Organization, Desert Research Institute, University of
Nevada, *Publication in Social Science and Humanities* (1966) 1:56-76, map 1. Reprinted with
permission of the publisher.

bows and arrows, fishing gear, bark and skin clothing. Some southern Paiute and western Shoshoni groups made and used a crude form of pottery.

In short, ethnographic Great Basin cultures were geared to a transhumant, foraging existence, exploiting available food resources. This adaptive pattern was well suited to Great Basin ecology, and the general pattern is of considerable antiquity— at least 10,000 to 12,000 years.

Great Basin aboriginal cultures were severely affected by intrusion into the American West. Between about 1780 and 1850, various central and southern Numic bands were subjected to slave raids by Spanish, Ute, and Navaho parties. Slaves were sold into New Mexico or southern California by raiders moving along the Old Spanish Trail (Malouf and Malouf 1945). With the advent of white migration across, and settlement in, the Great Basin after 1840, traditional aboriginal food resources were badly disrupted, and the ecological base of the aboriginal cultures largely destroyed by grazing, lumbering, and mining activities (Gould, Fowler, and Fowler 1972). Native peoples were forced onto reservations, or into ghetto colonies adjacent to towns and cities, where many of their descendants remain.

REFERENCES

Aikens, C. M. 1965. Excavations in southwest Utah. *Univ. Utah Anthropological Papers,* no. 76.

———. 1970. Hogup Cave. *Univ. Utah Anthropological Papers,* no. 93.

———. 1978. The Far West. In *Ancient Native Americans,* ed. J. D. Jennings, pp. 131-182. San Francisco: W. H. Freeman.

Aikens, C. M. et al. 1977. Excavations at Dirty Shame Rockshelter, southeastern Oregon. Tebiwa. *Miscellaneous Paper of Idaho State Univ. Museum of Natural History,* no. 4.

Albers, John P. 1967. Belt of Sigmoidal bending and right-lateral faulting in the western Great Basin. *Geol. Soc. Amer. Bulletin* 78(2):143-156.

Alcorn, J. R. 1940a. New and noteworthy records of birds for the state of Nevada. *Condor* 42:169-170.

———. 1940b. New and additional Nevada bird records. *Condor* 43:118-119.

———. 1941. Two new records for Nevada. *Condor* 43:294.

———. 1943. Additions to the list of Nevada birds. *Condor* 45:40.

———. 1946. The birds of Lahontan Valley, Nevada. *Condor* 48:129-138.

Allred, D. M.; Beck, D. E.; and Jorgensen, C. D. 1963. Biotic communities of the Nevada Test Site. *Brigham Young Univ. Science Bulletin, Biol. Series* 2(2).

Antevs, Ernst. 1948. Climatic Changes and pre-White man. In *The Great Basin, with emphasis on glacial and postglacial times. Bulletin, Univ. Utah,* 38(20):168-191.

———. 1952. Cenozoic climates of the Great Basin. *Geol. Rundschau* 40:94-108.

———. 1955. Geologic-climatic dating in the West. *American Antiquity* 20(4):317-335.

Armstrong, David M. 1977. Distributional patterns of mammals in Utah. *Great Basin Naturalist,* vol. 37(4):457-474.

Armstrong, Richard Lee. 1963. Geochronology and geology of the eastern Great Basin in Nevada and Utah. Ph.D. dissertation, Yale University.

———. 1968. Sevier Orogenic belt in Nevada and Utah. *Geol. Soc. Amer. Bulletin* 79:429-458.

Atwater, Tanya. 1970. Implications of plate tectonics for the Cenozoic tectonic evolution of western North America. *Geol. Soc. Amer. Bulletin* 81:3513-3536.

Austin, G. T., and Bradley, W. G. 1965. Bird records from southern Nevada. *Condor* 67:445-446.

———. 1968. Bird records for Clark County Nevada. *Great Basin Nat.* 28(2):61-62.

Austin, G. T., and Rea, A. M. 1976. Recent southern bird records. *Condor* 78:405-408.

Axelrod, Daniel. 1940. The Pliocene esmeralda flora of west-central Nevada. *J. Wash. Acad. Sci.* 30:163-174.

———. 1944. The alturus flora. *Carnegie Inst. Wash. Publ.* 553:263-284.

———. 1944. The Alvord creek flora (Oregon). *Carnegie Inst. Wash. Publ.* 553:225-262.

———. 1949. Eocene and Oligocene formations in the western Great Basin. *Geol. Soc. Amer. Bulletin* 60:1935.

———. 1950. Evolution of desert vegetation in western North America. *Carnegie Inst. Wash. Publ.* 590:217-306.

———. 1956. Mio-Pliocene floras from west-central Nevada. *Univ. Calif. Pub. Geol. Sci.* 33:1-316.

———. 1957. Late Tertiary floras and the Sierra Nevadan uplift. *Geol. Soc. Amer. Bulletin* 68:19-45.

———. 1958. Evolution of the Madro-Tertiary geoflora. *Bot. Rev.* 24:433-509.

———. 1960. Late Pliocene floras east of the Sierra Nevada. *Univ. Calif. Publ. Geol. Sci.* 39:1-118.

———. 1962. A Pliocene sequoiadendron forest from western Nevada. *Univ. Calif. Pub. Geol. Sci.* 39:195-268.

———. 1964. The Miocene Trapper Creek flora of southern Idaho. *Univ. Calif. Publ. Geol. Sci.* 51:1-180.

———. 1966. The Eocene Copper Basin flora of northeastern Nevada. *Univ. Calif. Publ. Geol. Sci.* 59:1-125.

———. 1976. History of the coniferous forests, California and Nevada. *Univ. Calif. Publ. Bot.* vol. 70.

Axelrod, D. I. and Ting, W. S. 1960. Late Pleistocene floras east of the Sierra Nevada. *Univ. Calif. Publ. Geol. Sci.* 39:1-118.

Axler, R. P.; Gersberg, R. M.; and Paulson, L. J. 1978. Primary productivity in Meromictic Big Soda Lakes, Nevada. *Great Basin Naturalist* 38(2):187-192.

Bailey, Vernon. 1936. The mammals and life zones of Oregon. *North American Fauna* 55. Washington, D.C.: Bureau of Biological Surveys, U.S. Department of Agriculture.

Baird, S. F., and Girard, C. 1852a. Characteristics of some new reptiles in the museum of the Smithsonian Institution. *Proc. Acad. Nat. Sciences Philadelphia* 6:68-70.

———. 1852b. Reptiles. Appendix C. In *An expedition to the Valley of the Great Salt Lake of Utah,* ed. H. Stansbury, pp. 336-353. Philadelphia: Lippincott, Gram Bo. & Co.

Baird, S. F. 1853. Birds. Appendix C. In *Capt. Howard Stansbury's report on exploration and surveys of the valley of the Great Salt*

Lake of Utah, pp. 314-325. Robert Armstrong, public printer.

———. 1859. Report on reptiles collected on the survey. Explorations and surveys for a railroad route from the Mississippi River to the Pacific Ocean. Reports of Lt. E. G. Beckwith, Third Artillery, upon explorations for a railroad route, near the 38th and 29th parallels of north latitude, by Lt. E. G. Beckwith, Third Artillery. *War Dept.* Wash. D.C. vol. 10, art. 3, pp. 17-20.

Banta, B. H. 1961. The variation and zoogeography of the lizards of the Great Basin. Ph.D. dissertation, Stanford University.

———. 1963. Preliminary remarks upon the zoogeography of the lizards inhabiting the Great Basin of the western United States. *Wasmann J. Biol.* 20(2):253-287.

———. 1965a. A distributional check list of the recent amphibians inhabiting the state of Nevada. *Occas. Pap. Bio. Soc. Nevada,* no. 7.

———. 1965b. A distributional check list of the recent reptiles inhabiting the state of Nevada. *Occas. Pap. Bio. Soc. Nevada,* no. 5.

———. 1965c. An annotated chronological bibliography of the herpetology of the state of Nevada. *Wasmann J. Biol.* 23(1&2).

———. 1967. Some miscellaneous remarks on recent Nevada lizards. *Occas. Pap. Biol. Soc. Nevada,* no. 16.

Banta, B. H., and Tanner, W. W. 1964. A brief historical resumé of herpetological studies in the Great Basin of the western United States. Part I: The reptiles. *Great Basin Nat.* 24(3):37-57.

Basinski, Paul. 1978. Personal communication.

Beatley, J. C. 1976. Vascular plants of the Nevada Test Site and central-southern Nevada Tech. Info. Center. Energy Research and Development Admin., Springfield, Va.

Becker, H. F. 1962. Reassignment of *Eopuntia* to *Cypeoacites. Bulletin Torrey Club* 89:319-330.

Beckwith, Lt. E. G. 1855. Reports of explorations for a route for the Pacific railroad of the line of the 41st parallel of north latitude, by Lt. E. G. Beckwith, 1854:1-132. In *Reports of explorations and surveys to ascertain the most practicable and economical route for a railroad from the Mississippi River to the Pacific Ocean, made under the direction of the Secretary of War, in 1853-54, according to Acts of Congress of March 3, 1853, May 31, 1854 and August 5, 1854,* vol. 2, 33d Cong., 2d sess. *Senate Executive Doc. No. 78,* Washington, D.C., Beverly Tucker, printer.

Bedwell, S. F. 1973. *Fort Rock Basin prehistory and environment.* Eugene: University of Oregon Books.

Beetle, A. A. 1960. A study of sagebrush. *Univ. Wyoming Agric. Exp. Sta. Bulletin* 368.

Behle, W. H. 1941. A collection of birds from the La Sal Mountain region of southeastern Utah. *Wilson Bulletin* 53:181-184.

———. 1943. Birds of Pine Valley Mountain region, southwestern Utah. *Bulletin, Univ. Utah* 34(2):1-85.

———. 1944. Checklist of the birds of Utah. *Condor* 46:67-87.

———. 1948. Birds observed in April along the Colorado River from Hite to Lee's Ferry. *Auk* 65:303-306.

———. 1955. The birds of the Deep Creek Mountains of central-western Utah. *Univ. Utah Biol. Ser.* 11(4):1-34.

Behle, W. H. 1960. The birds of southeastern Utah. *Univ. Utah Biol. Ser.* 12(1):1-56.

———. 1976. Mohave Desert avifauna in the Virgin River Valley of Utah, Nevada and Arizona. *Condor* 78:40-48.

———. 1978. Avian biogeography of the Great Basin and inter-mountain region. Great Basin memoirs no. 2. In *Intermountain biogeography: A symposium,* p. 55-80. Salt Lake City: University of Utah.

Behle, W. H., and Ghiselin, J. 1958. Additional data on the birds of the Uinta Mountains and basin of northeastern Utah. *Great Basin Nat.* 18:1-22.

Behle, W. H.; Bushman, J. B.; and Greenhalgh, C. M. 1958. Birds of the Kanab area and adjacent high plateaus of southern Utah. *Univ. Utah Biol. Ser.* 11(7):1-92.

Bendire, C. 1877. Notes on some of the birds found in southeastern Oregon, particularly in the vicinity of Camp Harney, from November 1874 to January 1877. *Proc. Boston Soc. Nat. Hist.* 19:109-149.

Berry, E. W. 1927. Flora of the Esmeralda formation in western Nevada. *Proc. U.S. Natl. Museum* 72(23):1-15.

Bettinger, R. L. 1977. Aboriginal human ecology in Owens Valley: Prehistoric change in the Great Basin. *American Antiquity* 42:3-17.

———. 1979. Multivariate statistical analysis of a regional subsistence-settlement model for Owens Valley. *American Antiquity* 44(3):455-470.

Billings, W. D. 1941. Quantitative correlations between vegetational changes and soil development. *Ecology* 22:448-456.

———. 1945. The plant associations of the Carson Desert region, western Nevada. *Butler Univ. Bot. Studies* 7:1-35.

———. 1949. The Shadscale Vegetation Zone of Nevada and eastern California in relation to climate and soils. *Amer. Midl. Nat.* 42:87-109.

———. 1950. Vegetation and plant growth as affected by chemically altered rocks in the western Great Basin. *Ecology* 31:62-74.

———. 1951. Vegetation zonation in the Great Basin of western North America. In *Les bases écologiques de la régénération de la végétation des zones arides. Int. Union Biol.* ser. B, no. 9:101-122.

———. 1954. Nevada trees. Agricultural Extension Service, *Univ. Nevada Bulletin* 94.

Billings, W. D. 1978. Alpine phytogeography across the Great Basin. *Great Basin Naturalist Memoirs* 2:105-117.

Billings, W. D. and Mark, A. F., 1957. Factors involved in the persistence of montane treeless balds. *Ecology* 38:140-142.

Birkeland, P. W. 1962. Multiple glaciation of the Truckee area, California. In *Geol. Soc. Sacramento and Sacramento Sect. Calif. Assoc. Eng. Geologists, Guidebook,* pp. 64-67.

Birman, J. H. 1964. Glacial geology across the crest of the Sierra Nevada. *Geol. Soc. Amer. Spec. Papers* 75.

Blackburn, W. H., and Tueller, P. T. 1970. Pinyon and juniper invasion in Black Sagebrush Communities in east-central Nevada. *Ecology* 51:841-848.

Blackwelder, Eliot. 1931. Pleistocene glaciation in the Sierra Nevada and Basin Ranges. *Geol. Soc. Amer. Bulletin* 42:865-922.

———. 1934. Supplementary notes on Pleistocene glaciation in the Great Basin. *Wash. Acad. Sci. Journal* 24:217-222.

Blanchard, C. J. 1907. National reclamation of arid lands. In *Ann. Report, Board Regents, Smithsonian Institute* (1905-1906), 1907, pp. 469-492.

Bortz, Louis C. 1959. Geology of the Copenhagen Canyon area, Monitor Range, Eureka County, Nevada. Master's thesis, University of Nevada.

Boulenger, G. A. 1884. Synopsis of the families of existing Lacertilia. *Annals and Magazine Nat. History,* ser. 5, 14:117-122.

————. 1885. *Catalogue of the lizards in the British Museum.* London: Taylor and Francis. Vol. 2.

————. 1896. *Catalogue of the snakes in the British Museum.* London: Taylor and Francis. Vol. 3.

Bradley, W. G. 1967. A geographical analysis of the flora of Clark County, Nevada. *J. Ariz. Acad. Sci.* 4:151-162.

Bradley, W. G., and Deacon, J. E. 1965. The biotic communities of southern Nevada. *Desert Research Institute,* no. 9. University of Nevada.

Brattstrom, B. H. 1954. Amphibians and reptiles from Gypsum Cave, Nevada. *Bulletin Southern Calif. Acad. Sci.* 53(1):8-12.

————. 1958. Additions to the Pleistocene herpetofauna of Nevada. *Herpetologica* 14(1):36.

Breese, C. R., Jr. 1968. A general limnological study of Big Soda Lake. Master's thesis, University of Nevada.

Brodkorb, William P. 1958. Birds from the Middle Pliocene of McKay, Oregon. *Condor* 60(4):252-255.

Broecker, W. S., and Orr, P. C. 1958. Radiocarbon chronology of Lake Lahontan and Lake Bonneville. *Geol. Soc. Amer. Bulletin* 69:1009-1032.

Brown, James E. 1971. Mammals on mountaintops: Nonequilibrium insular biogeography. *Amer. Nat.* 105:467-478.

————. 1978. The theory of insular biogeography and the distribution of boreal birds and mammals. In *Great Basin memoirs no. 2: Intermountain biogeogrpahy: A symposium,* pp. 209-227.

Brown, R. W. 1949. Pliocene plants from Cache Valley, Utah. *Wash. Acad. Sci. Journal* 39:224-229.

————. 1959. Some paleobotanical problematica. *J. Paleontol.* 33:122-124.

Brues, Charles T. 1927. Animal life in Hot Springs. *Quar. Rev. Biol.* 2(2):181-203.

————. 1928. Studies on the fauna of Hot Springs in the western United States and the biology of thermophilous animals. *Proc. Amer. Acad. Arts and Sciences* 63(4):139-228.

————. 1932. Further studies on the fauna of North American Hot Springs. *Proc. Amer. Acad. Arts and Sciences* 67(7):185-303.

Bryan, A. L. 1972. Summary of the archaeology of Smith Creek and Council Hall caves, White Pine County, Nevada. *Nevada Archaeological Survey Reporter* 6(1):6-8.

Burchfield, B. C., and Davis, G.A. 1972. Structural framework and evolution of the southern part of the Cordilleran orogen, western United States. *American J. of Science* 272:97-118.

Burt, C. E. 1933. Some lizards from the Great Basin of the West and adjacent areas, with comments on the status of various forms. *Amer. Midl. Nat.* 14(3):228-250.

Burt, William H. 1929. A new goose, *Branta,* from the Lower Pliocene of Nevada. *Univ. Calif. Dept. Geol. Sci. Bulletin* 18(6):221-224.

————. 1934. The mammals of southern Nevada. *Trans. San Diego Soc. Nat. Hist.* 7(36):375-428.

Bryson, R. A., and Hare, F. K. 1974. Climates of North America. In *World survey of climatology,* vol. 2. New York: Elsevier Scientific Pub. Co.

Cavender, T. M. 1969. An Early Pliocene *Salvelinus* from Nevada with comparative notes on recent species (Abst.). 49th Annual Meeting, Amer. Soc. Ichth. and Herpot., June 1969, pp. 29-30.

————. 1978. Taxonomy and distribution of the bull trout, *Salvelinus Confluentus* (Suckley) from the American Northwest. *Calif. Fish and Game* 64(3):139-174.

Chaney, R. W. 1940. Tertiary forests and continental history. *Geol. Soc. Amer. Bulletin* 51:469-488.

————. 1944a. A fossil cactus from the Eocene of Utah. *Amer. J. Bot.* 31:507-528.

————. 1944b. The Troutdale flora. *Carnegie Inst. Wash. Publ.* 553:323-351.

————. 1947. Tertiary centers and migration routes. *Ecol. Monographs* 17:139-148.

Chaney, R. W., and Axelrod, D. I. 1959. Miocene floras of the Columbia Plateau. *Carnegie Inst. Wash. Publ.* 617:1-237.

Christensen, E. M., and Johnson, H. B. 1964. Presettlement vegetation and vegetational change in three valleys in central Utah. *Brigham Young Univ. Sci. Bulletin, Biol. Ser.* 4:1-16.

Clark, John B.; Dawson, Mary; and Wood, Albert E. 1964. Fossil mammals from the lower Pliocene of Fish Lake Valley, Nevada. *Harvard Univ. Mus. Comp. Zool. Bulletin* 131(2):27-63.

Clements, F. E. 1920. *Plant indicators.* Washington, D.C.: Carnegie Institution.

Cline, Gloria Griffen. 1963. *Exploring the Great Basin.* Norman: University of Oklahoma Press.

Clokey, I. W. 1951. *Flora of the Charleston Mountains, Clark County, Nevada.* Berkeley: University of California Press.

Coogan, Alan H. 1964. Early Pennsylvanian history of Ely Basin, Nevada. *Amer. Assoc. Petroleum Geologists Bulletin* 48(4):487-495.

Cope, E. D. 1868. Observations on some vertebrata from western Nevada and northern lower California. *Proc., Acad. Nat. Sciences Philadelphia* 22:2.

————. 1872. On the Tertiary coal and fossils of Osino, Nevada. *Amer. Philosophical Soc. Proc.* 12:478-481.

————. 1875. Checklist of North American Batrachia and Reptilia with a systematic list of the higher groups, and an essay on geographic distribution based on the specimens contained in the U.S. Nat. Mus. *Bulletin U.S. Nat. Mus.* 1:1-104.

————. 1883a. On the fishes of the Recent and Pliocene lakes of the western part of the Great Basin and of the Idaho Pliocene Lake. *Acad. Nat. Science Philadelphia Proc.* 35:134-166.

————. 1883b. Notes on the geographic distribution of Batrachia and Reptilia in western North America. *Proc. Acad. Nat. Sciences Philadelphia* 35:10-35.

————. 1883c. Zoological geography of western North America. *Science* 1:21.

————. 1889. The Batrachia of North America. *Bulletin U.S. Nat. Mus.* 34:1-525.

————. 1892. A critical review of the characters and variations of the snakes of North America. *Proc. U.S. Nat. Mus.* 14:589-694.

————. 1893. The report of the Death Valley Expedition. *Amer. Nat.* 27:990-995.

————. 1896. The geographical distribution of Batrachia and Reptilia in North America. *Amer. Nat.* 30:886-902, 1003-1026.

————. 1900. The crocodilians, lizards and snakes of North America. *Annual Report, U.S. Nat. Mus. for year ending June 30, 1898* pp. 55-1270.

Cope, E. D., and Yarrow, H. C. 1875. Report upon the collections of fishes made in portions of Nevada, Utah, California, Colorado, New Mexico and Arizona, during the years 1871, 1872, 1873 and 1874. *Report of the United States Geographical Surveys west of the one-hundreth meridian (Wheeler Survey)* 5:635-703.

Cottam, W. P., and Evans, F. R. 1945. A comparative study of the vegetation of grazed and ungrazed canyons of the Wasatch Range, Utah. *Ecology* 26:171-181.

Cottam, W. P., and Stewart, G. 1940. Plant succession as a result of grazing and meadow desiccation by erosion since settlement in 1862. *J. Forest.* (Washington) 38:613-626.

Cottam, W. P. 1961. *Our renewable wild lands—A challenge.* Salt Lake City: University of Utah Press.

Critchfield, W. B., and Allenbaugh, G. L. 1969. The distribution of Pinaceae in and near northern Nevada. *Madrono* 19:12-26.

Cronquist, A.; Holmgren, A. H.; Holmgren, N. H.; and Reveal, J. L. 1972. *Intermountain flora—vascular plants of the intermountain West, U.S.A.* New York: Hafner.

Curry, H. Donald. 1939. Tertiary and Pleistocene mammal and bird tracks in Death Valley. *Geol. Soc. Amer. Bulletin* 50:1971-1972.

Dalley, G. F. 1976. Swallow shelter and associated sites. *Univ. Utah Anthropological Paper*, no. 96.

David, L. R. 1941. *Leptolepis Nevadensis*, A new cretaceous fish. *Paleontology* 15(3):318-321.

Davidson, P. 1919. A cestaciont spine from the Middle Triassic of Nevada. *Univ. Calif. Geol. Publ.* 11(4):433-435.

Davis, E. L. 1978. The ancient Californians. Ranchlabrean hunters of the Mohave Lakes country. *Natural History Museum Los Angeles County Science Series*, no. 29.

Davis, G. A. 1973. Subduction-obduction model for the Antler and Sonoma orogenies, western Great Basin area. *Geological Society of America* 5(7):592.

Deacon, James E.; Bradley, William G.; and Larsen, Karl M. 1964. Ecological distribution of the mammals of Clark Canyon, Charleston Mountains, Nevada. *Mammalogy* 45(3):397-409.

Deffeyes, K. S. 1959. Late Cenozoic sedimentation of tectonic development of central Nevada. Ph.D. dissertation, Princeton University.

Dott, R. H., Jr. 1955. Pennsylvanian stratigraphy of Elko and northern Diamond Ranges, northeastern Nevada. *Bulletin of Amer. Assoc. Petroleum Geologists* 39:2211-2305.

Driscoll, R. S. 1964. A relict area in the central Oregon Juniper Zone. *Ecology* 45:345-353.

Durrant, Stephen D. 1952. Mammals of Utah: Taxonomy and distribution. *Univ. Kansas Publ. Mus. Nat. Hist.* 6:1-549.

Dutton, C. E. 1880. *Geology of the high plateaus of Utah.* Washington, D.C.: Government Printing Office.

Eardley, A. J. 1962. *Structural geology of North America.* 2d ed. New York: Harper and Row.

Eastman, C. R. 1917. Fossil fishes in the collection of the United States National Museum. *U.S. Nat. Mus. Proc.* 52:235-304.

Edmonson, W. T. 1963. Pacific coast and Great Basin. In *Limnology in North America,* ed. David G. Frey, pp. 371-392. Madison: University of Wisconsin Press.

Elliott, R. R. 1973. *History of Nevada.* Lincoln: University of Nebraska Press.

Evernden, J. F., and James, G. T. 1964. Potassium-argon dates and

the Tertiary floras of North America. *Amer. J. Science* 262:945-974.

Fagan, J. L. 1974. Altithermal occupations of spring sites in the northern Great Basin. *Univ. Oregon Anthropological Papers,* no. 6.

Feth, J. H. 1964. Review and annotated bibliography of ancient lake deposits (Precambrian to Pleistocene) in the western states. *U.S.G.S. Bulletin* 1080.

Fisher, A. K. 1893. Report on the ornithology of the Death Valley expedition of 1891, comprising notes on the birds observed in southern California, southern Nevada and parts of Arizona and Utah. *North Amer. Fauna,* 7:7-158.

Flint, Richard F. 1971. *Glacial and Quaternary geology.* New York: John Wiley.

Flowers, S. 1934. Vegetation of the Great Salt Lake region. *Botan. Gaz.* 95:353-418.

———. 1973. *Mosses: Utah and the West.* Provo, Utah: Brigham Young University Press.

Fowler, C. S. 1977. Ethnography and Great Basin prehistory. *Desert Research Institute Publications in the Social Sciences* 12:11-48.

Fowler, D. D. 1966. Great Basin social organization. *Desert Research Institute Publications in the Social Sciences and Humanities* 1:56-76.

———. 1968. The archeology of Newark Cave, White Pine County, Nevada. *Desert Research Institute Publications in the Social Sciences,* no. 3.

———. 1973. Dated split-twig figurine from Etna Cave, Nevada. *Plateau* 46(2):54-63.

———. 1977. Models and Great Basin prehistory. *Desert Research Institute Publications in the Social Sciences* 12:3-10.

Fowler, D. D.; Madsen, D. B.; and Hattori, E. M. 1973. Prehistory of southeastern Nevada. *Desert Research Institute Publications in the Social Sciences,* no. 6.

Furlong, Eustace L. 1910. An aplodont rodent from the Tertiary of Nevada. *Univ. Calif. Dept. Geol. Bulletin* 5:397-403.

———. 1932. A new genus of otter from the Pliocene of the Northern Great Basin Province. *Carnegie Inst. Wash. Publ.* 418:93-104.

———. 1935a. New merycodonts from the Upper Miocene of Nevada. *Carnegie Inst. Wash. Publ.* 453:1-10.

———. 1935b. Pliocene antelopes of the Pronghoro type. *Science,* n.s. 82(2124):250-251.

———. 1943. Occurrence of the Pliocene antelope *Ilingoceros,* in Nevada. *Science* 97(2516):262.

Garman, S. 1884. The North American reptiles and Batrachians: A list of the species occurring north of the Isthmus of Tehuantepec, with references. *Bulletin, Essex Inst.* 16:1-46.

Gidley, James W. 1908. Notes on a collection of fossil mammals from Virgin Valley, Nevada. *Calif. Univ. Geol. Bulletin* 5:235-242.

Gilbert, Grove Karl. 1872. *1874 U.S. Geographical and Geological surveys west 100th meridian progress report.*

———. 1890. Lake Bonneville. *U.S.G.S. Mono.* no. 1.

Girard, C. 1852. A monographic essay on the genus *Phrynosoma.* In *Expedition to the Valley of the Great Salt Lake of Utah,* ed. H. Stansbury, pp. 354-365. Philadelphia: Lippincott, Gram Bo. & Co.

————. 1858. *Herpetology: U.S. exploring expedition, during the years 1838, 1839, 1840, 1841, 1842, under the command of Charles Wilkes, U.S.N.* vol. 20. Philadelphia: Lippincott.

Gould, R. A.; Fowler, D. D.; and Fowler, C. S. 1972. Diggers and doggers: parallel failures in economic acculturation. *Southwestern J. of Anthropology* 28(3):265-281.

Graham, A. E. 1965. The Trout Creek and Sucker Creek Miocene floras of southeastern Oregon. *Kent State Univ. Research Series* 9(6):9-147.

Grant, T. A. 1974. Minor folding in the Rabbit Hill formation, Eureka County, Nevada. Master's thesis, University of Nevada.

Hall, E. R. 1929. A second new genus of hedgehog from the Pliocene of Nevada. *Univ. Calif. Dept. Geol. Sci. Bulletin* 18(8):227-231.

————. 1930a. A new genus of bat (*Mystipterus*) from the later Tertiary of Nevada. *Univ. Calif. Dept. Geol. Sci. Bulletin* 19(14):319-320.

————. 1930b. Rodents and lagomorphs from the later Tertiary of Fish Lake Valley, Nevada. *Calif. Univ. Dept. Geol. Sci. Bulletin* 19(12):295-312.

————. 1940. An ancient nesting site of the white pelican in Nevada. *Condor* 42(1):87-88.

————. 1946. *Mammals of Nevada.* Berkeley: University of California Press.

Hall, E. R., and Kelson, Keith. 1959. *The mammals of North America,* vols. 1-2. New York: Ronald Press.

Hanford, F. S. 1903. The summer birds of Washoe Lake, Nevada. *Condor* 5:50-52.

Hanna, W. C. 1904. Nevada notes. *Condor* 6:47-48, 76-77.

Harner, R. F., and Harper, K. T. 1976. The role of area, heterogeneity, and favorability in plant species diversity of Pinyon-Juniper Ecosystems. *Ecology* 57:1254-1263.

Harrington, M. R. 1933. Gypsum Cave. *Southwest Museum Papers,* no. 8.

————. 1934. American horses and ancient men in Nevada. *Masterkey* 8(6):165-169.

Harrington, M. R., and Simpson, R. D. 1961. Tule Springs, Nevada, with other evidence of Pleistocene man in North America. *Southwest Museum Papers,* no. 18.

Haskill, 1872. Short descriptions of some of the valleys of Nevada. In *Preliminary report of the U.S. Geological Survey of Montana and portions of adjacent territories being a 5th annual report of progress,* F. V. Hayden. Washington, D.C.: Government Printing Office.

Hattori, E. M. 1980. The archeology of Falcon Hill, Winnemucca Lake, Nevada. Ph.D. dissertation, Washington State University.

Hattori, Eugene M. (in press). Late Holocene occurrence of Pacific Pont turtle (*Clemmys Marmorata*), from western Nevada.

Hay, O. P. 1907. A new fossil stickleback fish from Nevada. *U.S. Nat. Mus. Proc.* 32:271-273.

————. 1927. The Pleistocene of the western region of North America and its vertebrated animals. *Carnegie Inst. Wash. Publ.* 322B.

Haynes, C. V., Jr. 1969. The earliest Americans. *Science* 166(3906):709-715.

Hayward, C. L.; Kilpack, M. L.; and Richards, G. L. 1963. Birds of the Nevada Test Site. *Brigham Young Univ. Sci. Bulletin, Biol. Ser.* 3(1).

Hayward, C. L.; Cottam, C.; Woodbury, A. M.; and Frost, H. H. 1976. The birds of Utah. *Great Basin Nat. Memoir* 1.

Hecht, M. K. 1960. A new frog from an Eocene oil-well core in Nevada. *Amer. Mus. Novitates* 2006.

Heizer, R. F. 1951. Preliminary report on the Leonard Rockshelter site, Pershing County, Nevada. *American Antiquity* 17(2):89-98.

Heizer, R. F., and Baumhoff, M. A. 1961. The archeology of two sites at Eastgate, Churchill County, Nevada. *Univ. California Anthropological Records* 20(4).

————. 1965. Post glacial climate and archaeology in the desert west. In *The Quaternary of the United States,* eds. H. E. Wright, Jr. and D. Frey, pp. 697-708. Princeton: Princeton University Press.

————. 1965. post glacial climate and archaeology in the desert West. In *Quaternary of the United States,* eds. H. E. Wright, Jr. and D. Frey. pp. 697-708. Princeton: Princeton University Press.

Heizer, R. F., and Berger, R. 1970. Radiocarbon age of the gypsum culture. *Univ. California Archeological Research Facility Conts.* 7:13-18.

Heizer, R. F., and Napton, L. K. 1970. Archaeology and the prehistoric Great Basin lacustrine subsistence regime as seen from Lovelock Cave, Nevada. *Univ. California Archeological Research Facility Contribution,* no. 10.

Heizer, R. F. et al. 1968. Archeology of South Fork Rockshelter, Elko County, Nevada. *Univ. California Archaeological Survey Reports* 71:1-58.

Heller, A. A. 1909. The mountain mahogany. *Muhlenbergia* 5:62-63

————. 1910. The limber pine. *Muhlenbergia* 6:128-132.

————. 1 101912. *Pinus Ponderosa* and *Pinus Jeffreyi. Muhlenbergia* 1912. *Pinus Ponderosa* and *Pinus Jeffreyi. Muhlenbergia* 8:73-79.

Henshaw, H. W. 1874a. An annotated list of the birds of Utah. *Annals Lyceum Nat. Hist., New York* 11:1-14.

————. 1874b. An annotated list of birds of Utah. In *Geog. and Geol. Expl. & Survey West 100th Meridian,* George M. Wheeler, pp. 39-54. Washington, D.C.: Government Printing Office.

————. 1875. Report on ornithological collections. *Geog. and Geol. Survey West of 100th Meridian,* George M. Wheeler 5:131-507. Washington, D.C. Government Printing Office.

Henshaw, Paul C. 1942. A Tertiary mammalian fauna from the San Antonio Mountains near Tonopah, Nevada. *Carnegie Inst. Wash. Publ.* 530:77-168.

Herre, A. W. C. T. 1911. The desert lichens of Reno, Nevada. *Bot. Gaz.* 51:286-297.

Hibbard, C. W.; Ray, D. E.; Savage, D. E.; Taylor, D. W.; and Guilday, J. E. 1965. Quaternary Mammals of North America. In *The Quaternary of the United States,* eds. H. E. Wright, J. R. and D. G. Frey, pp. 509-525. Princeton: Princeton University Press.

Hickman, Terry J., and Duff, Donald A. 1978. Current status of cutthroat trout. Subspecies in the western Bonneville Basin. *Great Basin Nat.* 38 (2):193-202.

Hoffman, W. J. 1881. Annotated list of the birds of Nevada. *Bulletin, U.S. Geol. and Geog. Surv. Terr.* 6 (2):203-256.

Hotz, P. E., and Willden, Ronald. 1955. Lower Paleozoic sedimentary facies transitional between eastern and western types in the

Osgood Mountains quadrangle, Humboldt County, Nevada. *Geol. Soc. Amer. Bulletin* 66:1652-1653.

Houghton, John G. 1969. Characteristics of rainfall in the Great Basin. Sources of climatic data for Nevada. *Nevada Weatherwatch Spcl. Pub.,* no. 1. Reno: Desert Research Institute.

Houghton, John G.; Sakamoto, Clarence M.; and Gifford, Richard O. 1975. Nevada's weather and climate. *Nevada Bureau of Mines and Geology Specl. Pub.,* no. 2.

Houghton, Samuel G. 1976. *A trace of desert waters: The Great Basin story. Western lands and water series X.* Glendale, Calif.: Arthur A. Clark Co.

Howard, Hildegarde. 1935. A new species of eagle from a Quaternary Cave deposit in eastern Nevada. *Condor* 37 (4):206-209.

———. 1946. A review of the Pleistocene birds of Fossil Lake, Oregon. *Carnegie Inst. Wash. Publ.* 551:141-195.

———. 1952. The prehistoric avifauna of Smith Creek Cave, Nevada, with a description of a new gigantic raptor. *Bulletin So. Calif. Acad. Sciences* 51 (2):50-54.

———. 1958. An ancient cormorant from Nevada. *Condor* 60 (6):411-413.

———. 1964. A new species of the "pigmy goose," *Anabernicula*, from the Oregon Pleistocene, with a discussion of the genus. *Am. Mus. Novitates* 2200.

Howard, Hildegarde, and Miller, Alden. 1939. The avifauna associated with human remains at Rancho La Brea, California. *Carnegie Inst. Wash. Publ.* 514:41-48.

Hubbert, M. K., and Rubey, W. W. 1959. Role of fluid pressure in mechanics of overthrust faulting. *Geol. Soc. Amer. Bulletin* 70:115-166.

Hubbs, C. L., and Miller, R. R. 1948a. The zoological evidence: Correlation between fish distribution and hydrographic history in the desert basins of western U.S. Part 2, the Great Basin with emphasis on glacial and post-glacial times. *Bulletin Univ. of Utah* 38 (2) ; *Biol. Ser.* 10 (7):18-166.

———. 1948b. Two new relict genera of Cyprinid fishes from Nevada. *Univ. Mich. Mus. Zool. Occas. Papers,* no. 507.

———. 1972. Diagnoses of new cyprinid fishes of isolated waters in the Great Basin of western North America. *Trans. San Diego Soc. Nat. Hist.* 17 (8):101-106.

Hubbs, C. L.; Miller, R. R.; and Hubbs, L. C. 1974. Hydrographic history and relict fishes of the north-central Great Basin. *Memoir Calif. Acad. Sciences,* vol., 7.

Hutchinson, G. E. 1937. A contribution to the limnology of arid regions. *Conn. Acad. Arts Science Trans.* 33:47-132.

Jaeger, E. C. 1940. *Desert wild flowers.* Stanford: Stanford University Press.

———. 1957. *The North American deserts.* Stanford: Stanford University Press.

Jehl, Joseph R., Jr. 1967. Pleistocene birds from Fossil Lake, Oregon. *Condor* 69 (1):24-27.

Jennings, J. D. 1957. Danger Cave. *Univ. Utah Anthropol. Papers,* no. 27

———. 1978. *Prehistory of North America.* 2d. ed. New York: McGraw Hill.

Johnson, N. K. 1965. The breeding avifaunas of the Sheep and Spring Ranges in southern Nevada. *Condor* 67:93-124.

———. 1970. The affinities of the boreal avifauna of the Warner Mountains, Calif. *Occas. Pap. Biol. Soc. Nevada.* no. 22.

———. 1973. The distribution of boreal avifaunas in southeastern Nevada. *Occas. Pap. Biol. Soc. Nevada,* no. 36.

———. 1974. Montane avifaunas of southern Nevada: Historical change in species composition. *Condor* 76:334-337.

———. 1975. Controls of number of bird species on montane islands in the Great Basin. *Evolution* 29:545-563.

———. 1978. Patterns of avian geography and specialization in the intermountain region. *Great Basin Nat. Memoirs* 2:137-160.

Jones, J. C. 1925. The geologic history of Lake Lahontan. In *Quaternary climates, Carnegie Inst. Wash. Publ.* 352:1-49.

Jordan, D. S. 1907. The fossil fishes of California with supplementary notes on other species of extinct fishes. *Univ. Calif. Dept. Geol. Bulletin* 5(5):95-144.

———. 1908. Note on a fossil stickleback fish from Nevada. *Smithsonian Miscellaneous Collection* 52:117.

———. 1924. Description of a recently discovered fossil Sculpin from Nevada regarded as *Cottus beldingi. U.S. Nat. Mus. Proc.* 65:1-2.

———. 1925. The fossil fishes of the Miocene of southern Calif. *Stanford Univ. Publ., Univ. Series, Biol. Sci.* 4(1).

Jorgensen, Clive, and Hayward, C. Lynn. 1965. Mammals of the Nevada Test Site. *Brigham Young Univ. Science Bulletin Biol. Ser.* 6(3).

Kay, M., and Crawford, J. P. 1964. Paleozoic facies from miogeosynclinal to the eugeosynclinal belt in thrust slices, central Nevada. *Geological Society of America Bulletin* 75:425-454.

Kelleher, J. 1970. The ecological distributon of the summer birds of the Carson range (easter Sierra) in the area of the Lake Tahoe Nevada State Park. Master's thesis, University of Nevada.

Kellogg, Louise. 1910. Rodent fauna of the Late Tertiary beds at Virgin Valley and Thousand Creek, Nevada. *Calif. Univ. Dept. Geol. Bulletin* 6:421-437.

Kennedy, P. B. 1911. Alpine plants—I, II, III, IV, V. *Muhlenbergia* 7: 95, 103, 111, 121, 133.

———. 1912. Alpine Plants—VI, VII, VIII, IX, X, XI. *Muhlenbergia* 8:18, 25, 46, 59, 72, 95

Kimmel, B. L.; Gersberg, R. M.; Paulson, L. J.; Axler, R. P.; and Goldman, C. R. 1978. Recent changes in the meromictic status of Big Soda Lake, Nevada. *Limnol. Oceanogr.* 23(5):1021-1025.

King, Clarence. 1870. *U.S. geological explorations 40th parallel report.* 3:451-473.

———. 1878. *Systematic geology: U.S. geological explorations 40th parallel* (King), vol. 1.

King, Philip B. 1959. *The evolution of North America.* Princeton: Princeton University Press.

King, P. B. 1969a. *Tectonic map of North America.* U.S. Geological Survey. Reston, Va.

———. 1969b. The tectonics of North America—A discussion to accompany the tectonic map of North America. *U.S. Geological Survey Professional Paper* 628. U.S. Geological Survey, Reston, Va.

Klauber, L. M. 1930. New and renamed subspecies of *Crotalus confluentus* Say, with remarks on related species. *Trans. San Diego Soc. Nat. Hist.* 6 (3):95-144.

————. 1956. *Rattlesnakes: Their habits, life histories, and influence on mankind.* Berkeley: University of California Press.

Koch, David L. 1972. Life history information on the Cui-ui Lakesucker (Chasmistes Cujus, Cope, 1883) endemic to Pyramid Lake, Washoe County, Nevada. Ph. D. dissertation, University of Nevada.

Koenig, E. R.; Baker, J. R.; Paulson, L. J.; and Tew, R. W. 1971. Limnological status of Big Soda Lake, Nevada, October 1970. *Great Basin Naturalist* 31 (2):106-108.

Krochmal, A., and Krochmal, C. 1973. *A guide to the medicinal plants of the United States.* New York: Quadrangle Press.

LaMotte, R. A. 1936. The Upper Cedarville flora of northwestern Nevada and adjacent California. *Carnegie Inst. Wash Publ.* 416:21-68.

Langenheim, Ralph L., Jr., and Larson, E. R. 1973. Correlation of Great Basin stratigraphic units. *Nevada Bureau of Mines and Geology, Bulletin* 72.

LaRivers, Ira. 1942. Some new amphibian and reptile records for Nevada. *J. Ento. and Zool.* 34 (3):53-68.

LaRivers, Ira, and Trelease, T. J. 1952. An annotated checklist of the fishes of Nevada. *Calif. Fish and Game* 38 (1):113-123.

————. 1952. A key to Nevada fishes. *S. Calif. Acad. Science Bulletin* 51 (3):86-102.

————. 1953. A Lower Pliocene frog from western Nevada. *Paleontology* 27 (4):77-91.

————. 1962. Fishes and fisheries of Nevada. *Biol. Soc. of Nevada Memoir* 1.

————. 1964. A new trout from the Barstovian (Miocene) of western Nevada. *Biol. Soc. Nev. Occas. Papers* 3:1-4

————. 1966. Paleontological miscellanei I: A new Cyprinid fish from the Esmeralda (Pliocene) of southeastern Nevada. *Biol. Soc. Nev. Occas. Papers* 11:1-4.

————. 1966a. Paleontological miscellanei II: A new trout from the Esmeralda (Pliocene) of southeastern Nevada. *Biol. Soc. Nev. Occas. Papers.* 11:4-6.

————. 1966b. Paleontological miscellanei III: A new frog from the Nevada Pliocene. *Biol. Soc. Nev. Occas. Papers* 11:6-8.

————. 1978. *Algae of the western Great Basin.* Bioresources Center, Desert Research Institute, Univ. of Nevada System, no. 50008.

Larrison, Earl J. 1967. Guide to Idaho mammals. *J. Idaho Acad. Science,* vol. 7.

Larrison, E.J.; Tucker, J. L.; and Jollie, M. T. 1967. Guide to Idaho birds. *J. Idaho Acad. Sci.* 5:1-220.

Larson, E. R. 1978. Personal communication.

Lawson, C. L. 1973a. Charadriformes new to Nevada. *Western Birds* 4:77-82.

————. 1973b. Notes on Pelecaniformes in Nevada. *Western Birds* 4:23-30.

————. 1974. First Nevada record of chestnut-collared longspur. *Auk* 91:432.

————. 1975. Fish catching by a black phoebe. *Western Birds* 6:107-109.

————. 1977. Nonpasserine species new or unusual to Nevada. *Western Birds* 8:73-90.

Layton, T. N. 1979. Archaeology and paleo-ecology of pluvial Lake Parman, northwestern Great Basin. *J. of New World Archaeology* 3(3):41-56.

Leidy, J. 1868. Notice of some reptilian remains from Nevada. *Proc. Acad. Nat. Sciences Philadelphia* 20:117-178.

Linsdale, J. M. 1936. The birds of Nevada. *Cooper Ornithological Club, Avifauna,* no. 23.

————. 1938. Environmental responses of vertebrates in the Great Basin. *Amer. Midl. Nat.* 19(1):1-206.

————. 1940. Amphibian and reptiles in Nevada. *Proc. Amer. Acad. Arts and Sciences* 73 (8):197-257.

————. 1951. A list of the birds of Nevada. *Condor* 53 (5):228-249.

Linsdale, M. A.; Howell, J. T.; and Linsdale, J. M. 1952. Plants of the Toiyabe Mountains area, Nevada. *Wasmann J. Biol.* 10:129-254.

Lintz, Joseph, Jr. 1978. Personal communication.

Little, E. L., Jr. 1956. Pinaceae, Rutaceae, Meliaceae, Anacardiaceae, Tamaricaceae, Cornaceae, Oleceae, Bignoniaceae of Nevada. *Contr. toward a Flora of Nevada* 40:1-43.

Lloyd, R. M., and Mitchell, R. S. 1973. *A flora of the White Mountains, California and Nevada.* Berkeley: University of California Press.

Louderback, George D. 1907. General geological features of the Truckee Region east of the Sierra Nevada. *Geol. Soc. Amer. Bulletin* 18.

————. 1923. Basin range structure in the Great Basin. *Univ. Calif. Publ. Bulletin Dept. Geol. Sci.* 14 (10):329-376.

————. 1924. Period of scarp production in the Great Basin. *Univ. Calif. Publ. Bulletin Dept. Geol. Sci.* 15 (1):1-44.

————. 1926. Morphologic features of the basin range displacement in the Great Basin. *Univ. Calif. Publ. Bulletin Dept. Geol. Sci.* 16 (1):1-42.

Lovejoy, Earl M. P. 1969. Mount Rose, northern Carson Range, Nevada: New light on the Late Cenozoic tectonic history of the Sierra Nevada from a classic locality. *Geol. Soc. Amer. Bulletin* 80:1833-1842.

Lucas, F. A. 1900. Description of a new species of fossil fish from the Esmeralda formation. In *The Esmeralda Formation: A freshwater lake deposit,* H. W. Turner *U.S.G.S., 21st Annual Report,* pp. 223-226.

Lugaski, T.; Worley, D.; and Kleiner, E. 1972. The first confirmed occurence of white-tailed kites (Elanus leucurus) in Nevada. *Biol. Soc. Nev. Occas. Papers,* no. 28.

Lugaski, T. 1972. A new species of speckle dace from Big Smokey Valley, Nevada. *Biol. Soc. Nev. Occas. Papers,* no. 30.

————. 1977a. Additional notes and discussion of the relationship of *Gila esmeralda.* LaRivers, 1966, from the "Esmeralda" Formation, Nevada. *Biol. Soc. Nev. Occas. Papers,* no. 43.

————. 1977b. The occurrence of *Myzobdella moorei,* a leech, on the Tui-chuls, *Gila bicolor obesus,* in Pyramid Lake, Washoe County, Nevada. *J. Idaho Acad. Science* 13 (1):7-10.

————. 1978. *Fundulus lariversi,* a new miocene fossil cyprinodont fish from Nevada. *Wasmann J. Biology* 35 (2):203-211.

————. 1979. *Gila traini,* a New Pliocene Cyprinid fish from Jersey Valley, Nevada. *J. Paleontology* 53:1160-1164.

Lugaski, T.; Vreeland, P.; and Vreeland, H. 1978. Northern range extension of the Paramint rattlesnake *Crotalus mitchellii stephensi* in the Great Basin. *J. Idaho Acad. Science* 14 (2):32-34.

————. 1979. The integrated ecology of the Great Basin I: A

preliminary analysis of northwestern Stewart Valley. *J. Idaho Acad. Sci.* 15:21-28.

MacDonald, James Reid. 1948. A new Clarendonian fauna from northeastern Nevada. *Calif. Univ. Dept. Geol. Sci. Bulletin* 28 (7):173-194.

———. 1956a. A Blancan mammalian fauna from Wichman, Nevada. *J. Paleo.* 30 (1):213-216.

———. 1956b. A New Clarendonian mammalian fauna from the Truckee formation of Western Nevada. *J. Paleo.* 30 (1):186-202.

———. 1959. The Middle Pliocene mammalian fauna from Smiths Valley, Nevada. *J. Paleo.* 33 (5):872-887.

———. 1966. The Barstovian Camp Creek fauna from Elko County, Nevada. *L.A. Co. Mus. Nat. Hist. Quart.* 4 (3)18-22.

MacNeish, R.S. 1978. Late Pleistocene adaptations: A new look at early peopling of the New World as of 1976. *J. Anthropological Research* 34 (4):475-496.

Madsen, D. B., and Lindsay, L. 1977. Backhoe village. *Utah Antiquities Section, Selected Papers* 4(12).

Madsen, D. B. et al. 1979. New views on the Fremont. *American Antiquity* 44 (4):711-739.

Malouf, C., and Malouf, A. A. 1945. The effects of Spanish slavery on the Indians of the intermountain west. *Southwestern J. Anthropology* 1:378-391.

Martin, P. S. 1963. *The last 10,000 years.* Tucson: University of Arizona Press.

———. 1967. Prehistoric overkill. In Pleistocene extinction, the search for a cause, eds. P. S. Martin and H. E. Wright, Jr. *Proc. 7th. International congress of the Association for Quaternary Research* 6:75-120.

Mason, H. L., and Stockwell, W. P. 1945. A new pine from Mount Rose, Nevada. *Madrono* 8:61-63.

Mawby, John E. 1967. Fossil vertebrates of the Tule Springs site, Nevada. In H. M. Wormington and D. Ellis, *Pleistocene studies in southern Nevada. Nevada State Museum Antaro, Studies* 13:105-128.

———. 1968. *Megahippus* and *Hypohippus* (*Perissodactyla Mammalia*) from the Esmeralda formation of Nevada. *Paleo. Bios.,* no. 7.

———. 1968. *Megabelodon minor* (*Mammalia, Proboscidea*), A new species of mastadont from the Esmeralda formation of Nevada. *Paleo. Bios.,* no. 4.

Maxwell, John C. 1974. Early western margin of the United States. In *The geology of continental margins,* ed. C. A. Burk and C. L. Draker, pp. 831-832. New York: Springer-Verlag.

McConnell, W. J.; Clark, W. J.; and Sigler, W. F. 1957. *Bear Lake, Its fish and fishing.* Utah State Dept. Fish and Game, Idaho Dept. Fish and Game, and Utah State Agr. Coll.

McGinitie, D. 1933. The Trout Creek flora of southeastern Oregon. *Carnegie Inst. Wash. Publ.* 416:21-68.

———. 1941. A Middle Miocene flora from the central Sierra Nevada. *Carnegie Inst. Wash. Publ.* 534:1-178.

McKell, C. M., and Chilcote, W. W. 1957. Response of rabbitbrush following removal of competing vegetation. *J. Range Management* 10:228-230.

Medin, D. E. 1960. Physical site factors influencing annual production of true mountain mahogany, *Cercocarpus Montanus. Ecology* 41 (3):454-460.

Mehringer, P. J., Jr. 1977. Great Basin Late Quaternary environments and chronology. *Desert Research Institute Publications in the Social Sciences* 12:113-168.

Merriam, C. H. 1873. *Report on birds of the expedition,* pp. 670-715. Sixth Annual Report U.S. Geol. Surv. of the Territories embracing portions of Montana, Idaho, Wyoming and Utah, being a report of progress of the explorations for the year 1872 by F.V. Hayden, geologist. *Part III, Special Reports on Zoology and Botany.* Washington, D.C.: Government Printing Office.

Merriam, C. W., and Anderson, C. A. 1942. Reconnaissance surveys of the Roberts Mountains, Nevada. *Geol. Soc. Amer. Bulletin* 53:1675-1727.

Merriam, John C. 1902. Triassic Ichthyopterygia from California and Nevada. *Univ. Calif. Dept. Geol. Bulletin* 3 (4):63-108.

———. 1905. A Primitive Ichthyosaurian limb from the middle Triassic of Nevada. *Univ. Calif. Dept. Geol. Bulletin* 4 (2):33-38.

———. 1906. Preliminary note on a new marine reptile from the Middle Triassic of Nevada. *Univ. Calif. Dept. Geol. Bull.* 5 (5):75-79.

———. 1907. The occurrence of Middle Tertiary mammal-bearing beds in northwestern Nevada. *Science,* n.s. 26:380-382.

———. 1908. Triassic Ichthyosauria, with special reference to American forms. *Memoirs Univ. Calif.* 1 (1).

———. 1909. The occurrence of Strepsicerine antelopes in the Tertiary of north western Nevada. *Univ. Calif. Dept. Geol. Bulletin* 5:319-330.

———. 1910a. The skull and dentition of a primitive Ichthyosaurian from the Middle Triassic. *Univ. Calif. Dept. Geol. Bulletin* 5 (24):381-390.

———. 1910b. Tertiary mammal beds of Virgin Valley and Thousand Creek, in northwestern Nevada. Part I, geologic history. *Univ. Calif. Dept. Geol. Bulletin* 6:21-53.

———. 1911. Tertiary mammal beds of Virgin Valley and Thousand Creek in northwestern Nevada. Part II, vertebrate faunas. *Univ. Calif. Dept. Geol. Bulletin* 6 (11):119-304.

———. 1913. New Anchitheriine horses from the Tertiary of the Great Basin area. *Univ. Calif. Dept. Geol. Bulletin* 7:419-434.

———. 1914a. Correlation between the Tertiary of the Great Basin and that of the marginal marine province in California. *Science,* n.s. 40:643-645.

———. 1914b. The occurrence of Tertiary mammalian remains in northwestern Nevada. *Univ. Calif. Dept. Geol. Bulletin* 8:275-281.

———. 1915a. An occurrence of mammalian remains in a Pleistocene lake deposit at Astor Pass, near Pyramid Lake, Nevada. *Univ. Calif. Dept. Geol. Bulletin* 8:337-384.

———. 1915b. New species of *Hipparion* group from the Pacific Coast and Great Basin provinces of North America. *Univ. Calif. Dept. Geol. Bulletin* 9:1-8.

———. 1916a. *Hipparion*-like horses of the Pacific Coast and Great Basin provinces. *Geol. Soc. Amer. Bulletin* 27:171.

———. 1916b. Tertiary vertebrate fauna from the Cedar Mountain region of western Nevada. *Univ. Calif. Dept. Geol. Bull* 9:161-198.

———. 1917. Relationship of Pliocene mammalian faunas from the Pacific Coast and Great Basin provinces of North America. *Univ. Calif. Dept. Geol. Bulletin* 10:421-443.

————. 1918. Evidence of mammalian paleontology relating to the age of Lake Lahontan. *Univ. Calif. Dept. Geol. Bulletin* 10:517-521.

Merriam, John C., and Bryant, H. C. 1911. Notes on the dentition of *Omphalosaurus. Univ. Calif. Dept. Geol. Bulletin* 6 (14):329-332.

Meyer, S. E. 1978. Some factors governing plant distributions in the Mojave Intermountain Transition Zone. *Great Basin Nat. Memoirs* 2:197-208.

Mifflin, M. D., and Wheat, M. M. 1979. Pluvial lakes and estimated pluvial climates of Nevada. *Nevada Bureau of Mines and Geology Bulletin* 94.

Miller, R. R. 1943. The status of *Cyprinodon macularis* and *Cyprinodon nevadensis,* two desert fishes of western North America. *Univ. Mich. Mus. Zool. Occas. Papers,* 473:1-25.

————. 1946. Correlation between fish distribution and Pleistocene hydrography in eastern California and southwestern Nevada, with a map of the Pleistocene waters. *J. Geol.* 54 (1):43-53.

————. 1948. The Cyprinodont fishes of the Death Valley system of eastern California and southwestern Nevada. *Univ. Mich. Mus. Zool. Misc. Publ.* 68:1-155.

————. 1949. Desert fishes—clues to vanished lakes and streams. *Amer. Mus. Nat. Hist. Mag.* 58 (10):447-451.

————. 1950. Speciation in fishes of the genera *Cyprinodon* and *Empetrichthys,* inhabiting the Death Valley Region. *Evolution* 4 (2):155-163.

————. 1952. Bait fishes of the lower Colorado River from Lake Mead, Nevada to Yuma, Arizona, with a key for their identification. *Calif. Fish and Game* 38 (1):7-42.

————. 1958. Origin and affinities of the freshwater fish fauna of western North America. In Zoogeography. *Amer. Assoc. Adv. Sci. Publ.* 51:187-222.

Miller, R. R., 1965. Quaternary freshwater fishes of North America In *The Quaternary of the United States,* ed. H. E. Wright, Jr. and David G. Frey, pp. 569-581. Princeton: Princeton University Press.

Miller, R. R., and Hubbs, C. L. 1960. The spiny-rayed Cyprinid fishes (Plagopterini) of the Colorado River system. *Univ. Mich. Mus. Zool. Misc. Publ.* 115:1-39.

Monger, J. W. H.; Souther, J. G.; and Gabrielse, H. 1972. Evolution of the Canadian Cordillera: A plate-tectonic model. *American J. Science,* 272:577-602.

Morgan, W. J. 1968. Rises, trenches, great faults, and crustal blocks. *J. Geophysical Research* 73:1959-1982.

Morrison, Roger B. 1964. Lake Lahontan: Geology of southern Carson Desert, Nevada. *U.S.G.S. Prof. Papers,* vol. 401.

————. 1965. Quaternary geology of the Great Basin. In *The Quaternary of the United States,* eds. H. E. Wright, Jr. and David G. Frey, pp. 265-285. Princeton: Princeton University Press.

Morss, N. 1931. The Ancient culture of the Fremont River in Utah. Peabody Museum of Archaeology and Ethnology, *Harvard University Papers* 12(3).

Muller, S. W. 1949. Sedimentary facies and geologic structures in the basin and range province. In *Sedimentary facies in geologic history* (symposium), ed. C. R. Longwell. *Geol. Soc. Amer. Memoirs,* vol. 39.

Murphy, Michael A., and Gronberg, Eric C. 1970. Stratigraphy and correlation of the Lower Nevada Group (Devonian) north and west of Eureka, Nevada: *Geol. Soc. Amer. Bulletin* 81:127-136.

Nappe, L., and Klebenow, D. A. 1973. Rare and endangered birds of Nevada. *Agric. Exp. Station Publ.* R-95.

National Oceanic and Atmospheric Administration. 1975. *Local climatological data. Annual summaries for 1974,* pt. 2. Asheville, N.C.: National Climatic Center.

Nolan, T. B. 1928. A late Paleozoic positive area in Nevada. *American J. Science,* 5th series 16:153-161.

————. 1943. The basin and range province in Utah, Nevada, and California. *U.S. Geol. Survey Professional Paper* 197-D:141-196.

Nyquist, David. 1963. The ecology of *Eremichthys acros,* an endemic thermal species of Cyprinid fish from northwestern Nevada. Master's thesis, University of Nevada.

O'Connell, J. F. 1975. The prehistory of Surprise Valley. *Ballena Press Anthropological Papers,* no. 4.

Odum, E. P. 1971. *Fundamentals of ecology.* Philadelphia: W. B. Saunders.

O'Farrell, T. P., and Emery, L. A. 1976. *Ecology of the Nevada Test Site: A narrative summary and annotated bibliography.* Springfield, Va.: National Technical Information Services.

Oosting, H. J., and Billings, W. D. 1943. The red fir forest of the Sierra Nevada. *Ecol. Monogr.* 13:259-274.

Orr, Phil C. 1969. *Felis trumani,* a new radiocarbon dated cat skull from Crypt Cave, Nevada. *Santa Barbara Mus. Nat. Hist. Dept. Geol. Bulletin* 2.

Pack, H. J. 1930. Snakes of Utah. *Utah Agr. Exp. Station Bulletin* 221.

Parker, L. R. 1968. A reconnaissance of upper cretaceous plants from the Blackhawk formation in central Utah and their Paleoecological significance. Master's thesis, Brigham Young University.

Pippin, L. C. et al. 1979. Archeology at Eagle Lake, California. *Desert Research Institute Technical Report* 79003.

Potbury, S. S. 1935. The LaPorte flora of Plumas County, Calif. *Publ. Carnegie Inst. Wash.* 465:29-81.

Putnam, W. C. 1950. Moraine and shoreline relationships at Mono Lake, California. *Geol. Soc. Amer. Bulletin* 6:115-122.

Reed, Charles A. 1958. A fossorial mammal of unknown affinities from the Middle Miocene fauna of Nevada. *J. Mammalogy* 39 (1):87-91.

Reifschneider, O. 1964. *Biographies of Nevada botanists.* Reno: University of Nevada Press.

Remy, J., and Brenchley, J. 1861. *A journey to Great Salt Lake City.* London.

Richards, Gerald. 1962. Wintering habits of some birds at the Nevada Atomic Test Site. *Great Basin Nat.* 22:30-31.

Richardson, C. H., Jr. 1915. Reptiles of northwestern Nevada and adjacent territory. *Proc. U.S. Nat. Mus.* 48:403-435.

Richmond, G. M. 1960. Glaciations of the east slope of Rocky Mountain National Park, Colorado. *Geol. Soc. Amer. Bulletin* 71:1371-1382.

————. 1961. New evidence of the age of Lake Bonneville from the moraines in Little Cottonwood Canyon, Utah. *U.S.G.S. Prof. Papers* 424-D:127-128.

————. 1962. Quaternary stratigraphy of the LaSal Mountains, Utah. *U.S.G.S. Prof. Papers* 324.

————. 1964. Glaciation of Little Cottonwood and Bells Canyons, Utah. *U.S.G.S. Prof. Papers* 454-D:1-41.

Ridgeway, R. 1877. Ornithology. In *Clarence King's report of U.S. Geol. expl. of the 40th parallel*, pt. 3, 4:303-669. Washington, D.C.: Government Printing Office.

Roberts, R. J. 1964. Geology of the Antler Peak quadrangle. *U.S. Geological Survey, Professional Paper* 459-A.

————. 1972. Evolution of the Cordilleran Fold Belt. *Geol. Soc. Amer. Bulletin* 83:1989-2004.

Roberts, R. J.; Hotz, P. E.; Gilluly, James; and Ferguson, H. G. 1958. Paleozoic rocks of north-central Nevada. *Amer. Assoc. Petroleum Geologists Bulletin* 42:2813-2857.

Roberts, R. J., and Lehner, R. E. 1955. Additional data on the age and extent of the Roberts Mountain thrust fault, north-central Nevada. *Geol. Soc. Amer. Bulletin* 66:1661.

Robertson, J. H. 1947. Responses of range grasses to different intersites of competition with sagebrush (*Artemisia tridentata* Nutt.). *Ecology* 28:1-16.

Ruben, John. 1971. A Pliocene Colubrid snake (Reptilia: Colubridae) from west-central Nevada. *Paleo. Bios.* 13.

Russell, I. C. 1883. Sketch of the geological history of Lake Lahontan. *U.S.G.S. Annual Report* 3:189-235.

————. 1885. Geologic history of Lake Lahontan, A quaternary lake of northwestern Nevada. *U.S.G.S. Monogr.* 11.

————. 1897. *Glaciers of North America.* Boston: Ginn and Company.

Ryser, Fred A. 1963. Prothonotary warbler and yellow-shafted flicker in Nevada. *Condor* 65:334.

Salisbury, F. B. 1964. Soil formation and vegetation on hydrothermally altered rock material in Utah. *Ecology* 45:1-9.

Sampson, A. W. 1925. The foothill-montane-alpine flora and its environment. In *Flora of Utah and Nevada*, ed. I. Tidestrom. *Contr. from U.S. Nat. Herbarium* 25:24-31.

Schroeder, M. J., and Buck, C. C. 1970. *Fine WOeather U.S. Forest Service, Agriculture Handbook,* no. 360.

Schulze, E. D.; Mooney, H. A.; and Dunn, E. L. 1967. Wintertime photosynthesis of bristlecone pine (*Pinus aristata*) in the White Mountains of California. *Ecology* 48:1044-1047.

Selander, Robert K. 1965. Avian speciation in the Quaternary. In *The Quaternary of the United States*, ed. H. E. Wright, Jr., and D. G. Frey. pp. 527-542. Princeton: Princeton University Press.

Shantz, H. L. 1925. Plant communities in Utah and Nevada. In *Flora of Utah and Nevada*, ed. I. Tidestrom. *Contr. from U.S. Nat. Herbarium* 25:15-23.

Sharp, R. P. 1938. Pleistocene glaciation in the Ruby East Humboldt Range, northeastern Nevada. *J. Geomorphology* 1:296-323.

————. 1940. Geomorphology of the Ruby East Humboldt Range, Nevada. *Geol. Soc. Amer. Bulletin* 51:337-372.

Sharp, R. P., and Birman, J. H. 1963. Additions to classical sequence of Pleistocene glaciations, Sierra Nevada, Calif. *Geol. Soc. Amer. Bulletin* 74:1079-1086.

Shotwell, J. Arnold. 1961. Late Tertiary biogeography of horses in the northern Great Basin. *J. Paleont.* 35 (1):203-217.

————. 1967a. Late Tertiary geomyoid rodents of Oregon. *Oregon Univ. Mus. Nat. Hist. Bulletin* 9.

————. 1967b. *Peromyscus* of the Late Tertiary in Oregon. *Oregon Univ. Mus. Nat. Hist. Bulletin* 5.

————. 1970. Pliocene mammals of southeast Oregon and adjacent Idaho. *Oregon Univ. Mus. Nat. Hist. Bulletin* 17.

Shreve, F. 1942. The desert vegetation of North America. *Bot. Review* 8:195-246.

Shutler, M. E., and Shutler, R., Jr. 1963. Deer Creek Cave, Elko County, Nevada. *Nevada State Museum Anthropological Papers,* no. 11.

Shutler, R., Jr. 1961. Lost City, Peublo Grande de Nevada. *Nevada State Museum Anthropological Papers,* no. 5.

Sigler, William F., and Miller, R. R. 1963. *Fishes of Utah.* Utah State Dept. of Fish and Game, Salt Lake City.

Silberling, N. J. 1959. Pre-Tertiary stratigraphy and upper Triassic paleontology of the Union district, Shoshone Mountains Nevada. *U.S. Geological Survey Professional Paper,* no. 322.

Silberling, N. J. and Roberts, Ralph J. 1962. Pre-Tertiary stratigraphy and structure of northwestern Nevada. *Geol. Soc. Amer., Special Paper* 72:58.

Simms, S. R. 1977. A Mid-Archaic subsistence and settlement shift in the northeastern Great Basin. *Desert Research Institute Publications in the Social Sciences* 12:195-210.

Simpson, George G. 1961. *Horses. The story of the horse family in the modern world and through sixty million years of history.* Garden City, N.Y.: Doubleday.

Simpson, J. H. 1876. *Report of explorations across the Great Basin of the Territory of Utah for a direct wagon-route from Camp Floyd to Genoa, in Carson Valley, in 1859.* Washington, D.C.: U.S. Engineering Dept.

Simpson, R. D. 1976. A commentary on W. Glennan's article. *J. of New World Archeology* 1(7):63-66.

Slemmons, David B. 1967. Pliocene and Quaternary crustal movements of the Basin-and Range Province, USA. *J. of Geoscience, Osaka City University,* 10:1-11.

Slevin, J. R. 1928. The amphibians of western North America. *Ocas. Pap. Calif. Acad. Sciences* 17:1-152.

————. 1934. A handbook of reptiles and amphibians of the Pacific states including certain eastern species. *Calif. Acad. Sciences, Spec. Publ.*

Smith, E. R. 1941. The archeology of Deadman Cave, Utah. *Univ. Utah Bulletin* 32(4).

Smith, G. L. 1973. *A flora of the Tahoe Basin and neighboring areas.* San Francisco: University of San Francisco.

Smith, G. R.; Stokes, W. L.; and Horn, K. F. 1968. Some late Pleistocene fishes of Lake Bonneville. *Copeia* 4:807-816.

Smith, Gerald R. 1978. Biogeography of intermountain fishes. In *Great Basin memoirs no. 2: Intermountain biogeography: A symposium,* pp. 17-42.

Snyder, C. T.; Hardman, G.; and Zdendek, F. F. 1964. Pleistocene lakes in the Great Basin. *U.S.G.S. Misc. Geologic Investigations,* map 1-416.

Snyder, John Otterbein. 1917. The fishes of the Lahontan system of Nevada and northeastern California. *U.S. Bur. Fish. Bulletin* 35:33-86.

————. 1919. Three new whitefishes from Bear Lake, Idaho and Utah. *Bulletin U.S. Bur. Fish.* 36(1917-1918:)1-9.

————. 1921. Notes on some western fluvial fishes, described by Charles Girard in 1856. *Proc. U.S. Nat. Mus.* 59:23-28.

————. 1924. Notes on certain catostomids of the Bonneville

system, including the type of *Pantosteus virescens* Cope. *Proc. U.S. Nat. Mus.* 64:1-6.

Spurr, J. E. 1901. Origin and structure of the basin ranges. *Geol. Soc. Amer. Bulletin,* 12:217-270.

Stålfelt, M. G. 1960. *Plant ecology, plants, the soil, and man,* London: William Clowes & Sons. Translated by M. S. Jarvis and P. G. Jarvis, 1972. New York: John Wiley & Sons.

Stansbury, Howard. 1852. Exploration and survey of the valley of the Great Salt Lake of Utah, including a reconnaissance of a new route through the Rocky Mountains. *Special Session, U.S. Senate, March 1851, Exec.,* no. 3.

Stebbins, R. C. 1951. *Amphibians of western North America.* Berkeley: University of California Press.

———. 1954. *Amphibians and reptiles of western North America.* New York: McGraw-Hill.

Stejneger, L. H. 1893. Annotated list of the reptiles and Batrachians collected by the Death Valley expedition in 1891, with descriptions of new species. *North American Fauna* 7:159-229.

———. 1895. The poisonous snakes of North America. *Annual Report, U.S. Nat. Mus.* (1893):337-487.

Steward, J. H. 1937. Ancient caves of the Great Salt Lake Region. *Bureau of American Ethnology Bulletin,* no. 116.

———. 1938. Basin-Plateau aboriginal sociopolitical groups. *Bureau of American Ethnology Bulletin,* no. 120.

Stewart, J. H. 1972. Initial deposits in the Cordilleran geosyncline: Evidence of a late Precambrian (850 my) continental separation. *Geol. Soc. Amer. Bulletin* 83:1345-1360.

Stirton, R. A. 1929. Artiodactyla from the fossil beds of Fish Lake Valley, Nevada. *Univ. Calif. Dept. Geol. Sci. Bulletin* 18(11):291-302.

———. 1931. Castoridae from the Tertiary of Nevada. *Geol. Soc. Amer. Bulletin* 43(1):288.

———. 1932a. An Association of horn cores and upper molars of the antelope *Sphenophalos nevadanus* from the Lower Pliocene of Nevada. *Am. J. Sci.* 5th Series 24:46-51.

———. 1932b. Correlation of the Fish Lake Valley and Cedar Mountain beds in the Esmeralda formation of Nevada. *Science,* n.s. 76:60-61.

———. 1934. Phylogeny of North American Miocene and Pliocene Equidae. *Geol. Soc. Amer. Proc.* 1934:382-383.

———. 1935. A review of the Tertiary beavers. *Univ. Calif. Dept. Geol. Sci. Bulletin* 23(13):391-458.

———. 1936. Succession of North American continental Pliocene mammalian faunas. *Am. J. Sci.,* 5th series, 32:161-206.

Stock, Chester. 1921. Later Cenozoic mammalian remains from the Meadow Valley region, southeastern Nevada. *Am. J. Sci.,* 5th series, 2:250-264.

———. 1926. Anchitherine horses from Fish Lake Valley region, Nevada. *Univ. Calif. Dept. Geol. Sci. Bulletin* 16:43-60.

———. 1951. *Neohipparion leptode* (Merriam) from the Pliocene of northwestern Nevada. *Amer. J. Sci.,* 249:430-438.

Stokes, W. L. 1963. Triassic and Jurassic Periods in Utah. *Utah Geological and Mineralogical Survey Bulletin* 54:109-121.

Stokes, William L., and Condie, Kent C. 1961. Pleistocene bighorn sheep from the Great Basin. *J. Paleont.* 35(3):598-609.

Stokes, W. L.; Smith, G. R.; and Horn, K. F. 1964. Fossil fishes from the Stansbury Level of Lake Bonneville, Utah. *Proc. Utah Acad. Sci., Arts and Letters* 41:87-88.

Stutz, H. C. 1978. Explosive evolution of perennial *Atriplex* in Western America. *Great Basin Nat. Memoirs* 2:161-168.

Tanner, Vasco. 1936. A study of the fishes of Utah. *Utah Acad. Sci., Arts and Letters* 13:155-183.

Tanner, W. W. 1978. Zoogeography of reptiles and amphibians in the intermountain region. In *Great Basin Nat. memoir No. 2: intermountain biogeography: A symposium,* pp. 43-54.

Tanner, W. W., and Banta, B. H. 1977. The systematics of *Crotaphytus wislizeni,* the leopard lizard. Part III. The leopard lizards of the Great Basin and adjoining areas, with a description of a new subspecies from the Lahontan Basin. *Great Basin Nat.* 37:225-240.

Tanner, W. W., and Jorgensen, C. D. 1963. Reptiles of the Nevada Test Site. *Brigham Young Univ. Sci. Bulletin, Biol. Series* 3(3):1-31.

Taylor, E. H. 1941. A new anuran from the Middle Miocene of Nevada. *Univ. Kansas Science Bulletin* 27(4):61-64.

Taylor, W. P. 1912. Field notes on amphibians, reptiles and birds of northern Humboldt County, Nevada, with a discussion of some of the faunal features of the region. *Univ. Calif. Publ. Zool.* 7(10):319-436.

Thomas, C. E. et al. 1970. Annual climatological extremes in Nevada. *Univ. of Nevada, Dept. of Civil Engineering, Report,* no. 39.

Thomas, D. H. 1973. An empirical test for Steward's model of Great Basin settlement patterns. *American Antiquity* 38(2):155-176.

Tidwell, W. D. 1967. Flora of Manning Canyon shale. Part I: A lower-most Pennsylvanian flora from the Manning Canyon Shale, Utah, and its stratigraphic significance. *Brigham Young Univ. Geol. Studies* 14:3-66.

Tidwell, W. D.; Rushforth, Samuel R.; and Simper, Daniel. 1972. Evolution of floras in the intermountain region. In *Intermountain flora,* ed. A. Cronquist, A. Homgren, N. Holmgren, and J. Reveal, vol. 1, pp. 19-39. New York: Hafner.

Ting, I. P. 1961. An ecological study of *Dalea polyadenia.* Master's thesis, University of Nevada.

Tueller, P. T. 1973. Secondary succession, disclimax, and range condition standards in desert shrub vegetation. Arid Shrublands Proceedings of the Third Workshop, pp. 57-65.

Tueller, P. T.; Bruner, A. D.; Everett, R.; and Davis, J. B. 1974. *The ecology of Hot Creek Valley, Nevada and nonradiation effects on an underground nuclear detonation.* Springfield, Va.: National Technical Information Service.

Tueller, P. T., and Clark, J. E. 1976. *Nonradiation effects on natural vegetation from the Almendro underground nuclear detonation.* Springfield, Va.: National Technical Information Service.

Tuohy, D. R. 1974. A comparative study of late PaleoIndian manifestations in the Great Basin. *Nevada Archeological Survey Research Papers* 5:91-116.

———. 1980. The archeology of the Pyramid Lake region, Nevada. *Nevada State Museum Anthropological Papers,* no. 18.

Tuohy, D. R., and Rendell, D., eds. 1979. Archeological excavations in the Smith Creek region, eastern Nevada. *Nevada State Museum Anthropological Papers,* no. 17.

Turner, H. W. 1909. Contribution to the geology of the Silver Peak

Quadrangle, Nevada. *Geol. Soc. Amer. Bulletin* 20:223-264.

Van Denbergh, J. 1895. The species of the genus *Xantusia*. *Proc. Calif. Acad. Sciences Series 2* (5):523-534.

———. 1897. The reptiles of the Pacific Coast and Great Basin, an account of the species known to inhabit California, Oregon, Washington, Idaho and Nevada. *Occas. Pap. Calif. Acad. Sciences* 5:1-236.

———. 1922. The reptiles of western North America. Vol. I, Lizards. Vol. II, Snakes and turtles. *Occas. Pap. Calif. Acad. Sciences* 10:1-611, 617-1028.

Van Denbergh, J., and Slevin, J. R. 1918. The garter snakes of western North America. *Proc. Calif. Acad. Sciences Series 4* 8(6):181-270.

———. 1919. The gopher snakes of western North America. *Proc. Calif. Acad. Sciences Series 4* 9(6):197-220.

———. 1921. A list of the amphibians and reptiles of Nevada, with notes on the species in the collection of the academy. *Proc. Calif. Acad. Sciences Series 4* 11(2):27-38.

Van Rossem, A. J. 1931. Descriptions of new birds from the mountains of southern Nevada. *Trans. San Diego Soc. Nat. Hist.* 6:325-332.

Vasek, F. C. 1966. The distribution and taxonomy of three western junipers. *Brittonia* 18:350-372.

Vreeland, P.; Vreeland, H.; and Kleiner, E. F. 1976. *Sarcodes sanguinea* Torr., a Mycorrhizal species. *Amer. Midl. Nat.* 96:507-512.

Vreeland, H.; Vreeland, P.; and Lugaski, T. 1978a. Exploring the desert of the Great Basin. *Adventure Travel* 1:38-43.

———. 1978b. The integrated ecology of the Great Basin II: A preliminary analysis of southeastern Stewart Valley. *J. Idaho Acad. of Sciences* 14:42-50.

———. 1979. The integrated ecology of the Great Basin III: A preliminary analysis of Newark Summit, Eureka County, Nevada. *Northwest Science* 53:180-189.

Watters, D. R. 1979. On the hunting of "big game" by Great Basin aboriginal populations. *J. New World Archaeology* 3(3):57-64.

Webb, G. W. 1958. Middle Ordovician stratigraphy in eastern Nevada and western Utah. *Amer. Assoc. Petroleum Geologists Bulletin* 42:2335-2377.

Wedel, W. R. 1978. The prehistoric plains. In *Ancient native Americans*, ed. J. D. Jennings, pp. 183-220. San Francisco: W. H. Freeman.

Weide, M. L. 1974. North Warner subsistence network: A prehistoric band territory. *Nevada Archeological Survey Research Papers* 5:63-79.

Weide, M. L., and Weide, D. L. 1977. Time, space and intensity in Great Basin paleo-ecological models. *Desert Research Institute Publications in the Social Sciences* 12:79-112.

Welsh, J. E. 1959. Biostratigraphy of Pennsylvanian and Permian systems in southern Nevada. Ph.D. dissertation, University of Utah.

Wemple, E. M. 1906. New Cestraciont teeth from the west American Triassic. *Univ. Calif. Publ. Geol. Sci.* 5(4):71-73.

Went, F. W. 1948. Ecology of desert plants I. Observations on germination in Joshua Tree National Monument, California. *Ecology* 29:242-253.

———. 1949. Ecology of desert plants II. The effect of rain and temperature on germination and growth. *Ecology* 30:1-13.

———. 1964. Growing conditions of alpine plants. *Israel J. Bot.* 13:82-92.

West, N. E.; Tausch, R. J.; Rea, K. H.; and Tueller, P. T. 1978. Phytogeographical variation within juniper-pinyon woodlands of the Great Basin. *Great Basin Nat. Memoirs* 2:119-136.

Wheeler, H. E. 1939. *Helicoprion* in the Anthracolithic (Late Paleozoic) of Nevada and California and its stratigraphic significance. *J. Paleont.* 13(1):103-114.

Wheeler, H. E.; Scott, W. F.; and Thompson, T. L. 1949. Permian-Mesozoic statigraphy in northeastern Nevada (Abstract). *Geol. Soc. Amer. Bulletin* 60:1928.

Wheeler, S. M. 1942. *The archeology of Etna Cave, Lincoln County, Nevada.* Carson City: Nevada State Parks Dept.

Wheeler, S. S. 1967. *The desert lake.* Caldwell, Idaho: Caxton Printers.

———. 1971. *The Nevada Desert.* Caldwell, Idaho: Caxton Printers.

———. 1978. *The Black Rock Desert.* Caldwell, Idaho: Caxton Printers.

Whittaker, R. H. 1970. *Communities and ecosystems.* London: Macmillan.

Williams, J. E. 1978. Taxonomic status of *Rhinichthys osculus* (Cyprinidae) in the Moapa River, Nevada. *Southwestern Nat.* 23(3):511-518.

Wilson, Robert Warren. 1936. A Pliocene rodent fauna from Smiths Valley, Nevada. *Carnegie Inst. Wash. Publ.* 473:15-34.

———. 1938. Pliocene rodents of western North America. *Carnegie Inst. Wash. Publ.* 487:21-73.

Wolfe, J. A. 1964. Miocene floras from Fingerrock Wash, southwestern Nevada. *U.S.G.S. Prof. Papers* 454N:N1-N36.

Woodbury, A. M. 1931. A descriptive catalog of the reptiles of Utah. *Bulletin Univ. Utah* 21(5).

———. 1947. Distribution of pigmy conifers in Utah and northeastern Arizona. *Ecology* 28:113-126.

———. 1952. Amphibians and reptiles of the Great Salt Lake Valley. *Herpetologica* 8:42-50.

Woodbury, A. M.; Cottam, C.; and Sugden, J. W. 1949. Annotated Checklist of the Birds of Utah. *Bulletin Univ. Utah,* 39(16).

Woodbury, A. M., and Smart, Earl W. 1950. Unusual snake records from Utah and Nevada. *Herpetologica* 6:45-47.

Woods, W. F., and Brock, K. 1964. Interspecific transfer of Ca^{45} and P^{32} by root systems. *Ecology* 45:886-889.

Wormington, H. M., and Ellis, D., eds. 1967. Pleistocene studies in southern Nevada. *Nevada State Museum Anthropological Papers,* no. 13.

Yarrow, H. C., and Henshaw, H. W. 1878. Report upon the reptiles and batrachians collected during the years 1875, 1876 and 1877, in Calif., Arizona and Nevada. Appendix L of annual report, chief of engineers for 1878. *Geographical and geological surveys of territory of U.S. west of the 100th meridian under the direction of Lt. George M. Wheeler,* pp. 206-226.

Young, J. A., and Evans, R. A. 1974. Population dynamics of green rabbitbrush. *J. Range Management* 27:127-132.

Zweifel, R. G. 1956. Two pelobatid frogs from the Tertiary of North America and their relationships to fossil and recent forms. *Amer. Mus. Novitates* 1762.

RECENT NATIVE FISHES OF THE GREAT BASIN

LAHONTAN BASIN DRAINAGE

Family Salmonidae

Salmo clarki henshawi. Lahontan cutthroat trout, also known as Lahontan blackspotted trout, is found in the Lahontan drainage system of Nevada: the Humboldt, Truckee, Carson, and Walker lakes.

Salmo gairdneri regalis. The Tahoe rainbow trout or royal silver trout was originally described from Lake Tahoe-Nevada-California and is thought to be extinct.

Salmo gairdneri smaragdus. The pyramid rainbow trout or emerald trout, originally described from Pyramid Lake and now extinct.

Salvelinus confluentus. The bull trout recently has been found to exist in the streams of the Humboldt drainage. Little is known about its ecology and distribution.

Family Coregonidae

Prosopium williamsoni. The mountain whitefish is found throughout the western United States. Its distribution in the Lahontan drainage is confined to upper headwaters and Lake Tahoe.

Family Catostomidae

Catostomus platyrhynchus (Pantosteus lahontan). The mountain sucker is found throughout the Lahontan drainage area and numerous adjacent areas in the northern and eastern Great Basin.

Catostomus tahoensis. The Tahoe sucker is found in the Walker, Carson, Truckee, Susan, and Humboldt rivers of the Lahontan drainage, as well as in Walker, Pyramid, and Tahoe lakes.

Chasmistes cujus. The cui-ui lakesucker is found only in Pyramid Lake and is considered to be an endangered species. The cui-ui was a major food resource of the Indians in the area, but in the last one hundred years the decreasing water levels in the lake and Truckee River spawning grounds have reduced the population.

Family Cyprinidae

Richardsonius egregius. The Lahontan redshiner is found in the Walker, Carson, Truckee, Susan, Quinn, Reese, and Humboldt rivers and Walker, Tahoe, and Pyramid lakes of the Lahontan drainage.

Gila (Siphateles) bicolor obesus. The Lahontan tui-chub is found in the Walker, Carson, Truckee, Susan, Quinn, Humboldt, and Reese rivers, as well as the major lakes of the Lahontan drainage.

Gila bicolor newarkensis. The Newark Valley tui-chub is found in isolated springs and ponds of Newark Valley, Nevada. It has become extinct in many habitats due to the introduction of predatory game fish.

Gila bicolor euchila. The Fish Creek Springs tui-chub is found in the warm springs and creeks of Fish Creek (south of Newark Valley) Valley. It has become extinct in many of its habitats due to the introduction of predatory game fish.

Gila bicolor isolata. The Independence Valley tui-chub is found in the warm springs, pools, creeks, and marshes of that valley. It is being eliminated from many of the larger pools by introduced predatory game fish.

Gila alvordensis. The Alvord chub is found in Borax Lake of southeastern Oregon and the Virgin Valley-Thousand Creeks drainage of northwestern Nevada. Although limited in its distribution, it has maintained large populations.

Rhinichthys osculus robustus. The Lahontan speckled dace is found in the Walker, Carson, Truckee, Susan, Quinn, Humboldt, and Reese rivers and major lakes in the Lahontan drainage.

Rhinichthys osculus lariversi. The Big Smoky Valley speckled dace is found in a few isolated springs in Big Smoky Valley where it maintains moderate populations. Each population could easily become extinct if predatory game fish were introduced or if excessive water were removed from the springs.

Rhinichthys osculus religuus. The Grass Valley speckled dace was known from the creeks and springs of Grass Valley but is now considered extinct.

Rhinichthys osculus oligoporus. The Clover Valley speckled dace is found only in isolated warm springs of Clover Valley and due to introduction of predatory game fish and disruption of habitat by humans is facing extinction.

Rhinichthys osculus lethoporus. The Independence Valley dace is found in the same habitats as the Independence Valley chub and has also suffered from introduced predatory game fish.

Eremichthys acros. The Soldier Meadows dace is restricted to the warm springs found in Soldier Meadows. The ecology, distribution and physiology of this unique endemic fish is described by Nyquist (1963) and Davey (1966). For a long time the Soldier Meadows dace maintained a rather large population, but in recent years water diversion for irrigation and the use of water for recreation (swimming) has resulted in the elimination of several populations.

Family Cottidae

Cottus beldingi. The Belding sculpin (also known as the Lahontan Sculpin or Pahute Sculpin) is found throughout the Lahontan drainage in mountain streams in eastern Nevada.

WHITE RIVER-COLORADO RIVER DRAINAGE

Family Catostomidae

Catostomus clarki (Pantosteus intermedius). This sucker is found in the White River drainage and throughout tne Colorado River system.

Catostomus latipinnis. The flannelmouth sucker is found in the Colorado River drainage in great numbers but not in the White River.

Xyranchan texanus. The razorback sucker is found in the Colorado River system but not in the White River.

Family Cyprinidae

Ptychocheilus lucius. The Colorado squawfish is found throughout the Colorado River system but not in the White River. It is considered endangered over most of its range.

Gila cypha. This Colorado chub is found only in isolated areas of the Colorado River but not in the White River. It is considered endangered and may be extinct in many areas.

Gila robusta. The roundtail chub is found in the Colorado River system.

Gila elegans. The bonytail chub is found in the Colorado River system.

Gila robusta jordani. The White River roundtail chub is known from the Pahranagat Lakes of Pahranagat Valley in the White River drainage. It is considered to be endangered.

Rhinichthys osculus velifer. The White River speckled dace is known only from certain isolated springs, pools, and small lakes in the White River drainage. It has become extinct in several areas due to the introduction of predatory game fish.

Rhinichthys osculus moapae. Recently the Moapa speckled dace

has been tentatively described (Williams 1978) but is not well known.

Moapa coriacea. The Moapa dace is a relict of past pluvial times and is restricted to the springs and headwaters of the Moapa River. It is endangered and very scarce over most of its known range.

Lepidomeda albivallis. The White River spinedace occurs in cool springs, their outflows, and in the White River of the White River drainage. This species is very restricted and rare.

Lepidomeda altivelis. The Pahranagat spinedace occurred only in Pahranagat Valley of the White River drainage but is now extinct.

Lepidomeda mollispinis mollispinis. The Virgin River spinedace occurs in the Virgin River of the Colorado River drainage and is limited in numbers in many areas.

Lepidomeda mollispinis pratensis. The Big Spring spinedace was originally described from Big Spring near Panaca, Nevada, near the head of the eastern affluent of the Pluvial White River. It is now extinct.

Plagopterus argentissimus. The woundfin is known from the Virgin River of the Colorado River drainage. It is endangered and very scarce over most of its known range.

Family Cyprinodontidae

Crenichthys baileyi. The White River springfish is known from numerous springs, pools, outflows, and small lakes of the White River drainage and in the Moapa River, springs, and pools. In many areas it is very abundant, but it has also been eliminated from numerous areas by the introduction of predatory game fish and exotic tropical fish.

Crenichthys nevadae. The Railroad Valley springfish is found in springs in Railroad Valley. It is very abundant in the warm springs in which it occurs but can be easily eliminated by the introduction of species.

AMARGOSA-DEATH VALLEY SYSTEM

Family Cyprinidae

Rhinichthys osculus nevadensis. The Amargosa speckled dace is found throughout the springs and pools of the Amargosa River drainage. It has few associate fishes in this restricted environment, most notably the members of the genus *Cyprinodon*. Due to its restricted habitat and distribution, the fish is considered to be rare. Major habitat disruptions could easily eliminate local populations.

Cyprinodon nevadensis nevadensis. The pupfish is found in the lower Amargosa River drainage in the southern end of Death Valley. It occurs principally in Saratoga Springs but is also found in adjacent bodies of water.

Cyprinodon nevadensis amargosae. The Amargosa pupfish is found throughout the lower Amargosa River and several permanent springs near Tecopa and Shoshone, California.

Cyprinodon nevadensis calidae. This pupfish was found in the Tecopa Hot Springs, California, but is now considered extinct.

Cyprinodon nevadensis shoshone. This pupfish was found in Shoshone Springs, California, but is now considered extinct.

Cyprinodon nevadensis mionectes. This pupfish is found in the warm springs of Ash Meadows, Nevada. It has been known from Big Spring, Deep Spring, Eagle Spring, Point-of-Rocks Spring, Fairbanks

Spring, and others, but in several areas it has become extinct due to the introduction of predatory species.

Cyprinodon nevadensis pectoralis. This pupfish is found at Lovell Spring, Nevada, and adjacent localities. It is considered to be endangered.

Cyprinodon diabolis. The Devil's Hole pupfish is found only in Devil's Hole, Ash Meadows, Nevada, in relatively small numbers. This pupfish, more than any other, has been protected due to its unique habitat, which is totally dependent upon water supplies from adjacent areas in Ash Meadows. Its habitat is a single hole in the ground, in which the water level fluctuates with the surrounding groundwater. As groundwater is removed for irrigation, the water level in Devil's Hole drops. Recent court decisions have stopped nearby groundwater pumping, which has saved the Devil's Hole pupfish from extinction.

Cyprinodon salinus. The Salt Creek pupfish is found throughout the springs and tributaries of Salt Creek in Death Valley. The population fluctuates with the water supply, although the species is usually found in large numbers throughout its range.

Cyprinodon milleri. This recently described species of pupfish is found only in Cottonball Marsh, Death Valley.

Empetrichthys merriami. The Ash Meadows poolfish is confined to the springs and pools of Ash Meadows. Due to reduced population and habitat, the species may be extinct.

Empetrichthys latos latos. The Manse Ranch poolfish of Pahrump Valley, Nevada, is confined to this single locality.

Empeticichthys latos pahrump. The Pahrump Ranch poolfish is confined to this single locality in Pahrump Valley, Nevada.

Empetrichthys latos concavus. The Raycraft Ranch poolfish is confined to this single locality in Pahrump Valley, Nevada.

All the different subspecies of *Empetrichthys latos,* the poolfish, are considered to be endangered. The slightest change in habitat could cause their extinction.

LAKE BONNEVILLE SYSTEM

Family Salmonidae

Salmo clarki utah. The Utah cutthroat is restricted to small areas of the Bonneville basin. It has recently been found on the western edge of the Bonneville basin in several mountain streams but is extinct over most of its former range (Sigler and Miller 1963; Hickman and Duff 1978).

Prosopium gemmiferum. The Bonneville cisco.

Prosopium spilonotus. The Bonneville whitefish.

Prosopium abyssicola. The Bear Lake whitefish.

These three whitefish are confined to Bear Lake, Utah-Idaho, where they are found in great numbers. They occur nowhere else in the Bonneville system.

Prosopium williamsoni. The mountain whitefish occurs throughout the Bonneville basin and is especially abundant in the Logan, Blacksmith Fork, and Weber rivers.

Family Cyprinidae

Gila atraria. The Utah chub is native to the drainage of ancient Lake Bonneville and occurs in almost all bodies of water that fish can tolerate, usually in large numbers.

Gila copei The Leatherside chub is native to the eastern and southern Bonneville basin. It occurs in the Bear, Logan, and Weber rivers, the Provo River of Utah Lake drainage, and the Beaver and Sevier rivers of the Sevier River System.

Richardsonius balteatus hydrophlox. The Redside shiner is widespread throughout the Bonneville basin.

Iotichthys phlegethontis. The least chub is known from the streams around Salt Lake City, Utah, the ponds and swamps around the Great Salt Lake and Utah Lake, Beaver River, and the Provo River.

Rhinichthys osculus. The Western speckled dace is found throughout the many streams of Utah, with a unique form *Rhinichthys osculus adobe* known from the Sevier River.

Rhinichthys cataractae. The longnose dace is known from the northeastern Bonneville basin in the streams draining into Bear Lake, Great Salt Lake, and Utah Lake. They are also known from the Bear, Logan, Ogden, Weber, Jordan, and Provo rivers.

Family Catostomidae

Catostomus ardens. The Utah sucker is found throughout the Bonneville basin, especially in Bear and Utah lakes.

Catostomus platyrhynchus. The mountain sucker is found throughout the Bonneville basin drainage.

Chasmistes liorus. The June lakesucker is confined to Utah Lake and at one time was considered to be extinct but is found in a limited population.

Family Cottidae

Cottus bairdi semiscaber. The Bonneville mottled sculpin is found throughout the Bonneville basin drainage and the Snake River drainage above Shoshone Falls.

Cottus extensus. The Bear Lake sculpin is found only in the Bear Lake and its drainage.

Cottus echinatus. The Utah Lake sculpin was found only in Utah Lake and is now considered to be extinct.

3

THE NORTHERN GREAT BASIN REGION

DON McKENZIE

THE GREAT BASIN desert of the United States comprises about 305,710 square kilometers (km) within the five western states of Utah, Nevada, Idaho, California, and Oregon. This section will focus on the northern Great Basin of Oregon with specific reference to the Steens Mountain region (figure 3-1). The terms *high desert* and *cold desert* have been used in describing this region characterized by a landscape of expansive basins and fault-block ranges. Sagebrush (*Artemisia* sp.) is generally considered the index plant species, while jackrabbits and magpies are common faunal representatives.

UNIQUE FEATURES

The first human habitation of the region is radiocarbon dated at about 9000 years ago with the discovery of sagebark sandals in Fort Rock Cave near the northwest boundary of the northern Great Basin. Northern Paiute Indians roamed the region in search of food and shelter, and their artifacts are still quite commonly found.

Cattle Country

Today the northern Great Basin remains active in the cattle industry, with origins dating back to 1869 when John S. Devine first settled the region. His famous Whitehorse Ranch of the Alvord Basin remains a classic and productive spread that is also a weather-reporting station.

In 1872 Peter French settled the west side of the Steens Mountain, where he controlled vast expanses of land until his death, by gunshot wound, in 1897. Reminders of his days as a cattle king can still be observed in buildings and structures that dot the Harney Basin.

The effect of cattle grazing on the public lands currently is a topic of great interest and sometimes heated debate, especially regarding such areas as the Malheur National Wildlife Refuge and the Steens Mountain.

The raising and herding of domestic sheep is no longer a problem, but in 1901 there were about 140,000 head grazed on the Steens Mountain. This was reduced to 100,000 in the 1920s, and in 1972, sheep grazing was banned from the Steens Mountain.

Steens Mountain Auto Trips

A round trip of about 394 km can be made around the Steens Mountain, thus exposing the gentle west slope and the magnificent south scarp. One can begin in Burns, Oregon, by driving east on Oregon Highway 78 about 3 km, then south on Oregon Highway 205, crossing between Harney and Malheur lakes on the Malheur National Wildlife Refuge. Continuing south through the quaint settlement of Frenchglen into the Catlow Valley, the highway changes to a well-maintained rock-surfaced road. Continuing south along the Catlow Rim to Long Hollow and then east, one passes across the southern extension of the Steens Mountain to the community of Fields. Here one can turn left (north) and follow the road by the base of the east escarpment of the Steens Mountain, through the Alvord Basin, to Oregon Highway 78. Turning left (northwest), one returns to Burns.

The crest of the mountain can be explored by driving the Steens Mountain loop road from Frenchglen, about a 113 km round trip. The road is improved gravel to Fish Lake (29 km), where it changes to a rough, unimproved road that is passable usually in mid-July. This drive, coupled with short hikes, provides excellent views of the glaciated valleys of the Steens Mountain, along with spectacular views of the east scarp and Alvord Basin.

Few places in North America provide such an accessible view of vegetational zones, geologic features, and unique birds and mammals.

Malheur National Wildlife Refuge

President Theodore Roosevelt established the Malheur National Wildlife Refuge in 1908 for the chief purpose of providing nesting habitat for migrating birds. Today the

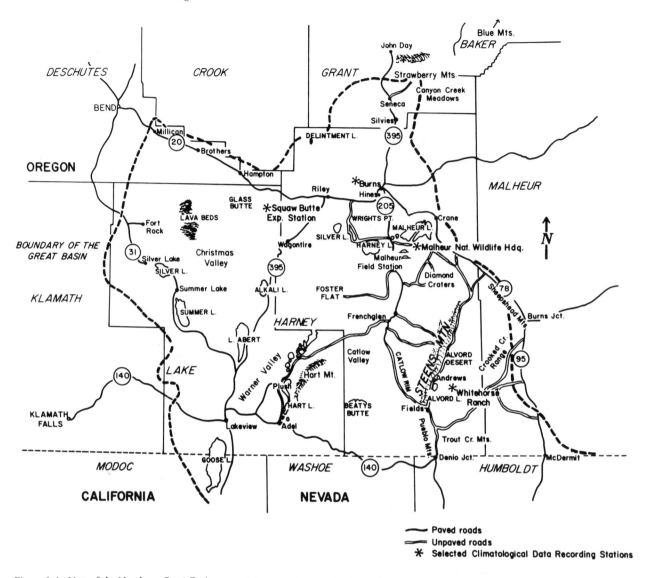

Figure 3-1. Map of the Northern Great Basin,
state and county lines, major highways and landmarks, and selected climatological data
recording stations. From *Oregon's Great Basin Country,* Denzel and Nancy Ferguson, Burns,
Oregon, 1978. Reprinted by permission of the authors, Malheur Environmental Field Station.

32,780 hectares (ha) of the refuge contain large, shallow marshes, small ponds, alkaline lakes, irrigated meadows and grasslands, alkali flats, and sagebrush uplands. This variety of habitats attracts large flocks of waterfowl during the fall and spring migration along the Pacific Flyway.

The refuge is located about 52 km south of Burns, Oregon, in Harney County. It lies within Harney Basin, which has internal drainage of water from two major sources, the Silvies River (north) and Donner and Blitzen River (southeast).

Much of the refuge is strictly managed by an expansive series of canals and dikes and by alternate flooding and farming of meadows. Cattle grazing is practiced, and certain sites show the effects of overuse.

Some 264 species of birds and 57 species of mammals have been recorded on the refuge, and the refuge headquarters maintains an excellent museum with displays of the local mounted birds and mammals, an egg collection, and a variety of publications, including maps, automobile tourguides, checklists of birds and mammals, and historical notes, among others.

Malheur Field Station

A consortium of twenty-two colleges and universities from Oregon and Washington operates a field station located in the Malheur National Wildlife Refuge about 6.4 km west of the refuge headquarters (figure 1). Field-oriented college credit courses are taught each summer by visiting and resident staff, and the station offers lodging, meals, and laundry facilities for numerous groups exploring the Great Basin region throughout the year. Laboratory and research space is available as is a small library, and reference herbarium. Complete information regarding use can be obtained by writing: Director, Malheur Field Station, P.O. Box 989, Burns, Oregon 97720.

Hart Mountain National Antelope Refuge

The refuge was established in 1936 as a management area for the protection of pronghorn herds. There are 111,291 ha of land, including a wide range of habitats. Warner Peak (2,458 meters elevation) marks the summit of Hart Mountain, a fault-block land-form characteristic of the northern Great Basin.

Vegetation types are varied, ranging from stands of ponderosa pine (*Pinus ponderosa*) to sagebrush and bitterbrush.

Warner Valley, on the west border of the refuge, lies at about 1361 m elevation and contains numerous lakes, sand dunes, dry playas, greasewood flats, and meadows with associated bird and mammal populations.

Besides the pronghorn, the region supports a fine herd of bighorn sheep that can generally be viewed as they range high in the mountains.

Diamond Craters

Located about 16 km south of Malheur Lake is a lava shield approximately 10 km in diameter. Here aspects of recent vulcanism are clearly illustrated by the presence of lava flows, craters and tubes, and scattered cinder cones.

Stages of primary ecological succession can be viewed, along with interesting plants and animal associations. The great horned owl (*Bubo virginianus*) uses lava crevices for nesting, and there is a population of melanistic western fence lizards (*Sceloporus occidentalis*) adapted to this dark lava habitat. Various stages of aquatic plant succession are nicely illustrated in one of the craters containing a pond (Malheur Maar), which is also the site of numerous nesting blackbirds.

Geothermal Activity

Hot springs, lakes, and small streams are common primarily in the Alvord Basin, Harney Lake, and Warner Valley regions. Many of these sites provide hot baths for local residents and knowledgeable travelers. Several sites have been leased from the Bureau of Land Management for exploration of potential commercial use of the geothermal activity.

Weather and Climate

The present climate of the northern Great Basin is thought to be quite different from that of the past based upon fossil flora collected from the Alvord Creek fossil beds (Hansen 1956). The annual precipitation was estimated to be nearly double that of the present, with the main distribution probably occurring, as now, in the winter.

Using Burns, Harney County, Oregon, as a reference station, the following discussion illustrates the more general climatological pattern of the northern Great Basin.

Burns is located in the northeast section of the high plateau area that comprises the major portion of the Great Basin region of Oregon. The crest of the Cascade Mountain range is about 217 km to the west and the Steens Mountain about 80 km to the southeast. An extension of the Blue Mountains to the north approaches within 80 km of Burns, and approximately 48 km east a number of low hills separate this area from the Malheur Valley. From the Blue Mountains in the north, to the Oregon southern border, and between the foothills of the Cascades in the west, to a chain of smaller mountains to the east, is a series of shallow valleys, each with its own small creeks. These are separated by gently rolling bench lands of somewhat higher elevation, which in turn are cut by rocky canyons and by numerous buttes or small mesas that rise from 152 m to 457 m above the general terrain. Well distributed over much of central Oregon south of Burns are a number of large, relatively shallow, landlocked lakes and marshes.

Burns, like most of the rest of the Great Basin area of the western plateau, has a semiarid climate. Maritime air moving inland from the Pacific Ocean gives up much of its precipitable moisture to the coastal and Cascade Mountain ranges that lie westward of this region. Burns's annual precipitation totals (table 3-1) are small and the humidities generally low, making for an abundance of sunshine and a rather wide range between daily maximum and minimum temperatures. The yearly average temperature is 7.8°C with a maximum summer average temperature of 32.2°C and minimum winter average temperature of −8.3°C. Extremes range from 41.1°C to −37.2°C. Nighttime frosts may occur in any month, and the growing season is 117 days a year. The average annual total precipitation for the entire high plateau region of Oregon is between 22.5 and 27.9 centimeters (cm), with about one-third falling in the form of snow. Thunderstorms occur each year and are occasionally accompanied by hail.

The other selected stations for climatological data occur in the region under discussion with the Malheur Refuge Headquarters and Squaw Butte Experimental Station west of the Steens Mountain and White Horse Ranch east of the Steens Mountain near the Alvord Basin.

There are two snow courses over the Steens Mountain over which a snow survey is conducted at least once a year in April (table 3-2). The Silvies Station is located at 2103 m elevation, and the Fish Creek Station is at 2408 m elevation.

Geology

The Steens Mountain of Harney County, Oregon, is the state's highest mountain within the northern Great Basin. It is an isolated block of volcanic rock extending approximately 80 km in a northeasterly direction and tilted to the west. The east escarpment is very steep, rising about 1.7 km in 4.8 km distance, while the west slope is gentle, extending about 32 km from the crest to the base.

Lund and Bentley (1976) presented an excellent account of the geology of the Steens Mountain, and Dr. Bruce Nolf of Central Oregon Community College (Bend, Oregon) has directed extensive field studies in the region. The following discussion is drawn principally from their work (figure 3-2).

Historical, Physiographic, and Structural Setting

The region of the Steens Mountain is within the northern extension of the Great Basin, which occupies Nevada and sections of adjacent states.

Tribes of Northern Paiute Indians hunted in the region and used the mountain for summer retreats (figure 3-3). They belonged to a band known as *Toso odo tuviwarai*, meaning "cold dwellers" (Stewart 1939). Arrowheads and chips are still found, and Indian caves have been located at the mouth of the Kiger Gorge, at Roaring Springs, and in the Catlow Valley (Hansen 1956).

The name *Snow Mountain* was first used to describe the mountain by trapper-explorer John Work in his journal of 1831. In 1860 Major Enoch Steen rode to near the summit of the mountain in pursuit of a band of Indians, thus lending his name to the present-day site.

Geographical survey work of the region was first reported by James Blake in 1873 and dealt with a preliminary survey of the Pueblo Mountains and southern Steens.

In 1877 Charles E. Bendire collected several eggs of Hepburn's (gray-crowned rosy) finch (*Leucosticte tephrocotis*). This work marked the first reported scientific endeavor on the mountain.

The first of many botanists to collect here was Thomas Howell in 1885.

The Steens Mountain block was uplifted and tilted gently toward the west along a set of faults that define the mountain's northeasterly trend. The major block is cut by numerous smaller faults, some with a northwesterly trend and others parallel to the principal northeasterly direction. Differential

TABLE 3-1

Climatological data for selected collecting stations, the northern Great Basin region, Oregon

STATION	ELEVATION (M)	AVERAGE TEMPERATURE (°C)	PRECIPITATION (CM)	DATES
Burns	1,262	7.8	25.4	1941-1977
Malheur Refuge Hdq.	1,252	8.1	22.4	1941-1977
Squaw Butte Exp. Station	1,425	7.6	24.4	1941-1977
White Horse Ranch	1,280	8.8	19.3	1970-1977

SOURCE: Climatological data. Oregon Annual Summary, 1971-1977. National Climatic Center.

NOTE: Elevations are rounded off to the nearest whole number. Temperatures and precipitations are rounded off to the nearest 0.1

TABLE 3-2

Average snow depths and water equivalence, Steens Mountain

STATION	ELEVATION (M)	AVERAGE SNOW DEPTH (CM)	AVERAGE WATER EQUIVALENCE (CM)	DATES RECORDED
Silvies	2,103	83.8	33.0	1963-1977
Fish Creek	2,408	175.3	66.0	1963-1977

SOURCE: Summary of snow survey measurements for Oregon, 1973. Oregon Soil Conservation Service.

NOTE: Elevations are rounded off to the nearest whole number. Numbers in centimeters are rounded off to nearest 0.1. All dates were recorded in April.

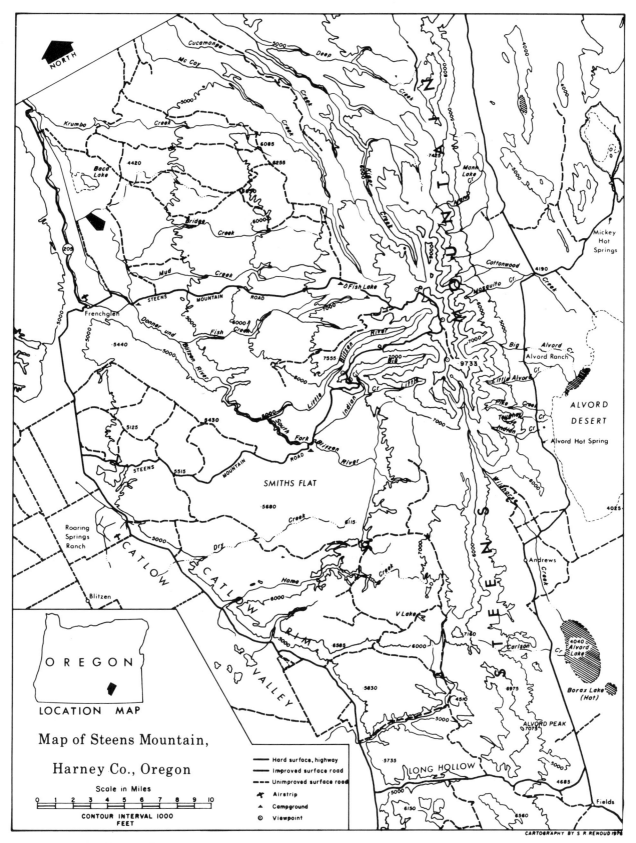

Figure 3-2. Topographic map of Steens Mountain.
Major landforms, landmarks, and roads are shown.

Figure 3-3. Catlow Rim near Roaring Springs Ranch.
Artemisia vegetational zone in the foreground, with Juniperus zone above. Caves near the base of
the scarp were once occupied by Paiute Indians. Photograph of the Oregon Highway Division.
Courtesy of State of Oregon Department of Geology and Mineral Industries.

displacement along northwesterly trending faults has divided the mountain into three distinct topographic units: northern, high, and southern Steens (Fuller 1931).

The northern Steens is about 40 km long and is bounded on the east by a steep, continuous scarp trending N 30° E. The scarp rises to a maximum of 975 m above the valley floor at the southern end; the north end is of lower elevation and merges with the surrounding hills southeast of Malheur Lake. This block slopes gently northwestward and is covered on the northwest by young sediments and volcanic rocks. A long, narrow valley is formed east of the scarp along a down-dropped fault block paralleling the northern Steens. Several shallow, intermittent lakes are found within this valley.

The central segment of the mountain is about 24 km long and forms the high Steens (figure 3-4). The mountain's highest elevation, 2967 m, is here attained, as it rises along a north-south scarp some 1744 m above the playa lake floor in the Alvord Desert. The abrupt eastern scarp is characterized by spectacular views of valleys and sharp ridges shaped by stream erosion and glaciation. At the mouths of Alvord Creek, Pike Creek, and Indian Creek are excellent examples of alluvial fans formed at the base of the scarp by streams originating in cirques high on the mountain. The high Steens block is bounded on the northeast and southwest by northwest trending faults. The high Steens rises as much as 557 m above the adjacent northern Steens block and has a regional slope toward the northwest of about 3°. The southern Steens block has a much gentler westward slope, having been tilted much less than the high Steens. Consequently the east edge of the high

Steens block stands as much as 958 m above the southern Steens block, but the west edge of the high Steens block is 305 m below the adjacent southern Steens block, and is covered by the young sediments of the Upper Blitzen Valley.

The Catlow Rim, a prominent fault scarp, rises as much as 620 m above the Catlow Valley and marks the west boundary of the southern Steens. The southern Steens are divided into two sections by another fault scarp trending about N 60° W, with the downfaulted block to the south. Smith Flat is the name given to the northern and larger section, a structural sag, which dictates the northwest-trending course of the Blitzen River. The other section south of Smith Flat is shaped by numerous faults, producing an irregular terrain through block tilting and subsequent stream erosion.

The south extension of the Steens Mountain merges with the Pueblo Mountains at a point northwest of Fields where a topographic break is marked by a saddle and by Long Hollow. This saddle is arbitrarily considered the southern end of the Steens Mountain.

The exposed bedrock of the Steens Mountain is chiefly lava flows with pyroclastics and intrusive bodies exposed low in the east scarp of the high Steens. Rocks as old as Oligocene (Walker and Repenning 1965) crop out at the base of the scarp.

Alvord Creek Beds

The light-colored tuffs that form interrupted outcrops for over 8 km in the lower 300 m of the scarp between Cotton-

Figure 3-4. High Steens viewed from the south.
Wildhorse Canyon and Little Wildhorse Canyon are in lower-left quarter of the photograph.
Conkling photo. Courtesy of the State of Oregon Department of Geology and Mineral Industries.

wood and Toughey creeks were termed the Alvord Creek beds by Fuller (in Lund and Bentley 1976). The color is chiefly white, but brownish and greenish forms of the altered stratified silicic tuffs are common. North of Big Alvord Creek are two thick andesite flows interlayered with volcaniclastic sediment, while south of Big Alvord Creek, a 61 m thick basalt sill is intruded into the beds. There is disagreement in the literature as to the age and distribution of these beds, but it appears that they are at the base of the sequence in the east scarp of the Steens Mountain.

Axelrod (1944) found fossil floral species of fir, spruce, pine, juniper, maple, aspen, cottonwood, willow, beech, Oregon grape, service berry, mountain mahogany, christmas berry, cherry, rose, mountain ash, sumac, madrona, chaparral, and pondweed. From this fossil record he concluded that the region was one of moderate topographic diversity and weather conditions at the time of deposition of these beds.

Pike Creek Volcanic Series

Pike Creek Canyon (figure 3-5) exposes a thick series of silicic flows, domes, and stratified tuffs with an aggregate thickness of over 460 m. Two tuff members interlayered with two rhyolite flows form the lower 305 m of the series. The lowermost tuff unit is intruded by at least five basalt sills, and the uppermost rhyolite is overlaid by a 12 m layer of tuff, which in turn is overlaid by two biotite-dacite flows. This series has been named the Pike Creek Formation and assigned an age of Oligocene and Miocene (Walker and Repenning 1965).

Steen Mountain Andesitic Series

This series is composed of an andesite flow of about 60 m thickness that is capped by a 6 m thick stratified tuff unit; a unit of substantial thickness, called the great flow, and a series of alternating thin layers of andesite breccias and platy flow andesite with a maximum combined thickness of over 183 m. The upper andesite series includes small cones of pyroclastic deposits.

The dominant unit in the Alvord Creek region is the great flow, reaching a maximum thickness of 274 m with joint columns up to 1.5 m in diameter and rising 90 m above the talus. About 1.6 km north, in the Cottonwood Creek Valley, the unit thins to about 152 m and is not traced beyond with certainty, while south of Alvord Creek the unit is exposed only in scattered outcrops. Some workers believe this unit represents a thick, intrusive sill.

Steens Mountain Basalt (Steens Basalt)

This unit forms the bulk of the mountain, comprising the upper 914 m of the east scarp and the rock exposed along glacial valleys on the back slope. It is a series of thin flood or plateau basalts and, except where covered locally by younger ash flows, is the bedrock on most of the western slope. Abundant feeder dikes for these flows are exposed in the eastern fault scarp. Most dikes trend approximately parallel to the fault.

The maximum original thickness of the Steens basalt is

Figure 3-5. Pike Creek flow formation at the mouth of Pike Creek Canyon.
Oregon Highway Division photograph. Courtesy of the State of Oregon Department of Geology and Mineral Industries.

unknown because of erosion at the top of the section. The average individual flow thickness is about 3 m, ranging from less than 30 cm to over 21 m. A conservative measurement of 1000 m thickness was made by Wilkerson (1958) at a section on the west rim of Wildhorse Canyon.

Most of the flows are composed of wholly crystalline rock, predominantly olivine basalt. Rock textures range from fine grained to coarsely porphyritic, with plagioclase crystals up to 4 cm long.

The formation is Middle to Late Miocene in age.

Ash-flow Tuffs (Danforth Formation)

Smith Flat has numerous small, isolated patches of Pliocene welded tuff scattered over the surface, while an extensive sheet veneers the Steens basalt on the lower slopes of the high Steens and the northern Steens. Rock composition is of glass shards, crystals, and fragments of pumice and other rock aggregated by welding. Where welding is intense, the rock is compact and glassy or porcelaneous in appearance, while less-welded rock is soft and easily eroded.

The present distribution of welded tuff suggests that it was originally distributed widely over low elevation portions of the Steens, but may have been thinner and less welded, or absent, on top of the high Steens. Erosion has subsequently removed much tuff. The welded tuffs filled and preserved a drainage system, which was already well developed in response to northwestward tilting of the Steens block. Therefore the onset of block faulting predates the oldest welded tuff (about 9.10^6

years B.P.) and postdates extrusion of the youngest Steens basalts (about 15.10^6 years B.P.). Movement along these faults probably continued at least episodically, into the Pleistocene, by which time sufficient elevation had been attained to allow glaciation. Recent displacement is indicated regionally by a few fresh fault scarps in surficial material.

GLACIATION

Erosional Features

Changes in surface configuration on the long, gentle back slope and the steep eastern scarp have been attributed primarily to the erosional agents of running water, glacial ice, mass movement, and wind. It is estimated that the east scarp has retreated 2.4 km from its original position (Williams and Compton 1953). The western slope has been eroded by glacial ice, which has sculptured deep U-shaped canyons and lake basins.

The upper region of the high Steens was marked by glacial activity in the Pleistocene, as were adjacent locations of the northern and southern Steens into which glaciers originating in the high Steens flowed. Valley glaciers originating close to the summit of the mountain flowed down former stream valleys and sculptured Kiger Gorge, Little Blitzen, Big Indian Creek, Little Indian Creek, Wildhorse Creek, and Little Wildhorse Creek canyons. A spectacular view of these canyon heads is possible by hiking short distances from the Steens Loop Road to viewpoints (figure 3-6).

Figure 3-6. Kiger Gorge showing the glacier-carved valley below and Steens Basalt flows of the canyon walls.
Oregon Highway Division photograph. Courtesy of the State of Oregon Department of Geology and Mineral Industries.

Glaciers on the east side of the Steens Mountain extended about halfway down the scarp, forming U-shaped canyons below which are stream valley features. Small cirque basins just below the rim of the mountain mark the position of small glaciers that extended a short distance down the scarp. The crest of the rim shifted westward as the glaciers on the east scarp eroded into the mountain. During the same period, west slope glaciers extended their valleys headward into the mountain, resulting in only thin walls separating the heads of Little Blitzen and Big Indian canyons from cirques east of the rim. A col, or gap in the mountain rim, is formed where the head of Kiger Glacier was close to the head of a glacier in the east scarp valley of Cottonwood Creek. Other examples of col formations are observed in the Big Nick on the east rim of Kiger Gorge and where the cirque wall of Little Wildhorse Glacier intersected the headwall of the glacier in Little Indian Canyon.

Lake basins on the high Steens are formed by glaciation and are of two types. Depressions dammed at the west end by moraines are typified by Fish Lake and other small lakes of the west slope. Glacially eroded depressions in the bedrock form Wildhorse Lake in the cirque of Wildhorse Canyon and a small lake in the cirque of Little Wildhorse Canyon.

Glacial Advances

Glaciation on the Steens Mountain has occurred in three advances, two major and one minor, according to Lund and Bentley (1976). Fish Lake advance was the first and formed a widespread ice cap of approximately 295 square km. It left an extensive mantle of till and around the lower margin ice was channeled along several major valleys. The ice deposited a set of moraines in the Little Blitzen Canyon that have since been severely eroded by running water, leaving only weathered boulders.

On the east side of the mountain, ice extended to about 762 m above the floor of Alvord Desert. Glacial cirques were left along the scarp of the high Steens, as shown at the head of Alvord Creek (figure 3-7).

The second major advance, named the Blitzen advance, covered less than 130 square km and was restricted mainly to canyons where cirques were formed on the rims and ground the headwalls. During the period, cirque glaciers formed on the east scarp.

The moraines of the Blitzen advance are relatively intact, with little weathering of the till. The grayish color distinguishes it from the weathered buff-colored till in the moraines of the Fish Lake advance, and there is no cementation and leaching of the till as is shown in the Fish Lake deposits.

Both the Fish Lake and Blitzen advances were in two stades, with two moraines at the mouth of the Little Blitzen Canyon formed during the Fish Lake stades and another pair formed during the Blitzen stades.

The post-Blitzen was the third glacial advance. It was of minor importance and is recorded by the pocket cirques along the east rim of the mountain and by tiered cirques on several canyon walls.

Figure 3-7. Small cirques around the upper edge of the large cirque at the head of Alvord Creek. The Alvord Desert is in the distance. Oregon Highway Division photograph. Courtesy of the State of Oregon Department of Geology and Mineral Industries.

VEGETATIONAL PATTERNS ON THE STEENS MOUNTAIN AND SURROUNDING REGION

There are five rather distinct horizontal zones of vegetation on the Steens Mountain. The most obvious plant species act as indicators of each zone and the species composition changes with increased elevation: Blitzen River Valley; *Artemisia* zone; *Juniperus* zone; *Populus tremuloides* zone; and alpine bunchgrass zone. The vegetational zones generally hold true, but patterns of vegetation may extend as fingers above or below the listed elevations as a result of changes in soil type, rock quality, exposure, and so forth (figure 3-8).

Blitzen River Valley

This region is about 1250 m elevation and forms the western drainage from the Steens Mountain. Malheur Lake, the state's largest inland marsh, covers between 20,235 ha and 27,115 ha at maximum water level and is about 32.2 km long and 19 km wide. The lake is better described as a shallow freshwater marsh with associated emergent aquatic plants such as hardstem bulrush (*Scirpus acutus*), Baltic rush (*Juncus balticus*), bur reed (*Sparganium eurycarpus*), and cattail (*Typha latifolia*). In areas of open water, sago pondweed (*Potamogeton pectinatus*), water milfoil (*Myriophyllus spicatum*), and horned pondweed (*Zannichellia pulustris*) dominate.

The area is greatly modified by an extensive irrigation canal system bordered by patches of willows (*Salix* sp.) and black cottonwood (*Populus trichocarpa*). Along the streams of the numerous canyons draining into the region are found alder (*Alnus sp.*), chokecherry (*Prunus* sp.), elderberry (*Sambucus* sp.), dogwood (*Cornus* sp.), and currant (*Ribes* sp.).

Lowland meadows are found throughout with sedges (*Carex* sp.), creeping spike-rush (*Eleocharis palustris*), water plantain (*Alisma* sp.), broad-leaved arrowhead (*Sagittaria cuneata*), and grasses like giant wildrye (*Elymus cinereus*) and meadow barley (*Hordeum brachyantherum*) dominating.

Raised land forms, such as buttes, provide rocky soils inhabited by big sage (*Artemisia tridentata*) and rabbitbrush (*Chrysothamnus* sp.) on north-facing slopes and spiny hopsage (*Atriplex spinosa*), saltbush (*Atriplex confertifolia*), greasewood (*Sarcobatus vermiculatus*), and horsebrush (*Tetradymia* sp.) on south-facing slopes. Cheat grass (*Bromus tectorum*), peppergrass (*Lepidium* sp.), and Indian rice grass (*Oryzopsis hymenoides*) are found in association with the other species on all slopes.

Intensive management of wildlife habitat is practiced in this region, along with agricultural pursuits of cattle ranching and haying.

Artemisia Zone

This zone is located at an elevation range of 1250 m to 1646 m. Big sagebrush dominates the lowlands of the mountains, especially on well-developed soils, while low sagebrush (*Artemisia arbuscula*) grows chiefly on rocky outcroppings and

ELEVATION meters	feet
3,048	(10,000)
2,896	(9,500)
2,743	(9,000)
2,591	(8,500)
2,439	(8,000)
2,286	(7,500)
2,134	(7,000)
1,982	(6,500)
1,829	(6,000)
1,676	(5,500)
1,524	(5,000)
1,372	(4,500)
1,220	(4,000)

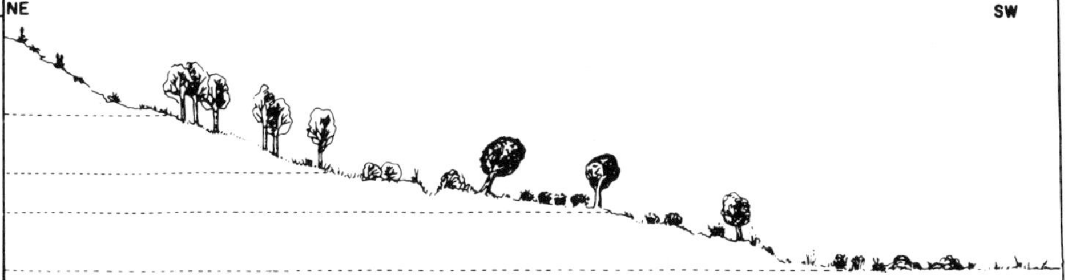

MACRO HABITAT BY ELEVATIONAL ZONES	ALPINE BUNCHGRASS ZONE 2,408-3,048/(7,900-10,000)	POPULUS tremuloides ZONE 1,951-2,408/(6,400-7,900)	JUNIPERUS ZONE 1,646-1,951/(5,400-6,400)	ARTEMISIA ZONE 1,250-1,646/(4,100 5,400)	BLITZEN RIVER VALLEY 1,250/(4,100)
MICRO HABITAT / DOMINANT VEGETATION: Meadow	Carex sp. Potentilla sp.	Salix sp. Western false hellebore	Quaking aspen	------	Giant wild rye grass Greasewood
Riparian	Short-leaved cinquefoil Snow buttercup	Quaking aspen Black cottonwood	Western juniper (lower) Salix sp. Black cottonwood (upper)	Salix sp. Western chokecherry	Salix sp. Western chokecherry
North Slope	Alpine sorrel	Quaking aspen	White fir Western juniper - Low sagebrush Green rabbitbrush	Low sagebrush	Big sagebrush
South Slope	Orange sneezeweed Steens Mt thistle	Low sagebrush	Mountain mahogany	Low sagebrush Gray rabbitbrush	Saltbrush Spiny hopsage
East Slope	Rose willow-herb	Low sagebrush	Western juniper Idaho fescue	Low sagebrush Gray rabbitbrush	Big sagebrush Low sagebrush
West Slope	Low sagebrush Sheep fescue	Low sagebrush	Western juniper Idaho fescue	Low sagebrush Gray rabbitbrush	Big sagebrush Western juniper
Rimrock (including summit outcroppings)	White shrubby cinquefoil Alpine prickly currant	Mountain mahogany	Western juniper Mountain mahogany Antelope bitterbrush	Western juniper	Western juniper
Talus Slopes	Orange sneezeweed Downy oatgrass Steens Mt thistle	Low sagebrush	Western juniper Big sagebrush	Big sagebrush	Big sagebrush
Alkaline Soils	Dwarf desert knotweed	------	-------	Low sagebrush	Greasewood Saltgrass
Frost Pockets (areas of cold air drainage)	Carex sp. Creeping sibbaldia Alpine sorrel Patterson's bluegrass	Carex sp Creeping sibbaldia	White fir	Quaking aspen	-------
Nivation Hollow	Creeping sibbaldia Tufted hairgrass Dwarf desert knotweed	Carex sp Creeping sibbaldia	Quaking aspen	-------	--------
AMPHIBIANS AND REPTILES	-------	Western toad	Pacific tree frog Western skink	Western fence lizard Western garter snake	Western spadefoot toad Pacific tree frog Western fence lizard Sagebrush lizard Gopher snake Western rattlesnake
BIRDS	American kestrel Horned lark White-throated swift Black rosy finch	White-crowned sparrow Dipper Rufous hummingbird House wren Common flicker	Valley quail Mountain bluebird Lazuli bunting	Marsh hawk Say's phoebe Sage thrasher	Turkey vulture Common raven Bobolink Sandhill crane Trumpeter swan
MAMMALS	Yellow-bellied marmot Northern pocket gopher Mountain mole Long-tailed meadow mouse Badger Mountain sheep	Little brown bat Belding ground squirrel Golden-mantled ground squirrel Least chipmunk Beaver Long-tailed weasel	Preble shrew Yellow-pine chipmunk Deer mouse Bushy-tailed woodrat Porcupine Coyote	Mountain cottontail Deer mouse Western jumping mouse Coyote Pronghorn	Black-tailed jackrabbit Great Basin pocket mouse Ord kangaroo rat Western harvest mouse Northern grasshopper mouse Muskrat Bobcat Mule deer

Figure 3-8. Schematic profile of south-facing slope of Steens Mountain.
The illustration indicates macrohabitats by elevational zones of dominant vegetation, microhabitats by selected site features, and index species of plants and vertebrate animals. Elevations are expressed in meters and feet. The vegetational zones elevate approximately 304.8 m (1000 ft.) in a southwestern direction. From "Vegetative Patterns on Steens Mountain, Harney County, Oregon," by Karl A. Urban, Blue Mountain Community College, Pendleton, Ore., 1973.

shallow alkaline soil above 1524 m elevation. In riparian habitats, willow (*Salix* sp.) and western chokecherry (*Prunus virginiana*)are found at the lower elevations. The most abundant grass is introduced cheat, while gray rabbitbrush (*Chrysothamnus nauseosus*) is common in lowlands and dry slopes of the gorges.

Western juniper (*Juniperus occidentalis*) is widespread at lower elevations, growing in association with basaltic fracture zones. Areas of cold air drainage and frost pockets produce conditions favorable for groves of quaking aspen (*Populus tremuloides*).

Juniperus Zone

This area extends from about 1646 m to 1951 m elevation and shows much effect of erosion. On many sites, the only dry, deep soil is found under quaking aspen, mountain mahogany (*Cercocarpus ledifolius*), and the dominant western juniper trees. The presence of the only grove of relict, white (Sierran) fir (*Abies concolor*) is found near the upper limit of this zone and is believed to be associated with frost pockets on north slopes near Muddy, Little Fir, and Big Fir creeks. They grow in very thick stands of trees from about forty to three hundred years old and appear to be reproducing and encroaching on local aspen groves. There is evidence of fire on the trunks of larger trees that date back about 150 years, possibly explaining the absence of more extensive stands on the Steens Mountain.

Between 1524 and 1676 m elevation, big sage drops out, and short sage comes into dominance. There is much evidence of overgrazing in seasonally wet meadows where grasses and sedges predominate.

In riparian areas western juniper is common at the lower elevations, while *Salix* sp. and black cottonwood are associated with higher elevations. Mountain mahogany is common on rimrock sites and fracture zones above 1768 m elevation. Green rabbitbrush (*Chrysothamnus viscidiflorus*) is locally abundant in the upper reaches of this zone.

Antelope bitterbrush (*Purshia tridentata*), although not abundant, is scattered in areas of rimrock where it appears to be overgrazed.

Populus tremuloides Zone

The aspen zone ranges above the *Juniperus* zone to an elevation of about 2408 m. Aspen at the lower portions are found primarily on north slopes and at the higher elevations on sites protected from severe climatic conditions. At middle elevations, the aspen groves are primarily associated with moist conditions along streams, around seeps, or in the numerous meadows. The meadows are further vegetated by grasses, sedges (*Carex* sp.), western false hellebore (*Veratrum californicum*), and *Salix* sp. Short sagebrush is the dominant shrub on sites without aspen. Sites with cold air drainage and nivation hollows show sedges and creeping Sibbaldia (*Sibbaldia procumbens*) as common associates. The higher elevation of this zone, about 2377 m, marks the upper limit of mountain mahogany stands.

Alpine Bunchgrass Zone

Above 2408 m elevation, the vegetation is of two distinct types. The lower 244 m is dominated by short sagebrush, while the upper area is predominantly a bunchgrass grassland. A striking feature of this area is the presence of large patches of bare ground caused by erosion. Gullies up to 1.8 m deep have been cut, and during the spring snowmelt freshets carry vast quantities of exposed soil down the mountain while wind erosion is most pronounced in the summer and fall.

Dry areas with deep soil are mainly inhabited by perennial grasses forming a typical subalpine grassland. This area is marked by erosion where stools of grass extend sometimes 15.2 cm above the surrounding bare ground. The Steens mountain thistle (*Cirsium peckii*) is a common endemic species in rocky areas above 1981 m elevation. Also at this elevation is found the alpine sorrel (*Oxyria digyna*), a common and widespread flower of talus slopes, and the orange sneezeweed (*Helenium hoopesii*) inhabiting disturbed sites. Downy oat grass (*Trisetum spicatum*) occasionally grows on rock outcroppings and talus slopes above 2377 m elevation. Alpine prickly currant (*Ribes montigenum*) is common in aspen woodlands of lower elevations and on rocky outcroppings and fracture zones at higher elevations, while shrubby cinquefoil (*Potentilla fruticosa*) is common on rimrock and rocky summit outcroppings. Sheep fescue (*Festuca ovina*) is locally abundant on sites above 2469 m elevation in the alpine scree zone.

Nivation hollows of the higher elevations provide favorable habitat for a variety of species, including creeping Sibbaldia, dwarf desert knotweed (*Polygonum heterosepalum*), short-leaved cinquefoil (*Potentilla brevifolia*), Patterson's bluegrass (*Poa pattersonii*), tufted hairgrass (*Deschampsia cespitosa*), and, above 2469 m elevation, snow buttercup (*Ranunculus eschscholtzii*).

Alvord Desert

Nonmountainous regions within the northern Great Basin are characterized by soils with high alkalinity, internal drainage of water, and sites with high geothermal activity. The Alvord Basin, east of the Steens Mountain, is an area of special interest because of the presence of one of North America's finest playas, the Alvord Desert. The desert proper is devoid of vegetation; however, the desert edge is primarily inhabited by desert saltgrass (*Distichlis sticta*) and greasewood; spiny hopsage and saltbush grow inland from the desert edge, as does gray rabbitbrush, cheatgrass, horsebrush, giantwild rye (*Elymus cinereus*), Russian thistle (*Salsola kali*), squirrel-tail barley (*Hordeum jubatum*), and thorny buffalo berry (*Shepherdia argentea*). Big sagebrush is restricted to sandier soils with better drainage above the desert edge community.

Mormon tea (*Ephedra viridis*), iodine bush (*Allenrolfea occidentalis*), and *Mirabilis bigelovi* var. *retrorsa* (a four-o'-clock) inhabit this area as the northernmost extent of their range. Several other plants are endemic to the northern Great Basin: Cusick's horsemint (*Agastache cusickii* var. *cusickii*), Steens Mountain paintbrush (*Castilleja steenensis*), Cusick's draba (*Draba sphaeroides* var. *cusickii*), sulfur buckwheat (*Eriogonum umbellatum* var. *glaberrimum*), and Davidson's penstemon (*Penstemon davidsonii* var. *praeteritus*).

The present-day desert flora that characterizes the northern Great Basin is the result of vast evolutionary change that has occurred since the Late Pleistocene when pluvial lakes receded from the region. During this period, subtropical flora replaced tropical vegetation with cooler and drier conditions. Analysis of fossil pollen indicates the presence of sagebrush in Early

Miocene (about 20 million to 25 million years ago). By Late Miocene, the landscape was dominated by forests of mixed deciduous hardwoods and conifers, while more xeric sites were occupied by chaparral and live oaks. Savanna and grasslands dominated in Early Pliocene, while chaparral continued to occupy the xeric areas, and cottonwoods and maples inhabited more mesic sites such as stream edges (Ferguson and Ferguson 1978).

Keeping in mind this rather remarkable shift in floral composition tends to help in understanding the present-day plant distributions and associations.

Vertebrate Animals

The wide diversity of vertebrate animals found in the northern Great Basin demonstrates their adaptability to a great variety of available habitats. Lakes, streams, canals, inland marshes, alkali and greasewood flats, irrigated grasslands, forested mountains, and subalpine environs provide refuge for local residents.

A complete classification and checklist of the vertebrates inhabiting the region is given in Appendix 3-B; therefore a brief discussion of only the most representative and/or unique forms will follow.

Fishes

The adaptations and distribution patterns of the fish fauna of this region have been shaped by geological activity of the past. Periods of uplift and faulting, fluctuations of pluvials, and changes in internal drainage patterns have affected the region that today is occupied by forty-one species of fishes. Four of the nine families found here—Ictaluridae (catfish), Poeciliidae (live bearers), Centrarchidae (sunfish), and Percidae (perch)—are introduced.

The Alvord chub (*Gila alvordensis*), a member of the Cyprinidae (minnows) family, is restricted to the southern Alvord Basin (Borax and Soap Lakes). This interesting species was not described until 1972.

Amphibians and Reptiles

Amphibians are poorly represented, with only one species of salamander, the long-toed salamander (*Ambystoma macrodactylum*), found in the northern border of the Great Basin. It prefers mountainous lakes and ponds, and during the breeding season the adults may be found under logs, rocks, or debris at the water's edge. Frogs and toads are represented by four families and five species. The Great Basin spadefoot (*Scaphiopus intermontanus*) is a nocturnal and secretive toad that spends most of its time below ground in burrows. During periods of spring rains, they become very active and search out ponds and pools for breeding. The western toad (*Bufo boreas*) is generally associated with higher elevations and habitats

about lakes, streams, and ponds. The Pacific treefrog (*Hyla regilla*) is wide ranging and found in vegetation surrounding springs, streams, ponds, and lakes or in crevices in rimrock. The spotted frog (*Rana pretiosa*) is native to permanent water habitats and is suffering from competition with the introduced bullfrog (*Rana catesbeiana*).

Reptiles are much more successful in adapting to the desert conditions of the region as demonstrated by the representative forms present: ten species of lizards (four families) and eight species of snakes (three families). The collared lizard (*Crotaphytus collaris*) frequents rocky gullies or canyons, especially where massive rocks or rock piles are available. This region marks the northern extent of distribution for this species. The leopard lizard (*Crotaphytus wislizeni*) is usually found in areas of sand, gravelly soil, or desert pavement. It is quite active and when caught will many times open the mouth, exposing a threatening black throat lining. The western fence lizard (*Sceloporus occidentalis*) is found in a wide range of habitats from talus outcrops to old buildings. A melanistic population inhabits the black lava flows of Diamond Craters.

Cnemidophorus tigris, the western whiptail, is the most active lizard of the region. It is swift of foot and at times demonstrates bipedal locomotion. In locations of mixed sagebrush, saltbrush, and greasewood with sandy soils (such as northern Alvord Desert region), it can be the most numerous lizard species.

The racer (*Coluber constrictor*) and striped whip snake (*Masticophis taeniatus*) are the most active and fast-moving members of the family Colubridae (colubrids). Both species are good climbers and when foraging hold their head and neck upright above the substrate. Two other colubrids, the gopher snake (*Pituophis melanoleucus*) and the night snake (*Hypsiglena torquata*), inhabit this region. The former is primarily a diurnal predator on rodents, lizards, and bird eggs, while the latter is nocturnal and feeds primarily on lizards and insects.

The western rattlesnake (*Crotalus viridis*) is a member of the family Viperidae (pit vipers) and is quite common in the northern Great Basin. It eats a variety of food, including rodents, birds, and lizards. It is a winter hibernator and shares its den with other species, such as gopher snakes, racers, and striped whip snakes. During warm periods, the western rattlesnake, along with the gopher snake, is readily collected from road surfaces, especially in the evening.

Birds

The Malheur National Wildlife Refuge is famous for the variety and numbers of birds that utilize the area either as migrants or residents. Some 145 species are known to nest in the northern Great Basin.

Avian numbers, variety, and activity fluctuate with the seasons. In mid- to late February, large flocks of snow geese (*Chen hyperborea*), white-fronted geese (*Anser albifrons*), whistling swans (*Olor columbianus*), pintails (*Anas acuta*), and lesser sandhill cranes (*Grus canadensis*) arrive. The spring waterfowl migration peaks in mid-March and early April. Numerous species of shorebirds arrive in April, and songbird numbers increase in late April and early May.

Most northern migrants move out by April, when local sandhill cranes and Canada geese (*Branta canadensis*) begin nesting. By June most of the resident species have completed their nesting, and in midsummer most waterfowl are in eclipse plumage, making field identification difficult. Refuge grainfields are favorite feeding sites for numerous flocks of geese, ducks, and sandhill cranes in the late summer. October is the peak of the fall migration, beginning with the arrival of southward migrants of northern shoveler (*Spatula clypeata*), canvasback (*Aythya valisineria*), American widgeon (*Mareca americana*), and whistling swan. Malheur Lake is a favorite feeding site and attracts large flocks of these fall migrants especially during the years of high production of the aquatic Sago pondweed (*Potamogeton pectinatus*).

Over three hundred species of birds are listed in appendix 3-B as possible sightings, but it should be emphasized that many are migratory, are found in locations bordering the Great Basin (Strawberry Mountains, north), or are restricted to specific habitats (for example, Boreal forests or Steens Mountain crest).

Some of the more interesting avian forms deserve special mention: The western grebe (*Aechmophorus occidentalis*) is a summer resident of lakes and marshes where colorful mating displays are performed. Several popular films have been produced utilizing local populations of this species. The white pelican (*Pelecanus erythrorhynchos*) forms extensive flocks on many of the larger bodies of water such as Malheur and Harney Lakes and lakes in the Warner Valley. Summer breeding colonies are abundant. The white-faced ibis (*Plegadis chihi*) is a summer nester, with the Malheur Refuge marking its northern-most range. During years of southern drought, their populations increase dramatically. Between ten and twenty pairs of trumpeter swans (*Olor buccinator*) reside on the refuge and nest primarily in the Blitzen River Valley. Huge flocks of snow geese utilize the region during peak migrations in the spring (March-May) and fall (September-November).

A large roosting colony of turkey vultures (*Cathartes aura*) is located on the P-Ranch near Frenchglen, Oregon. Here sometimes more than one hundred birds utilize a large fire tower and nearby tall poplar trees as a communal roost. The species nests locally. The golden eagle (*Aquila chrysaetos*) is commonly sighted soaring above basin lands or perched upon utility poles along the highways. Their massive stick nests are used for generations and are normally positioned on rimrock outcrops or canyon walls.

There is seasonal variation of the numerous hawk species inhabiting the northern Great Basin. Plentiful food and good habitat allow for large populations of red-tailed hawks (*Buteo jamaicensis*), rough-legged hawks (*Buteo lagopus*), marsh hawks (*Circus cyaneus*), and American kestrels (*Falco tinnunculus*). The prairie falcon (*Falco mexicanus*) and peregrine falcon (*Falco peregrinus*) are permanent residents on the endangered species list.

Each year, from about February to mid-May, sage grouse (*Centrocercus urophasianus*) converge upon well-established leks to carry out their mating displays. These predawn rituals can be viewed at Foster Flat, Catlow Valley, and Hart Mountain.

About 235 pairs of sandhill cranes nest in the Malheur region. This majestic summer resident arrives in February when large flocks can be seen feeding or resting in fields and on lake shores. The sight and sound of this species is a favorite of visitors to the Malheur National Wildlife Refuge.

The common resident, great horned owl (*Bubo virginianus*), is found in a variety of habitats ranging from willow thickets and old buildings to caves. They, like the golden eagle, use the same nesting site for generations.

The common nighthawk (*Chordeiles minor*) is a late-arriving summer (May-June) nester. It resides in the region, actively diving for insects, or spending many daylight hours perched upon fence posts, utility lines, or large rocks. Its southern fall migration is in September and November.

Yellow-headed blackbirds (*Xanthocephalus xanthocephalus*), red-winged blackbirds (*Agelaius phoeniceus*), and Brewer's blackbirds (*Euphagus cyanocephalus*) are highly territorial. All nest in the region prior to forming massive interspecific premigratory flocks in the fall.

Mammals

Some eighty-six species of mammals inhabit the northern Great Basin ranging from shrews (*Sorex* sp.) to bighorned sheep (*Ovis canadensis*). Many species, such as the snowshoe hare (*Lepus americanus*), are restricted to montane forests on the margins of the Great Basin at the northernmost extension of the area under discussion. The classification scheme, common, and scientific names used follow recent changes in nomenclature (Larrison 1976: 222-226).

The order Insectivora (shrews and moles) is represented by only five species, four shrews and one mole. Currently several research projects are being conducted on the small mammals of the Steens Mountain, with much valuable information regarding shrew numbers and distribution resulting from this work.

Fourteen species of bats (family Vespertilionidae) inhabit this region, with most confined to lower elevations and sites of high insect density.

Five species of rabbits and hares (family Leporidae) are common. The black-tailed jackrabbit (*Lepus californicus*) is most representative of a species that demonstrates cycles with extreme fluctuation in population size. It provides a good food source for predators (coyote) and scavengers (turkey vulture) during peak years.

The rodents (order Rodentia) are the most numerous (nine families, thirty-nine species) mammal inhabiting the northern Great Basin. Several species are not found in the Great Basin proper but are restricted to forested mountains bordering this region; these are the columbian ground squirrel (*Spermophilus columbianus*), red squirrel (*Tamiasciurus hudsonicus*), Douglas's squirrel (*Tamiasciurus douglasi*), and the northern flying squirrel (*Glaucomys sabrinus*).

The family Heteromyidae (pocket mice and kangaroo rats) is represented by six species: the little pocket mouse (*Perognathus longimembris*), Great Basin pocket mouse (*Perognathus parvus*), dark kangaroo mouse (*Microdipodops megacephalus*), Ord's kangaroo rat (*Dipodomys ordi*), chisel-toothed kangaroo rat (*Dipodomys microps*), and a species restricted to the western border of the Great Basin (Klamath region), Heerman's kangaroo rat (*Dipodomys heermanni*). The Great Basin pocket mouse is one of the most numerous rodents in arid regions of mixed sagebrush, butterbrush, and rabbitbrush, where they feed primarily on seeds. The nocturnal Ord kangaroo rat is the most common *Dipodomys* species present and has been trapped in greasewood hummocks about a mile onto the eastern edge of the Alvord Desert.

The family Castoridae is represented by the beaver, *Castor fiber*. The greatest population of beaver in the Great Basin is found at higher elevations where food trees are more available. In these locations, the beavers generally dam streams and build lodges, while at lower elevations they burrow in the banks of streams and canals and seldom construct lodges.

Cricetid mice and rats (family Cricetidae) have seven species inhabiting the region. The common deer mouse (*Peromyscus maniculatus*) is by far the most numerous Cricetid present, ranging into a variety of habitats. The northern grasshopper mouse (*Onychomys leucogaster*) is a highly predacious species of the sagebrush, where it feeds on other mice and arthropods. The desert wood rat (*Neotoma lepida*) is an animal of greasewood, hopsage, and sagebrush habitats, where it usually constructs large stick-pile nests. It is quite common in locations surrounding the Alvord Desert. The bushy-tailed woodrat (*Neotoma cinerea*) prefers rocky ledges, caves, or crevices, and sometimes old buildings where stick nests are constructed. Both species are referred to as "packrat" because of their habit of transporting a variety of material, ranging from rabbit skulls to pieces of broken glass, to their nests.

Voles (family Microtidae) are not well established in the arid Great Basin. The seven species present are restricted to meadows and marshes; and three species—Gapper's red-backed mouse (*Clethrionomys gapperi*), heather mouse (*Phenacomys intermedius*), and Richardson's water vole (*Arvicola richardsoni*)—are restricted to the mountainous region (Blue Mountains) north of the Great Basin.

The family Erethizontidae has a single species, the porcupine (*Erethizon dorsatum*). It prefers coniferous forests but is wide ranging. One has been captured lumbering across the Alvord Desert, over 4 miles from the nearest vegetation.

Economically the porcupine is an important pest because of the damage it inflicts primarily on trees.

There are eighteen species in five families of the order Carnivorea (Carnivores) found in this region.

The coyote (*Canis latrans*) is the most notable and wide-ranging member of the dog family (Canidae). It is generally appreciated by the biological community and despised and hunted by local ranchers.

In 1974 a gray wolf (*Canis lupus*) was killed in Malheur County, Oregon, and probably marked the extinction of this fine Canid in the northern Great Basin. The red fox (*Vulpes vulpes*) is also very rare, if not exterminated and the kit fox (*Vulpes velox*) shows a scattered distribution as an endangered species of Oregon.

Nine species of mustelids (family Mustelidae) are found here. A wolverine (*Gulo gulo*), long believed extinct in the northern Great Basin, was trapped and released in 1975 on the Steens Mountain. The badger (*Taxidea taxus*) has a wide distribution and is an excellent excavator of burrowing rodents such as ground squirrels, pocket gophers, and mice.

Of the three species of cats (family Felidae) residing in the northern Great Basin, only the bobcat (*Lynx rufus*) is common, with many pelts taken each year by hunters and trappers.

Two species of the deer family (Cervidae) are found here. The elk or wapiti (*Cervus elaphis*) is generally restricted to the mountain forest and foothills of the Blue Mountains north. However, there is a small herd on the Steens Mountain. Mule deer (*Odocoileus hemionus*) are wide ranging and abundant in forested regions, meadows, and open willow thickets.

The pronghorn family (Antilocapridae) is composed of a single member in North America, the pronghorn (*Antilocapra americana*). This swift artiodactyl prefers open sagebrush and grassland plains, with the greatest concentration in the Hart Mountain National Antelope Refuge of southeastern Oregon.

Bighorn sheep (*Ovis canadensis*) is the lone representative of the cattle family (Bovidae) in the northern Great Basin. Bighorn sheep have been reintroduced into Hart Mountain and the Steens Mountain, where numbers have increased to a level to allow limited hunting. A 1976 introduction of sixteen Hart Mountain bighorn sheep to the Pueblo Mountains appears less successful.

REFERENCES

American Ornithologist Union. 1957. *Checklist of North American birds.* Baltimore, Md.: Lord Baltimore Press.

Axelrod, D. L. 1944. "The Alvord Creek flora (Oregon)." In *Pliocene floras of California and Oregon,* ed. R. W. Chaney, no. 38, pp. 225-262. Washington, D.C.: Carnegie Institute.

Bailey, Vernon. 1936. The mammals and life zones of Oregon. Washington, D.C.: U.S. Department of Agriculture, Bureau of Biological Surveys.

Baldwin, Ewart M. 1959. *Geology of Oregon.* Ann Arbor, Mich.: Edwards Brothers.

Bond, Carl E. 1973. *Keys to Oregon freshwater fishes.* Corvallis: Oregon State University.

Ferguson, Denzel, and Ferguson, Nancy. 1978 *Oregon's Great Basin country.* Burns, Ore.: Gail Graphics.

Fuller, R. E. 1931. *The geomorphology and volcanic sequence of Steens Mountain in southeastern Oregon* 3:1. Seattle: University of Washington Publ. in Geol.

George, Tom, and Haglund, James. 1973. *Summary of snow survey measurements for Oregon.* Portland, Ore.: Soil Conservation Service.

Gilkey, Helen M., and Dennis, LaRea J. 1973. *Handbook of northwestern flowering plants.* Corvallis, Ore.: Oregon State University Bookstores.

Hall, E. Raymond, and Kelson, Keith R. 1959. *The mammals of North America.* New York: Ronald Press.

Hansen, Charles G. 1956. *An ecological survey of the vertebrate animals on Steens Mountain, Harney County, Oregon.* Ph.D. dissertation, Oregon State College.

Hitchcock, C. Leo, and Cronquest, Arthur. 1973. *Flora of the Pacific Northwest.* Seattle and London: University of Washington Press.

Larrison, Earl J. 1976. *Mammals of the Northwest.* Seattle, Wash. Seattle Audubon Society.

Lund, Ernest H., and Bentley, Elton. 1976. "Steens Mountain." In *The Ore Bin* 38(4):51-66. Portland, Ore.: State of Oregon Department of Geology and Mineral Industries.

Peck, Morton E. 1941. *A manual of the higher plants of Oregon.* Portland, Ore.: Binfords and Mort.

Peterson, Roger T. 1969. *A field guide to western Birds.* Boston: Houghton Mifflin.

Smith, Gerald R. 1978. "Biogeography of intermountain fishes." In *Great Basin naturalist memoirs,* 2:17-42. Provo, Utah: Brigham Young University.

Stebbins, Robert C. 1954. *Amphibians and reptiles of western North America.* New York: McGraw-Hill.

Stewart, Omer C. 1939. *The northern Paiute bands* 23:127-149.

Berkeley: University of California Anthropological Records.

Tate, James, Jr., and Kibbe, Douglas P. 1974. "Update Your Field Guides." in *American Birds,* 28(4):747-753. New York: National Audubon Society.

Urban, Karl A. 1973. *Vegetative patterns on Steens Mountain, Harney County, Oregon.* Pendleton, Ore. Blue Mountain Community College. Unpublished manuscript.

———. 1977. *Checklist of the vascular plants of Steens Mountain, Harney County, Oregon.* Pendleton, Ore. Blue Mountain Community College.

U.S. Department of Commerce. 1973. *Climatography of the U.S. No. 81, monthly normals of temperature, precipitation, and heating and cooling degree days,* 1941-70. Asheville, N.C.: National Climatic Center.

———. 1971-1977. *Climatological data: Oregon annual summary,* pp. 77-83. Asheville, N.C.: National Climatic Center.

U.S. Department of the Interior. Fish and Wildlife Service. 1973. *Birds of the Malheur National Refuge.* Portland, Ore.: U.S. Department of Interior.

———. 1977. *Mammals of Malheur National Wildlife Refuge, Oregon.* Portland, Ore. U.S. Department of Interior.

Verts, B. J. 1971. *Keys to the mammals of Oregon.* Corvallis, Ore. Oregon State University Bookstores.

Walker, G. W., and Repenning, C. A. 1965. *Reconnaissance geologic map of the Adel Quadrangle, Lake, Harney, and Malheur counties, Oregon.* Washington, D.C.: U.S. Geol. Survey Misc. Geol. Invest.

Wilkerson, W. L. 1958. The geology of a portion of the southern Steens Mountain, Oregon. Master's thesis, University of Oregon.

Williams, Howel, and Compton, R. R. 1953. Quicksilver deposits of the Steens and Pueblo Mountains, Southern Oregon. *U.S. Geol. Survey Bulletin* 995-B:19-77. Washington D.C.

Wright, A. H., and Wright, A. A. 1949. *Handbook of frogs and toads of the United States and Canada.* New York: Comstock.

———. 1957. *Handbook of snakes.* Ithaca, N.Y.: Comstock.

A CHECKLIST OF THE PLANTS OF THE NORTHERN GREAT BASIN REGION OF OREGON

KEY TO STATUS AND/OR SOURCE

+   Vascular plants of the Steens Mountain, Harney County, Oregon (K. A. Urban 1977).

A   Rare and endangered species from Oregon rare and endangered species checklist (K. A. Urban 1977).

B   Rare on Steens Mountain (K. A. Urban 1977).

C   Plant list from Steens Mountain (C. G. Hansen 1956).

D   Plants collected on the Steens Mountain previous to 1952 (C. G. Hansen 1956).

E   Plants with northernmost range in the Harney Basin or endemic to the northern Great Basin (Ferguson and Ferguson 1978).

Status and/or Source

    ISOETACEAE   The Quillwort Family
B   *Isoetes bolander's* Engelm. Bolander's quillwort
C   *Isoetes occidentalis* Hend. Lake quillwort

    EQUISETACEAE   The Horsetail Family
C   *Equisetum funstoni* Eat. California horsetail
D   *Equisetum hyemale* v. *Californicum* Milde. Western scouring rush
D   Equisetum paluster L. Marsh horsetail
C   *Equisetum praealtum* Raf. Prairie scouring rush

    OPHIOGLOSSACEAE   The Adders'-Tongue Family
C   *Botrychium simplex* Hitchc. Little grape-fern

    POLYPODIACEAE   The Common Fern Family
+   *Athyrium distentifolium* Tausch. Alpine lady fern
B   *Athyrium felix-femina* (L.) Roth. Lady fern

+   *Cystopteris fragilis* (L.) Bernh. Brittle bladder fern
A   *Pellaea breweri* D.C. Eat. Brewer's cliff brake
C   *Polystichum lonchitis* (L.) Roth. Mountain holly-fern
A/B   *Polystichum scopulinum* (D.C. Eat.) Maxon Christmas fern
C   *Woodsia oregona* Eat. Woodsia

    CUPRESSACEAE   The Cypress family
+   *Juniperus communis* L. v. *montana* Ait. Mountain juniper
+   *Juniperus occidentalis* Hook. Western juniper

    PINACEAE   The Pine Family
B   *Abies concolor* (Gord. & Gland) Lindl. v. *lowiana* Sierran or white fir

    EPHEDRACEAE   The Ephedra Family
D   *Ephedra nevadensis* Wats. Nevada Ephedra
D   *Ephedra viridis* Cov. Green Ephedra

    SALICACEAE   The Willow Family
+   *Populus tremuloides* Nichx. Quaking aspen
+   *Populus trichocarpa* T. & G. Black cottonwood
A/B   *Salix arctica* Pall. Arctic or ptarmigan willow
C   *Salix commutata* Bebb. Undergreen or variable willow
A/B   *Salix drummondiana* Barratt. Drummond willow
+   *Salix exigua* Nutt. v. *melanopsis* (Nutt.) Cronq. Dusky willow
+   *Salix lasiandra* Benth. v. *caudata* (Nutt.) Sudw. Whiplash willow
D   *Salix lemmonii* Bebb. Lemmon's willow
D   *Salix mackenziana* (Hook) Barr. Mackenzie Willow
D   *Salix nivalis* Hook. v. *nivalis* Snow willow

C   *Salix nivalis* Hook. v. *salimontana* (Rydb.) Schneid. Rocky Mountain willow

C   *Salix pseudocordata* (Ander) Rydb.

C   *Salix sitchensis* Sanson. Sitka willow

BETULACEAE   The Birch Family

+   *Alnus incana* (L.) Moench. Mountain alder

C   *Alnus tenuifolia* Nutt. Thin-leaf alder

C   *Betula fontinalis* Sarg. Water birch

+   *Betula occidentalis* Hook. Spring or western birch

URTICACEAE   The Nettle Family

+   *Urtica dioica* L. ssp. *gracilis* (Ait.) Seland. Stinging nettle

C   *Urtica dioica* L. v. *holosericea* (Nutt.) Hitchc. Hoary nettle

C   *Urtica dioica* L. v. lyallii (Wats.) Hitchc. Lyall nettle

SANTALACEAE   The Sandlewood Family

D   *Comandra pallida* DC Pale Comandra

POLYGONACEAE   The Buckwheat Family

D   *Chorizanthe watsoni* T. & G. Watson's spine-flower

+   *Eriogonum androsaceum* Benth. Rockjasmine buckwheat

A/B   *Eriogonum chrysops* Rydb. Golden buckwheat

A   *Eriogonum cusickii* M.E. Jones Cusick's buckwheat

+   *Eriogonum heracleoides* Nutt. Wyeth's buckwheat

+   *Eriogonum marifolium* T. & G. Mountain buckwheat

+   *Eriogonum microthecum* Nutt. v. *laxiflorum* Hook. Slenderbush

+   *Eriogonum ovalifolium* Nutt. v. *depressum* Blank. Dwarf buckwheat

+   *Eriogonum ovalifolium* Nutt. v. *nivale* (Canby) Jones. Snow buckwheat

+   *Eriogonum ovalifolium* Nutt. v. *ovalifolium* Oval-leaved buckwheat

C   *Eriogonum nidularium* Coville. Birds-nest eriogonum

B   *Eriogonum spergulinum* Gray. Spurry buckwheat

D   *Eriogonum sphaerocephalum* Benth. Round-headed Eriogonum or rock buckwheat

+   *Eriogonum strictum* Benth. ssp. *proliferum* (T. & G.) Stokes v. *anserinum* (Greene) Davis. Strict buckwheat

A/B   *Eriogonum umbellatum* Torr. v. *hausknechtii* (Dammer) Jones.

+   *Eriogonum vimineum* Dougl. v. *shoshonense* (Nels.) Stokes Broom buckwheat

E   *Eriogonum umbellatum* v. *glaberrimum* (Gandg.) Reveal sulfur buckwheat

+   *Eriogonum umbellatum* Torrv. *umbellatum* Sulfur buckwheat

B   *Eriogonum watsonii* T. & G. Watson's buckwheat

+   *Oxyria digyna* (L.) Hill. Alpine sorrel

D   *Oxytheca dendroides* Nutt. Northern Oxytheca

C   *Polygonum austiniae* Greene. Austin's knotweed

+   *Polygonum aviculare* L. Prostrate knotweed

+   *Polygonum bistortoides* Pursh. American bistort

C   *Polygonum coccineum* Muhl. Water smartweed

D   *Polygonum confertiflorum* Nutt. Closeflowered knotweed

C   *Polygonum douglasii* Greene. Douglas' knotweed

A   *Polygonum heterosepalum* Peck & Ownbey. Dwarf desert knotweed

C   *Polygonum kelloggii* Greene. Kellogg's knotweed

A   *Polygonum majus* (Moien.) Piper. Palouse knotweed

C   *Polygonum minimum* Wats. Broad-leaf or leafy knotweed

D   *Polygonum watsonii* Small. Water knotweed

C   *Rumex acetosella* L. Red sorrel

+   *Rumex crispus* L. Curly dock

C   *Rumex cuneifolius* Campd. Wedge-leaved dock

C   *Rumex occidentalis* Wats. Western dock

C   *Rumex paucifolius* Nutt. Alpine sorrel

+   *Rumex salicifolius* Weinm. Willow-leaved dock

CHENOPODIACEAE   The Goosefoot Family

E   *Allenrolfea occidentalis* (Wats) Kuntze. Iodine bush

+   *Chenopodium album* L. Lamb's quarters

C   *Chenopodium capitatus* (L.) Asch. Strawberry blite

B   *Chenopodium foliosum* (Moench.) Asch. Leafy goosefoot

C   *Chenopodium polyspermum* L.

+   *Chenopodium rubrum* L. Red goosefoot

C   *Grayia spinosa* (Hook) Collotzi. Spiny hopsage

+   *Monolepsis nuttalliana* (Schultes) Greene. Patata

D   *Monolepsis pusilla* Torr. Dwarf Monolepsis

C   *Nitrophila occidentalis* S. Wats. Borax weed

+   *Salsola kali* L. Russian thistle

+   *Sarcobatus vermiculatus* (Hook.) Torr. Greasewood

D   *Suaeda diffusa* Wats. Bushy blite

D   *Suaeda nigra* (Raf.) Macbr. Bushy Seablite

AMARANTHACEAE   The Amaranth Family

D   *Amaranthus graecizans* L. Prostrate pigweed, tumbleweed, or tumbleweed amaranth

NYCTAGINACEAE   The Four-O'Clock Family

E   *Mirabilis bigelovii* v. *retrorsa* (Heller) Munz

PORTULACACEAE   The Purslane Family

A   *Claytonia megarhiza* (Gray) Parry v. *bellidifolia* (Rydb.) Hitchc. Alpine spring beauty

A/B   *Claytonia nevadensis* Wats. Sierran spring beauty

+   *Lewisia pygmaea* (Gray) Robins v. *nevadensio* (Gray) Fosberg. Nevada Lewisia

+   *Lewisia rediviva* Pursh. Bitterroot

+   *Lewisia triphylla* (Wats.) Robins. Three-leaved Lewisia

+   *Montia chamissoi* (Ledeb.) Robins. and Fern. Water Montia

+   *Montia cordifolia* (Wats.) Pax and Hoffm. Heart-leaved Montia of miner's lettuce

C   *Montia fontana* Linna.   Water chickweed

+   Montia linearis (Dougl.) Greene.   Narrow-leaved miner's lettuce

+   *Montia perfoliata* (Donn.) Howell.   Miner's lettuce

+   *Spraguea umbellata* Torr. v. *caudicifera* Gray

CARYOPHYLLACEAE   The Pink Family

+   *Arenaria aculeata* Wats.   Needle-leaved sandwort

+   *Arenaria capillaris* Poir.   Thread-leaved sandwort

+   *Arenaria congesta* Nutt. v. *cephaloides* (Rydb.) Maguire Ball-headed sandwort

+   *Arenaria nuttallii* Pax v. *fragilis* (Mag. & Holm.) Hitch. Nuttall's sandwort

B   *Arenaria rubella* (Wahlenb.) J. E. Smith.   Boreal sandwort

+   *Cerastium arvense* L.   Common field mouse ear

+   *Cerastium berringianum* Cham. & Schlecht.   Alpine chickweed

C   *Cerastium viscosum* L.   Sticky chickweed or Cerastium

D   *Lychnis drummondii* (Hook) Wats.   Drummond Campion

D   *Sagina occidentalis* Wats.   Western pearlwort

C   *Sagina saginoides* (L.) Britt.   Alpine or arctic pearlwort

D   *Saponaria vaccaria* L.   Cow herb

D   *Silene antirrhina* L.   Sleepycat

B   *Silene douglassi* Hook v. *douglassii* Douglas' Silene

D   *Silene macounii* S. Wats.   Parry's Silene

+   *Silene menziesii* Hook. v. *menziesii* Menzies' Silene

+   *Silene oregana* Wats.   Oregon Silene

B   *Silene parryi* (Wats.) Hitch. & Mag.   Parry's Silene

D   *Silene scaposa* Robins.   Scapose Silene

D   *Spergula arvensis* L.   Stickwort, starwort or cornspurry

+   *Spergularia rubra* (L.) Presl.   Red sandspurry

C   *Stellaria crassifolia* Ehrb.   Thickleaved starwort

C   *Stellaria crispa* Cham. & Schlecht.   Crisped starwort

+   *Stellaria jamesiana* Torr.   Sticky chickweed

B   *Stellaria longifolia* Muhl.   Long-leaved chickweed

+   *Stellaria longipes* Goldie.   Long-stalked chickweed

+   *Stellaria media* (L.) Cyrill.   Common chickweed

NYMPHAEACEAE   The Waterlily Family

+   *Nuphar polysepalum* Engelm.   Indian pond lily or wakas

PAEONIACEAE   The Paeony Family

+   *Paeonia brownii* Dougl.   Wild paeony

RANUNCULACEAE   The Buttercup or Crowfoot Family

+   *Aconitum columbianum* Nutt. v. *columbianum* Columbian monkshood

+   *Actaea Rubra* (Ait.) Willd.   Western red baneberry

+   *Aquilegia formosa* Fisch.   Western columbine

B   *Caltha biflora* D.C.   Broad-leaved marshmarigold

+   *Caltha leptosepala* D. C. v. *leptosepala* marshmarigold

+   *Clematis ligusticifolia* Nutt.   Vine clematis

+   *Delphinium andersonii* Gray.   Anderson's larkspur

B   *Delphinium bicolor* Nutt.   Little larkspur

+   *Delphinium depauperatum* Nutt.   Slim larkspur

+   *Ranunculus alismaefolius* Geyer.   Plantain-leaved buttercup

+   *Ranunculus aquatilis* L.   White Water buttercup

B   *Ranunculus cymbalaria* Pursh.   Seaside buttercup

+   *Ranunculus eschscholtzii* Schlecht. v. *trisectum* (Eastw.) Benson Snow buttercup

+   *Ranunculus flammula* L.   Creeping buttercup

+   *Ranunculus glaberrimus* Hook. v. *ellipticus* Greene.   Sagebrush buttercup

C   *Ranunculus macounii* Britt.   Macoun's buttercup

+   *Ranunculus orthorhynchus* Hook. v. *platyphyllus* Gray. Straight-backed buttercup

+   *Ranunculus populago* Greene.   Mountain buttercup

+   *Ranunculus testiculatus* Granz.   Hornseed or horned buttercup

+   *Ranunculus uncinatus* D. Don v. *uncinatus* Little or wood buttercup

C   *Ranunculus verecundus* Robins.   Modest buttercup

+   *Thalictrum fendleri* Engelm.   Fendler's meadowrue

+   *Thalictrum occidentale* Gray.   Western meadowrue

B   *Thalictrum venulosum* Trel.   Veiny meadowrue

BERBERIDACEAE   The Barberry Family

+   *Berberis repens* Lindl.   Creeping barberry

PAPAVERACEAE   The Poppy Family

C   *Canbya aurea* Wats.   Golden Canbya

FUMARIACEAE   The Bleeding Heart Family

D   *Corydalis aurea* Willd.   Golden Corydalis

+   *Dicentra uniflora* Kell.   Steershead bleedingheart

BRASSICACEAE (CRUCIFERAE)   Mustard Family

+   *Alyssum alyssoides* L.   Pale alyssum

+   *Arabis divaricarpa* Nels.   Spreading pod rockcress

+   *Arabis drummondii* Gray.   Drummond's rockcress

B   *Arabis glabra* (L.) Bernh.   Towermustard

+   *Arabis holboellii* Hornem.   Hoelboell's rockcress

D   *Arabis lemmonii* Wats.   Lemmon's rockcress

C   *Arabis microphylla* Nutt.   Little-leafed rockcress

C   *Arabis puberula* Nutt.   Hoary rockcress

+   *Arabis sparsiflora* Nutt.   Sicklepod rockcress

+   *Barbarea orthoceras* Ledeb.   Wintercress

+   *Camelina microcarpa* Andrz.   Hairy falseflax

C   *Camelina sativa* (L.) Crantz.   Falseflax or gold of pleasure

+   *Capsella bursa-pastoris* (L.) Medic.   Shepherd's purse

+   *Cardamine oligosperma* Nutt.   Little western bittercress

+   *Descurainia pinnata* (Walt.) Britt.   Western tansy mustard

+   *Descurainia richardsonii* (Sweet) Schultz. Richardson's tansy mustard

+   *Descurainia sophia* (L.) Webb Flixweed

+   *Draba crassifolia* R. Grah. Thick-leaved Draba

D   *Draba cruciata* Pays. Steen Mountain willow grass

D   *Draba douglasii* Gray. Douglas' Draba

C   *Draba nemorosa* (L.) Woods. Draba

A   *Draba sphaeroides* Pays. v. *cusickii* (Robbins) C. L. Hitchc. Cusick's Draba

C   *Draba stenoloba* Ledeg. Alaskan whitlow-grass or slender Draba

D   *Erysimum parviflorum* Nutt. Small wallflower

+   *Lepidium perfoliatum* L. Peppergrass

C   *Lesquerella diversifolia* (Greene) Hitchc. Western bladder-pod

B   *Lesquerella occidentalis* Wats. Western bladderpod

+   *Phoenicaulis cheiranthoides* Nutt. Daggerpod mustard

D   *Physaria didymocarpa* (Hood) Gray. Common twinpod

D   *Physaria oregana* Wats. Oregon twinpod

+   *Polyctenium fremontii* (Wats.) Greene. Desert combleaf

+   *Rorippa curvisiliqua* (Hook.) v. *lyrata* (Nutt.) Peck bessey

+   *Rorippa sinuata* (Nutt.) Hitchc. Spreading yellowcress

+   *Rorippa nasturtium-aquaticum* (L.) Schinz. & Thell. Watercress

+   *Rorippa obtusa* (Nutt.) Britt. Blunt-leaved-yellow cress

+   *Sisymbrium altissium* L. Jim Hill tumblemustard

C   *Smelowskia fremontii* S. Wats. Fremont's Smelowskia

+   *Stanleya pinnata* (Pursh) Britt. Prince's plume

D   *Stanleya viridiflora* Nutt. Perennial Stanleya

D   *Streptanthus crassicaulis* Torr. Thick-stemmed Streptanthus

+   *Streptanthus cordatus* Nutt. Heart-leaved Streptanthus

B   *Thelypodium flexuosum* Robins. Spreading Thelypodium

+   *Thelypodium integrifolium* (Nutt.) Endl. Entire-leaved Thelypody

+   *Thelypodium laciniatum* (Hook.) Thick-leaved Thelypody

+   *Thelypodium sagitattum* (Nutt.) Endl. Slender Thelypody

+   *Thlaspi arvense* L. Fanweed

CAPPARIDACEAE   The Caper Family

D   *Cleome platycarpa* Torr. Golden spider flower or cleome

CRASSULACEAE   The Stonecrop Family

A   *Sedum debile* Wats. Weak-stemmed stonecrop

+   *Sedum lanceolatum* Torr. Lance-leaved stonecrop

+   *Sedum rosea* (L.) Scop. King's crown

B   *Sedum stenopetalum* Pursh. Wormleaf stonecrop

SAXIFRAGACEAE   The Saxifrage family

D   *Heuchera cusickii* R. B. L. Cusick's Heuchera

+   *Heuchera cylindrica* Dougl. v. *alpina* Wats. Roundleaf alumroot

C   *Heuchera ovalifolia* Nutt. Round-leaved alumroot

D   *Heuchera rubescens* Torr. Reddish Heuchera

+   *Lithophragma bulbifera* Rydb. Bulbous fringecup

+   *Lithophragma glabra* Nutt. Smooth fringecup

+   *Lithophragma parviflora* (Hook.) Nutt. Small-flowered fringecup

C   *Lithophragma rupicola* Greene. Larva fringecup

+   *Lithophragma tenella* Nutt. Slender fringecup

B   *Mitella breweri* Gray. Brewer's Mitella

B   *Mitella caulescens* Nutt. Leafy mitrewort

C   *Mitella pentandra* (Hook) Rydb. Alpine or five stemmed mitrewort

+   *Parnessia fimbriata* Konig. v. *intermedia* (Rydb.) Hitchc. Fringed grass of parnassus

C   *Saxifraga adscendens* L. Wedge-leaved saxifrage

C   *Saxifraga aprica* Greene.

+   *Saxifraga arguta* D. Don Brook saxifrage

A/B   *Saxifraga caespitosa* L. v. *emarginata* (Small) Rosend

C   *Saxifraga columbiana* (Piper) Hitchc. Columbia saxifrage

C   *Saxifraga debilis* Englm. Pygmy or weak saxifrage

+   *Saxifraga integrifolia* Hook. Bog saxifrage

D   *Saxifraga montanensis* (Small) Hitchc. Montana saxifrage

+   *Saxifraga oregana* Howell. Oregon saxifrage

+   *Saxifraga rhomboidea* Greene. Diamond-leaved saxifrage

GROSSULARIACEAE   The Gooseberry Family

+   *Ribes aureum* (Pursh) Golden. currant

+   *Ribes cereum* Dougl. Squaw currant

B   *Ribes inerme* Rydb. White-stemmed gooseberry

+   *Ribes lobbii* Gray. Lobb's gooseberry

+   *Ribes montigenum* McClatchie. Alpine prickly currant

C   *Ribes petiolare* (Dougl.) Jancz. Western black currant

ROSACEAE   The Rose Family

+   *Amelanchier alnifolia* Nutt. Serviceberry

C   *Amelanchier utahensis* Koehn. Utah serviceberry

+   *Cercocarpus ledifolius* Nutt. Mountain mahogany

C   *Crataegus columbiana* v. *piperi* (Britt.) Eggleston. Columbian hawthorn

B   *Fragaria virginiana* Duchesne. Wild strawberry

+   *Geum marcophyllum* Willd. Large-leaved avens

+   *Geum triflorum* Pursh v. *ciliatum* (Pursh) Fassett long-plumed avens

+   *Geum triflorum* Pursh v. *triflorum* Old man's whiskers

D   *Geum strictum* Soland.

+   *Holodiscus dumosus* (Hook.) Heller Creambush rockspirea

C   *Holodiscus dumosus* v. *glabrescens* (Greene) Hitchc. Gland ocean-spray

+   *Horkelia fusca* Lindl. v. *capitata* (Lindl.) Peck Pinewoods Horkelia

A/B   *Ivesia baileyi* Wats. Great Basin Ivesia

+   *Ivesia gordonii* (Hook) T. &. G. Gordon's Ivesia

C	*Potentilla biennis* Greene. Biennial cinquefoil
C	*Potentilla blaschkeana* Tuvca. Jeps.
B	*Potentilla brevifolia* Nutt. Short-leaved cinquefoil
+	*Potentilla breweri* Wats. Brewer's cinquefoil
+	*Potentilla concinna* Richards. v. *divisa* Rydb. Early cinque-foil
B	*Potentilla diversifolia* Lehm. Diverse-leaved Potentilla
C	*Potentilla drummondii* Lehm. Drummond's cinquefoil
C	*Potentilla fastigiata* Nutt.
+	*Potentilla fruticosa* L. Shrubby cinquefoil
+	*Potentilla glandulosa* Lindl. Sticky cinquefoil
D	*Potentilla glomerata* Nels.
+	*Potentilla gracilis* Dougl. Fan-leaved cinquefoil
D	*Potentilla nuttallii* Lehm. Nuttal's cinquefoil
C	*Potentilla pumila* (Rydb.) Fedde. Dwarf western cinquefoil
D	*Potentilla quinquefolia* Rydb. Snow or five-leaved cinque-foil
B	*Potentilla rivalis* Nutt. Brook cinquefoil
+	*Prunus emarginata* (Dougl.) Walp. Bitter cherry
+	*Prunus virginiana* L. v. *melanocarpa* (Nels.) Sarg. Western chokecherry
+	*Purshia tridentata* (Pursh) DC. Antelope bitterbrush
+	*Rosa nutkana* Presl. v. *hispida* Fern. Spalding's wildrose
C	*Rosa pisocarpa* Gray. Clustered or peafruit wildrose
+	*Rosa woodsii* Lindl. v. *ultramontana* (Wats.) Jeps. Wood's rose
+	*Sibbaldia procumbens* L. Creeping Sibbaldia
C	*Sorbus sitchensis* Roemer. Sitka mountain ash

FABACEAE (LEGUMINOSAE)   The Pea Family

D	*Astragalus collinus* Dougl. Hill or hillside milkvetch
+	*Astragalus conjunctus* Wats. Stiff milkvetch
+	*Astragalus curvicarpus* (Sheld) Macbr. Curvepod milkvetch
D	*Astragalus diurnus* S. Wats. Transparent milkvetch
+	*Astragalus filipes* Torr. Slim milkvetch
D	*Astragalus glareosus* Dougl. Wooly Milkvetch
D	*Astragalus howeleii* Gray. Howell's milkvetch
C	*Astragalus lectulus* S. Wats.
D	*Astragalus lentiginosus* Dougl. Freckled or specklepod milk-vetch
C	*Astragalus macregorii* (Rydb.) Tides.
+	*Astragalus malacus* Gray. Shaggy milkvetch
D	*Astragalus miser* Dougl. Weedy milkvetch
C	*Astragalus owyheensis* Nels.
+	*Astragalus purshii* Dougl. Pursh's or woolly-podded milk-vetch
C	*Astragalus salinus* Howell. Alkali milkvetch
D	*Astragalus sinuatus* Piper. Whited milkvetch
C	*Astragalus stenophyllus* T. & G. Threadstalk milkvetch
+	*Astragalus whitneyi* Gray. Balloon-podded milkvetch
+	*Lathyrus pauciflorus* Fern. Few-flowered peavine

D	*Lathyrus rigidus* White. Rigid or bushy peavine
D	*Lupinus albifrons* Benth. White-leaved lupine
C	*Lupinus brevicaulis* Wats. Short-stemmed lupine
+	*Lupinus caudatus* Kell. Tailcup lupine
C	*Lupinus corymbosus* Hel. Corymbed lupine
+	*Lupinus lepidus* Dougl. v. *lepidus* Prairie lupine
+	*Lupinus leucophyllus* Dougl. v. tenuispicus (Nels.) Smith Velvet lupine
C	*Lupinus lyallii* v. *lobbii* Gray. Low mountain lupine
D	*Lupinus nevadensis* Hel. Nevada lupine
D	*Lupinus polyphyllus* Lindl. Big, many, or large-leaved lupine
C	*Lupinus saxosus* Howell. Rock lotus
+	*Lupinus sericeus* Pursh v. *sericeus* Silky lupine
D	*Medicago lupulina* L. Black Medic(k) or hop clover
+	*Medicago sativa* L. Alfalfa
+	*Melilotus alba* Desr. White sweetclover
+	*Melilotus officinalis* (L.) Lam. Yellow sweetclover
B	*Onobrychis viciaefolia* Scop. Holy clover or sandfain
D	*Petalostemon ornatum* Dougl. Western prairie clover
+	*Thermopsis montana* Nutt. v. *montana* Mountain thermop-sis
C	*Trifolim beckwithii* Brewer. Beckwith's clover
+	*Trifolium cyathiferum* Lindl. Cup clover
C	*Trifolium fibriatum* Lindl. Springbank clover
+	*Trifolium longipes* Nutt. v. *hansenii* (Greene) Teps. Long-stalked clover
+	*Trifolium marcocephalum* (Pursh) Poiret. Big-headed clo-ver
+	*Trifolium multipedunculatum* Kennedy. Many-stalked clo-ver
C	*Trifolium repens* L. White or dutch clover
C	*Trifolium spinulosum* Dougl. Springbank clover
D	*Trifolium variegatum* Nutt. Few-flowered clover
+	*Trifolium wormskjoldii* Lehm. Springbank clover
+	*Vicia americana* Muhl. American vetch
+	*Vicia sativa* L. Common vetch

GERANIACEAE   The Geranium Family

+	*Erodium circutarium* (L.) L'Her. Filaree
+	*Geranium viscosissimum* F & M Sticky geranium

LINACEAE   The Flax Family

+	*Linum perenne* L. v. *lewisii* (Pursh) Eat. & Wright. West-ern blue flax

EUPHORBIACEAE   The Spurge Family

+	*Euphorbia glyptosperma* Engelm. Corrugate-seeded spurge
C	*Euphorbia serpyllifolia* Engelm. Thyme-leaf spurge

CALLITRICACEAE   The Water Starwort family

+	*Callitriche hermaphroditica* L. Northern water starwort
+	*Callitriche verna* L. Spring water starwort

LIMNANTHACEAE   The Meadow Foam Family
+   *Floerkea proserpinacoides* Willd. False mermaid

RHAMNACEAE   The Buckthorn Family
+   *Ceanothus velutinus* Dougl. v. *velutinus* Tobacco bush

MALVACEAE   The Mallow Family
+   *Iliamna rivularis* (Dougl.) Greene. Mountain hollyhock
+   *Malva neglecta* Wallr. Cheeseweed or buttonweed
D   *Sida hederacea* (Dougl.) Torr. Alkali-mallow
+   *Sidalcea oregana* (Nutt.) Gray. Oregon Sidalcea
D   *Sidalcea spicata* (Regel) Greene. Spiked Sidalcea
+   *Sphaeralcea grossulariaefolia* (H & A) Rydb. Gooseberry-leaved globemallow

HYPERICACEAE   The St. John's Wort Family
D   *Hypericum anagalloides* C. & S. Bog St. John's wort
B   *Hypericum formosum* H, B, K. v. *nortoniae* (Jones) Hitchc. Western St. John's wort
+   *Hypericum formosum* H, B, K. v. *scouleri* Western St. John's wort

VIOLACEAE   The Violet Family
+   *Viola adunca* Sm. Early blue violet
C   *Viola beckwithii* T. & G. Beckwith's violet
B   *Viola nephrophylla* Greene. Northern bog violet
+   *Viola nuttallii* Pursh v. *vallicola* (Nels.) Hitch. Nuttal's violet
B   *Viola palustris* L. Marsh violet
+   *Viola purpurea* Kell. Goosefoot violet

LOASACEAE   The Loasa or Blazingstar Family
+   *Mentzelia albicaulis* Dougl. White-stemmed blazingstar
B   *Mentzelia dispersa* Wats. Bushy Mentzelia
+   *Mentzelia laevicaulis* (Dougl.) T. & G. Great blazingstar
B   *Mentzelia veatchiana* Kell. Veatch's stick leaf

CACTACEAE   The Cactus Family
D   *Coryphantha vivipara* (Nutt) Britt. & Brown. Cushion or ball cactus

ELAEAGNACEAE   The Oleaster Family
D   *Elaeagnus utilis* Nels. Buffalo-berry
C   *Shepherdia argentea* (Pursh) Nutt. Thorny buffalo-berry

ONAGRACEAE   The Evening Primrose Family
+   *Boisduvalia densiflora* (Lindl.) Wats. Dense spikeprimrose
+   *Boisduvalia stricta* (Gray) Greene. Brook spikeprimrose
+   *Camissonia andina* (Nutt.) Raven. Obscure evening primrose
+   *Camissonia contorta* (Dougl.) Kearney. Contorted-pod evening primrose
A/B   *Camissonia pygmaea* (Dougl.) Raven. Dwarf evening primrose

+   *Camissonia scapoidea* Nutt. Naked-stemmed evening primrose
+   *Camissonia subacaulis* (Pursh) Raven. Stemless evening primrose
+   Camissonia tanacetifolia (T. & G.) Raven. Tansy-leaved primrose
C   *Circaea pacifica* A. & M. Western enchanter's nightshade
+   *Clarkia pulchella* Pursh. Ragged robin
B   *Clarkia rhomboidea* Dougl. Diamond-petaled Clarkia
C   *Epilobium adenocaulon* Haussk. Common willow herb
+   *Epilobium alpinum* L. Alpine willow herb
+   *Epilobium angustifolium* L. Fireweed
C   *Epilobium brevistylum* Barb. Short-styled willow herb
C   *Epilobium californicum* Haussk. California willow herb
C   *Epilobium ciliatum* Raf.
C   *Epilobium clavatum* Trel. Club fruited willow herb
C   *Epilobium fastigiatum* Piper. Smooth willow herb
+   *Epilobium glaberrimum* Barvey v. *glaberrimum* Smooth willow herb
B   *Epilobium glandulosum* Leh. Common willow herb
C   *Epilobium halleanum* Haussk. Hall's willow herb
C   *Epilobium hornemannii* Reich. Hornemann's willow herb
+   *Epilobium obcordatum* Gray. Rose willow herb
+   *Epilobium paniculatum* Nutt. Tall annual willow herb
C   *Epilobium ursinum* Parish. Hairy willow herb
+   *Epilobium watsonii* Barbey. Watson's willow herb
D   *Gayophytum caesium* T. & G. Racemed groundsmoke
D   *Gayophytum diffusum* T. & G. Spreading groundsmoke
C   *Gayophytum eriospermum* Coville. Spreading groundsmoke
+   *Gayophytum humile* Juss. Dwarf groundsmoke
C   *Gayophytum nutallii* T. & G. Dwarf groundsmoke
+   *Gayophytum ramosissimum* Nutt. Haristem groundsmoke
D   *Oenothera biennis* L. Common evening primrose
+   *Oenothera caespitosa* Nutt. Tufted evening primrose
+   *Oenothera hookeri* T. & G. Hooker's evening primrose

HALORAGACEAE   The Water Milfoil Family
E   *Myiophyllum spicatum* L. Spiked watermilfoil

HIPPURIDACEAE   The Mare's Tail Family
D   *Hippuris vulgaris* L. Common mare's tail

APIACEAE (UMBELLIFERAE)   The Parsley Family
C   *Angelica arguta* Nutt. Lyall's or sharptooth Angelica
D   *Berula erecta* (Huds.) Cov. Cut-leaved water parsnip or stalky Berula
+   *Cicuta douglasii* (DC) Coult. & Rose. Water hemlock
A/B   *Cymopterus bipinnatus* Wats. Hyaden's Cymopterus
B   *Cymopterus petraeus* M.E. Jones. Rock-loving Cymopterus
+   *Cymopterus terebinthinus* (Hock) T. & G. Turpentine Cymopterus

+   *Heracleum lanatum* Michx.   Cow parsnip

D   *Leptotaenia multifida* Nutt.   Lace-leaved Leptotaenia

+   *Ligusticum grayii* Coult. & Rose.   Fray's lovage

+   *Lomatium dissectum* (Nutt.) Math. & Const.   Fern-leaved Lomatium

+   *Lomatium donnellii* Coult. & Rose.   Donnell's Lomatium

C   *Lomatium gormani* (Howell) Coult. & Rose.   Gorman's Lomatium or desert parsley

D   *Lomatium grayi* C. & R.   Gray's Desert parsley

C   *Lomatium leptocarpum* (T. & G.) Coult. & Rose.   Slender-flowered or bicolor biscuit root

D   *Lomatium macrocarpum* (Nutt.) C. & R.   Large-flowered Lomatium

C   *Lomatium montanum* (Wats.) Coult. & Rose.   Cous biscuit-root

+   *Lomatium nevadense* (Wats.) Coult. & Rose   Nevada desert parsley

+   *Lomatium triternatum* (Pursh) Coult. & Rose.   Nine-leaved desert parsley

+   *Lomatium vaginatum* Coult. & Rose.   Boradsheath Lomatium

D   *Osmorhiza depauperata* Phil.   Blunt-flowered sweet root

+   *Osmorhiza occidentalis* (Nutt.) Torr.   Sweet cicely

+   *Perideridia bolanderi* (Gray) Nels. and Macbr.   Bolander's yampah

+   *Perideridia gairdneri* (H. & A.) Math.   Gaidner's yampah

+   *Sium sauve* Walt.   Hemlock water parsnip

+   *Sphenosciadium capitallatum* Gray.   Swamp woolly head

CORNACEAE   The Dogwood Family

   Cornus stolonifera Michx. v. *Occidentalis* (T. & G.) Hitchc.   Red osier dogwood

ERICACEAE   The Heath Family

B   *Arctostaphylos uva-ursi* (L) Spreng.   Bearberry or kinnick-innick

B   *Kalmia microphylla* (Hook) Heller.   Alpine laurel

B   *Vaccinium caespitosum* Michx.   Alpine huckleberry

PRIMULACEAE   The Primrose Family

+   *Dodecatheon alpinum* (Gray.) Greene.   Alpine shooting star

+   *Dodecatheon conjugens* Greene.   Desert shooting star

C   *Dodecatheon hendersonii* Gray.   Henderson's or broad-leaved shooting star

C   *Dodecatheon poeticum* Hend.   Narcissus shooting star

+   *Dodecatheon pulchellum* (Raf.) Merrill v. *watsonii* (Tidestron) C.L. Hitchcock.   Dark-throated shooting star

D   *Glaux maritima* L.   Sea milkwort

GENTIANACEAE   The Gentian Family

D   *Centaurium exaltatum* (Griseb.) Wight.   Western centaury

+   *Frasera speciosa* Dougl.   Giant Frasera or monument plant

+   *Gentiana affinis* Criseb.   Pleated gentian

D   *Gentiana calycosa* Griseb..   Explorer's or Mountain bog gentian

C   *Gentiana oregana* Engelm.   Oregon gentian

D   *Gentiana parryi* Engel.   Parry's gentian

B   *Swertia perennis* L.   Mountain Swertia

C   *Swertia radiata* (Kell.) Ktze.

APOCYNACEAE   The Dogbane Family

+   *Apocynum androsaemifolium* L.   Spreading dogbane

C   *Apocynum medium* v. *floribundum* Greene.   Woodson dogbane

ASCLEPIADACEAE   The Milkweed Family

+   *Asclepias fascicularis* Dcne.   Narrow-leaved milkweed

C   *Asclepias mexicana* Gav.   Narrow-leaf milkweed

CUSCUTACEAE   The Dodder Family

+   *Cuscuta occidentalis* Mills.   Western dodder

POLEMONIACEAE   The Phlox Family

+   *Collomia grandiflora* Dougl.   Large-flowered Collomia

+   *Collomia linearis* Nutt.   Narrow-leaved Collomia

C   *Collomia tenella* Gray.   Diffuse Collomia

+   *Collomia tinctoria* Kell.   Yellow-staining Collomia

+   *Eriastrum sparsiflorum* (Eastw.) Mason.   Few-flowered Eriastrum

D   *Gilia aggregata* (Pursh) Spreng.   Scarlet Gilia or skyrocket

+   *Gilia capillaris* Kell.   Smooth-leaved Gilia

C   *Gilia capitata* Hook.   Blue field Gilia

+   *Gilia congesta* Hook. v. *palmifrons* (Brand) Cronq.   Ball-head Gilia

D   *Gilia filifolia* Nutt.   Wooly-headed Gilia

C   *Gilia leptomeria* (Gray) M. & G.   Great Basin Gilia

D   *Gilia montana* Nels and Kenn.   Mountain Gilia

C   *Gilia ochroleuca* M.E. Jones.   Spicate Gilia

D   *Gilia tenerrima* Gray.   Delicate Gilia

A/B   *Gymnosteris nudicaulis* (H. & A.) Greene.   Large-flowered Gym nosteris

+   *Leptodactylon pungens* (Torr.) Nutt.   Lava or granite phlox

+   *Linanthastrum nuttallii* (Gray) Wean.   Nuttall's Linanthantrum

+   *Linanthus harknessii* (Curran) Greene.   Harkness' Linanthus

C   *Linanthus septentrionalis* H. L. Mason.   Northern Linanthus

+   *Navarretia breweri* (Gray) Greene.   Yellow-flowered Navarretia

+   *Navarretia divaricata* (Torr.) Greene.   Mountain Navarretia

D   *Navarretia intertexta* (Benth.) Hook.   Needle-leaved Navarretia

D   *Navarretia minima* (Nutt.) Gray.   Least Navarreta

C   *Navarretia propinqua* (Suksd.) Brand   Needle-leaved Navarretia

+   *Phlox diffusa* Benth.   Spreading Phlox

C   *Phlox gracilis* (Hook) Greene. Pink Microsteris

D   *Phlox lanata* Piper. Moss Phlox

+   *Phlox longifolia* Nutt. Long-leaved Phlox

C   *Polemonium elegans* Greene. Elegant Polemonium

+   *Polemonium micranthum* Benth. Annual Polemonium

+   *Polemonium occidentale* Greene. Western Polemonium

+   *Polemonium pulcherrimum* Hook. Skunk-leaved Polemonium

A/B   *Polemonium viscosum* Nutt. Sticky Polemonium

C   *Polemonium viviparum* L.

HYDROPHYLLACEAE   The Waterleaf Family

+   *Hesperochiron pumilus* (Griseb.) Porter. Dwarf Hesperochiron or centaur flower

+   *Hydrophyllum capitatum* Dougl. v. *alpinum* Wats. Alpine waterleaf

+   *Nemophila breviflora* Gray. Great Basin Nemophila

+   *Nemophila parviflora* Dougl. v. *austineae* (Eastw.) Brand. Small-flowered Nemophila

D   *Phacelia ciliata* Benth. Field Phacelia

D   *Phacelia glandulifera* Piper. Sticky Phacelia

+   *Phacelia glandulosa* Nutt. Glandular Phacelia

+   *Phacelia hastata* Dougl. v. *alpina* Alpine scorpionweed

C   *Phacelia heterophylla* Pursh. Varileaf Phacelia

+   *Phacelia humilis* (T.& G.) Low Phacelia

+   *Phacelia linearis* (Pursh) Holz. Thread-leaved Phacelia

+   *Phacelia lutea* (H. & A.) J.T. Howell. Yellow Phacelia

D   *Phacelia mutabilis* Greene. Varileaf or virgate Phacelia

+   *Phacelia sericea* (Grah.) Gray. Silky Phacelia

D   *Phacelia thermalis* Greene. Hot-spring Phacelia

BORAGINACEAE   The Borage Family

D   *Amsinckia intemedia* Fisch & Mey. Rancher's fireweed

C   *Amsinckia tessellata* A. Gray. Tessellate fireweed

+   *Coldenia nuttallii* Hook. Nuttall's Coldenia

C   *Cryptantha affinis* (Gray) Greene. Slender Cryptantha

D   *Cryptantha ambigua* (Gray) Greene. Obscure Cryptantha

+   *Cryptantha circumscissa* (H & A) Johnst. Mat Cryptantha

+   *Cryptantha echinella* Greene. Prickly Cryptantha

D   *Cryptantha humilis* (Greene) Pays. Low oaeocarya

+   *Cryptantha intermedia* (Gray) Greene. White desert forget-me-not

B   *Cryptantha nubigena* (Greene) Pays. Sierra Cryptantha

C   *Cryptantha simulans* Greene. Pinewoods Cryptantha

+   *Cryptantha torreyana* (Gray) Greene. Torrey's Cryptantha

+   *Cryptantha watsonii* (Gray) Greene. Watson's Cryptantha

D   *Hackelia diffusa* (Lehm.) Johnst. Diffuse stickseed

C   *Hackelia floribunda* (Lehm.) Many-flowered tickseed

+   *Hackelia micrantha* (Eastw.) J.L. Gentry. Blue stickseed

D   *Helioptropium curassavicum* L. Salt or seaside heliotrope

+   *Lappula redowskii* (Hornem) Greene. Western stickseed

+   *Lithospermum ruderale* Dougl. Wayside gromwell

+   *Mertensia ciliata* (Torr.) G. Don. Tall mountain bluebell

+   *Mertensia oblongifolia* (Nutt.) G. Don. Leafy bluebell

D   *Mertensia paniculata* (Piper) G. Don. Tailor pan bluebell

+   *Mertensia viridis* A. Nels. Green bluebell

C   *Myosotis laxa* (Lehm.) Small flowered forget-me-not

D   *Plagiobothrys hispidus* (Gray) Bristly popcorn flower

+   *Plagiobotrys scouleri* (H & A) Johnst. Scouler's Plagiobotrys

LAMIACEAE (LABIATAE)   The Mint family

C   *Agastache cusickii* (Greene) Heller. Cusick's giant-hyssop or horsemint

+   *Agastache urticifolia* (Benth.) Kuntze. Nettle-leaved horsemint

+   *Marrubium vulgare* L. Horehound

+   *Mentha arvensis* L. Common field mint

+   *Monardella odoratissima* Benth. Mountain Monardella

+   *Nepeta cataria* L. Catnip

+   *Scutellaria antirrhinoides* Benth. Skullcap

B   *Scutellaria nana* Gray. Dwarf skullcap

SOLANACEAE   The Nightshade Family

+   *Nicotiana attenuata* Torr. Coyote tobacco

+   *Solanum rostratum* Dunel. Buffalo bur

+   *Solanum triflorum* Nutt. Cut-leaved nightshade

SCROPHULARIACEAE   The Figwort or Snapdragon family

+   *Castilleja applegatei* Fern v. *fragilis* (Zeile) N. Holmgren Wavy-leaved paintbrush

D   *Castilleja camporum* (Green.) How. Blue Montain paintbrush

+   *Castilleja chromosa* A. Nels. Desert paintbrush

C   *Castilleja glandulifera* Pennell. Glandular Indian paintbrush

+   *Castilleja hispida* Benth. Harsh paintbrush

+   *Castilleja linaerifolia* Benth. Wyoming paintbrush

+   *Castilleja miniata* Dougl. v. *miniata* Thin-leaved paintbrush

D   *Castilleja nana* Estw.

C   *Castilleja psittacina* (Eastw.) Penn. Hairy or white paintbrush

D   *Castilleja rhexifolia* Rydb. Rhexia-leaved paintbrush

+   *Castilleja rustica* Piper. Rustic paintbrush

A   *Castilleja steenensis* Pennell. Steens Mountain paintbrush

C   *Collinsia sparsiflora* Fisch. and Mey. Few-flowered Collinsia or blue-eyed mary

+   *Collinsia parviflora* Lindl. Small-flowered blue-eyed mary

D   *Cordylanthus capitatus* Nutt. Clustered or Yakima birdbent

D   *Cordylanthus ramosus* Nutt. Bushy birdbeak

+   *Linaria dalmatica* (L.) Mill. Dalmatian toadflax

+   *Mimulus breviflorus* Piper. Short-flowered Mimulus

+   *Mimulus breweri* (Greene) Rydb. Brewer's monkeyflower

+   *Mimulus cusickii* (Greene) Piper. Cusick's monkeyflower

B   *Mimulus floribundus* Lindl.   Purple-stemmed monkeyflower

+   *Mimulus guttatus* DC v. *guttatus*   Common yellow monkey-flower

+   *Mimulus lewisii* Pursh.   Lewis's monkeyflower

C   *Mimulus longulus* Greene.

+   *Mimulus moschatus* Dougl. v. *moschatus*   Musk plant

+   *Mimulus nanus* H & A.   Dwarf monkeyflower

+   *Mimulus primuloides* Benth.   Primrose-leaved monkey-flower

B   *Mimulus suksdorfii* Gray.   Suksdorf's monkeyflower

+   *Mimulus tilingii* Regel v. *tiligii*   Large mountain monkey-flower

B   *Orthocarpus copelandii* Estw. v. *cryptanthus* (Piper) Keck   Copeland's owl clover

+   *Orthocarpus hispidus* Benth.   Hairy owl clover

B   *Orthocarpus luteus* Nutt.   Yellow owl clover

+   *Orthocarpus tenuifolius* (Pursh) Benth.   Thin-leaved owl clover

B   *Pedicularis attollens* Gray.   Little elephant's head

D   *Pedicularis centranthera* Gray.   Long-flowered Pedicularis

+   *Pedicularis groenlandica* Retz.   Pink elephant head

D   *Penstemon cusickii* Gray.   Cusick's Penstemon

A   *Penstemon davidsonii* Greene v. *praeteritis* Cronq.   Davidson's Penstemon

+   *Penstemon deustus* Dougl. v. *heterander* (T. & G.) Cronq.   Scabland Penstemon

C   *Penstemon fruticosus* (Pursh) Greene.   Shrubby or bush Penstemon

C   *Penstemon oreocharis* Greene.   Rydberg's Penstemon

B   *Penstemon pratensis* Greene.   White Penstemon

+   *Penstemon procerus* Doubl. v. *formosus* (A. Nels.) Cronq.   Small-flowered Penstemon

D   *Penstemon productus* Greene.   Tailed Penstemon

+   *Penstemon rydbergii* A. Nels.   Rydberg's Penstemon

+   *Penstemon speciosus* Dougl.   Showy Penstemon

+   *Scrophularia lanceolata* Pursh.   Lance-leaved figwort

+   *Verbascum thapsus* L.   Flannel mullein

+   *Veronica americana* Schwein.   American speedwell

+   *Veronica anagallis-aquatica* L.   Water speedwell

D   *Veronica arvensis* L.   Wall or common speedwell

+   *Veronica cusickii* Gray.   Cusick's speedwell

C   *Veronica peregrina* L.   Purslane speedwell

+   *Veronica serphyllifolia* L.   Thyme-leaved speedwell

+   *Veronica wormskjoldii* Roem & Schult.   American alpine speedwell

ORBANCHACEAE   The Broomrape Family

+   *Orobanche californica* Cham. and Schl. v. *grayana* (Beck) Cronq. Stout broomrape

+   *Orobanche corymbosa* (Rydb.) Ferris.   Flat-topped broom-rape

+   *Orobanche fasciculata* Nutt.   Clustered broomrape

C   *Orobanche pinorum* Geyer.   Pine broomrape

B   *Orobanche uniflora* L.   Broomrape

LENTIBULARIACEAE   The Bladderwort Family

+   *Utricularia minor* L.   Lesser bladderwort

PLANTAGINACEAE   The Plantain Family

+   *Plantago lanceolata* L.   English plantain

+   *Plantago major* L.   Nippleseed plantain

RUBIACEAE   The The Madder or Coffee family

+   *Galium aparine* L.   Cleavers or goose-grass

+   *Galium asperrimum* Gray.   Rough bedstraw

C   *Galium bifolium* Wats.   Thin-leaved or low mountain bedstraw

D   *Galium boreale* L.   Northern bedstraw

+   *Galium multiflorum* Kell.   Shrubby bedstraw

C   *Galium triflorum* Michx.   Sweetscented or fragrant bedstraw

+   *Kelloggia galioides* Torr.   Kelloggia

CAPRIFOLIACEAE   The Honeysuckle family

+   *Lonicera involucrata* (Rich.) Banks.   Black twinberry

+   *Sambucus cerulea* Raf.   Blue elderberry

C   *Sambucus glauca* Nutt.   Blue elderberry

+   *Sambucus racemosa* L. v. *melanocavpa* (Gray) McMinn.   Black elderberry

+   *Symphoricarpos oreophilus* Gray.   Mountain snowberry

C   *Symphoricarpos rotundifolius* Gray.   Round-leaved snowberry

VALERIANACEAE   The Valerian Family

+   *Plectritis macrocera* T. & G.   Popcorn flower

+   *Valeriana acutiloba* Rydb.   Downy-fruited Valerian

C   *Valeriana occidentale* Hel.   Western Valerian

C   *Valeriana sitchensis* Bong.   Sitka Valerian

DIPSACACEA   The Teasel family

+   *Dipsacus sylvestris* Huds.   Gypsy combs or teasel

LOBELIACEAE   The Lobelia Family

D   *Downingia elegans* (Dougl.) Torr.   Common or showy Downingia

D   *Downingia willamettensis* Peck.   Willamette Downingia

C   *Downingia bicornuta* Gray.   Double-horned Downingia

+   *Porterella carnulose* (H & A) Torr.   Porterella

ASTERACEAE (COMPOSITAE)   The Sunflower family

C   *Achillea lanulosa* Nutt.   Western yarrow

+   *Achillea millefolium* L.   Yarrow

C   *Agoseris alpestris* (Gray) Greene.   Alpine lake Agoseris

B   *Agoseris aurantiaca* (Hook.) Greene.   Orange Agoseris

+   *Agoseris glauca* (Pursh) Raf. v. *laciniata* (DC Eat.) Smiley   Pale Agoseris

+   *Agoseris grandiflora* (Nutt.) Greene.   Large-flowered Agoseris

+   *Agoseris heterophylla* (Nutt.) Greene.   Annual Agoseris

+   *Anaphalis margaritaceae* (L.) B. & H.   Pearly everlasting

+   *Antennaria alpina* (L.) Gaertn.   Alpine pussytoes

+   *Antennaria anaphaloides* Rydb.   Tall pussytoes

+   *Antennaria dimorpha* (Nutt.) T. & G.   Low pussytoes

+   *Antennaria luzuloides* T. & G.   Woodrush pussytoes

+   *Antennaria microphylla* Rydb.   Rosy pussytoes

D   *Antennaria rosea* Greene.   Rosy pussytoes

D   *Antennaria stenophylla* Gray.   Narrow-leaved pussytoes

+   *Antennaria umbrinella* Rydb.   Umber pussytoes

C   *Anthemis cotula* L.   Stinking mayweed

D   *Arnica alphina* (L.) Olin.   Alpine Arnica

+   *Arnica chamissonis* Less.   Leafy or streambank Arnica

+   *Arnica cordifolia* Hook.   Heart-leaved Arnica

+   *Arnica longifolia* DC Eat.   Long-leaved Arnica

+   *Arnica mollis* Hook.   Hairy or seepspring Arnica

D   *Arnica parryi* Gray.   Nodding Arnica

C   *Arnica rydbergii* Greene.   Rydberg's Arnica

B   *Arnica sororia* Greene.   Twin Arnica

+   *Artemisia arbuscula* Nutt.   Low sagebrush

C   *Artemisia cana* Pursh.   Silver or hoary sagebrush

+   *Artemisia dracunculus* L.   Tarragon

+   *Artemisia ludoviciana* Nutt. v. *ludoviciana*.   Louisiana sage

+   *Artemisia michauxiana* Bess.   Michaux mugwort

+   *Artemisia tridentata* Nutt.   Big sagebrush

B   *Artemisia tripartita* Rydb.   Three-tipped sagebrush

+   *Artemisia vulgaris* L.   Common sage or mugwort

+   *Aster alpigenus* (T. & G.) Gray. *haydenii* (Porter) Cronq.   Alpine Aster

C   *Aster canescens* (Pursh) Gray.   hoary Aster

+   *Aster eatonii* (Gray) Howell.   Eaton's Aster

+   *Aster foliaceous* Lindl. v. *parryi* (Eat.) Gray.   Leafy Aster

C   *Aster fremontii* Gray.   Fremont's Aster

C   *Aster frondosus* (Nutt.) T. & G.   Rayless alkali Aster

C   *Aster integrifolius* Nutt.   Thick-stemmed or entire Aster

D   *Aster occidentalis* Nutt. T. & G.   Western mountain Aster

D   *Aster oregonensis* (Nutt.) Cronq.   White-topped Aster

C   *Aster scopulorum* Gray.   Crag or lava Aster

+   *Balsamorhiza sagittata* (Pursh) Nutt.   Arrow-leaved balsamroot

+   *Balsamorhiza serrata* Nels. and Macbr.   Serrated-leaved balsamroot

+   *Bidens cernuua* L.   Nodding beggartick

+   *Blepharipappus scaber* Hook.   Scabland Blepharipappus

B   *Brickellia grandiflora* (Hook.) Nutt.   Large-flowered tassleflower

C   *Brickellia microphylla* (Nutt.) Gray.   Small-leaved Brickellia

+   *Chaenactis douglasii* (Hook.) H. & A.   Dusty maiden

+   *Chrysothamnus nauseosus* (Pall.) Britt.   Gray rabbitbrush

+   *Chrysothamnus viscidiflorus* (Hook.) Nutt. v. *lanceolatus* (Nutt.) Greene.   Green rabbitbrush

D   *Cirsium foliosum* (Hook.) T. & G.   Elk thistle

A   *Cirsium peckii* Henders.   Steens Mountain thistle

+   *Cirsium utahense* Petr.   Utah thistle

+   *Cirsium vulgare* (Savi) Tenore   Bull thistle

+   *Conyza canadensis* (L.) Cronq.   Conyza

+   *Crepis atrabarba* Heller.   Slender hawksbeard

D   *Crepis acuminata* Nutt.   Long-leaved or tapertip hawksbeard

D   *Crepis modocensis* Greene.   Siskiyou hawksbeard

+   *Crepis occidentalis* Nutt.   Western hawksbeard

A/B   *Dimeresia howellii* Gray.   Dimersia

C   *Erigeron bloomeri* Gray.   Scabland fleabane

C   *Erigeron canadensis* L.   Canadian fleabane or horseweed

+   *Erigeron compositus* Pursh.   Fern-leaved fleabane

D   *Erigeron concinnus* (Hook.) T. & G.   Shaggy Erigeron

+   *Erigeron corymbosus* Nutt.   Foothill daisy

C   *Erigeron divergens* T. & G.   Diffuse or spreading fleabane

+   *Erigeron eatonii* Gray v. *villosus* Cronq.   Eaton's daisy

C   *Erigeron eradiatus* (Gray) Piper.   Payless Erigeron

D   *Erigeron filifolius* (Hook.) Nutt.   Thread-leaved Erigeron

+   *Erigeron glabellus* Nutt.   Smooth daisy

C   *Erigeron leiomerus* Gray.   Smooth daisy

+   *Erigeron linearis* (Hook.) Piper.   Desert yellow fleabene

+   *Erigeron peregrinus* (Pursh) Greene ssp. *callianthemus* (Greene) Cronq. v. *eucallianthemus* Conq.   Subalpine daisy

D   *Erigeron peucephyllus* Gray.   Line-leaved fleabane or yellow daisy

C   *Erigeron philadelphicus* Linn.   Philadelphia daisy or fleabane

C   *Erigeron salsuginosus* (Rich) Gray.   Aster Erigeron

+   *Erigeron speciosus* (Lindl.) DC.   Showy fleabane

C   *Erigeron tegetarius* Cov.   Mat Erigeron

D   *Erigeron tener* Gray.   Slender daisy

D   *Eriophyllum integrifolium* (Hook) Smiley

+   *Eriophyllum lanatum* (Pursh) Forbes.   Woolly sunflower

+   *Eupatorium occidentale* Hook.   Western boneset

+   *Gnaphalium palustre* Nutt.   Cudweed

+   *Happlopappus acaulis* (Nutt.) Gray.   Cushion goldenweed

C   *Haplopappus bloomeri* Gray.   Rabbitbrush goldenweed

+   *Haplopappus carthamoides* (Hood.) Gray   large-flowered Haplopappus

+   *Haplopappus greenei* Gray.   Greene's goldenweed

C   *Haplopappus hirtus* Gray.   Sticky or hairy goldenweed

D   *Haplopappus lanuginosus* Gray.   Woolly goldenweed

+   *Haplopappus macronema* Gray.   Discoid goldenweed

B   *Haplopappus nanus* (Nutt.) Eat.   Dwarf goldenweed

+   *Haplopappus stenophyllus* Gray.   Narrow-leaved goldenweed

+   *Haplopappus suffruticosus* (Nutt.) Gray.   Shrubby goldenweed

A/B   *Haplopappus uniflorus* (Hook.) T. & G.   One-flowered goldenweed

+   *Helenium hoopesii* Gray.   Orange sneezeweed

+	*Helianthella uniflora* (Nutt.) T. & G. Little sunflower
+	*Helianthus annuus* L. Common sunflower
D	*Helianthus cusickii* Gray. Cusick's sunflower
C	*Hieracium albertinum* Farr. Western hawkweed
+	*Hieracium cynoglossoides* Arv. Touv. Hound's tongue hawkweed
C	*Hieracium rydbergii* Zahn. Rydberg's hawkweed
D	*Iva axillaris* Pursh. Deeproot
+	*Lactuca lulchella* (Pursh) DC. Blue-lettuce
+	*Lactuca serriola* L. Prickly lettuce
C	*Lagophylla ramosissima* Nutt. Slender or common hareleaf
+	*Layia glandulosa* (Hook.) H. & A. White Layix
+	*Lygodesmia spinosa* Nutt. Spiny skeleton plant
+	*Machaeranthera canescens* (Pursh) Gray. Hoary aster
B	*Machaeranthera grindeloides* (Nutt.) Keck & Cronq. (*Haplopappus nuttalii*) Gumweed Machaeranthera
+	*Machaeranthera shastensis* Gray v. *eradiata* (Gray) Cronq. & Keck. Shasta aster
C	*Madia dissitiflora* (Nutt.) T. & G. Slender or common tarweed or gumweed
+	*Madia glomerata* Hook. Clustered tarweed
+	*Madia gracilis* (J. E. Smith) Keck. Slender tarweed
+	*Malacothrix torreyana* Gray. Torrey's Malacothrix
C	*Matricaria matricarioides* (Loss) Porter. Native cardilleran weed or pineapple weed
+	*Microseris nutans* (Geyer) Schultz-Bip. Nodding Microseris
+	*Microseris troximoides* Gray. False Agoseris
D	*Psilocarphus elatior* Gray. Tall woolly-head
+	*Rudbeckia occidentalis* Nutt. Western coneflower
+	*Senecio canus* Hook. Woolly butterweed
+	*Senecio crassulus* Gray. Thick-leaved groundsel
B	*Senecio cymbalarioides* Buck. Alpine meadow groundsel
+	*Senecio fremontii* T. & G. Dwarf mountain butterweed
C	*Senecio howellii* Greene. Woolly groundsel
+	*Senecio integerrimus* Nutt. v. *exaltatus* (Nutt.) Cronq. Western groundsel or one-stemmed butterweed
B	*Senecio pseudaureus* Rydb. Streambank butterweed
+	*Senecio serra* Hook. Sawtooth butterweed
C	*Senecio suksdorfi* Greene. Cleft-leafed groundsel
+	*Senecio triangularis* Hook. Arrow-leaved groundsel
+	*Solidago canadensis* L. v. *salebrosa* (Piper) Jones. Meadow goldenrod
+	*Solidago decumbens* Greene. Dwarf goldenrod
C	*Solidago lepida* DC. v. *elongata* (Nutt.) Fern. Narrow goldenrod
C	*Solidago missouriensis* Nutt. Mountain goldenrod
C	*Stephanomeria exigua* Nutt. Small wire lettuce
+	*Stephanomeria tenuifolia* (Torr.) Hall v. *myrioclada* (Eat.) Cronq. Wire lettuce
C	*Tanacetum canum* Eat. Gray tansy

+	*Taraxacum officianale* Weber. Dandelion
+	*Tetradymia canescens* DC. Spineless horsebrush
+	*Tetradymia glabrata* Gray. Littleleaf horsebrush
A/B	*Townsendia parryi* Eat. Parry's Townsendia
+	*Tragopogon dubius* Scop. Salsify
+	*Wyethia amplexicaulis* Nutt. Mule's ears
+	*Wyethia helianthoides* Nutt. White Wyethia
+	*Xanthium strumarium* L. Cocklebur

ALISMATACEAE   The Water Plantain Family

C	*Alisma plantago aquatica* Linn. Water plantain
+	*Sagittaria cuneata* Sheld. Wapato

HYDROCHARITACEAE   The Frog's Bite Family

D	*Anacharis planchonii* (Casp.) n. comb. Rocky Mountain waterweed

JUNCAGINACEAE   The Arrowgrass Family

B	*Lilaea scilloides* (Poir.) Hanman. Flowering quillwort

ZANNICHELLIACEAE   The Horned Pondweed Family

C	*Zannichellia palustris* Linn. Horned pondweed

POTAMAGETONACEAE   The Pondweed Family

A/B	*Potamageton diversifolius* Raf. Diverse-leaved pondweed
+	*Potamogeton epihydrus* Raf. Ribbon-leaf pondweed
E	*Potamogeton pectinatus* L. Fennel-leaved or sago pondweed
+	*Potamogeton zosteriformis* Fern. Eel-grass pondweed

JUNCACEAE   The Rush Family

+	*Juncus articulatus* L. Jointed rush
C	*Juncus badium* (Suksd.) Hitchc.
+	*Juncus balticus* Willd. Baltic rush
+	*Juncus bufonius* L. Toad rush
+	*Juncus drummondii* E. Meyer. Drummond's rush
C	*Juncus ensifolius* Wiks. Dagger-leaf rush
+	*Juncus kelloggii* Engelm. Kellogg's rush
+	*Juncus mertensianus* Bong. Merten's rush
C	*Juncus regelii* Buch. Rengel's rush
C	*Juncus saximontanus* Nels. Rocky mountain rush
C	*Juncus sphaerocarpus* Nels. Toad rush
C	*Juncus uncialis* Greene. Short-stemmed dwarf rush
+	*Luzula campestris* (L.) DC. Sweep's rush
C	*Luzula divaricata* Wats. Spreading woodrush
+	*Luzula parviflora* (Ehrh.) Desr. Small-flowered woodrush
+	*Luzula spicata* (L.) DC. Spiked woodrush

CYPERACEAE   The Sedge Family

C	*Carex athrostochya* Olney. Slender-beaked sedge
+	*Carex atrata* L. v. *erecta* Boott. Blackened sedge
+	*Carex aurea* Nutt. Golden-fruited sedge
C	*Carex douglasii* Boott. Douglas sedge

C — *Carex epapillosa* Mack. Black-and-white sedge

+ — *Carex filifolia* Nutt. Thread-leaved sedge

C — *Carex gymnoclada* Holm. Holm's Rocky Mountain sedge

C — *Carex hoodii* Boott. Hodd's sedge

+ — *Carex illota* Bailey Sheep sedge

C — *Carex kelloggii* Boott. Kellogg's sedge

C — *Carex lanuginosa* Michx. Woolly sedge

C — *Carex microptera* Mack. Small-winged sedge

+ — *Carex multicostata* Mack. Many-ribbed sedge

C — *Carex nebraskensis* Dewey. Nebraska sedge

C — *Carex petasata* Dewey. Liddon's sedge

C — *Carex phaeocephala* Piper. Dunhead or mountain hare sedge

C — *Carex praegracilis* W. Boott. Clustered field sedge

C — *Carex preslii* Steud. Thick-headed sedge

C — *Carex pseudoscirpoidea* Hydb. Cronq. Western single-spiked sedge

+ — *Carex raynoldsii* Dewey. Raynold's sedge

+ — *Carex rostrata* Stokes. Beaked sedge

D — *Carex saximontana* Mack. Back's sedge

+ — *Carex scirpoidea* Michx. v. *pseudoscirproidea* (Rybd.) Cronq. Canadian single-spiked sedge

+ — *Carex scopulorum* Holm. Holm's Rocky Mountain sedge

C — *Carex sheldonii* Mack. Sheldon's sedge

+ — *Carex spectabilis* Dewey. Showy sedge

C — *Carex subnigricans* Stacey. Dark-alpine sedge

C — *Carex vallicola* Dewey. Valley sedge

C — *Carex vernacula* Bailey. Foetid sedge

C — *Carex vesicaria* L. Inflated sedge

C — *Cyperus inflexus* Muhl. Awned Cyperus

C — *Eleocharis acicularis* (L.) R. & S. Needle-spike-rush

+ — *Eleocharis palustris* (L.) R. & S. Creeping spike rush

C — *Scirpus americanus* Pers. American or three square bulrush

C — *Scirpus microcarpus* Presl. Small fruit bulrush

C — *Scirpus validus* Vahl. Tule, soft-stem or Am. Great Basin bulrush

+ — *Scirpus olneyi* Gray. Olney's bulrush

POACEAE (GRAMINEAE)   The Grass Family

D — *Agropyron intermedium* (Host) Beauv. Intermediate wheatgrass

C — *Agropyron spicatum* (Pursh) Sc. & Sm. Bluebunch wheat-grass

C — *Agrostis exarata* Trin. Spike bentgrass

C — *Agrostis hiemalis* (Walt) B.S.P. Rough bentgrass

C — *Agrostis palustris* (Huds.) Pers. Creeping bentgrass

C — *Agrostis rossiae* Vasey. Ros' bentgrass

+ — *Agrostis scabra* Willd. Winter bentgrass, rough hair-grass or tickle-grass

C — *Alopecurus aequalis* Sobol. Little meadow foxtail

+ — *Alopecurus alpina* Smith. Alpine meadow foxtail

+ — *Alopecurus pratensis* L. Meadow foxtail

+ — *Bromus carinatus* H & A. California brome

+ — *Bromus inermis* Leys. ssp. *pumpellianus* (Scribn.) Wagnon Pumpelly brome

C — *Bromus marginatus* Nees. Large mountain bromegrass

C — *Bromus socalenus* L. Ryebrome, chess or cheat

+ — *Bromus tectorum* L. Downy chess or cheatgrass

C — *Calamagrostis canadensis* (Michx) Beauv. Blue joint reed-grass

C — *Calamagrostis koelerioides* Vasey. Fire reedgrass

C — *Calamagrostis neglecta* (Ehrh.) G.M. & S. Slimstem reed-grass

+ — *Calamagrostis purpurescens* R. Br. Purple reedgrass

D — *Catabrosa aquatica* (L.) Beauv. Water whorl-grass

+ — *Cinna latifolia* (Trevir.) Griseb. Wood reedgrass

+ — *Danthonia intermedia* Vasey. Timber oatgrass

C — *Deschampsia atropurpurea* (Wahl) Scheele. Mountain hairgrass

+ — *Deschampsia caespitosa* (L.) Beauv. Tufted hairgrass

C — *Deschampsia elongata* (Hook) Munro. Slender hairgrass

E — *Distichlis stricta* (Torr.) Rydb. Saltgrass

+ — *Elymus cinereus* Scribn. & Merr. Giant wildrye

+ — *Elymus glaucus* Buckl. Western ryegrass

+ — *Festuca idahoensis* Elmer. Idaho fescue

+ — *Festuca scabrella* Torr. Buffalo bunchgrass

+ — *Festuca ovina* L. Sheep fescue

C — *Festuca viridula* Vasey. Green Festuca or mountain bunch-grass

C — *Glyceria elata* (Nash) Hitchc. Tall manna-grass

C — *Glyceria pauciflora* Presl. Few-flowered manna-grass

+ — *Hesperochloa kingii* (Wats.) Rydb. Spike fescue

+ — *Hordeum brachyantherum* Nevski. Meadow barley

C — *Hordeum jubatum* L. Meadow barley

+ — *Koeleria cristata* (L.) Pers. Junegrass

C — *Melica bulbosa* Geyer. Oniongrass

+ — *Melica fugax* Bolander. Little oniongrass

+ — *Melica spectablis* Scribn. Showy oniongrass

A/B — *Melica stricta* Bolander. Rock melic

C — *Muhlenbergia filiformis* (Thurb) Rydb. Pullup or slender muhly

+ — *Muhlenbergia richardsonis* (Trin.) Tydb. Mat muhly

+ — *Oryzopsis hymenoides* (R & S) Ricker. Indian ricegrass

+ — *Phleum alpinum* L. Timothy or Alpine timothy

C — *Poa ampla* Merr. Alkali bluegrass

D — *Poa arida* Vasey. Dryland bluegrass or prairie speargrass

C — *Poa canbyi* (Scribn.) Piper. Malpais bluegrass

+ — *Poa cusickii* Vasey v. *epilis* (Scribn.) Hitch. Skyline blue-grass

+ — *Poa incurva* Scribn. & Will. Curly bluegrass

D — *Poa juncifolia* Scribn. Alkali bluegrass

C — *Poa leibergii* Scribn. Leiberg's blue

C   *Poa longiligula* Scribn. & Will. Muttongrass

+   *Poa nervosa* (Hook.) Vasey. Wheeler's bluegrass

B   *Poa pattersonii* Vasey. Patterson's bluegrass

+   *Poa pratensis* L. Kentucky bluegrass

C   *Poa secunda* Presl. Sandberg's bluegrass

C   *Poa stenantha* Trin. Trinius' bluegrass

D   *Puccinellia lemmoni* (Vasey) Scribn. Lemmon's alkaligrass

C   *Sitanion hystrix* (Nutt.) J. G. Smith. Bottlebrush squirreltail

D   *Spartina gracilis* Trin. Alkali cordgrass

+   *Stipa comata* Trin. & Rupr. Needle and thread

+   *Stipa lettermanii* Vasey. Letterman's needle and thread grass

+   *Stipa occidentalis* Thurb. Western needlegrass

+   *Stipa thurberiana* Piper. Thurber's Stipa

+   *Trisetum spicatum* (L.) Richter. Downy oat grass

TYPHACEAE   The Cattail Family

+   *Typha latifolia* L. Common cattail

LILIACEAE   The Lily Family

+   *Allium acuminatum* Hook. Tapertip onion or Hooker's onion

C   *Allium ampetens* Torr. Slim-leaf onion

D   *Allium bisceptrum* (Wats.) Palmer's or patis onion

A/B   *Allium lemmonii* Wats. Lemon's onion

+   *Allium macrum* Wats. Rock onion

C   *Allium pleianthum* Wats. Many-flowered onion

C   *Allium punctum* Bend. Punctate onion

+   *Allium tolmei* Baker v. *tolmei* Tolmei's onion

D   *Brodiaea douglasii* S. Wats. Douglas's Brodiaea

+   *Brodiaea hyacinthina* (Lindl.) Baker. Hyacinth Brodiaea

B   *Calochortus bruneaunis* Nels. & McBr. Bruneau mariposa

D   *Calochurtus eurycarpus* S. Wats. Wide-flowered or big-pod mariposa

+   *Calochortus macrocarpus* Dougl. Green-banded mariposa

+   *Camassia quamash* (Pursh) Greene. Blue camas

B   *Erythronium grandiflorum* Pursh. Glacier lily

B   *Fritillaria atropurpurea* Nutt. Leopard lily

+   *Fritillaria pudica* (Pursh) Spreng. Yellow bell

+   *Lloydia serotina* (L.) Sweet alp lily

C   *Smilacina amplexicaulis* Nutt. Western solomon's seal

C   *Smilacina liliacea* (Greene) Wynd. Lily-leaved false solomon's seal

+   *Smilacina stellata* (L.) Desf. Starry solomon's seal

+   *Smilacina racemose* (L.) Desf. False solomon's seal

+   *Veratrum californicum* Durand. Western false hellebore

D   *Veratrum viride* Ait. American or green false-hellebore or Indian poke

+   *Zigadenus elegans* Pursh. Alpine death camas

+   *Zigadenus paniculatus* (Nutt.) Wats. Foothill death camas

B   *Zigadenus venenosus* Wats. Meadow death camas

IRIDACEAE   The Iris Family

D   *Iris missouriensis* Nutt. Western blue flag

+   *Sisyrinchium angustifolium* Mill. Blue-eyed grass

+   *Sisyrinchium inflatum* (Suksd.) St. John grass widow

ORCHIDACEAE   The Orchid Family

A/B   *Corallorhiza trifida* Chat. Early coral root

+   *Corallorhiza maculata* Raf. Spotted coral root

+   *Habenaria dilatata* (Pursh) Hook. White bog orchid

D   *Habenaria saccata* Greene. Slender bog orchid

C   *Habenaria unalaschensis* (Spreng) Wats. Alaskan rein-orchid

C   *Spiranthes romanzoffiana* Cham. & Schl. Hooded ladies-tresses

A CHECKLIST OF THE VERTEBRATE ANIMALS OF THE NORTHERN GREAT BASIN REGION OF OREGON

FISHES

Family-PETROMYZONTIDAE (Lampreys)
 Lampetra tridentata (Gairdner) Pacific lamprey
 Lampetra lethophaga (Hubbs) Pit-Klamath brook lamprey

Family-SALMONIDAE (Trout and relatives)
 Oncorhynchus nerka (Walbaum) Kokanee
 Salmo trutta (Linnaeus) Brown trout
 Salmo sp. (Richardson) Redbank trout
 Salmo clarki (Richardson) Cutthroat trout
 Salmo gairdneri (Richardson) Rainbow trout
 Salvelinus fontinalis (Mitchill) Brook trout
 Prosopium williamsoni (Girard) Mountain whitefish

Family-CYPRINIDAE (Minnows)
 Cyprinus carpio (Linnaeus) Carp
 Gila bicolor (Girard) Tui chub
 Gila alvordensis (Hubbs and Miller) Alvord chub
 Richardsonius balteatus (Richardson) Redside shiner
 Richardsonius egregius (Girard) Lahontan redside
 Ptychocheilus oregonensis (Richardson) Northern squawfish
 Hesperoleucus symmetricus (Baird and Girard) California roach
 Acrocheilus alutaceus (Agassiz) Chiselmouth
 Rhinichthys osculus (Girard) Speckled dace
 Rhinichthys cataractae (Valenciennes) Longnose dace

Family-CATOSTOMIDAE (Suckers)
 Catostomus macrocheilus (Girard) Largescale sucker
 Catostomus occidentalis (Ayres) Sacramento sucker
 Catostomus warnerensis (Snyder) Warner sucker
 Catostomus tahoensis (Gill and Jordan) Tahoe sucker
 Catostomus platyrhynchus (Cope) Mountain sucker

Catostomus columbianus (Eigenmann and Eigenmann) Bridgelip sucker

Family-ICTALURIDAE (Catfish-Introduced)
 Ictalurus punctatus (Rafinesque) Channel catfish
 Ictalurus natalis (Lesueur) Yellow bullhead
 Ictalurus melas (Rafinesque) Black bullhead
 Ictalurus nebulosus (Lesueur) Brown bullhead
 Pylodictis olivaris (Rafinesque) Flathead catfish

Family-POECILIIDAE (Livebearers-Introduced)
 Gambusia affinis (Baird and Girard) Mosquitofish

Family-CENTRARCHIDAE (Sunfishes-Introduced)
 Micropterus dolomieui (Lacepede) Smallmouth bass
 Micropterus salmoides (Lacepede) Large mouth base
 Lepomis gibbosus (Linnaeus) Pumpkinseed
 Lepomis macrochirus (Rafinesque) Bluegill
 Pomoxis nigromaculatus (Lesueur)
 Pomoxis annularis (Rafinesque) White crappie

Family-PERCIDAE (Perch-Introduced)
 Perca flavenscens (Mitchill) Yellow perch

Family-COTTIDAE (Sculpins)
 Cottus bairdi (Girard) Mottled sculpin
 Cottus pitensis (Bailey and Bond) Pit sculpin (may be extinct in Oregon)
 Cottus rhotheus (Smith) Torrent sculpin

AMPHIBIANS

Family-AMBYSTOMIDAE (Ambystomids)
 Ambystoma macrodactylum (Shaw) Long-toed salamander

Family-PELOBATIDAE (Spadefoot toads)
  Scaphiopus intermontanus (Baird) Great Basin spadefoot

Family-BUFONIDAE (Toads)
  Bufo boreas (Baird and Girard) Western toad

Family-HYLIDAE (Hylids)
  Hyla regilla (Baird and Girard) Pacific tree frog

Family-RANIDAE (Frogs)
  Rana pretiosa (Baird and Girard) Spotted frog
  Rana catesbeiana (Shaw) Bullfrog

REPTILES

Family-IGUANIDAE (Iguanids)
  Crotaphytus collaris (Say) Collared lizard
  Crotaphytus wislizeni (Baird and Girard) Leopard lizard
  Sceloporus occidentalis (Baird and Girard) Western fence lizard
  Sceloporus graciosus (Baird and Girard) Sagebrush lizard
  Uta stansburiana (Baird and Girard) Side-blotched lizard
  Phrynosoma platyrhinos (Girard) Desert honed lizard
  Phrynosoma douglassi (Bell) Short-horned lizard

Family-SCINCIDAE (Skinks)
  Eumeces skiltonianus (Baird and Girard) Western skink

Family-TEIIDAE (Teids)
  Cnemidophorus tigris (Say) Western whiptail

Family-ANGUIDAE (Alligator lizards)
  Gerrhonotus coeruleus (Baird and Girard) Northern alligator
    lizard

Family-BOIDAE (Boas)
  Charina bottae (Blainville) Rubber boa

Family-COLUBIDAE (Colubrids)
  Coluber constrictor (Linnaeus) Racer
  Masticophis taeniatus (Halloway) Striped whipsnake
  Pituophis melanoleucus (Stejneger) Gopher snake
  Thamnophis sirtalis (Linnaeus) Common garter snake
  Thamnophis elegans (Baird and Girard) Western terrestrial
    garter snake
  Hypsiglena torquata (Tanner) Night snake

Family-VIPERIDAE (Pit Vipers)
  Crotalus viridis (Rafinesque) Western rattlesnake

BIRDS

Family-GAVIIDAE (Loons)
  Gavia immer (Brunnich) Common loon
  Gavia arctica (Linnaeus) Arctic loon

Family-PODICIPEDIDAE (Grebes)
  Podiceps grisegena (Boddaert) Red-necked grebe
  Podiceps auritus (Linnaeus) Horned grebe
  Podiceps caspicus (Hablizl) Eared grebe
  Aechmophorus occidentalis (Lawrence) Western grebe
  Podilymbus podiceps (Linnaeus) Pied-billed grebe

Family-PELECANIDAE (Pelicans)
  Pelecanus erythrorhynchos (Gmelin) White pelican

Family-PHALACROCORACIDAE (Cormorants)
  Phalacrocorax auritus (Lesson) Double-crested cormorant

Family-ARDEIDAE (Herons and Bitterns)
  Ardea herodias (Linnaeus) Great blue heron
  Bubulcus ibis (Linnaeus) Cattle egret
  Casmerodius albus (Linnaeus) Great egret
  Leucophoyx thula (Molina) Snowy egret
  Hydranassa tricolor (Muller) Louisiana heron
  Nycticorax nycticorax (Linnaeus) Black-crowned night heron
  Ixobrychus exilis (Gmelin) Least bittern
  Botaurus lentiginosus (Rackett) American bittern

Family-THRESKIORNITHIDAE (Ibises)
  Plegadis chihi (Vieillot) White-faced ibis

Family-ANATIDAE (Swans, Geese, and Ducks)
  Olor columbianus (Ord) Whistling swan
  Olor buccinator (Richardson) Trumpeter swan
  Branta canadensis (Linnaeus) Canada goose
  Anser albifrons (Scopoli) White-fronted goose
  Philacte canagica (Sewastianov) Emperor goose
  Chen hyperborea (Pallas) Snow goose
  Chen caerulescens (Linnaeus) Snow goose (Blue and white
    phases)
  Chen rossii (Cassin) Ross' goose
  Anas platyrhynchos (Linnaeus) Mallard
  Anas rubripes (Brewster) Black duck
  Anas strepera (Linnaeus) Gadwall
  Anas acuta (Linnaeus) Pintail
  Anas crecca carolinensis (Gmelin) Green-winged teal
  Anas discors (Linnaeus) Blue-winged teal
  Anas cyanoptera (Vieillot) Cinnamon teal
  Anas penelope (Linnaeus) European widgeon
  Anas americana (Gmelin) American widgeon
  Spatula clypeata (Linnaeus) Northern shoveler
  Aix sponsa (Linnaeus) Wood duck
  Aythya americana (Eyton) Redhead
  Aythya collaris (Donovan) Ring-necked duck
  Aythya valisineria (Wilson) Canvasback
  Aythya marila (Linnaeus) Greater scaup
  Aythya affinis (Eyton) Lessor scaup
  Bucephala clangula (Linnaeus) Common goldeneye
  Bucephala islandica (Gmelin) Barrow's goldeneye
  Bucephalia albeola (Linnaeus) Bufflehead
  Clangula hyemalis (Linnaeus) Oldsquaw
  Malanitta douglandi (Bonaparte) White-winged scoter

Melanitta perspicillata (Linnaeus) Surf scoter
Oxyura jamaicensis (Gmelin) Ruddy duck
Lophodytes cucullatus (Linnaeus) Hooded merganser
Mergus merganser (Linnaeus) Common merganser
Mergus serrator (Linnaeus) Red-breasted merganser

Family-CATHARTIDAE (American vultures)
Cathartes aura (Linnaeus) Turkey vulture

Family-ACCIPITRIDAE (Hawks, Kites, Harriers, and Eagles)
Elanus leucurus (Vieillot) White-tailed kite
Accipiter gentilis (Linnaeus) Goshawk
Accipiter striatus (Vieillot) Sharp-shinned hawk
Accipiter cooperii (Bonaparte) Cooper's hawk
Buteo jamaicensis (Gmelin) Red-tailed hawk
Buteo lineatus (Gmelin) Red-shouldered hawk
Buteo swainsoni (Bonaparte) Swainson's hawk
Buteo lagopus (Pontoppidan) Rough-legged hawk
Buteo regalis (Gray) Ferruginous hawk
Aquila chrysaetos (Linnaeus) Golden eagle
Haliaeetus leucocephalus (Linnaeus) Bald eagle
Circus cyaneus (Linnaeus) Marsh hawk

Family-PANDIONIDAE (Ospreys)
Pandion haliaetus (Linnaeus) Osprey

Family-FALCONIDAE (Falcons)
Falco mexicanus (Schlegel) Prairie falcon
Falco peregrinus (Tunstall) Peregrine falcon
Falco columbarius (Linnaeus) Merlin
Falco tinnunculus (Linnaeus) American kestrel

Family-TETRAONIDAE (Grouse)
Dendragapus obscurus (Say) Blue grouse
Bonasa umbellus (Linnaeus) Ruffed grouse
Centrocercus urophasianus (Bonaparte) Sage grouse

Family-PHASIANIDAE (Quail, Partridges, and Pheasants)
Lephortyx californicus (Shaw) California quail
Oreortyx pictus (Douglas) Mountain quail
Phasianus colchicus (Linnaeus) Ring-necked pheasant
Alectoris graeca (Meisner) Chukar
Perdix perdix (Linnaeus) Gray partridge

Family-GRUIDAE (Cranes)
Grus canadensis (Linnaeus) Sandhill crane

Family-RALLIDAE (Rails, Gallinules, and Coots)
Rallus limicola (Vieillot) Virginia rail
Porzana carolina (Linnaeus) Sora
Gallinula chloropus (Linnaeus) Common gallinule
Fulica americana (Gmelin) American coot

Family-CHARADRIIDAE (Plovers)
189*Charadrius semipalmatus* (Bonaparte) Semipalmated plover
Charadrius alexandrinus (Linnaeus) Snowy plover

Charadrius vociferus (Linnaeus) Killdeer
Pluvialis dominica (Muller) American golden plover
Pluvialis squatarola (Linnaeus) Black-bellied plover

Family-SCOLOPACIDAE (Snipes, Sandpipers, and Other Shorebirds)
Arenaria interpres (Linnaeus) Ruddy turnstone
Capella gallinago (Linnaeus) Common snipe
Numenius americanus (Bechstein) Long-billed curlew
Numenius phaeopus (Linnaeus) Whimbrel
Bartramia longicauda (Bechstein) Upland sandpiper
Actitis macularia (Linnaeus) Spotted sandpiper
Tringa rolitaria (Wilson) Solitary sandpiper
Catoptrophorus semipalmatus (Gmelin) Willet
Tringa melanoleucas (Gmelin) Greater yellowlegs
Tringa flavipes (Gmelin) Lessor yellowlegs
Calidris canutus rufa (Linnaeus) Red knot
Calidris melanotos (Vieillot) Pectoral sandpiper
Calidris bairdii (Coues) Baird's sandpiper
Calidris minutilla (Vieillot) Least sandpiper
Calidris alpina (Linnaeus) Dunlin
Calidris alba (Pallas) Sanderling
Limnodromus scolopaceus (Say) Long-billed dowitcher
Calidris mauri (Cabanis) Western sandpiper
Limosa fedoa (Linnaeus) Marbled godwit

Family-RECURVIROSTRIDAE (Avocets and Stilts)
Recurvirostra americana (Gmelin) American avocet
Himantopus mexicanus (Miller) Black-necked stilt

Family: PHALAROPODIADAE (Phalaropes)
Steganopus tricolor (Vieillot) Wilson's phalarope
Phalaropus fulicarius (Linnaeus) Red phalarope
Lobipes lobatus (Linnaeus) Northern phalarope

Family: STERCORARIIDAE (Jaegers and Skuas)
Stercorarius parasiticus (Linnaeus) Parasitic jaeger

Family: LARIDAE (Gulls and Terns)
Larus argentatus (Pontoppidan) Herring gull
Larus californicus (Lawrence) California gull
Larus delawarensis (Ord) Ring-billed gull
Larus atricilla (Linnaeus) Laughing gull
Larus pipixcan (Wagler) Franklin's gull
Larus philadelphia (Ord) Bonaparte's gull
Xema sabini (Sabine) Sabine's gull
Sterna forsteri (Nuttall) Forster's tern
Sterna hirundo (Linnaeus) Common tern
Hydroprogene caspia (Pallas) Caspian tern
Childonias niger (Linnaeus) Black tern

Family: COLUMBIDAE (Pigeons and Doves)
Columba fasciata (Say) Band-tailed pigeon
Zenaida macroura (Linnaeus) Mourning dove

Family-CUCULIDAE (Cuckoos)
Coccyzus americanus (Linnaeus) Yellow-billed cuckoo

Family-TYTONIDAE (Barn Owls)
  Tyto alba (Scopoli) Barn owl

Family-STRIGIDAE (Owls)
  Otus asio (Linnaeus) Screech owl
  Otus flammeolus (Kaup) Flammulated owl
  Bubo virginianus (Gmelin) Great horned owl
  Nyctea scandiaca (Linnaeus) Snowy owl
  Glaucidium gnoma (Wagler) Pygmy owl
  Speotyto cunicularia (Molina) Burrowing owl
  Strix nebulosa (Forsiter) Great gray owl
  Asio otus (Linnaeus) Long-eared owl
  Asio flammeus (Pontoppidan) Short-eared owl
  Aegolius acadicus (Gmelin) Saw-whet owl

Family-CAPRIMULGIDAE (Poor-wills and Nighthawks)
  Phalaenoptilus nuttallii (Audubon) Poor-will
  Chordeiles minor (Foster) Common nighthawk

Family-APODIDAE (Swifts)
  Chaetura vauxi (Townsend) Vaux's swift
  Aeronautes saxatalis (Woodhouse) White-throated swift

Family-TROCHILIDAE (Hummingbirds)
  Archilochus alexandri (Boucier and Mulsant) Black-chinned
    hummingbird
  Selasphorus platycercus (Swainson) Broad-tailed hummingbird
  Selasphorus rufus (Gmelin) Rufous hummingbird
  Stellula calliope (Gould) Calliope hummingbird

Family-ALCEDINIDAE (Kingfishers)
  Megaceryle alcyon (Linnaeus) Belted kingfisher

Family-PICIDAE (Woodpeckers)
  Calaptes auratus (Linnaeus) Common flicker
  Dryocopus pileatus (Linnaeus) Pileated woodpecker
  Asyndesmus lewis (Gray) Lewis' woodpecker
  Sphyrapicus varius (Linnaeus) Yellow-bellied sapsucker
  Sphyrapicus thyroideus (Cassin) Williamson's sapsucker
  Dendrocopos villosus (Linnaeus) Hairy woodpecker
  Dendrocopos pubescens (Linnaeus) Downy woodpecker
  Dendrocopos albolarvatus (Cassin) White-headed woodpecker
  Picoides articus (Swainson) Black-backed three-toed woodpecker
  Picoides tridactylus (Linnaeus) Northern three-toed woodpecker

Family-TYRANNIDAE (Flycatchers)
  Tyrannus tyrannus (Linnaeus) Eastern kingbird
  Tyrannus verticalis (Say) Western kingbird
  Muscivora forficata (Gmelin) Scissor-tailed flycatcher
  Myiarchus cinerascens (Lawrence) Ash-throated flycatcher
  Sayornis saya (Bonaparte) Say's phoebe
  Empidonax traillii (Audubon) Willow flycatcher
  Empidonax hammondii (Xantus) Hammond's flycatcher
  Empidonax oberholseri (Phillies) Dusky flycatcher
  Empidonax wrightii (Baird) Gray flycatcher
  Empidonax difficilis (Baird) Western Flycatcher

  Contopus virens (Linnaeus) Eastern wood pewee
  Contopus sordidulus (Scalater) Western wood pewee
  Nuttallornis borealis (Swainson) Olive-sided flycatcher

Family-ALAUDIDAE (Larks)
  Eremophila alpestris (Linnaeus) Horned lark

Family-HIRUNDINIDAE (Swallows)
  Tachycineta thalassina (Swainson) Violet-green swallow
  Iridoprocne bicolor (Vieillot) Tree swallow
  Riparia riparia (Linnaeus) Bank swallow
  Stelgidopteryx ruficollis (Vieillot) Rough-winged swallow
  Hirundo rustica (Linneaus) Barn swallow
  Petrochelidon pyrrhonota (Vieillot) Cliff swallow
  Progne subis (Linnaeus) Purple martin

Family-CORVIDAE (Jays Magpies, and Crows)
  Perisoreus canadensis (Linnaeus) Gray jay
  Cyanocitta cristata (Linnaeus) Blue jay
  Cyanocitta stelleri (Gmelin) Steller's jay
  Aphelocoma coerulescens (Bose) Scrub jay
  Pica pica (Linnaeus) Black-billed magpie
  Corvus corax (Linnaeus) Common raven
  Corvus brachyrhynchos (Brehm) Common crow
  Gymnorhinus cyanocephala (Wied) Pinon jay
  Nucifraga columbiana (Wilson) Clark's nutcracker

Family-PARIDAE (Chickadees and Relatives)
  Parus atricapillus (Linnaeus) Black-capped chickadee
  Parus gambeli (Ridgway) Mountain chickadee
  Parus inornatus (Gambel) Plain titmouse
  Psaltriparus minimus (Townsend) Bushtit

Family-SITTIDAE (Nuthatches)
  Sitta carolinensis (Latham) White-breasted nuthatch
  Sitta canadensis (Linnaeus) Red-breasted nuthatch
  Sitta pygmaea (Vigors) Pigmy nuthatch

Family-CERTHIIDAE (Creepers)
  Certhia familiaris (Linnaeus) Brown creeper

Family-CINCLIDAE (Dippers)
  Cinclus mexicanus (Swainson) Dipper

Family-TROGLODYTIDAE (Wrens)
  Troglodytes aedon (Vieillot) House wren
  Troglodytes troglodytes (Linnaeus) Winter wren
  Thryomanes bewickii (Audubon) Bewick's wren
  Telmatodytes palustris (Wilson) Long-billed marsh wren
  Catherpes mexicanus (Swainson) Canon wren
  Salpinctes obsoletus (Say) Rock wren

Family-MIMIDAE (Mockingbirds and Thrashers)
  Mimus polyglottos (Linnaeus) Mockingbird
  Dumetella carolinensis (Linnaeus) Gray catbird
  Toxostoma rufum (Linnaeus) Brown thrasher
  Oreoscoptes montanus (Townsend) Sage thrasher

Family-TURDIDAE (Thrushes, Bluebirds, and Solitaires)
  Turdus migratorius (Linnaeus) American robin
  Ixoreus naevius (Gmelin) Varied thrush
  Catharus guttata (Pallas) Hermit thrush
  Catharus ustulata (Nuttall) Swainson's thrush
  Catharus fuscescens (Stephens) Veery
  Sialia mexicana (Swainson) Western bluebird
  Sialia currucoides (Bechstein) Mountain bluebird
  Myadestes townsendi (Audubon) Townsend's solitaire

Family-SYLVIIDAE (Gnatcatchers and Kinglets)
  Polioptila caerulea (Linnaeus) Blue-gray gnatcatcher
  Regulus satrapa (Lichtenstein) Golden-crowned kinglet
  Regulus calendula (Linnaeus) Ruby-crowned kinglet

Family-MOTACILLIDAE (Pipits)
  Anthus spinoletta (Linnaeus) Water pipit

Family-BOMBYCILLIDAE (Waxwings)
  Bombycilla garrulus (Linnaeus) Bohemian waxwing
  Bombycilla cedrorum (Vieillot) Cedar waxwing

Family-PTILOGONATIDAE (Silky Flycatchers)
  Phainopepla nitens (Swainson) Phainopepla

Family-LANIIDAE (Shrikes)
  Lanius excubitor (Linnaeus) Northern shrike
  Lanius ludovicianus (Linnaeus) Loggerhead shrike

Family-STURNIDAE (Starlings)
  Sturnus vulgaris (Linnaeus) Starling

Family-VIREONIDAE (Vireos)
  Vireo huttoni (Cassin) Hutton's vireo
  Vireo solitarius (Wilson) Solitary vireo
  Vireo olivaceus (Linnaeus) Red-eyed vireo
  Vireo gilvus (Vieillot) Warbling vireo

Family-PARULIDAE (Wood Warblers)
  Mniotilta varia (Linnaeus) Black-and-white warbler
  Vermivora peregrina (Wilson) Tennessee warbler
  Vermivora celata (Say) Orange-crowned warbler
  Vermivora ruficapilla (Wilson) Nashville warbler
  Parula americana (Linnaeus) Northern parula
  Dendroica petechia (Linnaeus) Yellow warbler
  Dendroica tigrina (Gmelin) Cape May warbler
  Dendroica caerulescens (Gmelin) Black-throated blue warbler
  Dendroica coronata auduboni (Linnaeus and Townsend) Yellow-rumped warbler
  Dendroica nigrescens (Townsend) Black-throated gray warbler
  Dendroica townseni (Townsend) Townsend's warbler
  Denfroica virens (Gmelin) Black-throated green warbler
  Dendroica occidentalis (Townsend) Hermit warbler
  Dendroica pensylvanica (Linnaeus) Chestnut-sided warbler
  Dendroica castanea (Wilson) Bay-breasted warbler
  Dendroica striata (Forster) Blackpoll warbler
  Dendroica palmarum (Gmelin) Palm warbler

  Seiurus aurocapillus (Linaeus) Ovenbird
  Seiurus noveboracensis (Gmelin) Northern waterthrush
  Oporornis tolmiei (Townsend) MacGillivray's warbler
  Geothlypis tirchas (Linnaeus) Common yellowthroat
  Icteria virens (Linnaeus) Yellow-breasted chat
  Wilsonia pusilla (Wilson) Wilson's warbler
  Setophaga ruticilla (Linnaeus) American redstart

Family-PLOCEIDAE (Weaver Finches)
  Passer domesticus (Linnaeus) House sparrow

Family-ICTERIDAE (Meadowlarks, Blackbirds, and Orioles)
  Dolichonyx oryzivorus (Linnaeus) Bobolink
  Sturnella neglecta (Audubon) Western meadowlark
  Xanthocephalus xanthocephalus (Bonaparte) Yellow-headed blackbird
  Agelaius phoeniceus (Linnaeus) Red-winged blackbird
  Icterus cucullatus (Swainson) Hooded oriole
  Icterus galbula bullockii (Linnaeus and Townsend) Northern oriole
  Euphagus cyanocephalus (Wagler) Brewer's blackbird
  Quiscalus quiscula (Linnaeus) Common grackle
  Molothrus ater (Boddaert) Brown-headed cowbird

Family-THRAUPIDAE (Tanagers)
  Piranga ludoviciana (Wilson) Western tanager
  Piranga rubra (Linnaeus) Summer tanager

Family-FRINGILLIDAE (Grosbeaks, Finches, Sparrows, and Buntings)
  Pheucticus lodovicianus (Linnaeus) Rose-breasted grosbeak
  Pheucticus malanocephalis (Swainson) Black-headed grosbeak
  Passerina cyanea (Linnaeus) Indigo bunting
  Passerina amoena (Say) Lazuli bunting
  Passerina ciris (Linnaeus) Painted bunting
  Hesperiphona verpertina (Cooper) Evening grosbeak
  Carpodacus purpureus (Gmelin) Purple finch
  Carpodacus cassinii (Baird) Cassin's finch
  Carpodacus mexicanus (Muller) House finch
  Pinicola enucleator (Linnaeus) Pine grosbeak
  Leucosticte tephrocotis (Swainson) Gray-crowned rosy finch
  Leucosticte atrata (Ridgeway) Black-rosy finch
  Acanthis flammea (Linnaeus) Common redpoll
  Spinus pinus (Wilson) Pine siskin
  Spinus tristis (Linnaeus) American goldfinch
  Spinus psaltria (Say) Lesser goldfinch
  Loxia curvirostra (Linnaeus) Red crossbill
  Chlorura chlorura (Audubon) Green-tailed towhee
  Pipilo erythrophthalmus (Linnaeus) Rufous-sided towhee
  Passerculus sandwichensis (Gmelin) Savannah sparrow
  Pooecetes gramineus (Gmelin) Vesper sparrow
  Chondestes grammacus (Say) Lark sparrow
  Amphispiza bilineata (Cassin) Black-throated sparrow
  Amphispiza belli (Cassin) Sage sparrow
  Junco hyemalis oreganus (Townsend) Dark-eyed junco
  Junco caniceps (Woodhouse) Gray-headed junco
  Seizella arborea (Wilson) Tree sparrow
  Seizella passerina (Bechstein) Chipping sparrow

Spizella pallida (Swainson) Clay-colored sparrow
Spizella breweri (Cassin) Brewer's sparrow
Zonotrichia querula (Nuttell) Harris' sparrow
Zonotrichia leucophrys (Foster) White-crowned sparrow
Zonotrichia atricapilla (Gmelin) Golden-crowned sparrow
Zonotrichia albicollis (Gmelin) White-throated sparrow
Passerella iliaca (Merrem) Fox sparrow
Melospiza lincolnii (Audubon) Lincoln's sparrow
Melospiza melodia (Wilson) Song sparrow
Plectrophenax nivalis (Linnaeus) Snow bunting

MAMMALS

Order-INSECTIVORA
  Family-SORICIDAE (Shrews)
    Sorex preblei (Jackson) Preble's shrew (for name change, see
      Larrison 1976)
    Sorex vagrans (Merriam) Vagrant shrew
    Sorex palustris (Baird) Water shrew
    Sorex merriami (Dobson) Merriam's shrew
  Family-TALPIDAE (Moles)
    Scapanus latimanus (True) Broad-footed mole

Order-CHIROPTERA
  Family-VESPERTILIONIDAE (Evening Bats)
    Myotis lucifugus (Thomas) Little brown bat
    Myotis yumanensis (Grinnel) Yuma brown bat
    Myotis evotis (Allen) Long-eared brown bat
    Myotis thysanodes (Miller) Fringed brown bat
    Myotis volans (Miller) Long-legged brown bat
    Myotis californicus (Audubon & Bachman) California brown
      bat
    Myotis subulatus (Merriam) Small-footed brown bat
    Lasionycteris noctivagans (LeConte) Silver-haired bat
    Pipistrellus hesperus (Allen) Western pipistrel
    Eptesicus fuscus (Young) Big brown bat
    Lasiurus cinereus (Beauvois) Hoarny bat
    Euderma maculatum (Allen) Spotted bat
    Plecotus townsendi (Grinnell) Townsend's big-eared bat
    Antrozous pallidus (Bailey) Pallid bat

Order-LAGOMORPHA
  Family-OCHOTONIDAE (Pikas)
    Ochotona princeps (Grinnell) Common pika
  Family-LEPORIDAE (Rabbit and Hare)
    Brachylagus idahoensis (Merriam) Pygmy rabbit
    Sylvilagus nuttalli (Bachman) Nuttall's cottontail
    Lepus americanus (Merriam) Snowshoe hare
    Lepus townsendi (Bachman) White-tailed jackrabbit
    Lepus californicus (Merriam) Black-tailed jackrabbit

Order-RODENTIA
  Family-SCIURIDAE (Marmots & Squirrels)
    Eutamias minimus (Hall & Hatfield) Least chipmunk
    Eutamias amoenus (Allen) Yellow pine chipmunk
    Marmonta flaviventris (Bangs) Yellow-bellied marmot

Ammospermophilus leucurus (Merriam) White-tailed antelope
  ground squirrel
Spermophilus townsendi (Merriam) Townsend's ground
  squirrel
Spermophilus richardsoni (A.H. Howell) Richardson's ground
  squirrel
Spermophilus beldingi (Hall) Belding's ground squirrel
Spermophilus columbianus (A.H. Howell) Columbian ground
  squirrel
Spermophilus lateralis (Merriam) Golden-mantled ground
  squirrel
Tamiasciurus hudsonicus (Bachman) Red squirrel
Tamiasciurus douglasi (Audubon and Bachman) Douglas'
  squirrel
Glaucomys sabrinus (Rhoads) Northern flying squirrel
Family-GEOMYIDAE (Pocket Gophers)
  Thomomys townsendi (Davis) Townsend's pocket gopher
  Thomomys talpoides (Merriam) Northern pocket gopher
Family-HETEROMYIDAE (Pocket Mice and Kangaroo Rats)
  Perognathus longimembris (Merriam) Little pocket mouse
  Perognathus parvus (Peale) Great Basin pocket mouse
  Microdipodops megacephalus (Merriam) Dark kangaroo mouse
  Dipodomys ordi (Merriam) Ord's kangaroo rat
  Dipodomys microps (Goldman) Chisel-toothed kangaroo rat
  Dipodomys heermanni (Merriam) Heerman's kangaroo rat
Family-CASTORIDAE (Beavers)
  Castor fiber (Nelson) Beaver
Family-CRICETIDAE (Cricetid Rats and Mice)
  Reithrodontomys megalotis (Baird) Western harvest mouse
  Peromyscus crinitus (Merriam) Canyon mouse
  Peromyscus maniculatus (LeConte) Common deer mouse
  Peromyscus truei (J. A. Allen) Pinyon mouse
  Onychomys leucogaster (Anthony) Northern grasshopper
    mouse
  Neotoma lepida (Tayler) Desert wood rat
  Neotoma cinerea (Hooper) Bushy-tailed wood rat
Family-MICROTIDAE (Voles)
  Clethrionomys gapperi (Merriam) Gapper's red-backed mouse
  Phenacomys intermedius (Merriam) Heather mouse
  Microtus montanus (Peale) Montane meadow mouse
  Microtus longicaudus (Merriam) Long-tailed meadow mouse
  Arvicola richardsoni (Rhoads) Richardson's water vole
  Lagurus curtatus (Cooper) Sagebrush vole
  Ondatra zibethicus (Lord) Muskrat
Family-MURIDAE (Old World Rats and Mice)
  Rattus norvegicus (Berkenhout) Norway rat
  Mus musculus (Rutty) House mouse
Family-DIPODIDAE (Jumping Mice)
  Zapus princeps (Preble) Western jumping mouse
Family-ERETHIZONTIDAE (Porcupines)
  Erethizon dorsatum (Brandt) Porcupine

Order-CARNIVORA
  Family-CANIDAE (Dogs)
    Canis latrans (Merriam) Coyote
    Canis lupus (Richardson) Gray wolf (probably extinct in
      northern Great Basin).

Vulpes vulpes (Desmarest) Red fox (probably extinct in
  northern Great Basin).
Vulpes velox (Goldman) Kit fox
Family-URSIDAE (Bears)
  Ursus americanus (J. Miller) Black bear
Family-PROCYONIDAE (Procyonids)
  Procyon lotor (Nelson and Goldman) Raccoon
Family-MUSTELIDAE (Mustelids)
  Martes americana (Rafinesque) Marten
  Martes pennanti (Rhoads) Fisher
  Mustela erminea (Bangs) Short-tailed weasel
  Mustela frenata (Hall) Long-tailed weasel
  Lutreola lutreola (Bangs) Mink
  Gulo gulo (Elliot) Wolverine
  Taxidea taxus (Mearns) Badger

Spilogale putorius (Merriam) Spotted skunk
Mephitis mephitis (Howell) Striped skunk
Family-FELIDAE (Cats)
  Felis concolor (Nelson & Goldman) Mountain lion
  Lynx lynx (Kerr) Lynx
  Lynx rufus (Merriam) Bobcat

Order-ARTIODACTYLA
  Family-CERVIDAE (Deer)
    Cervus elaphus (Bailey) Elk or Wapiti
    Odocoileus hemionus (Rafinesque) Mule deer
  Family-ANTILOCAPRIDAE (Pronghorns)
    Antilocapra americana (Bailey) Pronghorn
  Family-BOVIDAE (Cattle)
    Ovis canadensis (Douglas) Bighorn sheep

4

THE MOJAVE DESERT

PETER ROWLANDS, HYRUM JOHNSON, ERIC RITTER, AND ALBERT ENDO

THE MOJAVE DESERT is the smallest of the North American deserts and the most used and abused by humans. Since prehistoric times it has served as a crossroads of travel, trade, and migration. As early as 900 A.D., a trade network extended across the Mojave Desert to the Pacific and the Gulf of California. Lying across the major migration routes of settlers moving to California, the Mojave was exploited early by miners, cattlemen and others for the resources it contained. In recent years it has become a haven for retirees seeking a respite from the cold and snow of former homes in the East and Midwest. Today it is surrounded by metropolitan areas whose inhabitants increasingly utilize the desert for recreation. At the same time the Mojave has some areas so remote and little visited that they are recommended for wilderness status. (See figures 4-1 and 4-2.)

Lying between the Great Basin Desert to the north and the Sonoran Desert to the south, the Mojave has often been referred to as an ecotone incorporating vegetational elements from each of its neighbors. It does contain Great Basin vegetational elements and Sonoran vegetational elements, as well as endemic species of its own. Perhaps one indication of these interrelationships can be seen in the references to the Mojave Desert in other chapters of this book.

Crosswhite and Crosswhite in chapter 5 include a discussion of the geological history of the Mojave; Fowler and Koch in chapter 2 discuss the transitional zone between the Great Basin and Mojave deserts; Ohmart and Andersen in chapter 10 discuss the riparian ecosystems of the Mojave; Smith describes and illustrates the dune systems in chapter 11; Hadley in chapter 8 and Solbrig in chapter 9 discuss the adaptations of animals and plants to desert conditions; Bender lists the natural areas and other research areas and facilities within the Mojave Desert in chapter 13. It is indeed a pivotal desert.

PHYSICAL SETTING

For purposes of easy reference, the Mojave has been divided into the following regions:

Eastern Mojave
    southern Nevada section
    northwestern Arizona section
    California section
Northern Mojave
    Nevada section
    California section
Southwestern Mojave
Central Mojave
Southcentral Mojave

The location of each is shown in figure 4-3.

Eastern Mojave

Bradley and Deacon (1967) have described the southern Nevada section, which is part of the basin and range province with numerous isolated mountain ranges that run from north to south. Enormous bajadas slope from the flanks of the mountains, extending down to the intervening basins which may contain dry lake beds or playas. Elevation varies from 154 meters (m) at the Colorado River to 3652 m on Charleston Peak, the highest peak in the area. Major mountain ranges include the Spring Mountains and Sheep Range, which have a large limestone component. The limestone provides a substrate for calciphyte plant communities and supports a number of spectacular caverns. There are a number of springs but few permanent streams. The Moapa and Virgin rivers drain into Lake Mead on the Colorado River. Lake Mead is one of the largest lakes on the Colorado and provides outstanding opportunities for recreation. The valley of Fire State Park east of Las Vegas has spectacular red sandstone formations together with petrified trees and petroglyphs. Extensive forests of Joshua trees are found west of Searchlight, while Lake Mojave lies a few miles to the east.

The northwestern Arizona section follows the basin and range pattern, with northwest to southeast-trending mountain

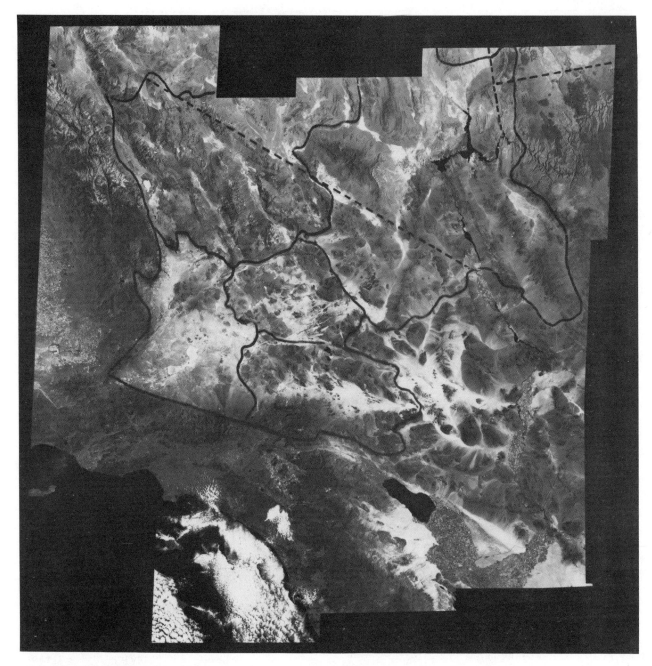

Figure 4-1. A digital mosaic of ten Landsat scenes.
Prepared by the Jet Propulsion Laboratories in Pasadena, California for the Bureau of Land
Management California Desert Project.

ranges such as the Cerbat and Black mountains interspersed with basins. Red Lake lying just east of the Cerbat mountains is a large, dry lake bed or playa. There are dry washes but no permanent streams in this section, although it is bounded on the north and west by the Colorado River. Extensive Joshua tree forests extend from Chloride to Pierce's Ferry. Elevations range from 155 m to 2246 m.

The California section includes some twenty mountain ranges varying in height from the Little Cow Hole Mountains at 430 m to the New York Mountains at 2296 m. Other large ranges include the Granite Mountains 2054 m and the Providence Mountains 2148 m with the impressive Mitchell Caverns. Valleys in the section include Ivanpah, Kelso, Lanfair, and Pinto.

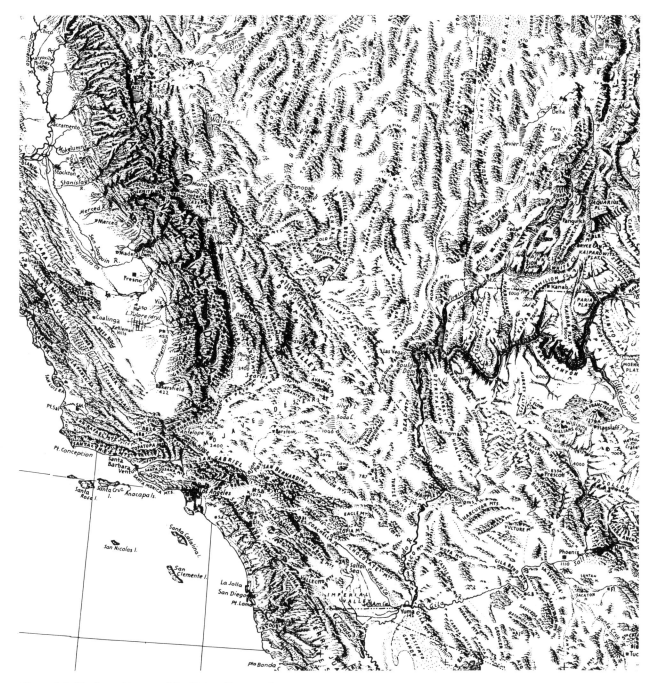

Figure 4-2. Physiographic map of the Mojave Desert region.
Modified from *Landforms of the United States* by Erwin Raisz, 1957.

Northern Mojave

The major feature of the California section is Death Valley with its tremendous relief ranging from −82 m at its lowest point to 3300 m at the top of Telescope Peak. The relief is very dramatic because one can stand at the lowest point and look directly up at the highest point just across the valley. Major mountain ranges include the Panamint Range, Amar-

gosa Range, Black Mountains, Funeral Mountains, Cotton-wood Mountains, and Grapevine Mountains. The general orientation of these ranges is from northwest to southeast. The areas proposed for wilderness status are in this section.

The Nevada section includes the Amargosa Desert, the Atomic Energy Commission Mercury Test Site, and Devil's Hole, home of the Devil's Hole pupfish.

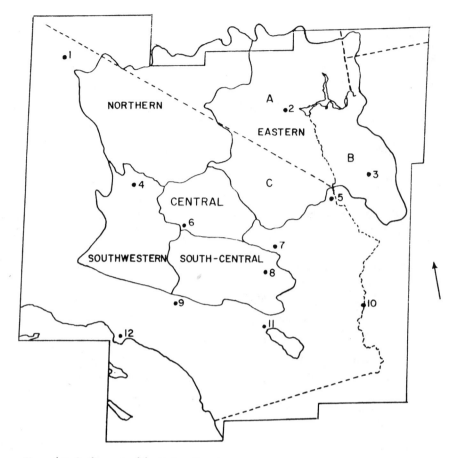

Figure 4-3. Outline map of the Mojave Desert.
The map shows the major regions of the Mojave Desert together with important points of
reference: Bishop (1), Las Vegas (2), Kingman (3), China Lake (4), Needles (5), Barstow (6),
Amboy (7), Twenty-Nine Palms (8), San Bernardino (9), Blythe (10), Indio (11), Los Angeles
(12). The eastern Mojave has been subdivided into a southern Nevada section (A), a northwestern Arizona section (B), and a California section (C).

Southwestern Mojave

This region is bounded to the south by the San Gabriel
Mountains and to the west by the Tehachapi Mountains. It is
an area of relatively slight relief with elevations ranging from
600 m to 800 m. Included in the region are Antelope Valley,
Edwards Air Force Base, the Mojave River, China Lake,
Trona Pinnacles, and the Desert Tortoise Preserve near Lancaster. The preserve was set aside to protect this unique desert
species.

South-central Mojave

This region is bordered to the south by the Little San Bernardino Mountains. It includes Joshua Tree National Monument, Twenty-nine Palms, Lucerne Valley, Apple Valley, and
the Eagle Mountains with their rich deposits of iron ore.

Central Mojave

Barstow is the major city in this region. Other points of
interest include the Calico Mountains, the NASA Goldstone
Tracking Station, and outstanding high desert scenery.

Figures 4-4 and 4-5 show transects across the Mojave
Desert in two different locations.

DRAINAGE

Following the last glacial period, the Mojave Desert had a
number of lakes and streams. The Owens River to the west,
the Mojave River to the south, and the Amargosa River to the
north and east formed an extensive drainage system with
several large lakes. Lakes associated with the Owens River
included Owens Lake, Searles Lake, Panamint Lake, Lake

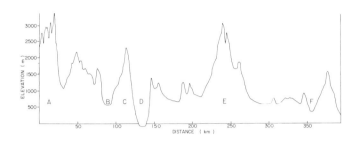

Figure 4-4. Profile diagram of the Mojave Desert at about 36°30′N.
The map begins at the crest of the Sierra Nevada (A) and continues due east across Panamint Valley (B), the Panamint Mountains (C), Death Valley (D), the Spring Range (E), and Lake Meade (F).

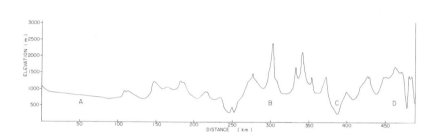

Figure 4-5. Profile diagram of the Mojave Desert beginning near Gorman at approximately 34°47′N, 118°47′W.
The map continues across the Antelope Valley (A), Clark Mountain (B), the Colorado River (C), and the Grand Wash Cliffs (D), ending at approximately 35°54′N, 113°38′W.

Manly, and Lake Rogers, the last two in Death Valley. Lakes associated with the Mojave River included Lake Manix, Soda Lake, and Silver Lake. The Amargosa River united with the Mojave River before flowing into Lake Manly in Death Valley. This system existed until about 10,000 to 8000 years ago.

As the climate became drier, the large lakes gradually disappeared. Their original sites are now marked by remnants such as Silver Lake, Soda Lake, Panamint Lake, Searles Lake, China Lake, and Owens Lake, many of which are dry lake beds or playas that are dry except during periods of heavy rainfall and flooding.

Today the Colorado River is the major river of the Mojave Desert. In addition to draining the eastern portion of the desert, it is also the source of water via Lake Mead and Lake Mojave, which increasingly is being utilized for industry and agricultural development. The Mojave River flows north out of the San Bernardino Mountains to the Mojave sink, where it disappears underground. In extremely wet seasons, it may flow through Soda Lake and reach Silver Lake. The Owens River flows out of the Sierra Nevadas into Owens Lake. The Amargosa River drains portions of western Nevada. It discharges into Death Valley only during periods of extreme high water.

CLIMATE

Precipitation

Bounded by the Sierra Nevadas, Tehachapi, San Gabriel, and San Bernardino mountains, the Mojave is shielded from abundant sources of moisture. Precipitation is low while evaporation is high.

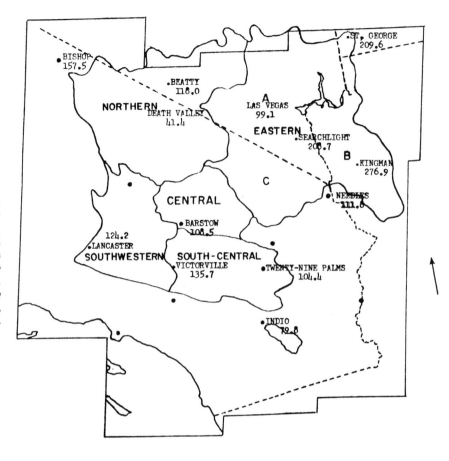

Figure 4-6. Mean annual precipitation in millimeters for selected stations in the Mojave Desert.
Data from P. Rowlands, Regional bioclimatology of the California desert. In *The California desert: An introduction to its resources and man's impact,* ed. J. Latting. *California Native Plant Society Special Publication* no. 5, 1980, Berkeley, Calif.

As is true of most other desert areas, precipitation is extremely variable from year to year and from place to place within the desert. Figure 4-6 shows annual precipitation figures for selected stations in the Mojave Desert.

The Mojave Desert is characterized by three kinds of precipitation patterns: winter storms, summer storms, and hurricanes or chubascos. Winter storms come from the Pacific Ocean, are generally widespread, may last for several days, and are relatively mild in terms of the amount of precipitation delivered, usually less than 20 millimeters (mm) in a twenty-four hour period. However, occasionally a severe storm may drop several times this amount. In 1969 one of the largest storms began on January 17 and ended January 27. Some desert stations recorded as much as 270 mm of precipitation from this single storm. Figure 4-7 shows the average annual amounts of winter rain received at selected stations.

Snow is also frequent in the Mojave Desert. The amount and frequency increase with the altitude and latitude of the reporting stations. Figure 4–8 shows the average amounts of snow received annually at several reporting stations. As with rainfall, the amounts of snow received are highly variable, and occasionally heavy amounts are reported. In 1937-1938 and 1948-1949 snow cover at Cima Dome was 1.2 to 1.5 meters (m). Huning (1978) states that snow is a very important

component of winter precipitation in the northern Mojave region and is common in the other regions at elevations above 1200 m. Winter precipitation is the most important component of the average annual precipitation (figure 4-9).

The summer rainfall period is July through September. Precipitation occurs as strong, localized thunderstorms that enter the desert from the south and east. The moisture comes from the Gulf of Mexico and the Gulf of California and is drawn into the desert by strong convectional currents. These storms can be very intense. On August 12, 1978, a thunderstorm dropped 97 mm of rain at Searchlight, Nevada. Summer precipitation is an important component of the annual total only in the eastern and central Mojave regions (figure 4-10).

Figure 4-11 summarizes the relationship between summer and winter rainfall for selected stations throughout the Mojave desert.

Chubascos are intense tropical storms that arise in Mexican waters and may move up the Gulf of California. Rarely they reach southern sections of the Mojave and produce some rainfall.

There is a direct relationship between elevation and the amounts of precipitation that fall within a region. Rowlands (1978) studied the effects of elevation on precipitation in the Mojave Desert regions of California, southern Nevada, north-

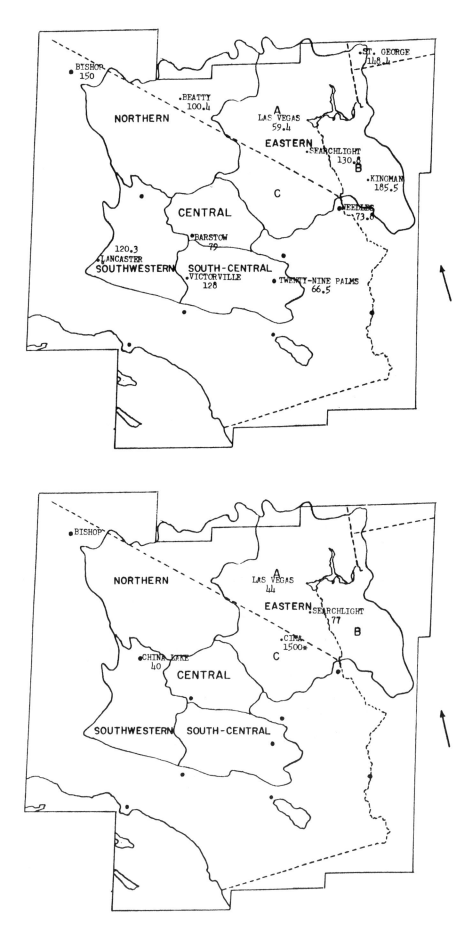

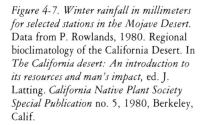

Figure 4-7. Winter rainfall in millimeters for selected stations in the Mojave Desert. Data from P. Rowlands, 1980. Regional bioclimatology of the California Desert. In *The California desert: An introduction to its resources and man's impact,* ed. J. Latting. *California Native Plant Society Special Publication* no. 5, 1980, Berkeley, Calif.

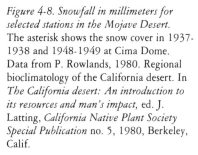

Figure 4-8. Snowfall in millimeters for selected stations in the Mojave Desert. The asterisk shows the snow cover in 1937-1938 and 1948-1949 at Cima Dome. Data from P. Rowlands, 1980. Regional bioclimatology of the California desert. In *The California desert: An introduction to its resources and man's impact,* ed. J. Latting, *California Native Plant Society Special Publication* no. 5, 1980, Berkeley, Calif.

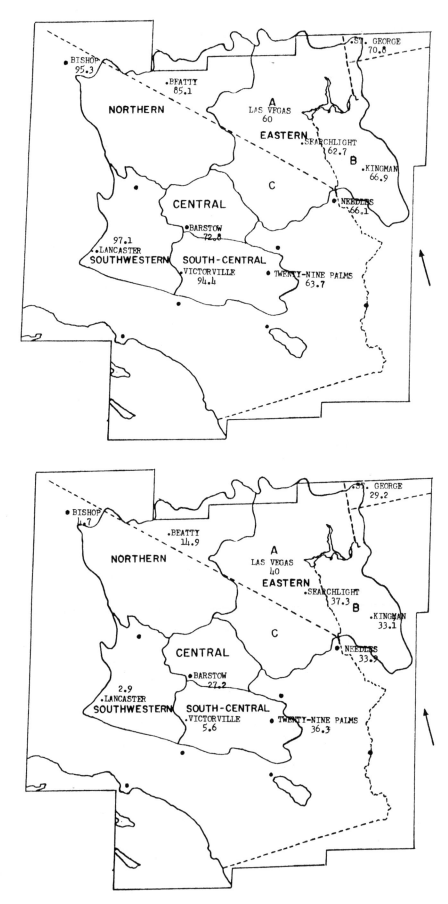

Figure 4-9. Percent annual precipitation falling as winter rain for selected stations in the Mojave Desert. Data from P. Rowlands, 1980. Regional bioclimatology of the California desert. In *The California desert: An introduction to its resources and man's impact,* ed. J. Latting, *California Native Plant Society Special Publication* no. 5, 1980, Berkeley, Calif.

Figure 4-10. Percent annual precipitation falling as summer rain for selected stations in the Mojave Desert. Data from P. Rowlands, 1980. Regional bioclimatology of the California desert. In *The California desert: An introduction to its resources and man's impact,* ed. J. Latting, *California Native Plant Society Special Publication* no. 5, 1980, Berkeley, Calif.

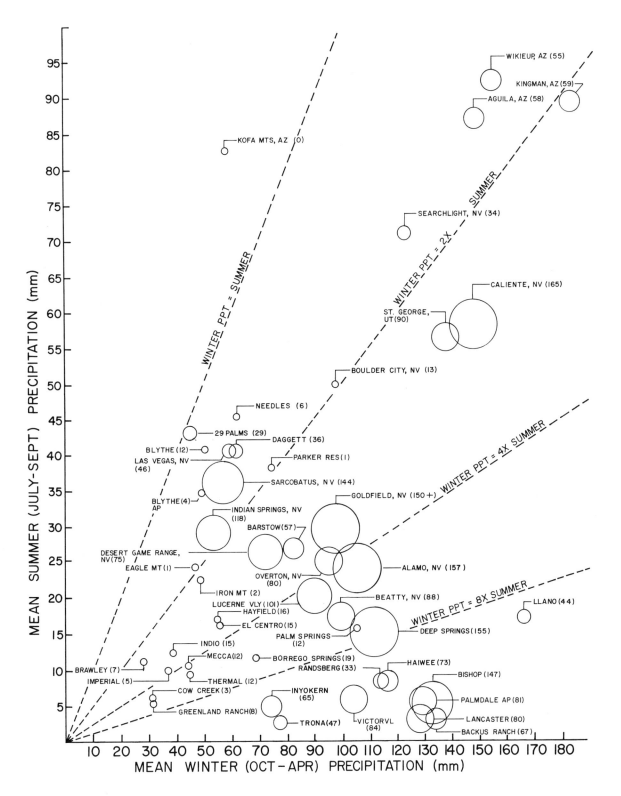

Figure 4-11. Scatter diagram of selected southwestern desert localities.
The illustration shows the seasonality of precipitation and the mean number of days per year
(number of parentheses) when the temperature falls below freezing. The size of the black circle is
proportional to the value of this variable. From P. Rowlands, 1980. Regional bioclimatology of
the California desert. In *The California desert: An introduction to its resources and man's impact.*
California Native Plant Society Special Publication no. 5, figure 3. Reprinted with permission.

western Arizona, and southwestern Utah. He found that precipitation increased with increasing elevation in all regions, with northwestern Arizona having a significantly higher rate than the others. This may be explained by the strongly bimodal precipitation pattern in northwestern Arizona, with summer precipitation being more important than in other Mojave desert regions.

Temperature

The Mojave Desert is an area of temperature extremes with mean minimum January temperatures ranging from $-2.4°C$ at Beatty, Nevada, to $47°C$ for a mean July maximum at Death Valley. Temperature means and extremes decrease with increasing elevation and latitude. Figure 4-12 shows mean January minimum and mean July maximum temperatures for selected stations.

There is a large daily temperature range, with a variation of $25°C$ between maximum and minimum temperatures in a twenty-four-hour period. The daily range is greater in summer than in winter.

An important factor affecting temperatures in closed basins is cold air inversion. Because cold air is heavier than warm air, it flows downslope during the night and settles into the bottoms of the basins with a layer of warmer air on top. Such temperature inversions may affect the distribution of vegetation in a localized area. Beatley (1974, 1975) described this for the Nevada Test Site where creosote bush (*Larrea tridentata*), which is less tolerant of low temperatures than shadscale (*Atriplex confertifolia*) is found occupying higher elevations on the bajada, while shadscale is found growing on the basin floor. A study of temperatures in the two areas revealed that the basin floor was 4 to 6° C colder than the bajada. Inversions are particularly important below 1000 m and are more intense during the winter.

Humidity

Humidity in desert areas is generally low and highly variable both during the year and during a twenty-four hour period. During the day it is highest in the early morning, lowest in late afternoon, and increases again during evening hours. Relative humidities of 10 percent or less are not unusual. Summer months show the greatest fluctuations in humidity.

Winds

Winds are common features of all desert areas and are particularly noticeable when they produce sand and dust storms. Localized "dust devils" are common during the summer moving across the desert floor. They sometimes reach heights of several hundred feet and develop considerable power.

The winds of spring and winter are generally stronger than those of summer. An important exception would be those associated with intense summer thunderstorms, which can be very violent. These strong, gusty winds often produce severe local dust storms.

Evapotranspiration

The combination of high temperatures, low humidity, and wind produces high evaporation rates from moist surfaces and affects the rate of transpiration in plants. The ratio between potential evapotranspiration and precipitation is frequently used as a measure of relative dryness. The higher the ratio the drier the area. Figure 4-13 shows evapotranspiration-precipitation ratios for selected stations in the Mojave Desert.

LAND FORMS

The major landforms of the Mojave Desert are hills and mountains, plains and alluvial fans, plateaus, badlands, pediments, river washes, playas, and sand dunes.

Mountains have local relief in excess of 150 m, 30 to 75 percent slopes and rugged summits. Soils are 10 to 60 inches deep, and rock outcrops cover 10 to 90 percent of the area.

Hills have local relief of less than 150 m, 20 to 50 percent slopes, and somewhat rounded outlines. Rock outcrops may cover from 0 to 80 percent of the area.

Plateaus are bounded by cliffs and have level to gently sloping surfaces, with 1 to 5 percent slope. Soil textures range from loamy sand to clay loam, with coarse fragments of gravel, cobbles, and stones.

Alluvial fans have surfaces with 2 to 25 percent slope and surface irregularities varying in depth 1 to 60 feet, with a frequency of five to fifty per mile.

Plains have surfaces with 0 to 3 percent slope with surface irregularities of 1 to 6 feet. This landform often is found adjacent to playa areas.

Playas have nearly horizontal surfaces with 0 to 1 percent slope, are poorly drained, usually saline, and covered with water after floods. Soils are fine-textured silty clay loam, silty clay, or clay. Vegetation is usually absent.

Pediments are rock plains with 10 to 20 percent slope, bedrock knobs, a thin layer of alluvium, and topographic irregularities 3 to 15 feet high occurring five to twenty per mile. Surface textures range from gravelly sand to sandy clay loam.

Badlands have a large number of gullies and ravines, slopes from 15 to 75 percent, with soil textures ranging from sand to clay.

River washes have braided channels 15 to 40 feet wide, 0 to 10 percent slope, 1 to 4 foot topographic irregularities

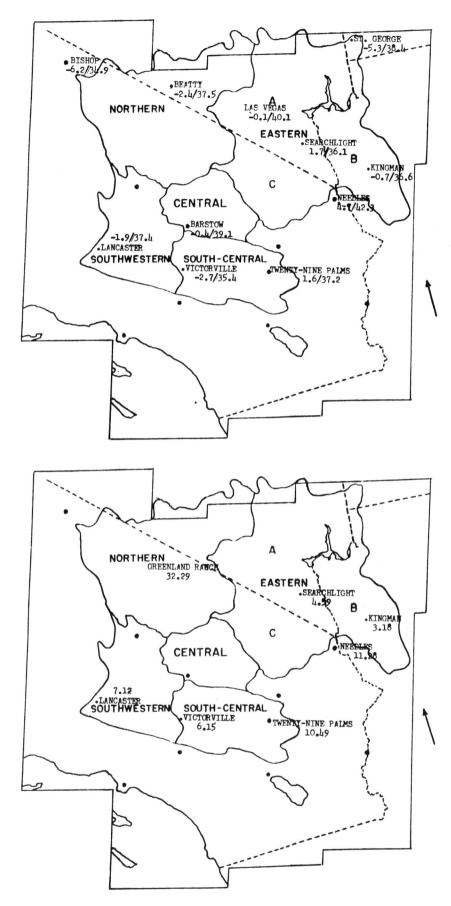

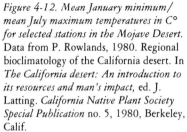

Figure 4-12. Mean January minimum/
mean July maximum temperatures in C°
for selected stations in the Mojave Desert.
Data from P. Rowlands, 1980. Regional
bioclimatology of the California desert. In
*The California desert: An introduction to
its resources and man's impact,* ed. J.
Latting. *California Native Plant Society
Special Publication* no. 5, 1980, Berkeley,
Calif.

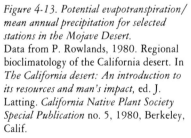

Figure 4-13. Potential evapotranspiration/
mean annual precipitation for selected
stations in the Mojave Desert.
Data from P. Rowlands, 1980. Regional
bioclimatology of the California desert. In
*The California desert: An introduction to
its resources and man's impact,* ed. J.
Latting. *California Native Plant Society
Special Publication* no. 5, 1980, Berkeley,
Calif.

averaging three to four per hundred feet. Surface textures are gravelly to stony sand or coarse sand.

Sand dunes have a cover of wind-blown sand, highly variable slopes, and various characteristic configurations.

Table 4–1 shows the approximate percentage coverage of each landform type in the Mojave desert.

TABLE 4-1
Coverage of each landform type, Mojave Desert

MAJOR LANDFORMS	PERCENT OF MOJAVE DESERT
Playas	2
Plains and alluvial fans	65
Hills and mountains	30
River washes	<1
Sand dunes	<1
Badlands	<1
Plateaus	<1

SOURCE: Prepared by Albert Endo.

SOILS

Soils of the Mojave desert show a wide range of properties. Many of the soils have high percentages of sand and coarse fragments, low organic matter, minimal soil horizon differentiation, concentrations of carbonates, and depths of several feet or more. In contrast, other soils have high percentages of silt and clay, higher concentrations of organic matter, well-developed soil horizons, and shallow depths.

Soils with high percentages of sand and coarse fragments are found on plains and alluvial fans and in river washes. Soils with high percentages of silt and clay include soils on playas, soils on old lacustrine deposits, soils with well-developed argillic horizons, and soils formed from volcanic and metamorphic parent materials.

Soils with well-developed desert pavement covered with desert varnish may be found on dissected and highly dissected alluvial fans. Vesicular surface crusts and clay-enriched subsurface horizons are usually found beneath the well-developed desert pavement. The age of desert pavement may vary from place to place, with some pavement soils possibly as old as pre-Wisconsin (Springer 1958).

Desert soils with organic matter concentrations are found mostly in localized areas under shrub canopies or at higher elevations with greater amounts of precipitation and vegetation. Soils under shrub canopies may have higher concentrations of essential elements (such as nitrogen) than intershrub soil.

Soils with shallow rooting depth of less than 20 inches occur on hills and mountains. Strongly cemented subsurface horizons of carbonates or silica may also limit rooting depth usually on more level terrain.

The following soil types have been identified for the Mojave Desert.

Calciorthids—soils with horizons with high concentrations of calcium carbonate

Camborthids—soils with minimal soil development

Cryoborolls—soils with dark surface horizons occurring at very high elevations with cold climatic conditions

Cryorthents—soils with no diagnostic horizons occurring at very high elevations with cold climatic conditions

Durargids—soils with horizons with increased clay content and silica cementation below

Durorthids—soils with silica cementation

Haplargids—soils with horizons with increased clay content

Haploxeralfs—soils with thick horizons of increased clay content and textures of sandy loam to clay loam

Haploxerolls—soils with dark-colored surface horizons, found at higher elevations

Haplustalls—soils with dark surface horizons and minimal development below

Natragids—soils with horizons with both increased clay content and concentrations of sodium

Paleargids—soils with horizons with increased clay content more than 35 percent

Paleorthids—soils with indurated horizons of calcium carbonate

Salorthids—soils with horizons with high concentrations of salts

Torrifluvents—alluvial soils with textures of sandy loam to clay with irregularity of organic matter content with depth

Torripsamments—sandy soil throughout

Torriorthents—soil with varying texture and with high percentages of coarse fragments

Xerochrepts—soils with minimal development in areas of higher precipitation

Table 4-2 shows the soil types associated with each of the major land forms.

PLANTS

Desert plant life as treated here is viewed from two perspectives: that of floristics and that of vegetation. In the first, emphasis is placed on the kinds of plants found in the desert and their distribution. In the second, more attention is given to the importance of certain species in terms of abun-

TABLE 4-2

Soil subgroups for major landforms, Mojave Desert

MAJOR LANDFORMS	SOILS SUBGROUPS
Playas	Typic Salorthids
	Typic Natragids
Plains and alluvial fans	Typic Torrifluvents
	Typic Torriorthents
	Typic Durorthids
	Typic Calciorthids
	Typic Camorthids
	Typic Salorthids
	Typic Durargids
	Typic Haplargids
	Typic Natragids
	Typic Haploxeralfs
	Calcic Haploxerolls
	Pachic Haploxerolls
Pediments	Typic Torripsamments
	Lithic Torriorthents
	Lithic Camorthids
	Typic Haplargids
Badlands	Typic Torriorthents
	Typic Calciorthids
	Typic Haplargids
Plateaus	Lithic Torriorthents
River washes	Typic Torripsamments
	Typic Torriorthents
Sand Dunes	Typic Torripsamments
Hills and mountains	Typic Torriorthents
	Lithic Torriorthents
	Xeric Torriorthents
	Typic Torripsamments
	Typic Cryorthents
	Typic Calciorthids
	Lithic Haplargids
	Xerollic Haplargids
	Xeralfic Haplargids
	Typic Durargids
	Typic Xerochrepts
	Typic Haplustolls
	Lithic Haplustolls
	Typic Cryoborolls

SOURCE: Prepared by Albert Endo.

dance, cover, and production and to the combinations in which these species occur.

The regions of the Mojave Desert are used as subdivisions for the flora. Species richness is highest in the northern Mojave, with over a thousand species, while the central Mojave with the least elevational range has the smallest number of species. Table 4-3 shows the number of plant species recorded for each of the regions.

TABLE 4-3
Vascular plant species, Mojave Desert regions

MOJAVE DESERT REGIONS	NUMBER OF VASCULAR PLANT SPECIES
Northern Mojave	1025
Eastern Mojave	
California section	754
Southern Nevada section	700-800
Central Mojave	458
South-central Mojave	777
Southwest Mojave	663

SOURCE: Prepared by Hyrum Johnson from unpublished data given in the California Desert Plan.

Records have been compiled on 113 plant families in the Mojave Desert. (See appendix 4.) However, more than half of all the species recorded are members of ten plant families (table 4-4). Native and introduced species numbers are indi-

TABLE 4-4
Ten most important plant families, California Desert

FAMILY	NATIVE SPECIES	INTRODUCED SPECIES
Asteraceae	270	17
Poaceae	139	47
Fabaceae	124	11
Polygonaceae	81	3
Scrophulariaceae	80	1
Polemoniaceae	67	0
Boraginaceae	63	0
Brassicaceae	61	17
Hydrophyllaceae	55	0
Chenopodiaceae	37	11

SOURCE: Prepared by Hyrum Johnson from unpublished data given in the California Desert Plan.

NOTE: Importance was determined by number of species and contribution to total desert flora.

cated. It is estimated that the vascular flora of the Mojave Desert is comprised of 1750 to 2000 species. Of the total species of plants found in the Mojave Desert only a few hundred make significant contributions to the vegetational

aspects. Vegetation types include shrubs, trees, succulents, and perennial and annual herbs.

Shrubs occupy large areas in the Mojave Desert. Shrub ground cover ranges from 1 percent to more than 40 percent. Creosote bush and burrobush are the most prevalent in the hotter and drier portions below 4000 feet, with blackbrush, shadscale, and sagebrush becoming important in higher, cooler areas of the desert.

Small tree woodlands consist of juniper and pinyon in the desert mountains and riparian woodlands along desert streams and washes. Joshua tree stands have been designated as woodlands by some authors.

Succulents include all members of the Cactaceae and Agavaceae. They may be the dominant vegetation type in localized areas but are also generally distributed throughout shrub communities in the desert. They do not occupy more than a few percent of the area of the Mojave Desert.

Herbaceous vegetation may be of two types: perennial grasslands and annual forbs and grasses. Perennial grasslands are widely distributed in the Mojave and become especially prominent following high rainfall years. The principal perennial grass species is big galleta grass, with desert needle grass being dominant on sandy sites. Black grama, bush muhley, and galleta are important species in the eastern Mojave. Grasslands occupy less than 5 percent of the Mojave.

Annual forbs and grasses are divided into winter and summer species, their growth and flowering being dependent upon seasonal rainfall. Winter annuals number more than a hundred species, while summer annuals consist of only ten to fifteen species. Spring displays of flowers are more frequent in the southwestern Mojave, while summer displays are more characteristic of the eastern Mojave, reflecting the differences in the seasonal rainfall patterns.

Vegetation

Although only a few studies have been published on the vegetation of the entire Mojave, a great deal has been published on the vegetation of its various regions. These can be pieced together to form a picture of the whole desert.

CALIFORNIA

Munz (1973), Munz and Keck (1959), Ornduff (1974), Knapp (1965), Kuchler (1964, 1967), Jaeger and Smith (1966), and Thorne (1976) have authored broad classifications and descriptions of California plant communities, including those of the Mojave Desert. On a more restricted scale Vasek and Barbour (1977) described the Mojave Desert scrub and transmontane coniferous vegetation in Barbour and Major (1977). Rowlands (1980) discussed the vegetation of the entire California desert. Regional floras are of importance with respect to the vegetation of the Mojave Desert. They include Kern County flora (Twisselman, 1967), which describes much of the desert vegetation and flora of the Indian Wells Valley and northern Antelope Valley of the western Mojave Desert; Prigge's (1975) flora of the Clark Mountains, and a flora of Inyo County (De Decker, in preparation). Site vegetational studies have been done by Hunt (1966) on the plant ecology of Death Valley, Randall (1972) on the vegetation of Saline Valley, Kornoeldge (1973) on the vegetation of Upper Covington Flats Joshua Tree National Monument, Leary (1977) on the plant communities of the Joshua Tree National Monument, Shefi (1971) on the plant communities of the Pinto Basin, Gaines and Logan (1955) on the vegetation of the Providence Mountains and surrounding desert. In a study of the vegetation dynamics of the Joshua tree, Rowlands (1978) did a desert-wide study of the vegetation of the upland Mojave Desert.

ARIZONA

Few published works are concerned exclusively with the Mojave Desert in northwestern Arizona. Rowlands (1978) studied sites near Congress Junction, Yucca, and Dolan Springs, VanDevender (1973) discussed the present vegetation of northwestern Arizona as it relates to the paleoecology of Late Pleistocene plant communities. Kearney and Peebles (1960), Nichol (1937), and Lowe (1964) treat the vegetation of the entire state.

NEVADA

Clokey (1951) published a flora of the Spring Mountains, Beatley (1976) a treatise on the vascular plants and plant ecology of the Nevada Test Site and south-central Nevada, and Bradley and Deacon (1967) described the biotic communities of southern Nevada. Although primarily concerned with the flora and vegetation of the Great Basin, Tidestrom (1925) and Conquist et al. (1972) discuss the transitional zones between the Mojave and Great Basin deserts.

Vegetation Classifications

Many authors have devised classification systems to facilitate the study of vegetation, and the Mojave Desert has not been spared. These systems will be considered with reference to the major regions of the Mojave Desert.

MOJAVE DESERT REGIONS IN CALIFORNIA

Table 4-5 summarizes eight different classifications of Mojave Desert vegetation of California. Because Rowlands's (1980) is the most inclusive, it has been used as a baseline, and the other classifications have been compared to it. Nomenclature follows that of Rowlands (1980). It is readily apparent that the authors disagree on the number of vegetation types and plant communities. The number varies from six to twenty-

seven, with only a few plant community types being agreed upon completely. The reasons for the lack of agreement vary from basic philosophic differences to minor disagreements on details. There are as many ideas as there are authors.

NORTHWESTERN ARIZONA

Lowe (1964) recognizes five association types of Mojave Desert vegetation in northwest Arizona: creosote bush, Joshua tree, blackbrush, saltbush, and bladder sage. Riparian tree species in this Mojave Desert region include desert willow (*Chilopsis linearis*), western honey mesquite (*Prosopis juliflora, torreyana*) and catclaw (*Acacia greggii*). Lowe and Brown (1973) refer to this vegetation as Mojave Desert scrub occurring at elevations of 240 and 1500 m with a mean annual precipitation of between 125 mm and 275 mm.

NEVADA

Beatley (1976) recognized the following vegetation types and associations in central Nevada and the Nevada Test Site.

I.   Mojave Desert
  A.   Bajadas
      Creosote bush, bursage
      Creosote bush, Anderson boxthorn, hopsage
      Creosote bush, shadscale

  B.   Mountains
      Shadscale, desert holly

  C.   Arroyos

  D.   Springs and seeps
      Ash, screwbean
      Shadscale
      Mesquite

II.   Transition desert
  A.   Upper Bajadas
      Blackbrush
      Creosote bush, hopsage, Anderson boxthorn

  B.   Lower bajadas
      Hopsage, Anderson boxthorn
      Rabbit thorn, hopsage
      Shockley Lycium, shadscale

III.   Great Basin Desert
      Shadscale
      Four winged saltbush
      Great Basin sagebrush
      Black sagebrush
      Sagebrush, pinyon, juniper
      Great Basin sagebrush, mountain mahogany
      White fir

TABLE 4-5
Selected classifications of California Desert vegetation

PLANT COMMUNITY	ROWLANDS (1980)	KÜCHLER (1977)	KNAPP (1965)	THORNE (1976)	JAEGER AND SMITH (1966)	KÜCHLER (1964)	MUNZ AND KECK (1959)	ORNDUFF (1974)
Sagebrush Scrub	x	x	x	x	x	x	x	x
Blackbrush Scrub	x	x	x	x				
Hopsage Scrub	x							
Shadscale Scrub	x	x	x	x	x	x	x	x
Desert Holly Scrub	x							
Mojave Saltbush-Allscale Scrub	x							
Allscale-Alkali Scrub } (Alkali Sink)	x	x	x }	x	x	x	x	x
Iodinebush-Alkali Scrub }	x	x	x }				x	x
Creosotebush Scrub	x	x[a]	x	x	x	x[a]	x	x
Succulent Scrub	x	x		x[b]				
Joshua Tree Woodland		x	x	x	x		x	x
Utah Juniper-One-leaf Pinyon Woodland[c]	x }							
California Juniper-One-leaf Pinyon Woodland[c]	x }	x	x	x	x	x	x	x
California Juniper-Four-leaf Pinyon Woodland[c]	x }							
White Fir Forest Enclave	x							
Subalpine Forest (inc. Bristlecone Pine)	x							
Foothill Paloverde-Saguaro Woodland	x					x		
Paloverde-Ironwood-Smoketree Woodland	x	x	x	x	x			
Mesquite Thicket	x		x					
Cottonwood-Willow-Mesquite Bottomland	x				x			
Cottonwood-Willow Streamside Woodland	x	x		x				
Washington Fan Palm Oasis	x	x	x	x				
Indian Rice Grass Scrub-Steppe	x							
Desert Needlegrass Scrub-Steppe	x							
Big Galleta Scrub-Steppe	x							
Galleta-Blue Grama Scrub-Steppe	x							
Saltgrass Meadow	x		x	x				
Calciphyte Saxicole Subscrub	x			x				
Non-Calciphyte Saxicole Subscrub	x			x				
Desert Psammophytic Scrub Complex	x	x	x		x			
Desert Urban					x			
Desert Rural					x			
Winter Annual Vegetation			x					
Summer Annual Vegetation			x					

SOURCE: Modified from Johnson, H.B. 1976. Vegetation and plant communities of southern California deserts—a functional view. In *Plant communities of southern California.* ed. by J. Latting. *California Native Plant Society Special Publ.* no. 2. table 2, pp. 133-13 reprinted with permission.

[a] Küchler subdivides the creosotebush Scrub type into Mojave Desert Creosotebush Scrub and Sonoran Desert Creosotebush Scrub and Sonoran Desert Creosotebush Scrub (Creosotebush-Bursage).

[b] Thorne subdivides the Succulent Scrub type into Stem-Succulent (Cacti and Yuccas) and semi Succulent (Agaves) types.

[c] These are referred to collectively by all other authors, save Rowlands (1980), as Pinyon-Juniper Woodland.

[d] Not Mojave desert types.

Bradley and Deacon (1967) in southern Nevada recognized the following desert vegetation types.

I. Terrestrial
  A. Zonal communities
    1. Desert shrub vegetation
      Creosote bush
      Blackbrush
    2. Woodland vegetation
      Juniper-pinyon
  B. Transzonal communities
    1. Shrub and woodland vegetation
      Desert riparian
      Saltbush

II. Hydric and Aquatic
    Desert spring and marsh
    Stream riparian

Fowler and Koch in chapter 2 of this book discuss the ecosystems found in the Great Basin, including the southern Nevada section of the Mojave Desert. They have used Whittaker's (1970) division of ecosystems as the basis for their discussion:

Desert shrub ecosystem
  Northern desert shrub zone
    Sagebrush community
    Rabbitbrush community
    Shadscale community
    Winter fat community
    Hopsage-blackbrush community
    Mat saltbrush community
    Gray molly community
  Salt desert shrub zone
    Greasewood community
    Greasewood-shadscale community
    Saline-alkaline community
    Rabbitbrush community
  Southern desert shrub zone
    Desert saltbush community
    Creosote bush community
    Joshua tree community
    Mesquite community
  Transition desert shrub zone
    Creosote bush-boxthorn community
    Hop sage-boxthorn community
    Boxthorn-shadscale community
    Sagebrush community
  Playa and dry lake zone

Refer to chapter 2 for discussions of each of these zones and communities.

Rowlands (1980) discussed the vegetational attributes of the California regions of the Mojave Desert in terms of vegetation types. A list of these vegetation types, together with a summary of the ranges of climatological variables associated with each, is shown in table 4-6. A brief summary of each type follows.

I. Desert Scrub Complex
  A. Great Basin Subcomplex
    1. Sagebrush Scrub
    Primary species: *Artemisia tridentata*—Big Sagebrush
      Artemisia nova
    Associated species: *Juniperus osteosperma*
      Pinus monophylla
      Oryzopsis hymenoides
      Chrysothamnus viscidiflores
      Coleogyne ramosissima
      Gutierrezia sarothrae
    Soil types: *A. tridentata*—deep, well-drained soil
      A. nova—shallow, heavier soils
    General location: Panamint Mountains
      White Mountains
      New York Mountains
      Inigo Mountains
    2. Blackbrush Scrub
    Primary species: *Coleogyne ramosissima*—Blackbrush
    Associated species: *Ambrosia dumosa*
      Atriplex confertifolia
      Ephedra nevadensis
      Lycium pallidum
    Soil types: rocky, well-drained soil
    General location: rocky bajadas above 1000 m
    3. Hopsage Scrub
    Primary species: *Grayia spinosa*—Hopsage
      Lycium pallidum
    Associated species: *Lycium andersoni*
      Lycium shockleyi
      Larrea tridentata
      Ambrosia dumosa
      Yucca brevifolia
    Soil types: sandy, loamy, moderately rocky
  B. Saline-Alkali Scrub
    1. Shadscale scrub
    Primary species: *Atriplex confertifolia*—Shadscale
    Associated species: *Atriplex canescens*
      Yucca brevifolia
      Garotia lanata
      Artemisia spinescens

TABLE 4-6
Vegetation types within the California Desert, with a summary of the ranges of climatological variables associated with each.

VEGETATIONAL CATEGORIES	MEAN ANNUAL PRECIPITATION (MM)		TEMPERATURE (°C) MEAN JANUARY MINIMA		MEAN JULY MAXIMA		POTE/PPT		APPROX. ELEVATION RANGE (x100M)	
	LL	UL	LL	UL	LL	UL	LL	UL	LL	UL
Desert Scrub Complex										
Great Basin Subcomplex										
Sagebrush Scrub	175	325	−12	−4	25	36	2	5	12	26(30)
Blackbrush Scrub	150	240	− 8	−4	29	37	3	7	10	20
Hopsage Scrub	150	240	− 8	−2	29	37	3	7	10	20
Saline-Alkali Scrub Subcomplex										
Shadscale Scrub	130	225	− 8	−4	31	37	3	7	10	18
Desert Holly Scrub	42	90	0	5	42	47	20	32	− .8	4
Mojave Saltbush-Allscale Scrub	110	150	− 1	1	37	40	6	8	6	10
Allscale-Alkali Scrub	82	170	− 5	5	36	43	8	20	− .8	12(18)
Mojave-Colorado Desert Scrub Subcomplex										
Creosote Bush Scrub	42	275	− 6	6	34	47	4	32	− .7	13
Cheesebush Scrub	42	275	−10	6	30	47	3	32	− .7	20
Succulent Scrub	150	275	− 8	−2	29	47	2	7	10	20
Xeric-Conifer Woodland/Forest Complex										
Xeric Conifer Woodland Subcomplex										
Utah Juniper—One-leaf Pinyon Woodland	175	375	−13	−4	23	36	1	4	15	30
California Juniper—One-leaf Pinyon Woodland	175	400	− 9	−2	34	38	1	4	12	18
California Juniper—Four-leaf Pinyon Woodland	225	400	− 9	−1	35	39	1	4	11	17
Desert Montane Forest Subcomplex										
White-Fir Forest Enclave	250	325	−10	−7	26	30	1.5	3	19	24
Subalpine Forest (inc. Bristlecone Pine)	370	440	−17	−12	15	21	0.5	1	29	35

SOURCE: Rowlands, P. 1980. Vegetational attributes of the California Desert. In *The California desert: An introduction to its resources and man's impact*, ed. J. Laiting. *California Native Plant Soc. Special Publ.* no. 5, table 1. Reprinted with permission.

NOTE: LL = lower limit; UL, upper limit.
 An adequate synecological analysis should result in substantial subdivision of these types.

TABLE 4-6 (continued)
Vegetation types within the California Desert, with a summary of the ranges of
climatological variables associated with each.

VEGETATIONAL CATEGORIES	MEAN ANNUAL PRECIPITATION (MM)		TEMPERATURE (°C)				POTE/PPT		APPROX. ELEVATION RANGE (x100M)	
			MEAN JANUARY MINIMA		MEAN JULY MAXIMA					
	LL	UL	LL	UL	LL	UL	LL	UL	LL	UL
Desert Microphyll Woodland Complex										
Paloverde Microphyll Woodland Subcomplex										
Foothill Paloverde-Sauaro Woodland	115	160	1	6	40	44	10	12	3	4
Blue Paloverde-Ironwood-Smoketree Woodland	80	160	1	6	40	44	10	20	0	8
Mesquite Microphyll Woodland Subcomplex	42	160	− 2	6	40	47	8	32	− .8	8
Streamside and Oasis Woodland Complex										
Streamside Woodland Subcomplex										
Cottonwood-Willow Mesquite Bottomland	80	160	− 4	6	35	42	5	17	0	10
Cottonwood-Willow-Streamside Woodland	125	250	− 7	1	30	38	3	9	8	20
Desert Oasis Woodland Subcomplex										
Washington Fan Palm Oasis	80	150	1	6	40	44	10	15	0	10
Desert-Semidesert Scrub-Steppe Complex										
Desert-Semidesert Scrub-Steppe Subcomplex										
Indian Rice Grass Scrub-Steppe	120	300	− 9	0	28	40	2	8	6	23
Desert Needlegrass Scrub-Steppe	120	250	− 9	−2	30	38	2	5	10	20
Big Galleta Scrub-Steppe	110−(80)	250	− 4	3	35	44	3	8(15)	3(0)	13
Galleta-Blue Gramma Scrub-Steppe	175	300	− 9	−3	28	36	2	4	12	23
Desert Alkali Grassland Subcomplex										
Saltgrass Meadow	42	120	− 5	5	38	47	8	32	− .8	10
Desert Saxicole Subscrub Complex										
Calciphyte Saxicole Subcomplex										
Calciphyte Saxicole Subscrub	100	300	− 9	0	26	38	2	10	6	24
Noncalciphyte saxicole Subcomplex										
Noncalciphyte Saxicole Subscrub	100	300	− 9	0	26	38	2	10	6	24
Desert Psammophytic (Sand Dune) Complex	42	150	− 4	6	37	47	7	32	0	10

Soil types: heavy, rocky soils on steep mountain slopes
General location: eastern Mojave desert region in California and southern Nevada; northern Mojave desert region: Death Valley
                    Owens Valley
                    Ingo Range
                    Panamint Range
                    Black and Funeral mountains

2. Desert Holly Scrub
Primary species: *Atriplex hymenelytra*—Desert Holly
Associated species: *Atriplex polycarpa*
                  Atriplex confertifolia
                  Tidestromia oblongifolia
Soil types: saline gravel fans
    high proportion of carbonates
General location: bottom of Death Valley

3. Mojave Saltbush-Allscale Scrub
Primary species: *Atriplex spinifera*—Mojave Salt-bush
Associated species: *Atriplex polycarpa*
                  Ceratoides lanata
                  Tetradymia glabrata
                  Yucca brevifolia
                  Soil types: mildly saline soils
General location: western Mojave desert region, Mojave to Barstow

4. Allscale-Alkali Scrub
Primary species: *Atriplex polycarpa*—Allscale
Associated species: *Suaeda fruticosa*
                  Atriplex confertifolia
                  Atriplex canescens
                  Atriplex parryi
Soil types: saline soils to 2.5 percent salt
General location: washes on Nevada test site
                    Saline valley salt sink

5. Iodinebush-Alkali Scrub
Primary species: *Allenrolfea occidentalis*—Iodine-bush
Associated species: *Suaeda torreyana*
                  Sporobolus airoides
                  Juncus cooperi
                  Salicornia utahensis
                  Atriplex spp.
Soil types: highly saline soils; groundwater salt content as high as 6 percent
General location: widely distributed in playas and desert sinks in western, northern, and eastern Mojave regions

C. Mojave-Colorado Desert Scrub
1. Creosotebush Scrub
Primary species: *Larrea*—Creosotebush

Associated species: *Ambrosia dumosa*
                  Encelia farinosa
                  Hymenoclea salsola
                  Lycium andersoni
                  Opuntia spp.
                  Yucca spp.
Soil types: wide range from finely divided, poorly drained soils to coarser soil types
General location: widely distributed throughout the Mojave Desert below 1200 m

2. Cheesebush Scrub
Primary species: *Hymenoclea salsola*—Cheesebush
Associated species: *Cassia armata*
                  Atriplex spp.
                  Ambrosia eriocentra
                  Brickellia incana
Soil types: sandy washes, gravel pans
General location: widely distributed in dry washes and also on disturbed sites

3. Succulent Scrub
Component species: *Yucca spp.*
                  Agave spp.
                  Nolina spp.
                  Echinocereus spp.
                  Ferocactus spp.
Soil types: bajada, alluvial fans with well-drained soil
General location: dependent on summer rains; prevalent in eastern Mojave regions in Arizona and California

II. Xeric Conifer Woodland/Forest
  A. Desert Montane Forest
    1. Limber pine woodland
Primary species: *Pinus flexilis*—Limber pine
Associated species: *Juniperus communis*
                  Pinus monophylla
                  Acer glabrum
                  Ribes cereum
                  Pinus longaeva
Soil types: shallow rocky soils
General location: above 2950 m in Panamint Mountains
                    Inyo mountains in California
                    Spring and Sheep ranges in Nevada

2. Great Basin Bristlecone Pine Woodland
Primary species: *Pinus longaeva*—Bristlecone Pine
Associated species: *Pinus flexilis*
Soil types: shallow rocky soils
General location: highest mountain peaks 2750-3500 m Panamint mountains Inyo Mountains in California
                    Spring and Sheep ranges in Nevada

3. White Fir-Pine Forest
Primary species: *Abies concolor*—White Fir
Associated species: *Populus tremuloides*
Juniperus osteosperma
Pinus flexilis
Acer glabrum
Soil types: granitic or limestone soils
General location: 1950-2400 m; Clark, Kingston, and
New York mountains in California
Spring, Sheep, and Virgin mountains
in Nevada

B. Xeric Conifer (Juniper-Pinyon) subcomplex
1. Utah Juniper-One Leaf Pinyon Woodland
Primary species: *Juniperus osteosperma*—Utah juniper
Pinus monophylla—One leaf pinyon
Associated species: *Yucca brevifolia*
Coleogyne ramosissima
Artemisia tridentata
Echinocarpus engelmannii
Yucca baccata
General location: between 1200 and 2200 m on desert
mountain ranges; eastern Mojave region
2. California Juniper-One Leaf Pinyon Woodland
Primary species: *Juniperus californica*—California Juniper
Pinus monophylla-One leaf pinyon
Associated species: *Yucca brevifolia*
Canotia holacantha (Arizona only)
Larrea tridentata
Quercus turbinella
Stipa speciosa
General location: Western Mojave region
eastern Mojave region to Granite
Mountains
northwestern Arizona section

III. Desert Microphyll Woodland
A. Mesquite Thickets
Primary species: *Prosopis pubescens*
Prosopis glandulosa
Associated species: *Atriplex spp.*
Suaeda spp.
Pluchea sericea
Tamarix spp.
Soil types: highly saline soils
General location: desert seeps; playas; floodplains;
below 800 m

IV. Streamside Woodland, Oasis Woodland
A. Streamside Woodland
1. Cottonwood-Willow-Mesquite Bottomland

Primary species: *Salix lasiolepis*
Salix exigua
Salix goodingii
Populus fremontii
Prosopis spp.
Associated species: *Tamarix spp.*
Primary species: *Salix lasiolepis*
Salix goodingii
Salix exigua
Populus fremontii
Populus macdougallii
Prosopis glandulosa
Prosopis pubescens
Associated species: *Tamarix spp.*
Phragmites australis
Pluchea sericea
Typha spp.
Carex spp.
Juncus spp.
Soil types: sandy; sandy loam with coarse fragments
General location: floodplains and bottomlands of
Colorado, Mojave, and Virgin rivers
2. Cottonwood-Willow Streamside Woodland
Typha spp.
Pluchea sericea
Soil types: sandy, sandy loam
General location: floodplains and bottomlands of
smaller streams; springs

B. Desert Oasis Woodland
Primary species: *Washingtonia filifera*
Associated species: *Pluchea sericea*
Atriplex polycarpa
Sporobolus airoides
Soil type: saline
General location: isolated stands

V. Desert and Semidesert Grassland
A. Desert-Semidesert Scrub Steppe
1. Indian Rice Grass Scrub-Steppe
Primary species: *Oryzopsis hymenoides*—Indian Rice
Grass
Stipa speciosa
Associated species: *Larrea tridentata*
Ambrosia dumosa
Artemisia tridentata
Yucca brevifolia
Soil types: sand sheets; sandy, well-drained soils
General location: western Mojave region above
1500 m
2. Desert Needle-Grass Scrub Steppe
Primary species: *Stipa speciosa*—Desert Needle
Grass

Associated species: *Yucca brevifolia*
Juniperus californica
Lycium andersonii
General location: Joshua Tree National Monument desert-facing slopes of mountains
3. Big Galleta Scrub-Steppe
Primary species: *Hilaria rigida*—Big Galleta Grass
Bouteloua erippoda
Muhlenbergia porteri
Associated species: *Juniperus osteosperma*
Yucca brevifolia
Juniperus californica
Stipa speciosa
Oryzopsis hymenoides
Soil type: sand sheets; sandy, well-drained soils
General location: eastern Mojave desert region
Joshua Tree National Monument
Mojave Desert region east of Lucerne Valley and Barstow
4. Galleta Scrub Steppe
Primary species: *Hilaria jamesii*—Galleta Grass
Bouteloua gracilis
Associated species: *Oryzopsis hymenoides*
Sitanion hystrix
Juniperus osteosperma
Yucca brevifolia
Chrysothamnus viscidiflorus
General location: above 1400 m northern and eastern Mojave desert
5. Saltgrass Meadow
Primary species: *Distichlis spicata*—Saltgrass
Sporobolus airoides
Associated species: *Allenrolfea occidentalis*
Juncus cooperi
Prosopis pubescens
Anemopsis californica
Soil type: saline
General location: near springs and seeps

VI. Desert Saxicole Complex
A. Calciphyte-Saxicole
1. Calciphyte Saxicole Subscrub
Primary species: *Astragalus funereus*
Astragalus panamintensis
Cercocarcus intricatus
Eriogonum spp.
Associated species: a very wide range of calcium-loving plants
Soil type: rock faces and crevices of calciferous outcroppings such as dolomite and limestone
General location: eastern and northern Mojave Desert regions in mountains: Pana-

mint, Inyo, Grapevine, Kingston, Providence, Sheep, Spring.
2. Noncalciphyte Saxicole Subscrub
Primary species: *Peucephyllum schottii*
Associated species: *Perityle emoryi*
Brickellia desertorum
Penstemon spp.
Cheilanthus spp.
Soil type: noncalciferous soils
General location: widely distributed in all regions of the Mojave Desert.

VII. Desert Psammophyte Complex
This is the vegetation type found on sand dunes. It is very diverse and is difficult to classify. Table 4-7 summarizes the perennial plant species for dune systems in California.

CREOSOTE BUSH

Creosote bush (*Larrea tridentata*) is the principal plant species of the Mojave Desert. It ranks first in ground cover and in geographic extent. The Mojave race differs from those found in the Chihuahuan and Sonoran deserts in chromosome number, it being a hexaploid while populations from the Sonoran Desert are generally tetraploids and those from the Chihuanhuan Desert are diploids (Yang 1970).

A distinctive feature of creosote bush is the frequent development of doughnut-shaped clones of extreme age (Johnson, Vasek, and Yonkers 1975; Vasek 1980). This growth pattern is clearly evident from aerial photos (figure 4-14). Clones may reach sizes greater than 25 m in diameter. The slow growth of these clones in a radial direction is evident from matched aerial photographs covering a span of thirty years, differences in clone size above and below the shoreline of Lake Cahuilla, which existed until about four hundred years ago, and observations and interpretations of the annual increment and direction of size increases (Vasek et al. 1975, Sternberg 1976; Vasek 1980). The last observations included morphological interpretations of the cloning process, as well as radiocarbon dating of remnant wood left behind as the clones expand outward. Isozyme analysis has shown that most large circular clumps with hollow centers are indeed clonal in origin and not the result of multiple seedling establishment (Sternberg 1976).

Although most of the wood from the center portions of the larger clones had decomposed and was unavailable for dating, some preserved remnants closer to the living parts of the clones were determined to be as old as 730 years. Extrapolated ages of some of the larger clones yield estimates of thousands of years. One particularly large clone studied by Sternberg (1976) and Vasek (1980) was estimated by Vasek to be 11,700 years old. Such age estimates are particularly interesting for Mojave Desert creosote bush since it appears to be a relative newcomer to the Mojave Desert (Wells and Hunziker 1976). These dates may represent the initial introduction (Johnson 1976).

The expanding clonal growth pattern also appears to occur in other Mojave Desert shrubs such as Mojave yucca (*Yucca schidigera*) (figure 4-15 and mesquite (*Prosopis juliflora*). Such a growth habit has important implications for the origin and dynamics of desert plant communities.

ANIMALS

Fowler and Koch (chapter 2 of this book) have provided extensive reviews of the literature for each of the groups of vertebrate animals for the Great Basin area including the southern Nevada section of the Mojave Desert. Bradley and Deacon (1967) reviewed the literature of the biota of southern Nevada. References that seem particularly useful include those of Allred et al. (1953, 1960, 1963) at the Nevada Test Site, Banta's studies of the reptiles of the Great Basin (1961, 1963, 1964, 1965, 1967), Hayward et al. (1963) on the birds of the Nevada Test Site, Tanner (1978) on the zoogeography of Great Basin reptiles, and Bradley (1964, 1966) on selected mammals.

Bradley and Deacon (1967) studying the biotic communities of southern Nevada reported a total of thirty species of reptiles, forty species of birds and forty-four species of mammals in the desert biotic communities (table 4-8). In considering the distribution of the various animal groups within the biotic communities, they reported that 77 per cent of the reptiles in southern Nevada are found in the creosote bush community with 36.7 percent being restricted to this community. Sixty percent are shared with the Blackbrush community, 60.2 percent of the mammals are found in the creosote bush community, with 27 percent being restricted to that community. The largest number of bird species was found in the desert spring and stream riparian communities (table 4-9).

Laudenslayer and Boyer (1980) studied the mammals of the California desert, which includes the California portions of the Mojave and Colorado desert, and reported ninety-four mammalian species representing seven orders. The order Rodentia made up half of the mammalian species reported. The order Chiroptera was second in number of species, with 21 species of bats being reported. England and Laudenslayer (1980) reported 427 species of birds in the California desert.

Ohmart and Andersen in appendix 10-A of this volume list the amphibians and reptiles occurring in the deserts of the southwest including the Mojave desert. In appendix 10-B the same authors list the birds found in the Mojave and other deserts of the southwest.

Checklists of animals in Death Valley National Monument at the northern border of the Mojave Desert list 6 species of fish, 3 species of amphibians, 36 species of reptiles, 53 species of mammals, and 258 species of birds. These checklists are reproduced in appendixes 4B and 4-C through the courtesy of the Death Valley Monument Natural History Association.

Checklists of animals in Joshua Tree National Monuments, which is located at the southern border of the Mojave Desert, list 2 species of amphibians, 37 species of reptiles, 46 species of mammals, and 219 species of birds. These checklists are reproduced in appendixes 4-D, 4-E, and 4-F through the courtesy of the Joshua Tree Monument Natural History Association.

Some species deserve special mention. The desert tortoise (*Gopherus agassiz,*) a unique desert species, is under increasing pressure from increasing use and development of the Mojave Desert. There are four major population centers of the tortoise: Fremont-Stoddard Valley, Ivanpah Valley, and Fenner-Che-mehuevi valleys. Minor populations are found in Lucerne, Johnson, and Shadow valleys. In some areas tortoises reach densities of 20 to 250 per square mile. Because they are grazers, they compete with cattle for food and are considered in making up grazing allotments. A desert tortoise preserve has been established near Lancaster, California to protect this sensitive species.

The desert bighorn sheep (*Ovis canadensis nelsoni*) is found in a number of mountain ranges in the Mojave Desert. Their populations have decreased over the years in most areas of their range and have been eliminated from some of the mountain areas. Poaching, declining habitat, and competition with other animals have been factors in their decline. This decline has occurred in the face of extensive efforts to manage their populations.

One of the animals believed to be an important factor in the decline of the desert bighorn is the feral burro. Burros have been part of the desert scene ever since prospectors began exploring. Some burros escaped; others were turned loose when the claim did not prove out, and they became wild or feral. Over the years their population has increased rapidly, and since 1976 it has doubled. Their population is increasing at the rate of 10 to 20 percent per year. The estimated number in the desert is approximately 8800 animals, with particularly high concentrations in the northern and eastern Mojave regions. This has resulted in severe overuse of the vegetation in some areas. They compete with bighorn sheep, mule deer, and livestock for food and water. Population control measures include shooting to remove excess animals and the Adopt-a-Burro program where the animals are rounded up and held for adoption.

Animals that are considered to be sensitive species because of limited distribution or specialized habitat requirements include the Panamint kangaroo rat, which is found only in the Panamint, Argus, and New York mountains, the yellow-eared pocket mouse, which is found only in a narrow band on the east slope of the Sierra Nevada in Kern County, California, and the golden eagle.

TABLE 4-7
Perennial plant species presence data for fourteen California Desert dune systems

PLANT SPECIES	NORTH		SAND DUNE LOCATION											SOUTH
	EUV	SAV	OLD	PVD	DUD	KED	BAD	WPD	DLD	CVD	RVD	CHV	COV	ALD
Ammobroma sonorae												*		*
Acamptopappus sphaerocephalus						*								
Acacia greggii						*								
Ambrosia dumosa		*												*
Aristida californica											*			*
Astragalus aginus						*								
A. lentiginosus var. borreganus										*				*
A. lentiginosus var. micans	*			*										
A. magdalenae var. piersonii														*
A. subulonum	*													
Atriplex canescens	*		*				*	*		*		*	*	
A. confertifolia	*													
A. hymenelytra		*												
A. parryi			*		*									
A. polycarpa	*	*						*	*			*	*	
A. truncata	*													
Baileya pauciradiata									*	*			*	
B. pleniradiata														*
Cercidium floridum														*
Chaetadelphia wheeleri	*													
Chilopsis linearis						*						*		*
Chrysothamnus teretifolius									*					
Coldenia plicatata	*			*	*	*			*	*	*	*		*
Croton californica						*								*
C. wigginsii												*		*
Dalea emoryii										*	*	*	*	*
D. fremontii	*													
D. mollisima											*			
D. polyadenia	*								*		*			
Distichlis spicata			*											
Encelia frutescens										*				*
Ephedra californica											*			
E. trifurca											*			*
Eriogonum deserticola														*
E. inflatum	*				*	*								*
Eriastrum densifolium	*					*								*

SOURCE: Rowlands, P. 1980. Vegetational attributes of the California Desert. In *The California desert: An introduction to its resources and man's impact*, ed. J. Latting, *California Native Plant Soc. Special Publ.* No. 5. Table 2 reprinted with permission.

TABLE 4-7 (continued)
Perennial plant species presence data for fourteen California Desert dune systems

PLANT SPECIES	NORTH					SAND DUNE LOCATION								SOUTH
	EUV	SAV	OLD	PVD	DUD	KED	BAD	WPD	DLD	CVD	RVD	CHV	COV	ALD
Fagonia pachyacantha						*								*
Haplopappus acradenius						*	*					*	*	
Helianthus nivens var. *tephrotes*														*
Heliotropium curassavicum			*											
Hesperocallis undulata					*	*				*	*	*		*
Hilaria rigida						*			*	*	*	*		*
Hymenoclea salsola	*							*						
Larrea tridentata	*	*		*	*	*	*		*	*	*	*	*	*
Lyrocarpa coulteri														*
Nitrophila occidentalis			*											
Oenothera avita ssp. *eurekensis*	*													
Olneya tesota												*		*
Opuntia echinocarpa						*								
Orobanche cooperi											*			*
Oryzopsis hymenoides	*					*								
Palafoxia arida ssp *arida*	*					*			*		*	*	*	*
P. arida ssp. *gigantea*														*
Panicum urvilleanum						*						*		*
Penstemon thurberi						*								
Petalonyx thurberi		*			*	*						*	*	*
Phragmites australis												*		
Pluchea sericea		*										*		
Prosopis pubescens		*												
Prosopis glandulosa		*				*						*	*	*
Rumex hymenosepalus						*								
Sarcobatus vermiculatus			*											
Sphaeralcea ambigua	*													
S. emoryi														*
Stanleya pinnata ssp. *inyoensis*	*													
Stipa speciosa												*		
Suaeda fruticosa		*												
Suaaeda torryana		*						*						
Swallenia alexandrea	*													
Tamarisk aphylla												*		*
Tidestromia oblongifolia					*									
Yucca brevifolia								*						

NOTE: Sources: Bureau of Land Management Desert Plan Staff, Dean (1978), Westec (1977). Dune Systems are arranged in the table more or less in order from north to south. EUV = Eureka Valley Dunes, SAV = Saline Valley Dunes, OLD = Olancha Dunes, PAV = Panamint Valley Dunes, DUD = Dumont Dunes, KED = Kelso Dunes, BAD = Barstow Area Dunes, WPD = Wilsona Ranch-Piute Butte Dunes, DLD = Dale Lake Dunes, CVD = Cadiz Valley Dunes, RVD = Rice Valley Dunes, CHV = Chuckwalla Valley Dunes, COV = Coachella Valley Dunes, ALD = Algodones Dunes.

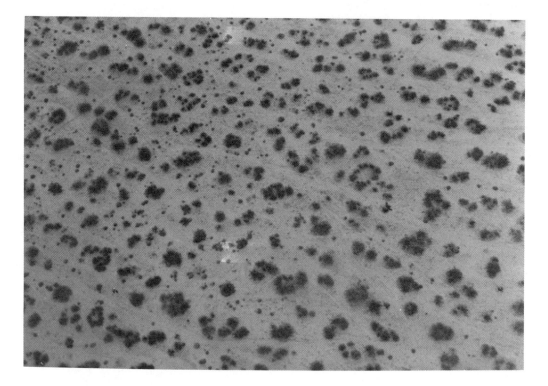

Figure 4-14. Aerial photograph showing doughnut-shaped clones of creosote bush, north of Lucerne Dry Lake. Photograph by H. Johnson.

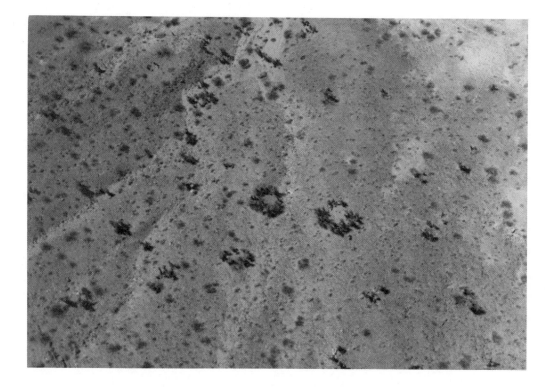

Figure 4-15. Aerial photograph showing doughnut-shaped clones of Mojave yucca. Fry Mountains, Lucerne Valley. Photograph by H. Johnson.

TABLE 4-8
A summary of the distribution of biota in the zonal biotic communities of southern Nevada

	DESERT COMMUNITIES			MONTANE COMMUNITIES					HYDRIC-AQUATIC COMMUNITIES				
	CR	BL	TOTAL	JP	FP	BR	PA	TOTAL	DS	SR	ST	LA	TOTAL
Vascular plants	256	185	311	258	275	75	13	414	21	36	3	0	50
Fish	0	0	0	0	0	0	0	0	20	0	21	17	41
Amphibians	0	0	0	0	0	0	0	0	7	7	7	3	9
Reptiles (total)	30	19	30	9	5	0	0	9	0	14	1	0	15
Turtles	1	1	1	0	0	0	0	0	0	0	1	0	1
Lizards	14	13	14	7	4	0	0	7	0	7	0	0	7
Snakes	15	5	15	2	1	0	0	2	0	7	0	0	7
Birds (total)	33	26	40	47	89	29	5	107	202	159	15	44	245
Permanent residents	8	6	9	12	15	7	0	24	22	18	0	2	26
Summer residents	6	8	11	13	31	18	0	24	19	20	3	0	28
Winter residents	10	7	10	3	4	0	0	5	65	40	7	27	71
Nonresidents	9	5	10	19	39	4	5	54	107	87	5	16	139
Mammals (total)	44	33	448	46	38	23	1	49	26	37	7	2	45
Insectivores	0	0	0	0	1	1	0	1	0	0	0	0	1
Bats	14	6	14	8	9	3	0	10	6	9	6	0	9
Rodents	16	14	18	22	15	10	1	22	10	18	1	1	19
Lagomorphs	2	2	2	3	3	2	0	3	2	2	0	0	2
Carnivores	9	6	9	9	7	4	0	9	4	6	0	1	9
Ungulates	3	5	5	4	3	3	0	4	4	2	0	0	5
Totals	363	263	429	360	407	127	19	579	276	253	54	70	405

SOURCE: The biotic communities of southern Nevada, W. Glen Bradley and James E. Deacon, *Nevada State Museum Anthropological Papers*, no. 13, part 4, October 1967. Table 4 reprinted with permission.

NOTE: Communities are grouped as desert, montane, and hydric-aquatic communities. Transzonal communities are not considered since they pass through two or more of the zonal communities. Code letters for the biotic communities are: CR = creosote bush, BL = blackbrush, JP = juniper-pinyon, FP = fir-pine, BR = bristlecone pine, PA = pseudo-alpine, DS = desert spring and marsh, SR = stream riparian, ST = stream, LA = lake.

TABLE 4-9
Total vascular plants and vertebrates found in biotic communities (in percentages)

	BIOTIC COMMUNITIES												
BIOTA	CR	BL	SA	DR	JP	RC	FP	BR	PA	DS	SR	ST	LA
Vascular plants	36.4	26.3	9.2	39.4	36.7	43.7	39.1	10.7	1.8	2.8	5.1	0.4	0
Fish	0	0	0	0	0	0	0	0	0	48.8	0	51.1	41.5
Amphibians	0	0	0	0	0	0	0	0	0	70.0	70.0	70.0	
Reptiles	77.0	48.8	20.2	56.5	23.1	15.4	12.8	0	0	0	35.9	2.5	0
Birds	11.4	8.9	4.1	19.0	16.2	12.8	30.8	10.0	1.7	69.5	54.9	5.2	15.2
Mammals	60.2	45.2	37.0	52.0	63.0	53.5	52.0	31.5	1.4	35.6	50.7	9.7	2.8
Total	31.4	22.7	9.7	34.0	31.1	33.7	35.2	11.0	1.6	23.8	21.9	4.7	6.0

SOURCE: The biotic communities of southern Nevada, W. Glen Bradley and James E. Deacon, *Nevada State Museum Anthropological Papers*, no. 13, part 4, October 1967. Table 4 reprinted with permission.

NOTE: Code letters are: CR = creosote bush, BL = blackbrush, SA = saltbrush, DR = desert riparian, JP = juniper-pinyon, RC = riparian and cliff face, FP = fir-pine, BR = bristlecone pine, PA = pseudo-alpine, DS = desert spring and marsh, SR = stream riparian, ST = stream, LA = lake.

ARCHAEOLOGY

Cultural developments proposed for the Mojave Desert during the earlier periods of human occupation are similar in many respects to those proposed for the adjoining Sonoran and Great Basin deserts. During the late periods, within the last several millenia or so, there are cultural patterns unique to the Mojave Desert, and the picture is further complicated by subregional variability. The salient characteristics of anthropological significance in the far western North American deserts include the examination of human societal adaptations to an often harsh and changing environment, cultural stability and persistence in some regions, and changes in other regions (cf. Aikens, 1978:71).

The location of the Mojave Desert between other larger deserts is significant in the development of cultural patterns due to the considerable human interaction that occurred. Proximity to the environmentally richer Central Valley of California, coastal areas, and the Southwest had a profound impact on the lifeways of the desert inhabitants as these populations expanded and interactions increased.

In prehistoric and early historic times, the Mojave Desert served as crossroads of travel, diffusion, and trade, due in some part to the predominance of the west to east-flowing Mojave River in an otherwise water-scarce area. In the last 1500 years influences from the advancing cultural developments in the American Southwest, Meso-America, and California coastal and Central Valley areas became evident. Here again variability is present on an east to west cline.

Development of a Cultural History

Early formal archaeological studies within the Mojave Desert beginning with the work of Malcolm Rogers (1929, 1939, 1966), Campbell (1931), Campbell and Amsden (1934), Campbell and Campbell (1935), Campbell et al. (1937), and Harrington (1933, 1957) concentrated on problems of cultural chronology and descriptive assessments of the kinds of cultural remains, their environmental setting and past to geologic and climatic events. Dating, the development of cultural chronologies, attempts to understand the relationship of prehistoric and historic human groups with their natural and cultural environment, recognition of processes of cultural change, and cultural resource management are some of the more important problems in today's archaeological work within the Mojave Desert. The predominant work to date has been oriented toward questions of sequential changes in technology and human settlement and subsistence patterns and the effects that environmental change, cultural intrusions, and internal developments played in these changes. While there is a proliferation of schemes, there is a continuing scarcity of data to back them up, particularly where earlier peoples are concerned. Among the more important chronologies are those of Rogers (1939), who established the working framework, and more recently Meighan (1959), Wallace (1962, 1978), Hester (1973), Bettinger and Taylor (1974), Taylor and Meighan (1978) and Warren and Crabtree (in press).

Early Man

When the subject of Early Man in the New World is broached, the Calico Site in the heart of the Mojave Desert stands out as being of great importance (Leakey et al. 1968, 1972; Schuiling 1979; Simpson 1979). The natural or cultural origin of the lithic materials of this Late Pleistocene site is the focus of considerable debate among scholars (Haynes 1969; Taylor and Payen 1979; Singer 1979; Payen, in press). Other proposed lithic industry sites of generally accepted human origin are the source of different controversies. Questions have been raised concerning their age (Pleistocene, Early Holocene, or later), function (quarry, big game hunting or campsite workshop), and cultural affiliation (pre projectile point or later; cf. Krieger 1962, 1964). These include the Lake Manly stone tools (Clements and Clements 1951), the Lake Manix lithic industry (Simpson 1958, 1960, 1961, 1965, 1976; Glennan 1976; and Alsoszatai-Petheo 1978), the Antelope Valley materials (Glennan 1971; Robinson et al. 1976), the core tool traditions of China Lake (Davis 1978) and Panamint Valley (Oavis 1970; Davis, Brott and Weide 1969), and the Malpais industry (Rogers 1939; Hayden 1976). These assemblages of materials have been characterized by stone "choppers" and bifaces, possible scrapers and cores and flakes. A recent work by Davis, Brown, and Nichols (1980) focuses on the problems involved in the Early Man debate in the California deserts and contiguous areas, the difficulties of identifying and interpreting stone reduction technology, dating quandaries, and complications of assessing geomorphic environmental associations.

Pluvial Lakes and Points

Most archaeologists, while noting the possibility of Pleistocene sites within the Mojave Desert, see the first evidence of human occupation in a series of distinct long-stemmed lanceolate projectile points (for example, at Lake Mojave), short-bladed and stemmed points (such as at Silver Lake), or fluted lanceolate points (such as at Clovis), all presumably used on darts or spears. In association are crescents and other less distinct flaked stone tools, such as spiked gravers, heavy core tools, specialized scrapers, and leaf-shaped knives all dating from terminal Pleistocene to Early Holocene times 9000 B.C. to approximately 5000 B.C. Human occupation and activities are viewed as adaptive responses by small groups to terminal Wisconsin warming and drying conditions (Antevs 1948, 1952, 1955; Mehringer 1977; Mifflin and Wheat 1979; Van Devender and Spaulding 1979).

As noted by Fowler (chapter 2 to this book), in the discussion of Great Basin archaeology, there is no overall agreement on the contemporaneity of fluted and nonfluted projectile points (also see Tuohy 1971). These and subsequent projectile point forms are the principal time markers within cultural sequences despite disagreement on the point taxonomies. Authors have assigned various designations to this time period depending on their interpretations of the evidence.

The major site locations identified for this period have been pluvial lakes (for example, Soda, Silver, Panamint, China, Ivanpah, Coyote, Troy, Manix) (Campbell et al. 1937; Adams 1938; Rogers 1939; Roberts 1940, 1951; Strong 1941; Brainerd 1953; Meighan 1954; Warren and True 1961; Simpson 1961, 1965; Warren and De Costa 1964; Heizer 1965; 1970; Woodward and Woodward 1966; Carter 1967; Warren 1967, 1970; Davis 1967, 1970, 1975, 1978; Davis et al. 1969; Ore and Warren 1971; Borden 1971; Warren and Ore 1978; Venner 1978). Findings around these lakes have led to the definition of a cultural tradition stemming from a pluvial lake adaptation (Bedwell 1973; Hester 1973). Tool inventories suggest big game hunting as well as lakeside subsistence. Certain recovery and recording biases no doubt have played a part in the lifeway constructs, and some scholars suspect that the subsistence economy and settlement pattern was more diverse, perhaps even seasonally and regionally varied (cf. Warren 1967; Hall and Parker 1975; Weide 1976; Heid, Warren, and Rocchio 1979).

Mojave Archaic

The apparent onset of increasing aridity in some regions of the desert west around 5000 B.C. coincides with observable changes in cultural material remains, reflecting more generalized hunting and gathering lifeways. This adaptive pattern, which included the first firm evidence of plant-food processing tools, has been termed Archaic by some authors (Jennings and Norbeck 1955; Shutler 1968; Hester 1973:125; Fowler, chapter 2). The inception of this lifeway pattern has been identified much earlier in parts of the Great Basin (Jennings 1957, 1958; Fowler, Chap. 2) than is apparent in the Mojave Desert. Authors variously contend that there was a period of virtual abandonment or drastic population decrease in the Mojave Desert until about 3000 B.C. (Wallace 1962:175, 1978:4; Kowta 1969) or that the earlier period, designated Lake Mojave or Mojave by many, continued until much more recently (Donnan 1964). Warren et al. (1980) have noted that full understanding of this question of sequence cannot be gained until the problems of chronology and classification of artifacts can be settled.

One of the proposed time markers for the initial Archaic period is the Pinto (Campbell and Amsden 1934; Amsden 1935) or Little Lake (Harrington 1957) projectile point. These dart or spear points are generally medium sized and shouldered with a basal notch (Warren in press). Silver Lake points apparently continue into this time frame as well (Harrington 1957; Davis 1969; Borden 1971).

The occurrence of these points coincides with the first substantiated evidence of milling tools, indicating that foraging and wild plant food processing was becoming an important subsistence activity in addition to big game hunting. The discovery of a series of approximately 6000-year-old human footprints in the muds along the Mojave river by Rector et al. (1979) may substantiate riparian area foraging as part of the subsistence routine. Although the evidence for major subsistence changes from earlier times is equivocal at best, widely scattered sites situated not only near lakes but also near springs and riparian areas and in uplands suggest an exploitive pattern less restrictive and localized than perhaps was the case earlier.

The similarity of Pinto-Little Lake points with proposed later period shouldered, "eared," or basally notched points (Elko and Humboldt series points) has caused confusion in the chronological ordering of cultural assemblages (Hall and Barker 1975:56-57). On the other hand Gypsum Cave points, which are triangular with a tapering stem and are named for a famous archaeological cave in southern Nevada (Harrington 1933), are generally accepted as distinct for this time frame. Those who classify points believe these Gypsum Cave specimens are associated with a change in cultural patterns dating from about 3000-2000 B.C. into the first millenium A.D. (Bettinger and Taylor 1974; Wallace 1978; Warren and Crabtree in press). An amelioration of climatic conditions is thought to have influenced population movements at this time. King (1976:27-28) suggested that cultural changes may have been the result of social changes influenced by the evolution of social systems in adjoining central and southern California and areas of the Southwest.

Site reports for this time period are numerous, reflecting a long history of interest in Mojave Desert archaeology (Harrington 1933, 1957; Smith et al. 1957; Wallace 1958; Shutler, Shutler, and Griffith 1950; Hunt 1960; Shutler 1961; Schroeder 1961; Lanning 1963: Hillebrand 1972, 1974; Panlaqui 1974). Plant foraging was becoming even more prominent as evidenced by a pronounced increase in milling tools compared to the frequency of flaked stone tools as in other areas of western North America. Other technological items present are large, triangular stone knife blades, flake scrapers in lesser quantities than from earlier times, an array of choppers, and heavy scraping tools and flaked stone drills. By this time rock art (petroglyphs and pictographs) of abstract geometric design or occasional naturalistic representations such as mountain sheep are found near water sources, caves, and in certain canyons (Smith et al. 1957; Rector 1976; Wallace 1978) (figure 4-16). These motifs have been interpreted variously as representing hunting, magic symbolism, or directional signs in conjunction with water sources. Archaeological testing of these hypotheses is only in the planning stages.

Although it may be a problem of sampling, cultural assemblages of this period are not as well defined for the Mojave

Desert as they are for the Great Basin. One of the better Mojave Desert collections, which includes many perishable items, was obtained from Newberry Cave (Smith et al. 1957). Materials from this shelter such as split twig mountain sheep figurines and other items suggest a ritual-religious association with hunting. Petroglyphs of mountain sheep, hunting scenes, shield figures, "medicine bags," and elaborate anthropomorphs from the Coso Range, a center of rock art in the Mojave Desert, may be the product of an elaborate and perhaps widespread sheep hunting cult in the western Mojave Desert with extensions throughout much of the Great Basin (Grant, Baird and Pringle 1968).

Toward the end of this little-known period, influences from the Southwest Basketmaker cultures become apparent within the eastern Mojave; an example is the split twig figurines. In the central and western Mojave, there is some evidence of influences from the developing Central Valley and coastal groups, particularly in terms of little understood trade networks such as obsidian and shell beads (Singer and Ericson 1977).

During this time habitation apparently lasted longer at certain preferred base camps or shelters as suggested by the trash accumulation. The cumulative effects of certain ecological changes prior to and during this time frame resulted in a number of areal and regional adaptive and technological stylistic shifts. For example, decreases in larger game animals, such as bighorn sheep, may have necessitated greater foraging.

Regional Diversity

During the last two millennia culture change becomes apparent. The archaeological record improves as the historic period is approached, population increase and involvement are apparent, diffusion and trade of material goods from surrounding regions are more evident, and increased numbers of desert areas exhibit not only use but distinctive use.

From about 0 A.D. until 1200 A.D. variability among the far western, central, and eastern regions of the Mojave Desert is most apparent. Robinson (1977) and Sutton (1979) commented on the marked increase in the number, kind, and complexity of archeological sites in the western region. Robinson (1977:47) noted that the relatively productive Antelope Valley region of the western Mojave is a highly favorable crossroads area for trade with nearby cultures. The result was the development of a complex and unique cultural system, a system efficient in subsistence matters and deeply involved in intercultural trade and diplomacy with groups to the north, northwest, and south. Some of the trade items appearing in sites include wealth or status items such as carved steatite and ornaments and beads of marine shells (*Haliotus* sp., *Mytilus* sp., *Tivella* sp., *Megathura* sp., and *Olivella* sp.).

Within the central Mojave Desert, occupation and activities centered on the resource-rich Mojave River and satellite camps at certain springs and ephemeral lakes (Sutton 1980). During

Figure 4-16. Photograph of abstract-geometric petroglyphs,
Cow Cove, San Bernardino County, California. Photograph by E. Ritter.

this period the bow and arrow was introduced, not only in the central region but throughout the Mojave Desert, as evidenced by small notched and stemmed projectile points of the Rose Spring and Eastgate series (Lanning 1963; Heizer and Baumhoff 1961). The effect of the bow and arrow, assuming increased hunting efficiency, must have been significant but is as yet poorly understood. Hall and Barker (1975:60) note that adverse effects on game populations in the north central Mojave Desert may correlate with the lexico-statistical indications of human movement out of this area about 900-1000 A.D. However, there are many conflicting linguistic interpretations. In general, the preceding cultural pattern appears to have changed little in the central region. Milling stones, incised stones and slate pendants, and both heavy and light-weight flaked stone tools continue to be in evidence. No doubt certain political and economic influences from East and West modifed adaptive patterns. The central Mojave appears to have been a transitional zone for cultural developments. Its inhabitants may have served as intermediaries or provided way stations for an increasing east-west trade network.

In the eastern Mojave region, Southwest Puebloan influences are greatest. Anasazi gray ware pottery is found in a number of eastern Mojave locations. The southwestern presence of pottery varies from numerous small, well-established settlements concentrated along the Muddy River in southern Nevada (Shutler 1961) to the occurrence of a few sherds of trade ware at small campsites in the central Mojave (Sutton 1980). Fowler (chapter 2 to this book) has discussed the southern Nevada and Utah situations. Rudy (1970) discussed the southwestern trade network, which extended across the Mojave Desert to the coast of California, Central Valley, and even down to the Gulf of California (Heizer 1941). Part of this trade and exploitation network included turquoise mining in the eastern Mojave (Eisen 1898; Rogers 1929; Murdock and Webb 1948; Sigleo 1974). Puebloan-type gray, black on gray, and black on white pottery sherds, grooved axes and crudely flaked axes, picks, and hammers all point to mining by or under the influence of Southwestern peoples. Southwestern influences have been reported on by Rogers (1939), Hunt (1960), Hunt and Hunt (1964), McKinney et al. (1971), Wallace and Wallace (1978), Smith (1963), Peck and Smith (1957), and Drover (1979). Aschmann (1958) and Warren et al. (1980) have noted that wetter than normal climate along with southwestern influences may have brought about agricultural practices as a supplement to hunting and gathering in favorable locations such as the Mojave sink.

Influences from the southwest also came from the lower Colorado River. Davis (1962), Donnan (1964), and Drover (1979) have excavated and surveyed in the eastern Mojave section of California and have seen extensions of the lowland Patayan culture of the Sonoran Desert beginning about 800 A.D. This influence is characterized by Colorado Buff and Tizon Brown ceramic wares.

Cultural Expansion

Beginning rather suddenly about 1000 A.D. there was a dramatic change in the archaeology of the Mojave Desert. This transformation is an expression of a much broader panwestern North American cultural modification, which persisted until the historic period. Within the Mojave Desert, this change was precipitated by the expansion of Numic (Shoshonean) speakers into the area, the result of climatic stress, population pressures, shifting social systems, or other yet-unknown factors. These groups are thought to be the predecessors of the historic Paiute, Kawaiisu, Panamint Shoshone, and Kitanemuk groups.

The diagnostic time markers of this period include the small desert side notched and triangular cottonwood projectile points and regionally distributed utilitarian brown ware ceramics. The variability of projectile points may represent different subperiods or spatial differences (Rector et al. 1979:16). This period also is characterized by the apparent lack of southwestern influences, particularly in terms of colonial expansion and significant trade.

Because the physical remains from this period are most frequently found in the Mojave Desert, at least in terms of surface manifestations, considerable archaeological work has been undertaken at these sites. Among the most important studies are the excavation reports of Riddell (1951); Meighan (1953); Wallace (1958); Hunt (1960); Davis (1962); Smith (1963); Donnan (1964); Wallace and Desautels (1960); Peck and Smith (1957); Wallace and Taylor (1959); McKinney et al. (1971); Davis (1970); Lathrap and Meighan (1951); Hillebrand (1972); Gearhart (1974); Rector et al. (1979); Drover (1979); and the surveys of Wallace (1964); True, Davis, and Sterud (1966); True, Sterud, and Davis (1967); Ritter (1976); Coombs (1979 a, 1979b); and Brooks et al. (1979); and Gallegos et al. (1979).

Drover (1979) has noted Yuman or Patayan Colorado River influences in sites from the Cronise Lakes area of the eastern Mojave in California resulting from local lacustrine intervals. Movements of Paiute peoples during the latest prehistoric period also is well documented (Goss 1968; King 1976:30-33). Population increased during this period, presumably as a response to a fuller utilization of natural resources, particularly plant foods such as *Prosopis juliflora* (Wallace 1962:178). Material remains from various sites suggest a high degree of cultural homogeneity. Milling tools are widespread, now including bedrock mortars and pestles. Flaked stone knife blades, drills, scrapers, basketry, marine shell beads, and incised slates are other period characteristics. Cremation became the mode of disposing of the dead replacing, in most regions, inhumation.

The settlement pattern was extremely diverse during this increasingly arid period with major settlements along sections of the Mojave River (Smith 1963; Rector et al. 1979; Coombs

1979b), which formed a corridor for continuing trade between the Colorado-River-based Mojave Indians and groups to the west and east (Farmer 1935; Davis 1961). Other important locations included springs (Ritter 1976), major playas, rock shelters, and areas of mesquite (figure 4-17) Within the western and eastern Mojave Desert, a pattern of seasonal movement or transhumance was practiced with group or subgroup movement corresponding to the seasonal ripening of various plant foods (such as *Pinus monophylla, Agave utahensis, Prosopis juliflora, Salvia* sp., and *Oryzopsis hymenoides*) and, to a lesser degree, the availability of game. Such patterns were observed or recorded through discussions with the historic Numic-, Tubatulabic-, and Takic-speaking people who occupied the Mojave Desert at the time of European contact. In many areas of the desert, this pattern did not occur rapidly but represents a relatively slow and sporadic evolution of changing and improving adaptation (Davis 1970).

New Directions and Trends

Within the last few years archaeological work in the Mojave Desert has been divided between studies that are academically oriented (such as Davis 1978; Drover 1979; Rector et al. 1979) and those geared more to the identification of resource variability and preservation protection management (such as Ritter 1976; Fowler et al. 1978; Coombs 1979a, 1979b; Gallegos et al. (1979) Greenwood and McIntyre 1979; Barker et al. 1979; Brooks et al. 1979). As an outgrowth of management or applied archaeology, primarily on federal lands, a number of cultural resource overviews that provide much important information on research directions and results have been completed by various federal agencies (Hall and Barker 1975; King 1976; King and Casebier 1976; Norwood and Bull 1979; Stickel, Weinman, Roberts 1980; Wallace 1977; Hauck et al. 1979; Warren et al. 1980; Greenwood and McIntyre 1980).

The nature of existing resources is another problem that earlier inventories addressed inadequately. Such surveys tended to be biased toward the large archaeological sites such as major middens, campsites, and structures, and toward what were thought to be the oldest sites, or toward those with the highest visibility such as rock art sites, intaglios, and alignments or geoglyphs (figures 4-18, 4-19).

Less visible site types and historic sites have become a recent

Figure 4-17. Photograph of Von Trigger Rock shelter, San Bernardino County, California. Photograph by E. Ritter.

Figure 4-18. Photograph of rock ring in the El Paso Mountains, Kern County, California. Photograph by E. Ritter.

Figure 4-19. Photograph of rock alignment, Eureka Valley. Inyo County, California. Photograph by E. Ritter.

focus of attention because of two major factors. One is the realization that a great deal of information was being lost. The second is an increasing tendency to focus on remnants of prehistoric and historic cultural systems and to examine all evidence of human adaptations to a changing environment, referred to as ecological archaeology. As a result, examination of cultural remains has become much more systematic, involving some sort of dispersed sampling or other carefully considered orientations. A few recent studies have been directed toward historic archaeology (Hickman 1977; Teague and Shenk 1977), a trend that undoubtedly will increase in the future.

Current Status of the Resource Base

Since a large part of the Mojave Desert is undergoing use planning by various agencies, a profusion of new information or surface-exposed site types and their distribution has been made available (Coombs 1979a, 1979b; Brooks et al. 1979; Gallegos et al. 1979). Systematic studies by the Bureau of Land Management California Desert Planning Staff over most of the California portion of the Mojave Desert have resulted in the recording of 1113 new sites at about the 1 percent inventory level. The site-type distribution based on the Bureau of Land Management classification scheme is presented in table 4-10. The general classes of features and artifacts found in association with these sites are noted in tables 4-11 and 4-

TABLE 4-10

Example of Mojave Desert (California) site types: Small-scale systematic sample, 1976-1978

SITE TYPE	NUMBER
Village	19
Temporary Camp	206
Shelter/cave	80
Milling station	57
Lithic scatter	301
Quarry	24
Pottery locus	15
Cremation locus	2
Intaglio rock alignment	4
Petroglyph/petrograph	36
Trail	3
Roasting pit	33
Isolated find	117
Cairn	6
Historic (pre-World War II)	173
Other	10
Multiple (historic and one of the above)	27
Total	1113

TABLE 4-11

Example of Mojave Desert (California) site features: Small scale systematic sample, 1976-78

FEATURE	NUMBER
Structural depression	3
Rock ring	49
Rock structure	34
Cairn/shrine	24
Roasting pit/fire affected	
Rock cluster	58
Hearth	10
Petroglyph/pictographs	36
Bedrock Mortars	6
Grinding sticks	50
Other	51
Combined	83
None	709
Total	1113

12. These data should be considered only approximate because a 1 percent sample is not very large. They do not represent the full range of archaeological remains in the Mojave Desert nor do they adequately reflect temporal variability. However, they do point out the heterogeneity and diversity of these resources at the prehistoric level.

The historic level, not addressed here, is even more complex.

Table 4-12

Example of Mojave Desert (California) site artifacts: Small-scale systematic sample, 1976-1978

ARTIFACTS	NUMBER
Projectile points	20
Flaked stone toole	82
Core–detritus	248
Milling tools	47
Other ground stone	0[a]
Tools	3
Ceramics	25
Bone tools	1
Perishables	0[a]
Ornaments	0[a]
Historic	142
Other	2
None	156
Multiple artifacts	387
Total	1113

[a]Found in other sites.

Based on a very limited set of data it is possible to project some important variables for site locations in the Mojave Desert (table 4-13). Correlations relate to one or more factors. These include cultural choices (economic, social, religious) for activity placement, natural phenomena (erosion, deposition, dryness, preservation), utilized resource distribution such as plants, animals, water resources, and stone materials, and landform constraints such as degree of slope.

The condition of the cultural resources is a concern to researchers, interpreters, managers, and the interested public. Since 1970 there has been a tremendous increase in the number of threats to archaeological sites. A study of Lyneis et al. (1980) from a sample of site records indicates that only 64 percent of those sites within the California desert are in good condition. Villages, historic sites, and shelter caves frequently are in poor condition, primarily because of vandalism and theft. Sites in the eastern Mojave region are in better condition that those in the western Mojave, which is closer to the large population centers.

Major threats to cultural resources include erosion, deposition, animal disturbance by cattle, rodents, and burros, damage by off-road vehicles, vandalism, and development. The fragile and irreplaceable nature of these cultural resources makes them particularly vulnerable to destructive land use. They are the key to extending our knowledge of the prehistoric and historic Mojave Desert cultures. It is hoped that human activities can be channeled to preserve the remains of the earlier peoples of the Mojave Desert.

TABLE 4-13
Important Mojave Desert archaeological site/environmental associations

Multiple landform areas

Major landform interfaces

Proximity to relict and modern fresh water sources

Floodplain proximity

Pediments and older alluvial fan surfaces

Limestone and Tertiary and Quaternary volcanic areas

Recent volcanic (lava) areas

Lower periphery of the present pinyon-juniper zones

Areas of Agave sp.

Areas of Prosopis sp.

Diversely vegetated low slope areas

Areas of stabilized or relict dune

Ore-bearing bodies

Surface transportation corridors

References

Adams, Robert H. 1938. Implements from Lake Mohave. *American Antiquity* 4(2):154-155.

Aikens, C. Melvin. 1978. Archaeology of the Great Basin. *Annual Review of Anthropology* 7:71-87.

Allred, C. M., and Beck, D. E. 1963. Comparative and ecological studies of animals exposed to radiation at the Nevada Test Site. In *Proceedings, National Symposium on Radioecology*, New York: Reinhold Publications.

Allred, D. M.; Beck, D. E.; and Jorgensen, C. D. 1963. Biotic communities of the Nevada Test Site. *Brigham Young Univ. Science Bulletin, Biology Series* 1 (2).

Alsoszatai-Petheo, J. A. 1978. *Early man in the Manix Basin.* Cerro Coso College. Calif: Ridgecrest.

Amsden, Charles A. 1935. The Pinto Basin artifact. *Southwest Museum Papers* 9:33-51.

Anderson, B. W., and Ohmart, R. D. 1976a. *A vegetation management study for the enhancement of wildlife along the lower Colorado River.* USDI, Bur. of Reclamation, Boulder City, Nev. Report for Contract 7-07-30-V0009.

———. 1976b. *An inventory of densities and diversities of birds and mammals in the Lower Colorado River Valley, 1975.* USDI, Bur. of Reclamation, Boulder City, Nevada. Report for Contract 7-07-30-V0009.

Anderson, E. R., and Prichard, D. W. 1951. Physical limnology of Lake Mead. *Navy Electronics Laboratory, Report* 258.

Antevs, Ernst. 1948. Climatic changes and pre-white man. *Univ. Utah Bulletin* 38(20):167-191.

———. 1952. Climatic history and the antiquity of man in California. *Univ. California Archaeological Survey Reports* 16:23-31.

———. 1955. Geologic-climatic dating in the West. *American Antiquity* 20(4):317-335.

Aschmann, Homer. 1958. Great Basin climates in relation to human occupance. *Univ. California Archaeological Survey Report* 42:23-40.

Axelrod, D. I. 1979. Age and origin of Sonoran Desert vegetation. *Occ. Papers of the Cal. Acad. of Sciences,* no. 132.

Bailey, H. P. 1966. *The climate of southern California.* Berkeley: University of California Press.

Banta, B. H. 1961. On the occurrence of *Hyla regilla* in the Lower Colorado River, Clark County, Nevada. *Herpetologica* 17 (2).

Banta, B. H., and Tanner, W. W. 1964. A brief historical resumé of herpetological studies in the Great Basin of the western United States. Part 1, The reptiles. *Great Basin Naturalist* 24 (2).

Barbour, M. G., and Major, J. eds. 1977. *Terrestrial vegetation of California.* New York: John Wiley.

Barker, James P.; Rector, Carol H.; and Wilke, Philip T. 1979. An archaeological sampling of the proposed Allen-Warner Valley Energy System, Western Transmission Line Corridors, Mojave Desert, Los Angeles and San Bernardino Counties, California and Clark County, Nevada. Report on file with the Archaeological Research Unit, University of California-Riverside, and Southern California Edison Company, Rosemead.

Bartholomew, G., and Hudson, J. 1961. Desert ground squirrels. *Scientific American* 205.

Beatley, J. C. 1974. Effects of rainfall and temperature on the distribution and behavior of *Larrea tridentata* (creosote bush) in the Mojave Desert of Nevada. *Ecology* 55(2):245-261.

———. 1975. Climate and vegetation pattern across the Mojave-Great Basin Desert transition of southern Nevada. *Amer. Mid. Nat.* 93:53-70.

———. 1976. *Vascular plants of the Nevada Test Site and central southern Nevada ecologic and geographic distributions.* Springfield, Va.: National Tech. Info. Serv.

Bedwell, Stephen F. 1973. *Fort Rock Basin prehistory and environment.* Eugene: University of Oregon Books.

Benson, L., and Darrow, R. A. 1944. A manual of southwestern desert trees and shrubs. *Ariz. Univ. Bulletin* 15(2) (*Biol. Sci. Bulletin* 6).

Bettinger, R. L., and Taylor, R. E. 1974. Suggested revisions in archaeology sequences of the Great Basin in interior southern California. *Nevada Archaeological Survey Research Papers* 5:1-26.

Borden, F. W. 1971. The use of surface erosion observations to determine chronological sequence in artifacts from a Mojave Desert site. *Archaeological Survey Association of Southern California Papers.*

Bradley, W. G. 1964. The vegetation of the Desert Game Range with special reference to the desert bighorn. *Transactions of the Desert Bighorn Council*, pp. 43-67.

Bradley, W. G. 1966. The status of the cotton rat in Nevada. *J. of Mammalogy*, 47 (2).

Bradley, W. G., and Deacon, J. F. 1966. Amphibian and reptile records for southern Nevada. *Southwest Naturalist* 11 (1).

———. 1967. The biotic communities of southern Nevada. *Nevada State Museum Anthropological Papers*, 13.

Brainerd, George W. 1953. A re-examination of the dating evidences for the Lake Mojave artifact assemblage. *American Antiquity* 18(3):270-271.

Brooks, Richard H.; Wilson, Richard; Brooks, Sheilagh. 1979. An archaeological inventory of the Owlshead/Amargosa-Mohave Basin planning units of the Southern California Desert Area. Report on file with the Bureau of Land Management, Riverside, Calif.

Brum, G. D. 1972. Ecology of the Saguaro (*Carnegiea gigantea*): Phenology and establishment in marginal populations. Master's thesis, University of California, Riverside.

———. 1973. Ecology of the Saguaro (*Carnegea gigantea*): phenology and establishment in marginal populations. *Madrono* 22:195-204.

Bureau of Land Management. 1976. Eastern Mojave Unit resource analysis step 11, physical profile—vegetation.

Burk, J. H. 1977. Sonoran Desert. In *The terrestrial vegetation of California* eds. M. G. Barbour and J. Major. New York: pp 869-889. New York: John Wiley.

Burt, W. H. The mammals of southern Nevada. *Trans. of the San Diego Society of Natural History* 7 (36).

Cable, D. R., and Martin, P. 1975. Vegetation responses to grazing, rainfall, site condition and mesquite control on semidesert range. *USDA Forest Serv. Res. Papers* RM-149, Rocky Mountain Forest and Range Exp. Sta.

Campbell, Elizabeth W. C. 1931. An archaeological survey of the Twenty-Nine Palms region. *Southwest Museum Papers* 7.

Campbell, Elizabeth W. C., and Amsden, Charles A. 1934. The Eagle Mountain site. *Masterkey* 8(6):170-173.

Campbell, Elizabeth W. C., and Campbell, William H. 1935. The Pinto Basin site. *Southwest Museum Papers* 9:21-31.

Campbell, Elizabeth W. C.; Campbell, William H.; Antevs, Ernst; Avery Amsden, Charles; Barbieri, Joseph A.; and Bode, Francis D. 1937. The archaeology of Pleistocene Lake Mohave. *Southwest Museum Papers* 11.

Carter, George F. 1967. A cross check on the dating of Lake Mojave artifacts. *Masterkey* 41(1):26-33.

Chew, R., and Butterworth, B. 1964. Ecology of rodents in Indian Cove, Joshua Tree National Monument, California. *J. Mammalogy* 45.

Clements, F. E. 1916. Plant succession: An analysis of the development of vegetation. *Carnegie Inst. Wash. Publ.* 242.

Clements, T., and Clements, L. 1951. Indian artifacts and collecting localities in Death Valley, California. *Masterkey* 25(4):125-128.

Clokey, I. W. 1951. Flora of the Charleston Mountains, Clark County, Nevada. *Univ. California Publ. in Botany* 24:1-274.

Coombs, Gary B. 1979a. *The archaeology of the northeast Mojave Desert.* Riverside, Calif.: Cultural Resources Publications.

———. 1979b. *The archaeology of the western Mojave.* Riverside, Calif: Cultural Resources Publications.

Cronquist, A.; Holmgren A. H.; Holmgren, N. H.; and Reveal, J. L. 1972. *Intermountain flora—Vascular plants of the intermountain West, U.S.A.* New York: Hafner.

Curtis, J. T. 1959. *The vegetation of Wisconsin. An ordination of plant communities.* Madison: University of Wisconsin Press.

Davis, Emma Lou. 1967. Man and water at Pleistocene Lake Mohave. *American Antiquity* 32(3):345-353.

———. 1970. Archaeology of the North Basin of Panamint Valley, Inyo County, California. *Nevada State Museum Anthropological Papers* 15:83-141.

———. 1975. The "exposed archaeology" of China Lake, California. *American Antiquity* 40(1):39-53.

———. 1978. The ancient Californians: Rancholabrean hunters of the Mojave Lakes Country. *Natural History Museum of Los Angeles County Science Series* 29.

Davis, Emma Lou; Brott, Clark W.; and Weide, David L. 1969. The western lithic co-tradition. *San Diego Museum Papers* 6.

Davis, Emma Lou; Brown, Kathryn H.; Nichols, Jacqueline. 1980. Evaluation of early human activities and remains in the California desert. Report on file with the Bureau of Land Management, Riverside.

Davis, James T. 1961. Trade routes and economic exchange among the Indians of California. *Univ. California Archaeological Survey Reports* 54.

———. 1962. The Rustler Rockshelter (SBr-288), A culturally stratified site in the Mohave Desert, California. *Univ. California Archaeological Survey Reports* 57:27-73.

Deacon, J. E.; Bradley, W. G.; and Larsen, K. M. 1964. Ecological distribution of the mammals of Clark Canyon, Charleston Mountains, Nevada. *J. of Mammalogy* 45 (3).

Dean, L. E. 1978. *The California desert sand dunes.* University of California, Riverside, Dept. of Earth Sciences Publ.

Decker, D. In preparation. A flora of Inyo County, California.

Donnan, Christopher B. 1964. A suggested culture sequence for the Providence Mountains. *Univ. California at Los Angeles Archaeological Survey, Annual Report 1963-1964* 1-26.

Drover, Christopher E. 1979. The late prehistoric human ecology of the Northern Mohave Sink, San Bernardino County, California. Ph.D. dissertation, University of California, Riverside.

Egler, F. E. 1979. Vegetation change—changed. (Review of J. Miles, *Vegetation dynamics.* Wiley Publication, New York, 1979) *Ecology* 60:1077.

Eisen, Gustav. 1898. Long lost mines of precious gems are found again, located in the remotest wilds of San Bernardino County, and marked by strange hieroglyphics. *San Francisco Call,* March 18.

Emberger, L. 1955. Une classification biogéographique des climats. *Rec. Trav. lab. Bot. Geol. Fac. Sc. de Montepellier* 7:3-43.

England, A. S., and Laudenslayer, W. F. Jr. 1980. Birds of the California desert. In *The California Desert: An introduction to natural resources and man's impact,* ed. J. Latting. *Calif. Native Plant Soc. Sp. Publ.,* 5.

Farmer, Malcolm F. 1935. The Mojave trade route. *Masterkey* 9(5):155-157.

Fisher, J. C. 1975. Impact of feral asses on community structure in the *Acamptopappus—Grayia* Community of the Panamint Mountains, Death Valley National Monument. Report to the USDI Nat. Park Serv. Death Valley Nat. Mon., Calif.

Fisher, J. C. 1977. Studies relating to the accelerated mortality of *Atriplex hymenelytra* in Death Valley National Monument. Master's thesis, University of California, Riverside.

Fowler, Don D.; Budy, Elizabeth; DeSart, Dennis; Bath, Joyce; and Smith, Alma. 1978. Class II cultural resources field sampling inventory along proposed IPP transmission line corridors, Utah-Nevada-California. *Desert Research Institute, Social Sciences Center, Technical Report,* 4, Reno.

Fox, K. 1977. Importance of riparian ecosystems: Economic considerations. In *Importance, preservation and management of riparian habitats: A symposium, USDA, Forest Service Gen. Tech. Rep.,* RM-43. Rocky Mtn. Forest and Range Experiment Stn., Fort Collins, Colo.

Gaines, D. 1974. Review of the status of the yellow-billed cuckoo in California: Sacramento Valley populations. *Condor* 76.

———. 1978. California yellow-billed cuckoo survey, 1977. Calif. Dept. Fish and Game. Project E-1-1, Job IV-1.4, Final Report.

Gaines, J. F., and Logan, R. F. 1956. The physical geography of the Providence Mountains area, eastern Mojave Desert, California. Unpublished report submitted to Environmental Protection Division, Research and Development Center, U.S. Army. Natick, Mass.

Gallegos, Dennis; Cook, John, Davis, Emma Lou; Lowe, Gary; Norris, Frank; and Thesken, Jay. 1980. Cultural resources inventory of the central Mojave and Colorado Desert regions, California. Report on file with Cultural Resources Publications, Riverside, Calif.

Gearheart, Patricia L. 1974. Shoshone shelter cave number two, A preliminary report. *Pacific Coast Archaeological Society Quarterly* 10(2):35-51.

Glennan, W. S. 1971. *A glimpse at the prehistory of Antelope Valley.* Rosamond: Kern-Antelope Historical Society.

———. 1976. The Manix Lake lithic industry: Early lithic tradition or workshop refuse? *J. of New World Archaeology* 1(7):43-62.

Goss, James A. 1968. Cultural-historical inferences from Utaztekan linguistic evidence. In *Utaztekan Prehistory,* ed. Earl H. Swanson, Jr. *Idaho State Museum, Occasional Papers* 22:1-42.

Gould, G., and Bleich, V. 1977. Amargosa vole study: Progress report. Calif. Dept. Fish and Game. Nongame wildlife investigations.

Grant, Campbell; Baird, James W.; and Pringle, J. Kenneth. 1968. Rock drawings of the Coso Range, Inyo County, California. *Maturango Museum Publication* 4, China Lake.

Greenwood, Roberta S., and McIntyre, Michael J. 1979. Class III cultural resource inventory: Victorville-McCullough transmission lines 1 and 2 right-of-way. Los Angeles Department of Water and Power Report.

———. 1980. Cultural resources overview for Edwards Air Force Base. Report on file with Edwards Air Force Base, Edwards.

Grime, J. P. 1979. *Plant strategies and vegetation processes.* Chichester, U.K.: John Wiley.

Hall, E. R. 1946. *The mammals of Nevada.* Berkeley: University of California Press.

Hall, E. R., and Kelson, K. 1959. *The mammals of North America.* New York: Ronald Press.

Hall, Matthew G., and Barker, James P. 1975. *Background to prehistory of the El Paso/Red Mountain Desert region.* Riverside, Calif.: Bureau of Land Management.

Hanes, T. L. 1971. Succession after fire in the chaparral of southern California. *Ecol. Monogr.* 41:27-52.

Harrington, Mark R. 1933. Gypsum Cave, Nevada. *Southwest Museum Papers* 8.

———. 1957. *A Pinto Site at Little Lake, California.* Southwest Museum Papers 17.

Hastings, J. R., and Turner, R. M. 1965. *The changing mile.* Tucson: University of Arizona Press.

Hauck, F. R.; Weder, D. G.; Drollinger, L.; and McDonald, A. 1979. A cultural resource evaluation in Clark County, Nevada. Report on file with the Bureau of Land Management, Las Vegas.

Hayden, Julian. 1976. Pre-altithermal archaeology in the Sierra Pinacate, Sonora Mexico. *American Antiquity* 41 (3):274-289.

Haynes, C. Vance. 1969. The earliest Americans. *Science* 166:709-715.

Heid, James; Warren, Claude N.; and Rocchio, Patricia. 1979. The western pluvial lakes tradition: A critique. Paper presented at the Society for California Archaeology Meeting, San Luis Obispo.

Heizer, Robert F. 1941. Aboriginal trade between the Southwest and California. *Masterkey* 15(5):185-188.

———. 1965. Problems in dating Lake Mohave artifacts. *Masterkey* 39(4):125-134.

———. 1970. Environment and culture, The Lake Mojave case. *Masterkey* 44(2):68-72.

Heizer, Robert F., and Baumhoff, Martin A. 1961. Wagon jack shelter. In *The archaeology of two sites at Eastgate, Churchill County, Nevada, Univ. Calif. Anthropological Records* 20(4):119-138.

Hendrickson, J., and Prigge, B. 1975. White fir in the mountains of eastern Mojave Desert of California. *Madrono* 23:164-168.

Hester, Thomas R. 1973. Chronological ordering of Great Basin prehistory. *Univ. California Archaeological Research Facility Contributions* 17.

Hickman, Patricia Parker. 1977. County Nodes, An anthropological evaluation of William Keys' Desert Queen Ranch. Joshua Tree National Monument, California. *National Park Service, Western Archeological Center Publications in Anthropology,* 7.

Hillebrand, Timothy S. 1972. The archaeology of the Coso locality of the Northern Mojave region of California. Ph.D. dissertation, University of California, Santa Barbara.

————. 1974. The Baird site. In *Excavation of two sites in the Coso Mountains of Inyo County, California. Maturango Museum Monograph,* 1:63-80.

Hoffman, W. J. 1881. Annotated list of the birds of Nevada. *Bulletin, U.S. Geological and Geographical Survey of the Territory,* 6 (2).

Holmgren, R. C., and Hutchings, S. S. 1972. Salt desert shrub response to grazing use. In *Wildland shrubs, their biology and utilization. Proc. Int. Symp.* Utah State University, Logan. pp. 153-165.

Howell, A. B. 1923. The influence of the southwestern deserts upon the avifauna of California. *Auk* 40.

Hubbs, C. L., and Miller, R. R. 1948. Correlation between fish distribution and hydrographic history in the desert basins of western United States. In *The Great Basin with emphasis on glacial and postglacial times. Bulletin, Univ. Utah,* 38 (20).

Humphrey, R. R. 1962. *Range ecology.* New York: Ronald Press.

Huning, J. R. 1978. A characterization of the climate of the California desert. USDI Bureau of Land Management California Desert Plan Program, Riverside. Report for Contract CA-060-CT7-2812.

Hunt, Alice P. 1960. Archaeology of the Death Valley Salt Pan. *Univ. Utah Anthropological Papers* 47.

Hunt, Alice P., and Hunt, Charles B. 1964. Archaeology of the Ash Meadows Quadrangle, California, and Nevada. Manuscript on file with the National Park Service, Western Archaeological Center, Tucson.

Hunt, C. B. 1966. *Plant ecology of Death Valley, California.* Geology. Serv. Prof. Pap. 509.

Jaeger, E. C., and Smith, A. C. 1966. *Introduction to the natural history of southern California.* Berkeley: University of California Press. 104 pp.

Jennings, Jesse D. 1957. Danger Cave. *Univ. Utah Anthropological Papers* 27.

————. 1978. Prehistory of Utah and the eastern Great Basin. *Univ. Utah Anthropological Papers* 98.

Jennings, Jesse D., and Norbeck, Edward. 1955. Great Basin prehistory: A review. *American Antiquity* 21(1):1-11.

Johnson, D. H.; Bryant, M. D.; and Miller, A. H. 1948. Vertebrate animals of the Providence Mountains area of California. *Univ. Calif. Publ. Zool.* 48.

Johnson, H. B. 1975. Gas exchange strategies in desert plants. In *Perspectives of biophysical ecology,* eds. D. M. Gates and R. B. Schmerl, New York: Springer-Verlag. pp. 105-120.

————. 1976. Vegetation and plant communities of southern California deserts—a functional view. In *Plant communities of southern California.* ed. J. Latting, *California Native Plant Society Special Publ.* 2.

Johnson, H. B.; Vesek, F. C.; and Yonkers, T. 1975. Productivity, diversity, and stability relationships in Mojave Desert roadside vegetation. *Bull. Torrey Botanical Club* 102(3): 106-115.

Johnson N. K. 1965. The breeding avifaunas of the Sheep and Spring ranges of southern Nevada. *Condor,* 67 (2).

————. 1973. The distribution of boreal avifaunas in southeastern Nevada. *Biol. Soc. Nevada Occasional Papers* 36.

Johnson, N. K., and Garrett, K. L. 1974. Interior bird species expand breeding ranges into southern California. *Western Birds* 5.

Jones, J. D.; Carter, D.; and Genoways, H. 1975. Revised checklist of North American mammals north of Mexico. *Occas. Papers, The Museum, Texas Tech. Univ.,* 28.

Jones, L. 1970. The whip-poor-will in California. *Calif. Birds* 2:33.

Kearney, T. H., and Peebles, R. H. 1960. *Arizona flora.* Berkeley: University of California Press.

King, Chester. 1976. Background to prehistoric resources of the east Mojave Desert region. In *Background to historic and prehistoric resources of the East Mojave Desert region,* Chester King and Dennis Casebier, pp. 19-34. Riverside: Bureau of Land Management.

King, Chester, and Casebier, Dennis G. 1976. *Background to historic and prehistoric resources of the East Mojave Desert region.* Riverside: Bureau of Land Management.

King, J. E., and VanDevender, T. R. 1977. Pollen analysis of fossil packrat middens from the Sonoran Desert. *Quaternary Research* 8:191-204.

King, Thomas F. 1975. *Fifty years of archeology in the California Desert: An archeological overview of Joshua Tree National Monument.* Tucson: National Park Service, Western Archeological Center.

Knapp, R. 1965. Die vegetation von Nord-und Mittelamerika und der Hawaii inseln. Stuttgart: G. Fisher Verlag.

Kornoelje, T. A. 1973. Plant communities of the Covington Flats area, Joshua Tree National Monument. Master's thesis, California State University, Long Beach.

Kowta, Makoto. 1969. The Sayles Complex: A late milling stone assemblage from Cajon Pass and the ecological implications of the scraper planes. *Univ. California Publications in Anthropology,* 6.

Krieger, Alex D. 1962. The earliest cultures in the western United States. *American Antiquity* 28(2):138-143.

————. 1964. Early man in the New World. In *Prehistoric man in the New World,* ed. J. D. Jennings and E. Norbeck, pp. 23-81. Chicago: University of Chicago Press.

Kuchler, A. W. 1967. Appendix: the map of the natural vegetation of California. In *The terrestrial vegetation of California.* ed. M. G. Barbour and J. Major, New York: Wiley Publications.

Küchler, A. W. 1964. Manual to accompany the map: Potential natural vegetation of the coterminous United States. *Amer. Geo. Soc. Special Publ.* 36.

Lamb, Sydney M. 1958. Linguistic prehistory in the Great Basin. *International J. American Linguistics* 24(2):95-100.

Lanning, Edward P. 1963. Archaeology of the Rose Spring site, Iny-372. *Univ. California Publications in American Archaeology and Ethnology* 49(3):237-336.

Lathrap, D. W., and Meighan, Clement W. 1951. An archaeological reconnaissance in the Panamint Mountains. *Univ. California Archaeological Survey Reports* 11:11-32.

Laudenslayer, W. F., Jr., and Boyer, K. B. 1980. Mammals of the California Desert. In *The California desert: An introduction to natural resources and man's impact,* ed. J. Latting. Calif. Native Plant Soc. Sp. Pub., 5.

Laudenslayer, W. F., Jr.; Cardiff, S. W.; and England, A. S. 1980. Checklist of birds known to occur in the California Desert. In *The California Desert: An introduction to natural resources and man's impact* ed. J. Latting. *Calif. Native Plant Soc. Sp. Pub.* 5.

Leakey, L. S. B.; Simpson, R. D. and Clements, T. 1968. Archaeological excavations in the Calico Mountains, California: Preliminary report. *Science* 160:1022-1023.

Leakey, L. S. B.; Simpson, R. D.; Clements, T.; Berger, R.; Witthoft, J.; and Schuiling, W. C., 1972. *Pleistocene man at Calico.* San Bernardino: County Museum Association.

Leary, P. J. 1977. Investigations of the vegetational communities of Joshua Tree National Monument California. Contribution National Park Service, No. CP5U/University of Nevada, Las Vegas, No. 19.

Linsdale, J. M. 1936. The birds of Nevada. *Pacific Coast Avifauna* 23.

———. 1940. Amphibians and reptiles in Nevada. *Proc. of the American Academy of Arts and Sciences* 73 (8).

———. 1951. A list of the birds of Nevada. *Condor* 54 (4).

Lloyd, R. M., and Mitchell, R. S. 1973. *A flora of the White Mountains, California and Nevada.* Berkeley: University of California Press.

Logan, R. F. 1968. Causes, climates and distribution of deserts. In *Desert biology: special topics on the physical and biological aspects of arid regions,* ed. G. W. Brown, New York: Academic Press. 1:21-50.

Lowe, C. H. 1964. Arizona landscapes and habitats. In *The vertebrates of Arizona,* ed. by C. H. Lower. Tucson: University of Arizona Press.

Lowe, C. H., and Brown, D. E. 1973. The natural vegetation of Arizona. *Arizona Resources Information System Cooperative Publication* 2.

Lyneis, Margaret M.; Weide, David L.; and von Till Warren, Elizabeth. 1980. Impacts: *Damage to cultural resources in the California Desert.* Riverside, Calif: Cultural Resources Publications.

Major, J. 1951. A functional, factorial approach to plant ecology. *Ecology* 32:392-412.

Major, J. 1977. California climate in relation to vegetation. In *Terrestrial vegetation of California,* eds. M. G. Barbour and J. Major, pp. 11-74. New York: John Wiley.

Margalef, R. 1968. *Perspectives in ecological theory.* Chicago: University of Chicago Press.

Martin, S. C. 1972. Some effects of continuous grazing on forage production. *Arizona Cattlelog* 28 (10).

Martin, S., and Cable, D. R. 1974. Managing semidesert grassland shrub ranges: Vegetation responses to precipitation, grazing, soil texture, and mesquite control. *USDA Forest Serv. Tech. Bul.* 1480.

McGinnies, W. G.; Goldman, B. J.; and Paylore, P. 1977. *Deserts of the world: An appraisal of research into their physical and biological environments.* Tucson: University of Arizona Press.

McKinney, Aileen; Hafner, Duane; and Gothold, Jane. 1971. A report on the China Ranch area. *Pacific Coast Archaeological Society Quarterly* 7(2):1-47.

McNaughton, S. J. Grazing as an optimization process. *Amer. Nat.* 113: 691-703.

McNaughton, S. J. 1969. Climates of organisms. (Review of *Ground level climatology,* ed. R. H. Shaw, Amer. Assoc. Adv. Sci., 1967) *Ecology* 50:526-528.

Mehringer, P. J., Jr. 1965. Late Pleistocene vegetation in the Mojave Desert of southern Nevada. *J. Ariz. Acad. Sci.* 3:172-188.

———. 1967. Pollen analysis of the Tule Springs site, Nevada. In *Pleistocene studies in southern Nevada,* ed. H. M. Wormington and D. Ellis. *Nevada State Museum Anthropological Papers,* 13:129-200.

———. 1977. Great Basin late Quaternary environments and chronology. In *Models and Great Basin perhistory,* ed. D. Fowler, *Desert Research Institution Publications in the Social Sciences,* 12:113-168.

Mehringer, P. J., and Ferguson, C. W. 1969. Pluvial occurrence of bristlecone pine (*Pinus aristae*) in a Mojave Desert Mountain Range. *J. Ariz Acad. Sci.* 5:284-292.

Meighan, Clement W. 1953. The Coville rock shelter, Inyo County, California. *Univ. Calif. Anthropological Records* 12(5).

———. 1954. The Lake Mojave site. University of California Archaeological Survey Manuscript 209.

———. 1959. California cultures and the concept of an Archaic stage. *American Antiquity* 24:289-305.

Meigs, P. 1957. Weather and climate. In *The North American Deserts,* ed. E. C. Jaeger, pp. 13-32. Palo Alto: Stanford University Press.

Meyer, S. E. 1976. An annotated checklist of the vascular plants of Washington County, Utah. Master's thesis, University of Nevada.

Mifflin, M. D., and Wheat, M. M. 1979. Pluvial lakes and estimated pluvial climates of Nevada. *Nevada Bureau of Mines and Geology Bulletin* 94.

Miles, J. 1979. *Vegetation dynamics.* New York: John Wiley.

Miller, A. H. 1940. A transition island in the Mojave Desert. *Condor* 42:161-163.

Miller, A. H., and Stebbins, R. C. 1964. *The lives of desert animals in Joshua Tree National Monument.* Berkeley: University of California Press.

Miller, R. R. 1948. The cyprinodont fishes of the Death Valley system of eastern California and southwestern Nevada *Miscellaneous Publications of the University of Michigan Museum of Zoology,* 68.

———. 1959. Origin and affinities of the freshwater fish fauna of western North America. *Zoogeography, American Association for the Advancement of Science, Publication* 51.

Miller, R. R., and Alcorn, J. R. 1946. The introduced fishes of Nevada with a history of their introduction. *Trans. of the American Fish Society,* 73.

Mueller-Dombois, D., and Ellenburg, H. 1974. *Aims and methods of vegetation ecology.* New York: John Wiley.

Muller, C. H. 1940. Plant succession in the *Larrea-Flourensia* climax. *Ecology* 21:206-212.

Munz. P. A. 1973. *A flora of southern California.* Berkeley: University of California Press.

Munz, P. A., and Keck, D. D. 1959. *A California flora.* Berkeley: University of California Press.

———. 1949, 1950. California plant communities. *El Aliso* 2:87-105, 199-202.

Murdock, Joseph, and Webb, Robert W. 1948. Minerals of California. *State of California Department of Natural Resources, Division of Mines Bulletin.*

National Oceanographic and Atmospheric Administration (NOAA). 1969. *Climatic summary of the United States—California.*

Nichol, A. A. 1937. *The natural vegetation of Arizona. Univ. of Arizona College of Agric., Agric. Exp. Sta. Tech. Bull.* 68.

Norton, B. E. 1975. *Effects of grazing on desert vegetation.* US/IBP Desert Biome Res. Memo. 75-50, Utah State University, Logan, Utah.

Norwood, Richard H., and Bull, Charles S. 1979. A cultural resource overview of the Eureka, Saline, Panamint and Darwin Region, east central California. Riverside, Calif.: Cultural Resource Publications.

Oakeshott, G. B. 1972. *California's changing landscape.* New York: McGraw-Hill.

Odum, E. P. 1971. *Fundamentals of ecology.* Philadelphia: W. B. Saunders.

Ohmart, R. D.; Deason, W. O.; and Burke, C. 1977. A riparian case history: The Colorado River. In *Importance, preservation and management of riparian habitat: A symposium,* R. R. Johnson and D. A. Jones, pp. 35-47. *USDA Forest Serv. Gen. Tech. Rep.* RM-43. Rocky Mt. For. and Range Exp. Stn., Fort Collins, Colo.

Ore, H. Thomas, and Warren, Claude N. 1971. Late Pleistocene-Early Holocene geomorphic history of Lake Mohave. *Geol. Soc. Amer. Bulletin* 82:2553-2562.

Ornduff, R. 1974. *Introduction to California plant life.* Berkeley: University of California Press.

Panlaqui, Carol. 1974. The Ray Cave site. In excavation of two sites in the Coso Mountains of Inyo County, California. *Maturango Museum Monograph* 1:1-62.

Parker, K. W. 1942. *General guide to the satisfactory utilization of the principal southwestern range grasses.* USDA Forest Serv. S.W. Forest and Range Exp. Sta. Note No. 104.

Pavlik, B. M. 1979. *The biology of endemic psammophytes, Eureka Valley, California and its relation to off-road vehicle impact.* USDI Bureau of Land Management, California Desert Plan Program, Riverside, CA. Report for Contract AC-060-C78-0000-49.

Payen, Louis A. (In press.) Artifacts or geofacts at Calico: Application of the Barnes test. In *Peopling of the New World,* eds. J. E. Ericson, R. E. Taylor and R. Berger. *Ballena Press Publications in Archaeology, Ethnology and History,* Socorro.

Peck, Stuart L. and Smith, Gerald A. 1957. The archaeology of Seep Spring. *San Bernardino County Museum Association, Scientific Series* 2.

Pilanka, E. R. 1974. *Evolutionary ecology.* New York: Harper and Row.

Pickett, S. T. A. 1976. Succession, an evolutionary interpretation. *Amer. Nat.* 110:107-119.

Prigge, B. A. 1975. Flora of the Clark Mountain Range. Master's thesis, California State University, Los Angeles.

Randall, D. C. 1972. An analysis of some desert shrub vegetation of Saline Valley, California. Ph. D. dissertation, University of California, Davis.

Ratliffe, R. D.: Westfall, S. E.; and Robarts, R. W. 1972. *More California poppy in stubble field than in old field.* USDA Forest Serv. Res. Note PSW-271.

Ratliffe, R. D., and Hubbard, R. L. 1975. *Clipping affects flowering of California poppy at two growth stages.* USDA Forest Serv. Res. Note PSW-303.

Ratliffe, R. D., and Westfall, S. E. 1976. Disturbance—not protection—may benefit culture of California poppy (*Eschcholzia californica* var. *peninsularis* (Greene). *Hort. Science* 11:210-212.

Raven, P. H. and Axelrod, D. I. 1976. Origin and relationships of the California flora. *Univ. Calif. Publ. Bot.* 72.

Rector, Carol H. 1976. Rock art of the east Mojave Desert. In *Background to historic and prehistoric resources of the East Mojave Desert region,* pp. 236-260. Riverside, Calif. Bureau of Land Management.

Rector, Carol H.; Swenson, James D.; Wilke, Philip J. 1979. Archaeological studies at Oro Grande, Mojave Desert, California. Report on file with University of California, Riverside.

Remsen, J. V., Jr., and Berry, K. H. 1976. Twenty-ninth winter bird population study: 56. Desert riparian. *Amer. Birds* 30.

Remsen, J. V. Jr.; Berry, K. H.; and Wessman, E. 1976a. Twenty-ninth winter bird population study: 51. Joshua Tree Woodland 2. *Amer. Birds.* 30.

———. 1976b. Twenty-ninth winter bird population study: 52. Catclaw-rabbitbrush desert wash. *Amer. Birds,* 30.

Remsen, J. V., Jr.; Cardiff, S.; and Hale, L. 1978. Avifaunal surveys in white fir and pinyon-juniper woodlands of the Kingston and New York Mountains. USDI Bur. of Land Management, Calif. Desert Plan Program, Riverside, Calif. Report for Purchase Order CA-060-PH7-1791.

Riddell, Harry S., Jr. 1951. The archaeology of a Paiute village in Owens Valley. *University of California Archaeological Survey Reports* 12:14-28.

Ritter, Eric W. 1976. Archaeology. In *Final environmental analysis record for proposed geothermal leasing in the Randsburg-Spangler Hills-So. Searles Lake areas, California.* Riverside, Calif.: Bureau of Land Management.

Roberts, Frank H. H., Jr. 1940. Recent developments in the problems of the North American Paleo Indian. *Smithsonian Miscellaneous Collections* 100:51-116.

———. 1951. Early man in California. In *The California Indians: A sourcebook,* eds. R. F. Heizer and M. A. Whipple, pp. 123-129. Berkeley: University of California Press.

Robinson, Roger W. 1977. The prehistory of Antelope Valley, California: An overview. *Kern County Archaeological Society Journal,* 1:43-49.

Robinson, Roger W.; Sutton, M. Q.; Eggers, A. V. D. 1976. Investigations at LAn-298: A re-evaluation of cultural traditions in Antelope Valley, California. Paper presented at the Society for California Archaeology Meeting, San Diego.

Robinson, T. W. 1965. Studies of evapotranspiration: Introduction, spread, and areal extent of salt cedar (*Tamarix*) in the western states. *U.S. Geological Survey Prof. Papers.*

Rogers, Malcolm J. 1929. Report of an archaeological reconnaissance in the Mohave Sink region. *San Diego Museum Papers* 1.

————. 1939. Early lithic industries of the lower basin of the Colorado River and adjacent desert areas. *San Diego Museum Papers* 2.

————. 1966. The ancient hunters . . . who were they? In *Ancient hunters of the Far West*, ed. R. F. Pourade, pp. 21–108. *San Diego Union Tribune*.

Rowlands, P. G. 1978. The vegetation dynamics of the Joshua tree (*Yucca brevifolia* engelm.) in the southwestern United States of America. Ph.D. dissertation, University of California, Riverside.

————. 1980. The vegetational attributes of the California deserts. In *The California desert: An introduction to its resources and man's impact*. ed. J. Latting. *California Native Plant Soc. Special Publ.* 5.

Ruby, Jay. 1970. Culture contact between aboriginal southern California and the Southwest. Ph.D. dissertation, University of California, Los Angeles.

Sampson, A. W. 1959. *Range management practices and principles*. New York: John Wiley.

Schroeder, A. H. 1961. The archaeological excavations at Willow Beach, Arizona, 1950. *University of Utah Anthropological Papers* 50.

Schuiling, Walter C., ed. 1979. Pleistocene man at Calico. *San Bernardino County Museum Quarterly* 26(4).

Seegmiller, R. 1977. Ecological relationships of feral burros and desert bighorn sheep, western Arizona. Master's thesis, Arizona State University.

Sellers, W. D. 1960. Precipitation trends in Arizona and western New Mexico, *Proc. 28th Ann. Snow Conf.*, Santa Fe, April, pp. 81–94.

Sharp, R. P. 1972. *Geology field guide to southern California*. Dubuque, Iowa: Wm. C. Brown.

Shefi, R. M. 1971. Plant communities of the Pinto Basin, Joshua Tree National Monument. Master's thesis, California State College, Long Beach.

Shimwell, D. W. 1971. *The description and classification of vegetation*. Seattle: University of Washington Press.

Shreve, F. 1942. The desert vegetation of North America. *Bot. Rev.* 8:195–246.

Shutler, Richard, Jr. 1961. Lost City, Pueblo Grande de Nevada. *Nevada State Museum Anthropological Papers* 5.

————. 1968. The Great Basin Archaic. In *Archaic prehistory in the western United States*, ed. C. Irwin-Williams. *Eastern New Mexico University, Contributions in Anthropology* 1(3):24–26.

Shutler, Richard, Jr.; Shutler, Mary E.; and Griffith, James S. 1960. Stuart rockshelter, A stratified site in southern Nevada. *Nevada State Museum Anthropological Papers* 3.

Sigleo, Ann Colberg. 1975. Turquoise mine and artifact correlation for Snaketown Site, Arizona. *Science* 189:456–460.

Simpson, Ruth D. 1958. The Manix Lake archaeological survey. *Masterkey* 32(1):4–10.

————. 1960. Archaeological survey of the eastern Calico Mountains. *Masterkey* 34(1):25–35.

————. 1961. Coyote Gulch. *Archaeological Survey Association of Southern California Papers* 5.

————. 1965. An archaeological survey of Troy Lake, San Bernar-

dino County, A preliminary report. *San Bernardino County Museum Association Quarterly* 12(3).

————. 1976. A commentary on W. Glennan's article. *J. of New World Archaeology* 1(7):63–66.

————. 1979. An overview of the major elements of the Calico lithic assemblage. In Pleistocene man at Calico, ed. Walter C. Schuiling. *San Bernardino County Museum Quarterly* 26(4):35–47.

Singer, Clay A. 1979. A preliminary report on the analysis of calico lithics. In Pleistocene man at Calico, ed. Walter C. Schuiling. *San Bernardino County Museum Quarterly* 26(4):55–65.

Singer, Clay A., and Ericson, Jonathon E. 1977. Quarry analysis at Bodie Hills, Mono County, California: A case study. In *exchange systems in prehistory*, eds. Timothy K. Earle and Jonathon E. Ericson, pp. 171–191. New York: Academic Press.

Smith, Gerald A. 1963. *Archaeological survey of the Mojave River area and adjacent regions*. San Bernardino: County Museum Association.

Smith, Gerald A.; Schuiling, W.; Martin, L.; Sayles, R.; and Jillson, P. 1957. Newberry Cave, California. *San Bernardino County Museum Association Scientific Series* 1.

Springer, M. E. 1958. Desert pavement and vesicular layer of some desert soils of the Lahontan Basin, Nevada. *Proc. Soil Sci. Soc. Am.* 22:63–66.

Stebbins, G. L., and Major, J. 1965. Endemism and speciation in the California flora. *Ecol. Monogr.* 35:1–35.

Stebbins, R. C. 1980. Off-road vehicle impacts on desert plants and animals, and BLM management prescriptions. In *The California Desert: An introduction to natural resources and man's impact.*, ed. J. Latting. *Calif. Native Plant Soc. Sp. Publ.* 5.

Sternberg, L. 1976. Growth forms of *Larrea tridentata*. *Madrono* 23:408–417.

Stickel, E. Gary, and Weinman-Roberts, Lois J. 1980. An overview of the cultural resources of the western Mojave Desert. Riverside, Calif.: Cultural Resources Publications.

Stoddart, L. A., and Smith, A. D. 1955. *Range management*. New York: McGraw-Hill.

Strong, William D. 1941. Review of early lithic industries of the lower basin of the Colorado River and adjacent desert areas, Malcolm J. Rogers. *American Anthropologist* 43:453–455.

Sutton, Mark Q. 1979. Some thoughts on the prehistory of the Antelope Valley. Paper presented at the Society for California Archaeology Meeting, San Luis Obispo.

————. 1980. Archaeological investigations at the Owl Canyon Site (SBr-3801). Report on file with the Bureau of Land Management, Barstow.

Szarek, S. K. 1979. Primary production in four North American deserts. *J. Arid Environments* 2:187–209.

Szarek, S. R., and Ting, I. P. 1977. The occurrence of crassulacean acid metabolism among plants. *Photosynthetica:* 11:330–342.

Tanner, W. W., and Jorgensen, C. D. 1963. Reptiles of the Nevada Test Site. *Brigham Young Univ. Science Bulletin, Biology Series* 3 (3).

Taylor, D. W. 1979. *Ecological life history studies of localized calciphytes in the Mojave Desert of California*. USDI Bureau of Land Management, California Desert Plan Program, Riverside CA Progress Report for contract CA-060-CT8-48.

Taylor, R. E., and Meighan, Clement W. eds. 1978. *Chronologies in New World archaeology.* New York Academic Press.

Taylor, R. E. and Payen, Louis A. 1979. The role of archaeometry in American archaeology: Approaches to the evaluation of the antiquity of *Homo sapiens* in California. In *Advances in archaeological method and theory,* ed. Michael B. Schiffer, 2:239–284. New York: Academic Press.

Teague, George A. and Shenk, Lynnette O. 1977. Excavations at Harmony Borax Works: Historical archaeology at Death Valley National Monument. National Park Service, *Western Archaeological Center Publications in Anthropology* 6.

Thompson, D. G. 1929. *The Mojave Desert Region, California: A geographic, geologic and hydrologic reconnaissance.* Water Supply Paper 578. Washington, D.C.: Government Printing Office.

Thorne, R. F. 1976. California plant communities. In *Plant communities of southern California.* ed. J. Latting. *Calif. Native Plant Soc. Special Publ.* 2:1–31.

Thornthwaite, C. W. 1948. An approach toward a rational classification of climate. *Geogr. Rev.* 38:55–94.

Thornthwaite, C. W., and Mather, J. R. 1957. Instruction and tables for computing potential evapotranspiration and the water balance. *Drexel Inst. Technol. Pub. Climatol.* 10:183–311.

Tidestrom, I. 1925. Flora of Utah and Nevada. *Contr. U.S. Nat. Herb.* 25.

Tomoff, C. S. 1977. *The spring avifauna of the Colorado Desert of southeastern California.* USDI, Bur. of Land Management, Riverside, Calif. Report for Contract CA-060-CT7-987.

True, D. L.; Davis, E. L.; and Sterud, E. L. 1966. Archaeological surveys in the New York Mountains region, San Bernardino County, California. *Univ. Calif. at Los Angeles Archaeological Survey, Annual Report* 1965–1966:247–274.

True, D. L.; Sterud, E. L.; Davis, Emma L. 1967. An archaeological survey at Indian Ranch, Panamint Valley, California. *Univ. California at Los Angeles Archaeological Survey, Annual Report* 1967:15–36.

Tuohy, Donald R. 1971. Review of the western lithic co-tradition, Emma Lou Davis, Clark W. Brott, and David L. Weide. *Amer. Anthropologist* 73:417–418.

Twisselman, E. 1967. A Kern County flora. *Wassman J. Biol.* 25: U.S. Dept. of Commerce Weather Bureau. 1952. *Climatic Summary of the United States Supplement for 1931 through 1952.* 11–4 Arizona; 11–4 California; 11–22 Nevada.

———. 1958. *Climatography of the United States,* number 70–26 Nevada, number 70–42 Utah; precipitation data from storage-gauge stations.

———. 1964. Climatography of the United States Number 86. *Decennial census of United States climate—climatic summary of the United States, supplement for 1951 through 1960;* 86–2 Arizona; 86–3 Calif.; 86–22 Nevada; 86–37 Utah.

VanDevender, T. R., 1973. Late pleistocene plants and animals of the Sonoran Desert: A survey of ancient packrat middens in southwestern Arizona. Ph.D. dissertations, University of Arizona.

———. 1977. Holocene woodlands in the southwestern deserts *Science* 198:189–192.

VanDevender, T. R., and Spaulding, W. G. 1979. Development of vegetation and climate in the southwestern United States. *Science* 204:701–710.

Vasek, F. C. 1979. Plant succession in the Mojave Desert. *Proc. Symposium on southern California desert communities* (In press).

———. Vasek, F. C., and Thorne, R. F. 1977. Transmontane coniferous vegetation. In *Terrestrial vegetation of California* ed. H. Barbour and J. Major, pp. 797–832. New York: John Wiley.

Vasek F. C. and Barbour, M.G. 1977. Mojave Desert scrub vegetation. In *Terrestrial vegetation of California,* ed. M. Barbour and J. Major, pp. 835–867. New York: John Wiley.

Vasek, F. C.; Johnson, H. B.; and Eslinger, D. H. 1975a. Effects of pipeline construction on Creosote Bush Scrub vegetation of the Mojave Desert. *Madrono* 23:1–64.

Vasek, F. C.; Johnson, H. B.; and Brum, G. D. 1975b. Effects of power transmission lines on vegetation of the Mojave Desert. *Madrono* 23(3):114–130.

———. 1980. Creosote bush: long-lived clone in the Mojave Desert. *Amer. J. Bot.* 67:246–255.

Venner, William T. 1978. An analysis of the Party Hill Bay Rock alignments. *Archaeological Survey Association of Southern California Journal* 2(1).

Vogl, R. J. and McHargue, L. T. 1966. Vegetation of California fan palm oases on the San Andreas fault. *Ecology* 47 (4): 533–540.

Wallace, William J. 1958. Archaeological investigations in Death Valley National Monument 1952–1957. *Univ. Calif. Archeological Survey Reports* 42:7–22.

———. 1962. Prehistoric cultural developments in the southern California deserts. *American Antiquity* 28:172–180.

———. 1964. An archaeological Reconnaissance in Joshua Tree National Monument, California. *J. of the West* 3(1):90–101.

———. 1977. Death Valley National Monument's prehistoric past: An archaeological overview. Report on file with the National Park Service, Western Archeological Center, Tucson.

———. 1978. Post-Pleistocene archeology, 9000 to 2000 B.C. In *Handbook of North American Indians: California,* ed. Robert F. Heizer, 8:25–37. Washington, D.C.: Smithsonian Institution.

Wallace, William J. and Desautels, R. J. 1960. An excavation at the Squaw Tank Site, Joshua Tree National Monument, California. *Archeological Research Associates Contributions to California Archaeology* 4(2).

Wallace, William J. and Taylor, E.S. 1959 A Preceramic Site at Saratoga Springs, Death Valley National Monument. *Archaeological Research Associates Contributions to California Archaeology* 3(2):1–13.

Wallace, William J. and Wallace, Edith. 1978. *Ancient peoples and cultures of Death Valley National Monument.* Ramona: Acoma Books.

Warren, Claude N. 1967. The San Dieguito complex: A review and hypothesis. *American Antiquity* 32(4):168–185.

———. 1970. Time and topography: Elizabeth W.C. Campbell's Approach to the prehistory of the California Deserts. *Masterkey* 44(1):5–14.

———. (in press) Pinto Points, period, and problems, eds. In *Archaeological papers in Memory of Earl H. Swanson,* eds. L. Hartin, D. Tuohy, and C. Warren. Idaho State University Press.

Warren, Claude N., and Crabtree, Robert. (in press) The prehistory of the southwestern Great Basin. In *Handbook of North American Indians: Great Basin.* Washington, D.C.: Smithsonian Institution.

Warren, Claude N. and DeCosta, John. 1964. Dating Lake Mohave artifacts and beaches. *American Antiquity* 30(2):206–209.

Warren, Claude N.; Knack, Martha; and von Till Warren, Elizabeth. 1980. A cultural resource overview for the Amargosa-Mojave Basin planning units. Riverside, Calif.: Cultural Resources Publications.

Warren, Claude N., and Ore, H. Thomas. 1978. Approach and process of dating Lake Mohave Artifacts. *J. of California Anthropology* 5(2):179–188.

Warren, Claude N., and True, Delbert L. 1961. The San Dieguito complex and its place in California prehistory. *Univ. of Calif. at Los Angeles Archaeological Survey, Annual Report 1960–61:*246–338.

Weaver, J. E., and Clements, F. E. 1938. *Plant ecology.* New York: McGraw-Hill.

Weaver, R. 1975. Status of the bighorn sheep in California. In *The wild sheep in modern North America,* ed. J. Trefethen. New York: Boone and Crockett Club in cooperation with Winchester Press.

Webb, R. H., and Wilshire, H. G. 1980. Recovery of soils and vegetation in a Mojave Desert ghost town. *J. Arid Environments* 3(4):291–303.

Weide, David. 1976. The altithermal as an archaeological "nonproblem" in the Great Basin. In *Holocene environmental change in the Great Basin. Nevada Archeological Survey Research Paper* 6:174–181.

Weinman-Roberts, L. 1980. History narrative overview. In *An overview of the cultural resources of the western Mojave Desert,* eds. E. Gary Stickel and Lois J. Weinman. Riverside, Calif.: Cultural Resources Publications.

Weinstein, M., and Berry, K. H. 1978. Forty-first breeding bird census: 164. *Desert Marsh. Amer. Birds* 32.

Wells, P. V. 1961. Succession in desert vegetation on the streets of a Nevada ghost town. *Science* 134:670–671.

———. 1976. Marcofossil analysis of woodrat (*Neotoma*) middens as a key to the Quaternary vegetational history of North America. *Quaternary Research* 6:223–248.

———. 1979. An equable glaciopluvial in the west: Pleniglacial evidence of increased precipitation on a gradient from the Great Basin to the Sonoran and Chihuahuan deserts. *Quaternary Research* 12:311–325.

Wells, P. V. and Berger, R. 1967. Late Pleistocene history of coniferous woodland in the Mojave Desert. *Science* 155:1640–1647.

Wells, P. V. and Hunziker, J. H. 1976. Origins of the Creosote bush (*Larrea*) in the deserts of southwestern North America. *Ann. Mo. Bot. Gardens* 63:843–861.

Went, F. W. 1948. Ecology of desert plants vol. 1. Observation on germination in the Joshua Tree National Monument, California. *Ecology* 29(3):242–253.

Westec Services. 1977. *Survey of sensitive plants of the Algodones Dunes.* Report prepared for USDI Bureau of Land Management, Riverside Calif.

Whittaker, R. H. 1975. *Communities and ecosystems,* 2d ed. New York: Macmillan.

Woodward, John A., and Woodward, Albert F. 1966. The carbon-14 dates from Lake Mohave. *Masterkey* 40:96–102.

Woodward, S. L., and Ohmart, R. D. 1976. Habitat use and fecal analysis of feral burros Equus asinus), Chemehuevi Mountains, California. *J. Range Management* 29.

Yang, T. W. 1970. Major chromosome races of *Larrea divaricata* in North America, *J. Ariz. Acad. Sci.* 6:41–45.

QUANTITATIVE SUMMARY OF VASCULAR PLANTS OF THE MOJAVE DESERT IN CALIFORNIA

FAMILY	NUMBER OF GENERA	NUMBER OF SPECIES	NUMBER OF TAXA	INTRODUCED TAXA
Lycopsida				
Sellaginella-ceae	1	3	3	
Sphenopsida				
Equisetaceae	1	2	2	
Pteropsida				
Filicae				
Aspidiaceae	3	5	5	
Blechraceae	1	1	1	
Polypodiaceae	1	1	1	
Pteridaceae	5	16	19	
Coniferae				
Cupressaceae	1	3	3	
Pinaceae	2	7	8	
Gnetae				
Ephedraceae	1	8	10	
Angiosper-meae				
Monocotyle-doneae				
Agavaceae	3	9	13	
Alismaceae	1	1	1	
Amaryllida-ceae	7	19	25	
Arecaceae	1	1	1	
Cyperaceae	7	36	36	2
Iridaceae	1	3	3	
Juncaceae	1	14	15	
Juncaginaceae	1	2	2	
Lemnaceae	1	6	6	

FAMILY	NUMBER OF GENERA	NUMBER OF SPECIES	NUMBER OF TAXA	INTRODUCED TAXA
Pteropsida (continued)				
Liliaceae	5	11	14	
Najadaceae	1	2	2	
Orchidaceae	2	3	3	
Poaceae	55	186	198	47
Potamogeton-aceae	1	1	1	
Ruppiaceae	1	2	2	
Sparganiaceae	1	1	1	
Typhaceae	1	3	3	
Zannichelli-aceae	1	1	1	
Dicotyledoneae				
Aceraceae	1	2	2	
Aizoaceae	4	5	5	3
Amarantha-ceae	2	9	9	3
Anacardi-aceae	2	3	4	
Apiaceae	15	32	36	2
Apocynaceae	3	6	7	1
Asclepiada-ceae	4	13	13	
Asteraceae	99	281	334	17
Berberidaceae	1	4	5	
Betulaceae	2	2	2	
Bignoniaceae	2	2	2	1
Boraginaceae	9	63	73	
Brassicaceae	30	78	94	17
Buxaceae	1	1	1	

FAMILY	NUMBER OF GENERA	NUMBER OF SPECIES	NUMBER OF TAXA	INTRODUCED TAXA
Pteropsida (*continued*)				
Cactaceae	9	23	41	
Callitricha-ceae	1	1	1	
Campanula-ceae	4	11	13	
Capparaceae	5	9	11	
Caprifoliaceae	2	4	4	
Caryophylla-ceae	8	14	17	1
Celastraceae	1	1	1	
Ceratophylla-ceae	1	1	1	
Chenopodi-aceae	16	48	56	11
Convolvula-ceae	5	10	13	2
Crassulaceae	3	5	6	
Crossosomata-ceae	2	2	3	
Cucurbitaceae	5	7	8	2
Datiscaceae	1	1	1	
Eleagnaceae	2	2	2	1
Ericaceae	1	2	2	
Euphorbi-aceae	9	38	40	2
Fabaceae	26	135	167	11
Fagaceae	1	7	7	
Gentianaceae	1	7	7	
Geraniaceae	2	3	3	
Hydrophylla-ceae	9	55	70	
Hypericaceae	1	1	1	
Juglandaceae	1	1	1	
Krameriaceae	1	2	4	
Lamiaceae	13	30	32	1
Lennoaceae	2	2	2	
Linaceae	1	3	3	
Loasaceae	3	32	38	
Loganiaceae	1	1	1	
Lythraceae	1	1	1	
Malvaceae	9	19	24	
Martyniaceae	1	1	1	
Moraceae	1	2	2	2

FAMILY	NUMBER OF GENERA	NUMBER OF SPECIES	NUMBER OF TAXA	INTRODUCED TAXA
Pteropsida (*continued*)				
Myrtaceae	1	1	1	1
Nyctaginaceae	7	24	31	
Oleaceae	3	6	7	
Onagraceae	7	37	60	
Orobancha-ceae	1	7	9	
Papaveraceae	10	17	20	1
Plantagina-ceae	1	3	5	1
Polemoni-aceae	9	67	86	
Polygonaceae				
Polygalaceae	1	1	1	
Portulacaceae	5	9	12	1
Primulaceae	3	4	4	1
Ranuncula-ceae	6	10	11	
Rosedaceae	2	2	2	1
Rhamnaceae	5	9	12	
Rosaceae	17	33	37	4
Rubiaceae	1	12	17	1
Rutaceae	1	1	1	
Salicaceae	2	8	11	
Santalaceae	1	1	1	
Saururaceae	1	1	1	
Saxifragaceae	6	12	13	
Scrophulari-aceae	13	81	94	1
Simarouba-ceae	1	1	1	
Solanaceae	7	27	32	4
Sterculiaceae	2	2	2	
Tamaricaceae	1	6	6	6
Ulmaceae	1	2	2	
Urticaceae	3	5	5	1
Verbenaceae	2	8	8	1
Violaceae	1	2	3	
Viscaceae	2	8	8	
Vitaceae	1	1	1	
Zygophylla-ceae	4	8	8	2
Total	575	1836	2179	156

A CHECKLIST OF BIRDS OF DEATH VALLEY NATIONAL MONUMENT IN CALIFORNIA

Death Valley National Monument covers about 3,000 square miles of western desert terrain. The habitat varies from the salt pan below sea level to the sub-alpine conditions found on the summit of Telescope Peak which rises to 11,049 feet. Vegetation zones include creosote bush and desert holly at the lower elevations and range up through shadscale, blackbrush, Joshua Tree, pinyon-juniper to limber and bristlecone pine woodlands. Annual precipitation varies from 1.71 inches below sea level to about 15 inches or more in the higher mountains which surround the valley. The valley floor and lower slopes are almost devoid of vegetation, yet where water is available an abundance is usually present. Natural ponds and springs as well as a golf course and date orchard at Furnace Creek Ranch attract a number of unexpected species of birds that prefer moist habitats.

A network of roads and trails provides access to all of the vegetation zones. Be sure to check with a park ranger for travel details before venturing off main routes of travel.

This list of 258 species on the main list plus **74** on the supplemental list represents observation and studies reported primarily since 1933.

Codes for relative abundance and seasonal occurrence are listed below.*

COMMON indicates the species can usually be seen during season indicated and in appropriate habitat. Generally more than 10 per day per locality.

UNCOMMON indicates the species cannot always be seen unless hunted for in the appropriate habitat. Generally less than 10 per day per locality.

RARE indicates generally that few species are observed annually. A flock may be a single occurance.

*The common names on this check-list and their order conform to the 1977 American Ornithologists Union listing.

SUPPLEMENTAL LIST

CASUAL or ACCIDENTAL indicates the species has been observed few times and not on an annual basis. All are confirmed sightings.

C — common   S — March — May
U — uncommon   S — June — August
R — rare   F — September — November
               W — December — February

	S	S	F	W
Grebes—Podicipedidae				
Horned Grebe	r	r		
Eared Grebe	c	u	c	r
Western Grebe	r		r	
Pied-billed Grebe	c	u	c	u
Cormorant—Phalacrocoracidae				
Double-crested Cormorant	r		r	
Herons & Bitterns—Ardeidae				
Great Blue Heron	u	u	u	u
Green Heron	u	r	u	
Cattle Egret	r		r	
Great Egret	u	r	u	
Snowy Egret	u	r	u	
Black-crowned Night Heron	u	u	u	r
Least Bittern	r	r	r	
American Bittern	r		r	r
Ibises—Threskiornithidae				
White-faced Ibis	r	r	r	
Waterfowl—Anatidae				
Canada Goose	u		u	u
White-fronted Goose	r		r	r
Snow Goose	r		r	r
Mallard	c	u	c	c
Gadwall	u		u	u
Pintail	c	u	c	c
Green-winged Teal	c	u	c	c
Blue-winged Teal	u	r	u	
Cinnamon Teal	c	r	c	r
American Widgeon	c		c	c
Northern Shovler	c		c	c
Wood Duck	r		r	r

	S	S	F	W
Redhead	u	r	u	r
Ring-necked Duck	r		r	r
Canvasback	r		r	r
Lesser Scaup	r		r	r
Common Goldeneye	r		r	r
Bufflehead	u		r	u
Ruddy Duck	c	c	c	c
Hooded Merganser			r	r
Red-breasted Merganser	u	r		

Vultures—Cathartidae

	S	S	F	W
Turkey Vulture	c	c	c	

Kites, Hawks & Eagles—

Accipitridae

	S	S	F	W
Goshawk			r	
Sharp-shinned Hawk	r		u	u
Cooper's Hawk	u		u	u
Red-tailed Hawk	c	u	c	c
Red-shouldered Hawk			r	
Swainson's Hawk	r		r	
Ferruginous Hawk	r		r	r
Golden Eagle	r	r	r	r
Marsh Hawk	u		u	r

Ospreys—Pandionidae

	S	S	F	W
Osprey	r		r	

Caracaras & Falcons—Falconidae

	S	S	F	W
Prairie Falcon	u	u	u	u
Peregrine Falcon	r		r	r
Merlin			r	r
American Kestrel	u	r	u	u

Quail, Partridge & Pheasants—Phasianidae

	S	S	F	W
California Quail	r	r	r	r
Gambel's Quail	c	c	u	u
Mountain Quail	u	u	u	u
Chukar	c	c	c	c

Rails, Gallinules & Coots—Rallidea

	S	S	F	W
Virginia Rail	r	r	r	r
Sora	u		u	r
Common Gallinule	r	r	r	
American Coot	c	c	c	c

Avocets & Stilts—Recurvirostridae

	S	S	F	W
Black-necked Stilt	c	u	c	
American Avocet	c	u	c	

Plovers—Charadriidae

	S	S	F	W
Semipalmated Plover	r		r	
Snowy Plover	r			
Killdeer	c	u	c	u
Mountain Plover			r	
Black-bellied Plover	r			

Sandpipers—Scolopacidae

	S	S	F	W
Marbled Godwit	r		r	
Long-billed Curlew	r		r	
Greater Yellowlegs	u		u	
Lesser Yellowlegs	r		r	
Solitary Sandpiper	r		u	
Willet	r		r	
Spotted Sandpiper	c	u	c	
Common Snipe	c		c	c
Short-billed Dowitcher			u	
Long-billed Dowitcher	c		u	
Western Sandpiper	c		u	
Least Sandpiper	u		u	r
Baird's Sandpiper			r	
Pectoral Sandpiper			r	
Dunlin	r		r	

Phalaropes—Phalaropodidae

	S	S	F	W
Wilson's Phalarope	c	r	c	
Northern Phalarope	u		u	

Gulls & Terns—Laridae

	S	S	F	W
California Gull	u		u	
Ring-billed Gull	u		u	
Franklin's Gull	r		r	
Bonaparte's Gull	r		r	
Forster's Tern	r			
Black Tern	r		r	

Pigeons & Doves—Columbidae

	S	S	F	W
Band-tailed Pigeon	r		r	
Rock Dove	r	r	r	r
White-winged Dove	r	r	r	
Morning Dove	c	c	c	r

Cuckoos & Roadrunners—Cuculidae

	S	S	F	W
Yellow-billed Cuckoo	r		r	
Roadrunner	c	c	c	c

Owls—Tytonidae

	S	S	F	W
Screech Owl	r	r	r	r
Great Horned Owl	u	u	u	u
Burrowing Owl	r	r	r	r
Long-eared Owl	r	r	r	r
Short-eared Owl	r		r	
Saw-whet Owl	r		r	r

Goatsuckers—Caprimulgidae

	S	S	F	W
Poor-will	u	u	u	r
Common Nighthawk	r			
Lesser Nighthawk	c	c	c	

Swifts—Apodidae

	S	S	F	W
Vaux's Swift	u		u	
White-throated Swift	c	c	c	

Hummingbirds—Trochilidae

	S	S	F	W
Black-chinned Hummingbird	u			
Costas Hummingbird	c	c	r	
Broad-tailed Hummingbird	u	u		
Rufous Hummingbird	u	u	u	
Calliope Hummingbird	r			

Kingfishers—Alcedinidae

	S	S	F	W
Belted Kingfisher	u		u	

Woodpeckers—Picidae

	S	S	F	W
Common Flicker	u	r	c	c
Lewis' Woodpecker	u		u	u
Yellow-bellied Sapsucker	u		u	u
Williamson's Sapsucker				r
Hairy Woodpecker	r	r	r	r
Ladder-backed Woodpecker	r	r	r	r

Tyrant Flycatchers—Tyrannidae

	S	S	F	W
Eastern Kingbird	r	r	r	
Western Kingbird	c	u	c	
Cassin's Kingbird	r	r	r	
Ash-throated Flycatcher	c	c	c	
Black Phoebe	u	r	u	u
Say's Phoebe	c	c	c	c
Willow Flycatcher	c		u	
Hammond's Flycatcher	u		r	
Dusky Flycatcher	u		u	
Gray Flycatcher	u	u	u	
Western Flycatcher	u	r	u	
Western Wood Pewee	c	u	c	
Olive-sided Flycatcher	u		r	
Vermilion Flycatcher	r	r	r	r

Larks—Alaudidae

	S	S	F	W
Horned Lark	u	u	u	u

Swallows—Hirundinidae

	S	S	F	W
Violet-green Swallow	c	u	c	
Tree Swallow	c		c	
Bank Swallow	u	r	u	
Rough-winged Swallow	c	u	u	
Barn Swallow	c	r	c	
Cliff Swallow	c	u	u	

Jays, Magpies & Crows—Corvidae

	S	S	F	W
Steller's Jay			r	r

	S	S	F	W
Scrub Jay	u	u	u	u
Common Raven	c	c	c	c
Common Crow	r	r	r	r
Pinyon Jay	u	u	u	u
Clark's Nutcracker	u	u	u	u

Titmice, Verdins & Bushtits—Paridae

	S	S	F	W
Mountain Chickadee	c	c	c	c
Plain Titmouse	u	u	u	u
Verdin	r	r	r	r
Bushtit	u	u	u	u

Nuthatches—Sittidae

	S	S	F	W
White-breasted Nuthatch	u	u	u	u
Red-Breasted Nuthatch	r		u	r

Creepers—Certhiidae

	S	S	F	W
Brown Creeper			r	r

Wrens—Troglodytidae

	S	S	F	W
House Wren	u	r	u	
Winter Wren			r	r
Bewick's Wren	u	u	u	r
Long-billed Marsh Wren	u		c	r
Canyon Wren	u	u	u	u
Rock Wren	c	c	c	c

Mockingbirds & Thrashers—Mimidae

	S	S	F	W
Mockingbird	c	c	c	u
Brown Thrasher			r	
Le Conte's Thrasher	u	u	u	u
Crissal Thrasher	r	r	r	r
Sage Thrasher	r	r	r	

Thrushes, Solitaires & Bluebirds Turdidae

	S	S	F	W
American Robin	c	u	c	c
Varied Thrush	r		r	r
Hermit Thrush	u	u	u	u
Swainson's Thrush	u		u	
Western Bluebird	r	r	r	
Mountain Bluebird	u	u	u	u
Townsend's Solitaire	u	u	u	r

Gnatcatchers & Kinglets—Sylviidae

	S	S	F	W
Blue-gray Gnatcatcher	c	c	c	r
Black-tailed Gnatcatcher	r	r	r	r
Golden-crowned Kinglet			r	r
Ruby-crowned Kinglet	c	r	c	c

Pipits & Wagtails—Motacillidae

	S	S	F	W
Water Pipit	c		c	c

Waxwings—Bombycillidae

	S	S	F	W
Bohemian Waxwing			r	r
Cedar Waxwing	c		c	c

Silky Flycatchers—Ptilogonatidae

	S	S	F	W
Phainopepla	r	r	r	r

Shrikes—Laniidae

	S	S	F	W
Loggerhead Shrike	u	u	u	u

Starling—Sturnidae

	S	S	F	W
Starling	c	c	c	c

Vireos—Vireonidae

	S	S	F	W
Bell's Vireo	r	r	r	
Gray Vireo	r	r		
Solitary Vireo	u	r	r	
Red-eyed Vireo	r		r	
Warbling Vireo	c		c	

Wood Warblers—Parulidae

	S	S	F	W
Black and white Warbler	r		r	
Tennessee Warbler	r		r	
Orange-crowned Warbler	c	r	c	
Nashville Warbler	u		u	
Virginia's Warbler	r	r	r	
Lucy's Warbler	u	u	u	
Northern Parula	r		r	

	S	S	F	W
Yellow Warbler	c	r	c	
Black-throated Blue Warbler			r	
Yellow-rumped Warbler	c	u	c	c
Black-throated Gray Warbler	u	u	u	
Townsend's Warbler	u		u	
Hermit Warbler	r		r	
Blackpoll Warbler	r		r	
Ovenbird	r		r	
Northern Waterthrush	r		r	
MacGillivray's Warbler	c	r	c	
Common Yellowthroat	c	u	u	
Yellow-breasted Chat	u	u	r	
Wilson's Warbler	c		c	
American Redstart	r		r	

Weaver Finches—Ploceidae

	S	S	F	W
House Sparrow	c	c	c	c

Blackbirds & Orioles—Icteridae

	S	S	F	W
Bobolink	r		r	
Western Meadowlark	u	r	c	c
Yellow-headed Blackbird	c	u	c	r
Red-winged Blackbird	c	c	c	r
Hooded Oriole	r	r		
Scott's Oriole	u	u		
Northern Oriole	c	u	c	
Rusty Blackbird			r	
Brewer's Blackbird	c	c	c	c
Brown-headed Cowbird	c	c	c	

Tanagers—Thraupidae

	S	S	F	W
Western Tanager	c	u	c	
Summer Tanager	r		r	

Grosbeaks, Finches, Sparrows, & Buntings—Fringillidae

	S	S	F	W
Rose-breasted Grosbeak	r		r	
Black-headed Grosbeak	c	c	c	
Blue Grosbeak	u	u	r	
Indigo Bunting	r		r	
Lazuli Bunting	u	r	u	
Evening Grosbeak	u		u	
Purple Finch	r		r	r
Cassin's Finch	u	u	u	u
House Finch	c	c	c	c
Gray-crowned Rosy Finch				r
Pine Siskin	u		u	r
American Goldfinch	r		r	r
Lesser Goldfinch	u	u	u	r
Red Crossbill	r	r	r	
Green-tailed Towhee	r	u	u	
Rufous-sided Towhee	u	u	u	u
Lark Bunting			r	
Savannah Sparrow	c		c	u
Grasshopper Sparrow	r			
Vesper Sparrow	u		u	
Lark Sparrow	u		u	
Black-throated Sparrow	c	c	c	c
Sage Sparrow	c	c	c	c
Dark-eyed Junco	c	c	c	c
Gray-headed Junco	r	r	r	r
Tree Sparrow			r	
Chipping Sparrow	u	u	u	r
Clay-colored Sparrow	r		r	
Brewer's Sparrow	u	u	u	
Black-chinned Sparrow	r	r		
Harris' Sparrow	r		r	r
White-crowned Sparrow	c		c	c
Golden-crowned Sparrow	r		r	r
White-throated Sparrow	r		r	r
Fox Sparrow	r		r	r
Lincoln's Sparrow	c		c	r
Swamp Sparrow	r		r	r
Song Sparrow	u		u	r
Lapland Sparrow			r	
Chestnut-collared Longspur			r	

SUPPLEMENTAL LIST

The following Species have been identified in Death Valley National Monument but are either of casual or accidental occurrence.

Common Loon
White Pelican
Little Blue Heron
Wood Stork
Whistling Swan
Fulvous Whistling Duck
Europeon Widgeon
Greater Scaup
Barrow's Goldeneye
White-winged Scoter
Common Merganser
Mississippi Kite
Broad-winged Hawk
Sandhill Crane
Whimbrel
Upland Sandpiper

Ruddy Turnstone
Sanderling
Red Phalarope
Sabines Gull
Ground Dove
Inca Dove
Barn Owl
Pygmy Owl
Black Swift
Chimney Swift
Anna's Hummingbird
Acorn Woodpecker
Downy Woodpercker
White-headed Woodpecker
Scissor-tailed Flycatcher
Wied's Crested Flycatcher

Olivaceous Flycatcher
Eastern Phoebe
Least Flycatcher
Purple Martin
Black-billed Magpie
Pygmy Nuthatch
Cactus Wren
Gray Catbird
Bendire's Thrasher
Rufous-backed Robin
Northern Shrike
Yellow-throated Vireo
Philadelphia Vireo
Prothonotary Warbler
Golden-winged Warbler
Magnolia Warbler
Cap May Warbler
Black-throated Green
  Warbler
Blackburnian Warbler
Yellow-throated Warbler
Chestnut-sided Warbler

Bay-breasted Warbler
Prairie Warbler
Palm Warbler
Connecticut Warbler
Mourning Warbler
Hooded Warbler
Canada Warbler
Painted Redstart
Orchard Oriole
Tricolored Blackbird
Common Grackle
Great-tailed Grackle
Scarlet Tanager
Hepatic Tanager
Painted Bunting
Dickcisal
Lawrence's Goldfinch
LeConte's Sparrow
Sharp-tailed Sparrow
Rufous-crowned Sparrow
Snow Bunting

A CHECKLIST OF FISHES, AMPHIBIANS, REPTILES, AND MAMMALS OF DEATH VALLEY NATIONAL MONUMENT

Death Valley National Monument covers about 3,000 square miles of western desert terrain. The habitat varies from the salt pan below sea level to the sub-alpine conditions found on the summit of Telescope Peak which rises to 11,049 feet. Vegetation zones include creosote bush, desert holly, and mesquite at the lower elevations and range up through shadscale, blackbrush, Joshua tree, pinyon-juniper to limber and bristlecone pine woodlands. Annual precipitation varies from 1.68 inches below sea level to about 15 inches or more in the higher mountains which surround the valley. The valley floor and lower slopes are almost devoid of vegetation; yet where water is available, an abundance is usually present. Natural ponds, springs, and small streams provide habitats for five species of fish; a sixth introduced species lives in the irrigation ditches at Furnace Creek.

A network of roads and trails provides access to all of the vegetation zones. Be sure to check with a park ranger for travel details before venturing off main routes of travel.

This list of 6 fishes, 3 amphibians, 36 reptiles, and 53 mammals represents observations and studies reported primarily since 1933.

FISHES

_____AMARGOSA PUPFISH, **Cyprinodon nevadensis amargosa.** Found in the Amargosa River northwest of Saratoga Springs.

_____SARATOGA PUPFISH, **Cyprinodon nevadensis nevadensis.** Found in Saratoga . Springs at the south end of Death Valley.

_____DEVILS HOLE PUPFISH, **Cyprinodon diabolis.** Found in Devils Hole, 37 miles east of Death Valley in western Nevada.

_____SALT CREEK PUPFISH, **Cyprinodon salinus.** Found in Salt Creek in the central part of Death Valley.

_____COTTONBALL MARSH PUPFISH, **Cyprinodon mil-**

leri. Found in Cottonball Marsh on the west side of central Death Valley.

_____WESTERN MOSQUITOFISH, **Gambusia affinis.** Introduced into Furnace Creek irrigation ponds and streams.

AMPHIBIANS

_____RED SPOTTED TOAD, **Bufo punctatus.** Common at Furnace Creek; also found in water areas at Saratoga Springs, Cottonwood Springs, and at springs in Johnson, Hanaupah, and Emigrant Canyons.

_____PACIFIC TREEFROG, **Hyla regilla.** Water areas throughout Death Valley: Scotty's Castle, Furnace Creek, Saratoga Springs, and at springs in Johnson and Hanaupah Canyons.

_____BULLFROG. **Rana catesbeiana.** Introduced around 1920 at Furnace Creek.

REPTILES

_____DESERT TORTOISE, **Gopherus agassizi.** Found in the flats and surrounding foothills from 1,500 to 3,500 feet; lives in burrows.

_____DESERT BANDED GECKO, **Coleonyx variegatus variegatus.** Nocturnal; around springs and well-watered places from valley floor to 3,500 feet.

_____DESERT IGUANA, **Dipsosaurus dorsalis.** In and around mesquite hummocks and other similar locations with fine sandy soil; in low canyons and washes up to 3,000 feet; Sand Dunes, Westside Road, north of Furnace Creek Campground.

_____CHUCKWALLA, **Sauromalus obesus.** Areas of large rocks and boulders on alluvial fans and in canyons; throughout Death Valley up to 5,000 feet; Towne Pass and Dantes View Road.

_____ZEBRA-TAILED LIZARD, **Callisaurus draconoides.** Sandy and gravelly areas near dunes and in washes; common on roads in morning in spring, summer, and fall.

_____MOJAVE FRINGE-TOED LIZARD, **Uma scoparia.** Sand dunes southeast of Saratoga Springs.

_____COLLARED LIZARD, **Crotaphytus collaris.** Among rocks in hilly areas and washes, on slopes; from 1,000 to 5,000 feet.

_____LEOPARD LIZARD, **Crotaphytus wislizenii.** Valley floor to 3,600 feet on alluvial fans, in canyons and washes with scattered vegetation.

_____DESERT SPINY LIZARD, **Sceloporus magister magister.** Rocky slopes and canyons from 3,500 to 7,000 feet around vegetation.

_____GREAT BASIN FENCE LIZARD, **Sceloporus occidentalis biseriatus.** Rocky areas over wide elevation range; rock outcrops, canyons, near springs.

_____SAGEBRUSH LIZARD, **Sceloporus graciosus.** From sagebrush through pinyon-juniper up to 10,500 feet.

_____DESERT SIDE-BLOTCHED LIZARD, **Uta stansburiana stejnigeri.** Throughout Death Valley below 5,000 feet in gravelly and rocky areas.

_____WESTERN BRUSH LIZARD, **Urosaurus graciosus graciosus.** Low desert in and around creosote bush and mesquite.

_____SOUTHERN DESERT HORNED LIZARD, **Phrynosoma platyrhinos calidiarum.** Sandy, gravelly areas; low desert to over 5,000 feet.

_____DESERT NIGHT LIZARD, **Xantusia vigilis vigilis.** In and near Joshua trees; under debris; near Dantes View, over 9,000 feet in Panamint Mountains.

_____WESTERN SKINK, **Eumeces skiltonianus skiltonianus.** Moist areas with good cover in pinyon-juniper.

_____WESTERN RED-TAILED SKINK, **Eumeces gilberti rubricaudatus.** Found in isolated populations in Hanaupah and Johnson Canyons in the Panamints and near Grapevine Peak in the Grapevine Mountains.

_____GREAT BASIN WHIPTAIL, **Cnemidophorus tigris tigris.** Dry sandy areas with sparse vegetation; rocky areas of upper washes; mesquite thickets and vegetated areas of Greenwater Valley and Harrisburg Flat.

_____PANAMINT ALLIGATOR LIZARD, **Gerrhonotus panamintinus.** Panamint and Grapevine Mountains above 3,500 feet.

_____WESTERN BLIND SNAKE, **Leptotyphlops humilis.** Nocturnal; under rocks, among roots on brush covered slopes; from below sea level to 4,000 feet.

_____DESERT ROSY BOA, **Lichanura trivigata gracia.** Low foothills and canyons below 4,500 feet; in sandy and gravelly habitats.

_____WESTERN LEAF-NOSED SNAKE, **Phyllorhynchus decurtatus perkinsi.** Nocturnal; sandy and gravelly soil; rocky foothills.

_____COACHWHIP (RED RACER), **Masticophis flagellum piceus.** Sandy mesquite hummocks, gravelly desert; rocky foothills.

_____STRIPED WHIPSNAKE, **Masticophis taeniatus.** Willow Creek in Black Mountains; Hunter Spring in Cottonwood Mountains.

_____DESERT PATCH-NOSED SNAKE, **Salvadora hexalepis hexalepis.** Rocky and sandy areas from lower slopes and washes up to Towne Pass.

_____DESERT GLOSSY SNAKE, **Arizona elegans eburnata.** Nocturnal; mostly in sandy areas, also gravel.

_____GREAT BASIN GOPHER SNAKE, **Pituophis melanoleucus deserticola.** From rock-strewn desert foothills into mountains.

_____CALIFORNIA KINGSNAKE, **Lampropeltis getulus californiae.** Panamint Mountains from Emigrant Canyon to Wildrose; Daylight Pass.

_____WESTERN LONG-NOSED SNAKE, **Rhinocheilus lecontei lecontei.** Nocturnal; Aquereberry Point, Towne Pass, Daylight Pass.

_____WESTERN GROUND SNAKE, **Sonora semiannulata.** Sandy or fine gravel soil to over 4,000 feet; Wildrose Canyon, Greenwater, Daylight Pass.

_____MOJAVE SHOVEL-NOSED SNAKE, **Chionactus occipitalis occipitalis.** Sandy areas of the low desert; sage flats.

_____UTAH BLACK-HEADED SNAKE, **Tantilla planiceps utahensis.** Nocturnal; Panamint Mountains.

_____CALIFORNIA LYRE SNAKE, **Trimorphodon vandenburghi.** From sea level to over 4,000 feet in rocky areas.

_____DESERT NIGHT SNAKE, **Hypsiglena Torquata.** Many hatbitats from below sea level to over 5,000 feet.

_____PANAMINT RATTLESNAKE, **Crotalus mitchelli stephensi.** Below sea level to over 7,000 feet usually in foothills and mountains.

_____MOJAVE DESERT SIDEWINDER, **Crotalus cerastes cerastes.** Mesquite hummocks; nocturnal; from below sea level to 4,500 feet.

MAMMALS

_____YUMA MYOTIS, **Myotis yumanensis.** Cave bat found in well-watered areas.

_____LONG-LEGGED MYOTIS, **Myotis volans.** Around well-watered areas.

_____CALIFORNIA MYOTIS, **Myotis californicus.** Roosts in caves and buildings.

_____SMALL-FOOTED MYOTIS, **Myotis subulatus.** Cave bat.

_____WESTERN PIPISTREL, **Pipistrellus hesperus.** Roosts in rock crevices.

_____SILVER-HAIRED BAT. **Lasionycteris noctivagans.** Around water in forested areas.

————HOARY BAT, **Lasiurus cinereus.** Roosts in trees, found around well-watered areas.

————PALLID BAT, **Antrozous pallidus.** Roosts in crevices in caves.

————MEXICAN FREE-TAILED BAT. **Tadarida mexicana.** Roosts in caves, crevices, and buildings.

————BIG FREETAIL BAT, **Tadarida macrotis.** Roosts in caves, mines, crevices, and buildings.

————RINGTAIL CAT, **Bassariscus astutus.** Nocturnal; rock terrain in the arid brush and tree areas.

————SPOTTED SKUNK, **Spilogale putorius.** Mountains surrounding Death Valley.

————BADGER, **Taxidea taxus.** Low desert into mountains.

————KIT FOX, **Vulpes velox.** Common through most of Death Valley; Sand Dunes and Furnace Creek; nocturnal.

————GRAY FOX, **Urocyon cinereoargenteus.** East side of Grapevine Mountains.

————COYOTE, **Canis latrans.** From salt flats into mountains; common around mesquite thickets.

————MOUNTAIN LION, **Felis concolor.** Surrounding mountains.

————BOBCAT, **Lynx rufus.** From sea level into mountains.

————CALIFORNIA GROUND SQUIRREL, **Citellus beecheyi.** Hunter Mountain area of Cottonwood Mountains.

————WHITETAIL ANTELOPE SQUIRREL, **Citellus leucurus.** Mesquite hummocks of valley floor to over 6,000 feet in mountains; common along roadsides.

————ROUNDTAIL GROUND SQUIRREL, **Citellus tereticaudus.** Low desert; mesquite thickets near Furnace Creek; common along roadsides.

————PANAMINT CHIPMUNK, **Eutamias panamintinus.** Pinyon-juniper belt of Grapevine and Panamint Mountains.

————PIGMY POCKET GOPHER, **Thomomys umbrinus oreocus.** Higher elevations in surrounding mountains; up to 10,000 feet of Telescope Peak.

————PANAMINT POCKET GOPHER, **Thomomys umbrinus scapterus.** Panamint and Grapevine Mountains.

————LITTLE POCKET MOUSE, **Perognathus longimembris.** Sage habitat at Harrisburg Flat.

————GREAT BASIN POCKET MOUSE, **Perognathus parvus.** Grapevine Mountains.

————LONGTAIL POCKET MOUSE, **Perognathus formosus mohavensis.** Grapevine Mountains.

————UTAH LONGTAIL POCKET MOUSE, **Perognathus formosus formosus.** Valley floor into mountains on east side of Death Valley.

————DESERT POCKET MOUSE, **Perognathus penicillatus.** Mesquite Flat.

————PANAMINT KANGAROO RAT, **Dipodomys panamintinus.** Northern Panamint Mountains between 6,000 and 7,000 feet.

————MERRIAM KANGAROO RAT, **Dipodomys merriami.** Dry sandy soil on the valley floor.

————GREAT BASIN KANGAROO RAT. **Dipodomys microps.** Harrisburg Flat in dry sandy soil with sparse vegetation.

————DESERT KANGAROO RAT, **Dipodomys deserti.** Dry locations on valley floor, especially around mesquite.

————SOUTHERN GRASSHOPPER MOUSE, **Onychomys torridus.** Throughout Death Valley below 5,500 feet.

————CACTUS MOUSE, **Peromyscus eremicus.** Higher elevations in Grapevine and Cottonwood Mountains.

————CANYON MOUSE, **Peromyscus crinitus.** Surrounding mountains and rocky canyons.

————DEER MOUSE, **Peromyscus maniculatus.** Valley floor and mountains.

————BRUSH MOUSE, **Peromyscus boylei.** Northern Panamint Mountains.

————PIÑON MOUSE, **Peromyscus truei.** Rocky areas in pinyon-juniper belt.

————DESERT WOODRAT, **Neotoma lepida.** From salt marshes into surrounding mountains.

————DUSKY-FOOTED WOODRAT, **Neotoma fuscipes.** Furnace Creek.

————BUSHYTAIL WOODRAT, **Neotoma cinerea.** Pinyon-juniper area of northern Panamint Mountains.

————CALIFORNIA VOLE, **Microtus californicus.** Northern Panamint Mountains; southern Black Mountains near Willow Creek.

————WESTERN HARVEST MOUSE, **Reithrodontomys megalotis.** Well-watered areas: Salt Creek, Furnace Creek, Hanaupah Canyon, Wildrose.

————HOUSE MOUSE, **Mus musculus.** In and around human dwellings.

————PORCUPINE, **Erethizon dorsatum.** Grapevine, Panamint, and Cottonwood Mountains.

————BLACKTAIL JACKRABBIT, **Lepus californicus.** Near valley floor and in mountains.

————MOUNTAIN COTTONTAIL, **Sylvilagus nuttalli.** Surrounding mountains.

————DESERT COTTONTAIL, **Sylvilagus auduboni.** Mesquite thickets on valley floor.

————MULE DEER, **Odocoileus hemionus.** Along eastern and western boundaries of the monument in Panamint, Cottonwood, and Grapevine Mountains.

————DESERT BIGHORN SHEEP, **Ovis canadensis nelsoni.** Throughout Death Valley at all elevations; inaccessible ridges and canyons, usually near water.

————BURRO, **Equus assinus.** Introduced in 1880's; Panamint and Cottonwood Mountains.

————HORSE, **Equus caballus. Introduced;** Hunter Mountain, Cottonwood Basin, Pinto Peak.

A CHECKLIST OF AMPHIBIANS AND REPTILES OF THE JOSHUA TREE NATIONAL MONUMENT

Joshua Tree National Monument, established in 1936, consists of approximately 577,000 acres of Mohave and Colorado Desert. Elevations within the Monument range from about 305 meters (1000 feet) in the Pinto Basin to 1772 meters (5814 feet) at the summit of Quail Mountain in the Little San Bernardino Mountains.

This checklist includes 37 species of reptiles and 2 species of amphibians.* Most reptiles are active during the spring, summer and fall months. Lizards are usually seen during the daytime hours, while many snakes limit their activities to the hours of darkness. Amphibians are commonly found only near springs, oases, and other areas where open water is likely to occur during the summer and fall rainy seasons.

For the purposes of this checklist, three broad habitat types have been recognized: low desert (305 to 914 meters), high desert (914 to 1280 meters) and pinyon woodland (1280 to 1772 meters). The status and habitat codes below refer only to Joshua Tree National Monument and not to the entire geographic range of a particular species.

Key:

Status	Habitat
C — Common	L — Low Desert
U — Uncommon	H — High Desert
R — Rare	P — Pinyon Woodland
P? — Possibly Occurs	S — Springs

*Revised January 1979, P. Knuckles

Reproduced through courtesy of Joshua Tree Natural History Association.

	Status	Habitat
AMPHIBIANS—CLASS AMPHIBIA		
True Toads—Family Bufonidae		
Red-spotted Toad		
Bufo punctatus	C	S
Treefrogs—Family Hylidae		
California Treefrog		
Hyla californae	C	S
REPTILES—CLASS REPTILIA		
Tortoises—Family Testudinidae		
Desert Tortoise		
Goperus agassizi	U	L, H
Geckos—Family Gekkonidae		
Banded Gecko		
Coleonyx variegatus	R	L, H, P
Iguanid Lizards—Family Iguanidae		
Desert Iguana		
Dipsosaurus dorsalis	C	L
Chuckwalla		
Sauromalus obesus	C	L, H
Zebra-tailed Lizard		
Callisaurus draconides	U	L
Mohave Fringe-toed Lizard		
Uma scopariea	R	L
Collared Lizard		
Crotaphytus collaris	C	L, H, P
Leopard Lizard		
Crotaphytus wislizenii	U	L
Desert Spiny Lizard		
Sceloporus magister	C	L, H

	Status	Habitat
Iguanid Lizards—Family Iguanidae (*continued*)		
Western Fence Lizard		
Sceloporus occidentalis	C	H, P
Side-blotched Lizard		
Uta stansburiana	C	L, H, P
Long-tailed Brush Lizard		
Urosaurus graciosus	R	L
Coast Horned Lizard		
Phyrnosoma coronatum	U	P
Desert Horned Lizard		
Phyrnosoma platyrhinos	U	L, H
Night Lizards—Family Xantusiidae		
Desert Night Lizard		
Xantusia vigilis	U	H
Skinks—Family Scincidae		
Gilbert's Skink		
Eumeces gilberti	U	L, H, P
Whiptail Lizards—Family Teiidae		
Western Whiptail		
Cnemidophorous Tigris	C	L, H, P
Slender Blind Snakes—Family Leptotyphlopidae		
Western Blind Snake		
Leptotyphlops humilis	R	L
Boas—Family Boidae		
Rosy Boa		
Lichanura trivirgata	U	L, H
Colubrid Snakes—Family Colubridae		
Spotted Leaf-nosed Snake		
Phyllorhynchus decurtatus	U	L
Coachwhip (Red Racer)		
Masticophis Flagellum	C	L, H, P

	Status	Habitat
Colubrid Snakes—Family Colubridae (*continued*)		
Striped Racer		
Masticophis lateralis	U	P
Western Patch-nosed Snake		
Salvadora hexalepis	C	L, H
Glossy Snake		
Arizona elegans	U	L, H
Gopher (Bull) Snake		
Pituophis melanoleucus	C	L, H, P
Common Kingsnake		
Lampropeltis getulus	U	L, H, P
Long-nosed Snake		
Rhinocheilus lecontei	R	L
Western Shovel-nosed Snake		
Chionactis occipitalis	U	L
Western Black-headed Snake		
Tantilla planiceps	R	P
California Lyre Snake		
Trimorphodon vandenburghi	R	L,H
Night Snake		
Hypsiglena torquata	R	L, H, P
Vipers—Family Viperidae		
Western Diamondback Rattlesnake		
Crotalus atrox	R	L
Red Diamond Rattlesnake		
Crotalus ruber	P?	L
Speckled (Mitchell's) Rattlesnake		
Crotalus mitchelli	C	H, P
Sidewinder (Horned) Rattlesnake		
Crotalus cerastes	C	L, H
Western Rattlesnake		
Crotalus viridis	U	L, H, P
Mohave Rattlesnake		
Crotalus scutulatus	U	H, P

A CHECKLIST OF THE MAMMALS OF THE JOSHUA TREE NATIONAL MONUMENT

Joshua Tree National Monument, established in 1936, consists of approximately 233,954 hectares (577,000 acres) of Mohave and Colorado Desert. Elevations within the monument range from about 305 meters (1000 feet) in the Pinto Basin to 1772 meters (5814 feet) at the summit of Quail Mountain in the Little San Bernardino Mountains.

For the purposes of this checklist, three broad habitat types have been recognized: low desert (305 to 914 meters), high desert (914 to 1280 meters) and pinon woodland (1280 to 1772 meters).* These correspond to the creosote bush, yucca and pinon belts described by Miller and Stebbins (*Lives of Desert Animals of Joshua Tree National Monument*, Univ. California Press, 1964. 452 pp.).

This checklist includes 46 species which have been reported from Joshua Tree National Monument and 12 additional species which may occur there but have not been verified. The status and habitat codes below refer only to Joshua Tree National Monument and not to the entire geographic range of a particular species.

Key:	
Status	**Habitat**
C Common	L Low Desert
U Uncommon	H High Desert
R Rare	P Pinon Woodland
P Possibly Occurs	

*Compiled by John E. Cornely, Department of Biological Sciences, Northern Arizona University, Flagstaff, Arizona 86001

Reproduced through courtesy of Joshua Tree Natural History Association.

	Status	Habitat
INSECTIVORES—ORDER INSECTIVORA		
Shrews—Family Soricidae		
Desert Shrew		
Notiosorex crawfordi		P
BATS—ORDER CHIROPTERA		
Leaf-nosed Bats—Family Phyllostomatidae		
California Leaf-nosed bat		
Macrotus californicus	U	L
Insectivorous Bats—Family Vespertilionidae		
Yuma Myotis		
Myotis yumanensis	P	
Long-eared Myotis		
Myotis evotis	P	
Fringed Myotis		
Myotis thysanodes	U	P
Long-legged Myotis		
Myotis volans	U	P
California Myotis		
Myotis californicus	C	H, P
Small-footed Myotis		
Myotis leibii	P	
Western Pipistrelle		
Pipistrellus hesperus	C	L, H, P
Big Brown Bat		
Eptesicus fuscus	C	L, H, P
Red Bat		
Lasiurus borealis	P	
Hoary Bat		
Lasiurus cinereus	U	P
Southern Yellow Bat		
Lasiurus ega	R	H
Spotted Bat		
Euderma maculata	R	L

	Status	Habitat
Insectivorous Bats—Family Vespertilionidae (*continued*)		
Townsend's Big-eared Bat		
Plecotus townsendii	P	
Pallid Bat		
Antrozous pallidus	C	L, H
Free-tailed Bats—Family Molossidae		
Brazilian Free-tailed Bat		
Tadarida brasiliensis	P	
Pocketed Free-tailed Bat		
Tadarida femorosacca	R	P
Big Free-tailed Bat		
Tadarida macrotis	P	
Western Mastiff Bat		
Eumops perotis	P	

LAGOMORPHS—ORDER LAGOMORPHA

	Status	Habitat
Hares and Rabbits—Family Leporidae		
Desert Cottontail		
Sylvilagus audubonii	C	L, H, P
Black-tailed Jackrabbit		
Lepus californicus	C	L, H, P

RODENTS—ORDER RODENTIA

	Status	Habitat
Squirrels—Family Sciuridae		
Merriam's Chipmunk		
Eutamias merriami	U	P
White-tailed Antelope Squirrel		
Ammospermophilus leucurus	C	L, H, P
California Ground Squirrel		
Spermophilus beecheyi	C	H, P
Round-tailed Ground Squirrel		
Spermophilus tereticaudus	U	L
Pocket Gophers—Family Heteromyidae		
Botta's Pocket Gopher		
Thomomys bottae	C	L, H, P
Heteromyids—Family Heteromyidae		
Little Pocket Mouse		
Perognathus longimembris	C	L, H, P
Long-tailed Pocket Mouse		
Perognathus formosus	U	L, H
Desert Pocket Mouse		
Perognathus penicillatus	U	L, H
San Diego Pocket Mouse		
Perognathus fallax	C	L, H, P
Spiny Pocket Mouse		
Perognathus spinatus	U	L, H
Chisel-toothed Kangaroo Rat		
Dipodomys micropus	R	H, P
Desert Kangaroo Rat		
Dipodomys deserti	U	L
Merriam's Kangaroo Rat		
Dipodomys merriami	C	L, H
New World Rats and Mice—Family Cricetidae		
Cactus Mouse		
Peromyscus eremicus	C	L, H

	Status	Habitat
New World Rats and Mice—Family Cricetidae (*continued*)		
Deer Mouse		
Peromyscus maniculatus	U	H, P
Canyon Mouse		
Peromyscus crinitus	C	L, H, P
Brush Mouse		
Peromyscus boylii	U	P
Pinon Mouse		
Peromyscus truei	C	P
Southern Grasshopper Mouse		
Onychomys torridus	U	H, P
White-throated Woodrat		
Neotoma albigula	U	H
Desert Woodrat		
Neotoma lepida	C	L, H, P
Dusky-footed Woodrat		
Neotoma fuscipes	U	P
Old World Rats and Mice—Family Muridae		
House Mouse		
Mus musculus	U	L, H

CARNIVORES—ORDER CARNIVORA

	Status	Habitat
Canids—Family Canidae		
Coyote		
Canis latrans	C	L, H, P
Kit Fox		
Vulpes macrotis	U	L,H
Gray Fox		
Urocyon cinereoargenteus	C	H, P
Procyonids—Family Procyonidae		
Ringtail		
Bassariscus astutus	P	
Weasels and Relatives—Family Mustelidae		
Long-tailed Weasel		
Mustela frenata	P	
Badger		
Taxidea taxus	U	L, H
Western Spotted Skunk		
Spilogale gracilis	R	P
Striped Skunk		
Mephitus mephitus	P	
Cats—Family Felidae		
Mountain Lion		
Felis concolor	R	P
Bobcat		
Felis rufus	C	L, H, P

EVEN-TOED UNGULATES—ORDER ARTIODACTYLA

	Status	Habitat
Deer and Relatives—Family Cervidae		
Mule Deer		
Odocoileus hemionus	U	P
Bovids—Family Bovidae		
Mountain Sheep		
Ovis canadensis	U	H, P

A CHECKLIST OF BIRDS OF THE JOSHUA TREE NATIONAL MONUMENT

This check list, consisting of 219 species, is based on random field observations dating from 1930 through 1977.* Some reports are sketchy and unconfirmed; therefore, the Visitor Center staff would appreciate receiving field notes from any observer concerning all birds seen in Joshua Tree National Monument.

This check list is an aid, not an absolute. It is subject to change and variations from year to year. The relative frequencies listed below apply only if the observer visits the appropriate habitat in the proper season.

Key:

SP– spring:	March, April, May
SU– summer:	June, July, August
FA– fall:	September, October, November
WI– winter:	December, January, February
A– abundant:	one or more per day
C– common:	one or more per week
O– occasional:	one or more per month
R– rare:	one or more per season
Cas– casual:	not seen every year
Acc– accidental:	out of its range
N– known to nest here	
N?– juveniles seen, but nesting unconfirmed	

*Revised January, 1978, P. & D. Knuckler

Reproduced through courtesy of Joshua Tree Natural History Association.

	SP	SU	FA	WI
Loons and Grebes				
Common Loon	Cas			
Eared Grebe	Cas		Cas	
Pied-billed Grebe	Cas	Cas		
Pelicans				
Brown Pelican		Cas	Cas	
White Pelican	Cas			
Geese and Ducks				
Canada Goose				R
Black Brant	Cas			
Snow Goose	Cas		Cas	Cas
Mallard—N	O	O	O	O
Blue-Winged Teal	Cas	Cas		
Cinnamon Teal	Cas		Cas	
Green-winged Teal	Cas	Cas	Cas	R
Redhead				Cas
Ring-necked Duck	Cas		Cas	R
Ruddy Duck	Cas		R	O
Common Merganser			Cas	
Hooded Merganser			Cas	
Vultures, Hawks, Eagles, & Falcons				
Turkey Vulture	C	O	C	R
Cooper's Hawk—N	R	R	O	O
Sharp-shinned Hawk	R		R	R
Harrier (Marsh Hawk)	Cas		R	R
Rough-legged Hawk				Cas
Ferruginous Hawk	R		R	
Red-tailed Hawk—N	C	C	C	C
Swainson's Hawk	R	R	R	R
Golden Eagle	O	O	O	O
Bald Eagle	Cas		Cas	
Osprey	Cas		Cas	

	SP	SU	FA	WI
Vultures, Hawks, Eagles, & Falcons (*continued*)				
Prairie Falcon—N	O	R	O	O
Peregrine Falcon	Cas		Cas	
Merlin (Pigeon Hawk)	Cas		Cas	
Kestrel (Sparrow Hawk)	C	O	O	O
Gallinaceous Birds				
Gambels's Quail—N	A	A	A	A
Mountain Quail—N	A	A	A	A
Chukar	Cas	Cas	Cas	
Herons and Ibises				
Common Egret	Cas	Cas	Cas	
Snowy Egret		Cas	Cas	
Great Blue Heron	Cas			
Black-Crowned Night Heron		Cas	Cas	
American Bittern			Cas	
Rails				
Virginia Rail			Cas	
Sora			Cas	
Common Gallinule			Cas	Cas
American Coot—N	C	C	C	C
Shorebirds				
American Avocet	Cas	Cas	Cas	
Black-necked Stilt		Cas	Cas	
Killdeer	Cas	Cas	R	Cas
Solitary Sandpiper	Cas		Cas	
Spotted Sandpiper	Cas			
Long-billed Dowitcher	Cas			
Least Sandpiper	Cas			
Western Sandpiper	Cas			
Gulls				
Ring-billed Gull	Cas			
Pigeons and Doves				
Band-tailed Pigeon	R	R		
White-winged Dove—N	O	O		Cas
Rock Dove	R			
Mourning Dove—	C	A	C	R
Ground Dove	R	R	R	
Cuckoos				
Yellow-billed Cuckoo		Cas		
Roadrunner—N	O	O	C	R
Owls				
Screech Owl	O	O	O	O
Great-horned Owl—N	O	O	O	O
Long-eared Owl	C	Cas	C	O
Barn Owl	Cas	Cas	Cas	
Burrowing Owl	R	R	R	O
Saw-whet Owl				Cas
Elf Owl—N	Cas	Cas	Cas	
Goatsuckers				
Poor-Will	C	C	C	R
Lesser Nighthawk—N?	O	O		
Swifts and Hummingbirds				
Vaux's Swift	O	R	O	

	SP	SU	FA	WI
Swifts and Hummingbirds (*continued*)				
White-throated Swift—N	A	C	C	C
Calliope Hummingbird	R			
Anna's Hummingbird—N	A	O	A	C
Black-chinned Hummingbird	Cas			
Costa's Hummingbird—N	C	C	R	O
Rufous Hummingbird—N	C	Cas	C	
Allen's Hummingbird	R		R	
Kingfishers				
Belted Kingfisher	Cas		Cas	
Woodpeckers				
Common Flicker	C	O	A	C
Gila Woodpecker		Cas		
Ladder-backed Woodpecker—N	A	C	C	C
Nuttall's Woodpecker	R		R	
Red-headed Woodpecker	Acc	Acc		
Acorn Woodpecker	Cas	Cas	Cas	
Lewis's Woodpecker	Cas		Cas	Cas
Yellow-bellied Sapsucker	R		R	R
Williamson's Sapsucker			Cas	Cas
Downy Woodpecker			Cas	
Flycatchers				
Vermilion Flycatcher	Cas	Cas		Cas
Western Kingbird—N	C	C	Cas	
Cassin's Kingbird—N?	R	Cas		
Ash-throated Flycatcher—N	O	O		
Black Phoebe	R	R	R	Cas
Say's Phoebe—N	C	O	C	O
Traill's Flycatcher	R	R	O	
Hammond's Flycatcher	R		R	
Dusky Flycatcher	Cas		Cas	
Gray Flycatcher	Cas			
Western Flycatcher	O	Cas	O	
Western Wood Pewee	O	R	O	
Olive-sided Flycatcher	R	Cas	Cas	
Larks				
Horned Lark—N	C	R	A	C
Swallows				
Barn Swallow	R		R	
Cliff Swallow—N	R			
Violet-green Swallow	O	Cas	O	R
Tree Swallow	R			
Rough-winged Swallow	R	O	Cas	
Jays, Crows and Ravens				
Steller's Jay	R	R		
Scrub Jay—N?	A	A	A	C
Pinyon Jay—N	R	O	O	R
Clark's Nutcracker		Cas	R	Cas
Common Raven—N	A	A	A	A
Chickadees and Titmice				
Mountain Chickadee	Cas	O	C	R
Plain Titmouse—N	C	C	C	C
Verdin—N	C	O	C	O

	SP	SU	FA	WI
Chickadees and Titmice (*continued*)				
Common Bushtit—N	C	O	C	C
Wrentit	Cas			
Nuthatches				
White-breasted Nuthatch			O	O
Red-breasted Nuthatch		R	O	
Pygmy Nuthatch		R	R	
Wrens				
House Wren	O	Cas	O	O
Winter Wren			R	R
Bewick's Wren—N?	C	O	C	O
Cactus Wren—N	A	C	A	A
Rock Wren—N	C	C	C	C
Canyon Wren—N	O	R	C	R
Long-billed Marsh Wren				Cas
Mockingbirds and Thrashers				
Mockingbird—N	A	O	A	Cas
Brown Thrasher	Acc		Acc	
Sage Thrasher—N	C	R	C	O
Bendire's Thrasher—N	O	R	Cas	
California Thrasher—N	R	O	O	R
Le Conte's Thrasher—N	C	O	O	O
Crissal Thrasher—N	R	Cas	O	O
Thrushes				
Robin	C		C	C
Varied Thrush			Cas	
Townsend's Solitaire	O	Cas	O	O
Hermit Thrush	O		O	O
Swainson's Thrush	O	Cas	R	Cas
Veery	Acc			
Western Bluebird—N	O	Cas	C	O
Mountain Bluebird—N	C		C	C
Gnatcatchers and Kinglets				
Blue-gray Gnatcatcher—N	C	C	C	R
Black-tailed Gnatcatcher—N	O	O	O	O
Golden-crowned Kinglet	Cas		Cas	
Ruby-crowned Kinglet	A		A	C
Pipits				
Water Pipit			R	Cas
Waxwings and Silky Flycatchers				
Cedar Waxwing	O	Cas	O	O
Phainopepla—N	A	A	A	A
Shrikes and Starlings				
Loggerhead Shrike—N	A	A	A	C
Starling	C	C	C	C
Vireos				
Gray Vireo—N	R	Cas	Cas	
Solitary Vireo	O		O	
Bell's Vireo	R			
Hutton's Vireo	R		R	
Warbling Vireo	O	Cas	O	
Wood Warblers				
Black & White Warbler	Acc			

	SP	SU	FA	WI
Wood Warblers (*continued*)				
Tennessee Warbler	Acc			
Orange-Crowned Warbler	A	O	A	
Nashville Warbler	O	Cas	O	
Virginia's Warbler	Acc		Acc	
Lucy's Warbler			Cas	
Yellow Warbler	O	Cas	R	
Yellow-rumped Warbler (Audobon's)	A	Cas	C	R
Hermit Warbler	R	Cas	Cas	
Townsend's Warbler	O		Cas	Cas
Black-throated Gray Warbler	O	Cas	O	
Yellow-throated Warbler	Acc		Acc	
Yellowthroat	R		Cas	
Yellow-Breasted Chat	R	Cas		
Maggillivray's Warbler	O	R	O	
Wilson's Warbler	A	R	C	
American Redstart	Acc	Acc	Acc	
Painted Redstart	Acc		Acc	
Weaver Finches				
House Sparrow—N	C	C	Cas	
Meadowlarks, Blackbirds and Orioles				
Western Meadowlark	O	Cas	C	O
Yellow-headed Blackbird	O	R	R	
Red-winged Blackbird	R		R	R
Rusty Blackbird	O	R	O	
Brown-headed Cowbird—N?	O	R	R	
Orchard Oriole	Acc	Acc		
Scott's Oriole—N	C	C	Cas	
Hooded Oriole—N	C	O	O	
Northern Oriole (Bullock's)	C	O	O	
Tanagers				
Western Tanager—N	C	O	O	
Summer Tanager	Cas			
Grosbeaks, Buntings and Finches				
Pyrrhuloxia	Acc			
Rose-breasted Grosbeak		Acc	Acc	
Black-headed Grosbeak	C	O	O	
Lazuli Bunting	O	R	O	
Purple Finch	O		O	
Cassin's Finch	R		Cas	Cas
House Finch—N	A	A	A	A
Pine Siskin	R	Cas	R	Cas
American Goldfinch	R		R	R
Lesser Goldfinch—N	C	O	C	O
Lawrence's Goldfinch—N	C		O	
Crossbills and Towhees				
Dickcissel	Acc			
Green-tailed Towhee	C	Cas	O	
Rufous-sided Towhee	O	R	O	O
Brown Towhee	R	R		
Sparrows and Juncos				
Savannah Sparrow	O		C	O
Grasshopper Sparrow			Cas	

Sparrows and Juncos (*continued*)	SP	SU	FA	WI
Vesper Sparrow			Cas	
Black-throated Sparrow—N	C	A	C	O
Sage Sparrow	C	R	A	C
Dark-eyed Junco (Oregon, etc.)	C		A	A
Rufous-crowned Sparrow	R			
Cassin's Sparrow	Acc			
Chipping Sparrow	O	R	O	O
Brewer's Sparrow—N	O	R	O	O

Sparrows and Juncos (*continued*)	SP	SU	FA	WI
Black-chinned Sparrow—N?	R	Cas	Cas	
Harris's Sparrow	Acc			
White-Crowned Sparrow	A		A	A
Golden-Crowned Sparrow	R		R	Cas
Fox Sparrow	R		R	
Lincoln's Sparrow	R		Cas	Cas
Song Sparrow	R		R	Cas

The records available on the following species are not completely satisfactory or need to be confirmed. If you observe any of these species, please record the sighting at the Twentynine Palms Visitor Center.

Horned Grebe
American Widgeon
Canvasback
Lesser Scaup
Bufflehead
White-tailed Kite
Harris Hawk
California Quail
Green Heron

White-faced Ibis
Black-bellied Plover
Long-billed Curlew
Marbled Godwit
Greater Yellowlegs
Sanderling
Wilson's Phalarope
Northern Phalarope
California Gull

Black Tern
Inca Dove
Common Nighthawk
Broad-tailed Hummingbird
Hairy Woodpecker
Wied's Crested Flycatcher
Bank Swallow
Purple Martin
Dipper
Sprague's Pipit
Bohemian Waxwing
Parula Warbler
Black-throated Blue Warbler
Ovenbird

Northern Waterthrush
Connecticut Warbler
Tricolored Blackbird
Evening Grosbeak
Blue Grosbeak
Varied Bunting
Painted Bunting
Red Crossbill
Abert's Towhee
Baird's Sparrow
Lark Sparrow
Yellow-eyed Junco
White-throated Sparrow

5

THE SONORAN DESERT

FRANK S. CROSSWHITE AND CAROL D. CROSSWHITE

FOR MAKING AVAILABLE unpublished manuscripts, we are indebted to Paul E. Damon, Muhammed Shafiqullah, and Daniel J. Lynch for geology of the Southern Basin and Range Province, and Kim Cliffton, Dennis Oscar Cornejo, and Richard S. Felger for sea turtles. Raymond Turner provided a large number of base maps of the Sonoran Desert upon which the rough data for the finished maps were compiled. For reading the manuscript in whole or in part and for contributing valued suggestions, we thank Gordon L. Bender, E. Lendell Cockrum, Richard S. Felger, Bernard Fontana, Charles T. Mason, Jr., James L. Patton, and Newell A. Younggren. Particular appreciation is extended to Nile Jones for making the satellite imagery available, to C. Allan Morgan for the photography, and to Russell Vogt for the cartography. For access to data in the files of the Arizona Game and Fish Department, we thank Robert A. Jantzen, director.

A semipermanent subtropical area of atmospheric high pressure off the Pacific Coast of North America intensified with increasing temperature in the Late Eocene, over 40 million years ago, to create increasing drought in the 23°-30° N latitude region to the east. Heat and drought have intensified to the present day, with minor trend reversals owing to the same climatic fluctuations that resulted in glaciation in the northern part of the continent in the Pliocene and Pleistocene.

What is now Baja California was wrenched from the mainland by tectonic rifting along the San Andreas system, and the Gulf of California was formed in the center of the present Sonoran Desert about 12 million years ago. Associated with this phenomenon was the plastic stretching, thinning, and faulting of the earth's crust inland to create the basin and range topography characteristic of most of the Sonoran Desert. Crustal thinning, subsidence, and erosion lowered the average

elevation of the region and consequently intensified the heat and drought.

The Sonoran Desert emerged as a true regional desert during the interglacials (Axelrod 1979). Fluctuation of sea level and subsidence and emergence of islands in the Gulf of California resulted in the isolation of organisms and fascinating patterns of endemism. Other patterns involving the plants and animals represent a large field of study still requiring much investigation. Critical study of life forms and adaptive strategies appears to be in its infancy. An insight into the subject of man in this desert is gained by studying the indigenous O'odham people.

Without question the Sonoran Desert is the most complex of North American deserts. The presence of the Gulf of California in the center of it increases the complexity. The desert is composed of seven regional divisions, each recognized on the basis of distinctive vegetation, which has responded to influences of elevation, latitude, geology, soil, and climate. Each of the seven divisions described by Shreve (1951) is of a level of complexity that can be compared to that of the Mojave Desert alone, and there is a growing tendency for the Mojave to be considered as an eighth division of the Sonoran Desert. Winter temperatures in the Mojave are much colder than those in the Sonoran, however, and on the whole Sonoran Desert vegetation would not tolerate the killing frosts there.

The Sonoran Desert differs from the Chihuahuan, Mojave, and Great Basin deserts in being subtropical. Indeed this is probably the most important cause of diversity, the entire spectrum between temperate and tropical being represented. Axelrod (1979) pointed out that the richest desert floras in the world as a whole are those of the margins of the tropics because their geographic areas have had dry climates for a much longer period of time than the cool (Mojave) or cold (Great Basin) deserts.

The Sonoran Desert, unlike many others, centers about water, be it seawater, river or lake water, or pumped groundwater coursing through irrigation canals. In one year in the

A Cooperative Project of the Boyce Thompson Southwestern Arboretum and the Department of Plant Sciences, University of Arizona.

Arizona Agricultural Experiment Station Contribution No. _____ (Received _____, 1980).

southern Arizona portion of the desert alone, enough water is pumped from the ground to cover a football field with a column of water over a thousand miles in height (Peirce 1979). It is incorrect to restrict discussion of the desert to its xeric elements alone because the food chains and ecologic interactions do not allow it.

The dominant geographic feature of the Sonoran Desert is the Gulf of California around which the desert is situated and to which most Sonoran Desert streams and rivers naturally flow. Indeed Shreve (1951) stated that as a geographic entity, the Sonoran Desert could best be described as the region immediately surrounding the upper two-thirds of the Gulf of California. Considerably more than half of the Sonoran Desert lies within fifty miles of a seacoast. Except for the portion that lies in Baja California, the desert is practically locked between the mountain ranges to the north and east and the water of the Gulf to the south and west. With this in mind, its boundaries are easy to distinguish. It is necessary only to make a critical division at the point of separation of the Mojave Desert from the Sonoran along a line between Needles and Indio California, and the dividing line between desert and thorn forest in southern Sonora. It is instructive to note that authors have sometimes chosen not to make these separations. For example, Paylore (1976) included the Mojave Desert in the Sonoran, and Dunbier (1968) included the entire coastal plain from southern Sonora across the Rio Mayo, Rio Fuerte, Rio Sinaloa, Rio Mocorito, and Rio Culiacan as a southward extension of the Sonoran Desert.

The Sonoran Desert consists of 310,000 square kilometers (km) (119,000 square miles) and has 3,800 km (2360 miles) of seacoast. About three-fourths of the land surface is in southern Arizona and western Sonora.

Literature pertaining to the Sonoran Desert is abundant, but there remain significant gaps of knowledge. Publications often deal with the Sonoran Desert within political boundaries rather than as a whole. Some of the maps prepared for this chapter represent the first attempts known to the authors to portray features such as geology and rainfall in any degree of detail for the entire desert.

The book by Dunbier (1968) on the geography, economy, and people of this desert integrates information from many sources and remains useful. An essential book for those planning detailed studies of the desert is the extensive bibliography of the Sonoran Desert edited by Paylore (1976). The complexity of this desert is indicated by the fact that a treatment of the plants alone by Shreve and Wiggins (1964) extends to 1740 printed pages in two volumes.

GEOLOGY

Surface geologic features of the Sonoran Desert are mapped in figure 5-1 according to rock type and in figure 5-2 according to time of rock formation. The oldest rocks of the Sonoran Desert were formed nearly 2 billion years ago and are rather common surface features in parts of the desert today. Together with various diabasic or granitic intrusions, they are probably the basement rock underlying many other rocks. The Early Precambrian rocks were probably laid down in a great elongated sinking trough (geosyncline) in the form of mudstones and sandstones composed of material weathered away from the primeval crust of the young earth. Basaltic lavas were locally interbedded with the sedimentary material. Thickness of deposition in this trough amounted to 20,000 feet or more. This great deposit is known today in its metamorphosed form—schist, gneiss, and amphibolite (Wilson 1962).

Metamorphosis of the older Precambrian rock appears to extend back 1.7 billion years. By studying potassium-argon isotope degradation of mica minerals of Precambrian rocks, Damon (1968) was able to derive a chronology of geologic revolutions, disturbances, or orogenies that had "reset the isotopic clocks" of these rocks. Evidence supported three major Arizonan Precambrian events: the Arizonan Revolution (previously questioned by geologists), the Mazatzal Revolution, and the Grand Canyon Disturbance. These three well-separated waves of activity occurred during the first 700 million years of geologic history of the area. There followed a quiescent period occupying the greater portion of 1 billion years. The long calm came to an end in what is now Arizona and Sonora with the occurrence of the Laramide Orogeny and subsequent Basin and Range Disturbance, events occurring in only the last 5 percent of the overall time period, with the last disturbance apparently not yet concluded.

In light of modern knowledge concerning plate tectonics and continental drift, much of the geology of the Sonoran Desert can be explained by the region's being in a zone of influence of the collision between continental and sea-floor plates. In addition to the geologic revolutions or disturbances, the Nevadan Revolution, before the Laramide Orogeny, was important to the West. When land that is now Sonoran Desert lay next to the ocean, it was possible for pieces of sea-floor crust to be subducted under the land crust. Being forced between molten mantle and hardened land crust, the subducted material melted, expanded, and induced igneous activity in the zone where it came to rest.

Igneous activity in what is now Sonoran Desert has ranged from intrusions of plutonic rock to extrusions of volcanic material. Faulting and mountain building are the characteristic disturbances associated with the igneous revolutions. Ore bodies containing metals from the mantle were placed during periods of upheaval. Noteworthy economically for the Sonoran Desert was the Laramide Orogeny at the end of the Cretaceous and beginning of the Cenozoic.

Erosion and deposition followed mountain building. Sedimentary rocks dating back to the Precambrian are well known. Some sedimentary series were rather widespread, characteristic, and of well-established stratigraphy. Very well known are

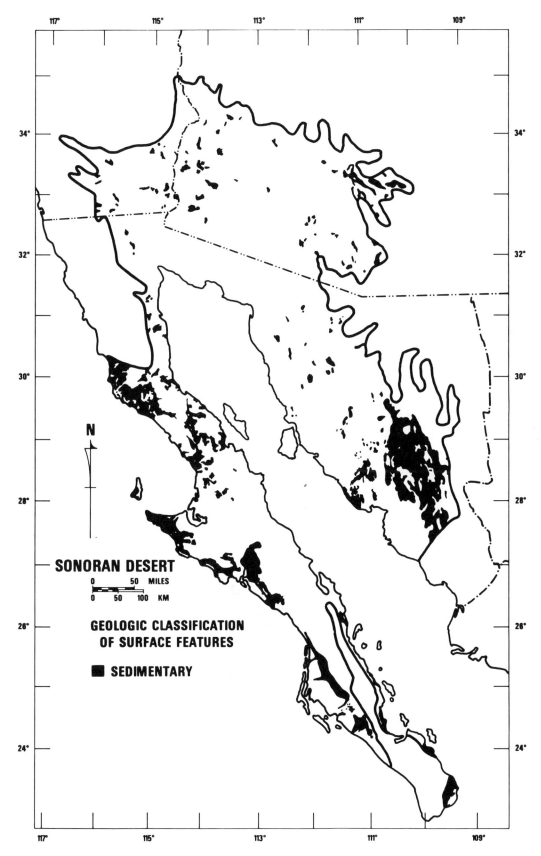

117°   115°   113°   111°   109°

N

SONORAN DESERT

0   50   MILES
0   50   100   KM

GEOLOGIC CLASSIFICATION
OF SURFACE FEATURES

■ SEDIMENTARY

Figure 5-1. Geology of the Sonoran Desert by rock types.
Compiled by the authors from United States and Mexican government surveys. Cartography by
Russell Vogt, University of Arizona graphics department.

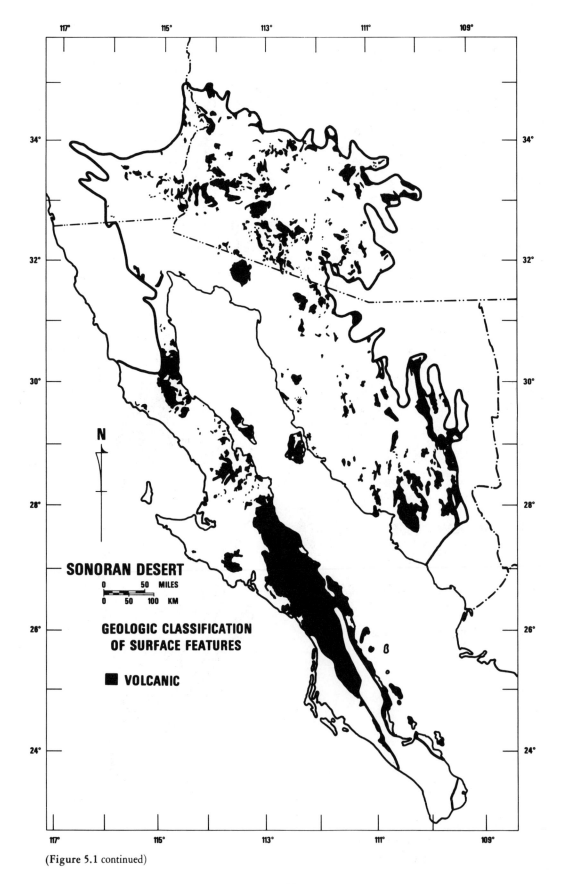

SONORAN DESERT

0   50 MILES
0   50   100 KM

GEOLOGIC CLASSIFICATION
OF SURFACE FEATURES

VOLCANIC

(Figure 5.1 continued)

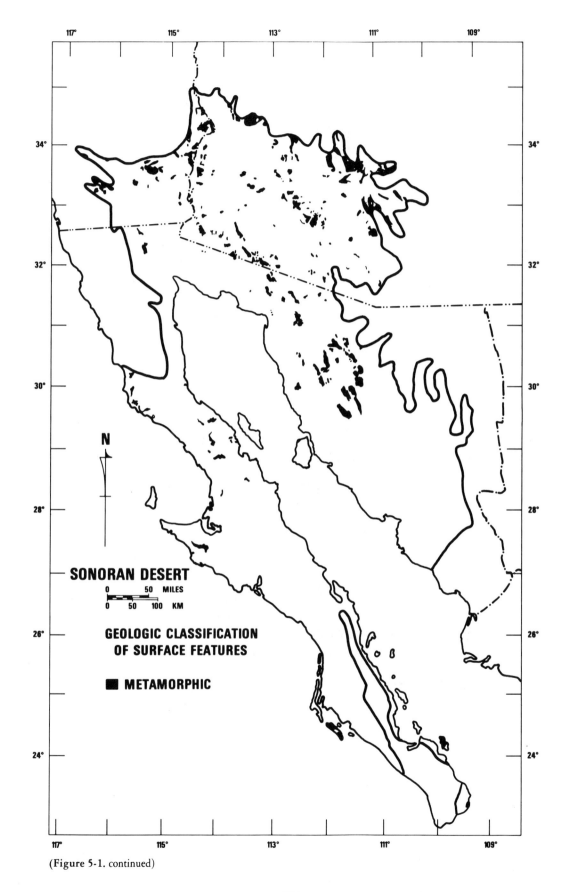

SONORAN DESERT

0   50   MILES

0   50   100   KM

**GEOLOGIC CLASSIFICATION
OF SURFACE FEATURES**

■ **METAMORPHIC**

(Figure 5-1. continued)

167

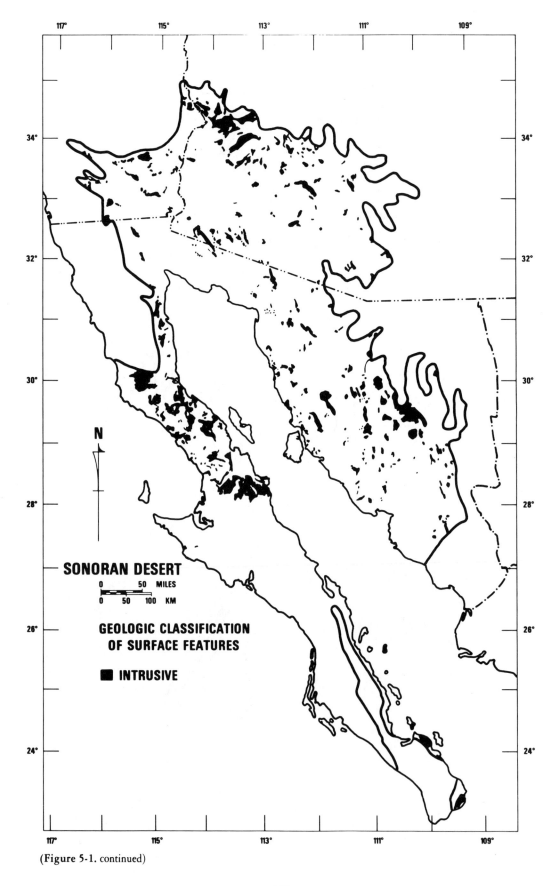

N

SONORAN DESERT

0   50   MILES
0   50   100   KM

GEOLOGIC CLASSIFICATION
OF SURFACE FEATURES

■ INTRUSIVE

(Figure 5-1. continued)

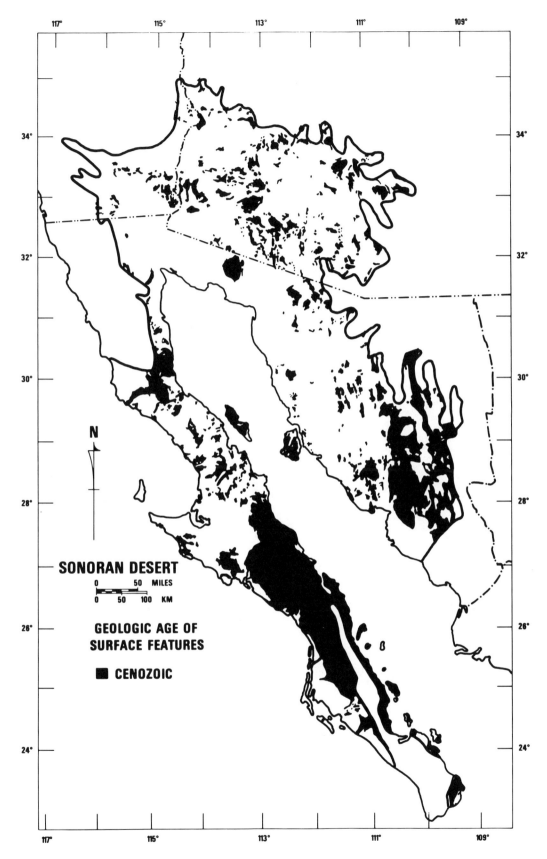

Figure 5-2. Geology of the Sonoran Desert by age of rocks.
Compiled by the authors from United States and Mexican government surveys. Cartography by
Russell Vogt, University of Arizona graphics department.

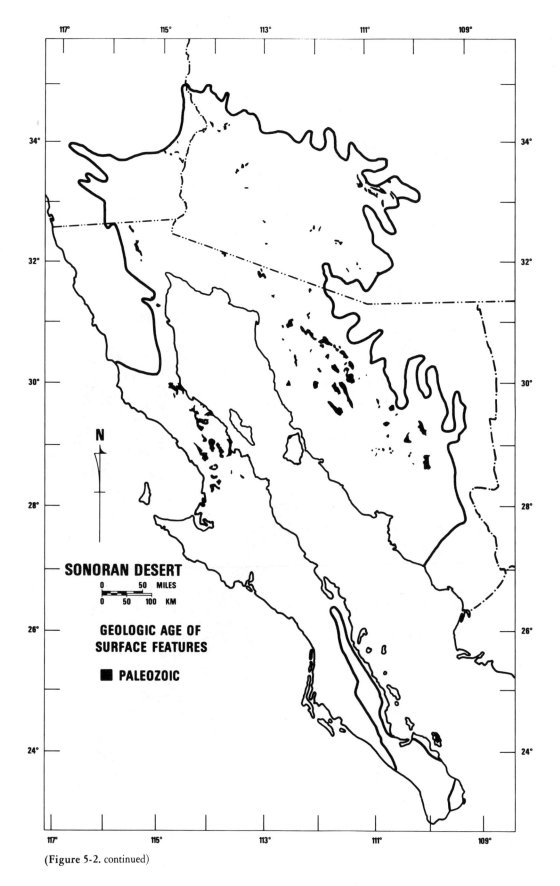

SONORAN DESERT

0   50 MILES
0   50   100 KM

GEOLOGIC AGE OF
SURFACE FEATURES

■ PALEOZOIC

(Figure 5-2. continued)

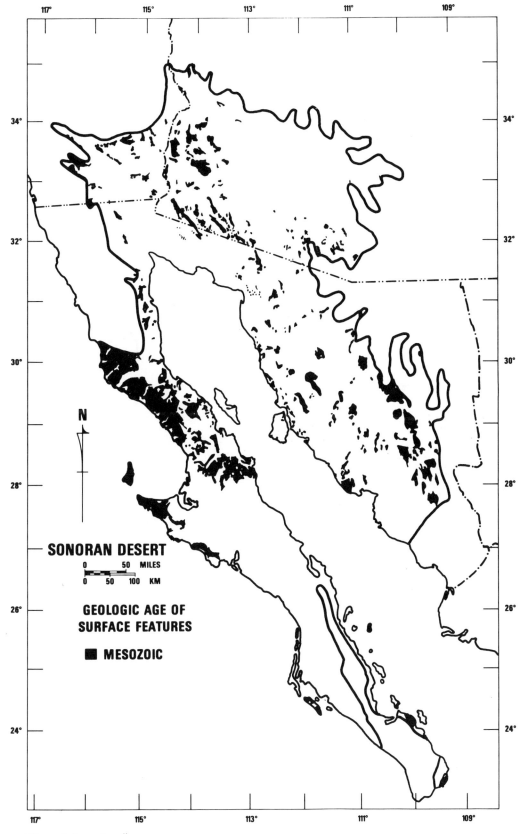

SONORAN DESERT

0   50 MILES

0 50 100 KM

GEOLOGIC AGE OF
SURFACE FEATURES

■ MESOZOIC

(Figure 5-2. continued)

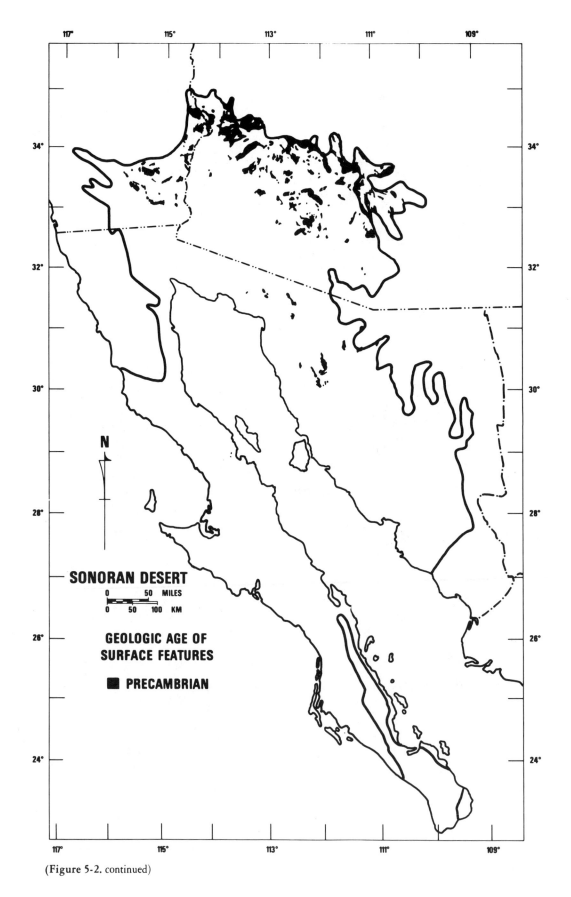

(Figure 5-2. continued)

the Younger Precambrian sediments known as the Apache Group, laid down on the Older Precambrian schist—Scanlan Conglomerate, Pioneer Shale, Barnes Conglomerate, Dripping Spring Quartzite and Mescal Limestone. Presumably these sediments were created when the mountain ranges that had arisen in the Arizonan or Mazatzal revolutions were beveled to a peneplain. Diabase was intruded under the Apache Group rocks. The nature of the trough into which the Apache Group sediments were deposited is not certain, but an ocean-continent interface is suggested by the Barnes unit, which consists of flattened ellipsoidal pebbles that must have traveled long distances on their way to the sea before being subjected to ocean wave buffeting. Although these pebbles are found within the Sonoran Desert, similar ones are to be found in the process of becoming flattened ellipsoids on Pacific Ocean beaches such as at Carlsbad, California.

Cambrian seas spread over the Apache Group rocks, and there was repeated flooding in Devonian, Mississippian, Pennsylvanian, and Permian times in what is now the Sonoran Desert. Great thicknesses of ocean sediments were sometimes interfingered with deltaic, floodplain, and sand-dune deposits. Most of these Paleozoic rocks, including limestone units up to 1000 feet thick or more, have eroded away or lie hidden. Where these limestones still exist, such as near Superior, Arizona, where they have been covered by a protective layer of dacitic lava, the various units can be identified on the basis of the abundant fossils of shellfish and other marine fossils that they contain.

Uplift occurred at the end of the Paleozoic. Sedimentary and volcanic rocks that formed in the Jurassic and Triassic are no longer present on the surface. Although erosion is cited as the most probable reason that the Paleozoic and Mesozoic sediments are largely absent, companies drilling for oil in the Sonoran Desert do so on the theory (unproven) that the sedimentary units were covered by a great overthrust of older rocks.

Baja California

As the Jurassic gave way to the Cretaceous, the Nevadan Revolution occurred. Cretaceous granite-like rocks of the California Batholith, together with related metamorphic rocks formed by intrusion of the batholith, are the basement rocks of present-day Baja California and the Pacific Coast of Mexico south to Jalisco, but the Gulf of California now cuts across these rocks. The pattern indicates that Baja California was part of the mainland of Mexico and that land now within the city limits of San Diego, California, once was slightly south of what is now Guaymas, Sonora (see diagrams in Gastil and Jensky 1973).

Similar evidence for major displacement in this region comes from finding conglomerates in southern California that were deposited in the Eocene and Oligocene and seem to be derived from source materials far away in southern Arizona

and Sonora (Doyle and Gorsline 1977, Sage 1973, Merriam 1968, Woodford et al. 1968).

Studies by Minch (1971) and by Abbott and Smith (1978) reinforce the concept of Baja California and the portion of California west of the San Andreas Fault having been literally wrenched northwestward. Composite satellite imagery (figure 5-3) clearly shows the Salton Trough (which connects the Gulf of California with the San Andreas Fault) to separate rifted land to the west from the original mainland to the northeast. Pliocene strike-dip displacement along the San Andreas system and ocean-floor spreading with subcontinental magma welling up into the rift are significant features of the chronology of Baja California and the Gulf. Rift chronology of this area has been summarized by Gastil et al. (1972, 1975), Gastil and Jensky (1973), and Gastil and Krummenacher (1977).

As a prelude to separation of Baja California from the mainland, there was a simple trench-arc system along the west coast of Mexico. A chain of volcanoes near what became the Gulf spread andesite and basalt eastward into what is now Sonora and westward into what is now Baja California (Karig and Jensky 1972). An ignimbrite flare-up occurred to the east. Rifting itself did not begin before the last 10 million or 15 million years.

There are currently active spreading centers in the Gulf of California region. De La Fuente Duch (1973) conducted aeromagnetic studies of the Colorado River delta to investigate a previously uncharted spreading center (designated Panga de Abajo). Such studies have been conducted with the expectation that spreading areas could be associated with exploitable geothermal energy fields.

Hamilton (1971) reported on Gemini and Apollo space photographs of Baja California. Some previously unrecognized structural features were seen. Young faults, monoclines, and warps created a stepped topography. This structural, rather than erosional, origin suggested that other older topographic steps such as in the Sierra Nevada may have had similar origins.

Sea-Level Changes and Ages of Desert Islands

During the Pleistocene the most important changes in Baja California probably involved sea-level changes rather than climate (Auffenberg and Milstead 1965). Such sea-level changes have been estimated to include drops of more than 100 meters and rises of up to 30 m (Flint 1971). These fluctuations had little effect other than to create or drown many islands in the Gulf of California (Durham and Allison 1960). Examples of such youthful islands given by Soulé and Sloan (1966), partly on the basis of faunistic data, are Tiburón, San Marcos, Coronados, San José, San Francisco, Espíritu Santo, and numerous shallow-water islands (also Felger and Lowe 1976). Other islands are considerably older and were apparently formed when the peninsula was torn away from mainland Mexico. This group includes Santa Catalina, San Diego, and Santa Cruz.

Salton Trough

Muffler and Doe (1968) studied composition and mean age of detritus of the Colorado River delta in the Salton Trough. To the north, the structural depression of the Gulf of California is filled with fine-grained sandstones and siltstones of the Colorado River delta. These are Late Cenozoic, average 20,000 feet in thickness, and derive from weathering and redeposition of Mesozoic sedimentary rocks of the Upper Colorado Basin states.

The present Salton Sea occurs in the bottom of what was once a dry lake, Lake Cahuilla. Radiocarbon dating of fish bones indicates that Lake Cahuilla dried up about three hundred years ago (Theilig et al. 1978). The Salton Sea was accidentally created by man in 1905 when floodwaters of the Colorado River washed out canal headgates south of Yuma,

Figure 5-3. Satellite imagery (NASA-ERTS) showing the northwesterly trending Salton Trough with agricultural fields of the Coachella Valley northwest of the Salton Sea (upper center) and those of the Imperial Valley and Baja California to the southeast.
The trough connects the Gulf of California to the southeast with the San Andreas Fault to the northwest. All of the land mass between the trough and the Pacific Ocean (lower left) was attached to the coast of Sonora fifteen million years ago before rifting.

Arizona, and the whole river changed course and began to flow into the Imperial Valley of southern California. The river cut through the land with a tremendous force, removing twice as much earth in a few months than had been excavated for the Panama Canal in ten years. This amounted to 450 million cubic yards. The Southern Pacific Railroad brought in train-loads of rock quarried wherever it could be obtained between New Orleans and the Pacific Coast, dumping carloads by the hundreds in the breech. By 1907 the river was returned to its normal channel.

Blake (1914), who had discovered the Salton Trough in 1853 during a U.S. government expedition for an overland route to the Pacific, discussed the deltaic silts and sands, sandhills, travertine deposits associated with Lake Cahuilla, silty lacustrine deposits, and volcanic buttes. The drying up of Lake Cahuilla is depicted as leaving a low, wide area at the head of the Gulf of California below sea level, flat and extending northwest about 200 miles beyond the present tidewater. The Salton Sea and the Salton Trough are shown by satellite imagery in figure 5-3. Figure 5-4 consists of satellite

Figure 5-4. Satellite imagery (NASA-ERTS) showing fine sediments of the Colorado River delta choking the head of the Gulf of California.
Orderly rows of dunes in the Gran Desierto (upper right) of Sonora contrast markedly with the rugged topography of Baja California (lower left).

imagery showing the connection of the trough with the head of the Gulf of California. The present delta of the Colorado is particularly prominent. The extensive sand dunes in figure 5-4 derive from Colorado River sediments that were washed down to the trough from the Upper Colorado Basin states. The construction of Hoover Dam in 1935 marked the end of development of the Colorado River delta as a major geologic entity (Sykes 1938).

Southern Basin and Range Province (SBRP)*

With the exception of Baja California, which is more an extension of southern California's complex physiography, the Sonoran Desert is essentially the southern part of Fenneman's (1928) Basin and Range Physiographic Province. This southern area is quite different from the regions that surround it, and it has come to be given the distinctive name Southern Basin and Range Province.

To the west of the SBRP, the Mojave Desert is characterized by fault systems that have lateral rather than vertical movement. To the northwest, the Great Basin is higher, altitudinally and latitudinally colder, with a significant amount of snow resulting in a different style of erosion, and with a greater frequency of internal drainage. The Colorado Plateau in the north is of much greater stability, having a crust roughly twice the thickness observed in the Sonoran Desert and a tectonic history that has not produced block faulting with subsidence of valleys. To the east are volcanic ash flows of the Sierra Madre Occidental in Mexico and other volcanic areas of New Mexico.

Damon et al. (1981) pointed out that nearly half of the production of copper in the United States comes from basin and range country, that the sediments filling the valleys represent important aquifers for agricultural and urban areas in the SBRP, and that knowledge of events that caused the peculiar physiography (and continue to influence it) give reason for reflection concerning ongoing tectonic events.

Government explorations in the mid-nineteenth century for railroad routes to connect California with the East resulted in the first geologic study of the basin and range country as a whole. These early geologists supposed that the topography had been sculptured by water and wind, but streams were often small, and external drainage from some valleys was low or nonexistent. Formation of this distinctive topography by erosion and deposition would have required enormous rainfall and large, raging rivers to transport huge quantities of material. There was a paradox here. The climate was arid. Low precipitation had resulted in desert weathering of many of the

* Special acknowledgment is made to Paul E. Damon, Muhammed Shafiqullah, and Daniel J. Lynch for making available unpublished material used in preparing this section.

mountains. The mountains were frequently seen to display a virtual checkerboard of bedrock geology, the presence and configuration of each mountain seeming to have less to do with its rock type than with deep-seated structure-controlling events.

Basin and range topography in the western United States was convincingly attributed to block faulting by Gilbert (1928), who established that valleys were down-dropped relative to the mountains. Thus it became clearly established that deep-seated structure-controlling events had produced the basin and range phenomenon. A good example is Picacho Peak in the southern Arizona portion of the SBRP. It rises from the landscape like a volcanic plug. Examination has proven it to be a faulted mountain block from the side of a volcano otherwise gone (Shafiqullah et al. 1976).

Older series of rocks that must have once been continuous over the entire landscape now appear as isolated outcroppings that are so deformed and out of place that early investigators experienced difficulty in arriving at the true explanation for the overall geology of the SBRP.

Sediments filling the valleys obscure the bedrock and structural features between mountains. Although there are isolated examples of sand covering the low areas between rather homogeneous mountain peaks that are obviously connected (for example, the Sierra del Rosario in northwestern Sonora, figures 5-4 and 5-5), deformation is usually so complete that a simple buried connection between mountains cannot explain the situation observed. Extrusive and sedimentary (conglomerate) layers of Cenozoic age usually cannot be traced beyond a given range of mountain. Correlating sediments of the Late Cenozoic was long difficult or impossible because of the scarcity or absence of diagnostic fossils.

Newer techniques of geologic investigation, borrowed from other sciences, have finally clarified the once-confusing situation. Over the last decade, potassium-argon dates have been derived for either extrusive, coarse-grained igneous, or metamorphic rocks (Damon et al. 1981). These age determinations have bracketed the time periods during which specific geologic processes were occurring and have also yielded dates for specific events.

Erosion and deposition rates, as well as the number of years before present when specific events of deformation took place in the SBRP, can be calculated from isotope studies of lava that has been faulted, eroded, tilted, or buried. Cores from deep drilling for oil, water, or minerals were obtained by Damon et al. (1981) and other workers for isotopic dating by the potassium-argon technique. In addition, studies of gravitational anomalies and the responses of artificial seismic waves have revealed deep bedrock structures where excavation would be impossible and drilling would be difficult.

The structural complexities of crust and mantle that caused the valleys to subside relative to the mountains in the SBRP were discussed by Damon et al. (1981). The thin, brittle crust is cooler and lighter in weight than the hot, dense mantle to

Figure 5-5. Satellite imagery (NASA-ERTS) showing the round Pinacate lava field (center) with volcanic craters in Sonora, and series of northwest trending mountains:
Rosario, Gila, Tinajas Altas, Cabeza Prieta, Sierra Pinta, Mohawk, Granite and Growler mountains. The trend lines are perpendicular to the direction of stretch in the earth's crust during third stage tectonic activity explained in the text. Volcanism shows no relationship with downfaulted valleys. Clouds are rolling in over the Gulf of California (bottom).

the interior of the earth and literally floats on its surface. The crust is subjected to gravitational and other stresses while floating, and when it breaks, the molten magma flows with each shift of the crust. Mountains, which are but thicker portions of the crust, are similar to icebergs in that they extend downward as well as upward.

That the mantle in the Basin and Range country is anomalously warmer and less dense than mantle material elsewhere is

inferred from studies by Pakiser (1963) showing low seismic wave velocities at the base of the crust and from heat flow statistics for the crust roughly double the values elsewhere (Warren et al. 1969). In addition, Langston and Helmberger (1974) have shown that the crust of the Basin and Range area is peculiarly thin, only half or two-thirds the thickness in some surrounding geologic provinces. Heat in the Magma Mine, located at the junction of the three subdivisions of the SBRP described by Damon et al. (1981), is so intense as to provoke serious discussion concerning the nearness of the mantle.

Topography and geologic deformation in the Basin and Range country can be explained by a theory that also accounts for the thinness observed in the crust (Damon et al., 1981). Shifting of the large plates that compose the outer layer of the earth results in areas where land masses and ocean floor collide and others where they are pulled apart. By theories of plate tectonics now accepted by geologists, the Basin and Range situation has resulted from a stretching of the crust in directions perpendicular to the mountain trend lines. Anomalies in the underlying mantle may be due to plastic movements in response to alterations in the crust.

The stability of the earth's outer layer depends on physical support. When the layer is stretched, support is removed and a section of the earth's crust may slump down and push into the mantle, which responds plastically and may push upward elsewhere. Splits in the crust (faults) form perpendicular to the direction of stretching.

As the crust of the SBRP thinned with stretching, the entire province was probably reduced in elevation. Although in structurally determined mountain and valley systems the valleys normally are considered subsidized graben elements and the mountain blocks upthrusted horst elements, when stretching has been prolonged, as in the case of the SBRP, heavy mountains may slowly sink into the mantle. Some mountains in the SBRP currently are sinking in relation to others (Damon et al. 1981).

Shreve (1951) noted that the SBRP mountains in southern California and extreme western Arizona were narrow, elongated, and dropped abruptly to the plain in what he considered true basin and range form. To the east in Arizona and Sonora, although the basin and range situation prevailed, the mountain shapes and slopes were seen to vary.

Damon et al. (1981) pointed out that the Sonoran Desert in reality covers only the western section of the SBRP. In the Lower Colorado-Gila vegetational division, the arid mountains indeed have sharp topographic features, have little soil and little vegetation, and are relatively narrow in comparison with the valleys. The mountain slopes rise sharply above the relatively flat valleys without significant accumulation of talus on the sides (Bryan 1925).

To the east, the Arizona Upland and Sonoran Foothills vegetational divisions occupy country of increasingly higher altitude and rainfall, which gradually gives way to an eastern mountain region still within the SBRP but outside of the desert. Figure 5-6 shows the more arid country southwest of the Phoenix-Tucson corridor and the less arid country to the east and north. To the east the mountains tend to be wider, and the valleys are high enough to support grassland and a variety of organisms uncommon to or absent from the more arid desert.

The increased height of the eastern mountains compared with those of the Sonoran Desert makes them more efficient at intercepting moisture-laden clouds, and the increased rainfall has resulted in more complete weathering. Thus these mountains have more rounded features than the rugged ones of the desert, have more soil, and have typical debris-covered talus slopes that increase their width. The Arizona Upland and Sonoran Foothills are intermediate.

Once the talus slopes achieve a favorable slope, debris becomes relatively stable. Melton (1965b) found that frictional properties of rock fragments resulted in slopes in southern Arizona less than 28.5 degrees being relatively stable, while slopes steeper than 36 degrees were ordinarily devoid of debris. Slopes in the intervening 7.5 degree range exhibited varying amounts of stability. In those mountains with stable slopes, increased availability of moisture to support a denser plant cover results not only from increased rainfall but from the water-holding capacities of the soils and talus.

The central mountain region that borders the Sonoran Desert to the north and northeast is somewhat intermediate between the Colorado Plateau and the SBRP. Some structural basins formed in the Late Cenozoic are present. Paleozoic and Mesozoic rocks similar to those still found on the Colorado Plateau have apparently eroded away in this region to expose Precambrian schist, gneiss, and granite.

Cenozoic Tectonic Chronology

The Laramide Orogeny was a 25-million-year period of folding and faulting in the crust with formation of both extrusive and intrusive rocks and placement of ore bodies, which are still being exploited. This period began late in the Cretaceous, lasted for about the first 15 million years of the Cenozoic, and ceased 50 million years ago. For 12 million years the SBRP was tectonically quiescent. Erosion and deposition were the most important geologic events. During this 12 million year period, the Laramide landscape was thoroughly eroded and gravels were deposited to form the lower units of the Cenozoic conglomerates.

Tectonic activity in the SBRP subsequent to the Laramide Orogeny has occurred for 38 million years and can be divided into three stages (Damon et al. 1981), the last of which has not yet ended (Shafiqullah et al. 1976). The first stage began with melting of rocks in the lower crust to form hot spots of molten magma. The crust naturally expanded as it melted and enormous pressures were created.

FIRST STAGE: THE IGNIMBRITE FLARE-UP

During the later Oligocene and earlier Miocene, great quantities of calc-alkalic volcanic material erupted in the SBRP as andesites and enormous masses of rhyolite and ash. This Mid-Tertiary Orogeny (Damon and Bikerman 1964) may

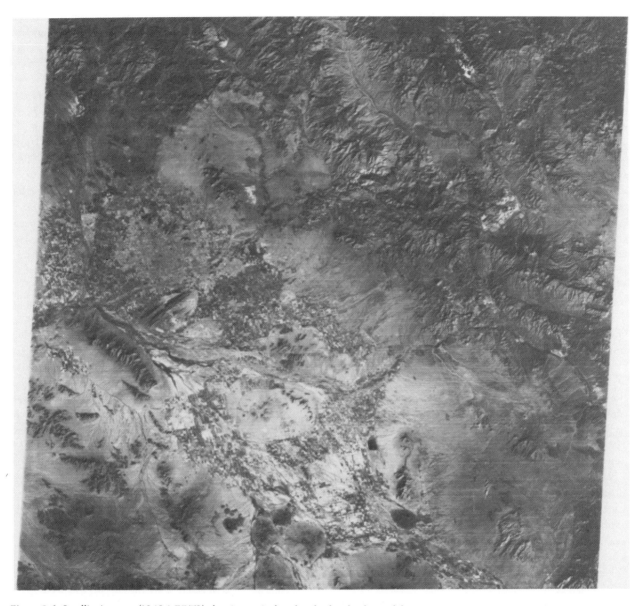

Figure 5-6. Satellite imagery (NASA-ERTS) showing agricultural and urban land use of the valleys in the Southern Basin and Range Province from Glendale and Phoenix (left center) to the outskirts of Tucson (lower right).

The highly arid mountains and valleys in the lower left quadrant contrast sharply with the mountains to the north and east which have higher rainfall.

have been the surface expression of the foundering Benioff Zone, which had slid under the earth's crust during a Laramide episode of collision between continental and ocean plates (Scarborough 1979, Coney and Reynolds 1977, Coney 1976). The piece of ocean crust that was subducted beneath the westward-moving North American plate may have partially melted, expanded, and created an episode of tectonic disturbance.

In the case of the Superstition Volcanic Field in central Arizona, violent pyroclastic activity resulted, and nearly a

thousand cubic miles of volcanic rock, chiefly rhyolite or rhyolitic tuff, became placed over an area of 3000 square miles (Stuckless and Sheridan 1971, Royse et al. 1971, Suneson 1976).

In the case of the Santa Catalina Mountains near Tucson, pressures were not relieved by violent eruption. As lower rock layers melted and expanded, the magma rose, and the overlying crust was metamorphosed to form a gneiss dome. When such intrusions cooled, they became masses of granite or similar coarse-grained crystalline rock. As a result, eroded mountains in the SBRP often show metamorphic rocks in association with intrusive ones.

Sheridan (1971, 1978) has discussed calderas in the Superstition-Superior Volcanic Field of central Arizona. About 29 million years ago, a ring of dacite volcanoes formed a series of mountains 2000 to 3000 feet high. Magma charged with dissolved gas built up and expanded under the mountain masses. Then tremendous explosions placed at least 2000 feet of ash, pumice, and fractured rock on the surface. The material spread laterally as a cloud of red-hot dust moving along the ground for miles in all directions, engulfing and covering the dacite cones. The volcanic material that was extruded compressed together to form a welded tuff.

Removal of great quantities of magma from the earth and placement on the surface caused a void that collapsed to form a huge pit or collapsed caldera, into which faulted blocks of steeply dipping rocks settled. As molten rock pushed upward in a later phase, the large, central portion of the collapsed volcanic rock became a resurgent dome, which is the present Superstition Mountain. Similar but smaller collapse structures in the same region are the Black Mesa, Florence Junction, and Haunted Canyon caldrons (Sheridan 1971, 1978). Other welded ash flows are common in the SBRP and have been particularly well studied in the southern Arizona portion at the Boyce Thompson Southwestern Arboretum and Picketpost Mountain (Nelson 1966), Chiricahua National Monument (Enlows 1955), the Bighorn and Ajo mountains (Watson 1968), the Tucson-Roskruge Mountains (Bikerman 1967; Eastwood 1970), and the Galiuro Mountains (Kreiger 1969).

Magmas associated with this first stage of SBRP Cenozoic tectonic activity, which cooled in the earth rather than erupting, formed granite-like rocks, some of which are exposed on the surface because of erosion. Examples include the Kitt Peak Granite in the Quinlan Mountains (Damon and Bikerman 1964) and the Ash Creek Monzonite and Deer Creek Granodiorite in the Rincon Mountains (Marvin et al. 1973). Intrusions that have cooled to form crystalline rock may exist under volcanic rocks of the first stage but lie so deep that they have not been discovered.

Damon et al. (1963) studied older Precambrian rocks of the Catalina Mountains that were heated by intrusive magma when the mountains were domed up during the first stage. Although these rocks were nearly 1500 million years old, the intense heat of the more recent intrusion had forced the argon-40 out of the rocks and essentially "reset the potassium-argon clocks." As soon as the rocks cooled, potassium-40 again started to decay to argon-40 and the potassium-argon dates calculated revealed the date of the more recent deformation and cooling.

Aside from major vertical faulting in the SBRP, nearly horizontal faults occurred in conjunction with doming of the first stage. Rocks apparently glided considerable distances due to gravity when the mountains were domed upward. Rocks near the Rincon Mountains in the eastern sector of Saguaro National Monument may have been deformed under a gravity glide sheet (Davis 1975).

SECOND STAGE: BLOCK ROTATION

Eruption of high silica volcanic material became of less importance about 24 million years ago in the SBRP (Damon and Mauger 1966, Damon 1971) but continued to a lesser degree up to 13 million years ago (Bikerman 1967). The division between the first and second stages of Cenozoic tectonic disturbance has been placed at the 24 m.y. B.P. date by Damon et al. (1981).

The second stage was a period of transition between extrusion of highly silicic lava of the first stage and extrusion of basaltic lava in the third stage. High-potassium rock was erupted during the early part of the second stage at what is now Picacho Peak (Shafiqullah et al. 1976) and elsewhere in the SBRP. Shafiqullah et al. have shown that these high potassium rocks are comparable to those associated with continental rifting elsewhere in the world, including the Rhine Graben in Europe and the East African Rift. Continental rifting in the SBRP probably began with the eruption of these unusual rocks and with crustal block rotation.

Perhaps the most important event of the second stage was the rotation of fault blocks of the crust. Evidence for such rotation comes from finding relatively youthful lava flows laid down on older ones that are tilted. Time when tilting occurred can be bracketed by determining potassium-argon values for the youngest volcanic rocks that are tilted and the oldest ones that still lie flat. Damon et al. (1981) presented potassium-argon dates for such unconformities in the SBRP, establishing the end of rotational tectonics in various localities. Most of the rotation was completed about 15 million years ago. Possible mechanisms for crustal block rotation have been proposed whereby nearly vertical faults at the surface curve within the crust and converge laterally at shallow depth or at the base of the crust.

Damon et al. (1981) believed that east-west crustal extension was the best explanation for block rotation, as well as for subsidence without rotation. The explanation for block rotation must account for the phenomenon's ceasing about 15 million years ago. What change could have occurred that caused block rotation to cease? One change in the earth's crust may have been the rehardening of the melted crust, which had been typical of the first stage. Since subsidence in the third stage replaced block rotation of the early phases of the second stage, if both processes were driven by the same east-west crustal extension, the change in response would have to be due to a change in the crust.

That there was a significant change in the crust during the second stage is implied by studies of Damon (1971) that showed a decrease in strontium isotope ratios, a decrease in extrusion of high-silica lava, and an increase of extrusion of low-silica basalt to be correlated. Change in the strontium ratio and the type of volcanic rock was thought to be due to rehardening of the lower crust after the melting of the first stage was reversed by diffusion of the heat. Damon and Shafiqullah (1976) believed that as the crust cooled and once again became brittle, deeper-seated magma of basaltic composition was able to break through to produce volcanic activity without mixing with crustal magma of high strontium isotope ratio and high silica content.

The second stage lasted 12 million years and ended 12 million years ago. By the end of the second stage, extrusion of rhyolite had completely ended, basaltic eruptions had become well established, crustal block rotation had ceased, and subsidence of basins had begun.

THIRD STAGE: THE BASIN AND RANGE DISTURBANCE

The third stage of SBRP Cenozoic tectonic activity is currently in progress and has been occurring during the last 12 million years. It apparently began with inception of the San Andreas transform margin of the western United States. Tensional shear stresses began literally to rip the SBRP apart, causing such deep fracturing of the crust that low-strontium isotope basalt of the upper mantle was extruded.

A system of subsidence structures formed that determined the directional trend of the mountains and valleys and their configuration. The down-faulting of valley grabens was followed by deposition of sediment into them. Deep well records, studies of gravitational anomalies, and seismic profiles have shown that the valley sediments are of great depths (Eberly and Stanley 1978; Aiken and Sumner 1974). Sediment depths of 3000 m (2 miles) are not uncommon.

Lynch (1976) has reported lava flows several million years old that have preserved desert mountain slopes, canyons, and alluvial fans that cannot be differentiated from ones developing under the present climate. This suggests that the climate of the Sonoran Desert has been rather arid during the third tectonic period when the basin and range topography was being formed.

In general in the SBRP, young volcanoes are not associated with surface faulting, and there is no evidence that they are directly related to block faulting or to basin subsidence (figure 5-6). The Pinacate Volcanic Field in northwestern Sonora near the Gulf (figure 5-6) results from third-stage volcanism and is composed of basalt. Volcanic cones still surrounding their vents indicate youthful age.

Ives (1964) studied the volcanic Pinacate region. A Papago legend was recorded to the effect that the most recent eruption of ash occurred after 700 A.D. The presence of prehistoric artifacts dated at 700 A.D. to 1300 A.D. in a sedimentary stratum including volcanic ash at Sonoyta seems to verify this legend. The main peaks at Pinacate, however, probably assumed their present size and form about 100,000 years ago (Ives 1964). Spectacular maar craters in the Pinacate field were blasted into the surface by explosions of steam created when basaltic magma mixed with groundwater.

Drainage

Rather than the streams of the SBRP determining the positions of the valleys, the down-faulted valleys determine the positions of the streams (figure 5-7). Streams may simply extend down the sides of a mountain, filling the adjacent valley. Often alluvial fans are produced at the edge of the mountain, less so under extreme desert conditions. A typical bajada in such country gradually slopes from mountain side to the center of the adjacent valley. The upper portion may be typically aggrading and is properly termed a fan, while the lower portion, if eroding, is termed a pediment. Drainage is clearly to the center of the valley. Local climate, rate of deposition, and amount of valley subsidence determine the course of events after water reaches the center of the valley. Drainage may be strictly internal, stream capture may occur, and a number of streams may become integrated into a regional drainage system.

Queen Creek in south-central Arizona represents a regional system that collects enormous quantities of water. In its upper reaches, it flows for much of the year. After rain in its watershed, it often becomes a raging river that has repeatedly washed out railroad bridges and trestles. For example, a single storm in 1954 produced a stream flow of 43,000 cubic feet per second (University of Arizona 1972). But this runoff sinks into the sediments of the desert valley before reaching the Gila River, which would be its potential outlet to the Colorado River and the gulf.

Earthquakes and Faults

Sumner (1976) mapped earthquake epicenters and sizes for the northeastern Baja California, southeastern California, northwestern Sonora, and California portions of the desert. DuBois (1979) discussed earthquakes in this region, particularly the disastrous earthquake of 1887, which had an epicenter in northern Sonora between Benson, Arizona, and Bavispe, Sonora. This earthquake had an estimated magnitude of 7.5 (DuBois 1979) and shook an area of 720,000 square miles (Sturgal and Irwin 1971). Partial disintegration of mountains as far away as Picketpost Mountain near Superior, Arizona, and the Sierra Estrella and South Mountains near Phoenix occurred as a result of this event, and the ground shook all the way to Mexico City and Albuquerque. A steep scarp extending for 50 km was created at La Cabellera Gorge, which can still

Figure 5-7. Satellite imagery (NASA-ERTS) showing the course of the Gila River bending first north of the Sierra Estrella Mountains near Phoenix (right center), then south of the Gila Bend Mountains (lower center) and flowing at lower left toward the Colorado River.
Mountain ranges appearing as arcs at upper left are the Harquahala, Harcuvar and Buckskin ranges. Mountains and valleys are clearly shown as structurally determined with drainage systems imposed later.

be seen. Fissures were produced in the San Pedro Valley more than 100 km away from the scarp.

Damon et al. (1981) used the 5 m maximum scarp height of the 1887 earthquake to calculate the number of such earthquakes that would be required for a typical basin block in the SBRP to subside to its present depth. Considering the distance from the bottom of the Picacho Basin, which subsided 15 million years ago, to the top of the adjacent Picacho Mountains, and adding an amount for erosion during the last 15 million years, it was calculated that nine hundred earthquakes similar to the one of 1887 would be necessary for that basin alone. But this is equivalent to only one every 17,000 years, a rate that cannot be disproven. Such a frequency applies only to a 5 km long scarp. By calculating total lengths of fault lines in the SBRP and by considering the effects of smaller quakes, even ones that can be recorded only by sensitive instruments, it might be possible to prove that the present rate of earthquake activity could be responsible for tectonic events of the third stage in the SBRP. This would verify that currently the SBRP is in the third stage and that tectonic events of this stage are not abating.

Recent tectonic activity in the Sonoran Desert is associated with the Gulf of California-San Andreas rift and northwest coast of Sonora on the one hand and with the northern and eastern sections of the SBRP on the other (Sumner 1976, DuBois 1979, Damon et al. 1980) according to earthquake data. Youthful fault scarps in western Sonora from the intense 1887 eathquake and in Chino Valley, Arizona (Soulé unpublished, in Damon et al. 1981), although within the SBRP, are outside of the Sonoran Desert portion and a quiescent center is inferred for much of the Sonoran Desert mainland.

Earth Fissures

Peirce (1979) discussed recent subsidence fissures and faults in the SBRP. Most spectacular cases of recent subsidence are thought to be related to removal of groundwater by pumping. The Picacho Basin has subsided between 7 and 12 feet since 1952, a rate too high to be accounted for by normal geologic events.

Eaton (1972) studied an area in the Salt River Valley where earth fissures had developed, and this had been attributed to groundwater withdrawal. Arching of recent alluvium and of quaternary deposits created a domed effect. By drilling, it was proven that a salt body lay under the dome. Plastic flow of salt was apparently responsible for causing the overlying sediments to become arched.

Erosion and Deposition

A number of tectonically disturbed prevolcanic sediments date from the early Oligocene in the SBRP (Scarborough 1979). These older Cenozoic sediments include the Whitetail Conglomerate at the Boyce Thompson Southwestern Arboretum (Sell 1968), redbeds in the Galiuro Mountains, redbeds near Yuma, the Sil Murk Formation near Gila Bend, and others. Since these are prevolcanic in age, they can be distinguished by the absence of volcanic clasts.

As Cenozoic volcanic activity occurred, conglomerates, fanglomerates, and tuffaceous sediments accumulated (Scarborough 1979). According to Damon et al. (1981), conglomerates appear to cover more than 25,000 square km (9600 square miles) in the Arizona portion of the SBRP alone, and in places the deposits are over 3000 m (9800 feet) in depth. These conglomerates, with poorly sorted fragments and angular boulders, indicate that rapid erosion and deposition occurred without movement of debris for great distances.

Damon et al. (1981) discussed rates of erosion and deposition in the SBRP during the Cenozoic. A 15-million-year old lava flow at the bottom of the Picacho Basin was covered with 3000 m of sediment, while another volcanic flow nearly 12

million years old in the Tucson Basin was covered with 2400 m of sediment. Both of the sites were therefore filled at the same rate of 200 m per million years.

In the case of mountains that are rising in relation to their adjacent valleys, the mountain-valley interface is essentially at the fault line. When vertical displacement along the fault has ceased, at least temporarily, erosion of the sides of the mountain block enlarges the size of the valley, and the fault may lie buried by sediment well out in the valley. Melton (1965a) calculated the rate at which mountains near Sacaton were becoming more narrow, a rate found to be about 1 km per side per million years. Shafiqullah et al. (1975) calculated that the Mogollon Rim, which separates the Colorado Plateau from the SBRP, is retreating by erosion at a maximum rate also of 1 km per million years.

CLIMATE

The Sonoran Desert is subtropical. Its high parameters of heat and sunlight derive from its lying a few degrees either side of 30° N latitude, only seven degrees away from the Tropic of Cancer, from its high primary surface solar heat load resulting from low cloud cover and low vegetative cover, and from its low topographic relief, which inhibits contact of the solar heat load with cold air aloft.

The heat and sunlight are important assets of this desert, which have drawn an influx of retired persons during the last decade, as well as the inferred southern migrations of biota during Pleistocene glaciation. The high parameters of heat and sunlight are assets shared with the tropics. Not shared is an abundance of water, a commodity that owes its geographic pattern either to evaporation from tropical oceans between 23° N and 23° S, and subsequent precipitation over the land, or to jet-stream exchange with colder polar air and consequent precipitation. In the intermediate desert-prone Horse Latitudes, tropical air having already ascended, cooled, lost its moisture, and increased in density, subsides and exerts a desiccating influence on the land.

Causes of aridity in the Sonoran Desert can be traced to the dominating effect of a high-pressure area that repeatedly forms off the Pacific Coast between 25° N and 39° N; the cold ocean currents running from north to south off the California and Baja California Coast, which to the south are often colder than the land and cause more of the moist air coming toward the continent to prematurely precipitate back into the ocean than to do so on the land; the rain-shadow effect resulting from the low elevation of the desert in relation to the surrounding mountains, which intercept incoming moisture; and isolation from both the moist air masses forming over tropical seas and from influence of the polar jet stream.

Rainfall in the Sonoran Desert is mapped in figure 5-8 from published data and regional maps of the government weather services of the United States and Mexico. Total rainfall

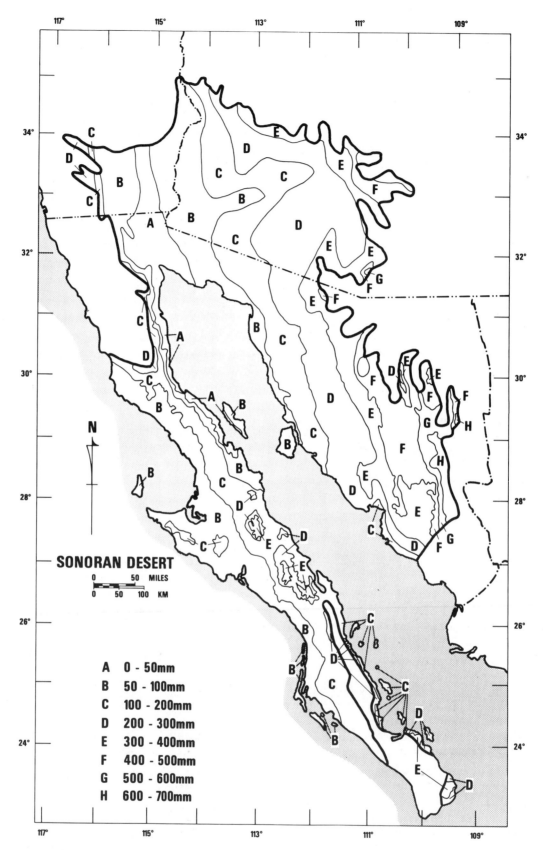

Figure 5-8. *Average annual precipitation in the Sonoran Desert.*
Compiled by authors from United States and Mexican government precipitation records.
Cartography by Russell Vogt, University of Arizona graphics department.

increases in the Sonoran Desert roughly from west to east, the isohyets running more parallel to the axis of the mountain systems than to the coastline of the Gulf or Pacific. The isohyets intersect the Gulf with such little deviation that the latter can have little more than a transient effect on precipitation. The space between the northwest-southeast trending Gulf and the north-south trending Arizona-Sonora mountains naturally narrows as one travels along the Gulf Coast. Consequently the band of relatively heavy precipitation that passes through the eastern reaches of the Sonoran Desert in Arizona is found again along the Gulf Coast between Guaymas and the southern limit of the desert. This creates the illusion to travelers that rainfall increases to the south in the Sonoran Desert or that the bands of rainfall become progressively displaced toward the coast as one travels south. Actually it is the coast itself that trends southeasterly toward the mountains.

Despite the obvious aridity of the Sonoran Desert, a number of recurrent phenomena bring moisture. Patterns of distribution and adaptation of the biota are intimately associated with this moisture, both with regard to the geographic pattern of its precipitation and to its subsequent patterns of flow or retention in the landscape.

Recurrent phenomena are (1) the Mediterranean syndrome, in this case winter rain deriving from kona and related storms forming near the Hawaiian Islands and moving inland when the Pacific high is to the north and the storm track is to the south; (2) the monsoon syndrome of summer thunderstorms resulting largely from Gulf of Mexico air responding to the seasonally recurrent Sonoran low centering on the average at about Gila Bend or Ajo in southern Arizona; (3) the tropical storm syndrome of torrential rain when one of the cordonazos de San Francisco reaches up the coast of Mexico into the Gulf of California and dissipate on the surrounding land masses; (4) the northern storm syndrome when the polar jet stream flows south over the desert; and (5) the coastal fog syndrome of Baja California.

Geology and climate are somewhat interrelated. Although the geologic sedimentation cycle involving weathering, erosion, and deposition clearly relates to precipitation, the opposite interaction of geology and precipitation—the effect of geologic events themselves on precipitation patterns—has been neglected in traditional discussions. Some hypothetical relations are apparent, however, in the Sonoran Desert. First, the ignimbrite flare-up of the first stage of SBRP Cenozoic tectonics, which resulted in rhyolitic mountains and gneiss-plutonic domes of significant height, must have increased precipitation and runoff in the desert due to the orographic effect. Second, the style of weathering, erosion, and deposition in the western part of the SBRP created much less soil and talus for vegetational cover on the mountains than the style in the east, and this resulted in intense heat load. Third, the stretching and thinning of the crust beginning 12 million years ago resulted in a general lowering of the topography, which may have made

the desert significantly hotter and drier than it had been before. Fourth, opening of the Gulf of California about this same time, and the drifting away of Baja California, interjected the ocean into the center of the desert and provided an entering wedge for tropical storms. Finally, drifting of the severed land northwesterly placed the inland portions of the desert to the leeward side of a huge mass that intercepted Pacific moisture.

In addition, Axelrod (1979) has pointed out that the plateau region of Mexico and its fringing mountains, as well as the southern Rocky Mountains, were uplifted chiefly during the Pliocene and Pleistocene, and blocking of ingress of moist marine air would have dated from the uplift. This probably decreased summer rainfall. Also Axelrod pointed out that the oceans gradually chilled during the Tertiary, restricting the northward advance of tropical storms.

Winter rain in the Sonoran Desert varies considerably from year to year. Even when it is abundant, it is more characteristic of the northern part of the desert than the southern. Winter rain derives from moist air masses moving in from the Pacific Ocean. Usually a definite storm track develops during any one season. When the storm track is high in latitude, British Columbia, Washington, and Oregon receive high precipitation. When the storm track is low, the chaparrals of southern California and central Arizona receive the winter waterings, which caused their development and maintain them as a distinctive vegetational type. In addition, the Lower Colorado-Gila and Arizona Upland divisions of the Sonoran Desert (and also the Mojave Desert) receive precipitation. Some rain may fall in Baja California, with concentration in the northern part. Winter precipitation comes from slow-moving clouds and is apportioned to the land relatively slowly so that the earth becomes well soaked, often to significant depth. Evaporation is relatively slow because of the temperatures of winter being cooler than those of summer.

Summer rain results from thunderstorm activity when a low pressure system draws moist air into the desert. As the desert land heats up in the summer and the air in contact with the surface becomes superheated, it has an enormous capacity for moisture. Convectional currents develop, and moist air is uplifted to great heights where it cools and precipitation occurs. One convectional thunderstorm covers a relatively small area, but numerous such storms occur each day during the season, generally in the late afternoon and early morning because of land-air temperature relations.

Thunderstorms begin to develop in Sonora in June and in southern Arizona in July. The storms are much more characteristic of the Arizona Upland and Sonoran Foothills than of the other divisions but occur also in the Sonoran Plains and in southern Baja California. Summer moisture seldom reaches to the desert portions of western Arizona and eastern California. The summer thunderstorms are violent and result in heavy runoff. Evaporation occurs quickly. Although the summer rainfall is often treated in the literature as not as effective as winter rainfall, a number of life forms clearly are adapted to exploit it in preference to that of winter.

Mean July temperatures and mean January temperatures for the mainland portions of the desert have been mapped by Dunbier (1968) in the Fahrenheit system. Mean July temperatures over 90° are restricted to the Lower Colorado Division and to thermal pockets downstreams from Altar and Hermosillo in Sonora. Mean January temperatures over 60° are encountered south of about 29.5° N latitude.

Catastrophic freezes in the Sonoran Desert have been analyzed by Bowers (1980) and Jones (1979). Such freezes, which occurred in 1913, 1937, 1949, 1962, 1971, and 1978, are a consequence of the desert's being in a zone of transition between the tropics and the midlatitude temperate belt of westerlies. During these rare events, the position of the jet stream is altered, and climate typical of the temperate zone temporarily replaces the normally subtropical climate of the Sonoran Desert. Durations of eighteen to twenty hours below 0° C (32° F) have been observed during these disastrous freezes. Reports on freeze damage to plants in the Sonoran Desert and a minisymposium on the disastrous freeze of 1978 were published in the journal *Desert Plants* (August 1979).

WATER

Water is perhaps the most precious of natural resources within the Sonoran Desert because it limits the extent to which the other resources can be utilized. When the desert is irrigated, agricultural productivity is extremely high because the additional water allows the fertile soils, the precious heat, and the abundant sunshine to be exploited better.

In the form of the salt waters of the Gulf and the fresh waters of the lakes, streams, and rivers, extensive habitats are provided within the desert for a rich variety of marine and aquatic organisms. The watercourses and their biota are inseparable assets of the deserts they flow through. With the exception of a few animals that obtain their moisture from vegetation, from metabolic processes, or from condensation or absorption, the land fauna consists of forms displaying obligate relationships with free sources of water—the rivers, streams, lakes, and waterholes of the desert.

The life cycles of plants relate directly to water and the numerous life forms in the Sonoran Desert seem to exploit every source of water conceivable. Even salty water is utilized by halophytes. Productivity in plants is particularly high along the watercourses, and a number of desert animals depend on the seeds, fruits, or other edible parts that such plants produce. Humans in the desert, both in prehistory and today, have devised ingenious ways to exploit the water resource.

Clyma and Young (1968) pointed out that irrigation as a modifier of the environment in the Sonoran Desert has existed for nearly 2500 years. The Hohokam Indians of central Arizona diverted river water for irrigating crops by about 300 B.C. (Haury 1976; Ayres 1971). By about 900 A.D., the irrigation system had evolved to a high state, and use of water was at its peak. Productivity was probably relatively high. Due to poor drainage and waterlogging of dense soils, accumulation of salts and alkali over the centuries eventually may have rendered the Hohokam fields unfit for agriculture. These prehistoric Indians seem to have vanished from central Arizona about 1450 A.D., probably because their agriculture failed (Ayres 1971).

Today in the Sonoran Desert, dams on the Colorado, Gila, Salt, Sonora, and Yaqui rivers halt floodwaters on their way to the sea and allow them to be apportioned to the fields in a rational manner. The valleys in the SBRP have large groundwater reserves. Unfortunately more water is pumped from the ground each year than is naturally replenished in most basins.

Floodwaters that cannot be held in reservoirs of the Salt and Verde rivers sometimes rage through a normally dry channel through the heart of the Phoenix metropolitan area, destroying bridges, homes, and other structures in the floodplain. Excess floodwaters are collected in Painted Rocks Reservoir downstream where some naturally recharge the groundwater reserves as a constant flow is released to the Colorado River.

Water from sewage in the Phoenix metropolitan area is intentionally infiltrated into the ground to replenish groundwater. Some alleviation of water shortages in central Arizona should result from completion of the Central Arizona Project about 1985, a series of canals and aqueducts that will deliver Colorado River water to central Arizona.

The larger rivers of the Sonoran Desert originate outside the desert in the mountains of Colorado, Arizona, and Sonora. Only the Colorado River has continuous flow on the surface, although the Gila probably did flow continuously before Coolidge and the Salt River dams were constructed. Although the Colorado River is thought of as continuously flowing, canals remove virtually all of its water just before it reaches the head of the Gulf. Not until the extremely heavy floods of 1980 did the river once again flow into the Gulf.

The major Sonoran Desert tributaries to the Colorado are the Bill Williams River, entering just above Parker Dam and draining the region from Congress Junction to the east side of the Hualpai Mountains, and the more extensive Gila River, which itself has several major tributaries. The Gila begins in southwestern New Mexico and, before reaching Phoenix, picks up two important tributaries from the south that originate in the vicinity of the Arizona-Sonora border: the San Pedro River, which flows north through Benson, and the Santa Cruz River, which flows north through Tucson. Phoenix marks the confluence of the Salt River with the Gila. The Salt drains the south flank of the mountain plateau that transverses central Arizona, receiving the waters of the Verde River, Tonto Creek, Cherry Creek, Cibecue Creek, Carrizo Creek, White River, and Black River. Most of these tributaries lie to the north of the Salt River. The Salt itself roughly parallels the Gila, with the Superstition and Pinal mountains between.

The rivers of Sonora flow during the rainy season. Disregarding present-day dams, the Yaqui would ordinarily discharge floodwaters to the Gulf; the Rio Sonora and Rio Magdalena would do so rarely; and the Rio Sonoyta, the only river arising within the Sonoran Desert, is landlocked. The Yaqui is a mighty system, the most important in Sonora. At the delta during wet periods, it once rivaled the Colorado (Shreve 1951). Through its tributaries, the Moctezuma, Bavispe, Haros, and Papagochic, it drains the western slope of the Sierra Madre Occidental north of 28° N latitude and even part of the east side of the Sierra Madre in Chihuahua. The Rio Sonora begins near the Arizona-Sonora border and flows south past Arizpe and Ures, at which point it leaves the hills to flow out on the plain. Just above Hermosillo, the Rio San Miguel joins it. From Hermosillo west to Kino Bay on the Gulf, the Rio Sonora becomes more an intermittent river with seasonal flow.

The Rio Magdalena is an important river, arising as the Rio Babasac west of Cananea and picking up several small tributaries as it flows to Santa Ana, where it reaches flatter ground and becomes merely a channel for floodwaters. Above Pitiquito the Rio Altar joins the Magdalena from the north. At about Pitiquito or a little farther at Caborca, the floodwaters from the summer rains usually sink into the sand. In the lower stretches the Magdalena is referred to as the Rio Concepción.

Ives (1936) described the Rio Sonoyta. Several washes in the Altar district of Sonora and in western Pima County, Arizona, converge to form this river, which then cuts through a pass in the Sierra de Nariz from which point it flows west-northwest for 70 km. Springs are present as the river passes east of the village of Sonoyta. Near Quitobaquito the river widens and turns sharply to the southwest at a point called Salada de Sonoyta, where floodwaters generally dissipate. The river then extends for 95 km through arid plains and retains a distinct channel up to the sand dunes bordering the Gulf Coast. Currently the river is incapable of discharging into the Gulf.

Fisher and Minckley (1978) studied chemical characteristics of a Sonoran Desert stream in flash flood. Peak concentrations of dissolved substances occurred at the leading edge of the initial flood wave and declined regularly during flooding. Damon et al. (1980) pointed out that although an individual flood in most parts of the SBRP might not flow as far as the sea, the salt that is carried eventually will be washed to the sea because rain falling on salt left by the previous flood will dissolve the salt and take it farther from the land.

Nevertheless, the SBRP does include evaporite deposits and closed drainage systems or playas. Damon et al. (1980) pointed out that subsiding structural basins tend to result in isolated drainage systems where stream flow is short and may extend only from the mountains into the nearby basin. Widely spaced

rainfall causes streams to be intermittent. Water spreads out as sheet flow when more sediment is present than can be carried. The water may evaporate rapidly in the arid air or infiltrate into the earth.

The Salton Sea represents a closed drainage system. It was the subject of a series of papers edited by MacDougal (1914). A biological survey of this body of water was published by Coleman (1926). Blaney (1955) reported on evaporation and stabilization of the Salton Sea water surface. Carpelan (1958) recorded physical and chemical characteristics. Arnal (1961) reported on the limnology, sedimentation, and micro-organisms. Ecology in relation to sport fishery was treated by Walker (1961). The ecology of the Salton Sea is rapidly changing.

Kim (1973) examined the Salton Sea ecosystem, pointing out that increasing salinity and pollution threaten the usefulness of this body of water. For years, sudden massive die-offs of millions of fish occurred due to toxic algae poisoning. The Salton Sea has no outlet and collects and concentrates large quantities of treated and untreated sewage. Agricultural chemicals and fertilizers combine with salts leached from the land by irrigation and runoff to add to the salinity of this inland sea.

Porter (1973) reported that development of geothermal brines in the Imperial Valley could produce 10 million kilowatts of power and over 2 million acre-feet of high-quality water per year, as well as possible mineral by-products. Several billion acre-feet of hot brine apparently exist in the Imperial Valley geothermal field. The hot brine should flash into a mixture of steam and water and flow to the surface when tapped by a deep well. Large-scale development of geothermal resources in the Imperial Valley would require bringing in seawater as a replacement fluid.

The headwaters of the Gulf of California once represented a unique habitat. The salt water was significantly diluted by the massive flow of fresh water from the Colorado River, a situation that presumably dated from the inception of the Gulf. A large number of organisms used the unique habitat for breeding or for other essential parts of their life cycles. As dams were built on the Colorado River and more and more water was diverted for agricultural and domestic use in California, one of the mightiest of North American rivers totally changed. And perhaps equally important, the once-unique Gulf habitat became significantly altered.

Limnology of the American Southwest and Middle America was treated by Cole (1963), and it was the general subject of desert limnology by the same author (Cole 1968). Stull and Kessler (1978) studied 23 lakes in Arizona and concluded that variance in chemical concentration could be accounted for by variables of climate and watershed morphology. Cole et al. (1967) described unusual monomixis in two saline Arizona ponds, and Cole and Minckley (1968) reported anomalous thermal conditions in a hypersaline inland pond. Adaptational biology of inhabitants of temporary desert ponds was treated by Belk and Cole (1975). Useful regional reports have appeared, such as an ecological reconnaisance of Quitobaquito Spring near the Arizona-Sonora border (Cole and Whiteside 1965) and a report on the aquatic biota of the Sonoita Creek basin (Minckley 1969a). Minckley (1965) reflected that human modification of the earth's surface had been particularly exaggerated in the arid Southwest and had included great changes in virgin aquatic environments. The aquatic fauna native to the Southwest has undergone significant contraction in overall numbers and in range.

SOILS

Soils are complex materials formed through the interaction of climate and vegetation with the parent geologic materials. Weathering of rock, accumulation of leaf litter, and infiltration and downward percolation of water all take different turns of events under desert conditions than under humid conditions of temperate or tropical areas (Fuller 1974). In the Sonoran Desert, however, some soil horizons are present that formed during more humid conditions of the Pleistocene. Although soils of the agricultural areas of the Sonoran Desert (such as the Imperial Valley and Salt River Valley) have been studied and mapped by soil scientists, very little work has been done on remote "nonproductive" lands that constitute the majority of the desert. Nevertheless there have been numerous observations on Sonoran Desert soils, and the repetitive nature of the basin and range physiography that characterizes most of the desert allows generalizations to be stated that apply to most areas with a high degree of reliability. It is not possible, however, to present an accurate and precise map of the soils of the Sonoran Desert.

In the parts of the Sonoran Desert with relatively higher rainfall, such as the Arizona Upland Division and the Sonoran Foothills Division, some chemical weathering of parent rock occurs when rain falls and rocks stay wet for a considerable period of time. As the saprolytic rock collects water, it "rots" into soil. As the wet soil rests on the parent rock, further chemical weathering of the rock is promoted. The vegetation growing on the soil eventually dies, decays, and usually becomes at least partially incorporated into the soil. Humic acids deriving from the vegetation further promote chemical decomposition of the rock. This type of weathering is more characteristic of the development of soils in humid regions, but it occurs, together with more arid processes, in some parts of the desert.

In the extremely arid portions of the Sonoran Desert, such as the Lower Colorado-Gila Division or Central Coast Division, disintegration of rock occurs not so much from chemical weathering but from stresses of heating and cooling, salt wedging, and growth of salt crystals in cracks (Rice 1976). The material weathered away is flushed from the parent rock by rain and accumulates in the valley as valley fill.

Thin Soils

Large parts of the Sonoran Desert fall into the "gravel, cobbles, and stones" syndrome of Fuller (1974), which he characterized as present in deserts on "stream outwash deposits, sediment plains, alluvial flood plains, volcanic plateaus, and alluvial fans and on consolidated upland such as foothills, denuded hills, lower mountains, steep upland, and residual upland mountains, most of which are highly dissected and subjected to erosion." Much of the land surface in such areas is devoid of soil. When scattered pockets of soil are present, they are usually shallow and so recent that climate and vegetation have not fully acted in their development. Such poorly developed soils, with little or no profile structure, are called entisols. They are also termed lithosols because of their poor horizon development on the original parent rock.

Crosswhite (1975) discussed stabilization of, coalescence of, and biological succession on such soil pockets on steep rhyolite slopes in the Arizona Upland Division of the Sonoran Desert. A variety of lichens produce acids that dissolve and break up the exposed parent rock. Roughness of the resulting surface derives from the effects of the acids and the physical configuration of the lichens. Downhill movement of soil weathered from the rock is impeded by the roughness. When a small amount of soil accumulates, resurrection plant (*Selaginella arizonica*), an extremely xerophytic fern relative, becomes established. At this point the continued presence of a soil pocket is much more greatly ensured than when lichens alone were present.

Several species of cacti germinate in the *Selaginella*-stabilized soil pockets, and the extensive lateral root systems of the cacti are efficient at holding more soil. In brecciated zones, or where small faults and fissures occur, the cactus roots penetrate crevices. If crevices are not present, cacti absorbing large amounts of water during rainy periods may become so heavy that they fall from the slope, ripping the soil pocket and the *Selaginella* from the parent rock. More often than not, however, roots penetrate the rock, if only where lichen acids and weathering permit. The next stage consists of germination and establishment of palo verde (*Cercidium*) trees as crevices expand, fill with soil, and succession continues. At this stage, soil pockets tend to coalesce, and vegetation in considerable quantity and diversity may occur on slopes so steep and with soil so thin that laws of gravity seem to be defied.

Microphylls falling from the leguminous *Cercidium* are rich in fixed nitrogen and become incorporated into the soil of the pocket. The organic material and nitrogen improve the soil so that plants adapted to the periodic absence of water find an extremely favorable habitat free from competition with plants not adapted to the habitat. Large quantities of water are soaked up by the organically enriched soil and by the prostrate creeping *Selaginella*. During times of extreme drought, however, the habitat is almost completely dry with the exception of a small amount of moisture in the rock itself and in the fissures.

Deep Soils

Many valleys and basins of the Sonoran Desert have deep soils of medium to coarse texture often referred to as red desert soils. The red color comes from iron oxides. Under the particular type of desert conditions prevailing in red desert soil genesis, iron oxides form rapidly from mineral compounds. Ferrugination of arid soils in Arizona has been shown to result in higher free Fe_2O_3 content than in the parent material from which the soils were derived (Fuller 1974).

Clays in the Sonoran Desert soils are generally thought to have formed during Pleistocene or other periods when the region had a humid climate. They do not form to any significant extent under truly arid regimes. Clays may blow into the desert from adjacent nonarid regions, however. Soils that formed when the Sonoran Desert was less arid have classical profile development, which frequently has been altered by subsequent arid regimes. For example, argillic (clay) layers may become impregnated with calcium carbonate (lime) because of the shallow penetration of water from summer thunderstorms.

The areas with deep soils are generally the basins of the SBRP. Characteristics of the deep soils depend in part on the geologic history of the particular basin and the type of drainage system that eventually evolved. In closed basins with salt deposits, the more soluble sodium chloride (table salt) is usually found at the center where evaporation occurred late. Typically sulfates surround the chlorides and carbonates surround the sulfates (Fuller 1974).

Agricultural soils in the Sonoran Desert are among the most fertile in the world because of their good texture and good drainage and also because they have not been leached of their basic nutrients to the extent of humid soils in temperate or tropical regions. Fertility and management of irrigated soils have been treated in a general way by Thorne and Peterson (1950). Fuller (1975) pointed out that the most effective management techniques for most Sonoran Desert agricultural soils involve simply turning organic material into the soil and periodically leaching to drive salts down below the root zone.

Preparation of Sonoran Desert land for agricultural use may involve ripping the subsoil layers to break up caliche hardpan and to minimize piping. Under continued agricultural use with turning of organic matter into the soil and periodic leaching with heavy irrigation, Sonoran Desert soils may take on characteristics quite different from those of their native desert aspect. But when these fields are converted for urban use, as has been done in the Phoenix-Tucson corridor (figure 5-5), and agricultural management practices have ceased, rapid reversion to desert soil with caliche and alkalinity problems may ensue.

The repeated cycles of wetting and drying that are characteristic of Sonoran Desert situations may frequently result in a vesicular structure forming as air is forced out. The physical effect of rain beating down on the soil frequently eradicates the vesicular structure on the surface and replaces it with a platy structure. If the carbonate content of the surface is high, cementation in the platy layer may occur.

Piping

Piping is essentially a tunneling form of erosion that has been observed in Sonoran Desert situations. It occurs when the subsoil erodes without removal of the upper layers. Some buildings and highways have faced loss of foundations because of this mechanism of erosion in Arizona (Fuller 1974). Piping may occur when irrigation is applied to soils that have not previously been subjected to such large amounts of water in their native soil condition.

When the rate of water infiltration at the surface exceeds the permeability factor of some soil layer below the surface and there is an erodable layer immediately above the retarding layer, piping is a distinct possibility. It may become a serious problem when there is a good outlet for lateral flow (Fletcher and Carroll 1948; Fletcher et al. 1954). The lateral flow may be enhanced by earth shrinking and cracking (Fuller 1974).

Calcareous Soils

The morphologic and genetic sequence of carbonate accumulation in desert soils was treated by Gile et al. (1966). Sonoran Desert soils are frequently calcareous. Management of calcareous soils for home gardening or for larger-scale agricultural use requires a different set of guidelines than soil management elsewhere. Phosphates in calcareous Arizona soils were treated by Fuller and McGeorge (1950) and Fuller (1953), reactions of nitrogenous fertilizers in calcareous soils by Fuller (1963), and the basic concepts of nitrogen phosphorus, and potassium in calcareous soils by Fuller and Ray (1965).

Breazeale and Smith (1930) described the formation of caliche by the dissolving, transporting, and precipitating of calcium carbonate. Caliche forms at the level of evaporation (or use of water by plants) of either downward-filtering surface water or upward-rising groundwater. As the upper soil surface erodes away, the caliche layer normally migrates downward at a comparable rate. In some areas, however, caliche forms on the surface of the soil and is often associated with algae (Breazeale and Smith 1930).

Cooley (1966) studied the effect of caliche on excavation costs for engineering projects in the Tucson area. Caliche was found to occur at the surface or near it in this area and to act as a preserver of ridges by effectively capping them to erosion. Significant increases in excavation costs were found when caliche was present. When some alkaline soils are wet, silica tends to dissolve and move in the horizon. The precipitation of silica or of silicate compounds is most pronounced in the harder layers of caliche.

Carbonate soil layers in Arizona were studied by Yesilsoy (1962). At 1 m depth the carbonate layer was found to be less than 2300 years old on the basis of organic material that happened to be locked into the layer when it formed. Another 0.05 m deeper, the layer was calculated to be 9800 years old, and at the 2 m depth, the layer was 32,000 years old. Such data suggest that continued erosion, deposition, and soil formation have occurred cyclically through the Pleistocene and that new caliche layers formed in response to altered conditions.

Organic Material

Organic material accumulating on the surface of the soil dries out quickly in the Sonoran Desert because of the extreme heat and low relative humidity. Because of the torrential nature of the summer thunderstorms and also due to stratification, compaction, and plating of the surface and other factors, the dried organic material typically is flushed from the surface with the heavy runoff and is carried with floodwaters into lakes or is deposited along the floodplains of the rivers and streams. As a result, the soils typical of this desert are extremely deficient in organic material. A notable exception, and one less characteristic of the other North American deserts, is the desert leaf litter which forms under microphyllous palo verde (*Cercidium*), and other leguminous trees. The thin elliptic microphylls adhere to each other and to the microtopography long enough to become incorporated into the soil.

Although microphyllous leguminous trees theoretically improve the soil, one study has shown one such tree, mesquite (*Prosopis*), to be associated with poor soil. Paulsen (1953) compared the surface soil properties under perennial grass in the Santa Rita Experimental Range with those under adjacent recently invaded mesquite. The soil occupied by mesquite proved to be chemically and physically deteriorated, coarser, lower in pore volume, higher in weight per unit volume, and generally soil that supported such grasses.

Algal and Lichen Soil Crusts

Shreve (1951) pointed out that crusts of algae and lichens on Sonoran Desert soils inhibited wind erosion and often represented an important stage in the stabilization of aeolian soils. Cameron (1958) found that blue-green algae and *Scytonema* are significant fixers of nitrogen in soils of the Sonoran Desert. Blue-green algae studied were ranked according to their nitrogen-fixing capacities. This ranged from 4 milligrams per culture to 326.

Algal and lichen floras in relation to nitrogen content of volcanic and arid range soils were discussed by Shields (1957). Fixation of nitrogen in desert soils was treated by Fuller et al. (1961). The role of algae in soil formation in general has been treated by Bolyshev (1964).

Other Microorganisms

Other soil microorganisms are important factors in Sonoran Desert soils as well. Perhaps most important are the nitrogen-fixing bacteria associated with root nodulation in legumes, both native and introduced. Pathogenic soil microorganisms may result in disastrous losses to cultivated crops. Some general studies of soil microorganisms have deal with survival of microorganisms in alkali soils (Greaves and Jones 1941), characteristics of the microflora of desert soils (Nicot 1960), studies on the rhizosphere microflora of desert plants (Mahmoud et al. 1964), and mycorrhiza in desert soils (Khudairi 1969).

Soil Temperature

Temperature is an important controlling factor in soil development in the desert. According to Fuller (1974) the rate of chemical reactions in the soil doubles with each 10°C rise in soil temperature. Turnage (1939) found the desert soils of Tumamoc Hill near Tucson to have smaller annual temperature ranges, a markedly higher minimum, a somewhat higher maximum, and a higher annual mean than humid soils.

There has been considerable recent attention paid to the subject of soil genesis as influenced by temperature and moisture at depth. The U.S. Forest Service is conducting long-term soil studies with buried probes adjacent to the weather station at the Boyce Thompson Southwestern Arboretum in central Arizona and at other locations. Soil studies adjacent to a weather station are of particular value in correlating climatic variables with soil development.

An example of soil temperature variations is seen in the study by Cable (1969) in the Santa Rita Experimental Range. Soil temperatures at the 3, 12, and 24 inch levels ranged from 40° to 60° F in winter and between 70° and 90° in summer. Diurnal variation at the surface was as high as 75° F but only one or two degrees at the 24-inch level. Maximum and minimum temperatures ranged from 141° to 29° F.

Soil Moisture

Livingston (1910) studied the relation of soil moisture to desert vegetation in the Tucson area. A plant's water supply was seen to depend on the nature of its root system and the moisture content of its soil, whereas the amount of water required by the plant mostly depended on its rate of transpiration and the evaporation potential of the air. Soil moisture was seen to depend on the rate at which water entered and left the soil. This classic survey was the basis for a considerable amount of work done by various investigators later in the Sonoran Desert using lysimeters and probes to measure responses of plants to soil moisture and temperature. A series of large cement lysimeters that were later converted to minigreenhouses still remains at the Boyce Thompson Southwestern Arboretum where they were used in experiments by Franklin J. Crider. These classic studies were instrumental in directing the course of events relating to the development of the U.S. Soil Conservation service in its formative years.

Lyford (1968) studied soil infiltration rates as affected by desert vegetation on the Atterbury Watershed near Tucson. Three times the amount of water penetration occurred under *Cercidium microphyllum* (little-leaf palo verde) and *Larrea tridentata* (creosote bush) than in adjacent open areas. Increased infiltration was attributed to' mulching by litter, increased biotic activity, and protection of the soil from trampling by animals.

Hendricks (1942) studied the effect of grass litter on infiltration of rainfall on granitic soils in a semidesert shrub-grass area using lysimeters. More effective penetration of water could be produced by limiting grazing of rangeland to allow grass litter to develop.

Shreve (1951) claimed that moisture at a depth of 2 m in soils of fine texture in the Sonoran Desert remains nearly constant throughout the year, changing from one year to the next only with exceptional periods of rain or drought.

In alluvial deposits, the soil materials are often separated by particle size due to the action of running water. Since the smaller particles remain suspended for the longest time and with reduced intensities of flow, typical stratification may result in thin layers of clay overlying silt, which lies on sand, which in turn rests on gravel. Such stratification, with the clay layer on top, inhibits infiltration of rain water so that runoff following a storm is characteristically high.

Desert Pavement

Desert pavements are common to all deserts of the world (Fuller 1974). Removal of fine material through wind and soil erosion may combine with an upward displacement of stones and gravel resulting from soil shrinkage, to produce a "pavement" of closely spaced stones and gravel. Such pavements are common in the Sonoran Desert.

Desert pavement has been found to consist of a surface layer of pebbles covered with what has been called desert varnish. Moore and Elvidge in chapter 12 of this book discusses its chemistry and origins.

Aeolian Features

Smith, in chapter 11 of this book, discusses the physical features and origins of sand dunes throughout the deserts of the Southwest.

VEGETATION

The Sonoran Desert is the richest of North American deserts in types of plant communities and in diversity of plant life forms. The richness of this desert derives from the bisea-

sonal rainfall, which varies from low in the west to high in the east and from falling during winter in the north and west to falling during summer in the south and east; a pattern of decline in the incidence of freezing weather, a pattern positioned at right angles to the pattern for total annual precipitation; the extreme topographic, geologic, and edaphic fractionation, which has promoted diversity of habitat; strong and perhaps fortuitous patterns of endemism with roots in a succession of vegetation types described by Axelrod (1979); and in-migration encouraged by the central location with respect to the present Mojave, Great Basin and Chihuahuan deserts, the North American Chaparral, the Sierra Madre Occidental, and the tropical Thorn Forest of Mexico.

Endemism is noteworthy in the Sonoran Desert in such genera as *Cercidium, Olneya, Jatropha, Bursera, Pachycormus, Simmondsia,* and in species of the Cactaceae and Fouquieriaceae families. Highly xeric as well as more mesic plants are represented in the flora. Leaf size and thickness are highly variable, although the number of microphyllous and sclerophyllous species is impressive. Aside from low-stature desert life forms, many trees are present along the drainage ways or blend in with the shrubs of the plains and bajadas. Both evergreen and deciduous plants are well represented. Stem-succulent plants, leaf succulents, and woody plants with swollen stems are abundant and varied in form and size. The biseasonal rainfall and mild winter temperatures allow growth of winter ephemerals in addition to those of summer.

Although the extent of the Sonoran Desert was determined by Shreve (1951) on vegetational grounds, the vegetation mapped was evidently responding to clear climatic forces. Hastings and Turner (1965) showed that Shreve's southern boundary proved to correlate with the line of average southern penetration of frost, his eastern and northern boundaries with the line beyond which there has been a complete day of freezing weather, and his western boundary with an isohyet of summer rainfall.

Studies of Desert Plants

Aridity as a factor in angiosperm evolution has been discussed by MacDougal (1909), Axelrod (1967, 1972) and others. Physiology of plants under drought was treated by Henckel (1964). MacDougal (1921) investigated the high-temperature extreme for plant growth, and Mooney et al. (1975) reported on photosynthetic adaptations to high temperature. Neales et al. (1968) treated physiological adaptation to drought in carbon assimilation and water loss of xerophytes.

Koller (1969) reviewed the physiology of dormancy and survival of plants in desert environments. Ehrler (1975) studied factors influencing transpiration of desert plants. Heat flux and the thermal regime of desert plants were reported on by Gibbs and Patten (1970) and Patten and Smith (1975).

Gates, Alderfer, and Taylor (1968) reported on leaf temperatures of desert plants. Orians and Solbrig (1977) discussed leaves and roots of desert plants. Mulroy and Rundell (1977) discussed adaptations of annual plants to desert environments. Seasonal rhythms relating to desert plant physiology have been treated by Adams and Strain (1969), Strain (1969), and Mooney et al. (1974). Nocturnal behavior of xerophytes was discussed by Gindel (1970) and diurnal physiologic patterns of selected xerophytes by Woodhouse and Szarek (1977).

The quarterly journal *Desert Plants* is devoted to the study of xerophytes and other plants within desert situations. It has been published by the Boyce Thompson Southwestern Arboretum at Superior, Arizona, since 1979. Its sponsoring institution lies within the Sonoran Desert, was founded in the 1920s, and has the distinction of being the oldest institution devoted to desert plant studies in the world.

Research Background

Axelrod (1979) detailed the age and origin of Sonoran Desert vegetation. From the Late Cretaceous into the Pliocene, the region of the present Sonoran Desert had a series of vegetation types from savanna to dry tropic forest, short-tree forest, woodland chaparral, and thorn forest. Semidesert vegetation was probably appearing on restricted dry sites by the Late Eocene and is documented by Axelrod for local basins by early Miocene.

Axelrod (1979) has shown that many taxa remaining from the progression of vegetation types were preadapted through physiology and life form potential to cope with increasing aridity during the Tertiary. The present vegetational associations in the Sonoran Desert were thought to have emerged only in the interglacials and especially after the Wisconsin phase of glacial chronology. The dominance of diverse life forms, the sources of the flora, the floristic imbalance, and relations to the Chihuahuan and Hidalgan desert floras were discussed by Axelrod (1979) in relation to the chronology.

VanDevender and Spaulding (1979) and Wells (1979) discussed Quaternary development of vegetation and climate in the southwestern United States. Other details were supplied by VanDevender (1977) and King and VanDevender (1977). Johnson (1968) treated evolution of desert vegetation in southern California. Turner (1959) has also considered evolution of southwestern desert vegetation. Axelrod's (1950) studies of Tertiary paleobotany and his (1958) monograph on evolution of the Madro-Tertiary geoflora are useful. Unfortunately the origin and development of desert plant communities is often obscured by gaps in the fossil record (Darrow 1961). This is due either to the weathering away of sedimentary rocks of key time periods or to their failure to form. The reduced incidence of fossil formation under desert conditions is also important.

Many plants of the Sonoran Desert have generic relatives in South America, although the number is not large when compared to that of the total flora of either region. Bray (1898) considered relations of Lower Sonoran vegetation in North America to the flora of the arid zones of Chile and Argentina. Johnston (1940) investigated the floristic significance of shrubs common to North and South American deserts, discussing a number of Sonoran Desert genera—*Allenrolfea, Larrea, Holacantha (Castela), Cercidium, Prosopis, Hoffmanseggia,* and *Krameria*—which have amphitropical distribution patterns. Axelrod (1950) questioned the hypothesis of an Early Tertiary connection of vegetation between North and South America and doubted the evolution and migration of these plants from a common central point. Raven (1963) thoroughly documented the disjunctive species and genera, Stebbins and Major (1965) showed that the genera in question were not of humid tropical relationship, and Cruden (1966) suggested long-range dispersal by means of birds to account for the phenomenon. The floristic disjunctions between the Sonoran Desert and the Monte Desert in Argentina were discussed by Solbrig (1972). Origin of the creosote bush deserts of southwestern North America was detailed by Wells and Hunziker (1977). Darrow (1961) discussed origin and development of the vegetational communities of the Southwest.

For many years, the headquarters for study of Sonoran Desert vegetation was located at the Desert Laboratory of the Carnegie Institution of Washington near Tucson, Arizona. McGinnies (1968) prepared a thorough bibliography of this classic work. The vegetation of the Sonoran Desert has been treated by Shreve (1951), Jaeger (1957), Dunbier (1968), and others. Harshberger (1911) is credited with first using the name Sonoran Desert and applying it to a vegetation map in essentially the present sense. Dice (1939) discussed vegetation of the region as the Sonoran Biotic Province. The concept of this region as a distinctive biogeographic province was strengthened by Dice (1943), Dasmann (1974), and Brown, Lowe, and Pase (1979). A two-volume definitive work on the vegetation and flora of the Sonoran Desert by Shreve and Wiggins (1964) includes all of the material from Shreve's (1951) classic paper. Hastings, Turner, and Warren (1972) have published a useful atlas of plant distributions in the Sonoran Desert.

Riparian vegetation within the Sonoran Desert has been treated by Bryan (1928), Haase (1972), Lacey et al. (1975), Pase and Layser (1977), and Rea (1979a). Brown et al. (1979) have treated coastal and marine environments on the shores of the Sonoran Desert region, partially following the classification by Ray (1975). Sonoran Desert vegetation classifications are being recorded as wildlife habitat data in the RUNWILD information storage and retrieval system (Patton 1978).

Lowe (1955) advocated modification of the eastern boundary of the desert in Arizona on the basis of distribution of plant and animal communities. Fish and Smith (1973), as well as Turner (1973), have demonstrated the value of remote sensing devices for defining desert areas and for distinguishing the natural boundaries. Shreve (1951) investigated the boundaries of the desert by means of ground surveys. He published vegetational analyses at zones of transition such as between desert and grassland (Shreve 1940, 1942), desert and chaparral (Shreve 1936), and desert and thorn forest (Shreve 1937).

It is essential to distinguish between the geographic distribution of Sonoran Life Zone vegetation in the classic sense of life zones and crop zones of the United States summarized by Merriam (1898) and the present use of the term *Sonoran Desert*. The Sonoran Life Zone includes not only the present Sonoran Desert but the Mojave, Great Basin, and Chihuahuan deserts, as well as extensive grassland, oak woodland, chaparral, and pinyon-juniper woodland. Merriam's definition of Lower Sonoran Life Zone is narrower, but encompasses the Mojave and Chihuahuan deserts in addition to the present Sonoran Desert. Brown, Lowe, and Pase (1979) distinguish the Sonoran Desert as subtropical and the Mojave and Chihuahuan deserts as temperate. In this sense, the Sonoran Desert would represent, at least in the United States, the subtropical division of the Lower Sonoran Life Zone.

Dunbier (1968) defined the Sonoran Desert so as to include much of the Sinaloan Thorn Forest of the west coast of Mexico, while Felger (1966) and Brown and Lowe (1977) in a map of biotic communities of the Southwest removed Shreve's (1951) foothills of Sonora division from the Sonoran Desert, classifying it as part of the thorn scrub of the Sinaloan biogeographic province. Since Shreve (1937) studied the Sinaloan Thorn Forest, since his boundary between Sonoran Desert and Thorn Forest is intermediate between the extremes of Dunbier (1968) and that of Felger (1966) or Brown and Lowe (1977), and since the natural boundary outlined on the basis of nontropical versus tropical conditions by Garcia-Castañeda (1978) corresponds to Shreve's southern boundary, the delimitation by Shreve (1951) is retained here.

The occurrence of the Sonoran Desert on either side of the U.S. and Mexican boundary has resulted in fractionation of its study. Significant information is to be found in publications based on the study of plants within political rather than natural boundaries. Thus important source books are the taxonomic treatment of the Arizona flora by Kearney and Peebles (1960), the summary of trees and shrubs of southwestern deserts by Benson (1980), the guide to southwestern trees by Little (1968), and the multivolume study of the trees and shrubs of Mexico by Standley (1920-1926).

The guide to the literature of the flowering plants of Mexico by Langman (1964) and the bibliography on native plants of Arizona (Schmutz 1978) are useful. Lehr (1978) has published an updated catalog of the flora of Arizona, which can be used in conjunction with the keys for identification by Kearney and Peebles (1960). The vegetation of the California portion of the Sonoran Desert has been treated by Parish (1930), Munz and Keck (1949), Jaeger (1957), and

Johnson (1976). Keys for identification of plants are available in the treatment of California flora by Munz and Keck (1959) and in the supplement by Munz (1968). The flora of southern California by Munz (1974) is useful. Collins (1976) published a key to trees and shrubs of the deserts of southern California. The natural vegetation of California was mapped by Küchler (1977).

The vegetation of Arizona has been treated by Shreve (1942), Nichol (1952), and Humphrey (1965). The map of natural vegetational communities of Arizona by Brown (1973) has been interpreted elsewhere in text supplied by Lowe and Brown (1973). A treatment of the natural vegetation of the state, private, and Bureau of Land Management lands in Arizona was published by the Arizona State Land Department (Sayers 1974). Other significant information is found in a treatment of Arizona's natural environment by Lowe (1964) and in a review of established natural areas by Smith (1974). Easily used descriptive works on native plants (Dodge 1963, 1973), weeds (Parker 1958), and poisonous plants (Schmutz and Hamilton 1979) are useful for rapid orientation to common Sonoran Desert plants by visiting college classes or investigators from outside the region.

Ochoterana (1945) outlined the geographic distribution of plants in Mexico, and Leopold (1950) summarized the vegetation zones. The treatment of the vegetation of Mexico by Gomez-Pompa (1965) and the map published by Flores Mata et al. (1971) are quite helpful. Rzedowski (1973) discussed geographic relationships of the flora of Mexican dry regions, and Garcia-Castañeda (1978) discussed geographic aspects of the vegetation of Mexico with respect to potential desertification processes. The vegetation of northwestern Mexico was discussed by Brand (1936), and the vegetation and flora of the Rio Bavispe region in northeastern Sonora was studied by White (1948). Felger (1980a) published on plants of the Gran Desierto. Gentry's (1942) study of Rio Mayo plants just south of the desert is useful.

Wiggins (1960) studied the origin and relationships of the land flora of Baja California and has produced a book (1980) detailing the species. Coyle and Roberts (1975) have prepared a useful illustrated guide to plants of the same region. The Vizcaíno portion was treated by Humphrey (1974) in conjunction with studies of the cirio or boojum tree (*Idria columnaris*). Kasheer (1961) summarized the flora of Baja California Norte. Moran and Lindsay (1949) reported on desert island vegetation off the Baja coast. Felger and Lowe (1976) treated vegetation of the Gulf and the islands off the Sonora coast. Plants of the Gulf of California region have been treated by Johnston (1924), Gentry (1949), Felger (1966), and Felger and Lowe (1976). Various aspects of the vegetation of Baja were summarized by Goldman (1916), Lindsay (1965), Aschmann (1967), Wiggins (1969), and Felger (1972). Plants of Baja California relationship in central Arizona were briefly considered by Kearney (1929).

There have been several detailed regional studies of Sonoran Desert vegetation in Arizona. Simmons (1966) prepared a floristic treatment of the Cabeza Prieta Game Refuge near the U.S.-Mexico boundary and later (1969) a survey of the environment of the desert bighorn sheep in this region. Steenbergh and Warren (1977) investigated the ecology of natural communities at Organ Pipe Cactus National Monument, and Warren (1979) summarized information on plant and rodent communities at this location. A checklist of the vascular flora of Tonto National Monument was published by Burgess (1965). The vegetation of flora of the White Tank Mountains were treated by Keil (1973).

Maps showing vegetation of the Phoenix and Tucson areas have been published by Turner (1974a, 1974b). Yield and value of wild-land resources of the Salt River watershed were considered by Cooper (1956). An ecological analysis of relictual desert vegetation in south-central Arizona was prepared by Ferguson (1950), and plant associations in the vicinity of the Desert Laboratory at Tucson were investigated by Spalding (1910) and Gibble (1950).

Specific ecological studies are available for individual plant species or situations in the Sonoran Desert. Wood and Nash (1976) evaluated the effects of copper smelter effluent on surrounding vegetation. Steenbergh (1972) presented evidence for lightning-caused destruction of vegetation. Productivity and flowering of desert ephemerals in relation to shrubs was considered by Halvorson and Patten (1975). Went (1942) studied the dependence of annual plants on shrubs in southern California deserts. Environmental relations of annual plants were discussed by Patten and Smith (1974) and by Smith and Patten (1974). Marks (1950) investigated vegetation and soil relations in the lower Colorado River region, and Wright (1965) evaluated homogeneity of vegetation stands.

Ecology of creosote bush was treated by Dalton (1961), its distributional relation to physical and chemical factors by Yang (1950) and Johnson (1961), its geographic response to moisture and temperature by Saunier (1967), factors affecting its susceptibility to burning by White (1968), and its recent expansion by Yang (1961). Welsh and Beck (1976) studied relationships between creosote bush and bush muhly grass. Woodell et al. (1969) reported the behavior of creosote bush in response to rainfall. Measurements of tissue water potential and photosynthate-related substances in creosote bush were reported by Strain (1970) and Oechel et al. (1972). Chew and Chew (1965) studied primary productivity of a creosote bush community. Papers on the biology and chemistry of the species in New World deserts have been edited by Mabry et al. (1977). Life history characteristics were treated by Valentine and Gerard (1968).

A literature review and annotated bibliography on jojoba (*Simmondsia chinensis*) was prepared by Sherbrooke and Haase (1974). Ecology of the species at its lower elevational limit was treated by Burden (1970), its natural history by Gentry (1958), and its physiological ecology by Al-Ani et al. (1972). Ehrler (1975) reported on transpiration rates in jojoba. Scarlett (1978) reviewed its natural history and cultivation. Haase and McGinnies (1972) have edited a symposium on jojoba and its uses, and the journal *Jojoba Happenings*, published by the University of Arizona Office of Arid Lands Studies, regularly features research papers on the plant.

Bibliographies on mesquite (*Prosopis*) were published by Bogusch (1950), Schuster (1969), and Schmutz (1978). Simpson (1977) edited a book of recent research papers on mesquite. North American species have been treated by Benson (1941), Johnston (1962a), and Burkart (1976). Effects of mesquite on herbaceous vegetation and soils were analyzed by Tiedemann (1970). Gavin (1973) prepared an ecological survey of a mesquite forest. Strain (1970) reported physiological details of tissue water potential and carbon dioxide levels in mesquite.

Bibliographies on the saguaro cactus (*Carnegiea gigantea*) have been published by Steenbergh (1974) and by Mitich and Bruhn (1975) and an ethnobotanical bibliography by Fontana (1980). Ecology of the saguaro has been monographed by Steenbergh and Lowe (1977) and its history reviewed by Mitich (1972). Ecology of saguaro forests was treated by Shantz (1937), Humphrey (1938), Howes (1954), and Lauber (1974). Alcorn et al. (1959) discussed pollination requirements of the saguaro.

An account of perennial vegetation associated with the organ-pipe cactus (*Stenocereus thurberi*) was published by Mulroy (1971). A study of perennation and proliferation in the fruits of chain-fruit cholla cactus (*Opuntia fulgida*) was completed by Johnson (1918). Tschirley (1963) prepared a physioecological study of the same species, and Tschirley and Wagle (1964) published on its growth rate and population dynamics.

Information on the cacti can be found in books on cacti of Arizona (Benson 1969a) and California (Benson 1969b) and in treatments of the family by Britton and Rose (1923) and Lamb and Lamb (1964). Individual discussions of senita cactus (*Lophocereus*) by Lindsay (1963) and Felger and Lowe (1967), of cardón (*Pachycereus*) by Moran (1968), and of pincushion cacti (*Mammillaria*) by Craig (1945) have been published. Articles concerning Sonoran Desert cacti frequently appear in the *Cactus and Succulent Journal* published by the Cactus and Succulent Society of America.

Patten and Dinger (1969) discussed CO_2 exchange patterns of cacti from different environments. Seasonal temperature acclimation of a prickly-pear cactus in south-central Arizona was treated by Nisbet and Patten (1974). Productivity and water stress in cacti has been discussed by Patten (1971, 1972, 1973). Dinger (1971) studied comparative physiological ecology of the cactus genus *Echinocereus*. Nobel (1977) reported on water relations and photosynthesis in the compass barrel cactus (*Ferocactus acanthodes*). Felger and Lowe (1967) investigated clinal variation in the surface-volume relationships of senita cactus (*Lophocereus schottii*). Cannon (1916) studied distribution of cacti with reference to soil temperature and soil moisture.

Century plants, mescal, maguey, or lecheguilla (*Agave*) of the southwestern United States were summarized by Breitung (1968), of Sonora by Gentry (1972), and of Baja California by Gentry (1978). Studies of southwestern soaptree or Spanish dagger (*Yucca*) were published by Webber (1953) and by McKelvey (1938-1947).

Many Sonoran Desert grass species are included in Humphrey's (1968) monograph of the desert grassland and in his (1970) treatment of Arizona range grasses. Copple and Pase (1967) published a useful vegetative key to local range grasses. Neuenschwander et al. (1975) and Sohns (1956) reviewed grasses of the genus *Hilaria*, and Pinckney (1969) studied factors affecting *Hilaria* distribution.

Saltbushes (*Atriplex*) and their allies were treated by Bidwell and Wooton (1925). Williams et al. (1974) studied improvement in growth of four-wing saltbush (*A. canescens*) by formation of vesicular arbuscular mycorrhizae. Sammis (1974) wrote a monograph on the microenvironment represented by bushes of desert hackberry (*Celtis pallida*). Humphrey (1932) studied morphology, physiology, and ecology of canescent borage (*Tiquilia canescens*). Ecology and taxonomy of burroweed or jimmyweed (*Aplopappus*) have been treated by Hall (1928), Humphrey (1937), Voth (1938), and Tschirley and Martin (1961).

Useful treatments of individual plant genera present in the Sonoran Desert include *Acacia* (Isely 1969), *Castela* (Moran and Felger 1968), *Condalia* (Johnston 1962b), *Ephedra* (Cutler 1939), *Eriogonum* (Reveal 1969), *Fouquieria* (Henrickson 1972), *Lycium* (Hitchcock 1932), *Nicotiana* (Goodspeed 1954), *Petalonyx* (Davis and Thompson 1967), *Sphaeralcea* (Kearney 1935), *Vauquelinia* Williams 1971) and *Washingtonia* (D. E. Brown et al. 1976).

LIFE FORMS AND SURVIVAL STRATEGIES

Development of diverse structural and behavioral syndromes among the species that compose the vegetation of the Sonoran Desert brings a plethora of distinct life forms into different sets of relationships with the environment. Using anthropocentric terminology, these life forms have come to be called strategies for survival in the desert. Sonoran Desert plants may be ephemeral, perennial or intermediate, woody or herbaceous. The leaves or stems may be succulent or not. Leaves may be deciduous or evergreen, broad or tiny. The deciduous leaves may fall in winter or summer. The stature of the plants may be treelike, shrublike, subshrubby, grasslike, or creeping on the ground. Plants may behave as parasites, may be salt tolerant, or may be phreatophytic. That there is no one

most successful strategy comes from observations of situations in the Sonoran Desert, such as in central Baja California, where the vegetation in a small region with similar soil and climate may consist of as many as five dominant species, each with a different life form.

MacDougal (1903) is credited with pioneering the study of Sonoran Desert vegetational life forms. He classified the common species on the basis of seasonal appearance and on characteristics of shoots and roots: ephemerals, perennials with subterranean tubers or thickened roots, deciduous perennials, spinescent perennials with reduced leaves, perennials with foliar protectants such as wax, resin, oil, or varnish, succulent perennials, and halophytes. Shreve (1951) recognized twenty-five major life forms for plants of the Sonoran Desert. He believed that the proliferation of life forms in the Sonoran Desert allowed a separation of activities among the species, in either time or space, so as to reduce or eliminate competition among them. The number of life forms in a particular vicinity varies from three or four in level fine-soil plains of the Lower Colorado-Gila Division up to fifteen or so in the Arizona Upland Division. The summary below is modified and updated from the classification by Shreve (1951).

Winter Ephemerals

Plants of this group exploit the moist soils resulting from precipitation due to Pacific air masses that move inland in winter. The geographic occurrence of these plants mirrors the pattern of winter precipitation in the region where soil temperatures allow germination, with concentration in northern Baja California, the Lower Colorado-Gila, the Arizona Upland, and the Sonoran Plains Divisions of the Sonoran Desert, but also in the Coastal Sage Scrub and Mojave Desert of California. The diversity of species and abundance of individuals relates directly to soil temperatures and to ancient and modern precipitation patterns.

The winter ephemerals germinate in winter, flower in late winter or early spring, and mature fruit in the arid foresummer of May and early June or earlier. These plants are shallowly rooted, relatively small, and short of stature because their brief life span does not allow either extensive growth or production of much woody tissue. The conditions exploited are present within a few inches of either side of the interface of soil with atmosphere. It is well known to residents of the northern part of the Sonoran Desert that there are good years and bad years for the spring wild flowers. When the Pacific storm track is to the south, the winter rain proceeds as a series of storms across the northern part of the Sonoran Desert from west to east, such as during February 1980. Seeds of winter ephemerals have lain in the soil through dry periods of the year, often for many seasons, before finally germinating. The mature ephemeral may show few, if any, adaptations of leaf or stem to desert conditions other than signs of rapid growth to exploit seasonal moisture. The major adaptations are to be seen in the seed and in the rapidity with which the life cycle is completed.

Because the winter ephemerals are grossly similar morphologically to mesic plants in the most obvious part of their life cycle, their strategy for survival frequently has been referred to as the avoidance of dry periods by persisting through this period in dormant heat-resistant and drought-resistant seeds. A fairly stable condition of soil moisture is necessary for these plants to complete their life cycles, and many have seeds that will not germinate unless the soil is drenched. Most commonly this resistance to germination derives from the relatively impervious nature of the seed coat or to the presence of germination inhibitors, which must be leached from the seed coats.

Plants that have been thought rare or virtually extinct quite unexpectedly have appeared in large quantities during a year with unusually optimum winter conditions of soil moisture and temperature. Most winter ephemerals of the Sonoran Desert, particularly those which are of most regular year-to-year occurrence, must complete a substantial portion of their life cycle under cool or cold conditions. In this respect, it is instructive to note that a relatively high percentage of Sonoran Desert winter ephemerals also occur in the Mojave Desert or are related to species of that desert. Examples of winter ephemerals include species of poppy (*Eschscholtzia*), bluebonnet (*Lupinus*), evening primrose (*Oenothera*), water-leaf (*Phacelia*), fiddleneck (*Amsinckia*), and Indian wheat (*Plantago*).

Summer Ephemerals

The moist soils of summer resulting from thunderstorm activity are also exploited by ephemerals. These plants are practically restricted to the regions of vegetation heavily influenced by the Sonoran monsoon: the Arizona upland, the central coast, and the Sonoran plains and foothills. The moisture for summer thunderstorm activity originates in the Gulf of Mexico. It is not unexpected, therefore, that the geographic pattern of the summer ephemerals is in direct opposition to that of the winter ephemerals. The Arizona upland and central coast are unique in having both groups in considerable abundance.

As opposed to the Mojave Desert relationships of the winter ephemerals, the summer ephemerals of the Sonoran Desert were shown by Shreve (1951) to be identical with or similar to summer active plants of western Texas, an area also influenced by Gulf of Mexico moisture in summer. The summer ephemerals have the basic drought-avoiding strategy of those of winter but are capable of more extensive, often rank, growth because of relatively high temperatures during the wet season in which they are active. Many of the species have distinctly weedy characteristics. Thunderstorm agriculture of endemic

peoples of the Sonoran Desert carries the summer ephemeral syndrome to its logical extreme by enhancing, controlling, and modifying it for production of food.

Examples of summer ephemerals include species of jimson weed (*Datura*), spurge (*Euphorbia*) four-o'clock (*Boerhaavia*), careless weed (*Amaranthus*), cocklebur (*Xanthium*), and devils-claw (*Proboscidea*).

Facultatively Perennial Ephemerals

This category includes plants that show ephemeral characteristics but that under favorable conditions may persist for two or three seasons. They become relatively dormant during unfavorable months and like true ephemerals may die if conditions are severe. Some, like desert penstemon (*Penstemon parryi*), respond to winter rain and others to summer rain (*Aplopappus*). There are a number, such as desert marigold (*Baileya multiradiata*), crownbeard (*Verbesina encelioides*), and globe mallow (*Sphaeralcea*), that respond to both.

The facultative ephemerals exploit the environment of and compete with the true ephemerals but have a headstart if and when climate permits. For these species to survive, they must produce seed synchronously with the true ephemerals, for once the factors conducive to seed production have passed, the life of the facultative ephemeral is very uncertain, and the likelihood of a future seed crop cannot be predicted.

The strategy of the facultative ephemeral is one of opportunism. By having a periodic headstart, it can produce a greater quantity of seed and compete favorably with species not having such an advantage. Although theoretically the individuals that persist from one season to the net have this advantage over ones that must germinate anew, the persistent plants carry over diseases and pests that also have a headstart. Although perennials in general should have such problems, the problems may be of greater importance in situations where a large stand of plants must be rapidly produced by ephemerals than where a relatively few seedlings need to be established by perennials. That facultative ephemerals may have little real advantage over true ephemerals is suggested by the few species that can be classified in the former category. A discussion of the merits of facultative versus true ephemerals is analagous to the seed-crop versus stub-crop debate of agriculturists.

Root Perennials

Plants of this life form have underground perennating organs that cannot be classified as bulbs. The above-ground portion of the plant dies back each year after the seasonally moist conditions of winter or summer have passed. No species in this group is known to respond to both winter and summer rain according to Shreve (1951), but the mechanism responsible for this phenomenon is not known. Although the annual dieback is most usually a response to xeric conditions, some species die back as a response to winter cold. Some such as grape ivy (*Cissus trifoliata*) are facultative evergreens during mild winters. Grape ivy is also a leaf succulent.

New stem growth arises from an underground rhizome or from a root stock consisting of a small amount of stem tissue associated with a massive root. Desert larkspur (*Delphinium scaposum*) and desert anemone (*Anemone tuberosa*) spring up from enlarged structures that consist chiefly of the largest persistent roots. Canaigre dock (*Rumex hymenosepalus*), maravilla (*Mirabilis multiflora*), wild cucumber (*Marah gilensis*), grape ivy, and some Cucurbitaceae have more grossly enlarged roots as perennating organs.

It is instructive to examine a genus of root perennial as an example of adaptation to desert conditions. Bemis and Whitaker (1969) discussed the four wild Sonoran Desert species of *Cucurbita* (the genus of pumpkins and squashes). These wild species share limited compatibility with *Cucurbita moschata*, a cultivated pumpkin, but are separated from the cultivated species by distinct biochemical and morphological differences. By development of tuberous roots, the desert *Cucurbita* species have adjusted from an annual to a perennial growth habit. Currently man is learning how to exploit this life form in the Sonoran Desert for agricultural purposes.

The strategy of the root perennial is similar to that of the ephemeral in avoiding the dry months or cold months, although the mechanism is quite different. The above-ground habitat of the root perennial is often usurped by trees, shrubs, or other vegetation when the root perennial has died back for the year. This would not be a major problem to the ephemeral because abundant seeds are dispersed widely each year. The root perennial, however, must occupy the one spot where its perennating organ lies. This has apparently created an additional growth form response, as root perennials with the largest and most permanent underground structures prove most frequently to be vines or spreading plants, which in turn seasonally proliferate, encroach on, and may grow over the other vegetation.

Bulb Perennials

Bulbs consist of a dormant apical meristem protected by fleshy leaves. Morphologically they are similar to the buds of trees and shrubs, which are destined to open and provide the new year's growth. In the strategy of the bulb perennial, the bud is protected from a hostile environment by resting in the ground until conditions are favorable.

All of the bulb perennials of the Sonoran Desert respond to winter rain. None responds to summer rain. Examples of these plants are desert lily (*Hesperocallis undulata*), covena (*Brodiaea*), blue sand lily (*Triteleiopsis palmeri*), mariposa lily (*Calochortus kennedyi*), and wild onion (*Allium*). These plants have a monopodial scapose growth habit and are uniformly of small or moderate stature, none proliferating over vegetation in the sense of many of the root perennials. The entire above-ground portions are very narrow and tend to rise up through other small plants to obtain sunlight. Usually they grow in open, sandy areas where competition is chiefly with small ephemeral plants on thin soils where large perennial plants are at a disadvantage.

The energy stored in the bulb is sufficient to give plants of this life form a headstart over ephemerals. The competitive advantage may be outweighed by the possibility of perennial vegetation usurping the living space of the immobile bulb when the above-ground portion has died back. Bulb perennials frequently occur outside the desert in grassy situations or where woody perennials are deciduous during the growth period of the bulb plants. The scarcity of grasses as dominant features of Sonoran Desert vegetation and the frequent dominance of nondeciduous woody perennials may combine to select against the success of bulb perennials. It is not unexpected, therefore, that relatively few bulb perennials are represented in the vegetation of the Sonoran Desert.

Perennial Grasses

Although grasses are well adapted to semiarid conditions, they rarely are of much dominance in the lower highly xeric reaches of the Sonoran Desert. An exception, big galleta grass (*Hilaria rigida*), which roots at the nodes and may be dominant in stabilizing sand dunes, has the life form of a perennial shrub rather than of the typical grass. In the upper reaches of the Sonoran Desert and particularly in the broad ecotonal region between true desert and oak woodland or chaparral, the grasses become important features of the vegetation. Some, like annual brome (*Bromus rubens*), are essentially winter ephemerals. The best range grasses, however, perennate from the base after seasonally dying back or after being grazed by native or domestic animals. Shreve (1951) pointed out that the strategy of perennial grasses was similar to that of bulb or root perennials except that the "inner basal part of the cluster of shoots and leaves and the upper part of the root system persist under ordinary unfavorable conditions." Perennial grasses undergo vegetative growth in the Sonoran Desert any time of year when moisture conditions are favorable, but they flower only in summer on the mainland. Examples include tobosa grass (*Hilaria mutica*, tanglehead grass (*Heteropogon contortus*), bush muhly (*Muhlenbergia porteri*), sacaton grass (*Sporobolus wrightii*), and salt grass (*Distichlis spicata*). Perennial species often flower in winter in Baja California (Felger 1980b). The grasses characteristically have fibrous root systems in the top foot of soil and must become dormant when moisture is low in that stratum. The length of time that dormancy can be maintained without death of the grass is such that grasses are severely restricted in distribution in the desert.

Leaf Succulent Caudiciforms

In this group of species, succulent leaves are crowded together on a caudex, a short, compact stem. Included are all the types of *Agave* (century plant, mescal, maguey, lecheguilla), *Dudleya* (siempreviva), and *Graptopetalum* (hen-and-chicks). These plants have extensive shallow roots that can harvest water from light rainfalls or from condensation of dew. The waxy covering on the succulent leaves reduces water loss and also permits condensation of dew into droplets of water that roll down the leaves of the rosette to soak into the soil immediately under the plant, where root hairs absorb it.

The moisture is stored in the leaf close to where it will be used in the photosynthetic process. Plants of this group may go for long periods of time without rain by relying on internal moisture. The waxier the leaf surface, the more apt it is to concentrate the rays of the sun to produce radiation burns of the epidermis that are harmful also to the underlying chlorophyll. For this reason the leaves may be oriented so as to avoid the direct rays of the sun. Plants of this life form are abundant in the Vizcaíno Division of the Sonoran Desert, where fog tempers the rays of the sun and atmospheric moisture is available for condensation.

The high moisture content of the leaves renders plants of this group relatively frost sensitive. Although radiation burn and freeze damage should render this survival strategy of limited geographic significance, which is true in a general way, the genus *Agave* has been able to radiate adaptively into a variety of habitats through subtle anatomic modifications. In the Sierra San Pedro Martir of Baja California, succulent species constitute 35 percent of the vegetation at the foot of the range but are not present at all at the top, 800 m higher Mooney and Harrison (1972).

Leaf Semisucculent Caudiciforms

This group includes all of the species of *Yucca*, variously referred to as Spanish dagger, palmilla, datilillo, and Joshua tree. The leaves vary in degree of succulence and in the massiveness, woodiness, and length of the trunk. Although moisture is stored to some extent by the roots, trunk, and leaves, the plants are distinctly less succulent than *Agave* and are most frequently more frost resistant. The leaves are characteristically stiffer and narrower than those of *Agave*.

The leaves in most instances are also less waxy than those of the leaf succulent caudiciforms, and condensation of atmospheric moisture is probably of lesser significance. Adventitious roots may form in parts of the trunk covered by old leaf bases, however, and may absorb water, which soaks into the fibers of these old leaves. Shreve (1951) treated the characteristics of

Yucca as intermediate between those of the dicotyledonous tree and those of the columnar cactus, observing that the "photosynthetic and transpiring surfaces are confined to the leaves, but are fixed in area so as to admit of no seasonal reduction."

Leaf Nonsucculent Caudiciforms

This life form consists of plants somewhat similar to *Yucca* but with long grasslike or ribbon-like leaves that are not succulent. Only beargrass (*Nolina*) and sotol (*Dasylirion*) are found in the Sonoran Desert. Species of beargrass, *Hesperaloe sonorensis,* and to a lesser extent of sotol, seem to have the life form of a grass combined with highly xeric evergreen leaves.

To this life form is also added the heart, or caudex, seen in *Yucca* or *Agave*. Some species develop a trunk and are almost palmlike in appearance. These leaf nonsucculent caudiciforms are important in the eastern and northern sections of the Sonoran Desert and in the zone of transition of desert with grassland, oak woodland, or chaparral. In the Sonoran Desert itself they tend to occur chiefly on north-facing slopes where conditions are relatively cool and moist during a much longer period of the year than elsewhere in the desert. The leaves, though somewhat grasslike, are thick, tough, and sharp. The survival strategy seems related both to that of grasses and to the leafy semisucculent caudiciforms. If a grass had tough evergreen leaves with sharp edges that resisted grazing animals, it would have to be classified in the same life form category as *Nolina* and *Dasylirion*.

Palms

This group was maintained as a separate Sonoran Desert life form by Shreve (1951). The plants would be considered arborescent monopodial caudiciforms, but since the taller species of *Nolina* and *Dasylirion*, here considered leaf nonsucculent caudiciforms, are also arborescent and monopodial in habit, the simpler designation as palms is used for this easily recognized morphologic type. Five genera of palms grow naturally in the Sonoran Desert. The leaves occur in rosettes at the tops of tall, unbranched, usually rather slender stems. Fibrous tissues in the stem retain some water for emergency use, but the palm life form is not at all succulent.

In the Sonoran Desert, palms grow only where water is rather freely available, along drainage ways or where moisture seeps from aquifers in hills or canyon bottoms. The roots are fibrous and do not extend to any great depth. Adventitious roots at the base of the trunk probably absorb water that runs down the trunk or pools up on the soil surface.

The palm strategy in the Sonoran Desert is to exploit water that is readily available near the soil surface throughout the year and to hold the habitat by size dominance. The habitat is one of the most favorable for plant growth in the desert, being essentially an oasis. Adaptations are necessary in the palm life form to allow it to monopolize such favorable habitats. Chief among these is the strong monopodial habit, which allows all of the plant's stem growth to extend upward to an exaggerated height where the rosette of leaves dominates the living space and intercepts the energy from the sun. Another adaptation is evident in the young plant. The wide trunk, with innumerable shallow roots and the dense shade of the spreading leaves, monopolizes the living space of a seep hole or moist spot and pushes out or shades out the competing vegetation. The numerous roots within a limited space extract moisture from the soil so efficiently that competitors are at a considerable disadvantage and are inevitably finally eliminated. Palms that are native to or now grow wild in portions of the Sonoran Desert include palma ceniza (*Brahea armata*), palma de taco (*B. brandegeei*), *Brahea roezlii*, datil (*Phoenix dactylifera*), palma (*Washingtonia robusta, W. filifera*), and cocotera (*Cocos nucifera*).

Leaf Succulent Noncaudiciforms

This group includes species that have succulent leaves attached to normal stems. The leaves are relatively small in comparison with those of the caudiciforms. The group includes only a few species and is of little significance to the Sonoran Desert as a whole. Species include a number of members of the Portulacaceae family such as *Talinum paniculatum*, rock purslane (*Calandrinia ambigua, C. ciliata*), *Calyptridium monandrum*, and purslane (*Portulaca lanceolata, P. suffrutescens*); other members are *Nitrophila occidentalis* in the Chenopodiaceae family and *Stylophyllum attenuatum* in the Crassulaceae family.

Degree of succulence varies greatly in this group, but the plants obviously belong to a single life form. Funneling of water to the base of the plant is not as significant as in the rosette-forming caudiciforms. Leaves store enough water to avoid wilting from periodic drought but not enough to survive in the more xeric regions of the Sonoran Desert. Many of the plants are annuals. The survival strategy of the leaf succulent noncaudiciform is to compete for moisture with other small, shallow-rooted plants but to survive periods of water stress by relying on water reserves of the succulent leaves.

Monopodial Stem Succulents

This life form is represented in the Sonoran Desert chiefly by the barrel cacti (*Ferocactus wislizenii, F. covillei, F. acanthodes, F. gracilis, F. chrysacanthus,* and others). The plant body is an unbranching barrel-shaped structure with massive quantities of succulent tissue. The popular analogy to a barrel of water, which neglects to mention the cellular structure of the succulent tissue, is nevertheless true in principle if not in detail. This moist tissue is protected from utilization by animals through the presence of numerous hard, sharp spines, often with fishhook curvature. These spines represent highly modified leaves, and true leaves in the classic sense are entirely absent.

Reproduction of monopodial stem succulents is by seed since no branches occur that could separate to form adventitiously rooted plants. The survival strategy is a textbook example of the storing of water from time of plenty to time of scarcity. Conservation of water is achieved through phylogenetic loss of leaves and development of a thick, waxy layer on the plant surface, which prevents dehydration of the succulent tissue. The striking feature of this life form is its discrete functional structure. Its failure to rise vegetatively to great heights or to branch to great widths or to proliferate above ground in any way severely restricts its capability of dominating relatively moist sites of optimum plant growth where competition is significant. Compensation is achieved through growth on relatively xeric sites where vegetative cover and potential competition is relatively low, through extreme lateral root proliferation with numerous root hairs that quickly extract moisture from shallow soil, and through minimizing photosynthetic demand and energy use by extremely slow, sometimes imperceptible, plant growth. The fruits of these plants are dry and not sweet, continuing the theme of low photosynthetic demand and water conservation.

Arborescent Oligopodial Stem Succulents

This group is represented by the immense columnar cacti, saguaro (*Carnegiea gigantea*), cardón pelón (*Pachycereus pringlei*), and cardón hecho (*P. pecten-aboriginum*). These plants grow where soil moisture conditions, although often higher than average for the Sonoran Desert, fluctuate considerably through the year. The seeds germinate in detritus under palo verde (*Cercidium*) trees or where protected by coastal fog or other overhead situations. Those germinating in the open soon die.

Seedlings are very susceptible to death from freezing weather or from radiation burn caused by intense sunlight. The life cycle depends on the presence of the nurse tree or surrounding vegetation until the extremely extensive water-harvesting root system has developed and the plant has passed the juvenile stage. This ordinarily takes about twenty-five years. The association with nurse trees that already occupy sites having favorable soil moisture, coupled with the extensively radiating water-harvesting roots, ensures a favorable water balance for arborescent oligopodial stem succulents. The notion that plants of this life form use very little water was questioned first by Blumer (1909), who discovered an exceptionally robust and healthy saguaro in a moist and nitrogen-rich situation near a stock tank. It flowered profusely, producing 300 to 400 blooms.

Juvenility ends with a burst of growth that extends the

plant up and away from the other vegetation. In the mature condition, it tolerates considerable cold and is damaged by intense sunlight only when plants are severely dehydrated. Using saguaro as an example, monopodial growth is characteristic to about the age of seventy-five years, usually sufficient time to ensure that the plant has risen above any potentially competing vegetation. Having risen above all other strata of the plant community, oligopodial growth commences, and a number of vigorous parallel trunks ascend to increase the photosynthetic capability of the plant. At the end of the juvenile stage, the root system is capable of providing more water than the plant has been actually utilizing. Photosynthesis and growth have proceeded deceptively slowly when the plant has been in the shade of other vegetation. With the onset of the mature growth phase, photosynthesis suddenly increases dramatically, the metabolic rate and use of energy both increase, and water demand suddenly intensifies.

These plants are the most massive of the entire Sonoran Desert. Although true xerophytes, they nevertheless use a considerable amount of energy and water, particularly in their mature reproductive phase when each plant produces a large number of juicy fruits with high sugar content. Reproduction is by seed, and severed branches produce adventitious roots only with difficulty or not at all. Although these plants store water like the monopodial stem succulents, they go much further in adaptive strategy.

The strategy of the arborescent oligopodial stem succulent is to gain a precarious foothold at a site with favorable soil moisture but with potentially high interspecific competition; thus the moisture is sufficient eventually to satisfy the demand of an enormous plant. Next, the strategy is to extend the root system and plant body slowly while under the protection of a canopy of vegetation, then to surge upward, creating a massive plant body with relatively higher demands of water and energy. The capability of storing a great quantity of water and energy allows the plant to compete effectively with and outlive the prior vegetation at the site. The root system takes water systematically from a much greater area than that occupied by the plant body. Since this is usually not readily apparent, the effect of the plant on surrounding vegetation is frequently underestimated. After a saguaro dies and decomposes, a new palo verde or other tree is apt to become established and the cycle is repeated.

Clustering Stem Succulents

These plants vary from low clustering cacti such as pincushion (*Mammillaria*) and hedgehog (*Echinocereus*) to taller organ-pipe (*Stenocereus thurberi*) and senita (*Lophocereus schottii*). The candelilla (*Pedilanthus macrocarpus*) in the Euphorbiaceae family is also included in the category. Rain contacting the clustered stems is naturally channelled, often by

ridges and valleys of the plant surface or by a waxy covering, to the base of the plant. Adventitious roots usually are formed at the base of each branch so that the number of roots under the cluster is extremely high. Branches that are not at ground level frequently have adventitious aerial roots and may even break off during violent storms to form new plants.

The survival strategy seems similar to that of the monopodial stem succulent modified by increase in photosynthetic area. The numerous stems arising from a common base push outward to enlarge the original living space after the plant has become established. The increased photosynthetic area allows a higher energy level and more active growth, as well as the development of juicy fruits with a high sugar content.

Cylindrocaulescent Shrubby Stem Succulents

Plants of this life form are represented by the cholla (*Cylindropuntia*) subdivision of the genus *Opuntia*, common throughout the Sonoran Desert, by octopus cactus (*Rathbunia*) in Sonora, and by *Machaerocereus gummosus* in Baja California. The plant body is narrow, elongate, and cylindrical. The photosynthetic stem surface area is greatly increased through branching, sometimes in a rambling or octopus-like manner. Branches root into the soil if they contact the ground, whether yet connected to the main plant or severed from it. Cholla spines stick tenaciously to flesh or fur so that the severed branches unintentionally may be dispersed by animals. The stems often detach quite readily and, once rooted, grow into new plants. The fruits of some species remain green, fall to the ground, and grow into new plants by vegetative proliferation of the ovary wall.

Chollas reproduce frequently by vegetative means, and the seeds often abort or are infertile. In this life form, vegetative growth represents potential reproduction since any portion of the stem can grow into a new plant. The plants therefore become colonial perennials. The survival strategy represents a variation of the basic theme of the clustering stem succulent, but with selection for a growth habit resulting in exaggerated increase in photosynthetic stem area by diffuse branching, octopus-like growth, or fragmentation and proliferation to form extensive clones. The large photosynthetic area and vigorous root system support very rapid plant growth in comparison with that of other cacti.

Rathbunia and *Machaerocereus* have fruits attractive to animals, but expenditure of water and energy on producing sweet fleshy fruits to attract animals as agents of dispersal is rendered unnecessary in species of cholla by the obligatory dispersal by animals that accidentally contact the plants. The water and energy that would have been expended on production of sweet juicy fruit is channeled into ever-increasing elaboration of new stems. This life form sacrifices genetic flexibility characteristic of sexual reproduction in favor of temporary fitness to the present environment. As long as the environment remains constant, the species of cholla that reproduce by vegetative means will thrive in the Sonoran Desert.

Platycaulescent Shrubby Stem Succulents

This flat-stemmed succulent plant group is represented by the prickly pear (*Platyopuntia*) subdivision of the genus *Opuntia*. These plants are most similar to the cylindrocaulescent shrubby stem succulent life form, differing chiefly in having pancake-shaped stem segments and in not being dispersed to much extent by animals. The stem segments root readily upon contact with the ground. Most prickly pears spread over the surface of the soil, with each of the stem segments individually rooted.

Prickly pears are most abundant on rolling hills and moderate slopes in the Sonoran Desert but may be more common in the region of transition with the life zone at immediately higher elevation to the north and east. The flat-stemmed life form develops a greater heat load than does the cylindrical-stemmed form, although the orientation of the flat segment in relation to the angle of the sun's rays can reduce the load on that segment to almost the value for the cylindrical stem. In general, prickly pear species are adapted to somewhat cooler and moister situations than all cholla species. The increased heat load is beneficial at higher and cooler elevations but detrimental in summer at lower, hotter elevations. It is significant that species of cacti are more abundant in the Sonoran Desert than in the Mojave Desert and that the commonest cacti of the Mojave are of the prickly pear type. Cannon (1916) found that root growth of cacti required both high soil temperature and moist soil conditions. Summer rain of the Sonoran Desert is conducive to growth of cactus roots, whereas winter rain of the Mojave Desert is not. The increased heat load on the flat-stemmed life form in winter could allow root growth when heat exchange between stem and root permits.

Examination of prickly pear clumps on hillsides usually reveals the presence of a deep but unobtrusive soil pocket most pronounced on the uphill side. The chain of decumbent rooted stem segments impedes downhill soil erosion. As soil accumulates around the plant and either partially or completely buries stem segments, new adventitious roots at a higher level grow into the soil and stabilize it. As the soil deepens around the plants, soil moisture conditions generally improve.

The survival strategy is similar to that of the cylindrocaulescent shrubby stem succulent modified for heat load and often modified through decumbency for habitat improvement through soil retention. Arborescent prickly pear species sacrifice decumbency for protection from cactus-eating animals such as the javelina (*Pecari tajacu*). Reproduction is mostly by seed. Some species have sweet fleshy fruits. Clonal reproduction may occur by the older central part of a plant's dying out and the ends of separately rooted branches remaining as individual plants.

Leafless Green-Stemmed Plants

In this category are the crown of thorn plant types: *Canotia holacantha* (a tree in the Celastraceae), *Holacantha emoryi* (a coarse shrub in the Simaroubaceae), and *Koeberlinia spinosa* (a tree in the Koeberliniaceae). Also included are jointfir or cañutillo (*Ephedra,* a shrubby gymnosperm) and even the desert mistletoe (*Phoradendron californicum*), which parasitizes leguminous microphyllous trees.

The cacti are also all leafless green-stemmed plants but are segregated in the present treatment as stem succulents. Palo verde (*Cercidium*) and smoketree (*Dalea spinosa*), which are drought-deciduous green-stemmed microphyllous perennials, represent a transition toward the present life form. Through genetic loss of leaves, plants in many phyletic lines have independently arrived at the green-stemmed leafless condition. The survival strategy is to eliminate water loss from the leaves by eliminating the leaves and carrying on photosynthesis in stem tissue. In its various modifications, this strategy is one of the most effective throughout the Sonoran Desert.

Low Leafy Softwood Bushes

These are the plants referred to as semishrubs by Shreve (1951), who wrote that they "vastly outnumber any other life form in the entire region" of the Sonoran Desert. The plants differ from the true woody shrubs, which "have hard wood, restricted branching, determinate growth and marked ability to withstand cold or dry periods," according to Shreve. These low softwood bushes include species of saltbush (*Atriplex*), wild buckwheat (*Eriogonum*), jimmyweed (*Aplopappus*), bursage (*Ambrosia*), and brittlebush (*Encelia*).

Cunningham and Strain (1969) and Walter (1931) found that variations in leaf quantity and structure in brittlebush (*Encelia farinosa*) are controlled by soil moisture and help adapt the species to a seasonally changing desert environment by altering the carbon dioxide exchange capacity and water status of the shrub through changes in resistance to carbon dioxide and water vapor diffusion. Such adaptation to a seasonally changing environment may be the key to the great success of this life form in the Sonoran Desert, which is noted for great seasonal fluctuations in soil moisture in many of its divisions. This is apparently the survival strategy that had been elusive to previous investigators. In this context the indeterminate patterns of branching and growth, as well as the periodic loss of leaves and stems—all factors inferred by Shreve to be important in characterizing this life form—are logical external manifestations of the phenomena observed by Cunningham and Strain (1969) and by Walter (1931).

Evergreen Hardwood Bushes

This is a somewhat heterogeneous assemblage, including both sclerophyllous forms such as jojoba (*Simmondsia chinensis*), yerba de la flecha (*Sapium biloculare*), jito (*Forchammeria watsoni*), Sonoran caper (*Atamisquea emarginata*) and *Jacquinia pungens,* as well as microphyllous forms such as creosote bush (*Larrea tridentata*) desert hackberry (*Celtis pallida*), and crucillo (*Condalia warnockii*).

It is instructive to examine jojoba as an example of the first category. Gentry (1958) summarized the natural history of this species. It grows on well-drained, coarse desert soils where annual precipitation is relatively high. Burden (1970) found biseasonal rainfall to be important for establishment and growth. Summer rains allow germination and seedling establishment, and winter rains recharge the soil with moisture for utilization by the deep root system. Al-Ani et al. (1972) found that the plant could photosynthesize at extremely high temperatures of 40° to 47° C. Photosynthesis also occurred under conditions of low leaf water potentials during dry periods. The plants maintained a favorable carbohydrate balance during drought periods by reducing carbon loss and maintaining carbon gain.

Creosote bush can be examined as an example of the second category. Leaf xeromorphy was treated generally in relation to physiological and structural influences by Shields (1950), and the significance of reduction in leaf size was treated generally by Thoday (1931). Runyon (1934) discussed organization of creosote bush with respect to drought. Changes in soil moisture were found to influence branching pattern, as well as density and color of leaves. Leaves persisting through intense drought were discovered to be dormant until favorable conditions returned. Hull et al. (1971) studied variations in thickness of cuticle and epidermal cell walls of the species. A low photosynthetic rate characteristic of dormancy seems to be governed by low tissue-water potential rather than high temperature. Dechel et al. (1972) concluded that the adaptive value of small leaves and the high stomatal resistance were important in the adaptive strategy.

Although it may not be safe to generalize too far from these studies, it seems likely that the evergreen plants with thick, leathery leaves are deep rooted, require winter rain to recharge the soil with moisture at great depth, and continue life processes and photosynthesis at high temperatures and during dry periods, whereas the evergreen microphyllous plants rely on reduced leaf size and water-conserving leaf morphology to survive but become more or less dormant during severe drought.

Drought-Deciduous Indurate-Base Perennials

In the Sonoran Desert, this category is restricted to ocotillo (*Fouquieria*). The plants have broad, thin leaves that fall from the stems when soil moisture becomes low. Leaves are produced whenever soil moisture becomes favorable and temperatures are not too low. This may happen only once or twice during the year elsewhere such as in ecotonal situations above the Sonoran Desert, but according to Shreve (1951) as many as six or seven sets of leaves may be produced in a single year in the Sonoran Desert.

When ocotillo is leafed out, its water requirements and rates of metabolism and photosynthesis are relatively high. When the leaves have fallen, the plants are virtually dormant, with all physiological processes occurring at a very slow rate. The drought resistance in this condition involves the presence of resin-filled cells in the tissues of the inner bark. The hard and slightly enlarged indurate base of this life form, from which the many wandlike branches arise, may be involved with resin dynamics, with equalizing the osmotic balance among stems, and with regulating the simultaneous nature of production and shedding of leaves.

The petiole of the primary leaf indurates to become a spine. The later leaves that are produced derive from buds on short, sunken stems in the axils of these spines. In this respect, the important perennating organs are well protected from most herbivorous animals. Ocotillo is characteristic of thin soils where competition is rather low. Its strategy is to exploit an environment unfavorable to other plants by surviving through complex precision-regulated physiological responses that allow repeated periods of dormancy alternating with periods of high metabolic and photosynthetic activity.

Drought-Deciduous Sarcocaulescent Perennials

Shreve (1951) considered this life form second only to that of the stem succulent "in giving the vegetation of the Sonoran Desert its characteristic physiognomy." These plants have stout and swollen trunks, which seem out of proportion with their other structures. Species include the cirio or boojum tree (*Idria columnaris*), the elephant tree (*Bursera microphylla*), fernleaf elephant tree (*B. filicifolia*), ciruelo (*Cyrtocarpa edulis*), and copalquín (*Pachycormis discolor*). Although the trunks of these plants are not truly succulent, they nevertheless represent a concentration of a large mass of internal moist tissue into a relatively well-circumscribed space, much as in the example of the barrel cactus.

Narrow structures with thin tissue layers that would be vulnerable to desiccation from wind or to damage by unfavorable climatic conditions are at a minimum. The chief among these, the leaves, are deciduous during periods of drought. The survival strategy seems somewhat intermediate between that of the stem succulent and that of the drought-deciduous tree. The typical sarcocaulescent perennial usually represents the greatest deviation toward the stem succulent condition to be found in its particular family or order. The convergence toward this life form along several phyletic lines is noteworthy.

Drought-Deciduous Large-Leaved Perennials

This group includes plumeria (*P. acutifolia*), lomboy (*Jatropha cinerea*), torote prieto (*J. cuneata*), and sangre de cristo (*J. cordata*). The species are all winter deciduous, with the behavioral characteristic of leafing out only if and when adequate summer rains occur. As emphasized by Shreve (1951), the life form is related behaviorally to the strictly winter-deciduous tree. The survival strategy represents a slight betterment in one sense over the winter-deciduous perennial in that water and energy are not needlessly expended if rains fail to materialize.

But if adequate summer rain does occur, the plants are not yet in leaf and cannot photosynthesize as quickly as the typical winter-deciduous perennial, which leafed out in spring. There is a parallel here with the ephemerals that have germination inhibitors in their seed coats, which ensure that germination occurs only after heavy rains. Whereas an ephemeral passes periods of heat and drought in the seed, the drought-deciduous large-leaved perennial must be capable of surviving for long periods as a dormant leafless plant. Just as the ephemeral would seem to be derived from the nondesert annual, so would the drought-deciduous large-leaved perennial seem to be derived from a nondesert type—the winter-deciduous life form so characteristic of perennials in temperate regions.

The uncertainty of leaf production and of maintaining an adequate carbon balance is clearly a negative factor with this life form. It is not unexpected, therefore, that species with developed compensating features might be found in the category. Some types of spurge (*Euphorbia colletioides, E. xanti*), as well as the chuparosa (*Justicia californica*), maintain a favorable carbon budget during drought stress through reduced photosynthetic activity carried out in stem tissues. In this respect they approach the category of the green-stemmed leafless perennial, a very effective strategy in the Sonoran Desert.

Green-Stemmed Microphyllous Perennials

This classification includes a number of very common, characteristic, and dominant plants of the Sonoran Desert. The smoke tree (*Dalea spinosa*) and species of palo verde (*Cer-*cidium microphyllum, C. floridum, C. sonorae*) fall here and are also drought deciduous. Other green-stemmed microphyllous species, although not strictly drought deciduous, have more leaves during the summer rainy season than during other times of year. Included are desert broom (*Baccharis sarothroides*), rush broom (*Bebbia juncea*), graythorn (*Condalia lycioides*), yerba del venado (*Porophyllum gracile*), and the various species of rhatany (*Krameria*).

The survival strategy involves lowering water loss by reducing leaf size and quantity, while maintaining a favorable carbon balance partially through photosynthesis in stem tissues. This life form represents a more flexible and xeric adaptation over the drought-deciduous large-leaved perennial (such as *Plumeria* or *Jatropha*) in that the leaf size is reduced. Also it is a more flexible adaptation over the winter-deciduous microphyllous perennial (such as *Prosopis*) in that the green stem and the leaves, which are usually present in some degree, contribute to the carbon balance throughout the year.

Winter-Deciduous Broad-Leaved Perennials

This group is represented by such plants as cottonwood (*Populus*), willow (*Salix*), walnut (*Juglans*), ash (*Fraxinus*), Soapberry (*Sapindus*), pochote (*Ceiba acuminata*), palo santo (*Ipomaea arborescens*), and trumpet tree (*Tabebuia palmeri*). All grow under conditions of good soil moisture, particularly along drainage ways, where water seeps down from the mountains or where abundant seasonal rainfall is present.

The first five types listed relate to the temperate deciduous forests outside the desert and are occupying sites within the Sonoran Desert that approach the former most closely. The winter-deciduous habit is probably retained from prior adaptation to dormancy during periods of intense winter cold elsewhere. In the Sonoran Desert the canyon bottoms that they often occupy are subject to cold air drainage, so the adaptation is not without some value.

The last three examples are of tropical derivation but occupy a geographic region in southern Sonora on the west coast of Mexico where vegetation often shows adaptation to summer rain. Their winter-deciduous character may be related more to genetic adaptation to previous winter-drought stress than to freezing weather.

Winter-Deciduous Microphyllous Perennials

This group includes mesquite (*Prosopis*), catclaw (*Acacia greggii*), sweet acacia (*A. farnesiana*), and *Lysiloma*. The original winter-deciduous nature of this life form is suspected to have arisen as part of the thorn forest syndrome of the west coast of Mexico, where precipitation generally occurs in summer and where plants have small leaves as a strategy relating to the prevalence of drying winds. Both *Prosopis* and *Acacia* have increased their abundance dramatically in the upper reaches of the Sonoran Desert in historic time and have extensively invaded parts of the Upper Sonoran Life Zone as well.

The strategy appears to be one of opportunism. The winter-deciduous nature has allowed extension into relatively cold habitats. The seeds are spread by animals. The plants exploit all available sources of groundwater. A closely regulated carbon budget and the advantage of the green stem—factors so important to palo verde—are foregone by mesquite and *Acacia* in favor of a high rate of photosynthesis sustained by groundwater and sun.

Parasites

The strategy of the parasite is simply to attach to another organism and obtain nutrition at its expense. Humphrey (1971) commented on parasitism in the Sonoran Desert by dodder (*Cuscuta*), a plant that spreads as a vine over other vegetation.

Cannon (1904) described germination and establishment of mistletoe near Tucson. Birds are important in dispersal, and the seeds germinate in February, March, and April before the onset of intense heat. The seeds, growing directly into the branch of the host, do not contact the soil and thus are not insulated to high temperatures as are those of nonparasitic plants. *Phoradendron villosum*, which parasitizes riparian trees such as ash and cottonwood, has thick, leathery sclerophylls. *Phoradendron californicum*, which parasitizes extremely desert-adapted leguminous trees and shrubs such as palo verde, catclaw, and mesquite, represents the essentially aphyllous adaptation to a more severe Sonoran Desert condition. A literature review of mistletoes was published by Gill (1961), and Schmutz (1978) has provided a bibliography of mistletoe in Arizona.

Root parasites include broom rape (*Orobanche*), which attaches to roots of bursage (*Ambrosia*), and sand root (*Ammobroma*), which also parasitizes bursage but can be found on roots of arroweed (*Pluchea*), wild buckwheat (*Eriogonum*), borage (*Tiquilia*), and *Dalea* as well.

Halophytes

This category includes plants that are tolerant of salt and extreme alkalinity. Some show considerable anatomic variation. For example, the saltcedar (*Tamarix*) has salt glands that actually excrete salt to allow it to modify the water that it utilizes. Waisel (1972) published a monograph on biology of halophytes. The treatment by Walter and Stadelmann (1974) is useful. Plant growth relationships on saline and alkaline soils have been discussed by Hayward and Wadleigh (1949) and

Hayward and Bernstein (1958). Physiology of salt tolerance has been discussed by Bernstein and Howard (1958) and Bernstein (1964). Inland halophytes of the United States were discussed by Ungar (1974), and factors affecting their distribution by Branson et al. (1967).

Plants often not considered to be halophytic may tolerate a moderate amount of salinity. Yermanos et al. (1967) studied the effects of soil salinity on the development of jojoba and found a considerable degree of tolerance.

Saline conditions are common in lower drainage-way situations in the Sonoran Desert. Some halophytes are also phreatophytes, and complex situations may exist. Gary (1965) reported on relationships of groundwater depth to soil texture and electrical conductivity in the alluvium of floodplain communities of saltcedar (*Tamarix*), arroweed (*Pluchea*), and mesquite (*Prosopis*) in the Salt River Valley.

Halophytes of the Sonoran Desert are many and include desert saltbush (*Atriplex polycarpa*), mangle dulce (*Maytenus phyllanthoides*), yerba reuma (*Frankenia palmeri*), and quelite salado (*Suaeda*). The strategy of the halophyte is to adapt physiologically in order to utilize salty water and thereby to be able to occupy territory where competition from ordinary plants is insignificant.

Phreatophytes

This category of plant depends on exploitation of abundant groundwater where a definite water table is present. Species include cottonwood (*Populus fremontii*), batamote (*Baccharis glutinosa*), saltcedar (*Tamarix*), arroweed (*Pluchea sericea*), and willow (*Salix gooddingii*). Turner (1974c) studied vegetation as a part of the Gila River Phreatophyte Project of the U.S. Geological Survey. Giluly (1971) discussed phreatophyte management. Horton (1973) published a phreatophyte bibliography and a guide for surveying phreatophyte vegetation (Horton et al., 1964) and discussed management problems relating to phreatophytes (Horton 1972). Nabhan and Sheridan (1977) discussed phreatophytes used as living fence rows for floodwater management in Sonora.

Phreatophytes take great quantities of water from the ground or nearby streams and put it up into the air by transpiration. Advocates of removal of these plants from waterways in the Sonoran Desert argue that removal would make the water saved available for beneficial use by humans. Wildlife enthusiasts and many conservationists, on the other hand, oppose removal of phreatophytes because they view riparian vegetation as a major wildlife habitat in the desert. When the vegetation is eliminated, these persons predict negative, possibly irreversible, effects on wildlife. Other papers dealing with phreatophytes include those by Gatewood et al. (1950), Haase (1972), Lacey et al. (1975), and Robinson (1958).

ENVIRONMENTAL RELATIONS

Plants of the Sonoran Desert respond to the factors of their environment in many ways. Although the factors interact one with another, the most important can be isolated for analysis. By far the most important is that dealing with water relations. Influences of heat, cold, soil, pH, site orientation, and elevation are also important. In addition, germination, establishment, survival, dormancy, and competition are the determinants of vegetation dynamics.

A pioneer in investigating environmental relations of Sonoran Desert plants was Spalding (1909), who examined factors that determine distribution of desert plants. Soil water was the primary factor affecting local distribution, but soil aeration and pH were found to play roles. Other factors found to be limiting were summer heat on south-facing slopes and winter cold on north-facing slopes. The latter was found to especially limit saguaro (*Carnegiea gigantea*).

Water Relations

Livingston (1906, 1910) studied water relations of Sonoran Desert plants near Tucson. He is credited with being the pioneer in this area of investigation. He reasoned that water influx and rate of removal determined soil moisture. Water availability to a plant was found to depend not only on the parameters of soil moisture but also on the soil structure. The driest soil in most of the Sonoran Desert was found to occur in June just before the summer rains and at the time of greatest evaporative power of the air. Livingston (1908, 1911) showed that the evaporative power of the air was important in regulating soil moisture and was an important factor in determining vegetation types in the Sonoran Desert. He discussed the relation between summer evaporation intensity and centers of plant distribution.

Livingston and Brown (1912) and Livingston and Shive (1914) subjected various Sonoran Desert plants to physiological study with regard to water relations. The Briggs and Shantz wilting coefficient was determined for various species at the Desert Laboratory. Microphyllous xerophytes were found not to undergo a diurnal moisture change, and their leaf moisture content was found often to be higher in the daytime than at night.

Other early studies of water relations dealing with Sonoran Desert plants involved relative transpiration in cacti (Livingston 1907), the water balance of succulent plants (MacDougal and Spalding 1910), and the reversibility of the water reaction in a desert fern (Cannon 1914).

Harris et al. (1916) found that osmotic concentrations of plant sap increased from winter annuals, to perennial herbs, to dwarf shrubs and half shrubs, and finally to arborescent woody perennials in Sonoran Desert plants. Klikoff (1966) reported

on competitive response to moisture stress in a winter annual of the Sonoran Desert and later discussed moisture stress in a vegetational continuum in the Sonoran Desert (Klikoff 1967).

Bryan (1928) reported on changes in vegetation brought about by changes in the water table. Judd (1971) attributed the death of the mesquite trees at Casa Grande National Monument near Coolidge, Arizona, to a lowering of the water table at that location. Martin (1976) analyzed soil moisture use by mesquite. Odening et al. (1974) studied the effect of decreasing water potential on net CO_2 exchange of desert shrubs.

McGinnies and Arnold (1939) reported relative water requirements of range plants of the Sonoran Desert, finding that trees and shrubs such as mesquite, jojoba, and catclaw acacia used about three times as much water as does the average perennial grass. Darrow (1935) studied transpiration rates of grasses and shrubs as related to environmental conditions and stomatal periodicity. Hastings (1960) recorded information on precipitation in relation to saguaro growth. Shantz (1948) summarized information on the water economy of plants, and Horton (1973) published a bibliography on evapotranspiration and water research as related to riparian and phreatophyte management.

Walter (1931) presented extensive data on the water relations of brittlebush (*Encelia farinosa*), barrel cactus (*Ferocactus wislizenii*), saguaro (*Carnegiea gigantea*), and Sonoran Desert ferns. Later (Walter 1960) he presented osmotic spectra for various life forms in the Arizona Upland Division of the Sonoran Desert. After fifty years of research into plant water relations, he presented a new approach to water relations of desert plants (Walter and Stadelmann 1974) and provided much information of value in Sonoran Desert studies. Ocotillo (*Fouquieria splendens*) and sangre de cristo (*Jatropha cardiophylla*) were classified as stenohydric xerophytes. Sclerophyllous and aphyllous xerophytes constituted another category. Foothills palo verde (*Cercidium microphyllum*) was a Sonoran Desert example of the aphyllous xerophyte. The sclerophyllous xerophyte category would include jojoba (*Simmondsia chinensis*). Brittlebush (*Encelia farinosa*) and triangle-leaf bursage (*Ambrosia deltoidea*) were presented as examples of malacophyllous xerophytes. Creosote bush (*Larrea tridentata*) was considered intermediate between the malacophyllous and sclerophyllous categories. Sonoran Desert ferns were given as examples of poikilohydric xerophytes. Solbrig, in chapter 9 of this book, discusses the adaptations of plants to desert conditions.

Influence of Heat and Cold

Lowe (1959) traced plants of the Sonoran Desert to tropical and subtropical sources that were already adapted to heat. Some that are winter deciduous due to adaptations to cope with winter drought may be preadapted to colder areas. This may explain growth of mesquite (*Prosopis velutina*) and *Acacia* species in colder habitats within the Sonoran Desert and their invasion of communities in the Upper Sonoran Life Zone. Investigators have found that characteristic Sonoran Desert plants such as jojoba (*Simmondsia chinensis*), creosote bush (*Larrea tridentata*), and most cacti can maintain metabolism, photosynthesis, and growth during extremely high temperatures as long as water is not limiting. Strain and Chase (1966) reported on the effect of past and prevailing temperatures on the carbon dioxide exchange capacities of some woody desert perennials.

Only the outer portions of plants facing the sun are subjected to intense solar radiation. McDonough (1964) found that a leafy creosote bush reduced solar load by 57 percent under the light-intercepting canopy. Dormant leafless desert shrubs reduced solar energy by 10 to 49 percent, depending on pattern of branching. Patten and Smith (1975) observed that small trees such as palo verde (*Cercidium*) moderate the effects of heat in their environment. Broad-leaved bursage (*Ambrosia*) proved to be slightly warmer than ambient air temperature, while small-leaved creosote bush proved to be slightly cooler. Cacti were found to heat up rapidly well above ambient air temperatures but to cool down quickly at night.

Gibbs and Patten (1970) reported that cacti in the Sonoran Desert dissipate heat through such mechanisms as critical stem-segment angle in relation to the sun, spine reflectivity, convection, temperature-dependent reradiation, and heat dissipation from apical atem to basal stem. Hull (1958) studied the effect of day and night temperatures on growth, foliar wax content, and cuticle development of mesquite.

Capon and Van Asdall (1967) found heat pretreatment to be an effective means of increasing germination of desert annual seeds in the Sonoran Desert. Shreve (1951) noted that seeds of many Sonoran Desert plants (such as palo fierro, *Olneya tesota;* and saguaro, *Carnegiea gigantea*) germinate immediately after forming and when very high temperatures prevail.

Shreve (1924) described the transition in vegetation from the north, where winter freezes are common, to the south at Puerto Libertad and Kino Bay where freezing weather is rare. Frost-sensitive species such as palo fierro (ironwood) were found in increasingly more open and flatter exposures. Saguaro eventually was replaced to the south by the more frost-sensitive but physiognomically similar cardón (*Pachycereus*).

Shreve (1911a) treated the influence of low temperature on the distribution of saguaro, the response of Sonoran Desert plants to cold air drainage (1912), and the role of winter temperatures in determining the distribution of plants. Patten (1973) discussed winter as a major determinant of desert vegetation patterns. Wiggins (1937) reported the effects of freeze on the pitahaya (*Stenocereus*). Turnage and Hinckley

(1938) wrote on the subject of freezing weather in relation to plant distribution in the Sonoran Desert.

Hastings and Turner (1965) believed that warming trends in climate over the last hundred years had significantly altered Sonoran Desert plant communities and was partially responsible for invasion of desert plants such as mesquite and *Acacia* into communities above the desert. Calder (1977) presented a lucid summary of fluctuations in heat and cold in the northern hemisphere, documenting a cold period from about 1750 to 1850 and a warm period from about 1850 to 1950. Within these trends, any one year may be an exception. The exceptional years provoke comment on minor vegetational effects, but the insidious effects of long-term trends and reversals are more important to the vegetation.

Recently Steenbergh and Lowe (1977) have reopened the subject of freeze damage and its effect on the dynamics of Sonoran Desert vegetation. Jones (1979) reported the effects of the 1978 freeze on native plants of Mexican portions of the Sonoran Desert, and a minisymposium on the severe freeze of 1978-1979 appeared in the periodical *Desert Plants* (August 1979). Some evidence indicates that the full effect of freeze damage to vegetation in the Sonoran Desert may not be evident until a number of years after the freezing event. This was found to be significant in the case of the saguaro by Steenbergh and Lowe (1977).

Influence of Soil

Shreve (1951) believed that the influence of soil on Sonoran Desert vegetation was manifold but was of utmost importance in exercising control of moisture availability. Near Tucson, Beutner and Anderson (1943) showed that protection of the soil surface by organic litter or by plants themselves was of great importance in allowing water to penetrate the soil and in producing a mulching effect to conserve water for plant growth. Musick (1975) commented on the barrenness of natural desert pavement in Yuma County, Arizona. Aside from the importance of surface permeability, it is apparent that soil depth, texture, salinity, and nutrient content are important factors governing vegetation.

Phillips (1963) discussed depth of roots in soil, finding roots of a woody legume, probably mesquite, over 175 feet below the surface at the site of a recent excavation for an open pit mine. Hastings and Turner (1965) pointed out that woody perennials in the Sonoran Desert are reduced in numbers on thin soils. Perennials with specialized survival strategies, however, can successfully colonize such soils and live undisturbed to an old age. Ocotillo (*Fouquieria splendens*) is characteristic of the thin soil habitat and can live to an age of one hundred or two hundred years (Darrow 1943) in spite of periodic conditions of extreme drought.

Marks (1950) reported on relations of soil and vegetation in the Lower Colorado River region whereas Yang and Lowe

(1956) were able to correlate major vegetation climaxes with soil characteristics in the Sonoran Desert. Kramer (1962) analyzed the distribution of saguaro in relation to soil characteristics and Bingham (1963) reported on vegetation-soil relationships in two stands of palo verde and saguaro vegetation. Phillips and McMahon (1978) related changes in plant species composition along a Sonoran Desert bajada to a gradient of soil particle size distribution and salinity.

The adaptive strategy of the halophyte has been discussed previously. It is also instructive to note vegetational responses to salt concentrations as they relate to soil patterns. Cannon (1908b) studied the vegetation in concentric rings around a salt spot in the floodplain of the Santa Cruz River near Tucson. No vegetation occurred in the center of the spot, but three distinct zones of vegetation radiated around it. Nearest the center, dwarf quelite salado (*Suaeda*) grew with species of saltbush (*Atriplex nuttallii*, *A. elegans*). In the second zone were desert saltbush (*Atriplex polycarpa*), tomatillo (*Lycium*), and quelite salado. The outer zone was of four-wing saltbush (*Atriplex canescens*), mesquite (*Prosopis*), and *Aplopappus* ("*Bigelovia*").

The soils of the Sonoran Desert, having developed under relatively dry conditions, characteristically are not excessively leached of potassium, phosphorus, and other important soil nutrients. A minisymposium relating to soils for cultivation of desert plants appeared in the periodical *Desert Plants* (November 1979) and was based primarily on experience in the Sonoran Desert. The chief deficiencies in Sonoran Desert soils from the standpoint of vegetation are water and available nitrogen.

The breakdown of organic material to yield humus (contributing to soil texture and cation exchange capacity) and ammonium ions is characteristic of soils of humid regions more than those of deserts. Dead plant material in the Sonoran Desert unfortunately may rest on the soil surface without significant decomposition before being washed away with runoff from rain. When decomposition does produce ammonium ions, they are lost to potential absorption by plant roots because of the rapid volatilization in dry soil or dry air.

As a result, leguminous plants with nitrogen-fixing bacteria in root nodules are significantly more dominant in Sonoran Desert vegetation than in other world vegetation types. Leguminous genera of major significance include *Cercidium*, *Prosopis*, *Acacia*, *Olneya*, *Lysiloma*, *Krameria*, *Caesalpinia*, *Calliandra*, *Eysenhardtia*, and *Pithecellobium*. The microphylls that fall from these leguminous plants are rich in nitrogen and also display increased resistance to being flushed away by runoff. The flat elliptic leaflets, which conform to the microtopography when wet, represent the closest approach to a true leaf litter in the Sonoran Desert. Since these plants also tend to occupy the moister sites of drainage ways and upper bajadas, where classic ammonification of soil can most closely be approached under arid conditions, they are major contributors

of fixed nitrogen to the soil and to the surrounding vegetation.

Nevertheless, a deficiency of fixed nitrogen in the soil can be an important factor for Sonoran Desert vegetation and the study of other potential sources becomes important. Green and Reynard (1932) found that dens of kangaroo rat (*Dipodomys spectabilis*) and pack rat (*Neotoma albigula*) contained larger quantities of soluble salts than did nearby soil of the same type outside the dens. Soil of the dens was particularly high in calcium, magnesium, bicarbonate, and nitrate.

Cameron (1962) reported that the soil algal flora of the Sonoran Desert was rather homogeneous. Following rain, an algal crust of oscillatoroid forms of blue-green algae often develops over large tracts of desert land. These soil algae are important because they add to the organic content of the soil and fix atmospheric nitrogen into a form that can be used by other plants.

Influence of pH

Alkaline soil (high pH) is so common through much of the Basin and Range Physiographic Province and indeed through such a large portion of the North American deserts that the presence of plants characteristic of high pH soil is inadvertently considered the norm. These alkaline soils are those in which a calcium carbonate (caliche) layer forms. Soils developing from granite and rhyolitic parent rock, or from any other acid igneous material, tend to be at least somewhat acid and are never known to contain caliche. The relation of caliche to desert plants was discussed by Shreve and Mallery (1933).

The influence of pH on vegetation is observed with textbook clarity on natural hillsides at the Boyce Thompson Southwestern Arboretum in the Arizona Upland Division of the Sonoran Desert where fortuitous geologic events have resulted in soils of different pH occurring at a single site. At that location, creosote bush (*Larrea tridentata*) occupies a high pH (alkaline) relictual alluvial hill (with caliche), which had been spared from elimination through weathering by being capped with a thin layer of a relatively recent basalt, which has now weathered away to reveal the old erosion surface.

At the Arboretum, jojoba (*Simmondsia chinensis*) and palo verde (*Cercidium microphyllum*) are dominant on an acid igneous (rhyolite) mountain, which was preserved from weathering by being capped with hard viscous lava of quartz latite composition. On lower slopes where the acid rhyolite has weathered away, a schist is exposed with a soil of nearly neutral pH on which buckwheat bush (*Eriogonum fasciculatum*) is very abundant. Creosote bush is never found on the rhyolite, quartz latite, or schist. Four-wing saltbush (*Atriplex canescens*) grows with the creosote bush but on none of the other soils.

Cable (1972) reported on revegetation trials of four-wing saltbush in two different areas in southern Arizona. It was shown that establishment was higher on a site with creosote bush and calcareous soil (pH 8 or above) than on a site where mesquite was dominant and soil was neutral and noncalcareous. Johnson (1961) found that creosote bush prefers shallow, sandy soils where a calcium carbonate layer typically is present and low moisture availability allows the plant to compete effectively against other species. Soils with low pH (acid) were found to prevent germination of creosote bush, and soils with much clay or moisture content were found to favor other plant species.

Influence of Site Orientation

Despite the relatively low latitude when compared to northern North America and despite the fact that the sun's rays are actually from the north during the extreme summer heat, the south-facing slope is much hotter and drier than the north-facing slope and the vegetation is often strikingly different on these orientations at any one site. The influence on the vegetation is probably exercised primarily in winter when the rays of the sun strike the south-facing slope more directly and warm it to the point where, in combination with the draining away of cold air, a truly subtropical, relatively frost-free situation usually prevails.

Although the sun's rays strike the north-facing slope rather directly in summer, the residual soil moisture is higher than on slopes with other orientations due to the less direct rays during the remainder of the year. Numerous reports indicate that saguaro (*Carnegiea gigantea*) is a plant of south-facing slopes (Steenbergh and Lowe 1977), as is palo fierro (*Olneya tesota*) or brittlebush (*Encelia farinosa*) in the Arizona portion of the desert (Shreve 1951), but that grasses are more abundant on the northern exposures (Cumming 1952). Blumer (1911) provided examples of Sonoran Desert plants that grow at one orientation at lower altitudes of a desert mountain range but at a different orientation at higher altitudes.

At the Boyce Thompson Southwestern Arboretum, vegetation on north- and south-facing slopes of the same rhyolite countryrock can be compared while standing at one point. The northern aspect there at Ayer Lake is dominated by hopbush (*Dodonaea viscosa*) jojoba (*Simmondsia chinensis*), ocotillo (*Fouquieria splendens*), mescal (*Agave chrysantha*), and compass barrel cactus (*Ferocactus acanthodes*). Other plants of north-facing slopes at the Arboretum include Arizona rosewood (*Vauquelinia californica*), sotol (*Dasylirion wheeleri*), bush penstemon (*Penstemon microphyllus*), and numerous grasses, lichens, mosses, ferns, and liverworts. The south-facing slope opposite the lake is dominated by foothills palo verde (*Cercidium microphyllum*), prickly pear (*Opuntia phaeacantha*), saguaro (*Carnegiea gigantea*), bursage (*Ambrosia deltoidea*), and brittlebush (*Encelia farinosa*).

Whitaker and Niering (1965) published a gradient analysis of south-facing slope vegetation of the Santa Catalina Mountains near Tucson. Haase (1970) studied environmental fluctuations on south-facing slopes at the same location. Solar radiation, evaporation, air temperature extremes, soil temperature extremes, soil moisture, and precipitation were measured on the south-southeast, south, south-southwest, and southwest slopes in the desert foothills. The sequence of warmest and driest exposures was south, south-southwest, southwest, and south-southeast. The driest exposures varied with the season. In the dry fall, south and southwest were driest, but in the dry spring, southwest was driest. During the wet winter, south was driest, but during the wet summer south-southeast was driest.

Influence of Elevation

The boundaries of the Sonoran Desert to the east and north are chiefly determined by the influence of elevation. Hastings and Turner (1965) thought that mean annual temperatures in Arizona may have risen by three to three and one-half degrees Fahrenheit since the 1870s and equated this to an elevational difference of about 1000 feet. Shreve (1922) described conditions under which quite dissimilar communities of plants and animals can be found at the same elevation in adjacent localities in the Santa Catalina Mountains. It is clear from several studies that within certain limits, changes in elevation can compensate for differences in site orientation, temperature, rainfall, or soil conditions.

Generally as elevation increases, temperature decreases and rainfall increases. Plants that grow at low elevations in the northern part of the Sonoran Desert often grow at higher elevations in the southern part. Cold air is heavier than warm air and therefore sinks from higher to lower elevations at night. For this reason the habitat on a higher slope may be warmer than one in a flat location where cold air pools. These low flatlands often are dominated by creosote bush (*Larrea tridentata*), which is relatively cold hardy. Also cold hardy are mesquite (*Prosopis velutina*) and riparian plants that grow in the low drainages. The most frost-sensitive species grow on slight slopes at elevations slightly above the flatlands. These include elephant tree (*Bursera microphylla*), brittlebush (*Encelia farinosa*), *Trixis californica*, and palo fierro (*Olneya tesota*). To the south, these may grow in the flatlands. The plants at highest elevations in the Sonoran Desert are similar to those of the north-facing slopes at lower elevations. On the north-facing slopes of the very highest elevations, scrub oak (*Quercus turbinella*), cedar (*Juniperus*), grasses, and various chaparral or scrub plants begin to appear, and there is a transition to the vegetation type elevationally above the desert.

Shreve (1922) showed that the upper limit of desert vegetation on mountains arising from desert plains is largely controlled by the height of the mountain. Thus forests tend to come down to lower elevations on higher mountains. The Pinal Mountains in Arizona and the Sierra Babiso in Sonora are, according to Shreve (1951), the only mountains with mesic vegetation that are surrounded by Sonoran Desert.

Brown (1978) noted chaparral elements on 19 of 21 desert

mountain ranges over 1300 m elevation in the Sonoran Desert of southern Arizona. These populations were assumed to be relictual. The typical vegetation altitudinally above Sonoran Desert was described as scrub. In Arizona this was described as interior chaparral. In California and Baja California it was described as coastal sagescrub or coastal chaparral and in Sonora as sinaloan thornscrub.

Germination, Establishment, and Survival

Fahn and Werker (1942) discussed mechanisms of seed dispersal. Seed reserves in desert soils were treated by Childs and Goodall (1973) and Goodall and Morgan (1974). Germination and seedling behavior of desert plants was reviewed by Went (1979). Tevis (1958a) reported on germination and growth of ephemerals induced by sprinkling a sandy desert and (Tevis 1958b) on a population of desert ephemerals germinated by less than 1 inch of rain.

During germination and the early life of desert plants, species that eventually will assume quite divergent life forms are faced with rather similar environmental situations. Germinating perennials, for example, must compete with ephemerals. Although there may indeed be major differences in germination and in the early phases of life cycles of the divergent plants that compose the Sonoran Desert vegetation, it is nevertheless instructive to examine some case studies of germination, establishment, and survival.

Shreve (1917) discussed the establishment of desert perennials. He believed that desert vegetation was relatively stable because the perennials were slow growing, long-lived, and had a low rate of establishment. Barton (1936) reported on experiments in germinating desert seeds at the Boyce Thompson Institute for Plant Research. Simpson (1977) compared breeding systems of dominant perennial plants of the Sonoran Desert near Tucson with those in Andalgala, Catamarca, Argentina.

Evenari (1949) summarized information on germination inhibitors. Many Sonoran Desert plants have proven to have germination inhibitors, which must be thoroughly leached from the seed coat before germination commences. This has been well documented for the winter ephemerals. Amen (1968) discussed the concept of seed dormancy. Went (1949) discussed effects of rain and temperature on germination and growth.

The leguminous trees of the Sonoran Desert, with the exception of palo fierro (*Olneya tesota*), have seeds with very hard seed coats that are virtually impervious to water. Commercial nurserymen germinate them by soaking in concentrated sulfuric acid for about one-half hour until the reticular vascular system can be seen, flushing them with cold water and planting them immediately. In nature these seeds are often carried along sandy washes by runoff water and abraded against the sand and gravel until the seed coat is worn away.

Went (1957) explained the abundance of such trees along sandy washes by referring to this phenomenon. Hastings and Turner (1965) referred also to bacterial and fungal decomposition of the seed coat.

Factors affecting germination of Sonoran Desert range plants were studied by Muhktar (1961). Maceration of the seed coat with sulfuric acid increased germination of catclaw acacia (*A. greggii*), mesquite (*Prosopis velutina*) and panic grass (*Panicum*) but decreased germination of bluestem grass (*Andropogon*) and lovegrass (*Eragrostis*). Scarification by mechanical means improved germination of all trees and grasses that were studied. Blue and far-red light inhibited panic grass germination, but red light promoted it. White light increased mesquite germination.

There is evidence that a number of seeds resistant to germination may be adapted to dispersal by means of passing through the digestive system of an animal. The stomach acid results in breakdown of the hard seed coat. Dispersal by animals theoretically would be selected for in desert areas because animals frequent places with moisture, and seed dispersal would tend to be concentrated in such areas. Martin (1980) believed that prehistoric animals were probably responsible for spreading mesquite around the Sonoran Desert.

The germination of mesquite in cow manure has been well studied. Glendening and Paulsen (1955) found as many as 1617 undigested mesquite seeds in a single cow chip. Tschirley and Martin (1960) found that germination of mesquite was improved by planting the entire fruit or just the seeds into manure. Possibly bacterial and fungal decay of the fruit assists breakdown of the seed coat. Martin (1948) reported that mesquite seed had been found to be viable after forty-four years. Paulsen (1950) investigated survival of mesquite seedlings.

Shreve (1911b) studied the establishment behavior of foothills palo verde (*Cercidium microphyllum*), documenting survival during the first ten years of life. Age classes of palo verde back to 1490 were delimited. McCleary (1973) discussed comparative germination and early growth of six species of *Yucca*. Barbour (1968) discussed germination requirements of creosote bush. Development of seedlings of tanglehead grass (*Heteropogon contortus*) as related to soil moisture and competition was treated by Glendening (1941). The effect of different osmotica on germination of alkali sacaton grass (*Sporobolus*) was discussed by Knipe (1971).

Kurtz and Alcorn (1960) reported on germination requirements of saguaro (*Carnegiea gigantea*). Booth (1964) investigated a disease of saguaro seedlings. Turner et al. (1966) studied the influence of shade, soil, and water on saguaro seedling establishment. Despain (1967) investigated survival of saguaro seedlings on soils of differing albedo, cover, and temperature at the Boyce Thompson Southwestern Arboretum. Steenbergh and Lowe (1969) discussed critical factors during the first years of life of the saguaro, and Shreve (1910) has reported the rate of establishment. Brum (1973) investigated both phenology and establishment of saguaro in marginal populations. McDonough (1964) treated germination responses of both saguaro and organ-pipe cactus (*Stenocereus thurberi*).

Dormancy

Most plants of the Sonoran Desert have definite periods of dormancy. For the ephemerals, the dormant period is passed in the seed. Mesquite and many riparian species become dormant in the winter. Bush penstemon (*P. microphyllus*), tomatillo (*Lycium*), and Mexican elderberry (*Sambucus*) are dormant in the summer. Ocotillo (*Fouquieria*) and cirio (*Idria*) are dormant whenever the soil is dry.

Some species, like jojoba, grow rapidly any time of year when soil moisture permits but maintain favorable carbon balance during times of drought because of the extremely deep root system and sclerophyllous leaves. According to Dechel et al. (1972), creosote bush has no set dormant period but may reduce its life functions severely during periods of water stress. Because of its inherent flexibility, the species was thought to be able to grow in regions with divergent seasonal rainfall and temperature. The species is successful not because it has no dormancy but rather because its dormant period is flexible.

Dormancy is a major key to survival of plants in the Sonoran Desert, and it is evident that most of the dominant life forms owe their success to key periods of dormancy in their life cycle. The strategy of dormancy allows the environment of a species to be divided into a fraction that has little or no effect on the plant and another fraction that becomes the new environment. In this manner adversity is eliminated within the dormancy capability of the species.

Competition

The proliferation of life forms in the Sonoran Desert divides the habitat and tends to reduce competition. A number of published studies have dealt with competition between range grasses and shrubby perennials. For example, Cable (1969) treated competition in semidesert grassland as influenced by root systems, growth habits, and extraction of soil moisture. Humphrey (1937) reported on competition between *Aplopappus* and grasses. Yeaton et al. (1977) treated competition and spacing in the Arizona Upland.

Another type of study has dealt with chemical substances elaborated by one Sonoran Desert plant to inhibit the germination or growth of another. These fall within the class of substances known as plant growth regulators (PGRs). Gray and Bonner (1948) reported on an inhibitor of plant growth from leaves of brittlebush (*Encelia farinosa*). Woods (1960) discussed biological antagonism due to phytotoxic root exudates. This type of activity has been proposed for roots of creosote bush. Bennett and Bonner (1953) isolated growth inhibitors from *Thamnosma montana*. Self and Bartels (1980) have recently studied native Sonoran Desert plants with germination-inhibiting PGRs growing at the Boyce Thompson Southwestern Arboretum.

Volatile substances produced by aromatic shrubs are often treated as inhibitors, and these have been treated by Muller et al. (1964). Many rank chemicals incompatible with basic life processes may be produced by desert vegetation, primarily to inhibit grazing animals and secondarily to act as inhibitors of competing vegetation.

Ecotypic Variation

This subject has been poorly studied in the Sonoran Desert. It is improbable that it does not exist in most of the commoner species, which seem to grow in more than one habitat. It seems to occur in jojoba. One naturally wonders if the mesquite, which has invaded the Upper Sonoran Life Zone, might not represent an ecotype adapted to cold. Yang and Lowe (1968) reported that creosote bush of the different North American deserts represented ecotypes with different chromosome numbers. Dina and Klikoff (1974) studied possible ecotypic fluctuation in carbohydrate balance in *Plantago insularis*, a winter annual of the Sonoran Desert.

Cryptogamic Plants

Walter (1931) reported on water relations of Sonoran Desert ferns. Hevly (1963) studied adaptation of Cheilanthoid ferns to desert environments and later (1965) treated taxonomy of the *Notholaena sinuata* complex of species. Tryon (1957) revised the entire genus *Notholaena*. Jones (1966) published an annotated bibliography of Mexican ferns.

McCleary (1959) treated the bryophytes of the desert region in Arizona and detailed (1951) the factors affecting the distribution of mosses in Arizona. Haring (1961) published a checklist of mosses of Arizona, and mosses of Sonora were treated by Bartram and Richards (1941).

Mosses display effective short-term survival capabilities when they are exposed to heat and desiccation (Dilks and Proctor 1974; Lange 1955; Norr 1974) but nevertheless cover only a small percentage of ground in the Sonoran Desert. Alpert (1979) suggested that the scarcity of these plants in the desert might be due more to an unfavorable carbon balance when the plants are desiccated than to direct physical effects of heat and desiccation. He showed that mosses in a natural Sonoran Desert situation at the Boyce Thompson Southwestern Arboretum, in the Arizona Upland, desiccate rapidly during daylight hours after a major summer rainfall. Taking water saturation of photosynthesis (at standard conditions of 15° to 25° C and 50 to 600 microEinsteins m^{-2} s^{-1} to lie normally between 1.5 and 3.0 g H_2O g^{-1}dw, the water content measured following the rain indicated that water became limiting for photosynthesis by the time four hours of daylight had passed. The water compensation point was reached by six to nine hours after sunrise. Thus, less than nine hours seemed available for net photosynthesis following rainfall in a typical Sonoran Desert habitat.

Desert lichens, algae, and fungi were treated generally by Friedmann and Galun (1974). Adaptations of desert lichens to drought and to extreme temperatures have been treated by Lange et al. (1975). Lichens of Arizona have been treated by Nash (1975a), with a separate treatment for those of Maricopa County (Nash 1975b). Fink (1909) discussed the lichens of the Desert Laboratory near Tucson. The monograph of *Parmelia* in North America by Berry (1941) is useful, as is the report on taxonomy and morphology of the lichen genus *Acarospora* and its phycosymbiont *Trebouxia* by Duewer (1971). Nebeker et al. (1977) discussed lichen-dominated vegetation of coastal Baja California.

Temperature tolerance of algae in dry soil has been reported by Trainor (1962, 1970). Cameron (1961) treated algae of the Arizona portion of the desert. Wien (1958-1959) studied the algae of irrigation waters of the Salt River Valley. Kidd and Wade (1965) reported on the algae of Quitobaquito Spring, and Sommerfeld et al. (1975) surveyed the phytoplankton of Canyon Lake in the Arizona Upland Division. Cameron (1964) treated the terrestrial algae of southern Arizona. Fletcher and Martin (1948) reported the effects of algae and molds in the rain crusts of desert soils. Cameron and Fuller (1960) and Fuller et al. (1960) discussed fixation of nitrogen in desert soils by algae, as did MacGregor and Johnson (1971). Nitrogen fixation by arid soil lichens was treated by Rodgers et al. (1966). Shields et al. (1957) and Cameron and Blank (1966) discussed aspects of alga and lichen-stabilized soil crusts. Snyder and Wullstein (1973) reviewed the role of desert cryptogams in nitrogen fixation.

Fungi that decay mesquite were detailed by Gilbertson et al. (1976). Hosts of Arizona rust fungi were listed by Gilbertson and McHenry (1969), and those of wood-rotting fungi were published by Gilbertson et al. (1974). A heart-rot fungus, *Fomes robustus*, that attacks cacti and other desert plants was discussed by Davidson and Mielke (1947). Lindsey and Gilbertson (1975) treated wood-inhabiting homobasidiomycetes on saguaro. The relation of *Poria carnegiea* to the decay of saguaro was discussed by Gilbertson and Canfield (1972) and Lindsey (1975). Levine (1933) had reported crowngall on saguaro. Alcorn and May (1962) believed that the bacterium *Erwinia carnegieana* was causing saguaro to decline.

Whaley (1964) studied representative samples of rhizosphere microflora of saguaro and palo verde in relation to antibiotic activity against the common Sonoran Desert soil-borne pathogens. Of 386 isolates, 147 showed antibiotic activity against *Fusarium oxysporum*, 117 against *Verticillium albo-atrum*, 109 against *Phymototrichum omnivorum*, and 85 against *Rhizoctonia solani*. All four pathogens studied were inhibited by 38 isolates. These data indicate a considerable degree of antagonism to pathogens by normal rhizosphere microflora. An increase in soil-borne diseases often results when large amounts of organic matter are added to soil in combination with watering. This increase may result from altering the normal balance of Sonoran Desert rhizosphere microflora.

Lamb (1964) studied the fungal flora of the slime flux of Sonoran Desert plants. Elbein (1956) found Myxobacteria species diversity to be much higher in saguaro-palo verde desert of the Tucson area than in nearby oak woodland or in pine forest. McLaughlin (1933) reported on a fusarium disease of Senita cactus.

Treatment of Vegetation by Geographic Divisions

The classic geographic treatment of Sonoran Desert vegetation is that of Shreve (1951). His delimitation of these divisions is shown in figure 5-9. Brown and Lowe (1977) have published a map of biotic communities south to Latitude 27° N. They eliminate the Sonoran Foothills Division following Felger (1966), and associate it with the Sinaloan Thorn Forest. Figure 5-10 compares their delimitations with those of Shreve (1951). The present treatment corresponds more closely with that of Shreve. Each division of the desert was referred to by Shreve by alternative descriptive names, which tend to emphasize the phytophysiognomic, floristic, and geographic unity of each division recognized. The names, modified from Shreve's (1951) terminology, are used in the present discussion (table 5-1).

Lower Colorado-Gila Division

At the head of the Gulf of California and on either side of its upper reaches lies the largest vegetational division of the Sonoran Desert, the Lower Colorado-Gila Division. This division occupies the northern Gulf coast of Baja California, all of the California portion of the Sonoran Desert (including El Centro, the Salton Sea, Indio, Blythe and Needles), about half of the Arizona portion (from Parker to Phoenix and Gila Bend, south to the Mexican border), and all of Sonora below 400 m elevation down to the valley of the Rio Concepción, including a wedge of land extending 150 km south of Caborca.

The north-south dimension of the Colorado-Gila Division is great enough, nearly 700 km, that some latitudinal diversification of vegetation is evident, with important dominant species of the south varying somewhat from those of the north. There are also east-west differences in the vegetation.

The parameters of heat and aridity of the Lower Colorado-Gila Division are extreme when compared with those of most of the other divisions, and this is reflected in the vegetation. Insolation is relatively high. Precipitation, although low in both summer and winter, is greater in winter than summer. Temperatures in all seasons are relatively high. Hastings and Turner (1965) commented on the relatively low number of life forms in the low plains, which constitute such a large part of this division.

This division is often described as the microphyllous desert because of the very small leaves of the characteristic species, or as the *Larrea-Franseria* region, because of the predominance of *Larrea tridentata,* called creosote bush, and *Ambrosia* (subgenus *Franseria*) *dumosa,* called white bursage. Shreve (1951) noted that in all of the larger intermountain plains, creosote bush and white bursage constituted 90 percent or more of the plant cover over thousands of hectares. This plant community is not restricted to the Sonoran Desert but occurs also in valleys of the Mojave Desert (Allred et al. 1963; Hastings and Turner; 1965). Dice (1939) treated the *Larrea-Franseria* association as a vegetation climax for the Lower Colorado-Gila, but Lowe (1959) pointed out that factors of climate and soil control distributions and that an ordered ecological succession is not present.

This division of the desert straddles the lower reaches of the Colorado and Gila rivers, the confluence of these rivers near Yuma being near the geographic center. For this reason, the revised designation Lower Colorado-Gila Division is used here in preference to the Lower Colorado Valley of Shreve (1951), a designation that has confused students. The California portion of the Lower Colorado Gila region is often referred to ambiguously as the Colorado Desert in scientific literature. Parish (1930) contrasted the vegetation of the Mojave Desert with the California portion of the Sonoran Desert. Migration of flora was seen to be impeded by the geographical barrier of the Chuckawalla Mountains. Similarities in species composition were evident, and 51 percent of the drought-adapted flowering plants that occur in the Mojave were said to also be found in the Colorado.

When compared to other Sonoran Desert divisions, the Lower Colorado-Gila has the lowest incidence of abrupt changes in topography, about seven-eighths of the land surface consisting either of relatively level plains of gravelly outwash, sand dunes, and the Colorado delta, or of gently inclined bajadas. Nevertheless, the highly variable edaphic and geologic features make for a slightly more diverse vegetation than the deficient rainfall would suggest. On the gravelly outwashes and sandy plains, the species diversity is relatively low, but in the hills and mountains, in the upper bajadas, and in the volcanic fields, the vegetation is richer. Although woody plant species in the Lower Colorado-Gila Division are not great in number, the ephemerals add to the diversity (Shreve 1937). Since summer thunderstorms rarely reach as far west as this division, the summer ephemerals are not of as much consequence as the ephemerals coming as a result of winter rains. Halvorson and Patten (1975) found that ephemerals were denser in areas where shrub density was reduced or shrub canopy growth was higher.

As a contrast to the creosote bush and bursage of the plains, four microphyllous leguminous trees are found along the flood plains of the rivers, in the washes, and with reduced stature on the rockier soils of the upper bajadas and foothills: mesquite (*Prosopis velutina*), blue palo verde (*Cercidium floridum*), foothills palo verde (*Cercidium microphyllum*), and palo fierro (*Olneya tesota*).

Although soils and geology exercise a significant degree of control over the stature, density, and species composition of the vegetation in the Lower Colorado-Gila Division, the correlations are less pronounced than in the moister divisions. The most clear-cut vegetational differences are observed when comparing plants of the granitic hills with those of the volcanic ones or the plants of the rocky slopes with those of the gravelly and sandy plains.

The dominance of creosote bush and bursage on soils that vary considerably in depth and texture, and their importance in even the vegetation of basaltic fields and volcanic hills, is

Table 5-1
Sonoran Desert Vegetation

PHYTOPHYSIOGNOMIC NAME	FLORISTIC NAME	GEOGRAPHIC NAME
Microphyllous	Larrea-Franseria	Lower Colorado-Gila
Crassicaulescent	Cercidium-Opuntia	Arizona Upland
Arbosuffrutescent	Olneya-Encelia	Sonoran Plains
Arborescent	Acacia-Prosopis	Sonoran Foothills
Sarcocaulescent	Bursera-Jatropha	Central Coast
Sarcophyllous	Agave-Franseria	Vizcaíno
Arbocrassicaulescent	Lysiloma-Machaerocereus	Magdalena

SOURCE: Modified from *Vegetation of the Sonoran Desert,* F. Shreve, Carnegie Institute, Washington, D.C., 1951.

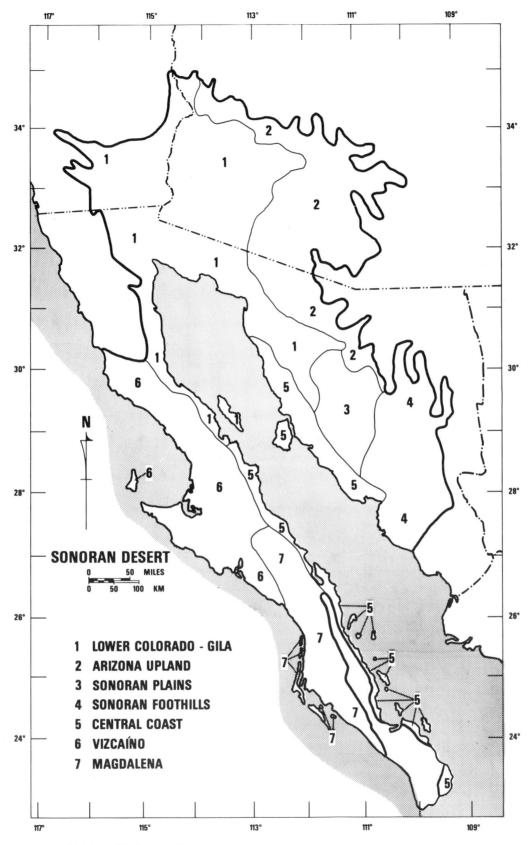

SONORAN DESERT

0   50   MILES
0   50   100   KM

1   LOWER COLORADO - GILA
2   ARIZONA UPLAND
3   SONORAN PLAINS
4   SONORAN FOOTHILLS
5   CENTRAL COAST
6   VIZCAÍNO
7   MAGDALENA

Figure 5-9. Divisions of the Sonoran Desert
based on data taken from *Vegetation of the Sonoran Desert* by F. Shreve, Carnegie Institution of
Washington, Washington, D.C. 1951.

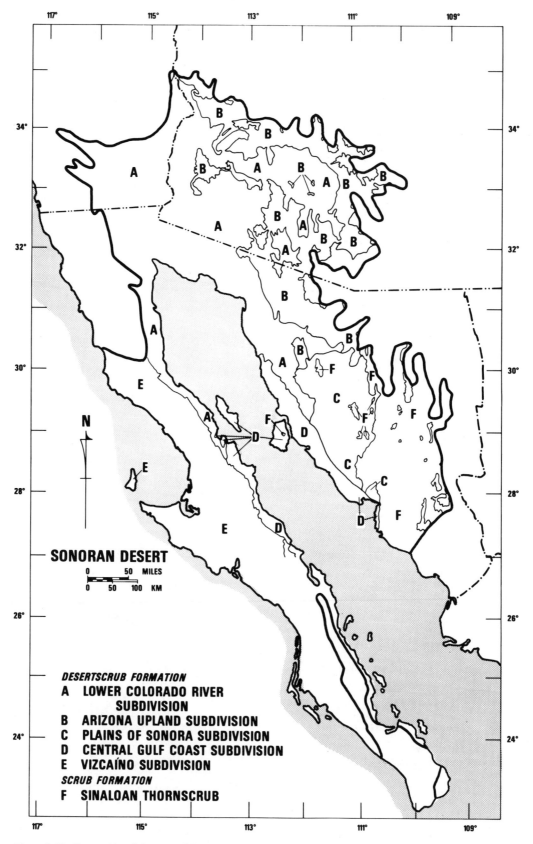

Figure 5-10. *Geographic subdivisions of the Sonoran Desert,*
suggested by D. E. Brown and C. H. Lowe in *Biotic communities of the Southwest,* USFS *Rocky*
Mountain and Range Experimental Station General Technical Bulletin 1PM-41, 1977.
Superimposed on the outline determined by F. Shreve (1951).

remarkable. Shreve (1951) believed that this was due to the fact that no other shrubs had evolved biological and physiological adaptations sufficient to allow them to compete successfully with creosote bush and bursage in areas of such extremely low rainfall. Yang and Lowe (1956) correlated the fine soils of lower slopes and valleys with *Larrea-Franseria* (creosote bush-bursage) vegetation and light rocky soils of higher slopes with *Cercidium-Carnegiea* (palo verde-saguaro) vegetation. They concluded that this was a prime example of edaphic determination of distinctly different and major vegetation types under the same climate. Halvorson and Patten (1974) analyzed water-potentials of plants characteristic of the Lower Colorado-Gila Division. Triangle-leaf bursage (*Ambrosia deltoides*) had the lowest water potential and responded markedly to seasonal and diurnal fluctuations in soil moisture and in evaporative power of the air. Foothills palo verde responded the least. Buckwheat bush (*Eriogonum fasciculatum*) showed a high degree of response to changes in soil moisture, while creosote bush, white rhatany (*Krameria grayi*), and jojoba (*Simmondsia chinensis*) showed moderate responses.

VALLEY PLAINS AND BAJADAS

The plains and lower bajadas of the Lower Colorado-Gila Division are among "the most unfavorable habitats of the Sonoran Desert from the standpoint of water relations" (Shreve 1951). Vegetation gradually becomes more abundant in passing from the plains through the lower bajadas to the upper bajadas as moisture conditions become slightly more favorable. Not only does the rainfall increase somewhat from the valley floor toward the adjacent mountain, but the coarser soils of the bajadas allow better infiltration of the precipitation. The rockier nature of the bajadas provides better conditions for germination and establishment and for anchorage and drainage, particularly for cacti but also for the other perennials.

On any given plain, the stand of creosote bush and bursage seems remarkably uniform in spacing, density, and stature. Vegetative cover is usually 10 percent of the land surface or less. Cover is always very low, down to about 3 percent, when rainfall is under 100 mm. The spacing of creosote bush is often more regular than that of bursage, probably due to the presence of chemical inhibitors in the roots of creosote bush that reduce competition by other plants. White bursage seems to grow better on deep, sandy loams than on deep clay loams that suffice for creosote bush. Densities of creosote bush and bursage seem not to compensate one for the other. When rainfall is low, the densities of each are low. When bursage is less abundant because of edaphic factors, creosote bush seems not to increase to fill the void.

Of seventeen species commonly found in the plains and lower bajadas, only four—creosote bush, white bursage, branching cholla (*Opuntia ramosissima*), and ground cholla (*Opuntia wrightiana*)—are more characteristic of these areas than of other sites. Species more common to washes but that also occur on the open ground of the plains and lower bajadas are mesquite, blue palo verde, graythorn (*Condalia lycioides*), tomatillo (*Lycium andersonii*), burrobrush (*Hymenoclea monogyra*), and *Encelia frutescens*.

Plants of the upper bajadas that are found to a lesser extent on the plains and lower bajadas are triangle-leaf bursage, ocotillo (*Fouquieria splendens*), white rhatany, hedgehog cactus (*Echinocereus engelmannii*), and *Opuntia echinocarpa*. In addition, big galleta grass (*Hilaria rigida*), more common on stabilizing sand dunes, and four-wing saltbush (*Atriplex canescens*, more characteristic of alkaline flats, occur to a minor degree on the plains and lower bajadas.

Throughout these plains and lower bajadas, plants clearly are intimately associated with even the tiniest drainage features. When the drainage pattern is reticulate, the plants seem to be scattered over the entire surface of the land, although close inspection usually reveals correlation with drainage. When the pattern is dendritic, the vegetation coalesces in linear bands and is more clearly seen to follow the drainage ways. Glendenning (1949) pointed out that some species occur in this division only where moisture is most readily available in drainage ways. An example is smoke tree (*Dalea spinosa*).

Twenty-six species of woody plants are found in association with margins of drainage ways in the plains and lower bajadas (Shreve 1951). Of these, eight are more or less restricted to southern latitudes of the division, partly because of sensitivity to frost: yerba de la flecha (*Sapium biloculare*), heart-leaf bursage (*Ambrosia cordata*), hinds nightshade (*Solanum Hindsianum*), *Stegnosperma halimifolium*, *Phaulothamnus spinescens*, shrub mallow (*Horsfordia alata*), *Berginia virgata*, and copperleaf (*Acalypha californica*).

Only two species, desert hackberry (*Celtis pallida*) and catclaw acacia (*Acacia greggii*), have northerly patterns in this group of drainage-way plants. Drainage-way species most common in sand of the washes are rush broom (*Bebbia juncea*), smoke tree, and two species of burrobrush (*Hymenoclea monogyra* and *H. pentalepis*).

The remaining twelve species are characteristic of drainage ways throughout the plains and lower bajadas of this division: mesquite, blue palo verde, palo fierro, desert broom (*Baccharis sarothroides*), tomatillo, graythorn, desert willow (*Chilopsis linearis*), large-leaf ragweed (*Ambrosia ambrosioides*), chuparosa (*Jacobinia californica*), hummingbird bush (*Anisacanthus thurberi*), *Encelia frutescens*, and creosote bush.

As the lower bajada gives way to the upper, palo verde and palo fierro begin to grow away from the washes. Ocotillo may become common on thin soils, catclaw acacia and *Opuntia echinocarpa* usually become more abundant, and saguaro (*Carnegiea gigantea*) and compass barrel cactus (*Ferocactus acanthodes*) are encountered. Often triangle-leaf bursage grows with or replaces white bursage, and although creosote bush is still present, it is no longer the dominant plant that it was on the plains and lower bajadas. Although trees and shrubs are more abundant along the washes in the upper bajadas than along those in the lower bajadas, they are not as noticeable as when they stood in stark contrast to the sparse creosote bush and bursage at lower elevations.

ACTIVE AND STABILIZED SAND DUNES

Sand areas are quite common in the Lower Colorado-Gila Division, comprising about one-seventh of its surface area (Shreve 1951). Successional series can be observed from very active dunes to moderately active ones with creosote bush and bursage, to ones being stabilized by big galleta grass, to fairly level sand plains where algae and lichens also help to keep the sand from being blown out.

Aside from creosote bush and bursage, six perennials are characteristically found on the moderately active dunes: four-wing saltbush, borage (*Tiquilia palmeri*), dye bush (*Dalea emoryi*), cañutillo (*Ephedra trifurca*), desert buckwheat (*Eriogonum deserticola*), and sandpaper plant (*Petalonyx thurberi*). Establishment and persistence of plants on these active dunes are both difficult processes. Germination is difficult in the dry surface sand.

Once a plant is established, it can easily die when winds blow the sand away from its roots. The opposite peril, of being covered by drifting sand, is also very real. If the plant can grow fast enough to extend beyond the sand as it builds up, it can become healthier than other plants of the same species because the deepening levels of sand become a reservoir for moisture.

Big galleta grass grows in clumps on dunes that are beginning to stabilize. It seems to be the most important binder of sand in the stabilization process. It occurs only on dunes that have passed their most active phases. As big galleta becomes established, *Tiquilia*, *Ambrosia*, and *Dalea* become more abundant, and the stabilization process seems assured.

Stabilization eventually leads to leveling as the blowing sand settles around the bases of the shrubs and grasses. The level, sandy surfaces would continue to be subjected to the peril of being blown out if it were not for the presence of a surface layer of blue-green algae and *Lecidia* and *Acarospora* ground lichens that develop.

MOUNTAINS OF METAMORPHIC AND INTRUSIVE IGNEOUS ROCKS

Rocks of a metamorphic or intrusive nature generally support a greater abundance of vegetation than volcanic ones with similar precipitation under the arid regime of the Lower Colorado-Gila Division. This is due in part to the more permeable nature and saprolytic tendencies of the crystalline rocks. The rock beneath the soil is also usually highly fractured. Bryan (1925) discussed the vegetation of the granitic mountains. MacDougal (1912) referred to mountains of the Lower Colorado as being "biological islands."

Most abundant on south-facing slopes are saguaro, organ-pipe cactus (*Stenocereus thurberi*), buckhorn cholla (*Opuntia acanthocarpa*) and smaller cacti, the shrubs brittlebush (*Encelia farinosa*) yerba de la flecha, mallow (*Hibiscus denudatus*), and *Trixis californica,* as well as the elephant tree (*Bursera microphylla*).

Conspicuous on the north-facing slopes, but also present on those with a southern aspect, are palo fierro, foothills palo verde, ocotillo, jojoba, and torote prieto (*Jatropha cuneata*). Less conspicuous but also characteristic of the north-facing slopes are catclaw acacia, desert mescal (*Agave deserti*), turpentine bush (*Aplopappus laricifolius*), cañutillo (*Ephedra viridis*), white bursage, shrubby bedstraw (*Galium stellatum*), and tree beargrass (*Nolina bigelovii*).

ERODED MOUNTAINS OF EXTRUSIVE IGNEOUS ROCKS

Small perennial plants and grasses are scattered over the older volcanic mountains and hills. All of these areas in the Lower Colorado-Gila Division lie in the zone where rainfall is limited to 150 mm or less. From a distance these arid mountains seem devoid of vegetation. Tanglehead grass (*Heteropogon contortus*) and tobosa grass (*Hilaria mutica*) are characteristic. Desert mescal is sometimes impressive, particularly when the flower stalk is present. Brittlebush grows on the warm south-facing slopes where cold air drains away in winter. Ocotillo grows on the slopes but may be more abundant on the alluvium outwashed below. Organ-pipe cactus and palo fierro may grow with ocotillo. Cacti are represented by the large compass barrel and by the small beehive cactus (*Echinomastus johnsonii*).

Seven species of low shrubs complete the inventory of characteristic plants: white bursage, *Fagonia californica*, rhatany (*Krameria*), desert lavender (*Hyptis emoryi*), *Trixis californica,* brickell bush (*Brickellia atractyloides*), and pygmy cedar (*Peucephyllum schottii*). In the canyon bottoms, palo verde and palo fierro may grow if water retention and soil depth are adequate.

FIELDS AND HILLS OF LESS ERODED EXTRUSIVE IGNEOUS ROCKS

Abundance of vegetation on recent volcanics relates to the degree of weathering of the lava and to the physical placement of cinders and massive lava flows in relation to the previous substrate. When lava forms a thin layer over a preexisting soil, runoff may be channelled to cracks where it seeps into the underlying soil. In such cases the lava inhibits evaporation, and trees or shrubs germinating in the cracks find a favorable environment.

As volcanic beds weather, clays that have a natural fertility are produced. Volcanic cinders insulate underlying soil pockets and provide an excellent mulch, which retards evaporation.

When massive lava has no soil and few cracks, vegetation is sparse or entirely absent even if cinders are present. That plants occur rather sporadically in most areas of recent volcanic activity is due to the slow decomposition of the parent rock and the unreliable occurrence of suitable soils and fissures under the lava or cinders of the surface.

Frost-sensitive plants that are common in the Pinacate lava flow of northern Sonora but are absent or much less common in more northern volcanic areas are elephant tree, torote prieto, and hinds nightshade. Shrubs that are abundant on recent volcanics in the lower Colorado-Gila division are creosote bush, brittlebush, desert lavender, rhatany, whitethorn acacia (*Acacia constricta*), broom rush, *Trixis californica*, desert hackberry, white bursage, tomatillo, and *Encelia frutescens*. Characteristic trees are foothills palo verde and palo fierro. Subshrubs are represented by yerba del venado (*Porophyllum gracile*) and *Dyssodia porophylloides*. Cacti and cactus-like plants of thin soils are ocotillo, hedgehog cactus, and buckhorn cholla. Tanglehead grass is also characteristic.

RIVERS

The mighty Colorado, although important for irrigation, has only a minute effect on the natural vegetation of the Sonoran Desert. It enters the desert about 15 miles north of Needles, California, at an elevation of 490 feet (150 m), falling steadily to sea level at the Gulf about 215 miles (358 m) south. Where surrounding terrain is nearly level with the river, there is riparian vegetation of cottonwood and a floodplain periodically inundated and swamplike. The riparian and floodplain vegetation is rarely more than 3 km wide (Shreve 1951).

Haase (1972) surveyed the floodplain vegetation along the lower Gila River in southwestern Arizona, primarily using aerial photography. Six communities were described: *Typha, Tamarix-Pluchea, Prosopis, Suaeda-Allenrolfea, Atriplex,* and *Larrea-Prosopis.* The introduced *Tamarix pentandra* has come in on over 50 percent of the 6622 hectares surveyed. Aldous and Shantz (1924) found desert saltbush (*Atriplex polycarpa*) to represent dominant vegetation of certain bottom lands along the Gila River.

Arizona Upland Division

Hastings and Turner (1965) treated the large number of life forms in the vegetation of the Arizona upland as remarkable and due to the heterogeneity of the environment. This second-largest division is in the south-central Arizonan and north-central Sonoran portion of the desert where hot and dry conditions through much of the year alternate with seasonal precipitation. Elevations range between 150 and 950 m. Rainfall is distinctly biseasonal and ranges from under 100 mm to over 300 mm per year. A considerably greater percentage of the surface area of the Arizona Upland Division consists of hills and mountains when compared with the Lower Colorado-Gila Division. In the Arizona Upland, vegetational cover may be so full that it may be difficult for a person to see more than a few hundred feet (Hastings and Turner 1965).

The intermittent nature of the dry and wet conditions allows both the deep-rooted woody habit and the succulent habit to be effective adaptations. Although creosote bush (*Larrea tridentata*) retains an importance where caliche is present, foothills palo verde (*Cercidium microphyllum*) becomes abundant through most of the division, except the lower bajadas. Mesquite (*Prosopis velutina*), palo fierro (*Olneya tesota*), and ocotillo (*Fouquieria splendens*) are characteristic. Cacti include over a dozen species of *Opuntia* (cholla and prickly pear), fishhook barrel (*Ferocactus wislizenii*), saguaro (*Carnegiea gigantea*), and many smaller species.

The noteworthy number of cacti and other succulents with fleshy stems has given rise to the name crassicaulescent desert. The simultaneous abundance of leguminous trees and cacti, symbolized by the palo verde and cholla-prickly pear association, has given rise to the designation of this division as the *Cercidium-Opuntia* region.

There is a definite abundance of green-stemmed plants in the Arizona upland. Both *Cercidium* and *Opuntia* fall into this category. Cannon (1908b) studied chlorophyll-bearing tissues in the stems of plants at the Desert Laboratory in the Arizona upland. Chlorophyll was found to persist through the life of cacti, ocotillo, rhatany (*Krameria*), palo verde, and probably zizyphus (*Condalia*). Chlorophyll capable of functioning was found 6.6 mm deep in the stem of saguaro, a distance 165 times greater than in leaves of ordinary plants.

The dominance shared by microphyllous trees and by cacti is remarkable because the root systems are so different. Cannon (1911) studied the root system of the microphyllous trees, the cacti, and some of the subdominant species, dividing them into three categories: generalized, with taproot and laterals, such as mesquite, creosote bush, and bursage; taproot, such as *Ephedra;* and subsurface lateral, which includes most cacti. The generalized root system, found in weeds and annuals, was seen as subject to considerable environmental modification. The taproot system was specialized to allow plants to draw water from a depth where the water source was constant. The shallow subsurface lateral system allowed plants to absorb water from light showers or other unpredictable precipitation. Only succulent plants capable of storing water against eventual use could make complete use of such a system.

The flora of the Arizona Upland is extremely diverse. Many of the elements of the Lower Colorado-Gila Division are present, but a large number of species whose water requirements are too great for them to be of much importance in the former division become numerous here. Some species of the Arizona Upland extend into the higher mountains past the edge of the Sonoran Desert. Some Sonoran Desert species are found only in the upper reaches of the Arizona Upland Division.

Although biseasonal rainfall allows both winter and summer ephemerals to pass their life cycles, the winter ephemerals are more numerous and are responsible for the desert-in-bloom phenomenon in spring. Although all of the Arizona Upland Division does experience rain in both summer and winter, a slightly higher incidence of winter rain in the north gradually gives way to a preponderance of summer rain in the south.

Plant cover in the Arizona Upland usually represents 20 to 60 percent of the land surface but may extend down to about 10 percent in its lower reaches or up to 80 percent or more in some of its upper reaches. Although the Arizona upland division is notable for its trees, they rarely occupy more than one-eighth of its entire surface in any particular situation and much less when the entire surface area of the division is considered.

Valleys where acid igneous alluvium is not a factor generally have creosote bush as a strong dominant, with white bursage (*Ambrosia dumosa*) and triangle-leaf bursage (*Ambrosia deltoidea*) also important. On coarse, rocky slopes and gravelly outwash plains, triangle-leaf bursage tends to become more dominant, with various cacti and four microphyllous leguminous trees—foothills palo verde (*Cercidium microphyllum*), blue palo verde (*C. floridum*), mesquite (*Prosopis velutina*), and palo fierro (*Olneya tesota*)—being scattered over the surface. Where trees are most abundant, large cacti are also usually common, and the shrub cover is somewhat less dense than in some other situations. The upper bajadas, hills, and mountains have a rich flora of trees, shrubs, and other perennials.

VALLEY PLAINS AND LOWER BAJADAS

Creosote bush is dominant over the plains and lower bajadas of this division. The percentage cover of the land for this community in the Arizona Upland varies from 10 to 20 percent, roughly double its average in the Lower Colorado-Gila Division. Close associates of *Larrea* in the finer soils of the valleys include mesquite, both whitethorn acacia (*A. constricta*) and catclaw acacia (*A. greggii*), and four species of cholla: the pencil cholla (*Opuntia arbuscula*), the staghorn cholla (*Opuntia versicolor*), the cane cholla (*Opuntia spinosior*), and the chain-fruit cholla (*Opuntia fulgida*).

Hastings and Turner (1965) characterized desert saltbush (*Atriplex polycarpa*) vegetation as of wide extent in the Arizona Upland and occurring essentially in pure stands. Shantz and Piemesel (1924) found this community in areas where surface drainage was impeded. It has been correlated with alkaline soil, and Hastings and Turner have pointed out that it occurs as a community in the Lower Colorado region and in the Mojave Desert as well.

Grasses occur locally where soils are heavier and more retentive of water. Characteristic species are tobosa (*Hilaria mutica*) and bush muhly (*Muhlenbergia porteri*). The latter

species grows where it is protected by shrubs, but because it is a preferred winter forage, it is scarce where grazing pressure has been significant.

To the south in the valleys of the Arizona upland, particularly in Mexico, mesquite, palo verde, and grasses seem to become slightly more frequent. Plants that enter the Arizona Upland only in the plains include gum bush (*Bumelia occidentalis*) and two species of crucifixion thorn (*Holocantha emoryi* and *Koeberlinia spinosa*).

Several small shrubs seem to occur where there is a presence of gravel on the soil surface, either because of physical surface germination requirements or because of differences in water relations of the underlying soil solution. These plants include white bursage, triangle-leaf bursage, canescent rhatany (*Krameria grayi*), desert zinnia (*Zinnia grandiflora*), and jimmyweed (*Aplopappus*).

Plants more characteristic of upper bajadas but that are encountered to a lesser degree in the plains and lower bajadas are ocotillo, barrel cactus, saguaro, desert hackberry (*Celtis pallida*), and graythorn (*Condalia lycioides*). Plants of the valley plains and lower bajadas increase in density and stature along drainage ways (figure 5-11) but not in such a pronounced manner as in the Lower Colorado-Gila Division. Minor drainage ways characteristically are bordered with catclaw acacia (*A. greggii*) and whitethorn (*A. contricta*). As the washes become larger and coalesce, mesquite, blue palo verde, large-leaf ragweed (*Ambrosia ambrosioides*), and desert willow (*Chilopsis linearis*) become common.

DRAINAGE WAYS

As streams fall from the upper elevations of the mountains down into the Arizona Upland Division, a combination of abundant moisture with the hot temperatures of the desert floor results in numerous ecotonal situations. In the upper reaches of desert canyons, species of chaparral or oak woodland may grow beside riparian zone plants of the Sonoran Desert. Such situations are the most mesic within the entire Sonoran Desert.

Since the streams rarely flow during more than half of the year and since average annual rainfall is greatly variable both in the desert and in the higher mountains where the streams arise, the placement of vegetation depends largely on the nature of the floodplain and its water-retaining characteristics. Thus even the most riparian species depend less on moisture of the stream banks themselves and more on moisture reserves of the floodplain than would be usual in a truly mesic situation.

Water relations are favorable for broad-leaf vegetation in canyon situations where stream channels are clogged with large boulders and the floodplains, though narrow, are silty. The trees in the streamway or on its banks, with roots extending into the floodplain, are cottonwood (*Populus fremontii*), Arizona ash (*Fraxinus velutina*), Arizona walnut (*Juglans major*),

Figure 5-11. Association of vegetation with drainage features in the Arizona Upland Division.
Photograph by C. Allan Morgan.

Arizona sycamore (*Platanus wrightii*), canyon hackberry (*Celtis reticulata*), and elderberry (*Sambucus mexicana*). Shrubs are not as numerous as trees on the stream bank itself in these situations, perhaps because the rapidly moving floodwaters more easily dislodge them in these upper reaches. Such streambank shrubs are practically limited to seepwillow (*Baccharis glutinosa*) and true willow (*Salix gooddingii*).

On the floodplain away from the stream bank, mesquite is the commonest tree, while the characteristic shrubs and other perennials are desert hackberry, catclaw acacia, large-leaf bursage, tomatillo (*Lycium*), and graythorn.

In ecotonal situations at the very upper edge of the desert, various species of oak (*Quercus emoryi, Q. arizonica, Q. oblongifolia,* and *Q. hypoleucoides*) may be associated with the river-bank trees, while scrub oak (*Q. turbinella*) may mix with mesquite in the floodplain.

Sandy floodplains are ones that are built up little above the streambed itself. As Shreve (1951) pointed out, these are often little more than seldom used sections of the streambed. The dominant species are burrobrush (*Hymenoclea monogyra*) and desert broom (*Baccharis sarothroides*), with large-leaf bursage and desert willow (*Chilopsis linearis*) common. Mesquite and catclaw acacia may also be present. Canaigre dock (*Rumex hymenosepalus*), with underground tubers, dies back above ground in summer and is characteristic of these sandy areas. Within historic time, saltcedar (*Tamarix*) has replaced native shrubs in many cases.

A third type of floodplain occurs generally at lower elevations where floodwaters move more slowly. In the deep silty and clayey soils that develop, mesquite forms nearly pure stands, often of several miles extent. Once these were virtual riverine forests, but they have been converted to agricultural fields or cut down for firewood frequently. Gavin (1973) prepared an ecological profile of a typical mesquite forest. He enumerated the reasons mesquite forests are cut down as the need to convert floodplains to agricultural croplands, improvement of grazing land, and elimination of phreatophytes to save water. When allowed to revert to their previous condition, second-growth mesquite forests at first have an extensive admixture of graythorn, desert hackberry, catclaw acacia, and other species. Shreve (1951) noted that disturbed mesquite forests had more vines than any other area of the Sonoran Desert, listing virgin's bower (*Clematis drummondii*), Arizona grape (*Vitis arizonica*), snail-seed (*Cocculus diversifolius*), climbing milkweed (*Funastrum*), and spiny gourd (*Echinopepon wrightii*).

Bryan (1928) reported that the Santa Cruz River once had extensive cienegas. As groundwater level lowered, the vegetation changed from bulrush (*Scirpus*), which used surface water, to sacaton grass (*Sporobolus*) and cottonwood (*Populus*), which used shallow water, to mesquite, which uses water deep in the ground. Arroyo cutting was attributed to a drier climate and overgrazing.

Rea (1979a) described the vegetation of the fertile banks of the lower reaches of the major rivers. Between the river and the mesquite floodplain, cottonwood and true willow once created a lush green ribbon along the rivers, with seepwillow and arroweed (*Pluchea sericea*) occupying the openings. Cane breaks of carrizo (*Phragmites communis*) also occurred. There were meadows of saltgrass (*Distichlis spicata*) and sacaton grass (*Sporobolus wrightii*). With the impoundment of the rivers by artificial reservoirs, this vegetation has greatly declined.

UPPER BAJADAS

A relatively rich vegetation exists in the upper bajadas of the Arizona Upland Division. About half of the surface area of the valleys can be referred to this classification. Density, abundance, diversity, and cover are high, and individual plants are usually robust in stature. Perennials that are ubiquitous throughout these upper bajadas are foothills palo verde, triangle-leaf bursage, mesquite, desert hackberry, graythorn, tomatillo, staghorn cholla, Engelmann prickly pear (*Opuntia phaeacantha*), and saguaro. This upper bajada vegetation was designated "paloverde, bursage and cacti desert" by Nichol (1952), "the paloverde-saguaro association" by Yang and Lowe (1956), and "paloverde-triangle leaf bursage range" by Humphrey (1960).

Creosote bush is more abundant on flatter areas with deeper and finer soils than on steeper slopes with shallow rocky soils. The opposite is true for ocotillo. Creosote bush is conspicuously absent from acid igneous rocks.

Through the broad, central portion of the Arizona Upland Division, the following larger perennials are also characteristic

of the upper bajadas: catclaw acacia, desert hackberry, jojoba (*Simmondsia chinensis*), cañutillo (*Ephedra trifurca*), adelia (*Forestiera neomexicana*), barrel cactus, and chain-fruit cholla. Palo fierro occurs in areas with warm winters. Whitethorn acacia may be present but avoids acid igneous alluvium.

Smaller perennials are very common between these larger plants and include triangle-leaf bursage, brittlebush (*Encelia farinosa*), feather duster (*Calliandra eriophylla*), Christmas cholla (*Opuntia leptocaulis*), desert zinnia, rhatany (*Krameria*), paperflower (*Psilostrophe cooperi*), bush muhly grass (*Muhlenbergia porteri*), sangre de cristo (*Jatropha cardiophylla*), and yerba del venado (*Porophyllum gracile*).

Stratification of vegetation in the upper bajadas is clear. Brome grass (*Bromus rubens*), pincushion cactus (*Mammillaria microcarpa*), and other small plants occur at ground level. Triangle-leaf bursage (*Ambrosia deltoidea*) rises to a height of about a foot. Jojoba rises to a height of 2 feet on south-facing slopes and to 4 feet on north-facing ones. Foothills palo verde (*Cercidium microphyllum*) rises to 10 or 20 feet, while the massive saguaro (*Carnegiea gigantea*) towers to 30 or 40 feet. The dominant strata most frequently are the low bush layer of bursage and the higher tree level of palo verde.

In the upper bajadas of the northwestern section of the Arizona Upland, a number of interesting species are present from adjacent juniper woodlands or from the Mojave Desert to the west. Examples include joshua tree (*Yucca brevifolia*), Utah juniper (*Juniperus utahensis*), crucifixion tree (*Canotia holacantha*), scrub oak (*Quercus turbinella*), banana yucca (*Yucca baccata*), beavertail cactus (*Opuntia basilaris*), bladder sage (*Salazaria mexicana*), goldenhead (*Acamptopappus sphaerocephalus*), and turpentine broom (*Thamnosma montanum*).

At the southern end of the Arizona Upland Division in Sonora, the vegetation of the upper bajadas varies somewhat from the more typical facies. Palo verde, saguaro, cholla, and prickly pear are still common and dominant, but crucillo (*Condalia warnockii*), *Jatropha cordata*, smooth mimosa (*M. laxiflora*), *Croton sonorae*, *Caesalpinia pumila*, and heart-leaf bursage (*Ambrosia cordifolia*) also become important.

Frost-sensitive species that come into the upper bajadas of the Arizona upland division only in the section from Bend, Arizona, to Caborca, Sonora, include organ-pipe cactus (*Stenocereus thurberi*), senita cactus (*Lophocereus schottii*), torote prieto (*Jatropha cuneata*), and elephant tree (*Bursera microphylla*).

HILLS AND MOUNTAINS

Vegetation of hills and mountains in the Arizona Upland Division includes most of the species of the upper bajadas and numerous additional ones; the perennial species alone number about 350. A list of the most characteristic species is very similar to the list for the upper bajadas. On south-facing slopes saguaro (figure 5-12), teddy bear cholla (*Opuntia bigelovii*), Engelmann prickly pear, foothills palo verde, blue palo verde, and brittlebush are common.

On north-facing slopes, jojoba, scrub oak, Arizona rosewood (*Vauquelinia californica*), turpentine bush, lip fern (*Cheilanthes*), cloak fern (*Notholaena*), cliff fern (*Pellaea*), and resurrection plant (*Selaginella*) are common. In the northern section of the Arizona Upland, bush penstemon (*Penstemon microphyllus*), shrubby bedstraw (*Galium stellatum*), rhyolite bush (*Crossosoma bigelovii*), buckwheat bush (*Eriogonum fasciculatum*), and beargrass (*Nolina*) are noteworthy.

It is thought that perennial grasses were once more common in the upper reaches of the Arizona Upland Division, but due to overgrazing and subsequent increase of cholla and mesquite, which compete for moisture, they reestablish with great difficulty (Humphrey 1958). *Bromus rubens*, an introduced annual grass, can respond to the occasionally heavy winter rains to cover nearly all of the soil surface between perennials in some parts of the Arizona Upland.

Sonoran Plains Division

On either side of the mid-lower stretch of the Rio Sonora and mostly west of the Rio San Miguel lies a division of the desert that seems topographically much like an extension of the Lower Colorado-Gila Division but that has considerably more rainfall and a flora that becomes increasingly richer from north to south. As Dunbier (1968) pointed out, this division represents the southern terminus of the basin and range physiographic province. Landforms rising above the gently sloping plains comprise only about one-tenth of the total

Figure 5-12. A south-facing slope in the Arizona Upland Division dominated by palo verde, saguaro and ocotillo.
Photograph by C. Allan Morgan.

surface area (figure 5-13). Summer temperatures are lower and winter temperatures are higher than those of the Lower Colorado-Gila Division. Freezing weather is less common, and the extreme lows are higher. Rainfall varies from 250 to nearly 400 mm. Dunbier (1968) pointed out that vegetation of this division "must be as drought resistant as that of the Lower Colorado Valley, while at the same time able to withstand greater desiccating heat during the long periods of drought."

The Sonoran Plains Division has been referred to as an "arbosuffrutescent desert" because of the dominance of both trees and shrubs. The alternative designation, *Olneya-Encelia* region, is taken from a representative tree, palo fierro (*Olneya tesota*), and a representative shrub, brittlebush (*Encelia farinosa*). It is informative to examine *Olneya* and *Encelia* as being particularly adapted to the Sonoran Plains. Although these species are present in the northern divisions of the desert in Arizona, where they occupy the relatively frost-free micro-

habitats there, in the Sonoran Plains they grow in only average situations, including many exposed and flat situations. The pods of *Olneya* ripen from May to mid-June and fall to the ground as they mature. Many are quickly eaten by rodents, but those remaining germinate quickly if early rain allows it. The somewhat later seasonal advent of rain in those divisions of the desert north of the Sonoran Plains is detrimental to reproduction in the species there.

The dominant trees—palo fierro, foothills palo verde (*Cercidium microphyllum*), and mesquite (*Prosopis velutina*)—show their greatest abundance and vigor in this division. Palo estribo (*Cercidium sonorae*) and palo brea (*C. praecox*) become frequent, and the northern ocotillo (*Fouquieria splendens*) may be replaced by Palo Adán (*Fouquieria macdougalii*). The saguaro (*Carnegiea gigantea*) becomes less common, and the numbers of organ-pipe cactus (*Stenocereus thurberi*) and senita cactus (*Lophocereus schottii*) increase. The relatively frost-

Figure 5-13. The Sonoran Plains Division near Hermosillo.
Photograph by C. Allan Morgan.

sensitive octopus cactus (*Rathbunia alamosensis*) is not uncommon. Arizona's staghorn cholla (*Opuntia versicolor*) is replaced by a southern relative, *Opuntia thurberi*. Chain-fruit cholla (*O. fulgida*), pencil cholla (*O. arbuscula*), and Christmas cholla (*O. leptocaulis*) remain important, however.

The Sonoran Plains is an area of transition between northern and southern elements. Dunbier (1968) characterized the Sonoran Plains as the region where "easily recognized features of the desert—low stature, open spacing, and diversity of life forms—are still prevalent but are beginning to wane." Although creosote bush (*Larrea tridentata*), white bursage (*Ambrosia dumosa*), and triangle-leaf bursage (*A. deltoidea*) are still present in local situations, they lack the importance they display in the north. The aspect of the vegetation takes on the appearance of a woodland, with numerous trees and intervening shrubbery. Columnar cacti are generally more common than prickly pear and cholla. Species diversity increases greatly from north to south in the Sonoran Plains Division, as more and more southern elements are encountered.

NORTHERN REGION

Two distinct heights of vegetation (trees and shrubs) are present, and intermediates are not found. Open woodlands consisting of foothills palo verde, palo fierro, mesquite, and blue palo verde (*Cercidium floridum*) seem to alternate with stands of creosote bush, the latter probably representing areas where there is a carbonate (caliche) layer in the soil. Brittlebush, ocotillo, and organ-pipe cactus are characteristic of the open woodlands. The creosote bush stands are rarely as pure in composition as in the northern parts of the Sonoran Desert, usually including plants of palo verde, palo fierro, ocotillo, and brittlebush.

Various trees and shrubs found intermixed with the creosote bush stands or in the open woodlands include elephant tree (*Bursera microphylla*), desert hackberry (*Celtis pallida*), graythorn (*Condalia lycioides*), Christmas cholla, sangre de cristo (*Jatropha cardiophylla*), whitethorn acacia (*A. constricta*), *Caesalpinia pumila*, tomatillo (*Lycium brevipes*), and rhatany (*Krameria parvifolia*).

The vegetation on volcanic hills is similar to that of the open woodland, with foothills palo verde and ironwood retaining dominance. Species of deeper soils such as mesquite and *Acacia occidentalis* are replaced by plants of rockier and thinner soils, such as ocotillo, Engelmann prickly pear (*Opuntia phaeacantha*), barrel cactus (*Ferocactus covillei*), saguaro (*Carnegiea gigantea*), hedgehog cactus (*Echinocereus engelmannii*), and desert lavender (*Hyptis emoryi*).

Limestone hills have a quite different vegetation in which palo verde, palo fierro, and cacti are virtually absent. The characteristic plants are ocotillo, whitethorn acacia, kidneywood (*Eysenhardtia orthocarpa*), yellow bells (*Tecoma stans*),

crucillo (*Condalia warnockii*), *Caesalpinia pumila*, *Bursera laxiflora*, and *Croton sonorae*.

SOUTHERN REGION

Density, stature, and diversity of the trees and shrubs increase. Creosote bush becomes less common and is replaced by jito (*Forchammeria watsoni*), a sclerophyllous tree with a light, spongy trunk. The two-tier vegetation is broken by the addition of larger shrubs and smaller trees. Cacti such as organpipe, saguaro, senita, pencil cholla, and barrel become less common. Chain-fruit cholla is still present, while cardón hecho (*Pachycereus pecten-aboriginum*), octopus cactus (*Rathbunia alamosensis*), and *Opuntia thurberi* appear.

Grasses, ephemerals, and small perennials respond to summer rains and often occupy much of the land surface not covered by trees and large shrubs. Brittlebush becomes less abundant, but the individuals become larger. The following species become important: palo santo (*Ipomaea arborescens*), tree ocotillo, *Acacia willardiana*, espino (*Acacia cymbispina*), *Randia thurberi*, palo de asta (*Cordia parvifolia*), lantana (*L. horrida*), Sonoran caper (*Atamisquea emarginata*), and pochote (*Ceiba acuminata*).

DRAINAGE WAYS

The streamways of the Sonoran Plains are studded with mesquite, blue palo verde, palo fierro, catclaw acacia (*A. greggii*), and *Acacia occidentalis*. Various shrubs that grow with these trees to make virtually closed thickets are desert hackberry, adelia (*Forestiera neomexicana*), large-leaf bursage (*Ambrosia ambrosioides*), whitethorn acacia, tomatillo (*Lycium*), graythorn, paperfruit (*Mascagnia macroptera*), hummingbird bush (*Anisacanthus wrightii*), desert broom (*Baccharis sarothroides*), lantana, smooth mimosa, and honey bush (*Aloysia lycioides*).

In the streamway thickets, the large number of vines is remarkable. These are represented by over a dozen genera. Some of the species are queen's wreath (*Antigonon leptopus*), virgin's bower (*Clematis drummondii*), passion flower (*Passiflora mexicana*), blood-leaf (*Iresine interrupta*), snail-seed (*Cocculus diversifolius*), and wild bean (*Phaseolus atropurpureus*).

Sonoran Foothills Division

Felger and Lowe (1976) stated that this area is a broad ecotone between desert scrub (that is, the Sonoran Desert) and thorn scrub (the Sinaloan Thorn Forest). Although this is true in a general way, it is necessary to make a subjective decision as to whether the region is Sonoran Desert with thorn forest elements or Thorn Forest with desert elements. The weight of the evidence available supports Shreve's placement of the region as the southernmost facies of the Sonoran Desert.

This division, together with the Arizona Upland, contains areas where rainfall is at its highest for the Sonoran Desert. Rainfall has been known to exceed 500 mm in some areas, but high parameters of evaporation and transpiration, as well as long periods without rain, result in desert conditions. The division consists of a large heartland at 27° to 29° N latitude on either side of the Rio Yaqui, with fingers extending north in the valleys of the Rio Sonora, Rio Moctezuma, and Rio Bavispe.

Although this is clearly the least desert part of the Sonoran Desert, the long periods of drought, high evaporation, and species composition cannot be dismissed as being nondesert in character. Many of the species of the Arizona Upland are present, such as desert hackberry (*Celtis pallida*), catclaw acacia (*A. greggii*), ocotillo (*Fouquieria splendens*), desert hopbush (*Dodonaea viscosa*), brittlebush (*Encelia farinosa*), and organ-pipe cactus. Three-fourths of the precipitation falls in summer, in strong contrast to the regime of the Arizona Upland. The Sonoran Foothills are often referred to as an arborescent desert because of the preponderance of trees. The alternative name, *Acacia-Prosopis* region, alludes to the importance of espino (*Acacia cymbispina*) and mesquite (*Prosopis velutina*). Other *Acacia* species such as sweet acacia (*A. farnesiana*), feather acacia (*A. pennatula*), and catclaw acacia (*A. greggii*) are also important. Mesquite is dominant to the north, while espino and mauto (*Lysiloma divaricata*) are dominant to the south.

Elevations range from about 1000 m down to sea level. Grasses are rather abundant at the upper elevations, particularly in the north. Trees are normally abundant, and shrubs may grow in dense thickets. Cacti, although of less importance, nevertheless occur, even in heavy soils of the Yaqui delta. Palms (*Brahea roezlii*) are sometimes frequent along washes.

Shreve (1951) painted a picture of the Sonoran foothills division as having "an important position in the vegetation of Mexico," being bounded to the north by grassland, to the south by thorn forest, and to the east by the Sierra Madre Occidental.

NORTHERN REGION

Dunbier (1968) described the heavy areas of vegetation of the precipitous cuestas and coarse outwashes as denser than all but the most propitious situations in the divisions of the Sonoran Desert to the north, describing them as desert forests. In the northern parts of the Sonoran Foothills the characteristic species are mesquite, palo estribo (*Cercidium sonorae*), desert hackberry, catclaw acacia, kidneywood (*Eysenhardtia orthocarpa*), sangre de cristo (*Jatropha cardiophylla*), ocotillo, desert hopbush, brittlebush, organ-pipe cactus, cholla (*Opuntia fulgida, O. thurberi*), crucillo (*Condalia warnockii*), palo adán

(*Fouquieria diguetii* and coyotillo (*Karwinskia humboldtiana*). Thorn forest species are largely absent from the northern reaches of the Sonoran Foothills, but several characteristic Arizona upland species are absent as well, such as creosote bush, foothills palo verde, blue palo verde, saguaro, and palo fierro.

Beds of dark brown basalt between the northern and southern regions support a unique vegetation because of several factors. The rainfall is retained because of the irregular surface of the lava beds, and it soaks into the water-retaining clay. The mulching, shading, and wind-breaking effect of the lava inhibits evaporation and excessive transpiration. The basalt area, perhaps because of color and topography, is apparently warmer than surrounding regions. The vegetation on these lavas includes palo estribo, kidneywood, sangre de cristo, espino, palo santo (*Ipomaea arborescens*), *Randia obcordata*, tree ocotillo, prickly pear, smooth mimosa (*M. laxiflora*), and heart-leaf bursage (*Ambrosia cordifolia*).

SOUTHERN REGION

Vegetation in the southern part of the Sonoran Foothills consists of denser stands of trees and shrubs, ordinarily covering 75 percent or more of the land surface. Communities with the most open desert aspect have cover of about 50 percent, while those with the most pronounced thorn forest aspect have cover of 100 percent.

Thorn forest from the south comes into the floodplains, but desert plants intervene in the uplands. Even these truly desert elements are of increased stature, and it is obvious that the southern terminus of the desert is near. Broadleaf trees become more frequent in the thorn forest element than anywhere else within the limits of the Sonoran Desert.

The truly desert communities can be walked in, but the thorn forest element is impenetrable in most situations. In this area of transition, the desert communities are dominated by mesquite, blue palo verde, palo estribo, espino, tree ocotillo, kidneywood, elephant tree (*Bursera microphylla*), *Piscidia mollis*, coyotillo, organ-pipe cactus, graythorn (*Condalia lycioides*), *Pithecellobium sonorae*, *Bursera laxiflora*, *Guaiacum coulteri*, *Janusia gracilis*, tomatillo, and smooth mimosa.

The delta of the Rio Yaqui may once have supported considerable stands of mesquite, but now much is agricultural land producing rice, wheat, and other crops. Shreve and Wiggins (1964) published a photograph of the northern edge of the delta showing abundant cacti, including organ-pipe, chain-fruit cholla, octopus cactus, cardón hecho, pincushion cactus (*Mammillaria*), and barrel cactus (*Ferocactus*). Along the coast, mangrove vegetation (*Rhizophora mangle, Avicennia germinans*) and (*Laguncularia racemosa*) and salt-tolerant plants such as mangle dulce (*Maytenus phyllanthoides*), fourwing saltbush (*Atriplex canescens*), quelite salado (*Suaeda*), *Stegnosperma*, and *Phaulothamnus* occur.

The thorn forest element along the streamways is dominated by espino, heart-leaf bursage, palo santo, octopus cactus (*Rathbunia alamosensis*), cardón hecho, large-leaf bursage (*Ambrosia ambrosioides*), *Pithecellobium sonorae*, *Zizyphus sonorensis*, palo de asta (*Cordia sonorae*), *Randia echniocarpa*, soapberry (*Sapindus saponaria*), desert hackberry, *Lantana*, *Opuntia thurberi*, and *Plumbago scandens*.

Central Coast Division

Felger (1966) studied the ecology of the Gulf Coast and islands off the coast of Sonora. The vegetation of the coast of Baja California from Angel de la Guardia Island to La Paz and on an isolated coastal spot northeast of San José is similar to, but not identical with, that of the mainland across the Gulf from the Rio Concepción south nearly to the Rio Yaqui. This coastal area is the driest division of the entire Sonoran Desert. Long periods of time may pass without a trace of rain. When it does come, it may be either in summer or winter.

The vegetation of the mesas along the Gulf of California was discussed by MacDougal (1904), who remarked upon the preponderance of perennials with lactiferous sap and ones with volatile or resinous exudates. Also found to be abundant in this classic survey were microphyllous spinose forms, which became deciduous during drought.

The division is often referred to as being a sarcocaulescent desert because of the presence of trees with swollen trunks (figure 5-14) such as elephant tree (*Bursera microphylla*), *Bursera hindsiana*, lomboy (*Jatropha cinerea*), and cirio or boojum tree (*Idria columnaris*). The alternative name, *Bursera-Jatropha* region, also refers to these peculiarly structured trees.

Figure 5-14. The sarcocaulescent "swollen trunk" syndrome characteristic of many plants in the Central Coast Division and Vizcaíno Division.
Photograph by C. Allan Morgan.

Plants that grow with the sarcocaulescent trees and that are usually more important in the vegetation are palo fierro (*Olneya tesota*), blue palo verde (*Cercidium floridum*), ocotillo (*Fouquieria splendens*), and mesquite (*Prosopis*). Cacti are represented by a large number of cholla species (*Opuntia bigelovii, O. clavellina, O. cholla, O. ramosissima,* and *O. tesajo*) and by cardón pelón (*Pachycereus pringlei*). Vegetation is most abundant and vigorous along the drainage ways.

Felger and Lowe (1976) recognized four communities associated directly with the saline coastal areas: (1) seagrass meadow in subtidal zones, consisting of eelgrass (*Zostera marina*) and ditch-grass (*Ruppia maritima*); (2) mangrove scrub in quiet bays and lagoons consisting of *Avicennia germinans, Laguncularia racemosa,* and *Rhizophora mangle;* (3) salt scrub on salt flats, beaches, and mangrove margins, consisting of succulent halophytes (*Allenrolfea, Batis, Salicornia*), shrubby halophytes (*Atriplex, Tricerma* and *Suaeda*), and grass halophytes (*Jouvea, Monanthochloe* and *Sporobolus*); and (4) coast scrub of yerba reuma (*Frankenia palmeri*).

SONORAN REGION

The land vegetation within reach of the salt spray consists of four-wing salt bush (*Atriplex canescens*) and yerba reuma. A little way from the salt spray are found torote prieto (*Jatropha cuneata*), *Errazurizia megacarpa*, shrubby spurge (*Euphorbia misera*), tomatillo (*Lycium fremontii*), white bursage (*Ambrosia dumosa*), quelite salado (*Suaeda ramosissima*), and *Stegnosperma halimifolium*.

On the bajadas away from the coast, the halophytes are absent, and blue palo verde, mesquite, ocotillo, elephant tree, palo fierro, cardón pelón, *Bursera hindsiana*, brittlebush, creosote bush, whitethorn acacia (*A. constricta*), desert lavender (*Hyptis emoryi*), prickly pear, catclaw acacia (*A. greggii*), and jojoba are seen. The cirio is found on the mainland only between Puerto Libertad and El Desemboque.

The larger streamways have tree vegetation of palo fierro, mesquite, catclaw acacia, and *Bursera hindsiana*. The sandy washes have shrub vegetation of burrobrush (*Hymenoclea pentalepis*), large-leaf bursage (*Ambrosia ambrosioides*), chuparosa (*Jacobinia californica*), rush broom (*Bebbia juncea*), *Baccharis, Aplopappus,* and *Lagascea*.

Near Guaymas the vegetation is enriched, and there is a noteworthy endemic element. Near the coast northwest of Guaymas, on thin soil of dry slopes, guapilla (*Hechtia montana*), palo adán (*Fouquieria diguetii*), elephant tree, *Acacia willardiana*, torote prieto, and lomboy predominate.

In the vicinity of San Carlos Bay, the enriched vegetation along streamways is dominated by mesquite, mauto (*Lysiloma divaricata*), soapberry (*Sapindus saponaria*), *Pithecellobium sonorae*, mangle dulce (*Maytenus phyllanthoides*), paperfruit (*Mascagnia macroptera*), and large-leaf bursage (*Ambrosia ambrosioides*).

DESERT ISLANDS

Felger and Lowe (1976) reported on the vegetation of desert islands in the Gulf of California: Tiburón, San Esteban, San Pedro Nolasco, San Pedro Mártir, Dátil (Turners Island), Alcatraz, Patos, and Cholludo. They considered eight taxa to be island endemics. On San Esteban were *Echinocereus grandis* and *Mammillaria estebanensis* in the Cactaceae, *Agave dentiens* in the Agavaceae, and *Lyrocarpa linearifolia* in the Cruciferae. On San Pedro Nolasco were *Echinocereus websterianus, Mammillaria multidigitata,* and *M. tayloriorum*, all in the Cactaceae. On San Lorenzo were also found *Echinocereus grandis* and *M. estebanensis*. An undescribed *Mammillaria* related to *M. dioica* of Baja California was found on Dátil and Cholludo. In addition, endemics occur on other islands in the Gulf, such as *Mammillaria insularis* on San Marcos and Smith Island, *Ferocactus gatesii* on Smith Island, and *Ferocactus johnstonianus* and *Lyrocarpa linearifolius* on Angel de la Guarda. Endemics on islands in the southern part of the Gulf include *Ferocactus diguetii* on Santa Catalina, Monserrat, Danzante, San Diego, and Cerralvo, *Mammillaria cerralboa* on Cerralvo, and *Opuntia brevispina* on Espíritu Santo.

BAJA CALIFORNIA REGION

The narrow ribbon of Central Coast Division vegetation in Baja California extends through nearly six degrees of latitude. Predominant light winter rains to the north gradually give way to a prevalence of scant summer rain in the south. Although only a few kilometers from the Pacific side of the peninsula where moist Pacific air is indicated by epiphytic Heno Pequeño (*Tillandsia recurvata*), the Gulf Coast generally receives little or no moisture from the Pacific. Shreve (1951) emphasized that the "influence of the Pacific Ocean on the vegetation of the Gulf Coast could scarcely be less if the ocean were 1000 kilometers away instead of 70 to 100."

Gradually from north to south the vegetation changes from creosote bush, ocotillo, foothills palo verde, and other species from the lower Colorado-Gila division and eventually gives way to torote prieto, palo adán, elephant tree, lomboy, and *Bursera hindsiana*. Near Los Angeles Bay, cirio comes near the coast along the rocky streamways. Together with Copalquín (*Pachycormus discolor*), it is occasional on south-facing slopes.

The vegetation of uplands and streamways is often strikingly different. Aside from the plants mentioned previously, the drier upland sites have vegetation of cardón pelón, jojoba, candelilla (*Pedilanthus macrocarpus*), rhatany (*Krameria parvifolia*), mescal (*Agave sobria*), magdalena bursage (*Ambrosia magdalenae*), and cholla.

Along the streamways can be found mesquite, palo blanco (*Lysiloma candida*), *Lantana*, paperfruit, burrobrush, and rush broom. Near the coast itself appear mangrove (*Rhizophora mangle*), mangle dulce (*Maytenus phyllanthoides*), iodine bush (*Allenrolfea*), and yerba reuma. Near the southern end of the peninsula, volcanic soils support desert vegetation, while thorn forest occurs on granite.

Vizcaíno Division

This division consists of the central third of the Pacific side of Baja California. To the north is an extensive area of transition between desert and chaparral and to the south lies the Magdalena Division. Winter rainfall, scant as it is, becomes even more undependable to the south. In occasional years rainfall may be more abundant. Summer storms are very rare. Long periods may pass without rain in the interior. The Pacific air, however, is humid, and the epiphytic heno pequeño (*Tillandsia recurvata*) is to be found.

The most conspicuous aspect of the Vizcaíno Division is the abundance of leaf-succulent plants such as maguey (*Agave shawii*), siempreviva (*Dudleya*), flor de sol (*Mesembryan-*

themum), and datilillo (*Yucca valida*). The genus *Agave* is unusually diverse in Baja California, where it is represented by twenty-five different taxa (Gentry 1978). The swollen leaves of many plants in the Vizcaíno Division have given rise to the name sarcophyllous desert. The commonest shrub, goosefoot bursage (*Ambrosia [Franseria] chenopodifolia*), is responsible, together with *Agave,* for the floristic designation of this as the *Agave-Franseria* Region. On deep soils, datilillo may become more common than maguey. Datilillo has the appearance of joshua tree of the Mojave Desert. Due to its woodier and taller nature than *Agave,* it requires a deep root system. Maguey, being typical of the century plant group, has a shallow but extensive root system that can harvest water from even light rains.

The Vizcaíno Division seems to be the true homeland for spectacular plants that have given Baja California a reputation as a botanical wonderland (figure 5-15). Such plants, which can be dominant in the vegetation, include cirio (*Idria colum-*

Figure 5-15. The Vizcaíno Division with tall spires of cirio and ground rosettes of maguey.
Photograph by C. Allan Morgan.

naris), copalquín (*Pachycormus discolor*), palo adán (*Fouquieria diguetii*), cardón pelón (*Pachycereus pringlei*), datilillo, maguey, guayacán (*Vizcainoa geniculata*), and palma ceniza (*Brahea armata*). The plants of this region have been ably treated in *The Boojum and its Home* by Humphrey (1974) and in *Field Guide to Common and Interesting Plants of Baja California* by Coyle and Roberts (1975).

Much of the Vizcaíno Division is an area of extreme aridity and/or alkalinity and is largely devoid of plants of unusual physiognomy. In such areas the dominants include goosefoot bursage, magdalena bursage (*Ambrosia magdalenae*), desert saltbush (*Atriplex polycarpa*), creosote bush (*Larrea tridentata*), tomatillo (*Lycium*), *Viguieria deltoidea,* and pitahaya agría (*Machaerocereus gummosus*). Near the coast the windward slopes lack large plants but are covered with yellow lichen and short perennials. The leeward slopes have trees and larger shrubs.

Humphrey (1974) studied the associates of cirio in the Vizcaíno Division in considerable detail on sites with different soil characteristics. In coarse, sandy areas derived from granite, two low-growing densely branched shrubs, magdalena bursage and goosefoot bursage, as well as the leaf-succulent desert mescal, were the most abundant associates. Desert mescal also proved to be a dominant plant on soils derived from shale and on Cerrito Blanco volcanic ash, El Arenoso clay, El Ciprés granite, and Agua Dulce granite. Goosefoot bursage was also common on Cerrito Blanco volcanic ash, San Fernando limestone (with *Encelia californica*), San Fernando rhyolite (with magdalena bursage), and Arroyo del Rosario clay loam. White bursage was common on shale soils and on Agua Dulce granite. Buckwheat bush was a dominant on Cerrito Blanco volcanic ash and El Ciprés granite. Estafiate (with maguey) was the dominant plant on El Arenoso clay. Saltbushes (*Atriplex*) were dominants on Las Arrastras clay and on basaltic plains from Rosario to San Borja. These data indicate the complexities of soil-vegetation relations within the habitat of a single Sonoran Desert plant species.

CEDROS ISLAND

Cedros Island lies off the Pacific Coast of Baja California at about 28° N latitude. A considerable endemic element is present, including such species as *Ferocactus chrysacanthus*, *Cochemiea pondii* and *Mammillaria goodridgei* in the Cactaceae and *Penstemon cerrosensis* in the Scrophulariaceae.

NORTHERN REGION

Shreve (1936) described the transition from Sonoran Desert to chaparral at the head of the northern region, finding that in traveling north, density increased, the dominants became greater in stature and more uniform, and the evergreen (chaparral) shrubs increased as deciduous shrubs decreased.

The most common shrub of the chaparral-desert transition is goosefoot bursage, a facultatively evergreen species that is abundant in loam soil. This is replaced on heavy clay soil by estafiate (*Ambrosia camphorata*).

In the northern desert below the transition, the characteristic vegetation consists of goosefoot bursage or estafiate (depending on soil), maguey, *Viguieria deltoidea*, brittlebush (*Encelia farinosa*), ocotillo (*Fouquieria splendens*), jojoba (*Simmondsia chinensis*), buckwheat bush (*Eriogonum fasciculatum*), cañutillo (*Ephedra californica*), and cholla.

The vegetation becomes somewhat more xeric farther inland, with goosefoot bursage being replaced by white bursage (*Ambrosia dumosa*) and creosote bush being dominant or nearly so. Other species that become important are desert saltbush, foothills palo verde (*Cercidium microphyllum*), graythorn (*Condalia lycioides*), and canescent borage (*Tiquilia canescens*).

SOUTHERN REGION

Gradually species that were rare in the north become common in the southern region. In relatively optimum sites with clay or loam soil on the plains and bajadas, cirio, cardón pelón, maguey, desert mescal (*Agave deserti*), magdalena bursage, *Viguieria deltoidea, Encelia frutescens,* and *Opuntia clavellina* are the chief dominants. Species of clear secondary importance are datilillo, copalquín, goosefoot bursage, lomboy (*Jatropha cinerea*), creosote bush, candelilla (*Pedilanthus macrocarpus*), senita cactus (*Lophocereus schottii*), ocotillo, tomatillo, pitahaya agría, and desert saltbush.

Shreve (1951) considered it remarkable that the contrasting physiognomies of cirio, cardón pelón, maguey, copalquín, and datilillo simultaneously could reach their optimum development under the same climatic and edaphic conditions. In the rainy season, copalquín often becomes covered with parasitic dodder (*Cuscuta veatchii*). On the coast this tree is generally more gnarled and prostrate than elsewhere.

On the Vizcaíno Plain, datilillo becomes the dominant plant with the shrubs such as magdalena bursage, tomatillo, and *Encelia frutescens* important. Many of the previously mentioned species are still present, but cirio, cardón pelón, maguey, desert mescal, copalquín, and candelilla are either absent or uncommon. East of the Vizcaíno plain lie volcanic mesas with vegetation of torote prieto (*Jatropha cuneata*), palo adán, tomatillo, magdalena bursage, brittlebush, lomboy, creosote bush, pitahaya agría, and cholla.

Magdalena Division

In the southern one-third of Baja California, with definite Pacific influence and south of the Vizcaíno Division, lies the Magdalena Division. Characteristic plants of the Vizcaíno Division such as cirio, capalquín, datilillo, and maguey are notably absent or uncommon, but otherwise the Magdalena Division seems closely related floristically to the Vizcaíno. The Magdalena has large trees such as palo blanco (*Lysiloma candida*), mesquite (*Prosopis torreyana*), palo hierro (*Prosopis palmeri*), Baja palo verde (*Cercidium peninsulare*), *Bersera laxiflora*, and palo adán (*Fouquieria diguetii*).

Another notable group of plants consists of large cacti such as organ-pipe (*Stenocereus thurberi*), pitahaya agría (*Machaerocereus gummosus*), cardón pelón (*Pachycereus pringlei*), and cholla (*Opuntia cholla*). Because of the nearly equal importance of trees and large cacti, this is often referred to as an arbocrassicaulescent desert on physiognomic grounds, or as the *Lysiloma-Machaerocereus* region floristically.

On the Pacific side, rain comes only in late winter, but interior regions may also receive moisture from summer thunderstorms. Rain can often be very uncertain, however; periods of three to five years without precipitation have been recorded at some sites. Rain that does fall is most pronounced in April and May.

In the northern section, vegetation similar in physiognomy to that of the Central Coast of Sonora is found, although many species are Baja Californian ecological equivalents of those of the mainland. Creosote bush, torote prieto (*Jatropha cuneata*), desert saltbush (*Atriplex polycarpa*), lomboy (*Jatropha cinerea*), senita cactus (*Lophocereus schottii*), and elephant tree (*Bursera microphylla*) are the same as in Sonora. Shreve (1951) pointed out that in the Magdalena Division, *Opuntia cholla* replaces *O. fulgida*, pitahaya agría replaces octopus cactus (*Rathbunia alamosensis*), palo adán replaces ocotillo (*Fouquieria splendens*), and cardón pelón replaces saguaro (*Carnegiea gigantea*).

Volcanic mesas in the north differ from the nonvolcanic areas in having a pronounced scarcity of cacti. Valleys draining the volcanic mesas have an abundance of palo blanco, palo hierro, *Bursera odorata*, organ-pipe cactus, palo adán and cardón pelón.

On sandy plains that extend inland from coastal dunes, the caterpillar cactus (*Machaerocereus eruca*) is an unusual endemic. It grows prostrate on the ground, rooting in the soil as it grows, with the older parts dying. The plant wanders over the surface of the soil by means of new apical growth. The spines are wider than in its closest relative, probably because the prostrate nature of the plant renders it susceptible to solar energy burn on clear days, the wider spines promote shading of the epidermis and underlying chlorophyll. The stout, tough spines probably also give protection from the chewing and biting animals of the ground zone. The plant is a unique example of a truly mobile terrestrial rooted plant.

Vegetation of the sandy plains consists chiefly of cholla, brittlebush (*Encelia farinosa*), magdalena bursage, cardón pelón, lomboy, palo adán, creosote bush, and the endemic *Euphorbia magdalenae*. The Magdalena Plain in the southern region is clearly dominated by mesquite, tomatillo, and cholla, with abundant palo adán, pitahaya agría, cardón pelón, and *Euphorbia californica*.

Magdalena and Santa Margarita islands, off the Pacific Coast of Baja California, and part of the Magdalena region, have significant endemic elements. These include *Agave margaritae*, *Opuntia santamaria*, *O. pycnantha* and *Echinocereus barthelowanus* on Magdalena Island and *Agave margaritae* and *Opuntia pycnantha* on Santa Margarita.

Digitized Classification in the Sonoran Desert

The digitized classification of biotic communities proposed by Brown, Lowe, and Pase (1979), although compatible with computer programming and useful as a standardized shorthand for designating areas on maps, lacks the versatility of the spoken or written word. Concepts that seem logical in nature may become fragmented in the system. For example, Shreve (1951) showed that the Sonoran Desert logically included frost-resistant desert plants, often with northerly distributions, and frost-sensitive plants, generally of southerly distributions.

The digitized classification separates warm temperate desertlands as a 1153 series and tropical subtropical desertlands as 1154 series. The entire Sonoran Desert must be assigned to one or the other classification, and the obvious choice is 1154. But as the classification system progresses and plant communities must be assigned one way or the other with regard to a large number of characteristics, cumulative compromises are disturbing. Specific communities do not always have the general characters of the larger classes they fall within. As long as the user realizes that the system is more utilitarian than philosophical, those who devised it are to be complimented.

FAUNA

With the exception of invertebrates, the fauna of the Sonoran Desert are summarized in appendixes 5-A, 5-B, 5-C, and 5-D. Disregarding the estimated 800 species of marine fishes of the Gulf but including the marine mammals, there are 1014 principal taxonomic units (species plus additional subspecies) of vertebrates associated with the Sonoran Desert. Included are 137 families, 422 genera, and 735 species. Subspeciation is of great significance among the terrestrial mammals and reptiles, of moderate significance among the freshwater amphibians and fishes, and of less significance among the aerial bats and birds.

During glacial maxima, desert animals were presumably pushed south to refugia in central Mexico (Chihuahuan elements) and on either side of the Gulf of California (Sonoran elements). During interglacials, Sonoran and Chihuahuan animals probably met on the Deming Plain as they do today (Hubbard 1973; Findley 1969; Lowe 1955). MacMahon (1979) pointed out that bird species of these two deserts overlap greatly, mammals to some extent, but reptiles hardly at all. Thus degree of faunal similarity observed today could be a result of past ability of species to reunite populations during interglacials. Biogeography of the Pleistocene was treated by Deevey (1949) and southwestern animal communities in the Late Pleistocene by Martin (1961).

MacMahon (1979), Stuart (1970), Hagmeier (1966), and Hagmeier and Stutz (1964) have addressed the question of faunal boundaries of the North American deserts. Mac-Mahon's (1979) Sonoran Desert limits are unusually large, extending the desert all the way from the Arizona-New Mexico boundary to southern Nevada. Stuart (1970) segregated the desert fauna of much of Baja California under the heading San Lucan in a coordinate rank with the Sonoran Desert. Hagmeier (1966) and Hagmeier and Stutz (1964) recognized a Yaquinian transition between Sonoran and Chihuahuan and also attributed part of the Sonoran to the "Mohavian."

MacMahon's (1979) concept of the Sonoran Desert as extending all the way to the Rio Mayo was based on the presence there of some typical Sonoran Desert mammals such as the black tailed jackrabbit, Harris's antelope squirrel, round-tailed ground squirrel, valley pocket gopher, Bailey's pocket mouse, and badger, as well as many typical Sonoran Desert reptiles and amphibians (Bogert and Oliver 1945). Although the fauna of the Sonoran, Mojave, and Chihuahuan deserts are similar to some extent, more species of animals can be seen in a shorter time in the Sonoran than in the Mojave or Chihuahuan deserts (MacMahon 1979), and the Sonoran has much more regional differentiation.

As opposed to Pleistocene history, life zones govern animal distributions through the effects of local environmental factors. Merriam (1898) described life zones of the United States largely on the basis of studies in Arizona. Cockerell (1900) discussed lower and middle Sonoran zones at an early date. Hall and Grinnell (1919) discussed life zone indicators in California, and Daubenmire (1938) published a well-known critique and discussion. The life zones have been discussed well for Arizona by Lowe (1964) in *Vertebrates of Arizona*.

Swarth (1929) divided southern Arizona into two faunal areas: the eastern plains extending from the Santa Catalina, Santa Rita, and Rincon mountains to the New Mexico boundary, and the western desert area, extending from Tucson to the Colorado River. Phillips (1939) wrote on the faunal areas of Arizona based on bird distributions, describing four such areas: Pima, Yuma, Mogollon, and Navajo. Sellers (1960) produced a map based on environmental factors, which is now widely used for delimiting faunal areas of the state.

Contemporary biota of the Sonoran Desert has been treated by Lowe (1959) and the biotic communities of the sub-Mogollon region of the inland Southwest by the same author (Lowe 1961). Origins and affinities of vertebrates of the Sonoran Desert and the Monte Desert of Argentina were detailed by Blair et al. (1976). The biotic provinces of Mexico were discussed by Goldman and Moore (1945) and the faunal

areas of Baja California Norte by Bancroft (1926). Leopold (1959) published a book on the wildlife of Mexico and Tinker (1978) one on Mexican wilderness and wildlife.

Extensive studies of Sonoran Desert animals are present in the Arizona Game and Fish Department reports—for example W-53-R-25, 26, 27 (1974-1977) on small-game species and W-53-R-28 (1977–1978) on big-game animals. General faunal surveys such as that for unit 37-B (FW-11-R-8, Job 1, 1975), which includes the Boyce Thompson Southwestern Arboretum and Picketpost Mountain, often include extensive lists of nongame mammals, birds, reptiles, and amphibians. Other useful publications include classic studies of mammals and birds of the Lower Colorado Valley (Grinnell 1914) and of vertebrates of Organ Pipe Cactus National Monument (Huey 1942). Dickerman (1954) published an ecological survey of the Three-Bar Game Managment Unit in central Arizona.

Habitat Considerations

Species composition of Sonoran Desert animal communities depends not only on adaptational biology (Hadley, chap. 8) but on vagaries of cover, temperature, humidity, and food availability, the factors of the habitat. The number of species showing biological adaptations to the desert is large, but the number with less pronounced adaptations reflects the diversity of the habitat. Factors of the habitat are valuable to the species in ameliorating the basic environment. For example, were it not for the holes made in giant cacti by the gila woodpecker, the elf owl probably could not survive in the Sonoran Desert.

A better knowledge of animals of the desert can be gained by studying ill-adapted species. The Mexican funnel-eared bat requires a sheltered place in the daytime with a rather constant high humidity, which puts it at a disadvantage in the desert (Mitchell 1967). McNab (1973) theorized that cold weather limited vampires from extending their range north into the Sonoran Desert because the cost of thermoregulation would require a blood meal beyond the holding capacity of the bat.

Sonoran Desert plants frequently have toxic or indigestible compounds that render them unavailable or deleterious as food for animals. Examples of such plant substances are oxalates in cacti and liquid wax and simmondsin in jojoba. Once the javelina and white-throated woodrat evolved adaptations to cope with the oxalates and once Bailey's pocket mouse could cope with jojoba compounds (Rosenzweig and Winakur 1969; Sherbrooke 1976; Sherwood 1980), the animals suddenly could utilize a habitat factor not available to other species.

The microphyllous and open nature of dominant trees and shrubs of the Sonoran Desert raises questions as to the effectiveness of such species in ameliorating temperature extremes and incidentally creating habitat for wildlife. Lowe and Hinds (1971) measured radiation flux at ground level under a typical species, palo verde (*Cercidium*), finding that in winter when leaf cover was low, incoming radiation was nevertheless cut to less than half that on open ground and that net outgoing radiation at night was in a similar ratio. For a typical 1 kg homiotherm, heat loss at night by the animal in the open was calculated to be 2.6 times what it would have been under the palo verde. On open ground, the 1 kg homiotherm would have to increase its metabolic rate by 6.2 times to balance the heat loss. Calculations indicate that homiotherms smaller than 1 kg or poikilotherms would die under such conditions in the open but could survive through typical days and nights in the ameliorated environment of the Palo Verde. In areas with colder winters, however, heat trapped under the canopy of such a microphyllous plant typically would be insufficient to give such nighttime protection in winter.

Favorable habitat temperatures allow poikilotherms and small homiotherms to be more abundant and active during a greater portion of the year in the Sonoran Desert than in typical situations in surrounding regions. In studying body weights of Sonoran Desert rodents, Reichman and Van De Graaff (1973) found that heavier species were most affected by high temperatures. Interrelationships were found among low ambient air temperature, body weight, and inactivity. The rodent species lightest in weight was the least active in winter.

Mammals

High species diversity of mammals of the Sonoran Desert is logically explained by the study of McCoy and Connor (1980). When 21 marine species and 25 aerial ones are added to the inventory of terrestrial species, the mammalian fauna of this desert includes 30 families, 72 genera, 141 species, and 343 principal taxonomic units (species plus additional subspecies). Mammals of the Sonoran Desert are listed in appendix 5-A. The most useful reference to Sonoran Desert mammals is the classic two-volume work by Hall and Kelson (1959).

Cockrum (1960) detailed the recent mammals of Arizona and summarized them in a checklist (Cockrum 1963). Drabek (1967) prepared a thesis on mammals of the Boyce Thompson Southwestern Arboretum and surrounding area in the Arizona upland division. Hatfield (1942) treated mammals of south-central Arizona. The treatment of mammals of Deep Canyon by Ryan (1968) is a fine representative treatment for a southern California portion of the desert.

Huey (1964) and Nelson (1922) reported on mammals of Baja California. Burt (1932) described mammals from islands in the Gulf of California, and Lowe (1955) published an evolutionary study of the desert island faunas. Burt (1938) reported on the faunal relationships and geographic distribution of mammals of Sonora.

Mearns (1907) published a classic paper on mammals of the Mexico-U.S. boundary. Olin (1959) treated mammals of the southwestern deserts, and Findley (1969) discussed biogeography of southwestern boreal and desert mammals.

Dice and Blossom (1937) discussed mammalian ecology in southwestern North American with special attention to the colors of desert mammals. Chew and Chew (1970) reported on energy relationships of the mammals of a creosote bush community. MacMahon (1976) presented a functional analysis of North American mammal communities, reporting on species and guild similarities.

ARTIODACTYLA

The importance of this group to humans and to the modern sportsmen should not be underestimated. Present in the Sonoran Desert are bighorn (Bovidae: *Ovis*), pronghorn (Antilocapridae: *Antilocapra*), deer (Cervidae: *Odocoileus*), and javelina (Tayassuidae: *Tayassu*).

Desert bighorns seem well adapted to the Sonoran Desert, though less suited to cope with changes caused by civilization. Sheep can go without water for periods up to six days, even during June, the hottest and driest month. Population dynamics of the bighorn are monitored by the Desert Bighorn Council, which publishes a journal, and by the Arizona Desert Bighorn Sheep Society. Present status of the bighorn in the Arizona portion of the desert is summarized in a report by the Arizona Game and Fish Department (1979), in Baja California by Alvarez (1976), and in Sonora by Valverde (1976). Overgrazing on bighorn ranges in the Sonoran Desert was analyzed by Gallizioli (1977). Russo (1956) treated the life history of bighorns in Arizona, and Simmons (1969) studied their behavior in the Cabeza Prieta Game Range.

Introduction of domestic sheep into the Sonoran Desert brought the sheep bot fly into contact with the native bighorn. Due to the different anatomy of the desert bighorn skull as compared with that of domestic sheep, larger bot larvae become trapped in cranial sinuses off the moist nasal passages, die, and cause massive infection and osteonecrosis (Bunch et al. 1978). Sheep so affected frequently die from secondary meningitis, pneumonia, or encephalitis. Thus man has unintentionally eliminated the bighorn in many desert mountain ranges and greatly reduced their populations elsewhere.

Haury et al. (1950), in reporting on the stratigraphy and archaeology of Ventana Cave in Papaguería, found remains of bighorns that had likely been used as food by prehistoric desert people. The absence of such remains in the older levels suggested that the animals had not migrated into the Sonoran Desert until after 8000 B.C.

That bighorns represented important food for indigenous people of the Sonoran Desert is attested to by Lumholtz (1912) and numerous early explorers. An observer of the seventeenth century (translation, Manje 1954) encountered a huge pile of horns of this species near the present-day Pima Indian village of Blackwater, which he estimated contained in excess of 100,000 horns. From such evidence he judged that the species represented the chief source of food for the local Indians.

McKusick (1976a) found evidence that reliance on bighorns for food may have increased in late prehistory in central Arizona.

Mearns (1907) reported large heaps of horns in the Tule and Granite mountains. Similar heaps were reported by early explorers in northern Sonora. Greene and Mathews (1976) estimated that bighorn sheep meat represented only 6 percent of total meat used by prehistoric Hohokam from 300 B.C. to 1100 A.D. at the Snaketown site in the lower Colorado-Gila Division. The Pima (Akimel O'Odham) Chief Antonio Azul revealed a taboo against bringing home the horns of the bighorn sheep, although the meat was utilized (Russell 1908). Haury (1976) believed that if the Hohokam also had such a taboo, then the estimate of bighorns, providing 6 percent of the meat used by these prehistoric people would be low since remains would have been concentrated outside the villages where excavations were conducted.

Bighorn sheep pictographs were pecked through desert varnish on basalt boulders along the trail into Ventana Cave (Haury et al. 1950), and a complete bighorn sheep was artistically carved onto a bone hairpin discovered in prehistoric Snaketown. Haury (1976) believed that the importance of the bighorn to the Hohokam was demonstrated by their art, "for it was sculptured in stone, modeled in clay, carved in bone and painted on pottery." Aschmann (1959) indicated that the bighorn was also of importance to the Cochimí of Baja California.

The Sonoran pronghorn is classified as endangered and has been specially studied (Carr 1971, 1972, 1973; Phelps 1974). Its habitat was described by Hornaday (1908), Lumholtz (1912), and Monson (1968). Halloran (1957) noted that southern stock had been genetically swamped by introduction of pronghorns from northern Arizona. Brown and Webb (1979) reported on the status of the peninsular pronghorn, detailing its distribution on the Llano del Berrendo (="Antelope Plain") and adjacent portions of the Vizcaíno Desert.

Deer have been treated by Rue (1978). Lumholtz (1912) recorded a Papago (Tohono O'Odham) claim that a race of deer in the western part of the desert did not need to drink free water. Although mule deer thus far studied by scientists obtain some water from eating vegetation, they generally need free water (Elder 1956), taking four to eleven quarts during a typical visit to a water hole. Limitations on mule deer populations in the Sonoran Desert relate to water availability in summer (Clark 1953).

White tailed deer apparently have always been absent from the California and Baja California portions of the Sonoran Desert. In a study of over 350 deer bones excavated from prehistoric Snaketown, none were from white-tailed deer. This evidence indicates that they were likely not present in the area during the period 300 B.C. to 1100 A.D.

Greene and Mathews (1976) calculated that mule deer provided 82 percent of the meat eaten by the Hohokam at Snaketown. Castetter and Bell (1942) found that venison was also the most important meat for the Papago (Tohono O'odham) and second only to black-tailed jackrabbit for the Pima (Akimel O'odham).

Javelina apparently were not utilized for food by the prehistoric Hohokam at Snaketown (Greene and Mathews 1976), although Russell (1908) reported them as a food source for the Pima (Akimel O'odham) in the same general region over 800 years later. The possibility that the species migrated north during this interval deserves investigation. Once the ability to subsist on cacti high in oxalates became genetically fixed, the species might have dispersed through *Opuntia* habitat with incredible speed, utilizing a food resource that nonadapted animals could not. Subspecies patterns indicate separate northward radiations through the Chihuahuan and Sonoran deserts from the southern center of diversity of the genus after Baja California had rifted from the mainland. A tendency for prickly pears to be rather treelike in certain areas, with a definite trunk, may be related to natural selection since the taller-growing individuals would escape being utilized for food. The dynamics of *Opuntia*-javelina interactions represent an area for future research.

Zervanos and Hadley (1973) published on the adaptational biology and energy relationships of javelina, and later Zervanos and Day (1977) compared water and energy requirements of captive and free-living javelinas. Research on the influence of aridity on reproduction of the species in Texas was published by Low (1970), and seasonal movements and activity patterns of the species were reported by Bigler (1974). The recent status of the javelina in Arizona is stable (Arizona Game and Fish Department 1979). Javelina have also been studied by Knipe (1957), Sowls (1958), Neal (1959), and Eddy (1961).

PINNIPEDIA

Included here are harbor seals (Phocidae: *Phoca*), elephant seals (Phocidae: *Mirounga*), and sea lions (Otariidae: *Zalophus*). These animals have been utilized by man in the Sonoran Desert, including the indigenous Indians. Pacific coastal waters and islands off the Baja California coast provide breeding areas for the California sea lion. Two kinds of earless seals are also found there. The California harbour seal extending from San Ignacio Lagoon, San Gerónimo, and San Martín islands, north to the boundary. The northern elephant seal occurs from the San Benitos Islands north to the boundary. Populations have recouped from near extinction in the 1890s to nearly ten thousand in 1958 (Huey 1964).

CARNIVORA

Included in this order are jaguars, mountain lions, ocelots and jaguarundis (Felidae: *Felis*), bobcats (Felidae: *Lynx*), badgers (Mustelidae: *Taxidea*), skunks (Mustelidae: *Spilo-*

gale, Mephitis, Conepatus), sea otters (Mustelidae: *Enhydra*), ringtails (Procyonidae: *Bassariscus*), raccoons (Procyonidae: *Procyon*), coatimundis (Procyonidae: *Nasua*), coyotes (Canidae: *Canis*), and foxes (Canidae: *Vulpes, Urocyon*)

As in biotic communities around the world, carnivores are less numerous than herbivores because of the normal configuration of the food pyramid. Carnivora of the Sonoran Desert have rarely been used as food by either prehistoric or recent man either because of preference or taboo.

Many of the Sonoran Desert carnivores, particularly the cats, are rare or extinct in the northern parts of their range. This may be due to several factors. They have been widely hunted north of the Mexican border for sport. A documented increase in aridity during the first half of this century probably reduced population numbers of prey. Urbanization and agriculturization in the northern part of the desert eliminated prey in river valleys and at water sources. And the characteristic of predators ranging far afield may have exaggerated the ranges of some species that typically are more southern elements.

Carnivores of Yuma County, Arizona, were treated by Halloran and Blanchard (1954). Jaguarundi in Arizona were discussed by Little (1938). Pocock (1941) treated the ocelot. Details concerning habits, characteristics, and classification of mountain lions were summarized by Young (1946) and Young and Goldman (1946). Habits and characteristics of bobcats were detailed in a book by Young (1958). Food habits of desert skunks have been summarized by Reichman et al. (1979), and use of cactus as protection by hooded skunks has been documented by Reed and Carr (1949).

Formerly sea otters were found off the Pacific coast of Baja California, including Cedros Island. Their extinction in this area occurred when the value of their pelts was realized. Ringtails occur through much of the desert, with endemic subspecies on Gulf islands. Studies of growth and development of ringtails were published by Richardson (1942). Taylor (1954) studied their food habits and life history. Range of the coatimundi in the United States was reported by Taber (1940). The status in Arizona was delineated by Wallmo and Gallizioli (1954) and in the United States by Kaufmann et al. (1967). Gilbert (1973) published a book on the species.

Books on the coyote have been published by Dobie (1949), Young and Jackson (1951), and Leydet (1977). Longevity was treated by Manville (1953) and the food habits by Sperry (1941) and Murie (1951). The kit fox was studied by Waitman and Roest (1977).

CETACEA

This order of marine mammals includes the right whales (Balaenidae: *Eubalaena*), fin-backed whales (Balaenopteridae: Balaenoptera, Sibbaldus, Megaptera), gray whales (Eschrichtidae: *Eschrichtius*), dolphins (Delphinidae: *Delphinus, Tursiops, Lagenorhynchus*), killer whales (Delphinidae: Or-

cinus, Pseudorca), blackfish (Delphinidae: *Globicephala*) porpoises (Delphinidae: *Phocoena*), pygmy sperm whales (Kogiidae: *Kogia*), and sperm whales (Physeteridae: *Physeter*).

Although these mammals are not terrestrial, the use of various parts of them is well documented for aboriginal peoples of the Sonoran Desert, including the Cochimí, Seri, and HiacheD O'odham. Utilization usually resulted when the animals accidentally beached and could not return to the sea, a recurrent phenomenon.

Warm, quiet bays such as Scammon's Lagoon and San Ignacio Lagoon provide calving areas for the gray whale. Plankton and krill-eating whales such as the fin-backed whale, sei whale, little piked whale, and Pacific right whale have the southern limits of their range off the Baja coast.

RODENTIA

This order contains by far the largest number of mammals in the Sonoran Desert, including the porcupines (Erethizontidae: *Erethizon*), rice rats (Cricetidae: *Oryzomys*), harvest mice (Cricetidae: *Reithrodontomys*), white-footed mice (Cricetidae: *Peromyscus*), pigmy mice (Cricetidae: *Baiomys*), grasshopper mice (Cricetidae: *Onychomys*), cotton rats (Cricetidae: *Sigmodon*), Wood rats (Cricetidae: *Neotoma*), muskrats (Cricetidae: *Ondatra*), beaver (Castoridae: *Castor*), pocket mice (Heteromyidae: *Perognathus*), kangaroo rats (Heteromyidae: *Dipodomys*), spiny pocket mice (Heteromyidae: *Liomys*), pocket gophers (Geomyidae: *Thomomys*), chipmunks (Sciuridae: *Eutamias*), antelope squirrels (Sciuridae: *Ammospermophilus*), and ground squirrels (Sciuridae: *Spermophilus*).

Because these mammals are easily collected and make good laboratory subjects, they have been frequent subjects for study of ecology, physiology and population biology. Prakash and Ghosh (1975) edited a collection of 23 articles concerning rodents in desert environments. Brown (1975b) discussed geographic ecology of desert rodents. Rosenzweig et al. (1975) analyzed patterns of food, space, and diversity. Whitford (1976) discussed temporal fluctuations in density and diversity of desert rodent populations. Price (1976) analyzed the role of microhabitat in structuring desert rodent communities. French et al. (1966) reported on periodicity of desert rodent activity. Brown and Lieberman (1973) discussed resource utilization and coexistence of seedeating rodents in sand-dune habitats. Bateman (1967) reported on home-range studies of a desert nocturnal rodent fauna at the Boyce Thompson Southwestern Arboretum.

Lemen (1976) discussed relations of body size and seed size in desert rodents. Reichman (1976) reported on the effects that rodents have on the germination of desert annuals. Smith and Jorgensen (1975) cataloged data on reproduction in forty-one species of North American rodents. Integumentary modifications of desert rodents in North America were discussed by

Quay (1964). Van De Graaff (1973) related details of osteology of selected desert rodents to their habits and habitats.

The reactions of desert rodents in dry heat were discussed by Schmidt-Nielsen (1964b). Physiological adaptations of desert rodents were discussed by Ghobrial and Nour (1975). The relation of oxygen consumption to temperature in desert rodents was reviewed by Dawson (1955). Pulmonary and other evaporative loss of water by desert rodents was treated by Schmidt-Nielsen and Schmidt-Nielsen (1950a, 1950b). Yousef et al. (1974) compared tritiated water turnover rates in rodents of desert and mountain. MacMillen (1972) reviewed water economy of nocturnal desert rodents. Johnson and Groepper (1970) discued bioenergetics of North American rodents.

Thyroid function and metabolic rate of desert rodents received treatment by Yousef and Johnson (1972), and corticosterone plasma levels of desert rodents were reported by Vanjonack et al. (1975). A comparison of body fat content and metabolic rate in rodents of desert and mountain was published by Scott et al. (1972).

Reichman and Van De Graaff (1973) correlated body weight of Sonoran Desert rodents with activity patterns. Vorhies (1945) investigated microclimates and water requirements of selected Sonoran Desert rodents in a classic study. Ecological aspects of the diets of Sonoran rodents were recently treated by Reichman (1978). Price (1978) reported on seed-dispersion preferences of coexisting Sonoran Desert rodent species. Seasonal reproductive patterns of some species of Sonoran Desert rodents were treated by Reichman and Van De Graaff (1973). Population ecology, water relations, and social behavior of a southern California semidesert rodent fauna were reported by MacMillen (1964). Stamp and Ohmart (1979) discussed rodents of desert scrub and riparian woodland habitats in the Sonoran Desert. Plant and rodent communities of Organ Pipe Cactus National Monument were carefully described by Warren (1979).

Porcupines normally live in habitats more mesic than the Sonoran Desert but have made well-documented invasions into the desert. Destructive girdling of trees and limbs at the Boyce Thompson Southwestern Arboretum presented a difficult control problem in the 1930s and 1940s. Porcupines in southwestern Arizona were discussed by Monson (1948) and Reynolds (1957). The animals have been recorded in the Sonora portion of the desert as far south as Kino Bay at the mouth of the Rio Sonora (Leopold 1959).

Peromyscus represents a complex assortment of geographically replacing species and subspecies in North America (Osgood 1909; Hall and Kelson 1959). In the Sonoran Desert a number of endemic taxa are associated with specific physiographic features or are restricted to islands in the Gulf of California. Lawlor (1971), in studying the evolution of *Peromyscus* on islands at the northern end of the gulf, concluded

that the chief factor allowing divergence of the island populations had been time since formation of the islands.

Peromyscus eremicus has often been studied as a representative desert species. Ryan (1968) pointed out that distribution of *P. eremicus* and related species often showed correlation with local soil types. Reichman (1978) studied the food habits of *P. eremicus*, and MacMillen (1965) reported on aestivation. Lewis (1972) discussed high winter population density and low summer density in the species. McNab and Morrison (1963) compared body temperature and metabolism in subspecies of *Peromyscus* from arid and mesic environments. Murie (1961) compared metabolic characteristics of desert populations of *Peromyscus* with ones from mountain and coast.

Pearson (1960) analyzed the oxygen consumption and bioenergetics of harvest mice. Water economy and salt balance of the western harvest mouse were studied by MacMillen (1964). Schmidt-Nielsen and Haines (1964) studied water balance in the grasshopper mouse as an example of a carnivorous desert rodent. Diet of this mouse was treated by Flake (1973) and Bailey and Sperry (1929).

Cotton rats enter the desert along watercourses and in agricultural regions. They can undergo population explosions at which time "they usually are the most numerous mammals and are active both day and night. Sometimes they are serious pests in agricultural areas" (Hall and Kelson 1959). Interlocking runways of these animals typically occur on ditch banks and margins of cultivated fields. Rea (1979a) reported that among all of the game animals of the Pima Indians (Akimel O'odham), they considered the cotton rat prime food. The rat is associated with Pima fields allowed to remain fallow a year or two with rank growth of grass and weeds. That these rats are relatively recent immigrants at least to the central Arizona portion of the desert seems possible since none of their bones occur in the excavations at Snaketown (Greene and Mathews 1976; Olsen 1976) where from 300 B.C. to 1100 A.D. the Hohokam apparently utilized a wide range of fauna as food, including *Neotoma*, *Dipodomys*, *Thomomys*, *Spermophilus*, and *Peromyscus* in addition to larger game.

Seven species of *Neotoma* occur in the Sonoran Desert, 4 of them isolated island endemics. Adaptations to arid environments in *Neotoma* were discussed by Lee (1963). Bergman's Rule and climatic adaptation of wood rats were treated by Brown and Lee (1969). Food habits were discussed by Ryan (1968), MacMillen (1964), Schmidt-Nielsen (1964), Finley (1958), and Jaeger (1948). In addition, Vorhies and Taylor (1940) made a special study of the white-throated wood rat. Spencer and Spencer (1941) made further reports of food habits, and Hill (1942) discussed the "cactus paved runways" of the species.

The muskrat is a northern element unknown from Mexico except along the lower Colorado River at the extreme northern tip of Baja California. Halloran (1946) described a muskrat house on the lower Colorado River. According to Rea (1979a, 1979b), beavers occurred in the Sonoran Desert "in numbers all the way from the small Gila tributary streams of the uplands down to the Colorado River." Lagoons, marshes, or *cienegas* created by the actions of the beaver provided habitat for a number of animals.

The genus *Perognathus* is composed of a rather large number of taxa of arid and semiarid situations (Merriam 1889; Osgood 1900; Hall and Kelson 1959; Patton 1967). Chew et al. (1965) reported on circadian rhythms in the metabolic rate. Patton (1967) correlated chromosome variation with morphologic evolution. Reichman and Oberstein (1977) analyzed selection of seed types. Other studies have been performed by Rosenzweig and Winakur (1969), Sherbrooke (1976), and Sherwood (1980). Populations of *Perognathus* were stranded on desert islands in the Gulf or off the Pacific Coast of Baja California. These now represent island endemics.

Stock (1974) reported data on chromosome evolution in *Dipodomys* and discussed the taxonomic and phylogenetic implications. Shroder and Rosenzweig (1975) reported on competition in habitat between species of *dipodomys*. Conservation of water by *Dipodomys* was treated in a classic paper by Howell and Gersh (1935) and by succeeding authors. Van De Graaff and Balda (1973) discussed the importance of green vegetation for reproduction in *Dipodomys*. Yousef et al. (1970) studied the energy expenditure of running kangaroo rats.

Mullen (1971) reported on energy metabolism and body-water turnover rates of two free-living kangaroo rats *D. merriami* and *D. microps*. Carpenter (1966) compared thermoregulation and water metabolism between *D. merriami* and *D. agilis*. A comparative study of growth and development of *D. merriami* and *D. deserti* was published by Butterworth (1961). Lockard and Lockard (1971) reported on seed preference and buried seed retrieval by *Dipodomys deserti*.

Reichman and Oberstein (1977) analyzed the selection of seed types by *Dipodomys merriami*. Bradley and Mauer (1971) treated reproduction and food habits of the species. Responses of *D. merriami* to heat were reported by Yousef and Dill (1971a) and daily cycles of hibernation by Yousef and Dill (1971b). Reynolds (1950) found that *D. merriami* tended to decrease in numbers with an increase in perennial grasses. Numbers of these rats proved to be greater in grazed areas than in areas where livestock had been excluded. Vorhies and Taylor (1922) described the life history of *D. spectabilis*.

Reichman (1975) determined diets of *Dipodomys* and *Perognathus* species and compared actual foods utilized with the resources available in specific habitats of the Sonoran Desert. Reichman and Van De Graaff (1975) studied the association between reproduction and ingestion of green vegetation by *Perognathus* and *Dipodomys* near Tucson. Wondal-

leck (1978) analyzed forage-area separation and overlap in Heteromyid rodents south of Tucson. Heteromyid rodents have also been treated by Reichman (1978), Reichman et al. (1975, 1979), Schmidt-Nielsen (1964), and Hall and Linsdale (1929).

The Heteromyid genus *Liomys* (spiny pocket mice) is found from Mexico to central America. The painted spiny pocket mouse of eastern Sonora has been collected near the Arizona-Sonora boundary. Among recent studies on *Liomys,* the paper by Pinkham (1973) on the evolutionary significance of locomotor patterns in the genus is important.

Classification of North American Sciuridae was detailed by Howell (1938). Since desert ground squirrels are active in the daytime, they may be subjected to rather intense water-balance problems in desert situations (Bartholomew and Hudson 1961). These animals are omnivorous and also store food. The diet has been discussed by Walker (1968), Hudson (1962), and Bartholomew and Hudson (1961). Temperature regulation in desert ground squirrels was discussed by Hudson (1962, 1967) and by Hudson et al. (1972). Thyroid function in desert ground squirrels was treated by Hudson and Wang (1969). The energetic cost of running in the antelope ground squirrel was analyzed by Yousef et al. (1973).

Neal (1964) studied *Spermophilus harrisi* and *S. tereticaudus,* species that are sympatric and appeared to violate Gause's Principle of Competitive Exclusion. It was found that *S. harrisi* preferred rocky soils and hills, whereas *S. tereticaudus* preferred sandy soils of flatlands. Drabek (1970) discussed behavior and ecology of *S. tereticaudus neglectus* in the Arizona Upland Division.

The peninsula chipmunk is very rare, occurring in an area less than 25 miles in diameter in Baja California and stands "in strange contrast with its northern relatives by living in lava-bound palm-cactus associations" (Huey 1964). The status of this chipmunk has been discussed by Callahan (1975).

LAGOMORPHA

This order is represented by cottontail and brush rabbits (Leporidae: *Sylvilagus*) and jackrabbits (Leporidae: *Lepus*). Madsen (1974) pointed out that lagomorphs are the chief grazing competitors of domestic cattle in the Sonoran Desert. They also compete for water. Population studies of rabbits in the Sonoran Desert, examining the three commonest species, were published by Hungerford et al. (1973, 1974). Survival of the antelope jackrabbit and black-tailed jackrabbit in the hot and dry environment of the desert was studied by Schmidt-Nielsen et al. (1965).

Taylor, Vorhies, and Lister (1935) pointed out that jackrabbits feed on valuable range vegetation and originally were in a state of fluctuating balance with the vegetation before man introduced grazing livestock. Sex ratios, age classes, and reproduction of black-tailed jackrabbits were discussed by Lechleitner (1959). The effects of thermal conductance on the water economy of the antelope jackrabbit were detailed by Dawson and Schmidt-Nielsen (1966). Vorhies and Taylor (1933) contrasted the antelope jackrabbit of what they considered Sonoran grassland with the black-tailed jackrabbit of what they treated as typical desert. Leopold (1959) characterized the black-tailed jackrabbit in Mexico as found "on the most desolate stretches of cactus desert or on adjoining mesquite-grasslands that have been so overgrazed" that the ground is almost bare, with the antelope jackrabbit frequenting brushy watercourses.

At prehistoric Snaketown on the Gila River, where a brushy watercourse was certainly present, the jackrabbits used for food from 300B.C. to 1100A.D. proved to consist of both black-tailed and antelope jackrabbits. The black-tailed was represented by 938 bones, or 55 percent of all animal bones excavated. The antelope jackrabbit was represented by 149 bones, or only 7 percent of the total. Greene and Mathews (1976) calculated that jackrabbits provided slightly more than 5 percent of the meat for the Snaketown inhabitants even though their bones collectively accounted for 62 percent of those excavated. Leopold (1959) estimated that the number of wild rabbits killed for food each year in Mexico probably numbers in the millions.

Sowls (1957) published details of reproduction in the desert cottontail in Arizona. Hinds (1973) investigated seasonal physiological responses to ambient temperatures. Stout (1970) reported on the breeding biology near Phoenix. In a study area north of Phoenix, Turkowski and Reynolds (1974) found the desert cottontail to consume forty-six different kinds of plants as well as arthropods.

CHIROPTERA

In the Sonoran Desert are found vespertilionid bats (Vespertilionidae: *Myotis, Pipistrellus, Eptesicus, Lasiurus, Dasypterus, Euderma, Plecotis, Pizonyx, Antrozous*), free-tailed bats (Mollosidae: *Tadarida, Eumops*), and leaf-nosed bats (Phyllostomidae: *Mormoops, Macrotis, Choeronycteris, Leptonycteris*). Regional treatments include those for Baja California (Huey 1964), Mexico (Villa-R. 1966), Arizona (Cockrum 1964), and America (Barbour and Davis 1969). As Hoffmeister (1970) reported for Arizona, species compositions change seasonally; only fifteen of the twenty-eight species of the state remain through the winter.

Desert insectivorous species seem to possess better renal adaptations for water conservation than do bats of more mesic areas (Geluso 1978). Ross (1961) studied prey specificity of insectivorous bats. Information on pollination by bats has been presented by Howell (1973).

Of special interest in the Sonoran Desert are associations of bats with cacti. The long-nosed bat pollinates the saguaro cactus (Cockrum and Hayward 1962; Howell 1973). Carpenter (1969) found that nectar comprised up to three-fourths of the diet of this species. Villa-R. (1966) observed *Choeronycteris mexicana* consuming fruits of pitahaya cactus (*Stenocereus*) and garambulla cactus (*Myrtillocactus*) and saw both *Leptonycteris* and *Choeronycteris* pollinating cactus flowers near Guaymas in the Central Coast Division. Cross and Huibregtse (1964) reported on the use of holes in the saguaro cactus as roosting sites for the big brown bat.

Perhaps the most unusual member of the normally insectivorous Vespertilionidae is the fish-eating bat (*Pizonyx vivesi*), limited to coasts and coastal islands of central Baja California and central Sonora. Burt (1932) first commented on the fish diet. Reeder and Norris (1954) described distribution and habits. Novick and Dale (1971) discussed foraging behavior. The bat catches fish with the sharp claws of the hind feet (Carpenter 1969).

INSECTIVORA

This order is represented in the Sonoran Desert only by the desert shrew (Soricidae: *Notiosorex*). Lindstedt (1976) treated its physiological ecology. The species is the smallest desert mammal. Its food habits were treated by Fisher (1941) and Hoffmeister and Goodpaster (1962). Armstrong and Jones (1972) summarized life history data for the species.

MARSUPIALIA

Only the opossum (Didelphidae: *Didelphis*) is found in the Sonoran Desert. The Mexican subspecies is native in the southern part of the desert (Hall and Kelson 1959,) and the Virginia subspecies has been weakly introduced in the northern part (Cockrum 1960, 1963).

Birds

The avian fauna of the Sonoran Desert includes 62 families, 231 genera, and 365 species. The numerous birds of the Gulf of California swell the total considerably. Birds of the Sonoran Desert are summarized in appendix 5-B. Bird guano has been mined from islands in the Gulf. Although Gulf birds are often important in the desert region immediately surrounding the Gulf, their occasional presence in more distant regions of the desert highlights the overpowering presence of the gulf in the center of the desert. For example, Vorhies and Phillips (1937) documented a virtual invasion of brown pelicans into Arizona in 1936. The magnificent frigatebird, common over coastal waters, has been sighted at scattered points inland in southern Arizona. The parasitic jaeger and pomarine jaeger have been collected rather far north along the lower Colorado River (Phillips et al. 1964). Disoriented red-billed tropicbirds have been found deep within the Arizona desert more than once,

(Vorhies 1934). Catastrophes or errors in navigation have brought large numbers of eared grebes down in Arizona as well. One case reported by newspapers involved large numbers found living and dead scattered around the desert near Tucson in 1956.

Birds are attuned to the desert by complex physiological means. Serventy (1971) treated the biology of desert birds. Bartholomew (1960) reviewed the physiology. Dawson and Bartholomew (1968) discussed temperature regulation and the water economy. Gordon (1934) discussed the drinking habits, Salt (1964) treated respiratory evaporation, and Schmidt-Nielsen et al. (1958) pointed out the significance of extrarenal salt secretion. Skadhauge (1972) dealt with salt and water excretion in xerophilic birds, and Bartholomew and Dawson reviewed respiratory water loss in birds of the Southwestern United States. Lasiewski et al. (1971) reported on cutaneous water loss in the poor-will. Cade (1964) treated water and salt balance in granivorous birds. Bartholomew (1971) discussed the water economy of seedeating birds that survive without drinking. Emery et al. (1972) treated production of concentrated urine by avian kidneys.

Desert birds in dry heat were discussed by Dawson and Schmidt-Nielsen (1964). The relation of metabolism to the development of temperature regulation in birds has been treated by Kendeigh (1939) and the reactions of fowls to hot temperatures by Yeates et al. (1941). Lasiewski and Dawson (1967) discussed the relation between standard metabolic rate and body weight in birds. The metabolic significance of differential absorption of radiant energy by black and white birds was pointed out by Heppner (1970).

Reichman et al. (1979) pointed out that birds frequenting the desert can often seasonally or even daily move out of it, or to oasis-like situations within it, to avoid stresses that other desert animals experience. Species diversity of desert birds correlates strongly with vegetational parameters, particularly with diversity of plant shape and life form (Tomoff 1974; Raitt and Maze 1968; Hensley 1954). The endangered nature of the southwestern riparian avifauna was discussed by Haight and Johnson (1977).

Hubbard (1973) discussed avian evolution in the arid lands of North America. Hensley (1954) discussed ecological relations of breeding bird populations in the Arizona desert biome. Van Rossem (1936) treated birds in relation to faunal areas of south-central Arizona. Russell et al. (1973) dealt with the population structure, foraging behavior, and daily movement of certain Sonoran Desert birds.

Short (1974) found that a predominance of nesting behavior among birds in southern Sonora took place during the June through September rainy season because the thorn forest vegetation is particularly lush at this time. Anderson and Anderson (1946) noted that large, pure stands of creosote bush in the northern part of the desert rarely attract large numbers of birds but that many birds do use the plants when they grow

intermixed with or at the edge of cactus, mesquite (*Prosopis*), or catclaw (*Acacia*) associations.

The birds of Arizona have been treated by Swarth (1914), Brandt (1951), Phillips and Monson (1963), and Phillips et al. (1964). Special studies of the birds of Saguaro National Monument were published by Swarth (1920), of the Tucson region by Vorhies et al. (1935), and of Papaguería by Sutton and Phillips (1942). Birds of California have been treated by Grinnell (1915), Willet (1933), Grinnell and Miller (1944), Miller (1951), and Brown et al. (1979).

Field guides to the birds of Mexico have been published by Edwards (1972) and by Peterson and Chalif (1973). Birds of Sonora were treated by Van Rossem (1931, 1945), Alden (1969), and Russell et al. (1972). Van Rossem (1932) published on the birds of Tiburón Island off the coast of Sonora. The avifauna of Baja California was treated by Grinnell (1928), Stager (1960), and Short and Crossin (1967).

At prehistoric Snaketown in the Lower Colorado-Gila Division, a number of birds were utilized for food by the Hohokam people from 300 B.C. to 1100 A.D. (McKusick 1976a). These included the Canada goose, white-fronted goose, snow goose, mallard, pintail, green-winged teal, lesser scaup, and ruddy duck. In addition, bones of red-winged blackbirds and hooded orioles were excavated in sufficient quantity to suggest that the Hohokam may have killed them for their feathers, if not their food value.

The roadrunners are desert birds par excellence. The greater roadrunner occurs through most of the desert, but the lesser roadrunner is found only in the Sonoran portion. Roadrunners subsist chiefly on a diet of arthropods and the diurnal whiptail lizard. As the birds grow older, they feed chiefly on whiptails (Bryant 1916; Ohmart 1973). Observations on breeding adaptations of roadrunners were made by Ohmart (1973). Calder (1968) reported on diurnal activity. Thermoregulation was treated by Lowe and Hinds (1969). Energy conservation by hypothermia and absorption of sunlight were detailed by Ohmart and Lasiewski (1971). Evidence that urine enters the rectum and ceca was presented by Ohmart et al. (1970). The urine concentration was compared to that of a carnivorous sea bird by Calder and Bentley (1967).

Predatory birds can live and reproduce in harsh desert conditions because they obtain enough moisture from their carnivorous diet (Bartholomew and Cade 1963; Dawson and Bartholomew 1968). Nasal salt secretion in falconiform birds has been treated by Cade and Greenwald (1966). Calder and Bentley (1967) compared urine concentration of pelicans with that of carnivorous birds characteristic of xeric portions of the desert.

Both the golden eagle and the bald eagle apparently were once more common in the Sonoran Desert than they are today. Howard (1947) presented evidence that the golden eagle has decreased in size since the Pleistocene. Glinski and Ohmart

(1977) reported on the habitat and diet of the black hawk. Whaley (1979) prepared a thesis on ecology of Harris's hawk in Arizona. The birds have to learn to avoid cholla cactus, and young ones have been found dead, accidentally impaled on the spines. Ligon (1968) reported on the biology of the elf owl. Bradshaw and Howard (1960) reported on mammal skulls recovered from owl pellets in Sonora.

Quail are ground birds requiring varying types of vegetative cover in upland situations. They suffer predation by a large variety of animals. Their eggs are relatively easy prey, which are eaten by gila monsters (Bogert and del Campo 1956). Water relations of southwestern quail were discussed by Vorhies (1928). Carey and Morton (1971) compared salt and water regulation in California quail and Gambel's quail. Lowe (1955) discussed Gambel's quail in relation to the water supply on Tiburón Island, off the coast of Sonora. Bartholomew and Dawson (1958) compared body temperatures in California quail and Gambel's quail, and Gorsuch (1934) treated the life history of Gambel's quail in Arizona. Beck, Engen, and Gelfand (1974) found an inverse relationship, accounted for by predatory dynamics, whereby Gambel's quail were crepuscular in activity at a Sonoran Desert water hole, whereas Raptorial species tended to come to the water hole at midday.

Tomlinson (1972) reviewed the literature on the endangered *Colinus virginianus ridgwayi* ("masked bobwhite"). Ornithologists discovered the race in 1884 at about the time of its decline due to drought and overgrazing. By 1900 the bird was extinct in Arizona. During the 1940s, with the increase of the Mexican cattle industry, it became threatened in its only other habitat, middle Sonora.

Since doves and pigeons are largely granivorous (Tomoff 1974), they must consume green vegetation or drink water (usually twice a day) to compensate for the dryness of the seed. Fluid is regurgitated to their young (Salt and Zeuthen 1960). Bartholomew and Dawson (1954) published on the body temperature and water requirements of mourning doves, and MacMillen (1962) reported on minimum water requirements of the species. Smyth and Bartholomew (1966) discussed effects of water deprivation and sodium chloride on the blood and urine of the mourning dove. MacMillen and Trost (1967) treated water economy and salt balance in white-winged and Inca doves. Willoughby (1966) discussed water requirements of the ground dove. Cade (1965) outlined relations between raptors and columbiforms at a desert waterhole. The daily movement of these birds between water source and xeric habitat sometimes occurs in large numbers and can be striking.

Hummingbirds have oil globules in the cones of the retina of the eye, which filter out colors other than red-orange-yellow. In the Sonoran Desert, hummingbirds live on nectar from red, orange, or yellow flowers, which have a tubular morphology conforming to the beaks of the birds. The flowers often require the hummingbirds for pollination. Hummingbird plants may be widespread in the desert (for example, *Fouquieria splendens,* the ocotillo) and characteristic of open areas with thin soil, or it may be narrowly endemic (for example, *Penstemon cedrosensis*) on open hillsides or cliff faces.

The important feature of hummingbird-pollinated plants is that they are characteristic of open, sunny areas. Visual perception in animals favors recognition of colors in the red end of the color spectrum ("hot colors") in open sunny areas and recognition of colors in the blue-violet band ("cold colors") in shady areas (Crosswhite and Crosswhite 1980). Flowers of Sonoran Desert hummingbird plants typically are thrust out into intense sunlight of the sky (for example, *Fouquieria splendens*) on spreading, relatively leaf-reduced stems, well within efficient flight paths of the hummingbirds (Crosswhite and Crosswhite 1980), whereas blue-flowered bee-pollinated plants tend to occur at lower light intensities, and the flowers are often closely appressed to a leafy background (for example, *Lycium*).

Introductions by man of red- and yellow-flowered plants into the Sonoran Desert such as tree tobacco (*Nicotiana glauca*) and several species of *Aloe* have resulted in nectar sources for hummingbirds being available through the year. As these plants become more common, increasing numbers of hummingbirds are overwintering, failing to migrate.

Hummingbirds obtain concentrated energy in the form of high-calorie nectar, which is quickly used in formation of ATP (adenosine triphosphate) to power the extremely high energy level of the birds. Protein comes from eating insects that inhabit flowers of the nectar plants. Lasiewski (1963) discussed oxygen consumption of torpid, resting, active, and flying hummingbirds. Body temperature, heart rate, breathing rate, and evaporative water loss in hummingbirds were teated by Lasiewski (1964). Hummingbirds and their flowers were subjects of a book by Grant and Grant (1968).

A number of Sonoran Desert birds are candidates for further study. The purple martin has a diminutive desert race common through the Sonoran Desert and associated with woodpecker holes in saguaro. Austin (1976) discussed behavioral adaptations of the verdin to the desert. Taylor (1971) reported on food habits. Information on the cactus wren has been published by Anderson and Anderson (1973) and Austin (1974). It nests in cholla cactus. The phainopepla feeds on berries of desert mistletoe (*Phoradendron californicum*). The loggerhead shrike may impale its prey on sharp thorns or spines of desert vegetation while consuming it. The starling is a European species that has invaded the Sonoran Desert in recent years by the thousands. Bell's vireo has been reduced in numbers and in distribution in the desert by cowbirds replacing vireo eggs with their own and letting the vireos raise the cowbird young. The red-eyed cowbird lays its eggs largely in nests of hooded orioles and summer tanagers, letting the "foster parents" raise the young. This has resulted in a reduction of the oriole and tanager populations.

Although the California condor (*Gymnogyps californicus*) was sighted by early explorers in the Sonoran Desert, there are no recent records. The presence of bones of this bird in caves in the Grand Canyon (Phillips et al. 1964) suggests that it may have once been common in Arizona.

Formerly the thick-billed parrot (*Rhynopsitta pachyrhyncha*) was found in Arizona in the wild state (Phillips et al. 1964). Thick-billed parrot bones have been excavated from prehistoric Snaketown in the Lower Colorado-Gila Division (McKusick 1976a), and the strata from which they came are dated to 200-350 A.D. by Haury (1976). Burials of parrots and macaws by prehistoric Indians in Arizona are known, suggesting that the birds were kept in captivity, perhaps for their colorful feathers (McKusick 1976).

Reptiles and Amphibians

In the Sonoran Desert there are 24 families of reptiles and amphibians in 65 genera, 146 species and 222 principal taxonomic units. These are summarized in appendix 5-C.

General treatments of reptiles and amphibians include those of Stebbins (1966) for western North America, Brown (1974) for the western United States, and Klauber (1939) for the arid Southwest. Cope's (1866) report on Reptilia and Batrachia of the Sonoran provinces was a cornerstone built upon by subsequent workers.

The Baja California peninsula and the islands off the shores of Baja and Sonora have intrigued herpetologists. A high degree of endemism exists in many of these insular faunas. Murphy (1975), in studying reptiles of Santa Catalina Island, noted that of ten reptiles all were endemic. This is one of the deep-water islands formed during rifting of Baja California from Sonora. The longer time of separation (as compared to more recent shallow-water coastal islands) is reflected in this high rate of endemism. Reptiles and amphibians of Baja have been treated by Murray (1955), Tevis (1944), Van Denburgh and Slevin (1921), Linsdale (1932), and Van Denburgh (1896). In addition to species of Baja, Schmidt (1922) considered those of neighboring islands. Soulè and Sloan (1966) discussed biogeography and distribution of reptiles and amphibians on Gulf coastal islands. Mosauer (1936b) considered reptiles of sand dune areas of the Vizcaíno Division and of northwestern Baja California. Bostic (1971) investigated herpetofauna of western Baja California in relatively poorly known regions of coastal desert, particularly in the Vizcaíno Division. The evolution of the Baja California herpetofauna was discussed by Savage (1960).

Herpetofauna of the State of Sonora has been treated by Bogert and Oliver (1945), Langebartel and Smith (1954), and Zweifel and Norris (1955). Smith and Smith (1973) produced a book analyzing the literature on Mexican herpetofauna.

Lowe (1963) provided an annotated checklist of the amphibians and reptiles of Arizona and delimited an eastern boundary of the Sonoran Desert in the United States with the aid of herpetological data (Lowe 1955). Gates (1957) did a special study of the herpetofauna of the Wickenburg area in central Arizona, whereas Ortenburger and Ortenburger (1926) observed reptiles and amphibians of Pima County. Kauffield (1943) and Coues (1875) studied reptiles and amphibians of the state in general. Fossil reptiles and amphibians from Arizona were treated by Brattstrom (1955).

California reptiles and amphibians were considered by Grinnell and Camp (1917). Reptiles of a sand dune area in the California portion of the Lower Colorado-Gila Division were described by Mosauer (1935). Brattstrom (1965) reported on Cenozoic amphibians and reptiles of California.

Effects of temperature and thermoregulation have been of special interest in study of poikilothermic reptiles and amphibians. Cloudsley-Thompson (1972) discussed temperature regulation in desert reptiles, as well as temperature and water relations of reptiles generally (1971). Desert reptiles in dry heat were investigated by Schmidt-Nielsen and Dawson (1964). Andreev (1948) studied the adaptations of reptiles to high temperatures of deserts. Thermoregulation in reptiles as a factor in evolution was reported on by Bogert (1949a). Bogert and Cowles (1944) studied thermal requirements of desert reptiles.

Reptilian thermal reponses (Grant 1939), thermal tolerance (Cowles 1939), and toleration of solar heat (Mosauer 1936a) have all been considered in the literature. Lowe et al. (1971) studied supercooling in reptiles and other vertebrates. The effect of solar radiation through the body wall of living vertebrates, especially desert reptiles, was discussed by Porter (1966). Norris (1967) interrelated color adaptation in desert reptiles with thermal relationships. Thermal acclimation in Anurans as a function of latitude and altitude was treated by Brattstrom (1968). Studies of water relations have treated cutaneous water loss in reptiles (Bentley and Schmidt-Nielsen 1966) and osmoregulation in amphibia (Heller, 1965). The biology of desert amphibians and reptiles, including many Sonoran Desert forms, was broadly treated by Mayhew (1968).

Adaptability of lizards with regard to temperature has been studied by Atstatt (1939) in relation to temperature and light induced color changes; by Cole (1943) in reference to heat toleration in respect to adaptive coloration; by Licht (1965) in reference to effects of temperature on heart rates of resting and active lizards; and again by Licht (1964) in reference to thermoregulation and thermal physiological adjustments. Tucker (1967) investigated the role of the cardiovascular system in thermoregulation and in oxygen transport. Soulè (1963) considered aspects of thermoregulation in lizards of Baja California. Lizard diet in relation to dentition was the object of a study by Hotton (1955).

LIZARDS

There are 9 families, 23 genera, 69 species, and 125 principal taxonomic units of lizards in the Sonoran Desert (appendix 5-C). General treatments of lizards can be found in the works by Smith and Taylor (1950) for Mexico and by Smith (1946) for the United States. Pianka (1965, 1967) discussed species diversity and ecology of flatland desert lizards in western North America. Asplund (1967) investigated the ecology of lizards in the cape region of Baja. Experimental determination of lizard home ranges was discussed by Jorgensen and Tanner (1963).

Lizards are abundant both in density and in species diversity in the Sonoran Desert. Arboreal, rock dwelling, detritus dwelling, digging, sand swimming, burrowing, insectivorous, carnivorous, herbivorous, diurnal, nocturnal: the range of habits and niches reflects the success of this group in radiating into the desert environment. Behavioral strategies are the primary means enabling them to retain relative stasis in body temperature under a desert regime. Pioneering studies of behavioral thermoregulation were done by Bogert (1949) at the Boyce Thompson Southwestern Arboretum.

Sonoran Desert lizards include gila monsters (Helodermatidae: *Heloderma*), banded geckos (Gekkonidae: *Coleonyx*), leaf-toed geckos (Gekkonidae: *Phyllodactylis*), barefoot geckos (Gekkonidae: *Anarbylus*), skinks (Scincidae: *Eumeces*), whiptails (Teiidae: *Cnemidophorus*), alligator lizards (Anguidae: *Gerhonotus*), legless lizards (Anniellidae: *Anniella*), worm snakes (Amphisbaenidae: *Bipes*), and 14 genera in the Iguanidae: false iguanas (*Ctenosaura*), desert iguanas (*Dipsosaurus*), chuckwallas (*Sauromalus*), earless lizards (*Holbrookia*), zebra-tailed lizards (*Callisaurus*), fringe-toed lizards (*Uma*), leopard or collared lizards (*Crotaphytus*), rock lizards (*Petrosaurus*), spiny lizards (*Sceloporus*), utas (*Uta*), brush and tree lizards (*Urosaurus*), horned lizards (*Phrynosoma*), island lizards (*Sator*), and night lizards (*Xantusia*).

Bogert and del Campo (1956) published a definitive monograph on the gila monster. Funk (1966) commented on the species at the western extremity of its range. Food habits were discussed by Pianka (1966), Bogert and del Campo (1956), Stahnke (1950, 1952), Shaw (1948), and Hensley (1949). The gila monster can climb bushes and small trees (personal observation) to search for food or to avoid flood. It seizes its prey in a firm bite and chews venom into wounds made by the teeth. The venom has been investigated by Styblova and Kornalik (1967) and by Loeb et al. (1913). Tyler (1956) reported on an autoantivenin in the gila monster as a natural autoantibody. Edwards and Dill (1935) looked at properties of gila monster blood.

Population dynamics of the banded gecko were studied by Parker (1972), social behavior by Greenberg (1943), and daily emergence by Evans (1967). Murphy (1974) described a new genus and species of gecko, Switak's barefoot gecko (*Anarbylus switaki*), near San Ignacio in Baja California.

The desert iguana runs swiftly over hot sands, lifting forelegs off the ground in a bipedal running stance. Nasal salt glands help maintain an electrolyte balance. *Dipsosaurus dorsalis* has a range closely corresponding to that of creosote bush (Norris 1953), being chiefly herbivorous (Minnich and Shoemaker 1970) and feeding especially on flowers of creosote bush (Norris 1953; Pianka 1971a), although the juveniles feed on insects. The species climbs bushes and branches to reach selected food. Physiological studies abound for this desert species, especially with regard to temperature tolerance and thermoregulation (Dawson and Bartholomew 1958; Weathers 1971; Templeton, 1960; DeWitt 1963, 1967a 1967b; McGinnis and Dickson 1967a, 1967b) or to water and electrolyte balance (Philpott and Templeton, 1964; Minnich 1970a, 1970b; Templeton et al. 1972; Chan et al. 1970; Minnich and Shoemaker 1970). Other aspects appearing in publications include treatments of ecology (Norris 1953), feeding habits (Cowles 1946), hibernation (Moberly 1963), growth response to photoperiodic stimulation (Rath 1966; Mayhew 1965c), social behavior (Carpenter 1961), and reproduction (Mayhew 1971; Lisk 1967).

Food habits of the chuckwalla were reported by Hansen (1974), Nagy (1973), Shaw (1945), and Mayhew (1963c). Respiration, including panting and pulmonary inflation, was studied by Templeton (1964, 1967), Boyer (1967), and Schmidt-Nielsen et al. (1966). Water and electrolyte budgets were investigated by Nagy (1972) and Norris and Dawson (1964). Ecology (Berry 1974; Johnson 1965), energy and nitrogen budgets (Nagy and Shoemaker 1975), and blood properties (Dill et al. 1935) have also been treated. Soulè and Sloan (1966) noted the occurrence of deep-water island forms of chuckwallas on shallow-water islands near the peninsular coast and thought their occurrence there most likely to be due to human agency, since chuckwallas were a common food of the Indians of the Viscaíno division (Aschmann 1967).

Pianka (1970) reported on food habits of southwestern earless lizards, whereas Pianka and Parker (1972) reported on those of zebra-tailed lizards. Heifetz (1941) reviewed the genus *Uma*. Stebbins reported on ecology (1944) and adaptations of the nasal passage for sand burrowing (1943). Norris (1958) considered evolution and systematics. Various aspects of reproduction were studied by Mayhew (1960, 1961, 1965a). Ecological studies of the collared lizard have been made by Fitch (1956), with social behavior observed by Greenberg (1945). Respiration was studied by Templeton and Dawson (1963). The collared lizard's physiological responses to temperature were reported by Dawson and Templeton (1963). The genus *Crotaphytus* was treated by Robison and Tanner (1962). Mackay (1975) studied home range behavior in *Petrosaurus*. The ecology of *Sceloporus* was studied by

Parker and Pianka (1973). Mayhew studied the biology (1963a), reproduction (1963b), and temperature preferences (1926b) of the granite spiny lizard.

The majority of Sonoran Desert *Uta* species are confined to desert islands in the Gulf. Soulè (1964) studied evolution and population genetics of these island Utas. Tinkel (1967) reported on the life history and demography of *Uta stansburiana.* Energy requirements were covered by Alexander and Whitford (1968). Various aspects of reproduction have been treated by Cuellar (1966), Hahn and Tinkle (1965), Hahn (1967), Ferguson (1966), and Christiansen (1965).

Reproductive cycles in *Urosaurus ornatus* were investigated by Asplund and Lowe (1964). Vance (1953) reported on respiratory metabolism and temperature acclimation. Seasonal variation in diet was studied by Asplund (1964). Dominance shifts were studied by Carpenter and Grubits (1960).

Taxonomy and distribution of *Phrynosoma* have been discussed by Reeve (1952). Pianka and Parker (1975) reviewed the ecology. Lynn (1963) investigated comparative behavior, while aggressive behavior was reported by Lowe and Woodin (1954) and display behavior by Lynn (1965). Blount (1929) and Smith (1941) treated reproduction. Color changes, including thermoregulatory and protective aspects, were studied by Redfield (1918), Parker (1938), Norris and Lowe (1964), and Norris (1967). Various aspects of temperature control were discussed by Heath (1962a, 1962b, 1964, 1965). Weese (1917) investigated urine, and Whitford and Whitford (1973) investigated combat. Mexican *Phrynosoma* were treated by Terron (1932), and Bryant (1911) reported on those of California.

Since *Phrynosoma* eats ants, the feces usually consist of little more than remains of the hard parts of ants (Norris 1949; Reeve 1952). Although *Phrynosoma* may be eaten by such snakes as *Arizona elegans* or *Crotalus,* the *Phrynosoma* horns function so as to puncture the intestines of the snakes, resulting in death of the latter (Miller and Stebbins 1964; Klauber 1956). *Phrynosoma* also twists when captured, presenting sharp horns to the animal that has captured it. Defensive squirting of blood from the corner of the eye by *Phrynosoma* has been investigated by Burleson (1942), Hay (1892), Cutter and Heath (1966). Blood on a predator but not on the horned lizard may confuse the predator. Horned lizards also gulp air to puff themselves up, or they flatten out and feign death.

Studies on the reproduction of *Xantusia* have been conducted by Bartholomew (1950, 1953), Cowles and Burleson (1945), and Miller (1948, 1955). Life history has been studied by Miller (1951) and population ecology by Zweifel and Lowe (1966). Felger (1965) described the habitat of *X. vigilis* in Sonora. Lowe (1966) investigated territorial behavior. Cannibalism in captive *X. vigilis* was reported by Heimlich and Heimlich (1947).

Burt (1931) wrote a monograph on *Cnemidophorus.* Asplund (1970) considered metabolic scope of body temperatures within the genus. Interspecific ecological relationships were investigated by Echternacht (1967). Reproduction in some *Cnemidophorus* is by parthenogenesis, males being unknown. Clonal populations thus can arise with no interbreeding possible, perpetuating distinct groups not easily separable on morphologic grounds. Parthenogenesis in the genus has been treated by Parker (1979a, 1979b), Wright (1967), Taylor et al. (1967), Lowe and Wright (1966), Zweifel (1965), Maslin (1962), and Capocaccia (1962). Goldberg and Lowe (1966) studied the reproductive cycle of *C. tigris.* Benes (1966) investigated progressive color discrimination. Food habits of *Cnemidophorus* have been investigated by Pianka (1970), Bostic (1966), and Milstead (1965).

Smith and Smith (1977) treated the anomalous two-legged worm lizard, which resembles an earthworm. Although it is found in southern Baja California , it may extend to other areas as well. Its secretive habits and lack of collecting do not allow the range and habits to be fully understood.

SNAKES

The snake fauna of the Sonoran Desert includes 6 families, 26 genera, 53 species, and 97 principal taxonomic units (appendix 5-C). Included here are the worm snakes (Leptotyphlopidae: *Leptotyphlops*), boas (Boidae: *Lichanura, Constrictor*), coral snakes (Elapidae: *Micruroides*), sea snakes (Hydrophidae: *Pelamis*), rattlesnakes (crotalidae: *Crotalus*), and 18 genera in the Colubridae: regal king snakes (*Diadophis*), leaf-nosed snakes (*Phyllorhynchus*), racers and whipsnakes (*Masticophis*), patch-nosed snakes (*Salvadora*), rat snakes (*Elaphe*), glossy snakes (*Arizona*), king snakes (*Lampropeltis*), long-nosed snakes (*Rhinocheilus*), water snakes (*Thamnophis*), ground snakes (*Sonora*), shovel-nosed snakes (*Chionactis*), sand snakes (*Chilomeniscus*), hook-nosed snakes (*Ficimia*), black-headed snakes (*Tantilla*), vine snakes (*Oxybelis*), lyre snakes (*Trimorphodon*), night snakes (*Hypsiglena*), and eridiphas (*Eridiphas*).

General works that include treatments of Sonoran Desert snakes are those by Wright and Wright (1957) and by Smith and Taylor (1945). Fowlie (1965) published a volume on the snakes of Arizona. Klauber's (1956) two-volume treatment of rattlesnakes is classic. He also treated Leptotyphlopidae (Klauber 1940) and shovel-nosed snakes (Klauber 1951). Norris and Kavanau (1966), as well as Warren (1953), also discussed shovel-nosed Snakes; and Murphy (1957) reported on Leptotyphlopidae. Tanner (1966) treated the night snakes of Baja California.

In poisonous snakes, the injection of a foreign substance to subdue prey is a strategy that attracts human attention, often of necessity where activities of man and dangerously venomous species widely coexist. Zeller (1948) studied enzymes of snake venoms and reported on the venoms and their biological significance. Russell (1960) reported on snake venom poisoning in southern California. Lethal and hemorrhagic properties of North American snake venoms were discussed by Emery and Russell (1963). Arnold (1973) produced a book on treatment of bites and stings of venomous animals, including snakes. Minton (1957) reported on venoms in general and also on venoms of desert animals (1968).

Among Sonoran Desert snakes, venomous species are in the minority. Among Colubrids, only four species are poisonous— the vine snake, two species of lyre snake, and the night snake— and these are not dangerous to man. Other families of snakes with venomous members in the Sonoran Desert are the cobra family (Elapidae), sea snake family (Hydrophidae), and pit viper family (Crotalidae).

Treatment for rattlesnake bite has been varied and often controversial. Arnold (1973) published a guide to first-aid for rattlesnake bite, with references to technical articles. Work by Russell (1962, 1967a, 1967b) has also dealt with first aid and emergency treatment of snakebite.

TURTLES

The turtle fauna, including marine species, encompasses 4 families, 10 genera, and 11 species (appendix 5-C). Included here are mud turtles (Chelydridae: *Kinosternon*), sliders (Testudinidae: *Pseudemys*), desert tortoises (Testudinidae: *Gopherus*), green turtles (Cheloniidae: *Chelonia*), loggerheads (Cheloniidae: *Caretta*), ridleys (Cheloniidae: *Lepidochelys*), hawksbills (Cheloniidae: *Eretmochelys*), leatherbacks (Dermochelidae: *Dermochelys*,) and softshells (Trionychidae: *Trionyx*).

Definitive treatments of the turtles and tortoises of the United States and Baja California can be found in the *Handbook of Turtles* by Carr (1952). Two species of mud turtles, the yellow and Sonora, range into the Sonoran Desert. Southern Baja California is the range of a subspecies of the pond slider. The yellow box turtle primarily is a species of open plains, ranging into central Sonora in the Sonoran Plains Division. An apparently recent (around 1900 A.D.) introduction to the Sonoran Desert, the Texas softshell now inhabits the Gila and Lower Colorado River drainages.

The desert tortoise has been extensively studied, including life history and habits (Nichols 1953; Miller 1932, 1955; Woodbury and Hardy 1948; Grant 1960); food habits (Nichols 1953; Burge and Bradley 1976; Hansen et al. 1976); growth (Patterson and Brattstrom 1972; Jackson et al. 1976, 1977; Bogert 1937); reproduction (Householder 1950; Booth 1958; Lee 1963; Stuart 1954); physiology (Schmidt-Nielsen and Bentley 1966; Dantzler and Schmidt-Nielsen 1966); behavior in captivity (Nichols 1957); and in regard to livestock grazing (Berry 1978). Recent investigations have been encouraged by efforts of the Desert Tortoise Council.

Aschmann (1967) reported that Indians of the Vizcaíno Division regularly caught sea turtles with harpoons and tridents from their tule balsas off both coasts of the peninsula. The Seri of coastal Sonora harvested them by diving and were estimated to use sea turtle for 25 percent of their diet by McGee (cf. Aschmann 1967) when he visited them in 1895 and 1896.

The Pacific green turtle is found in the Gulf as well as off the Pacific coast of Baja California in water of moderate depth with an abundance of marine food plants, seldom coming on shore in the Sonoran Desert, the nesting beaches lying farther to the south in Michoacán. From original use by Seri and other indigenous people for food, for shell vessels (Aschmann 1967), or for medicine (Felger and Moser 1974), the green turtle suddenly became a commercial meat animal harvested in large quantities for sale in urban markets of Mexico and the United States. Felger et al. (1976) described harvesting of winter-dormant turtles by Mexican fishermen, a rediscovered technique very detrimental to existing gulf turtle populations. During the 1970's intensive harvesting of all sizes of turtles by increasingly efficient methods, along with discovery and subsequent exploitation of nesting beaches for both eggs and animals, have decimated green turtle populations (Cliffton et al. 1980). Now an endangered species, green turtles are still threatened by poaching and by depredations of man, pigs, and dogs in nesting areas.

Other sea turtles that are used for meat, leather, and oil are the Pacific loggerhead, the olive ridley, and the Pacific hawksbill, which is also the source of commercial tortoise-shell or carey. All three are rapidly declining in numbers because of excessive commercial demand (Cliffton et al. 1980).

The largest and least known of the sea turtles is the Pacific leatherback, a pelagic animal occurring as a visitor along coasts of the Gulf and Pacific. Although not used for food, it is sometimes taken for oil. Recently discovered (1976) nesting beaches in Guerrero have been subjected to extensive egg harvesting (Cliffton et al. 1980).

AMPHIBIANS

The amphibian fauna includes 5 families, 6 genera, 14 species, and 15 principal taxonomic units (appendix 5-C). Present are true frogs (Ranidae: *Rana*) burrowing treefrogs (Hylidae: *Pternohyla*), treefrogs (Hylidae: *Hyla*), toads (Bufonidae: *Bufo*), narrow-mouthed toads (Microhylidae: *Gastrophryne*), and spadefoot toads (Pelobatidae: *Scaphiopus*). Adaptations of amphibians to arid environments have been studied by Bentley (1966). Nitrogen excretion in arid-adapted or dehydrated amphibians has been investigated by McClanahan (1975) and Balinsky et al. (1961). McClanahan (1964) also studied osmotic tolerances of muscles of desert dwelling toads. A volume on the frogs and toads of the United States and Canada (Wright and Wright 1949) gives detailed information for species of the northern Sonoran Desert. Smith and Smith (1976) compiled a book on Mexican amphibians. Amphibia of California were treated by Storer (1925). Low (1976) discussed the evolution of desert amphibian life histories. Adaptations of Aneurans to desert scrub were treated by Blair (1976).

Lowe (1964) noted the presence of marked geographic variation in local populations of leopard frogs in Arizona. Collins and Lewis (1947) discussed overwintering of tadpoles and breeding season variation in the *Rana pipiens* complex in Arizona. Jameson et al. (1966) published on the systematics and distribution of the pacific treefrog.

Skin secretions of *Bufo alvarius* contain a white toxic substance that can paralyze or even kill animals that try to eat them (Stebbins 1966). Musgrave and Cochran (1929) commented on the poisonous nature. The mating call and its significance was discussed by Blair and Pettus (1954), while Cole (1962) published notes on distribution and food habits. McClanahan and Baldwin (1969) studied the rate of water uptake through the integument of *Bufo punctatus*. Tevis (1966) investigated unsuccessful breeding at this toad's limit of ecological tolerance.

Mayhew (1965b) and McClanahan (1967) have studied adaptations of *Scaphiopus couchii* to desert environments. Its life history was investigated by Strecker 1908), and its distribution in the California part of the Lower Colorado-Gila Division was treated by Mayhew (1962a). Shoemaker et al. (1969) analyzed seasonal changes in body fluids in a field population. The western spadefoot was studied by Ruibal et al. (1969). Temperature adaptation and evolutionary divergence within the species was investigated by Brown (1966, 1967b). Jasinski and Gorbman (1967) studied hypothalmic neurosecretion under different environmental conditions in *S. hammondi*. High temperature tolerance of the eggs was determined by Brown (1967a).

Fishes

Aside from the 21 families, 54 genera, and 83 species of fishes known from fresh waters of the Sonoran Desert (appendix 5-D), more than 800 species occur in the Gulf of California. Desert fishes were treated by Deacon and Minckley (1974). Somero (1975) treated enzymic mechanisms of eurythermality in desert and estuarine fishes. Heath (1967) studied the ecological significance of temperature tolerance in Gulf of California shore fishes. Lowe et al. (1967) discussed experimental catastrophic selection and tolerances to low oxygen concentration in native Arizona freshwater fishes. Hubbs (1959) discussed vertebral deformities in fishes inhabiting warm springs. Bersell (1973) studied vertical distribution of

fishes relative to physical, chemical, and biological features in two central Arizona reservoirs.

Minckley (1965) discussed native fishes as natural resources in arid lands of the Southwest. The fishery of the Colorado River was treated by Dill (1944) and the effect of channelization on it by Beland (1953). John (1964) discussed the survival of fish in intermittent streams in southern Arizona. Schaut (1939) treated the general subject of fish catastrophes during drought.

The relation of man to the changing fish fauna of the American Southwest was discussed by Miller (1961). Endangerment of desert fishes was discussed by Deacon (1968) and Minckley and Deacon (1968). Johnson and Lew (1970) reported on chlorinated hydrocarbon pesticides recovered from fishes of southern Arizona. James (1968) discussed copepod infection of fishes in the Salt River Basin of southern Arizona.

Miller (1959) discussed origin and affinities of the freshwater fish fauna of western North America. Follett (1960) treated origin and affinities of the freshwater fishes of Baja California. Miller (1949) pointed out that desert fishes could represent clues to vanished lakes and streams. Fish remains from an archaeological site along the Verde River in central Arizona were treated by Minckley (1968), as were those from Snaketown, a prehistoric site in the Lower Colorado-Gila Division (Minckley 1976; Olsen 1976).

Freshwater fishes of Mexico were treated by Meek (1904) and Beltrán (1934). Those of Baja California were discussed by Follett (1960) and those of Sonora by Branson et al. (1960). Fishes of California were treated by Shapovalov et al. (1959) and Kimsey and Fisk (1960). Fishes of the Salton Sea were enumerated by Walker (1961) and the bait fishes of the lower Colorado River by Miller (1952). The fishes of Arizona have been listed by Miller and Lowe (1964) and Minckley (1973). Studies of fishes of the Arizona Upland Division include those by Barber and Minckley (1966) for Aravaipa Creek and Stout et al. (1970) for Cave Creek.

FISHES OF THE GULF

Oceanography of the Gulf of California has been detailed by Roden (1964), Parker (1964), Brusca (1973), Thomson et al. (1969), and Hendrickson (1974). The 800 species of fish that occur in the Gulf (Thomson et al. 1979) can be classified into five distinct communities: pelagic fishes of inshore and offshore surface waters, true deep sea fishes, fishes of offshore soft-bottomed shelves, fishes of sandy shores and estuaries, and reef fishes.

Thomson et al. (1979) have published a book on the ecology and taxonomy of reef fishes of the Gulf of California. A total of 271 species in 39 families inhabit the rocky shores of the gulf. Chief among these are sea basses (Serranidae), gobies (Gobiidae), clinids (Clinidae), morays (Muraenidae), grunts (Haenulidae), damselfishes (Pomacentridae), wrasses

(Labridae), tube blennies (Chaenopsidae), and clingfishes (Gobiesocidae).

Gulf fishes not ordinarily found on reefs can be identified using the *Gulf of California Fishwatcher's Guide* (Thomson and McKibbin 1976). This is a very diverse group of over five hundred species with quite variable characteristics. Included are sharks, stingrays, marine catfishes, totoabas, sailfish, marlins, sea bass, yellowtails, swordfish, amberjacks, sheepheads, skipjacks, tunas, pompanos, barracudas, mullets, halibuts, turbots, corvinas, flounders, soles, mackerels, herrings, anchovies, snappers, grunts, groupers, grunions, and many more.

Sharks are frequent in the Gulf. Tiburón Island is translated as "Shark Island." Of thirty-eight species of sharks known from the Gulf, about half are common enough or have habits that result in their being harvested by fishermen using nets, hooks, or harpoons. The dorsal fins and lower lobe of the tail of the sharks are dried and sold for making shark-fin soup. The shark meat is made into fillets and dried for three to five days on bushes or on racks made of ocotillo branches. When salted and dried, the meat is sold throughout Mexico as dried codfish (Harris 1972).

The jaws of the shark are dried and sold to tourists as novelties. Skin of many of the sharks is made into leather. Before the advent of synthetic vitamins, the shark livers were saved for making shark-liver-oil. The shark fishermen often cut down and dry the shovelnose guitarfish (*Rhonobatos productus*), trimming the head and pectoral fins to make the creature look like a human figure—a "Diablo del Mar" sold to tourists as a novelty, a twisted caricature.

The totoaba (*Cynoscion macdonaldi*) is a benthic sandy-bottom fish of the gulf, which reaches six feet in length and a weight of two hundred pounds. It has the distinction of being the only endangered marine fish in the world, with the possible exception of the coelacanth (Hendrickson 1979). Millions once migrated to near the mouth of the Colorado each spring to spawn. They trapped schools of smaller fish against the shore and fed on them. Indigenous people of the desert utilized the totoaba as food and commercial fishing by non-Indians finally developed to the point where 2,000 tons were harvested in a single year. Since construction of dams on the Colorado River, the waters at the head of the Gulf have become saltier and the totoaba faces possible extinction (Hendrickson 1979).

NATIVE FISHES OF FRESHWATER

The native freshwater fishes of the Sonoran Desert belong to only five families: the pupfishes (Cyprinodontidae), minnows (Cyprinidae), suckers (Catostomidae), catfishes (Ictaluridae), and livebearers (Poeciliidae). All others are introduced.

The desert pupfish (*Cyprinodon macularius*) is an excellent example of a freshwater species with salt tolerance sufficient for it to move across the Gulf of California from one freshwater basin to another. The species has been known to occupy various mineralized springs and lakes, including the Salton Sea. It is the only freshwater species dispersed in Sonora, Arizona, California, and Baja California portions of the desert.

Salinity requirements of *Cyprinodon macularius* were treated by Kinne (1965). Barlow (1958) studied daily movement of desert pupfish in shore pools of the Salton Sea and reported high salinity mortality. Sweet and Kinne (1964) discussed effects of various temperature-salinity combinations on the body form of newly hatched desert pupfish. Lowe and Heath (1969) discussed behavioral and physiological responses to temperature. Cox (1966) reported on a behavioral and ecological study of the species in Quitobaquito Spring. Cowles (1934) treated the ecology and breeding behavior. Barlow (1961) reported on the social behavior. Lowe and Heath (1969) reported a heat tolerance of 44.6°C in the desert pupfish, apparently the highest yet recorded for fishes. Younger fish selected warmer microenvironments than did older ones. They displayed successful behavioral thermoregulation when frequenting water near the lethal temperature limit. Cox (1972) studied the food habits.

The genus *Gila* in the lower Colorado River basin was discussed by Rinne (1969). The bonytail (*Gila elegans*) once was present in the Sonoran Desert sections of the Colorado, Gila, and Salt rivers. The small eyes and reduced or embedded scales were suggested by Minckley (1973) to be adaptations to the high silt loads, "which characterized the remarkably-erosive, turbid Colorado River system prior to construction of dams." The verde trout (*G. robusta*) is found at riffles of streams in the northern part of the Sonoran Desert, feeding like true trout, and caught by means of dry flies. Native chubs of the Sonoran Desert are geographically replacing units, including the Gila chub (*G. intermedia*), the Sonoran chub (*G. ditaenia*), and the Yaqui chub (G. purpurea). Miller (1949) discussed *G. ditaenia*.

The spikedace is endemic to the Gila River sytem. Its biology was treated by Barber et al. (1970). The woundfin is adapted to live in the swift parts of silty streams and was once found in the Gila and lower Colorado rivers. The longfin dace is common in the Yaqui, Magdalena, Sonoyta, Gila, and Bill Williams River systems. Minckley and Barber (1971) treated the biology of the longfin dace in streams of the Sonoran Desert. The speckled dace is common at elevations above the Sonoran Desert, infrequent downstream in the desert. John (1963) discussed the effect of torrential rains on the reproductive cycle of the species. The loach minnow is endemic to the Gila River basin. The Sonoran shiner occurs in Mexican tributaries to the Gulf and the Mexican stoneroller in the Rio Yaqui drainage.

The Colorado River squawfish may now be extinct within the boundaries of the desert due to interference of spawning runs by dams (Minckley 1973). This large fish probably reached a length of six feet and nearly one hundred pounds (Minckley 1973). It was "top carnivore" of the Colorado and

Gila rivers and was used for food by prehistoric Indians (Olsen 1976) and early settlers. The species was called "salmon" in Arizona and occurred in such numbers that it was used not only for human food but as fertilizer and for feeding hogs. Branson (1966) discussed the species in the Salt River in the Arizona Upland division, and Vanicek and Kramer (1969) treated the life history.

The razorback sucker is a very large Catostomid, which has frequently been reported as food of prehistoric Indians. The conspicuous hump on the back apparently functions to orient the fish to an optimum feeding angle in swift water. The fish is present in the Colorado River and was once common in the river's Sonoran Desert tributaries. Douglas (1952) discussed spawning of the species.

Hubbs and Miller (1953) reported on hybridization between *Catostomus* and *Xyrauchen*. *Pantosteus* also hybridizes with *Catostomus* in the Sonoran Desert (personal observation). Smith (1966) treated distribution and evolution of North American *Pantosteus*, treating the taxon as a subgenus of *Catostomus*. Amin (1969) reported on helminths associated with suckers of the Gila River sytems, and Koehn (1967) reported on blood proteins of Catostomids.

The only instance of catfish occurring natively on the Pacific coast of North America is in the drainage of the Rio Yaqui where the Yaqui catfish occurred. This distribution undoubtedly resulted from the drainage pattern whereby the Yaqui drains an area east of the Continental Divide.

Native topminnows in the Sonoran Desert are represented by the Gila topminnow and the Yaqui topminnow. Moore et al. (1970) discussed an all-female form of *Poeciliopsis* in northwestern Mexico. Livebearers among the native fishes of Arizona were treated by Minckley (1969). Attempted reestablishment of the Gila topminnow was discussed by Minckley (1969). Heath (1962) discussed maximum temperature tolerance as a function of acclimation in the species. Gila topminnows are considered an endangered species because they can not tolerate competition from the introduced *Gambusia affinis*. Relations of Gila topminows with *Gambusia* were discussed by Minckley et al. (1977). Ayer Lake at the Boyce Thompson Southwestern Arboretum is maintained as a sanctuary for the Gila topminnow in cooperation with the Arizona Game and Fish Department.

INTRODUCED FISHES

Man has introduced fish species to Sonoran Desert waters for decades. Early introductions were of food fishes such as carp, buffalo, and catfish, followed by sport fishes such as bass, crappie, bluegill, and walleye. Bait fishes such as the golden shiner, red shiner, and flathead minnow have escaped or were intentionally released. Aquarium introductions include the variable platyfish, green swordtail, guppy, sailfin molly, and convict chichlid. The threadfin shad was introduced as a forage

fish for larger game fish to eat. Mosquitofish have been widely introduced to control mosquito larvae. Introduced species, together with reason introduced and date of establishment, are summarized in appendix 5-D.

Fish introductions result in unpredictable consequences. The effects on other fishes are subtle but numerous. The threadfin shad introduction was dramatic as the species underwent a population explosion (Haskell 1959; Beers and McConnell 1966; Johnson 1970). Striped bass, white bass, and yellow bass were introduced to recycle shad biomass into trophy-size game fish. But in some situations the shad have now all been eaten, and a food source for the bass is lacking.

Gambusia affinis introduction inevitably results in extinction of the Gila topminnow. Krumholz (1948) discussed reproduction in *Gambusia* and use of the fish in mosquito control. Winkler (1973) studied the ecology and thermal physiology of the species in a hot spring in southern Arizona. Otto (1973) compared a population of *Gambusia affinis* from the Sonoran Desert with one from northern Utah with regard to critical thermal maxima.

Tilapia have recently been promoted by the Food and Agricultural Organization of the United Nations as a protein source. The objects of numerous fish-farming experiments, three species have been introduced in the Sonoran Desert. They are prolific and tend to become stunted in smaller bodies of water. As a result, sterile forms are sometimes introduced to avoid overpopulation. *Tilapia* are mouthbrooders from Africa, holding the eggs in the mouth until they hatch. The very young fish return to the mouth of the mother at night or when under attack by a predator. McConnell (1966) discussed *Tilapia* as sportfish and Hoover and Amant (1970) described their establishment in Imperial County, California.

Invertebrates

The importance of invertebrates in the desert is underscored by some of the roles they play in food chains as detritovores, as pollinators, as sources of moisture to larger animals, and as economically important pests and parasites. They are also sources of basic information on metaboic and physiological processes. Knowledge of invertebrate fauna lags behind that of vertebrate animals because of the number of organisms involved. One investigator recorded 1160 species of invertebrates in 245 families and 796 genera in Deep Canyon at the western edge of the Sonoran Desert (Edney 1974).

Reichman et al. (1979) reviewed food selection and consumption in desert arthropods. MacMahon (1979) briefly considered some arthropods of North American deserts. *Larrea* as a habitat component for desert invertebrates was reported on by Shultze (1975). Chemical defenses of arthropods were described by Roth and Eisner (1962) and Eisner and Meinwald (1966). Wallwork (1972) studied the distribution patterns and population dynamics of desert soil microarthropods in southern California. Edney (1974) wrote a review article on desert arthropods and also wrote on their water relations (1967). Adaptations of arthropods to arid environments were discussed by Cloudsley-Thompson (1975), who also presented data on thermoregulation in terrestrial invertebrates (1970), on arthropods in dry heat (1964b), and on lethal temperatures and the mechanism of heat death in desert arthropods (1962; Cloudsley-Thompson and Crawford 1970). Hadley (1970a) published on micrometeorology and energy exchange in desert arthropods. Shaw and Stobbart (1972) studied water balance and osmoregulatory physiology of desert and xeric arthropods.

INSECTS

General treatments of insects include those of Essig (1958) for western North America, Borror and White (1970), and Truxal (1960) for Baja California. Crawford (1979) discussed desert Collembola, Thysanurans, crickets, cockroaches, beetles, and social insects. Reichman et al. (1979) considered the feeding habits of various desert insects. Thermoregulation and flight energetics of desert insects were reviewed by Heinrich (1975). Beard (1963) published on insect toxins and venoms. The utilization of metabolic water by insects was studied by Fraenkel and Blewett (1944). The effect of coloration in thermoregulation for diurnal desert insects was discussed by Hamilton (1975). Chewing insects found with *Larrea* were investigated by Werner and Olsen (1973). Hurd and Linsley (1975a, 1975b) reported on insects associated with creosote bush. Insects associated with mesquite were treated by Riazance and Whitford (1974). Aschmann (1967) listed insects known as food for Sonoran Desert peoples.

Hymenopterans (bees, wasps and ants) are prominent throughout the desert. For example, carpenter bees pollinate flowers of desert willow and bore in stalks of century plant and sotol; bumblebees pollinate flowers of *Lycium;* leaf-cutter bees pollinate *Penstemon* and trim neat circles of tender leaves to line nests. Many solitary and semisocial bees are present in the desert, as indicated from data of Hurd and Linsley (1975a) recording eighty-four species of bees in eight families found in association with creosote bush (*Larrea*). Wasps that evoke comment by visitors are the large orange and black tarantula hawks (*Pepsis*) and the wingless female velvet ants (*Mutillidae*).

Hunt and Snelling (1975) published a checklist of the ants of Arizona. The ants of Deep Canyon were treated by Wheeler and Wheeler (1973). Ant venoms, attractants, and repellants were reported on by Cavill and Robertson (1965). Ecology of foraging ants was investigated by Carroll and Janzen (1973). Wheeler (1907) described colonies of fungus-growing ants in a wash near the Carnegie Desert Laboratory defoliating an *Acacia greggii* (catclaw). Such ants are thought to assist plants in seed dispersal and possibly to have an effect on plant distributions. Tevis (1958) studied seed utilization by *Vero-*

messor pergandei (harvester ant) in the Coachella Valley of California. Overall numbers of seeds taken by the ants from an acre in twelve months was not major in relation to total number of seeds produced. However, composition of the vegetation could be influenced by selective foraging by the ants. Seeds from only three species of plants made up 90 percent of the seeds collected but less than 8 percent of total seeds produced in the area.

Werner (1973) reported the leaf-cutter ant (*Acromyrmex versicolor*) to gather leaves of creosote bush. Reichman et al. (1979) thought that mesquite (*Prosopis*) was probably the most important leaf material for this ant because the nests are usually near mesquites. It collects the new leaves of mesquite in spring and later the flowers. Also used are leaves and petals of jimmyweed (*Aplopappus*), catclaw acacia (*A. greggii*), blue palo verde (*Cercidium floridum*), devil's claw (*Proboscidea*), zinnia, *Allionia, Tidestromia, Euphorbia, Opuntia*, grasses, as well as feces of birds, grasshoppers and moths. Many of the leaves and petal pieces are eaten, but others are combined with feces and dirt as a base for cultivation of fungi.

Flies are well represented among the desert insect fauna. Cole (1969) authored a book on the flies of western North America that treats Sonoran Desert species in part. Some interesting groups are characteristic of desert regions. Larvae of the syrphid genus *Volucella*, rat-tail maggots, inhabit the soupy mass of decaying tissues within dead cacti. The families Apioceridae and Nemestrinidae are poorly known large flies occurring mainly in arid regions, including the Sonoran Desert. Beeflies (Bombyllidae) are fuzzy-bodied small- to medium-sized flies, found frequently in hot, arid areas. Cole (1969) considered many southwestern United States species to be Sonoran in distribution or allied to species that are dominant on the plains of northern Mexico. Some flies (Syrphidae, Asilidae) mimic venemous bees and wasps. A southwestern mydas-fly (*Mydidae*) is a mimic of the large red-winged tarantula Hawk (a wasp, *Pepsis*). An unusual fly of the family Neriidae, *Odontoloxozus longicornis* (cactus fly), is associated with decaying cacti. Some species of *Drosophila* also utilize the mesic, nutrient-rich environment represented by rotting succulent cactus tissue.

Moths and butterflies (*Lepidoptera*) are conspicuous members of nocturnal and diurnal insect faunas during warm months. Klots (1951) generally treated many of the butterflies that extend into the northern Sonoran Desert. The pipe-vine swallowtail can be seen in areas where the larval food-plant, *Aristolochia watsoni*, is abundant. Snout-nosed butterflies are also locally common, for larvae feed on leaves of widely distributed desert hackberry (*Celtis*), as do the larvae of the hackberry butterfly. The great purple hairstreak extends its range into Arizona from a more southerly distribution. Metalmarks (*Riodinidae*) are a small family of butterflies that occur primarily in the West. Giant skippers (*Megathymidae*) occur where *Yucca* grows, for larvae bore in the stems and roots.

Among the most conspicuous of moths are the large sphinx moths (*Sphingidae*). Occasionally diurnal, such as the lined morning sphinx (*Celerio lineata*) and clearwing sphinxes, these nectar-feeding insects probe tubular flowers with long probosces. Larger-sized moths are sometimes mistaken for hummingbirds. Noctuid and Geometrid moths are probably the most abundant moths to congregate about lights at night, but many other macro- and micro-*Lepidoptera* are found.

Beetles comprise the largest insect order and are abundant both in species and numbers in the Sonoran Desert. Arnett (1960-1968) produced a volume on the beetles of the United States that is helpful for the northern part of the desert. Many beetles are quite showy and conspicuous. Long-horned beetles (*Cerambycidae*) such as the large brown palo verde root borer (*Prionus*) and the flightless cactus-eating beetle (*Moneilema*) are frequently encountered on summer evenings. Linsley (1958) studied the geographic origins and phylogenetic affinities of western American Cerambycids. He believed that 22 percent of all North American Cerambycid beetles were a Sonoran element "centering in southwestern United States and Mexico."

Blister beetles (*Meloidae*) can be seasonally very abundant as they feed on vegetation. Often conspicuously colored with iridescent metallic sheen or with red-and-black or black, blister beetles can release the irritating vesicant cantharidin when handled. Werner et al. (1966) treated the Meloids of Arizona. Metallic wood-boring beetles (*Buprestidae*) are noticeable when the large, wary adults buzz from tree to tree during warm months. Larvae burrow in wood (mostly dead wood) or stems and leaves of desert trees and shrubs. The Scarabeidae have some very robust members such as the rhinoceros beetle (*Xyloryctes*) and the golf-ball-sized *Dynastes*.

Tenebrionids are desert beetles. They are usually black. Some, like *Eleodes*, can release an offensive-smelling substance when disturbed. Cloudsley-Thompson (1979) theorized that the black color of Tenebrionids renders them conspicuous against light desert substrates as an aposematic strategy; beetles are hard (some being called iron-clad beetles) and distasteful. Cloudsley-Thompson (1964a) reported on the function of the subelytral cavity in desert Tenebrionids. The control of water loss was investigated by Ahearn (1970). Ahearn (1971) also studied ecological factors affecting population sampling of desert Tenebrionid beetles.

Weevils (Curculionidae) are often common. The large, black agave weevil causes death in century plants. Larvae burrow in soft heart tissues, spreading bacterial rot with their feeding activities. Seed weevils (Bruchidae) are very abundant when legume seeds are ripe. Riazance and Whitford (1974) reported 26 to 57 percent of the seed crop in populations of mesquite to be damaged by *Bruchus prosopis*. A Bostrychid stem borer was found to kill small leaves and stems of mesquite and to be capable of significaatly lowering biomass productivity.

Homopterans are plant feeders, present sometimes in great numbers when suitable host plants are available. Aphid populations increase especially in spring when ephemerals and spring annuals are plentiful in the vegetation. Lac insects of several species can be found on cacti and desert shrubs such as creosote bush. Mealybugs and cochineal insects (once used as sources of crimson dye) are also found, sometimes abundantly, on cacti. The desert cicada (*Diceroprocta apache*) is often very abundant. It was studied with regard to temperature response by Heath and Wilkin (1970).

Hemipterans are present in both aquatic and terrestrial habitats. Plant-feeding members of the order (Lygaeids, Mirids, Tingids, Pyrrhocorids, and Pentatomids) contain species that can cause damage to crop plants. Some coreids feed on plant seeds, including many legumes, injecting digestive saliva to liquefy portions of the endosperm. Others feed almost exclusively on cacti, causing necrotic spots to form in subepidermal tissues because of their feeding. Large infestations can cause plant mortality. Reduviid bugs are predaceous, subsisting mostly on other arthropods, but some assassin bugs take only blood meals from vertebrate hosts. Cone-nosed bugs (*Triatoma* sp., also known as kissing bugs or hualapai tigers) inhabit wood rat (*Neotoma*) nests, feeding on the wood rats. Occasionally they feed on human blood, causing very painful and uncomfortable symptoms, which sometimes result in a general systemic toxic reaction to the injected saliva.

Isoterans are important desert detritivores. According to Nutting et al. (1974) and Haverty and Nutting (1974), dead wood from mesquite (*Prosopis*), palo verde (*Cercidium*), catclaw acacia (*A. greggii*), and *Opuntia* makes up about 97 percent of the dead wood available to termites in the Sonoran Desert. The various termites (*Heterotermes aureus, Gnathamitermes perplexus, Paraneotermes simplicicornis,* and *Amitermes* spp.) divide this resource by mode of attack or species preference so that potential interspecific competition is reduced. Collins (1973) reported that two subterranean termites of the Sonoran Desert, *Heterotermes aureus* and *Gnathamitermes perplexus,* can survive at temperatures up to 50°C.

Orthopterans of Arizona have been treated by Ball et al. (1942). Orthopterans in relation to creosote bush were investigated by Otte and Joern (1975), Hurd and Linsley (1975a), and Rehn (1958). Desert grasshoppers have been studied in relation to plants of overgrazed land (Ball 1936), species richness in relation to plant diversity (Otte 1975), and effects of solar radiation on their temperature and activities (Pepper and Hastings 1952). The desert cockroach or sand roach (*Arenivaga*) lives in areas of sand and sand dunes. It consumes decaying roots and other vegetation, particularly the micorrhizae associated with mesquite roots (Hawke and Farley 1973). *Arenivaga* of southern California sand dunes were studied by Friauf and Edney (1969). Their behavior and ecology were investigated by Hawke and Farley (1973). Edney summarized work on *Arenivaga* in a review paper (1974). Other orthopterans (walking sticks, crickets, katydids, and mantids) are seasonally common in areas where there is suitable food.

ARACHNIDS

A general treatment of Arachnids is present in Savory (1977). Scorpions hunt for invertebrate (and occasionally vertebrate) prey at night. Venom injected into seized animals through the caudal stinger acts to subdue them in most cases, but Hadley and Williams (1968) recorded that scorpions (*Hadrurus*) studied in northwestern Sonora uusually immobilized prey by pedipalps alone because stinging actually increased the mobility of captured mice and lizards. The bark scorpion (*Centruroides sculpturatus*) possesses neurotoxic venom that can be dangerous to humans. Stings from this species can be serious (and sometimes fatal) in children or in adults with health problems such as hypertension. Scorpion venom has been investigated by Adam and Weiss (1959), Glenn et al. (1962), Whittemore et al. (1963), Watt (1964), Potter and Northey (1962), Patterson (1960), and Master et al. (1963). Stahnke (1966) studied scorpion behavior. Water relations of *Hadrurus arizonicus* were reported on by Hadley (1970b). Hadley and Hill (1969) discussed oxygen consumption by *Centruroides sculpturatus*. Taxonomic studies of *Hadrurus* were completed by Williams (1970) and of *Vejovis* by Williams (1976) for Baja California and by Gertsch and Soleglad (1966).

Pseudoscorpions occur in cryptic habitats, where they are predaceous on other invertebrates. They are commonly found in the unique microcosm presented by rotting cacti. Lee (1979) reported on maritime pseudoscorpions of Baja California. Muma (1951) studied solpugids of the southwest and published on their feeding behavior (1966). Mites (Acari) exist in desert environments as external parasites on vertebrate and invertebrate animals, as well as free-living predators and herbivores. Herbivores inhabit soils and organic debris, vegetation, and aquatic habitats. Tuttle and Baker (1968) published a book on the spider mites of the southwestern United States.

General treatments of spiders (Araneae) include those by Gertsch (1979) and Levi et al. (1968). Chew (1961) investigated the ecology of spiders in a desert community. Of Sonoran Desert spiders, the only species that can seriously harm humans are the black widow (*Latrodectus*) and the brown recluse (*Loxosceles*). Venoms of these spiders have been studied by Denny and Dillaha (1964), Keegan et al. (1960), Smith and Russell (1967), and McCrone (1964). *Latrodectus* venom is neurotoxic and potentially dangerous to fatal. *Loxosceles* venom can cause a necrotic lesion that is very slow to heal. A historically interesting account of the black widow was produced by Thorp and Woodson (1945).

Some of the more conspicuous spiders likely to be encountered in the Sonoran Desert are funnel weavers (Agelenidae), wolf spiders (Lycosidae), huntsman spiders (Heteropodidae),

crab spiders (Thomisidae), and jumping spiders (Salticidae). The monogeneric Diguetidae has species limited to the Sonoran Desert and to the Monte Desert of Argentina.

Largest desert spiders are members of the family Theraphosidae, commonly known as tarantulas. Covered with brownish or blackish hairs, these largely nocturnal spiders forage in warm weather, retreating to burrows or other protected sites during daytime and the cold of winter. Tarantulas have been observed eating ants (Reichman et al. 1979) and hunting other insects and arthropods, perhaps occasionally small mammals and lizards. Venom is not dangerous to man. Tarantula thermal relations, water loss, and oxygen consumption were detailed by Seymour and Vinegar (1973).

OTHER ARTHROPODS

Centipedes ordinarily subsist on a diet of invertebrates, but the large Sonoran Desert *Scolopendra heros* has been observed feeding on lizards and toads. Easterla (1975) recorded a *Scolopendra heros* carrying a freshly killed long-nosed snake (*Rhinocheilus*). During daytime hours and cold months, centipedes retreat to crevices, soil under rocks or under dead wood, or to burrows they dig in protected places. Centipede bites from small species can be painful, from large ones more so. *Scolopendra heros*, because of its size and amount of venom, can inflict a very painful bite.

Warburg (1965a) studied microclimate in isopod habitats of southern Arizona and also compared water relations and internal body temperatures of isopods from xeric versus mesic habitats (1965b). The fairy shrimp (Anostraca) and the tadpole shrimp (Notostraca) have desiccation-resistant eggs that can survive for years in dry desert soil. Zoogeography of Arizona Anostraca was discussed by Belk (1974). Tadpole shrimps were investigated by Horne (1968), Horne and Beyenbach (1971), and Carlisle (1968). Cole (1968) discussed not only these crustaceans but the typical crustacean freshwater fauna of cladocerans, copepods, and ostracods as well. Belk and Cole (1975) considered the adaptational biology of desert temporary pond inhabitants. Brusca (1973) treated marine crustaceans found in Gulf intertidal zones.

ECOLOGY OF THE INDIGENOUS O'ODHAM PEOPLE

People of the Sonoran Desert in southern Arizona and northern Sonora who call themselves O'odham realize that they are all ethnically and linguistically related even though they are members of distinct political tribes, the Pima and Papago. According to the O'odham dictionary by Saxton and Saxton (1969:156), they partition themselves by habitat (*akimel*, "river"; *tohono*, "desert"; *hiach-eD*, "sand") as Akimel O'odham, Tohono O'odham, and Hiach-eD O'odham but differentiate O'odham compatriots of eastern Sonora and adjacent Chihuahua on the basis of diet as Chuhwi Ko'adam

("Jackrabbit Eating People"). On the other hand, the Chuhwi Ko'adam refer to themselves as O'odham and northern compatriots as Papawi Ko'adam ("Tepary Bean Eating People"). Habitat and diet diversity underscore ecotypic differentiation and make for incipient ethnogenesis.

O'odham life is so linked to plants, animals, climate, soils, and other physical factors of the Sonoran Desert that these people are considered aboriginal residents par excellence for this desert (Fontana 1974). These people have been a significant factor in the overall ecology of a large part of the Sonoran Desert. They have themselves differentiated into three ecologically defined facies in the north that are closely correlated with rainfall patterns and water availability. A brief survey of relationships of the northern O'odham within their desert homeland may contribute to a better understanding of the Sonoran Desert and to the general subject of humans in arid lands.

Origin of the O'odham

That the O'odham might represent an ethnic group that had migrated into the Sonoran Desert from the southeast is a logical possibility suggested by a clear geographic distribution to the southeast in Mesoamerica of all eight groups most closely related by language to them: Pima Bajo, northern Tepehuan, southern Tepehuan, Yaqui, Mayo, Cora, Huichol, and Tarahumara (Saxton and Saxton 1969:188). Indeed the Akimel ("river") facies of the O'odham, although now confined to a tiny spot of Sonoran Desert adjacent to the Gila and Salt rivers in central Arizona, nevertheless was spread over a wide area of ecological transition above and below the Sonoran Desert boundary at the time the Spanish padres encountered these people (Spicer 1962).

Nevertheless, there is evidence cited by Haury (1950) from Ventana Cave in the Papaguería of southern Arizona that an endemic facies of a desert culture extending back 10,000 years may have evolved essentially in situ into the O'odham. Elsewhere Haury (1976) has stressed his belief that the O'odham descended from the broken remnants of the Hohokam culture, which he believed in turn had migrated north to the Gila and Salt rivers of central Arizona about 300 B.C. from Mesoamerica, merging with the indigenous facies of the desert Cochise culture and flourishing over many hundreds of years, only to decline and break up in the fifteenth century. Such an origin of the O'odham from the Hohokam is not proven and remains as speculative as when proposed earlier by Gladwin et al. (1937). A theoretical transition from Hohokam to O'odham would be easier to envision by postulating divergence of a Hohokam desert subgroup long before the river Hohokam culture itself broke up. This was done by Haury (1950).

Such a theory unfortunately ignores the fact that the O'odham, except for occupation of the San Miguel and Sonoran rivers from Cucurpe to Opodepe by possibly intrusive Opata (Spicer 1962), were essentially in distributional unity with, and not unlikely enjoyed an ethnic oneness with, the Sonoran Desert facies of the Pima Bajo. The Pima Bajo had a very respectable history deep down in Sonora and adjacent Chihuahua and a population large enough in the seventeenth century that the Jesuit Father Olinano baptized 4350 of them "within a short time" (Spicer 1962). It is instructive to note here that the O'odham word for enemy (*Ohb*) relates to their name for an Opata tribesman (*Ohbadi*) and that when O'odham came into contact with the more northerly Apache, a people who proved to be unfriendly, they referred to them as "the enemy" (*Ohb;* cf.: *Opata, Ohbadi, Apache*).

Differentiation within the O'odham

Today the northern O'odham are notably homogeneous in that they speak a common language (O'odham) but display differentiation in that the language has nine major dialects, eight of which (S-ohbmakan, Totoguani, Ge Aji, Ahngam, Ko-Lohdi, Huhuwosh, Gigimai, and Huhhu'ula) are referred to by Americans as "Papago" with the ninth dialect ("Pima") being spoken by both Kohadk Papago and the river Pima (Saxton and Saxton 1969).

The O'odham divide themselves into Akimel O'odham (River People), Tohono O'odham (Desert People), and Hiach-eD O'odham (Sand People). The central and currently the largest assemblage both geographically and demographically is the true Papago (Tohono O'odham), from which both Sand Papago (Hiach-eD O'odham), and Pima (Akimel O'odham) can be easily distinguished as subgroups specialized to moisture conditions drier and wetter respectively than those of the Tohono O'odham.

At first glance the cultural complexity seems to increase in the sequence Hiach-eD O'odham → Tohono O'odham → Akimel O'odham, and it is tempting to postulate ethnogenesis in this sequence. The speculation that the Hiach-eD O'odham may have reverted to a lower cultural level to survive in the sand as a refuge from pressures of other Indians or the Spaniards might be expressed by the hypothetical sequence Hiach-eD O'odham ← Tohono O'odham → Akimel O'odham. This model would presume strong consolidation of the Tohono O'odham as a people with subsequent differentiation of facies adapted to drier and wetter environmental conditions.

But the Tohono O'odham, with a dual subsistence economy and a complex saguaro harvest and crop cycle, as well as a degree of vigor that has preserved them essentially intact to the present, might have benefitted from a cultural and actual heterosis suggested by the sequence Hiach-eD O'odham → Tohono O'odham ← Akimel O'odham ←→ Pima Bajo. Such attempts at analysis are superficial in that they ignore the obvious time depth required for ethnogenesis. The real sequence of events may never have followed the unidirectional course of a straight line, and cultural trends need not have followed linguistic lines. When faced with the need to account for the cultural time depth in the region—the San Pedro Cochise culture of the desert and the Hohokam culture of the Gila and tributaries—the inevitable conclusion is that we stand only on the brink of knowledge.

The pancultural ethnic and linguistic unity of the O'odham over a large geographic range and presumably over a respectable time depth, with clear ecotypic cultural differentiation, seems remarkable when compared with Yuman and Puebloid patterns. The Puebloid people were linguistically diverse but relatively uniform culturally, apparently welcoming groups to live among them and share their technology. Frank Cushing, an anthropologist with the Bureau of American Ethnology, demonstrated that an Anglo willing to accept Zuñi culture could become one of them. The O'odham, on the other hand, seem from observations by Russell to have been obsessed with maintaining the purity of their heredity, killing babies born of American or Mexican fathers, babies born of loose women, and even nonuniform babies, for example, one with six toes (Russell 1975).

Apparently no stigma is attached to babies born of a union between people of different O'odham subgroups or of O'odham with linguistically related people such as Yaquis. Nor is there indication that grown half-breeds from Seri or Yuma country were killed if they came among the O'odham. Grown people were killed, however, if it was thought that somehow they threatened the well-being of the O'odham by being possessed by alien spirits, by being "bad medicine men" or sorcerers.

Opportunities for Research

In the future researchers will be able to construct fascinating monographs on phases of O'odham ecology as the archaeological record is extended and interpreted, as snatches of history recorded by early Spanish priests (Spicer 1962) or by O'odham calendar-stick keepers (Lumholtz 1971; Hall 1907; Kilcrease 1939; Underhill 1938) are subjected to workshop analysis and correlation, as ethnobotanical inventories (Castetter and Underhill 1935; Russell 1975) are subjected to detailed analysis and correlated with advancing plant knowledge, as oral legends and ritual speeches (Saxton and Saxton 1973; Russell 1975; Underhill et al. 1979) are interpreted, and as ecohistoric ordinations of ethnographic continuity, of adaptation, and acculturation over the centuries are achieved. Some of the patterns are already obvious but researchers in this field are confronted with a bewildering array of stereotype terms and concepts that require close scrutiny.

The Piman Concept

The terms *Pima* and *Piman* are applied to Indians in different senses and are easily confused. *Piman*, although ostensibly an adjective derived from *Pima*, is used in the sense of "Pima-like" and has arbitrarily been assigned the generic connotation of designating that larger group of peoples that includes the Sonoran Desert O'odham and their nondesert close relatives. These close relatives historically were the first of the larger Piman group to be encountered by early Spaniards. The explorers and missionaries erroneously believed that the response *pimatc* ("I don't know") elicited from these Indians, in response to questions, represented the name of the tribe (Russell 1975). The Spaniards quickly recorded them as Pima and their homeland as the Pimería.

Today these Mexican Pima Indians live above the Sonoran Desert in the region of the Sonoran-Chihuahuan boundary near the Sonoran Foothills vegetational division. When compared ecologically with the other remaining Pima of the Gila and Salt rivers in central Arizona, these people seem so different that if the strong linguistic and ethnic tie were not known, they doubtless would be dismissed as not closely related.

Details of how the terminology *Pima* became associated with only a segment of the natural O'odham group is in order. As the Spaniards penetrated the northern part of the Sonoran Desert, they followed river systems and naturally first encountered and became most familiar with the Akimel O'odham rather than with either Tohono or Hiach-eD O'odham. If the O'odham encountered were not already Akimel O'odham, they may have been converted quickly to an Akimel phase by being taught irrigation agriculture as part of the missionary reduction program. The sites chosen for missions were on rivers where wheat and other crops could be grown by irrigation using techniques perhaps already being used by the Akimel O'odham for their traditional crops. It was to these river-dwelling Indians, perhaps not much different from the desert river facies of the previously encountered Pima tribe, that the *Pima* designation was also extended.

Because the Akimel O'odham were clearly related in the eighteenth century by language and customs to the southern Pima, they were called Pima Alto or Upper Pima. The originally encountered *Pima* was then differentiated as the Pima Bajo or Lower Pima on latitudinal grounds (Fontana 1974). The Upper Pima (simply *Pima* to residents of the United States) are now confined to lower elevations than the Lower Pima of today, and the Lower Pima who have retained an ethnic identity now live at the upper elevations. To avoid confusion, the Upper Pima are now sometimes referred to as the River Pima and the Lower Pima as the Mountain Pima for their habitat in the Sierra Madre.

Within historic time the Upper Pima had a nondesert contingent, and the Lower Pima had a Sonoran Desert contingent (figure 5-16). Spanish padres encountered the Pima Bajo in a wide swath stretching on either side of the Sonoran Desert from near present-day Hermosillo east past Maicoba into present-day Chihuahua (Spicer 1962). A thousand of these Indians left their homeland willingly, but perhaps under

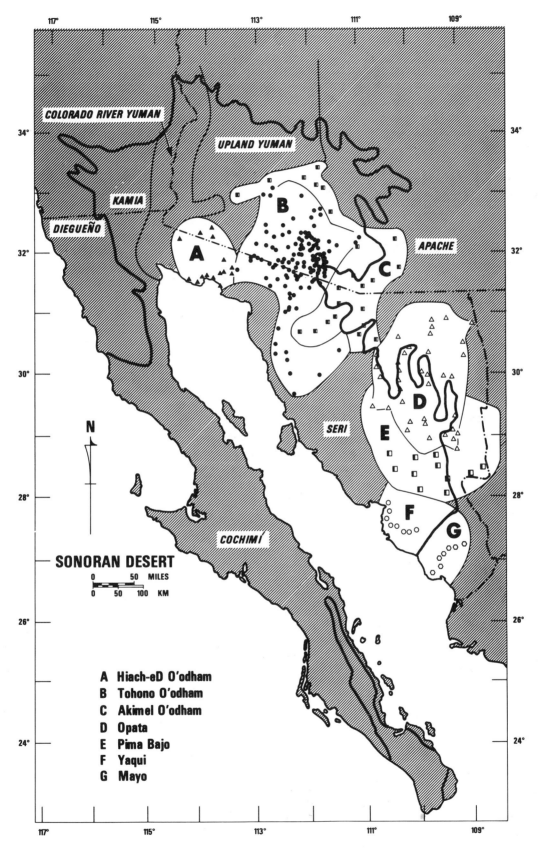

COLORADO RIVER YUMAN

UPLAND YUMAN

KAMIA

DIEGUEÑO

APACHE

A

B

C

SERI

D

E

N

F

G

SONORAN DESERT

0   50   MILES

0   50   100 KM

COCHIMÍ

A Hiach-eD O'odham
B Tohono O'odham
C Akimel O'odham
D Opata
E Pima Bajo
F Yaqui
G Mayo

Figure 5-16. Historic geographic patterns of the O'odham in relation to other tribes.
Compiled by the authors from the ethnological literature. Cartography by Russell Vogt,
University of Arizona Graphics Department.

pressure from the Opata, to settle in Sinaloa to be close to the Spaniards after Cabeza de Vaca passed through their country in 1540 and before 1619 when a missionary was finally assigned to the Pima Bajo country. Spicer (1962) traced the geographic distribution in 1750 of both lower (desert) and upper (mountain) groups of the Lower Pima in Sonora and adjacent Chihuahua.

The Lower Pima are clearly the closest relatives of the O'odham of the Sonoran Desert, and their geographic separation from these latter Indians by the intervening Opata deserves comment. Spicer (1962) thought that the Opata may have pushed into Sonora prior to 1600 from the northeast. If so, the Opata eventually may have severed the northern Pima from the southern Pima by taking over the region from Cucurpe to Opodepe on the Rio San Miguel and Rio Sonora. Dirst (1979) discussed such a hypothesis in the light of new archaelogical methods for studying shifting frontiers.

If intervention by the Opata was the opening wedge for differentiation, subsequent decimation of the easternmost groups of the Akimel O'odham or their forced migration due to Apache attack, and subsequent Mexicanization of the westernmost (desert) Pima Bajo, should have effectively obliterated any cultural intercourse of the O'odham with their Pima Bajo compatriots. At the end of the seventeenth century, there is a casual mention in Spanish records of "chieftains from as far north as the Gila Valley" being present at a religious festival at Remedios in 1698 (Russell 1975), indicating an apparent willingness to travel up to 200 miles. Whether Pima Bajo were willing to travel a somewhat shorter distance north to attend such an event at Remedios is not known.

It is known that the Pima Bajo who had migrated to live among the Spaniards in Sinaloa continued communication and cultural contact with the people of their ancestral homeland (Spicer 1962), passing on information concerning the military strength of the Spanish garrisons, the benefits of mission life, and a willingness to cooperate with the bearers of European culture. It has been speculated that the O'odham knew a great deal more about the Spaniards by the time the padres approached them than the Europeans would know about the O'odham for a long time.

Decline of the O'odham in Mexico

Although the Akimel, Tohono, and Hiach-eD facies of the O'odham were once common to both the northern (United States) and the southern (Mexican) sections of the Sonoran Desert, the O'odham in general have survived less well physically and culturally in Mexico. That O'odham who preferred to keep their traditional ways and to live apart from the Spaniards migrated north to avoid the conquest is not known but is suspected.

It is well documented that the O'odham of the north were sufficiently familiar with the strengths and habits of the

Spaniards to submit and be docile. Their peaceful characteristic, except when fighting Apaches, has been frequently noted in the literature. Castetter and Underhill (1935) thought that due to the inhospitality of the desert homeland of the Tohono O'odham, these people "did not suffer the thorough conquest which befell the other Mexican tribes and have kept many of their ancient customs intact." Although most O'odham have now adopted the industrial culture of the United States, they retain a few customs that some fondly remember and try to perpetuate.

Present Status of the O'odham in Southern Arizona and Adjacent Sonora

The Hiach-eD O'odham are essentially extinct as an identifiable group; the individuals have either been killed by a posse of Mexicans (Childs 1954), rounded up and taken to Caborca, Sonora, or escaped to Arizona. The Akimel O'odham today survive only in small villages on the Gila and Salt rivers of central Arizona. Fontana (1974) has traced them to aboriginal occupancy on the San Pedro River, Santa Cruz River, Rio San Miguel, Rio Magdalena, Rio Concepción, Rio Altar, and possibly between the Rio San Miguel and Rio Sonora.

The Tohono O'odham historically lived on either side of the international boundary, and although any one family had definite summer and winter quarters of a semipermanent nature, these Indians seemed to pass at will between north and south. They are thought to have adjusted the boundaries of their homeland with various fluctuations of climatic cycles. Crosswhite (1980) discussed the movement of Tohono O'odham into the coastal lowlands of Mexico about 1850 following a decade of cold weather, which had included three feet of snow one year in their previously subtropical Papaguería. The movement south into Mexico was reversed by the return of favorable weather and by the threat of yellow fever in the Mexican lowland.

During the last 150 years, the Mexican contingent of the Tohono O'odham has decreased in numbers through attrition, including losses from warfare, through migration into the United States, and through assimilation into the mestizo population of Mexico. The great majority of Tohono O'odham now live in cities such as Tucson, Phoenix, or Los Angeles or in the Papaguería of Arizona east of Yuma, south of Casa Grande, and west of Tucson. The tribe is guaranteed the tenancy of land by the U.S. government.

Correlation of the O'odham Subgroups with Rainfall

The Tohono O'odham traditionally occupied mid-longitudes of the Sonoran Desert where precipitation, although deficient for usual agricultural techniques, was about average for the desert. The Hiach-eD O'odham lived to the west where rainfall was lower and the habitat was severe. The Akimel O'odham lived to the east where rainfall was more plentiful and where rivers flowed through much of the year from snowmelt in the mountains and from biseasonal rainfall from winter Pacific moisture and summer moisture from the Gulf of Mexico.

Central Tohono O'odham

The Desert People were semiagricultural. This was not because they could not master the full concept of agriculture but rather because their homeland would support growth of crops during only a portion of the year. Indeed to master the technology of water-harvesting agriculture and also to master the details of saguaro fruit technology would seem to place these Indians at a higher technical level than has generally been appreciated.

RUNOFF AGRICULTURE

The Tohono O'odham depended on channelled runoff water from short, intense summer thunderstorms for a type of farming distinct in methodology from ordinary dry-land agriculture and equally different from normal irrigation agriculture. They depended on a summer harvest of saguaro (*Carnegiea gigantea*) cactus fruit and on an autumn harvest of beans, corn, and squash. Mesquite beans and other wild foods were also gathered. Today the Tohono O'odham are the familiar Papago who have adapted to Anglo industrial culture. Few continue to farm and harvest saguaro fruit by traditional methods.

NAWAIT RITUAL AND AN EPIC CYCLE OF SAGUARO AND CROP SPEECHES

Much can be learned of the central Tohono O'odham by studying their integrated cycle of harvesting saguaro fruit and making a portion into wine drunk in the *nawait* ritual as part of formal preparations for planting agricultural crops. The Tohono O'odham ceremonially called to the supernatural through saguaro wine communion as an intermediary to ask for rain for their agricultural fields (Waddell 1973). Speeches relating to this ritual, although archaic and difficult to comprehend, seem to comprise an epic cycle dealing with the natural history of the saguaro and its reaching to the sky in supplication to the supernatural for rain, followed by an account of gathering wild food plants by the O'odham and their subsequent agricultural acculturation (Crosswhite 1980). The speeches of this cycle occur in the order Running Speech, Seating Speech, and Mockingbird Speeches.

The ritual Running Speech seems to detail the life cycle of the saguaro: its rising into the sky during intervals of growth,

its contact with the sky and the rainhouse of the supernatural, its touching the supernatural being who resided there, and the subsequent breathing by the supernatural on the land to cause rain. The ritual Seating Speech refers to preagricultural plant gathering, which was made possible by the saguaro's having succeeded in asking for rain. The ritual Mockingbird Speeches describe drinking of wine from the saguaro, sending of rain by the supernatural, and subsequent "de temporal" growing of corn by the O'odham to the east, north, and south but remarkably only growth of wild food plants in the western sector. Indeed the westerly Hiach-eD O'odham were hardly serious agriculturists because Gulf of Mexico moisture of the Sonoran monsoon rarely reached as far west as their homeland.

ANNUAL SAGUARO HARVEST AND CROP CYCLE

Harsh areas of the Sonoran Desert were made more habitable to the Tohono O'odham by their developing or assimilating techniques of saguaro fruit harvest, fruit-processing technology, hermetic sealing, food storage technology, and by their growing desert-adapted races of corn, beans, and squash with rapid life cycles, using water harvested from the brief summer thunderstorms. Were it not for the fruit harvest in June and July, the Tohono O'odham would not have had adequate food reserves for proper nutrition during the period when crops were planted in July and before they were harvested in September and October (Crosswhite 1980). Wild food plants, including mesquite beans, were also important.

The Tohono O'odham combined a knowledge of ecology with a large amount of symbolism to develop an annual saguaro harvest and crop cycle notable for its ritualism and its effectiveness in keeping the people on schedule with gathering and planting while providing a sense of fulfillment in a difficult subsistence economy (Crosswhite 1980). The promise of wine from saguaro fruit may have made easier the hard work of fruit harvesting, processing, and storing, which resulted in eleven other food products aside from wine: syrup, jam, dehydrated pulp, animal food seed mix, seed flour, seed oil, pinole, atole, snack foods, soft drinks, and vinegar.

RESPECT AND EMULATION OF THE SAGUARO

The Tohono O'odham demonstrated profound respect for the saguaro cactus, successfully emulating some of its qualities to survive on the desert as it did. The Tohono O'odham scarcely transgressed the distributional limits of the saguaro to the north, east, south, or west. They channelled water to their fields from a great expanse of desert just as the extensive water-harvesting roots of the saguaro channelled water to the plant from a great expanse of land. Just as the saguaro stored water from time of plenty to time of scarcity, the Tohono O'odham learned to store food from time of plenty to time of scarcity, both in storage facilities and in adipose tissue. Underhill (1979) stated that these people likened the seedling of a saguaro under a nurse plant to a fat child cared for by a mother. Both Tohono O'odham and saguaro were adapted to the warmer south-facing sites in the desert, indicating a subtropical nature that may have flourished during the warm cycle of the last 10,000 years, particularly the warm period from 500 B.C. to 1500 A.D. (Crosswhite 1980). The Tohono O'odham began annual preparations for planting crops by waiting until the seeds of saguaro were about to fall to the ground and subtropical germination prerequisites of heat and moisture were imminent.

FURTHER DETAILS OF SAGUARO SIGNIFICANCE

The Tohono O'odham may have believed according to the ritual Running Speech that the saguaro had a supernatural anthropomorphic dimension coming from its contact with the man who resided at the rainhouse where winds, clouds, and seeds were located (Crosswhite 1980). Contact with wine made from the saguaro was definitely used in calling for rain by the Tohono O'odham, but whether they thought its use also conferred on them some of the qualities of the saguaro is not known. Was the ability to learn from and emulate the saguaro enhanced by drinking its wine? The overall significance of the saguaro to the Tohono O'odham can be seen from their annual calendar, which began the new year with ripening of saguaro fruit and was kept on schedule by reference to phenology of the saguaro, crops, and wild plants: the progression of the seasons in the environment.

In their traditional culture, Tohono O'odham depended on bacterial necrosis of saguaro resulting from *Erwinia carnegieana* for a natural process of rotting to eliminate the fleshy parts of old and fallen saguaros. This allowed the wooden skeletal ribs to fall away to be used as construction materials and for household purposes. This bacterial rot also allowed the callous "boots" from dead saguaros to be separated out and used as containers or as household implements.

ENVIRONMENTAL BASIS OF THE INDIGENOUS CALENDAR

Whereas the Aztecs and others regulated their calendar by reference to the sun and its equinoxes, the O'odham based their calendar on phenological observations of the environment. This type of calendar would be more useful to a people attuned to the environment because the alternative solar calendar might often be off schedule with regard to phenology. When phenology itself is the basis of the calendar, it is self-regulated for planting, hunting, and gathering purposes. The following outline of the environmentally based calendar is taken from Crosswhite (1980) and was modified from Lumholtz (1971), Russell (1975), Underhill (1939), and Saxton and Saxton (1969).

1. *Hahshani Mashad* (June). The "Saguaro [harvest] month," sometimes referred to as the "hot month." This is the time to set up the cactus camp and begin to harvest, process, and eat the fruit—a joyous time but one with hard work.

2. *Jukiabig Mashad* (July). The "rainy month." Now is the time to finish the saguaro harvest, to start the rain by means of a *nawait* ritual, and then to plant the seeds of beans, corn, and cucurbits.

3. *Shopol Eshabig Mashad* (August). The "short planting month." This is the last chance of the year to get crops into the ground. If the *nawait* ritual at one village has already started the rains, now other villages should have ceremonies to ask for the rains to continue.

4. *Washai Gak Mashad* (September). The "dry grass month." When the rains stop, the desert grass turns brown and the late crops ripen.

5. *Wi'ihanig Mashad* (October). The "month of persisting [vegetation.]" Certain food plants characteristically surviving seasonal drought and the onset of cold weather are harvested now. It is autumn and frost might be seen. This has also been called the month of winds, light rain, light frost, or "when cold touches mildly."

6. *Kegh S-hehpijig Mashad* (November). The "month when it is really getting nice and cold." Also translated as the fair cold, pleasant cold, or low cold month. This would be a good time to go hunting.

7. *EDa Wa'ugad Mashad* (December). The inner bone [backbone of winter] month." This is the dead of winter. It is also referred to as the month of great cold or the month when leaves [for example, of mesquite] fall.

8. *Gi'ihodag Mashad; Uhwalig Mashad* (January to early February). The time when animals "have lost their fat," then "go into heat" and mate. There is not a lot of activity in the desert, and time seems to hang heavy. This was once a good time for the Tohono O'odham to go south to work in Mexico or travel north to work among the Akimel O'odham of the Gila River.

9. *Kohmagi Mashad* (February). The "gray month." This is the time when the landscape is at a bleak climax of gray. The trees are without leaves, but the flowers are already coming on the cottonwood trees.

10. *Chehdagi Mashad* (March). The "green month."
Leaves are finally coming on the cottonwood and
mesquite trees, and there are abundant green herbs
and grasses. A saguaro ritual is held.

11. *Oam Mashad* (April). The "yellow-orange
month." It is spring, and the beautiful yellow and
orange flowers of desert poppy, brittlebush, desert
marigold, and other wild flowers make the desert
colorful and happy, but food stored by humans is
beginning to run low.

12. *Kai Chukalig Mashad* (May). The month when
saguaro "seeds are turning black" in the developing
fruit. This is the optimistic name; the pessimistic
name was "painful month." The flowers of spring
have disappeared. Hunger pangs were once a real
possibility. A parent might have chosen painful sacrifice to make sure the children were well nourished.
After Father Kino introduced wheat, the Akimel
O'odham harvested grain during this month, and it
became a good time for the Tohono O'odham to
demonstrate friendship by helping them with harvesting and singing.

SPECIAL ASSOCIATIONS WITH CROP PLANTS

Although the Tohono O'odham have been strongly linked
to saguaro, they also have clearly been associated with a special
crop plant, the tepary bean (*Phaseolus acutifolius* var. *latifolius*). This bean has an astonishingly quick lifecycle adapted to
the "de temporal" summer thunderstorm agriculture of the
Tohono O'odham. Mathews (1951) thought that the common
name Papago for these people became associated with them
from *pavi coatam*, "the bean eating people," while Castetter
and Underhill (1935) suggesting *papawi o'otam* ("bean people") *papawi* being the plural of *pawi* ("tepary bean"). The
name *tepary* itself is said to derive from the O'odham phrase *t
pawi* ("it is a pawi"). Aside from their close association with
saguaro and tepary, the Tohono O'odham were masters at
growing a locally adapted race of maize or Indian corn (*Zea
mays*), which likely was their most important crop plant.
According to Spicer (1962), the Pima Bajo used the term
Papabotas to refer vaguely to Piman speakers to the north and
northwest, probably to the entire O'odham in what is now
southern Arizona and adjacent Sonora.

TRANSHUMANCE PATTERNS REQUIRED BY RUNOFF AGRICULTURE

The Tohono O'odham were discussed by Fontana (1974)
under the heading "Two-Village People." In the traditional
way of doing things, these families had a winter home, a
summer home, and a cactus camp. The central Papaguería
consisted of broad valleys with alluvial soil, which were
bordered by small, steep mountains (Bryan 1925). Water was
found in winter only in such hills so the winter home, or "the
well," was located there (Castetter and Underhill 1935).
Thunderstorms in summer brought rainfall and runoff to the
valleys so they could be lived in and farmed. This summer
home was known as "the fields." As part of the annual pattern
of transhumance, moving from winter well site to summer
fields, the Tohono O'odham camped in impressive stands of
saguaro for about three weeks to harvest and prepare the fruit.

After food that could be eaten or stored had been provided
by the saguaro harvest and after the *nawait* ritual had brought
rain, it was time to plant maize, tepary, and *cucurbita* seeds.
Underhill (1946) stated that by tradition, crops could not be
planted until after the saguaro wine ceremony. The fields of
the Tohono O'odham characteristically were located at mouths
of washes in the flatlands where relatively fertile, silty soil had
been deposited by runoff floodwaters. These desert washes
were dry ordinarily but flowed briefly from violent summer
thunderstorms.

COLLECTION OF WILD FOODS CONCURRENTLY WITH FARMING

Valleys used for summer farming also yielded wild foods,
particularly mesquite beans, which were not generally found in
the vicinity of the winter quarters or cactus camp. Fortunately
these ripened just as the Tohono O'odham had finished the
annual saguaro harvest and had settled into the valleys for the
seasonal farming. Mesquite beans and wild food plants growing in remote valleys or in ones not suitable for semipermanent
summer farming were harvested by setting up a mesquite camp
of brief duration. This was not nearly the elaborate affair of
the cactus camp. Whereas saguaro harvest camps were occupied by the entire family for a rather long period each year,
mesquite harvesting and processing sites were probably used
only by women who walked to and from the summer village.
A mild, nutritious mesquite drink that must have made this
hard work more sociable was aptly named "women's wine."

TRADITIONAL AGRICULTURE

Castetter and Underhill (1935) described agronomy of the
Tohono O'odham, observing that ditches "are dug from the
wash to the fields situated alongside, the ditch ending in the
middle of the field so that the water will flow over it."
Preparation of new fields generally involved leveling, stone
removal, terracing, and then channelling the runoff.

When a prepared field had been soaked and was starting to
dry on the surface, it was seeded. Underhill (1951) described
planting preparations: "Grandfather stands at the end of the
field with the corn in a little bag of deerskin and sings the corn

song. Then he speaks to the corn . . . then he gives the bag to Mother."

Holes were made with ironwood (*Olneya tesota*) planting sticks with sharpened fire-hardened points. Such holes were generally 3 feet apart and 6 inches deep. A man generally laid out the fields and made the holes. A woman followed behind "carrying the seeds in a small basket; she pushed a little loose earth into the hole with her bare toes, dropping in four seeds, covered them with a single foot movement" (Castetter and Underhill 1935).

Aside from normal crops, a little tobacco might be secretly grown away from the food plants to preserve its mystical nature. Although cotton was known, it is unlikely that the small fields of the Tohono O'odham permitted growing this crop. This limitation did not extend to the Akimel O'odham who had mastered river irrigation technology. Indeed early explorers differentiated the Desert People, who wore animal skin clothing, from the River People, who wore cotton clothing.

While families occupied summer villages, the women collected wild food plants and did the cooking, boys hunted small game, and "the men attended to what agriculture was possible" (Castetter and Underhill 1935). Through the growing season, men kept water channels clean and functionable, kept terraces intact, and used a sharp-edged piece of mesquite wood to hoe weeds in the fields.

After a growing season of 100 days or less, the crops were harvested in October. "After being flailed and winnowed, then thoroughly dehydrated on the roof of the house, they were stored in jars or baskets. The Papago usually hid their food for safekeeping" (Castetter and Underhill 1935).

After European crops had been adopted and the government had dug wells in the valleys, the Tohono O'odham eventually practiced some winter farming. This was most pronounced at San Xavier on the Santa Cruz River where Tohono O'odham had replaced much of the endemic Akimel O'odham population, which locally had been largely devastated prior to the twentieth century by Apache warfare.

By 1935, Castetter and Underhill thought that the new wells had not affected Tohono O'odham agronomy to a significant degree: "The government wells scattered through the reservation are used for watering stock, and to make it possible for residents to remain the year around, but they do not affect agriculture."

CULTURAL CHANGE AND THE RELICTUAL NATURE OF THE SAGUARO HARVEST AND CROP CYCLE

Gradually the Tohono O'odham tended to give up the winter home with a southern exposure in the little mountains in favor of the flatlands of the valleys even though cold air settles there at night. They adopted sturdier and warmer houses, began wearing cotton and other store-bought clothing, and a few began to grow winter wheat. Fewer and fewer continued practicing the traditional runoff agriculture of summer. The Tohono O'odham link with saguaro through perpetuation of the *nawait* ritual, which was at the heart of the saguaro harvest and crop cycle, preserves a relictual subtropical element of their culture, which for many may merely remind them of their older agricultural and fruit-processing economy.

DEMOGRAPHICS

Fontana (1974) reported that the two-village transhumance system of the Tohono O'odham "was at the core of the way of life of more than 4,000 O'odham at the time of Spanish contact." Thackery and Leding (1929) estimated the number of Tohono O'odham families in the 1920s to be 1200. Since the O'odham lived in extended families whereby a number of miscellaneous relatives might live as one family and sometimes under one roof, the Thackery and Leding statistic would seem to indicate that Tohono O'odham demographics fared well in historic time despite Apache warfare and disease. Today there are probably 12,000 to 15,000 of these people, 5000 to 6000 of whom live in the traditional Papaguería.

SUBSISTENCE SOURCES AND RATIOS OF DEPENDENCE

It has been estimated that about one-fifth of the food supply of the Tohono O'odham came from cultivating 0.1 to 1.0 hectare family-sized garden plots (Fontana 1974). In the first half of the twentieth century, Tohono O'odham probably harvested and stored 600,000 pounds of saguaro fruit per year in addition to the fruit that they ate raw during the three-week saguaro harvest (Crosswhite 1980). This quantity would easily account for another one-fifth of their food supply and likely provided trade products that were significant in their obtaining another one-fifth of their food supply from the Akimel O'odham or others (figure 5-17).

Abundant records indicate that saguaro syrup, saguaro jam, saguaro seeds, and saguaro seed meal prepared by the Tohono O'odham were common items of commerce in the Sonoran Desert for as far back as history has been recorded. When W. H. Emory collected the first botanical specimen of saguaro, the specimen from which the species was scientifically described, he encountered saguaro fruit products in local O'odham Indian commerce. Crosswhite (1980) compared changing dollar values for saguaro syrup from the time American money began to be used in the Sonoran Desert up to the present.

Although the calculation that one-fifth of Tohono

O'odham food came from trade might seem high, it should be kept in mind that the value of the saguaro fruit products that these people offered was increased by their being packaged in reusable clay pottery containers, which were further enhanced in value by being decorated with painted designs. Such an advanced marketing technique is not at all dissimilar to the modern flour mill practice of packaging flour in printed muslin bags for sale to rural people who sew their own clothes (including the Tohono O'odham) and is not dissimilar to the marketing technique of U.S. manufacturers of detergent and oatmeal who package cups or wash cloths in their products.

The value of saguaro fruit products was increased by offering them in conjunction with ritualistic group singing, with its inseparable religious value to the village being sung for. If the occasion required, ritualistic singing with little or no transfer of saguaro products was a great enough item to trade for grain. Transfer of food among individual O'odham was commonplace and to a certain degree expected, indeed even required, by custom. An old saying of early Anglo settlers was that when one of these people had food, all of them had food. Although this was generally true at the village level, the custom of one village's trading and singing for another and being paid in food had the effect of making food sharing a pan-O'odham phenomenon.

The extent of saguaro product commerce is emphasized by the fact that a considerable number of clay pots used by the Akimel O'odham were believed by Russell (1975) to have accumulated from the Tohono O'odham's having brought them "filled with cactus sirup to exchange for grain."

With three-fifths of the food supply of the Tohono O'odham accounted for by the annual saguaro harvest and crop cycle and by the trade resulting from this cycle, the remaining two-fifths came from general hunting and gathering that was similar to that which characterized the less agricultural Hiach-eD O'odham (figure 5-17). It is reasonable to estimate that plant and animal food was about equally represented in this general hunting and gathering portion of their economy and that mesquite alone may have accounted for as much as half of such additional plant material. Extensive lists of the plants and animals utilized by the Tohono O'odham are given by Castetter and Underhill (1935) at a time when older informants were alive and could remember traditional details.

Aside from the traditional subsistence patterns of the Tohono O'odham, these people adopted the raising of livestock, which had been introduced to them by Father Kino and succeeding Europeans. Cattle and chickens became particularly important in this latter phase of their economy. Feeding of saguaro seeds to chickens as a high-protein supplement became common with the Tohono O'odham and is a noteworthy example of their blending of traditional and new culture. (This practice proved quite effective and won at least one 4-H ribbon for a Tohono O'odham youngster.)

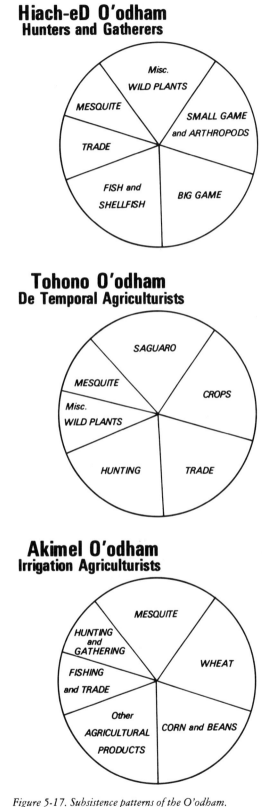

Figure 5-17. *Subsistence patterns of the O'odham.* Compiled by the authors from the ethnological literature.

USE OF NATIVE MATERIALS FOR HOUSING

The house or *kih* of the traditional Tohono O'odham was usually described as a dome-shaped brush structure. Although there is no thorough published review of the details and variations of its construction, it probably frequently resembled the structure described by Lumholtz (1971) as a dome-shaped grass hut with a framework of mesquite posts with two to four forked uprights in the middle supporting the dome-shaped roof "made of saguaro ribs, surmounted with greasewood twigs and some large coarse grass called sacate colorado." This sacaton grass (*Sporobolus*) and "greasewood" (creosote bush) covering was kept in place "by hoops of ocotillo inside and outside, placed at intervals of eight or ten inches," with the top of the house covered by earth. A small *huhulga kih* was provided as a temporary residence for menstruating women and was the site where they usually accomplished time-consuming tasks such as weaving baskets.

When cattle began to be raised by the Tohono O'odham, a stockade-like fence of ocotillo (*Fouquieria splendens*) stems or saguaro ribs had to be built around the house to keep the livestock from eating the grass thatch of the structure. This fence may have also helped to discourage skunks, gila monsters, and, depending on the tightness of the stockade, even snakes. It also provided something of an enclosed house yard for domestic work. Crosswhite (1980) has speculated on the effect of changing climatic cycles on the availability of saguaro ribs for construction purposes.

Although ocotillo stems were cut from living plants by the Tohono O'odham, saguaros were not killed, probably for religious reasons or out of respect. The skeletal ribs from dead saguaros were readily used, however, and the abundance of these was higher when freeze death of saguaro was significant in cold periods about 1550 to 1650 A.D., again from 1750 to 1850, and during the recent cold trend since about 1950.

Cooking by the Tohono O'odham was not ordinarily done in the house used for sleeping but rather in a brush kitchen usually some distance away. This kitchen was round and surrounded on all sides, except for the doorway, by walls of dried plant material—often saguaro ribs or other posts supporting a dense wind break of grass stems or corn stalks. This kitchen had no roof and was suited for use on sunny winter days and through much of the remaining times of the year.

During the intense heat of summer, cooking and other daytime household chores were likely to be pursued under the *watto*, a ramada with upright mesquite posts and roof but no sides. Thus aside from the *huhulga kih*, the domestic housing of the Tohono O'odham usually consisted of a cluster of three similar structures: a complete sleeping *kih*, a modified roofless *kih* for kitchen work, and a modified sideless *kih* for household activities and summer shade. To this architectural array was added the house yard with stockade fence, apparently in historic time.

BASKETRY

The Tohono O'odham relied on basketry for a variety of purposes (Shreve 1943; Kissell 1972; DeWald 1979). A plaited basket with a twill weave was made from split sotol (*Dasylirion wheeleri*), often incorrectly referred to as agave in the literature. The long ribbon-like leaves were cut from the plant, split into strips, and scraped against a grooved stone to remove prickles and to smooth the surface. Such a *huari* basket was apparently common to all Piman peoples (Brugge 1956). It was used when a watertight basket was not desired, particularly to store dried products. The spaces between weaving elements allowed this type of basket to be used also for sifting dry products or for straining liquids. The same basket weave was used for making sleeping mats when animal skins were not placed directly on the ground. Small plaited baskets were used for storing fetishes and even enemy scalps. Plaited sotol baskets were probably more characteristic of the Tohono O'odham than of the Akimel O'odhham because the former are known to have traded these baskets to the Akimel O'odham (Russell 1975).

Another type of basket, very similar to one made by the Akimel O'odham, was a very tightly coiled kind made from strips of willow bark woven over a foundation of split cattail rods. Black designs were woven into the basket using devil's claw (*Proboscidea*) fiber. These willow baskets had the important quality of swelling when wet and were used when a shallow watertight vessel was needed. Notable uses of willow baskets were for gathering moist saguaro fruit and for distributing saguaro wine. The willow basket has generally been replaced by the metal or plastic bucket. Coils of willow splints were obtained in trade from the Akimel O'odham, although some willow could be found by the Tohono O'odham where it grew at permanent watering spots. Most such sites were controlled by the Akimel O'odham, by unfriendly Yumans, or by warlike Apaches, however. After upstream diversion of the Gila River by Anglos, after impoundment of the river waters by the U.S. government near San Carlos, and after assertion of Anglo or Mexican ownership to most permanent watering places where towns could be built, the sources of willow for Tohono O'odham basketry disappeared.

The advent of the commercial bucket at the same time willow became unavailable made watertight baskets unnecessary. The type of coiled basket that has survived with the Tohono O'odham utilizes local materials, beargrass (*Nolina*), soapweed (*Yucca*), and devil's claw (*Proboscidea*). Such a basket, when tightly coiled, resembles a willow basket but apparently is not nearly as watertight. These baskets are used now as trays, sewing baskets or decorative containers for dry materials so they need not be watertight. The result is that the sewing element (*Yucca*) woven over the split beargrass coils is applied in an open coarse pattern to save time, to make a pattern, and to expose the beargrass foundation. Somewhat larger baskets with such coarse stitching were probably previously used only as granary baskets (Kissell 1972).

POTTERY

Materials and methods of Tohono O'odham pottery manufacture were examined by Fontana et al. (1962). Clay deposits were considered tribal property, and no techniques, forms, or designs were the exclusive property of individuals, as was sometimes the case with Puebloid people. Pottery making was the domain of women, although men might help gather the clay or carve the wooden paddles used in the paddle and anvil technique of manufacture. Underhill (1939) observed that Tohono O'odham women would not "sit down" to pottery making until middle aged or "too old for the heavy work of corn grinding and food gathering."

Pottery making was a summer occupation because drying of green pots was less uniform in winter and the summers were more comfortable (Fontana et al. 1962). Various shapes of vessels were used for holding water, for boiling beans, for storing corn, for fermenting wine, and so forth and were referred to by names denoting utilization rather than shape.

Often more pots were made than needed for family use with an intent to trade or, with the advent of money, to sell the excess. Trading trips were made to the Akimel O'odham of the Gila River, to Mexico, and to the Yuma Indians of the Colorado River (Underhill 1939). Grain or willow splints were received from the Akimel O'odham, and in later years payment by various people could be made by cash, salt, flour, sugar, or coffee (Fontana et el. 1962).

Underhill (1939) noted that a water olla was worth enough beans or corn to fill two basket bowls. In early days of Tucson, Tohono O'odham women walked the streets with large burden baskets filled with clay pots for sale to townspeople. Such ollas were an inseparable part of the scenery at Arizona dude ranches in the first half of the twentieth century.

Unfortunately clay pots tend to break easily and become relatively useless for most intended functions. The hard work of making replacement pots finally was alleviated by the advent of metal cookware and metal water barrels. Water hauled by horse and wagon in a metal barrel or two revolutionized Tohono O'odham life by replacing a large number of water-storage ollas in the household and by ending the frequent need for long daily treks for water.

Pottery making by the Tohono O'odham is now nearly an extinct art, and the clay pots in the O'odham villages and Anglo dude ranches are rapidly decreasing in number by the inevitable attrition of breakage. Although the Tohono O'odham generally remember that beans tasted better when boiled in a clay pot (Fontana et al. 1962), they have nevertheless acceded to the forces of Anglo acculturation and now eat food cooked in metal pots.

SALT GATHERING AND SALT TRADE

Caravans by foot to the Gulf of California in the vicinity of the present-day Puerto Penasco to collect salt were a traditional O'odham event. The journey is thought to extend back into antiquity and had a religious significance discussed by Underhill et al. (1979). Some of this salt was used in cooking and for domestic purposes, but much of it was traded to the Akimel O'odham and others. One suspects that prehistorically the Tohono O'odham must have kept inland Indians well supplied with this commodity because records show that by 1860 they were selling very large quantities of salt to the residents of Tubac and Tucson; the superintendent of the Sonoran Mining Company at that date stated that he bought some 20,000 pounds of salt from them annually (Russell 1975).

ADDITIONAL ETHNOGRAPHIC STUDIES

Numerous ethnographic publications dealing with the Tohono O'odham have appeared. Treatments dealing with general aspects of their culture were published by Lumholtz (1971), Joseph et al (1949), and Underhill (1939). The role of women in life was studied (Underhill 1979), as was Tohono O'odham religion (Underhill 1946). Music and dance have been rather thoroughly documented by Densmore (1929), Gunst (1930), Chesky (1943), Underhill (1973), and Haefer (1977). The omnibus Viikita ceremony, with elements of thanksgiving and with either feigned or real drinking of saguaro wine and with hand-carried parade floats, has been described by Jones (1971), Chesky (1942), Hayden (1937), Davis (1920), and Mason (1920). An analysis of Tohono O'odham communities was presented by Jones (1969). Dictionaries of the O'odham language have been published by Saxton and Saxton (1969) and Mathiot (1973). Archaeological studies in Papagueria include those of Scantling (1939), Withers (1944), Haury (1950), Goodyear and Dittert (1973), Goodyear (1975a, 1975b), and Bruder (1975, 1977).

Western Hiach-Ed O'odham

The Hiach-eD O'odham were nomadic hunters and gatherers, their population in any one place being ephemeral. They roamed widely through the sand dunes at the head of the Gulf of California east of the Colorado River in a manner not dissimilar to that of unrelated coastal Seri to the south. They traded with the Cocopa (Fontana 1974) or Yuma (Childs 1954) but mainly moved about looking for wild food plants and game animals.

Lumholtz (1971) described the Hiach-eD O'odham as intelligent, healthy, and ablebodied although his informants told him that a large number had died about 1850 from a disease that included vomiting blood. This may have been an epidemic of yellow fever that affected other cultural facies of the O'odham about the same time. Lumholtz concluded that the existence of these people depended on their knowing the few places in the mountains where rain filled *tinajas* and knowing the few places along the coast where potable water could be found by digging.

Heaps of stones were often positioned to show the way to *tinajas* in the mountains surrounding the sand. Often a series of *tinajas* occurred one above the other. The lower ones seemed to dry up first, but water could be scooped from the upper to the lower. Sometimes as many as eight stone tanks naturally occurred one above the other, and access to the upper ones could be precipitous (Lumholtz 1971). It has been stated that the Hiach-eD O'odham jealously guarded their water supply from the guzzling horses, mules, and burros of strangers. Lumholtz (1971) believed that they were merciless to strangers, making travel on the roads dangerous and "nobody could follow them into the sand dunes to their principal retreat at Pinacate."

Literature on the Hiach-eD O'odham is meager. Father Kino and Father Kappus, traveling with a military escort under Juan Manje, encountered them in 1694. The first group seen were described as nearly naked people who covered parts of their body with jackrabbit fur, were poor and hungry, living on roots, locusts, and shellfish. The next group were similar and "lived by eating roots of wild sweet potatoes, honey, mesquite beans and other fruits. They travelled about naked; only the women had their bodies half covered with hare fur" (Manje 1954). In 1701 the Spaniards encountered a band of these Hiach-eD O'odham in the Sierra del Pinacate, counting 50 persons described as poor and naked, subsisting on roots, locusts, lizards, and some fish.

The Hiach-eD O'odham were already nearly extinct as an endemic group when Thomas Childs (born 1870) married a woman whose father belonged to that band. His sketch of the "Sand Indians" (Childs 1954) reveals much information that would otherwise have been lost to history. The same can be said of the publication by Lumholtz (1971) of ethnographic material on these people. He traveled through the homeland of the Hiach-eD O'odham in 1910 with two of them, one an old medicine man, as guides.

At the time of Lumholtz's 1910 expedition, the Hiach-eD O'odham had not lived in their sand dune country for 40 to 50

years, and they existed only as a few families who had taken up residence elsewhere. Lumholtz described one Hiach-eD O'odham guide as being "not very good in following tracks, but knew how to trap wild animals, and the rapid and dexterous manner in which he prepared them for eating reminded me of the Australian savages. His ideas about property were not highly developed."

On archaeological evidence Ezell (1954) concluded that the plainware pottery of these people was Yuman and the decorated types Piman. Fontana (1974) stated that the plainware was obtained by trade with the Cocopa. Childs (1954) stated that the Sand People made their own pottery, while Ezell (1954) suggested that they made their own utility pottery in a [borrowed] Yuman tradition while obtaining their decorated pottery in trade with O'odham in Sonora. If the Hiach-eD O'odham are defined as a non pottery-making, nonagricultural people who built no permanent houses, then ones who did any of these things are considered Areneños (Hiach-eD-like Tohono O'odham) by Fontana (1974).

Ezell (1954) believed that the nucleus of the Sand People in historic time was O'odham ("Papago") but cited a statement by Alberto Celaya that Seris, Tiburones, and Pima Bajo Indians had joined them. If the Hiach-eD O'odham did indeed receive recruits seeking a refuge first from Spanish conquest and later from encroaching civilization, as believed by Ezell, just as surely there is evidence of movement of O'odham from the Hiach-eD facies to the Tohono facies.

Childs (1954) "lived with and around them" all his life and described a period in history when older Hiach-eD O'odham remembered when the people had spent their entire existence as a nomadic people. Child's own experiences with them were during a period when they seasonally visited the dunes, Pinacate, and ocean to relive their history. Since Childs's father-in-law was Hiach-eD O'odham and his mother-in-law Tohono O'odham, he was in a good position to compare the dialects. He stated that the Hiach-eD people laughed at the speech of the Tohono and Akimel people, and vice-versa, and that the other O'odham thought that the Hiach-eD people talked with an accent, "like Chinamen." Childs stated that the Hiach-eD people killed only to eat and that if one had meat, they all had meat. He could tell when food was scarce because the Hiach-eD O'odham gathered around him in greater numbers. Since he was practically considered one of them, he was expected to share.

Tohono O'odham villages mentioned most frequently in connection with the Hiach-eD O'odham are Sonoita, Quitobaquito, and Quitovac. Indians in this region were considered to be Areneños by Fontana (1974) but not true Hiach-eD O'odham because they practiced some Tohono O'odham agriculture, made their own pottery, and lived in houses. There is ample historic evidence for migration from the Hiach-eD country east. A Hiach-eD informant of Lumholtz (1971) revealed that "the great annual feast, now given at Quitovac, was removed from the Pinacate region at least by 1840 because the old men who had charge of the ceremonial objects of the feast had died, and it was decided that the latter should be taken care of at Quitovac."

It is of interest that in the present *nawait* ritual of the Tohono O'odham, in which the invited villages sit at prescribed points of the compass according to long-standing tradition, the host village always takes the position to the west. The Mockingbird Speeches of this ritual clearly distinguish the dominant hunting and gathering economy of the Hiach-eD O'odham to the west from the more agricultural economy of the O'odham to the north, east, and south.

Childs (1954) noted that the O'odham still had the rain feast, with people coming from all around—Tucson, Sacaton, and Sells—to ask for rain. Childs's Hiach-eD informant stated that the rain feast used to be held by the Hiach-eD O'odham "down near Chibo Tanks" in the region of the sand dunes, an area traversed by various O'odham salt pilgrimages, which had a religious connotation and the people occupying the positions in the rain ceremony to the north, east, and south can only be speculated on.

Whether the rain feast of the Hiach-eD O'odham was attended by other O'odham people is not known. That ceremonial centers drawing O'odham from great distances have existed, however, is clear from the literature. The transcription by Kilcrease (1939) of the covered wells calendar stick makes it clear that Santa Rosa seemed to have become a ceremonial center for a large area by 1919. The unearthing by Russell (1975) of seventeenth-century records that "chieftains from as far north as the Gila Valley" were present at ceremonies in Remedios, Sonora, in 1698 is noteworthy. Ferdon's (1967) suggestion that prehistoric sites commonly referred to in Arizona as "Hohokam ball courts" were forerunners of O'odham ceremonial grounds, although speculative, was apparently based on the fact that whole villages were invited to participate in O'odham ceremonial events, and a large space was needed. Regardless of Hohokam theories, pan-O'odham ceremonies would have contributed to the remarkable ethnic and linguistic unity of the O'odham in the face of clear ecotypic differentiation patterns over a large geographic range, the eastern section of which had 5 times as much rain as the western section of the Hiach-eD O'odham.

HIACH-ED O'ODHAM SUBSISTENCE PATTERNS

The Sand People had abundant seafood resources, knew of a number of tasty wild food plants (Felger 1980), killed both large and small game animals, utilized some food from small birds, large lizards, and insects, and possibly (although this point is debated) grew a very small quantity of agricultural crops. Lumholtz (1971) stated that they made an annual journey to Yuman to trade baskets and sea shells for maize, tepary beans, and squash.

Childs (1954) related how two Hiach-eD O'odham men, Caravajales (sometimes referred to in the literature as a "hermit" because of reverting to the Hiach-eD facies of his youth) and José Augustin, caught fish at the Gulf in the traditional way. These fish were totoaba, some over 250 pounds in weight. The Indians waited until the moon changed from full to new and the tides were high. As the tide rushed in, the large fish came in to eat the small fish. As the tide broke, the fish were left stranded where they could easily be speared and then strung on a rope. The fish spear often consisted of the retrorsely barbed underside of the tail of the sting ray (*Dasyatis*) lashed to a willow branch. An incipient fisheries management system is suggested by Fontana's (1974) mention that sometimes the Hiach-eD O'odham built low stone enclosures for the fish to become stranded in.

Surveyors for the Southern Pacific Railroad observed these people in 1854 (Gray 1856). It was noted that all of them had very short front teeth, but the cause could not be determined. Childs (1954) solved the mystery by reporting that the Hiach-eD O'odham "ate their Missmire Clams raw and opened them with their teeth." The teeth were worn down from abrasion in opening the clam shells. These clams lived in the sand and were very easy to gather. Oysters lived out on coral reefs, and since the Hiach-eD O'odham had no boats, this food source was difficult to obtain and then only at low tide. Fontana (1974) stated that they ate sea turtle, fish, shrimp, clams, and oysters, both fresh and dried.

Hiach-eD O'odham killed mountain sheep with bows and arrows, especially in the large craters of the Pinacates (Lumholtz 1971). They refrained from killing these animals unless needed for food and then only in the quantity required (Childs 1954). When all other food failed, the Hiach-eD O'odham would ask I'itoi, the legendary Elder Brother, to send an old mountain sheep ram that I'itoi had no further use for. They would send a man to sit and wait with bow and arrow at a large rock on the east side of the Pinacate. An old ram invariably would appear when needed and would not run away, as if it had been sent by I'itoi.

Mountain sheep were not particularly difficult to kill and were one of the most important food reserves of the Hiach-eD O'odham. The sheep would often stand motionless at a distance, and the Indians would climb stealthily into the mountains to surprise them. Lumholtz (1971), however, found places where the sheep could go that were too precipitous for man. Once mountain sheep were killed, the horns would be added to shrinelike piles scattered over the desert (Fontana 1974).

Hiach-eD O'odham hunters could approach near enough to mule deer and antelope to kill them with bow and arrow (Lumholtz 1971). Pronghorn antelope grazed on flat llanos.

When among creosote bush, the antelope were easily spotted from the color, which though light reddish-brown above, presented a white aspect from the legs and underside. At a distance they were hard to distinguish from cholla cactus (*Opuntia fulgida*), however, according to Lumholtz (1971). Antelope were chased until the animals stopped a distance away. The hunter would then lie down and wait for an animal to move, then chase again, lie down again, and continue the cycle until the animal tired, lay down, and could be killed (Childs 1954).

Packrats were flushed from their nests by the Hiach-eD hunter's setting fire to the cactus spines that lined their burrows. When the rodents appeared, they were shot with bow and arrow. In summer jackrabbits (*Lepus*) were run down by foot in the sand (Lumholtz 1971), but cocopa could outrun the Hiach-eD people in the sand according to Childs (1954). Lumholtz described the preparation of a jackrabbit for eating by an elderly Hiach-eD O'odham man. The legs were skillfully broken with a stick in a few seconds, and the hair was singed from the body by holding it over the fire. He had first cut off the ears, which he reserved for himself, "for they are considered a great delicacy."

Meat of both large game and small game was dried for future use (Lumholtz 1971), usually being cut up and strung on ropes for this process (Childs 1954). Quails and probably doves were eaten but buzzards, hawks, and eagles were taboo as food (Fontana 1974). Quails being ground birds, were trapped in a saguaro rib box propped up by a stick leading to an Indian boy hiding behind a bush. The lizards that were eaten belonged to the family Iguanidae (Fontana 1974). The "locusts" referred to by Spanish explorers as Hiach-eD O'odham food were probably the *ma-kum* referred to by Greene (1936) and Russell (1975) that were processed by various O'odham into a tasty snack chip munched while drinking saguaro wine. These sphinx moth larvae (*Celerio lineata*) were processed by Hiach-eD O'odham by removing the heads and inner organs, placing them between two ollas, and roasting them by covering them with hot coals (Childs 1954).

Felger (1980) has made a rather complete listing of the food plants that these people probably used. The sand root (*Ammobroma sonorae*) was one of the most important foods of the Hiach-eD O'odham. This large fleshy parasite of *Franseria, Dalea, Coldenia, Eriogonum*, and *Pluchea* bushes is endemic to the sand dunes at the head of the Gulf of California. The plant is about three feet long, subterranean, without leaves, and "looks like a root covered with scales which grow thicker toward the top" (Lumholtz 1971). When eaten raw, the plant is tenderer and more succulent than a radish, satisfying both hunger and thirst. When toasted on coals of a campfire, it tastes somewhat like sweet potatoes. Unfortunately the plant generally can be found only in March and April when it protrudes slightly from the ground and produces

flowers. Some of the Hiach-eD O'odham proficient at gathering the plant were said to be able to find it during any time of year. Some were said to live chiefly by eating this plant.

Pink or yellow flower buds of certain species of cholla cactus (*Opuntia*) were gathered, allowed to dry slightly, hit with a stick to remove the spines, and roasted. This food, *chi'odima*, had a pleasing somewhat sour taste, but residual spines could be a problem if the food was not carefully prepared (Childs 1954). Lumholtz thought that the Hiach-eD O'odham roamed as far east as Quitobaquito and Santo Domingo to gather mesquite beans, saguaro fruit, and pitahaya fruit. He observed that bedrock mortars from 6 to 10 inches deep used for pounding mesquite beans were seen near all of the *tinajas* and that frequently the pestles were found nearby. Childs (1954) stated that when mortars in large boulders had been well used, the huge boulders were turned over and new pounding holes for mesquite preparation were made.

The Hiach-eD people toasted, ground, and mixed with water the beans of palo fierro to make pinole, a nutritious beverage (Lumholtz 1971). The seeds of Indian wheat (*Plantago*) were eaten, either uncooked or toasted, and ground into pinole as well. Evening primrose (*Oenothera*) growing on sand dunes was abundant after scant winter rains, some of the clumps being over 5 feet in diameter and having more than 100 flowers (Lumholtz 1971). These juicy plants were boiled and eaten.

Other plant food utilized by the Hiach-eD O'odham recorded by Fontana (1974) consisted of wild century plant (*Agave*), stalks and seeds of careless weed (*Amaranthus palmeri*), the bulb of covena (*Brodiaea capitata*), the tender new growth of lambsquarters (*Chenopodium murale*), and patata (*Monolepis nuttalii*). Childs (1954) thought that the Hiach-eD O'odham smoked tobacco "from time immemorial."

Childs (1954) described a single Hiach-eD O'odham agricultural site lying "about 12 miles east of the Soda on the Gulf, and about 8 miles west of Papago Tanks" where watermelons, pumpkins, beans, corn, and wheat were grown. Since Childs received much of his information from the Hiach-eD O'odham "hermit" Caravajales and Lumholtz (1971) recorded that Caravajales lived at Los Papagos and grew a little maize, a few squashes, and watermelons, the garden plot referred to by Childs must have been that of Caravajeles. The exact location of this Hiach-eD O'odham agricultural site cannot be determined from these references. Elsewhere, however, Lumholtz (1971) spoke of a single agricultural site in Hiach-eD O'odham country, stating that the majority of the people had no agriculture whatsoever. This site, where beans, maize, and squash were grown, was identified as Suvuk by Lumholtz and according to his map was slightly east of all mapped Hiach-eD O'odham camping spots and west of all other sites mapped as sites of agricultural (Thono or Akimel) O'odham. Because this was an agricultural site, it is placed on the present map as a Tohono O'odham location, the westernmost. (figure 5-16).

OTHER DETAILS

According to Fontana the Hiach-eD people did not have houses. They piled up boulders one or two courses high and slept within the enclosure as a windbreak. Clumps of grass were placed against the roofless shelters to deaden the force of the wind and penetration of the cold further. Lumholtz (1971) claimed that in winter, they erected grass huts, but he did not state that he ever saw one. Such a rumored structure may have been nothing more than a stone sleeping ring with grass clumps piled around it.

The Hiach-eD O'odham wore their hair long and periodically plastered it with mud to kill lice (Childs 1954). Men, but apparently not women, might keep long hair out of the eyes by tying it with badger hair plaited into a cord or ribbon (Lumholtz 1971). Although the Hiach-eD people were frequently referred to as practically naked, they did wear clothing, although not as much as the overdressed Spaniard (Fontana 1974).

Lumholtz (1971) stated that the Hiach-eD O'odham clothing came from the skin of mountain sheep, antelope and muledeer. Childs (1954) mentioned a preference for antelope skin because it was tanned more easily than deer skin. Early explorers mentioned rabbit fur. In reality, the clothing may have depended on whichever kind of animals the people had been eating.

In preparing the skins, "The hair was removed with a bone taken from the lower foreleg of the animal" (Lumholtz 1971) or with a whale rib bone found lying on the coast (Childs 1954). The skin was smeared with brains of the animal killed to soften it and rubbed with a rock for the same purpose. The root of *Jatropha* was crushed and steeped in water to make a solution for tanning the leather.

Hiach-eD O'odham men wore a breech cloth held up by cord made from badger hair and occasionally wore a skin shirt. Women apparently wore a short skirt but were naked above (Lumholtz 1971). This skirt had fringes hanging down (Childs 1954) and apparently consisted of nothing more than these fringes (Fontana 1974) hanging from buckskin strings (Lumholtz 1971). The leather clothing was often colored with mineral or plant dyes to make it red or yellow (Childs 1954). Sandals and straps were made from sea lion skin, the sea lions being killed where they rested on shoreline rocks along the Gulf (Fontana 1974) by striking them on the nose with rocks (Lumholtz 1971).

The Hiach-eD people constructed baskets from torote prieto (*Jatropha*), willow (*Salix*), and bulrush (*Scirpus*). They made carrying nets but not burden baskets. They traveled to the Colorado River for true willow (*Salix*) for making bows (Lumholtz 1971) or made them from desert willow [*Chilopsis?*], according to Fontana (1974). Arrows were made from arrow-weed (*Pluchea sericea*) and plumed with hawk feathers (Childs 1954).

Akimel O'odham of the East and Peripheral Rivers

Although the Akimel O'odham were once distributed in a broad ecotone on either side of the eastern boundary of the Sonoran Desert (figure 5-16), they also extended downstream (west) along the Gila River and Rio Concepción. Akimel O'odham who survived into the twentieth century live on the Gila in central Arizona between Coolidge and Phoenix and on the Salt River north of Mesa. The settlement on the Salt was made in the nineteenth century from the Gila at the request of Mormon settlers who wanted a buffer of friendly Indians between them and the Apache.

The map of Akimel O'odham sites in historic time (figure 5-16) reveals that the Gila River contingent represented a westerly arm of the total distribution of the Akimel facies. These Gila River Indians repeated a tradition to Russell (1975) that they had come from the east, and indeed some of their ancestors even in historic time lived east of the Sonoran Desert along the San Pedro River. That the Akimel O'odham may have extended their range onto the Gila about 1450 is suggested (but not proven) by the Akimel O'odham legend recorded by Russell (1975) detailing the order and manner of their laying siege to the Salado ("Hohokam") pueblos beginning with the Casa Grande near present-day Coolidge. The relation of the Akimel O'odham to the true pit-house (as opposed to pueblo) dwelling Hohokam people is uncertain. Haury (1976) concluded that the Hohokam were declining culturally when the Salado built their pueblos among them and were further weakened during the Salado occupation. Haury's contention that the Akimel O'odham represent the remnants of the true Hohokam (not Salado) culture is unproven, however.

SUBSISTENCE PATTERNS

The Akimel O'odham could have lived on their agricultural products alone if the Gila River contingent is at all characteristic. These people had numerous mesquite recipes, however, and it is suggested that one-fifth of their food came from this river floodplain tree. Although an additional one-fifth of their food probably came from hunting, fishing, and gathering, the plants and animals exploited mostly lived in the river-desert econtone and in the enhanced riverine environment of the Akimel O'odham canals, fencerows, and fields. Wheat probably represented one-fifth of their food, corn and beans another fifth, with various agricultural products comprising the remaining one-fifth.

PRE-EMINENCE OF AGRICULTURE

Although Hiach-eD and Tohono O'odham people experienced periodic food shortages, the Akimel O'odham, who characteristically practiced irrigation agriculture in the rich riverine environments, often had food surpluses. Much of the annual excess was traded to the Tohono O'odham. But a little-known fact is that the Akimel O'odham fed the great quantities of people passing through the Sonoran Desert when California was being settled in the nineteenth century. The Gila villages were on the direct route and became a major stopping point for transcontinental travel because the inhabitants were friendly and the food was abundant and good.

J. Ross Brown passed through the Gila villages in 1864 and recorded his observations on Akimel O'odham agriculture (Brown 1869). Travelers passing through the villages had been fed for some time by 1858 when the Overland Mail stagecoach line was established with a stop there. The successful nature of Akimel O'odham agriculture is indicated by the fact that in the first year of this transportation company's existence, it purchased 100,000 pounds of surplus wheat from these Indians. By 1861 the annual purchase was up to 300,000 pounds of wheat, 50,000 pounds of corn, 20,000 pounds of beans, and large quantities of dried and fresh pumpkin (Brown 1869).

In 1862 the entire California column of the U.S. Army, charged with protecting the western United States from the Confederacy, about 1000 men strong, received their food "for many months" (Brown 1869) from the Akimel O'odham. In that year to the U.S. government alone, these Indians sold over 1 million pounds of wheat, aside from substantial quantities of "pinole, chickens, green peas, green corn, pumpkins and melons."

Brown (1869) commented on the productivity of the Akimel O'odham fields. Wheat was planted in December and January, harvested before the summer rains, and the same fields planted to corn, pumpkins, melons, and other vegetables for a second crop. Tobacco and cotton were planted "when Mesquite leafs out" in early March. Brown (1869) correctly suggested that silt from the irrigation waters renewed the fertility of the fields. He commented that in 300 years of recorded history, the same fields were known to produce bountiful crops each year without the addition of manure or fertilizer other than the silt of the irrigation water.

Rea (1979a) described Akimel O'odham irrigation practices. By means of weirs or brush dams across the Gila River, they diverted water into canals, which ran sometimes for miles through level fertile terraces of the floodplains. A succession of fields, canals, ditches, and fence rows increased the biological productivity of the land near the river. This enhanced riverine environment not only allowed the farming of crops on a regular basis but provided habitat for game, encouraged edible wild plants, and provided pastures for grazing stock.

The Akimel O'odham waited until floodwaters turned "half way muddy, half way clear" (Rea 1979a) before using the river's water for irrigation. The water chosen helped renew the fertility of the fields without leaving a muddy deposit of

adobe, which would have cracked on drying. With the introduction of winter wheat to the Southwest by Father Kino, the agriculture of the Akimel O'odham increased and prospered. The impact of winter wheat on Akimel O'odham agriculture is indicated by the fact that although one of Russell's (1975) informants in 1902 gave the traditional "saguaro harvest month" (June) as the beginning of the O'odham year, another already used "wheat harvest month" for May and considered this the beginning of the Akimel O'odham year. The river customarily flooded each year from snowmelt in the mountains in late winter and spring, just in time to provide water to crops which had been planted in winter.

UTILIZATION OF MESQUITE

Rea (1979b) considered velvet mesquite (*Prosopis velutina*) to be the most important native plant in the ecology of the Akimel O'odham of the Gila River, calling it their "tree of life." This mesquite is characteristic of the riparian zone and floodplains of rivers and streams of the country occupied by Akimel O'odham, occurring in large *bosques* or riverine forests. One bosque was reported by Rea (1979b) to extend for as far as 6 miles near the mouth of the Santa Cruz River. The importance of mesquite to the Akimel O'odham can be seen from their recognition of a "mesquite leaves month" when the tree leafs out in spring and a "mesquite flowers month" when it blooms as spring is about to change into summer.

Although mesquite flowers could be eaten in May and the immature long, thin green string-bean like pods could be eaten in June, the real harvest was not until August when the mature pods fell to the ground. These seed pods were gathered when thoroughly dry in enormous quantities and stored in granaries for later use. The stored pods eventually were crushed with a stone pestle in a wooden mortar made out of a 2-foot log cut from a mesquite or cottonwood trunk. Flour was sifted out from the crushed pod fragments by means of a sifting basket. This flour consisted of the dried sweet pulp of the pods after the seeds and fiber had been discarded.

A loaf or cake 6 inches or more thick was made by moistening repeated increments of mesquite meal. When the cake was thoroughly dry, it was extremely hard and would keep through the year. To eat this *chuuk* it was necessary to break pieces off by hitting the cake with a stone.

Other Akimel O'odham mesquite recipes were also researched by Rea (1979b). A pudding made from mesquite flour and water looked like butterscotch but was richer. Rea thought that this *vihuk hidut* tasted like carob. Thick wheat tortillas were layered like pancakes with dried mesquite pods in between, covered with water, and cooked slowly for several hours until almost dry. The pods were discarded and the mushy *chuchumit* eaten. Another dish, *hawhawhi wichoda*, was made by boiling dried mesquite pods until soft, cooling, then adding whole-wheat dumplings, and cooking until the dish thickened. *Vihok wongim chui* consisted of wheat pinole sweetened with mesquite flour. *Vaa'o* was a sweet drink prepared by soaking crushed mesquite pods in cold water.

HUNTING, FISHING, AND GATHERING

Big game such as mule deer and bighorn sheep were hunted, but these apparently had become less common in Akimel O'odham time than during Hohokam occupation of the region. Although calendar-stick records of the Akimel O'odham indicate that they sometimes went into the Sierra Estrella to gather saguaro cactus fruit and that they explored as far north as Prescott, the floodplain of the river yielded cottontail rabbit (*Sylvilagus auduboni*), jackrabbit (*Lepus californicus*, and *L. alleni*), raccoon (*Procyon lotor*), Gambel's quail (*Callipepla gambelii*), cotton rat (*Sigmodon hispidus*), and pack rat (*Neotoma albigula*) used as food (Rea, 1979a). Habitat for these small animals was greatly enhanced by fence rows built around the fields. These consisted of two parallel fence lines with a 6-foot width of spiny brush stacked head high in between. New branches were added at the top as the lower layers deteriorated. Often fish could be gathered up from the fields where they were left stranded from irrigation water.

Wild greens or "spinaches" growing on the banks of canals included *awpon* (*Monolepis nuttaliana*) in winter, *onk iivak* (*Atriplex wrightii*) in spring, *chuhugia* (*Amaranthus palmeri*) in summer, and *ku'ukpark* (*Trianthema portulacastrum*) in summer. Berries of *Lycium* and *Ziziphus* were produced by bushes that grew in the fence rows.

Volunteer wild food plants coming up in the fields included devil's claw (*Proboscidea parviflora*), patota (*Monolepsis nuttalliana*), and goosefoot (*Chenopodium*). Rea (1979a) pointed out that biological activities, productivity, and diversity increased due to the edge effect in the ecotone where broadleaved river woods and low desert scrub came together in complex associations with Akimel O'odham fields, fence rows, and canals.

OTHER DETAILS

The housing arrangement of the Akimel O'odham was similar to that of the Tohono O'odham except that riverine plants were used in the construction. The standard *kih* had the four forked mesquite posts, on which rested substantial willow or cottonwood poles. According to Rea (1979b), "Thin willow poles made a framework for the walls and roof, which was thatched with arrowweed, willow, cattails, or even corn stalks, covered with a thick layer of adobe mud." Plaited mats were sometimes spread out over the exterior of the house (Fontana, 1974).

The Akimel people plaited sleeping mats from carrizo cane (*Phragmites communis*) of river cane brakes. Large granaries placed on the roof of the *kih* or the *watto* were constructed from arrowweed (*Pluchea sericea*) and willow (*Salix gooddingii*). Large granary storage baskets, so large that the weaver had to stand within them, were made from coils of wheat straw sewn with strips of mesquite or willow bark. Russell (1975) presented a monograph on the Akimel O'odham, with particular attention to their material culture. Apparently most, if not all, of the material culture of the Tohono O'odham was characteristic of the Akimel O'odham as well; the significant differences lay in the degree of utilization of resources beyond the floodplain. Ezell (1961) traced the hispanic acculturation of the Gila River contingent. Shaw (1974) presented the remembrances of an Akimel O'odham woman.

Conclusions

Although there are significant differences in subsistence patterns among the three O'odham subgroups of southern Arizona and adjacent Sonora, the overall cultural and linguistic unity is nevertheless impressive. The historic patterns of distribution of the three facies and perhaps of incipient ethnogenesis can be explained by reference to the pattern of increasing rainfall from west to east and by the nature of the desert rivers arising in the mountains of the east and flowing west.

Life-styles of the historic O'odham have been heavily influenced by the ecology of the Sonoran Desert. It is not surprising, therefore, that their subsistence patterns often relate to those of prehistoric cultures in the same regions. Where prehistoric and historic similarities are greatest, the common ties seem to be of nothing more than an ecological nature. Dissimilar ethnic groups could easily have converged independently by ecologically induced acculturation even though well separated in time and not cross-conversant. Those impressive features of the Hohokam culture that relate the least to the environment are exactly those features missing from the culture of the O'odham. The statistical study of this subject might represent a fruitful research project.

Of the three O'odham subgroups, the one that has survived with the least change is the Tohono facies. The Akimel people, on the routes traversed by bearers of European culture, changed more readily. They adopted culture of winter wheat using irrigation from spring snowmelt in the mountains. The Hiach-eD facies, apparently of a relictual nature, may have waxed and waned with both historic and prehistoric cycles of acculturation and of conquest, but the habit of retreating into the Pinacate where they could not be followed may have bought time and allowed them to keep their ancient ways longer. This O'odham facies could not cope with the revolutionary changes that occurred in historic time other than by hiding in the sand and trying to avoid them, and was forced into extinction as a recognizable subculture.

If agricultural acculturation of the O'odham occurred upon an advancing frontier, then the legend common to the Tohono

O'odham of corn (Zea mays) being brought to them from the east and of their searching for suitable planting spots would suggest that the Hiach-eD people might have been those who could not find favorable moist areas and continued their old way of life. The ritual speeches of the *nawait* ceremony suggest that the proto-Tohono facies may have been existing at the Hiach-eD cultural level. If so, the important contribution of the proto-Hiach-eD may have been the transmittal of hunting and gathering fundamentals to produce a people who became enriched through agricultural acculturation to become the strong Tohono O'odham subculture that has been successful over a large expanse of the basin and range province in the Sonoran Desert.

References

Abbott, P. L., and Smith, T. E. 1978. Trace-element comparison of clasts in Eocene conglomerates, southwestern California and northwestern Mexico. *J. Geol.* 86:753-762.

Adam, K. R., and Weiss, C. 1959. Scorpion venom. *Z. Tropenmed. Parasitol.* 10:334-339.

Adams, M. S., and Strain, B. R. 1969. Seasonal photosynthetic rates in stems of *Cercidium floridum* Benth. *Photosynthetica* 3:55-62.

Adolph, E. F. 1943. Physiological fitness for the desert. *Fed. Proc.* 2(3):158-164.

Ahearn, G. A. 1970. The control of water loss in desert Tenebrionid beetles. *J. Exptl. Biol.* 53:349-359.

———. 1971. Ecological factors affecting population sampling of desert Tenebrionid beetles. *Amer. Midl. Nat.* 86:385-406.

Aiken, C. L. V., and Sumner, J. S. 1974. *A geophysical and geological investigation of potentially favorable areas for petroleum exploration in southwestern Arizona*. Phoenix: Arizona Oil and Gas Commission.

Al-Ani, H. A.; Strain, B. R.; and Mooney, H. A. 1972. The physiological ecology of diverse populations of the desert shrub *Simmondsia chinensis*. *J. Ecology* 60:41-57.

Alcorn, S. M.; McGregor, S. E.; Butler, G. D.; and Kurtz, E. B. 1959. Pollination requirements of the saguaro (*Carnegiea gigantea*). *Cactus Succulent J.* 31(2):39-41.

Alcorn, S. M., and May, C. 1962. Attrition of a saguaro forest. *Plant Dis. Rept.* 46(3):156-158.

Alden, P. 1969. *Finding the birds in western Mexico—a guide to the states of Sonora, Sinaloa, and Nayarit*. Tucson: University of Arizona Press.

Aldous, A. E., and Shantz, H. L. 1924. Types of vegetation in the semi-arid portion of the United States and their economic significance. *J. Agr. Res.* 28:99-127.

Alexander, C. E., and Whitford, W. G. 1968. Energy requirements of *Uta stansburiana*. *Copiea* 4:678-683.

Alpert, P. 1979. Desiccation of desert mosses following a summer rainfall. *Bryologist* 82(1):65-71.

Alvarez, T. 1976. Status of desert bighorns in Baja California. *Desert Bighorn Council Transact.* 1976:18.

Amen, R. D. 1968. A model of seed dormancy. *Bot. Rev.* 34:1-31.

Amin, O. 1969. Helminth fauna of suckers (Catostomidae) of the Gila River system, Arizona. *Amer. Midl. Nat.* 82:188-196, 429-443.

Anderson, A. H., and Anderson, A. 1946. Notes on use of creosote bush by birds. *Condor* 48:179.

———. 1973. *The cactus wren*. Tucson: University of Arizona Press.

Anderson, R. E. 1971. Thin skin distension in Tertiary rocks of southeastern Nevada. *Geol. Soc. Amer. Bull.* 82:43-58.

Andreev, S. F. 1948. Adaptations of reptiles to high temperatures of deserts. *Uch. Zap. Biol. Fak. Univ. Chernovitsy* 1:109-118.

Arizona Game and Fish Department. 1979. *Arizona big game investigations, 1977-1978*. F. A. Project W-53-R-28. Phoenix.

Armstrong, D., and Jones, J. 1972. *Notiosorex crawfordi*. *Mammalian Species* 17:1-5.

Arnal, R. E. 1961. Limnology, sedimentation, and microorganisms of the Salton Sea, California. *Bull. Geol. Soc. Amer.* 72:427-478.

Arnett, R. H.: Jr. 1960-1968. *The beetles of the United States*. Ann Arbor, Mich.: American Entomological Institute.

Arnold, R. E. 1973. *What to do about bites and stings of venomous animals*. New York: Collier Books.

Aschmann, H. 1959. Reprinted as H. Aschmann. 1967. q.v.

———. 1967. *The Central desert of Baja California: demography and ecology*. Riverside, Calif.: Manessier Publishing Company.

Asplund, K. K. 1964. Seasonal variation in the diet of *Urosaurus ornatus* in a riparian community. *Herpetologica* 20:91-94.

———. 1967. Ecology of lizards in the Relictual Cape flora, Baja California. *Amer. Midl. Nat.* 77:462-475.

———. 1970. Metabolic scope and body temperatures of whiptail lizards (*Cnemidophorus*). *Herpetologica* 26:403-410.

Asplund, K. K., and Lowe, C. H. 1964. Reproductive cycles of the iguanid lizards *Urosaurus ornatus* and *Uta stansburiana* in southeastern Arizona. *J. Morph.* 115:27-33.

Atstatt, S. R. 1939. Color changes as controlled by temperature and light in the lizards of the desert regions of southern California. *Univ. Calif. (Los Angeles) Publ. Biol. Sci.* 1:237-276.

Auffenberg, W., and Milstead, W. W. 1965. Reptiles in the quaternary of North America. In H. E. Wright and D. G. Frey (eds.), *The Quaternary of the United States*, pp. 557-568. Princeton: Princeton University Press.

Austin, G. T. 1974. Nesting success of the cactus wren in relation to nest orientation. *Condor* 76:216-217.

———. 1976. Behavioral adaptations of the verdin to the desert. *Auk* 93:245-262.

Axelrod, D. I. 1950. Studies in late Tertiary paleobotany. *Carneg. Inst. Wash. Publ.* 590:1-323.

———. 1958. Evolution of the Madro-Tertiary geoflora. *Bot. Rev.* 24:453-509.

———. 1967. Drought, diastrophism and quantum evolution. *Evolution* 21:201-209.

———. 1972. Edaphic aridity as a factor in angiosperm evolution. *Amer. Nat.* 106:311-320.

———. 1979. Age and origin of Sonoran desert vegetation. *Occ. Pap. Calif. Acad. Sci.* 132:1-74.

Ayres, J. E. 1971. Man, the desert farmer. In *Hydrology and water resources in Arizona and the Southwest*, pp. 373-379. *Amer. Water Resources Assoc., Ariz. Sect. Proc. 1971 Meetings*, Tempe, Arizona.

Bailey, V. E., and Sperry, C. C. 1929. Life history and habits of grasshopper mice, genus *Onychomys*. *U.S.D.A. Techn. Bull.* 145:1-20.

Balinsky, J. B.; Cragg, M. M.; and Baldwin, E. 1961. The adaptation of amphibian wastes nitrogen excretion to dehydration. *Comp. Biochem. Physiol.* 3:236-244.

Ball, E. D. 1936. Food plants of some Arizona grasshoppers. *J. Econ. Entomol.* 29:679-684.

Ball, E. D. et al. 1942. The grasshoppers and other Orthoptera of Arizona. *Ariz. Agr. Exp. Sta. Techn. Bull.* 93:257-373.

Bancroft, G. 1926. The faunal areas of Baja California del Norte. *Condor* 28:209-215.

Banks, R. C. 1964. The mammals of Cerralvo Island, Baja California. *Trans. San Diego Soc. Nat. Hist.* 13:397-404.

Barber, W. E., and Minckley, W. L. 1966. Fishes of Aravaipa Creek, Graham and Pinal counties, Arizona. *Southwest Nat.* 11:313-324.

Barber W. E.; Williams, D. C.; and Minckley, W. L. 1970. Biology of the Gila spikedace, *Meda fulgida*, in Arizona. *Copeia* 1970:9-18.

Barbour, M. G. 1968. Germination requirements of the desert shrub *Larrea divaricata*. *Ecology* 49:915-923.

Barbour, R. W., and Davis, W. H. 1969. *Bats of America*. Lexington: University of Kentucky Press.

Barlow, G. W. 1958. High salinity mortality of desert pupfish, *Cyprinodon macularius*, in the field and in the aquarium. *Amer. Midl. Nat.* 65:339-358.

Bartholomew, G. A. 1950. The effects of artificially controlled temperature and day length on gonadal development in a lizard, *Xantusia vigilis*. *Anat. Rec.* 106:49-60.

————. 1953. The modification by temperature of the photoperiodic control of gonadal development in the lizard, *Xantusia vigilis*. *Copeia* 1953:45-50.

————. 1960. The physiology of desert birds. *Anat. Rec.* 137:338.

————. 1963. Behavioral adaptations of mammals to the desert environment. *Proc. 16th Inter. Congr. Zool.* Washington, D.C. 3:49-52.

————. 1964. The roles of physiology and behavior in the maintenance of homeostasis in the desert environment. *Symp. Soc. Exp. Biol.* 18:7-29.

————. 1971. The water economy of seed-eating birds that survive without drinking. *Int. Orn. Congr.* 15.

Bartholomew, G. A., and Cade, T. J. 1963. The water economy of land birds. *Auk* 80:504-539.

Bartholomew, G. A., and Dawson, W. R. 1968. Temperature regulation in desert mammals. In G. Brown (ed.), *Desert biology*, vol. 1, pp. 395-421. New York: Academic Press.

Bartholomew, G. A., and Hudson, J. W. 1961. Desert ground squirrels. *Sci. Amer.* 205(5):107-116.

Barton, L. 1936. Germination of some desert seeds. *Contrib. Boyce Thompson Inst.* 8:7-11.

Bartram, E. B., and Richards, D. 1941. Mosses of Sonora. *Bryologist* 44:59-65.

Bateman, G. C. 1967. Home range studies of a desert nocturnal rodent fauna. Ph.D. dissertation, University of Arizona.

Beard, R. L. 1963. Insect toxins and venoms. *Ann. Rev. Entomol.* 8:1-18.

Beck, B. B.; Engen, C. W.; and Gelfand, P. W. 1974. Behavior and activity cycles of Gambel's quail and raptorial birds at a Sonoran Desert waterhole. *Condor* 75:466-470.

Beers, G. D., and McConnell, W. J. 1966. Some effects of threadfin shad introduction on black crappie diet and condition. *J. Ariz. Acad. Sci.* 4:71-74.

Beland, R. D. 1953. The effect of channelization on the fishery of the lower Colorado River. *Calif. Fish and Game* 39:137-139.

Belk, D. 1970. Zoogeography of the Arizona Anostraca with a key to the North American species. Ph.D. dissertation, Arizona State University.

Belk, D., and Cole, G. A. 1975. Adaptational biology of desert temporary-pond inhabitants. In N. F. Hadley (ed.), *Environmental physiology of desert organisms*, pp. 207-226. Stroudsburg, Pa.: Dowden, Hutchinson & Ross.

Beltrán, E. 1934. *Lista de peces Mexicanos*. Secretaría de Agricultura y fomento. Mexico. D.F.: Instituto Bio Técnico.

Bemis, W. P., and Whitaker, T. W. 1969. The xerophytic *Cucurbita* of northwestern Mexico and southwestern United States. *Madroño* 20:33-41.

Benes, E. S. 1966. Progressive color discrimination in the lizard, *Cnemidophorus tigris (Teidae)*. Ph.D. dissertation, University of California.

Bennett, E. L., and Bonner, J. 1953. Isolation of plant growth inhibitors from *Thamnosma montana*. *Amer. J. Bot.* 40:29-33.

Benson, L. 1941. The mesquite and screw-beans of the United States. *Amer. J. Bot.* 28:748-754.

————. 1969a. *The cacti of Arizona*. 3rd ed. Tucson: University of Arizona Press.

————. 1969b. *Native cacti of California*. Stanford: Stanford University Press.

————. 1980. *Trees and shrubs of the southwestern deserts*. Tucson: University of Arizona Press.

Bentley, P. J. 1966. Adaptation of amphibia to arid environments. *Science* 152:619-623.

Bentley, P. J., and Schmidt-Nielsen, K. 1966. Cutaneous water loss in reptiles. *Science* 151:1547-1549.

Bequaert, J. C., and Miller, W. B. 1973. *The mollusks of the arid Southwest*. Tucson: University of Arizona Press.

Bernstein, L. 1964. Salt tolerance of plants. *U.S.D.A. Agr. Inform. Bull.* 283:1-23.

Bernstein, L., and Howard, H. E. 1958. Physiology of salt tolerance. *Ann. Rev. Plant Physiol.* 9:25-46.

Berry, E. L. 1941. A monograph of the genus *Parmelia* in North America, north of Mexico. *Ann. Mo. Bot. Gard.* 28:31-146.

Berry, K. H. 1974. The ecology and social behavior of the chuckwalla, *Sauromalus obesus obesus* Baird. *Univ. Calif. Publ. Zool.* 101:1-60.

————. 1978. Livestock grazing and the desert tortoise. *Proc. Forty-third North American Wildlife Conference: 505-519.*

Bersell, P. O. 1973. *Vertical distribution of fishes relative to physical, chemical and biological features in two central Arizona reservoirs.* Tempe: Arizona State University.

Beutner, E. L., and Anderson, D. 1943. The effect of surface mulches on water conservation and forage production in some semi-desert grassland soils. *Agronomy J.* 35:393-400.

Bidwell, G. L., and Wooton, E. O. 1925. Salt bushes and their allies in the United States. *U.S.D.A. Bull.* 1345:1-40.

Bigler, W. J. 1974. Seasonal movements and activity patterns of the collared peccary. *J. Mammalogy* 55(4):851-855.

Bikerman, M. 1967. Isotopic studies in the Roskruge Mountains, Pima County, Arizona. *Geol. Soc. Amer. Bull.* 78:1029-1036.

Bingham, S. B. 1963. Vegetation-soil relationships in two stands of the Cercidium-Carnegiea community of the Sonoran desert. Master's thesis, University of Arizona.

Blair, W. F. 1976. Adaptation of anurans to equivalent desert scrub of North and South America. In D. W. Goodall (ed.), *Evolution of desert biota*, pp. 195-222. Austin: Univ. of Texas Press.

Blair, W. F.; Hulse, A. C.; and Mares, M. A. 1976. Origins and affinities of vertebrates of the North American Sonoran Desert and the Monte Desert of northwestern Argentina. *J. Biogeogr.* 3:1-18.

Blair, W. F., and Pettus, D. 1954. The mating call and its significance in the Colorado River toad (*Bufo alvarius* Girard). *Texas J. Sci.* 8:72-77.

Blake, W. P. 1914. The Cahuilla Basin and Desert of the Colorado. In D. T. MacDougal (ed.), *The Salton Sea. Carneg. Inst. Wash. Publ.* 193:1-12, Washington, D.C.

Blaney, H. F. 1955. Evaporation from and stabilization of Salton Sea water surface. *Trans. Amer. Geophys. Union* 36:633-640.

Bligh, J. 1972. Evaporative heat loss in hot arid environments. In G. M. O. Maloiy (ed.), *Comparative physiology of desert animals*, pp. 357-369. New York: Academic Press.

Blount, R. F. 1929. Seasonal cycles of the interstitial cells in the testis of the horned toad (*Phrynosoma solare*); seasonal variation in the number and morphology of the interstitial cells and the volume of the interstitial tissue. *J. Morph. Physiol.* 48:317-344.

Blumer, J. C. 1909. Observations on cacti in cultivation. *Plant World* 12(7):162-164.

———. 1911. Change of aspect with altitude. *Plant World* 14(10):236-248.

Bodenheimer, F. S. 1953. Problems of animal ecology and physiology in deserts. In *Proc. Int. Symp. Desert Res. Jerusalem* 205-229.

———. 1954. Problems of physiology and ecology of desert animals. In *Biology of deserts*, pp. 162-167. London: Institute of Biology.

Bogert, C. M. 1937. Note on the growth rate of the desert tortoise, *Gopherus agassizii. Copeia* 1937:191-192.

———. 1949. Thermoregulation in reptiles, a factor in evolution. *Evolution* 3:195-211.

Bogert, C. M., and del Campo, R. M. 1956. The gila monster and its allies. *Bull. Am. Mus. Nat. Hist.* 109:1-238.

Bogert, C. M., and Cowles, R. B. 1944. A preliminary study of desert reptiles. *Bull. Amer. Mus. Nat. Hist.* 83:265-296.

Bogert, C. M., and Oliver, J. A. 1945. A preliminary analysis of the herpetofauna of Sonora. *Bull. Amer. Mus. Nat. Hist.* 83:303-425.

Bogusch, E. R. 1950. A bibliography of mesquite. *Texas J. Sci.* 2:528-538.

Bolyshev, N. N. 1964. Role of algae in soil formation. *Sov. Soil Sci.* 6:630-635.

Booth, J. A. 1964. An investigation of a saguaro seedling disease. Ph.D. dissertation, University of Arizona.

Booth, K. 1958. Development of eggs and young of desert tortoise. *Herpetologica* 13:261-263.

Borror, D. J., and White, R. E. 1970. *A field guide to the insects of America north of Mexico.* Boston: Houghton Mifflin.

Bostic, D. L. 1966. Food and feeding behavior of the teiid lizard, *Cnemidophorus hyperythrus beldingi. Herpetologica* 22:23-31.

———. 1971. Herpetofauna of the Pacific coast of north central Baja California, Mexico, with a description of a new subspecies of *Phyllodactylus xanti. San Diego Soc. Nat. Hist. Trans.* 16:237-264.

Bowers, J. E. 1981. A chronology of catastrophic freezes in the Sonoran Desert. *Desert Plants* 2:(4)232-237.

Boyer, D. R. 1967. Interaction of temperature and hypoxia on respiratory and cardiac responses in the lizard, *Sauromalus obesus. Comp. Biochem. Physiol.* 20:437-447.

Bradley, W. C., and Mauer, R. A. 1971. Reproduction and food habits of Merriam's kangaroo rat, *Dipodomys merriami. J. Mammal.* 52:497-507.

Bradley, W. G., and Yousef, M. K. 1942. Small mammals in the desert. In M. K. Yousef, S. M. Horvath and R. W. Bullard (eds.). *Physiological adaptations: Desert and mountain.* New York: Academic Press.

Bradshaw, G. and Howard, B. 1960. Mammal skulls recovered from owl pellets in Sonora, Mexico. *Journ. Mammal.* 41:282-283.

Brand, D. B. 1936. Notes to accompany a vegetation map of northwest Mexico. *Univ. New Mex. Bull. Biol. Ser.* 4(4):5-27.

Brandt, H. 1951. *Arizona and its bird life.* Cleveland, Ohio: Bird Research Foundation.

Branson, B. A. 1966. Histological observations on the sturgeon chub, *Hybopsis gelida* (Cyprinidae). *Copeia* 1966:872-876.

Branson, B. A.; McCoy, C. J.; and Sisk, M. E. 1960. Notes on the freshwater fishes of Sonora with an addition to the known fauna. *Copeia* 1960:217-220.

Branson, F. A.; Miller, R. F.; and McQueen, I. S. 1967. Geographic distribution and factors affecting the distribution of salt desert shrubs in the United States. *J. Range Managm.* 20:287-296.

Brattstrom, B. H. 1965. Body temperatures of reptiles. *Amer. Midl. Nat.* 73:376-422.

———. 1968. Thermal acclimation in anuran amphibians as a function of latitude and altitude. *Comp. Biochem. Physiol.* 24:93-111.

Bray, W. L. 1898. The relation of the Lower Sonoran Zone in North America to the flora of the arid zone of Chile and Argentina. *Bot. Gaz.* 26:121-147.

Breazeale, J. F., and Smith, H. V. 1930. Caliche in Arizona. *Ariz. Agr. Exp. Sta. Bull.* 131:419-430.

Breitung, A. J. 1968. The agaves. *Cactus and Succulent Soc. Yearbook* 1968:1-107.

Brinck, P. 1956. The food factor in animal desert life. In K. G. Wingstroud (ed.). *Bertil Hanstrom: Zoological papers in honour of his 65th birthday, November 20th 1956.* pp. 120-137. Sweden: Zool. Inst. Lund.

Britton, N. L., and Rose, J. N. 1919-1923. The Cactaceae. *Carneg. Inst. Wash. Publ.* 248.

Brown, D. E. 1973. *The natural vegetative communities of Arizona.* Phoenix: Ariz. Game and Fish Dept.

———. 1978. The vegetation and occurrence of chaparral and woodland flora on isolated mountains within the Sonoran and

Mojave deserts in Arizona. *J. Arizona-Nevada Acad. Sci.* 13:7-12.

Brown, D. E.; Carmony, N. B.; Lowe, C. H.; and Turner, R. M. 1976. A second locality for native California fan palms (*Washingtonia filifera*) in Arizona. *J. Ariz. Acad. Sci.* 11(1):37-41.

Brown, D. E., and Lowe, C. H. 1977. Biotic communities of the Southwest. *U.S.F.S. Rocky Mtn. Forest and Range Exp. Sta. Gen. Techn. Rep.* RM-41.

Brown, D. E.; Lowe, C. H.; and Pase, C. P. 1979. A digitized classification system for the biotic communities of North America, with community (series) and association examples for the Southwest. *J. Arizona-Nevada Acad. Sci.* 14:1-16.

Brown, D. E., and Webb, P. M. 1979. A preliminary reconnaissance of the habitat of the peninsular pronghorn. *J. Arizona-Nevada Acad. Sci.* 14(1):30-32.

Brown, G. W. (ed.). 1968. *Desert biology.* Vol. 1. New York: Academic Press.

———. 1974. *Desert biology.* Vol. 2. New York: Academic Press.

Brown, H. A. 1966. Temperature adaptation and evolutionary divergence in allopatric populations of the spadefoot toad, *Scaphiopus couchii. Copiea* 1966:365-370.

———. 1967a. High temperature tolerance of the eggs of a desert anuran, *Scaphiopus hammondi. Copiea* 1967:365-370.

———. 1967b. Embryonic temperature adaptations and genetic compatibility in two allopatric populations of the spadefoot toad, *Scaphiopus hammondi. Evolution* 21:742-761.

Brown, J. H. 1975a. A preliminary study of seed predation in desert and montane habitats. *Ecology* 56(4):987-992.

———. 1975b. Geographical ecology of desert rodents. In M. L. Cody and J. M. Diamond (eds.). *Ecology and evolution of communities.* pp. 315-341. Cambridge, Mass.: Belknap Press.

Brown, J. H., and Lee, A. K. 1969. Bergman's rule and climatic adaptation in woodrats (*Neotoma*). *Evolution* 23:329-338.

Brown, J. H., and Lieberman, G. L. 1973. Resource utilization and coexistence of seed-eating desert rodents in sand dune habitats. *Ecology* 54:788-797.

Brown, J. R. 1869. *Adventures in Apache country.* New York: Harper and Brothers.

Brown, V. 1974. *Reptiles and amphibians of the West.* Happy Camp, Calif.: Naturegraph Publishers.

Brown, V.; Weston, Jr. H.; and Buzzell, J. 1979. *Handbook of California birds.* Happy Camp, Calif.: Naturegraph.

Bruder, J. S. 1975. Historic Papago archaeology. *Ariz. State Univ. Anthropol. Res. Paper 9.*

———. 1977. Changing patterns in Papago subsistence strategies: Archaeology and ethnohistory compared. *Kiva* 42 (3-4):233-256.

Brugge, D. M. 1956. Pima Bajo basketry. *Kiva* 22(1):7-11.

Brum, G. D. 1973. Ecology of the saguaro (*Carnegiea gigantea*): phenology and establishment in marginal populations. *Madroño.* 22:195-204.

Brusca, R. C. 1973. *A handbook to the common intertidal invertebrates of the Gulf of California.* Tucson: University of Arizona Press.

Bryan, K. 1925. The Papago country, Arizona. *U.S. Geol. Surv. Water Supply Paper* 499. Washington, D.C.: Government Printing Office.

———. 1928. Changes in plant associations by change in ground-water levels. *Ecology* 9(4):474-478.

Bryant, H. C. 1911. The horned lizards of California and Nevada of the genera *Phrynosoma* and *Anota. Univ. Calif. (Berkeley) Publ. Zool.* 9:1-84.

Bryant, Harold C. 1916. Habits and food of the roadrunner in California. *Univ. Calif. Publ. Zool.* 17:21-58.

Bullock, T. H. 1955. Compensation for temperature in the metabolism and activity of poikilotherms. *Biol. Rev. Cambridge Phil. Soc.* 30:311-342.

Bunch, T. D.; Paul, S. R. and McCutchen, H. 1978. Chronic sinusitis in the desert bighorn. *Desert Bighorn Council Transact.* 1978:16.

Burden, J. D. 1970. Ecology of *Simmondsia chinensis (Link.) Schneid.* at its lower elevational limits. Master's thesis, Arizona State University.

Burge, B. L., and Bradley, W. G. 1976. Population density, structure and feeding habits of the desert tortoise, *Gopherus agassizi*, in a low desert study area in southern Nevada. *Proc. Desert Tortoise Council Symp.* 1976:51-74.

Burgess, R. L. 1965. A checklist of the vascular flora of Tonto National Monument. *J. Ariz. Acad. Sci.* 3:213-223.

Burkart, A. 1976. A monograph of the genus *Prosopis* (Leguminosae subfam. Mimosoideae). *J. Arnold Arboretum* 57:219-249, 450-525.

Burleson, G. L. 1942. The source of blood ejected from the eye by horned toads. *Copeia* 1942:246-248.

Burt, C. E. 1931. A study of the teiid lizards of the genus *Cnemidophorus* with special reference to their phylogenetic relationships. *U.S. Nat. Mus. Bull.* 154:1-286.

Burt, W. H. 1932. Descriptions of heretofore unknown mammals from islands in the Gulf of California, Mexico. *Trans. San Diego Soc. Nat. Hist.* 7:161-182.

———. 1938. Faunal relationships and geographic distribution of mammals in Sonora, Mexico. *Misc. Publ. Mich. Mus. Zool.* 39:1-77.

Buterworth, B. B. 1961. A comparative study of growth and development of the kangaroo rats, *Dipodomys deserti* Stephens and *Dipodomys merriami* Mearns. *Growth* 25:127;-139.

Buxton, P. A. 1922. Heat, moisture and animal life in deserts. *Proc. Roy. Soc. Ser. B.* 96:123-131.

———. 1923. *Animal life in deserts: A study of the fauna in relation to the environment.* London: Arnold..

———. 1955. *Animal life in deserts.* London: Arnold.

Cable, D. R. 1969. Soil temperature variations on a semi-desert habitat in southern Arizona. *U.S.F.S., Rocky Mtn. Forest and Range Exp. Sta. Research Note* RM-128:1-4

———. 1972. Fourwing saltbush revegetation trials in southern Arizona. *J. Range Managm.* 25(1):150-153.

Cade, T. J. 1964. Water and salt balance in granivorous birds. In M. J. Wayner (ed.), *Thirst*, pp. 237-256. Oxford:Pergamon Press.

———. 1965. Relations between raptors and columbiform birds at a desert water hole. *Wilson Bull.* 77:340-345..

Cade, T. J., and Greenwald, L. 1966. Nasal salt secrretion in falconiform birds. *Condor* 68:338-350.

Cailleux, A., and Wuttke, K. 1964. Morphoscopie des sables quartzeux dans l'ouest des Etats-unis D'Amerique du Nord. *Bol. Paranaense Geografia* 10-15:79-87.

Calder, N. 1977. *The weather machine.* Harmondsworth, Middlesex: Penguin Books.

Calder, W. A. 1968. The diurnal activity of the roadrunner, *Geococcyx californianus. Condor* 70:84-85.

Calder, W. A., and Bentley, P. J., 1967. Urine concentrations of two carnivorous birds, the white pelican and the road runner. *Comp. Biochem. Physiol.* 22:607-609.

Callahan, J. R. 1975. Status of the peninsula chipmunk. *J. Mammal.* 56(1):266-269.

Cameron, R. E. 1958. Fixation of nitrogen by algae and associated organisms in semi-arid soils: Identification and characterization of soil organisms. Master's thesis, University of Arizona.

———. 1961. *Algae of the Sonoran Desert in Arizona.* Ph.D. dissertation, University of Arizona.

———. 1962. Species of *Nostoc* Vaucher occurring in the Sonoran Desert in Arizona. *Trans. Amer. Microscop. Soc.* 81:379-384

———. 1964. Algae of southern Arizona. Vol. 2. Algal flora (Exclusive of blue-green algae). *Rev. Algol.,* n.s.7:151-177.

Cameron, R. E., and Blank, G. B. 1966. Desert algae: Soil crusts and diaphanous substrata as algal habitats. *Nat. Aeronautics and Space Adm. Techn. Rept.* 32-971.

Cameron, R. E., and Fuller, W. H. 1960. Nitrogen fixation by some algae in Arizona soils. *Soil. Sci. Soc. Amer. Proc.* 24:353-356.

Cannon, W. A. 1904. Observations on the germination of *Phoradendron villosum* and *P. californicum. Bull. Torr. Bot. Club* 31: 435-443.

———. 1908a. On the electric resistance of solutions of salt plants and solutions of alkali soils. *Plant World* 11:10-14.

———. 1908b. The topography of the chlorophyll apparatus in desert plants. *Carneg. Inst. Wash. Publ.* 98(1):1-43.

———. 1911. The root habits of desert plants. *Carneg. Inst. Wash. Publ.* 131:1-96.

———. 1914. A note on the reversibility of the water reaction in a desert liverwort. *Plant World* 17:261-265.

———. 1916. Distribution of the cacti with especial reference to the role played by the root response to soil temperature and soil moisture. *Amer. Nat.* 50:435-442.

Capon, B., and Van Asdall, W. 1967. Heat pre-treatment as a means of increasing germination of desert annual seeds. *Ecology* 48(2):305-306.

Capocaccia, L. 1962. Probable natural parthenogenesis in lizards of the genus *Cnemidophorus. Natura* 53:109-110.

Carey, C., and Morton, M. L. 1971. A comparison of salt and water regulation in California quail (*Lophortyx californicus*) and Gambel's quail (*Lophortyx gambelii*). *Comp. Biochem. Physiol.* 38A:75-101.

Carlisle, D. B. 1968. *Triops* (Entomostraca) eggs killed only by boiling. *Science* 161:179-180.

Carpelan, L. H. 1958. The Salton Sea. Physical and chemical characteristics. *Limnol. Oceanogr.* 3:373-386.

Carpenter. C. C. 1961. Patterns of social behavior in the desert iguana, *Dipsosaurus dorsalis. Copeia* 1961:396-405.

Carpenter, C. C., and Grubits III, G. 1960. Dominance shifts in the tree lizard (*Urosaurus ornatus*—Iguanidae). *Southwest Nat.* 5:123-128.

Carpenter, R. E. 1966. A comparison of thermoregulation and water metabolism in the kangaroo rats, *Dipodomys agilis* and *Dipodomys merriami. Univ. Calif. Publ. Zool.* 78:1-36.

———. 1969. Structure and function of the kidney and the water balance of desert bats. *Physiol. Zool.* 42:288-302.

Carr, A. 1952. *Handbook of turtles.* Ithaca, N.Y.: Cornell University Press, (Comstock Publ. Assoc.).

Carr, J. N. 1971, 1972, 1973. Endangered species investigations—Sonoran pronghorn. Proj. W-53-21, 22, 23, WP-7, Job 1. *Ariz. Game and Fish Dept.*

Carroll, C. R., and Janzen, D. H. 1973. Ecology of foraging ants. *Ann. Rev. Ecol. and Syst.* 4:231-258.

Castetter, E. F. and Bell, W. H. 1942. Pima and Papago Indian Agriculture. *Inter-Americana Studies.* No. 1. Albuquerque.

Castetter, E. F., and Underhill, R. M. 1935. Ethnobiological studies in the American southwest, vol. 2. The ethnobiology of the Papago Indians. *Univ. N. Mex. Bull.* 275.

Causey, N. B. 1975. Desert Millipedes (Spirostreptidae; Spirostreptida) of the southwestern United States and adjacent Mexico. *Occas. Pap. Mus. Texas Tech. Univ.* 35:1-12.

Cavill, G. W. K., and Robertson, P. 1965. Ant venoms, attractants and repellants. *Science* 149:1337-1345.

Chaffee, R. R. J., and Roberts, J. C. 1971. Temperature acclimation in birds and mammals. *Ann. Rev. Physiol.* 33:155.

Chan, K. K. O.; Callard, I. P.; and Jones, I. C. 1970. Observations of the water and electrolyte composition of the iguanid lizard *Dipsosaurus dorsalis* (Baird and Girard), with special reference to the control by the pituitary gland and the adrenal cortex. *Gen. Comp. Endocrin.* 15:374-387.

Chew, R. M. 1961. Ecology of spiders in a desert community. *J. New York Entomol. Soc.* 69:5-41.

———. 1961. Water metabolism of desert-inhabiting vertebrates. *Biol. Rev.* 36(1):1-31.

———. 1965. Water metabolism of mammals. *In* W. V. Mayer and R. G. Van Gelder (Eds.) *Physiological mammalogy.* Vol. 2, pp. 43-178. New York: Academic Press.

Chew, R. M., and Chew, A. E. 1965. The primary productivity of a desert shrub (*Larrea tridentata*) community. *Ecol. Monogr.* 35:355-375.

Chew, R. M., and Chew, A. E. 1970. Energy relationships of the mammals of a desert scrub (*Larrea tridentata*) community. *Ecol. Monogr.* 40:1-21.

Chew, R. M., and Damman, A. E. 1961. Evaporative water loss of small vertebrates, as measured with an infrared analyzer. *Science* 133:384-385.

Chew, R. M.; Lindberg, R. G.; and Hayden, P. 1965. Circadian rhythm of metabolic rate in pocket mice. *J. Mammal.* 46:477-494.

Chesky, J. 1942. The wiikita. *Kiva* 8(1):3-5.

———. 1943. *The nature and function of Papago music.* Master's thesis, University of Arizona.

Childs, S., and Goodall, D. W. 1973. Seed reserves of desert soils. *US/IBP Desert Biome Research Memorandum* RM 73-5. Logan, Utah.

Childs, T. 1954. Sketch of the "Sand Indians." *Kiva* 19 (2-4):27-39

Christiansen, J. L. 1965. Reproduction in Uta stansburiana stansburiana Baird and Girard. Master's thesis, University of Utah.

Clark, E. D. 1953. A study of the behavior and movement of the Tucson Mountains mule deer. Masters thesis, University of Arizona.

Cliffton, K.; Cornejo, D.; and Felger, R. 1980. Sea turtles of the

Pacific Coast of Mexico. *Poc. World Conf. Sea Turtle Conservation* (in press).

Cloudsley-Thompson, J. L. 1962. Lethal temperatures of some desert arthropods and the mechanism of heat death. *Entomol. Exp. Appl.* 5:270-280.

———. 1964a. Terrestrial animals in dry heat: Arthropods. In D. B. Dill (ed.), *Handbook of physiology*, 451-456. Washington, D.C.: American Philosophical Society.

———. 1964b. On the function of the subelytral cavity in desert Tenebrionidae. *Entomol. Month. Mag.* 100:148-151.

———. 1970. Terrestrial invertebrates. In G. C. Whittow (ed.), *Comparative physiology of thermoregulation.* vol. 1, pp. 15-77. New York: Academic Press.

———. 1971. *The temperature and water relations of reptiles.* Watford, Herts,: Merrow,

———. 1972. Temperature regulation in desert reptiles. In G. M. O. Maloiy (Ed.), *Comparative physiology of desert animals*, pp. 39-59. New York: Academic Press.

———. 1975. Desert expansion and the adaptive problems of the inhabitants. In N. F. Hadley (ed.), *Environmental physiology of desert organisms*, pp. 255-268. Stroudsburg, Pa.: Dowden, Hutchinson & Ross.

———. 1979. Adaptive functions of the colors of desert animals. *J. Arid Envir.* 2:95-104.

Cloudsley-Thompson, J. L., and Chadwick, M. J. 1964. *Life of deserts.* Philadelphia: Dufour.

Cloudsley-Thompson, J. L., and Crawford, C. S. 1970. Lethal temperatures of some arthropods of the southwestern United States. *Entomol. Month. Mag.* 106:26-29.

Clyma, W., and Young, R. A. 1968. Environmental effects of irrigation in the central valley of Arizona. *Amer. Soc. Civil Engin., Natl. Meet. Environm. Engin., Chattanooga, Tenn.*

Cockerell, T. D. A. 1900. The Lower and Middle Sonoran Zones in Arizona and New Mexico. *Amer. Natl.* 34:285-293.

Cockrum, E. L. 1960. *The recent mammals of Arizona: Their taxonomy and distribution.* Tucson: University of Arizona Press.

———. 1964. Recent mammals of Arizona. In C. H. Lowe, (ed.), *The vertebrates of Arizona*, pp. 249-259. Tucson: University of Arizona Press.

Cockrum, E. L. and Hayward. 1962. Cited by B. Villa-R. q.v. Cole, C. J. 1962. Notes on the distribution and food habits of *Bufo alvarius* at the easten edge of its range. *Herpetologica* 18:172-175.

Cole, F. R. 1969. *The flies of western North America.* Berkeley: University of California Press.

Cole, G. A. 1963. The American Southwest and Middle America. In D. G. Frey (ed.), *Limnology in North America*, pp. 393-434. Madison: University of Wisconsin Press.

———. 1968. Desert limnology. In G. W. Brown (ed.), *Desert biology*, vol. 1, pp. 423-486. New York: Academic Press.

Cole, G. A., and Minckley, W. L. 1968. "Anomalous" thermal conditions in a hypersaline inland pond. *J. Ariz. Acad. Sci.* 5:105-107.

Cole, G. A., and Whiteside, M. C. 1965. An ecological reconnaissance of Quitobaquito Spring, Arizona. *J. Ariz. Acad. Sci.* 3:159-163.

Cole, G. A.,; Whiteside, M. C.; and Brown, R. J. 1967. Unusual monomixis in two saline Arizona ponds. *Limnol. Oceanogr.*

12:584-591.

Cole, L. C. 1943. Experiments on toleration of high temperatures in lizards with reference to adaptive coloration. *Ecology* 24:94-108.

Coleman, G. A. 1926. A biological survey of the Salton Sea. *Calif. Fish Game* 17:218-227.

Collins, B. J. 1976. *Key to trees and shrubs of the deserts of southern California.* Northridge, Calif.: California State University Foundation.

Collins, J. P., and Lewis, M. A. 1979. Overwintering tadpoles and breeding season variation in the *Rana pipiens* complex in Arizona. *Southwestern Nat.* 24:371-373.

Collins, K. J., and Weiner, J. S. 1968. Endocrinological aspects of exposure to high environmental temperatures. *Physiol. Rev.* 48:785.

Collins, M. S. 1973. High temperature tolerance in two species of subterranean termites from the Sonoran Desert in Arizona. *Env. Ent.* 2(6):1122-1123.

Coney, P. J. 1976. Plate tectonics and the Laramide Orogeny. *N. Mex. Geol. Soc. Spec. Publ.* 6:5-10.

Coney, P. J., and Reynolds, S. J. 1977. Cordilleran Benioff zones. *Nature* 270:403-405.

Cooley, D. B. 1966. Geological environment and engineering properties of caliche in the Tucson area, Arizona. Master's thesis, University of Arizona.

Cooper. 1956. Yield and value of wild-land resources of the Salt River watershed, Arizona. Master's thesis, University of Arizona.

Cope, E. D. 1866. On the reptilia and batrachia of the Sonoran province of the Nearctic region. *Proc. Acad. Nat. Sci. Phila.* 18:300-314.

Copple, R. F. and Pase, C. P. 1967. A vegetative key to some common Arizona range grasses. *U.S.F.S. Rocky Mtn. Forest Range Exp. Sta. Res. Paper* RM-27:1-72.

Coues, E. 1875. Synopsis of the reptiles and batrachians of Arizona, with critical and field notes, and an extensive synonomy. *U.S. Geogr. Geol. Explor. Surv. West 100th Meridian* 5:585-633.

Cowles, R. B. 1934. Notes on the ecology and breeding habits of the desert minnow, *Cyprinodon macularius.* Baird and Girard. *Copeia* 1934:40-42.

Cowles, R. B. 1939. Possible implications of reptilian thermal tolerance. *Science* 90:465-466.

———. 1946. Note on the arboreal feeding habits of the desert iguana. *Copeia* 1946:172-173.

Cowles, R. B., and Burleson, G. L. 1945. The sterilizing effect of high temperature on the male germ-plasm of the Yucca night lizard, *Xantusia vigilis. Am. Nat.* 79:417-435.

Cox, T. J. 1966. A behavioral and ecological study of the desert pupfish *(Cyprinodon macularius)* in Quitobaquito Springs, Organ Pipe Cactus National Monument, Arizona. Ph.D. dissertation, University of Arizona.

———. 1972. The food habits of the desert pupfish *(Cyprinodon macularius)* in Quitobaquito Springs, Organ Pipe National Monument, Arizona. *J. Ariz. Acad. Sci.* 7:25-27.

Coyle, J., and Roberts, N. C. 1975. *A field guide to the common and interesting Plants of Baja California.* La Jolla, Calif.: Natural History Publishing Co.

Craig, R. T. 1945. *The mammillaria handbook.* Pasadena, Calif.: Abbey Garden Press.

Crawford, C. S. 1976. Feeding-season production in the desert millipede *Orthoporus ornatus*. *Oecologia* 24:265-276.

————. 1979. Desert detritovores: A review of life history patterns and trophic roles. *J. Arid Envir.* 2:31-42.

Cross, S. P., and Huigbregtse, W. 1964. Unusual roosting site of *Eptesicus fuscus. J. Mammal.* 45:628.

Crosswhite, F. S. 1966. Revision of *Penstemon* section Chamaeleon (Scrophulariaceae). *Sida* 2:339-346.

————. 1972. Studies of *Simmondsia chinensis* at the Boyce Thompson Southwestern Arboretum. In E. F. Haase and W. G. McGinnies (eds.), *Jojoba and its uses,* pp. 5-10. Tucson: University of Arizona Office of Arid Lands Studies.

————. 1975. *Relations of precipitation and vegetation at the Boyce Thompson Southwestern Arboretum.* Superior, Ariz.: Boyce Thompson Southwestern Arboretum.

————. 1890. The annual saguaro harvest and crop cycle of the Papago, with reference to ecology and symbolism. *Desert Plants* 2(1):2-61.

Crosswhite, F. S., and Crosswhite, C. D. 1981. Geography of a pollinator-correlated character in *Penstemon. Desert Plants* 2:000-000 (in press).

Cruden, R. W. 1966. Birds as agents of long-distance dispersal for disjunct plant groups of the temperate western hemisphere. *Evolution* 20:517-563.

Cuellar, O. 1966. Delayed fertilization in the lizard *Uta stansburiana. Copeia* 1966:549-552.

Cumming, K. J. 1952. The effect of slope and exposure on range vegetation in desert grassland and oak woodland areas of Santa Cruz County, Arizona. Master's thesis, University of Arizona.

Cunningham, G. L., and Strain, B. R. 1969. An ecological significance of seasonal leaf variability in a desert shrub. *Ecology* 50(3):400-408.

Cutler, H. C. 1939. Monograph of the North American species of the genus *Ephedra. Ann. Mo. Bot. Gard.* 26:373-428.

Cutter, W. L. 1959. An instance of blood-squirting by *Phrynosoma solare. Copeia:* 176.

Dalton, P. D. 1961. Ecology of the creosotebush *Larrea tridentata* (DC.) Cov. Ph.D. dissertation, University of Arizona.

Damon, P. E. 1964. *Correlation and chronology of ore deposits and volcanic rocks.* Annual report to U.S. Atomic Energy Commission.

————. 1968. Application of the potassium-argon method to the dating of igneous and metamorphic rock within the basin ranges of the Southwest. In S. R. Titley (ed.), *Southern Arizona Guidebook vol. 3,* pp. 7-20. Tucson: Arizona Geological Society.

————. 1971. The relationship between late Cenozoic volcanism and tectonism and orogenic periodicity. In K. K. Turekian (ed.), *The Late Cenozoic glacial ages,* pp. 15-35. New York: John Wiley.

Damon, P. E., and Bikerman, M. 1964. K-Ar dating of post-Laramide plutonic and volcanic rocks within the Basin and Range Province of southeastern Arizona and adjacent areas. *Ariz. Geol. Soc. Dig.* 7:63-78.

Damon, P. E.; Erickson, R. C.; and Livingston, D. E. 1963. K-Ar dating of basin and range uplift, Catalina Mountains, Arizona. In *Nuclear Geophysics,* pp. 113-121. NAS-NRC Publication 1075, Woods Hole Conference.

Damon, P. E., and Mauger, R. C. 1966. Epirogeny and orogeny viewed from the Basin and Range Province. *Trans. Soc. Mining Engin.* 235:99-112.

Damon, P. E.; Shafiqullah, M.; and Lynch, D. J. 1981. The valleys and mountains of southern and western Arizona. In T. Smiley, J. D. Nations, T. L. Pewe, and J. P. Schafer (eds.), *Landscapes of Arizona: Geological Story.* Tucson: University of Arizona Press.

Dantzler, W. H., and Schmidt-Nielsen, B. 1966. Excretion in freshwater turtle (*Pseudemys scripta*) and desert tortoise (*Gopherus agassizii*). *Am. J. Physiol.* 210:198-210.

Darrow, R. A. 1935. A study of the transpiration rates of several desert grasses and shrubs as related to environmental conditions and stomatal periodicity. Master's thesis, University of Arizona.

————. 1943. Vegetative and floral growth of *Fouquieria* splendens. *Ecology* 24(3):310-322.

————. 1961. Origin and development of the vegetational communities of the Southwest. In *Bioecology of the arid and semiarid lands of the southwest,* pp. 30-47. Las Vegas, N.M.: American Association for the Advancement of Science. Southwestern and Rocky Mountain Division.

Dasmann, R. 1974. Biotic provinces of the world—further development of a system for defining and classifying natural regions for the purpose of conservation. *I.U.C.N. (Morges, Switzerland) Occ. Papers* 9:1-57.

Daubenmire, R. F. 1938. Merriam's life zones of North America. *Quart. Rev. Biol.* 13:327-332.

Davidson, R. W., and Mielke, J. L. 1947. *Fomes robustus,* a heart-rot fungus on cacti and other desert plants. *Mycologia* 39:210-217.

Davis, E. H. 1920. The Papago ceremony of Vikita. *Indian Notes and Monographs* 3(4):158-178.

Davis, G. H. 1975. Gravity-induced folding off a gneiss dome complex, Rincon Mountains, Arizona. *Geol. Soc. Amer. Bull.* 86:979-990.

Davis, W. S., and Thompson, H. J. 1967. A revision of *Petalonyx* (Loasaceae) with a consideration of affinities in subfamily Gronovivideae. *Madroño* 19:1-18.

Dawson, T., and Schmidt-Nielsen, K. 1966. Effect of thermal conductance on water economy in the antelope jack rabbit, *Lepus alleni. J. Cell. Physiol.* 67:463-471.

Dawson, W. R. 1955. The relation of oxygen consumption to temperature in desert rodents. *J. Mammal.* 36:543-553.

Dawson W. R., and Bartholomew, G. A. 1958. Metabolic and cardiac responses to temperature in the lizard *Dipsosaurus dorsalis. Physiol. Zool.* 31:100-111.

————. 1968. Temperature regulation and water economy of desert birds. In G. W. Brown (ed.), *Desert biology,* vol. 1, pp. 357-394. New York: Academic Press.

Dawson, W. R. and K. Schmidt-Nielsen. 1964. Terrestrial animals in dry heat: Desert birds. In D. B. Dill (ed.), *Handbook of Physiology. Sec. 4. Adaptation to the Environment.* pp. 481-492. Washington, D.C.: American Physiological Society.

Deacon, J. E. 1968. Endangered non-game fishes of the West: Causes, prospects, and importance. *Proc. 48th Ann. Conf. West. Assoc. Game Fish Comm.* pp. 534-549.

Deacon, J. E., and Minckley, W. L. 1974. Desert fishes. In G. W. Brown (ed.), *Desert biology,* vol. 2, pp. 385-488. New York: Academic Press.

Dechel, W. C.; Strain, B. R.; Odening, W. R. 1972. Tissue water potential, photosynthesis, C^{14}-labelled photosynthate utilization, and growth in the desert shrub *Larrea divaricata* Cav. *Ecol. Monogr.* 42:127-141.

Deevey, E. 1949. Biogeography of the Pleistocene. Part I. Europe and North America. *Bull. Geol. Soc. Amer.* 60:1315-1416.

De La Fuente Duch, M. F. F. 1973. Aeromagnetic study of the Colorado River delta area, Mexico. Master's thesis, University of Arizona.

Denny, W. F., and Dillaha, C. J. 1964. Hemotoxic effect of *Loxosceles reclusus* venom. *In vivo* and *in vitro* studies. *J. Lab. Clin. Med.* 64:291-296.

Densmore, F. 1929. Papago music. *Bull. Bur. Amer. Ethnol.* 90:229.

Despain, D. G. 1967. The survival of saguaro (*Carnegiea gigantea*) seedlings on soils of differing albedo, cover, and temperature. Master's thesis, University of Arizona.

DeWald, T. 1979. *The Papago Indians and their basketry.* Tucson: By the author.

DeWitt, C. B. 1963. Behavioral thermoregulation in the desert Iguanid, *Dipsosaurus dorsalis.* Ph.D. dissertation, University of Michigan.

———. 1967a. Precision of thermoregulation and its relation to environmental factors in the desert iguana, *Dipsosaurus dorsalis. Physiol. Zool.* 40:49-66.

———. 1967b. Behavioral thermoregulation in the desert iguana. *Science* 158:809-810.

Dice, L. R. 1939. The Sonoran biotic province. *Ecology* 20:118-129.

———. 1943. *The biotic provinces of North America.* Ann Arbor: University of Michigan Press.

Dice, L. R., and Blossom, P. M. 1937. Studies of mammalian ecology in southwestern North America with special attention to the colors of desert mammals. *Carneg. Inst. Wash. Publ.* 485:1-129.

Dickerman, R. W. 1954. An ecological survey of the Three-Bar Game Management Unit located near Roosevelt, Arizona. Master's thesis, University of Arizona.

Dilks, T. J. K., and Proctor, M. C. F. 1974. The pattern of recovery of bryophytes after desiccation. *J. Bryology* 8:97-115.

Dill, D. B.; Adolph, E. F.; and Wilber, D. G. (eds.). 1964. Adaptation to environment. In *Handbook of physiology.* Washington, D.C.: American Physiological Society.

Dill, D. B.; Bock, A. V. and Edwards, H. T. 1933a. Mechanisms for dissipating heat in man and dog. *Amer. J. Physiol.* 104:36-43.

Dill, D. B.; Edwards, H. T.; Bock, A. V.; and Talbot, J. H. 1935. Properties of reptilian blood. Vol. 3. The chuckwalla (*Sauromalus obesus* Baird). *J. Cell. Comp. Physiol.* 6:37-42.

Dill, W. A. 1944. The fishery of the lower Colorado River. *Calif. Fish and Game* 30:309-201.

Dina, S. J., and Klikoff, L. G. 1974. Carbohydrate cycle of *Plantago insularis* var. *fastigiata,* a winter annual from the Sonoran Desert. *Bot. Gaz.* 135(1):13-18.

Dinger, B. E. 1971. Comparative physiological ecology of the genus *Echinocereus* (*Cactaceae*) in the Piñaleno Mountains, Arizona. Ph.D. dissertation, Arizona State University.

Dirst, V. 1979. A prehistoric frontier in Sonora. Master's thesis, University of Arizona.

Dobie, J. F. 1949. *The voice of the coyote.* Boston: Little, Brown.

Dodge, N. N. 1963. *One hundred desert wildflowers in natural color.* Globe, Ariz.: Southwest Monuments Association.

———. 1973. *Flowers of the Southwest desert.* 8th ed. Globe, Ariz.: Southwest Monuments Association.

Douglas, P. A. 1952. Notes on the spawning of the humpback sucker, *Xyrauchen texanus* (Abbott). *Calif. Fish and Game* 38:149-155.

Downton, W. J. S. 1971. Check list of C$_4$ species. In M. D. Hatch, C. B. Osmond, and R. O. Slayter (eds.), *Photosynthesis and photorespiration,* pp. 554-556. New York: Wiley-Interscience.

Doyle, L. J., and Gorsline, D. J. 1977. Poway conglomerate in northwestern Baja California derived fron Sonora. *Am. Assoc. Pet. Geol. Bull.* 61:903-917.

Drabek, C. M. 1967. Ecological distribution of the mammalian fauna of the Desert Biology Station Area. Master's thesis, University of Arizona.

———. 1970. Ethoecology of the round-tailed ground squirrel, *Spermophilus tereticaudus.* Ph.D. dissertation, University of Arizona.

DuBois, S. M. 1979. Earthquakes. *Fieldnotes Ariz. Bur. Geol. Mineral Techn.* 9:1-9.

Duewer, E. A. H. 1971. The taxonomy and morphology of the lichen genus *Acarospora* and its *Phycosymbiont Trebouxia.* Ph.D. dissertation, University of Arizona.

Dunbier, R. 1968. *The Sonoran Desert, Its geography, economy and people.* Tucson: University of Arizona Press.

Durham, J. W., and Allison, E. C. 1960. Geologic history of Baja California and its marine faunas. *Syst. Zool.* 9:47-91.

Easterla, D. A. 1975. Giant desert centipede preys upon snake. *Southwestern Nat.* 20:411.

Eastwood, R. L. 1970. A geochemical-petrological study of Mid-Tertiary volcanism in parts of Pima and Pinal counties, Arizona. Ph.D. dissertation, University of Arizona.

Eaton, G. P. 1972. Deformation of quaternary deposits in two intermontane basins in southern Arizona, U.S.A. *24th Int. Geol. Congress Sect. 3:*607-616.

Eberly, L. D., and Stanley, T. B. 1978. Cenozoic stratigraphy and geologic history of southwestern Arizona. *Geol. Soc. Amer. Bull.* 89:921-940.

Echternacht, A. C. 1967. Ecological relationships of two species of the lizard genus *Cnemidophorus* in the Santa Rita Mountains of Arizona. *Amer. Midl. Nat.* 78:448-459.

Eddy, T. A. 1961. Foods and feeding patterns of the collared peccary in southern Arizona. *J. Wildl. Managm.* 25(3):248-257.

Edney, E. B. 1960. The survival of animals in hot deserts. *Smithson. Inst. Ann. Rept.* 1960:407-425.

———. 1966. Animals of the desert. In E. S. Hills (ed.). *Arid lands,* pp. 181-218. London: Methuen.

———. 1967. Water balance in desert arthropods. *Science* 156:1059-1066.

———. 1968. The effect of water loss on the haemolymph of *Arenivaga* sp. and *Periplaneta americana. Comp. Biochem. Physiol.* 25:149-158.

———. 1974. Desert arthropods. In G. W. Brown, Jr. (ed.), *Desert biology,* vol 2. pp. 311-383. New York: Academic Press.

Edwards, E. P. 1972. *A field guide to the birds of Mexico.* Sweet Briar, Va.: By the author.

Edwards, H. T., and Dill, D. B. 1935. Properties of reptilian blood.

Vol. 2. The gila monster (*Heloderma suspectum* Cope). *J. Cellular Comp. Physiol.* 6:21-35.

Ehrler, W. L. 1975. Environmental and plant factors influencing transpiration of desert plants. In N. F. Hadley (ed.), *Environmental physiology of desert organisms*, pp. 52-66. Stroudsburg, Pa.: Dowden, Hutchinson & Ross.

Eisner, T., and Meinwald, J. 1966. Defensive secretions of arthropods. *Science* 153:1341-1350.

Elbein, A. D. 1956. Studies of some of the myxobacteria isolated from soils of the Tucson area. Master's thesis, University of Arizona.

Elder. H. B. 1956. Watering patterns of some desert game animals. *J. Wildl. Managm.* 20:368-378.

Elder, J. B. 1953. Utilization of man-made waterholes by wildlife in southern Arizona. Master's thesis, University of Arizona.

Emery, J. A., and Russell, F. E. 1963. Lethal and hemorrhagic properties of some North American snake venoms. In H. L. Keegan and W. V. Macfarlane (eds.), *Venomous and poisonous animals and noxious plants of the Pacific area*, pp. 409-413. New York: Pergamon Press.

Emergy, N.; Poulson, T. L.; and Kinter, W. B. 1972. Production of concentrated urine by avian kidneys. *Am. J. Physiol.* 223:180-187.

Enlows, H. E. 1955. Welded tuffs of the Chiricahua Monument, Arizona. *Geol. Soc. Amer. Bull.* 66:1215-1246.

Essig, E. O. 1958. *Insects and mites of western North America.* New York: Macmillan.

Evans, K. J. 1967. Observations on the daily emergence of *Coleonyx variegatus* and *Uta stansburiana. Herpetologica* 23:217-222.

Evenari, M. 1949. Germination inhibitors. *Botan. Rev.* 15:153-194.

Eymann, J. L. 1953. A study of the sand dunes in the Colorado and Mojave deserts. Master's thesis, University of Arizona.

Ezell, P. H. 1954. An archaeological survey of northwestern Papaguería. *Kiva* 19(2-4):1-26.

———. 1961. The Hispanic acculturation of the Gila River Pima. *Mem. Amer. Anthrop. Assoc.* No. 90.

Fahn, A., and Werker, E. 1942. Anatomical mechanisms of seed dispersal. In T. T. Kozlowski (ed.), *Seed biology.* pp. 151-221. New York: Academic Press.

Felger, R. S. 1965. *Xantusia vigilis* and its habitat in Sonora, Mexico. *Herpetologica* 21:146-147.

———. 1966. Ecology of the Gulf Coast and Islands of Sonora, Mexico. Ph.D. dissertation, University of Arizona.

———. 1972. Fog desert. *Pacific Discovery* 25(3).

———. 1980a. Vegetation and flora of the Gran Desierto, Sonora, Mexico. *Desert Plants* 2:87-114.

———. 1980b. Personal communication. Felger, R. S.; Cliffton, K. and Regal, P. J. 1976. Winter dormancy in sea turtles: Independent discovery and exploitation in the Gulf of California by two local cultures. *Science* 191:283-285.

Felger, R. S., and Lowe, C. H. 1967. Clinal variation in the surface-volume relationships of the columnar cactus *Lophocereus schottii* in northwestern Mexico. *Ecology* 48(4):530-536.

———. 1976. The island and coastal vegetation and flora of the northern part of the Gulf of California. *Contrib. Sci. Nat. Histo Mus. Los Angeles County* 285:1-59.

Felger, R. S., and Moser, M. B. 1974. Seri Indian pharmacopoeia. *Econ. Bot.* 28:414-436.

Fenneman, N. F. 1928. Physiographic divisions of the United States. *Ann. Assoc. Amer. Geographers* 18:261-353.

Ferdon, E. N., Jr. 1967. The Hohokam "ball court": An alternative view of its function. *Kiva* 33:1-14.

Ferguson, C. W. 1950. An ecological analysis of Lower Sonoran Zone relic vegetation in south-central Arizona. Master's thesis, University of Arizona.

Ferguson, G. W. 1966. Effect of follicle-stimulating hormone and testosterone proprionate on the reproduction of the side-blotched lizard, *Uta stansburiana. Copeia* 1966:495-498.

Fink, B. 1909. The composition of a desert lichen flora. *Mycologia* 1:87-103.

Findley, J. S. 1969. Biogeography and southwestern boreal and desert mammals. *Kans. Mus. Nat. Hist. Misc. Publ.* 51:113-128.

Finley, R. B. 1958. The wood rats of Colorado; distribution and ecology. *Univ. Kans. Publ. Mus. Nat. Hist.* 10:213-552.

Fish, E. B., and Smith, E. L. 1973. Use of remote sensing for vegetation inventories in a desert shrub community. *Progr. Agr. Ariz.* 25(3):3-5.

Fisher, H. I. 1941. Notes on shrews of the genus *Notiosorex. J. Mammal.* 22:262-269.

Fisher, S. G., and Minckley, W. L. 1978. Chemical characteristics of a desert stream in flash flood. *J. Arid. Envir.* 1:25-33.

Fitch, H. S. 1956. An ecological study of the collared lizard (*Crotaphytus collaris*). *Univ. Kans. Publ. Mus. Nat. Hist.* 8:215-274.

Flake, L. D. 1973. Food habits of four species of rodents on a shortgrass prairie in Colorado. *J. Mammal.* 54:636-647.

Fletcher, J. E. and Carroll, P. H. 1948. Some properties of soils associated with piping in Southern Arizona. *Soil. Sci. Soc. Amer. Proc.* 13:545-547.

Fletcher, J. E.; Harris, K.; Peterson, H. B.; and Chandler, V. N. 1954. Piping. *Trans. Amer. Geophys. Union* 35:358-263.

Fletcher, J. E., and Martin, W. P. 1948. Some effects of algae and molds in the rain-crust of desert soils. *Ecology* 29:95-100.

Flint, R. F. 1971. *Glacial and quaternary geology.* New York: John Wiley.

Flores Mata, G.; Lopez, J. Jimenez; Sanchez, X. Madrigal; Ruiz, F. Moncayo; and Takaki, F. Takaki. 1971. Memoria del mapa de tipos de vegetacion de la Republica Mexicana. Secretaría de Recursos Hidáulicos. Subsecretaría de Planeacion. Dirección General Estudios. Dirección de Agrologia. Mexico City (map and manual).

Follett, W. I. 1960. The tresh-water fishes—their origins and affinities. In *The biogeography of Baja California and adjacent seas. Syst. Zool.* 9:212-232.

Fontana, B. L. 1974. Man in arid lands: The Piman Indians of the Sonoran Desert. In G. W. Brown (ed.), *Desert biology,* vol. 2, pp. 489-528. New York: Academic Press.

———. 1980. Ethnobotany of the saguaro, an annotated bibliography. *Desert plants* 2:63-78.

Fontana, B. L.; Robinson, W. J.; Cormack, C. W.; and Leavitt, E. E. 1962. *Papago Indian pottery.* Seattle: University of Washington Press.

Fowler, L. E. 1979. Hatching success and nest predation in the green sea turtle, *Chelonia mydas,* at Tortuguero, Costa Rica. *Ecology* 60:946-955.

Fowlie, Jack A. 1965. *The snakes of Arizona.* Fallbrook, Calif.: Azul Quinta Press.

Fraenkel, G., and Blewett, M. 1944. The utilization of metabolic water in insects. *Bull. Entomol. Res.* 35:127-139.

Freckman, D. W., and Mankau, R. 1977. Distribution and trophic structure of nematodes in desert soils. *Ecol. Bull.* 25:511-514.

Freckman, D. W.; Sher, S. A.; Mankau, R. 1974. Biology of nematodes in desert ecosystems. *US/IBP Desert Biome Research Memorandum.* RM 74-35. Logan, Utah.

French, N. R.; Maza, B. G.; and Aschwanden, A. P. 1966. Periodicity of desert rodent activity. *Science* 854:1194-1195.

Friauf, J. J., and Edney, E. B. 1969. A new species of *Arenivaga* from desert sand dunes in southern California (Dictyoptera, Polyphagidae). *Proc. Ent. Soc. Wash.* 71:1-7.

Friedmann, E. I., and Galun, M. 1974. Desert algae, lichens, and fungi. In G. W. Brown (ed.), *Desert biology,* vol. 2, pp. 165-212. New York: Academic Press.

Fry, F. E. J. 1947. Effects of the environment on animal activity. *Publ. Ontario Fisheries Res. Lab.* 68:1-62.

———. 1967. Responses of vertebrate poikilotherms to temperature. In A. H. Rose (Ed.), *Thermobiology,* pp. 375-409. New York: Academic Press.

Fuller, W. H. 1953. Effect of kind of phosphate fertilizer and method of placement on phosphorus absorption by crops grown on Arizona calcareous soil. *Ariz. Agr. Exp. Sta. Tech. Bull.* 128:235-255.

Fuller, W. H. 1963. Reactions of nitrogenous fertilizers in calcareous soils. *J. Agr. Food Chem.* 11:188-193.

———. 1974. Desert soils. In G. W. Brown (ed.), *Desert biology,* vol. 2, pp. 31-101. New York: Academic Press.

———. 1975. *Soils of the desert Southwest.* Tucson: University of Arizona Press.

Fuller, W. H.; Cameron, R. E.; and Raica, N. 1960. Fixation of nitrogen in desert soils by algae. In *Seventh Int. Congr. Soil Sci. Trans.* vol. 2, pp. 617-624. Madison, Wis.

Fuller, W. H.; Cameron, R. E.; and Raica, N. 1961. Fixation of nitrogen in desert soils by algae. *Trans. Seventh Int. Congr. Soil Sci.* 2:617-624.

Fuller, W. H., and McGeorge, W. T. 1950. Phosphates in calcareous soils. *Soil Sci.* 71:45-49, 315-323, 441-460.

Fuller, W. H., and Ray, H. 1965. Basic concepts of nitrogen, phosphorus and potassium in calcareous soils. *Ariz. Agr. Exp. Sta. Bull.* A-42:1-30.

Funk, R. S. 1966. Notes about *Heloderma suspectum* along the western extremity of its range. *Herpetologica* 22:254-258.

Gallizioli, S. 1977. Overgrazing on desert bighorn ranges. *Desert Bighorn Council Transact.* 1977:21-22.

Garcia-Castañeda, F. 1978. Marco geographico de la desertificacion en Mexico. In Medellín-Leal, F. (ed.), *La Desertificacion en Mexico,* pp. 35-54. Mexico. Universidad Autonoma de San Luis Potosi.

Gary, H. L. 1965. Some site relations in three flood-plain communities in central Arizona. *J. Ariz. Acad. Sci.* 3(4):209-212.

Gastil, R. G., and Jensky, W. 1973. Evidence for strike-slip displacement beneath the Trans-Mexican volcanic belt. *Stanford Univ. Publ. Geol. Sci.* 13:181-190.

Gastil, R. G., and Krummenacher, D. 1977. Reconnaissance geology of coastal Sonora between Puerto Lobos and Bahia Kino. *Geol. Soc. Am. Bull.* 88:189-198.

Gastil, R. G., Phillips, R. P., and Rodriguez-Torres, R. 1972. The reconstruction of Mesozoic California. *24th Int. Geol. Congr. Montreal Sect.* 3:217-229.

Gastil, R.G.; Phillips, R. P.; and Allison, E. C. 1975. Reconnaissance geology of the state of Baja California. *Geol. Soc. Am. Mem.* No. 140:1-170.

Gates, D. M.; Alderfer, R.; and Taylor, E. 1968. Leaf temperatures of desert plants. *Science* 195:994-995.

Gates, G. O. 1957. A study of the herpetofauna in the vicinity of Wickenburg, Maricopa County, Arizona. *Trans. Kansas Acad. Sci.* 60:403-418.

Gatewood, J. S.; Robinson, T. W.; Colby, B. R.; Hem, J. D.; and Halpenny, L. C. 1950. Use of water by bottom-land vegetation in lower Safford Valley, Arizona. *U.S. Geol. Surv. Water Supply Paper* 1103:1-210.

Gavin, T. A. 1973. An ecological survey of a mesquite bosque. Master's thesis, University of Arizona.

Geluso, K. N. 1978. Urine concentrating ability and renal structure of insectivorous bats. *J. Mammal.* 59:312-323.

Gentry, H. S. 1942. Rio Mayo Plants. *Carneg. Inst. Wash. Publ.* 527:1-328.

———. 1949. Land plants of the California Gulf region. Allen Hancock Foundation. *Pacific Expeditions* 13:81-180.

———. 1958. The natural history of jojoba (*Simmondsia chinensis*) and its cultural aspects. *Econ. Bot.* 12:261-294.

———. 1972. The agave family in Sonora. *U.S.D.A. Agr. Handb.* 399:1-195.

———. 1978. The agaves of Baja California. *Occ. Pap. Calif. Acad. Sci.* San Francisco.

Gertsch, Willis J. 1979. *American Spiders.* New York: Van Nostrand Reinhold.

Gertsch, W. J., and Soleglad, M. 1966. The scorpions of the *Vejovis boreus* groups (Subgenus Paruroctonus) in North America (Scorpionida, Vejovidae). *Am. Mus. Novit.* 2278:1-54.

Ghobrial, L. I., and Nour, T. A. 1975. The physiological adaptations of desert rodents. In I. Prakash and P. K. Ghosh (Eds.), *Rodents in desert environments.* The Hague: Junk.

Gibble, W. P. 1950. Nineteen years of vegetational change in a desert habitat. Master's thesis, University of Arizona.

Gibbs, J. G., and Patten, D. T. 1970. Heat flux and plant temperatures in a Sonoran Desert ecosystem. *Oecologia* 5:165-184.

Gilbert, B. 1973. *Chulo.* New York:Alfred A. Knopf.

Gilbert, R. L.; Burdsall, H. H.; and Canfield, E. R. 1976. Fungi that decay mesquite in southern Arizona. *Mycotaxon* 3:487-551.

Gilbertson, R. L., and Canfield, E. R. 1972. *Poria carnegiea* and decay of saguaro cactus in Arizona. *Mycologia* 64:1300-1311.

Gilbertson, R. L.; Martin, K. J.; and Lindsey, J. P. 1974. Annotated check list and host index for Arizona wood-rotting fungi. *Ariz. Agr. Exp. Sta. Techn. Bull.* 209:1-48.

Gilbertson, R. L.; and McHenry, J. 1969. Check list and host index for Arizona rust fungi. *Ariz. Agr. Exp. Sta. Techn. Bull.* 186:1-40.

Gile, L. H.: Peterson, F. F.; and Grossman, R. B. 1966. Morphological and genetic sequences of carbonate accumulation in desert soils. *Soil Sci.* 101:347-360.

Gill, L. S. 1961. The mistletoes. A literature review. *U.S.D.A. Techn. Bull.* 1242:1-87.

Gilles, R., (ed.). 1979. *Mechanisms of osmoregulation in animals:*

Maintenance of cell volume. New York:John Wiley.

Giluly, . 1971. Wildlife versus irrigation. *Science News* 99(11):184-185.

Gindel, I. 1970. The nocturnal behavior of xerophytes. *New Phytologist* 69:399-404.

Gladwin, H. S.; Haury, E. W.; Sayles, E. B.; and Gladwin, N. 1937. *Excavations at Snaketown. Material culture.* Medallion Papers, no. 25. Globe, Ariz.: Gila Pueblo Archeological Foundation.

Glendening, G. E. 1941. Development of seedlings of *Heteropogon contortus* as related to soil moisture and competition. *Bot. Gaz.* 102:684-698.

Glendening, G. E., and Paulsen, H. A. 1955. Reproduction and establishment of velvet mesquite as related to invasion of semidesert grasslands. *U.S.D.A. Forest Service Tech. Bull.* 1127.

Glenn, W. G.; Keegan, H. L.; and Whittemore, Jr., F. W. 1962. Intergeneric relationships among various scorpion venoms and antivenoms. *Science* 135:434-435.

Glinski, R. L. and Ohmart, R. D. 1977. The population, habitat and diet of the black hawk in Arizona and New Mexico. *Proc. 21st Ann. Meeting Ariz. Acad. Sci.:*25.

Goldberg, S. R., and Lowe, C. H. 1966. The reproductive cycle of the western whiptail lizard (*Cnemidophorus tigris*) in southern Arizona. *J. Morphol.* 118:543-548.

Goldman, E. A. 1916. Plant records of an expedition to lower California. *Contrib. U.S. Natl. Mus.* 16:309-371.

Goldman, E. A., and Moore, R. T. 1945. The biotic provinces of Mexico. *J. Mammal.* 26:347-360.

Gomez-Pompa, A. 1965. La vegetación de México. *Bol. de la Soc. Bot. Mex.* 29:76-120.

Goodall, D. W., and Morgan, S. 1974. Seed reserves in desert soils. *US/IBP Desert Biome Research Memorandum* RM 74-16. Logan, Utah.

Goodspeed, T. H. 1954. *The genus Nicotiana.* Waltham, Mass.: Chronica Botanica.

Goodyear, A. C. 1975a. Hecla II and III: An interpretive study of archaeological remains from Lakeshore Project, Papago Indian Reservation, south central Arizona. *Ariz. St. Univ. Anthrop. Res. Paper 9.*

————. 1975b. *The historical and ecological position of protohistoric sites in the Slate Mountains, South Central Arizona.* Columbia:Institute of Archaeology and Anthropology, University of South Carolina.

Goodyear, A. C., and Dittert, A. E. 1973. Hecla I: A preliminary report on archaeological investigations at the Lakeshore project, Papago Reservation, South Central Arizona. *Ariz. St. Univ. Anthrop. Res. Paper.* 4.

Gordon, S. 1934. The drinking habits of birds. *Nature* 133: 436-437.

Gorsuch, D. M. 1934. Life history of the Gambel quail in Arizona. *Univ. Ariz. Bull.* 5:1-89.

Grant, C. 1960. Differentiation of the southwestern tortoises (genus *Gopherus*), with notes on their habits. *Trans. San Diego Soc. Nat. Hist.* 12:441-448.

Grant, K. A., and Grant, V. 1968. *Hummingbirds and their flowers.* New York:Columbia University Press.

Grant, R. L. 1939. The thermal responses of reptiles. Master's thesis, University of California.

Gray, A. B. 185. On the *Ammobroma sonorae. Proceed. AAAS*

9:233-236.

Gray, R., and Bonner, J. 1948. An inhibitor of plant growth from the leaves of *Encelia farinosa. Amer. J. Bot.* 35:52-57.

Greaves, J. E., and Jones, L. W. 1941. The survival of microorganisms in alkali soils. *Soil. Sci.* 52:359-364.

Greeley, R.; Womer, M. B.; Papson, R. P.; and Spudis, P. D. (Eds.). 1978. *Aeolian features of southern California.* Washington, D.C.:Government Printing Office.

Green, R. A., and Reynard, C. 1932. The influence of two burrowing rodents, *Dipodomys spectabilis* (kangaroo rat) and *Neotoma albigula albigula* (pack rat) on desert soils in Arizona. *Ecology* 13:73-80.

Greenberg, B. 1943. Social behavior of the western banded gecko, *Coleonyx variegatus* Baird. *Physiol. Zool.* 16:110-122.

————. 1945. Notes on the social behavior of the collared lizard. *Copeia* 1945:225-230.

Greene, J. L., and Mathews, T. W. 1976. Faunal study of unworked mammalian bones. In E. W. Haury, *The Hohokam: Desert farmers and craftsmen,* pp. 367-373. Tucson:University of Arizona Press.

Greene, R. A. 1936. The composition and uses of the fruit of the giant cactus (*Carnegiea gigantea*) and its products. *J. Chem. Educ.* 13:309-312.

Grinnell, J. 1914. An account of the mammals and birds of the lower Colorado Valley with special reference to the distributional problems presented. *Univ. Calif. Publ. Zool.* 12:51-294.

————. 1915. A distributional list of the birds of California. *Pacif. Coast Avifauna* 11:1-217.

————. 1928. A distributional summation of the ornithology of Lower California. *Univ. Calif. Publ. Zool.* 32:1-300.

————. 1935. A revised life-zone map of California. *Univ. Calif. Publ. Zool.* 40:327-330.

Grinnell, J., and Camp, C. L. 1917. A distributional list of the amphibians and reptiles of California. *Univ. Calif. Publ. Zool.* 17:127-208.

Grinnell, J., and Miller, A. H. 1944. The distribution of the birds of California. *Pacif. Coast Avifauna* 27.

Gunn, D. L. 1942. Body temperature of poikilotherms. *Biol. Rev.* 17:293-314.

Gunst, M. L. 1930. Ceremonies of the Papago and Pima Indians, with special emphasis on the relationship of the dance to their religion. Master's thesis, University of Arizona.

Haase, E.F. 1970. Environmental fluctuations on south-facing slopes in the Santa Catalina Mountains of Arizona. *Ecology* 51(6):959-974

————. 1972. Survey of floodplain vegetation along the lower Gila River in southwestern Arizona. *J. Ariz. Acad. Sci.* 7(2):75-81.

Haase, E. F., and McGinnies, W. G. (Eds.) 1972. *Jojoba and its uses.* Tucson:University of Arizona, Office of Arid Lands Studies.

Hadley, N. F. 1970a. Micrometeorology and energy exchange in two desert arthropods. *Ecology* 51:434-444.

————. 1970b. Water relations of the desert scorpion, *Hadrurus arizonensis. J. Exptl. Biol.* 53:547-558.

————. 1972. Desert species and adaptation. *Amer. Sci.* 60:338-347.

————. 1975. *Environmental physiology of desert organisms.* Stroudsburg, Pa.:Dowden, Hutchinson & Ross.

Hadley, N. F., and Hill, R. D. 1969. Oxygen consumption of the scorpion *Centruroides sculpturatus*. *Comp. Biochem. Physiol.* 29:217-226.

Hadley, N. F., and Williams, S. C. 1968. Surface activities of some North American scorpions in relation to feeding. *Ecology* 49:726-34.

Haefer, J. R. 1977. Papago music and dance. *Occ. Pap. Navajo Comm. Coll.*

Hagmeier, E. M. 1966. A numerical analysis of the distributional patterns of North American mammals. Vol. 2. Re-evaluation of the provinces. *Syst. Zool.* 15:279-299.

Hagmeier, E. M., and Stutz, C. D. 1964. A numerical analysis of the distributional patterns of North American mammals. *Syst. Zool.* 13:125-155.

Hahn, W. E. 1967. Estradiol-induced vitellinogenesis and concomitant fat mobilization in the lizard *Uta stansburiana*. *J. Exptl. Zool.* 158:79-86..

———. Estradiol-induced vitellinogenesis and concomitant fat mobilization in the lizard *Uta stansburiana*. *Comp. Biochem. Physiol.* 23:83-93.

Hahn, W. E., and Tinkle, D. W. 1965. Fat body cycling and experimental evidence for its adaptive significance to ovarian follicle development in the lizard *Uta sansburiana*. *Jo. Exper. Zool.* 158:79-86.

Haight, L. T., and Johnson, R. R. 1977. The endangered southwest riparian avifauna. *Proc. 21st Ann. Meeting Ariz. Acad. Sci.:*31.

Hall, E. R., and Kelson, K.R. 1959. The mammals of North America. 2 vols. New York: Ronald Press.

Hall, E. R., and Linsdale, J. M. 1929. Notes on the life history of the kangaroo mouse (*Microdipodops*). *J. Mammal.* 10:298-305.

Hall, F. G. 1922. The vital limits of exsiccation of certain animals. *Biol. Bull. Mar. Biol. Lab Woods Hole* 42:52-58.

Hall, H. M. 1928. The genus *Haplopappus*, a phylogenetic study in the Compositae. *Carneg. Inst. Wash. Publ.* 389:1-391.

Hall, H. M., and Grinnell, J. 1919. Life-Zone indictors in California. *Calif. Acad. Sci.*, Sec. 4.(9):37-67.

Hall, S. M. 1907. The story of a Pima record rod. *Out West* 26:413-423.

Halloran, A. F. 1946. A muskrat house on the lower Colorado River. *J. Mammal.* 27:88-89.

———. 1957. A note on the Sonoran pronghorn. *J. Mammal.* 38(3):423.

Halloran, A. F., and Blanchard, W. E. 1954. Carnivores of Yuma County, Arizona. *Amer. Midl. Nat.* 51:481-487.

Halvorson, W. L. and Patten, D. L. 1975. Productivity and flowering of desert ephemerals in relation to Sonoran Desert shrubs. *Amer. Midl. Nat.* 93(2):311-319.

Hamilton, W. 1971. Recognition on space photographs of structural elements of Baja California. *U.S. Geol. Surv. Prof. Paper* 718:1-26.

Hamilton, W. J. 1975. Coloration and its thermal consequences for diurnal desert insects. In N. F. Hadley (ed.), *Environmental physiology of desert organisms*, pp. 67-89. Stroudsburg, Pa.:Dowden, Hutchinson & Ross.

Hansen, R. M. 1974. Dietary of the chuckwalla, *Sauromalus obesus*, determined by drug analysis. *Herpetologica* 30:120-123.

Hansen, R. M., Johnson, M. K., and Van Devender, T. R. 1976. Foods of the desert tortoise, *Gopherus agassizii*, in Arizona and Utah. *Herpetologica* 32:247-251.

Haring, I. M. 1961. A checklist of the mosses of the state of Arizona. *Bryologist* 64:212-240.

Harris, D. 1972. Vagabundos del mar: Shark fishermen of the Sea of Cortez. *Oceans* 5(1):60-71.

Harris, J. A.; Lawrence, J. V. and Gortner, R. A. 1916. The cryoscopic constants of expressed vegetable saps as related to local environmental conditions in the Arizona deserts. *Physiological Researches* 2(1):1-49

Harshberger, J. W. 1911. *Phytogeographic Survey of North America. Vegetation der Erde.* Vol. 13. New York:G.E. Stechert.

Haskell, W. L. 1959. Diet of the Mississippi threadfin shad, *Dorosoma petenense atchafalayae* in Arizona. *Copeia* 1959:298-302.

Hastings, J. R. 1960. Precipitation and saguaro growth. *Arid Lands Colloquia. University of Arizona:* 30-38.

Hastings, J. R., and Turner, R. M. 1965. *The changing mile.* Tucson:University of Arizona.

Hastings, J. R., Turner, R. M. and Warren, D. K. 1972. *An Atlas of Some Plant Distributions in the Sonoran Desert.* Institute of Atmospheric Physics. Univ. of Ariz. Tucson.

Hatfield, D. M. 1942. Mammals from south-central Arizona. *Bull. Chicago Acad Sci.* 6:143-157.

Haury, E. W. 1950. *The stratigraphy and archaeology of Ventana Cave.* Tucson: University of Arizona Press.

———. 1976. *The Hohokam, desert farmers and craftsmen.* Tucson: University of Arizona Press.

Haury, E. W.; Bryan, K.; Colbert, E. H.; Gabel, N. E.; Tanner, C. L.; and Buehrer, T. E. 1950. See Haury, E. W. 1950.

Haverty, M. I., and Nutting, W. L. 1974. Natural wood-consumption rates and survival of a dry-wood and a subterranean termite at constant temperatures. *Ann. Ent. Soc. Am.* 67:158-157.

Hawke, D. D., and Farley, R. D. 1973. Ecology and behavior of the desert burrowing cockroach, *Arenivaga* sp (Dictyoptera, Polyphagidae). *Oecologia* 11:263-279.

Hay, O. P. 1892. On the ejection of blood from the eyes of horned toads. *Proc. U.S. Natl. Mus.* 15:375-378.

Hayden, J. D. 1937. The Vikita ceremony of the Papago. *Southwest. Monuments Monthly Rept. (Suppl. for April)*:263-277.

Hayward, H. E., and Bernstein, L. 1958. Plant-growth relationships on salt-affected soils. *Bot. Rev.* 24:584-635.

Hayward, H. E., and Wadleigh, C. H. 1949. Plant growth on saline and alkali soils. *Advances in Agronomy* 1:1-38.

Heath, J. E. 1962a. Temperature regulation and diurnal activity in horned lizards. Ph.D. dissertation, University of California.

———. 1962b. Temperature-independent morning emergence in lizards of the genus *Phrynosoma*. *Science* 138:891-892.

———. 1964. Head-body temperature differences in horned lizards. *Physiol. Zool.* 37:273-279.

———. 1965. Temperature regulation and diurnal activity in horned lizards. *Univ. Calif. Publ. Zool.* 64:97-136.

———. 1966. Venous shunts in the cephalic sinuses of horned lizards. *Physiol. Zool.* 39:30-35.

Heath, J. E., and Wilkin, P. J. 1970. Temperature responses of the desert cicada *Diceroprocta apache* (Homoptera, Cicadidae). *Physiol. Zool.* 43:145-154.

Heath, W. G. 1962. Maximum temperature tolerance as a function of constant temperature in the gila Topminnow (*Poeciliopsis occidentalis*. Ph.D. dissertation, University of Arizona Press.

————. 1967. Ecological significance of temperature tolerance in Gulf of California shore fishes. *J. Ariz. Acad. Sci.* 4:172-178.

Heifetz, W. 1941. A review of the lizards of the genus *Uma. Copeia* 1941:99-111.

Heimlich, E. M. and Heimlich, M. G. 1947. A case of cannibalism in captive *Xantusia vigilis. Herpetologica* 3:149-150.

Heinrich, Bernd. 1975. Thermoregulation and flight energetics of desert insects. In N. F. Hadley (ed.). *Environmental physiology of desert organisms.* New York: Halsted Press.

Heller, H. 1965. Osmoregulation in Amphibia. *Arch. Anat. Micros. Morph. Exp.* 54:471-490.

Henckel, P. A. 1964. Physiology of plants under drought. *Ann. Rev. Plant Physiol.* 15:363-386.

Hendricks, B. A. 1942. Effect of grass litter on infiltration of rainfall on granitic soils in a semi-desert shrub-grass area. *U.S.F.S. Southwest. Forest and Range Exp. Sta. Research Note* 96:1-3.

Hendrickson, J. R. 1974. Study of the marine environment of the northern Gulf of California. *Nat. Tech. Info. Serv. Publ.* N74-16008:1-95.

————. 1979. Totoaba: Sacrifice in the Gulf of California. *Oceans* (September): 14-18.

Henrickson, J. 1972. A taxonomic revision of the Fouquieriaceae. *El Aliso* 7:439-537.

Hensley, M. M. 1949. Mammal diet of *Heloderma. Herpetologica* 5:152.

————. 1954. Ecological relations of the breeding bird population of the desert biome in Arizona. *Ecol. Monogr.* 24:185-207.

Heppner, F. 1970. The metabolic significance of differential absorption of radiant energy by black and white birds. *Condor* 72:50-59.

Hevly, R. H. 1963. Adaptations of Cheilanthoid ferns to desert environments. *J. Ariz. Acad. Sci.* 2:164-175.

————. 1965. Studies of the sinuous cloak-fern (*Notholaena sinuata*) complex. *J. Ariz. Acad. Sci.* 3:2-5-208.

Heyne, B. 1814. *Transact. Linn. Soc.* 1818:213-215.

Hill, J. E. 1942. Cactus paved runways of *Neotoma albigula. J. Mammal.* 23:213-214.

Hinds, D. S. 1973. Acclimatization of thermoregulation in the desert cottontail, *Sylvilagus auduboni. J. Mammal.* 54:708-728.

Hitchcock, C. L. 1932. A monographic study of the genus *Lycium* of the western hemisphere. *Ann. Mo. Bot. Gard.* 19:179-374.

Hoff, C. C., and Riedesel, M. L. (eds.). 1969. *Physiological systems in semi-arid environments.* Albuquerque: University of New Mexico Press.

Hoffmeister, D. F. 1970. The seasonal distribution of bats in Arizona: A case for improving mammalian range maps. *Southwest. Nat.* 15(1):11-12.

Hoffmeister, D. F., and Goodpaster, W. W. 1962. Life history of the desert shrew *Notiosorex crawfordi. Southwest. Nat.* 7:236-252.

Holmes, W. N. 1965. Some aspects of osmoregulation in reptiles and birds. *Arch. Anat. Micr. Morph. Exp.* 54:491-514.

Hooke, R. L.; Yang, H-Y.; and Weiblen, P. W. 1969. Desert varnish, and electron probe study. *J. Geol.* 77(3):275-288.

Hoover, F. G. and St. Amant, J. A. 1970. Establishment of *Tilapia mossambica* Peters in Bard Valley, Imperial County, California. *Calif. Fish and Game* 56:70-71.

Hornaday, W. T. 1908. *Camp-fires on desert and lava.* New York: Charles Scribner's Sons.

Horne, F. 1968. Survival and ionic regulation of *Triops longicaudatus* in various salinities. *Physiol. Zool.* 41:180-186.

Horne, F., and Behenbach, K. W. 1971. Physiological properties of hemoglobin in the branchippod crustacean *Triops. Amer. Jour. Physiology* 220:1875-1881.

Horton, J. S. 1972. Management problems in phraeatophytes and riparian zones. *J. Soil Water Conserv.* 27:57-61.

————. 1973. An abstract bibliography—evapotranspiration and water research as related to riparian and phreatophyte management. *U.S.F.S. Misc. Publ.* 1234:1-192.

Horton, J. S.; Robinson, T. W.; and McDonald, H. R. 1964. Guide for surveying phraeatophyte vegetation. *U.S.D.A. Agr. Handbook.* 266:1-37.

Hotton, N. 1955. A survey of adaptive relationships of dentition to diet in the North American Iguanidae. *Amer. Midl. Nat.* 53:88-114.

Householder, V. H. 1950. Courtship and coition of the desert tortoise. *Herpetologica* 6:11.

Howard, H. 1947. An ancestral golden eagle raises a question in taxonomy. *Auk* 64:287-291.

Howell, A. B., and Gersh, I. 1935. Conservation of water by the rodent *Dipodomys. J. Mammal.* 16:1-9.

Howell, A. H. 1938. Revision of the North American ground squirrels. *N. Amer. Fauna* 56:1-256.

Howell, D. J. 1973. Bats and pollen: Physiological aspects of the syndrome of chiropterophily. *Comp. Biochem. Physiol.* 48A:263-276.

Howes, P. G. 1954. *The giant cactus forest and its world.* New York: Duell, Sloan and Pearce.

Hubbard, J. P. 1973. Avian evolution in the arid lands of North America. *Living Bird* 12:155-196.

Hubbs, C. 1959. High incidence of vertebral deformities in two natural populations of fishes inhabiting warm springs. *Ecology* 40:154-155.

Hubbs, C. L., and Miller, R. R. 1953. Hybridization in nature between the fish genera *Catostomus* and *Xyrauchen. Pap. Michigan Acad. Sci. Arts Letter.* 38:207-233.

Hudson, J. W. 1962. The role of water in the biology of the antelope ground squirrel, *Citellus leucurus. Univ. Calif. Publ. Zool.* 64:1-56.

————. 1964. Water metabolism in desert mammals. In M. H. Wayner (ed.), *Thirst,* pp. 211-233. New York: Pergamon.

————. 1964. Temperature regulation in the round-tailed ground squirrel, *Citellus tereticaudus. Ann. Acad. Sci. Fennicae* 71:16.

————. 1967. Variations in the patterns of torpidity of small homeotherms. In K. C. Fisher et al. (eds.), *Mammalian hibernation,* vol. 3, pp. 30-46. New York: American Elsevier.

Hudson, J. W., and Bartholomew, G. A. 1964. Terrestrial animals in dry heat: Estivators. In D. B. Dill (ed.), *Handbook of physiology. Sect. 4. Adaptation to the environment,* pp. 541-550. Washington, D.C.: American Physiological Society.

Hudson, J. W.; Deavers, D. R.; and Bradley, S. R. 1972. A comparative study of temperature regulation in ground squirrels with special reference to the desert species. In G. M. O. Maloiy (ed.), *Comparative physiology of desert animals,* pp. 191-213. New York: Academic Press.

Hudson, J. W., and Wang, L. C. 1969. Thyroid function in desert ground squirrels. In C. C. Hoff and M. L. Riedesel (eds.),

Physiological systems in semi-arid environments. Albuquerque: University of New Mexico Press.

Huey, L. M. 1942. A vertebrate faunal survey of the Organ Pipe Cactus National Monument, Arizona. *Trans. San Diego Soc. Nat. Hist.* 9:355-375.

———. 1951. The kangaroo rats (*Dipodomys*) of Baja California, Mexico. *Trans. San Diego Soc. Nat. Hist.* 11:205-256.

———. 1964. The mammals of Baja California, Mexico, *Trans. San Diego Soc. Nat. Hist.* 13:85-168.

Hull, H. M. 1958. The effect of day and night temperature on growth, foliar wax content, and cuticle development of velvet mesquite. *Weeds* 6:133-142.

Hull, H. M.; Shellhorn, S. J.; and Saunier, R. E. 1971. Variations in creosotebush (*Larrea divaricata*) epidermis. *J. Ariz. Acad. Sci.* 6:196-205.

Humphrey, R. R. 1932. The morphology, physiology and ecology of *Coldenia canescens. Ecology* 13:153-158.

———. 1937. Ecology of the borrowed. *Ecology* 18:1-9.

———. 1960. Forage production on Arizona ranges V: Pima, Pinal and Santa Cruz countries. *Ariz. Agr. Exp. Sta. Bull.* 302.

———. 1965. Arizona natural vegetation. *Ariz. Agr. Exp. Bull.* A-45: map only.

———. 1968. *The desert grassland.* Tucson: University of Arizona Press.

———. 1970. *Arizona range grasses.* Tucson: University of Arizona Press.

———. 1971. Comments on an epiphyte, a parasite and four independent spermatophytes of the central desert of Baja California. *Cactus and Succulent J.* 43(3): 99-104.

———. 1974. *The boojum and its home.* Tucson: University of Arizona Press.

Humphrey, Z. 1938. *Cactus forest.* New York: E. P. Dutton.

Hungerford, C. R.; Lowe, C. H.; Madson, R. L. 1973. Population studies of the desert cottontail (*Sylvilagus auduboni*) and black-tailed jackrabbit (*Lepus californius*) in the Sonoran Desert. *US/IBP Desert Biome Research Memorandum* RM 73-20:1-15. Logan, Utah.

———. 1974. Population studies of the desert cottontail (*Sylvilagus auduboni*), blacktailed jackrabbit (*Lepus californicus*) and Allen's jackrabbit (*Lepus alleni*) in the Sonoran Desert. *US/IBP Desert Biome Research Memorandum* RM 74-23:73-93. Logan, Utah.

Hunt, J. H., and Snelling, R. R. 1975. A checklist of the ants of Arizona. *J. Ariz. Acad. Sci.* 10:20-23.

Hurd, P. D. and Linsley, E. G. 1975a. The principal *Larrea* bees of the *Smithsonian Contr. Zool.* 193:1-74.

———. 1975b. Some insects other than bees associated with *Larrea tridentata* in the southwestern United States. *Proc. Ent. Soc. Wash.* 77:100-120.

Inman, D. L.; Ewing, G. C.; and Corliss, J. B. 1966. Coastal sand dunes of Guerrero Negro, Baja California, Mexico. *Geol. Soc. Amer. Bull.* 77(8):787-802.

Isely, D. 1969. Legumes of the United States. Vol. 1. Native acacia. *Sida* 3:365-386.

Ives, R. L. 1936. Desert floods in the Sonoyta Valley. *Amer. J. Sci.,* ser. 5., 32:349-360.

———. 1959. Shell dunes of the Sonoran shore. *Amer. J. Sci.* 257(6):449-457.

———. 1964. The Pinacate Region, Sonora, Mexico. *Calif. Acad.* *Sci. Occ. Papers* 47:1-43.

Jackson, D. G., Jr.; Trotter, T. H.; Trotter, M. W.; and Trotter; J. A. 1976. Accelerated growth rate and early maturity in *Gopherus agassizi. Herpetologica* 32:139-145.

———. 1977. Further observations of growth and sexual maturity in captive desert tortoise. *Desert Tortoise Council Second Annual Symposium,* Las Vegas, Nev.

Jackson, S. P. 1951. The Clever Coyote. Part I. Its history, life habits, economic status, and control. In S. P. Young and H. H. T. Jackson, *The clever coyote,* pp. 3-226. Washington, D. C.: Wildlife Management Institute.

Jaeger, E. C. 1948. Who trims the creosote bush? *J. Mammal.* 29:187-188.

———. 1957. *The North American deserts.* Stanford: Stanford University Press.

James, A. E. 1968. *Learnea* (copepod) infection of three native fishes from the Salt River Basin, Arizona. Master's thesis, Arizona State University.

Jameson, D. L.; Mackey, J. P.; and Richmond; R. C. 1966. The systematics of the Pacific tree frog, *Hyla regilla. Proc. Calif. Acad. Sci.* Ser. 4, 33:551-620.

Jasinski, A., and Gorbman, A. 1967. Hypothalamic neurosecretion in the spadefoot toad, *Scaphiopus hammondi,* under different environmental conditions. *Copeia* 1967:271-279.

John, K. R. 1963. The effect of torrential rains on the reproductive cycle of *Rhinichthys osculus* in the Chiricahua Mountains, Arizona. *Copeia* 1963:286-291.

———. 1964. Survival of fish in intermittent streams of the Chiricahua Mountains, Arizona. *Ecology* 41:112-119.

Johnson, Albert W. 1968. The evolution of desert vegetation in western North America. In G. W. Brown (ed.), *Desert biology,* Vol. 1, pp. 101-140. New York: Academic Press.

Johnson, C. R. 1965. The diet of the Pacific fence lizard, *Sceloporus occidentalis occidentalis* (Baird and Girard) from northern California. *Herpetologica* 21:114-117.

Johnson, D. E. 1961. Edaphic factors affecting the distribution of creosotebush, *Larrea tridentata (DC.)* Cov. in the desert grassland sites of southeastern Arizona. Master's thesis, University of Arizona.

Johnson, D. R., and Groepper, K. L. 1970. Bioenergetics of North Plains rodents. *Amer. Midl. Nat.* 84:537-548.

Johnson, D. S. 1918. The fruit of *Opuntia fulgida,* a study of perennation and proliferation in the fruits of certain Cactaceae. *Carneg. Inst. Wash. Publ.* 269:1-62.

Johnson, D. W., and Lew, L. 1970. Chlorinated hydrocarbon pesticides in representative fishes of southern Arizona. *Pesticide Monitoring J.* 4:57-61.

Johnson, H. B. 1976. Vegetation and plant communities of southern California deserts—a functional view. In J. Latting (ed.), *Plant communities of southern California,* pp. 125-152. Berkeley: California Native Plant Society.

Johnson, J. E. 1970. Age, growth, and population dynamics of threadfin shad, *Dorosoma petenense* (Günther), in central Arizona reservoirs. *Transact. Amer. Fish Soc.* 99:739-753.

Johnson, M. E., and Snook, H. J. 1927. *Seashore animals of the Pacific Coast.* New York: Macmillan. Reprint ed., New York: Dover, 1967.

Johnston, I. M. 1924. Expedition of the California Academy of

Sciences to the Gulf of California in 1921. *Proc. Calif. Acad. Sci.* 4(12):951-1218.

————. 1940. The floristic significance of shrubs common to North and South American deserts. *J. Arnold Arboretum (Harvard Univ.)* 21:356-363.

Johnston, M. C. 1962a. The North American mesquites. *Prosopis* section *Algarobia* (Leguminosae). *Brittonia* 14:72-90.

————. 1962b. Revision of *Condalia* including *Microrhamnus* (Rhamnaceae). *Brittonia* 14:332-368.

Jones, G. N. 1966. *An annotated bibliography of Mexican ferns.* Urbana, Ill.: University of Illinois Press.

Jones, R. D. 1969. An analysis of Papago communities, 1900-1920. Ph.D. dissertation, University of Arizona.

————. 1971. The wi'igita of Achi and Quitobac. *Kiva* 36(4):1-29.

Jones, W. 1979. Effects of the 1978 freeze on native plants of Sonora, Mexico. *Desert Plants* 1:33-39.

Jorgensen, C. B., and Tanner, W. W. 1963. The applications of the density probability function to determine the home ranges of *Uta stansburiana* and *Cnemidophorus tigris. Herpetologica* 19:105-115.

Joseph, A.; Spicer, R.; and Chesky, J. 1949. *The desert people.* Chicago: University of Chicago Press.

Judd, B. I. 1971. The lethal decline of mesquite on the Casa Grande National Monument. *Great Basin Nat.* 31:153-159.

Karig, D. E., and Jensky, W. 1972. The proto-Gulf of California. *Earth and Planet. Sci. Letter* 17:169-174.

Kasheer, T. 1961. *Flora of Baja Norte.* Glendale, Calif.: La Siesta Press.

Kauffield, C. F. 1943. Field notes on some Arizona reptiles and amphibians. *Amer. Midl. Nat.* 29:342-359.

Kaufmann, J. H.; Lanning, D. V.; and Poole, S. E. 1967. Current status and distribution of the coati in the United States. *J. Mammal.* 57:621-637.

Kearney, T. H. 1929. Plants of lower California relationship in central Arizona. *J. Wash. Acad. Sci.* 19:70-71.

————. 1935. The North American species of *Sphaeralcea* subgenus *Eusphaeralcea. Univ. of Calif. Publ. Bot.* 19:1-128.

Kearney, T. H., and Pheebles, R. H. 1960. *Arizona flora.* 2d ed. Berkeley: University of California Press.

Keegan, H. L.; Hedden, R. A.; and Whittemore Jr., F. W. 1960. Seasonal variation in venom of black widow spiders. *Am. J. Trop. Med. Hyg.* 9:477-479.

Keil, D. J. 1973. Vegetation and flora of the White Tank Mountains Regional Park, Maricopa County, Arizona. *J. Ariz. Acad. Sci.* 8:35-48.

Kendeigh, S. C. 1939. The relation of metabolism to the development of temperature regulation in birds. *J. Exptl. Zool.* 82:419-438.

Khudairi, A. K. 1969. Mycorrhiza in desert soils. *Bioscience* 19:598-599.

Kidd, D. E., and Wade, W. E. 1965. Algae of Quitobaquito: A spring-fed impoundment in Organ Pipe Cactus National Monument. *Southwest Nat.* 10:227-233.

Kilcrease, A. T. 1939. Ninety five years of history of the Papago calendar stick. *Southwest. Monuments Monthly Rept.* (suppl. for April):297-310.

Kinne, O. 1964. The effect of temperature and salinity of marine and brackish water animals. Vol. 2. Salinity and temperature-salinity relations. In: H. Barnes (ed.), *Oceanography and marine biology,* vol. 2, pp. 281-339.

————. 1965. Salinity requirements of the fish, *Cyprinidon macularius. U.S. Public Health Serv. Publ.* 999-WP-25:187-192.

King, J. E., and Van Devender, T. R. 1977. Pollen analysis of fossil packrat middens from the Sonoran Desert. *Quat. Res.* 8(2):191-204.

Kim, J. 1973. Ecosystem of the Salton Sea. In W. C. Ackerman, G. F. White, and E. B. Worthington (eds.), *Man-made lakes, their problems and environmental effects,* pp. 601-605. Washington, D.C.: American Geophysical Union.

Kimsey, J. B., and Fisk, L. O. 1960. Keys to the freshwater and anadromous fishes of California. *Calif. Fish and Game* 46:453-479.

Kirmiz, J. P. 1962. *Adaptation to desert environment.* London: Butterworths.

Kissell, M. L. 1972. *Basketry of the Pima and Papago Indians.* Glorieta, N.M.: Rio Grande Press.

Klauber, L. M. 1939. Studies of reptile life in the arid Southwest. *Bull. Zool. Soc. San Diego.* 14:1-100.

————. 1940. The worm snakes of the genus *Leptophlops* in the United States and northern Mexico. *Trans. San Diego Soc. Nat. Hist.* 9:87-162.

————. 1951. The shovel-nosed snake, *Chionactis,* with descriptions of two new subspecies. *Trans. San Diego Soc. Nat. Hist.* 11:141-204.

————. 1956. *Rattlesnakes.* 2 vols. Berkeley: University of California Press.

Klikoff, L. G. 1966. Competitive response to moisture stress of a winter annual of the Sonoran Desert. *Amer. Midl. Nat.* 75:383-391.

————. 1967. Moisture stress in a vegetational continuum in the Sonoran Desert. *Amer. Midl. Nat.* 77:128-137.

Klots, A. B. 1951. *A field guide to the butterflies.* Boston: Houghton Mifflin.

Knipe, O. D. 1971. Effect of different osmotica on germination of alkali sacaton (*Sporobolus airoides* Torr.) at various moisture stresses. *Bot. Gaz.* 132:109-112.

Knipe, T. 1957. The javelina in Arizona: A research and management study. *Ariz. Game and Fish Dept. Wildl. Bull.* 2:1-96.

Koehn, R. K. 1967. Blood proteins in natural populations of Catostomid fishes of western North America. Ph.D. dissertation, Arizona State University.

Koller, D. 1969. The physiology of dormancy and survival of plants in desert environments. In *Dormancy and Survival.* pp. 449-469. Cambridge: University Press.

Kramer, R. J. 1962. The relationship of saguaro (*Cereus giganteus* Engelm.) to certain soil characteristics. Master's thesis, Arizona State University.

Krieger, M. H. 1969. Ash-flow tuffs in the northern Galiuro Mountains, Pinal County, Arizona. Abstract. *Geol. Soc. Amer. Spec. Paper* 121:523.

Krumholz, L. A. 1948. Reproduction in the western mosquitofish, *Gambusia affinis affinis* (Baird and Girard), and its use in mosquito control. *Ecol. Monogr.* 18:1-43.

Kuchler, A. W. 1977. Natural vegetation of California. Map. In M. G. Barbour and J. Major (eds.). *Terrestrial vegetation of California.* New York: John Wiley.

Kurtz, E. B., and Alcorn, S. M. 1960. Some germination requirements of saguaro cactus seeds. *Cactus Succulent J.* 32(3):72-74.

Lacey, J. R.; Ogden, P. R.; and Foster, K. E. 1975. Southern Arizona riparian habitat: Spatial distribution and analysis. *University of Arizona. Office of Arid Lands Studies Bull.* 8:1-148.

Laddell, W. S. S. 1953. The physiology of life and work in high ambient temperatures. *Proc. Int. Symp. Desert Res. Jerusalem.* 1953:187-204.

Lamb, E., and Lamb, B. 1974. *Colorful cacti of the American deserts.* New York: Macmillan.

Lamb, J. W. 1964. The fungal flora of the slime flux of certain plants. Master's thesis, University of Arizona.

Lange, O. L. 1955. Untersuchungen über die Hitzresistenz der Moose in Bezeihung zu ihrer Verbreitung. II. Die Resistenz stark ausgetrockneter Moose. *Flora, Jena* 142:381-399.

Lange, O. L; Schulze, E.-D.; Kappen, L.; Buschbom, U.; and Evenari, M. 1975. Adaptations of desert lichens to drought and extreme temperatures. In N. F. Hadley (ed.), *Environmental physiology of desert organisms*, pp. 20-37. Stroudsburg. Pa.: Dowden, Hutchinson & Ross.

Langebartel, D. A., and Smith, H. M. 1954. Summary of the Norris collection of reptiles and amphibians from Sonora, Mexico. *Herpetologica* 10:125-136.

Langlois, J. P. 1902. Thermal regulation in poikilotherms. *J. Physiol. Path. Gen.* 4:249-256.

Langman, I. K. 1964. *A selected guide to the literature on the flowering plants of Mexico.* Philadelphia: University of Pennsylvania Press.

Langston, G. A., and Helmberger, D. V. 1974. Interpretation of body and Rayleigh waves from NTS to Tucson. *Bull. Seism. Soc. Amer.* 64:1919-1929.

Lasiewski, R. C. 1964. Body temperature, heart and breathing rate, and evaporative water loss in hummingbirds. *Physiol. Zool.* 37:212-223.

Lasiewski, R. C., and Bartholomew, G. A. 1969. Condensation as a mechanism for water gain in nocturnal poikilotherms. *Copeia* 1969:405-407.

Lasiewski, R. C.; Bernstein, M. H.; and Ohmart, R. D. 1971. Cutaneous water loss in the roadrunner and poor-will. *Condor* 73:470-472.

Lasiewski, R. C., and Dawson, W. R. 1967. A re-examination of the relation between standard metabolic rate and body weight in birds. *Condor* 69:13-23.

Lauber, P. 1974. *Life on a giant cactus.* Champaign, Ill.: Garrard Publishing Company.

Lawlor, T. E. 1971. Evolution of *Peromyscus* on northern islands in the Gulf of California, Mexico. *Trans. San Diego Soc. Nat. Hist.* 16(5):91-124.

Lechleitner, R. R. 1959. Sex ratio, age classes, and reproduction of the black-tailed jack rabbit. *J. Mammal.* 40:63-81.

Lee, A. K. 1963. The adaptations to arid environments in wood rats of the genus Neotoma. *Univ. Calif. Publ. Zool.* 64:57-96.

Lee, H. H. 1963. Egg-laying in captivity by *Gopherus agassizi* Cooper. *Herpetologica* 19:62-65.

Lee, V. F. 1979. The maritime pseudoscorpions of Baja California, Mexico (Arachnida: Pseudoscorpionida). *Occ. Pap. Calif. Acad. Sci.* 131:1-38.

Lehr, J. H. 1978. *Catalogue of the flora of Arizona.* Phoenix: Desert Botanical Garden.

Lemen, C. A. 1976. Body size and seed size in desert rodents. *Bull. Ecol. Soc. Amer.* 57:22.

Leopold, A. S. 1950. Vegetation zones of Mexico. *Ecology* 31:507-518.

———. 1959. *Wildlife of Mexico.* Berkeley: University of California Press.

Levi, Herbert W.; Levi, Lorna R.; and Zim, Herbert S. 1968. *A guide to spiders and their kin.* New York: Golden Press.

Levine, M. 1933. Crown gall on saguaro (*Carnegiea gigantea.*) *Bull. Torr. Bot. Club* 60:9-16.

Lewis, A. W. 1973. Seasonal population changes in the cactus mouse, *Peromyscus eremicus.* Ph.D. dissertation, Arizona State University.

Leydet, F. 1977. *The coyote.* Norman: University of Oklahoma Press.

Licht, P. 1964. The relation between thermoregulation and physiological adjustments to temperature in lizards. Ph.D. dissertation, University of Michigan.

———. 1965. Effects of temperature on heart rates of lizards during rest and activity. *Physiol. Zool.* 38:129-137.

Ligon, J. D. 1968. The biology of the elf owl, *Micranthene whitneyi. Misc. Publ. Mus. Zool. Univ. Mich.* 136:1-70.

Lindsay, G. 1963. The genus *Lophocereus. Cactus Succ. J.* 35:176-192.

———. 1965. Desert plants of Baja California. *Pacific Discovery* 18(4).

Linsdale, J. M. 1932. Amphibians and reptiles from lower California. *Univ. Calif. Publ. Zool.* 38:345-386.

Lindsey, J. P. 1975. The biology of *Poria carnegiea* in southern Arizona. Ph.D. dissertation, University of Arizona.

Lindsey, J. P., and Gilbertson, R. L. 1975. Wood-inhibiting Homobasidiomycetes on saguaro in Arizona. *Mycotaxon* 2:83-103.

Lindstedt, S. L. 1976. Physiological ecology of the smallest desert mammal, *Notiosorex crawfordi. Proc. 20th Ann. Meeting AAAS. J. Ariz. Acad. Sci. Suppl.* 11:155.

Linsley, E. G. 1958. Geographical origins and phylogenetic affinities of the Cerambycid beetle fauna of western North America. In C. L. Hubbs (ed.), *Zoogeography*, pp. 229-320. Washington, D.C.: American Association for the Advancement of Science.

Lisk, R. D. 1967. Neural control of gonad size by hormone feedback in the desert iguana, *Disposaurus dorsalis dorsalis. Gen. Comp. Endocrinol.* 8:258-266.

Little, E. L. 1936. A record of the jaguarundi in Arizona. *J. Mammal.* 19:500-501.

———. 1968. Southwestern trees. *U.S.D.A. Agr. Handbook* 9:1-109.

Livingston, B. E. 1906. The relation of desert plants to soil moisture and to evaporation. *Carneg. Inst. Wash. Publ.* 50:1-78.

———. 1907. Relative transpiration in cacti. *Plant World* 10:110-114.

———. 1908. Evaporation and plant habitats. *Plant World* 11(1):1-9.

———. 1910. Relation of soil moisture to desert vegetation. *Bot. Gaz.* 50:241-256.

———. 1911. A study of the relation between summer evaporation intensity and centers of plant distribution in the United States. *Plant World* 14:205-222.

Livingston, B. E., and Brown, W. H. 1912. Relation of the daily march of transpiration to variations in the water content of foliage leaves. *Bot. Gaz.* 53:309-330.

Livingston, B. E., and Shive, J. W. 1914. The relation of atmospheric evaporating power to soil moisture content at permanent wilting in plants. *Plant World* 17:81-121.

Lockard, R. B., and Lockard, J. S. 1971. Seed preference and buried seed retrieval of *D. deserti. J. Mammal.* 52:219-222.

Loeb, L. et al. 1913. The venom of Heloderma. *Carneg. Inst. Wash. Publ.* 177:1-244.

Low, B. S. 1976. The evolution of amphibian life histories in the desert. In D. W. Goodall (ed.), *Evolution of desert biota,* pp. 149-195. Austin: University of Texas Press.

Low, W. A. 1970. The influence of aridity on reproduction of the collared peccary (*Dictoyles tajacu*) Linn. in Texas. Ph.D. dissertation, University of British Columbia.

Lowe, C. H. 1955. An evolutionary study of island faunas in the Gulf of California, Mexico, with a method for comparative analysis. *Evolution* 9:339-344.

———. 1955. Gambel quail and water supply on Tiburon Island, Sonora, Mexico. *Condor* 57:244.

———. 1955. The eastern limit of the Sonoran Desert in the United States with additions to the known herpetofauna of New Mexico. *Ecology* 36:343-345.

———. 1959. Contemporary biota of the Sonoran Desert: Problems. *Univ. of Ariz. Arid Lands Colloquia* 1958-1959:54-74.

———. 1961. Biotic communities in the sub-Mogollon region of the inland Southwest. *J. Ariz. Acad. Sci.* 2:40-49.

———. 1963. An annotated check list of the amphibians and reptiles of Arizona. In C. H. Lowe (ed.), *The vertebrates of Arizona.* Tucson: University of Arizona Press.

———. 1964a. Arizona landscapes and habitats. In C. H. Lowe (ed.), *The Vertebrates of Arizona,* pp. 3-132. Tucson: University of Arizona Press.

———. 1964b. *Arizona's natural environment.* Tucson: University of Arizona Press.

———. 1968. Fauna of desert environments. In W. G. McGinnies, B. J. Goldman, and P. Paylore (eds.), *Deserts of the world,* pp. 569-645. Tucson: University of Arizona Press.

Lowe, C. H., and Brown, D. E. 1973. The Natural vegetation of Arizona. *Ariz. Res. Inf. Syst. Coop. Publ.* 2:1-53.

Lowe, C. H., and Heath, W. G. 1969. Behavioral and physiological responses to temperature in the desert pupfish, *Cyprinodon macularius. Physiol. Zool.* 42:53-59.

Lowe, C. H.; Hinds, D. S., and Halpern, E. A. 1967. Experimental catastrophic selection and tolerances to low oxygen concentration in native Arizona freshwater fishes. *Ecology* 48:1013-1017.

Lowe, C. H., and Hinds, D. S. 1969. Thermoregulation in desert populations of roadrunners and doves. In C. C. Hoff and M. Riedesel (eds.), *Physiological systems in semiarid environments.* Albuquerque: University of New Mexico Press.

———. 1971. Effect of paloverde (*Cercidium*) trees on the radiation flux at ground level in the Sonoran Desert in winter. *Ecology* 52(5):916-922.

Lowe, C. H.; Lardner, P. J.; and Hapern, E. A. 1971. Supercooling in reptiles and other vertebrates. *Comp. Biochem. and Physiol.* 39A:125-135.

Lowe, C. H., and Wright, J. W. 1966. Evolution of parthenogenetic species of *Cnemidophorus* (Whiptail lizards) in western North America. *J. Ariz. Acad. Sci.* 4:81-7.

Lowe, C. H.; Wright, J. W.; and Norris, K. S. 1966. Analysis of the herpetofauna of Baja California, Mexico. Vol. 4. The Baja California striped whiptail, *Cnemidophorus labialis,* with key to the striped-unspotted whiptails of the Southwest. *J. Ariz. Acad. Sci.* 4:121-127.

Lowe, C. H., and Woodin, W. H. 1954. Aggressive behavior in *Phrynosoma. Herpetologica* 10:48.

Lumholtz, C. 1912. Republished as Lumholtz, C. 1971. q.v.

———. 1971. *New trails in Mexico.* Glorieta, N.M.: Rio Grande Press.

Lund, D. 1977. *All about tarantulas.* Neptune, N. J.: T. F. H. Publications.

Lyford, F. P. 1968. Soil infiltration rates as affected by desert vegetation. Master's thesis, University of Arizona.

Lynch, D. J. 1976. A buried mid-Tertiary desert in south-western Arizona. Abstract. *J. Ariz. Acad. Sci.* 11:85.

Lynn, R. T. 1963. Comparative behavior of the horned lizards, genus Phrynosoma, of the United States. Ph.D. dissertation, University of Oklahoma.

Lynn, R. T. 1965. A comparative study of display behavior in *Phrynosoma* (Iguanidae). *Southwestern Nat.* 10:25-30.

Mabry, T. J.; Hunziker, J. H.; and DiFeo, D. R. (eds.). *Creosote bush: Biology and chemistry of* Larrea *in New World deserts.*

McCleary, J. A. 1951. Some factors affecting the distribution of mosses in Arizona. Ph.D. dissertation, University of Michigan.

———. 1959. The Bryophytes of the desert region in Arizona. *Bryologist.* 62:58-62.

———. 1973. Comparative germination and early growth studies of six species of the genera *Yucca. Amer. Midl. Nat.* 90:503-508.

McClanahan, L. 1964. Osmotic tolerance of the muscles of two desert-inhabiting toads, *Bufo cognatus* and *Scaphiopus couchi. Comp. Biochem. Physiol.* 12:501-508.

———. 1967. Adaptations of the spadefoot toad, *Scaphiopus couchii. Comp. Biochem. Physiol.* 12:501-508.

———. 1975. Nitrogen excretion in arid-adapted amphibians. In N. F. Hadley (ed.). *Environmental physiology of desert organisms,* pp. 106-116. Stroudsburg, Pa.: Dowden, Hutchinson & Ross.

McClanahan, L., and Baldwin, R. 1969. Rate of water uptake through the integument of the desert toad, *Bufo punctatus. Comp. Biochem. Physiol.* 28:381-389..

McConnell, W. J. 1966. Preliminary report on the Malacca *Tilapia* hybrid as a sport fish in Arizona. *Prog. Fish-Cult.* 28:40-45.

McCoy, E. D., and Connor, E. F. 1980. Latitudinal gradients in the species diversity of North American mammals. *Evolution* 34(1):193-203.

McCrone, J. D. 1964. Comparative lethality of several *Latrodectus* venoms. *Toxicon* 2:201-203.

McDonough, W. T. 1964. Reduction in incident solar energy by desert shrub cover, *Ohio J. Sci.* 64(4):250-251.

MacDougal, D. T. 1903. Some aspects of desert vegetation. *Plant World* 6:249-257.

———. 1904. Delta and desert vegetation. *Bot. Gaz.* 38:44-63.

———. 1909. The origin of desert floras. In V. M. Spalding,

Distribution and movements of desert plants. *Carneg. Inst. Wash. Publ.* 113:113-119, Washington, D. C.

———. 1912. North American deserts. *Geogr. J.* 39:116.

———. 1914. The Salton Sea. *Carneg. Inst. Wash. Publ.* 193:1-182.

———. 1921. A new high temperature record for growth. *Science* 53:370-372.

MacDougal, D. T., and Spalding, E. S. 1910. The water balance of succulent plants. *Carneg. Inst. Wash. Publ.* 141:1-77.

McGinnies, W. G. 1968. Vegetation of desert environments. In W. G. McGinnies, B. J. Goldman and P. Paylore, *Deserts of the world*, pp. 381-566. Tucson: University of Arizona Press.

McGinnies, W. G., and Arnold, J. F. 1939. Relative water requirement of Arizona range plants. *Ariz. Agr. Exp. Sta. Techn. Bull.* 80:167-246.

McGinnis, S. M., Dickson, L. L. 1967a. Thermoregulation in the desert iguana *Dipsosaurus dorsalis. Science* 156:1757-1759.

———. 1967b. Behavioral thermoregulation in the desert iguana. *Science* 158:810.

MacGregor, A. N., and Johnson, D. E. 1971. Capacity of desert algae crusts to fix atmospheric nitrogen. *Proc. Soil Sci. Soc. Amer.* 35:843-844.

Mackay, William P. 1975. The home range of the banded rock lizard *Petrosaurus mearnsi* (Iguanidae). *Southwestern Nat.* 20(1):113-120.

McKelvey, S. D. 1938-1947. *Yuccas of the southwestern United States.* 2 vols. Jamaica Plain, Mass.: Arnold Aboretum of Harvard University.

McKusick, C. 1976a. Unpublished study of archaelogic site U:14:8. Cited by Haury, E. W., 1976. q.v.

———. 1976b. Avifauna. In E. W. Haury, *The Hohokam: Desert farmers and craftsmen*, pp. 374-377. Tucson: University of Arizona Press.

McLaughlin, A. M. 1933. A fusarium disease of *Cereus schottii.* Master's thesis, University of Arizona.

MacMahon, J. A. 1976. Species and guild similarity of North American desert mammal faunas: A functional analysis of communities. In D. W. Goodall, (ed.), *Evolution of desert biota*, pp. 134-148. Austin: University of Texas Press.

MacMahon, J. A. 1979. North American deserts: Their floral and faunal components. In D. W. Goodall, R. A. Perry, and K. M. W. Howes (eds.), *Arid-land ecosystems: Structure, functioning and management*, vol. 1, pp. 21-82. Cambridge: At the University Press.

MacMillen, R. E. 1962. The minimum water requirements of mourning doves. *Condor* 64:165-166.

———. 1956. Population ecology, water relations, and social behaviour of a southern California semidesert rodent fauna. *Univ. Calif. Publ. Zool.* 1-66.

———. 1964. Water economy and salt balance in the western harvest mouse, (*Reithrodontomys megalotis. Physiol. Zool.* 37(1):45-56.

———. 1965. Aestivation in the cactus mouse, *Peromyscus eremicus. Comp. Biochem. Physiol.* 16:227-248.

———. 1972. Water economy of nocturnal desert rodents. In G. M. O. Maloiy (ed.), *Comparative physiology of desert animals.* New York: Academic Press.

MacMillen, R. E., and Trost, C. H. 1967. Nocturnal hypothermia in the Inca dove, *Scardafella inca. Comp. Biochem. Physiol.* 23:

McNab, B. 1968. The influence of fat deposits on the basal metabolism in desert homiotherms. *Comp. Biochem. Physiol.* 26:337-343.

McNab, B. K. 1973. Energetics and the distribution of vampires. *J. Mammal.* 54(1):131-144.

McNab, B. K., and Morrison, P. 1963. Body temperature and metabolism in subspecies of *Peromyscus* from arid and mesic environments. *Ecol. Monogr.* 33:63-82.

Madsen, R. L. 1974. The influence of rainfall on the reproduction of Sonoran Desert lagomorphs. Master's thesis, University of Arizona.

Mahmoud, S. A.; El-Fadl, M.; and El-Mofty, M. K. 1964. Studies on the rhizosphere microflora of desert plants. *Filia Microbiol. (Prague).* 9:1-8.

Maloiy, G. M. O. (ed.). 1972. *Comparative physiology of desert animals.* New York: Academic Press.

Manje, J. M. 1954. *Unknown Mexico and Sonora.* Translated by H. Karns. Tucson: Arizona Silhouettes.

Manville, R. H. 1953. Longevity of the coyote. *J. Mammal.* 34:390.

Marks, J. B. 1950. Vegetation and soil relations in the lower Colorado desert. *Ecology* 31(2):176-193.

Martin, P. S. 1961. Southwestern animal communities in the Late Pleistocene. *New Mexico Highlands Univ. Bull.* 1961:56-66.

———. 1980. Personal communication.

Martin, R. C. 1976. Soil moisture use by velvet mesquite (*Prosopis juliflora*). Master's thesis, University of Arizona.

Martin, S. C. 1948. Mesquite seeds remain viable after 44 years. *Ecology* 29:393.

Marvin, R. F.; Stearn, T. W.; Creasey, S. C.; and Mehnert, H. H. 1973. Radiometric ages of igneous rocks from Pima, Santa Cruz, and Cochise Counties, southeastern Arizona. *U.S. Geol. Surv. Bull.* 1379:1-27.

Maslin, T. P. 1962. All-female species of the lizard genus *Cnemidophorus*, Teiidae. *Science* 135:212-213.

Mason, J. A. 1920. The Papago harvest festival. *Amer. Anthropol.* 22(1):13-15.

Master, R. W.; Rao, S.; and Soman, P. D. 1963. Electrophoretic separation of scorpion venoms. *Biochem. Biophys. Acta* 71:422-428.

Mathews, M. M. 1951. *A dictionary of Americanisms.* Chicago: University of Chicago Press.

Mathiot, Madeleine. 1973. A dictionary of Papago usage. *Language Science Monographs* 8. (1-2).

Mayhew, W. W. 1960. Testes changes in three populations of the sand lizard *Uma* in southeastern California. *Anat. Record* 137:380.

———. 1961. Photoperiodic response of female fringe-toed lizards. *Science* 134:2104-2105.

———. 1962a. *Scaphiopus couchi* in California's Colorado Desert. *Herpetologica* 18:153-161.

———. 1962b. Temperature preferences of *Sceloporus oructti. Herpetologica* 18:217-233.

———. 1963a. Biology of the granite spiny lizard, *Sceloporus oructti. Am. Midl. Nat.* 69:310-327.

———. 1963b. Reproduction in the granite spiny lizard, *Sceloporus orcutti. Am. Midl. Nat.* 69:310-327.

————. 1963c. Some food preferences of captive *Sauromalus obesus. Herpetologica* 19:10-16.

————. 1965a. Reproduction in the sand-dwelling lizard *Uma inornata. Herpetologica* 21:39-55.

————. 1965b. Adaptations of the amphibian, *Scaphiopus couchii,* to desert conditions. *Amer. Midl. Nat.* 74:95-109.

————. 1965c. Growth response to photoperiodic stimulation in the lizard *Dipsosaurus dorsalis. Comp. Biochem. Physiol.* 14:209-216.

————. 1968. Biology of desert amphibians and reptiles. In G. W. Brown, Jr. (ed.), *Desert biology,* vol. 1, 195-356. New York: Academic Press.

————. 1971. Reproduction in the desert lizard *Dipsosaurus dorsalis. Herpetologica* 27:57-77.

Mearns, E. A. 1907. Mammals of the Mexican boundary of the United States. *U.S. Natl. Mus. Bull.* 56.

Meek, S. E. 1904. The freshwater fishes of Mexico north of the Isthmus of Tehuantecpec. *Publ. Field Columb. Mus.* 93., *Zool. Ser.* 5:1-252.

Melton, M. A. 1965a. The geomorphic and paleoclimatic significance of alluvial deposits in southern Arizona. *J. Geol.* 73:1-38.

————. 1965b. Debris-covered hillslopes of the southern Arizona desert—consideration of their stability and sediment contribution. *J. Geol.* 73(5):715-729.

Merriam, C. H. 1889. Revision of North America pocket mice. *N. Amer. Fauna* 1:1-36.

————. 1898. Life zones and crop zones of the United States *U.S.D.A. Biol. Surv. Bull.* 10:1-79.

Merriam, R. 1968. Geologic reconnaissance of northwest Sonora. *Stanford Univ. Publ. Geol. Sci.* 11:287.

————. 1969. Source of sand dunes of southeastern California and northwestern Sonora, Mexico. *Geol. Soc. Amer. Bull.* 80(3):531-534.

Miller, A. H. 1951. An analysis of the distribution of the birds of California. *Univ. Calif. Publ. Zool.* 50:531-644.

Miller, A. H., and Stebbins, R. C. 1964. *The lives of desert animals in Joshua Tree National Monument.* Berkeley: University of California Press.

Miller, L. 1932. Notes on the desert tortoise (*Testudo agassizii*). *Trans. San diego Soc. Nat. Hist.* 7:187-208.

Miller, M. R. 1948. The seasonal histological changes occurring in the ovary, corpus luteum and testis of the viviparous lizard, *Xantusia vigilis. Univ. Calif. Publ. Zool.* 47:197-224.

————. 1951. Some aspects of the life history of the yucca night lizard, *Xantusia vigilis. Copeia* 1951:114-120.

————. 1955. Cyclic changes in the thyroid and interrenal glands of the viviparous lizard, *Xantusia vigilis. Anat. Rec.* 123:19-32.

Miller, R. R. 1949. Desert fishes—clues to vanished lakes and streams. *Nat. Hist. N.Y.* 58:447-451, 475-476.

————. 1952. Bait fishes of the lower Colorado River, from Lake Mead, Nevada, to Yuma, Arizona, with a key for their identification. *Calif. Fish and Game* 38:7-42.

————. 1959. Origin and affinities of the freshwater fish fauna of western North America, In C. L. Hubbs, (ed.), *Zoogeography,* pp. 187-222. Washington, D.C.:American Association for the Advancement of Science.

————. 1961. Man and the changing fish fauna of the American Southwest. *Pap. Michigan Acad. Sci. Arts Lett.* 46:365-404.

Miller, R. R., and Lowe, C. H. 1964. An annotated check-list of the fishes of Arizona. In C. H. Lowe (ed.), *The vertebrates of Arizona,* pp. 133-151. Tucson: University of Arizona Press.

Milstead, W. W. 1965. Changes in competing populations of whiptail lizards (*Cnemidophorus*) in southwestern Texas. *Amer. Midl. Nat.* 73:75-80.

Minch, J. A. 1971. Early Tertiary paleogeography of the northern peninsular ranges, Baja California, Mexico. *Geol. Soc. Amer. Cordilleran Section. Abstract with Programs* 3(2):164.

Minckley, W. L. 1965. Native fishes as natural resources. In J. L. Gardner (ed.), *Native plants and animals as resources in arid lands of the southwestern United States,* pp. 48-60. *AAAS Committee on Desert and Arid Zones Research, Contribution* 8.

————. 1969a. Aquatic biota of the Sonoita Creek basin, Santa Cruz County, Arizona. *Ecol. Stud. Leafl.* 15:1-8.

————. 1969b. Attempted re-establishment of the Gila topminnow within its former range. *Copeia* 1969:193-194.

————. 1973. *Fishes of Arizona,* Phoenix: Arizona Game and Fish Department.

————. 1976. Fishes. In E. W. Haury, *The Hohokam: Desert farmers and craftsmen,* p.379. Tucson: University of Arizona.

Minckley, W. L., and Barber, W. E. 1971. Some aspects of the biology of the longfin dace, a cyprinid fish characteristic of streams in the Sonoran Desert. *Southwest. Nat.* 15.

Minckley, W. L., and Deacon, J. E. 1968. Southwestern fishes and the enigma of "endangered species." *Science* 159:1424-1432.

Minckley, W. H.; Rinne, J. N.; and Johnson, J. E.; 1977. Status of the Gila topminnow and its co-occurrence with mosquitofish. *U.S.D.A. For. Serv. Res. Papers.* R.M. -198:1-8.

Minnich, J. E. 1970a. Water and electrolyte balance of the desert iguana, *Dipsosaurus dorsalis,* in its natural habitat. *Comp. Biochem. Physiol.* 35:921-933.

Minnich, J. E., and Shoemaker, V. H. 1970b. Diet, behavior and water turnover in the desert iguana, *Disposaurus dorsalis. Amer. Midl. Nat.* 84:496-509.

Minton, S. A. 1957. Venoms. *Amer. J. Trop. Med.* 6:145-151, 1097-1107.

————. 1968. Venoms of desert animals. In G. W. Brown (ed.), *Desert biology,* vol. 1, pp. 487-516. New York: Academic Press.

Mitchell, G. C. 1967. Population study of the funnel-eared bat (*Natalus stamineus*) in Sonora. *Southwest, Nat.* 12(2):172-175..

Mitich, L. W. 1972. The saguaro—a history. *Cactus Succulent J.* 44(3):118-129.

Mitich, L. W., and Bruhn, J. G. 1975. The saguaro—a bibliography. *Cactus Succulent J. Yearbook,* Suppl. 47:56-64.

Moberly, W. R. 1963. Hibernation in the desert iguana, *Dipsosaurus dorsalis. Physiol. Zool.* 152-160.

Monson, G. 1948. Porcupines in southwestern Arizona *J. Mammal.* 29:182.

————. 1968. The desert pronghorn. *Desert Bighorn Council Trans.* 1968:63-69.

Mooney, H. A., and Harrison, A. T. 1972. The vegetational gradient on the lower slopes of the Sierra San Pedro Martir in northwest Baja California. *Madrono* 21:439-445.

Mooney, H. A.; Björkman, O.; and Troughton, J. 1975. Phtosynthetic adaptation to high temperature. In *Environmental physiology of desert organisms,* pp. 138-151. Stroudsburg, Pa.: Dowden, Hutchinson & Ross.

Moore, W. S.; Miller, R. R.; and Schultz, R. J. 1970. Distribution, adaptation and probable origin of an all-female form of Poeciliopsis (Pisces: Poeciliidae) in northwestern Mexico. *Evolution* 24:789-795.

Moran, R. 1968. Cardon. *Pacific Discovery* 21(2):2-9.

Moran, Reid and Felger, Richard 1968. *Castela polyandra*, a new species in a new section; Union of *Holacantha* with *Castela* *(Simaroubacae). Trans. San Diego Mus. Nat. Hist.* 15:3-40. 1968.

Moran, R. and G. Lindsay. 1949. Desert islands of Baja California. *Desert Plant Life* 22 (December).

Mosauer, W. 1935. The reptiles of a sand dune area and its surroundings in the Colorado Desert, California. A study in habitat preference. *Ecology* 16:13-27.

———. 1936a. The toleration of solar heat in desert reptiles. *Ecology* 17:56-66.

———. 1936b. The reptilian fauna of sand dune areas of the Vizcaíno Desert and of northwestern lower California. *Occ. Pap. Mus. Zool. Univ. Michigan* 329:1-21.

Muffler, L. J. P., and Doe, B. R. 1968. Composition and mean age of detritus of the Colorado River delta in the Salton Trough, southeastern California. *J. Sedim. Petr.* 38(2):384-399.

Muhktar, H. A. M. 1961. Factors affecting seed germination of some important desert plants. Master's thesis, University of Arizona.

Muller, C. H.; Muller, W. H.; and Haines, B. L. 1964. Volatile growth inhibitors produced by aromatic shrubs, *Science* 143:471-473.

Mulroy, T. W. 1971. Perennial vegetation associated with the organ pipe cactus in Organ Pipe Cactus National Monument, Arizona. Master's thesis, University of Arizona.

Mulroy, T. W., and Rundell, P. W. 1977. Annual plants: Adaptation to desert environments. *Bioscience* 27:109-115.

Muma, M. H. 1951. The arachnid order Solpugida in the United States. *Bull. Amer. Mus. Nat. Hist.* 97:35-144.

———. 1966. Feeding behavior of North American Solpugida (Arachnida). *Florida Entomologist* 49:199-216.

Munz, P. A. 1968. *Supplement to a California flora.* Berkeley: University of California Press.

———. 1974. *A flora of Southern California.* Berkeley: University of California Press.

Munz, P. A., and Keck, D. D. 1949. California plant communities. *El Aliso* 2:87-105.

Munz, P. A., and Keck, D. D. 1963. *A California flora.* Berkeley: University of California Press.

Murie, A. 1951. Coyote food habits on a southern cattle range. *J. Mammal.* 32:291-295.

Murie, M. 1961. Metabolic characteristics of mountain, desert and coastal populations of *Peromyscus.* Ecology 42:723-740.

Murphy, R. W. 1974. A new genus and species of eublepharine gecko (Sauria: Gekkonidae) from Baja California, Mexico. *Proc. Calif. Acad. Sci.* 4th ser. 40:87-92.

———. 1975. *Proc. Calif. Acad. Sci.* 4th Ser. 40:93-107.

Murray, K. F. 1955. Herpetological collections from Baja California. *Herpetologica.* 11:13-48.

Musgrage, M. E., and Cochran, D. M. 1929. *Bufo alvarius,* a poisonous toad. *Copeia* 1929:96-99.

Musick, H. B. 1975. Barrenness of desert pavement in Yuma County, Arizona. *J. Ariz. Acad. Sci.* 10(1):24-28.

Myers, G. S. 1951. *Stanford Ichthyol. Bull.* 4:11-21.

Nabhan, G. P., and Sheridan, T. E. 1977. Living fencerows of the Rio San Miguel, Sonora, Mexico: Traditional technology for floodplain management. *Human Ecology* 5:97-111.

Nagy, K. A. 1972. Water and electrolyte budgets of a free-living desert lizard, *Sauromalus obesus. J. Comp. Physiol.* 79:39-62.

———. 1973. Behavior, diet and reproduction in a desert lizard, *Sauromalus obesus. Copeia* 1973:93-102.

Nagy, K. A., and Shoemaker, V. H. 1975. Energy and nitrogen budgets of a free-living desert lizard, *Sauromalus obesus. Physiol. Zool.* 48:252-262.

Nash, T. H. 1975a. Catalog of the lichens of Arizona. *Bryologist* 78:7-24.

———. 1975b. Lichens of Maricopa County, Arizona. *J. Ariz. Acad. Sci.* 10:119-125.

Neal, B. J. 1959. A contribution on the life history of the collared peccary in Arizona. *Amer. Midl. Nat.* 6(1):177-190.

———. 1964. Comparative biology of two southwestern ground squirrels *Citellus harrisii* and *Citellus tereticaudus.* Ph.D. dissertation, University of Arizona.

Neales, T. F.; Patterson, A. A.; and Hartney, V. J. 1968. Physiological adaptation to drought in carbon assimilation and water loss of xerophytes. *Nature* 219:469-472.

Nebeker, G. T.; Nash, T. H.; and Moser, T. J. 1977. Lichen dominated systems of coastal Baja California. *Proc. 21st Ann. Meeting AAs. Ariz. Acad. Sci.*

Nelson, E. W. 1922. Lower California and its natural resources. 1*Mem. Nat. Acad. Sci.* 16:1-194.

———. 1966. The geology of Picketpost Mountain. Master's thesis, University of Arizona.

Neuenschwander, L. F.; Sharrow, S. H. and Wright, H. A. 1975. Review of tobosa grass (*Hilaria mutica*). *Southwest. Nat.* 20:255-263.

Nichol, A. A. 1952. The natural vegetation of Arizona. *Ariz. Agr. Exp. Sta. Techn. Bull.* 127:189-230.

Nichols, U. G. 1953. Habits of the desert tortoise, *Gopherus agassizii, Herpetologica* 9:65-69.

———. 1957. The desert tortoise in captivity. *Herpetologica* 13:141-144.

Nicot, J. 1960. Some characteristics of the micoflora of desert soils. In D. Parkinson and J. S. Ward, (eds.), *Int. symp. on ecology of soil fungi.* pp. 94-97. Liverpool: Liverpool University Press.

Nisbet, R. A., and Patten, D. T. 1974. Seasonal temperature acclimation of a prickly-pear cactus in south-central Arizona. *Oecologia* 15:345-352.

Nishida, K. 1963. Studies on stomatal movement of crassulacean plants in relation to the acid metabolism. *Physiol. Plant* 16:281-298.

Nobel, P. S. 1977. Water relations and photosynthesis of a barrel cactus, *Ferocactus acanthodes,* in the Colorado desert. *Oecologia* 27(2):117-133.

Norr, M. 1974. Trockenresistenz bei Moosen. *Flora, Jena* 163:371-378.

Norris, K. S. 1949. Observations on the habits of the horned lizard *Phrynosoma m'calli. Copeia* 1949:176-180.

———. 1953. The ecology of the desert iguana *Dipsosaurus dorsalis. Ecology* 34:265-287.

———. 1958. The evolution and systematics of the iguanid genus

Uma and its relation to the evolution of other North American desert reptiles. *Bull. Amer. Mus. Nat. Hist.* 114:253-326.

————. 1967. Color adaptation in desert reptiles and its thermal relationships. In *Lizard ecology: A symposium,* pp. 162-229. Columbia: University of Missouri Press.

Norris, K. S., and Dawson, W. R. 1964. Observations on the water economy and electrolyte excretion of chuckwallas (Lacertilia, *Sauromalus*). *Copeia* 1964:638-646.

Norris, K. S., and Kavanau, J. L. 1966. The burrowing of the western shovel-nosed snake, *Chionactis occipitalis* Hallowell, and the undersand environment. *Copeia* 1966:650-664.

Norris, K. S. and C. H. Lowe. 1964. An analysis of background color-matching in amphibians and reptiles. *Ecology* 45:565-580.

Norris, R. M., and Norris, K. S. 1961. Algodones dunes of southeastern California. *Geol. Soc. Amer. Bull.* 72(4):605-620.

Novick, A., and Dale, B. A. 1971. Foraging behavior in fishing bats and their insectivorous relatives. *J. Mammal.* 52(4):817-818.

Nutting, W. L.; Haverty, M. J.; and Le Fage, J. P. 1974. Colony characteristics of termites as elated to population density and habitat. *US/IBP Desert Biome Research Memorandum* RM 74-28. Logan, Utah.

Ochoterana, I. 1945. Outline of the geographic distribution of plants in Mexico. In *Plants and plant science in Latin America,* pp. 261-265. Waltham, Mass.: Chronica Botanica Co.

Odening, W. R.; Strain, B. R.; and Oechel, W. C. 1974. The effect of decreasing water potential on net CO_2 exchange of intact desert shrubs. *Ecology* 55:1086-1095.

Oechel, W. C.; Strain, B. R.; and Odening, W. R. 1972. Tissue water potential, photosynthesis, C^{14}-labeled photosynthate utilization, and growth in the desert shrub *Larrea divaricata* Cov. *Ecol. Monogr.* 42:127-141.

Ohmart, R. D. 1973. Observations on the breeding adaptations of the road-runner. *Condor* 73:140-149.

Ohmart, R. D., and Lasiewski, R. C. 1971. Roadrunners: Energy conservation by hypothermia and absorption of sunlight. *Science* 172:67-69.

Ohmart, R. D.; McFarland, L. Z.; and Morgan, J. P. 1970. Urographic evidence that urine enters the rectum and ceca of the roadrunner (*Geococcyx californicus*) (Aves). *Comp. Biochem. Physiol.* 35:487-489.

Olin, G. 1959. *Mammals of the Southwest deserts.* 2d ed. Globe, Ariz.: Southwest Monuments Association.

Olsen, S. J. 1976. Micro-vertebrates. In E. W. Haury, *The Hohokam: Desert farmers and craftsmen,* p. 378. Tucson: University of Arizona Press.

Orians, G. H., and Solbrig, O. T. 1977. A cost-income model of leaves and roots with special reference to aid and semiarid areas. *Amer. Nat.* Ill(980):677-690.

Ortenburger, A. L., and Ortenburger, R. D. 1926. Field observations on some amphibians and reptiles of Pima County, Arizona. *Proc. Oklahoma Acad. Sci.* 6:101-121.

Osgood, W. H. 1900. Revision of the pocket mice of the genus *Perognathus. N. Amer. Fauna* 18:1-72.

————. 1909. Revision of the mice of the American genus *Peromyscus. N. Amer. Fauna* 28:1-285.

Otte, D., and Joern, A. 1975. Insect territoriality and its evolution: Population studies of desert grasshoppers on creosote bushes. *J. Anim. Ecol.* 44:29-54.

Otto, R. G. 1973. Temperature tolerance of the mosquitofish, Gambusia affinis (Baird and Girard). *J. Fish Biol. Calif. Fish and Game* 5(5):575-585.

Pakiser, L. C. 1963. Structure of the crust and upper mantle in the western United States. *J. Geophysical Res.* 68:5747-5756.

Parish, S. B. 1930. Vegetation of the Mohave and Colorado deserts of southern California. *Ecology* 11:481-499.

Parker, E. D. 1979a. Phenotypic consequences of parthenogenesis in *Cnemidophorus* lizards, Vol. 1. Variability in parthenogenetic and sexual populations. *Evolution* 33:1150-1166.

————. 1979b. Phenotypic consequences of parthenogenesis in *Cnemidophorus* lizards Vol. 2. Similarity of *C. tessellatus* to its sexual parental species. *Evolution* 33:1167-1179.

Parker, G. H. 1938. The color changes in lizards, particularly in *Phrynosoma. J. Exp. Biol.* 15:48-73.

Parker, K. F. 1958. Arizona ranch, farm, and garden weeds. *Ariz. Agr. Ext. Serv. Circ.* 265:1-288.

————. 1972. *An illustrated guide to Arizona weeds.* Tucson: University of Arizona Press.

Parker, R. H. 1964. Zoogeography and ecology of macroinvertebrates of Gulf of California and continental slope of western Mexico. *Amer. Assoc. Petrol. Geol. Mem.* 3:331-376.

Parker, W. S. 1972. Aspects of the ecology of a Sonoran Desert population of the western banded gecko, *Coleonyx variegatus* (Sauria, Eublepharinae). *Amer. Midl. Nat.* 88:209-224.

Parker, W. S., and Pianka, E. R. 1973. Notes on the ecology of the iguanid lizard, *Sceloporus magister. Herpetologica* 29:143-152.

Pase, C. P., and Layser, E. 1977. Classification of riparian habitat in the Southwest. In *Importance, preservation and management of riparian habitat. U.S.F.S. Rocky Mtn. Forest and Range Exp. Sta. Gen. Techn. Rept.* RM-43:5-9.

Patten, D. T. 1971. Productivity and water stress in cacti. *US/IBP Desert Biome Research Memorandum* RM 71-12. Logan, Utah.

————. 1972. Productivity and water stress in cacti. *US/IBP Desert Biome Research Memorandum* RM 72-17. Logan, Utah.

————. 1973. Winter: A major determinant of desert vegetation patterns. *J. Ariz. Acad. Sci.* 8:9.

Patten, D. T., and Dinger, B. E. 1969. CO_2 exchange patterns of cacti from different environments. *Ecology* 50:686-688.

Patten, D. T., and Smith, E. M. 1974. Phenology and function of Sonoran Desert annuals in relation to environmental changes. *US/IBP Desert Biome Research Memorandum* RM 74-12. Logan, Utah.

————. 1975. Heat flux and the thermal regime of desert plants. In N. F. Hadley (ed.), *Environmental physiology of desert organisms,* pp. 1-19. Stroudsburg, Pa.: Dowden, Hutchinson & Ross.

Patterson, R. A. 1960. Action of scorpion venom. *Amer. J. Trop. Med. Hyg.* 9:410-413.

Patterson, R., and Brattstrom, B. 1972. Growth in captive *Gopherus agassizi. Herpetologica* 28:169-171.

Patton, D. R. 1978. RUNWILD—a storage and retrieval system for wildlife habitat information. *U.S.F.S. Rocky Mtn. Forest and Range Exp. Sta. Gen. Techn. Rept.* RM-51:1-8.

Patton, J. L. 1967. Chromosome studies of certain pocket mice, genus *Perognathus* (Rodentia: Heteromyidae). *J. Mammal.* 48:27-37.

Paulsen, H. A. 1950. Mortality of velvet mesquite seedlings. *J. Range Managm.* 3:281-286.

———. 1953. A comparison of surface soil properties under mesquite and perennial grasses. *Ecology* 34:727-732.

Paylore, P. (ed.). 1976. The Sonoran Desert, a retrospective bibliography. *Arid Lands Abstracts.* 8. Tucson: Office of Arid Lands Studies, University of Arizona.

Pearcy, R.; Björkman, O.; Harrison, A.; and Mooney, H. 1971. Photosynthetic performance of two species with C₄ photosynthesis in Death Valley, California. *Carneg. Inst. Year Book* 70:540-550.

Pearson, O. P. 1960. The oxygen consumption and bioenergetics of harvest mice. *Physiol. Zool.* 33:152-160.

Peirce, H. 1979. Subsidence-fissures and faults in Arizona. *Fieldnotes Ariz. Bur. Geol. Mineral Techn.* 9:1-2, 6.

———. Pepper, J. H., and Hastings, E. 1952. The effects of solar radiation on grasshopper temperature and activities. *Ecology* 33:96-103.

Peterson, R. T. 1969. *A field guide to western birds.* Boston: Houghton Mifflin.

Peterson R. T., and Chalif, E. L. 1973. *A field guide to Mexican birds.* Boston: Houghton Mifflin.

Phelps, J. S. 1974. Endangered species investigations—Sonoran pronghorn. Proj. W-53-24, WP-7, Job 1. *Ariz. Game and Fish Dept.*

Phillips, A.; Marshall, J.; and Monson, G. 1964. *The Birds of Arizona.* Tucson: University of Arizona Press.

Phillips, A. R. 1939. The faunal areas of Arizona, based on bird distribution. Master's thesis, University of Arizona.

Phillips, A. R., and Monson, G. 1963. An annotated check list of the birds of Arizona. In *The vertebrates of Arizona.* Tucson: University of Arizona Press.

Phillips, O. L., and McMahon, J. A. 1978. Gradient analysis of a Sonoran Desert bajada. *Southwest. Nat.* 23(4):669-680.

Phillips, W. S. 1963. Depth of roots in soil. *Ecology* 44(2):424.

Philpott, C. W., and Templeton, J. R. 1964. A comparative study of the histology and fine structure of the nasal salt secreting gland of the lizard, *Dipsosaurus. Anat. Rec.* 148:394-395.

Pianka, E. R. 1965. Species diversity and ecology of flatland desert lizards in western North America. Ph.D. dissertation, University of Washington.

———. 1966. Convexity, desert lizards, and spatial heterogeneity. *Ecology* 47:1055-1059.

———. 1967. On lizard species diversity: North American flatland deserts. *Ecology* 48:333-351.

———. 1970. Comparative autecology of the lizard *Cnemidophorus tigris* in different parts of its geographic range. *Ecology* 51:703-720.

———. 1971. Comparative ecology of two lizards. *Copeia* 1971:129-138.

———. 1975. Niche relations of desert lizards. In M. L. Cody and J. M. Diamond (eds.), *Ecology and evolution of communities,* pp. 292-314. Cambridge: Belknap Press.

Pianka, E. R., and Parker, W. S. 1972. Ecology of the iguanid lizard *Callisaurus draconoides. Copeia* 1972:493-508.

———. 1973. Notes on the ecology of the iguanid lizard, *Sceloporus magister. Herpetologica* 29:143-152.

———. 1975. Ecology of horned lizards: A review with special reference to *Phrynosoma platyrhinos. Copeia* 1975:141-162.

Pinckney, F. C. 1969. Factors affecting the distribution of Hilaria species in Arizona. Master's thesis. University of Arizona.

Pinkham, C. F. A. 1973. The evolutionary significance of locomotor patterns in the Mexican spiny pocket mouse, *Liomys irroratus. J. Mammal.* 54(3):742-746.

Pocock, R. I. 1941. The races of the ocelot and the margay. *Field Mus. Nat. Hist. Publ.* 511. Zool. Ser. 27:319-369.

Porter, L. R. 1973. Geothermal resource investigations. *J. Amer. Soc. Civil Engin., Hydraulics Div.* 99 (HY11):2097-2111.

Porter, W. 1966. Solar radiation through the body wall of living vertebrates with emphasis on desert reptiles. Ph.D. dissertation, University of California, Los Angeles.

Potter, J. M., and Northey, W. T. 1962. Immunological evaluation of scorpion venoms. *Am. J. Trop. Med. Hyg.* 11:712-716.

Prakash, I., and Ghosh, P. K. (eds.). *Rodents in desert environments.* The Hague: Junk.

Price, M. V. 1976. The role of microhabitat in structuring desert rodent communities. Ph.D. dissertation, University of Arizona.

———. 1978. Seed dispersion preferences of coexisting desert rodent species. *J. Mammal.* 59:624-626.

Proffitt, J. M. 1977. Cenozoic geology of the Yerington district, Nevada. *Geol. Soc. Amer. Bull.* 88:247-266.

Quay, W. B. 1964. Integumentary modifications of North American desert rodents. In A. G. Lyne and B. F. Short (eds.), *Biology of skin and hair growth,* pp. 59-74. New York: American Elsevier.

Raitt, R. J., and Maze, R. L. 1968. Densities and species composition of breeding birds of a creosotebush community in southern New Mexico. *Condor* 70:193-205.

Ranson, S. L., and Thomas, M. 1960. Crassulacean acid metabolism. *Ann. Rev. Plant Physiol.* 11:81-110.

Rath, L. 1966. The effect of light quality on growth of the desert iguana, *Dipsosaurus dorsalis.* Master's thesis, University of California, Riverside.

Raven, P. H. 1963. Amphitropical relationships in the floras of North and South America. *Quart. Rev. Biol.* 38:131-177.

Ray, G. C. 1975. A preliminary classification of coastal and marine environments. *I.U.C.N. (Morges, Switzerland) Occas. Papers* 14:1-26.

Rea, A. 1979a. The ecology of Pima fields. *Environment Southwest* 484. San Diego Society Natural History.

———. 1979b. Velvet mesquite. *Environment Southwest* 486. San Diego Society Natural History.

Redfield, A. C. 1918. The physiology of the melanophores of the horned toad *Phrynosoma. J. Exp. Zool.* 26:275-333.

Reed, C. A., and Carr, W. H. 1949. Use of cactus as protection by hooded skunk. *J. Mammal.* 30:79-80.

Reeder, W. G., and Norris, K. S. 1954. Distribution, type locality, and habits of the fish-eating bat, *Pizonyx vivesi. J. Mammal.* 35(1):81-87.

Reeve, W. L. 1952. Taxonomy and distribution of the horned lizards, genus *Phrynosoma. Univ. Kans. Sci. Bull.* 34:817-960.

Rehn, J. A. G. 1958. The origin and affinities of Dermaptera and Orthoptera of western North America. In C. L. Hubbs (ed.), *Zoogeography,* pp. 253-298. *Am. Assoc. Adv. Sci. Publ.* 51.

Reichman, O. J. 1975. Relation of desert rodent diets to available resources. *J. Mammal.* 56:731-751.

———. 1976. Dispersion patterns of seeds in desert soils. *US/IBP Desert Biome Research Memorandum* RM 76-20, Logan, Utah.

————. 1978. Ecological aspects of the diets of Sonoran Desert rodents. *Mus. Northern Ariz. Res. Paper*, ser. 20. Flagstaff, Ariz.

Reichman, O. J., and Oberstein, D. 1977. Selection of seed distribution types by *Dipodomys merriami* and *Perognathus amplus*. *Ecology* 56:636-643.

Reichman, O. J.; Prakash, I.; and Roig, V. 1979. Food selection and consumption. 1979. In D. W. Goodall, R. A. Perry, and K. M. W. Howes (eds.), *Arid-land ecosystems: Structure, functioning and management*, vol. 1, pp. 681-716. Cambridge: At the University Press.

Reichman, O. J., and Van de Graaff, K. M. 1973. Seasonal activity and reproductive patterns of five species of Sonoran Desert rodents. *Amer. Midl. Nat.* 90(1):118-126.

————. 1975. Association between ingestion of green vegetation and desert rodent reproduction. *J. Mammal.* 56:503-506.

Reveal, J. L. 1969. A revision of the genus Eriogonum (Polygonaceae). Ph.D. dissertation, Brigham Young University.

Reynolds, H. G. 1950. Relation of Merriam kangaroo rats to range vegetation in southern Arizona. *Ecology* 31(3):546-563.

————. 1957. Porcupine behavior in the desert-shrub type of Arizona. *J. Mammal.* 38(3):418-419.

Riazance, J., and Whitford, W. G. 1974. Studies of wood borers, girdlers and seed predators of mesquite. *US/IBP Desert Biome Research Memorandum* RM 74-30, Logan, Utah.

Rice, A. 1976. Insolation warmed over. *Geology* 4(1):61-62.

Richards, S. A. 1970. The biology and comparative physiology of thermal panting. *Biol. Rev.* 45:223-264.

Richardson, W. B. 1942. Ring-tailed cats (*Bassariscus astutus*): their growth and development. *J. Mammal.* 23:17-26.

Rinne, J. N. 1969. Cyprinid fishes of the genus Gila from the Lower Colorado River Basin. Master's thesis, Arizona State University.

Robinson, T. W. 1958. Phreatophytes. *U.S. Geol. Survey Water Supply Paper* 1423:1-84.

Robison, W. G., and Tanner, W. W. 1962. A comparative study of the species of the genus *Crotaphytus* Holbrook (Iguanidae). *Brigham Young Univ. Sci. Bull., Biolo. Ser.* 2:1-31.

Roden, G. I. 1964. Oceanographic aspects of Gulf of California. In T. H. van Andel and G. G. Shore, Jr. (eds.), *Marine Geology of the Gulf of California*, pp. 50-58. Amer. Assoc. Petrol. Geol. Mem. 3.

Rodgers, R. W.; Lange, R. T.; and Nicholas, D. J. D. 1966. Nitrogen fixation by lichens of arid soil crusts. *Nature* 209:96-97.

Rosenzweig, M. L.; Smigel, B. W.; and Kraft, A. 1975. Patterns of food, space, and diversity. In I. Prakash and P. K. Ghosh (eds.), *Rodents in desert environments*, pp. 241-268. The Hague: Junk.

Rosenzweig, M. L., and Winakur, J. 1969. Population ecology of desert rodent communities: Habitat and environmental complexity. *Ecology* 50:558-572.

Ross, A. 1961. Notes on food habits of bats. *Journ. Mammal.* 42:66-71.

Roth, L. M., and Eisner, T. 1962. The chemical defenses of arthropods. *Ann. Rev. Entomol.* 7:107-136.

Royse, C. F.; Sheridan, M. F.; and Peirce, H. W. 1971. Geologic guidebook 4, Highways of Arizona, Arizona Highways 87, 88, and 188. *Bulletin 184, Arizona Bureau of Mines*.

Rue, L. L. 1978. *The deer of North America*. New York: Crown Publishers.

Ruibal, R.; Tevis, L.; and Roig, V. 1969. The terrestrial ecology of the spadefoot toad *Scaphiopus hammondi*. *Copeia* 1969:571-584.

Runyon, E. H. 1934. The organization of the creosotebush with respect to drought. *Ecology* 15:128-138.

Russell, F. 1908. Republished as Russell, F. 1975. q.v.

————. 1975. *The Pima Indians*. Tucson: University of Arizona Press.

Russell, F. E. 1960. Snake venom poisoning in southern California. *Calif. Med.* 93:347-350.

————. 1962. Poisoning. *Am. J. Med. Sci.* 243:159-161.

————. 1967a. First aid for snake venom-poisoning. *Toxicon* 4:285-289.

————. 1967b. Injuries by venomous animals. *National Clearinghouse for Poison Control Centers Bulletin* (Jan.-Feb.).

Russell, S. M.; Gould, P. J.; and Smith, E. L. 1973. Population structure, foraging behavior and daily movements of certain Sonoran Desert birds. *US/IBP Desert Biome Research Memorandum* RM 72-31:1-6, Logan, Utah.

Russo, J. P. 1956. The desert bighorn sheep in Arizona. *Ariz. Game and Fish Dept. Rept.*

Ryan, R. M. 1968. *Mammals of Deep Canyon, Colorado Desert, California*. Palm Springs: Desert Museum.

Rzedowski, J. 1973. Geographical relationships of the flora of Mexican dry regions. In *Vegetation and vegetational history of northern Latin America*, pp. 61-72. New York: Elsevier.

Sage, O. 1973. Paleocene geography of the Los Angeles region. *Stanford Univ. Publ. Geol. Sci.* 13:348-357.

Salt, G. W. 1964. Respiratory evaporation in birds. *Biol. Rev. Cambridge Phil. Soc.* 39:113-136.

Salt, G. W., and Zeuthen, E. 1960. The respiratory system. In A. J. Marshall (ed.), *Biology and comparative physiology of birds*, vol. 1, pp. 363-409. New York: Academic Press.

Sammis, T. W. 1974. The Microenvironment of a desert hackberry plant (*Celtis pallida*). Ph.D. dissertation, University of Arizona.

Saunier, R. E. 1967. Geographic variability of creosotebush (*Larrea tridentata*) in response to moisture and temperature stress. Ph.D. dissertation, University of Arizona.

Savage, J. M. 1960. Evolution of a peninsular herpetofauna. *Syst. Zool.* 1960:184-212.

Savory, T. 1977. *Arachnida*. London: Academic Press.

Saxton, D., and Saxton, L. 1969. *Dictionary: Papago and Pima to English; English to Papago and Pima*. Tucson: University of Arizona.

Sayers, S. J. 1974. Natural vegetation of state, private and Bureau of Land Management lands in Arizona. *Ariz. State Land Dept. Arizona Landmarks* 4(4):1-45.

Scantling, F. H. 1939. Jackrabbit Ruin. *Kiva* 5(3):9-12.

Scarborough, R. B. 1979. Cenozoic history and uranium in southern Arizona. *Fieldnotes Ariz. Bur. Geol. Mineral Techn.* 9:1-3, 14-15.

Scarlett, P. L. 1978. *Jojoba in a nutshell*. Carpinteria, Calif.: Jojoba International Corporation.

Schaut, G. G. 1939. Fish catastrophes during droughts. *J. Amer. Waterworks Assoc.* 31:771-822.

Schmidt, K. P. 1922. The amphibians and reptiles of Lower California and the neighboring islands. *Bull. Amer. Mus. Nat. Hist.* 46:607-707.

Schmidt-Nielsen, B. 1958. The resourcefulness of nature in physiological adaptation to the environment. *Physiologist* 1(2).

Schmidt-Nielsen, B., and O'Dell, R. 1961. Structure and concentrating mechanism in the ammalian kidney. *Amer. J. Physiol.* 200:1119-1124.

Schmidt-Nielsen, B., and Schmidt-Nielsen, K. 1950a. Pulmonary water loss in desert rodents. *Amer. J. Physiol.* 162:31-36.

———. 1950b. Evaporative water loss in desert rodents in their natural habitat. *Ecology* 31:75-85.

Schmidt-Nielsen, K. 1963. Osmotic regulation in higher vertebrates. *Harvey Lectures* ser. 58:53-93.

———. 1964a. *Desert animals: Physiological problems of heat and water.* Oxford: Oxford University Press.

———. 1964b. Terrestrial animals in dry heat: desert rodents. In D. B. Dill (ed.), *Adaptation to the environment: Textbook of physiology,* pp. 493-507. Baltimore: Williams and Wilkins.

———. 1965. The jack rabbit, a study in its desert survival. *Itvalradets Skrifter* 48:125-142.

———. 1972. Recent advances in the comparative physiology of desert animals. In G. M. O. Maloiy (ed.), *Comparative physiology of desert animals,* pp. 371-382. New York: Academic Press.

Schmidt-Nielsen, K., and Bentley, P. J. 1966. Desert tortoise *Gopherus agassizzi:* Cutaneous water loss. *Science* 154:911.

Schmidt-Nielsen, K.; Crawford, E. C.; and Bentley, P. J. 1966. Discontinuous respiration in the lizard *Sauromalus obesus. Federation Proc.* 25:596.

Schmidt-Nielsen, K., and Dawson, W. R. 1964. Terrestrial animals in dry heat: Desert reptiles. In D. B. Dill (ed.), *Handbook of physiology. Section 4: Adaptation to the environment,* pp. 467-480. American Physiological Society.

Schmidt-Nielsen, K., and Haines, H. B. 1964. Water balance in a carnivorous desert rodent, the grasshopper mouse. *Physiol. Zool.:* 37:259-265.

Schmidt-Nielsen, K.; Jorgensen, C. B.; and Osaki, H. 1958. Extrarenal salt excretion in birds. *Amer. J. Physiol.* 193:101-107.

Schmidt-Nielsen, K.; Taylor, C. R.; and Shkolnik, A. 1971. Desert snails: Problems of heat, water and food. *J. Exptl. Biol.* 55:385-398.

Schmutz, E. M. 1978. *Classified bibliography on native plants of Arizona.* Tucson: University of Arizona Press.

Schmutz, E. M., and Hamilton, L. B. 1979. *Plants that poison: An illustrated guide for the American Southwest.* Flagstaff, Ariz.: Northland Press.

Schroder, G. D., and Rosenzweig, M. L. 1975. Perturbation analysis of competition and overlap in habitat utilisation between *Dipodomys ordii* and *Dipodomys merriami. Oecologia* 19:9-28.

Schuster, J. L. 1969. Literature on the mesquite (*Prosopis* L.) of North America, an annotated bibliography. *Int. Center for Arid and Semi-Arid Land Stud. Spec. Rept.* 26:1-84.

Scott, I. M.; Yousef, M. K.; and Bradley, W. G. 1972. Body fat content and metabolic rate of rodents: Desert and mountain. *Proc. Soc. Exp. Biol. Med.* 141:818.

Self, J., and Bartels, P. 1980. Germination inhibition by extracts from desert plants. Boyce Thompson Southwestern Arboretum. Manuscript.

Sell, J. D. 1968. Correlation of some Post-Laramide Tertiary units Globe (Gila County) to Gila Bend (Maricopa County), Arizona. In S. R. Titley (ed.), *Southern Arizona guidebook III,* pp. 69-74. Tucson: Arizona Geol. Society.

Sellers, W. D. 1960. The climate of Arizona. In *Arizona climate.* Tucson: University of Arizona Press.

Serventy, D. L. 1971. Biology of desert birds. In D. S. Farner and J. R. King (eds.), *Avian biology,* vol. 1, pp. 287-339. New York: Academic Press.

Seymour, R. S. and Vinegar, A. 1973. Thermal relations, water loss and oxygen consumption of a North American tarantula. *Comp. Biochem. Physiol.* 44A:83-96.

Shafiqullah, M.; Lynch, D. J.; Damon, P. E.; and Peirce, H. W. 1976. Geology, geochronology, and geochemistry of the Picacho Peak area, Pinal County, Arizona. *Ariz. Geol. Soc. Dig.* 10:305-324.

Shafiqullah, M.; Peirce, H. W.; and Damon, P. E. 1975. Rate of structural and geomorphic evolution of the southern margin of the Colorado Plateau in Arizona. *Geol. Soc. Amer. Cordilleran Sect. Abstracts with Programs* 7(3):373-374.

Shantz, H. L. 1937. The saguaro forest. *Natl. Geog. Mag.* 71:515-532.

———. 1948. Water economy of plants. *Leafl. Santa Barbara Bot. Gard.* 1:31-54.

Shantz, H. L., and Piemesel, R. L. 1924. Indicator significance of the natural vegetation of the southwestern desert region. *J. Agr. Res.* ˄8:721-801.

Shapovalov, L., et al. 1959. A revised check-list of the freshwater and anadromous fishes of California. *Calif. Fish and Game* 45:159-180.

Shaw, A. M. 1974. *A Pima past.* Tucson: University of Arizona Press.

Shaw, C. E. 1945. The chuckwallas, genus *Sauromalus. Trans. San Diego Soc. Nat. Hist.* 10:296-306.

———. 1948. A note on the food habits of *Heloderma suspectum* Cope. *Herpetologica* 4:145.

Shaw, J., and Stobbart, R. H. 1972. The water balance and osmoregulatory physiology of the desert locust (*Schistocerca gregaria*) and other desert and xeric arthropods. In *Comparative physiology of desert animals.* New York: Academic Press.

Sherbrooke, W. C. 1976. Differential acceptance of toxic jojoba seed by four Sonoran desert Heteromyid rodents. *Ecology* 57:596-602.

Sherbrooke, W. C., and Haase, E. F. 1974. Jojoba: A wax-producing shrub of the Sonoran Desert—Literature review and annotated bibliography. *University of Arizona. Office of Arid Lands Res. Inf. Paper* 5:1-141.

Sheridan, M. F. 1971. *Superstition wilderness guidebook.* By the author, Arizona State University.

———. 1978. The superstition cauldron complex. In D. M. Burt and T. L. Péwé (eds.), *Guidebook to the geology of central Arizona. Ariz. Bur. Geol. Mineral Techn. Spec. Paper 2.*

Sherwood, S. 1980. Unpublished studies at Boyce Thompson Southwestern Arboretum. Superior, Ariz.

Shields, L. M. 1950. Leaf xeromorphy as related to physiological and structural influences. *Bot. Rev.* 16:399-447.

———. 1957. Algal and lichen floras in relation to nitrogen content of certain volcanic and arid range soils. *Ecology* 38:661-663.

Shields, L. M.; Mitchell, C.; and Drouet, F. 1957. Alga- and lichen-stabilized surface crust as a soil nitrogen source. *Amer. J. Bot.* 44:489-498.

Shoemaker, V. H.; McClanahan, L.; and Ruibal, R. 1969. Seasonal changes in body fluids in a field population of spadefoot toads. *Copeia* 1969:585-591.

Short, L. L. 1974. Nesting of southern Sonoran birds during the summer rainy season. *Condor* 76:21-32.

Short, L. L., and Crossin, R. S. 1967. Notes on the avifauna of northwestern Baja California. *San Diego Soc. Nat. Hist.* 14:283-299.

Shreve, F. 1910. The rate of establishment of the giant cactus. *Plant World* 13(10):235-240.

———. 1911a. The influence of low temperatures on the distribution of the giant cactus. *Plant World* 14:136-146.

———. 1911b. Establishment and behavior of the paloverde. *Plant World* 14:289-296.

———. 1912. Cold air drainage. *Plant World* 15:110-115.

———. 1917. The establishment of desert perennials. *J. Ecol.* 5:210-216.

———. 1922. Conditions indirectly affecting vertical distribution on desert mountains. *Ecology* 3:269-274.

———. 1924. Across the Sonoran desert. *Bull. Torr. Bot. Club* 51:283-293.

———. 1936. The transition from desert to chaparral in Baja California. *Madroño* 3:257-264.

———. 1937a. Lowland vegetation of Sinaloa. *Bull. Torr. Bot. Club* 64:605-613.

———. 1937b. Plants of the sand. *Carneg. Inst. Wash. News Serv. Bull.* 4(10):91-96.

———. 1940. The edge of the desert. *Assoc. Pacif. Coast Geogr. Yearbook* 6:6-11.

———. 1942a. Grasslands and related vegetation in northern Mexico. *Madroño* 6:190-198.

———. 1942b. The vegetation of Arizona. In T. H. Kearney and R. H. Peebles, *Flowering plants and ferns of Arizona*, pp. 10-23. *U.S.D.A. Misc. Publ.* 423.

———. 1951. *Vegetation of the Sonoran Desert.* Washington, D.C.: Carnegie Institute.

Shreve, F., and Mallery, T. D. 1933. The relation of caliche to desert plants. *Soil Sci.* 35:99-112.

Shreve, F., and Wiggins, I. L. 1964. *Vegetation and flora of the Sonoran Desert.* 2 vols. Stanford: Stanford University Press.

Shreve, M. B. 1943. Modern Papago basketry. Master's thesis, University of Arizona.

Shultze, J. C. 1975. *Larrea* as a habitat component for desert invertebrates. In T. J. Mabry and J. Hunziker (eds.), *Larrea and its role in desert ecosystems.* Stroudsburg, Pa.: Dowden, Hutchinson & Ross.

Simmons, N. M. 1966. Flora of the Cabeza Prieta Game Range. *J. Ariz. Acad. Sci.* 4:93-104.

———. 1969. The social organization, behavior, and environment of the desert bighorn sheep on the Cabeza Prieta Game Range, Arizona. Ph.D. dissertation, University of Arizona.

Simpson, B. B. 1977. Breeding systems of dominant perennial plants of two disjunct warm desert ecosystems. *Oecologia* 27(3):203-226.

———. (ed.). *Mesquite: Its biology in two desert shrub ecosystems.* Stroudsburg, Pa.: Dowden, Hutchinson & Ross.

Skadhauge, E. 1972. Salt and water excretion in xerophilic birds. *Symposia of the Zoological Soc. London* 31:113-129.

Smith, C. F. 1941. Birth of horned toads. *Copeia* 1941:114.

Smith, D. S., and Russell, F. E. 1967. Structure of the venom gland of the black widow spider *Latrodectus mactans.* A preliminary light and electron microscope study. In F. E. Russell and P. R. Saunders (eds.), *Animal toxins*, pp. 1-15. Oxford: Pergamon Press.

Smith, E. L. 1974. *Established natural areas of Arizona.* Phoenix: Arizona Office of Economic Planning and Development.

Smith, G. R. 1966. Distribution and evolution of the North American catostomid fishes of the subgenus *Pantosteus* genus *Catostomus. Misc. Publ. Mus. Zool. Univ. Michigan* 29:1-132.

Smith, H. D., and Jorgensen, C. J. 1975. In I. Prakash and P. K. Ghosh (eds.), *Rodents in desert environments.* The Hague: Junk.

Smith, H. M. 1946. *Handbook of lizards.* Ithaca: Cornell University Press.

Smith, H. M., and Smith, R. B. 1973. *Analysis of the literature exclusive of the Mexican axolotl.* Augusta, W. Va.: Eric Lundberg.

———. 1976. *Synopsis of the herpetofauna of Mexico.* Vol. 3: *Source analysis and index for Mexican reptiles.* North Bennington, Vt.: John Johnson.

———. 1976. *Synopsis of the herpetofauna of Mexico.* Vol. 4: *Source analysis and index for Mexican amphibians.* North Bennington, Vt.: John Johnson.

———. 1977. *Synopsis of the herpetofauna of Mexico.* Vol. 5: *Guide to Mexican amphisbaenians and crocodilians. Bibliographic addendum II.* North Bennington, Vt.: John Johnson.

Smith, H. M., and Taylor, E. H. 1945. An annotated checklist and key to the snakes of Mexico. *U.S. Natl. Mus. Bull.* 187:1-239.

———. 1950. An annotated checklist and key to the reptiles of Mexico exclusive of the snakes. *U.S. Natl. Mus. Bull.* 199:1-253.

Smith, L. D. 1968. Factors affecting susceptibility of creosotebush (*Larrea tridentata (D.C.) Cov.*) to burning. Ph.D. dissertation, University of Arizona.

Smyth, M., and Bartholomew, G. A. 1966. Effects of water deprivation and sodium chloride on the blood and urine of the mourning dove. *Auk* 83:597-602.

Snyder, J. M., and Wullstein, L. H. 1973. The role of desert cryptogams in nitrogen fixation. *Amer. Midl. Nat.* 90:257-265.

Sohns, E. R. 1956. The genus *Hilaria* (Gramineae). *J. Wash. Acad. Sci.* 46:311-321.

Solbrig, O. T. 1972. The floristic disjunctions between the Monte in Argentina and the Sonoran Desert in Mexico and the United States. *Mo. Bot. Gard. Annals* 59:218-223.

Somero, G. N. 1975. Enzymic mechanisms of eurythermality in desert and estuarine fishes: Genetics and kinetics. In N. F. Hadley (ed.), *Environmental physiology of desert organisms*, pp. 168-187. Stroudsburg, Pa.: Dowden, Hutchinson & Ross.

Sommerfeld, M. R.; Gisneros, R. M.; and Olsen, R. D. 1975. The phytoplankton of Canyon Lake, Arizona. *Southwest. Nat.* 20:45-53.

Soulè, M. 1963. Aspects of thermoregulation in nine species of lizards from Baja California. *Copeia* 1963:107-115.

Soulè, M. 1964. The evolution and population phenetics of the side-blotched lizards (*Uta stansburiana* and relatives) on the islands in the Gulf of California, Mexico. Ph.D. dissertation, Stanford University.

Soulè, M., and Sloan, A. J. 1966. Biogeography and distribution of the reptiles and amphibians on islands in the Gulf of California, Mexico. *Trans. San Diego Soc. Nat. Hist.* 14:137-156.

Sowls, L. K. 1957. Reproduction in the Audubon cottontail in Arizona. *J. Mammal.* 38(2):234-243.

———. 1958. Experimental feeding and measurement of water

consumption of captive javelina. *Ariz. Crop. Wildl. Res. Unit Report* 8:14-16.

Spalding, V. M. 1909. Distribution and movements of desert plants. *Carneg. Inst. Wash. Publ.* 113:1-144.

———. 1910. Plant associations of the Desert Laboratory domain and adjacent valley. *Plant World* 13:31-42, 56-66, 86-93.

Spencer, D. A., and Spencer, A. L. 1941. Food habits of the white-throated wood rat in Arizona. *J. Mammal.* 22:280-284.

Sperry, C. C. 1941. Food habits of the coyote. *U.S. Dept. Interior. Fish and Wildl. Serv. Wildlife Research Bull.* 4:1-70.

Spicer, E. H. 1962. *Cycles of conquest.* Tucson: University of Arizona Press.

Spoehr, H. A. 1919. The carbohydrate economy of cacti. *Carneg. Inst. Wash. Publ.* 287:1-79.

Stager, K. E. 1960. The composition and origin of the avifauna. *Syst. Zool.*, Baja Calif. Symp. 179-183.

Stahnke, H. L. 1950. The food of the gila monster. *Herpetologica* 6:103-106.

———. 1952. A note on the food of the gila monster, *Heloderma suspectum. Herpetologica* 8:64-65.

———. 1966. Some aspects of scorpion behavior. *Bull. So. Calif. Acad. Sci.* 65:65-80.

Stamp, N. E., and Ohmart, R. D. 1979. Rodents of desert shrub and riparian woodland habitats in the Sonoran Desert. *Southwest. Nat.* 27(2):279-289.

Standley, P. C. 1920-1926. Trees and shrubs of Mexico. 5 vols. *Contrib. U.S. Natl. Herb.* 23:1-1721.

Stebbins, G. L. and J. Major. 1965. Endemism and speciation in the California flora. *Ecol. Monogr.* 35:1-35.

Stebbins, R. C. 1943. Adaptations in the nasal passages for sand burrowing in the saurian genus *Uma. Amer. Nat.* 77:38-52.

———. 1944. Some aspects of the ecology of the iguanid genus *Uma. Ecol. Monographs* 14:311-332.

———. 1966. *A field guide to western reptiles and amphibians.* Boston: Houghton Mifflin.

Stebbins, R. C., and Robinson, H. B. 1946. Further studies of a population of the lizard *Sceloporus graciosus gracilis. Univ. Calif. Publ. Zool.* 48:149-168.

Steenbergh, W. F. 1972. Lightning-caused destruction in a desert plant community. *Southwest. Nat.* 16:419-429.

———. 1974. *The saguaro, giant cactus: A bibliography.* San Francisco: U.S. National Park Service, Western Region.

Steenbergh, W. F., and Lowe, C. H. 1969. Critical factors during the first years of life of the saguaro (*Cereus giganteus*) at Saguaro National Monument, Arizona. *Ecology* 50(5):825-834.

———. 1977. Ecology of the saguaro, vol. 2. *National Park Service Sci. Monogr. Ser.*, No. 8.

Steenbergh, W. F., and Warren, P. L. 1977. Preliminary ecological investigation of natural community status at Organ Pipe Cactus National Monument. *U.S.D.I. Cooperative National Park Resources Studies Unit, Univ. Ariz. Tech. Rept.* 3:1-152.

Stock, A. D. 1974. Chromosome evolution in the genus *Dipodomys* and its taxonomic and phylogenetic implications. *J. Mammal.* 55:505-526.

Storer, T. I. 1925. A synopsis of the amphibia of California. *Univ. of Calif. (Berkeley) Publ. Zool.* 27:1-342.

Stout, G. A. 1970. The breeding biology of the desert cottontail in the Phoenix region, Arizona. *J. Wildlife Managm.* 34:47-51.

Stout, G. A.; Bloom, E. C.; and Glass, J. K. 1970. The fishes of Cave Creek, Maricopa County, Arizona. *J. Ariz. Acad. Sci.* 6:109-113.

Strain, B. R. 1969. Seasonal adaptations in photosynthesis and respiration in four desert shrubs growing in situ. *Ecology* 50:511-513.

———. 1970. Field measurements of tissue water potential and carbon dioxide exchange in the desert shrubs *Prosopis juliflora* and *Larrea divaricata. Photosynthetica* 4:118-122.

Strain, B. R., and Chase, V. C. 1966. Effect of past and prevailing temperatures on the carbon dioxide exchange capacities of some woody desert perennials. *Ecology* 47:1043-1045.

Strecker, J. K. 1908. Notes on the life history of *Scaphiopus couchii* Baird. *Proc. Biol. Soc. Wash.* 21:199-206.

Stuart, G. R. 1954. Observations on reproduction in the tortoise *Gopherus agassizii* in captivity. *Copeia* 1954:61-62.

Stuart, L. C. 1970. Fauna of Middle America. In R. C. West (ed.), *Handbook of Middle American Indians*, vol. 1, pp. 316-362. Austin: University of Texas Press.

Stuckless, J. S., and Sheridan, M. F. 1971. Tertiary volcanic stratigraphy in the Goldfield and Superstition mountains, Arizona. *Geol. Soc. Amer. Bull.* 82:3235-3240.

Stull, E. A., and Kessler, S. J. 1978. Major chemical constituents of Arizona lakes. *J. Arizona-Nevada Acad. Sci.* 13:57-61.

Sturgal, J. R., and Irwin, T. D. 1971. Earthquake history of Arizona, 1850-1966. *Ariz. Geol. Soc. Dig.* 9:1-37.

Styblova, Z., and Kornalik, F. 1967. Enzymatic properties of *Heloderma suspectum* venom. *Toxicon* 5:139-140.

Sumner, J. R. 1976. Earthquakes in Arizona. *Fieldnotes, Ariz. Bur. Mines* 6:1-5.

Suneson, N. H. 1976. Geology of the northern portion of the Superstition-Superior Volcanic Field, Arizona. Master's thesis.

Sutton, G. M., and Phillips, A. R. 1942. June bird life of the Papago Indian Reservation, Arizona. *Condor* 44:57-65.

Swarth, H. S. 1914. A distributional list of the birds of Arizona. *Pacif. Coast Avifauna* 10:1-133.

———. 1920. *Birds of the Papago Saguaro National Monument and the neighboring region, Arizona.* U.S. Dept. Interior. Natl. Park Service.

———. 1929. The faunal areas of southern Arizona: A study in animal distribution. *Proc. Calif. Acad. Sci.* 18(12):267-383.

Sweet, J. G., and Kinne, O. 1964. The effects of various temperature-salinity combinations on the body form of newly hatched *Cyprinodon macularius* (Teleostei). *Helgolander Wiss. Meeresunters.* 11:49-69.

Sykes, G. 1938. End of a great delta. *Pan-American Geol.* 69:241-248.

Szarek, S. R. 1974. Physiological mechanisms of drought adaptation in *Opuntia basilaris* Engelm. & Bigel. Ph.D. dissertation, University of California, Riverside.

Szarek, S. R.; Johnson, H. B.; and Ting, I. P. 1973. Drought adaptation in *Opuntia basilaris.* Significance of recycling carbon through crassulacean acid metabolism. *Plant Physiol.* 53:539-541.

Szarek, S. R., and Ting, I. P. 1974. Seasonal patterns of acid metabolism and gas exchange in *Opuntia basilaris. Plant Physiol.* 54:76-81.

Taber, F. W. 1940. Range of the coati in the United States. *J. Mammal.* 21:11-14.

Tanner, W. W. 1966. The night snakes of Baja California. *Trans. San Diego Nat. Hist. Soc.* 14:189-196.

Taylor, H. L.; Walker, J. M.; and Medica, P. A. 1967. Males of three normally parthenogenetic species of teiid lizards (genus *Cnemidophorus*). *Copeia* 1967:737-743.

Taylor, W. K. 1971. A breeding biology study of the Verdin, *Auriparus flaviceps* (Sundevall) in Arizona. *Amer. Midl. Nat.* 85:289-328.

Taylor, W. P. 1954. Food habits and notes on life history of the ring-tailed cat in Texas. *J. Mammal.* 35(1):55-63.

Taylor, W. P.; Vorhies, C. T.; and Lister, P. B. 1935. The relation of jack rabbits to grazing in southern Arizona. *J. Forestry* 33:490-498.

Templeton, J. R. 1960. Respiration and water loss at the higher temperatures in the desert iguana, *Dipsosaurus dorsalis*. *Physiol. Zool.* 37:300-306.

———. 1964. Cardiovascular responses during buccal and thoracic respiration in the lizard, *Sauromalus obesus*. *Comp. Biochem. Physiol.* 11:31-43.

———. 1967. Panting and pulmonary inflation, two mutually exclusive responses in the chuckwalla, *Sauromalus obesus. Copeia:* 224-225.

Templeton, J. R., and Dawson, W. R. 1963. Respiration in the lizard *Crotaphytus collaris*. *Physiol. Zool.* 36:104-121.

Templeton, J. R.; Murrish, D. E.; Randall, E. M.; and Mugaas, J. N. 1972. Salt and water balance in the desert iguana, *Dipsosaurus dorsalis*. I. The effect of dehydration, rehydration, and full hydration. *J. Vergleich. Physiol.* 76:245-254.

Terron, C. C. 1932. The Mexican horned lizards. *An. Inst. Biol. Mex.* 3:95-121.

Tevis, L. 1944. Herpetological notes from Lower California. *Copeia* 1944:6-18.

———. 1958a. Germination and growth of ephemerals induced by sprinkling a sandy desert. *Ecology* 39:681-688.

———. 1958b. Interrelations between the harvester ant *Veromessor pergandei* Mayr and some desert ephemerals. *Ecology* 39:695-704.

———. 1966. Unsuccessful breeding by desert toads (*Bufo punctatus*) at the limit of their ecological tolerance. *Ecology* 47:766-775.

Thackery, Frank A., and Leding, A. R. 1929. The giant cactus of Arizona: The use of its fruit and other cactus fruits by the Indians. *J. Heredity* 20(9):400-414.

Theilig, E.; Womer, M.; and Papson, R. 1978. Geological field guide to the Salton trough. In R. Gleeley, M. B. Womer, R. P. Papson, and P. D. Spudis, *Aeolian features of southern California*, pp. 100-159. Washington, D.C.: Supt. of Documents.

Thoday, D. 1931. The significance of reduction in the size of leaves. *J. Ecol.* 19:297-303.

Thomson, D. A.; Findley, L. T.; and Kerstitch, A. N. 1979. *Reef fishes of the Sea of Cortez*. New York: John Wiley.

Thomson, D. A., and McKibbin, N. 1976. *Gulf of California fishwatcher's guide*. Tucson, Ariz.: Golden Puffer Press.

Thomson, D. A.; Mead, A. R.; and Schreiber, J. F. (eds.). 1969. Environmental impact of brine effluents on Gulf of California. *O.S.W. Res. and Dev. Progr. Rept.* 387:1-196.

Thorne, D. W., and Peterson, H. B. 1950. *Irrigated soils, their fertility and management*. New York: McGraw-Hill.

Thorp, R. W., and Woodson, W. D. 1945. *The black widow spider*. Chapel Hill: University of North Carolina Press.

Tiedemann, A. R. 1970. Effect of mesquite (*Prosopis juliflora*) trees on herbaceous vegetation and soils in the desert grassland. Ph.D. dissertation, University of Arizona.

Ting, I. P.; Johnson, H. B.; Yonkers, T. A.; and Szarek, S. R. 1973. Gas exchange and productivity of *Opuntia* spp. *US/IBP Progress Rept. (Desert Biome)* RM 73-12:1-22.

Ting, I. P.; and Szarek, S. R. 1975. Drought adaptation in crassulacean acid metabolism plants. In N. F. Hadley (ed.), *Environmental physiology of desert organisms*, pp. 152-167. Stroudsburg, Pa.: Dowden, Hutchinson & Ross.

Tinkle, D. W. 1967. The life and demography of the side-blotched lizard, *Uta stansburiana*. *Misc. Publ. Mus. Zool. Univ. Mich.* 132:1-182.

Tinker, B. 1978. *Mexican wilderness and wildlife*. Austin: University of Texas Press.

Tomlinson. R. E. 1972. Review of literature on the endangered masked bobwhite. *U.S. Bureau of Sport Fisheries and Wildlife, Resource Publ.* 108.

Tomoff, C. S. 1974. Avian species diversity in desert scrub. *Ecology* 55:396-403.

Trainor, F. R. 1962. Temperature tolerance of algae in dry soil. *Phycol. Soc. Amer. News Bull.* 15:3-4.

———. 1970. Survival of algae in a desiccated soil. *Phycologia* 9:111-113.

Truxal, F. S. 1960. The entomofauna with special reference to its origins and affinities. *Syst. Zool.* 9:165-170.

Tryon, A. 1957. A revision of the fern genus *Pellaea* section Pallaea. *Ann. Mo. Bot. Gard.* 44:125-193.

Tschirley, F. H. 1963. A physio-ecological study of jumping cholla (*Opuntia fulgida* Engelm.) Ph.D. dissertation, University of Arizona.

Tschirley, F. H., and Martin, S. C. 1960. Germination and longevity of velvet mesquite seed in soil. *J. Range Managm.* 13:94-97.

———. 1961. Burroweed on southern Arizona range lands. *Ariz. Agr. Exp. Sta. Techn. Bull.* 146:1-34.

Tschirley, F. H., and Wagle, R. F. 1964. Growth rate and population dynamics of jumping cholla (*Opuntia fulgida* Engelm.). *J. Ariz. Acad. Sci.* 3:67-71.

Tucker, V. A. 1967. The role of the cardiovascular system in oxygen transport and thermoregulation in lizards. In W. W. Milstead (ed.), *Lizard ecology—A symposium*, pp. 258-269. Columbia: University of Missouri Press.

Turkowski, F. J., and Reynolds, H. G. 1974. Annual nutrient and energy intake of the desert cottontail, *Sylvilagus auduboni* under natural conditions. *US/IBP Desert Biome Research Memorandum* RM 47-24, Logan, Utah.

Turnage, W. V. 1939. Desert subsoil temperatures. *Soil Sci.* 47:195-199.

Turnage, W. V., and Hinckley, A. L. 1938. Freezing weather in relation to plant distribution in the Sonoran Desert. *Ecol. Monogr.* 8:529-550.

Turner, R. M. 1959. Evolution of the vegetation of the Southwestern desert region. In *Univ. of Ariz. Arid Lands Colloquia, 1958-59*, pp. 46-53.

———. 1973. *Dynamics of distribution and density of phraeatophytes and other arid-land plant communities*. Tucson: U.S. Dept. Interior, Geol. Surv.

————. 1974a. Map showing vegetation in the Phoenix area, Arizona. *U.S. Geol. Survey Misc. Invest. Series,* map I-845-I.

————. 1974b. Map showing vegetation in the Tucson area, Arizona. *U.S. Geol. Survey Misc. Invest. Series,* map I-844-H.

————. 1974c. Quantitative and historical evidence of vegetation changes along the upper Gila River, Arizona. *U.S. Geol. Surv. Prof. Paper* 655-H.

Turner, R. M.; Alcorn, S. M.; Olin, G.; and Booth, J. A. 1966. The influence of shade, soil, and water on saguaro seedling establishment. *Bot. Gaz.* 127 (2/3):95-102.

Tuttle, D. M., and Baker, E. W. 1968. *Spider mites of southwestern United States.* Tucson: University of Arizona Press.

Tyler, A. 1956. An auto-antivenin in the gila monster and its relation to a concept of natural auto-antibodies. In E. Buckley and N. Porges, (eds.), *Venoms,* pp. 65-74. Washington, D.C.: American Association for the Advancement of Science.

Underhill, R. M. 1938. A Papago calendar record. *Univ. N. Mex. Bull.* 322, *Anthrop. Series* 2(5).

————. 1939. *Social organization of the Papago Indians.* New York, Columbia University Press.

————. 1946. *Papago Indian religion.* Columbia University Press.

————. New York: 1951. *People of the crimson evening.* Riverside, Calif.: U.S. Indian Service.

————. 1979. *Papago woman.* New York: Holt, Rinehart and Winston.

Underhill, R. M.; Bahr, D. M.; Lopez, B.; Pancho, J.; and Lopez, D. 1979. *Rainhouse and ocean. American tribal religions.* vol. 4. Flagstaff: Museum of Northern Arizona.

Ungar, I. A. 1962. Influence of salinity on seed germination in succulent halophytes. *Ecology* 43:763-764.

Ungar, I. A. 1974. Inland halophytes of the United States. In R. S. Reimold and W. H. Queen (eds.), *Ecology of Halophytes,* pp. 235-306. New York: Academic Press.

University of Arizona. 1972. Water. In *Arizona, Its people and resources,* pp. 107-136. Tucson: University of Arizona Press.

Ushakov, B. P. 1958. On the conservation of protoplasm of the species of poikilothermal animals. *Zool. Zh.* 37:693-706.

————. 1959. Thermostability of the tissue as one of the diagnostic characters in poikilothermal animals. In *Proc. 15th Inter. Congr. Zool. London,* pp. 1046-1051.

Valentine, K. A., and Gerard, J. B., 1968. Life-history characteristics of the creosotebush, *Larrea tridentata. New Mex. Agr. Exp. Sta. Bull.* 526.

Valverde, J. M. 1976. The bighorn sheep of the state of Sonora. *Desert Bighorn Council Transact.* 1976:25.

Vance, V. J. 1953. Respiratory metabolism and temperature acclimatization of the lizard, *Urosaurus ornatus linearis* (Baird), Master's thesis, University of Arizona.

Van De Graaf, K. M. 1973. Comparative developmental osteology in three species of desert rodents, *Peromyscus eremicus, Perognathus intermedius* and *Dipodomys merriami. J. Mammal.* 54:3.

Van De Graaff, K. M., and Balda, R. P., 1973. Importance of green vegetation for reproduction in the kangaroo rat, *Dipodomys merriami merriami. J. Mammal.* 54:509-512.

Van Denburgh, J. 1896. *Proc. Cal. Acad. Sci., ser. 2,* 5:1004-1008.

Van Denburgh, J., and Slevin, J. R. 1921. A list of the amphibians and reptiles of the peninsula of Lower California, with notes on the species in the collection of the academy. *Proc. Calif. Acad. Sci., ser. 4,* 11:49-72.

Van Devender, T. R. 1977. Holocene woodlands in the southwestern deserts. *Science* 198 (4313):189-192.

Van Devender, T. R., and Spaulding, W. G. 1979. Development of vegetation and climate in the southwestern United States. *Science* 204:701-710.

Vanicek, C. D., and Kramer, R. H. 1969. Life history of the Colorado squawfish, *Ptychocheilus lucius,* and the Colorado chub, *Gila robusta,* in the Green River in Dinosaur National Monument, 1964-1966. *Transact. Amer. Fish Soc.* 98:193-208.

Vanjonack, W. J.; Scott, I. M.; Yousef, M. K.; and Johnson, H. D. 1975. Corticosterone plasma levels in desert rodents. *Comp. Biochem. Physiol.*

Van Rossem, A. J. 1931. Report on a collection of land birds from Sonora, Mexico. *Transact. San Diego Soc. Nat. Hist.* 6:237-304.

————. 1932. The avifauna of Tiburon Island, Sonora, Mexico, with descriptions of four new races. *Transact. San Diego Soc. Nat. Hist.* 7:119-150.

————. 1936. Notes on birds in relation to the faunal areas of south-central Arizona. *Transact. San Diego Soc. Nat. Hist.* 8:122-148.

————. 1945. A distributional survey of the birds of Sonora, Mexico. *Louisiana St. Univ. Mus. Zool. Occ. Pap.* 21:1-379.

Victor, J. B. 1979. *Tarantulas.* New York: Dodd, Mead.

Villa-R., B. 1966. *Los Murciélagos de México.* Instituto de Biología, Univ. Nac. Autónoma de México.

Villee, C. A. 1977. *Biology.* Philadelphia: W. B. Saunders.

Vorhies, C. T. 1928. Do southwestern quail require water? *Amer. Nat.* 62:446-452.

————. 1934. *Condor* 36: 85-86.

————. 1945. Water requirements of desert animals in the Southwest. *Ariz. Agr. Exp. Sta. Techn. Bull.* 107:487-525.

Vorhies, C. T.; Jenks, R.; and Phillips, A. R. 1935. Bird records from the Tucson region, Arizona. *Condor* 37:243-247.

Vorhies, C. T., and Phillips, A. R. 1937. *Condor* 39:175.

Vorhies, C. T., and Taylor, W. P. 1922. Life history of the kangaroo rat. *Bull. U.S. Dept. Agr.* 1091:1-39.

————. 1933. The life histories and ecology of jack rabbits, *Lepus alleni alleni* and *Lepus californicus* sp., in relation to grazing in Arizona. *Ariz. Agr. Exp. Sta. Techn. Bull.* 49:1-112.

————. 1940. Life history and ecology of the white-throated wood rat, *Neotoma albigula albigula* Hartley, in relation to grazing in Arizona. *Ariz. Agr. Exp. Sta. Techn. Bull.* 86:455-529.

Voth, A. 1938. Summary of investigations dealing with burroweed (*Aplopappus fruticosus*) and its ecological aspects. Master's thesis, University of Arizona.

Waddell, J. O. 1973. The place of the cactus wine ritual in the Papago Indian ecosystem. *Ninth Intl. Congr. Anthrop. Ethn. Sci.*

Waisel, Y. 1972. *Biology of halophytes.* New York: Academic Press.

Waitman, J. and Roest, A. 1977. A taxonomic study of the kit fox, *Vulpes macrotis. J. Mammal.* 58(2):157-164.

Walker, B. W. 1961. The ecology of the Salton Sea, California, in relation to the sportfishery. *Calif. Fish & Game Dept. Fish Bull.* 113:1-204.

Walker, B. W.; Whitney, R. R.; and Barlow, G. W. 1961. The fishes of the Salton Sea. In B. W. Walker, *The ecology of the Salton Sea, California, in relation to the sport fishery. Calif. Fish & Game Dept. Fish Bull.* No. 113:77-91.

Walker, E. P. 1968. *Mammals of the world.* Baltimore: Johns-Hopkins University Press.

Wallmo, O. C., and Gallizioli, S. 1954. Status of the coati in Arizona. *J. Mammal.* 35:48-54.

Wallwork, J. A. 1972. Distribution patterns and population dynamics of the micro-arthropods of a desert soil in southern California. *J. Anim. Ecol.* 41:291-310.

Walter, H. 1931. *Die Hydratur der Pflanze und ihre Physiologisch-ökologische Bedeutung (Untersuchungen über den Osmotischen Wert).* Jena: Fisher.

———. 1960. *Einführung in die Phytologie.* Vol. 3: *Grundlagen der Pflanzenverbreitung. Part 1: Standortslehre (analytisch-ökologische Geobotanik).* 2d ed. Stuttgart: Ulmer.

Walter, H. and Stadelmann, E. 1974. A new approach to the water relations of desert plants. In G. W. Brown, (ed.), *Desert biology,* vol. 2, 213-310. New York: Academic Press.

Warburg, M. R. 1965a. The microclimate in the habitats of two isopod species in Southern Arizona. *Amer. Midl. Nat.* 73:363-375.

———. 1965b. Water relations and internal body temperature of isopods from mesic and xeric habitats. *Physiol. Zool.* 38: 99-109.

Warren, J. W. 1953. Notes on the behavior of *Chionactis occipitalis. Herpetologica.* 9:121-124.

Warren, P. L. 1979. *Plant and rodent communities of Organ Pipe Cactus National Monument.* Master's thesis, University of Arizona.

Warren, R. E.; Sclater, J. C.; Vaquier, V.; and Roy, R. E. 1969. A comparison of terrestrial heat flow and transient geomagnetic variations in the southwestern United States. *Geophysics* 34:463-478.

Watson, B. N. 1968. Intrusive volcanic phenomena in southern and central Arizona. *Ariz. Geol. Soc. Guidebook III:*147-153.

Watt, D. 1964. Biochemical studies on the venom from the scorpion, *Centruroides sculpturatus. Toxicon* 2:171-180.

Weathers, W. W. 1971. Some cardiovascular aspects of temperature regulation in the lizard *Dipsosaurus dorsalis. Comp. Biochem. and Physiol.* 40A:503-515.

Webber, J. M. 1953. Yuccas of the Southwest. *U.S.D.A. Agr. Monogr.* 17:1-97.

Weese, A. O. 1917. The urine of the horned lizard. *Science* 46:517-518.

Wells, P. V. 1979. An equable glaciopluvial in the West: Pleniglacial evidence of increased precipitation on a gradient from the Great Basin to the Sonoran Desert, U.S.A., and Chihuahuan Desert, Mexico. *Quat. Res.* 12(3):311-325.

Wells, P. V., and Hunziker, Juan H. 1977. Origin of the creosote bush (*Larrea*) deserts of southwestern North America. *Ann. Mo. Bot. Gard.* 63(4):843-861.

Welsh, R. G. and Beck, R. F. 1976. Some ecological relationships between creosotebush and bush muhly. *J. Range Managm.* 29:472-475.

Went, F. W. 1942. The dependence of certain annual plants on shrubs in southern California deserts. *Bull. Torr. Bot. Club* 69:100-114.

———. 1949. Ecology of desert plants II: The effect of rain and temperature on germination and growth. *Ecology* 30:1-13.

———. 1957. *Experimental control of plant growth.* Waltham, Mass.: Chronica Botanica.

Went, F. W. 1979. Germination and seedling behavior of desert plants. In D. W. Goodall, R. A. Perry, and K. M. W. Howes,

(eds.), *Arid-land ecosystems: Structure, functioning and management, vol. 1,* pp. 477-489. Cambridge: At the University Press.

Werner, F. G. 1973. Foraging activity of the leaf-cutter ant, *Acromyrmex versicolor,* in relation to season, weather and colony condition. *US/IBP Desert Biome Research Memorandum* RM 73-28, Logan, Utah.

Werner, F. G.; Enns, W. R.; and Parker, F. H., 1966. The Meloidae of Arizona. *Tech. Bull. Ariz. Agr. Exp. Sta.* 175:1-96.

Werner, F. G., and Olsen, A. R. 1973. Consumption of *Larrea* by chewing insects. *US/IBP Desert Biome Research Memorandum* RM 73-32. Logan, Utah.

Whaley, J. W. 1964. Physiological studies of antagonistic actinomycetes from the rhizosphere of desert plants. Ph.D. dissertation, University of Arizona.

Whaley, W. H. 1979. The ecology and status of the Harris' hawk (*Parabuteo unicinctus*) in Arizona. Master's thesis, University of Arizona.

Wheeler, G. C., and Wheeler, J. 1973. Ants of Deep Canyon. In *Philip L. Boyd Deep Canyon Research Center Reports.* Riverside: University of California.

Wheeler, W. M. 1907. The fungus-growing ants of North America. *Am. Mus. Nat. Hist. Bull.* 23:669-807.

Whitaker, R. H., and Niering, W. A. 1965. Vegetation of the Santa Catalina Mountains, Arizona. II: A gradient analysis of the south slope. *Ecology* 46(4):429-452.

White, L. D. 1968. Factors affecting susceptibility of creosotebush (*Larrea tridentata (DC.) Cov.*) to burning. Ph.D. dissertation University of Arizona.

White, S. S. 1948. The vegetation and flora of the region of the Rio de Bavispe in northeastern Sonora, Mexico. *Lloydia* 2:229-302.

Whitford, W. B., and Whitford, W. G. 1973. Combat in the horned lizard, *Phrynosoma cornutum. Herpetologica* 29:191-192.

Whitford, W. G. 1976. Temporal fluctuations in density and diversity of desert rodent populations. *Journ. Mammal.* 57:351-369.

Whittemore, F. W., Jr.; Keegan, H. L.; and Borowitz, J. L. 1963. Venom collection and scorpion colony maintenance. *Bull. World Health Org.* 28:505-511.

Whittow, G. C. (ed.). 1971. *Comparative physiology of thermoregulation.* Vol. 2: *Mammals.* New York: Academic Press.

Wien, J. 1958-1959. The study of the algae of irrigation waters. *1st and 2nd Ann. Prog. Repts.* Arizona State University. Manuscript.

Wiggins, I. 1937. Effects of the January freeze upon the pitahaya in Arizona.

———. 1960. The origin and relationship of the land flora. In *The biogeography of Baja California and adjacent seas. Syst. Zool.* 9: 148-165.

———. 1969. Observations on the Vizcaíno Desert and its biota. *Proc. Calif. Acad. Sci.* 36. (11).

———. 1980. *Flora of Baja California.* Stanford, Calif.: Stanford University Press.

Willet, G. 1933. A revised list of the birds of southwestern California. *Pacif. Coast Avifauna* 21:1-204.

Williams, K. B. 1971. Ecological and morphological variations of *Vauquelinia californica (Torr.)* Sarg. populations in Arizona. Ph.D. dissertation, University of Arizona.

Williams, S. C. 1970a. A systematic revision of the giant hairy-scorpion genus *Hadrurus* (Scorpionida: Vejovidae). *Occ. Pap. Calif. Acad. Sci.* 87:1-62.

———. 1970b. Scorpion fauna of Baja California, Mexico: Eleven new species of *Vejovis* (Scorpinida: Vejovidae). *Calif. Acad. Sci. Proc.* 37:275-331.

Williams, S. E.; Wollum, A. G.; and Aldon, E. F. 1974. Growth of *Atriplex canescens* (Pursh) Nutt. improved by formation of vesicular-arbuscular mycorrhizae. *Soil Sci. Soc. Amer. Proc.* 38:962-965.

Willoughby, E. 1966. Water requirements of the ground dove. *Condor* 68:243-248.

Wilson, E. D. 1962 *A resume of the geology of Arizona.* Tucson: Arizona Bureau of Mines, University of Arizona.

Winkler, P. 1973. The ecology and thermal physiology of *Gambusia affinis* from a hot spring in southern Arizona. Ph.D. dissertation, University of Arizona.

Winogradov, B. S. 1948. Adaptations of animals to life in deserts. *Zhirotnij mir USSR* 2:17-62.

Withers, A. M. 1944. Excavations at Valshni Village, a site on the Papago Indian Reservation. *Amer. Antiquity* 10(1):33-47.

Wondolleck, J. T. 1978. Forage-area separation and overlap in Heteromyid rodents. *J. Mammal.* 59(3):510-518.

Wood, C. W., and Nash, T. N. 1976. Copper smelter effluent effects on some Sonoran Desert vegetation. *Ecology* 57:1311-1316.

Woods, F. W. 1960. Biological antagonisms due to phytotoxic root exudates. *Bot. Rev.* 26:546-569.

Woodbury, A. M., and Hardy, R. 1948. Studies of the desert tortoise, *Gopherus agassizii. Ecol. Monogr.* 18:145-200.

Woodell, S. J. R.; Mooney, H. A.; and Hill, A. J. 1969. The behavior of *Larrea divaricata* (Creosote bush) in response to rainfall in California. *J. Ecol.* 57:37-44.

Woodford, A. O.; Welday, E. E.; and Merriam, R. 1968. Siliceous tuff casts in the upper Paleogene of southern California. *Geol. Soc. Amer. Bull.* 79:1461-1468.

Woodhouse, R. M., and Szarek, S. R. 1977. Ecophysiological studies of Sonoran Desert plants. *Ariz. Acad. Sci. Proc. 21st Ann. Meet:* 18.

Wright, A. H., and Wright, A. A. 1949. *Handbook of frogs and toads.* Ithaca: Cornell University Press.

———. 1957. *Handbook of snakes.* 2 vols. Ithaca: Cornell University Press.

Wright, J. W. 1967. A new uniparietal whiptail lizard (genus *Cnemidophorus*) from Sonora, Mexico. *J. Ariz. Acad. Sci.* 4:185-193.

Wright, R. A. 1965. An evaluation of the homogeneity of two stands of vegetation in the Sonoran Desert. Ph.D. dissertation, University of Arizona.

Yang, T. W. 1950. Distribution of *Larrea tridentata* in the Tucson area as determined by certain physical and chemical factors of habitat. Master's thesis, University of Arizona.

———. 1957. Vegetational, edaphic, and faunal correlations of the western slope of the Tucson Mountains and adjoining Avra Valley. Ph.D. dissertation, University of Arizona.

———. 1961. The recent expansion of creosotebush (*Larrea divaricata*) in the North American Desert. *W. Res. Acad. Nat. Hist. Spec. Publ.* 1:1-11.

Yang, T. W., and Lowe, C. H. 1956. Correlation of major vegetation climaxes with soil characteristics in the Sonoran Desert. *Science* 123(3196):542.

———. 1968. Chromosome variation in ecotypes of *Larrea divaricata* in the North American desert. *Madroño* 19:161-164.

Yeates, N. T. M.; Lee, D. H. K.; and Hines, H. J. G. 1941. Reactions of fowls to hot atmospheres. *Proc. Roy. Soc. Queensland* 53:105-128.

Yeaton, R. I.; Travis, J.; and Gilinsky, E. 1977. Competition and spacing in plant communities: The Arizona Upland association. *J. Ecol.* 65(2):587-596.

Yermanos, D. M.; Francois, L. E.; and Tammadoni, T. 1967. Effects of soil salinity on the development of jojoba. *Econ. Bot.* 21:69-80.

Yesilsoy, M. 1962. Characterization and genesis of a Mohave sandy loam profile. Master's thesis, University of Arizona.

Young, S. P. 1946. *The puma—Mysterious American cat.* New York: Dover.

———. 1958. *The bobcat of North America.* Washington, D.C.: Wildlife Management Institute.

Young, S. P., and Goldman, E. A. 1946. *The puma.* New York: Dover.

Young, S. P., and Jackson, H. H. T. 1951. *The clever coyote.* Washington, D.C.: Wildlife Management Institute.

Yousef, M. K. 1975. Thyroid and adrenal function in desert animals: A view on metabolic adaptation. In N. F. Hadley, (ed.), *Environmental physiology of desert organisms,* pp. 188-206. Stroudsburg, Pa.: Dowden, Hutchinson & Ross.

Yousef, M. K., and Dill, D. B. 1971a. Responses of Merriam's kangaroo rat to heat. *Physiol. Zool.* 44:33.

———. 1971b. Daily cycles of hibernation in the kangaroo rats, *Dipodomys merriami. Cryobiology* 3:122.

Yousef, M. K.; Horvath, S. M.; and Bullard, R. W. (eds.). 1972. *Physiological adaptations: Desert and mountain.* New York: Academic Press.

Yousef, M. K., and Johnson, H. D. 1972. Thyroid function and metabolic rate of desert rodents. In S. W. Tromp and J. J. Bouma (eds.), *Biometeorology.* 5:134. Amsterdam: Swets Zeitlinger N.V.

Yousef, M. K.; Johnson, H. D.; Bradley, W. G.; and Seif, S. M. 1974. Tritiated water turnover rate in rodents: desert and mountain. *Physiol. Zool.* 47:153.

Yousef, M. K.; Robertson, W. D.; Dill, D. B.; and Johnson, H. D. 1970. Energy expenditure of running kangaroo rats, *Dipodomys merriami. Compar. Biochem Physiol.* 36:387-393.

———. 1973. Energetic cost of running in the antelope ground squirrel *Ammospermophilus leucurus. Physiol. Zool.* 46:139-147.

Zeller, E. A. 1948. Enzymes of snake venoms and their biological significance. *Advam. Enzymol.* 8:459-495.

Zervanos, S. M., and Day, G. I. 1977. Water and energy requirements of captive and free-living collared peccaries. *J. Wildl. Managm.* 41(3):527-532.

Zervanos, S. M., and Hadley, N. F. 1973. Adaptational biology and energy relationships of the collared peccary (*Tayassu* tajacu). *Ecology* 54:759-774.

Zweifel, R. G. 1965. Variation in and distribution of the unisexual lizard *Cnemidophorus tessellatus. Amer. Mus. Novit.* 2235:1-49.

Zweifel, R. G., and Lowe, C. H. 1966. The ecology of a population of *Xantusia vigilis,* the desert night lizard. *Amer. Mus. Novit.* 2247:1-57.

Zweifel, R. G., and Norris, K. S. 1955. Contribution to the herpetology of Sonora, Mexico. *Amer. Midl. Nat.* 54:230-249.

MAMMALS OF THE SONORAN DESERT

FAMILY	COMMON NAME	SCIENTIFIC NAME	PENINSULA	ISLAND ENDEMIC	N. MAINLAND	S. MAINLAND	COMMENT
Bovidae	Desert bighorn	Ovis canadensis mexicana			x		Arizona, Sonora
		O. c. nelsoni			x		Southern California
		O. c. cremnobates	x				Northern Baja
		O. c. weemsi	x				Central Baja
Antilocapridae	Pronghorn	Antilocapra americana mexicana			x		
	Sonoran pronghorn	A. a. sonoriensis				x	
	Peninsular pronghorn	A. a. peninsularis	x				Vizcaíno Division
Cervidae	Mule deer	Odocoileus hemionus eremica			x		
		O. h. fuliginata	x				Northern Baja
		O. h. peninsularis	x				Southern Baja
	Tiburón mule deer	O. h. sheldoni		x			Tiburón Island
	Cedros mule deer	O. h. cerrosensis		x			Cedros Island
	White-tailed deer	O. virginiana couesi			x	x	
Tayassuidae	Javelina	Tayassu tajacu sonoriensis			x	x	
Equidae	Burro	Equus assinus			x		Introduced
Hominidae	Man	Homo sapiens	x		x	x	Post Pleistocene immigrant
Phocidae	California harbor seal	Phoca vitulina geronimensis	x				
	Elephant seal	Mirounga angustirostris	x				
Otariidae	California sea lion	Zalophus californicus	x			x	
Felidae	Jaguar	Felis onca arizonensis			x		
		F. o. hernandesii				x	
	Mountain lion	F. concolor azteca			x	x	
		F. c. browni	x		x		
		F. c. improcera	x				Southern Baja
	Ocelot	F. pardalis sonoriensis			x	x	
	Jaguarundi	F. yagouaroundi			±	±	
	Bobcat	Lynx rufus baileyi			x		
		L. r. escuinapae				x	
	Peninsular bobcat	L. r. peninsularis	x				

FAMILY	COMMON NAME	SCIENTIFIC NAME	PENINSULA	ISLAND ENDEMIC	N. MAINLAND	S. MAINLAND	COMMENT
Mustelidae	Badger	Taxidea taxus berlandieri			x		
		T. t. sonoriensis				x	
		T. t. infusca	x				
	Western spotted skunk	Spilogale gracilis gracilis			x		
		S. g. martirensis	x				Central Baja
		S. g. microdon	x				Magdelena Division
		S. g. lucasana	x				Cape region
	Striped skunk	Mephitis mephitis estor			x		
	Hooded skunk	M. macroura			x	x	Arizona, Sonora
	Hog-nosed skunk	Conepatus mesoleucus venaticus			x		
	Sonoran hog-nosed skunk	C. m. sonoriensis				x	
	Sea otter	Enhydra lutris nereis	x				
Procyonidae	Ringtail	Bassariscus astutus arizonensis			x		
		B. a. yumanensis			x		
	San José ringtail	B. a. insulicola		x			San José Island
	Espíritu Santo ringtail	B. a. saxicola		x			Espíritu Santo Island
	Peninsular ringtail	B. a. palmarius	x				
	Raccoon	Procyon lotor pallidus			x		
		P. l. mexicanus				x	
		P. l. grinnelli	x				Southern Baja
	Coatimundi	Nasua narica molaris			x	x	
Canidae	Coyote	Canis latrans mearnsi			x	x	
	Peninsular coyote	C. l. peninsulae	x				
	Tiburón coyote	C. l. jamesi		x			Tiburón Island
	Kit fox	Vulpes macrotis	x		x		
	Gray fox	Urocyon cinereoargenteus scottii			x		
		U. c. madrensis				x	
	Peninsular gray fox	U. c. peninsularis	x				
Balaenidae	Pacific right whale	Eubalaena sieboldii	x				
Balaenopteridae	Fin-backed whale	Balaenoptera physalus	x			x	
	Sei whale	B. borealis	x				
	Little piked whale	B. acutirostrata	x				
	Blue whale	Sibbaldus musculus	x				
	Hump-backed whale	Magaptera novaeangliae	x			x	
Eschrichtidae	Gray whale	Eschrichtius gibbosus	x			x	
Delphinidae	Pacific dolphin	Delphinus bairdii	x			x	
	Gill's bottle-nosed dolphin	Tursiops gillii	x			x	
	Pacific bottle-nosed dolphin	T. nuuanu	x			x	
	Pacific white-sided dolphin	Lagenorhynchus obliquidens	x			x	
	Pacific killer whale	Orcinus rectipinna	x			x	
	False killer whale	Pseudorca crassidens	x			x	
	Scammon's blackfish	Globicephala scammoni	x			x	
	Pacific Harbor porpoise	Phocoena vomerina	x			x	
Kogiidae	Pygmy sperm whale	Kogia breviceps	x			x	
Physeteridae	Sperm whale	Physeter catodon	x			x	
Erethizontidae	Porcupine	Erethizon dorsatum couesi			x		South to Kino Bay
Cricetidae	Lamb's rice rat	Oryzomys couesi lambi				x	Guaymas region
	Baja California rice rat	O. peninsulae	x				Cape region
	Sonoran harvest mouse	Reithrodontomys burti				x	
	Western harvest mouse	R. megalotis megalotis	±		x		extreme northern Baja

FAMILY	COMMON NAME	SCIENTIFIC NAME	PENINSULA	ISLAND ENDEMIC	N. MAINLAND	S. MAINLAND	COMMENT
Cricetidae	Peninsular harvest mouse	R. m. peninsulae	x				Northern Baja
	Fulvous harvest mouse	R. fulvescens fulvescens			x		
		R. f. tenuis				x	
	Canyon mouse	Peromyscus crinitus stephensi			x		
		P. c. disparilis			x		Gran Desierto
		P. c. delgadilli			x		Pinacate lava field
		P. c. palidissimus		x			Gonzaga Bay Island
	Merriam's mouse	P. merriami merriami			x		
		P. m. goldmani				x	
	Angel Island mouse	P. guardia guardia		x			Angel de la Guarda Island
		P. g. interparietalis		x			South San Lorenzo Island
		P. g. mejiae		x			Méjia Island
	San Esteban mouse	P. stephani		x			San Esteban Island
	Turners Island Mouse	P. collatus		x			Turners Island
	Pemberton's deer mouse	P. pembertoni		x			San Pedro Nolasco Island
	Dickey's Deer Mouse	P. dickeyi		x			Tortuga Island
	False Canyon mouse	P. pseudocrinitus		x			Coronados Island
	Burt's deer mouse	P. caniceps		x			Monserrate Island
	Deer mouse	P. maniculatus sonoriensis			x		
		P. m. gambelii	x				Northern Baja
		P. m. coolidgei	x				
		P. m. hueyi		x			Gonzaga Bay Island
	Magdalena deer mouse	P. m. magdalenae	x				Magdalena division
	Margarita deer mouse	P. m. margaritae		x			Margarita Island
	Gerónimo deer mouse	P. m. geronimensis		x			Gerónimo Island
	Natividad deer mouse	P. m. dorsalis		x			Natividad Island
	San Roque deer mouse	P. m. cineritius		x			San Roque Island
	Slevin's mouse	P. slevini		x			Santa Catalina Island
	Santa Cruz mouse	P. sejugis		x			Santa Cruz Island
							San Diego Island
	White-footed mouse	P. leucopus arizonae			x		
	Brush mouse	P. boylii glasselli		x			San Pedro Nolasco Island
	Northern pygmy mouse	Baiomys taylori paulus				x	near Ciudad Obregón
	Southern grasshopper mouse	Onychomys torridus torridus					Central mainland
		O. t. pulcher			x		Southern California
		O. t. perpallidus			x		Southern Arizona
	Yaqui grasshopper mouse	O. t. yakiensis				x	
	Hispid cotton rat	Sigmodon hispidus arizonae			x		
		S. h. cienegae			x		
		S. h. eremicus			x		Lower Colorado River
		S. h. major				x	
	White-throated wood rat	Neotoma albigula albigula			x		
		N. a. venusta			x		
		N. a. mearnsi			x		Tinajas Altas
		N. a. sheldoni			x		Pinacate lava field
		N. a. melanura				x	
	Tiburón wood rat	N. a. seri		x			Tiburón Island
	Turners Island wood rat	N. varia		x			Turners Island
	Desert Wood rat	N. lepida lepida			x		
		N. l. grinnelli			x		
		N. l. devia			x		
		N. l. harteri			x		Southern Arizona

FAMILY	COMMON NAME	SCIENTIFIC NAME	PENINSULA	ISLAND ENDEMIC	N. MAINLAND	S. MAINLAND	COMMENT
Cricetidae		N. l. flava			x		Tinajas Altas
		N. l. auripila			x		Cabeza Prieta Mountains
		N. l. bensoni			x		Pinacate lava field
		N. l. aureotunicata			x		Puerto Peñasco
		N. l. felipensis	x				San Felipe region
		N. l. intermedia	x				Northern Baja
		N. l. egressa	x				El Rosario region
		N. l. molagrandis	x				Vizcaíno Division
		N. l. insularis	x				Angel de la Guarda Island
		N. l. ravida	x				Central Gulf Coast
		N. l. pretiosa	x				Magdalena Division
		N. l. arenacea	x				San José del Cabo
	San Marcos wood rat	N. l. marcosensis		x			San Marcos Island
	Carmen wood rat	N. l. nudicauda		x			Carmen Island
	Danzante wood rat	N. l. latirostra		x			Danzante Island
	San José wood rat	N. l. perpallida		x			San José Island
	San Francisco wood rat	N. l. abbreviata		x			San Francisco Island
	Espíritu Santo wood rat	N. l. vicina		x			Espíritu Santo Island
	San Martín wood rat	N. martinensis		x			San Martín Island
	Bryant's wood rat	N. bryanti		x			Cedros Island
	Bunker's wood rat	N. bunkeri		x			Coronados Island
	Sonoran wood rat	N. phenax				x	
	Muskrat	Ondatra zibethicus pallidus			x		Southern Arizona
		O. z. bernardi			x		Southern California
Castoridae	Beaver	Castor canadensis repentinus			x		Colorado River
		C. c. frondator			x		
Heteromyidae	Silky pocket mouse	Perognathus flavus flavus			x		
		P. flavus sonoriensis				x	
	Little pocket mouse	P. longimembris bombycinus	x		x		
		P. l. venustus	x				San Augustín
	Kino Bay pocket mouse	P. l. kinoensis				x	Kino Bay region
	Pima pocket mouse	P. l. pimensis			x		Casa Grande region
	Arizona pocket mouse	P. amplus jacksoni			x		Northern Arizona Upland
		P. a. taylori			x		Southern Arizona Upland
		P. a. rotundus			x		Lower Colorado-Gila
	Long-tailed pocket mouse	P. formosus mesembrinus	x		x		Southern California, Northern Baja
		P. f. infolatus	x				Central Baja
	Bailey's pocket mouse	P. baileyi baileyi			x	x	
		P. b. domensis			x	±	
		P. b. hueyi	x		x		Northern Baja, Northwest Sonora
		P. b. extimus	x				
	Monserrate pocket mouse	P. b. fornicatus		x			Monserrate Island
	Tiburón pocket mouse	P. b. insularis		x			Tiburón Island
	Desert pocket mouse	P. penicillatus penicillatus			x		Colorado River region
		P. p. angustirostris	x				Northern Baja
		P. p. pricei			x	x	
	Seri pocket mouse	P. p. seri		x			Tiburón Island
	Sand mouse	P. arenarius arenarius	x				Comondú region
		P. a. sublucidus	x				La Paz region
		P. a. ambiguus	x				Central Baja

FAMILY	COMMON NAME	SCIENTIFIC NAME	PENINSULA	ISLAND ENDEMIC	N. MAINLAND	S. MAINLAND	COMMENT
Heteromyidae	Magdalena sand mouse	P. a. albulus		x			Magdalena Island
	Margarita sand mouse	P. a. ammophilus		x			Santa Margarita Island
	Cerralvo pocket mouse	P. a. siccus		x			Cerralvo Island
	Sinaloan pocket mouse	P. pernix rostratus				x	
	Rock pocket mouse	P. intermedius intermedius			x		
		P. i. lithophilus			x		Puerto Libertad, Sonora
		P. i. phasma			x		Tinajas Altas
	Black Mountain pocket mouse	P. i. nigrimontis			x		Black Mountain, Tucson
	Pinacate pocket mouse	P. i. pinacate			x		Pinacate lava field
	Turners Island pocket mouse	P. i. minimus		x			Turners Island
	San Diego pocket mouse	P. fallax inopinus	x				San Bartolomé
	Anthony's pocket mouse	P. anthonyi		x			Cedros Island
	Spiny pocket mouse						Southern California,
		P. spinatus spinatus	x		x		Northern Baja
		P. s. prietae	x				Central Gulf Coast
		P. s. peninsulae	x				Southern Baja
	Magdalena pocket mouse	P. s. magdalenae		x			Magdalena Island
	Margarita pocket mouse	P. s. margaritae		x			Margarita Island
	Méjia pocket mouse	P. s. evermanni		x			Méjia Island
	Angel pocket mouse	P. s. guardiae		x			Angel de la Guarda Island
	San Marcos pocket mouse	P. s. marcosensis		x			San Marcos Island
	Coronados pocket mouse	P. s. pullus		x			Coronados Island
	Carmen pocket mouse	P. s. occultus		x			Carmen Island
	Danzante pocket mouse	P. s. seorsus		x			Danzante Island
	San José pocket mouse	P. s. bryanti		x			San José Island
	San Francisco pocket mouse	P. s. latijugularis		x			San Francisco Island
	Espíritu Santo pocket mouse	P. s. lambi		x			Espíritu Santo Island
	Goldman's pocket mouse	P. goldmani				x	
	Ord's kangaroo rat	Dipodomys ordii			x		
	Agile kangaroo rat	D. agilis plectilis	x				N. Vizcaíno region
	Baja California kangaroo rat	D. peninsularis peninsularis	x				Vizcaíno Division
		D. p. australis	x				Magdalena Division
		D. p. eremoecus	x				Central Gulf Coast
		D. p. pedionomus	x				Northern Baja
	Santa Catarina kangaroo rat	D. paralius	x				Northern Vizcaíno Division
	San Quintín kangaroo rat	D. gravipes	x				Northern Vizcaíno Division
	Banner-tailed kangaroo rat	D. spectabilis perblandus			x		Arizona, Sonora
	Merriam's kangaroo rat	D. merriami merriami			x	x	
		D. m. regillus			x		Arizona-Sonora border
		D. m. simiolus			x		Southern California, Northwest Sonora
		D. m. arenivagus	x				Laguna Salada region
		D. m. trinidadensis	x				Northern Baja
		D. m. semipallidus	x				Northern Vizcaíno Division
		D. m. platycephalus	x				Southern Vizcaíno Division
		D. m. brunensis	x				Central Gulf Coast

FAMILY	COMMON NAME	SCIENTIFIC NAME	PENINSULA	ISLAND ENDEMIC	N. MAINLAND	S. MAINLAND	COMMENT
Heteromyidae		D. m. llanoensis	x				Magdalena Division
		D. m. melanuris	x				Cape region
	Tiburón kangaroo rat	D. m. mitchelli		x			Tiburón Island
	Margarita kangaroo rat	D. m. margaritae		x			Margarita Island
	San José kangaroo rat	D. insularis		x			San José Island
	Desert kangaroo rat	D. deserti deserti	x		x		Northeast Baja
		D. d. arizonae			x		South-central Arizona
		D. d. sonoriensis			x	±	
	Painted spiny pocket mouse	Liomys pictus sonoranus				x	
Geomyidae	Valley pocket gopher						41 endemic subspecies in
		Thomomys bottae	x		x	x	Sonoran Desert
	Magdalena pocket gopher	T. bottae magdalenae		x			Magdalena Island
Sciuridae	Cliff chipmunk	Eutamias dorsalis			±	±	
	Peninsula chipmunk	E. merriami meridionalis	x				See Callahan 1975
	Harris's antelope squirrel	Ammospermophilus harrisii harrisii			x	±	
		A. h. saxicola			x	±	
	White-tailed antelope squirrel	A. leucurus leucurus	x				Northern Baja
		A. l. canfeldae	x				Central Baja
		A. l. extimus	x				Southern Baja
	Spotted ground squirrel	Spermophilus spilosoma canescens			±		
	Rock squirrel	Spermophilus variegatus			x	x	
	Baja California rock squirrel	S. atricapillus	x				
	Round-tailed ground squirrel	S. tereticaudus tereticaudus	±		x		California
		S. tereticaudus neglectus			x	x	Arizona, Sonora
Leporidae	Vizcaíno brush rabbit	Sylvilagus bachmani exiguus	x				Central Baja
	Cape brush rabbit	S. b. peninsularis	x				Southern Baja
	Cedros Island brush rabbit	S. b. cerrosensis		x			Cedros Island
	Pale brush rabbit	S. mansuetus		x			San José Island
	Holzner's Cottontail	S. floridanus holzneri			x		also central Mexico
	Desert Cottontail	S. auduboni arizonae			x		
	Goldman's desert cottontail	S. a. goldmani				x	
	Peninsula cottontail	S. a. confinis	x				
	Black rabbit	Lepus insularis		x			Espíritu Santo Island
	Black-tailed jackrabbit	L. californicus deserticola			x		
		L. c. eremicus				x	
		L. c. martirensis	x				Central Baja
		L. c. xanti	x				Southern Baja
	Magdalena Island rabbit						Magdalena and Margarita
		L. c. magdalenae		x			islands
	Carmen Island rabbit	L. c. sheldoni		x			Carmen Island
	Antelope jackrabbit	L. alleni alleni			x	±	
		L. alleni palitans			x		
	Tiburón Island rabbit	L. alleni tiburonensis		x			Tiburón Island
Vespertilionidae	Yuma myotis	Myotis yumanensis yumanensis			x		
	Lamb's myotis	M. y. lambi	x				San Ignacio region
	Cave myotis	M. velifer velifer				x	

FAMILY	COMMON NAME	SCIENTIFIC NAME	PENINSULA	ISLAND ENDEMIC	N. MAINLAND	S. MAINLAND	COMMENT
Vespertilionidae	Arizona Cave myotis	M. v. brevis			x		
	Cape Cave myotis	M. v. peninsularis	x				Cape region
	Arizona myotis	M. occultus			x	x	
	Long-eared myotis	M. evotis evotis	x		x		
		M. e. auriculus				x	
	Fringed myotis	M. thysanodes			x		Uninhabited buildings
	Long-legged myotis	M. volans volans	x				
		M. v. interior			x		
	California myotis	M. californicus californicus	x			x	
	Stephens' myotis	M. c. stephensi			x		
	Western pipistrelle	Pipistrellus hesperus hesperus	x		x		Northern Baja
		P. h. australis	x		x	x	Southern Baja
	Big brown bat	Eptesicus fuscus pallidus	x		x	x	Northern Baja
	Peninsula brown bat	E. f. peninsulae	x				Southern Baja
	Red bat	Lasiurus borealis teliotis	x		x	x	Roosts in vegetation
	Hoary bat	L. cinereus			x		Roosts in vegetation
	Southern yellow bat	Dasypterus ega xanthinus	x		x	x	
	Spotted bat	Euderma maculata			x		
	Lump-nosed bat	Plecotis townsendii pallescens			x		
	Fish-eating bat	Pizonyx vivesi	x			x	
	Pallid bat	Antrozous pallidus pallidus			x	x	
	Small pallid bat	A. p. minor	x				Southern Baja
	Pacific pallid bat	A. p. pacificus	x				Northern Baja
Molossidae	Mexican free-gailed bat	Tadarida brasiliensis mexicana	±		x	x	Northern Baja
	Pocketed free-tailed bat	T. femorosacca	x		x	x	
	Big free-tailed bat	T. molossa	x		x	x	
	Western mastiff bat	Eumops perotis californica			x		
	Underwood's mastiff bat	E. underwoodi sonoriensis			±	x	
Phyllostomidae	Peters' leaf-chinned bat	Mormoops megalophylla			±	x	Hot, humid caverns
	California leaf-nosed bat	Macrotis californicus	x		x	x	
	Mexican long-tongued bat	Choeronycteris mexicana	x		±	x	Pollinates cacti
	Long-nosed bat	Leptonycteris nivalis			±	x	Pollinates cacti
Soricidae	Desert shrew	Notiosorex crawfordi	x		x		
Didelphidae	Virginia opossum	Didelphis marsupialis virginiana			x		Weakly introduced
	Mexican opossum	D. m. californica				x	

BIRDS OF THE SONORAN DESERT

FAMILY	COMMON NAME	SCIENTIFIC NAME	PENINSULA	N. MAINLAND	S. MAINLAND
Gaviidae	Common loon	Gavia immer	x	x	x
	Arctic loon	G. arctica	x	±	x
	Red-throated loon	G. stelata	x	±	x
Podicipitidae	Horned grebe	Podiceps auritus	x	±	
	Least grebe	P. dominicus	x		x
	Eared grebe	P. caspicus	x	±	x
	Western grebe	Aechmophorus occidentalis	x	x	x
	Pied-billed grebe	Podilymbus podiceps	x	x	x
Diomedeidae	Black-footed albatross	Diomeda nigripes	x		
Procellariidae	Wedge-tailed shearwater	Puffinus pacificus	x		x
	Pink-footed shearwater	P. creatopus	x		
	Sooty shearwater	P. griseus	x		x
	Manx shearwater	P. puffinus	x		x
Hydrobatidae	Least petrel	Halocyptena microsoma	x		x
	Leach's petrel	Oceanodroma leucorhoa	x		
	Ashy petrel	O. homochroa	x		
	Galapagos petrel	O. tethys	x		x
	Black petrel	Loomelania melania	x		x
Phaëthontidae	Red-billed tropicbird	Phaëthon aethereus	x	rare	x
Pelecanidae	White pelican	Pelecanus erythrohynchos	x	x	x
	Brown pelican	P. occidentalis	x	x	x
Sulidae	Blue-footed booby	Sula nebouxii	x	rare	x
	Blue-faced booby	S. dactylatra	x		
	Red-footed booby	S. sula			x
	Brown booby	S. leucogaster	x	rare	x
Phalacrocoracidae	Double-crested cormorant	Phalacrocorax auritus	x	x	x
	Olivaceous cormorant	Ph. olivaceus		rare	x
	Brandt's cormorant	Ph. penicillatus	x		x
	Pelagic cormorant	Ph. pelagicus	x		
Fregatidae	Magnificent frigatebird	Fregata magnificens	x	rare	x

FAMILY	COMMON NAME	SCIENTIFIC NAME	PENINSULA	N. MAINLAND	S. MAINLAND
Aredeidae	Great blue heron	Ardea herodias	x	x	x
	Green heron	Butorides virescens	x	x	x
	Little blue heron	Florida caerulea	x	rare	x
	Louisiana heron	Hydranassa tricolor	x	rare	x
	Black-crowned night heron	Nycticorax nycticorax	x	x	x
	Yellow-crowned night heron	N. violacea	x	rare	x
	Reddish egret	Dichromanassa rufescens	x	±	x
	Common egret	Casmerodius albus	x	x	x
	Snowy egret	Leucophoyx thula	x	x	x
	Cattle egret	Bubulcus ibis			x
	Tiger bittern	Heterochus mexicanus			x
	Least bittern	Ixobrychus exilis	x	x	x
	American bittern	Botaurus lentiginosus	x	x	x
Ciconiidae	Wood stork	Mycteria americana	x	±	x
Threskiornithidae	White-faced ibis	Plegadis chihi	x	±	x
	Roseate spoonbill	Ajaia ajaia	x	±	x
Anatidae	Whistling swan	Olor columbianus	x	±	
	Canada goose	Branta canadensis	x	x	
	Black brant	B. nigricans	x		
	White-fronted goose	Anser albifrons	x	x	x
	Snow goose	Chen hyperborea	x	x	x
	Black-bellied tree duck	Dendrocygna autumnalis		x	x
	Fulvous tree duck	D. bicolor	x	rare	x
	Mallard	Anas platyrhynchos	x	x	x
	Gadwall	A. strepera	x	x	x
	Pintail	A. acuta	x	x	x
	Green-winged teal	A. carolinensis	x	x	x
	Blue-winged teal	A. discors	x	x	x
	Cinnamon teal	A. cyanoptera	x	x	x
	American widgeon	Mareca americana	x	x	x
	Shoveler	Spatula clypeata	x	x	x
	Wood duck	Aix sponsa		rare	x
	Redhead	Aythya americana	x	x	x
	Ring-necked duck	A. collaris	x	±	x
	Canvasback	A. valisineria	x	±	x
	Greater scaup	A. marila	x	±	x
	Lesser scaup	A. affinis	x	x	x
	Common goldeneye	Bucephala clangula	x	rare	x
	Bufflehead	B. albeola	x	x	x
	White-winged scoter	Melanitta deglandi	x		
	Surf scoter	M. perspicillata	x	rare	x
	Common scoter	Oidemia nigra	x		
	Ruddy duck	Oxyura jamaicensis	x	x	x
	Masked duck	O. dominica			x
	Hooded merganser	Lophodytes cucullatus	x	x	
	Common merganser	Mergus merganser	x	x	x
	Red-breasted merganser	M. serrator	x	x	x
Cathartidae	Turkey vulture	Cathartes aura	x	x	x
	Black vulture	Coragyps atratus		x	x
Accipitridae	Cooper's hawk	Accipiter cooperi	x	x	x
	White-tailed hawk	Buteo albicaudatus		rare	x
	Ferruginous hawk	B. regalis	x	±	
	Red-tailed hawk	B. jamaicensis	x	x	x
	Red-shouldered hawk	B. lineatus	x		x
	Swainson's hawk	B. swainsoni	x	x	x
	Broad-winged hawk	B. platypterus			x

FAMILY	COMMON NAME	SCIENTIFIC NAME	PENINSULA	N. MAINLAND	S. MAINLAND
Accipitridae	Gray hawk	B. nitidus		x	x
	Harris's hawk	Parabuteo unicinctus	x	x	x
	Great black hawk	Hypomorphnus urubitinga			x
	Common black hawk	Buteogallus anthracinus		x	x
	Marsh hawk	Circus cyaneus	x	x	x
	Golden eagle	Aquila chrysaëtos	x	x	x
	Bald eagle	Haliaëtus leucocephalus	x	±	
Pandionidae	Ospreay	Pandion haliaëtus	x	x	x
Falconidae	Laughing falcon	Herpetotheres cachinnans			x
	Crested caracara	Polyborus cheriway	x	±	x
	Prairie falcon	Falco mexicanus	x	x	x
	Peregrine falcon	F. peregrinus	x	x	x
	Bat falcon	F. albigularis			x
	Merlin	F. columbarius	x	x	x
	American kestrel	F. sparverius	x	x	x
Phasianidae	California quail	Lophortyx californica	x		
	Gambel's quail	L. gambelii	x	x	x
	Elegant quail	L. douglasii			±
	Bobwhite	Colinus virginianus			±
Guidae	Sandhill crane	Grus canadensis	x	x	x
Rallidae	Clapper rail	Rallus longirostris	x	x	x
	Virginia rail	R. limicola	x	x	x
	Sora rail	Porzana carolina	x	x	x
	Black rail	Laterallus jamaicensis	x		
	Purple gallinule	Porphyrula martinica		±	x
	Common gallinule	Gallinula chloropus	±	x	±
	American coot	Fulica americana	x	x	x
Haematopodidae	Oystercatcher	Haematopus ostralegus	x		x
Charadriidae	Black-bellied plover	Squatarola squatarola	x	rare	x
	Semipalmated plover	Charadrius semipalmatus	x	±	x
	Snowy plover	Ch. alexandrinus	x	x	x
	Wilsons plover	Ch. wilsonia	x		x
	Killdeer	Ch. vociferus	x	x	x
	Mountain plover	Eupoda montana	x	±	x
Scolopacidae	Whimbrel	Numenius phaeopus	x	x	x
	Long-billed curlew	N. americanus	x	x	x
	Marbled godwit	Limosa fedosa	x	x	x
	Lesser yellowlegs	Totanus flaviceps	x	x	x
	Greater yellowlegs	Tringa melanoleuca	x	x	x
	Solitary sandpiper	T. solitaria	x	x	x
	Spotted sandpiper	Actinitis macularia	x	x	x
	Willet	Catoptrophorus semipalmatus	x	x	x
	Wandering tattler	Heteroscelus incanus	x		x
	Surfbird	Aphriza virgata	x		x
	Ruddy turnstone	Arenaria interpres	x	x	x
	Black turnstone	A. melanocephala	x		x
	Short-billed dowitcher	Limnodromus griseus	x	x	x
	Long-billed dowitcher	L. scolopaceus	x	x	x
	Common snipe	Capella gallinago	x	x	x
	Knot	Calidris canutus	x	x	x
	Sanderling	Crocethia alba	x	x	x
	Western sandpiper	Ereunetes mauri	x	x	x
	Least sandpiper	Erolia minutilla	x	x	x
	Baird's sandpiper	E. bairdi	x	x	
	Pectoral sandpiper	E. melanotos	x	x	x
	Dunlin	E. alpina	x	x	x

FAMILY	COMMON NAME	SCIENTIFIC NAME	PENINSULA	N. MAINLAND	S. MAINLAND
Scolopacidae	Stilt sandpiper	Micropalama himantopus		x	x
Recurvirostridae	Black-necked stilt	Himantopus mexicanus	x	x	x
	American avocet	Recurvirostris americana	x	x	x
Phalaropidae	Red phalarope	Phalaropus fulicarius	x	x	x
	Wilson's phalarope	Steganopus tricolor	x	x	x
	Northern phalarope	Lobipes lobatus	x	x	x
Stercorariidae	Parasitic jaeger	Stercorarius parasiticus	x	rare	rare
	Pomarine jaeger	S. pomarinus	x	rare	rare
Laridae	Heermann's gull	Larus heermanni	x		x
	Ring-billed gull	L. delawarensis	x	x	x
	Herring gull	L. argentatus	x	x	x
	California gull	L. californicus	x	x	x
	Western gull	L. occidentalis	x	x	x
	Glaucous-winged gull	L. glaucescens	x	x	x
	Laughing gull	L. atricilla	x	x	x
	Franklin's gull	L. pipixcan		x	x
	Bonaparte's gull	L. philadelphia	x	x	x
	Kittiwake	Rissa tridactyla	x		
	Sabine's gull	Xema sabini	x	x	
	Black tern	Chidonia niger	x	x	x
	Gull-billed tern	Gelochelidon nilotica	x	x	x
	Caspian tern	Hydroprogne caspia	x	x	x
	Common tern	Sterna hirundo	x	x	x
	Forster's tern	S. forsteri	x	x	x
	Sooty tern	S. fuscata	x		x
	Least tern	S. albifrons	x	x	x
	Royal tern	Thalasseus maximus	x		x
	Elegant tern	Th. elegans	x		x
Rynchopidae	Black skimmer	Rynchops nigra			x
Alcidae	Xantus's murrelet	Endomychura hypoleuca	x		x
	Cassin's auklet	Ptychoramphus aleuticus	x		
	Rhinoceros auklet	Cerorhinca monocerata	x		
Columbidae	Rock dove	Columba liria	x	x	x
	Red-billed pigeon	C. flavirostris			x
	Mourning dove	Zenaidura macroura	x	x	x
	White-winged dove	Z. asiatica	x	x	x
	Inca dove	Scardafella inca		x	x
	White-fronted dove	Leptotila verreauxi			x
	Ruddy quail dove	Geotrygon montana			x
Psittacidae	Green parakeet	Aratinga holochlora			x
	Mexican parrotlet	Forpus cyanopygialis			x
	White-fronted parrot	Amazona albifrons			x
	Pacific parrot	A. finschi			x
Cuculidae	Black-billed cuckoo	Coccyzus erythropthalmus			x
	Yellow-billed cuckoo	C. americanus	x	x	x
	Squirrel cuckoo	Piaya cayana			x
	Groove-billed ani	Crotophaga sulcirostris		rare	x
	Greater roadrunner	Geococcyx californicus	x	x	x
	Lesser roadrunner	G. velox			x
Strigidae	Common screech-owl	Otus asio	x	x	x
	Guatemalan screech-owl	O. guatemalae			±
	Vinaceous screech-owl	O. vinaceus			±
	Great horned owl	Bubo virginianus		x	x
	Ferruginous pygmy-owl	Glaucidium brasilianum		x	x
	Elf owl	Micranthene whitneyi	x	x	x
	Burrowing owl	Speotyto cunicularia	x	x	x

FAMILY	COMMON NAME	SCIENTIFIC NAME	PENINSULA	N. MAINLAND	S. MAINLAND
Strigidae	Wood owl	Ciccaba virgata			x
	Short-eared owl	Asio flammeus	x		x
Caprimulgidae	Lesser nighthawk	Chordeiles acutipennis	x	x	x
	Tucuchillo	Caprimulgus ridgwayi		rare	x
	Common poor-will	Phalaenoptilus nuttallii	x	x	x
Apodidae	Vaux's swift	Chaetura vauxi	x	x	x
	White-throated swift	Aeronautes saxatilis	x	x	x
Trochilidae	Broad-billed hummingbird	Cyananthus latirostris		±	x
	Xantus's hummingbird	Hylocharis xantusii	x		
	Violet-crowned hummingbird	Amazilia violiceps		±	x
	Plain-capped starthroat	Heliomaster constantii			x
	Ruby-throated hummingbird	Archilochus colubris			x
	Black-chinned hummingbird	A. alexandri	x	x	x
	Anna's hummingbird	Calypte anna	x	x	x
	Costa's hummingbird	C. costae	x	x	x
	Allen's hummingbird	Selasphorus sasin	x	x	x
Trogonidae	Elegant trogon	Trogon elegans		±	x
Alcedinidae	Belted kingfisher	Ceryle alcyon	x	x	x
	Green kingfisher	Chloroceryle americana		±	x
Momotidae	Russet-crowned motmot	Momotus mexicanus			x
Picidae	Gilded flicker	Colaptes chrysoides	x	x	x
	Lewis's Woodpecker	Asyndesmus lewis	x	x	x
	Gila woodpecker	Centurus uropygialis	x	x	x
	Yellow-bellied sapsucker	Sphyrapicus varius	x	x	x
	Nuttall's woodpecker	Dendrocopus nuttallii	x		
	Ladder-backed woodpecker	D. scalaris	x	x	x
Cotingidae	Rose-throated cotinga	Platypsaris aglaiae		rare	x
Tyrannidae	Eastern phoebe	Sayornis phoebe		x	x
	Black phoebe	S. nigricans	x	x	x
	Say's phoebe	S. saya	x	x	x
	Vermilion flycatcher	Pyrocephalus rubinus	x	x	x
	Cassin's kingbird	Tyrannus vociferans	x	x	x
	Western kingbird	T. verticalis	x	x	x
	Tropical kingbird	T. melancholicus		x	x
	Thick-billed kingbird	T. crassirostris		±	x
	Sulphur-bellied flycatcher	Myiodynastes luteiventris		x	x
	Social flycatcher	Myiozetes similis			x
	Great kiskadee	Pitangus sulphuratus			x
	Ash-throated flycatcher	Myiarchus cinerascens	x	x	x
	Nutting's flycatcher	M. nuttingi		rare	x
	Wied's flycatcher	M. tyrannulus		x	x
	Olivaceous flycatcher	M. tuberculifer		x	x
	Traill's flycatcher	Empidonax traillii	x	x	x
	Least flycatcher	E. minimus		x	x
	Dusky flycatcher	E. oberholseri	x	x	x
	Gray flycatcher	E. wrightii	x	x	x
	Beardless flycatcher	Camptostoma imberbe		x	x
Alaudidae	Horned lark	Eremophila alpestris	x	x	x
Hirundinidae	Purple martin	Progne subis	x	x	x
	Cliff swallow	Petrochelidon pyrrhonota	x	x	x
	Barn swallow	Hirundo rustica	x	x	x
	Rough-winged swallow	Stelgidopteryx ruficollis	x	x	x
	Bank swallow	Riparia riparia	x	x	x
	Tree swallow	Iridoprocne bicolor	x	x	x
	Mangrove swallow	I. albilinea			x
	Violet-green swallow	Tachycineta thalassina	±		±

FAMILY	COMMON NAME	SCIENTIFIC NAME	PENINSULA	N. MAINLAND	S. MAINLAND
Corvidae	Common raven	Corvus corax	x	x	x
	White-necked raven	C. cryptoleucus		x	x
	American crow	C. brachyrhynos	±	±	±
	Sinaloa crow	C. sinaloae			x
	Magpie jay	Calocitta formosa			x
	Beechey's jay	Cissilopha beecheii			x
	Scrub jay	Aphelocoma coerulescens	x	x	
Paridae	Verdin	Auriparus flaviceps	x	x	x
Chamaeidae	Wrentit	Chamaea fasciata	±		
	Marsh wren	Telmatodytes palustris	x	x	x
	Cactus wren	Campylorhynchus brunneicapillus	x	x	x
	Sinaloa wren	Thryothorus sinaloa			x
	Happy wren	Th. felix			x
	Bewick's wren	Thryomanes bewicki	x	x	x
	Rock wren	Salpinctes obsoletus	x	x	x
	Canyon wren	Catherpes mexicanus	x	x	x
Mimidae	Gray thrasher	Toxostoma cinereum	x		
	Bendire's thrasher	T. bendirei	x	x	x
	Curve-billed thrasher	T. curvirostre	x	x	x
	California thrasher	T. redivivum	x		
	LeConte's thrasher	T. lecontei	x	x	x
	Crissal thrasher	T. dorsale	x	x	x
	Blue mockingbird	Melanotis caerulescens			x
	Northern mockingbird	Mimus polyglottos	x	x	x
	Sage thrasher	Oreoscoptes montanus	±	±	±
Turdidae	Rufous-backed robin	Turdus rufopalliatus		rare	x
	American robin	T. migratorius	±	±	±
	Varied thrush	Ixoreus naevius	x	±	
	Hermit thrush	Hylocichla guttata	x	x	x
Sylviidae	Blue-gray gnatcatcher	Polioptila caerulea	x	x	x
	Black-capped gnatcatcher	P. nigriceps			x
	Black-tailed gnatcatcher	P. melanura	x	x	x
	Ruby-crowned kinglet	Regulus calendula	x	x	x
Motacillidae	Water pipit	Anthus spinoletta	x	x	x
	Sprague's pipit	A. spragueii		x	x
Bombycillidae	Cedar waxwing	Bombycilla cedrorum	x	x	x
Ptilogonatidae	Phainopepla	Phainopepla nitens	x	x	x
Laniidae	Loggerhead shrike	Lanius ludovicianus	x	x	x
Sturnidae	Starling	Sturnus vulgaris	x	x	x
Vireonidae	Mangrove vireo	Vireo pallens			x
	Golden vireo	V. hypochryseus			x
	Bell's vireo	V. bellii	x	x	x
	Red-eyed vireo	V. olivaceus		x	x
Parulidae	Orange-crowned warbler	Vermivora celata	x	x	x
	Nashville warbler	V. ruficapilla	x	x	x
	Lucy's warbler	V. luciae	x	x	x
	Yellow warbler	Dendroica petechia	x	x	x
	Mangrove warbler	D. erithachorides	x		x
	Magnolia warbler	D. magnolia			x
	Myrtle warbler	D. coronata			x
	Townsend's warbler	D. townsendi	x	x	x
	Louisiana waterthrush	Seiurus motacilla			x
	Common yellowthroat	Geothlypis trichas	x	x	x
	Belding's yellowthroat	G. beldingi	x		
	Yellow-breasted chat	Icteria virens	x	x	x

FAMILY	COMMON NAME	SCIENTIFIC NAME	PENINSULA	N. MAINLAND	S. MAINLAND
Parulidae	Wilson's warbler	Wilsonia pusilla	x	x	x
	Rufous-capped warbler	Basileuterus rufifrons			x
Ploceidae	House sparrow	Passer domesticus	x	x	x
Icteridae	Red-eyed cowbird	Tangarius aeneus		x	x
	Brown-headed cowbird	Molothrus ater	x	x	x
	Boat-tailed grackle	Cassidix mexicanus		x	x
	Brewer's blackbird	Euphagus cyanocephalus	x	x	x
	Bullock's oriole	Icterus bullocki	x	x	x
	Wagler's oriole	I. wagleri			x
	Scott's oriole	I. parisorum	x	x	x
	Hooded oriole	I. cucullatus	x	x	x
	Streak-backed oriole	I. pustulatus			x
	Tricolored blackbird	Agelaius tricolor	x		
	Red-winged blackbird	A. phoeniceus	x	x	x
	Yellow-headed blackbird	Xanthocephalus xanthocephalus	x	x	x
	Eastern meadowlark	Sturnella magna	x	x	x
Thraupidae	Scrub euphonia	Euphonia affinis			x
	Summer tanager	Piranga rubra	x	x	x
Fringillidae	Cardinal	Richmondena cardinalis	x	x	x
	Pyrrhuloxia	Pyrrhuloxia sinuata	x	x	x
	Yellow grosbeak	Pheuticus chrysopeplus			x
	Blue grosbeak	Passerina caerulea	x	x	x
	Indigo bunting	P. cyanea		±	x
	Lazuli bunting	P. amoena	x	±	x
	Varied thrush	P. versicolor	x	±	x
	Dickcissel	Spiza americana		±	x
	House finch	Carpodacus mexicanus	x	x	x
	Blue-black grassquit	Volatinia jacarina			x
	Lesser goldfinch	Spinus psaltria	x	x	x
	Lawrence's goldfinch	S. lawrencei	x	x	x
	Red-tailed towhee	Chlorura chlorura	x	x	x
	Spotted towhee	Pipilo maculatus	x	x	
	Brown towhee	P. fuscus	x	x	x
	Abert's towhee	P. aberti	x	x	x
	Rusty-crowned sparrow	Melozone kieneri			x
	Savanna sparrow	Passerculus sandwichensis	x	x	x
	Lark bunting	Calamospiza melanocorys	x	x	x
	Vesper sparrow	Pooecetes gramineus	x	x	x
	Rufous-winged sparrow	Aimophila carpalis		x	x
	Rufous-crowned sparrow	A. ruficeps	x	x	
	Botteri's sparrow	A. botteri		±	x
	Black-throated sparrow	Amphispiza bilineata	x	x	x
	Sage sparrow	A. belli	x	x	x
	Lark sparrow	Chondestes grammacus	x	x	x
	Chipping sparrow	Spizella passerina	x	x	x
	Black-chinned sparrow	S. atrogularis	x	x	x
	Clay-colored sparrow	S. pallida	x	±	x
	Brewer's sparrow	S. breweri	x	x	x
	Junco	Junco hyemalis	x	x	x
	White-crowned sparrow	Zonotrichia leucophrys	x	x	x
	Golden-crowned sparrow	Z. atricapilla	x	rare	
	White-throated sparrow	Z. albicollis		rare	x
	Fox sparrow	Passerella iliaca	x	x	
	Lincoln's sparrow	Melospiza lincolni	x	x	x
	Swamp sparrow	M. georgiana		x	x
	Song sparrow	M. melodia	x	x	x

REPTILES AND AMPHIBIANS OF THE SONORAN DESERT

FAMILY	COMMON NAME	SCIENTIFIC NAME	PENINSULA	ISLAND ENDEMIC	N. MAINLAND	S. MAINLAND	COMMENT
Helodermatidae	Reticulate gila monster	Heloderma suspectum suspectum			x		
	Banded gila monster	H. s. cinctum			x	x	
Gekkonidae	Banded gecko	Coleonyx variegatus variegatus			x		
	Bogert's banded gecko	C. v. bogerti			x		Southern Arizona
	Abbott's banded gecko	C. v. abbotti	x		x		
	San Marcos banded gecko						San Marcos and Santa Inés islands
		C. v. slevini		x			
	Sonoran banded gecko	C. v. sonoriensis				x	Also Tiburón Island
	Peninsular banded gecko	C. v. peninsularis	x		x		
	Leaf-toed gecko	Phyllodactylus xanti xanti	x				
	Sloan's leaf-toed gecko	P. x. sloani	x				Vizcaíno Division
		P. unctus	x				Southern Baja
		P. tuberculosus	x				
	Sonoran leaf-toed gecko	P. homolepidurus				x	
	San Pedro Nolasco leaf-toed gecko	P. h. nolascoensis		x			San Pedro Nolasco Island
	Switak's barefoot gecko	Anarbylus switaki	x				
Iguanidae	False iguana	Ctenosaura hemilopha	x			x	
	San Esteban iguana	C. h. conspicua		x			San Esteban Island
	Cerralvo iguana	C. h. insulana		x			Cerralo Island
	Desert iguana	Dipsosaurus dorsalis dorsalis	x		x	x	
	Catalina iguana	D. d. catalinensis		x			Santa Catalina Island
	Western chuckwalla	Sauromalus obesus obesus			x		Lower Colorado-Gila Division
	Arizona chuckwalla	S. o. tumidus			x		
	Sonoran chuckwalla	S. o. townsendi				x	Also Tiburón Island
	Peninsula chuckwalla	S. australis	x				
	Espíritu santo chuckwalla	S. ater ater		x			On 5 islands

FAMILY	COMMON NAME	SCIENTIFIC NAME	PENINSULA	ISLAND ENDEMIC	N. MAINLAND	S. MAINLAND	COMMENT
Iguanidae	Santa Catalina chuckwalla	S. a. klauberi		x			Santa Catalina Island
	San Marcos chuckwalla	S. a. shawi		x			San Marcos Island
	Angel Island chuckwalla	S. hispidus		x			On 7 islands
	Carmen Island chuckwalla	S. slevini		x			On 3 islands
	San Esteban chuckwalla	S. varius		x			On 3 islands
	Southwestern earless lizard	Holbrookia texana scitula			x		
	Lesser earless lizard	H. maculata			x	x	
	Zebra-tailed lizard	Callisaurus draconoides ventralis			x		Eastern part
		C. d. gabbii			x		Western part
		C. d. draconoides				x	
		C. d. inusitatus				x	
	Angel Island zebra-tail	C. d. splendidus		x			Angel de la Guarda Island
	Carmen Island zebra-tail	C. d. Carmenensis	x				Also 4 islands
		C. d. rhodostictus	x				Also Méjia Island
		C. d. crinitus	x				
	Coachella Valley fringe-toed lizard	Uma inornata			x		Coachella Valley only
	Colorado Desert fringe-toed lizard	U. notata notata			x		Western part
	Cowles's fringe-toed lizard	U. n. rufopunctata			x		Eastern part
	Long-nosed leopard lizard	Crotaphytus wislizenii wislizenii	x		x		Also Tiburón Island
		C. w. copeii	x				Vizcaíno Division
	Collared lizard	C. collaris	x		x		
	Angel collared lizard	C. c. insularis		x			Angel de la Guarda Island
	Banded rock lizard	Petrosaurus mearnsi mearnsi	x				
	Angel Island rock lizard						Méjia and Angel de la Guarda islands
		P. m. sleveni		x			
		P. thalassinus	x				Also Danzante and Espíritu Santo islands
		P. repens	x				Vizcaíno Division
	Desert spiny lizard	Sceloporus magister magister			x	x	Also Tiburón Island
	Yellow-backed spiny lizard	S. m. uniformis			x		Lower Colorado-Gila Division
		S. m. rufidorsum	x				Also Cedros and Coronados islands
	Santa Catalina spiny lizard	S. m. lineatulus		x			Santa Catalina Island
	Sonora spiny lizard						Also Tiburón, San Pedro
		S. clarki			x	x	Nolasco islands
	Granite spiny lizard	S. orcutti lickii	x				
	Desert side-blotched lizard	Uta stansburiana stejnegeri	x		x	x	
		U. s. elegans	x				Southern Baja
		U. s. hesperis	x				Northwest Baja
		U. mearnsi			x		
		U. thalassina	x				Cape region
	San Pedro Nolasco uta	U. nolascensis		x			San Pedro Nolasco Island
	San Pedro Martín uta	U. palmeri		x			San Pedro Martín Island
	San Benito uta	U. stellata		x			San Benito Island
	Cedros Island uta						Cedros and Natividad islands
		U. concinna		x			
	San Martín uta	U. martinensis		x			San Martín Island
	Santa Catalina uta	U. squamata		x			Santa Catalina Island

FAMILY	COMMON NAME	SCIENTIFIC NAME	PENINSULA	ISLAND ENDEMIC	N. MAINLAND	S. MAINLAND	COMMENT
Iguanidae	Carmen Island uta	U. mannophorus		x			On three islands
	Western brush lizard	Urosaurus graciosus graciosus			x		Lower Colorado-Gila Division
	Arizona brush lizard	U. g. shannoni			x		
	Tree lizard	U. ornatus			x	x	
		U. o. symmetricus			x		Lower Colorado-Gila Division
		U. o. schottii				x	Also Tiburón Island
	Cape Urosaurus	U. nigricaudus	x				Cape region
	Small-scaled lizard	U. n. microscutatus	x				North of Cape region
	Flat-tailed horned lizard	Phrynosoma M'calli			x		Sand dunes
	Mountain short-horned lizard	P. douglassi hernandesi			x		Arizona
	Southern Desert horned lizard	P. platyrhinos calidiarum			x		
		P. p. goodei				x	
		P. coronatum coronatum	x				Cape region
		P. c. jamesi	x				Central Baja
		P. c. frontale	x				Northwest Baja, Cedros Island
		P. c. blainvillii	x				Northeast Baja
	Cedros horned lizard	P. cerrosense		x			Cedros Island
	Regal horned lizard	P. solare			x	x	Also Tiburón Island
	Santa Cruz sator	Sator angustus		x			Santa Cruz and San Diego islands
	Cerralvo sator	S. grandaevus		x			Cerralvo Island
Xantusiidae	Desert night lizard	Xantusia vigilis vigilis	x		x		
	Vizcaíno night lizard	X. v. wigginsi	x				Vizcaíno Division
	Cape night lizard	X. gilberti	x				Cape region
	Arizona night lizard	X. arizonae			x		
Scincidae	Great Plains skink	Eumeces obsoletus			x		Arizona
Teiidae	Great Basin whiptail	Cnemidophorus tigris	x		x		Lower Colorado-Gila Division
	Southern whiptail	C. t. gracilis			x		Also Tiburón Island
	Coastal whiptail	C. t. multiscutulatus	x				
	San Pedro Nolasco whiptail	C. t. bacatus		x			San Pedro Nolasco Island
		C. t. rubidus	x				Also 3 islands
	San Lorenzo whiptail	C. t. canus		x			On 3 islands
	Santa Catalina whiptail	C. t. catalinensis		x			Santa Catalina Island
	San Francisco whiptail	C. t. celeripes		x			San José and San Francisco islands
	Angel Island whiptail	C. t. dickersonae		x			Partida Norte and Angel de la Guarda islands
	San Esteban whiptail	C. t. estebanensis		x			San Esteban Island
	San Pedro Martír whiptail	C. t. martyris		x			San Pedro Martír Island
	Sonoran whiptail	C. t. aethiops				x	Also Tiburón and San Esteban islands
	Orange-throated whiptail	C. hyperythrus hyperythrus	x				Southern Baja
	Vizcaíno whiptail	C. h. schmidti	x				Vizcaíno Division
	San José whiptail	C. h. schmidti		x			San José Islands
	Monserrate whiptail	C. h. pictus		x			Monserrate Island
	Cedros whiptail	C. h. beldingi	x				Also Cedros Island
	Carmen whiptail	C. h. caeruleus		x			Carmen Island

FAMILY	COMMON NAME	SCIENTIFIC NAME	PENINSULA	ISLAND ENDEMIC	N. MAINLAND	S. MAINLAND	COMMENT
Teiidae	Chihuahuan whiptail	C. exsanguis			x		Arizona
	Baja California striped whiptail	C. labialis	x				
	Red-backed whiptail	C. burti xanthonotus			x		
		C. maximus	x				Southern Baja
		C. gadovi				x	
	Cerralvo whiptail	C. ceralbensis		x			Cerralvo Island
Anguidae	Arizona alligator lizard	Gerhonotus kingii			x		Mostly east of desert
	Southern alligator lizard	G. multicarinatus	x				Bostic, 1971
	Coronados alligator lizard	G. m. nana		x			Coronados Island
	San Lucan alligator lizard	G. paucicarinatus	x				Cape region
	Cedros alligator lizard	G. cedrosensis		x			Cedros Island
Anniellidae	Silvery legless lizard	Aniella pulchra pulchra	x				Northern baja
	Gerónimo legless lizard	A. geronimensis		x			Gerónimo Island
Amphisbaenidae	Two-legged worm lizard	Bipes biporus	x				Southern Baja
Leptotyphlopidae	Western worm snake	Leptotyphlops humilis	x		x	x	
	Slevin's worm snake	L. h. sleveni	x				Also on Carmen and Cerralvo Islands
	Colorado Desert worm snake	L. h. cahuilae			x		Lower Colorado-Gila Division
		L. h. levitoni		x			Santa Catalina Island
		L. h. lindsayi		x			Santa Catalina Island
Boidae	Desert rosy boa	Lichanura trivirgata gracia	x		x		Also on Angel de la Guarda Island
	Mexican rosy boa	L. t. trivirgata	x			x	Also on Tiburón Island
	Common boa	Constrictor constrictor imperator				x	
Colubridae	Regal king snake	Diadophis punctatus regalis			x		
	Western leaf-nosed snake	Phyllorhynchus decurtatus perkinsi	x		x		Also on Angel de la Guarda Island
	Clouded leaf-nosed snake	Ph. d. nubilis			x		Arizona, Sonora
	Sonoran leaf-nosed snake	Ph. d. ssp.				x	
	Baja California leaf-nosed snake	Ph. d. decurtatus	x				Vizcaíno Division
	Monserrate leaf-nosed snake	Ph. d. arenicola		x			Monserrate Island
	Pima leaf-nosed snake	Ph. browni browni			x		
	Maricopa leaf-nosed snake	Ph. b. lucidus			x		
	Red racer	Masticophis flagellum piceus	x		x	x	Also on 8 islands
	Sonora whipsnake	M. bilineatus bilineatus			x	x	Also on Tiburón Island
	Ajo Mountain whipsnake	M. b. lineolatus			x		Organ Pipe, N.M.
	San Esteban whipsnake	M. b. slevini		x			San Esteban Island
	California striped whipsnake	M. lateralis	x				
	Espíritu Santo whipsnake	M. l. barbouri		x			Espíritu Santo Island
	Cape whipsnake	M. aurigulus	x				Cape region
	Desert patch-nosed snake	Salvadora hexalepis hexalepis	±		x		
	Big Bend patch-nosed snake	S. h. deserticola				x	
	Baja California patch-nosed snake	S. h. klauberi	x				Also on San José Island

FAMILY	COMMON NAME	SCIENTIFIC NAME	PENINSULA	ISLAND ENDEMIC	N. MAINLAND	S. MAINLAND	COMMENT
Colubridae	Green rat snake	Elaphe triaspis			x	x	
	Baja California rat snake	E. rosaliae	x				
	Arizona glossy snake	Arizona elegans noctivaga			x	x	
	Desert glossy snake						Lower Colorado-Gila Division
		A. e. eburnata			x		
	Sonoran gopher snake	Pituophis melanoleucus affinis			x	x	
		P. m. annectans	x				
		P. m. bimaris	x				
		P. m. vetebralis	x				
	Yuma kingsnake	Lempropeltis getulus yumanensis			x		
	Black kingsnake	L. g. nigritus				x	
		L. g. ssp.	x				Southern Baja
		L. g. californiae	x				Northern Baja
	Santa Catalina kingsnake	L. g. catalinensis		x			Santa Catalina Island
	Western long-nosed snake	Rhinocheilus lecontei lecontei			x		
	Sonoran long-nosed snake	R. l. antonii				x	
	Water snake	Natrix valida	x				Cape region
	Checkered garter snake	Thamnophis marcianus			x		
	Black-necked garter snake	Th. cyrtopsis			±	x	
	Mexican garter snake	Th. eques			±		
		Th. digueti	x				Southern Baja
	Western terrestrial garter snake	Th. elegans	x				
	Western ground snake	Sonora semiannulata	x		x	x	
		S. mosaueri	x				
	Colorado Desert shovel-nosed snake	Chionactis occipitalis annulata			x		Lower Colorado-Gila Division
	Tucson shovel-nosed snake	Ch. o. klauberi			x		Arizona Upland Division
	Sonora shovel-nosed snake						Northern and central Sonora
		Ch. palarostris palarostris					
		Ch. p. organica			x		Organ Pipe, N.M.
	Banded sand snake	Chilomeniscus cinctus	x		x	x	
	Cape sand snake	Ch. stramineus stramineus	x				Cape region
		Ch. s. esterensis	x				Estero Salina
	Cerralvo sand snake	Ch. savagei		x			Cerralvo Island
	Espíritu Santo sand snake	Ch. punctatissimus		x			Espíritu Santo Island
	Desert hook-nosed snake	Ficimia quadrangularis				x	
	Western black-headed snake	Tantilla planiceps	x		x		
	Mexican vine snake	Oxybelis aeneus auratus				x	
	Sonora lyre snake	Trimorphodon lambda			x	x	
		T. lyrophanes	x				
	Spotted night snake	Hypsiglena torquata ochrorhyncha			x	x	Also 3 islands
	San Diego night snake	H. t. klauberi	x				Also Coronados Island
	Desert night snake	H. t. deserticola			x		Southern California
		H. t. venusta	x				Gulf coast
	Santa Catalina night snake	H. t. catalinae		x			Santa Catalina Island
	Partida night snake	H. t. gularis		x			Partida Island
	Cape night snake	H. t. ochrorhyncha	x				Cape region
	Tortuga night snake	H. t. tortuguensis		x			Tortuga Island
	San Martín night snake	H. t. martinensis		x			San Martín Island

FAMILY	COMMON NAME	SCIENTIFIC NAME	PENINSULA	ISLAND ENDEMIC	N. MAINLAND	S. MAINLAND	COMMENT
Colubridae	Slevin's Eridiphas	Eridiphas slevini	x				Also Cerralvo Island
Elapidae	Arizona coral snake	Micruroides euryxanthus euryxanthus			x		Also Tiburón Island
	Sonoran coral snake	M. e. ssp.				x	
Hydrophidae	Pelagic sea snake	Pelamis platurus	x				
Crotalidae	Western diamondback	Crotalus atrox			x	x	Also Santa Cruz and San Pedro Mártir islands
	Rattleless rattlesnake	C. catalinensis		x			Santa Catalina Island
	Sonora sidewinder	C. cerastes cercobombus		x			Arizona, Sonora
	Colorado Desert sidewinder	C. c. laterorepens			x		Lower Colorado River region
	Baja California rattlesnake	C. enyo enyo	x				Vizcaíno Division
	Cerralvo rattlesnake	C. e. cerralvensis		x			Cerralvo Island
		C. e. furvus	x				Northern Baja
	Cedros rattlesnake	C. exsul		x			Cedros Island
	San Lucan speckled rattlesnake	C. mitchellii mitchellii	x				Southern Baja
	Southwestern speckled rattlesnake	C. m. pyrrhus			x		
	Angel Island speckled rattlesnake	C. m. angelensis		x			Angel de la Guarda Island
	El Muerto speckled rattlesnake	C. m. muertensis		x			El Muerto Island
	Black-tailed rattlesnake	C. molossus molossus			x	x	Also Tiburón Island
	San Esteban rattlesnake	C. m. estebanensis		x			San Esteban Island
	Red diamond rattlesnake	C. ruber ruber	x				Northern Baja
	Cape red rattlesnake	C. r. lucasensis	x				Southern Baja
	Mojave rattlesnake	C. scutulatus			x		
	Tiger rattlesnake	C. tigris			x	x	
	Tortuga Island rattlesnake	C. tortugensis		x			Tortuga Island
	Southern Pacific rattlesnake	C. viridis helleri	x				Northern Baja
	Coronados rattlesnake	C. v. caliginis		x			Southern Coronado Island
Chelydridae	Yellow mud turtle	Kinosternon flavescens			x		
	Sonora mud turtle	K. sonoriense			x	x	
Testudinidae	Pond slider	Pseudemys scripta	x			x	
	Yellow box turtle	Terrapene ornata luteola			±		Central Sonora
	Desert tortoise	Gopherus agassizi			x	x	Also Tiburón Island
Cheloniidae	Pacific green turtle	Chelonia mydas agassizi	x			x	Marine
	Pacific loggerhead	Caretta caretta gigas	x			x	Marine
	Olive ridley	Lepidochelys olivacea	x			x	Marine
	Pacific hawksbill	Eretmochelys imbricata squamata	x			x	Marine
Dermochelidae	Pacific leatherback	Dermochelys coriacea schlegeli	x			x	Marine
Trionychidae	Texas softshell	Trionyx spiniferus emoryi			x		
Ranidae	Leopard frog	Rana pipiens			x	x	
	Bullfrog	R. catesbeiana			x		
Hylidae	Burrowing treefrog	Pternohyla fodiens			x	x	
	Canyon treefrog	Hyla arenicolor			x	x	
	California treefrog	H. californiae	x				Northern Baja
	Pacific treefrog	H. regilla regilla	x				Northern Baja
	Peninsula Desert treefrog	H. r. deserticola	x				Southern Baja
Bufonidae	Colorado River toad	Bufo alvarius			x	x	

FAMILY	COMMON NAME	SCIENTIFIC NAME	PENINSULA	ISLAND ENDEMIC	N. MAINLAND	S. MAINLAND	COMMENT
Bufonidae	Woodhouse's toad	B. woodhousei			x		
	Red spotted toad	B. punctatus	x		x	x	Also Tiburón Island
	Great Plains toad	B. cognatus			x	x	
	Sonoran green toad	B. retiformis			x	x	
Microhylidae	Sinaloa narrow-mouthed toad	Gastrophryne olivacea mazatlanensis			x	x	
Pelobatidae	Couch's spadefoot toad	Scaphiopus couchi	x		x	x	Southern Baja
	Western spadefoot toad	S. hammondi	x		x		Northern Baja

FRESHWATER FISHES OF THE SONORAN DESERT

FAMILY	COMMON NAME	SCIENTIFIC NAME	SALT TOLERANT	INTRODUCED	NATIVE NORTH	SOUTH	COMMENT AND/OR YEAR OF INTRODUCTION
Elapidae	Machete	Elops affinis	x				Colorado River
Clupeidae	Threadfin shad	Dorosoma petenense		x			Forage fish, 1953
Characidae	Banded tetra	Astyanax fasciatus mexicanus		x			Bait fish, 1950
Salmonidae	Rainbow trout	Salmo gairdneri	±				Northern Vizcaíno Division
Cyprinidae	Carp	Cyprinus carpio		x			Food fish, 1885
	Goldfish	Carassius auratus		x			Ornamental, 1939
	White amur	Ctenophayngodon idellus		±			Food fish, 1970
	Golden shiner	Notemigonus chrysoleucus		x			Bait fish, 1930s
	Bonytail	Gila elegans			x		
	Verde trout	G. robusta			x		
	Gila chub	G. intermedia			x		
	Yaqui chub	G. purpurea				x	
	Sonora chub	G. ditaenia				x	
	Spikedace	Meda fulgida			x		
	Woundfin	Plagopterus argentissimus			x		
	Colorado River squawfish	Ptychocheilus lucius			x		
	Longfin dace	Agosia chrysogaster			x	x	
	Speckled dace	Rhinichthys osculus			x	x	
	Loach minnow	Tiaroga cobitis			x		
	Red shiner	Notropis lutrensis		x			Bait fish, 1950s
	Sonoran shiner	N. formosus		x		x	Introduced in north
	Mexican stoneroller	Camptostoma ornatum pricei			±	x	
	Flathead minnow	Pimephales promelas		x			Bait fish, 1952
Catostomidae	Bigmouth buffalo	Ictiobus cyprinellus		x			Food fish, 1918
	Black buffalo	I. niger		x			Food fish, 1918
	Smallmouth buffalo	I. bubalus		x			Food fish, 1952
	Razorback sucker	Xyrauchen texanus			x		
	Flannelmouth sucker	Catostomus latipinnis			x		
	Gila sucker	C. insignis			x		

FAMILY	COMMON NAME	SCIENTIFIC NAME	PENINSULA	ISLAND ENDEMIC	N. MAIN-LAND	S. MAIN-LAND	COMMENT
Catostomidae	Sonora sucker	C. sonoriensis				x	
	Yaqui sucker	C. bernardini				x	
	Gila mountain sucker	Pantosteus clarki			x		
Ictaluridae	Flathead catfish	Pilodictis olivaris		x			Game fish, 1940s
	Channel catfish	Ictalurus punctatus		x			Game fish, 1913
	Yaqui catfish	I. pricei				x	
	Blue catfish	I. furcatus		x			Game fish, 1972
	Black bullhead	I. melas		x			Accidental introduction, 1904
	Yellow bullhead	L. natalis		x			Accidental introduction, early 1900s
	Brown bullhead	I. nebulosus		x			Accidental introduction, 1910
Cyprinodontidae	Desert pupfish	Cyprinodon macularius	x				
		Fundulus parvipinnis	x				Probable prototype of F. lima
		F. lima	?			x	Only Baja endemic species
Centropomidae	Snook	Centropomus nigrescens	x				Baja California
Mugilidae	Striped mullet	Mugil cephalus	x				Colorado River
		M. curema	x				Baja California
		Agonostomus monticola	x				Rio Yaqui
Lutjanidae	Snapper	Lutjanus argentiventris	x				Baja California
Sciaenidae	Croaker	Micropogon megalops	x				Colorado River
		Cynoscion xanthulus	x				Colorado River
		C. macdonaldi	x				Colorado River
Cottidae	Sculpin	Leptocottus armatus australis	x				Baja California
Gasterosteidae	Stickleback	Gasterosteus aculeatus microcephalus	x				Baja California
Eleotridae	Spotted sleeper	Eleotris picta	x				Colorado River
		Gobiomorus maculatus	x				Rio Yaqui
		Dormitator latifrons	x				Rio Yaqui
Gobiidae	Goby	Awaous transandeanus	x				Rio Yaqui
	Longjaw mudsucker	Gillichthys mirabilis	x	x			Bait fish, 1952
Poeciliidae	Mosquitofish	Gambusia affinis		x			Mosquito control, 1926
	Gila topminnow	Poeciliopsis occidentalis occidentalis			x		
	Yaqui topminnow	P. o. sonoriensis				x	
	Variable platyfish	Xiphophorus variatus		x			Aquarium introduction, 1963
	Green swordtail	X. helleri		x			Aquarium introduction, 1952
	Sailfin molly	Poecilia latipinna		x			Aquarium introduction, 1952
	Mexican molly	P. mexicana		x			Aquarium introduction, 1968
	Guppy	Lebistes reticulatus		x			Aquarium introduction, 1960s
Percichthyidae	Striped bass	Morone saxatilis		x			Game fish, 1959
	White bass	M. chrysops		x			Game fish, 1960
	Yellow bass	M. mississipiensis		x			Game fish, 1930
Centrarchidae	Smallmouth bass	Micropterus dolomieui		x			Game fish, 1942
	Largemouth bass	M. salmoides		x			Game fish, 1940s
	Warmouth bass	Chaenobryttus gulosus		x			Accidental, 1950s
	Green sunfish	Ch. cyanellus		x			Game fish, 1926
	Bluegill	Lepomis macrochirus		x			Game fish, 1934
	Redear sunfish	L. microlophus		x			Game fish, 1940s
	Pumpkinseed	L. gibbosus		x			Food fish, 1950
	White crappie	Pomoxis annularis		x			Game fish, 1934
	Black crappie	P. nigromaculata		x			Game fish, 1936

FAMILY	COMMON NAME	SCIENTIFIC NAME	SALT TOLERANT	INTRODUCED	NATIVE NORTH	SOUTH	COMMENT AND/OR YEAR OF INTRODUCTION
Centrarchidae	Sacramento perch	Archoplites interruptus		x			
Percidae	Walleye	Stizostedion vitreum		x			Game fish, 1960s
	Yellow perch	Perca flavescens		x			Game fish, 1930s
Chichlidae	Convict cichlid	Cichlasoma nigrofasciatum		x			Aquarium introduction, 1969
	Mozambique mouthbrooder	Tilapia mossambica		x			Food fish, 1960s
	Nile mouthbrooker	T. niloctica		x			Food fish, 1964
	Zilli's tilapia	T. zilli		x			Food fish, 1973

6

THE CHIHUAHUAN DESERT

FERNANDO MEDELLÍN-LEAL

THE CHIHUAHUAN DESERT, the largest division of the Great North American Desert, lies over the Tropic of Cancer, which crosses it in its southern portion, trapped among mountain chains. Beautiful, strong, and tough, it has been divided in two parts—a small one to the north and a larger one to the south—by a river and by history. It has served as a natural barrier not only for the fauna and flora that inhabit the surrounding regions but also for two types of cultures. Many years ago the Chihuahuan Desert held back the impetus and marked the limit of the Aztec Empire and the Tarascan Kingdom. Several centuries later, the capricious history overturned and it was from the north that the impulsive expansionism like a demolishing wave arrived, but once more the Chihuahuan Desert marked the limit.

ANTECEDENTS

The first bibliographic reference to the Chihuahuan Desert may be the report of Ramos-Arizpe on the reigning conditions in the province of Coahuila of which he was deputy before the Cortes de Cádiz and published in Spanish in 1812 (Ramos-Arizpe 1812). After him, there are tales of daring travelers, such as the Englishman W. H. Hardy in 1829, the Italian J. C. Beltrami in 1830, the North American A. M. Guilliam in 1847 (cited by Glantz 1964), and the Mexican J. T. de Cuéllar in 1869 (cited by Tavera-Alfaro 1964).

There are abundant bibliographies on the Chihuahuan Desert, and some key works have excellent bibliographic reviews. Among the most important are those of the Universidad Nacional Autónoma de México investigators published during 1936 and 1937 in volumes 7 and 8 of the *Anales* del Instituto de Biología about the portion of the Chihuahuan Desert corresponding to the state of Hidalgo; Shreve's (1942) paper about the North American desert vegetation; the works edited by Beltrán (1955a, 1955b, 1964), by Gonzáles-Cortés (1959, 1963), and by Marroquín et al. (1964); Jaeger's classic on North American deserts (1957); Rzedowski's botanical papers (1957, 1962, 1965, 1966, 1973, 1975, 1978); the monumental *Deserts of the World* (McGinnies, Goldman, and Paylore 1970); and the excellent bibliographies on the geology of the Mexican arid zones by Aranda (1973, 1975).

The approach to and the quality of what has been written about the Chihuahuan Desert varies enormously. Most of the works are not exclusively dedicated to it and their completeness depends on the scale of the investigations, which may be excessively regional or may cover the entire world. Many of the studies focus on a national level on both sides of the border. Few papers are dedicated exclusively to the Chihuahuan Desert. Of these, the ones edited in a volume by Wauer and Riskind (1977) and, particularly, Morafka's work (1977b) deserve to be pointed out because they are the most recent and are very complete.

*I wish to express my gratitude to my wife, María de los Angeles, for all the support given to me for the elaboration of this work. Without her help it would not have been possible to complete it. I also thank Dr. Jerzy Rzedowski and Ing. Fausto García-Castaneda for the documents and critical comments they furnished relative to some portions of the work; Raul Grande-López, for his collaboration in writing the section on soils; Nicolás Vásquez-Rosillo, biologist, for his critical review of the section on fauna; Lucio Gallegos-Juárez, geological engineer, for making the translation from Spanish into English; Raul Martinez-de-la-Rosa, civil engineer, for the excellent drawings, particularly the maps of communication and tourism; and Rebeca Montejano-Díaz and María del Carmen Vargas-Olvera, chemist, who had the enormous patience to type the manuscript several times. To all of them I express my deepest gratitude. They are all members of the Instituto de Investigación de Zonas Desérticas staff. Any errors are my responsibility.

NAME

According to Morafka (1977b:60) the origin of the name Chihuahuan Desert may be traced to the end of the first half of the nineteenth century when R. B. Hinde in 1843 spoke vaguely of a southwestern "Chihuahua Region"; however, it was not until 1943 that L. R. Dice actually confined the "Chihuahuan Desert" as a biogeographic province, and since then it has been known by that name. (Morafka 1977b:60) In 1911 Harshberger delimited vegetationally the "Chihuahuan Desert Region" and called it that (Shreve 1942:210). Shreve in 1942 also called it Chihuahuan Desert.

The Chihuahuan Desert gets its name from a Mexican state, Chihuahua, which is derived from a Tarahumara word meaning "in the workshop" or "place of a workshop."

LOCATION AND LIMITS

The Chihuahuan Desert is located in a slightly inclined way bearing southeast-northwest between the 20° and 35° N parallels and the meridians 98.45 and 109.15 west of Greenwich. The border line is quite sinuous and includes portions of Arizona, New Mexico, and Texas in the United States and the states of Coahuila, Nuevo León, Tamaulipas, San Luis Potosí, Guanajuato, Zacatecas, Aguascalientes, Durango, Chihuahua, and Sonora in the Mexican Republic, and in a discontinuous way, in its most southern part, two relatively small zones of the Mexican states, Querétaro and Hidalgo.

The limit proposed here is as arbitrary as all previous ones proposed. All authors have their own concept of what they consider a desert; they consider their definition merely as a working tool in order to present a series of facts in a graphic way.

In figures 6-1 and 6-2, the map of Raisz (1959) was used primarily for the part corresponding to the Mexican Republic, which, according to Morafka (1977b:39), includes 80 percent of the Chihuahuan Desert. The maps of the U.S. Geological Survey were used for the part corresponding to the United States, with slight changes because of the different plottings used on the original maps. The line used on the basic map is purposely thick and covers a belt approximately 10 kilometers (km) in width.

The Chihuahuan Desert as presented here, is limited by the following geographic characters:

Starting out from Albuquerque, considered here as the northern end, and following a clockwise direction, the Chihuahuan Desert is limited by the southern foothills of the Rocky Mountain system, the High Plains of New Mexico and Texas, the Sierra Madre Occidental, and mountains of the central highland of Mexico, and foothills mainly to the east of the Sierra Madre Occidental.

Regarding the two small southern portions of the Chihuahuan Desert, the Querétaro's forms part of a depression between the towns of Vizarrón and Peñamiller. The portion corresponding to Hidalgo is an intermountain valley limited by the Sierras de Zimapán, Juárez, Actopan, Tula, and Nopala in that state.

Within the limits but not plotted on the map are islands or spots that can hardly be considered desertic but cannot be excluded since they are subject to the strong influence of the surrounding desert. They cover, generally, the highest parts of the mountains, particularly but not exclusively in New Mexico, Texas, and Coahuila, where they form a mosaic with the surrounding desert.

In establishing the Chihuahuan Desert limits, I have taken many aspects into account, among them the distribution of several plant and animal species, types of soil, and weather data, all in combination with other elements, particularly topographic ones. The elaboration of this map, which I have divided into two figures for clarity, has been developed little by little throughout my twenty years of working in this zone. In some small portions, the limit I have proposed has not been evaluated systematically but has resulted from a somewhat intuitive reaction stemming from my long field experience. This limit differs from the limits proposed by many other authors (Shreve 1942; Miranda 1955; Jaeger 1957; Comité Mexicano de Zonas Aridas 1963; McGinnies, Goldman and Paylore 1970; Johnston, quoted by Henrickson and Straw 1976; Findley 1977; Morafka 1977a, 1977b). Also its limits differ from the ones I have proposed on other occasions (Medellín-Leal quoted by Aranda 1973; Medellín-Leal and Gómez-Gonzáles in press). The other authors generally mark the complete or partial limits of this desert on blank maps, without referring to geographic coordinates, political limits, towns, or other points. Those maps serve the purposes of the papers with which they have been associated, but readers have difficulty when they try to compare them with maps that contain other types of information, such as means of communication or tourism. Therefore the basic map presented in figures 6-1 and 6-2 and that in figure 6-3 compare the oldest proposed limit (Shreve 1942:212), the latest published one (Morafka 1977b:62), and the one presented here. The map of limit comparisons (figure 6-3) is curious because apparently both Shreve (1942) and Morafka (1977b) are stricter about their concept of desert, and their limits extend over a smaller area than the one I have proposed. Other authors are less conservative, and some of the limits they use include a larger area than the one I propose here (Miranda 1955; Jaeger 1957).

Shreve (1942:210-215) says that the Chihuahuan Desert is difficult to delimit due, among other things, to the existence of a great elasticity in the meaning of the word *desert* and to the presence of large ecotones. The transition is not only gradual but frequently regressive. The problem becomes complicated because the surrounding vegetation varies enough with the latitude and exposure in such a way that it is not comparable along the limit. Shreve believes that in many cases, it is only the density of the vegetation that varies when the limit is reached, and he suggests using the mountain ranges as a possible important demarcation line. On the other hand he mentions that those who plot maps of delimitation suffer from a partial dependence on their predecessors, and this is the reason why the limits that different authors make coincide at least in part.

Miranda (1955:102-108) believes that the precise limits of the Chihuahuan Desert cannot be marked until enough phytogeographic and phytoecologic works exist. These can be used instead of the meager meteorological data due to the insufficient net of meteorological stations. According to Miranda, any limit given should be conventional. Although plants could serve as indicators, their utility is more relative than absolute because their xeroplastic capacity induces confusion.

Morafka (1977b:59-60) proposes standardization of the term *desert* as a technical tool to establish the bases for a correct delimitation, and he indicates that this must be the objective of contemporaneous biographers.

CAUSES OF THE CHIHUAHUAN DESERT

Morafka (1977b:16) points out that the genetic causes of this desert are, in order of importance, the continentality, the tropicality, and the mountain shadow. Medellín-Leal and Gómez-González (in press) agree with these causes, but the relative importance they give them is just the opposite of Morafka's. The cold marine currents that are important as causes of the coastal deserts have no direct influence over the Chihuahuan Desert. This desert is clearly an independent climatic unit, which logically and due to its size and variable geomorphology has subdivisions among which can be found some symmetrical climatic parallels with other North American deserts (Morafka 1977b:21).

EXTENSION OF THE CHIHUAHUAN DESERT

The Chihuahuan Desert is the largest of the North American deserts. Henrickson and Straw (1976:1) say that the whole area of the Chihuahuan Desert is approximately 507,000 square km, but minus the least arid portions, it would be reduced to 378,000 square km. Morafka (1977b:27) establishes an approximate area of 450,000 square km for the same desert.

GEOMORPHOLOGY AND PHYSIOGRAPHY

The Chihuahuan Desert is boxed in by the Rocky Mountains, Sierra Madre Oriental, the many mountain ranges of central Mexico, and the Sierra Madre Occidental. It is formed mainly by a series of alluvial plains with an average altitude of 1400 meters (m). Although the core of the desert is relatively old from the earth's historical point of view, the main factors of geomorphology were originated by the rising of the sea bottom during the Laramide revolution (end of Cretaceous, beginning of the Tertiary, Paleocene) when parts of the Rocky Mountains, the Sierra Madre Oriental, and most of the geormorphologic structures of all the eastern portion of the Chihuahuan Desert were formed (Dunbar 1955:364-368). Subsequently the great volcanic activity occurring in North America from the transverse mountain ranges of central Mexico in the Tertiary and originated some recent lava flows (Dunbar 1955:424-429). This constitutes the major characteristic of the Chihuahuan Desert: basal core, predominantly formed toward the east by limestone rocks of marine sedimentary type and toward the west almost exclusively by rocks of igneous origin.

Although the Chihuahuan Desert has an average altitude of 1400 m, it is not a continuous plain. Rather it shows various landscapes (figure 6-4).

The Chihuahuan Desert represents an enormous geomorphologic structure with a descendant inclination from Zacatecas in the southwest, where it reaches an average altitude higher than 2000 m, toward Texas in the northeast, where it reaches only 250 m above sea level. There is another less slanted inclination from west toward the east at the southern edge. The desert drops—sometimes abruptly, sometimes gently, but generally in a stairway form—from 2000 m in Zacatecas until it reaches an area with an elevation of 900 m in San Luis Potosí.

From the southeast in San Ciro, San Luis Potosí (SLP), toward the northwest in Albuquerque, the ascents and descents are gentler and the difference in altitudes less. These features are shown in the two vertical profiles (figure 6-5).

The two small southern islands of the Chihuahuan Desert have purposely not been shown on these sections. The one belonging to Querétaro consists of a depression with its lower part at approximately 1700 m; and the one belonging to Hidalgo drops a little from a bit higher than 2000 m in the southeast to about 1500 m in the northwest. Both regions, relatively small in area, are surrounded by higher mountains.

According to the classification of Raisz (1959), somewhat modified in the map of physiographic provinces (figure 6-6), the Chihuahuan Desert spreads over eight physiographic provinces:

1. Basin and ranges (includes the subregion called Coahuila Upland by Raisz).

2. Upland with basins of the Sierra Madre Occidental

3. Central Highland (name and extension modified from Raisz, who called it Central Mesa and shows it smaller)

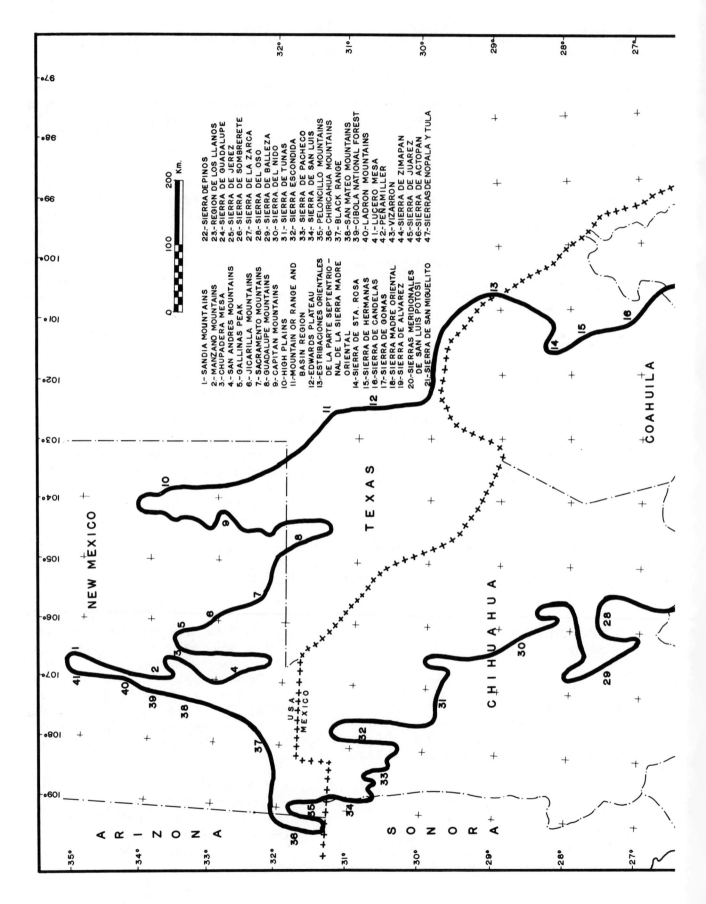

1.- SANDIA MOUNTAINS
2.- MANZANO MOUNTAINS
3.- CHUPADERA MESA
4.- SAN ANDRES MOUNTAINS
5.- GALLINAS PEAK
6.- JICARILLA MOUNTAINS
7.- SACRAMENTO MOUNTAINS
8.- GUADALUPE MOUNTAINS
9.- CAPITAN MOUNTAINS
10.- HIGH PLAINS
11.- MOUNTAIN OR RANGE AND
     BASIN REGION
12.- EDWARDS PLATEAU
13.- ESTRIBACIONES ORIENTALES
     DE LA PARTE SEPTENTRIO-
     NAL DE LA SIERRA MADRE
     ORIENTAL
14.- SIERRA DE STA. ROSA
15.- SIERRA DE HERMANAS
16.- SIERRA DE CANDELAS
17.- SIERRA DE GOMAS
18.- SIERRA MADRE ORIENTAL
19.- SIERRA DE ALVAREZ
20.- SIERRAS MERIDIONALES
     DE SAN LUIS POTOSI
21.- SIERRA DE SAN MIGUELITO

22.- SIERRA DE PINOS
23.- REGION DE LOS LLANOS
24.- SIERRA DE GUADALUPE
25.- SIERRA DE JEREZ
26.- SIERRA DE SOMBRERETE
27.- SIERRA DE LA ZARCA
28.- SIERRA DEL OSO
29.- SIERRA DE BALLEZA
30.- SIERRA DEL NIDO
31.- SIERRA DE TUNAS
32.- SIERRA ESCONDIDA
33.- SIERRA DE PACHECO
34.- SIERRA DE SAN LUIS
35.- PELONCILLO MOUNTAINS
36.- CHIRICAHUA MOUNTAINS
37.- BLACK RANGE
38.- SAN MATEO MOUNTAINS
39.- CIBOLA NATIONAL FOREST
40.- LADRON MOUNTAINS
41.- LUCERO MESA
42.- PEÑAMILLER
43.- VIZARRON
44.- SIERRA DE ZIMAPAN
45.- SIERRA DE JUAREZ
46.- SIERRA DE ACTOPAN
47.- SIERRAS DE NOPALA Y TULA

0   100   200
Km.

NEW MEXICO

TEXAS

USA
MEXICO

ARIZONA

SONORA

CHIHUAHUA

COAHUILA

324

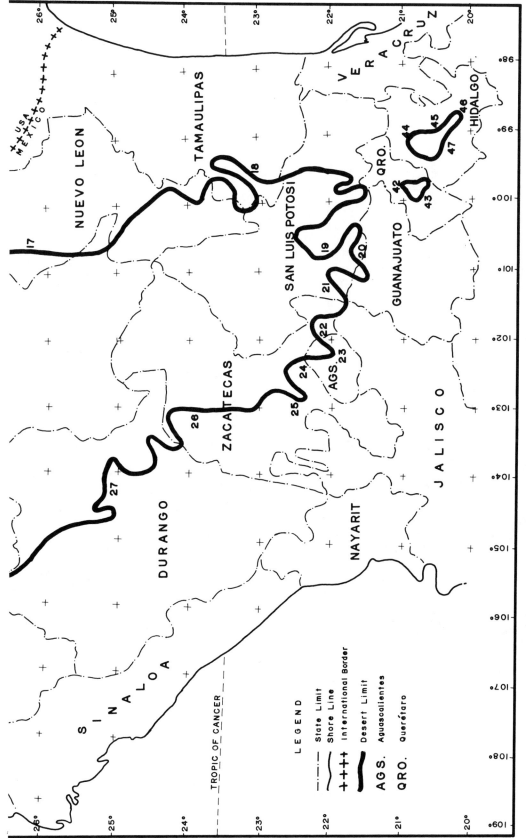

Figure 6-1. Map of the Chibuabuan Desert showing state limits, shoreline, international border, and desert limit.

The basic map for the figures in this chapter is taken from *A biogeographical analysis of the Chihuahuan Desert* by D. J. Morafka, published in 1977, W. Junk, The Hague, Netherlands. Reprinted with the permission of the publisher.

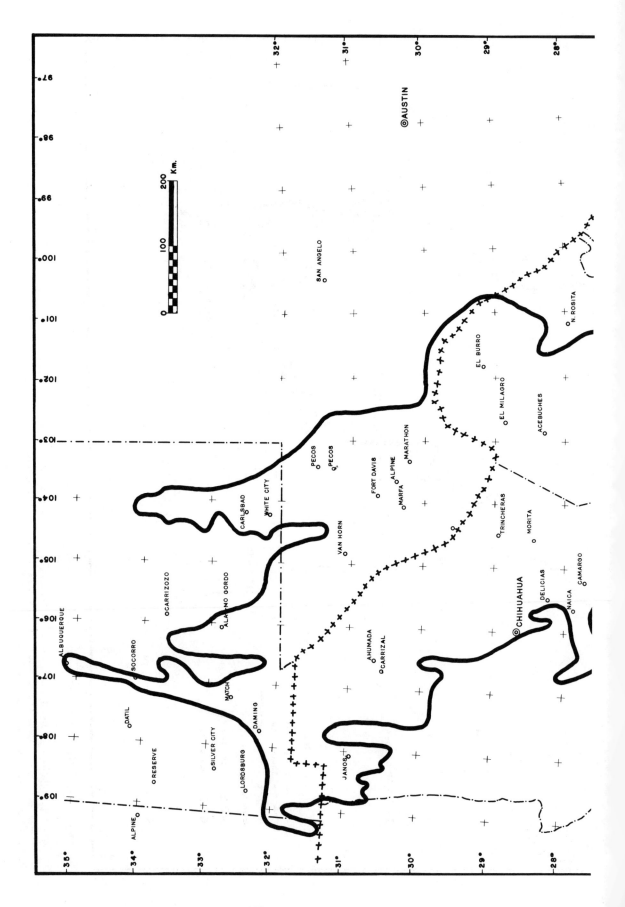

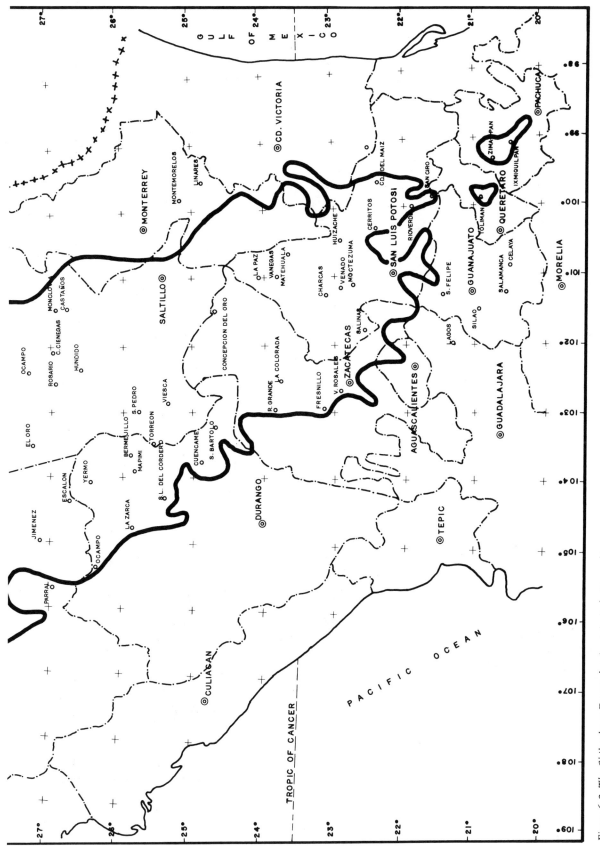

Figure 6-2. The Chihuahuan Desert showing principal cities and towns.

327

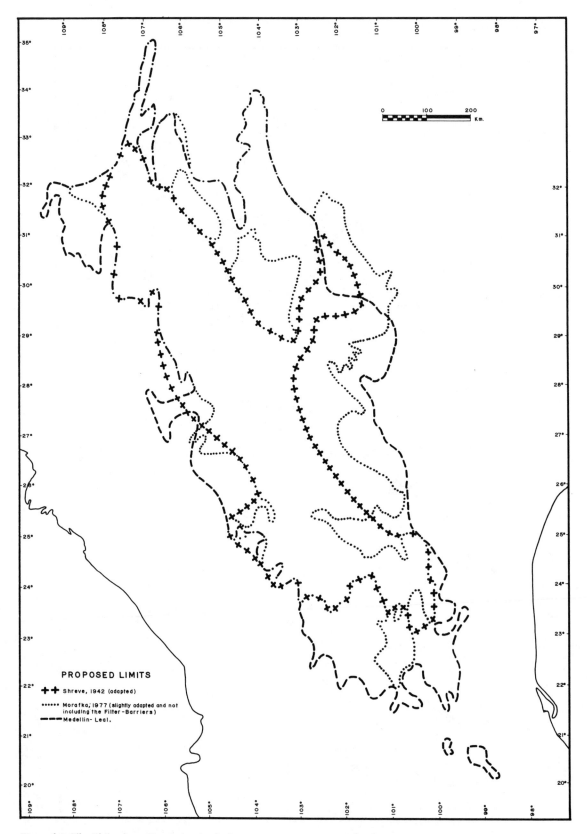

Figure 6-3. The Chihuahuan Desert showing limits
proposed by F. Shreve (adapted from *The desert vegetation of North America, Botanical Review*
8 [4]:195-246, 1942 and used with permission) and by D. J. Morafka (slightly adapted and not
including the filter-barriers, from *A biogeographical analysis of the Chihuahuan Desert*,
published in 1977, W. Junk, The Hague, Netherlands and used with permission), and by F.
Medellín-Leal here.

328

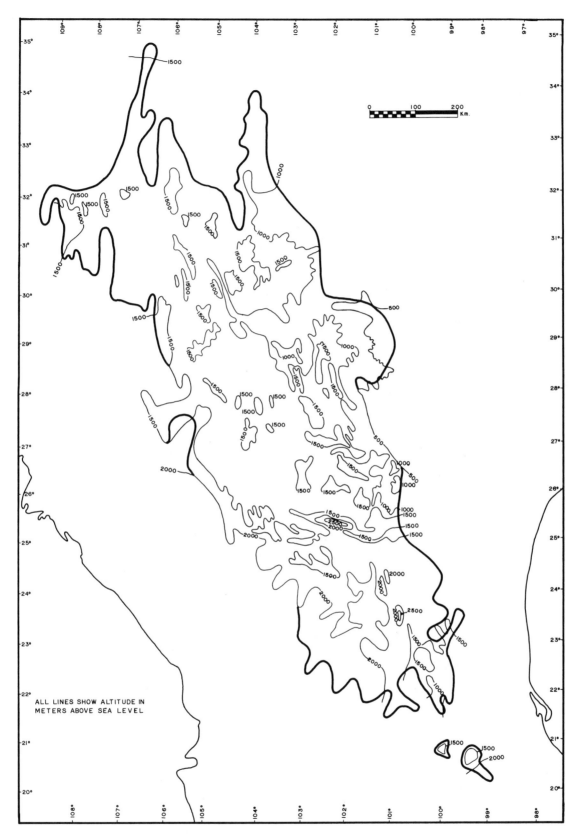

Figure 6-4. The topography of the Chihuahuan Desert.
Simplified from J. L. Tamayo (*Atlas geográfico general de México,* Instituto de Investigaciones
Económicas, México, 1962).

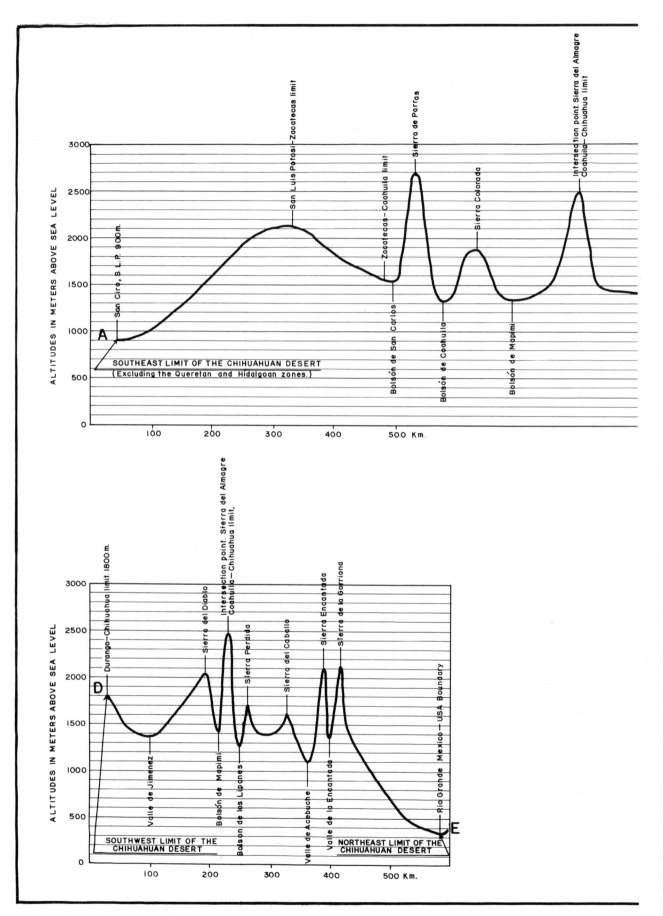

Figure 6-5. Vertical sections of the Chihuahuan Desert.
Prepared by Fernando Medellín-Leal.

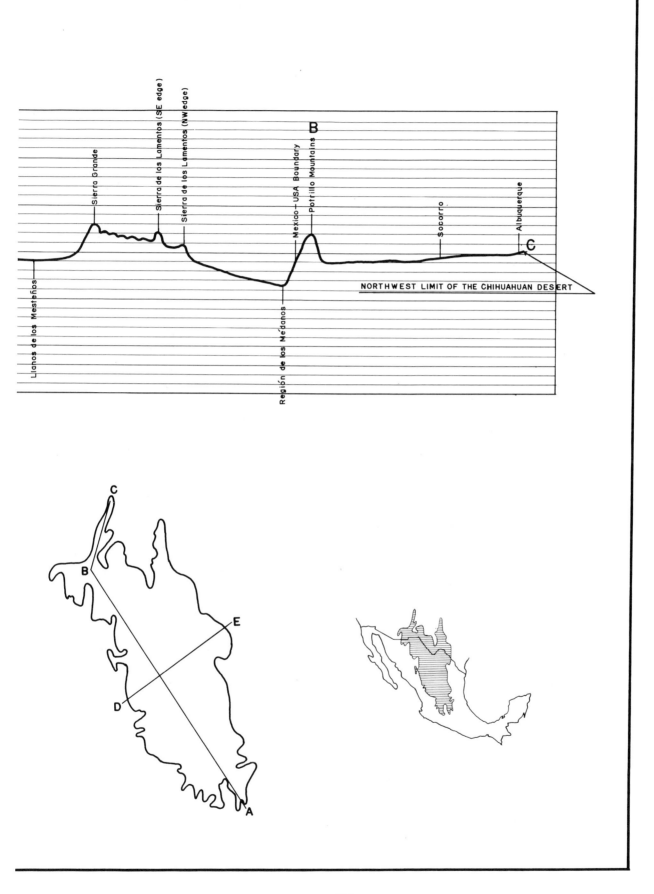

NORTHWEST LIMIT OF THE CHIHUAHUAN DESERT

Llanos de los Mesteños

Sierra Grande

Sierra de los Lamentos (SE edge)

Sierra de los Lamentos (NW edge)

Región de los Médanos

Mexico-USA Boundary

Potrillo Mountains

B

Socorro

Albuquerque

C

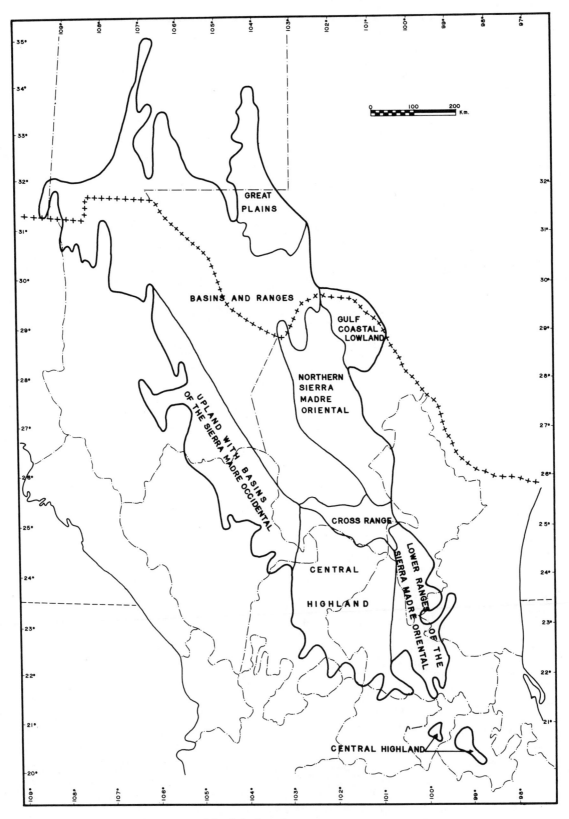

Figure 6-6. The physiographic provinces of the Chihuahuan Desert.
Slightly modified from E. Raisz (*Landforms of Mexico,* map prepared for the Geography Branch
of the Office of Naval Research, Cambridge, Mass., 1959) and used with permission and from
L. A. Heindl (Groundwater in the Southwest: A perspective, in *Ecology of Groundwater in the
Southwestern and Rocky Mountain Division of the American Association for the Advancement of
Science,* ed. J. E. Fletcher and G. L. Bender, Arizona State University, Tempe, 1965.)

4. Lower ranges of the Sierra Madre Oriental
5. Cross Range
6. Northern Sierra Madre Oriental
7. Gulf Coastal Lowland
8. Great Plains.

Basin and Ranges

This region extends from New Mexico and Texas to Chihuahua and Coahuila. It is formed by mountains that follow a general northwest-southeast direction and reach altitudes greater than 2000 m. Its origin in the northwest is mainly igneous although there are sedimentary traces in the southeast. In the southeast the origin of the mountains is from generally marine Cretacic sedimentary folds, but there are also igneous, intrusive, and extrusive traces, as well as other sedimentary rocks from other geologic periods.

The basins are clastic refills between ranges, originated by their erosion. Currently most of the basins are endorheic or arheic. Both cases form the numerous bolsones in this region. Other basins are exhorheic and form part of the present Rio Grande Basin. Many testimonial mountains, specially in Chihuahua, indicate a more active drainage and a larger basin than now exist in more humid seasons, perhaps during the Pleistocenical glaciations.

The Guadalupe Mountains in New Mexico and Texas and the Sierra de los Alamitos in Coahuila remained emerged during long epochs of the Mesozoic.

Upland with basins of the Sierra Madre

The oriental part of the Sierra Madre Occidental is formed by foothills that sometimes follow a northwest-southeast direction and sometimes a definite north-south direction. These mountain ranges form the western limit of the Chihuahuan Desert. The basins are generally alluvial or colluvial refills originated within the same Sierra Madre Occidental.

Central Highland

Raisz (1959) calls this province Central Mesa. Although he says a more proper name for it would be Central Basin, I have followed García-Castañeda's criteria (1978:46) and called it Central Highland. The region extends in the Chihuahuan Desert over part of the states of Coahuila, Durango, Zacatecas, Aguascalientes, San Luis Potosí, Guanajuato, Querétaro, and Hidalgo and continues in regions that do not form part of the Chihuahuan Desert. Most of this province in the desert zone is formed by dissected mountains or mountain chains of igneous origin that follow west-northwest and southeast or a definite west-east direction. Its valleys are refilled

detritus that originated in the same mountains. Sometimes they are not really valleys but old mountains eroded to the base, as can be seen in several regions in Zacatecas. Small isolated portions are formed by medium-altitude mountains of limestone origin. This occurs in Aguascalientes, Querétaro, and Hidalgo. The limestone origin of these regions explains, partially at least, the southern extension of the desert since they are much more permeable carstic zones, which hardly retain water at the surface.

Lower Ranges of the Sierra Madre Oriental

These are the western foothills of the Sierra Madre Oriental. Of marine sedimentary origin, mostly Cretaceous, the hills that form them have more or less gentle shapes with wide colluvial cones and enormous alluvial fans. The valleys are slightly undulated, sometimes flat in great extensions, sometimes endorheic or arheic and therefore with bolsones. The inferior part of the zone drains into the Gulf of Mexico through the superior parts of the Rio Panuco Basin.

Cross Range

It has the same origin and similar features as do the lower ranges of the Sierra Madre Oriental. It is a bit higher and extends in an east-west direction from Saltillo to Nazas, almost connecting both Sierra Madres and forming the Sierra de Parras. Some authors (Johnston in Henrickson and Straw 1976; Morafka 1977b) do not consider it part of the Chihuahuan Desert, but I believe that it should be included.

Northern Sierra Madre Oriental

It extends from the region of the Davis Mountains and the Big Bend in Texas toward the southeast, in the Sierra del Burro in Coahuila and Nuevo León. In its most northern part, it presents considerable vulcanism remains, and the mountain ranges are abrupt and irregular. Farther to the southeast, it is formed primarily by limestone rocks of marine sedimentary origin, mainly Cretaceous. Even so, the mountains continue to be abrupt and splashed here and there by rocks of igneous origin. This is a very rough region, which forms numerous internal islands of nondesert but is subject to the ecologic pressure from the desert.

Gulf Coastal Lowland

A structure of Tertiary origin toward the interior of the continent, it is formed by low mountains that descend little by little. As altitude is lost, they become an alluvial plain that reaches the Gulf of Mexico, although the desert does not extend all the way there. All of the transitions here are slow and long, and both the climate and the ecology become gradually different, while the actual river bed of the Rio Grande divides it all the way to its outlet.

Great Plains

The only portion of this enormous physiographic province that forms part of the Chihuahuan Desert has been named indistinctly Pecos Valley or Toyah Basin and forms the southern end of the Great Plains. The surface, with an average altitude of 750 m, is almost absolutely flat, although it is slightly higher toward the north than to the south.

The Chihuahuan Desert mountain systems and other physiographic features are shown in figure 6-7.

HYDROGRAPHY AND DRAINAGE

The Continental Divide is located beyond the Chihuahuan Desert toward the west and crosses only a small portion of it in the northwest between the northwest of the state of Chihuahua and the southwest of the state of New Mexico. Only the northern and southern ends of the Chihuahuan Desert have a well-developed hydrographic system that will allow the drainage of these dry lands (figure 6-8).

In the southeast part, the San Luis Potosí and Rioverde valleys and the two small separated portions of the desert, Hidalgo and Querétaro, are part of the Pánuco system, which is contained in a very extensive basin to the east of the republic and therefore drains out in the Gulf of Mexico. Most of the Pánuco Basin is located outside of the Chihuahuan Desert, touching it with some of its tributaries most remote from its outlet, which are present as intermittent or permanent streams with very little flow.

The portion in Hidalgo is drained mainly by the Río Tula that flows into the Río Moctezuma, which joins much lower and outside of the Sierra Madre to the Río Tamuin and becomes the actual Río Pánuco.

The part in Querétaro is drained by the Río Extóraz, which joins the Moctezuma. The Río Santa María and Rioverde drain tiny parts of the San Luis Potosí and Cerritos-San Ciro valleys, both in the southern end of the desert. The Santa María-Rioverde system originates outside of the desert and forms the Río Tamuin, which joins with the Río Moctezuma to form the Río Pánuco. Although the southern end of the Chihuahuan Desert forms part of the higher portion of the Río Pánuco Basin, the natural influence of this hydrographic system over the landscape would be practically nil except for some small crop areas that have developed through time, particularly in the water meadows of the same rivers.

At its beginning the Río Santa María is partially supplied by the dregs originating from the rain that occurs in the higher part of the Sierras de San Miguelito and Meridionales of the San Luis Potosí, which form a natural limit between the states of San Luis Potosí and Guanajuato. However, its main source of supply is made up from the numerous thermal springs found in Ojocaliente, La Labor del Río, Villa de Reyes, and Gogorrón in the state of San Luis Potosí and the Municipio de San Felipe in the State of Guanajuato. In times of extraordinary rain, the northern portion of the San Luis Potosí Valley, which normally constitutes a more or less closed basin, may overflow toward the Río Santa María because the divide between the northern and southern portions of the San Luis Potosí Valley is vey low, not well defined, and in case of flooding may overflow.

The Río Santa María follows a very sinuous course, boxed in among the hills of igneous rhyolitic origin. As it goes down toward the east, it collects the water of numerous intermittent tributaries. A bit to the south of San Ciro, SLP it joins the Rio Verde outside the desert. The Rio Verde also has a mixed origin. Its main tributary is formed by the Río Santa Catarina, which originates in a series of small, intermittent streams in the high part of the Sierra de Alvarez, which is not desertic. This mountain, made up of Cretaceous reef limestones and partially by Tertiary deposits, is highly permeable and presents a well-developed Carstical system. It originates in its eastern flank with some springs of the vauclusian fountain type, the most important formed by the Laguna de la Media Luna. At one time, they were a tourist attraction. Currently they are being used exhaustively for irrigation of dubious importance at a microregion level. The drainage in this zone does not seem to work very efficiently since the northern portion of the Cerritos-San Ciro Valley contains soils with a high concentration of salts, which make them useless for agriculture.

The Río Extóraz, which could be considered the natural drainage of the small desertic portion of Querétaro, marks its northern limit, although it is actually boxed in among very abrupt hills formed by limestone rocks of marine origin. The main population center of this small region is Tolimán, which lies beside a river bed that is wide, deep, erratic, and almost always dry. Water flows only as flash floods almost immediately after the desert heavy rains appear in the high part of the hills surrounding it. On some occasions the water may reach a depth of 2 m in a bed that at some points measures more than 50 m in width. Its force is considerable. It can carry along anything in its way and after about two hours leave no more than a few isolated puddles, which quickly evaporate. The people of Tolimán have not built a bridge to the only road on the other side of the river bed. Few crops, mostly corn and fruit trees, are grown, and no irrigation exists.

The Valle del Mezquital, Valle de Actopan, or Chihuahuan Desert portion of the state of Hidalgo was originally drained by the Río Tula of dendritic type, originating in the drainage of the Actopan, Juárez, Nopala, and Zimapán mountains and in several springs, some of the thermal type, the best known being the Tzindejé. The river, placid at the edge of the town of Ixmiquilpan, runs softly through the valley but is later boxed in among deep ravines—some formed in sedimentary limestone rocks, others in rhyolitic igneous rocks—until it flows into the Río Moctezuma, which is also boxed in a deep gully

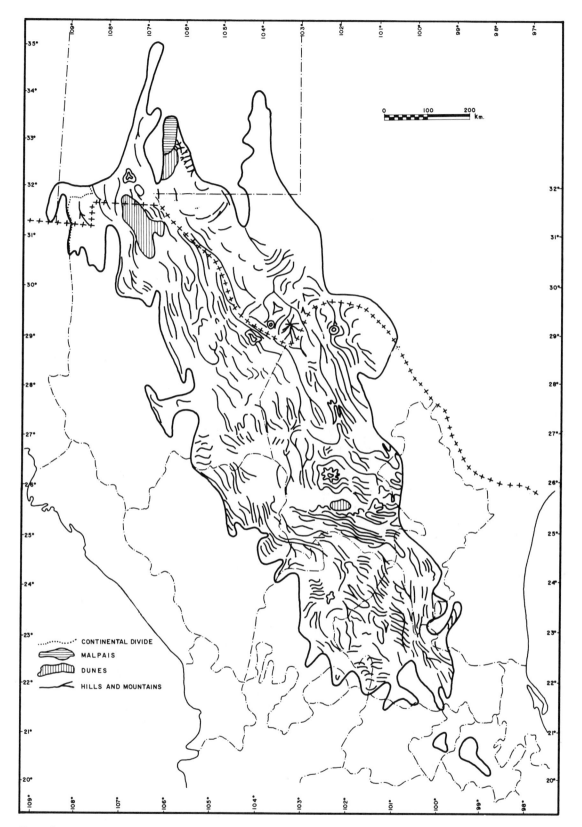

Figure 6-7. Mountain systems and other geomorphological features of the Chihuahuan Desert.
Simplified mainly from E. Raisz (*Landforms of Mexico,* map prepared for the Geography Branch
of the Office of Naval Research, Cambridge, Mass., 1959) and from several other minor sources.
Used with permission.

335

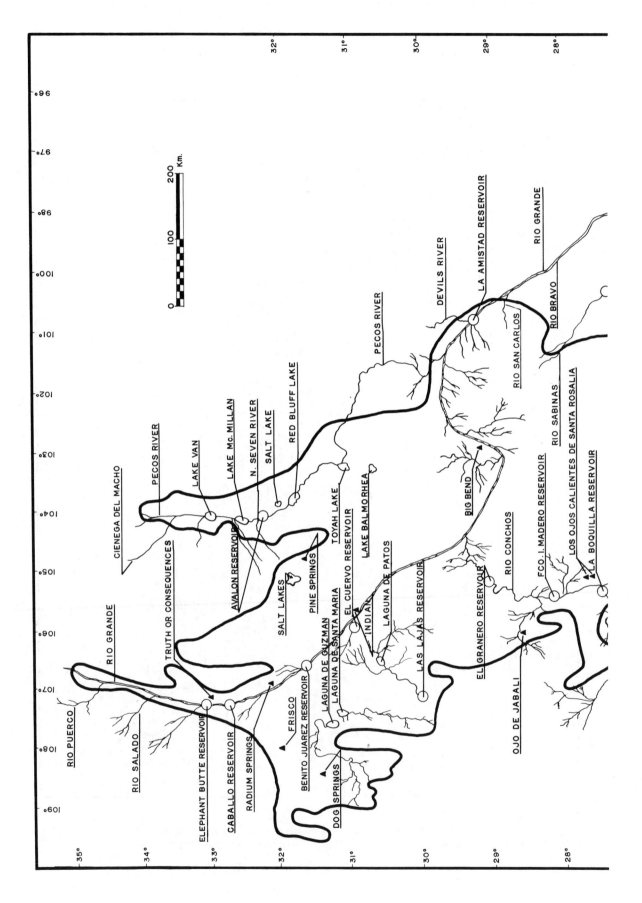

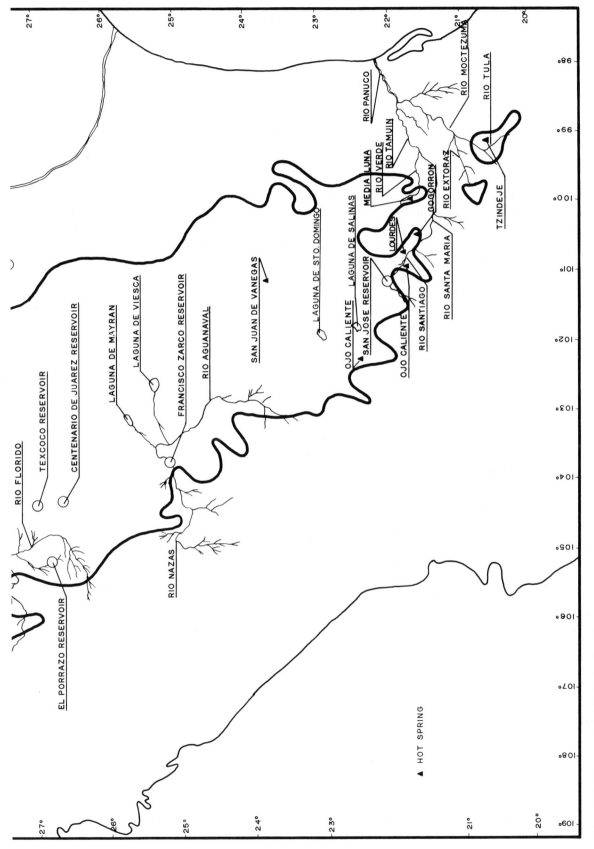

Figure 6-8. Hydrography of the Chihuahuan Desert.
Modified from L. A. Heindl (Ground water in the Southwest: A perspective, in *Ecology of groundwater in the Southwestern and Rocky Mountain Division of the American Association for the Advancement of Science,* ed. J. E. Fletcher and G. L. Bender, Arizona State University, Tempe, 1965 from S. A. Arbingast et al (*Atlas of Texas,* University of Texas, Austin, 1967), and the Secretaría de Recursos Hidráulicos (*Región Hidrológica Número 24 Poniente,* 1968 and *Región Hidrológica Numéro 34,* 1970).

337

opened through limestone rocks. Since the end of the nineteenth century, this natural system has been greatly altered. In order to avoid flooding in Mexico City, which is located at the bottom of a naturally closed basin, a cut and a tunnel system were built to connect the Mexico City Basin and the Valley del Mezquital.

In summary, the extreme southeast end of the Chihuahuan Desert belongs wholly to the Pánuco Basin, of dendritic type drainage, but since this basin is exceedingly large, it touches the desert only in a marginal way, and its influence over it is practically insignificant for it covers only tiny parts of the Central Highland physiographic province.

Something quite different occurs to the north of the Chihuahuan Desert. A real system of drainage does exist that belongs almost totally to the desert. Here the most important river is the one that on both sides of the border receives different names of Spanish origin. In the United States, it is called Grande, which means "big," and in Mexico it is called Bravo, which means "brave." The Rio Grande is born in the most northern portion in the Rocky Mountains and is supplied by the melting of glaciers. Its course begins near Albuquerque, New Mexico. There its bed is at approximately 1500 m above sea level, gently descending to the elevation of Socorro by following a slightly sinuous route of a general north-northeast, south-southwest direction. It subsequently describes a wide curve to the southeast, where it reaches the most western portion of the straight southern limit of the state of New Mexico with the most western portion of the state of Texas. From there to the city of El Paso, Texas, it forms an irregular natural limit with Texas. Forming a single population center with El Paso is the Mexican city of Juárez, Chichuahua. From there to its outlet in the Gulf of Mexico, the river forms the sinuous, and in part capricious, border between the United States and Mexico. First heading through a long stretch with a northwest-southeast direction up to a latitude of 29° N and 103° longitude west of Greenwich, a bit southeast of Big Bend it turns and heads toward the north-northeast, advancing almost one more degree in latitude and dropping about half a degree in longitude afterward. Although always following a twisted course, it advances in an almost west-east direction, dropping one more degree of longitude toward the east. At this point it joins the Pecos River. Here the Rio Grande again takes a northwest-southeast direction, enters the Gulf coastal lowland, and leaves the desert.

Due to its length, the Rio Grande crosses many flat and mountainous landscapes, and its river bed crosses deposits of igneous origin, mainly Tertiary or Quaternary; deposits of marine sedimentary origin, mostly Mesozoic; continental and marine deposits probably of Tertiary origin; over mosaics of all the aforementioned and, of course, over the degraded products of such rocks. In some parts it flows gently presenting a relatively wide and peaceful surface, and in some parts it

narrows and tends to form rapids of moderate intensity. Dams and other works developed by the governments of both countries have stabilized some parts of its course, which tends to change more frequently than the people of both could bear for the good of diplomatic relations. New international agreements have tried to establish a more solid approach in marking the border, avoiding incidents provoked by the whims or changes of nature regarding the Rio Grande.

Throughout its course, the Rio Grande has been subject to the building of reservoirs, which are used as sources of potable water, to generate electricity, as recreational spots, and for irrigation. In New Mexico the most important ones are the Elephant Butte Reservoir, Butte Dam, and Caballo Dam near Socorro. Shared by Texas and the Mexican Republic, the most spectacular are the Amistad Dam Reservoir (*Amistad* means "friendship") and the Falcon Dam Reservoir, binational projects of multiple use and a notable example of what can be done with bilateral collaboration. Only the Amistad Dam Reservoir is in what I have designated as desert.

Throughout its length, the Rio Grande receives a great number of tributaries, but only two can be considered as its major ones. The first to join the Rio Grande is the Río Conchos, born on the eastern slope of the Sierra Madre Occidental far to the south of the Rio Grande in the southwest part of the state of Chihuahua and outside of the desert. The other important tributary of the Rio Grande is the Pecos River, partially originated in the Rocky Mountains in the north, also outside of the desert, and partially with contributions from the mountains limiting the Pecos River Basin toward the east. Once the Pecos River reaches the limits that I have marked here, it follows an irregular but more or less north-south direction to the southern limit of New Mexico with Texas.

Starting in the desert and part of the Rio Grande Basin are several streams, such as Río San Carlos, Río Sabinas, and Río Salado, and many other intermittent ones that drain the eastern edge of the Chihuahuan Desert mainly in the states of Coahuila and Nuevo León. They are quite important to the population living there, but in the general hydrological scope of the Chihuahuan Desert are not important because they leave it very soon. Nevertheless they have played an essential role in the speciation of many aquatic animals and in the formation of endemisms (Contreras-Balderas 1977; Miller 1977; Minckley 1977).

In summary, the Rio Grande Basin forms the natural drainage of a great part of the north of the Chihuahuan Desert, which includes the physiographic province of basin and ranges, part of the upland with basins of the Sierra Madre Occidental, the northern Sierra Madre Oriental and the gulf coastal lowland provinces.

Both exorheic basins, the Rio Grande's and the Pánuco's, finally flow into the Gulf of Mexico and therefore belong to the Atlantic slope. The rest of the Chihuahuan Desert is made of a great majority of closed basins, endorheic and arheic, as a result of the complicated topography and not because the hills forming it are very high.

The drainage in these basins is of the centripetal dendritic type, and although the number of streams or rivers is considerable, most of them are of an intermittent type. These show a very relative seasonal importance as a real drainage, although they are of great importance for the life they sustain. Sooner or later they end up forming part of some salty lagoon, usually of an intermittent type and, if evaporation is extreme, of a bolson. Examples of these lagoons are the Laguna de Guzmán supplied by the short river of the same name, the Laguna de Santa María (the river supplying it is called Santa María, and the Laguna de Patos supplied by the Río del Carmen. These three small lagoons are located in the northern portion of the state of Chihuahua. Another small salt lagoon is Laguna de Alamos in the eastern part of the state of Chihuahua supplied by the river of the same name, which is born in the Sierra del Pino in the state of Coahuila, and two more that are called the same (Salt Lake), one in the southeast of New Mexico within the Potash Basin and another in Texas in the Bolson del Diablo. There are so many lagoons of this type throughout the Chihuahuan Desert that it is not possible to mention all of them. Many are called salada or salinas which means "salt lake" (figure 6-8).

It is important to mention two rivers that are born outside the desert in the eastern flank of the Sierra Madre Occidental, which actually modify the landscape. The Río Nazas, located farthest to the north, begins in the state of Durango, penetrates the desert at the town of Nazas (from which it takes its name) and follows an irregular direction but more or less southwest-northeast. The much longer Río Aguanaval is born in the state of Zacatecas a bit to the west of Fresnillo. It follows a south-north direction slightly inclined toward the east but at the elevation of Sierra de Guadalupe, and within the state of Durango it changes direction and heads toward the northwest. The Río Nazas and the Río Aguanaval join at the town of Lerdo just before the limits of Durango and Coahuila. They used to flow into an enormous closed basin that extended from Torreón to Saltillo through the state of Coahuila, forming the former lagoon known as Mayrán. This lagoon has been dried and through a complicated and ingenious system of dams and canals has become the most important agricultural region in the Mexican part of the Chihuahuan Desert.

Since this development is relatively recent, it has originated a whole series of human actions and reactions. Unfortunately the drying of the Laguna de Mayrán and its conversion to the Bolson de Coahuila, or Bolsón de Parras, has brought some problems of desertification. Incipient dunes of loose sand are formed at the unsheltered places of vegetation where the irrigation system cannot be extended. Some of the worst dust and sand storms in the Chihuahuan Desert occur between Torreón and Saltillo. Sometimes it is impossible to distinguish the highway from the surrounding lands because of blowing sand.

The bolsones are closed intermountain basins. They are generally dug by the wind in a very irregular way and are always surrounded by elevations of different heights. Sometimes they are related to a drainage system, usually of the dendritic centripetal type, but often it is not easy to distinguish if there is one or several intermittent supplying streams. At other times they are arheic basins without horizontal drainage, and the small precipitation is absorbed in situ. Besides being endorheic or arheic basins, the bolsones are actual evaporating pans in which a great deal of alluviums carried in from the surrounding mountain are accumulated.

These alluviums are rich in many kinds of salts. The bolsones generally appear as great extension plains; occasionally the ground is finely textured, sometimes rocky but rich in salts. Because they are useless for agricultural purposes, bolsones have been undervalued, but if we consider the great geological age of some of them, we can imagine the high potential that exists for industrial exploitation of the salts accumulated there.

Besides the Laguna de Mayrán, now converted to Bolsón de Coahuila, the most important ones are the Bolsón de Mapimí, which is the largest and most famous and which extends over part of the states of Chihuahua and Durango; the White Sands Bolson in New Mexico; Bolsón del Diablo in Texas; Bolsón de los Lipanes in Chihuahua; Bolsón de Monclova and the Bolsón de Viesca in Coahuila; Bolsón de Cedros and Bolsón San Carlos in Zacatecas; and Bolsón del Salado shared by Zacatecas and San Luis Potosí. There exist many more of less extent, some of them shown in figure 6-9.

METEOROLOGICAL DATA

As Reitan and Green have mentioned (1970:24), there exists quite a difference between the meteorological data published on both sides of the border that separates the countries sharing the Chihuahuan Desert. Several maps presented here regarding meteorologic data and climatic systems should be subject to a full critical review by specialists. They have been the hardest to synthesize because many of the data obtained for Mexico and the United States do not match. One of the problems is that the measures for Mexico are given in the metric system and for the United States mostly in the British system. Frequently, even after having made the necessary conversions, one finds that a line that bears a certain direction, after passing from one state to another, not only has a different value but sometimes forms a right angle with the first line. A great deal of criteria have been applied, but it is difficult to determine which is correct.

Most of the meteorological stations found in the Chihuahuan Desert are actually thermopluviometric stations. As a

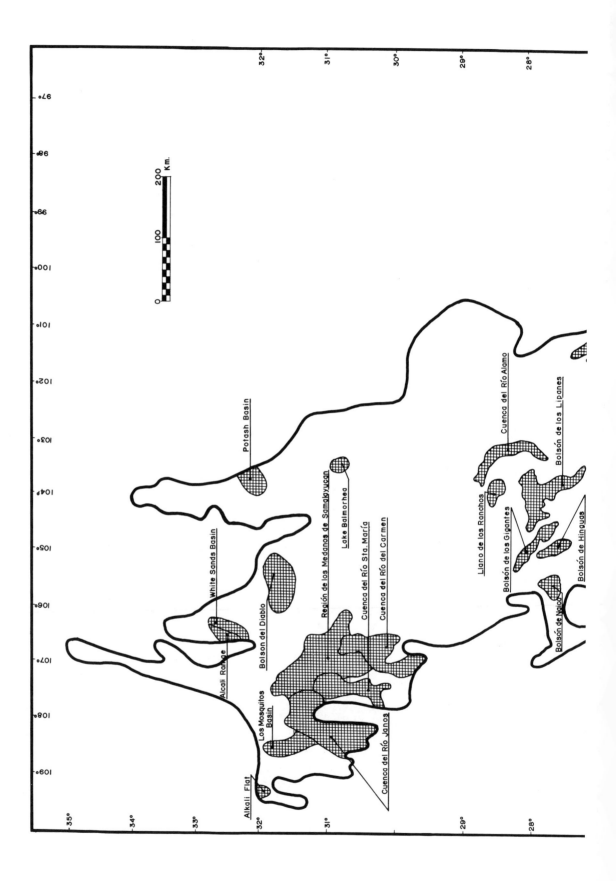

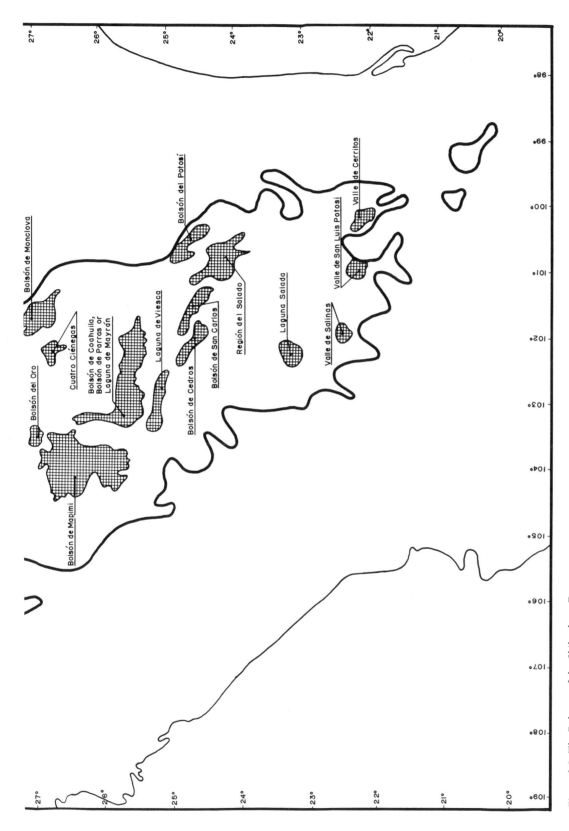

Figure 6-9. The Bolsones of the Chihuahuan Desert.
Modified from L. A. Heindl (Ground water in the Southwest: A perspective, in *Ecology of groundwater in the Southwestern and Rocky Mountain Division of the American Association for the Advancement of Science*, ed. J. E. Fletcher and G. L. Bender, Arizona State University, Tempe, 1965), and the Secretaría de Recursos Hidráulicos (*Región Hidrológica Número 24 Poniente*, 1968 and *Región Hidrológica Número 34*, 1970, and from S. A. Arbingast et al, *Atlas of Texas*, University of Texas, Austin, 1967).

result, most of the data we have are of temperature and precipitation. Those related to other meteorologic variables are scarce.

Temperature

Generally the Chihuahuan Desert is divided in two zones regarding temperature. In the warmer zone, the annual average temperature ranges from 22° to 28°C (figure 6-10).This zone is located in the Gulf Low Plains region (warmest zone, 22°C) and neighboring regions, follows a more or less parallel direction to the Sierra Madre Oriental, and reaches the southern part of the Chihuahuan Desert (annual average temperature 18°C). The colder zone includes the northwest region of the desert, and its limit is more or less inclined in a southwest-northeast direction. Its highest value is marked by the 18°C isotherm. It covers parts of the zone near the northern end of the eastern slope of the Sierra Madre Occidental, its continuation in the USA and the high mountains of New Mexico and Texas.

The daily and annual fluctuations in temperature are considerable. In some places, extreme values of fluctuation of 24°C may be reached in one day (Soto-Mora and Jáuregui 1965:20), and 60°C may be reached in a year (Morafka 1977b:19). The variation between the average temperatures in the coldest month and the hottest is also considerable—from 7 to 20°C (Morafka 1977b:19)—with the resulting effects on the relative humidity, the surface rock weathering, life, and human activities, including the consumption of energy. Frosts in this arid land do not occur in the lower parts of the Coastal Lowland region, but they do occur in the rest of the Chihuahuan Desert. They are more frequent in places of higher latitude and altitude.

For most of the desert, they usually occur from the beginning of October until the end of March. But early frosts occur occasionally in southern regions in September or as late as early May, with resulting harm to agriculture.

Precipitation

Precipitation is also subject to modifications imposed by latitude, topography, and continental relation. Generally it also follows a gradient that diminishes in centripetal form. At the margins of the Chihuahuan Desert, some places have an annual precipitation of 550 mm, which is above the old criterion of 250 mm as a limit for the desert. As McGinnies, Goldman, and Paylore (1970:3) point out, such a criterion is of little value. There is a minimal average precipitation below 200 mm annually in the middle of the desert. This zone may be the most arid of the Chihuahuan Desert. It is located on the limits that separate the states of Coahuila and Durango and in other smaller zones (figure 6-11).

Such lines represent only the average of precipitation received and measured by climatic or thermopluviometric sta-

tions. Precipitation, as well as temperature, shows great annual fluctuations, with high variations from one season to another. San Luis Potosí Valley, for example, which has an annual average of 360 mm precipitation, shows fluctuations from 119 mm in some years to 700 mm in others.

The precipitation rate in the Chihuahuan Desert is very common to other places with predominant rainfall in summer. February, March, April, and usually most of May are the driest months. The rainy season usually starts at the end of May, establishes itself in June, and reaches its height in July. It diminishes but does not stop in August, allowing for a short period known as *canicula* ("dog days").

The rainfall increases in September, which sometimes is the month of major precipitation depending on cyclonic disturbances originating in the Caribbean, their number, intensity, and the route followed by them. Finally the rainfall begins to diminish in October and ends about the middle of the month. November is usually a dry month, but December and January present another period of rains, though they average less than a tenth of the summer rains. This rainfall generally is the result of the advance of cold continental polar masses of a north-south direction that collide with warm and humid winds originating in the Gulf of Mexico.

TYPE OF PRECIPITATION

During the summer the Chihuahuan Desert storms are usually isolated showers. Scarce but very heavy, they may cause flash floods.

The precipitation peaks tend to show a direct relation with the solar cycles of eleven years and with the intensity of the cyclones originated in the Caribbean. Hence the years 1944, 1955, and 1966 were extremely wet. But since every rule has an exception, 1977 was rather dry.

Usually the summer rains are isolated showers scattered from June to October with two peaks of concentration, one in July and another in September, and a period of relative dryness in August. However, this rule also has exceptions. The period between autumn 1973 and autumn 1974 was one of the driest of the last decade for the southern region of the Chihuahuan Desert: no water at all fell from October 1973 to August 1974. It was necessary to import water from more humid places for indispensable needs. In September 1974, it did not stop raining for 22 continuous days, causing floods and drownings.

The year 1958 was one of unequaled floodings for the northern end of Chihuahua, although according to statistics, it was not supposed to be a rainy year. The only similar precedent had been registered in 1904. On the other hand, 1952 and 1953 were quite dry.

The winter precipitation occurs in the form of cold drizzles that are persistent but light. The bad weather originates in the Oriental Coast of Mexico, which Mexicans call *Nortes* ("Norths") due to their place of origin. Not all *Nortes* reach

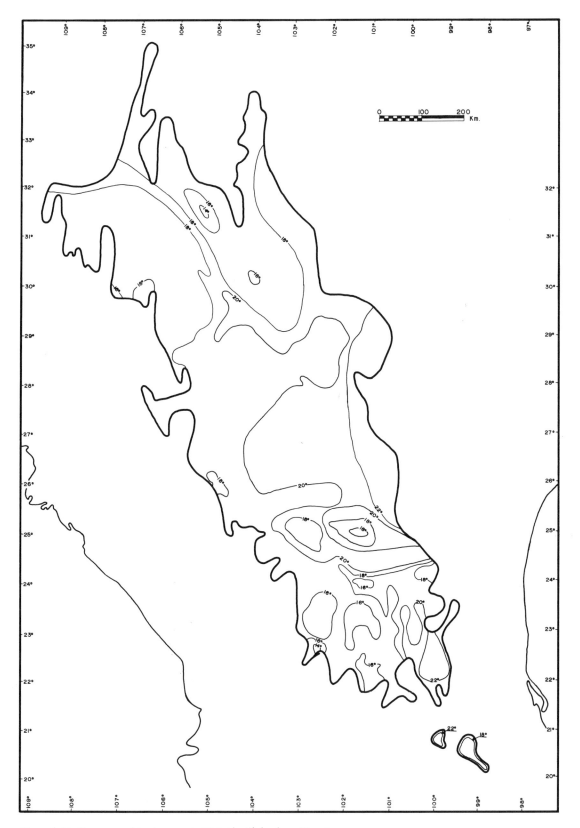

Figure 6-10. Mean annual temperature in C° in the Chihuahuan Desert.
Adapted and simplified from H. Gannett (Map of the United States showing mean precipitation
in *Irrigation principles and practices,* O. W. Israelsen, Wiley, New York, 1950, figure 3) and
from CETENAL (now Dirección de Estudios del Territorio Nacional, México, D.F.)

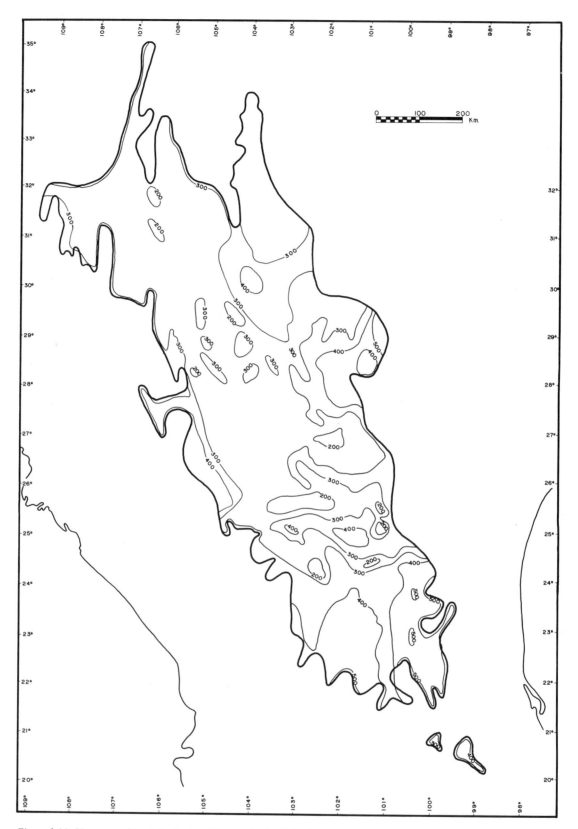

Figure 6-11. Mean annual precipitation in millimeters in the Chihuahuan Desert.
Adapted and simplified from H. Gannett (Map of the United States showing mean precipitation,
in *Irrigation principles and practices,* O. W. Israelsen, Wiley, New York, 1950, figure 3) and
from CETENAL (now Dirección de Estudios del Territorio Nacional, México, D.F.).

enough intensity to pass the Sierra Madre Oriental and drop some water on the desert, but when they do, the bad weather lasts only a few days, mostly in January.

SNOW

When the humidity coming from the *Nortes* is sufficient and the polar continental masses are severe, snow may occur. The snow gradient in the Chihuahuan Desert has to do with the latitude and altitude. In New Mexico, Texas, and Chihuahua, there are ten to twenty snowfalls annually, depending on the altitude and slope exposure. This number drops along with decreased latitude, but it is locally compensated by the altitude. In the southern parts of the desert with an altitude above 1500 m but less than 2500 m, snowfalls appear approximately every twenty years. This includes Querétaro and Hidalgo where the last two snowfalls occurred in January 1947 and January 1967.

Above 2500 m, the frequency of snowfall in the southern part (high parts of Sierra de Zacatecas and San Miguelito) is approximately one every ten years. The last occurred in January 1977. This is not a phenomenon confined to deserts but occurs in places as far away as the transversal volcanic axis located at least 100 km south of the most southern Chihuahuan Desert limit. In the areas below 1000 m, snowfalls either never occur or are so unusual tha they are not mentioned in bibliographic data.

HAIL

Hail does not occur as frequently as in other deserts. However, the strong hailstorms that generally occur together with some isolated drizzles at the end of winter or beginning of spring cause considerable damage due to their intensity. The annual crops, be they winter or summer, are usually not affected since they have either been harvested or not yet established. An opposite situation occurs with perennial or semiperennial crops that are seriously damaged (mostly fruits and alfalfa). The natural vegetation, particularly succulent plants, apparently suffers great temporal damage also.

In the hailstorm that occurred March 26, 1978, in San Luis Potosi Valley, covering a diameter of about 60 km, most of the leaves of young century plant (*Agave* spp.) specimens were completely destroyed. Practically all the pulp was eliminated. Only the fibers running longitudinally along the leaves of these plants remained connected to the plant, but as an entangled and hanging long hair. The stem joints of some large prickly pear cactus (*Opuntia* spp.) were pierced through by hail pellets, despite the fact that some of them were 5 cm in thickness. Many pellets did not pass through but were incrusted inside the stem joints. Many of these plants, which normally do not lose their leaves during winter, were practically defoliated. However, by the middle of spring and the beginning of summer, the most damaged ones, although they still showed signs of the occurrence, had recovered their initial strength, usually by a speedy replacement of their lost parts.

Throughout the Chihuahuan Desert hail occurs in a particular spot with an average frequency of four times per year.

Relative Humidity

The relative humidity is low on average and fluctuates in the different regions between 40 and 60 percent (Morafka 1977b:19).

Winds

The wind system in the Chihuahuan Desert is quite variable from one place to another because of the topography and the multiple directions presented by the numerous mountain ranges. The greatest wind intensities occur between February and March, although their sudden occurrence in June or July is not unusual. In winter the winds are very cold. Depending on the locality (the bolsones, for example, or places that are unsheltered by vegetation), the wind may carry a great amount of solid particles (suspended sand or clay), which create dust storms, locally known as *viento chivero* ("goat's wind") or *viento negro* ("black wind"). The first name derives from the popular belief that this kind of wind can kill goats. *Viento negro* obstructs visibility and darkens the sky greater than a black cloud.

Besides these storms, during the first months of the year, a great number of convection whirlwinds, which also contribute to blowing the soil occur in the valley bottoms.

Solar Radiation

Solar radiation according to Morafka (1977b:63-64) averages 500 (+) langley.

CLIMATE

Dregne (1977) includes the Chihuahuan Desert in his map "Status of Desertification in the Hot Arid Regions" and designates it as "hot." Morafka (1977b:13-25), based on an analysis of the meteorological variables, compared his results with the opinion of other authors and designates it as "warm." The difference between the two is that Dregne distinguishes only between the arid polar regions as cold deserts and the rest of them as hot. Morafka says that the arid polar regions should not be considered as real deserts, so for the other arid lands he gives a wide classification between cold and hot. The Chihuahuan Desert is found in an intermediate situation and is designated as warm.

The following maps presented here are based on two different criteria: Meigs's (quoted by McGinnies, Goldman, and Paylore 1970) and Köppen's modified by García (1964).

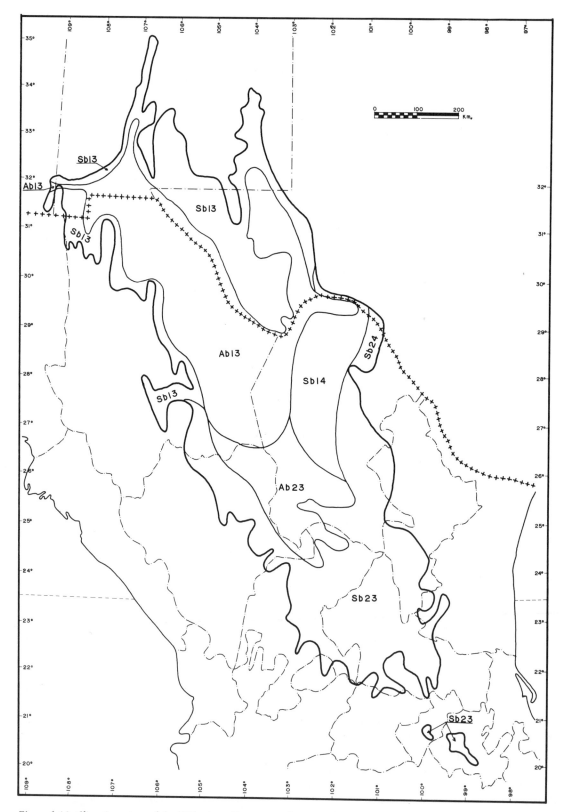

Figure 6-12. Climatic regions of the Chihuahuan Desert according to Meig's system and based on W. G. McGinnies, B. J. Goldman and P. Paylore (Deserts of the World, University of Arizona Press, Tucson, 1970) and on F. Medellín-Leal and Gómez-González (Management of natural vegetation in the semi-arid ecosystems of Mexico, in Management of semi-arid ecosystems, edited by B. H. Walker, Elsevier Scientific Publishing Co., Amsterdam, 1980). Used with permission.

Figure 6-12 has been developed according to Meigs's system (McGinnies, Goldman, and Paylore 1970:xxi, xxvi). It is different in details from the one presented for the same region by the above-mentioned authors and also different from the one elaborated by Medellín-Leal and Gómez-González (in press). One reason is that the scale used by the authors mentioned above is different than the one I use here. The second reason is that more data have been used to develop figure 6-12. According to this map and following Meigs's criteria (quoted by McGinnies, Goldman, and Paylore; 1970), six types of climate occur in the Chihuahuan Desert: Ab13, Ab23, Sb13, Sb14, Sb23, and Sb24.

Type Ab13 represents an arid climate with rain mostly in the summer, with an average temperature in the coldest month between 0° and 10°C, and with an average temperature in the hottest months between 20° and 30°C. This type of climate covers a minimal part of the state of Arizona, two southern portions of New Mexico, a strip along the Rio Grande Basin, part of the Pecos River Basin in Texas, and almost all of the desertic portion of Chihuahua and the northwest of Coahuila.

Type Ab23 represents an arid climate, with rain mostly in the summer, an average temperature of the coldest month between 10° and 20°, with an average temperature of the hottest month between 20° and 30°C. It is located southeast of type Ab13, practically in the middle of the desert, and covers two very small portions of the southeast end of Chihuahua, the southwest of Coahuila, small portions east of Durango, and the high western extreme of the state of Zacatecas.

Type Sb13 is a semiarid climate, with rain mostly during summer, an average temperature of the coldest month between 0° and 10°C, and an average temperature in the hottest month between 20° and 30°C. This type is found as a relatively narrow strip surrounding the more arid climates of type A. It is located following a clockwise direction in the eastern part of the desertic zone of Chihuahua that begins north of Parral, occupies the minimum part of the state of Sonora that corresponds to the Chihuahuan Desert and occupies a minimal part of the state of Arizona. In the Chihuahuan Desert it continues as a narrow strip surrounding type Ab13 in the high part of the Rio Grande Basin in New Mexico, it widens later, covering the south-central part of New Mexico and the extension of this region toward Texas, and finally thins out surrounding type Ab13 in the high part of the Pecos River Basin. The type Sb13 mainly bounds Ab13 type and also the northern ends of the Sb14 and Sb23, and in one point it touches the northwest end of type Ab23.

Type Sb14 represents a semiarid climate, with rain mostly in summer. The average temperature of the coldest month is between 0° and 10°C, and the average temperature of the hottest month goes above 30°C. It occupies a small portion of the Rio Grande Basin in Texas and practically all of the middle of the state of Coahuila. It limits to the northwest with the Sb13 climatic type, to the northeast with the Sb24 climatic

type, to the southeast with the Sb23 type, to the south with Ab23, and to the west with Ab13.

Type Sb23 represents a semiarid climate, with rain mostly in the summer. The average temperature during its coldest month is between 10°C and 20°C, and the average temperature of the hottest month is between 20°C and 30°C. Type Sb23 occupies all of the southern part of the Chihuahuan Desert, taking up about one third of it, to the southeast and south of Coahuila, the southeast of Tamaulipas that corresponds to the Chihuahuan Desert, all of the desertic part of San Luis Potosí, the middle and east of Zacatecas, all of the desertic portion of Aguascalientes, the western end of the desert in Durango, and the southwestern limiting strip of the Chihuahuan Desert south of Parras, as well as all of the isolated portions of the desert that correspond to the states of Querétaro and Hidalgo. This climatic type limits to the northwest with the Sb13 type and at one point with the Ab223 and Sb14.

Type Sb24 corresponds to a semiarid climate, with rain mostly in the summer. The average temperature of the coldest month is between 10° and 20°C, and the average temperature of the hottest month surpasses 30°C. This, the hottest region of the desert, occupies a small portion of the Rio Grande Basin north of the state of Coahuila and a tiny portion of the state of Texas. This climatic type only limits to the west with the type Sb14.

All the desert is bounded by more humid zones with different climates, whatever the classifications used, but those other limiting climates are not mentioned here.

The second climatic map (figure 6-13) is based on Köppen's system modified by García (1964). Six main types of climate are presented: BWhw(e), BWkw(e), BS₀hw(e), BS₀kw(e), BS₁hw(e), and BS₁kw(e).

Type BWhw(e) corresponds to an extremely hot arid climate, where the annual fluctuation of the monthly temperature average is more than 7°C. The average temperature of the hottest month, as well as the annual average temperature, surpasses 18°C. Precipitation occurs predominantly in the summer; hence the driest season, although it may present scarce precipitaion, is winter. The annual precipitation average expressed in centimeters (cm) is below the value of the temperature expressed in °C degrees (this value = T) or up to T + 14. This climate type extends throughout the Chihuahuan Desert, occupying practically all of the middle of it and spreading over part of the Rio Grande Basin in Texas and part of the Pecos River Basin in the south of New Mexico, as well as all of the oriental region of the states of Chihuahua and Durango, the north, west, and south of Coahuila (except the Sierra de Parras), and the southernmost end of the state of Zacatecas. It also includes a small isolated portion toward the north-center of the state of Chihuahua. Since it occupies the part that may be considered the middle of the desert, it is surrounded by the other climatic types. Because of its scale,

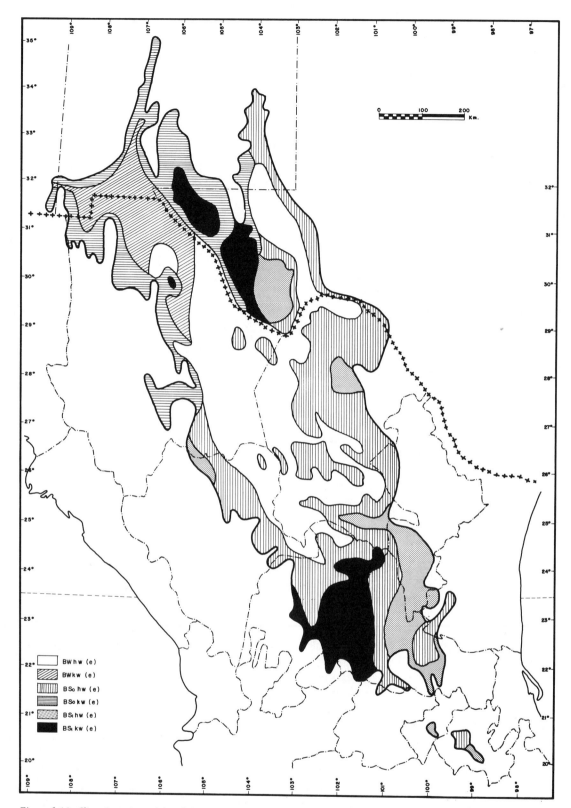

Figure 6-13. *Climatic regions of the Chihuahuan Desert. Köppen System modified by Garcia (1964).*

Adapted and simplified from S. Pettersen, (*Introduction to Meteorology,* copyright 1958, McGraw Hill, New York), from CETENAL (now Dirreción de Estudio del Territorio Nacional, and the Universidad Autonoma de México, México, D.F. 1970. Used with permission.

Legend:
- BWhw (e)
- BWkw (e)
- BSₒhw (e)
- BSₒkw (e)
- BSₗhw (e)
- BSₗkw (e)

this type is not represented on the map. The region where the climatic type BWhw(e) is found presents a rather irregular topography with very different altitudes that have an influence over the temperature as well as over the other meteorological variables. It causes a geat variety of changes in the climate over relatively small portions of land, producing a sort of mosaic of a very complicated design.

Type BWkw(e) is a warm but extremely arid climate, where the annual fluctuation of the average temperature is higher than 7°C. The average temperature of the hottest month surpasses 18°C, but the annual temperature average is less than 18°C. Precipitation occurs predominantly in the summer; therefore winter is the driest period, although some precipitation in the form of light rain and snow may occur. The annual precipitation average expressed in cm is located below T or up to T + 14. This climatic type extends over part of the southwest quadrant of New Mexico and the most western peak of Texas, occupying part of the Rio Grande Basin. Around the southwestern part of New Mexico, it extends toward Arizona and through it to other arid zones of North America. Toward the south it occupies the most nordic portion of the state of Chihuahua, where it includes a portion of land that also presents climate type BHhw(e). It also limits toward the east over the states of Texas and Chihuahua. As for the rest, it is totally surrounded, at least concerning the Chihuahuan desert, by the $BS_0kw(e)$ climatic type.

$BS_0hw(e)$. There is a $P/T < 22.9$ relation in this type where P = annual average precipitation in mm and T = annual average temperature in °C; therefore it belongs to the driest of semiarid hot climates where the annual fluctuation of the monthly average temperature is greater than 7°C. The annual average temperature, as well as the average temperature of the hottest month, surpasses 18°C. Most of the precipitation occurs during summer, therefore the dry season is winter. The $BS_0hw(e)$ type is found limiting the Chihuahuan Desert from the highest part of the Pecos River Basin in the state of New Mexico, part of the Rio Grande Basin in Texas, the Big Bend Basin, almost all of the middle of Coahuila, especially the Serranías del Burro, the lower half of Sierra de Parras, and all of the longitudinal middle of San Luis Potosí, from the north to the south, the small region of the Chihuahuan Desert that corresponds to the southwest of Tamaulipas, the extension of this region toward San Luis Potosí, considerable parts of north-central Zacatecas, and the west of the desertic portion of Durango, as well as the northern part of Valle del Mezquital or desert zone of the state of Hidalgo and many high parts of mountains, splashing here and there in a mosaic form, the states of Chihuahua, Durango and Coahuila. This climatic type is found limiting mostly with the BWhw(e) type and with the other semiarid types.

Type $BS_0hw(e)$ presents the same $P/T < 22.9$ relation as the previous one, but in this case it represents the driest of the semiarid warm climates where the annual fluctuation of the

monthly average temperature is higher than 7°C. The annual average temperature does not reach 18°C, but the average temperature of the hottest month surpasses that. Precipitation occurs fundamentally during summer, and although there might be precipitation during winter, this is the driest season. Type $BS_0kw(e)$ is located as a strip of variable thickness limiting the Chihuahuan Desert in its northwest end over the states of Chihuahua, a tiny portion of the northeast end of Sonora, New Mexico, and Texas. As in the case of BWkw(e) type, it goes over into Arizona, where it also extends to other deserts. Type $BS_0kw(e)$ partially limits with all of the other climatic types described here.

The relation in climatic type $BS_1hw(e)$ is $P/T < 22.9$. It represents the least dry of the semiarid hot climates where the annual fluctuation of the average monthly temperature is higher than 7°C. The annual average temperature, as well as average temperature of the hottest month, surpasses 18°C. Precipitation occurs mostly in the summer, and the driest season is winter, although some precipitation may occur during it. The $BS_1hw(e)$ type is located in the southeastern extreme of the mountain mass, which includes the Santiago Mountains, the Chalk Mountains, and Christmas Mountains, and over part of the Big Bend National Park in southwest Texas; a small portion of the middle of Coahuila and the Estribaciones Occidentales de la Sierra Madre Oriental located within the limits of the Chihuahuan Desert, which includes the superior half of the Sierra de Parras; also in Coahuila, and portions of the states of Nuevo León and Tamaulipas as well as a small fraction of the states of Chihuahua and Durango, the entire Querétaro Desert zone and the most southern strip of the Chihuahuan Desert, located southeast of the Hidalgo desert region. This climatic type limits only with the other semiarid climates.

Type $BS_1kw(e)$ also shows a relation $P/T < 22.9$. It represents the least dry of the semiarid warm climates with an annual fluctuation of the monthly average tempeature higher than 7°C. The annual average temperature does not reach 18°C, but the average monthly temperature of the hottest month surpasses the 18°C. Precipitation occurs mostly during summer; therefore although there might be some precipitation, the driest season is winter. The $BS_1kw(e)$ type is found only in the higher altitudes of the Chihuahuan Desert. It spreads over the mountain masses, which include the Guadalupe Mountains in New Mexico and Texas, the David Mountains, and the eastern part of the Santiago and Chalk mountains in Texas. Afterwards, it is found throughout the whole Chihuahuan Desert, splashed here and there over the high parts of the mountains, sometimes in minimum extensions (not marked on the map) and appearing later in the south on the limits of Zacatecas and San Luis Potosí, covering the western part of this last state and the southeastern part of the western portion of the desert zone of Zacatecas. It also covers the totality of the Chihuahuan Desert part corresponding to the state of Aguascalientes. This climatic type only limits with other semiarid climates and not with the arid ones under Köppen's concept.

Meigs (1957:13-32) and Morafka (1977b:21-25) present a comparative analysis of the climate of North American deserts. A brief summary of them indicates the principal meteorological differences between the Chihuahuan Desert and other North American arid lands. The Chihuahuan Desert, with an annual temperature average of 19.2°C is 3° to 4°C colder than the Sonoran Desert, which extends more or less in the same latitudinal range but closer to the coast. It presents the same average temperature of the Mojave, which is more nordic but generally lower in altitude; and it is 7°C warmer than the Great Basin.

The annual thermic variation of up to 60°C (−10 to 50°) is greater in the Chihuahuan Desert than in the others; 50°C in the Sonoran (0° to 50°), 40°C (5° to 45°) in the Mojave, and 55°C in the Great Basin (−15° to 40°). The annual thermic variation of the monthly average between the hottest and the coldest is smaller in the Chihuahuan Desert than in the others. It is only 14.3°C while in the Sonoran it is 20.5°C, in the Mojave it is 23.7°C, and in the Great Basin, 27.5°C. Perhaps both characteristics are due to the differences in the precipitation.

The average rainfall in the Chihuahuan Desert is slightly more abundant than in the other arid zones. The rain occurs in quick, strong showers or thunderstorms. In the arid lands with winter rains, these usually occur in relatively persistent drizzles. But although the rain in general is more abundant, its presence is more variable because the annual average deviation reaches up to 60 percent in the Chihuahuan Desert against a deviation of only 30 percent in the Great Basin.

Snow is more frequent and abundant in some parts of the Great Basin, where it can reach 60 cm monthly from the middle of November to the beginning of April. In the Chihuahuan Desert each snowfall reaches a maximum value on 7 cm, and the snow usually melts in a few hours or at least in a few days in the bottom of the valleys, although, of course, it could be more persistent in the high mountains. Neither Meigs nor Morafka gives figures for the snowfalls in the Sonoran Desert, although Meigs says it is rather scarce.

Fog, relative humidity average, and solar radiation are almost the same in all North American deserts, but fog is more frequent in the Sonoran, the relative humidity average is near half (25 percent) in the Mojave, and the solar radiation is slightly less in the Great Basin.

Other meteorological variables are not reported, so they cannot be compared.

Aridity Indexes

Figures 6-14 and 6-15 show the aridity indexes calculated with two different criteria. Figure 6-14, which uses the Budyko ratio, is taken with slight modifications from the map prepared by Henning and Flohn (1977a). Introduced into climatology by M. I. Budyko and is called the radiational index dryness, which refers to the ratio between net radiation and precipitation at the surface.

Figure 6-15 uses the aridity index devised by Emberger, modified by Stretta and Mosiño (1963). According to these modifications and Emberger's Aridity Index, I_A is obtained when the following formula is applied:

$$I_A = \frac{(M + m)(M - m)(m + 45)}{P}$$

where:

I_A = Index of aridity
P = Annual average precipitation in mm
M = maximum average temperature of the hottest month in °C
m = minimum average temperature of the coldest month in °C

I_A ranges from less than 18 to more than 1000; less than 18 corresponds to an extremely humid climate and 500 to more than 1000 to extremely desertic climates. The values from 18 to 52 include climates that can be classified in several degrees within the group of humid ones. Values between 53 and 118 indicate semiarid climates and those from 118 to 500 arid climates, according to Emberger's classification modified by Stretta and Mosiño (1963; Valera-Adam 1963; Soto-Mora and Jáuregui 1965).

GEOLOGY

The oldest rocks found in the Chihuahuan Desert correspond to the latter part of the Paleozoic, particularly Permian. Although they do not occupy great areas, they are particularly important in the semiarid mountains of southern New Mexico and the southwest end of Texas. In the Guadalupe Mountains, for example, according to Dunbar (1955:279-308), the most important Permian deposit in the world is found.

This situation suggests that the northern part of the Chihuahuan Desert was probably the first to emerge when the Laramide Revolution occurred. Plenty of evidence indicates that the rest of the desert was generally covered by the sea during the Mesozoic (Rzedowski 1966:17-18), although some parts of Coahuila might have remained as islands or peninsulas during long periods of such geologic era (Sierra de Alamitos). Triassic and Jurassic rocks outcrop in a large part of the hills splashed throughout the desert, but the outcrops are not large enough to be represented on a map as small as the one presented in figure 6-16. The predominating rocks can be classified by three main types and several secondary ones, considering importance and extension: (1) marine sedimentary

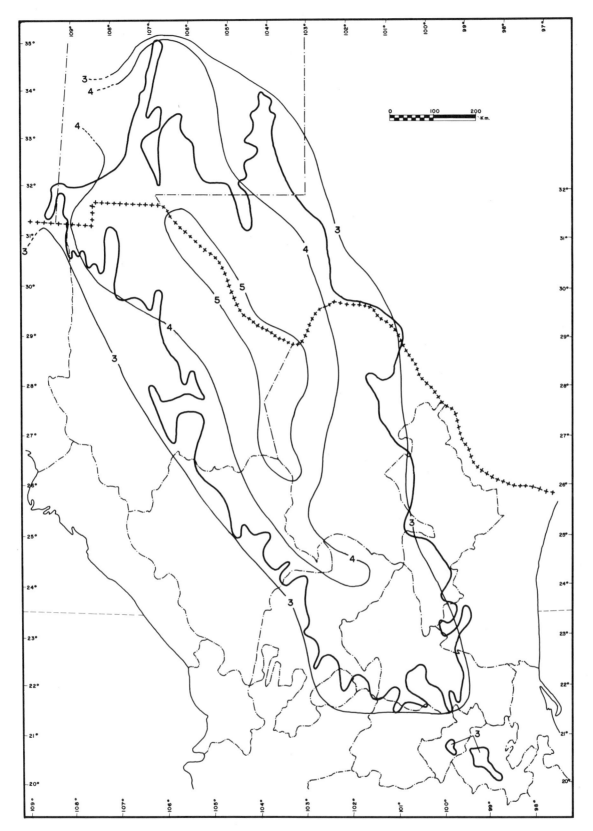

Figure 6-14. Climate aridity index of the Chihuahuan Desert
(Budyko-ratio). Slightly modified from D. Henning and H. Flohn (*Climate aridity index,*
UN Conference on Desertification, Nairobi, UNEP-FAO-UNESCO, 1977).

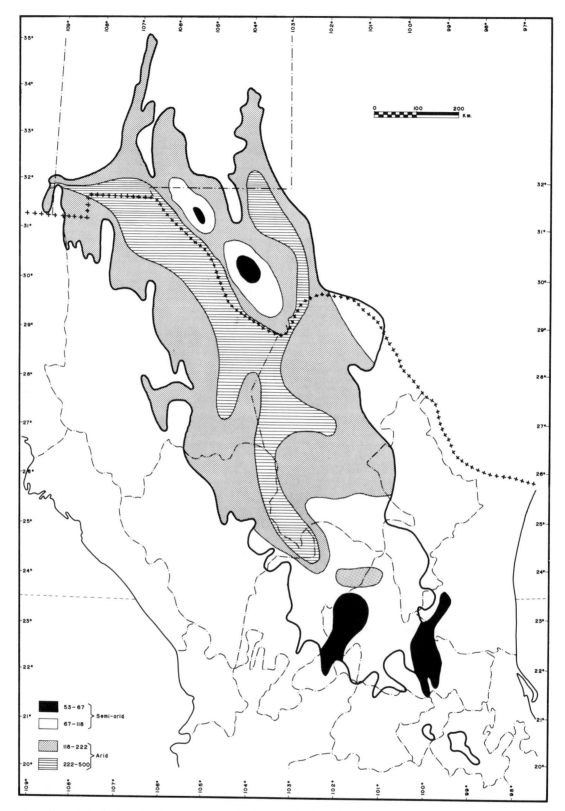

Figure 6-15. Distribution of Emberger's aridity index in the Chihuahuan Desert.
Modified by E. P. Stretta and P. A. Mosiño (*Distribución de las zonas áridas de la República Méxicana*, Ingeniería Hidráulica en México, 1963) and adapted from modifications and additions from C. Soto-Mora and E. Jáuregui (*Isotermas extremas e índice de aridez en la República Mexicana*, Instituto de Geografía, UNAM, 1965).

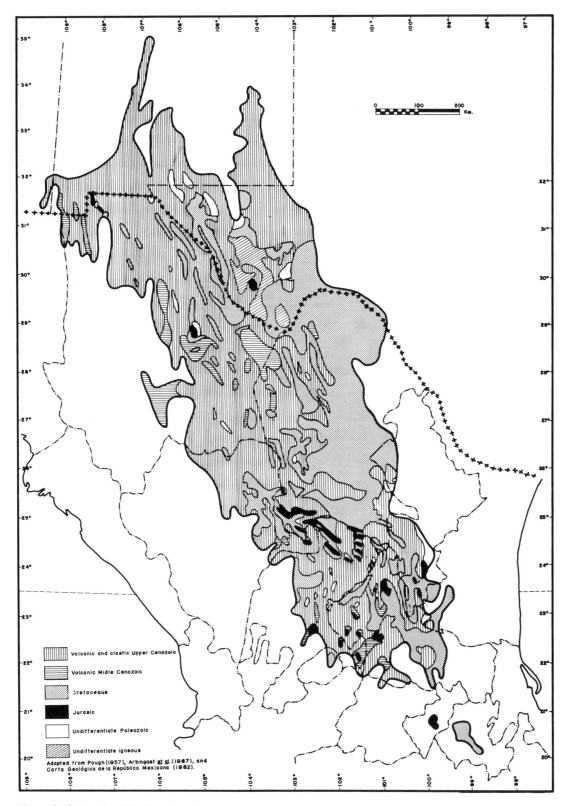

Figure 6-16. *Superficial lithology of the Chihuahuan Desert.*
Adapted from F. H. Pough (*A field guide to rocks and minerals,* Houghton Mifflin, Boston,
1957), S. A. Arbingast et al (*Atlas of Texas,* University of Texas, Austin, 1967), and Carta
Geológica de la República Mexicana (*Cartas Geológicas y Mineras* 5, Instituto de Geología,
1962).

cretaceous rocks predominantly to the east and northeast of the desert; (2) rocks of extrusive igneous origin (mainly rhyolites) of Eocene-Miocene origin; and (3) rocks derived from the degradation of the two former ones. The secondary rocks are the granites and the plutonic rocks from the beginning of the Tertiary (Rogers et al. 1964:57-63) exposed in some isolated points (for example, Salinas and north of Zacatecas) and the lava currents, mainly basalts from the Quaternary, that splash the desert here and there.

In short, we know practically nothing of the geological history of the zone occupied by the Chihuahuan Desert regarding the Precambrian and most of the Paleozoic. From the end of Paleozoic and during all the Mesozoic, almost all of this region was below the sea, leaving some portions as small islands or peninsulas connected to firm land to the north (Rzedowski 1966:17-18; Morafka 1977b:29-31). The Laramide Revolution marks the end of the Mesozoic and the beginning of the Cenozoic. During it, the emergence of all of the northern portion of the Mexican republic and a great part of southwestern United States occurred. The Sierra Madre Oriental was formed, and possibly the Chihuahuan Desert also. They were established due to the barrier created by the same Sierra Madre Oriental to the humid winds from the east. From the Eocene to the Miocene, the enormous effusion of lava that closed off humid winds coming from the south and west was produced during the formation of the southern sierras of the desert, the Sierra Madre Occidental, and other mountains. During the Tertiary, the communication with South America was produced and interrupted, an important event because it explains the origin of most of the flora and part of the fauna of this desert. Wells (1977) disagrees. He thinks it is most convenient to explain the South American origin of some floristic elements by means of another kind of phenomena such as the dispersion by migratory birds. Wells, based on his work in the Big Bend area, believes the Chihuahuan Desert is less than 11,000 years old.

During the end of the Tertiary and beginning of the Quaternary, the volcanic activity that reached its highest level during the Eocene to Miocene slowly diminished. Meanwhile the characteristic erosion of an arid zone refilled the bottom of the intermountain valleys with great amounts of clastic deposits. The more humid epoch, originated by the glaciations occurring farther north, led to a reduction in the extension of the desert but not its disappearance. A number of lakes surely existed at the time and perhaps drainage was more active, particularly through the fluvial system of the Rio Grande, which must have been larger. That might explain, partially at least, why the valleys with salty or gypsum soils are not as frequent as they apparently should be because of the age of this desert. After glaciations, which affected only the periphery of the Chihuahuan Desert, the dryness apparently increased, producing the bolsones and closed basins, favoring the eolian erosion over the hydraulic one, and producing a senile land-

scape with few or no signs of rejuvenation (Morafka 1977b:29-31, 173-211). However, it is important to remember Wells's opinion (1977).

Hydrogeology

Probstein and Alvarez (1973) have pointed out the great importance that the desalination of the brackish water is having in the development of the arid zones. They consider as brackish waters those whose content in salts and dissolved solids is less than 15 percent of that of sea water. As reference points, the total content of salts or dissolved solids (TDS) in sea water is approximately 35,000 parts per million (ppm); potable water has less than 500 ppm. According to these definitions, the Rio Grande water, as well as that of Pecos River, should be considered brackish. The Rio Grande shows a maximum of 10,200 ppm where the main anions, in order of importance, are sodium and calcium and the main cations, also in order of importance, are chloride and sulphates. A similar situation occurs in the lower part of the Pecos River, while in the high part of the river, a maximum of 6800 ppm of TDS occurs, but in this case the importance of the anions and cations is inverted. In the upper part of the Pecos River the calcium is more abundant than the nitrogen and the sulphates more abundant than the chlorides. There is little available information for the other surface waters of the Chihuahuan Desert.

The information available regarding underground water covers only the U.S. portion of the Chihuahuan Desert, the northeast end of Durango, and the semiarid zone of San Luis Potosí. In New Mexico and the southwestern end of Texas, 70 percent of underground water has a total dissolved solids (TDS) content between 1000 and 3000 ppm, and 30 percent of the aquifers have between 3000 and 10,000 ppm TDS. Sodium and calcium are the principal anions, and sulphates and chloride are the cations, in that order of importance. For the Valle de Caballos in the northeast end of Durango, there are no figures, but they are assumed to be similar to those found in San Luis Potosí. In the state of San Luis Potosí Probstein and Alvarez (1973), based on Villalobos and Díaz-de-León (1971), show that the brackish waters have a TDS content between 1500 and 5000 ppm. Calcium is the most important anion, with sodium next in importance, while the cations, in order of importance, are the sulphates and the chlorides. According to Villalobos and Díaz-de-León (1971) and Díaz-de-León (1977), the underground waters in the state of San Luis Potosí vary from potable to very brackish depending on the zone of origin. Those originated and trapped in igneous formations are less salty than the ones of sedimentary formations. Thermal springs, some of which show a certain radioactivity, are present. Unfortunately they have not been studied thoroughly (Stretta et al. 1964).

SOILS*

For the classification of arid and semiarid zone soils, the morphological and genetic characteristics are taken as a base, as well as soil formation factors identified under this concept initially by Jenny (1941). On the other hand, the evolution in the knowledge of soil in its dynamic concept as a natural body leads to a better identification of its properties.

In soil inventories performed in Mexico, with the object of establishing a more adequate use, there is a noticeable lack of definition in the soil units of arid and semiarid zones. The same is true to a lesser degree in the United States, although such soils continue to represent sources of rich potential. The identification of these soils demands a knowledge of their intrinsic characteristics, their behavior regarding natural and cultivated vegetation, and the variables imposed by natural phenomena and man.

The reconnaissance of the Chihuahuan Desert soils is focused on their classification through systems that allow grouping them by similar nature in order to give them cartographic expression.

Identifying soils for classification is a difficult task and more complex than the one applied to individual plants or animals, because no net limits exist for individuals in soils, and changes occur gradually. Consequently the taxonomy cannot be rigid.

For the identification of the Chihuahuan Desert soils, it has been necessary to refer to three systems of classification that usually have been used:

1. The superior categories of orders, suborders, and large groups that consider the family, series, type, and phase of soils. This system was revised in the United States by Thorp and Smith in 1949 (quoted by Ortiz-Villanueva 1973:181).

2. The seventh approximation system and subsequents, which is the systematic grouping proposed by the U.S. Department of Agriculture (1960) and considered as official in the United States since 1965 (Ortiz-Villanueva 1973:180).

3. The FAO-UNESCO system, based on preliminary projects started in 1964, to be used in the elaboration of the world map of soils (Ortiz-Villanueva 1973:180)

Because of the great surface covered by the Chihuahuan Desert, only the soil units that were established by Grande-López are presented here. The blank spaces on the soil map in figure 6-17 correspond to units yet to be identified.

Aridisoles

These are soils of places with little precipitation. They have an ochric epipedon and one or several additional diagnostic horizons: cambic, argylic, natric, calcic, gypsic, or salic, associated with duripans. With a low content of humus, its brown color is replaced by gray when the content of salt intensifies, and it accumulates on the surface.

The distribution of these soils is conditioned to the topography, and there exists a marked relation with the basal rock as an effect of the very low weathering. When they are red, these soils are associated with a larger oxidation of iron components.

Salt Soils

These are halomorphic soils, belonging to the aridisoles, of poorly drained regions. They belong to the suborder Orthid, large group Salorthid. When there is base leaching, Natralbolls and Natrustolls climatic phases occur, which present organic matter dissolution. When the humus content is low, the Natriboralfs, Natrustalfs, and Natrargid climatic phases occur, and when the humus and calcium content are high, Salorthidic Calciustolls occur. Those of the Argid suborder possess a developed argylic horizon. The salt soils are not presented in the map.

Lithic Soils

These are thin mountain soils with 8 to 20 percent slope and even larger, with two phases: the infertile (districal) and the fertile (eutrical).

Fluvents

Alluvial soils occupying the flat regions of floods and zones of fluvio-lacustrine sedimentation adjacent to the elevations, they include the fertile, infertile, and calcaric phases.

Molisoles

These include a molic epipedon, as well as cambic, argylic, or natric horizons. The suborder Ustols, which are the Molisoles located between the Udols and the Aridisoles, have a moderate proportion of organic matter and brown or walnut stain. When not well developed, they are called Haplustolls. If there is clay leaching, they are Argiustolls, and Calciustolls when they are derived from limestone rock.

---

*The section was written by Raúl Grande-López of the Laboratory of Soils of the Instituto de Investigatión de Zonas Desérticas de la Universidad Autónoma de San Luis Potosí.

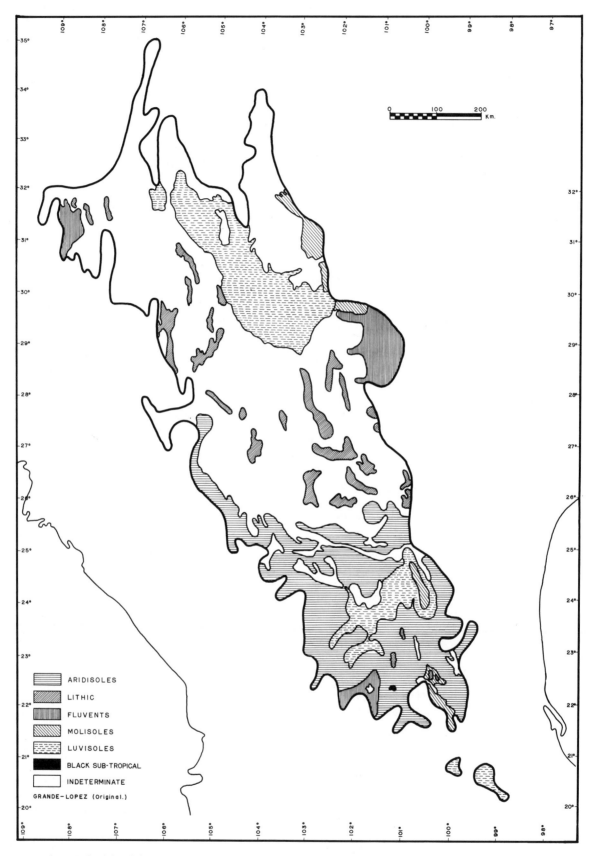

Figure 6-17. Soils of the Chihuahuan Desert.
Original by Raúl Grande-López, Instituto de Investigacion de Zonas Deserticas.

Luvisoles

These soils show a downward movement of clay in their morphology with the resulting accumulation. The climatic phases are Hapludalf, Haploxeralf, equivalent to Brown Forest Podsolized Soils, Rodoxeralf (terra-rosa), Haplustalfs (laterite podsolic soils), and Hapludults (red-yellow podsolic soils).

Black Subtropical Soils

These soils have been located in the San Luis Potosí Valley (Boulaine 1964). They lie over limestone rock and compare to the Vertisoles, being black soils with expandable clays in a two-to-one ratio. The most outstanding characteristic is their dark color, despite the low content of organic matter (2 percent maximum), and their slightly alkaline pH. The possibility of MnO_2 as the reason for the coloring is considered, although it is also possible (Segalen 1964) to attribute it to bridges of calcium, magnesium, iron, and manganese cations among the organic matter and the two-to-one clay, probably because under such conditions a very small part of organic matter is enough to produce the black color.

VEGETATION

The number of primary vegetation formations or climaxes in the Chihuahuan Desert is relatively reduced. They are greatly influenced by such factors as mother rock, drainage, and local edaphic characteristics. The historical geology and human factors must also be considered.

The communities that I consider to be primary, and which have been used basically to elaborate the map of vegetation shown in figure 6-18, are only six:

1. Microphyllous desertic brushwood.
2. Rosettophyllous desertic brushwood.
3. Crassicauleous brushwood.
4. Desert grassland.
5. Sclerophyllous brushwood.
6. Aciculifolious-squamifolious Low Forest.

Under the existing conditions in the Chihuahuan Desert, the adaptation gradient to the aridity diminishes progressively from vegetation type 1 to type 6.

Microphyllous Desertic Brushwood

This type represents the background scenery of the entire Chihuahuan Desert. It is not exclusive to this desert but is found in practically all of the other North American deserts,

with the characteristic monotony mentioned by McGinnies, Goldman, and Paylore (1970:16). It is widely ubiquitous and found from the sea level (although the lowest limit of the Chihuahuan Desert is located at a 250 m level) to approximately 2000 m.

The type of vegetation I have called mezquital in a previous work (Medellín-Leal and Gómez-González in press) is included. It tolerates all of the climates mentioned for this desert in the corresponding section and all mother rocks, drainage, and soils, except perhaps the gypsic type.

I have watched it develop with relative strength in zones that may remain flooded for more than a month at a time (Palma de la Cruz and other floodable zones of the bottom of the San Luis Potosí Valley) and in relatively salty soils (around El Huizache, SLP), although this situation is not very frequent. The wideness of its geographic extension can be explained in two ways. This is either the most evolved vegetation type and therefore the best adapted to the general condition of extraordinary variation of the desert (which in the extreme annual thermic fluctuation reaches 60°C; snowfalls; variations in rainfall from 100 to 700 mm of precipitation in different years for the same place; floods; prolonged droughts; different textures and chemical characteristics of the soil), or it is a highly evolved type and therefore is plastic enough to occur everywhere.

The most frequent biological form in this type of brushwood is the shrubbiness or subshrubbiness one of small leaves, generally with a production of different resins. Most of the shrubs conserve their leaves through the year. Prickle forms are very common, and aphyllia is a characteristic of many of its elements. All of the biological forms of the desert are found in some degree. Most of the plants are entomophilous.

The dominant layer is made up of plants with a 0.5 to 2 m height, depending greatly on local characteristics such as exposure, latitude, mother rock, drainage, temperature, and humidity, although they are of the same species.

The most characteristic species of this layer are *Larrea divaricata* and/or *Flourensia cernua*, which in extreme cases may be the only phanerogamic inhabitants of the community. These two elements sometimes exclude themselves mutually and sometimes are found associated. *Koeberlinia spinosa* is most commonly found along with them, as well as *Holacantha emoryi* and different species of *Ephedra* and *Acacia*. They may be relatively far apart from each other or sometimes very intertwined, forming actual cushions of mutual protection against grazing.

In the subshrubbish layer it is very common to find the creeping species of *Opuntia*, *Zinnia*, and other compositae, which in favorable periods of high precipitation and good temperature form a colorful flower rug over the desert soil.

In this brushwood there may exist a higher layer, up to 8 m in height, where different species of *Yucca*, *Prosopis*, or *Myrtillocactus* stand out. Although they dominate the land-

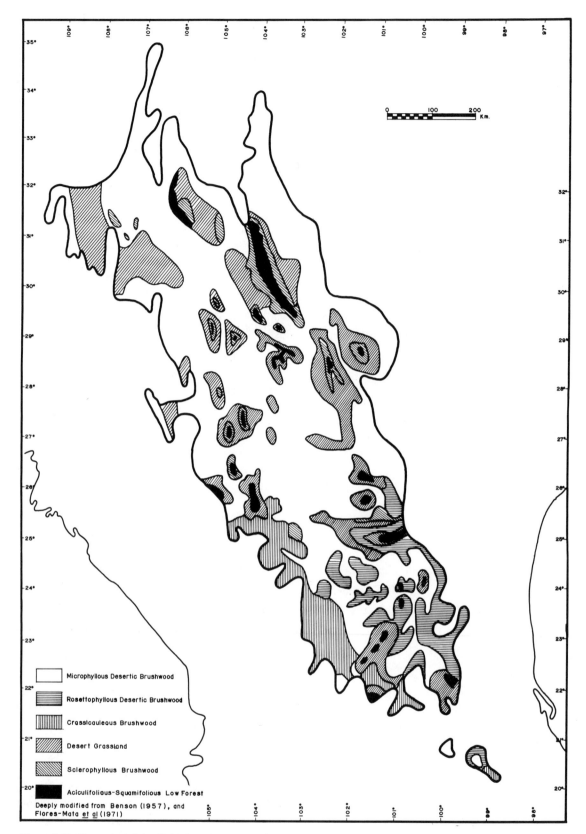

Figure 6-18. *Vegetation of the Chihuahuan Desert.*
Deeply modified from L. Benson (*Plant classification*, Heath, Boston, 1957) and G. Flores-Mata
et al (*Mata de tipos de vegetacion de la República Mexicana*, Secretaría de Recursos Hidráulicos,
1971).

358

scape physiognomy, they do not greatly alterate the relative floristic composition. The herbaceous layer is well represented by a large number of annual plants and by many perennial plants, including small cacti belonging to a great number of species and some grass species. The creeping and climbing types are also very abundant, especially those with bulbs, rhizomes, or tubers. Such abundance of cryptophytes is believed to correspond to a secondary indicator of desertification, caused by continuous stamping, since this brushwood is subject to an intense disorderly grazing by all kinds of cattle.

It is not unusual to find an epiphyte layer formed almost exclusively by *Tillandsia recurvata* over *Prosopis,* which in some cases make it appear gray.

This type of brushwood is subject to a disorderly and intense exploitation. It is common at the bottom of valleys and even on the sides of some hills to clear them for agricultural purposes. This agriculture may have variable success and very frequently also has problems of desertification due to accumulation of salts if irrigation is used.

Although this type of brushwood does not represent an important source of construction material, the same cannot be said regarding energy sources since it supplies most of the available combustible material. People of less means continue to use it on both sides of the border.

The microphyllous desertic brushwood practically borders with all of the other vegetation types of the desert, since it is highly ubiquitous. I have observed near Galeana, Nuevo Leon, a forest of *Pinus* sp. of a more mesophilous type, whose undergrowth was almost exclusively formed by *Larrea.*

Rosettophyllous Desertic Brushwood

This type is more specifically tied to conditions of aridity. However, such aridity is not of a climatic type since it is mostly found in semiarid climates. It is due to a lack of water caused by excellent drainage since it is located on the sides and high parts of alluvial fans of limestone hills between 500 and 2600 m altitude. It is better represented in the southeast part of the Chihuahuan Desert, perhaps indicating that low temperatures may represent a limiting factor—possibly not the average during the coldest month but the extreme minimum temperature. However this would not explain why it is not more frequent in the east, particularly in the lower elevations of the Gulf Coastal Lowland. Perhaps the drainage or the slope is not great enough. The mother rock is almost exclusively of the limestone type, sometimes marly. The soils are shallow, immature, from gray to black, and with a great amount of gravel. The bare rock forms part of the great extensions of the landscape. The biological form predominates in this type of vegetation and, as the name suggests, it is of rosetted leaves with or without a well-visible and developed stem. Plants of the most diversified groups and families characterize this type

of brushwood: *Selaginella* (*Selaginellaceae*) and other ferns within the Pteridophytes; *Echeveria* and related ones (Crassulaceae); *Leuchtenbergia* (Cactaceae) within the dicotyledons; and *Agave, Hersperaloe, Yucca* (Agavaceae), *Hechtia* (Bromeliaceae), *Dasylirion,* and *Nolina* (Liliaceae) among the monocotyledons. These are just a few of all of the species, genera, and families.

The dominating layer is shrubbish or subshrubbish with a height that fluctuates between 0.5 to 2 m. The elements of this layer sometimes form pure strands of one species covering considerable areas of land and practically all of the available soil surface. These species include *Agave lecheguilla, A. striata, A. stricta,* and *Hechtia glomerata.* The rosetted biological form is not exclusive of this type of brushwood. What is impressive here is the accumulation and frequency of the species representing it. Besides rosetted plants, the dominant layer exhibits with less richness almost all the biological forms that exist in arid zones. The most colorful are the barrel-shaped Cactaceae, such as *Echinocactus* and *Ferrocactus.* Also included are species of genera from other families, such as *Acacia, Bursera, Cassia, Ephedra, Euphorbia, Fouquieria, Karwinskia, Krameria, Larrea, Leucophyllum, Mimosa, Mortonia, Parthenium, Sophora,* and *Zinnia.*

Most of these plants are entomophilous and very frequently associated with a particular insect species. Within the rosettophyllous desert brushwood, there is sometimes a higher straight layer formed by plants that are also generally rosettophyllous, such as several species of *Yucca* and *Dasylirion.* The herbaceous layer is formed by plants that are mostly rosetted, many of them belonging to the Cactaceae: *Leuchtembergia, Pelecyphora, Roseocactus,* and others. They are found semi-buried in many cases. The herbaceous annuals vary in frequency according to rainfall. In their best periods they are briefly very colorful. A great number of composites and grasses form part of this group.

From the point of view of usefulness to humans, the Rosettophyllous desertic brushwood is highly exploited in a disorderly way. Some of the plants are used as combustibles or as construction material (mostly *Yucca* and *Dasylirion*). *Agave lecheguilla* and *Yucca carnerosana* are intensely exploited in the production of hard fibers, *Euphorbia antisyphilitica* for wax production, and *Parthenium argentatum* for the production of rubber. Others are used as food for humans, either the whole plant or some of its parts: these include *Yucca* (floral parts, fruits), *Echinocactus visnaga* (the whole plant in the production of sweets), and *Ferrocactus* (floral buttons). Lately this type of vegetation has been subject to looting because many of the numerous species are quite beautiful. I have seen this occur in San Luis Potosí and Zacatecas at the south end of the Chihuahuan Desert, and Burleson (1977:654) reports the same for the north. This brushwood borders the microphyllous desertic brushwood, the crassicauleous brushwood, the desert grassland, and the sclerophyllous brushwood.

Crassicauleous Brushwood

This type of vegetation is formed by fleshy plants, more specifically large cacti so characteristic of the Mexican landscape. It is present at altitudes between 1000 and 2500 m and is a semiarid climatic type. It is different from the rosettophyllous type in that it is found on the sides of hills formed by rocks of igneous origin or on well-drained soils originated from such mother rock. The soil may be very superficial or very deep; it can be sandy and from brown to reddish brown in color or gray to black if it derives from basalts in which the amount of calcium is poor. Its distribution is limited in part by the drainage and in part by the acidity of the soil. Although as a vegetational type it may be considered as one of the primaries, there are serious doubts regarding its original distribution, and it is believed that a great part of its actual distribution has been greatly influenced by human activities. I consider it as the first step toward less harsh and more humid conditions, at the same level as the desert grassland.

The predominant biological form is that of fleshy plants, including cacti, in the form of large prickly pears, spheres, columns, and so on, and *Agave* in the form of large century plants.

The main layer, if not the only one, is composed of numerous species of *Opuntia* and *Agave*, from 2 to 5 m in height, although for a given locality the size is more or less constant and the layer includes only one, or very few, species.

In some cases, the plants are mixed with *Myrtillocactus* or with *Lemaireocereus*. This situation is more common toward the extreme southeast part of the desert. Sometimes also *Yucca* spp., *Prosopis* sp., and the introduced *Schinus molle* blend with this brushwood. All the dominating plants are of zoophyllous pollination.

The shrubs are represented by species of *Acacia, Aloysia, Baccharis, Celtis, Eysenhardtia, Lycium, Rhus, Verbesina,* and others. The subshrubbish layer between 0.5 and 2 m is formed by species of *Adolphia, Dalea,* and others.

The herbaceous layer, less than 0.5 m in height, is very colorful or totally inconspicuous, according to the season. Many composites and some grasses are particularly notable. Since this type of vegetation is usually subject to intense exploitation for grazing as well as for harvesting of fruit and other parts of the body of the Cactaceae used in human and animal feeding, it is very hard to decide which plants of the herbaceous layer are original and which are secondary. Creeping and cryptophyte types that are indictors of a strong desertification process are well represented.

The Crassicauleous brushwood is in contact in some places with the rosettophyllous desertic brushwood but usually does not blend with it, and the ecotones are brief and quite clear due to the mother rock type (Rzedowski 1955). Its largest contacts are with the desert grassland and eventually with the sclerophyllous brushwood. Because it represents a step toward higher humidity conditions, currently and because of desertification, it is spreading toward the south and to the west over regions that in other times were not part of the Chihuahuan Desert.

Desert Grassland

The desert grassland as a type of primary vegetation has frequently been described for the Chihuahuan Desert in two main categories: climatic grassland and edaphic grassland (Rzedowski 1966:167-174; 1975). However, in the Chihuahuan Desert the edaphic factor also has a strong influence in the distribution of at least the rosettophyllous desertic brushwood and the crassicauleous brushwood. The edaphic factor is also considered by other authors (Burgess and Northington 1977; Henrickson 1977; Johnston 1977; Morafka 1977b:39-58; Pinkava 1977; Powell and Turner 1977). Therefore in the following description the usual distinction is not made, and they are considered as a more or less direct result of the soil characteristics. The predominant biological form is grass shape (Rzedowski 1966:166). The pollination is autogamous or anemophilous.

Having established this clarification I propose a new interpretation of the grassland: (1) desert grassland originated mostly by physical characteristics in the soil and where the chemical characteristics are secondary and (2) desert grassland caused mostly by chemical characteristics of the soil.

Desert grassland originated mostly by physical characteristics of the soil. Since this type of vegetation has been exploited for several centuries for pasture or for cultivation, assuming that a carpet of grasses over the soil is a guarantee of harvest (Medellín-Leal and Gómez-González in press), the plotting currently is more theoretic than real and its description is closer to an ideal than to a tangible truth. One can only infer what existed in those places currently used by cattlemen or what existed in places of the Chihuahuan Desert where most of the rain-fed agriculture is being carried out.

From the altitudinal and climatic point of view, this first type of desert grassland would occupy a similar situation to that of the crassicauleous brushwood, and therefore it must also be considered as a transitional type of vegetation toward more humid regions. The present edaphic-physical factor that characterizes it is that it develops over soils of a depth between 0.5 and 1 m, limited in their bottom part by igneous mother rock, and in the hills by low ridges and slopes of the same, or by a horizon of calcarean or waterproof ferric induction in the valleys. Apparently this waterproof horizon facilitates the establishment of the grassland from New Mexico to San Luis Potosí. In places where the indurative horizon is interrupted, frequently the grassland is splashed by other types of vegetation. The chemical characteristics of the soil are highly variable and are not the ones that cause the existence of this type of grassland. This type would be formed mostly by only one layer of approximately .7 to 1 m in height, homogeneous, with a dominance of several species of *Aristida, Bouteloua, Heteropogon, Muhlenbergia, Stipa,* and some others. But at present, in places that one would imagine that such a primary community would exist, either the soil is gone due to erosion and the indurative horizon is exposed in large extensions, or the land is covered by inefficient agriculture usually of temporal type where gramineans such as corn, wheat, rye and barley are cultivated. Finally in the best of the cases, there is a grassland where there may be a great number of invading shrubs and trees, mostly of the following genera—*Acacia, Adolphia, Agave, Asclepias, Berberis, Dalea, Juniperus, Larrea, Opuntia, Parthenium, Pinus, Quercus, Sophora,* and *Yucca*—and an enormous number of composites and other undesired herbs. Among them are some toxic to cattle, such as *Aplopappas* spp., *Asclepias subverticillata, Baileya multiradiata, Notolaena sinuata var cochisensis, Peganum harmala, Psilostrophe gnaphalodes,* and *Xanthium* spp.

Desert grassland caused by the chemical characteristics of the soil owes its origin primarily to factors of salt accumulation and therefore is found in old lakes and at the bottom of bolsones. The flora tend to be scarce, although the cover of the soil in some cases can be great if the type of existing grass is rigid and difficult for cattle to eat. This occurs in some floodable places without a great accumulation of soluble salts where the very low grassland, approximately 0.25 m high, is composed mostly of *Buchloe dactyloides* or *Sporobolus wrightii.* In soils with more salts, which usually present a certain alkalinity, the grassland is formed by *Sporobolus wrightii* and *S. nealleyi,* when there is minor saltiness. It changes to *Distichlis spicata, Eragrostis obtusiflora,* and *Sporobolus pyramidatus* with intermediate saltiness, or *Spartina spartinae* when saltiness is greater and a high water table exists. With these grasses there are other halophytic species blending more or less and pertaining to the following genera: Atriplex (Chenopodiaceae), *Sesuvium* (Aizoceae), *Suaeda* (Chenopodiaceae), and others (Rzedowski 1966:173; Henrickson 1977). In the case of gypsophilous grasslands, the soil covering is more scarce. In this case the main dominants are *Bouteloua breviseta, B. chasei,* and *Muhlenbergia purpussi,* sometimes along with *Coldenia hispidissima, Dalea filiciformis,* and *Flaveria oppositifolia* (Gomez-Gonzales 1973; Burgess and Northington 1977; Powell and Turner 1977).

The desert grassland as a traditional type is in contact with all of the other types of vegetation present in the Chihuahuan Desert.

Sclerophyllous Brushwood

This type develops at altitudes greater than 1000 m with climates that can hardly be considered arid. It represents the second step toward greater humidity and is always found on slopes and high parts of hills that splash the desert. It sometimes makes contact at higher elevations with pine forests and other types of woods that, although they may be located as spots within the desert, are not desertic any more and therefore are not considered here. Neither the mother rock nor the soil is of great influence for its existence, which apparently depends more on humidity and low temperatures present in one part of the year. This type of vegetation usually has just one shrubby layer 1 to 3 m in height where the predominating biological form has flat coriaceous leaves. Pollination may be anemophilous or entomophilous. It is mostly formed by shrubbish species, including *Quercus, Amelanchier, Arbutus, Arctostaphylos,* and *Rhus.*

This type is abundantly exploited by man, mostly for fuel, and it often forms mosaics with the other types of vegetation, but most of the grassland is related to topography, altitude, and shelter.

Aciculifolious-Squamifolious Low Forest

This type probably also represents a second step toward more humid climates, or perhaps it is already a third step after the sclerophyllous brushwood. It blends in its higher elevations with humid vegetation. The dominant biological type is small trees with needle or scale-form leaves, of anemophilous pollination, belonging to the genera *Pinus* and *Juniperus.* It usually constitutes open communities, sometimes very exploited by man, for fuel or for the edible seeds of many of the *Pinus* species. This type of vegetation sometimes blends in with the rest of vegetation types of the Chihuahuan Desert but more often does it with the desert grassland and with the sclerophyllous brushwood.

Historical Character of the Vegetation Origin

Of the many authors who have concerned themselves with the historical phytogeography of the Chihuahuan Desert, Rzedowski has been the most active. Several of his publications (1962, 1965, 1966, 1973, 1975, 1978) provide a good resumé of previous investigations.

It is not my intention to summarize them here but to make a correlation between them and the vegetation types currently found in the Chihuahuan Desert, although as Rzedowski (1973) mentions, the general lack of data concerning Mexican historical geology leaves this field open to a wide discussion.

Authors have not agreed on the geological period when the uplifting of Mexican High Plateau and Chihuahuan Desert occurred. However I believe that such uplifting started during the Laramide Revolution at the end of the Mesozoic and beginning of the Cenozoic and that it still has not ended.

When the uplifting occurred, there began to exist a virgin field that had the possibility of being colonized. Besides the existence of real aridity conditions, the age of which is also still under discussion (Rzedowski 1962, 1973; Morafka 1977a; Wells 1977), the initial colonization must have developed over rocks in different states of fragmentation but without an evolved soil. Taking this base, the theory developed here regarding the origin of the present Chihuahuan Desert vegetation is as follows.

As the first step toward the formation of the actual vegetation, colonization started from the previously emerged places with elements preadapted to a certain degree of aridity. This first colonization would be by paleotropical species belonging to genera such as *Acacia, Aristida, Artemisia, Atriplex, Cassia, Dodonaea, Ephedra, Frankenia, Jatropha, Lycium, Menodora, Mimosa, Notholaena, Peganum, Solanum,* and *Suaeda,* which today are represented by one or several species in all of the types of vegetation.

Perhaps together with these plants, the paleoendemisms shared by some of the arid zones of North America originated as a second step, among them *Fouquieria* (Fouquieriaceae), and *Mortonia* (Celastraceae), with several species, and the monotypical genera such as *Adolphia infesta* (Rhamnaceae) from the desert grassland and crassicauleous brushwood; *Chilopsis linearis* (Bignoniaceae), which is found along gullies and streams of the microphyllous desertic brushwood; *Holocantha emoryi* (Simaroubaceae) from the microphyllous desertic brushwood; *Lindleyella mespiloides* (Rosaceae) of the Rosettophyllous desertic brushwood. As an exclusive paleoendemism of the Chihuahuan Desert I mention only *Orthosphenia mexicana* (Celastraceae) of the rosettophyllous desertic brushwood but, as Johnston (1977:344-356) has pointed out, the number of endemic plants in the Chihuahuan Desert is very high.

Simultaneously with this formation of paleoendemisms, the desert grassland formation must have initiated as part of the second step—not as an endemism at a specific generic level but as an endemism at a level of all vegetation. This endemism is clearly represented by the gypsophyllous grasslands, which apparently are exclusive of the Chihuahuan Desert (Rzedowski 1975). Among them are species of the genera *Bouteloua, Muhlenbergia,* and sometimes *Sporobolus;* plus species of *Dalea* (Leguminosae), *Flaveria* (Compositae), and many others.

Of the other types of grasslands where the elements are not exclusively of the desert but that evolved specifically due to arid conditions, are representatives of the Gramineae *Andropogon, Aristida, Bouteloua, Erioneuron, Hilaria, Lycurus,* and *Muhlenbergia,* for the most common grassland; in the alluvial bolsones with accumulation of salts, the grasses *Distichlis, Eragrostis, Hilaria,* and *Sporobolus,* and in the floodable grasslands with few salts, species of *Buchloe, Hilaria, Panicum, Muhlenbergia,* and *Paspalum.*

The third step would be marked by the first communication during the Tertiary with the southern part of the American continent, which brought a first contribution of neotropical elements already adapted to the desert; among those are species

of *Castela* (Simarubaceae), *Coldenia* (Boraginaceae), *Condalia* (Rhamnaceae), *Flourensia, Franseria, Gutierrezia* (Compositae), *Larrea* (Zygophyllaceae); *Prosopis* (Leguminosae), *Sanvitalia* (Compositae), and many others, all of them from the microphyllous desertic brushwood. Here it is convenient to remember that Wells (1977) holds quite a different opinion. For him the origin of the present Chihuahuan Desert occurred less than 11,500 years ago after the Wisconsinian glacial; therefore these South American elements arrived at the Chihuahuan Desert in an accidental way until the Holocene and not during the Tertiary.

A fourth step toward the complete colonization of the Chihuahuan Desert would be represented by the occupation of the hills of marine sedimentary origin of limestone constitution by elements of neotropical origin but not highly adapted at the beginning to aridity, although with certain preadaptations to the relative lack of water. Among these elements there were species of *Agave, Echeveria, Hechtia,* and *Yucca.*

A fifth step would be represented by the colonization of the consolidated lava effusions and by mesophilous elements, which little by little were adapted to the arid conditions. Such elements, also of neotropical origin, give rise to the formation of neoendemisms where we find an enormous interspecific variety that would explain or justify putting them in this point of the sequence. Among them are representatives of the genera *Agave, Bursera* (Burseraceae), *Hechtia, Opuntia,* and other genera of the Cactaceae and many other representatives of the cassicauleous brushwood.

A sixth step would be a new South American contribution of mountain flora, especially Compositae, such as *Baccharis, Brickellia,* and *Zinnia,* which would give the final touches to all types of local vegetation.

The adaptations of *Juniperus* (Cupressaceae), *Pinus* (Pinaceae), and *Quercus* (Fagaceae) could be considered as a bannister along this stairway to total vegetation, but perhaps with a strong impulse during the Quaternary glaciations to conditions of relative aridity in the types of vegetation of the sclerophyllous brushwood and aciculifolious-squamifolious low forest.

The historical background of the Chihuahuan Desert vegetation has been presented here as a series of steps, but possibly several phenomena occurred simultaneously.

The dominant elements of the evolved vegetation types in situ—the desert grassland, sclerophyllous brushwood, and the aciculifollious-squamifolious low forest—are predominantly autogamous or anemophilous in their pollination, while the pantropicals and neotropicals are basically zoophilous in theirs.

FAUNA

When one writes about the Chihuahuan Desert fauna, one would like to do it from an ecological and integrated point of view. Unfortunately the faunistic studies, although abundant (Lowe 1970), are very dispersed and correspond to very different dates. Most studies do not give an ecological approach and usually do not cover all of the desert but refer only to a certain political-administrative unit.

Nevertheless this situation seems to be changing. The Pleistocene and Present mollusks are studied by Metcalf (1977); the Pleistocene-Holocene fossil herpetofauna by Devender and Worthington (1977); the Pleistocene higher vertebrates by Harris (1977); the fishes by Contreras-Balderas (1977), Miller (1977), and Minckley (1977); the herpetofauna by Axtell (1977), Conant (1977), Degenhardt (1977), Milstead (1977), Morafka (1977a, 1977b), and Scudday (1977); the avifauna by Barlow (1977), Hunt (1977), Raitt and Pimm (1977), Wauer (1977), Wauer and Ligon (1977), and Webster (1977) though at species level there is no one endemic bird in the Chihuahuan Desert (Phillips 1977:617); and the mammals by Baker (1977), Conley, Nichols, and Tripton (1977), Findley and Caire (1977), Hailey (1977), Packard (1977), Schmidly (1977), and Villa-Ramírez (1977). However, several of these works take into account only the northern part of the Chihuahuan Desert.

Fish

Miller (1977) lists 107 native species of fish in the Chihuahuan Desert, of which 15 percent are undescribed. They include 16 families and 40 genera. Their distribution in the desert is indicated and their origins discussed. The information is incomplete because large areas of the desert streams and rivers have not been sampled. Contreras-Balderas (1974) discussed the changes in and decline of fish habitat in the Chihuahuan Desert. He pointed out that isolated basins have far fewer fish species than are found in major drainages.

Herpetofauna

Morafka (1977) listed 57 species of herpetofauna as being present in true Chihuahuan Desert ecosystems. Their distribution, together with that of other species on the Mexican Plateau, was used to delimit the boundaries of the desert. Conant (1977) listed 19 species of semiaquatic reptiles and amphibians as occurring in streams within the Chihuahuan Desert. He discussed changes in habitat and riparian fauna and their causes. Axtell (1977) studied the geological history of desert playas and their influence on distribution of modern species of amphibians, reptiles and fish. (See also Ohmart and Anderson, appendix 10-A of this book.) Appendix 6-A is a checklist of amphibians and reptiles of the Big Bend National Park in Texas.

Mammals

Packard (1977) estimated that 101 species of mammals representing 55 genera, 20 families, and 7 orders are found in the Chihuahuan Desert south of the U.S.-Mexico boundary.

He listed them according to the trophic levels and types of habitat in which they are found. Schmidley (1977) listed 119 species, 60 genera, 24 families, and 8 orders of mammals found in the Chihuahuan Desert. He discussed the factors governing the distribution of these mammals in the desert, including physiographic barriers, plant-animal relationships, climate, soil, water, food, competition, and human activities. Baker (1977) listed 10 species of Chihuahuan Desert mammals whose existence is threatened largely due to increasing pressure from the activities of man. Appendix 6-C is a checklist of mammals of the Big Bend National Park in Texas.

Birds

Webster (1977), studying the southern half of the Chihuahuan Desert, lists 39 species of birds as breeding in the desert. Raitt and Pimm (1977), studying the Jornado del Muerto plain near the northern edge of the Chihuahuan Desert, lists 16 species of birds in grassland in summer and winter with 7 different species during the two seasons. (See also Ohmart and Anderson, appendix 10-B of this book.) Appendix 6-B is a checklist of birds of the Big Bend National Park in Texas.

Files of the Instituto de Investigacion de Zonas Deserticas list 880 species of vertebrate and invertebrate animals. Many are not found exclusively in the Chihuahuan Desert, and the list is woefully incomplete, particularly concerning invertebrates.

OCCUPATION OF THE CHIHUAHUAN DESERT BY MAN

Apparently the first evidence of the Chihuahuan Desert occupation by man was from 23,000 to 20,000 B.C. It has been called Sandia culture because its remains have been found in the Sandia Mountains, near Albuquerque. This one and other cultures like the Folsom and the San Juan flourished between 23,000 and 10,000 B.C. primarily in the high part of the Rio Grande Basin. Like other primitive cultures of the world, most human occupation did not move away from the course of the river. The desert was marginally exploited under the form of hunting and gathering. The Cochise culture flourished from 15,000 to 500 B.C. Little is known of what happened between 500 B.C. and 500 A.D. From 500 to 600 A.D. the groups known as Anasazi, which already had some semidomestic animals (turkey and dog) and were semiagricultural, occupied the high part of the Rio Grande Basin once more, giving birth later to the Pueblo culture about 800 A.D. which through the desert, brought a greater demographic pressure to it. The marginal forest wood was razed not just for construction material and mine timber but also as fuel on large scale (Beltran 1956). Large cattle exploitations were established in the semiarid climates. European culture invaded and in many cases devastated the desert. This situation continues to this day.

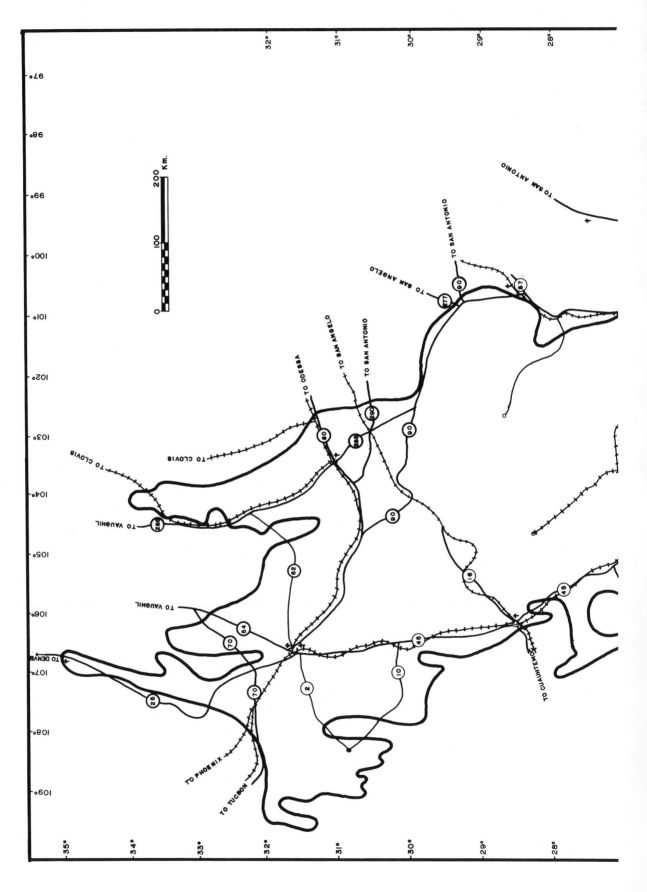

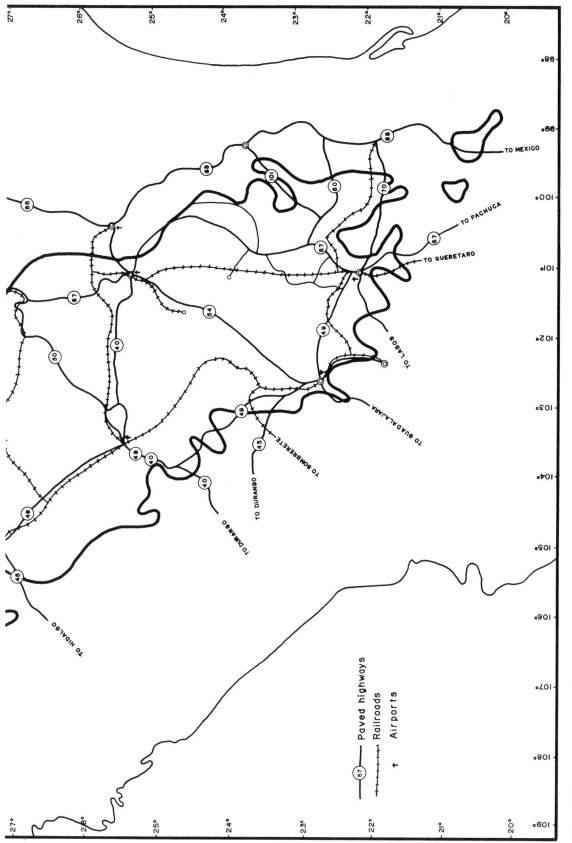

Figure 6-19. Communications in the Chihuahuan Desert.
Compiled from many sources by Raúl Martínez-de-la Rosa. Instituto de Investigación de Zonas Deserticas.

Paved highways
Railroads
Airports

TO MEXICO
TO PACHUCA
TO QUERETARO
TO LAGOS
TO GUADALAJARA
TO SOMBRERETE
TO DURANGO
TO DURANGO
TO HIDALGO

COMMUNICATIONS

I have compared many maps of successive dates supplied by the tourist departments and chambers of commerce from states in Mexico and the United States and have found that communications are spreading more and more, so a map soon becomes obsolete. Nevertheless a map is included here (figure 6-19).

The communications are quite efficient, as much in infrastructure as in services. The U.S. roads are generally excellent and the same can be said for the Mexican roads, particularly those following a north-south route.

TOURISM

The tourist attraction of the Chihuahuan Desert is relative, for it depends more on the different interests of people visiting it or expecting to cross it than on the desert itself. In general, for the average citizen of both countries, it represents more of a nuisance than an attraction. For most young people the main interest is the possibility of crossing over the border, but actually the experience is not very pleasant because neither country shows its best face near the border. The points of greater interest (figure 6-20) are almost all peripheral, and it is necessary to cross very monotonous landscapes in order to get to them.

Apparently, the attraction of the desert for the desert itself increases with one's age and maturity. It is then that the enormous solitude and the relaxation that may be found become attractive, not in the towns but precisely in the lack of them—in a sunset, in a starry sky without any light on the horizon, in the lack of most modern services, in the color of a spring or a summer fleetingly covering the ground with multicolors, in the changing color of naked mountains according to the time of day. They are priceless attractions. For scientists interested in the desert, every step toward the interior of it reserves a new surprise, and each plant; each animal, each rock may have its own enchantment.

SPECIAL PHENOMENA AND EVENTS

A large number of meteorites apparently fall in the Valle de Allende located between Jiménez and Parral in the state of Chihuahua (Clarke et al., 1970:17). Another important phenomenon is the existence of what has been called the "zone of silence." It is a region located where the limits of the states of Chihuahua, Coahuila, and Durango coincide, about 150 km north-northwest of the city of Torreón near a town called Ceballos (figure 6-20).It is said that in this zone the radiophonic communication sets do not work adequately. This phenomenon, according to newspapers, was discovered in 1968, though I do not know of any real scientific study about it. However, popular and journal fantasy has given it enormous circulation, at least in Mexico. The fact that on July 11, 1970, the remains of an Athena-type missile launched by the United States fell in this region has fired the imagination of pseudo-scientific and science-fiction writers who have related it to space phenomena and the possible existence of unidentified flying objects (Melendez 1973).

The other events have more to do with social aspects and the incredible differences that culture may present in the twentieth century. At present, while the laws of both countries are enlightened, it cannot be denied that the Indian populations on both sides of the border have not reached the state of cultural integration that should be desired.

DESERTIFICATION

A recently cited publication by Medellín-Leal (1978) provides the first study of this phenomenon in Mexico. As Dregne (1977) has already mentioned, the zones in the most danger in the Chihuahuan Desert are not in the center of it but on the periphery (figure 6-21).On both sides of the border the causes are slightly different but convergent. In Mexico the main cause of desertification is found in the strong social differences that still persist in the country and apply tremendous pressure on the autochthonous population and are caused by the asymetrical transfer of resources. An example is mining exploitation, which destroys raw materials and does not leave investments and infrastructure for later development that might allow an adequate management of the region when the minerals have been exhausted. On the other side of the border the social competition and the possibility of developing the desert for living and enjoyment influences desertification. Some aspects that contribute to this problem are country houses for city people, underground water being wasted in swimming pools, and the possibility of fun with dune buggies, cycles, and other off-road vehicles.

CONCLUSIONS

Throughout these pages the author tried to introduce a picture of what the Chihuahuan Desert is presently and what science knows of it. What is the conclusion? The wise reader must already have it in mind. Large and beautiful as it is, everything or almost everything has begun to be investigated. However in every field, meteorology, climatology, geology, etc., there exists large, sometimes enormous, holes in knowledge. We may say together with Del-Río, one of the most famous Spanish-Mexican wise men that "everything here which seems known also turns out to be new" (quoted by Arnáiz-y-Freg, 1965:197). For the future it will be necessary to try to integrate the knowledge more. The apparent and colorful aspects are already investigated. Now it is science's turn to extract from the Chihuahuan Desert its most intimate secrets, which may turn out to be the most important.

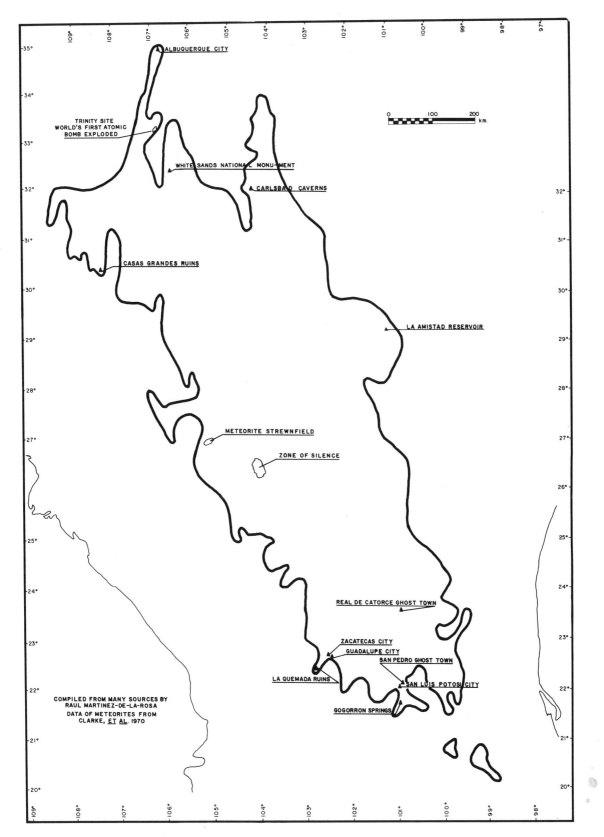

Figure 6-20. Tourism, special phenomena, and events in the Chihuahuan Desert.
Compiled from many sources by Raúl Martínez-de-la Rosa, Instituto de Investigacion de Zonas
Deserticas. Data on meteorites from R. S. Clarke et al (The Allende Mexico Meteorite Shower,
Smithsonian Contributions to the Earth Sciences 5:1-53, 1970).

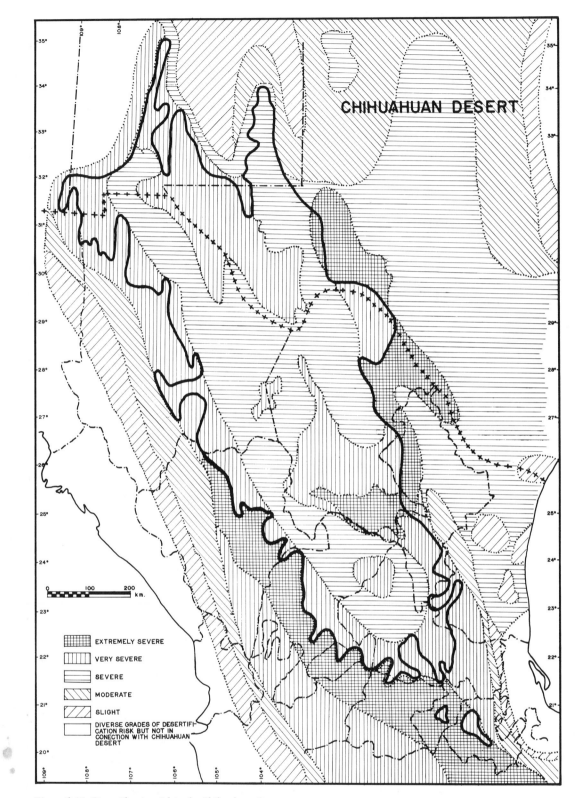

CHIHUAHUAN DESERT

EXTREMELY SEVERE

VERY SEVERE

SEVERE

MODERATE

SLIGHT

DIVERSE GRADES OF DESERTIFI-
CATION RISK BUT NOT IN
CONECTION WITH CHIHUAHUAN
DESERT

Figure 6-21. Desertification risk in the Chihuahuan Desert.
Modified from H. E. Dregne (*Status of desertification in the hot arid regions.* Map for the UN
Conference on Desertification, Nairobi, UNEP-FAO-UNESCO, 1977) and F. García-Castañeda
(*Marco geográfico* de la desertificación en México, in *La desertificación en México,* F. Medellín-
Leal, Universidad Autonoma, San Louis Potosi, 1978, used with the permission of the Instituto
de Investigacion de Zonas Deserticas).

REFERENCES

Alessio-Robles, V. comp. 1942. *Discursos, memorias e informes de Jose Miguel Ramos-Arizpe.* Mexico. UNAM.

Aranda, J. J. 1973. *Reporte del Servicio Social efectuado en el Instituto de Investigación de Zonas Desérticas sobre documentación geológica de las zonas áridas de México.* San Luis Potosí; Universidad Autónoma, Tesis Ingeniero Geólogo.

————. 1975. *Bibliografía geológica de las zonas áridas de San Luis Potosí.* Biblioteca de Historia Potosina, Serie Cuadernos 32. San Luis Potosí: Academia de Historia Potosina.

Arbingast, S. A. et al. 1967. *Atlas of Texas.* Austin: University of Texas, Bureau of Business Research.

Arnáiz-y-Freg, A. 1965. Don Andrés Manuel Del-Río y su ilustre magisterio en México. *Ciencia, Mex.* 23(5):196-200.

Axtell, R. W. 1977. Ancient playas and their influences on the recent herpetofauna of the northern Chihuahan Desert. In R. H. Wauer and D. H. Riskind (eds.), *Transactions of the Symposium on the Biological Resources of the Chihuahuan Desert Region. U.S. and Mexico. Trans. and Proc. Series* 3:493-512. Washington, D.C.: National Park Service.

Baker, R. H. 1977. Mammals of the Chihuahuan Desert region—future prospects. In R. H. Wauer and D. H. Riskind (eds.), *Transactions of the Symposium on the Biological Resources of the Chihuahuan Desert Region. U.S. and Mexico. Trans. and Proc. Series* 3:221-225. Washington, D.C.: National Park Service.

Barlow, J. C. 1977. Effects of habitat attrition on vireo distribution and population density in the northern Chihuahan Desert. In R. H. Wauer and D. H. Riskind (eds.), *Transactions of the Symposium on the Biological Resources of the Chihuahuan Desert Region. U.S. and Mexico, Trans. and Proc. Series* 3:591-596. Washington, D.C.: National Park Services.

Beltrán, E. 1956. El Virrey Revillagigedo y los bosques de San Luis Potosí. *Rev. Soc. Mex. Hist. Nat.* 17:121-132.

————. (ed.). 1955a. *Los recursos naturales de México. I. Estado actual de las investigaciones forestales.* Mexico: IMRNRAC.

————. 1955b. *Mesas redondas sobre problemas de las zonas aridas de México.* Biblioteca Central de la Ciudad Universitaria. 24 a 28 de enero de 1955. México: IMRNRAC.

————. 1964. *Las zonas áridas del centro y noreste de México y el aprovechamiento de sus recursos.* México: IMRNRAC.

Benson, L. 1957. *Plant classification.* Boston: Heath.

Benson, N. L., trans. 1950. *Report that Dr. Miguel Ramos de Arizpe priest of Borbon and deputy in the present general and special Cortes of Spain for the province of Coahuila. One of the four eastern interior provinces of the kingdom of Mexico presents to the August Congress on the natural political and civil condition of the provinces of Coahuila, Nuevo León. Nuevo Santander and Texas of the four eastern interior provinces of the kingdom of Mexico.* Austin: University of Texas Press.

Boulaine, J. 1964. Curso de pedología de las zonas subtropicales. Mimeographed. Chapingo: Colegio de Post-graduados. Curso semestral.

Burgess, T. L., and Northington, D. K., 1977. Desert vegetation in the Guadalupe Mountains region. In R. H. Wauer and D. H. Riskind (eds.), *Transactions of the Symposium on the Biological Resources of the Chihuahuan Desert Region. U.S. and Mexico. Trans. and Proc. Series* 3:229-242. Washington, D.C.: National Park Service.

Burleson, B. 1977. Second keynote address. In R. H. Wauer and D. H. Riskind (eds.). *Transactions of the Symposium on the Biological Resources of the Chihuahuan Desert Region. U.S. and Mexico. Trans. and Proc. Series* 3:651-658. Washington, D.C.: National Park Service.

Carta geológica de la República Mexicana. 1962. Compilada en 1960 por el Comité de la Carta Geológica de México. *Cartas Geológicas y Mineras 5.* Mexico: UNAM, Instituto de Geología.

CETENAL [Comisión de Estudios del Territorio Nacional]. 1970. *Carta de climas a escala 1:500 000 basada en el sistema de Köppen modificado por E. García.* Mexico: CETENAL-Instituto de Geografía de la UNAM.

Clarke, R. S. et al. 1970. The Allende México, meteorite shower. *Smithsonian Contributions to the Earth Sciences* 5:1-53.

Comité Mexicano de Zonas Aridas. 1963. *Informe Nacional. México. Conferencia Latinoamericana para el Estudio de las Zonas Aridas. Buenos Aires. Septiembre de 1963.* Mexico: El Comité.

Conant, R. 1977. Semiaquatic reptiles and amphibians of the Chihuahuan Desert and their relationships to drainage patterns of the region. In R. H. Wauer and D. H. Riskind (eds.), *Transactions of the Symposium on the Biological Resources of the Chihuahuan Desert Region. U.S. and Mexico. Trans. and Proc. Series* 3:455-491. Washington, D.C.: National Park Service.

Conley, W. J., Nichols, D., and Tripton, A. R. 1977. Reproductive strategies in desert rodents. In R. H. Wauer and D. H. Riskind (eds.), *Transactions of Symposium on the Biological Resources of the Chihuahuan Desert Region. U.S. and Mexico. Trans. and Proc. Series* 3:193-215. Washington, D.C.: National Park Service.

Contreras-Balderas, S. 1977. Speciation aspects and man-made community composition changes in Chihuahuan Desert fishes. In R. H. Wauer and D. H. Riskind (eds.), *Transactions of the Symposium on the Biological Resources of the Chihuahuan Desert Region. U.S. and Mexico. Trans. and Proc. Series* 3:405-431. Washington, D.C.: National Park Service.

Cserna, E. G., and Bello-Barradas, A. 1963. Geología de la parte central de la Sierra de Alvarez, Municipio de Zaragoza, Estado de San Luis Potosí. *Univ. Nal. Autón. México. Inst. Geología.* 71 (2):23-63.

Degenhardt, W. G. 1977. A changing environment: Documentation of lizards and plants over a decade. In R. H. Wauer and D. H. Riskind (eds.), *Transactions of the Symposium on the Biological Resources of the Chihuahuan Desert Region. U.S. and Mexico. Trans. and Proc. Series* 3:533-555. Washington, D.C.: National Park Service.

Díaz-de-León, E. 1977. *Composición de aguas en el Estado de San Luis Potosí II.* San Luis Potosí: Universidad Autónoma, IIZD.

Dodge, N. and Zim, H. S. 1955. *The American Southwest.* New York: Golden.

Dregne, H. E. 1977. *Status of desertification in the hot arid regions.* Map. presented at the United Nations Conference on Desertification, Nairobi, August 29-September 9, 1977. A/CONF. 74/31. Nairobi: UNEP-FAO-UNESCO.

Dunbar, C. O. 1955. *Historical Geology.* New York: Wiley.

Findley, J. S. and Caire, W. 1977. The status of mammals in the northern region of the Chihuahuan Desert. In R. H. Wauer and D. H. Riskind (eds.), *Transactions of the Symposium on the Biological Resources of the Chihuahuan Desert Region. U.S. and*

Mexico. Trans. and Proc. Series 3:127-139. Washington, D.C.: National Park Service.

Findley, R. 1977. *Desiertos de América*. Mexico: Diana.

Flores-Mata, G. et al. 1971. *Mapa de tipos de vegetación de la República Mexicana*. Mexico: Secretaría de Recursos Hidráulicos.

Gannett, H. 1950. Map of the United States showing mean precipitation. In O. W. Israelsen. *Irrigation principles and practices*. 2d ed. New York: Wiley.

García-Castañeda, F. 1978. Marco geográfico de la desertificación en México. In F. Medellín-Leal (ed.). *La desertificación en México*, pp. 35-54. San Luis Potosí: Universidad Autónoma, IIZD.

García, E. 1964. *Modificaciones al sistema de clasificación climática de Köppen*. México: Ed. de la autora.

Glantz, M. 1964. *Viajes en México. Crónicas extranjeras (1821-1855)*. México: Secretaría de Obras Públicas.

Gómez-González, A. 1973. *Ecología del pastizal de* Bouteloua chasei. Chapingo: Tésis Maestría Colegio Postgraduados.

González-Cortés, A. 1959. *Los recursos naturales de México. II. Estado actual de la investigación del suelo y agua*. México: IMRNRAC.

———. 1963. Bibliografía. In Comité Mexicano de Zonas Aridas, *Informe Nacional México. Conferencia Latinoamericana para el Estudio de las Zonas Aridas. Buenos Aires. Septiembre de 1963*, pp. 51-52. México: El Comité.

Hailey, T. L. 1977. Past, present and future status of the desert bighorn in the Chihuahuan Desert region. In R. H. Wauer and D. H. Riskind (eds.), *Transactions of the Symposium on the Biological Resources of the Chihuahuan Desert Region. U.S. and Mexico. Trans. and Proc. Series* 3:217-220. Washington, D.C.: National Park Service.

Harris A. H. 1977. Wisconsin Age environments in the northern Chihuahuan Desert: Evidence from the higher vertebrates. In R. H. Wauer and D. H. Riskind (eds.), *Transactions of the Symposium on the Biological Resources of the Chihuahuan Desert Region. U.S. and Mexico. Trans. and Proc. Series*, 3:23-52. Washington, D.C.: National Park Service.

Heindl, L. A. 1965. Groundwater in the Southwest: A perspective. In J. E. Fletcher and G. L. Bender (eds.), *Ecology of groundwater in the southwestern United States*, pp. 4-26. Symposium held at Arizona State University, 37th Annual Meeting of Southwestern and Rocky Mountain Division of the AAAS, April 18-19 1961, Tempe: Arizona State University

Henning, D. and Flohn, H. 1977a. *Climate aridity index (Budyko-Ratio)*. Map presented at the United Nations Conference on Desertification, Nairobi, August 29-September 9, 1977. A/CONF.74/31 Nairobi: UNEP-FAO-UNESCO.

———. 1977b. Climate aridity index map. Explanatory note. United Nations Conference on Desertification, Nairobi. August 29-September 9, 1977. A/CONF.74/31. Nairobi: UNEP-FAO-UNESCO.

Henrickson, J. 1977. Saline habitats and halophytic vegetation of the Chihuahuan Desert Region. In R. H. Wauer and D. H. Riskind (eds.), *Transactions of the Symposium on the Biological Resources of the Chihuahuan Desert Region. U.S. and Mexico. Trans. and Proc. Series*, 3:289-314. Washington, D.C.: National Park Service.

Henrickson, J. and Straw, R. M. 1976. *A gazetteer of the Chihu-ahuan Desert region: A supplement to the Chihuahuan Desert flora*. Los Angeles: California State University.

Hollon, W. E. 1966. *The great American desert*. New York: Oxford University Press.

Hunt, W. G. 1977. The significance of wilderness ecosystems in western Texas and adjacent regions in the ecology of the peregrine. In R. H. Wauer and D. H. Riskind (eds.), *Transactions of the Symposium on the Biological Resources of the Chihuahuan Desert Region. U.S. and Mexico. Trans. and Proc. Series*, 3:609-616. Washington, D.C.: National Park Services.

Jaeger, E. C. 1957. *The North American deserts*. Stanford: Stanford University Press.

Jenny, H. 1941. *Factors of soil formation: A system of quantitative pedology*. New York: McGraw-Hill.

Johnston, M.C. 1977. Brief resume of botanical, including vegetational, features of the Chihuahuan Desert region with special emphasis on their uniqueness. In R. H. Wauer and D. H. Riskind (eds.), *Transactions of the Symposium on the Biological Resources of the Chihuahuan Desert Region. U.S. and Mexico. Trans. and Proc. Series* 3:335-359. Washington, D.C.: National Park Services.

Josephy, A. M. 1964. American Indians. In *Collier's Encyclopedia* 12:642-695.

Lowe, C. H. 1970. Fauna of desert environments with desert disease information. In W. G. McGinnies, B. J. Goldman, and P. Paylore, *Deserts of the world*, pp. 567-645. Tucson: University of Arizona Press.

McGinnies, W. G.; Goldman, B. J. and Paylore, P. 1970. *Deserts of the world*. Tucson: University of Arizona Press.

Marroquín, J. S. et al. 1964. *Estudio ecológico dasonómico de las zonas áridas del Norte de México*. Mexico: INIF.

Medellín-Leal, F., ed. 1978. *La desertificación en México*. San Luis Potosí: Universidad Autónoma, IIZD.

Medellín-Leal, F., and Gómez-González, A. 1980. Management of natural vegetation in the semi-arid ecosystems of Mexico. In B. H. Walker (ed.), *Management of semi-arid ecosystems*. Amsterdam: Elsevier.

Meigs, P. 1957. Weather and climate. In E. C. Jaeger *The North American deserts*, pp. 13-32. Stanford: Stanford University Press.

Meléndez, J. 1976. La zona del silencio ¿Un "Triángulo de las Bermudas" en México? *Sucesos para todos*. 2233:14-15.

Metcalf, A. L. 1977. Some Quaternary molluscan faunas from the northern Chihuahuan Desert and their paleoecological implications. In R. H. Wauer and D. H. Riskind (eds.), *Transactions of the Symposium on the Biological Resources of the Chihuahuan Desert Region. U.S. and Mexico. Trans. and Proc. Series*. 3:53-66. Washington, D.C.: National Park Service.

Miller, R. R. 1977. Composition and derivation of the native fish fauna of the Chihuahuan Desert region. In R. H. Wauer and D. H. Riskind (eds.), *Transactions of the Symposium on the Biological Resources of the Chihuahuan Desert Region. U.S. and Mexico. Trans. and Proc. Series*. 3:365-381. Washington, D.C.: National Park Service.

Milstead, W. W. 1977. The black gap whiptail lizards after twenty years. In R. H. Wauer and D. H. Riskind (eds.), *Transactions of the Symposium on the Biological Resources of the Chihuahuan*

Desert Region. U.S. and Mexico. Trans. and Proc. Series. 3:523-532. Washington, D.C.: National Park Service.

Minckley, W. L. 1977. Endemic fishes of the Cuatro Ciénegas Basin, northern Coahuila, Mexico. In R. H. Wauer and D. H. Riskind (eds.), *Transactions of the Symposium on the Biological Resources of the Chihuahuan Desert Region. U.S. and Mexico. Trans. and Proc. Series.* 3:383-404. Washington, D.C.: National Park Service.

Miranda, F. 1955. Formas de vida vegetales y el problema de la delimitación de las zonas áridas de México. In E. Beltrán (ed.), *Mesas redondas sobre problemas de las zonas aridas de México. pp. 83-109. Biblioteca Central de la Ciudad Universitaria, 24-28 de enero de 1955. México: IMRNRAC.*

Morafka, D. J. 1977a. Is there a Chihuahuan Desert? A quantitative evaluation through a herpetofaunal perspective. In R. H. Wauer and D. H. Riskind (eds.), *Transactions of the Symposium on the Biological Resources of the Chihuahuan Desert Region. U.S. and Mexico. Trans. and Proc. Series.* 3:437-454. Washington, D.C.: National Park Service.

————. 1977b. *A biogeographical analysis of the Chihuahuan Desert.* The Hague: W. Junk.

Ortiz-Villanueva, B. 1973. *Edafología.* Mexico: Patena.

Packard, R. L. 1977. Mammals of the southern Chihuahuan Desert: An inventory. In R. H. Wauer and D. H. Riskind (eds.), *Transactions of the Symposium on the Biological Resources of the Chihuahuan Desert Region. U.S. and Mexico. Trans. and Proc. Series.* 3:141-153. Washington, D.C.: National Park Service.

Pettersen, S. 1958. *Introduction to meteorology.* New York: Mc-Graw-Hill.

Phillips, A. R. 1977. Summary of avian resources of the Chihuahuan Desert region. In R. H. Wauer and D. H. Riskind (eds.), *Transactions of the Symposium on the Biological Resources of the Chihuahuan Desert Region. U.S. and Mexico. Trans. and Proc. Series.* 3:617-620. Washington, D.C.: National Park Service.

Pinkava, D. J. 1977. Vegetation and flora of the Cuatro Ciénegas Basin, Coahuila, Mexico. In R. H. Wauer and D. H. Riskind (eds.), *Transactions of the Symposium on the Biological Resources of the Chihuahuan Desert Region. U.S. and Mexico. Trans. and Proc. Series.* 3:327-333. Washington, D.C.: National Park Service.

Pough, F. H. 1957. *A field guide to rocks and minerals.* Boston: Houghton Mifflin.

Powell, M. A., and Turner, B. L. 1977. Aspects of the plant gypsum outcrops of the Chihuahuan Desert. In R. H. Wauer and D. H. Riskind (eds.), *Transactions of the Symposium on the Biological Resources of the Chihuahuan Desert Region. U.S. and Mexico. Trans. and Proc. Series.* 3:315-325. Washington, D.C.: National Park Service.

Probstein, R. F., and Alvarez, J. M. 1973. *Desalting in the arid regions. Science and Man in the Americas.* AAAS-CONACYT Meeting, Mexico City, June 20-July 4, 1973; deserts and Arid Lands Symposium June 25-June 27. Massachusetts Institute of Technology. Department of Mechanical Engineering. *Fluid Mechanics Laboratory Pub.* 73-5.

Raisz, E. 1959. *Landforms of Mexico.* Cambridge, Mass.: Map prepared for the Geography Branch of the Office of Naval Research.

Raitt, R. J., and Pimm, S. L. 1977. Temporal changes in northern Chihuahuan Desert birds communities. In R. H. Wauer and D. H. Riskind (eds.), *Transactions of the Symposium on the Biological Resources of the Chihuahuan Desert Region. U.S. and Mexico. Trans. and Proc. Series.* 3:579-589. Washington, D.C.: National Park Services.

Ramos-Arizpe, Miguel. *Memoria ... sobre el estado natural, político y civil de su dicha Provincia (Coahuila), y los del Nuevo Reyno de León, Nuevo Santander, y los Texas, con exposición de los defectos ... de sus gobiernos y de las reformas que necesitan.*

Reitan, C. H. and Green, C. R. 1970. Appraisal of research on weather and climate of desert environments. In W. G. McGinnies, B. J. Goldman, and P. Paylore. *Deserts of the World,* pp. 19-92. Tucson: University of Arizona Press.

Rogers, C. L. et al. 1964. Rocas plutónicas del Norte de Zacatecas y áreas adyacentes, México. *Geo. y Met.* 9:57-63.

Rzedowski, J. 1955. Notas sobre la flora y la vegetación del Estado de San Luis Potosí. II. Estudio de diferencias florísticas y ecológicas condicionadas por ciertos tipos de sustrato geológico. *Ciencia, Mex.* 15:141-158.

————. 1957. Vegetación de las partes áridas de los estados de San Luis Potosí y Zacatecas. *Rev. Soc. Mex. Hist. Nat.* 18:49-101.

————. 1962. Contribuciones a la fitogeografía florística e histórica de México. I. Algunas consideraciones acerca del elemento endémico en la flora mexicana. *Bol. Soc. Bot. Méx.* 27:52-65.

————. 1965. Relaciones geográficas y posibles orígenes de la flora de México. *Bol. Soc. Bot. Mex.* 29:121-177.

————. 1966. Vegetación del Estado de San Luis Potosí. *Acta Cient. Potos.* 5(1-2):5-291.

————. 1973. Geographical relations of the flora of Mexican dry regions. In A. Graham, (ed.), *Vegetation and vegetational history of northern Latin America,* pp. 61-72. Amsterdam: Elsevier.

————. 1975. An ecological and phytogeographical analysis of the grasslands of Mexico. *Taxon* 24(1):67-80.

————. 1978. *Vegetación de México.* México: Limusa.

Schmidly, D. J. 1977. Factors governing the distribution of mammals in the Chihauhuan Desert Region. In R. H. Wauer, and D. H. Riskind, (eds.), *Transactions of the Symposium on the Biological Resources of the Chihuahuan Desert Region. U.S. and Mexico, Trans. and Proc. Series* 3:163-192.

Schreyer, R. 1978. The treasure of the Anasazi. *Edge* 1(2):38-43.

Scudday, J. F. 1977. Some recent changes in the herpetofauna of the northern Chihuahuan Desert. In R. H. Wauer and D. H. Riskind, (eds.), *Transactions of the Symposium on the Biological Resources of the Chihuahuan Desert Region. U.S. and Mexico, Trans. and Proc. Series* 3:513-522. Washington, D.C.: National Park Service.

Secretaría de Recursos Hidráulicos. 1968. *Región Hidrológica Número 24 Poniente. Zona alta de la cuenca del Río Bravo incluyendo la subcuenca del Río Conchos.* Boletín Hidrológico 29 (1-2).

————. 1970. *Región Hidrológica Número 34. Cuencas cerradas del Norte.* Boletín Hidrológico 33.

Segalen, P. 1964. Suelos de la zona intertropical. Chapingo: Colegio de Post-graduados. Series de Apuntes 4. Mimeographed.

Shreve, F. 1942. The desert vegetation of North America. *Bot. Rev.* 8(4):195-246.

Soto-Mora, C. and E. Jáuregui. 1965. *Isotermas extremas e índice de aridez en la República Mexicana.* México: UNAM, Instituto de Geografía.

Stretta, E. J. P., and Mosiño, P. A. 1963. *Distribución de las zonas áridas de la República Mexicana.* México: Ingeniería Hidráulica en México.

Stretta, E. J. P., et al. 1964. Geoquímica y radiactividad de las aguas de Lourdes. Mpio. de Sta. María del Río, S.L.P. *Geo. y Met.* 8:53-74.

Tamayo, J. L. 1962. *Atlas geográfico general de México.* Mexico: Instituto de Investigaciones Económicas.

Tavara-Alfaro, X. 1964. *Viajes en México. Crónicas mexicanas.* Mexico: Secretaría de Obras Públicas.

U.S. Department of Agriculture. 1960. *Soil classification. A comprehensive system. 7th. Approximation.* Conservation Service. Washington D.C.: Government Printing Office.

Valera-Adam, J. 1963. Clima. In *México. Informe Nacional presentado a la Conferencia Latinoamericana para el Estudio de las Zonas Aridas. Buenos Aires. Septiembre 1963,* pp. 4-5 Mexico: Comité Mexicano de Zonas Aridas.

Van Devender, T. R., and Worthington, R. D. 1977. The herpetofauna of Howell's Ridge Cave and the paleoecology of the northwestern Chihuahuan Desert. In R. H. Wauer and D. H. Riskind, (eds.), *Transactions of the Symposium on the Biological Resources of the Chihuahuan Desert Region. U.S. and Mexico,* 3:85-106. *Trans. and Proc. Series,* Washington, D.C.: National Park Service.

Villa-Ramírez, B. 1977. Major game mammals and their habitats in the Chihuahuan Desert Region. In R. H. Wauer and D. H. Riskind, (eds.), *Transactions of the Symposium on the Biological Resources of the Chihuahuan Desert Region. U.S. and Mexico, Trans. and Proc. Series,* 3:155-161, Washington, D.C.: National Park Service.

Villalobos, C. I., and Díaz-de-León, E. 1971. *Composición de aguas en San Luis Potosí.* San Luis Potosí: Universidad Autónoma, IIZD.

Wauer, R. H. 1977. Changes in breeding avifauna within the Chisos Mountains system. In R. H. Wauer and D. H. Riskind, (eds.), *Transactions of the Symposium on the Biological Resources of the Chihuahuan Desert Region. U.S. and Mexico. Trans. and Proc. Series,* 3:597-608. Washington, D.C.: National Park Service.

Wauer, R. H., and Ligon, J. D., 1977. Distributional relations of breeding avifauna of four southwestern mountain ranges. In R. H. Wauer, and D. H. Riskind, (eds.), *Transactions of the Symposium on the Biological Resources of the Chihuahuan Desert Region. U.S. and Mexico. Trans. and Proc. Series,* 3:567-578. Washington, D.C.: National Park Service.

Wauer, R. H., and Riskind, D. H., (eds.). 1977. *Transactions of the Symposium on the Biological Resources of the Chihuahuan Desert Region. United States and Mexico.* Sul Ross University, Alpine, Texas, 17-18 October 1974. *Transactions and Proceedings Series,* 3.

Webster, J. D. 1977. The avifauna of the southern part of the Chihuahuan Desert. In R. H. Wauer, and D. H. Riskind, (eds.), *Transactions of the Symposium on the Biological Resources of the Chihuahuan Desert Region. U.S. and Mexico, Trans. and Proc. Series,* 3:559-566. Washington, D.C.: National Park Service.

Wells, P. V. 1977. Post-glacial origin of the Present Chihuahuan Desert less than 11,500 years ago. In R. H. Wauer, and D. H. Riskind, (eds.), *Transactions of the Symposium on the Biological Resources of the Chihuahuan Desert Region. U.S. and Mexico. Trans. and Proc. Series,* 3:67-83. Washington, D.C.: National Park Service.

Woodbury, R. B. 1959. Pre-Spanish human ecology in the southwestern deserts. *University of Arizona, Arid Lands Colloquia 1958-1959:* 82-92.

A CHECKLIST OF THE AMPHIBIANS AND REPTILES OF THE BIG BEND NATIONAL PARK

Most common and scientific names used follow Raun and Gehlbach — Amphibians and Reptiles in Texas: Taxonomic Synopsis, Bibliography, and County Distribution Maps, 1972, Dallas Mus. Nat. Hist., Bull. 2, 132 pp. Subspecies have been ignored since they are primarily of academic interest and are often difficult to determine from field observations.

Big Bend National Park, in the southern Trans-pecos of southwestern Texas, was established in 1944 and consists of over 708,000 acres (over 1,100 sq. mi.) within the Chihuahuan Desert. The herpetofauna (amphibians and reptiles) is diverse and spectacular, but one must look for these animals in the proper place at the right time. Most kinds (species) of herptiles are nocturnal and retreat underground into burrows, rocky crevices, or under stones during the day to "beat the heat". One of the best ways to observe (not disturb or collect!) the park's herpetofauna (a total of 65 species) is to cruise the roads slowly on warm, humid, windless, moonless nights.

Ten species of amphibians and 55 species of reptiles are presently recorded for Big Bend National Park. Six additional species are listed as hypothetical. Unfortunately, a few species have been extirpated and no longer occur in the park because of habitat loss and human abuse. For amphibians in the park, frogs and toads are the only group (10 species). For reptiles in the park, snakes are the largest group (30 species), followed by lizards (21 species), and turtles (4 species). Although some species occur throughout the park, others are restricted to specific habitats and life-zones found at the different elevations. Five species (4 kinds of rattlesnake and 1 kind of copperhead) are poisonous and are dangerous to humans.

Any observations of rare, hypothetical, or unlisted amphibians and reptiles should be reported to: Chief Park Interpreter, Panther Junction, Big Bend National Park, Texas 79834.

CLASS AMPHIBIA

ORDER ANURA — Frogs and Toads

_____ Couch's Spadefoot (**Scaphiopus couchi**)
Subterranean (animal burrows) or fossorial, in desert from foothills (lower Green Gulch and Panther Jct.) to lowlands (Rio Grande); common, toxic skin, nocturnal — rainy.

_____ Western Spadefoot (**Scaphiopus hammondi**)
Subterranean (animal burrows) or fossorial in shortgrass plains, status unclear; a single park record from Panther Junction, well established north of park; toxic skin, nocturnal — rainy.

_____ Madrean Cliff Frog (**Syrrhophus guttilatus**)
Under moist rocks, sotols, and agaves, Chisos Mountains; tiny, uncommon, nocturnal — rainy.

_____ Canyon Treefrog (**Hyla arenicolor**)
Rocky streams and pools, under boulders and streambanks, sometimes in trees, Chisos Mountains and its foothills; uncommon, primarily nocturnal.

_____ Green Toad (**Bufo debilis**)
Secretive, burrowing, low hot deserts, only Tornillo Creek floodplain drainage, possibly Terlingua Creek drainage; very uncommon, nocturnal — heavy rainfall.

_____ Red-spotted Toad (**Bufo punctatus**)
Burrowing, desert, Rio Grande floodplain to foothills (i.e. Green Gulch); common, nocturnal.

_____ Texas Toad (**Bufo speciosus**)
Burrowing, desert, Rio Grande floodplain to Panther Junction; common, nocturnal.

_____ Woodhouse's Toad (**Bufo woodhousei**)
Burrowing, floodplains, only Rio Grande floodplain (Rio Grande Village and Santa Elena Canyon); rare, nocturnal — heavy rainfall.

_____ Leopard Frog (**Rana pipiens**)
Permanent pools, streams, springs, ponds, and tinajas, aquatic vegetation, Rio Grande floodplain to mountain foothills; common, nocturnal and diurnal.

_____ Great Plains Narrow-mouthed Toad (**Gastrophryne olivacea**)
Very secretive, burrowing, pools, damp crevices & burrows, under rocks, bark, & vegetation, floodplains and desert, Rio Grande to foothills (Panther Junction); uncommon, tiny, toxic skin, nocturnal — heavy rainfall.

CLASS REPTILIA

ORDER TESTUDINATA — Turtles

_____ Yellow Mud Turtle (**Kinosternon flavescens**)
Floodplains and desert, shallow (often temporary) muddy pools and tanks, park lowlands (Rio Grande

to Dagger Flat); uncommon, musky odor, mainly nocturnal.

_____ Big Bend Slider **(Chrysemys gaigeae)**
Sloughs, tanks, ponds, and rivers with a muddy bottom and aquatic vegetation, only Rio Grande floodplain; uncommon, diurnal and nocturnal.

_____ Western Box Turtle **(Terrapene ornata)**
Nonaquatic, prairies, only at Panther Junction (possibly escapes), well established north of the park; very rare, diurnal.

_____ Spiny Softshell **(Trionyx spiniferus)**
Rivers, only Rio Grande (rarely adjacent pools); uncommon, diurnal and nocturnal.

ORDER SQUAMATA (Suborder Lacertilia) — Lizards

_____ Mediterranean Gecko **(Hemidactylus turcicus)**
Human habitations: buildings, screens, often near lights, known only from Rio Grande Village (also across the river at Boquillas, Mexico); has voice, originally introduced from Old World, walks on smooth walls and ceilings, very rare, nocturnal.

_____ Texas Banded Gecko **(Coleonyx brevis)**
Rocky deserts, under stones, sotols, and agaves, or in rocky crevices, from lowlands to foothills (sparingly to Panther Junction and Green Gulch); has voice, fairly common, nocturnal.

_____ Reticulated Gecko **(Coleonyx reticulatus)**
Desert rock outcrops, lowlands, a single park record, known only from Presidio and Brewster Counties, Texas; has voice, large, can change pattern intensity, extremely rare, nocturnal.

_____ Greater Earless Lizard **(Cophosaurus texanus)**
Heat-loving, hot desert lowlands, Rio Grande floodplain to foothills (sparingly to Panther Junction); no ear opening, common, diurnal.

_____ Collared Lizard **(Crotaphytus collaris)**
Open deserts with boulders, rocks, and sparse vegetation, lowlands to Chisos Mt. foothills (i.e. Pine Canyon, Panther Junction, lower Green Gulch); sometimes bipedal (runs upright), uncommon, diurnal.

_____ Leopard Lizard **(Crotaphytus wislizeni)**
Low, hot deserts and arid flatlands of creosotebush, only recorded at Tornillo Flat, base of Rosillos Mts., between Panther Junction and Rio Grande Village, near Boquillas, and San Vicente; sometimes bipedal, extremely rare, diurnal.

_____ Texas Horned Lizard **(Phrynosoma cornutum)**
Grasslands, plains, and deserts, recent records only from Panther Junction (escapes?) and Nine Point Draw on Upper Tornillo Flat; well established north of park, extremely rare, diurnal.

_____ Round-tailed Horned Lizard **(Phrynosoma modestum)**
Desert flats with sparse vegetation, Rio Grande to foothills; often near ant mounds, fairly common, diurnal.

_____ Desert Spiny Lizard **(Sceloporus magister)**
Lowlands, desert washes and arroyos with ample vegetative cover (mesquite thickets, prickly pear, etc.), use animal burrows, climb trees, Rio Grande floodplain sparingly to Chisos Mts. foothills (about 4000 ft.; very rare at Panther Junction); very wary, fairly common in proper habitat, diurnal.

_____ Canyon Lizard **(Sceloporus merriami)**
Rock loving, canyon walls, cliffs, boulders, rocky arroyos, and rarely adobe buildings, Rio Grande floodplain well into Chisos Mts. to at least 5500 ft. ele.; unwary, common, diurnal.

_____ Crevice Spiny Lizard **(Sceloporus poinsetti)**
Rocky faces and boulders having crevices, sometimes buildings, rocky highlands of Chisos Mts. and foothills, a few records from the river canyons; wary, wedges itself in crevices, common, diurnal.

_____ Eastern Fence Lizard **(Sceloporus undulatus)**
Trees, shrubs, yuccas, or other types upright vegeta-

tion, sometimes logs, boulders, buildings, or signs, readily climbs, Chisos Mts. and foothills down to at least Giant Dagger Flats; fairly common, diurnal.

_____ Side-blotched Lizard **(Uta stansburiana)**
Lowland desert flats, sandy areas, hardpan, fine gravel, Rio Grande to foothills (3500 - 4000 ft. elev.; at least to Grapevine Hills & K-Bar Ranch); sometimes active in winter, burrows, common, diurnal.

_____ Tree Lizard **(Urosaurus ornatus)**
Rock faces, boulders, trees, Chisos Mts. and foothills (rarely to Panther Junction); climbs, uncommon, diurnal.

_____ Short-lined Skink **(Eumeces brevilineatus)**
Likes moisture, under rocks, agaves, sotols, prickly pears, boards, and logs near springs, seeps, or pools, parkwide except for possibly the highest parts of the Chisos Mts.; very secretive, a skulker, semi-fossorial, uncommon, diurnal.

_____ Great Plains Skink **(Eumeces obsoletus)**
Hides under sotols, agaves, prickly pears, stones, or in animal burrows, parkwide; secretive, a skulker, very uncommon, diurnal.

_____ Little Striped Whiptail **(Cnemidophorus inornatus)**
Grasslands, intermediate park elevations — K-Bar Ranch, Govt. Spring, Lone Mt., Persimmon Gap, Panther Jct., and Giant Dagger Flat; uncommon, diurnal.

_____ Rusty-rumped Whiptail **(Cnemidophorus scalaris)**
Grassy rocky deserts, shrublands, Chisos Mts. and foothills; common, diurnal.

_____ Checkered Whiptail **(Cnemidophorus tesselatus)**
Rocky barren areas, only Rio Grande floodplain (mainly canyon areas such as Santa Elena Canyon and Mariscal Canyon); parthenogenic (males extremely rare), very rare, diurnal.

_____ Western Whiptail **(Cnemidophorus tigris)**
Barren desert flats, animal burrows, Rio Grande floodplain to mountain foothills (rare at Panther Junction); common, diurnal.

_____ Texas Alligator Lizard **(Gerrhonotus liocephalus)**
Forests, brushlands, only Chisos Mts.; powerful jaws, slow moving, prehensile tail, very uncommon, mainly diurnal — sometimes nocturnal.

ORDER SQUAMATA (Suborder Serpentes) — Snakes

_____ Texas Blind Snake **(Leptotyphlops dulcis)**
Loose soil, burrowing, moist areas, under rocks, sotols, and agaves, only Chisos Mts.; vestigial eyes, food: termites and ants, uncommon, nocturnal — rainy.

_____ Western Blind Snake **(Leptotyphlops humilis)**
Loose soil, burrowing, desert flats, under rocks, sotols, and yuccas, Rio Grande floodplain to mountain foothills; vestigial eyes, food: termites and ants, fairly rare, nocturnal — rainy.

_____ Glossy Snake **(Arizona elegans)**
Deserts, creosotebush flats, burrowing, Rio Grande floodplain to foothills (up to Panther Jct. and lower Green Gulch); uncommon, nocturnal — rainy.

_____ Ringneck Snake **(Diadophis punctatus)**
Rocky moist areas, mountains, under rocks, logs, sotols, and agaves, mainly Chisos Mts. and foothills — rarely other park mountains and lowlands along Rio Grande (at canyons); uncommon, nocturnal and diurnal.

_____ Great Plains Rat Snake **(Elaphe guttata)**
Floodplains, thick riparian growth, cane thickets, mammal burrows, crevices, only Rio Grande floodplain; secretive, rare, nocturnal and diurnal.

_____ Common (Baird's) Rat Snake **(Elaphe obsoleta)**
Forested uplands, rocky wooded canyons, shrublands, upland meadows, only Chisos Mts. (above 4000 ft. ele.); rare, diurnal and nocturnal.

_____ Trans-pecos Rat Snake **(Elaphe subocularis)**
Animal burrows, rocky crevices, canyons, deserts, creosotebush flats, rocky terrain with agave, ocotillo, cactus or sotol, Rio Grande floodplain to Chisos Mts. foothills

(below 5000 ft.); secretive, uncommon, nocturnal (hot nights).

_____ Western Hook-nosed Snake **(Ficimia cana)**
Burrowing, rocky outcrops, mountains, Chisos Mts. and foothills (two lowland records — one at Boquillas Canyon Road and one at Tornillo Flat); secretive, "anal popping", rare, nocturnal.

_____ Night Snake **(Hypsiglena torquata)**
Deserts, rocky areas, creosotebush or lechuguilla associations, under rocks, boards, or sotols, Rio Grande floodplain to Chisos Mts. foothills (2 records for Basin); rear-fanged, common, nocturnal.

_____ Common (Sonora) Kingsnake **(Lampropeltis getulus)**
Deserts, Rio Grande floodplain to Chisos Mts. foothills; secretive, rare, nocturnal — rainy.

_____ Gray-banded Kingsnake **(Lampropeltis mexicana)**
Rocky outcrops, mountains, Chisos Mts. and foothills (known from rocky lowlands outside the park); very secretive, two main color phases: **alterna** morph (the one occuring in park) and **blairi** morph, extremely rare; probably the rarest park snake and one of the rarest in the United States, nocturnal.

_____ Milk Snake **(Lampropeltis triangulum)**
Rocky areas, Chisos Mts. and foothills (down to Panther Jct.); secretive, very rare, nocturnal — rainy.

_____ Coachwhip **(Masticophis flagellum)**
Deserts, lowlands, shrublands, Rio Grande floodplain to Chisos Mts. foothills; swift, sometimes climbs, common, diurnal — heat of day.

_____ Striped Whipsnake **(Masticophis taeniatus)**
Rocky outcrops, brushlands, animal burrows, parkwide except for possibly the highest parts of the Chisos Mts.; very alert, swift, readily climbs trees, uncommon, diurnal.

_____ Plain-bellied Water Snake **(Natrix erythrogaster)**
Aquatic, ponds, rivers, only Rio Grande between Hot Springs and Boquillas Canyon; secretive, very alert, very rare, nocturnal.

_____ Bullsnake **(Pituophis melanoleucus)**
Deserts, shrublands, woodlands, sometimes burrows, parkwide; hisses, vibrates tail, sometimes climbs, uncommon, diurnal and nocturnal.

_____ Long-nosed Snake **(Rhinocheilus lecontei)**
Burrowing, deserts, creosotebush flats, under rocks, sotols, or in mammal burrows, Rio Grande floodplain to Chisos Mts. foothills (2 records for Basin); uncommon, nocturnal — rainy.

_____ Big Bend Patch-nosed Snake **(Salvadora deserticola)**
Desert lowlands, desert scrublands, creosotebush flats, Rio Grande floodplain to Chisos Mts. foothills (up to Panther. Jct.); uncommon, diurnal.

_____ Mountain Patch-nosed Snake **(Salvadora grahamiae)**
Mountain slopes, plateaus, rocky canyons, open woodlands, scrublands, Chisos Mts. and foothills (down to Panther Jct.); uncommon, diurnal.

_____ Western Ground Snake **(Sonora semiannulata)**
Burrowing, scant vegetation, desert lowlands, river bottoms, rocky hillsides, creosotebush flats, Rio Grande floodplain to Chisos Mts. foothills (up to 5000 ft. in Green Gulch); secretive, numerous color phases and patterns, uncommon, nocturnal — windy hot dry nights.

_____ Mexican Black-headed Snake **(Tantilla atriceps)**
Burrowing, deserts, grasslands, shrublands, woodlands, in crevices, under rocks, sotols, agaves, and yuccas, Giant Dagger Flats (probably at lower elevations) to foothills and Chisos Mts.; secretive, rear-fanged, common, nocturnal.

_____ Red (Big Bend) Black-headed Snake **(Tantilla rubra)**
Burrowing, mountain habitat where pinyon, juniper, oak, agave, sotol, yucca, and nolina dominate, only Chisos Mts. (Basin, upper Green Gulch, and Panther Pass) and Volcanic Dike Overlook; two color-phases (collared and non-collared), secretive, rear-fanged, possibly does not surface during dry years, known from only three Texas counties, extremely rare, nocturnal — following heavy rainfall.

_____ Black-necked Garter Snake **(Thamnophis cyrtopsis)**
Semiaquatic, rock-inhabiting, canyons, pools, mountains, springs, tinajas, parkwide (down to Rio Grande) in the proper habitat but mainly Chisos Mts. and foothills; live-bearing, common, nocturnal or diurnal.

_____ Checkered Garter Snake **(Thamnophis marcianus)**
Semiaquatic, floodplains, permanent water, rivers, ponds, lakes, springs, pools, streams, only Rio Grande floodplain and its major tributaries (Tornillo and Terlingua Creeks); live-bearing, uncommon, nocturnal or diurnal.

_____ Lyre Snake **(Trimorphodon biscutatus)**
Rock-inhabiting, mountains, canyons, outcrops, Chisos Mts. and foothills (lowest records at Panther Jct., Volcanic Dike Overlook, and Grapevine Hills); secretive, rear-fanged, extremely rare, nocturnal.

_____ Copperhead **(Agkistrodon contortrix)**
Rocky habitat: canyons, outcrops, mountains, and cane patches, parkwide (in the proper habitat); pit viper (poisonous), live-bearing, rare, nocturnal.

_____ Western Diamondback Rattlesnake **(Crotalus atrox)**
Arid and semiarid regions: deserts, grasslands, brushlands, river bottoms, rocky canyons, Rio Grande floodplain to Chisos Mts. foothills (in Green Gulch up to about 4500 ft.); very aggressive, pit viper (poisonous), live-bearing, common, mostly nocturnal.

_____ Rock Rattlesnake **(Crotalus lepidus)**
Rocky terrain: ledges, canyons, mountains, outcrops, slides, parkwide but most common in Chisos Mts. and foothills; pit viper (poisonous), live-bearing, very uncommon, nocturnal — sometimes diurnal.

_____ Black-tailed Rattlesnake **(Crotalus molossus)**
Rock-inhabiting, mountains, canyons, outcrops, parkwide but most common in Chisos Mts. and foothills; pit viper (poisonous), live-bearing, common, nocturnal — sometimes diurnal.

_____ Mojave Rattlesnake **(Crotalus scutulatus)**
Low hot deserts, creosotebush flats, grasslands, Rio Grande floodplain to about 4000 ft. (up to Panther Junction); aggressive, pit viper (very poisonous), live-bearing, uncommon, nocturnal.

HYPOTHETICAL SPECIES

_____ Tiger Salamander **(Ambystoma tigrinum)**
Status uncertain; known only from a couple records from Panther Junction — possibly fish bait escapes; well established north of park.

_____ Bullfrog **(Rana catesbeiana)**
Collected outside the park at Lajitas, Texas; reported calling along Rio Grande in park.

_____ Texas Tortoise **(Gopherus berlandieri)**
Recently several found at Basin and Rio Grande Village; probably releases or escapes from park visitors; native to south Texas.

_____ Western Hognose Snake **(Heterodon nasicus)**
A single park record from Panther Junction (possibly an escape); native to the grasslands north of the park.

_____ Plains Black-headed Snake **(Tantilla nigriceps)**
An unconfirmed record from the Basin; one specimen of questionable identity from Rio Grande Village; native to grasslands north of park.

_____ Prairie Rattlesnake **(Crotalus viridis)**
One old park record; probably extirpated from park when grasses seriously depleted in earlier years; found north of park in grasslands.

A CHECKLIST OF THE BIRDS OF BIG BEND NATIONAL PARK

This list includes the names of 398 species of birds that have been reported for Big Bend National Park and vicinity. Thirty-seven of these are in a "hypothetical" list: those that have not been authenticated by a specimen, photograph, a sighting by more than one individual or party, or have not been reported during the last twenty years. Common names used are from the American Ornithologists' Union Check-list of North American Birds, 5th edition, 1957, as currently supplemented.

c = common   u = uncommon   s = sporadic
f = fairly common   r = rare   acc. = accidental

W = WATER AREAS as Rio Grande and tributaries, and isolated ponds.

R = RIPARIAN AREAS as river floodplain, cottonwood groves, mesquite thickets, and deciduous vegetation up into Chisos Mountains. Examples are Rio Grande Village, Cottonwood Campground, Hot Springs, Santa Elena Canyon picnic area, Dugout, lower Window Trail, and Old Ranch.

D = SHRUB DESERT habitat from lowlands into the Chisos Mountains. Examples are Rio Grande Village, Panther Junction, Castolon, Dagger Flat, and along River Road.

G = GRASSLANDS from the lowlands into the high Chisos Mountains. Examples are Panther Junction vicinity, Green Gulch, and Window Trail.

M = MOUNTAIN WOODLANDS of pinyon, junipers, and oaks. Examples are upper Green Gulch, Lost Mine Trail, and South Rim Trail.

B = BOOT CANYON and similar highland areas, generally above 6,000 feet.
If no locality symbol is shown the birds can be expected throughout.

Summer = breeding birds including permanent residents and those which arrive as early as March and remain as late as October.

After Breeding = species that are not known to nest within the park but arrive in spring, summer, or fall and may remain through December.

Winter = permanent residents and those which may arrive as early as September, remain all winter, and may stay as late as April.

Migrant = those species that pass through the park only in spring and/or fall from March to May and August to November.

Please report any observations of rare, unlisted or hypothetical birds, or those found at times other than designated, to the Chief Park Naturalist, Big Bend National Park, Texas 79834.

SPECIES	SUMMER	AFTER BREEDING	WINTER	MIGRANT	NOTES
LOON, Common				acc.	
GREBE, Eared			rW	uW	
Least		acc.	sW		
Pied-billed	uW		uW	fW	
PELICAN, White				acc.	
CORMORANT, Olivaceous				acc.	
HERON, Great Blue	rW		fW	fW	
Green	rW	fW	rW	rW	
Little Blue				acc.	
Louisiana		uW		rW	
Black-crowned Night		acc.	acc.	rW	
Yellow-crowned Night		uW		rW	
EGRET, Cattle				uW	
Great				uW	
Snowy				rW	
BITTERN, Least	rW			rW	
American				uW	
STORK, Wood				acc.	
IBIS, White-faced			rW	rW	
White			acc.		
SWAN, Whistling			rW		
GOOSE, Canada				rW	
White-fronted				acc.	
Snow				rW	
TREE DUCK, Fulvous				acc.	
DUCKS					
Mallard			uW	uW	
Mexican		rW	rW		
Black			acc.		
Gadwall		uW	uW	fW	
Pintail		rW	uW	uW	
Green-winged Teal		uW	cW	cW	
Blue-winged Teal		rW	rW	cW	
Cinnamon Teal		rW	uW	fW	
American Wigeon		rW	uW	fW	
Northern Shoveler				fW	
Wood		uW			
Redhead				rW	
Ring-necked			uW	uW	
Canvasback			sW		
Lesser Scaup			rW	rW	
Common Goldeneye			acc.		
Bufflehead				uW	
Ruddy				uW	
MERGANSER, Hooded				rW	
Common				rW	
VULTURE, Turkey	c		r	c	
Black	uW		uW	fW	

SPECIES	SUMMER	AFTER BREEDING	WINTER	MIGRANT	NOTES
KITE, White-tailed			acc.	acc.	
Swallow-tailed		acc.		acc.	
Mississippi		rR		r	
HAWKS					
Goshawk		rM	acc.	f	
Sharp-shinned	rB		f	f	
Cooper's		rM	f	f	
Red-tailed	f		f	f	
Red-shouldered				r	
Broad-winged				u	
Swainson's	rD	fD		f	
Zone-tailed	uRM			u	
White-tailed				acc.	
Rough-legged			acc.		
Ferruginous			u	u	
Gray				acc.	
Harris'		rD	rD	rD	
Black			r	r	
Marsh		r	u	f	
OSPREY				u	
CARACARA				acc.	
EAGLE, Golden	u		f	f	
Bald			acc.		
FALCON, Prairie	r		r	u	
Peregrine	r		r	u	
Aplomado	acc.				
Merlin			r	r	
American Kestrel	uM		c	c	
QUAIL, Scaled	cRD		cRD		
Gambel's	rRD		rRD		
Montezuma					reintroduced M
CRANE, Sandhill			rW	u	
RAIL, King		acc.		acc.	
Virginia				rW	
Sora	rW		uW	fW	
Yellow			acc.		
GALLINULE, Purple				acc.	
Common				rW	
COOT, American	rW		uW	fW	
SHOREBIRDS					
Semipalmated Plover				acc.	
Killdeer	uW		fW	fW	
American Woodcock				acc.	
Common Snipe			uW	uW	
Willet				rW	
Upland Sandpiper				rW	
Long-billed Dowitcher				rW	
CURLEW, Long-billed				u	
Whimbrel				acc.	
SANDPIPER, Spotted		fW	cW	cW	
Solitary				fW	
Baird's				rW	
Least			uW	fW	
Western			rW	rW	
YELLOWLEGS, Greater				uW	
Lesser			acc.	uW	
AVOCET, American				uW	
STILT, Black-necked				rW	
PHALAROPE, Wilson's				rW	
GULL, Ring-billed				rW	
Laughing				acc.	
Franklin's				r	
TERN, Forster's				rW	
Least				acc.	
PIGEON, Band-tailed	fB			sB	
DOVE, Rock		acc.			
White-winged	fRM		uRM		
Mourning	cRDM		fRD		
Ground	uR		sR	u	
Inca	uR		rR	rR	
White-fronted		acc.			
CUCKOO, Yellow-billed	fR				
ROADRUNNER	fRD		fRD		
ANI, Groove-billed			rD	rR	
OWL, Barn				rR	
Screech	f		f		
Flammulated	fB				
Great Horned	f		f		
Pygmy		rM			
Elf	fRDM				
Burrowing	rD			rD	
Long-eared	rB		rR	rD	
Short-eared			rD		
Saw-whet			rB		

SPECIES	SUMMER	AFTER BREEDING	WINTER	MIGRANT	NOTES
WHIP-POOR-WILL	fB				
POOR-WILL	f		r	c	
NIGHTHAWK, Common		uD		u	
Lesser	cRD			c	
SWIFT, Chimney				rR	
				c	
White-throated	cRM		uRM	c	
HUMMINGBIRD, Lucifer	f				
Ruby-throated		r	r		
Black-chinned	cRD	cRM		c	
Costa's		acc.		acc.	
Anna's		r			
Broad-tailed	cM			f	
Rufous		cM		u	
Allen's		acc.		acc.	
Calliope				rM	
Rivoli's	uB				
Blue-throated	cB				
White-eared		rM			
Broad-billed	rR	r			
KINGFISHER, Belted		rW	rW	fW	
Green		rW		acc.	
FLICKER, Common	uM		c	c	
WOODPECKER, Golden-fronted				r	
Red-headed		acc.			
Acorn	cM		cM		
Lewis'			r	rM	
Ladder-backed	cRD		cRD		
SAPSUCKER, Yellow-bellied			f	u	
Williamson's		sM		s	
KINGBIRD, Eastern				r	
Tropical		rR			
Western	uR			f	
Cassin's	rR			u	
Thick-billed		acc.	acc.		
FLYCATCHER, Scissor-tailed				uD	
Sulphur-bellied		acc.		acc.	
Great Crested		rR			
Wied's Crested				rR	
Ash-throated	c		sR	c	
Olivaceous		r		r	
Yellow-bellied		r			
Willow	rR			uR	
Least				f	
Hammond's				u	
Dusky			rR	f	
Gray				u	
Western	cB			r	
Coues'		rM		rM	
Olive-sided		rM		f	
Vermilion	fR		fR	fRD	
PHOEBE, Eastern			fR		
Black	fR		fR	r	
Say's	fRDM		cRD	cRD	
WOOD PEWEE, Western		u		c	
HORNED LARK			rD	uD	
SWALLOW, Violet-green	uM			f	
Tree			rW	uW	
Bank			acc.	uW	
Rough-winged	fW		uW	cW	
Barn	u			f	
Cliff	cW			u	
Cave	rW			r	
MARTIN, Purple				r	
JAY, Blue			s	acc.	
Steller's			acc.		
Scrub			sM		
Mexican	cM		cM		
Piñon			acc.		
RAVEN, Common	f		f		
White-necked	rD			rD	
NUTCRACKER, Clark's			sM		
TITMOUSE, Tufted	cM		cM		
VERDIN	fRD		fRD		
BUSHTIT	cM		cM		
NUTHATCH, White-breasted	uM		uM		
Red-breasted			sM	s	
Pygmy			sB		
CREEPER, Brown			sRM	s	
WREN, House			u	f	
Winter				rR	
Bewick's	cM		s	f	
Carolina		rM	rR		
Cactus	cD		cD		
Long-billed Marsh			fW	fW	
Short-billed Marsh			acc.		
Canon	c		c		
Rock	u		f	c	

SPECIES	SUMMER	AFTER BREEDING	WINTER	MIGRANT	NOTES
MOCKINGBIRD	cD		cD	c	
CATBIRD, Gray		acc.	sR		
THRASHER, Brown			uR	uR	
Long-billed				acc.	
Curve-billed	rD		fD	cD	
Crissal	uDM		uD	u	
Sage			sD	uD	
ROBIN, American		rM	fRM		
Rufous-backed		acc.			
THRUSH, Aztec		acc.			
Wood				r	
Hermit			fRM	c	
Swainson's				u	
Gray-cheeked				acc.	
BLUEBIRD, Eastern			u		
Western			fM	r	
Mountain			s	r	
SOLITAIRE, Townsend's			fM	fM	
GNATCATCHER, Blue-gray		cM	sDM	f	
Black-tailed	cD		cD		
KINGLET, Golden-crowned			uM	rRM	
Ruby-crowned			c	f	
PIPIT, Water			fW	f	
Sprague's				r	
WAXWING, Cedar			f	u	
PHAINOPEPLA	rG	r	uD	u	
SHRIKE, Loggerhead	uD		cD	c	
STARLING				r	
VIREO, Black-capped	rM				
White-eyed				acc.	
Hutton's	fM		uM		
Bell's	cR				
Gray	uM		r	u	
Yellow-throated				r	
Solitary	rB	uM		f	
Yellow-green		acc.			
Red-eyed				rR	
Philadelphia				r	
Warbling	sB			u	
WARBLER, Black-and-white			acc.	u	
Prothonotary				acc.	
Worm-eating				r	
Golden-winged				acc.	
Blue-winged				acc.	
Orange-crowned			f	f	
Nashville				f	
Virginia's				u	
Colima	cB				
Lucy's				acc.	
Northern Parula		r		uR	
Olive		rM			
Yellow				fR	
Magnolia				acc.	
Cape May				acc.	
Black-throated Blue			acc.	r	
Yellow-rumped			cR	c	
Black-throated Gray			acc.	u	
Townsend's			fM	f	
Black-throated Green				rM	
Hermit				u	
Blackburnian				acc.	
Grace's				r	
Chestnut-sided				rR	
Blackpoll				acc.	
Palm				acc.	
Kentucky				acc.	
MacGillivray's				f	
Red-faced		acc.			
Hooded				r	
Wilson's				c	
Canada				acc.	
Rufous-capped	sRM	sRM			
OVENBIRD				rR	
WATERTHRUSH, Northern				u	
Louisiana				acc.	
YELLOWTHROAT, Common	fR		cR	cR	
CHAT, Yellow-breasted	cR			uR	
REDSTART, American				f	
Painted	uB				
SPARROW, House	f			f	
MEADOWLARK, Eastern			uG	uG	
Western			fG	fG	

SPECIES	SUMMER	AFTER BREEDING	WINTER	MIGRANT	NOTES
BLACKBIRD, Yellow-headed		uR	r	f	
Redwinged	sW			uW	
Rusty			rW		
Brewer's			r	c	
ORIOLE, Orchard	fR	rG			
Hooded	uR	r		r	
Scott's	cDM	f			
Northern			u	u	
Black-vented		acc.			
GRACKLE, Great-tailed	rR		rR	uR	
COWBIRD, Brown-headed	c		c	c	
Bronzed	uR				
TANAGER, Western		rRM		f	
Scarlet				acc.	
Hepatic	uM				
Summer	cR				
CARDINAL	fR		fR	r	
PYRRHULOXIA	cD		cRD		
GROSBEAK, Rose-breasted			r		
Black-headed	cM			u	
Blue	fRG			u	
Evening			acc.		
BUNTING, Indigo		s		u	
Lazuli				r	
Varied	fG		s		
Painted	cR	rG		u	
DICKCISSEL		rG		u	
FINCH, Cassin's		sM	sM		
House	c		c	c	
SISKIN, Pine			uM	u	
GOLDFINCH, American			u	f	
Lesser	uRM		f		
CROSSBILL, Red		sM	sM		
TOWHEE, Green-tailed			fRD	f	
Rufous-sided	fM		f	u	
Brown	cG		cG		
LARK BUNTING			sG	f	
JUNCO					
Dark-eyed			f		
Gray-headed			fM		
SPARROW, Savannah			uDG	u	
Grasshopper		sG		r	
Baird's		rG		r	
LeConte's				r	
Vesper			fG	f	
Lark		uG	r	u	
Rufous-crowned	cG		cG		
Cassin's	sG	sG	sG		
Black-throated	cD		cD		
Sage			sD		
Chipping		r	f	c	
Clay-colored			s	u	
Brewer's			uG	f	
Field			sG	r	
Black-chinned	fM		sG		
Harris'			r		
White-crowned			c	c	
Golden-crowned			acc.		
White-throated			u	u	
Fox			rR		
Lincoln's			fRG	f	
Swamp			fR	r	
Song			uR	u	
LONGSPUR, Chestnut-collared				r	

Hypothetical

Anhinga	Black Swift	Tropical Parula
Masked Duck	Coppery-tailed Trogon	Golden-cheeked Warbler
Red-breasted Merganser	Hairy Woodpecker	Yellow-throated Warbler
Chukar	Downy Woodpecker	Bay-breasted Warbler
Turkey	Rose-throated Becard	Pine Warbler
Spotted Rail	Kiskadee Flycatcher	Mourning Warbler
Black Rail	Buff-breasted Flycatcher	Black-headed Oriole
Red Knot	Beardless Flycatcher	Common Grackle
Pectoral Sandpiper	Common Crow	Pine Grosbeak
Black Tern	Bridled Titmouse	Purple Finch
Red-billed Pigeon	Dipper	Olive Sparrow
Black-billed Cuckoo	Veery	Yellow-eyed Junco
Barred Owl		

A MAMMALS FIELD LIST OF BIG BEND NATIONAL PARK

Most common and scientific names used follow Jones, Carter, and Genoways — Checklist of North American Mammals North of Mexico, 1973, Occas. Papers, Mus., Texas Tech. Univ., 12:1-14. Subspecies have been ignored since they are primarily of academic interest and are often difficult to determine from field observations.

Big Bend National Park, in the southern Trans-Pecos of southwestern Texas, was established in 1944 and consists of over 708,000 acres (over 1,100 sq. mi.) in the Chihuahuan Desert. The mammalian fauna is diverse and spectacular, but one must look for these animals in the proper place at the right time. Most kinds (species) of mammals "beat the heat" by being nocturnal, usually retreating in underground burrows or rocky crevices during the day.

Seventy-five species of mammals have been recorded within Big Bend National Park. Regrettably, a few species have been extirpated and no longer occur in the park because of habitat loss and human abuse. Rodents are the largest group in the park (28 species), followed by bats (19 species), carnivores (17), even-toed ungulates (5), rabbits (3), and opossums and shrews each one. Although some kinds occur throughout the park, others are restricted to specific habitats and life-zones found at the different elevations.

Any observations of rare or unlisted mammals should be reported to: Chief Park Interpreter, Panther Junction, Big Bend National Park, Texas 79834.

ORDER MARSUPIALIA — Opossums

_____ Opossum **(Didelphis virginiana)**
No recent records, evidently extirpated from park; old records from Pine Canyon, Pinnacle Spring, and Rosillos Mts., nocturnal.

ORDER INSECTIVORA — Shrews

_____ Desert Shrew **(Notiosorex crawfordi)**
Arid conditions, probably throughout park; known only from Burro Mesa, K-Bar Ranch, and the Pinnacles Trail to Boot Canyon; extremely rare and secretive, mainly nocturnal.

ORDER CHIROPTERA — Bats

_____ Ghost-faced Bat **(Mormoops megalophylla)**
Canyons, cliffs, and caves, throughout park; uncommon, nocturnal.

_____ Mexican Long-nosed Bat **(Leptonycteris nivalis)**
Caves in Chisos Mts. in daytime, ranges to Rio Grande at night; pollen and nectar feeder, especially on century plants; uncommon to rare, nocturnal.

_____ Yuma Myotis **(Myotis yumanensis)**
Cliffs, canyons, caves, and buildings, mainly lowlands along Rio Grande, rarely in Chisos Mts.; uncommon, nocturnal.

_____ Cave Myotis **(Myotis velifer)**
Caves, mines, and buildings, lowlands; uncommon, nocturnal.

_____ Fringed Myotis **(Myotis thysanodes)**
Caves and canyon crevices, throughout park; uncommon, nocturnal.

_____ Long-legged Myotis **(Myotis volans)**
Rock crevices, trees, and caves, only Chisos Mts.; rare, nocturnal.

_____ California Myotis **(Myotis californicus)**
Canyon crevices, mines, caves, buildings, and trees, throughout park; uncommon, nocturnal.

_____ Small-footed Myotis **(Myotis leibii)**
Canyon crevices and caves, possibly throughout park; known only from a single specimen from Pine Canyon; very rare, nocturnal.

_____ Western Pipistrelle **(Pipistrellus hesperus)**
Canyons, cliffs, crevices, caves, mines, and buildings, throughout park; abundant, a tiny bat that flies before dark, crepuscular and nocturnal.

_____ Big Brown Bat **(Eptesicus fuscus)**
Canyons, cliffs, crevices, trees, and buildings, throughout park; uncommon, nocturnal.

_____ Red Bat **(Lasiurus borealis)**
Trees, rarely caves, status uncertain; only park record supposedly from cave in Chisos Mts., known from riparian areas surrounding park; nocturnal.

_____ Hoary Bat **(Lasiurus cinereus)**
Trees, throughout park; uncommon, nocturnal.

_____ Spotted Bat **(Euderma maculatum)**
Canyon and cliff crevices, mainly in lowlands; rare,

North America's most spectacular and colorful bat, nocturnal.

—— Townsend's Big-eared Bat (**Plecotus townsendii**)
Caves, mines, cliffs, and rock shelters, throughout park; fairly common, nocturnal.

—— Pallid Bat (**Antrozous pallidus**)
Buildings, cliffs, canyons, crevices, rocky outcrops, and rarely mines, throughout park; will land on ground after insects, abundant, nocturnal.

—— Brazilian Free-tailed Bat (**Tadarida brasiliensis**)
Caves, buildings, and canyons, throughout park; common, nocturnal.

—— Pocketed Free-tailed Bat (**Tadarida femorosacca**)
Canyon and cliff crevices, only lowlands; rare, nocturnal.

—— Big Free-tailed Bat (**Tadarida macrotis**)
Cliff crevices and canyons, throughout park; uncommon, nocturnal.

—— Western Mastiff Bat (**Eumops perotis**)
Cliff and canyon crevices, buildings, lowlands; rare, North America's largest bat (1½ ft. wingspan), nocturnal.

ORDER LAGOMORPHA — Hares and Rabbits

—— Blacktail Jackrabbit (**Lepus californicus**)
Grassland, brushland, and desert, Rio Grande floodplain to mountain foothills (up to Basin, 5400 ft. ele.); common, crepuscular and nocturnal.

—— Eastern Cottontail (**Sylvilagus floridanus**)
Brushland, forest edge, and weedy areas, Chisos Mts.— 4700 ft. ele. and above; uncommon, crepuscular and nocturnal.

—— Desert Cottontail (**Sylvilagus audubonii**)
Plains, deserts, foothills, low valleys, creosote flats, and brushlands, Rio Grande floodplain to about 4700 ft. ele.; fairly common, crepuscular and nocturnal.

ORDER RODENTIA — Rodents

—— Rock Squirrel (**Spermophilus variegatus**)
Boulder-strewn slopes, cliffs, and rocky canyons, throughout park but most common in Chisos Mts.; uncommon, diurnal.

—— Mexican Ground Squirrel (**Spermophilus mexicanus**)
Campgrounds, lawns, and grasslands, Bermuda grass community and sandy soil, known only from Rio Grande Village, perhaps other areas along Rio Grande; rare, diurnal.

—— Spotted Ground Squirrel (**Spermophilus spilosoma**)
Desert flats and foothills, between 2200 and 4000 ft. ele.; uncommon, wary, diurnal.

—— Texas Antelope Squirrel (**Ammospermophilus interpres**)
Rocky canyons, boulder-strewn slopes, bluffs, and desert flats, sparse vegetation, Rio Grande to at least 6100 ft. ele., most common in Chisos Mts.; uncommon, secretive, diurnal.

—— Botta's Pocket Gopher (**Thomomys bottae**)
Dry, rocky flats and mountain slopes and valleys, Chisos Mts. down to about 2000 ft. ele.; common, diurnal or nocturnal.

—— Yellow-faced Pocket Gopher (**Pappogeomys castanops**)
Easily worked and deep, fine-textured soils, only along Rio Grande and its tributaries; common, diurnal or nocturnal.

—— Merriam's Pocket Mouse (**Perognathus merriami**)
Desert, grassland, short or sparse vegetation, sandy or gravelly soil, mainly Chisos Mt. foothills — 1840 to 4000 ft. ele.; uncommon, nocturnal.

—— Desert Pocket Mouse (**Perognathus penicillatus**)
Sandy desert floors, sparse vegetation, Lower Sonoran Life Zone, Rio Grande to around 3500 ft. ele.; common, nocturnal.

—— Nelson's Pocket Mouse (**Perognathus nelsoni**)
Rocky slopes having sparse vegetation, upper grassland, mountain foothills from 2300 to 5500 ft. ele.; uncommon, nocturnal.

—— Ord's Kangaroo Rat (**Dipodomys ordii**)
Sandy soil, floodplains, creosote flats, Rio Grande floodplain and Upper Tornillo Creek; rare, nocturnal.

—— Merriam's Kangaroo Rat (**Dipodomys merriami**)
Sandy or rocky soil, desert, sparse vegetation, lowlands from Rio Grande to 4000 ft. ele.; common, nocturnal.

—— Beaver (**Castor canadensis**)
Rivers, streams, and lakes with wooded banks, only along Rio Grande; uncommon, nocturnal.

—— Western Harvest Mouse (**Reithrodontomys megalotis**)
Grassland, desert, weedy areas, dense ground cover, between 2500 and 6700 ft. ele.; uncommon, nocturnal or sometimes diurnal.

—— Fulvous Harvest Mouse (**Reithrodontomys fulvescens**)
Canyon grasslands, known only from Pine Canyon, Juniper Canyon, and Green Gulch, perhaps throughout Chisos foothills; very rare, mainly nocturnal.

—— Cactus Mouse (**Peromyscus eremicus**)
Desert, creosote flats, grama grass-lechuguilla and prickly pear associations, sandy, clay, and rocky soil, Rio Grande floodplain to 5500 ft. ele.; common, nocturnal.

—— Deer Mouse (**Peromyscus maniculatus**)
Desert plains, arroyos, creek and river floodplains, lower foothills, Rio Grande to 3500 ft. ele.; rare, nocturnal.

—— White-footed Mouse (**Peromyscus leucopus**)
Wooded or brushy areas, riparian habitat, Rio Grande floodplain; rare, nocturnal.

—— Brush Mouse (**Peromyscus boylii**)
Rocky areas with pinyon, juniper, oak, and brush, mesic forest, mountains -- 4500 ft. ele. and above; uncommon, nocturnal.

—— White-ankled Mouse (**Peromyscus pectoralis**)
Rocky areas, pinyon-oak association, grassy meadows, mountains — 4000 ft. ele. and above; uncommon, nocturnal.

—— Rock Mouse (**Peromyscus difficilis**)
Exposed rocky slopes, rockpiles, boulders, highest parts of Chisos Mts. — 6300 ft. ele. and above; rare, nocturnal.

—— Southern Grasshopper Mouse (**Onychomys torridus**)
Desert, grassland, sandy or gravelly soil; known only from a single specimen from Gano Spring at Burro Mesa; vagrant or possibly extirpated, nocturnal.

—— Southern Plains Woodrat (**Neotoma micropus**)
Desert brushland, river cane and mesquite thickets, creosote flats, prickly pear and cholla flats, deep sand, Rio Grande floodplain to 3500 ft. ele.; uncommon, nocturnal.

—— White-throated Woodrat (**Neotoma albigula**)
Woodland, brushland, pinyon-juniper and grass-cactus associations, rocky soils, mountains—between 3700 and 7000 ft. ele.; fairly common, nocturnal.

_____ Mexican Woodrat (**Neotoma mexicana**)
Mountain cliffs, rock slides, mesic habitat, only the rocky slopes of Mt. Emory; rare, nocturnal.

_____ Hispid Cotton Rat (**Sigmodon hispidus**)
Grassland, weed patches, moist areas, Rio Grande floodplain to about 3850 ft. ele.; uncommon, nocturnal or diurnal.

_____ Yellow-nosed Cotton Rat (**Sigmodon ochrognathus**)
Mountain meadows and grasslands, grassland-woodland association, Stipa grass, dense vegetation, Chisos Mts. — between 5000 and 7000 ft. ele.; uncommon, nocturnal or diurnal.

_____ House Mouse (**Mus musculus**)
Man's dwellings, sometimes fields; one old questionable park record, evidently this introduced pest has never established itself within the park; mainly nocturnal.

_____ Porcupine (**Erethizon dorsatum**)
Forest, sometimes brushland and desert, Rio Grande floodplain to at least 5400 ft. ele. (Basin); rare, mainly nocturnal.

ORDER CARNIVORA — Carnivores

_____ Black Bear (**Ursus americanus**)
Mainly mountain forest, status uncertain; originally fairly common in Chisos Mts., today mainly strays along Rio Grande from Mexico; vagrant, nocturnal or diurnal.

_____ Raccoon (**Procyon lotor**)
Rio Grande and its tributaries, riparian habitat, lowlands; rare, nocturnal.

_____ Coati (**Nasua nasua**)
Floodplains and open woodlands, lowlands and foothills; vagrant, diurnal or nocturnal.

_____ Ringtail (**Bassariscus astutus**)
Canyons, cliffs, and rocky ledges, often near water, throughout park; fairly common, nocturnal.

_____ Long-tailed Weasel (**Mustela frenata**)
Evidently no habitat preference, possibly throughout park; extremely rare, nocturnal or diurnal.

_____ Badger (**Taxidea taxus**)
Desert and grassland, Rio Grande lowlands to Chisos foothills; uncommon, mainly nocturnal.

_____ Western Spotted Skunk (**Spilogale gracilis**)
Brushlands, along streams, grasslands, rocky outcrops, throughout park; rare, nocturnal.

_____ Striped Skunk (**Mephitis mephitis**)
Woodlands, brushlands, grasslands, and rocky areas, throughout park; common, nocturnal.

_____ Hooded Skunk (**Mephitis macroura**)
Streams, rocky ledges, desert and lowlands (mainly in riparian habitat); evidently only one park record from Tornillo Creek; vagrant or extremely rare, nocturnal.

_____ Hog-nosed Skunk (**Conepatus mesoleucus**)
Open woodlands, brushy or rocky areas, Rio Grande up to at least 5700 ft. ele.; uncommon, nocturnal.

_____ Coyote (**Canis latrans**)
Desert, grassland, open woodland, brushland, and rocky areas, Rio Grande up to 5500 ft. ele., rarely higher; very common, nocturnal or diurnal.

_____ Gray Wolf (**Canis lupus**)
Wilderness desert, grassland, and woodland; only two recent park records, three recently confirmed records outside park boundaries, nearly extirpated; vagrant, nocturnal or diurnal.

_____ Kit Fox (**Vulpes macrotis**)
Desert and grassland, from Rio Grande to 5000 ft. ele. (Chisos foothills); rare, nocturnal.

_____ Gray Fox (**Urocyon cinereoargenteus**)
Desert, shrubland, open woodland, and rocky areas, Rio Grande to at least 5500 ft. ele., probably parkwide; climbs trees; uncommon, nocturnal.

_____ Mountain Lion (**Felis concolor**)
Rugged mountains, canyons, woodlands, and deserts, prefers wilderness, throughout park but prefers mountains (Chisos); uncommon, tame, secretive, mainly nocturnal; a necessary predator to keep the deer herds healthy, climbs trees.

_____ Ocelot (**Felis pardalis**)
Brushland, riparian thickets, thorn scrub, and rocky areas; one recent sight record at Rio Grande Village, old record from Persimmon Gap; vagrant, nocturnal.

_____ Bobcat (**Lynx rufus**)
Desert, brushland, woodland, rocky outcrops and canyons, throughout park; uncommon, mainly nocturnal.

ORDER ARTIODACTYLA — Even-toed Ungulates

_____ Collared Peccary or Javelina (**Dicotyles tajacu**)
Desert, brushland, woodland openings, arroyos, canyons, cacti, often near waterholes, throughout park but most common in mountain foothills, rarer in high parts of Chisos Mts.; common, travel in bands, crepuscular and nocturnal.

_____ Mule Deer (**Odocoileus hemionus**)
Desert, grassland, shrubland, Rio Grande floodplain to mountain foothills (up to about 5000 ft. ele. — the lower parts of the Chisos Mts.); common, crepuscular and nocturnal.

_____ White-tailed Deer (**Odocoileus virginianus**)
Mountain meadows, forests, canyons, and shrublands, only Chisos Mts. down to about 4000 ft ele.; common, crepuscular and nocturnal.

_____ Pronghorn (**Antilocapra americana**)
Grasslands, plains, and semidesert, park lowlands to mountain foothills (most records are between Persimmon Gap and park headquarters); rare, small bands, mainly diurnal.

_____ Mountain Sheep (**Ovis canadensis**)
Mountain slopes, canyons, ridges, ledges, rocky desert, rugged terrain; formerly at Boquillas Canyon, Mariscal Mt., and Mesa de Anguila; extirpated, now returning, recent records from lower Tornillo Flat, Dagger Flat, fossil bone exhibit, tunnel, Panther Pass, and Blue Creek Canyon; extremely rare, diurnal.

7

THE ARCTIC DESERT

DAVID M. HICKOK, JOSEPH C. LaBELLE, AND ROBERT G. ADLER

THERE ARE MANY misconceptions about the North American Arctic. Few people have actually been there, and it lies far from large centers of population. Many conceive of the Arctic as an externally frozen wasteland. Although dominated by rigorous cold for most of the year, the arctic summer offers comfortably warm temperatures and long periods of continuous daylight. The vista of the short summer's fragile green tundra, however, is soon replaced by a vast white blanket of snow for nine months.

Four major physiographic regions make up the North American Arctic (figure 7-1) The Alaskan Coastal Plain consists of flat terrain graduating into the foothills of the Brooks Range to the south. The Beaufort Sea of the Arctic Ocean defines its northern limits. The Continental Canadian Coastal Plain extends from the Mackenzie River lowlands east to the highly glaciated Labrador peninsula. Treeline defines its southern boundary and the coastline its northern extent. The Canadian Arctic Archipelago consists of an extensive group of glaciated islands bounded by Canda to the south, Greenland to the east, and the Arctic Ocean to the north and west. Last is Greenland, dominated by its continental ice sheet and bounded by the sea that surrounds it.

The arctic region, with all of its unique features, has no single characteristic to delineate its boundaries. Scientists define the Arctic according to their field of study. Astronomers and cartographers of the Renaissance period derived latitude 66° 33″ as the Arctic Circle, defined as the line above which the sun would not rise on the winter solstice or set on the summer solstice. This line is widely used as the general boundary of the Arctic today. Geologists, in their study of the earth's surface, establish the arctic limit by the occurrence of continuous permafrost. This zone of permanently frozen ground is also of great importance to botanists. Plant growth and water penetration are determined by its surface extent. Oceanographers and marine biologists confine their boundaries to the sea. Sea temperatures of 0°C define arctic waters. Climate also greatly influences the form and structure of the arctic landscape. One obvious result is the northern limit of trees, beyond which the climate is too severe for tree survival.

Pewe (1974) defined a polar desert as a glacier-free terrestrial area where the mean annual temperature for the warmest month of the year is less than 10°C. Bovis and Barry (1974) outlined the north polar desert according to the Budyko Radiational Index of Dryness, the Thornthwaite-Mather Moisture Index, and the Turc Evaporation Formula. Each produces slightly different boundaries for the desert.

The most meaningful and useful boundaries to man are climatic factors. In terms of temperature and duration of arctic cold, the effects of climate are most profound. This intensely harsh environment molds the landscape and life upon it. Solid rock is weathered by high winds and frost action that produces extensive boulder-strewn landscapes. Solar heating produces meltwater and soil moisture, giving rise to a soggy, marshy surface layer overlying the permafrost zone. These processes create land features that are recognizable throughout the year, including patterned ground, thaw lakes, and ice mounds called pingos. Plants and animals are limited to the species that have adapted to this environment. This paucity of life forms, combined with the desolation and solitude one feels in the Arctic, give it the tranquil appearance of a desert.

REGIONAL PHYSIOGRAPHY

Alaska's Arctic Slope

The arctic foothills of the Brooks Range are characterized by irregular buttes, knobs, mesas, east-trending ridges, and intervening rolling tundra plains; maximum elevations reach 1050 meters (m). (See figure 7-2.) Most streams crossing the foothills have north-trending courses from sources in the Brooks Range. The Colville River, the longest in arctic Alaska, has a structurally controlled east-trending course for more than 350 kilometers (km) through the foothills before it turns north and crosses the coastal plain directly to the sea. Most streams

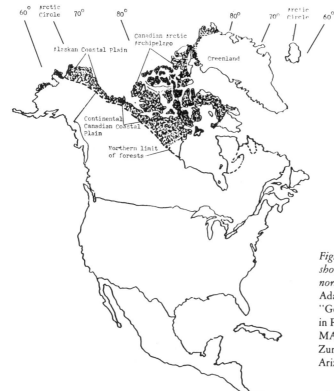

Figure 7-1. North American Arctic Desert showing major physiographic regions and northern limit of forests.
Adapted by permission, Troy L. Péwé, "Geomorphic Processes in Polar Deserts" in POLAR DESERTS AND MODERN MAN, Terah L. Smiley and James H. Zumberge, editors, Tucson: University of Arizona Press, copyright, 1974.

Figure 7-2. Arctic slope near mouth of the Kellik River.
Photograph by C. D. Evans. Courtesy of the Arctic Environmental Information and Data Center, University of Alaska.

east of the Colville River have broad, braided courses, and many of their upper valleys contain moraine-dammed lakes. A few thaw lakes also occur.

No modern glaciers exist in the foothills, but a considerable amount of Pleistocene glacial debris mantles the surface. Frost features including ice-wedge polygons, stone stripes and nets, and solifluction sheets and lobes are common.

Alaska's arctic coastal plain is a smooth surface rising from the shores of the Chukchi and Beaufort seas to a height of approximately 180 m at its southern boundary where it meets the arctic foothills of the Brooks Range. West of the Colville River to the Kuk River, the plain is essentially flat except for occasional pingos and a section of active and stabilized sand dunes that originated during the Pleistocene. East of the Colville River, the White Hills produce scattered low relief.

Due to extensive flat terrain and the continuous occurrence of impermeable permafrost, drainage on the coastal plain is very poor and marshes occur in most low areas. West of the Colville River, rivers meander sluggishly in valleys 15 to 90 m deep, whereas to the east, rivers cross the coastal plain in braided channels and are building large deltas into the Beaufort Sea.

The western part of the coastal plain is more than half covered by oriented thaw lakes that have their long axes aligned just west of north (figure 7-3). Many former lakes have drained, leaving behind marshy basins; numerous overlapping active and drained lake basins are in evidence.

Continental Canadian Arctic Region

The Mackenzie Lowland is a flat surface of almost no relief, lying almost entirely below 330 m elevation. The Mackenzie Delta consists of complex river channels, lakes, and low, marshy islands, which are forested almost to the shores of the Beaufort Sea. East of the delta and about 60 m above it, the coastal plain is uniformly flat and unvarying, with numerous thaw lakes. West of the delta, terrain rises sharply to the adjacent Richardson Mountains on the southwest and to the narrow coastal plain, which melds with Alaska's coastal plain on the west.

The Canadian Shield begins in the west as a series of escarpments running southeastward from Darnley Bay through Great Bear Lake, Great Slave Lake, Lake Athabasca, and Lake Winnepeg. The shield is continuous to the Atlantic

Figure 7-3. Arctic coastal plain.
Photograph by J. C. LaBelle. Courtesy of the Arctic Environmental Information and Data Center, University of Alaska.

Ocean's shoreline in the east and extends into the arctic islands and Greenland. The part of the Canadian Shield on the continent is known as the Laurentian Upland Province and is subdivided into the Keewatin Section, the Ungava Section, and the Hudson Bay Lowland.

The topography of the Keewatin Section is gentle and undulating, rising to the divide at 750 m between the Mackenzie River and Hudson Bay drainages. Extensive Pleistocene glacial deposits mantle the land, interrupting drainage patterns and creating numerous lakes, some of which are now drained. The northern coastal slopes are gentle, with north-projecting peninsulas rising above the lowlands. South-facing scarps rising to 600 m are found in the Coppermine area. The large portion of the shield west of Hudson Bay, known as the Barren Lands, is the flattest, unvarying part of the entire shield, with elevations rising from sea level up to approximately 300 m. Melville Peninsula is a smooth plateau about 450 m high, while Boothia peninsula consists of high rocky uplands.

The Ungava Section consists of both the Ungava and Labrador peninsulas and is of moderate elevation with typical shield topography. Rocky plateaus rise abruptly from the east and north coasts with deep, rugged valleys and mountains rising to a maximum of about 1700 m. Lakes occupy fault troughs along the eastern border. The central uplands of the region drain radially into the watersheds of the St. Lawrence River, Ungava Bay, Hudson Bay, and Labrador Sea.

The Hudson Bay Lowland is a broad, flat coastal plain, superimposed on the Canadian Shield. It consists primarily of postglacial marine sands and clays, rising to elevations below 150 m.

All of the Canadian mainland arctic region, with the exception of the far northwest corner, was glaciated during the Pleistocene, leaving behind an ice-scoured surface with extensive glacial debris. Features related to frost action have since modified much of the surface.

Canadian Arctic Archipelago

This province contains all of the Canadian high arctic islands lying north of the North American mainland. Topography is highly varied, with lowlands, plains, plateaus, and mountains. Ice caps are found throughout the belt of high mountains running from eastern Labrador through Baffin Island to Ellesmere Island, with the coastline deeply cut by fjords.

The rugged relief of the entire range is due to glacial erosion, which in some cases extends all the way down to sea level. On Baffin Island, rugged mountains rise within a few tens of kilometers of the eastern coastline to elevations as high as 3000 m. Alpine ranges reach elevations of about 3000 m on Ellesmere Island and 2100 m on Axel Heiberg Island. The ranges become flat-topped on their western sides where they

grade through plateaus and uplands to the lowlands. The eastern uplands slope generally southwest (elevations from 300 to 1200 m) and are generally rough, rocky hills with flat tops. The western uplands of Victoria and Melville islands are characterized by ridge and escarpment topography with elevations as high as 600 m.

Baffin, Ellesmere, and Devon islands have plateaus composed of flat-lying or tilled sedimentary rocks with elevations as high as 900 m. On Ellesmere Island, the plateau above 1500 m is covered by an ice cap, but most plateau elevations do not exceed 1100 m. The plateaus are often dissected into cliffs, canyons, and inlets.

On Banks, Victoria, and Melville islands, the plateaus are lower, with elevation between approximately 300 and 750 m, with one exception to 1000 m on southwestern Melville Island. The surface of these plateaus is gently tilted and warped, with steep sea cliffs 150 to 300 m high. On Banks and Victoria islands, a lower plateau surface with rolling topography is found with elevations less than 300 m.

Here and there on the northwestern and southern portions of the islands are lowlands and plains with maximum elevations of about 100 m. These lowlands are generally basins amid higher topography, with poor drainage and many lakes. They are often mantled with extensive glacial deposits, especially eskers, drumlins, raised beaches, and marine deposits. The lowlands grade to the west into uplands and plateaus. In the extreme northwestern part of the province is a band of coastal plains.

All of the Canadian high arctic islands were covered by the Pleistocene ice sheet except the islands farthest to the northwest. Glacial erosion has modified the surface nearly everywhere, abetted by landforms of glacial deposition. Further modification of the land has taken place due to frost processes.

Greenland

Greenland, the largest island in the world (total area 2,186,000 square km), is principally part of the Canadian Shield. Five-sixths of the island is covered with snow and ice from the Greenland Ice Cap, which rises to elevations of 3000 m or more at the center. The surface is depressed by the weight of the ice; the lowest part beneath the central ice cap possibly is below sea level.

Near the coast are found many nunataks—mountain tops that protrude through the ice cover. The coast is typified by numerous cliffs, fjords, glaciers, and islands. There are a few areas of dwarf trees along the southwestern coast, but otherwise the terrain of rocks and gravel sustains only shrubs, grasses, mosses, and lichens.

The ice-free coastal sections vary considerably physiographically. On the west coast south of Disko Bay, the shield resembles that of eastern Baffin Island, although it is somewhat higher in elevation. Around Disko Bay and Scoresby Sound are basalt plateaus. In the north and northwest are mountain ranges and sedimentary plateaus closely related to similar features on northern Ellesmere Island. East Greenland consists of high arctic mountains reaching altitudes of more than 3000 m, cut by fjords and glaciers, with more ice-free land north of 70 degrees north than south of it.

The ice-free regions of Greenland were all covered by glacier ice during the Pleistocene, as evidenced by the universal extent of glacial erosional and depositional land forms. In most places frost processes have somewhat modified the older glacial surface.

GLACIAL HISTORY

During the Pleistocene epoch, beginning approximately 1 million years ago, four major glacial advances occurred in North America that covered the northern portion of the continent with ice to depths exceeding 3 km. Northern Canada was completely covered with ice except along the very northwest portions of both the continent and the arctic islands. Greenland was entirely covered. Alaska's arctic slope was mostly unaffected by ice, except in some of the valleys of the southern foothills into which some of the large Brook Range glaciers extended. Each glacial stage reached southward to a different degree, but the northern part of the continent was essentially inundated by ice each time.

The causes of continental glaciation are varied and complex, and researchers are uncertain of why it happens. Several climatic factors are known to contribute to glacial advance, such as a cooling trend and an increase in annual precipitation. If the climate cools sufficiently, summer will not be long or warm enough to melt all of the winter snowfall. If annual precipitation increases sufficiently, more snow falls in winter than can be melted during summer. Either condition leads to continually deepening accumulations of perennial snow, which finally metamorphose into ice through a number of interrelated processes. When the ice becomes thick enough (about 60 m) to flow downhill, it becomes a glacier.

Several times during the Pleistocene, climatic conditions in central Quebec became such that seasonal snowfall failed to melt totally. The formation of expanding permanent snowfields soon led to a radially moving glacial ice mass. As the ice cap expanded, inflowing moist air from the south and west interacted with cold air at the ice margins increasing precipitation that fell as snow on ice margins. This process, in which the expanding ice cap created conditions for its own further expansion, continued rapidly until a great continental ice sheet covered almost all of the northern part of the continent. In adjacent areas, expanded mountain glacier systems inundated most northern mountain ranges, especially in the Rocky Mountain Cordillera. The formation and expansion of each ice sheet

took many thousands of years, followed by a relatively stable glacial maximum period, and then a declining period of glacial retreat and disappearance when the climate reverted to a warm interglacial stage. Four episodes of continental glaciation, interspersed with warm interglacial stages, occurred during the Pleistocene (table 7-1).

TABLE 7-1
Pleistocene Epoch in North America

GLACIAL EPISODES	INTERGLACIAL EPISODES
Nebraskan	
	Aftonian
Kansan	
	Yarmouth
Illinoisan	
	Sangamon
Wisconsin	

Four times the incorporation of as much as 50 million cubic km of ice into the several continental ice sheets of the world robbed the oceans of their normal access to surface drainage waters, until sea level was lowered more than 100 m below the present position. Much of the continental shelves was exposed and, in unglaciated areas, populated by animals and men. Between present Siberia and Alaska several thousand square kilometers of exposed shelf became the Bering Land Bridge, connecting the two continents and allowing the interchange of animals and humans between Asia and North America.

Approximately 22,000 years ago the climate in the Northern Hemisphere began to warm again, and the Wisconsin ice front began to retreat northward. Like all other glacial retreats, the decline was not continuous but was interrupted by halting readvances, marking substages of the Wisconsin glaciation. Eventually ice disappeared from the lowland areas, leaving lingering highland ice caps that lasted for several thousand years. In the final stages, two lingering ice caps flowed outward from areas on either side of the Hudson Bay. As these shrunk even further, the sea entered the Hudson Bay basin between the ice domes, causing them to calve directly into the sea and waste away more quickly. Finally the ice sheet disappeared almost entirely, leaving behind only the smaller ice caps that dominate the higher mountain regions of the cordillera and arctic islands and the Greenland ice cap. All of these shrunk to smaller sizes than exist today during the height of the Hypsithermal warm interval, 5000 to 6000 years ago. Since that time the climate has fluctuated somewhat, leading at times to minor readvances of mountain glacier systems. Currently we are in a period of uncertain climatic tendency, with most glaciers retreating somewhat while others advance. Many glaciologists feel we are merely in another interglacial period and that time will bring about another continental extension of the glaciers.

Isostatic Rebound

At the height of the glacial age, the earth's crust in northern Canada had to support the weight of ice approximately 3 km thick. This forced the crust to subside somewhat to compensate for the great load. When the ice finally retreated and disappeared, removing the load, the crust began to rebound or readjust to its former position. Although the ice has been essentially gone for thousands of years, the slow rebound process still continues, raising the central dome of Canada a few centimeters per year. It has been estimated that an additional 260 m of uplift must occur before equilibrium is again established. When this happens, Hudson Bay will be drained almost to the Hudson Strait.

PERMAFROST

Formation and Distribution

Most of the arctic region is covered by continuous permafrost, which reaches depths as great as 600 m near Prudhoe Bay in Alaska and Melville Island in the Canadian Arctic. Permafrost is any earth material, such as soil or bedrock, that has remained below freezing for more than one season. It forms whenever summer thaw fails to reach as deeply into the ground as winter freezing. This usually occurs in regions, such as the Arctic, where a lengthy winter of intense cold is not compensated for by the short, relatively cool summer. Permafrost may or may not be in a frozen state, depending on whether the ground contains moisture and whether any present moisture is saline, depressing its freezing point. Only the ground temperature defines the presence of permafrost.

Much of the permafrost in the Arctic has been in existence for thousands of years, although its character continually changes to match the changing climatic conditions. Locally permafrost is newly forming in some areas while degrading in others. Permafrost is generally deepest and oldest in regions that were ice free during the Pleistocene epoch; the coverage of the thick continental ice sheet over much of the northern part of the continent tended to insulate the ground from extreme cold temperatures, preventing the formation of new permafrost and possible degrading older permafrost. For these reasons, permafrost is usually older and deeper under Alaska's arctic slope, which was never ice covered during the Pleistocene, than it is beneath the glaciated Canadian Shield.

The thickness of permafrost is affected by differences in climate, vegetative cover, topography, amount of groundwater, heat flow in the earth, and the type of sediment or bedrock involved. Thaw bulbs, free of permafrost, often form around and beneath deep-water bodies, such as rivers and lakes that are sufficiently deep to prevent freezing to the bottom in winter, because of the thawing of the relatively warm water body. Even here, though, permafrost usually exists at greater depths, depending on the size and depth of the water body. Relict permafrost has also recently been found to exist on the continental shelf of the Beaufort Sea. It remains from the time when much of the shelf was exposed above sea level during the height of the Pleistocene.

The upper soil layers in a permafrost region undergo a seasonal freeze-thaw cycle. The depth of this active layer may vary from season to season, depending on climatic variations. Whenever the active layer fails to freeze all the way down to the permafrost layer in winter, a zone of unfrozen talik exists between the frozen active layer and the permafrost.

The depth of the summer thaw depends upon the character of the surface sediments. Fine-grained sediments, such as silt, retain much more water and ice than does gravel or coarse sand, so silty soils generally thaw only to depths of tens of centimeters, whereas dry gravelly soils may thaw to depths greater than 2 m.

Frost Processes and Land Forms

The operation of the seasonal freeze-thaw cycle upon permanently frozen grounds creates a fascinating array of topographic features characteristic of the Arctic. Frost features are especially notable in the coastal plains and basins where thick, ice-rich soils cover the region, but some are common even to exposed bedrock terrain.

ICE WEDGE POLYGONS

Polygonal features, also known as patterned ground, are seen in many places in the tundra country (figure 7-4). They are initially produced by contraction of the ground during low winter temperatures. As the ground contracts, cracks form, delineating polygonal shapes. The next summer the cracks fill with water and then freeze during the following winter, forming ice wedges in the cracks. During ensuing seasons, more water is added and frozen to the wedges, which grow and expand, pushing soil material up and out of the cracks. This produces small soil ridges parallel to the crack edges.

On well-drained land surfaces, summer waters flow through the crack systems above the ice wedges, eroding the crack edges and causing the soil ridges to collapse back into the cracks. This process leaves the polygons themselves higher than the cracks, forming high-centered polygons. On poorly drained, marshy land surfaces, standing water in the cracks freezes to the ice wedges, pushing even more soil out onto the soil ridges. This causes the crack system to stand higher than the polygons, forming low-centered polygons.

Generally polygon patterns form completely at random; however, along retreating shorelines of draining lakes and

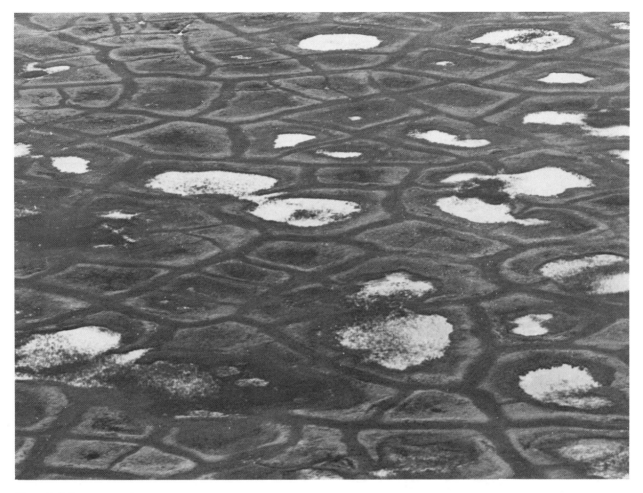

Figure 7-4. Low center polygons.
Photograph by J. C. LaBelle. Courtesy of the Arctic Environmental Information and Data Center, University of Alaska.

meandering rivers, lateral temperature gradients exist in the abandoned shore regions, resulting from newly aggrading permafrost. This causes a tendency for polygons to form parallel to the shoreline.

THERMOKARST TOPOGRAPHY

Differential thawing of ground ice in permafrost soils often produces an uneven surface of mounds; holes; tunnels; caves; short, water-cut valleys; lakes; and circular lowlands. Thaw can be started by any process that upsets the temperature balance of the permafrost, such as destruction of the insulating vegetative cover, erosion, or change in microclimate. An example of a natural thermokarst feature is a thaw lake; an example of a man-made thermokarst feature is the water-filled trench that results from driving a tractor over tundra, locally destroying the vegetative cover (figure 7-5).

THAW LAKES

Thaw lakes are a very common feature in the permafrost regions, covering more than half of the land surface in such regions as the flat Alaskan arctic coastal plain. They occur with less frequency on slopes or hilly regions. A thaw lake begins its existence as low-centered polygons where water pools in a low, marshy area. Pooling water thaws the permafrost below, adding more water to form a small pond covering several polygons. This process continues as thaw extends the edges of the pond, often joining other local ponds to form a larger lake. In this manner, the lake enlarges until it eventually thaws through to lower ground. Then the lake drains to the lower level, leaving behind a drained lake basin. This process may take hundreds of years. Many overlapping lakes and drained basins can be seen in the Arctic. Areas of never-thawed higher ground between basins are known as initial surface residuals and usually lie 3 to 5 m above the basins. The residuals, with their better drainage, are characterized by high-centered polygons, while the marshy basins have low-centered polygons.

In some areas, such as Alaska's arctic coastal plain where they are a characteristic feature, thaw lakes become elongated perpendicular to the prevailing wind direction, forming oriented lakes. In Alaska they form with their long axes pointing north-northwest, normal to the east-northeast prevailing wind. Wind-driven waves cause underwater sedimentary shelves to form on the windward and leeward shores of a thaw lake. The shelves dampen wave action on those shores, preventing or slowing further thaw extension in the east-west direction, while the lake continues its thaw extension on the unprotected northern and southern shores. Long, oval lakes are formed that are seldom deeper than about 3 m.

Teshekpuk Lake, the largest lake in arctic Alaska, was formed by the coalescence of several oriented lakes but is somewhat deeper than usual because it formed in a structurally controlled depression. Some researchers feel that series of

offshore lagoons paralleling Alaska's arctic coast behind barrier islands also formed this way.

BEADED STREAMS

An unusual type of stream forms in permafrost regions. Drainage waters, flowing along high-centered polygon crack systems, enlarge the polygon intersections through thaw erosion, forming small pools. The pools may be 1 to 3 m deep and a few meters wide. Connected by the stream running through the crack system, they appear like a string of beads (figure 7-6).

PINGOS

Rounded hills, as much as 30 m high and several hundred meters across, often having summit craters filled with water, are found in many places in the arctic plains. Some of the largest and most impressive pingos occur in the Mackenzie

Figure 7-5. Rolligon tracks on tundra.
Photograph by J. C. LaBelle. Courtesy of the Arctic Environmental Information and Data Center, University of Alaska.

River delta, and some have even been found submerged offshore on the continental shelf. They are essentially huge frost heaves that form when thaw lakes drain. After a lake drains away, permafrost begins to aggrade toward the center of the basin from the sides. As the freezing center reaches the middle of the basin, it can no longer expand laterally and is forced upward into an ice-cored mound. Often as the hill expands upward, a tension crack forms on the top, exposing the top of the ice core to thaw, which creates a small summit lake (figure 7-7).

FROST MOUNDS

Features similar to pingos but much smaller (usually no more than 1 m high and 3 to 5 m across) are often found in low, marshy basins. Like pingos, they also have ice cores and probably form by similar processes. The ice core of the large one is often connected to the permafrost below, but the smaller ones are usually limited to the active layer.

STRING BOGS

In some muskeg areas of the Arctic, marshes have relatively dry ridges of peat with ice lenses and vegetation running across them. They are several meters wide and are often interspersed with small, shallow ponds. The ridges take one of three forms: linear ridges transverse to the slope of the area, irregular ring-shaped ridges, or netlike patterns of ridges. Ridges formed transverse to a slope and usually have a steep face downslope and a gentler face upslope.

REGIONAL GEOLOGY

Alaska's Arctic Slope

The major structural features of the southern foothills are open folds of Mesozoic beds that become steeper and deeper toward the south. Most of the region is underlain by east-trending belts of Cretaceous rocks that were deposited from

Figure 7-6. Beaded Stream.
Photograph by J. C. LaBelle. Courtesy of Arctic Environmental Information and Data Center, University of Alaska.

Figure 7-7. Pingo south of the Colville River Delta.
Photograph by J. C. LaBelle. Courtesy of the Arctic Environmental Information and Data Center,
University of Alaska.

source areas in the uplifted Brooks Range to the south. Many of these, along with some earlier Paleozoic rocks to the south, were later displaced north by shallow thrust faults. Characteristic rocks in the foothills include Cretaceous sandstone, conglomerate, shale, siltstone, limestone, and chert. Less common rocks, exposed especially in the southern foothills, include Paleozoic and Mesozoic basic intrusive rocks, Mississippian to Cretaceous sedimentary rocks, and some Tertiary conglomerate and sandstone that crop out in the northern part of the foothills. Coal beds are widely distributed throughout the area, and rocks rich in oil shale crop out continuously in the thrust belt of the southern foothills. Small amounts of oil and gas have been found in the foothills.

Nearly all of the arctic coastal plain is covered by Quaternary unconsolidated deposits, which overlie gently south-dipping, Late Mesozoic sandstone, conglomerate, shale, and Tertiary beds of conglomerate, sandstone, and siltstone. Extensive coal fields occur near the surface of the coastal plain and crop out at several points along the western coast and along many riverbanks.

Deeper in the subsurface, Paleozoic rocks that represent the buried extensions of rock units in the Brooks Range and foothills thin out and truncate against the Barrow Arch, a regional structural high that parallels much of Alaska's arctic coast. Nonconformities at the base of the Cretaceous sequence and other units, as well as many widespread fault systems, truncate progressively older rock units northward. Each older unit, including the old pre-Mississippian basement rocks,

terminates against one or another of the nonconformities or faults. These conditions provide the impermeable stratigraphic traps atop the Barrow Arch where the great oil and gas reservoirs of the Prudhoe Bay field are found.

Continental Canadian Arctic Region

The broad Mackenzie lowland plain is underlain principally by gently folded Paleozoic strata with fewer beds of Mesozoic and Tertiary age. Most of the rock types are the result of marine shelf deposition and include sandstone, shale, conglomerate, limestone, and dolomite. Overlying the gently dipping sedimentary rocks are glacial sediments and drift left from Pleistocene continental glaciation. The Mackenzie River valley is filled with alluvial sediments. Occurrences of gypsum, salt, bituminous shale, oil, and gas are found in the plains, and some coal has been mined from Cretaceous beds north of Aklavik. Significant oil and natural gas accumulations have been found in the Mackenzie delta and its offshore extension.

The Canadian Shield extends around Hudson Bay in a horseshoe shape that forms the backbone of Canada. The shield consists almost entirely of Precambrian igneous and metamorphic rocks—mainly granite, gneiss, basaltic lava, greenstone, schists, quartzites, and crystalline limestone. The areal extent of the granite may be as much as 80 percent of the shield. The shield is a peneplain surface, which was uplifted to about its present position in Pliocene time. This warping uplift gave the shield its saucer-like shape, sloping generally inward toward the Hudson Bay Lowland, filled with mainly Ordovician, Silurian, and Devonian sediments, mainly limestones and shales.

The hard core of highly metamorphosed ancient rocks was folded in Precambrian times by mountain-building movements, pierced by granite intrusions, and then eroded to its present low, rounded topography regardless of differing hardness of underlying rocks. Exceptions to this are the rugged, glaciated mountain ranges of gneiss, rising to more than 1500 m along the coast of Labrador.

The surface of the shield has been dissected by stream action, scoured and smoothed, and the valleys filled by glacial debris. Particularly in the interior of the shield, the rock surface has been covered by the glacial overburden. Two orogenic belts have been identified: the west, north-trending Great Bear belt and the adjacent, eastern Yellow Knife belt. Minerals associated with the shield include gold, nickel, copper, iron, and uranium.

Canadian Arctic Archipelago

The arctic islands contain regions of the Canadian Shield, sedimentary basins and plains, and the Innuitian fold belt of mountains. Shield sections are composed of similar igneous and metamorphic rocks as the continental shield section, with the same peneplain-type surface except along the eastern mountainous coast of Baffin Island, where heavily glaciated mountains of gneiss rise to more than 2100 m, dissected by glacial valleys. In the northern part of the shield monadnocks—remnants of a former highland rising as isolated rock masses surrounded by younger rocks—rise from 460 to 700 m.

Sedimentary basins are scattered through the arctic islands, separated by exposed fingers of the Canadian Shield. These basins consist of flat-lying marine sedimentary rocks deposited mainly during the Paleozoic era but including sediments of all ages, some as young as Tertiary. The sediments in the southern part are thin and low-lying Paleozoic lowlands; toward the north they form the thicker plateaus of the Mesozoic and Tertiary uplands.

The arctic coastal plain, a narrow strip adjacent to the Arctic Ocean, extends from Banks Island to Meighen Island. It is composed of unconsolidated quartz sand and thin beds of gravel dipping gently northwest. These sediments contain abundant unaltered wood of late Tertiary or Pleistocene age.

The Innuitian fold belt extends from Prince Patrick Island to northernmost Ellesmere Island, where the mountains rise to elevations of 3000 m. This has been a tectonically active region from the Paleozoic era into the Tertiary. Five distinct fold belts have been mapped. The folds of the Parry Islands and Ellesmere Island belts are very regular, composed of rocks of the Carboniferous to Jurassic periods, with some possibility as old as Precambrian. Intensely folded sedimentary strata of Ordovician to early Devonian ages in many places are overlain by younger, flat-lying beds. Ordovician and Silurian shell fossils are found extensively in carbonate rocks, sandstones, and shales. Gypsum of Permo-Carboniferous age is found in domes, layers, and folds in several locations. Lavas, granitic intrusions, and metamorphics like gneiss, slate, quartzite, and schists are also found. The Sverdrup Basin, a tectonic element underlying the northwestern part of the region, is of particular economic importance. Extensive oil and gas exploration is being carried out in the region, and low-grade coal is found in the area.

Greenland

Greenland is probably a peneplain extension of the Canadian Shield, formed in two domes separated by a deep saddle, possibly below sea level located beneath the ice cap. These domes consist of gneiss and granite with some remnants of ancient sedimentary rocks and volcanic deposits.

The southern and western coast are typical of the Precambrian Canadian Shield. On Disko Island and Nugssuak Peninsula are some younger, flat-lying Cretaceous and Tertiary sediments and volcanic basalts, conglomerates, sandstone, and shale interbedded with volcanic tuffs. There is some faulting in these sediments.

The northern and eastern coasts are bordered by mountain ranges of younger sediments, folded and intruded in places by igneous rocks. These ranges are extensions of the Innuitian fold belt of northern Ellesmere Island. From Peary Land to northwestern Greenland, high fjords cut into the intensely folded schists, slates, sandstones, and limestones of late Paleozoic through Mesozoic age. The folding occurred in post-Silurian time, probably in the Caledonian orogeny of Mesozoic time. To the southwest, in Inglefield Land and Hayes Peninsula, sediments are thinner, flat-lying Precambrian and Paleozoic formations that terminate to the south into the shield. Rock formations later than Precambrian age probably do not extend very far south under the ice cap.

Along the coast of eastern Greenland are high mountains cut by fjords and glaciers. From Cape Farewell to Kangerdlugsuak at latitude 68° N is a narrow coastal fringe of igneous and metamorphic rocks composed of gneiss, schist, quartzite, and granite. There are also some small patches of Cretaceous sedimentary rocks. From 68° N to 70° N are found large Tertiary basalt flows, forming flat-lying plateaus. These are counterparts of the west coast basalts at Disko Island.

From 70° N to Peary Land are three belts of mountains parallel to the coast and perpendicular to the mountains of Peary Land. The first belt is a thick, folded belt of Precambrian to Mesozoic sediments, including quartzite, shale, limestone, and dolomite. The second belt, to the west, is composed of igneous and metamorphic rocks, and the third belt, northwest of the second, is composed of Caledonian-folded Cambrian and Ordovician sedimentary rocks.

The mineral potential of Greenland is similar to that of arctic Canada. There is a high potential shield area of known metal deposits and folded belts of sedimentary rocks with known coal deposits. Known mineral occurrences include copper, iron, lead, zinc, chromium, nickel, rare earths, graphite, asbestos, marble, and soapstone.

CLIMATE

The polar regions and the equatorial belt are the climatic engine of the world. The arctic engine is run by the polar heat sink interacting with the equatorial heat source. The general circulation pattern over the entire earth's surface depends on cold air from the Arctic moving southward to cool areas that normally would be warmer.

Equilibrium between these warm and cold areas keeps the climate in motion. Motion is further generated by a complex series of atmospheric and terrestrial factors interacting with each other. The sun's rays provide the energy to run this dynamic engine. The arctic region receives low angles of incidence by the sun's rays. The high-altitude arctic atmosphere therefore receives less heat energy than do the lower latitudes, producing climatic phenomena peculiar to polar regions.

Low temperature is the dominant feature of the arctic climate, and temperature is often used to define the arctic region in terms of climate—that is, the area where the monthly mean temperature during the warmest month rises above freezing but never above 10°C. Solar energy input is the dominant factor causing seasonal temperature differences. The angle of the sun's rays and the tilt of the earth combine to regulate that energy. Temperature is also related to location of land and open water.

Land-water juxtaposition gives the arctic region its unique climatic pattern. The Arctic has often been misconceived as the coldest area of the world. In fact the proximity of the Arctic Ocean moderates extreme temperatures. The coldest temperatures of the Northern Hemisphere occur in subarctic continental interiors least affected by oceanic influences. The real harshness of the arctic climate derives from the duration of continual high winds and low temperatures.

A region's annual precipitation depends not only on global weather patterns but also on its interrelationship with the air temperature and terrain of the surrounding area. Cold arctic temperatures limit the capacity of the air to hold substantial amounts of moisture. Precipitation throughout the region except at higher elevations is low, usually less than 10 inches. This is why the Arctic can be classified as a desert, a desert of frozen land. In summer the thawed tundra creates anomalous vast reservoirs of water and wetlands. The majority of the region's precipitation falls in the form of snow. Accumulation is slight but covers the ground surface for about nine months without melting.

Land contours interact with both the sun's energy and ocean influences to establish local climate conditions. Mountains are the best example of this. Their southern slopes will absorb more radiant energy, postponing freezing in fall and encouraging early warming in spring. Windward sides of mountains not only affect precipitation but also incoming and outgoing solar radiation by promoting accumulations of clouds. The warming effect of air coming down off the leeward side of mountains can be seen visually in the northern displacement of treeline on the leeward side of the Western Cordillera near the Mackenzie River. Downslope winds have also had a pronounced ameliorating effect on temperature in the coastal areas of Ellesmere Island, Baffin Island, and Greenland.

Air and Sea Interactions

In the Arctic, climatic patterns are accounted for by the relative locations of open water, ice, and land. With the exception of the interior of Greenland, all North American arctic regions are influenced by water areas. The differences of thermal properties are a dominant factor in the behavior of heat transport. Sea ice has the greatest effect on arctic heat transfer. The retreating polar sea ice in the summer results in open areas of water along the coastal zones throughout most of the region. This is accompanied by an increase in atmospheric circulation, bringing warmer temperatures and more moisture to these high latitudes. It is only for a brief two-month period that these open waters have an effect. When the polar ice expands and these open areas freeze, the ocean influence decreases. Cold dry conditions now dominate the region, expanding southward to the midlatitudes by winter. The more the ice cover expands southward, the colder the air temperatures will be in these lower regions. More snow will be received, and frost penetrates as far south as Florida and southern California.

Relative to the continents, the oceans act as a heat sink in summer and heat source in the winter. Ice covering the Arctic Ocean in winter prevents this area from becoming a heat source; therefore the temperature characteristics of the underlying arctic seas cannot be transmitted to the overlying air. As a result of these processes and cold ocean currents, the tundra has extended southward to James Bay and Newfoundland.

Heat is transported by atmospheric as well as oceanic circulation. Warm ocean currents retain heat from lower latitudes and affect coastal climate. A pertinent example is the North Atlantic Gulf Stream, which brings warm water as far north as the Barents Sea, located above the Arctic Circle, and the Soviet arctic port of Murmansk on the north shore of the Kola Peninsula, which is open year round.

High Latitude Winds

The most persistent climate element throughout the year in the Arctic is wind. Wind affects the activities of man and animals, scours and transports snow, exposes food for grazing, and at the same time creates barriers to travel (figure 7-8). It can also help to increase travel. Certain animals and man can transverse a hard snow crust with ease. Combined with other elements, the wind can make winter activities in this region almost impossible.

Differences in temperature determine different atmospheric pressure. Air movement is caused when cold, dense air of a high-pressure area travels to a low-pressure area of warmer air.

The greater these differences are, the faster the airflow. Approximately 600 miles north of Alaska's Arctic is an area where dense cold air piles up, resulting in a downdraft. Moving southward, this air flows to areas of lower pressure usually found in the mid-latitudes. The corridor effect further gives direction to the continual flow of air moving southward by deflecting it to the right in the Northern Hemisphere.

The result of radiational cooling over North America and Siberia in winter helps pull the high-pressure belt inland over these areas. At the same time a low-pressure trough dominates the high Pacific and Atlantic latitudes. In the high Pacific, the ocean is warmer relative to land, creating the Aleutian Low. The Atlantic trough is caused by the warming effect of the Gulf Stream extending to this area, commonly called the Icelandic Low. Basic air flow moves from the northwest throughout the Arctic until it meets this easterly in the mid-

Figure 7-8. Windblown snow formations.
Photograph by C. D. Evans. Courtesy of Arctic Environmental Information and Data Center, University of Alaska.

latitudes, resulting in large cyclonic weather systems moving eastward in these latitudes throughout the Northern Hemisphere. Although these are the predominating winds of the high latitudes, winds from other directions frequently interrupt this general flow.

The smooth terrain of much of the arctic lowlands offers less resistance to air flow. The Arctic coastal winds are strong and persistent. Speeds of 10 to 15 knots are common, with extremes of 30 to 50 knots during winter months, especially along Alaska's Arctic coast. As these winds travel into the interior regions, their duration and strength diminishes. The morphology of this land topography determines the rate of wind speed.

Wind is probably the most important transporting agent of this region. It not only transports snow but also much of the fine-grain, eroded material. The hardness of snow grains increases with lowering temperatures, becoming more abrasive to exposed rock surfaces. Local dust clouds from windblown material of glacial outwashes, deltas, and terraces are common in spring. Wind has oriented thaw lakes in the Arctic coastal lowland in a prevailing north-west, south-east direction.

Wind channeling, a local phenomenon, can double or triple wind speed. In order to carry an equal volume of air in an equal amount of time, the speed of the wind must increase when traveling through a narrower space. The most common occurrences are in valleys and mountain passes, especially in the Brooks Range.

A special feature that has a great crippling effect on man's activities is a whiteout. This condition can be created in an area of wind channeling where this drastically increased wind speed blows snow to an extent that all sense of depth and orientation are lost (figure 7-9).

Combined with cold temperatures, wind has a strong effect on the human body or any other body warmer than its surrounding air. Wind chill indicates a rate of heat loss greater than that which the body can replace. Improper clothing hastens the lowering of body temperature. The greater the wind and the lower the temperature, the faster the cooling.

Solar Radiation

The amount of heat energy supplied by the sun to the earth depends upon the angle of the sun's rays striking the earth. The higher the latitude, the smaller the angle; consequently the high latitudes of the polar region receive the smallest amount of heat energy. Most of the earth's heat energy supply

Figure 7-9. Whiteout.
Photograph by J. C. LaBelle. Courtesy of Arctic Environmental and Information Center, University of Alaska.

is distributed throughout the latitude 15 degrees north and south of the equator where the duration of high angles of the sun's rays are the greatest.

During the time of the summer solstice in the Northern Hemisphere, the arctic regions receive a greater allotment of heat energy. The earth's axis is tilted toward the sun, increasing the angles of the sun's rays and duration of time the sun's energy affects the earth's surface. During the winter solstice, the exact opposite happens; darkness prevails, and little or no solar energy is absorbed.

Adding to this heat supply deficiency is that much of the energy never reaches the surface. During the summer the existence of extensive cloud cover absorbs or reflects much of this energy back into space. In the winter the cloud cover is gone, but the extensive amount of ice and dry snow is highly reflective, further decreasing the small amount of energy already rationed to the Arctic. Solar radiation is affected greatly by the presence of ice and snow. It is the main factor producing the long, cold winters and short, cool summers that characterize arctic seasons. The arrival of the ice and snow cover increases solar radiation reflection from about 25 percent to about 80 percent.

Soil

Soil is an important requirement of plant life. Soil formation is an extremely slow process because soil is frozen for much of the year. The shallow layer available in summer lacks distinct horizons and is confined to a shallow surface layer of only a few inches before continuous permafrost begins. Like the arctic biota, these soils are relatively young, originating in the Quaternary. The slow plant succession induces slow accumulation of humus for the next stages of succession. Eventually the plant community develops the best soil that the piece of land can support. Tedrow (1974) divided the soil zones of northern polar lands into three zones: polar desert, subpolar desert, and tundra.

Polar desert soil has free drainage with a surface desert pavement. It varies from alkaline to acid and is sometimes saline. The upper mineral horizon contains small amounts of organic matter, the A horizon is gray to yellow and the solum is brown to yellow. It thaws to a depth of 60 to 120 cm (Tedrow 1974).

Subpolar desert soil is highly variable depending on location. Hummocky ground soil has a covering layer of mineral-organic matter 3 to 5 cm thick. The mineral horizon is gray to yellow. pH is near neutral. The soil may become very hard and dry by the end of summer, with a salty crust developing (Tedrow 1974).

Tundra soils occur where there is water available during the summer. They vary from acid to alkaline. When dry, a salty crust may develop. They thaw to a depth of 30 to 40 cm (Tedrow 1974).

BIOLOGICAL ENVIRONMENT

Together climate, topography, nutrients, the amount of solar energy, and water are the major influences molding the environment of all living things in the Arctic. In the Arctic, temperature is the most apparent determinant of the environment but not the most important. In such a fragile environment, the smallest alteration of a physical or biological factor or process can have a great effect on the others. Short summers and long, dark winters dominated by low temperatures cause biological, biochemical, and chemical processes to occur slowly. These sensitive plants and animals thus require a longer time span to repair any damage inflicted on them. Man is a significant part of the ecosystem. His most recent entry into the Arctic in search of fossil fuel has expanded his awareness of the delicate balance of this ecosystem.

Adaptation is the key to survival in the Arctic. Each plant and animal must either adapt to the strenuous physical conditions or become extinct. Organisms have adapted differently to the polar climatic conditions. Musk-oxen evolved a thick coat of hair as protection. The less-well-protected ground squirrel hibernates in a sheltered burrow. As the arctic winter approaches, great flocks of birds migrate south in search for a more hospitable habitat capable of supplying their food demands.

Plant life is the basis for the arctic food chain. Many unique features have evolved in such a short history. Most obvious are their small size and closeness to the ground surface, enabling them to be insulated from extreme cold by snow. In a less obvious instance, the Labrador tea plant has adapted to the dry climate by developing leaves that curl up on its underside for water conservation. Such minute features can be essential for survival.

LIFE IN THE ARCTIC

Conditions in the Arctic were once much different than they are today. Fossils indicate that a warmer climate existed about 100 million years ago. Archaeologists have found giant evergreen fossils along the Sagavanirktok River in arctic Alaska. Fossils from millions of years before that time show that much of the arctic land, including the Brooks Range of Alaska, was covered by the sea. As the land emerged, plant and animal life began to occupy it. Life developed and adapted with the changing climate in order to survive. The mid-Mesozoic was dominated by a warm climate, and throughout the world dinosaurs roamed the earth. The Arctic was characterized by rain forests of large, broad-leaved trees and swamps populated by turtles and alligators. The formation of abundant fossil fuels in arctic regions today is attributed to these ancient tropical forests and swamplands of the mid-Cenozoic. As millions of years passed, the climate grew cooler, and the leaved trees were succeeded by evergreens. This marked the beginnings of the boreal forest of today's subarctic region. It was in this climate that the modern-day Arctic fox, polar bears, and wolves evolved. The recent extinction of relatively modern animals, including the woolly mammoth, is not only due to fluctuating climate; it was also caused by the more recent presence of man.

The Pleistocene glaciations of the Northern Hemisphere produced the most profound changes on arctic ecosystems. Although thick layers of ice covered most lands of the northern latitudes, large portions of glaciated land were also ice free. Ice-free areas along Alaska's Arctic coast, extending down into the Mackenzie River Valley and the interior of Alaska and along the Yukon River Valley, are known as the Bering Refuge. Plants and animals were able to continue their existence there. Although these areas provided a separate sanctuary unattached to other ice-free land, many plants of the refuge were not only unique to their area, but similar or related fauna and flora were also found in more temperature regions.

As Pleistocene glaciers receded and exposed more and more land, plants and animals dispersed throughout the region by winds and currents, and animals moved to newly available habitats. This recent "ice age" produced probably the youngest biomes on earth, and their biota is the least developed. The simplicity of the arctic environment and its unstable characteristics show that it is in a youthful stage. Plants and animals are still responding to the changes in the Pleistocene climate. Over time, a more complex and stable system should develop unless climate changes substantially.

Arctic conditions have made it possible for only a few highly specialized species to survive. As elsewhere, arctic biota must have suitable habitat, nutrients on which to live and grow, and must be able to reproduce in those conditions. Arctic biotic communities are dynamic. If one part of the ecosystem changes, it affects a number of organisms, which, in turn, must adapt or relocate. The growth and decline of biotic populations significantly change the ecosystem. Equilibrium is achieved by both plant and animal species.

Animals

Different factors determine population increase or decrease, the most significant being reproduction and death rates. Fauna and flora in warm, hospitable areas at lower latitudes generate more species, inducing more competition for survival. The harsh arctic environment limits the number of species in the region. Those that can adapt attain large populations due to lack of interspecies competition. At times the environment becomes more hospitable for a species to increase its population, and this in turn affects an entire portion of the food chain. For example, if more vegetation is available for the Arctic hares, their reproduction rate will increase, and predators of Arctic hares, such as the owl and the Arctic fox, will also have an abundance of food and will increase their populations.

Animals that normally do not prey on Arctic hares may use hare as a supplement to their diet. A poor growing season with less plant life available may lower hare population due to starvation and reduce its reproduction rate. In turn, foxes and owls will be forced to find other sources of food, contributing to the boom-or-bust cycle common to arctic ecosystems.

Surging populations many times cause mass migration of the species in search of more food and space. Lemmings, like many other small creatures with high reproductive rates, will go through such a surge under certain conditions. Common to herbivores, this migration may become a stampede, complete with abnormal behavior. A mass lemming migration throughout lowland areas means that animals will move in all directions, swimming across small water bodies but perishing in the larger ones. In uneven mountainous terrain, such as the coasts of Labrador and Greenland, they swarm down valleys toward forage and, faced with a cliff face or fjord wall, are forced by stampede momentum to jump. This is not an act of self-destruction but a reaction to severe overcrowding. Migrations may also be responses to flooding of a habitat during spring breakup, overuse of available food, or modification of habitat. Caribou that overgraze an area of slow-growing lichens may cause a drop in range-carrying capacity for many years. Overpopulation of various species can also occur when predator populations drop.

An outstanding example of animal adaptability in the Arctic is the musk-ox (figure 7-10). Confined to northern Canada and Alaska, some islands of the Canadian archipelago, and some coastal areas of Greenland, musk-ox have adapted perfectly to the extreme cold. Their hair is longer, denser, and woolier than that of similar species in warmer climates. The cold, dry climate exposes them to little moisture, which can soak their fur, freeze when temperatures drop, and limit mobility due to ice weight, making them easy prey for predators. In summer, burrowing animals, such as lemmings and shrews, concentrate in well-drained areas, such as pingos and streambanks, to avoid the inhospitable wet tundra. These rodents remain active all winter long, using snow for insulation and feeding on buried plant life. The physiology of large polar animals such as the polar bear helps regulate their heat. Their large size and bulk increases the amount of heat being produced without great loss to the environment, and their short legs and tail, ears, and snout further cut down on surface heat loss. Twenty four species of mammals have been reported for the north slope and coastal plain region of Alaska (Appendix 7-A).

Pitelka (1974) listed 185 species of birds for the north slope of Alaska and 151 species for the Barrow region including 22 species which regularly breed in the Barrow region. Numerous species of waterfowl from throughout the world summer in the coastal lowlands of the region. The most abundant birds are shorebirds, many of which also migrate south in winter to escape the cold. Most characteristic of tundra

Figure 7-10. Musk Ox.
Photograph by Jo Keller, USFW. Courtesy of Arctic Environmental Information and Data
Center, University of Alaska.

birds are species of ptarmigan. An obvious adaptation is that their legs are feathered down to their claws. In winter their plumage changes from the earth-blending colors of brown, black, and gray to snow white for camouflage against potential prey. Some predators, such as the fox and weasel, also turn white in winter to evade detection by their prey.

Reptiles and amphibiana are conspicuously absent in arctic regions. In North America only the wood frog, *Rana sylvatica*, is known to exist north of the arctic circle. (Hodge 1976).

Plants

Vegetation tundra represents the most prevalent plant life of the Arctic. Tundra begins where the treeline ends, but defining the exact location of treeline is difficult since it changes, depending on climate, soil conditions, and site exposures (figure 7-11).

Tundra is comprised of some 900 species, dominated by mosses, lichens, herbs, grasses, sedges, and shrubs. Low-lying tundra ranges in height from a few centimenters to a few meters above the ground surface. To survive, tundra flora

expend only the amount of energy provided by available sunlight. Almost all of these plants survive freezing temperatures because they can chemically bind water molecules to proteins. As the insulating snow cover melts in spring, tundra responds quickly to the long daily periods of sunlight. Represented mainly by perennial plants, tundra flora essentially reproduces by means other than seeds. Multitudes of bright colors combine with an array of fragrances to transform the bleak winter landscape into a vibrant blanket that attracts insects for pollination.

Summer is the time of maximum plant activity. The arctic coastal plain is comprised of wet tundra and a mosaic of small lakes. Although precipitation is low, a great deal of water is available from these unfrozen surfaces. Poor drainage has formed muskeg and swamp areas, and these wetlands mostly support different species of sedges. Secondary species of cottongrass, lousewort, and buttercup accumulate in wetter areas, while heather and purple mountain saxifrage thrive in higher and drier habitats, such as the ridges between the polygons.

Mountain, foothill, and floodplain areas support high brush communities. Relatively well drained and with a deeper permafrost layer, these habitats support a wider variety of plant species. The river community, for example, contains such species as horsetail, alpine bluegrass, and dwarf fireweed. Many are destroyed by spring floodwaters and breakup ice, which also create constant change in the habitat. In more stable higher areas, such species as willows, mosses, lichens, and herbs may not be disturbed for several decades.

MAN IN THE ARCTIC

Asiatic

Over thousands of years arctic aborigines have learned to live in harmony with their surroundings. Through wise use of the resources available to them, these nomadic people adapted their lives to the rhythms of the land. At some time, almost all of the land in the Arctic has been inhabited by Eskimos or Indians, the only exception being a few islands in the Canadian Arctic Archipelago.

About 25,000 years ago, the retreating Pleistocene ice sheets allowed man to move northward from central Asia, probably in pursuit of such now-extinct animals as the mammoth. They ventured northeastward across an intercontinental land bridge to Alaska, and over thousands of years these ancestors of America's native peoples used this route to migrate to every corner of the American continent. About 11,000 to 15,000 years ago, oceans rose, reclaimed the land bridge, and formed what is currently the Bering and Chukchi seas and the Bering Strait, which connects them.

About 2500 years ago a migration of people made their way into the North American Arctic from Asia. They leap-frogged their way beyond the Aleutian chain and instead of heading south as had previous groups, they headed north to

Figure 7-11. Moist tundra,
sometimes called "tussock grass tundra" is found on well-drained upland areas. It covers the grazing grounds of both the Arctic and Porcupine caribou herds and is important as summer range. Photograph by J. C. LaBelle. Courtesy of the Arctic Environmental Information and Data Center, University of Alaska.

the Arctic coastline. Having already developed appropriate hunting techniques, they flourished in the austere arctic environment and later became known as Eskimos. They called themselves Inuit or Inupiat—the Real People.

Two cultural groups of Eskimos lived in the arctic region. The Dorset culture, after a thousand years of successful occupation of the Arctic, began to spread across the northern rim of North America. Archaeologists have located Dorset sites in areas of Newfoundland, Southhampton Island, Ellesmere Island, and Greenland. They favored coastal location due to the availability of sea mammals, birds, and fish. About 800 years ago, the Thule Eskimos of arctic Alaska (ancestors of present-day Inuit) migrated eastward. They invaded Dorset communities as far east as Greenland and displaced or absorbed them. Thus, the principal roots of native arctic peoples can be found in the Thule culture, whose subsistence economy was based on sea mammal hunting.

Euro-American

The arctic region escaped exploration and conquest by Westerners until relatively recently. Various enticements lured Europeans to the Arctic, but by far the most significant was the search for new lands to the west and trade routes to the Far East. This quest brought them to both the northeast and northwest Arctic. Later personal and national prestige prompted explorations in search of the North Pole. Today the demand for new mineral and fuel resources makes economics the principal goal of exploration in the Arctic.

Vikings

About 1200 years ago the prosperous Scandinavian people began feeling the effects of increased population. In search of food and more land, the Vikings set sail and raided and pillaged the coasts of western Europe. Their forays also took them northward and westward, and by the ninth century they had reached the Fiaerol Islands. Here they established their first permanent colony. To this day these islands remain a dependency of Denmark.

But the Fiaerols were only the Vikings' first stop. About one hundred years later they sailed around the relatively ice-free waters of the south Barents Sea to the south coast of the Kola Peninsula. During this same period, they also landed in Iceland. Many of their compatriots followed and established a flourishing colony based on fishing. Iceland became the base for voyages and explorations still farther west.

By 985 Eric the Red and his followers had reached Greenland. The somewhat milder climate prevailing a thousand years ago allowed these Norsemen to establish an agrarian colony on the western coast. Simultaneously the Inuit were working their way eastward and had reached the northern Greenland coast. Conflict with the Dorset and Inuit people,

climatic cooling trends, and a shrinking link with Norway brought about the end of the Greenland colony about the year 1415. Their homes and animal sheds of stone remain today in the sheltered fjords of this great island.

Search for the Northeast Passage and Exploration of Siberia

The desire to find a sea route from the Far East to Europe launched the search for a Northeast Passage. English and Dutch merchants began their efforts in the mid-1500s. For a bit more than a century, they continued their attempts, but ice and climatic conditions hindered all efforts. However, an important commercial link between England and Russia was established in the form of the Muscovy Company and a trade route between London and Archangel.

Russian exploration coincided with European efforts in North America. To control its fringe lands and expand territory, Cossacks pushed the Russian empire eastward into Asia. Finally afer a series of land and sea expeditions, Russia determined the vast extent of its Siberian territory. Maps and general outlines of its arctic lands and coastline were made available for the first time. In 1741 Vitus Bering extended the Russian empire beyond the Asian continent by discovering and claiming the Aleutian Islands and parts of the Alaska mainland. In 1867, after a century of resource exploration, mainly furs, the Russians sold their Alaskan territory to the United States for $7.2 million.

Sweden's A.E. Nordenskiold was the first to navigate completely the finally discovered Northeast Passage, but it was clearly impractical for commercial trade purposes. Ice-choked waters along almost the entire Russian coast were too difficult for the sailing vessels of the eighteenth and nineteenth century, so trade-hungry Europeans looked westward for a short route to the fabled east.

Search for the Northwest Passage

English merchants were the primary seekers of a Northwest Passage trade route. The voyages of Frobisher, Davis, Hudson and Baffin by 1617 had delineated all the major entrances and channels to the eastern Arctic of North America. Between 1616 and 1818, increasingly severe climate and ice conditions that restricted the use of wooden sailing ships practically halted explorations in the Arctic. The period later became known as the little ice age. In the late 1700s Sammuel Hearne of the Hudson's Bay Company and Alexander Mackenzie of the Northwest Company reached the North American Arctic coast by following the major river drainages of that region. The companies rapidly established overland trading routes and outposts. Climatic conditions had moderated by 1819 when

William Parry ventured through Lancaster Sound and Bering Strait. He was finally stopped by heavy pack ice at McClure Strait.

Significant land explorations were made by Sir John Franklin. From 1820 to 1825 he mapped almost the entire Arctic coast of Canada, from Hudson Bay to the Mackenzie River. In 1845 he set out from Great Britain with two steamships and was never again heard from after entering Lancaster Sound. His loss encouraged about forty separate search expeditions over the next fourteen years. Though fruitless, these expeditions provided more insight into the geography of the Arctic Archipelago. Maps and island outlines made during these and other expeditions paved the way for discovery of the Northwest Passage by Roald Amundsen of Norway in 1903-1906. As with the Northeast Passage, ice conditions made the route impractical for trade.

Expeditions to the North Pole

National prestige motivated most countries involved in the race to the North Pole. Many different techniques were tried, including deliberately driving a ship (the *Fram*) into the ice sheet near the New Siberian Islands in hopes that it would drift to the pole. Unfortunately the northward drift shifted, and the ship headed south. Other unsuccessful attempts were made, but in 1909 Robert Peary, an American, led an expedition by dog team that reached the geographic North Pole. Peary attributed his success to adaptation of Eskimo styles of travel and dress and the use of dog power, rather than manpower, for pulling sledges. The pole has since been overflown by aircraft, approached by submarines cruising beneath the ice cap, and battered by icebreaker ships.

Present Search for Resources

Arctic exploration today is more intense than ever. The immense potential of oil and natural gas discoveries has made the Arctic a new center of economic and resource development for North America. In addition to resource development, strategic importance holds a high priority in the region. The Arctic acts as a buffer zone between the United States and the Soviet Union (figure 7-12).

The increasing demand for oil and gas has resulted in a mass importation of modern technology into this region, as well as numerous problems. The inhospitable environment and high labor and transportation costs have made exploration and development in the region extremely expensive. The impact on sensitive arctic environmental systems has brought about continuing battles between environmentalists and developers. Perhaps most important, however, is the cultural impact on indigenous peoples. The arrival of white explorers and traders brought unforeseen changes to aboriginal ways of life. The

Figure 7-12. Petroleum camp at Prudhoe Bay, Alaska.
Photograph by T. L. Péwé.

most notable example is the Alaskan Eskimo. Abrupt changes in the last ten years, brought about by the invasion of a technological society, have made it extremely difficult for the Eskimos to maintain their traditional life-style and culture and to cope with the problems associated with modern life.

REFERENCES

Alaska. University Arctic Environmental Information and Data Center. 1975. Alaska regional profiles, Arctic region.

Bovis, M. J., and Barry, R. G. 1974. A climatological analysis of north polar desert areas. In *Polar deserts and modern man.* Tucson: University of Arizona Press.

Clark, T. H., and Stearn, C. W. 1960. The great ice age. In *The geological evolution of North America: A regional approach to historical geology.* New York: Ronald Press.

Embleton, C., and King, A. M. 1975. *Glacial and periglacial geomorphology.* 2 vols. New York: John Wiley.

Hodge, R. P. 1976. *Amphibians and reptiles in Alaska, the Yukon and Northwest Territories.* Anchorage; Alaska Northwest Publishing.

MacDonald, S. O. 1980. *Checklist of mammals of Alaska.* Fairbanks: University of Alaska Museum.

Mellor, M. 1964. Snow and ice on the earth's surface. In *Cold regions science and engineering, Part II: Physical science, Section C: the physics and mechanics of ice.* Hanover, N.H.: Cold Regions Research and Engineering Laboratory.

Péwé, T. L. 1974. Geomorphic processes in polar deserts. In T. L. Smiley and J. H. Zumberge (eds.). *Polar deserts and modern man.* Tucson: University of Arizona.

————. 1975. Quaternary geology of Alaska. *U.S. Geological Survey Professional Paper* 835.

Pitelka, F. A. 1974. An avifaunal review for the Barrow region and north slope of arctic Alaska. *Arctic and Alpine Research 6 (2):161-184.*

Pruitt, W. O. Jr., Lent, P. C. and Carl, E. 1960. Terrestrial mammal investigations Ogotoruk creek-Cape Thompson and vicinity. *Bull. Dept. Biol. Sci.* University of Alaska, Fairbanks.

Sater, J. E. 1969. *The arctic basin.* Washington, D.C.: Arctic Institute of North America.

Smiley, T. L. and Zumberge, J. H. 1974. *Polar Deserts and modern man.* Tucson: University of Arizona Press.

Stearns, R. S. 1965. *Selected aspects of the geology and physiography of the cold regions. Cold regions science and engineering, Part 1, environmental; section A, general.* Hanover, N.H.: Cold Regions Research and Engineering Laboratory.

Sugden, D. E., and John, B. S. 1976. *Glaciers and landscape, a geomorphological approach.* New York: John Wiley.

Tedrow, J. C. F. 1974. Soils of the high arctic landscape. In T. L. Smiley and J. H. Zumberge (eds.) *Polar deserts and modern man.* Tucson: University of Arizona.

Thornbury, W. D. 1966. *Principles of geomorphology.* New York: John Wiley.

Washburn, A. L. 1973. *Periglacial processes and environment.* New York: St. Martin's Press.

MAMMALS OF THE ARCTIC NATIONAL WILDLIFE REFUGE

The Arctic National Wildlife Refuge lies in the extreme northeast corner of Alaska. Biomes within its boundaries include arctic polar desert, subarctic mountains and subarctic taiga. Terrestrial ecosystems found within the refuge include wet tundra, moist tundra, alpine tundra, low brush muskeg and bog, high brush and upland spruce-hardwood forest.

The definition of arctic desert used in this book includes all of these ecosystems except the upland spruce-hardwood forest. Physiographically it includes the Alaskan coastal plain and the north slope of Alaska.

The list includes all the mammals currently known to exist on the refuge. Those found in arctic polar desert and the north slope of Alaska are indicated with an asterisk*.

MAMMALS OF THE ARCTIC NATIONAL WILDLIFE REFUGE

*Masked shrew	*Sorex cinereus*
*Dusky shrew	*Sorex monticolus*
*Arctic shrew	*Sorex arcticus*
Pigmy shrew	*Sorex hoyi*
*Showshoe hare	*Lepus americanus*
*Alaska marmot	*Marmota broweri*
*Arctic ground squirrel	*Spermophilus parryi*
Red squirrel	*Tamiasciurus hudsonicus*
Beaver	*Castor canadensis*
*Red-backed vole	*Clethrionomys rutilus*
Meadow vole	*Microtus pennsylvanicus*
*Tundra vole	*Microtus oeconomus*
Yellow-cheeked vole	*Microtus xanthognathus*
*Singing vole	*Microtus miurus*
Muskrat	*Ondatra zibethicus*
*Brown lemming	*Lemmus sibiricus*
*Collared lemming	*Dicrostonyx torquatus*
*Porcupine	*Erethizon dorsatum*
Bowhead whale	*Balaena mysticetus*
Narwhal	*Monodon monoceros*
Beluga whale	*Delphinapterus leucas*
Coyote	*Canis latrans*
*Gray wolf	*Canis lupus*
*Arctic fox	*Alopex lagopus*
*Red fox	*Vulpes vulpes*
Black bear	*Ursus americanus*
*Brown (Grizzly) bear	*Ursus arctos*
*Polar Bear	*Ursus maritimus*
Marten	*Martes americana*
*Short-tailed weasel or ermine	*Mustela erminea*
*Least weasel	*Mustela nivalis*
Mink	*Mustela vison*
*Wolverine	*Gulo gulo*
*River otter	*Lutra canadensis*
*Lynx	*Felis lynx*
Pacific walrus	*Odobenus rosmarus*
Larga seal	*Phoca larga*
Ringed seal	*Phoca hispida*
Bearded seal	*Erignathus barbatus*
Hooded seal	*Cystophora cristata*
*Moose	*Alces alces*
*Caribou	*Rangifer tarandus*
*Muskox	*Ovibos moschatus*
Dall sheep	*Ovis dalli*

SOURCE: U. S. Department of the Interior, Arctic National Wildlife Refuge, Fairbanks, Alaska.

NOTE: Revised 1/81. Taxonomic nomenclature according to S. O. Mac-Donald (1981) *Checklist to the Mammals of Alaska*, University of Alaska Museum, Fairbanks, Alaska.

8

ANIMAL ADAPTATIONS

NEIL F. HADLEY

NORTH AMERICAN DESERTS are among the most extreme terrestrial habitats in the world. Although climatic conditions vary in each according to geographical location, all are characterized by low rainfall and sparse vegetation. Water is the dominant controlling factor for biological processes in these desert ecosystems and largely determines the adaptive strategies of resident organisms (Noy-Meir 1973, 1974). Low precipitation, however, is only part of the problem. Desert rainfall is also erratic and unreliable. An animal prepared for a long period of drought may suddenly be faced with a flood because the parched soil surface cannot handle runoff from a summer thunderstorm. In low-latitude deserts, high temperatures are coupled with limited moisture availability. The addition of the temperature factor means that animals must reconcile the antagonistic demands of evaporative cooling and the need to maintain sufficient body water to carry out essential biological functions. Temperature, like rainfall, is a feature of extremes, with daytime highs often being replaced by stressful nighttime low temperatures that place further demands on the adaptive flexibility of an animal. Winds are another factor that contribute to the harshness of the desert environment. During hot, dry summer months, dust-laden winds can increase the rate of convective heat gain, hasten the rate of dehydration, and abrade the surfaces of animals that are unable to find shelter.

How animals maintain an economic water balance and a tolerable energy balance under desert conditions has been heavily researched during the past fifteen years. A vast literature on the ecophysiology of desert animals has accumulated as a result of these investigations, and today there exist a number of reviews and symposia in which these data have been synthesized and summarized. Among the more recent of these are *Desert Biology* (Brown 1968, 1974), *Physiological Adaptations: Desert and Mountain* (Yousef, Horvath, and Bullard 1972), *Comparative Physiology of Desert Animals* (Maloiy 1972), *Environmental Physiology of Desert Organisms* (Hadley 1975), *Man and Animals in Hot Environments* (Ingram and Mount 1975), *Evolution of Desert Biota* (Goodall 1976), and *Convergent Evolution in Warm Deserts* (Orians and

Solbrig 1977). In addition, a number of excellent surveys of adaptations of specific taxa (such as arthropods, amphibians, and reptiles) appear in physiological monographs or in texts on environmental physiology.

A further synthesis of the existing extensive literature is neither practical nor necessary. Instead I have chosen to present here data from recently published studies that employed innovative research techniques or addressed problems that were previously unstudied. Coverage centers on three basic areas of ecophysiology—thermal, water, and metabolic relations—with emphasis placed on investigations of free-roaming animals, the biochemical and/or genetic basis for observed adaptations, and those studies having applied potential. Investigations performed on North American desert animals are highlighted, but studies conducted on animals from desert regions worldwide are included where concepts illustrated are applicable.

TEMPERATURE RELATIONS

Thermal Tolerance

The upper lethal temperatures of desert-adapted animals, which range between 40° and 50°C, are partly responsible for their ability to function at high ambient temperatures. Arthropods are among the most tolerant species, often surviving temperatures approaching 50°C (Cloudsley-Thompson 1970; Edney 1974). Slightly lower thermal tolerances have been reported for a number of desert lizards and snakes (Templeton 1970). Many factors affect the determination of upper lethal temperature, including length of exposure, rate of heating, humidity, and previous thermal history of the animal. Because these factors have seldom been controlled in experimental designs, it is difficult to say if upper lethal temperatures of desert animals are actually higher than those of their nondesert counterparts. Apparently no significant difference exists between active body temperatures or lethal temperatures in desert versus nondesert birds. Seasonal acclimatization to temperature probably occurs to a limited extent in all groups but does not surpass behavioral control of temperature in importance.

Hyperthermia

Desert species that can tolerate high body temperatures capitalize on this ability to conserve water and energy. The body temperatures of many desert birds and mammals rise several degrees above normothermic levels during daytime hours. The elevated body temperatures preserve the direction of heat flow away from the body as long as possible. Once ambient temperature surpasses body temperature and the direction of heat flow is reversed, hyperthermia reduces the rate of heat gain because of the narrowed temperature gradient between the environment and the animal. Critical body water supplies are conserved because of the reduced demand for evaporative cooling. Most of the accumulated body heat is dissipated during the cooler evening hours through radiative and convective heat transfer. The animal conserves some energy as well since this behavior reduces the number of calories normally expended in evaporative cooling.

The periodicity and extent of hyperthermy are dependent upon the animal's size, hydration state, and metabolism. This adaptation is of particular value in larger animals, which have a larger heat capacity per unit surface area and consequently a higher ratio of heat storage potential to heat exchange with the environment than do smaller animals (Ingram and Mount 1975). Hydration state influences the extent of departure from normothermic levels. Under conditions of dehydration, body temperatures not only rise above normal values during the day but also drop to lower levels at night. For large mammals, such as the camel and eland, the low body temperature at the beginning of the heating period further delays the onset of evaporative cooling. In some instances the rate of heat gain is such that the animal does not acquire a body temperature that necessitates sweating or panting. When free water is plentiful, these same species may dispense with hyperthermy and simply use evaporative cooling to maintain body temperatures within a fairly narrow range (Schmidt-Nielson 1964; Taylor 1970a, 1970b). Smaller mammals, such as desert rodents, and desert birds are less tolerant of hyperthermia and typically seek alternative ways of avoiding temperature extremes. An exception among birds is the large-bodied ostrich, which exhibits an increase of 4°C in body temperature at high ambient temperature (Crawford and Schmidt-Nielsen 1967).

Thermoregulatory Control

The physiological basis for the relaxation of limits for thermoregulatory homeostasis is poorly known. It has been suggested that possibly dehydration reduces the effectiveness of evaporative cooling through the reduction of fluid secretion onto the evaporative surface or that dehydration causes a shift in the "set-point" of body temperature. Thermosensitivity and thermoregulation are active research areas, which can benefit from additional studies on thermolabile desert species. There are applied benefits to be gained as well. High temperatures typically result in diminished feeding by cattle and sheep (Ingram and Mount 1975). Other problems caused by high temperatures include a high level of disease and parasite infestation, reduced growth rates, reduced milk yields, and a decline in fertility. Native breeds do better in these categories but do not exhibit the productivity of selected animals raised in temperate climates (Ingram and Mount 1975). A better understanding of thermal control mechanisms is needed to develop domestic strains that not only survive but are highly productive in hot, dry climates.

Cardiovascular Adjustments

The cardiovascular system plays an important role in regulating the rates of heat gain and heat loss in desert animals. In those species that have adopted heat storage strategies, there are often special arrangements of blood vessels that influence heat transfer so that temperature-sensitive regions of the body are protected against extensive temperature changes. Such systems, which have been elegantly described for running antelopes, operate as follows. Arterial blood supplying the brain passes through hundreds of small arteries (carotid rete mirabile) that lie in a venous pool (cavernous sinus). Venous blood cooled by evaporation from the walls of the nasal passages drains through the cavernous sinus. This cool venous blood in turn cools the warm arterial blood en route to the brain. As a result of this heat exchange, brain temperatures remain several degrees below body core temperatures. This mechanism also exists in goats and sheep (Taylor 1974) and in the large running bird, *Rhea americana* (Kilgore et al. 1973).

A countercurrent heat exchange system was also proposed (Heath 1966) to explain head-body temperature differences in the horned lizard (*Phrynosoma coronatum*). In this case, higher head temperatures were maintained by the transfer of heat from warm blood draining the head into cooler blood entering the head by the internal carotid artery. When a preferred temperature was reached, blood draining the head was shunted into the external jugular, bypassing the heat exchange system and thus eliminating the head-body temperature differential. Head-body temperature differences subsequently have been reported for a number of lizards and snakes; however, a countercurrent system may not always be responsible (Pough and McFarland 1976). Differential blood shunting is also the basis for lizards' gaining heat more rapidly than they cool and thus being able to remain near or at their preferred temperature for longer periods (White 1973).

Heat produced during exercise places additional demands on an animal's thermoregulatory processes. Excess heat must be dissipated with a minimum of energy and water loss. In the brown bat, (*Eptesicus fascus*) vasodilation in the wing is an important mechanism for dissipating heat generated during flight (Kluger and Heath 1970). Cardiovascular adjustments are also employed by some desert insects to maintain thermal stability. The sphinx moth (*Manducta sexta*), relies on circulatory heat transfer to regulate thoracic temperature during flight over a wide range of ambient temperatures (Heinrich 1970, 1971) (figure 8-1). Heat generated in the thorax during flight was transferred to the blood pumped from the comparatively cool abdomen and eventually dissipated when blood returned to the abdomen. Active cooling by the circulatory system, however, is only effective in large insects, which generate a marked difference between thoracic and ambient temperatures (Heinrich 1975). In all cases, physiological regulation of body temperature is insufficient without appropriate behavioral adjustments of activity patterns.

Evaporative Heat Loss

Once ambient temperature reaches or exceeds body temperature, evaporative cooling is the only means available for dissipating heat. Deployment of evaporative cooling by desert animals is naturally restricted because of limited water supplies. This is especially true for small animals. Because of their relatively large surface area to body weight ratio, small animals tend to absorb more heat than do large ones. It is estimated that if a 30 gram (gm) mouse used primarily evaporation to keep cool, it would lose water amounting to about 20 percent of its body weight each hour (Taylor 1974). This rate obviously could not be sustained for any period of time without free water being available for rehydration. Evaporative cooling is a viable strategy for larger animals and is used effectively in a variety of desert vertebrates.

Evaporative cooling occurs when water is vaporized from the surface of the skin and respiratory tract. A certain amount of obligatory heat loss results from normal respiratory activities; however, this rate of loss can be augmented by increased ventilation rates associated with high ambient temperatures. Panting occurs in a number of desert reptiles, birds, and mammals. It is particularly important in reptiles and birds because these groups lack sweat glands. In some bird species, panting is supplemented by gular flutter, a rapid flexing of the hyoid apparatus that moves air over the highly vascularized skin of the throat or gular area (Lasiewski 1972). Certain species pant and gular flutter at frequencies that are harmonious with the natural resonant frequency of the respiratory system and, hence, are thought to benefit further from reduced energy expenditure.

Evaporative heat loss at the skin surface may involve the vaporization of water passively transported through the skin, sweat produced from special glands, and/or the evaporation of moisture from external or internal sources that has been spread onto the surface. The evaporation of water that diffuses through the skin increases when an animal vasodilates in response to heat; thus the surface temperature of the epidermis partially determines the amount of heat lost in this way (Ingram and Mount 1975). This mechanism may contribute to heat dissipation in desert reptiles, birds, and mammals that lack sweat glands. Sweating is restricted to mammals and, in arid regions, is effectively used by only larger species such as the camel and donkey. Because heat dissipation by sweating is costly in terms of water usage, its onset is usually delayed as long as possible by these species through the heat storage-heterothermy strategy. Besides quickly depleting body water, excessive sweating also results in loss of electrolytes, which may lead to osmotic imbalance and possibly fatigue. Hales (1974), reviewed the advantages and disadvantages of sweating versus panting and concluded that panting is a more efficient heat-dissipation mechanism from both an energetic and water economy standpoint for animals in hot, dry environments, regardless of their body size.

The efficiency of heat loss resulting from the evaporation of moisture spread onto the surface depends upon the source of the moisture. In a 55°C environment, ornate box turtles

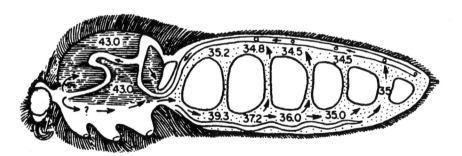

Figure 8-1. The circulatory system of a sphinx moth and the body temperatures observed during flight at 35°C.
From Thermoregulation and flight energetics of desert insects by B. Heinrich, in *Environmental physiology of desert organisms*, ed. by N. F. Hadley, Dowden, Hutchinson & Ross, Stroudsburg, Pa. Figure 6, page 101 used with the permission of the publisher.

(*Terrapene ornata*) produced copious amounts of saliva, which they spread over their heads and forelimbs. The spreading of saliva onto the fur of the rat is also an effective evaporative heat-loss mechanism (Hainsworth 1967), but its use in small desert rodents has to be considered an emergency measure. Desert turkey vultures reduce body temperatures by excreting on their legs at high ambient temperatures (Hatch 1970), a practice that utilizes moisture otherwise destined for elimination. When an external moisture source is applied to the surface, however, an animal can achieve effective cooling without increasing energy expenditure or further depleting body water. Species lacking sweat glands often benefit from this practice. An excellent example is house finches in the deserts of southern California, which bathe extensively at water holes during the middle of hot days, thereby behaviorally achieving a form of cutaneous evaporative cooling.

Thermal Consequences of Surface Coloration

The surface color of desert organisms and its relation to radiant energy and resulting thermal loads continues to attract the attention of environmental physiologists. This interest has been generated by the seemingly disproportionate number of desert species with black or dark coloration and the apparent thermal problems created by dark pigmentation, and by recent findings that radiative heating of light and dark coats may be different in the presence of forced convection.

Our understanding of the thermal consequences of surface coloration in desert arthropods is based largely on studies of tenebrionid beetles. This group includes both black and white or partially white species whose surface activity patterns often coincide with periods of high solar insolation. Measurements made under natural and artificial conditions have generally indicated that black beetles absorb more incident radiation than do white individuals and hence exhibit higher body-subelytral temperatures and greater rates of heat gain (Hadley 1970; Edney 1971; Hamilton 1973). Hamilton proposed that black coloration might be thermally advantageous in that it allows dark individuals to maximize absorption of radiation during early morning and late afternoon hours, thereby extending the time for surface activity. To facilitate the increase in heat gain at the beginning and end of the day, the black Namib Desert tenebrionid, *Onymacris plana*, exhibits elevated elytral transmittance to shortwave infrared radiation that predominates at times of low sun angle. The visible and ultraviolet radiation, approaching the maximum at mid-morning, are, in contrast, absorbed by the elytra, which can acquire temperatures in excess of 60°C (Henwood 1975). These high surface temperatures, however, facilitate heat loss by convection, allowing beetles to remain active longer into the midday period. In the Sonoran Desert, the tenebrionid beetle, *Cryptoglossa verrucosa*, exhibits color phases that range from black at high humidity to whitish-blue at low humidity. The morphol-

ogical basis for these color phases are wax filaments that radiate from the tips of miniature tubercles to cover the integument at low humidity (Hadley 1979). The accumulated wax meshwork decreases cuticular transpiration and has the potential to reduce the radiant heat load through increased reflectance.

The assumption that animals with dark-colored integuments or coats acquire greater solar heat loads has been substantiated by most studies conducted on vertebrates. This can mean additional thermal problems for desert species, especially small endotherms, which face the dual problem of rapid heat gain and the need to dissipate excess heat generated by metabolic processes, as well as that acquired from the environment. Nevertheless there are times when such coloration can provide thermal benefits for these same species. At low air temperatures, the increased absorption of radiant solar energy by dark surfaces reduces the metabolic cost of maintaining a constant core temperature (Hamilton and Heppner 1967, Heppner 1970, Ohmart and Lasiewski 1971). During summer heat loads, high surface temperatures are believed to enhance heat dissipation by convection and radiation and also to reduce the rate of heat gain from the environment because of the reduced temperature gradient (Marder 1973). A recent report indicates it may be necessary to reevaluate the entire relationship between coat color and radiation load. Using black and white pigeons, it was found that at wind speeds greater than 3 m s^{-1} black erected plumages acquired lower radiative heat loads than did white plumages (figure 8-2). The lower heat load transmitted to the skin by erected black plumages may be functionally significant during times of flight and in nesting individuals exposed to full sunlight during windy, hot afternoons (Walsberg et al. 1978).

Adaptations to Cold

Adaptations of desert animals are usually equated with high temperatures and limited water availability. In many arid regions, however, nighttime and/or seasonal cold can create equally stressful conditions for resident species. Low temperatures are frequently encountered in high latitude deserts or in deserts with mountains within their boundaries. Adaptation to cold by animals in these habitats requires a complex of adjustments that range from simple behavioral avoidance to more spectacular phenomena such as supercooling and torpor. Emphasis in this coverage will be on the latter two responses.

Freezing resistance and supercooling, typically associated with high latitude and polar species, have also been observed in a number of desert arthropods and reptiles. In the Sonoran Desert, *Drosophila nigrospiracula* is capable of supercooling to between −5° and −6°C, a feature that allows the species to be surface active and breed throughout winter months (Lowe et al. 1967). Supercooling points between approximately −3° and −12°C have been reported for a variety of desert tene-

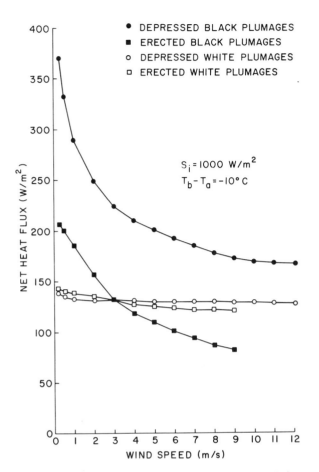

- DEPRESSED BLACK PLUMAGES
- ERECTED BLACK PLUMAGES
- DEPRESSED WHITE PLUMAGES
- ERECTED WHITE PLUMAGES

$S_i = 1000 \text{ W/m}^2$

$T_b - T_a = -10°C$

Figure 8-2. Projected thermal behavior of the experimental plumages under extreme desert conditions.
Positive heat flux values indicate a net body heat gain. From Animal coat color and radiative heat gain: A re-evaluation, by G. E. Walsberg, G. S. Campbell, and J. R. King, *Journal of Comparative Physiology* 126:211-222 (1978). Figure 5, page 217 used with the permission of Springer-Verlag Publishers, Secaucus, N.J.

brionid beetles, scorpions, and the desert locust (Cloudsley-Thompson and Crawford 1970; Cloudsley-Thompson 1973; Crawford and Riddle 1975). The physiological or biochemical basis for this capacity in desert species is unclear. Increased levels of cryoprotectants such as glycerol and sorbitol in the hemolymph, typically observed in insects habitating extreme cold environments, were not noted in these investigations. A substantial seasonal depression of the supercooling point of hemolymph of the scorpion *Paruroctonus aquilonalis,* which enables it to avoid winter freezing, is apparently associated with cessation of feeding in the fall (Riddle and Pugach 1976). Supercooling limits and freezing point temperatures ranged between $-3.89°$ and $-7.44°C$ for forty-five reptilian species, including a number of desert lizards and snakes. Ability to supercool to this level would provide protection against freezing cold likely to be experienced by lizards such as *Sceloporus jarrovi* that overwinter in shallow rock crevices (Congdon

1977). For many species, however, the supercooling capacity is greater than that necessary to survive climatic extremes and may be a taxonomic rather than an adaptive feature (Cloudsley-Thompson 1973).

Desert birds and mammals respond to cold by coupling behavioral adjustments with physiological and morphological mechanisms for reducing heat loss and/or increasing heat production. Smaller species utilize a variety of shelters that provide escape from excessive environmental stress. These periods of inactivity are often accompanied by torpor, which involves a drop in body temperature from normothermic levels as a result of decreased metabolic rates. Torpor can occur over a several hour period on a daily basis at any time of year or for prolonged periods during a specific season such as winter. The latter pattern is frequently observed in desert mammals and results in a savings of energy and possibly water. The occurrence and energetic consequences of torpor in desert species are discussed by Borut and Shkolnik (1974).

Seasonal acclimatization to cold has been reported for a few desert mammals that remain active throughout the year. The collared peccary (*Tayassu tajacu*) in the Sonoran Desert increases the density of its pelage in winter. This results in reduced convective heat loss, but the improved insulation cannot fully compensate for the colder air temperatures. Hence winter peccaries must also increase their metabolism to augment heat production (figure 8-3). Behavioral adjustments also play an important role in the adaptive process. Winter peccaries shift their major activity periods to daytime, spending much time basking, and then huddle together in a shelter at night to reduce heat loss (Zervanos and Hadley 1973; Zervanos 1975). Seasonal acclimatization to cold was also observed in two species of jackrabbits collected from desert habitats outside of Tucson, Arizona. The black-tailed jackrabbit (*Lepus californicus*) reduces heat loss during winter by increasing the insulative value of its fur by approximately 40 percent over summer values. The antelope jackrabbit (*L. alleni*) in contrast, relies more on increasing heat production than reducing heat loss. Its metabolism at thermoneutrality increases by 18 percent, which decreases the lower critical temperature by 3°C. The winter increase in fur insulation for *L. alleni* is estimated at only 26 percent; however, even this relatively small increase provides some thermal benefits as evidenced by the reduced metabolic rates in winter animals compared to rates for summer animals exposed to below thermoneutral temperatures (Hinds 1977).

WATER RELATIONS

Water Loss

Animals in warm deserts face a high dehydration potential due to the increased drying power of the air and, for many species, the heavy demands placed on limited water supplies by evaporative cooling. Despite these problems, observed water-loss rates for xeric species are typically low, especially when compared to rates reported for their mesic relatives (Hadley 1972). Water is lost through a variety of avenues in both the liquid and vapor state. Integumentary and respiratory transpiration are generally paramount, with the former predominating at low to middle ambient temperatures and the latter predominating at higher temperatures. Additional water-loss pathways include defecation, elimination of nitrogeneous waste products, extrarenal salt removal, emergency cooling mechanisms (salivation), and miscellaneous species-specific activities (such as defensive secretions, external digestion, and egg laying). Although these additional pathways are considered to be minor, at times they can surpass transpiratory water loss in importance.

Integumentary Transpiration

Integumentary transpiration is generally low in desert-adapted species. The ability to reduce the outward diffusion of water is especially important to desert arthropods due to their relatively high surface area and limited body water (Edney 1974). Experimental evidence indicates that several layers of the insect and arachnid cuticle actively participate in the waterproofing process. For many species, the principal barrier is a layer or layers of lipids associated with the epicuticle. Lipids that impregnate other layers of the cuticle, the chitin-protein complex of the endocuticle, and even epidermal cells that are responsible for secreting and maintaining the cuticle may contribute to integumentary impermeability (Locke 1974; Ebeling 1974, 1976).

Recently the chemical composition of epicuticular lipids has been correlated with cuticular permeability in a number of desert insects and arachnids (Hadley 1977, 1978a, 1978b; Hadley and Jackson 1977; Toolson and Hadley 1977, 1979). These investigations have demonstrated a number of patterns or trends regarding lipid composition that appear to be adaptive in terms of improved waterproofing. Hydrocarbons are often the most abundant epicuticular lipid constituent and are likely a major contributor to the water barrier. Hydrocarbon molecules are usually long chain (twenty to over forty carbon atoms) and completely saturated, two features that should enhance the cuticle's impermeability based on studies of plasma membranes and artificial bilayers (table 8-1). The lower cuticular permeability of the desert scorpion *Hadrurus arizonensis* when compared to the montane scorpion *Uroctonus apacheanus* is attributed in part to the higher surface densities of lipids and hydrocarbons in *H. arizonensis* (Toolson and Hadley 1977). Seasonal differences in permeability are also linked to cuticular lipid composition. Area-specific water loss rates for summer scorpions (*Centruroides sculpturatus*) are significantly lower than for winter scorpions. An increased percentage of long-chain epicuticular hydrocarbons in summer animals probably accounts for much of this increased imperme-

ability (Toolson and Hadley 1979). Summer beetles (*Eleodes armata*) and winter beetles acclimated to 35°C for periods of five and ten weeks exhibited greater quantities of hydrocarbons and a higher percentage of long-chain components than did winter beetles or appropriate controls (Hadley 1977).

Recent investigations of integumentary transpiration in desert vertebrates have produced findings that are contrary to earlier assumptions. Amphibia are widely acknowledged to have a very permeable skin, thus making them ill adapted for survival in dry atmospheres (Loveridge 1976). Evaporative water loss rates, however, for two African tree frogs of the genus *Chiromantis* and several members of the South American genus *Phyllomedusa* are comparable to those recorded for desert reptiles (Shoemaker and Nagy 1977). The basis for this reduction in phyllomedusine frogs is the presence of a lipid film secreted onto the skin surface (Shoemaker and Mc-Clanahan 1975; Blaylock et al. 1976; McClanahan et al. 1978). In *Phyllomedusa sauvagei*, the skin lipids are mainly wax esters and are effective in reducing integumentary water loss up to ambient temperatures of 35°C. Between 35° and 38°C, there is a marked increase in permeability that corresponds to a phase change in the wax ester molecules (Mc-Clanahan et al. 1978). No skin lipids or associated glands are found in *Chiromantis*. Instead their reduced evaporative water loss is believed to be correlated with a dense layer of chromatophores that is essentially unbroken throughout the dorsal skin (Drewes et al. 1977). Other amphibian species in arid environments form cocoons while burrowed in the soil that effectively reduce water loss (Lee and Mercer 1967; Mc-Clanahan et al. 1966).

Rates of cutaneous water loss in reptiles and birds are apparently higher than previously thought, despite their lack of sweat glands. Still these rates are low in comparison to more mesic species. The morphological and/or physiological bases for reduction in cutaneous transpiration are poorly understood in most cases. Water loss rates for a scaleless gopher snake (*Pituophis melanoleucus catenifer*) are comparable to those for a normal individual (Licht and Bennett 1972). The scaleless snake even lacked the outer superficial dermal layer and possessed a much thinner keratin layer. Lipids, although organized differently, may contribute to reduced cutaneous transpiration. The molted skin of a black rat snake exhibited a higher permeability following extraction of its lipids with an organic solvent. Furthermore, chemical analyses of these lipids indicated a composition similar to lipids extracted from the epicuticle of insects (Roberts, unpublished). In mammals, the minimization of cutaneous water loss is complicated by the presence of sweat glands in some species and the secretion of large volumes of fluid onto the surface for the purpose of evaporative cooling. Where low transepidermal water loss exists, the barrier responsible is uncertain. Hattingh (1972) found a poor correlation between epidermal thickness and "transepidermal water loss" in nine mammalian skins and

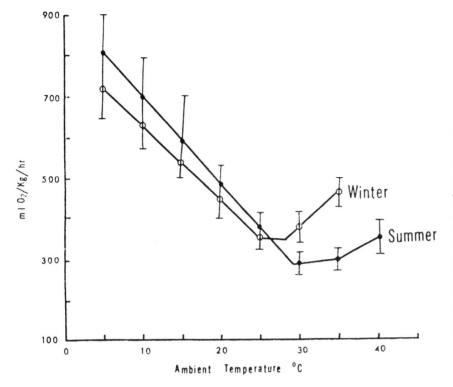

Figure 8-3. Relation of oxygen to ambient temperature of five summer- and five winter-acclimatized peccaries. Points represent means and vertical lines one standard deviation. Average weight of summer animals was 20.2 kg and winter animals 20.1 kg. Figure 1, page 366 reprinted with permission from *Comparative Biochemistry and Physiology,* vol. 50A, Stam M. Zervanos, "Seasonal effects of temperature on the respiratory metabolism of the collared peccary (*Tayssu tajacu*) copyright 1975, Pergamon Press, Ltd., New York.

TABLE 8-1
Comparative summary of lipid-hydrocarbon/body weight ratios, hydrocarbon composition, and cuticular permeability of desert tenebrionid beetles

	ELEODES ARMATA	CRYPTOGLOSSA VERRUCOSA	CENTRIOPTERA MURICATA	CENTRIOPTERA VARIOLOSA	PELECYPHORUS ADVERSUS
Sample size	20	31	168	28	27
Sample weight (gm)	16.54	19.57	64.17	12.10	21.45
Total lipid (mg)	29.73	24.99	18.98	1.90	5.07
Total Hydrocarbon (HC) (mg)	28.27	23.56	17.40	1.77	4.97
% hydrocarbon	95.1	94.2	91.7	93.2	94.3
Hydrocarbon/body weight (mg/gm)	1.71	1.20	0.27	0.15	0.23
Hydrocarbon size range (quantities 0.5%)	28br-44br	23-49br	23br-37br	25-37br	25-41br
% n-alkanes	7.7	78.5	55.3	53.1	48.8
% branched	92.3	21.5	44.7	46.9	51.2
Cuticular water loss (mg · gm^{-1} · hr^{-1})					
25°C	0.8	0.9		0.8	
30°C	1.4	1.2	1.1	1.0	
35°C	2.0	1.4	1.6	1.6	
40°C	2.5	1.6	2.2	2.6	

SOURCE: F. Hadley, Cuticular permeability of desert tenebrionid beetles. correlations with epicuticular hydrocarbon composition, N. F. Hadley, *Insect Biochemistry* 8:17-22, 1978.

NOTE: All extractions with 100 percent hexane.

concluded that some other integumentary feature or process was responsible for the barrier function.

Respiratory Transpiration

Respiratory transpiration is often the predominate source of water loss at high ambient temperatures. It is not surprising, therefore, to find in desert animals modifications of their respiratory activities or the respiratory apparatus itself that minimize water loss without sacrificing metabolic efficiency. The very nature of the respiratory system in desert arthropods is effective in reducing transpiration. Both the tracheal system of insects and the book lung system of most arachnids communicate to the outside by means of narrow spiracular openings that in xeric forms are often sunken below the cuticle surface or are partially covered by scales or hairs (Hadley 1972). Muscular control of spiracular opening and closing further limits water loss when conditions dictate conservation (Edney 1974). Ventilatory movements of desert locusts decreased in rate and amplitude when this species became dehydrated. Similar benefits in water economy are gained by diapausing insect larvae and pupae that show cyclic release of carbon dioxide.

Some desert rodents achieve a reduction in respiratory water loss by expiring air at a temperature lower than body temperature. This phenomenon is based on a modified water and heat countercurrent exchange mechanism. During inspiration, water evaporates from the moist nasal mucosa into the dry, hot air. This removes heat from the nasal area and decreases its surface temperature. During expiration, the warm, moist air from the lungs passes over the cool nasal surface, is cooled, and part of the moisture in the air recondenses. This process is repeated during the next ventilatory cycle. As a result, expired air is exhaled at a temperature lower than body temperature, thus containing less water when saturated. Nasal passages must be long and narrow in order for sufficient heat exchange to occur. In larger mammals, nasal passages are too wide for adequate contact between inspired air and the moist nasal mucosa. A similar method of water recovery prevails in some small desert reptiles and birds, but here too nasal morphology limits the efficiency of the mechanism (Schmidt-Nielsen et al. 1970).

Excretory Water Loss

A most important feature of desert animals is their ability to eliminate toxic end products and excess electrolytes in a minimum volume of fluid. A number of factors are responsible for this ability, paramount among these being specialized excretory structures, the biochemical capacity to excrete nitrogenous wastes in the form of purines, and the functional assist provided by extrarenal structures such as the gut, cloaca, and salt glands. Although many of these features are not unique to

desert animals, they are more effectively utilized by these species than they are by more mesic forms.

The principal excretory structures of insects and arachnids are the Malpighian tubules and the rectal segment of the hindgut. Their functions are described in detail by Phillips (1977). Significant in terms of desert adaptation is the elimination of nitrogenous wastes as either uric acid or guanine and the ability to produce urine that is hyperosmotic to the hemolymph. Uric acid and guanine are essentially nontoxic and very insoluble compounds that can be eliminated in a crystalline form, thus minimizing the concomitant loss of water. This process contributes to the production of hypertonic urine, which in some desert arthropods can approach values observed in xeric-adapted mammals. To some extent, similar processes operate in producing "dry" fecal materials, which in insects is eliminated along with urine as part of their excreta (Borut and Shkolnik 1974). The extent of water reabsorption depends largely on external and internal moisture conditions. Under desiccating conditions or when partially dehydrated, desert species typically extract a higher percentage of water from their feces.

Recent studies of water transport across the gut of the desert scorpion (*Hadrurus arizonensis*) illustrate the role of the ileum in water conservation (Ahearn and Hadley 1976, 1977). Using isolated perfused ilea, it was found that high levels of sodium ions in the lumen favored the transport of water from the gut into the hemolymph, whereas hemolymph uptake of water was inhibited when luminal potassium ion levels were high. The ileal water absorption mechanism appears to be adaptive, for it was also shown that ileal concentrations of sodium surpassed potassium in dehydrated and starved scorpions. Such a ratio would favor the retention of body water during a time when conservation is critical (figure 8-4). In freshly fed scorpions, excessive hydration could be reduced by the inhibitory interaction of potassium on water transport into the hemolymph.

The ability to minimize water loss associated with excretory function in vertebrates is closely linked to renal structure, especially the architecture of the kidney nephron. Amphibian and reptilian nephrons lack the loop of Henle, and, hence, these groups cannot produce a hypertonic urine. (The lizard, *Amphibolurus maculosus*, which inhabits dry salt lakes in South Australia, can void a urine that is hyperosmotic to the plasma, the urine being concentrated in the cloaca and not the kidney as it is in birds and mammals.) Amphibians can adjust to water deprivation, however, by reducing urine output through reduced glomerular filtration and increased reabsorption so that urine-plasma ratios approach one (Shoemaker and Nagy 1977). Although most amphibia excrete nitrogen as urea or ammonia, certain xeric-adapted members of the genera *Phyllomedusa* and *Chiromantis* eliminate a significant fraction of their nitrogenous wastes in the form of urate salts (Loveridge 1970; Shoemaker and McClanahan 1975; Drewes et al.

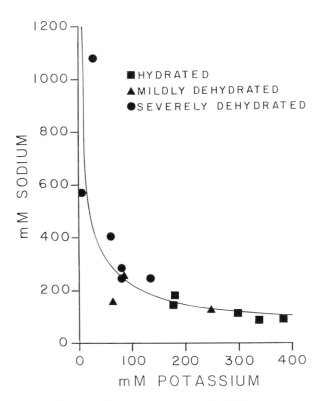

Figure 8-4. Effect of hydration state of scorpion (Hadrurus arizonensis) *on ileal sodium and potassium concentrations.*
Each point represents ion concentrations at an individual ileum. Curve drawn through data was calculated from power function: $y = a X^b$, where $a = 2,401$ and $b = -0.52$ ($4 = 0.84$). Figure 3, page R202 from Water transport in perfused scorpion ileum by G. A. Ahearn and N. F. Hadley, *American Journal of Physiology* 233 (5): R193-R207 was reprinted with permission of the publisher.

1977). The latter process effectively eliminates excess electrolytes while conserving body water. Similar relationships between habitat xerism and renal function (glomerular filtration rate, tubular reabsorption of water, uricotelism) have also been described for reptiles. Nonrenal structures are also important in reptilian osmoregulation. The cloaca, bladder, and possibly the intestine are sites of water reabsorption, whereas salt glands in desert lizards eliminate about half of the dietary potassium and most of the sodium (Shoemaker and Nagy 1977).

Birds and mammals can produce urine that is hyperosmotic to their blood and thus rely on renal function to maintain water and osmotic balance under xeric conditions. The structural feature responsible is the loop of Henle, which establishes and maintains the osmotic gradient necessary for water uptake by the collecting ducts. Because the avian kidney contains a mixture of reptilian- and mammalian-type nephrons, the maximum urine concentration possible exceeds plasma concentration usually by only a factor of two (Braun and Dantzler 1972). Some desert species exhibit slightly higher urine-to-plasma ratios, a fact that can be correlated with differences in

renal morphology (Johnson and Ohmart 1973). Still, these values are well below maximum values reported for xeric-adapted mammals. The use of conventional osmotic urine-to-plasma ratios for comparing renal concentrating abilities is misleading, however, since nitrogenous wastes and a high percentage of sodium, potassium, and ammonium ions are eliminated as urate salts that precipitate and do not contribute to measured osmotic pressure (McNabb and McNabb 1975). Postrenal processes also help maintain water and osmotic balance. Salt glands are present in the roadrunner, the desert partridge, and several xerophilic hawks. These structures assist in the excretion of excess salts with relatively small expenditure of water (Dawson 1976). Both the coprodeum and large intestine are capable of absorbing water following the uptake of solutes, thereby reducing the water content of urinary and fecal components (Skadhauge 1976).

The kidney's contribution to the water economy of desert mammals has been firmly established. Small desert rodents appear to possess the most efficient kidney based on reports of urine-to-plasma ratios exceeding twenty-five times and urea concentrations some eight hundred times higher than blood values (MacMillan and Lee 1967). Such concentrating abilities, when they occur, are usually strongly correlated with a high percentage of nephrons containing long loops of Henle and also increased medullary thickness. Reduced urine-to-plasma ratios in larger desert mammals indicate that their kidneys cannot achieve comparable urinary electrolyte concentrations. Instead species such as the camel, donkey, gazelle, and peccary reduce glomerular filtration and urinary flow rates. For example, GFRs of fully hydrated camels average 66 ml min^{-1} per 100 kg but drop by two-thirds to 22 ml min^{-1} after ten days without water (Siebert and MacFarlane 1971). Urine flow rates also exhibit a marked decrease. Species that lack the ability to reduce urine volume when water intake is restricted are unlikely to be successful in desert habitats.

Dehydration Tolerance

Despite efficient mechanisms for restricting water loss, high temperatures and desiccating atmospheric conditions can produce severe temporary water deficits in desert species. It is essential that changes in osmotic pressure and body fluid composition that result from dehydration do not exceed levels incompatible with biological functioning. Desert arthropods as a group are quite tolerable of dehydration. Survival following losses of body water amounting to more than 30 percent of their original wet weight has been reported (Edney 1974, Hadley 1974). Hemolymph osmotic pressure is regulated during dehydration in most insects despite a reduction in hemolymph volume (Edney 1968; Wall 1970; Okasha 1973). Patterns of hemolymph osmoregulation were recently examined in three desert arthropods: a tenebrionid beetle

(*Eleodes hispilabris*), a vejovid scorpion (*Paruroctonus aquilonalis*), and a spirostreptid millipede (*Orthoporus ornatus*) Riddle et al. 1976). During desiccation, beetles regulated hemolymph osmolality and scorpions tolerated increased osmolality; the millipede response was intermediate to these two strategies. Mechanisms for ionic regulation during desiccation and subsequent rehydration are poorly known for desert species.

Among the vertebrates, amphibians and reptiles in general tolerate much wider variations in body water content and electrolyte concentrations than can birds or mammals (Shoemaker and Nagy 1977). Desert toads may lose approximately 60 percent of their body water while burrowed (McClanahan 1967), and desert lizards can survive dehydration up to 50 percent of their body mass (Munsey 1972). Increased urea and electrolyte concentration in body fluids of burrowed toads and frogs provide for more favorable concentrations gradients between the animal and the surrounding soil. The desert lizard (*Sauromalus obsesus*) became dehydrated, but fractional fluid volumes and plasma electrolyte concentrations were maintained at normal levels (Nagy 1972). Birds tolerate body mass loss during dehydration that varies from less than 30 percent of initial mass in house finches to approximately 50 percent in the California quail; during this time some increase in ionic and osmotic concentrations can occur (Dawson 1976). Dehydration tolerance in mammals varies greatly, ranging from 10 to 12 percent loss of body weight in man to approximately 30 percent in the camel (Schmidt-Nielsen 1964). In the camel, only a small percentage of the water lost is removed from the plasma (unlike in man). As a result, blood volume is maintained, blood viscosity does not change greatly, and circulation is maintained more easily.

Water Uptake

Lost body water eventually must be replaced if an animal is to maintain overall water balance. Replenishment can be an acute problem, especially during hot summer months when evaporative water loss rates are high and usual sources of water are greatly diminished or even absent. Most species drink when free water is available and can quickly replenish deficits that may have occurred over relatively long periods of time. Camels have justly deserved reputations for the ability to rehydrate rapidly; however, the champion drinker is the bedouin goat, which reportedly replenished a water loss that amounted to 70 percent of its initial body weight within two minutes (Shkolnik et al. 1972). Oral water uptake is also possible in the absence of free-standing water, particularly in some coastal deserts where advective fogs are frequent (Louw 1972). Recent observations of tenebrionid beetles in the Namib Desert indicate that the direct collection of fog by several species is important to their water economy and that in certain cases elaborate behavior patterns have evolved for its procurement

(Hamilton and Seely 1976; Seely and Hamilton 1976). Other species never drink free water, even when available, but instead depend upon preformed water in their diet plus the small amount of water obtained from the oxidation of food. Independence of drinking is possible even when the moisture content of food is low, such as in "dry" seeds (less than 10 percent water), provided the species has the necessary mechanisms for effectively reducing water loss and is able to find appropriate shelter from temperature-humidity extremes. Small body size is not necessarily a prerequisite. Large African ungulates feed on grass that by day contains only 1 percent preformed water but over 30 percent at night due to hygroscopic absorption. By grazing only at night, these ungulates take advantage of both increased water availability and milder climatic conditions, thus maintaining water balance without having to drink (Taylor 1968).

The uptake (sorption) of water vapor from subsaturated atmospheres has been the subject of much investigation and speculation. Its occurrence among arthropods is restricted to insects, ticks, and mites and in truly desert forms is found only in adult females of the desert cockroach (*Arenivaga*) (Edney 1966; O'Donnell 1977), the thysanuran *Ctenolepisma terebrans* (Edney 1971), and larvae of tenebrionid beetles (Couchie and Crowe 1975). None of these insects are capable of flight, a feature shared by all species capable of direct atmospheric vapor sorption regardless of their habitat (Edney and Nagy 1976). The site(s) of uptake and the mechanisms involved are not completely understood for most species. Early emphasis suggested the cuticle as the site of absorption and even led to elaborate models explaining possible mechanisms. The possibility of cuticular sorption, however, has largely been discarded in view of recent investigations showing the mouth and/or anus to be the critical structures involved (Wharton and Richards 1978).

No desert vertebrate is capable of absorbing atmospheric water vapor directly, although some amphibians and reptiles may gain water from vapor by having it first condense on the integument. Uptake of water through the integument resulting from contact with moist soil or from simply sitting in standing water is, however, a primary avenue of water gain in desert anurans. Species such as the spadefoot toad (*Scaphiopus couchii*), which remains burrowed for ten months of the year, rapidly replenish lost body water upon emergence following summer rains (McClanahan 1967). Although rehydration takes place across the entire integument, certain regions such as the ventral pelvic patch absorb water more rapidly than do other areas. In newly metamorphosed and juvenile *S. couchii* toadlets, this specialized pelvic region accounts for 5 percent of the total surface area. The rate of rehydration through this region is 322 mg per square cm per hour, fifteen times greater than the mean rate through the remaining integument (Jones 1978). The absence of a specialized ventral region having similar function in adult toads is believed linked to the fact that adults usually do not expose themselves to situations in which water is preferentially available to the ventral surface. Moisture uptake through the integument also takes place while the toad is burrowed, provided the osmotic pressure of body fluids is higher than soil water potential. By elevating blood osmotic pressure through urea storage, desert toads can maintain a favorable gradient for continuous moisture uptake from the surrounding soil even when soil water tensions reach very low levels (Shoemaker et al. 1969).

METABOLIC RELATIONS

Metabolic function is an important adaptive feature of desert species that perhaps has not received adequate recognition. Many investigators have reported that metabolic rates of desert animals are often lower than predicted rates based on weight-metabolism equations established for the specific taxon. Among vertebrates such relationships have been described for reptiles, birds, and mammals. Two thermophilic Sonoran Desert lizards, *Dipsosaurus dorsalis* and *Crotaphytus collaris*, exhibited standard metabolic rates (measured at 37°C) that were approximately one-quarter to one-third of predicted levels (Templeton 1970). A similar response was found in lacertid lizards. *Acanthodactylus boskianus*, the most arid adapted of four species investigated, had a metabolic rate that deviated the most (57 percent) from the expected rate. This trend is not as clear-cut in either birds or mammals due to interordinal differences that influence the relationship between body weight and metabolism (passerine birds have higher weight-specific metabolic rates than do birds of comparable size belonging to other orders). Still, low metabolic rates are characteristic of a number of xeric birds including caprimulgids, owls, doves, and quail. Metabolism of twenty-two species of desert rodents was approximately 10 percent lower than rates observed for fifty-two nondesert species (Hart 1971). Furthermore the lower boundary of the thermoneutral zone of desert rodents was approximately 3°C higher than in the nondesert species. Reduced metabolism is also characteristic of larger desert mammals, especially when the animal is partially dehydrated (Borut and Shkolnik 1974). Desert invertebrates are no exception to this metabolic pattern, although the low metabolic rates observed in many cases reflect an inherent feature of the group (such as snails and scorpion) rather than a response to heat and aridity. Still, desert-adapted snails and scorpions exhibited lower metabolic rates than did their nondesert counterparts, and these rates were altered by humidity and starvation in manner that is adaptive.

The adaptive significance of reduced metabolic levels is multifaceted and depends in part on the thermoregulatory status of the animal. In all desert species, a reduction in the rate of gas exchange results in reduced respiratory transpiration, which can be a major component of total water loss at the high ambient temperatures common in desert regions. A more important consequence of lowered metabolism, however, is the resultant energetic benefits. This is especially true for desert invertebrates and ectothermic vertebrates. During extreme hot dry summer months, these species are often faced with an unpredictable food supply. To find sufficient food to satisfy water and nutrient needs, animals with high energy demands would have to be surface active for long periods that would likely extend into the more stressful times of the day. Because of lowered metabolic rates, nutrient reserves are utilized more slowly, permitting animals to remain in burrows or under cover longer where climatic conditions are more moderate. No desert species shows this better than do scorpions, which not only exhibit extremely low metabolic rates under normal conditions but depress these rates further during starvation. For birds and mammals, lower metabolic rates also mean reduced internal heat production. Heat generated from metabolic processes adds to the total heat load on an animal and subsequently must be dissipated if the animal is to maintain thermal balance. A reduction in this heat gain component will result in a savings of both water and calories that are normally expended during evaporative cooling.

Most information on metabolic performance of desert and nondesert animals is based on laboratory studies conducted under confined and essentially artificial conditions. Although these studies provide useful data on basal or resting metabolism under such conditions, they tell nothing about an animal's metabolic response in nature where temperatures and activity levels fluctuate. A number of telemetry and radioisotopic techniques have been devised to monitor metabolism in free-roaming animals. One of the most promising of these is the doubly labeled water (tritium and oxygen-18) method, which is based on turnover rates of hydrogen and oxygen of body water. Many of the initial field applications of this technique were performed on small desert mammals (Mullen 1970, 1971a, 1971b; Mullen and Chew 1973). Recently these investigations have been extended to include xeric-adapted lizards (Nagy 1975; Bennett and Nagy 1977; Congdon 1977) and even desert scorpions (King and Hadley 1979). Field metabolic rates of the scorpions were approximately three times higher than laboratory-recorded values, reflecting the higher activity levels of unrestrained animals. Validation experiments indicate that at least in vertebrates, metabolic data obtained using the doubly labeled water technique are within 10 percent of metabolic data obtained using conventional laboratory procedures. A larger error was reported in the scorpion validation tests (King and Hadley 1979); however, the difference apparently reflects the need for technique refinement rather than an inherent problem in technique theory when applied to invertebrates, for an error of less than 10 percent resulted when the doubly labeled water technique was used to measure respiratory CO_2 in locusts. Assuming that costs for the oxygen-18 can be reduced and that additional facilities for analyzing this isotope become available, the technique appears to have a bright future in investigations of energy utilization by field animals.

REFERENCES

Ahearn, G. A., and Hadley, N. F. 1976 Functional roles of luminal sodium and potassium in water transport across desert scorpion ileum. *Nature* 261:66-68.

——. 1977. Water transport in perfused scorpion ileum. *Amer. J. Physiol.* 233:R198-R207.

Bennett, A. F., and Nagy, K. A. 1977. Energy expenditure in free-ranging lizards. *Ecology* 58:697-700.

Blaylock, L. A.; Ruibal, R.; and Platt-Aloia, K. 1976. Skin structure and wiping behavior of phyllomedusine frogs. *Copeia* 1976:283-295.

Bligh, J. 1972. Evaporative heat loss in hot arid environments. In *Comparative physiology of desert animals*, ed. G. M. O. Maloiy, pp. 357-369. New York: Academic Press.

Borut, A., and Shkolnik, A. 1974. Physiological adaptations to the desert environment. In *Environmental physiology* eds. A. C. Guyton, D. Horrobin, and D. Robertshaw, pp. 185-229. London: Butterworths.

Braun, E. J., and Dantzler, W. H. 1972. Function of mammalian-type and reptilian-type nephrons in kidney of desert quail. *Amer. J. Physiol.* 222:617-629.

Braysher, M. L. 1976. The excretion of hyperosmotic urine and other aspects of the electrolyte balance of the lizard *Amphibolurus maculosus*. *Comp. Biochem. Physiol.* 54A:341-345.

Brown, G. ed. 1968. *Desert biology*. Vol. 1. New York: Academic Press.

——. 1974. *Desert biology*. Vol. 2. New York: Academic Press.

Buscarlet, L. A.; Proux, J.; and Gerster R. 1978. Utilisation due double marquage HT¹⁸O dans une étude de bilan metabolique chez *Locusta migratoria migratorioides*. *J. Insect Physiol.* 24:225-232.

Chaffee, R. J., and Roberts, J. C. 1971. Temperature acclimation in birds and mammals. *Ann. Rev. Physiol.* 33:155-202.

Cloudsley-Thompson, J. L. 1970. Terrestrial invertebrates. In *Comparative physiology of thermoregulation*, ed. G. C. Whittow, 1:15-77. New York: Academic Press.

——. 1973. Factors influencing the supercooling of tropical Arthropoda, especially locusts. *J. Nat. Hist.* 7:471-480.

Cloudsley-Thompson, J. L., and Crawford, C. S. 1970. Lethal temperatures of some arthropods of the southwestern United States. *Entomol. Month. Mag.* 106:26-29.

Congdon, J. D. 1977. Energetics of the montane lizard *Sceloporus jarrovi*: a measurement of reproductive effort. Ph.D. dissertation, Arizona State University.

Coutchie, P.A., and Crowe, J. H. 1975. Absorption of water vapor from subsaturated air by tenebrionid beetle larvae. *Am. Zool.* 15:802.

Crawford, C. S., and Riddle, W. A. 1975. Overwintering physiology of the scorpion *Diplocentrus spitzeri*. *Physiol. Zool.* 48:84-92.

Crawford, E. C., and Schmidt-Nielsen, K. 1967. Temperature Regulation and evaporative cooling in the ostrich. *Amer. J. Physiol.* 212:347-353.

Dawson, W. R. 1976. Physiological and behavioural adjustments of birds to heat and aridity. *Proc. 16th Internat. Ornithol. Congr.* Canberra City: Australian Acad. Science.

Drewes, R. C.; Hillman, S. S.; Putnam, R. W; and Sokol, O. M. 1977. Water, nitrogen and ion balance in the African treefrog *Chiromantis petersi* Boulenger (Anura: Rhacophoridae), with comments on the structure of the integument. *J. Comp. Physiol.* 116:257-267.

Duvdevani, I. and Borut, A. 1974. Oxygen consumption and evaporative water loss in four species of *Acanthodactylus* (Lacertidae). *Copeia* 1974:155-164.

Ebeling, W. 1974. Permeability of insect cuticle. In *The physiology of insecta*, ed. M. Rockstein, pp. 271-343. New York: Academic Press.

——. 1976. Insect integument: a vulnerable organ system. In *The insect integument*, ed. H. R. Hepburn, pp. 383-400. Amsterdam: Elsevier.

Edney, E. B. 1966. Absorption of water vapour from unsaturated air by *Arenivaga* sp. (Polyphagidae, Dictyoptera). *Comp. Biochem. Physiol.* 19:387-408.

——. 1968. The effect of water loss on the haemolymph of *Arenivaga* sp. and *Periplaneta americana*. *Comp. Biochem. Physiol.* 25:149-158.

——. 1971a. Some aspects of water balance in tenebrionid beetles and a thysanuran from the Namib Desert of Southern Africa. *Physiol. Zool.* 44:61-76.

——. 1971b. The body temperature of tenebrionid beetles in the Namib desert of southern Africa. *J. Exp. Biol.* 55: 253-272.

——. 1974. Desert arthropods. In *Desert Biology*, ed. G. W. Brown, 2:311-394. New York: Academic Press.

Edney, E. B., and Nagy, K. A. 1976. Water balance and excretion. In *Environmental physiology of animals*, eds. J. Bligh, J. L. Cloudsley-Thompson, and A. G. Macdonald, pp. 106-132. New York: John Wiley.

Goodall, D. W., ed. 1976. *Evolution of desert biota*. Austin, Texas: University of Texas Press.

Hadley, N. F. 1970. Micrometeorology and energy exchange in two desert arthropods. *Ecology* 51:434-444.

——. 1972. Desert species and adaptation. *Amer. Sci.* 60:338-347.

——. 1974. Adaptational biology of desert scorpions. *J. Arachnol.* 2:11-23.

——. 1975. *Environmental physiology of desert organisms*. Stroudsburg, Pa.: Dowden, Hutchinson & Ross.

——. 1977. Epicuticular lipids of the desert tenebrionid beetle, *Eleodes armata*: seasonal and acclimatory effects on composition. *Insect Biochem.* 7:277-283.

——. 1978a. Cuticular permeability of desert tenebrionid beetles: correlations with epicuticular hydrocarbon composition. *Insect Biochem.* 8:17-22.

——. 1978b. Cuticular permeability and lipid composition of the black widow spider, *Latrodectus hesperus*. *Symp. Zool. Soc. Lond.* 42:429-438.

——. 1979. Wax secretion and color phases of the desert tenebrionid beetle *Cryptoglossa verrucosa* (LeConte). *Science* 203:367-369.

Hadley, N. F., and Jackson, L. L. 1977. Chemical composition of the epicuticular lipids of the scorpion, *Paruroctonus mesaensis*. *Insect Biochem.* 7:85-89.

Hainsworth, F. R. 1967. Saliva spreading, activity and body temperature regulation in the rat. *Amer. J. Physiol.* 212:1288-1292.

Hales, J. R. S. 1974. Physiological responses to heat. In *Environmen-*

tal physiology, eds. A. C. Guyton, D. Horrobin, and D. Robertshaw, pp. 107-162. London: Butterworths.

Hamilton, W. J. III. 1973. *Life's color code*. New York: McGraw Hill.

———. 1975. Coloration and its consequences for diurnal desert insects. In *Environmental physiology of desert organisms*, ed. N. F. Hadley, pp. 67-89. Stroudsburg, Pa: Dowden, Hutchinson & Ross.

Hamilton, W. J. III, and Heppner, F. 1967. Radiant solar energy and the function of black homeotherm pigmentation: an hypothesis. *Science* 155:196-197.

Hamilton, W. J. III, and Seely, M. K. 1976. Fog basking by the Namib Desert beetle, *Onymacris unguicularis*. *Nature* 262:284-285.

Hart, J. S. 1971. Rodents. In *Comparative physiology of thermoregulation*, ed. G. C. Whittow, vol. 1, pp. 1-149. New York: Academic Press.

Hatch, D. E. 1970. Energy conserving and heat dissipating mechanisms of the turkey vulture. *Auk* 87:11-124.

Hattingh, J. 1972. The correlation between transepidermal water loss and the thickness of epidermal components. *Comp. Biochem. Physiol.* 43A: 719-722.

Heath, J. E. 1966. Venous shunts in the cephalic sinuses of horned lizards. *Physiol. Zool.* 39:30-35.

Heinrich, B. 1970. Thoracic temperature stabilization by blood circulation in a free-flying moth. *Science* 168:580-582.

———. 1971. Temperature regulation of the sphinx moth, *Manduca sexta*. II. Regulation of heat loss by control of blood circulation. *J. Exp. Biol.* 54:153-166.

———. 1975. Thermoregulation and flight energetics of desert insects. In *Environmental physiology of desert organisms* ed. N. F. Hadley, pp. 90-105. Stroudsburg, Pa: Dowden, Hutchinson & Ross.

Henwood, K. 1975. Infrared transmittance as an alternative thermal strategy in the desert beetle *Onymacris plana*. *Science* 189: 993-994.

Heppner, F. 1970. The metabolic significance of differential absorption of radiant energy by black and white birds. *Condor* 72:50-59.

Hinds, D. S. 1977. Acclimatization of thermoregulation in desert-inhabiting jackrabbits (*Lepus alleni* and *Lepus californicus*). *Ecology* 58:246-264.

Ingram, D. L. and Mount, L. E. 1975. *Man and animals in hot environments*. New York: Springer-Verlag.

Jackson, D.C., and Schmidt-Nielsen, K. 1964. Countercurrent heat exchange in the respiratory passages. *Proc Nat. Acad. Sci. USA* 51:1192-1197.

Johnson, O. W., and Ohmart, R. D. 1973. The renal medulla and water economy in vesper sparrows (*Pooecetes gramineus*). *Comp. Biochem. Physiol.* 44A:655-661.

Jones, R. M. 1978. Rapid pelvic water uptake in *Scaphiopus couchi* toadlets. *Physiol. Zool.* 51:51-55.

Kilgore, D. L., Jr.; Bernstein, M. H., and Schmidt-Nielsen, K. 1973. Brain temperature in a large bird, the rhea. *Amer. J. Physiol.* 225:739-742.

King, W. W., and Hadley, N. F. 1979. Water flux and metabolic rates of freeroaming scorpions using the doubly-labeled water technique. *Physiol. Zool.* 52:176-189.

Kluger, M. J., and Heath, J. E. 1970. Vasomotion in the bat wing: A thermoregulatory response to internal heating. *Comp. Biochem. Physiol.* 32:219-226.

Lasiewski, R. C. 1972. Respiration function in birds. In *Avian biology*, eds. D. S. Farner and J. R. King, vol. 2, pp. 287-342. New York: Academic Press.

Lee, A. K., and Mercer, E. H. 1967. Cocoon surrounding desert-dwelling frog. *Science* 157:87-88.

Licht, P. and Bennett, A. F. 1972. A scaleless snake: Tests of the role of reptilian scales in water loss and heat transfer. *Copeia* 1972:702-707.

Locke, M. 1974. The structure and formation of the integument in insects. In *The Physiology of Insecta*, ed. M. Rockstein, pp. 123-213. New York: Academic Press.

Louw, G. N. 1972. The role of advective fog in the water economy of certain Namib Desert animals. In *Comparative physiology of desert animals*, ed. G. M. O. Maloiy, pp. 297-314. New York: Academic Press.

Loveridge, J. P. 1968. The control of water loss in *Locusta migratoria migratorioides* R. and R. *J. Exp. Biol.* 49:1-29.

———. 1970. Observations on nitrogenous excretion and water relations of *Chiromantis xerampelina* (Amphibia, Anura). *Arnoldia Rhodesia* 5:1-6.

———. 1976. Strategies of water conservation in southern African frogs. *Zool. Afric.* 11:319-333.

Lowe, C. H.; Heed, W. B.; and Halpern, E. A. 1967. Supercooling of the saguáro species *Drosophila nigrospiracula* in the Sonoran Desert. *Ecology* 48:984-985.

Lowe, C. H.; Lardner, P. J.; and Halpern, E. A. 1971. Superwoling in reptiles and other vertebrates. Comp. Biochem. Physiol. 39A:125-135.

McClanahan, L. L. 1967. Adaptations of the spadefoot toad, *Scaphiopus couchi*, to desert environments. *Comp. Biochem. Physiol.* 20:73-99.

McClanahan, L. L; Shoemaker, V. H.; and Ruibal, R. 1976. Structure and function of the cocoon of a ceratophryd frog. *Copeia* 1976: 179-185.

McClanahan, L. L; Stinner, J. N.; and Shoemaker, V.H. 1978. Skin lipids, water loss, and energy metabolism in a South American tree frog (*Phyllomedusa sauvagei*). *Physiol. Zool.* 51:179-187.

MacMillen, R. E., and Lee, A. K. 1967. Australian desert mice: Independence of exogenous water. *Science* 158:383-385.

McNabb, R. A., and McNabb, F. M. 1975. Urate excretion by the avian kidney. *Comp. Biochem. Physiol* 51A:253-258.

Maloiy, G. M. O., ed. 1972. *Comparative physiology of desert animals*. Symp. Zool. Soc. Lond., 31. New York: Academic Press.

Marder, J. 1973. Body temperature regulation in the brown-necked raven (*Corcus corax ruficollis*)—II. Thermal changes in the plumage of ravens exposed to solar radiation. *Comp. Biochem. Physiol.* 45A:431-440.

Mullen, R. K. 1970. Respiratory metabolism and body water turnover rates of *Perognathus formosus* in its natural environment. *Comp. Biochem. Physiol.* 32:259-265.

———. 1971. Energy metabolism and body water turnover rates of two species of free-living kangaroo rats, *Dipodomys merriami* and *Dipodomys microps*. *Comp. Biochem. Physiol.* 39A:379-390.

———. 1971. Energy metabolism of *Peromyscus crinitus* in its natural environment. *J. Mammal.* 52:633-635.

Mullen, R. K., and Chew, R. M. 1973. Estimating the energy metabolism of freeliving *Perognathus formosus:* A comparison of direct and indirect methods. *Ecology* 54:633-637.

Munsey, L. D. 1972. Water loss in five species of lizards. *Comp. Biochem. Physiol.* 43A:781-794.

Nagy, K. A. 1972. Water and electrolyte budgets of a free-living desert lizard, *Sauromalus obesus. J. Comp. Physiol.* 79:39-62.

———. 1975. Water and energy budgets of free-living animals: measurement using isotopically labeled water. In *Environmental physiology of desert organisms*, ed. N. F. Hadley, pp. 227-245. Stroudsburg, Pa.: Dowden, Hutchinson & Ross.

Noy-Meir, I. 1973. Desert ecosystems: Environment and producers. *Ann. Rev. Ecol. Syst.* 4:25-51.

———. 1974. Desert ecosystems: Higher trophic levels. *Ann Rev. Ecol. Syst.* 5:195-214.

O'Donnell, M. J. 1977. Uptake of water vapour by the desert cockroach *Arénivaga*. In Comparative physiology water, ions and fluid mechanics, eds. K. Schmidt-Nielsen, L. Bolis, and S. H. P. Maddrell, pp. 115-121. Cambridge: At the Press.

Ohmart, R. D., and Lasiewski, R. C. 1971. Roadrunners: energy conservation by hypothermia and absorption of sunlight. *Science* 172:67-69.

Okasha, A. Y. K. 1973. Water relations in an insect, *Thermobia domestica*, III. Effects of desiccation and rehydration on the haemolymph. *J. Exp. Biol.* 58:385-400.

Orians, G. H., and Solbrig, O. T. 1977. Convergent evolution in warm deserts. Stroudsburg, Pa.: Dowden, Hutchinson & Ross.

Phillips, J. E. 1977. Excretion in insects: function of gut and rectum in concentrating and diluting the urine. Fed. Proc. 36:2480-2486.

Pough, F. H., and McFarland, W. N. 1976. A physical basis for head-body temperature differences in reptiles. *Comp. Biochem. Physiol.* 53A:301-303.

Riddle, W. A. 1977. Comparative respiratory physiology of a desert snail *Rabdotus schiedeanus*, and a garden snail, *Helix aspersa, Comp. Biochem. Physiol.* 56A:369-373.

———. 1978. Respiratory physiology of the desert grassland scorpion *Paruroctonus utahensis. J. Arid. Environ.* 1:243-251.

Riddle, W. A., Crawford, C. S.; and Zeitone, A. M. 1976. Patterns of hemolymph osmoregulation in three desert arthropods. *J. Comp. Physiol.* 112:295-305.

Riddle, W. A., and Pugach, S. 1976. Cold hardiness in the scorpion, *Paruroctonus aquilonalis.* Cryobiology 13:248-253.

Schmidt-Nielsen, K. 1964. *Desert animals: Physiological problems of heat and water.* London: Oxford University Press.

Schmidt-Nielsen, K.; Hainsworth, F. R. and Murrish, D. W. 1970. Counter current heat exchange in the respiratory passages: effect on water and heat balance. *Resp. Physiol.* 9:263-276.

Seely, M. K., and Hamilton, W. J., III. 1976. Fog catchment sand trenches constructed by tenebrionid beetles, *Lepidochora*, from the Namib Desert. *Science* 193:

Shoemaker, V. H.; McClanahan, L. L.; and Ruibal, R. 1969. Seasonal changes in body fluids in a field population of spadefoot toads. *Copeia* 1969:585-591.

Shoemaker, V. H., and McClanahan, L. L., Jr. 1975. Evaporative water loss, nitrogen excretion and osmoregulation in phyllomedusine frogs. *J. Comp* Physiol. 100:331-345.

Shoemaker, V. H., and Nagy, K. A. 1977. Osmoregulation in amphibians and reptiles. *Ann. Rev. Physiol.* 39:449-471.

Shkolnick, A.; Borut, A.; and Choshnick, J. 1972. Water economy of the beduin goat. *Symp. Zool. Soc. London* 31:229-

Siebert, B.D., and MacFarlane, W. V. 1971. Water turnover and renal function of dromedaries in the desert. *Physiol. Zool.* 44:225-240.

Skadhauge, E. 1976. Cloacal absorption of urine in birds. *Comp. Biochem: Physiol.* 55A:93-98.

Sturbaum, B. A., and Riedesel, M. L. 1977. Dissipation of stored body heat by the ornate box turtle, *Terrapene ornata. Comp. Biochem. Physiol.* 58A:93-97.

Taylor, C. R. 1968. Hygroscopic food: A source of water for desert antelopes? *Nature* 219:181-182.

———. 1970a. Strategies of temperature regulation: Effect on evaporation in East African ungulates. *Amer. J. Physiol.* 219:1131-1135.

———. 1970b. Dehydration and heat: Effects on temperature regulation of East African ungulates. *Amer. J. Physiol.* 219:1136-1139.

———. 1974. Exercise and thermoregulation. In *Environmental physiology*, eds. A. C. Guyton, D. Horrobin, and D. Robertshaw, pp. 163-184. London: Butterworths.

Taylor, C. R., and Lyman, C. P. 1972. Heat storage in running antelopes: independence of brain and body temperatures. *Amer. J. Physiol.* 222:114-117.

Templeton, J. R. 1970. Reptiles. In *Comparative physiology of thermoregulation*, ed. G. C. Whittow, vol. 1, pp. 204-249. New York: Academic Press.

Toolson, E. C., and Hadley, N. F. 1977. Cuticular permeability and epicuticular lipid composition in two Arizona vejovid scorpions. *Physiol. Zool.* 50:323-330.

———. 1979. Seasonal effects on cuticular permeability and epicuticular lipid composition in *Centruroides sculpturatus Ewing 1928* (Scorpiones: Buthidae). *J. Comp. Physiol.* 129:319-325.

Wall, B. J. 1970. Effects of dehydration and rehydration on *Periplaneta americana. J. Insect Physiol.* 16:1027-1042.

Walsberg, G. E.; Campbell, G. S.; and King, J. R. 1978. Animal coat color and radiative heat gain: A re-evaluation. *J. Comp. Physiol.* 126: 211-222.

Wharton, G. W., and Richards A. G. 1978. Water vapor exchange kinetics in insects and acarines. *Ann. Rve. Entomol.* 23:309-328.

White, F. N. 1973. Temperature and the Galapagos marine iguana—insights into reptilian thermoregulation. *Comp. Biochem. Physiol.* 45A:505-513.

Yousef, M. K., Horvath, S. M.; and Bullard, R. W., eds. 1972. *Physiological adaptations: Desert and mountain.* New York: Academic Press.

Zervanos, S. M. 1975. Seasonal effects of temperature on the respiratory metabolism of the collared peccary (*Tayassu tajacu*). *Comp. Biochem. Physiol.* 50A:365-371.

Zervanos, S. M., and Hadley, N. F. 1973. Adaptational biology and energy relationships of the collared peccary (*Tayassu tajacu*). *Ecology* 54:759-774.

9

PLANT ADAPTATIONS

OTTO T. SOLBRIG

IN SPITE OF the attention that the specialized morphologies and behaviors of plants from arid lands have attracted since the time of Schimper (1903) as attested by the existence of a large number of articles, reviews, and books (Kearny and Schantz 1911; Maximov 1929; Stocker 1928, 1968; Walter 1950; Evenari and Tadmor 1950), an accepted generalized theory regarding the dynamics of the evolution and physiological behavior of desert plants is still lacking. A generalized theory probably will not be available for some time because there still is incomplete knowledge regarding some basic physiological processes—for example, mineral absorption and utilization, photosynthetic mechanisms, and energy apportionment.

In this chapter I assess the characteristics of desert plants in order to detect general patterns. I look at desert plants as forming part of an ecological system, each individual behaving in such a way as to maximize its chances of survival and reproduction. I assume that natural selection favors survival and not physiological efficiency (Slobodkin and Rapopport 1974; Solbrig 1980) and that adaptations can best be understood if the plant is considered as a system that functions so as to allocate energy to maximize survival and reproduction.

In order to grow, survive, and reproduce, a plant needs to harvest energy and materials from the environment and allocate that energy and those materials to its various structures and functions in some precise way. The mechanisms by which a plant harvests energy and materials from the environment and apportions them to the various functions and structures constitute its adaptation to that environment. In most environments there is more than one possible adaptation, but they are always a small number as compared to the set of all known ways that plants can harvest and apportion energy. This is attested by the fact that only a limited number of species (in relation to all available species) is capable of growing in a given environment even when cultivated free from competition.

Green land plants have a relatively simple system of nutrition. All green plants manufacture carbohydrates from atmospheric carbon dioxide and soil water utilizing solar energy. In

addition, all plants require a number of major and minor minerals, which they obtain from the soil. All land plants also require soil water to absorb mineral ions, and to dissipate excess heat from leaf surfaces. Mineral availability is not uniform in the soil column nor is it uniform from site to site, both in microscale and in a larger geographical context. The same is true for water availability, which in addition may show large fluctuations over time. Finally, although carbon dioxide (CO_2) concentration in air is fairly uniform, sunlight shows a regular daily and yearly cycle, which is a function of latitude. Although light cannot be stored (as can water and minerals), it can be intercepted and effectively withdrawn by vegetation even when it is not directly utilized. This creates a vertical light gradient superimposed on the daily and yearly light cycles. Consequently although all green plants harvest the same limited number of resources, the differences in abundance and availability of each of these resources in time and space are very great, resulting in a large but limited number of ways of partitioning these resources.

Three other important aspects of adaptation are the modes of reproduction, seed dispersal, and predator defense. Since these processes also make energetic and material demands on the plant, they can be considered as part of the pattern of allocation. In addition, the type of breeding system that the species possesses has important evolutionary implications.

Resource availability varies in time and space. The morphological structures and physiological processes required for harvesting light, water, or minerals when abundant may differ from the structures and processes required when they are in low supply. This may create situations that allow the coexistence in time and space of plants with different requirements.

Finally plants are subjected to stresses that may cause death, such as floods, droughts, fires, or excessive predation. Species must have the ability of regenerating after these catastrophic events, either by regrowth from individual plant fragments or, as is more often the case, from special reproductive structures, primarily sexually produced seeds, but also asexually generated bulbils, bulbs, corms, and so forth. Formation of such structures diverts energy from maintenance and growth, thereby reducing the plant's ability to harvest resources but ensuring its survival in case of a catastrophic event. If potentially catastrophic events are frequent and predictable, such as fires in tropical and subtropical savannas, species may evolve special structures that reduce the effect of the disturbance, among them thick bark or specially protected buds. These structures are also energetically costly but presumably less so than reestablishment.

DESERT ENVIRONMENT

Two principal types of deserts are: rain shadow deserts and high pressure deserts. This description will be based primarily on the latter but most statements apply to both types.

The world-wide circulation of air results in the formation of high pressure zones at middle latitudes in both hemispheres. Within these belts, cold and dry air descend from high altitudes. As the air descends, its temperature is increased through compressional heating, rendering it incapable of any precipitation. This dry air permits the penetration of solar radiation with minimum scatter and little heat absorption by the atmosphere. As a result, the ground receives a large proportion of the incident solar radiation at that latitude, resulting in very high soil surface temperatures. In turn, the hot ground reradiates long infrared waves that are partly absorbed by the lower layers of the atmosphere, which thereby become greatly heated. Much of the radiation absorbed by the ground during the day is dissipated at night through the dry, clear air, into outer space. At night both ground and air temperatures drop rapidly, resulting in a great range of daily temperatures.

A second important diagnostic feature of desert regions is insufficient precipitation to support a vegetation dense enough to cover most of the ground surface during all of the year. Insufficient precipitation results from the existence of masses of descending air that are adiabatically heated. Consequently the saturation point of air is seldom reached, and dew, frost, and fog are lacking except on rare occasions following a rainstorm. As a result of latitudinal shifts of the high pressure cells, the poleward boundaries of the desert belts are penetrated occasionally by migratory front disturbances characteristic of higher latitudes. These bring soaking frontal rains. The equatorial boundaries are penetrated during summer months by humid tropical air that brings torrential but short-lived convectional rain.

Consequently the principal features of the physical environment of deserts that affect plants are: high solar radiation, low soil moisture most of the year, irregular and unpredictable precipitation, and relatively high daily and yearly fluctuations in temperature.

Mineral availability varies according to the parent rock and does not differ from other regions. In general leaching is low or nonexisting; nitrogen (N) levels are low (West and Skujins 1978) but calcium (Ca) and sodium (Na) and other cations are high, sometimes being present in toxic amounts. Nevertheless low soil water availability makes access to minerals difficult. Atmospheric carbon dioxide (CO_2) levels are normal.

Plant life in desert regions means adjusting to low soil moisture levels, high insolation, and varying mineral levels during most of the year, punctuated by periods of precipitation where soils reach or surpass field capacity. Because water is such an important aspect of desert life, the first aspect to be considered is water economy.

WATER ECONOMY OF DESERT PLANTS

Plant responses to water stress have been reviewed by Slatyer (1967), Hsiao (1973), and Lange, Kappen, and Schulze (1976). Since water is central to all metabolic functions, water deficit manifests itself in a pervasive way throughout the plant, from lowered photosynthesis, to changes in enzyme levels.

Water moves from the soil to the plant roots, through the xylem to the leaves, and from there to the atmosphere at a rate that is determined by the water potential drop in the soil-plant-atmosphere system, and the resistance encountered (figure 9-1). It is a purely physical process using exterior energy sources. The movement of water in the soil-plant-atmosphere system is driven by evaporation (=transpiration) of water from leaf surfaces. Transpiration can be viewed as a simple diffusion process. It can be expressed in terms of the energy budget of the leaf (Bannister 1976):

$$E_L = \frac{(S_L - C_L)}{L}$$

where E_L = transpiration; S_L = total radiation balance; C_L = convective heat loss; and L = latent heat of vaporization of water. Thus transpiration increases with net increase of radiation (S_L) and when convective heat losses (C_L) are low. Convective heat losses depend in part on leaf size, increasing with decreased leaf diameter.

Water loss by evaporation from leaf surfaces must be replaced by water absorbed from the soil by roots. Movement of water in the soil-plant system follows potential gradients, moving from soil into plant only when the water potential in the plant is more negative than that in the soil. Since the water potential in the atmosphere is very low (in excess of −300 bars in a typical desert day), a plant theoretically could extract soil water until that value is reached. However, angiospermous plants cannot withstand such internal water deficits and close their stomates (interrupting the stream of water vapor from leaf to atmosphere) long before that point is reached. For most

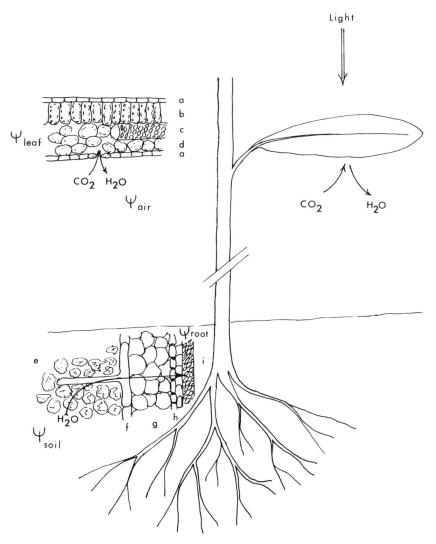

Figure 9-1. Flow of water from soil to atmosphere.
Water moves along gradients of water potential toward increasingly more negative potentials, from soil (ψ soil) to root (ψ root) to leaf (ψ leaf) to air (ψ air). When soil dessicate and their potentials go below values of -20 bars, it becomes increasingly more difficult for the plant to maintain even greater negative potentials in roots and leaves without incurring permanent damage. Original prepared by O. T. Solbrig.

Note: (a) leaf epidermis, (b) palisade parenchyma, (c) leaf vessel, (d) spongy parenchyma, (e) soil, (f) root epidermis with root hair, (g) cortex, (h) endodermis, (i) root xylem vessel.

plants of mesic regions that point lies between -10 and -15 bars; occasionally it is somewhat lower. Many desert plants can withstand much greater water deficits.

The supply of water to the leaf must be sufficient to replace the water molecules that are lost through the stomates; otherwise severe water stress will follow. According to Hsiao (1973) the loss of tissue water has these principal effects:

1. A lowering of the chemical potential (Ψ) or activity of cellular water. This change in potential activity of

water is not very large under the water deficits that plants tolerate (from 0.993 at $\Psi = -10$ bars to 0.978 at $\Psi = -30$ bars). Consequently it is unlikely that the effects on plant metabolism of water stress are due in major part to a reduced chemical potential of water.

2. Changes in turgor pressure. Normally cells of plants from mesic areas have turgor pressures (Ψ_p) in the order of 5 to 10 bars. Changes in soil water potential of 10 bars or more will lower the turgor pressure in

such plants to zero. Changes in turgor pressure can affect stomatal functioning and loading of phloem, as well as compression of macromolecular complexes against the cell wall. It is not surprising, therefore, that many desert plants that are subjected to high water stresses have high osmotic potentials (Breckle 1974).

3. Changes in molecular concentration and spatial relations in membranes and organelles. The change in concentration of molecules as water is lost from the cell is bound to affect biochemical reactions. So are spatial relations of enzymes bound to membranes inside the cell and inside cellular organelles. Vieira da Silva (1976) presents direct experimental evidence of the effect of spatial changes on metabolism, including disruption of compartmentation in the cells and liberation of destructive enzymes. However, there is still some question regarding the triggering mechanism of these reactions.

4. Changes in macromolecular structures as a result of water stress have been postulated, but there is insufficient experimental evidence in favor of a major role of this effect.

Although some microorganisms (Levitt 1972) can grow at water deficits of −200 to −300 bars, most drought-tolerant species of vascular plants cannot tolerate deficits of over −40 bars. However, occasional desert species, such as *Larrea tridentata,* function at values as low as −70 bars (Mabry, Hunziker, and DiFeo 1977).

According to Hsiao (1973), water stress primarily causes turgor pressure reduction, which in turn retards growth. If stress is extreme, cellular disruption and death follow. In order to maintain cell turgor pressure and to tolerate extreme water deficits, sufficient solutes must be accumulated to produce an osmotic potential higher than the environment.

Some authors (Itai and Benzioni 1976) believe that the observed change in growth results from disruption in hormonal balances, especially the kinetin-abscissic acid ratio. Drought-tolerant species also possess greater resistance to drought-induced membrane alteration and decompartmentation, and perhaps also possess different hormonal controls.

Breckle (1974) studied the water and salt relations of *Atriplex confertifolia* (Torr. & Frem.) S. Wats. and *Ceratoides lanata* Nevski from Utah. She found that in *C. lanata* the osmotic potential of the leaf cells remains almost constant around −30 bars but that in *A. confertifolia* it can drop to nearly −200 bars during the summer drought period. While *C. lanata* limits the uptake of salts at the roots, *A. confertifolia* allows salt to flow through the plant and subsequently secretes the salts through specialized bladder hairs in the leaves. The high osmotic potentials of these two species are an adaptation

to the low soil water potentials of the salty soils in which they grow.

But not all desert plants acquire characteristics to tolerate extreme water stresses. An alternative strategy is to escape water shortages by surviving the stress period in the drought-tolerant, dormant-seed stage. Other plants escape the drought as drought-tolerant, dormant underground structures, such as bulbs or corms. Finally, some plants escape water stress by dropping their leaves and nonwoody aerial tissue in times of drought. Although these behaviors avoid water stress damage, they also entail loss of biomass and a lag in productivity once the rains set in. This loss is greatest for ephemeral species and least for species that drop only their leaves. These last species require special morphological structures to reduce stem transpiration or the ability of stem tissues to tolerate dehydration. They also must expend metabolic reserves for maintenance respiration; the energy expenditure is less than that of drought-active plants but higher than seeds or underground tissues.

A third general strategy intermediate between drought tolerance and drought avoidance is to maintain reasonably high water potentials in tissues through mechanisms that reduce transpiration water loss or by gaining access to richer soil water supplies, or a combination of both. Such plants are sometimes called drought-resistant plants (Shantz 1927; Maximov 1929). Reduced transpirational loss as a result of stomatal closure, while conferring effective drought resistance, results in a shutdown of photosynthesis and therefore may lead to drought-induced starvation.

A fine example of a drought-resistant plant is the creosote bush (*Larrea tridentata*) from the Sonoran Desert and the jarillas, *L. cuneifolia* and *L. divaricata,* from the Monte Desert in Argentina (figure 9-2). These species rely on a variety of methods to harvest soil water, as well as different adaptations to reduce transpiration in times of water stress. For example, roots of *Larrea* spp. are found not only in surface layers (0-20 centimeters) that saturate during the rainy season but also at intermediate and deep layers (50-100 cm) that retain some moisture during the dry season (Solbrig, unpublished data). Furthermore, these species appear to be able to utilize water that condenses in the underside of rocks that cool more rapidly than surrounding soil at night (Stark and Love 1969; Syvertsen et al. 1975). There is also some indication that the leaves of *Larrea* spp. may be able to take up water present on leaf surfaces (Spalding 1904; Stark and Love 1969; Wallace and Romney 1972), although it is doubtful that this source of humidity is significant in the economy of the plant. Species of *Larrea* respond to increased drought by shedding tissue, both leaves and branches (Stark and Love 1969; Strain and Chase 1966; Morello 1955). Size and number of new leaves decrease with increased water stress (Morello 1955; Wallace and Romney 1972). In addition, *Larrea* spp. show very low cuticular transpiration rates (Oppenheimer 1960). Of several species of desert plants measured by Bamberg et al. (1973) in

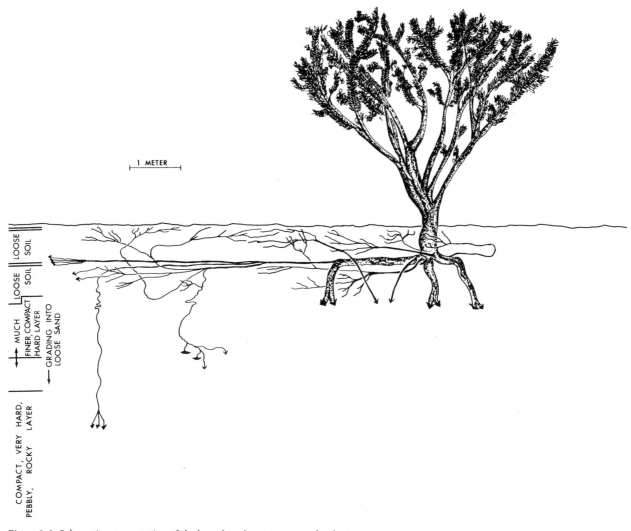

Figure 9-2. Schematic representation of the branch and root structure of a plant of Larrea cuneifolia *from Andalgala, prov. Catamarca, Argentina.*

situ in Rock Valley, Nevada, *L. tridentata* was clearly the most efficient in restricting transpirational water loss. Finally *Larrea* spp. can tolerate some of the lowest levels of tissue water potential among desert plants. Dawn water potentials of −50 to −70 bars have been measured repeatedly, and values of up to −115 bars have been recorded (Bamberg et al. 1973; Cunningham and Burk 1973; Halvorson and Patten 1974; Odening, Strain, and Oechel 1974; Oechel, Strain, and Odening 1972; Syvertsen et al. 1975.

ENERGY AND CARBON CAPTURE

A plant that is capable of harvesting a greater amount of carbon and energy than its neighbors has a selective advantage inasmuch as it has more photosynthate available for growth,

reproductive structures, and/or defense against predators. This statement assumes that photosynthate (carbon and energy) are the principal factors that limit growth. Although there is good evidence that photosynthate availability limits growth in certain situations, the important role of minerals, especially nitrogen, cannot be discounted.

Photosynthesis requires CO_2, which diffuses into the leaf through the stomata, the opening of which depends on leaf hydrature. When the stomata close, the rate of CO_2 diffusion into the leaf—and hence photosynthesis—is reduced or stopped. The rate at which water vapor is lost can be regulated by partial closure of the stomata but always at the cost of a reduced CO_2 flux. Leaf cells can also acquire characteristics (small cells, thick cell walls, high osmotic potential, biochemical changes) that allow them to endure higher levels of dessication (Boyer and Bowen 1970) and thereby prolong the time stomata remain open. However, these characteristics reduce the potential rates of photosynthesis when soil water is plentiful (Orians and Solbrig 1977). Consequently, there are inevitable trade-offs between the ability of desert plants to possess high potential photosynthetic rates to utilize the high light energy of desert areas and their ability to function under conditions of water stress.

The rate of photosynthesis is affected by a variety of environmental factors. (For detailed reviews on the subject see Stocker 1968; Mooney 1972; Mooney and Gulmon 1979.) Light intensity and duration during the day and year affect the rate of photosynthesis. Desert areas are characterized by clear air and low nubosity, and consequently desert plants receive a higher percentage of the radiation incident at the periphery of the atmosphere at each particular locality of any vegetation type. Another characteristic of the desert's light environment is the steepness of the light increase at sunrise and decrease at sunset. That is, both dawn and dusk are of short duration.

A second factor affecting the rate of photosynthesis is the flux of atmospheric carbon dioxide (CO_2). Since the concentration of CO_2 in air is fairly constant, the CO_2 flux is given by the concentration of CO_2 at the chloroplast and the resistances encountered in its path, following Ohm's law,

$$CO_2 \text{ flux} = \frac{CO_{2_{CHL}} - CO_{2_{air}}}{r_b + r_s + r_m},$$

where $CO_{2_{CHL}}$ = CO_2 concentration at the chloroplast; r_b = boundary layer resistance, r_s = stomatal resistance, and r_m = mesophyll resistance. The two principal factors in this equation are the concentration of CO_2 at the chloroplast and the stomatal resistance.

Stomatal resistance (r_s) is a function of the number of stomata per unit of surface and the size and morphology of the stomata. By varying the aperture of the stomata, the plant can control in part the value of the stomatal resistance (r_s). The number of stomata per unit surface and their morphology is species specific, although number of stomata per unit surface is also affected by the conditions encountered by the leaf during development. Stomatal resistance is therefore a variable resistance with values ranging from close to infinity at one end to a minimum value that depends on species at the other. The boundary layer resistance is a function of leaf shape, especially its width (Gates 1962) and wind velocity. In narrow-leaved desert plants, the value of the boundary layer is usually low and approaches zero or a very small value with wind speeds of 50 cm min^{-1} or higher.

The mesophyll resistance is the sum of the remaining factors that impede the free movement of CO_2 molecules from the substomatal cavity to the site of carboxylation. The exact physical factors involved are not well understood, but they include movement across cell walls, cell and chloroplast membranes, transport through a liquid phase in the cytoplasm, and

the CO_2 affinity of the carboxylating enzyme (Yocum and Lommen 1975).

The majority of green plants incorporate CO_2 into the photosynthetic pathway by combining a molecule of CO_2 with one of ribulose-diphosphate to produce phosphoglyceric acid, which is then further metabolized into glucose and a new molecule of ribulose-diphosphate, which repeats the cycle (Zelitch 1971). The enzyme responsible for the incorporation of CO_2, ribulose-diphosphate-carboxylase, has a dual function in that in the presence of oxygen (O_2) it also functions as an oxygenase that oxidizes glycolate and releases CO_2, a process known as photorespiration. Photorespiration, by increasing the internal concentration of CO_2, reduces its flux from the atmosphere.

Certain plants, known as C_4 plants, have a special CO_2 concentration mechanism. In such plants photosynthesis takes place primarily in specialized cells (bundle sheet cells) covered by parenchymatous cells that incorporate CO_2 into phosphoenol-pyruvate to give oxalacetate, catalyzed by the enzyme phospho-enol-pyruvate carboxylase or PEP carboxylase. Oxalacetate is reduced to malate, which is then transported to the bundle sheets, where it serves as a source of CO_2. PEP carboxylase has a greater affinity for CO_2 than Ru-D-P carboxylase, and since the bundle sheets are not in contact with air, photorespiration is reduced to zero or a very low value (Hatch and Osmond 1976).

Given the low concentrations of CO_2 in air (0.03 percent) but the much greater concentration differences in water vapor between leaf mesophyll and air (which can be as high or higher than 50 percent in desert regions), many more molecules of water vapor are lost than molecules of CO_2 gained per unit time. For an ordinary C_3 plant the ratio of H_2O vapor lost to CO_2 gained is in the order of 400 to 1000 (Kluge and Ting 1978; Mooney. 1972; Shantz and Piemeisel 1927). Because of the greater carboxylation efficiency of C_4 plants the ratio in these plants is on the order of 300 (Shantz and Piemeisel 1927; Black 1973). The advantages of C_4 photosynthesis in warm deserts in terms of water use efficiency is obvious.

C_4 plants differ in many biochemical and physiological features from C_3 plants, such as lack of oxygen inhibition, high light saturation values, and higher absolute photosynthesis in the range of 30 to 45 °C as contrasted to C_3 plants that have a large range but with the bulk having optimal temperatures in the range of 15 to 25 °C (Hesketh and Baker, 1969; Larcher 1969; Mooney 1972). This difference in optimal temperature, together with the other physiological characteristics, makes C_4 plants well suited for hot, bright desert environments, particularly in summer (Teeri 1979). However, evergreen or primarily winter-active C_3 desert plants can be as productive on a yearly basis as C_4 plants (Szarek 1979).

Caldwell et al. (1977) compared the productivity of two monospecific communities dominated by a C_3 plant (*Ceratoides lanata*) and a C_4 plant (*Atriplex confertifolia*) in Utah. Summers are dry while spring is wet in these communities. During the spring the photosynthesis-transpiration ratio of the two species was similar, while during the summer months the C_4 species *A. confertifolia,* had a superior photosynthesis-transpiration ratio. Since maximum photosynthesis occurred during the wet spring *C. lanata,* by starting growth slightly earlier in the season, had about the same annual ratio of carbon fixation to transpirational loss.

The opening of the stomata is dependent on leaf hydrature. Consequently whenever the plant is under water stress, stomata will close. Since evaporation is the principal mechanism for dissipation of heat, stomatal closure results in an increase in leaf temperature. The extent of the increase depends on a number of factors: air temperature and humidity, wind speed, incident energy flux, and leaf boundary resistance, which in turn is a function of leaf shape, especially its diameter (Gates 1962; Taylor 1975). Of these factors leaf shape is the only plant variable. Since energy flux and air temperature are likely to be high in deserts in times of drought, desert plants that are exposed to periods of water stress tend to have narrow leaves (Gates et al. 1968; Orians and Solbrig 1977; Givnish 1979) although the relation is complex. Narrow leaves mean decreased leaf surface, which must be compensated by an increase in the number of leaves, thereby increasing the unit cost of production of photosynthate per unit surface (Orians and Solbrig 1977; Givnish 1979). The increased cost must be balanced against the benefit of future photosynthesis. Clearly no photosynthesis takes place while the stomata are closed, while respiration proceeds. Assuming that the plant behaves in an economically optimal way, the added construction costs must be balanced by an equal gain in net photosynthate production (Orians and Solbrig 1977; Solbrig 1980).

Heat load is affected by the angle of the leaf in relation to the sun. Leaves held horizontally intercept the greatest light flux but also experience the greatest heat load. The opposite is true for leaves that are held vertically. Since deserts are characterized by the steepness of the light increase and decrease at sunrise and sunset, vertically held leaves still will intercept significant amounts of light while minimizing heat load during the hot midday hours. An extreme example is provided by *Larrea cuneifolia* of the Monte Desert of Argentina. This plant, locally known as the compass plant, produces all of its leaves in one plane—that is, held vertically in a north-south direction—maximizing morning and evening light intersection.

In extreme environments over long periods of time, soil water potential may be too low for any plant root to function. In such environments, as for example in parts of the Atacama Desert in Chile, survival during the drought period is possible only by a decoupling of the organism's water budget from the environment. One way is for the plant to survive as seeds. Seeds require a minimum of gas exchange and therefore can conserve some moisture. The other adaptation is succulence coupled with CAM (crassulacean acid metabolism) photosynthesis.

Many desert plants are succulent. Over 300 species in 109

genera, belonging to 18 different families (Kluge and Ting 1978), are known to have the special type of photosynthetic mechanism, known as CAM. The most important of these families are the Cactaceae and Bromeliaceae in the New World deserts and the Euphorbiaceae, Crassulaceae, and Aizoaceae in the Old World. All CAM species are succulent, although not all succulent plants have CAM photosynthesis. What characterizes CAM photosynthesis is an ability to incorporate CO_2 into organic acids in the dark (during the night)— primarily malate and aspartate with oxalacetate functioning as an intermediate. According to Kluge and Ting (1978) the reaction catalyzed by PEP carboxylase, is at follows:

$$\text{Phosphoenolpyruvate} + CO_2 \rightarrow \text{oxalacetate} \begin{array}{l} \nearrow \text{malate} \\ \searrow \text{aspartate} \end{array}$$

In turn, malate and aspartate (and all other organic acids produced in smaller amounts such as citrate and isocitrate) serve as CO_2 donors during the daytime. In this way, during the day ordinary C_3 photosynthesis can proceed with the stomata closed using the CO_2 donated by the organic acids. CAM plants can incorporate atmospheric CO_2 at night, when temperature and water vapor deficits are such that water vapor losses will be minimized. In addition CO_2 produced by respiration can be fixed by CAM plants even with the stomata closed. Some CAM plants can photosynthesize during the day in a regular way following the C_3 Calvin pathway, as well as recirculating respiratory carbon, while others apparently cannot. Those species that rely exclusively on the CAM pathway benefit from the very high water use efficiency of this system but also show the lowest net rates of any photosynthetic system (Mooney 1972; Kluge and Ting 1978). This does not apply to plants with a mixed CAM-C_3 system, which may show high productivity rates (Bloom 1979). However such plants normally do not grow in deserts.

Cactaceae and other succulents growing in desert regions can survive extensive drought periods by becoming almost totally decoupled from the environment. Succulents, especially those with succulent stems such as columnar cactii, store water in their tissues. During drought periods, when soil water potentials are very negative, such plants could lose water to the soil by diffusion. In order to prevent such losses, these plants slough off their active roots while the primary and secondary roots become suberized. Such plants can survive extensive droughts, extending for more than a year, with their stomata closed and their roots nonfunctioning, recirculating respiratory CO_2, and using light energy for maintenance and with a minimal loss of water through the cuticle.

In summary, the inevitable trade-off between the ability to sustain high photosynthetic rates and high water use efficiency in desert plants results in a tendency toward two diametrically opposite viable solutions. The first is acquiring the ability to photosynthesize at a high rate, at the cost of a very low water use efficiency. This solution is exemplified by desert ephemerals that have very high photosynthetic rates, as well as very high transpiration rates. Plants with this photosynthetic strategy can function only during periods of high moisture availability. The second solution is acquiring the ability to utilize water efficiently, at the cost of a low photosynthetic capacity. Succulents with CAM photosynthesis are an extreme example of this solution. Not all desert plants can be easily classified into these two classes; many are intermediate in both water use efficiency and photosynthetic ability, especially in areas such as the Sonoran Desert where droughts are not extreme. However as aridity increases, the survival of plants with intermediate strategies becomes increasingly more precarious, and consequently the percentage of ephemerals and succulents increases (Orians and Solbrig 1977).

MINERAL NUTRITION

The very high photosynthetic rates of desert ephemerals are not an automatic consequence of lowered leaf resistances, although low resistance to CO_2 flux is a necessary requirement. A well-developed photosynthetic apparatus as well as a high flux of CO_2 is also needed. An important additional factor is a high level of carboxylating enzyme (Mooney 1972), which in turn necessitates high nitrogen availability (Mooney and Gulmon 1979).

Mineral requirements of desert plants are apparently no different from those of mesic plants, nor is the distribution of soil minerals in desert soils inherently different from the pattern encountered in humid regions. However, since plants can only absorb minerals that are in solution, low soil moisture in desert soils presents a special problem of mineral availability even when minerals are abundant. Of the soil minerals, nitrogen is probably the most important one for the economy of the plant.

Total nitrogen in desert soils tends to be low. West and Skujins (1978) gives figures of 250 to 850 gm^{-2} for U.S. deserts. Much of the nitrogen is absorbed to the clay fraction where it is largely unavailable for plants. Leaching of nitrogen is low in desert soils due to lack of rain. Nitrogen tends to be concentrated in the upper part of the profile. Horizontal patterns are very striking, with spots of nitrogen accumulation and availability that coincide with the vegetation (Charley and West 1975; Tiedemann and Klemmendson 1973).

The low availability of minerals and of nitrogen in particular has an effect on photosynthesis through its effect on enzyme levels especially carboxylase (Mooney and Gulmon 1979). These authors have addressed themselves to the question of why different species have different content of carboxylase in their leaves. Since photosynthetic rate is directly proportional to leaf nitrogen (figure 9-3), it could be assumed that increasing leaf nitrogen is always advantageous. However, as Mooney and Gulmon (1979) show, increasing the content of carboxylating enzyme is favored only while the net marginal gain ($dP/$

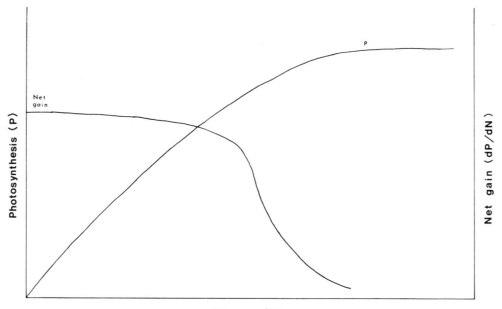

Figure 9-3. Hypothetical rate of photosynthesis versus leaf nitrogen content in a desert plant.
The net gain in photosynthesis per unit N investment (dP/dN) falls to zero with light saturation. From Environmental and evolutionary constraints on the photosynthetic characteristics of higher plants by H. N. Mooney and S. L. Gulman, in *Topics in Plant Population in Biology*, edited by O. T. Solbrig et al, Columbia University Press, New York, 1979.

$dE - dC/dE$, where P = photosynthesis, E = enzyme, and C = cost of leaf formation) is positive (figure 9-4). Factors affecting leaf cost are the actual construction costs in carbon and energy, the cost of root needed to supply water to the leaf and to extract nitrogen from the soil, and the cost of defending a leaf against predators, which increases with nitrogen content of the leaf. One factor that increases leaf cost is aridity, since it takes a greater investment of energy and materials per unit leaf surface to provide the same amount of water and nitrogen to a xerophytic than a mesophytic leaf. We therefore expect ephemeral desert plants that live in the wet part of the year to have a higher content of carboxylating enzyme than xerophytic perennials. There are insufficient data to test this prediction, but in a study of the Death Valley desert ephemeral *Camissonia claviformis,* Mooney and collaborators (1976) found very high photosynthetic rates and carboxylating enzymes. Another example is represented by *Encelia californica,* a winter-active deciduous shrub from the coastal sage association in California, and *E. farinosa* from the Sonoran Desert, which have higher photosynthetic rates and carboxylating enzyme levels than most C_3 plants measured to date (Ehrelinger and Bjorkman 1978). Ephemerals may also invest less in predator defense, relying on their unpredictability in time and space (Rhoades and Cates 1976). If so, this would also cut down the leaf cost for ephemeral and increase the net marginal gain.

The implications of mineral nutrition on desert plant adaptations have just begun to be investigated. Still needed are better baseline data on the distribution of minerals in desert soil, on mechanisms of absorption of minerals and data regarding the possibility of different strategies of mineral absorption, and on the effect of different levels of mineral nutrition on production and fitness.

PLANT FORM AND STRUCTURE AS ADAPTATIONS TO DESERT CONDITION

Natural selection acts on phenotypes and only indirectly on individual plant structures or processes such as leaf shape or photosynthetic performance. Nevertheless, photosynthetic processes furnish the necessary carbon and energy for growth and maintenance of the plant body. The form and general adaptive strategy of the species are therefore constrained by the type of leaf, root, and stem structure it possesses, and in turn, plant form restricts the organ morphology and physiology that the species can have.

The desert environment is characterized by high insolation, low air humidity, and uneven water availability. In subtropical high-pressure deserts such as the Sonora, Chihuahua, and Mojave deserts in North America and the northern Monte and Atacama deserts in South America, average to maximum

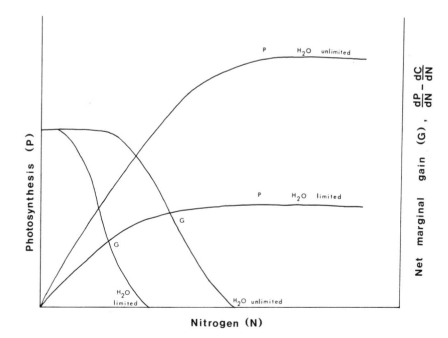

Figure 9-4. Hypothetical light-saturated rates of photosynthesis of plants versus leaf nitrogen in water limited and unlimited habitats.
The net marginal gain varies in those two habitats reaching zero earlier in water limited habitats. From Environmental and evolutionary constraints on the photosynthetic characteristics of higher plants, H. N. Mooney and S. L. Gulman, in *Topics in Plant Population in Biology,* edited by O. T. Solbrig et al, Columbia University Press, New York, 1979.

summer day temperature can be in excess of 30°C, and winter frosts are rare; in more temperate rain shadow deserts and in high elevation deserts such as the central basin deserts of North America or those of central Soviet Union, air temperatures are much lower. The problem is how to maximize the conversion of light energy into chemical energy given prevailing humidity and temperature regimes. To understand the basic life forms of desert plants, soil, water, and topography also need to be taken into account.

Surface flow of water derived from rainfall within the desert is very sporadic but quite often torrential. On flat surfaces, rainfall leads to unconcentrated flow over large areas, with thin and extensive sheets of moving water known as sheetflood. Sudden storms produce abrupt and sometimes violent flooding of streams—the well-known flash floods. As water flows down streambeds, it is lost at varying rates by evaporation and infiltration. These losses are usually sufficiently high so that little water reaches terminal basins and very seldom do they leave the desert areas.

Underground water remains available for much longer intervals of time than surface water. Following a rainfall, the soil column saturates from the top down as the water moves downward under the pull of gravity, although some water is retained near the surface by capillarity and related factors. The amount held varies inversely with the size of the pore spaces of the soil materials, being minimum for coarse-grained deposits and maximum for the fine-grained materials. Below the zone of soil moisture at most places there is a comparatively dry zone extending to depths of 5 to 100 m, followed by a lower zone saturated with water, referred to as the water table if the water is free. Within the dry zone, local and discontinuous bodies of perched water may be held up by impervious beds. A special case of this is the water stored in the channel fill of water courses trenched in bedrock.

The yearly amount and distribution of rainfall in desert areas varies greatly. There are areas, such as the Atacama Desert in northern Chile, where the average yearly rainfall is below 10 mm, while other areas such as Tucson, Arizona, receive an average in excess of 300 mm a year. All desert regions, however, share two characteristics: precipitation is concentrated in one or two short seasons, and precipitation is very unpredictable from year to year.

The depth of the saturated layer of soil following a rainfall depends on amount of rainfall, soil particle size, and runoff, which is determined primarily by slope and the rate of percolation. The finer the soil particle size, the more water is retained in the upper layers, and the slower the rate of percolation. After cessation of rainfall, water continues to move in the soil column, but at a slower rate, depending on the soil conductivity. Water also evaporates from the upper soil layers. Since more water is retained in the upper layers of fine-textured soil, they become alternatively wetter (during rainfall) and drier (during drought) than coarser-textured soils. However, if sufficient water accumulates to saturate the water column for considerable depths, more total water can be retained in the fine soil.

Uplands are drier than basins because on hilltops the soil is very shallow and formed by coarse elements with very little water retention capacity. Flat areas in valley bottoms where soils tend to be composed of fine-textured elements retain more of the rainwater but become saturated only in their upper layers. As this superficial water is lost through evaporation, these soils become very dry during the arid season. Water that moves to lower horizons by conduction and percolation is available to plants over longer periods. Alluvial fans, particularly in their middle and upper portions, retain a larger portion of rainfall at intermediate and low depths because their semicoarse texture permits more percolation.

Soil water availablility varies therefore according to topography, even when rainfall is uniform, which is seldom the case. Rainfall, especially summer rainstorms, is notoriously local and unpredictable. High, rocky areas are inherently driest; low-lying areas with fine-textured soil have good water retention in the upper layers of the soil but tend to become very dry during drought; hillsides and alluvial fans are very diverse in water availability due to the great diversity of microtopographic features; streambeds and drainage basins have the wettest soils. These differences in water availability are reflected in the distribution of plants with different life forms.

In effect, areas where environmental factors take extreme values will be inhabited by species that have specialized life forms, while areas with relatively more moderate environments will be inhabited by species that are less specialized but presumably better competitors and more productive. The theoretical reason for this trade-off is given by Orians and Solbrig (1977). Briefly a species cannot be both highly productive under conditions of high and low water availability due to the inevitable coupling of CO_2 flux (needed for production) and water vapor loss.

The life form of a plant—whether annual, perennial, herbaceous, woody, or succulent—and the characteristics of its roots, stems, and leaves are presumed to be adaptations to special conditions within a desert. The diversity of life forms in certain areas can be considerable (for example, in the Sonoran desert) or they may be fairly low (for example, in the Atacama Desert in Chile or the central Asian steppes). In general, life-form diversity decreases with decreasing rainfall and increases with an increase in topography, but no good quantitative studies have been conducted. Life forms can be classified into four major forms that represent four major classes of adaptation or strategies. According to Solbrig and Orians (1977) the four principal desert life forms are the drought evaders, the drought-enduring evergreen plants, the phreatophytes, and the succulents.

The drought evaders are mostly ephemeral species, although some are perennial. Their major adaptation is an ability to predict accurately the wet season and to restrict their carbon and energy harvest and their reproductive activities to the wet part of the year. Since they are not restricted by water, since they grow only when water is relatively abundant, they can exploit fully the favorable light and temperature conditions of the desert. For two annuals of the North American deserts, *Camissonia claviformis* and *Machaeranthera (Haplopappus) gracilis,* photosynthetic rates of approximately 70 and 60 milligrams per square decimeter per hour have been measured (Mooney et al 1976; Szarek and Woodhouse 1979), which are among the highest photosynthetic rates in any known C₃ plant. Desert ephemerals show special physiological adaptations to the desert environment as Szarek (Szarek and Woodhouse 1979) points out, but they are not of the type one usually associates with desert plants. Instead they are adaptations that enable these plants to function efficiently in the high light and high temperature of the desert.

The ephemeral growth habit represents an extreme adaptation to drought—avoidance. They are most frequent in areas with extreme drought, such as the Atacama Desert in Chile and Death Valley in California, and on microtopographic scale where soils are driest, (fine-textured soils). Ephemerals spend much of their life cycle in the dormant seed stage and show elaborate mechanisms of seed dormancy and staggered germination. Cohen (1967) has presented theoretical models that predict seed behavior in desert ephemerals.

Drought-enduring evergreen plants are mostly shrubs, although some small trees also belong to this category. Because these plants function throughout the year and are opportunistic in their carbon-fixing strategies, these plants show a great variety of morphological features that can be interpreted as devices to cut down transpiration. For example, many drought-enduring evergreen shrubs have leaves covered with a thick layer of waxes or resins, presumably to cut down cuticular water loss. A fine example is the cresote bush (*Larrea tridentata*) of the Sonoran Desert. These resins can also function as antiherbivore substances (Orians et al. 1977). Another common specialization is found in the form and shape of stomata that often are either sunken or raised in such a way as to increase the diffusion path of the gases. Another common adaptation is the presence of a thick indumentum of hairs. Although these hairs increase the boundary layer resistance somewhat, their primary function appears to be to reflect incident light, thereby cutting down heat load (Ehrelinger and Bjorkman 1978; Ehrelinger, Bjorkman, and Mooney 1976). Evergreen desert shrubs often are aphyllous (for example, *Cassia aphylla* of the Monte Desert of Argentina) or have ephemeral leaves present only during the more humid part of the year (such as *Cercidium microphyllum* of the Sonoran Desert). Those that are leafy have small, narrow leaves or finely divided compound leaves as in species of *Acacia*. All these specialized adaptations decrease the rate of evaporation during periods of water stress and reduce the heat load at times when stomata are closed. However, even with these adaptations, these plants could not function were it not for additional physiological and biochemical cell characteristics that enable

these plants to function even when they are subjected to water potentials that can be as low as −70 bars (Mabry, Hunziker, and DiFeo 1977). The ability of drought-enduring evergreen plants to sustain very low cell water potentials enables them to extract water held in intermediate layers (50-100 cm) in desert plants (Solbrig and Orians 1979). They can therefore photosynthesize during most of the year at a low rate. However, because of the inevitable trade-offs, the maximal photosynthetic rates of these plants are lower than those of drought-deciduous shrubs. This life form is dependent on some soil water for most of the year.

For evolutionists and ecologists interested in adaptation, drought-enduring evergreen plants are the most interesting among desert plants because of the large number of morphological and physiological specialized characterizatics. Nevertheless they represent an intermediate strategy that is neither as specialized to function under drought conditions as are succulents nor as productive under humid conditions as are ephemerals. Representing an intermediate situation between these extremes, it is not surprising that drought-enduring evergreen species are so diverse and hard to define.

Drought-enduring evergreen shrubs constitute the most common life form in semidesert regions, their abundance and importance decreasing as drought increases. Within an area, they can be found in all types of soils and topography but are most common on hillsides where soil diversity and number of microniches are higher.

Phreatophytes essentially are not different from drought-enduring evergreen plants in their basic life strategy. Drought-enduring evergreen shrubs can photosynthesize throughout most of the year, extracting water from intermediate soil layers held with relatively high matrix forces. By virtue of their long roots, phreatophytes can extract perched water below streambeds. This water can be plentiful or not, depending on the underlying substratum and the water flow during the wet season; water storage will be highest in large streambeds underlaid by impervious rock and minimal under small drainage ditches flowing over sandy soils. Consequently phreatophytes vary in their structure from those found exclusively in the first type of environment, such as *Populus grandifolia* of the Sonoran Desert, which shows no special morphological adaptation, to species such as *Cercidium floridum* of the Sonoran desert or *Bulnesia retama* from the Argentine Monte, which share many characteristics with desert shrubs. Basically phreatophytes tend to be trees or large shrubs, with deep roots found along streambeds (figure 9-5). Ohmart chapter 10 discusses riparian ecosystems.

An example of reduced transpiration is offered by *Prosopis glandulosa*, a phreatophytic leguminous tree native to the western Sonoran Desert. In studies performed in Death Valley, California, Mooney (1977) showed that as the vapor pressure deficit in the air increases, a point is reached where stomata close. Stomatal closure also results, of course, in a drastic reduction of photosynthesis. Since vapor pressure deficit in the

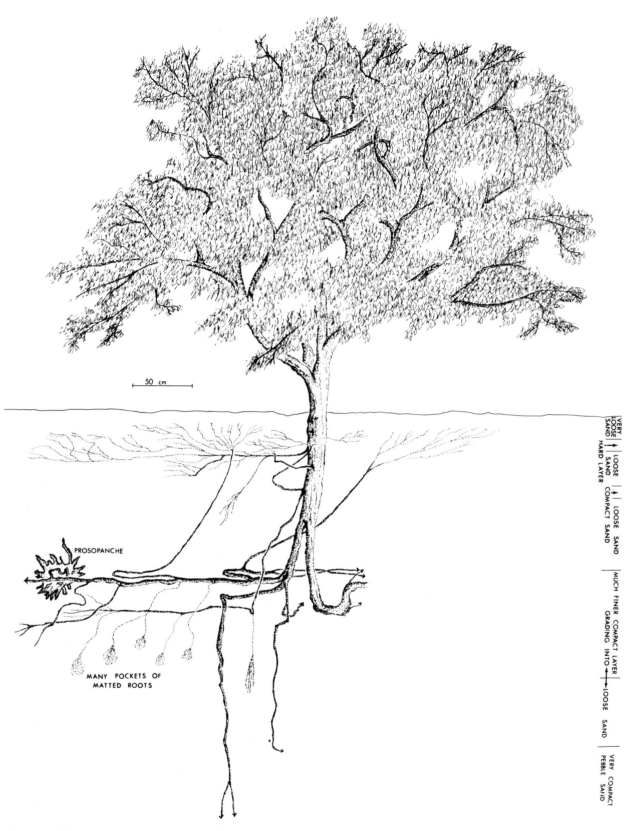

Figure 9-5. Diagramatic representation showing the development of both a superficial and a deep system of roots.
The diagram shows the tree of *Prosopis chilensis* growing in Adalgala, Catamarca.

air is lowest in early morning and maximal at noon, the midday closure of the stomata in *Prosopis glandulosa* and species with similar mechanisms enhances water use efficiency considerably, because as carbon is allowed to diffuse into the leaf only during periods of lowest potential water loss, the ratio of carbon dioxide fixed to water vapor lost is increased.

Drought-enduring succulents plants possess special morphological and physiological adaptations—most notably CAM photosynthesis and succulent stems that allow them to maintain an active metabolism even when no absorption of water by the roots can take place. Energy for maintenance is obtained by reutilizing respiratory carbon, which allows succulents to photosynthesize during the entire year but results in very low rates of carbon gain compared to other desert plants, even under the most favorable conditions. Succulents obtain their water from superficial soil layers after heavy rains, swelling as they store it in their tissues. During dry periods, permeable root hairs are shed, thereby reducing water loss to the soil, but new roots must be regrown before water can be taken in.

An example of a succulent desert plant is *Agave desertii*, a common plant of the Sonoran and Mojave deserts of the United States and Mexico. The plant has shallow roots, and during the dry season it opens its stomata only at night (Nobel 1977). However, as soon as the soil water potential at 10 cm depth becomes less than -3 bars, stomata open during the day and the stomatal resistance is less than 20 sec cm^{-1}. Such a soil water potential would correspond to a rainfall of approximately 7 mm. Transpiration ratios (mass of water transpired/mass of CO_2 fixed) of *Agave desertii* shows extremely low values of 25 for the entire year.

Desert succulents with CAM photosynthesis are truly adapted to growth in the desert because they are able to survive in an active metabolic state an entire year, and sometimes more, without soil water. They pay the penalty of being the least productive of all desert plants. Furthermore, they face the problem of heat dissipation during hot and dry seasons. Spines, reflective hairs, waxes, specialized morphologies, and orientations of branches and stems represent some of the many adaptations shown by these plants to decrease heat load. Some of these structures, especially spines, and alkaloids and other substances that make tissues unpalatable decrease herbivore damage.

Seedling establishment presents a particularly difficult problem for these plants. The soil surface in desert regions is a particularly inhospitable thermal microclimate, with maximum noon temperatures and minimum night temperatures. But given the low growth rate of succulents, seedlings of many species spend many years with all of their biomass close to the surface. Many species, such as the saguaro (*Carnegia gigantea*) in Arizona, cannot become established in the open ground but require the shade of other so-called nurse plants (Steenbergh and Lowe 1977). Another problem faced by succulents, especially the large columnar species, is an inability to withstand frost because of their high water content.

REFERENCES

Bamberg, S.; Wallace, A.; Kleinkopf, G.; and Vollmer, A. 1973. Plant productivity and nutrient interrelationships of perennials in the Mojave Desert. *IBP Desert Biome Report* RM73-10:1-52.

Bannister, P. 1976. *Introduction to physiological plant ecology.* New York: John Wiley.

Black, C. C. 1973. Photosynthetic carbon fixation in relation to net CO_2 uptake. *Ann. Rev. Plant Phys.* 24:253-286.

Bloom, A. 1979. Crassulacean acid metabolism in *Mesembryanthemum crystallinum.* Ph.D. dissertation, Stanford University.

Boyer, J. S., and Bowen, B. L. 1970. Inhibition of oxygen evolution in chloroplasts isolated from leaves with low water potentials. *Plant Physiol.* 45:612-615.

Breckle, S. W. 1974. Wasser- and Salzverhältnisse bei Halophyten der Salzsteppe in Utah/USA. *Ber. Deutsch. Bot. Ges.* 87:589-600.

Caldwell, M. M.; White, R. S.; Moore, R. T.; and Camp. L. B. 1977. Carbon balance, productivity, and water use of cold-winter desert shrub communities dominated by C_3 and C_4 species. *Oecologia* 29:275-300.

Charley, J. L., and West, N. E. 1975. Plant-induced soil chemical patterns in some shrub-dominated semi-desert ecosystems in Utah. *J. Ecol.* 63:945-963.

Cohen, D. 1967. Optimizing reproduction in a randomly varying environment. *J. Theoret. Biol.* 16:1-14.

Cunningham, G. L., and Burk, J. H. 1973. The effect of carbonate deposition layers ("caliche") in the water status of *Larrea divaricata. Am. Midl. Natur.* 90:474-480.

Ehrelinger, J. R., and Björkman, O. 1978. A comparison of photosynthetic characteristics of *Encelia* species possessing glabrous and pubescent leaves. *Plant Physiol.* 62:185-190.

Ehrelinger, J.; Björkman, O.; and Mooney, H. 1976. Leaf pubescence: Effects on absorptance and photosynthesis in a desert shrub. *Science* 192:376-377.

Evenari, M.; Shanan, L.; and Tadmor, N. 1950. *The Negev: The challenge of a desert.* Cambridge: Harvard University Press.

Gates, D. M. 1962. *Energy exchange in the biosphere.* New York: Harper and Row.

Gates, D. M.; Alderfer, R.; and Taylor, S. E. 1968. Leaf temperature of desert plants. *Science* 159:994-995.

Givnish, T. 1979. On the adaptive significance of leaf form. In *Topics in plant population biology,* ed. O. T. Solbrig et al., pp. 375-407. New York: Columbia University Press.

Halvorson, W. T., and Patten, D. T. 1974. Seasonal water potential changes in Sonoran Desert shrubs in relation to topography. *Ecology* 55:173-177.

Hatch, M. D., and Osmond, C. B. 1976. Compartmentation and transport in C_4 photosynthesis. In *Transport in Plants,* ed. C. R. Stocking and V. Heber. *Encyclopedia of Plant Physiology* 3:144-184.

Hesketh, J., and Baker, D. 1969. Relative rates of leaf expansion in seedlings of species with differing photosynthetic rates. *J. Ariz. Acad. Sci.* 5:216-221.

Hsiao, T. C. 1973. Plant responses to water stress. *Ann. Rev. Plant Physiol.* 24:519-570.

Itai, C., and Benzioni, A. 1976. Water stress and hormonal response. In *Water and plant life,* ed. O. Lange et al., pp. 225-240. Berlin: Springer.

Kearney, T. H., and Schantz, H. L. 1911. The water economy of dry land crops. *Yearbook of agriculture.* 10:331-366.

Kluge, M., and Ting, I. P. 1978. *Crassulacean acid metabolism.* Berlin: Springer.

Lange, O. L.; Kappen, L.; and Schulze, E. D. 1976. *Water and plant life.* Berlin: Springer.

Larcher, W. 1969. The effect of environmental and physiological variables on the carbon dioxide gas exchange of trees. *Photosynthetica* 3:167-198.

Levitt, J. 1972. *Responses of plants to environmental stresses.* New York: Academic Press.

Mabry, T.; Hunziker, J. H.; and DiFeo, D. R., Jr. 1977. *Creosote bush.* Stroudsburg, Pa.: Dowden, Hutchinson & Ross.

Maximov, N. A. 1929. *The plant in relation to water.* New York: Macmillan.

Mooney, H. A. 1972. Carbon balance of plants. *Ann. Rev. Ecol. Syst.* 3:315-346.

Mooney, H. A.; Ehleringer, J.; and Bery, J. 1976. High photosynthetic capacity of a winter annual in Death Valley. *Science* 194:322-323.

Mooney, H. A.; Simpson, B. B.; and Solbrig. O. T. 1977. Phenology, morphology, physiology. In: *Mesquite,* ed. B. B. Simpson, pp. 26-43. Stroudsburg, Pa.: Dowden, Hutchinson & Ross.

Mooney, H. A., and Gulmon, S. L. 1979. Environmental and evolutionary constraints on the photosynthetic characteristics of higher plants. In *Topics in plant population biology,* ed. O. T. Solbrig et al., pp. 316-337. New York: Columbia University Press.

Morello, J. 1955. Estudios botánicos en las regiones áridas de la Argentina. I. Ambiente, morfologia y anatomia de cuatro arbustos resinosos de follaje permanente del Monte. *Rev. Agron. Noroeste Arg.* 1:301-370.

Nobel, P. 1976. Water relations and photosynthesis of a desert CAM plant, *Agave deserti. Plant Physiol.* 58:576-582.

———. 1977. Water relations of flowering of *Agave deserti. Bot. Gaz.* 138:1-6.

Odening, W. R.; Strain, B. R.; and Oechel, W. C. 1974. The effects of decreasing water potentials on net CO_2 exchange of intact desert shrubs. *Ecology* 55:1086-1095.

Oechel, W. C.; Strain, B. R.; and Odening, W. R. 1972. Tissue water potential, photosynthesis, ^{14}C-labeled photosynthate utilization and growth in the desert shrub *Larrea divaricata. Ecol. Monogr.* 42:127-141.

Oppenheimer, H. R. 1960. Adaptation to drought: Xerophytism. In *Plant-water relationships in arid and semi-arid conditions,* pp. 105-138. Paris: UNESCO.

Orians, G. H., and Solbrig, O. T. 1977. A cost-income model of leaves and roots with special reference to arid and semi-arid areas *Amer. Nat.* 1:677-690.

Rhoades, D. F., and Cates, R. G. 1976. Towards a general theory of plant anti-herbivore chemistry. In *Biochemical interactions between plants and insects,* ed. J. W. Wallace and R. L. Man, pp. 168-213. New York: Plenum Press.

Shantz, H. L. 1927. Drought resistance and soil moisture. *Ecology* 8:145-157.

Shantz, H. L., and Piemeisel, L. N. 1927. The water requirements of plants at Akron, Colorado. *J. Agr. Res.* 34:1093-1190.

Schimper, A. F. W. 1903. *Plant geography upon a physiological basis.* Oxford: Clarendon Press.

Slatyer, R. O. 1967. *Plant-water relationship.* New York: Academic Press.

Slobodkin, L. B., and Rapoport, A. 1974. An optimal strategy of evolution. *Quart. Rev. Biol.* 49:181-200.

Solbrig, O. T. 1979. Life forms and vegetation patterns in desert regions. In *Arid land plant resources,* ed. J. R. Goodin and D. K. Northington, pp. 82-95. Lubbock: Texas Technical University.

Solbrig, O. T. 1980. Energy, information and plant evolution. (in preparation).

Solbrig, O. T., and Orians, G. H. 1977. The adaptive characteristics of desert plants. *Am. Scientist* 65:412-421.

Spalding, V. M. 1904. The creosote bush (*Covillea tridentata*) in its relation to water supply. *Bot. Gaz.* 38:122-138.

Stark, N., and Love, L. D. 1969. Water relations of three warm desert species. *Israel J. Bot.* 18:175-190.

Steenbergh, W. F., and Lowe, C. H. 1977. Ecology of the saguaro: II. Reproduction, germination, establishment, growth, and survival of the young plant. *National Park Service Monograph 8.*

Stocker, O. 1928. Der Wasseraushalt aegyptischer Wuesten-und Salzpflanzen. *Bot. Abhandl.* 13:1-200.

———. 1968. Physiological and morphological changes due to water deficiency. *Arid Zone Res.* 15:63-104.

Strain, B. R., and Chase, V. C. 1966. Effect of past and prevailing temperatures on the carbon dioxide exchange capacities of some woody desert perennials. *Ecology* 47:1043-1045.

Syvertsen, J. P.; Cunningham, G. L.; and Feather, T. V. 1975. Anomalous diurnal patterns of stem xylem water potentials in Larrea tridentata. *Ecology* 56:1423-1428.

Szarek, S. R. 1979. Primary production in four North American deserts: Indices of efficiency. *J. Arid Env.* 2:187-209.

Szarek, S. R., and Woodhouse, R. M. 1979. Ecophysiological studies of Sonoran Desert plants V. *Oecologia* 41:317-328.

Taylor, S. E. 1975. Optimal leaf form. In *Perspectives in biophysical ecology,* ed. D. M. Gates and R. B. Schmerl. New York: Sprinrer-Verlag.

Teeri, J. A. 1979. The climatology of the C_4 photosynthetic pathway. In *Topics in plant population biology,* ed. O. T. Solbrig et al., pp. 356-374. New York: Columbia University Press.

Tiedemann, A. R., and Klemmendson, J. U. 1973. Nutrient availability in desert grassland soils under mesquite (*Prosopis juliflora*) trees and adjacent open areas. *Soil Sci. Soc. Am. Proc.* 37:107-111.

Vieira da Silva, J. 1976. Water stress, ultrastructure and euzymatic activity. In *Water and plant life,* ed. O. L. Lange, L. Kappen, and E. D. Schulze. Berlin: Springer.

Wallace, A., and Romney, E. M. 1972. Radioecology and ecophysiology of desert plants at the Nevada test site. *USAEC Report* T10-25954.

Walter, H. 1950. *Einführung in die Physiologie.* I. *Die Grundlagen des Pflanzenlelbens.* Stuttgart: Ulmer.

West, N. E., and Skujins, J. 1978. *Nitrogen in desert ecosystems.* Stroudsburg, Pa.: Dowden, Hutchinson & Ross.

Yocum, C. S., and Lommen, P. W. 1975. Mesophyll resistances. In *Perspectives of biophysical ecology,* ed. D. M. Gates and R. B. Schmerl. New York: Springer.

Zelitch, I. 1971. *Photosynthesis, photorespiration and plant productivity.* New York: Academic Press.

10

NORTH AMERICAN DESERT RIPARIAN ECOSYSTEMS

ROBERT D. OHMART AND BERTIN W. ANDERSON

DESERT RIPARIAN ECOSYSTEMS have received little attention from ecologists. To our knowledge, no book on North American deserts has ever treated desert riparian areas as a distinct ecological system. In the late 1960s, Lowe (1968) appraised our overall knowledge of world deserts and ranked North American deserts as the best studied faunistically, but more than a decade later we still have little real knowledge of the total value of desert riparian ecosystems. To illustrate the importance of these ecosystems in North American deserts, we will pose a number of questions about desert riparian habitats and address them in this chapter. What are desert riparian ecosystems? Are they a separate ecological entity, and, if so, how do they differ from the remaining landscape, either physiographically or floristically and faunistically? How important are these areas to the desert fauna? Approximately what percentage of the arid Southwest is comprised of riparian habitats? What is their importance to man? Have these habitats been modified by man and, if so, to what extent? What is their current status and what does the future hold for them? Ultimately if these habitats are lost or almost completely eliminated, what would be the ecological impact on our North American deserts as a whole?

Desert regions of North America are characterized by aridity, high ambient temperatures, and generally a paucity of vegetation. Winter rains are usually gentle and widespread, whereas summer rains, if any, are convective in nature, usually of short duration, and result in highly localized and intensive precipitation patterns. Only a small amount of the summer rains penetrates the ground surface since the sparse vegetation and erosion pavement on the desert floor do little to slow surface runoff. Further, the high rate of evapotranspiration

typically exceeds precipitation by a substantial percentage (Logan 1968).

As precipitation collects and drains from the desert floor, its action cuts drainage ways. Consequently the desert regions are topographically characterized by water courses, termed washes or arroyos, that drain the adjacent desert uplands and eventually converge to form larger transport systems, which empty into primary or permanently flowing rivers. These rivers have their headwaters located in high mountain areas where they also drain high elevational watersheds.

Desert riparian ecosystems are comprised of these drainages, their attendant vegetation, and the fauna supported by these riparian plant assemblages. The drainage system itself may have permanently flowing water, be intermittent, or seldom (if ever) flow. Nevertheless, the available soil moisture is higher in these alluvial floodplains than in the adjacent desert uplands and hence supports a flora distinctly different from that in the adjacent desert. A working definition is, "A riparian association of any kind is one which occurs in or adjacent to drainageways and/or their floodplains and which is further characterized by species and/or life-forms different than that of the immediately surrounding non-riparian climax" (Lowe 1964:62).

Riparian vegetation is frequently termed phreatophytic, which denotes a collective group of plant species that have their roots located in perennial groundwater or in the capillary fringe above the water table. The term has a negative connotation among water managers and refers to those plant species that transpire large quantities of water from the groundwater table. Consequently phreatophytes are commonly viewed as undesirable, and their removal has been viewed as positive because it constitutes water salvage or a reduction in water loss.

PHYSICAL CONSIDERATIONS

The fluvial geomorphology of desert landscapes has been studied in some detail, but the paucity of long-term records of our changing landscape under natural and man-related conditions leaves many questions unresolved. In general, fluvial

*We want to express our gratitude to a number of individuals who helped make writing this chapter possible. Luanne Cali relentlessly typed the numerous copies of the manuscript; Kenneth V. Rosenberg and Scott Terrill worked many hours in preparing the avifaunal lists; Laurie Vitt helped in preparing the reptile and amphibian lists, and Jane Durham, Susan Cook, and Cindy Zisner provided invaluable editing help. We are extremely grateful for the help given by these people.

systems are considered open, being characterized by input, cycling, and output of energy and materials, with component variables considered to be self-regulated (Cooke and Warren 1973). Schumm and Lichty (1965) have classified ten component variables (including time, initial relief, geology, and climate) of fluvial systems as either independent or dependent, with the interrelationships varying according to time span. They list three time spans: cyclic (encompassing an erosion cycle), graded (when grade and condition of dynamic equilibrium exists), and steady (a fraction of graded time). Of the ten variables, vegetation is the most germane to our discussion and is classified as dependent on cyclic time and independent relative to graded and steady time. The length and periodicity of cyclic time primarily dictates the amount of available soil moisture in flumes, which in turn exerts a strong influence on riparian plant species composition.

Surface Runoff

Water entering small drainage systems originates primarily from surface runoff from the adjacent desert upland and secondarily from percolated groundwaters. The size of the watershed and the amount of precipitation dictate the amount of surface flow and percolated waters that enter the drainage during each erosion cycle. In general, the amount and type of vegetational ground cover and the slope of the terrain are directly related to the percentage of water that will enter the drainage system as surface flow, or as percolated water, the latter being highest when the watershed has a good perennial grass cover and shallow slopes.

Good watersheds have a high roughness coefficient. The force of falling raindrops is reduced before they hit the soil, and the vegetation retards the flow of the surface water, allowing more time for the water to penetrate the soil. This slower, decreased surface water flow reduces the erosion of topsoil. The relationship among rainfall, watersheds, geology, and surface versus percolated water is by no means simple. For a discussion of desert drainage basins, their constituent parts, and an in-depth analysis of some of these systems, see Melton (1958) and Schumm and Hadley (1961).

Surface flows carry varying amounts of sediment, minerals, and organic materials that precipitate or are filtered out in the soils of the water course. The deposited soils are termed alluvial and are rejuvenated with new soils, minerals, and nutrients each time there is adequate rainfall to produce surface runoff. These alluvial soils are generally restricted to stream fans and floodplains along stream bottoms (Denny 1965). The alluvial soils along permanent rivers can be enriched either locally by small flood pulses from adjacent uplands or along the entire reach of the river by floods from the melting snowpack in the upper watershed.

Drainage systems, whether having permanent or intermittent flow, transport enough surface and subsurface moisture to support a flora that is different in growth form and species composition from the adjacent uplands (figure 10-1). In very small drainages the soil moisture level may be too low to

Figures 10-1A, B, C. *Three types of riparian ecosystems in the Sonoran Desert.* Originals prepared by Lauren Porzer.

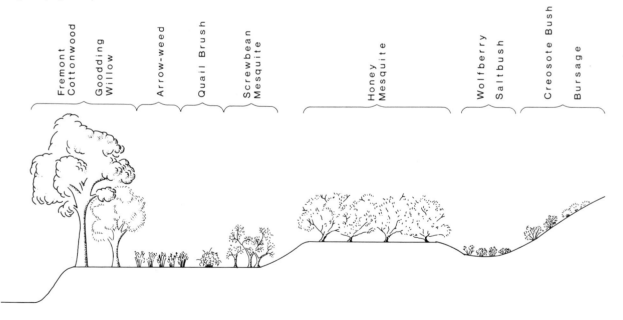

Figure 10-1A. *A permanent or primary drainage showing the position of the riparian species on the first and second terraces.*

support riparian species but will have higher moisture levels than will the adjacent upland. These drainages contain upland desert species that are usually more robust than their desert conspecies.

Soils

Desert uplands adjacent to riparian areas usually have poorly developed, shallow soils. Soil moisture varies greatly throughout the annual cycle but generally is very low. In contrast, the riparian soils, which are primarily alluvial in nature, are much deeper and highly variable in texture. The gentle surface runoff from winter rains transports more finely textured soils and small organic materials, such as leaves and rabbit and rodent droppings, whereas the heavier runoff from summer rains transports both coarse and finely textured soils and larger organic materials. The nature of alluvial soils is amply described by Dregne (1968:309):

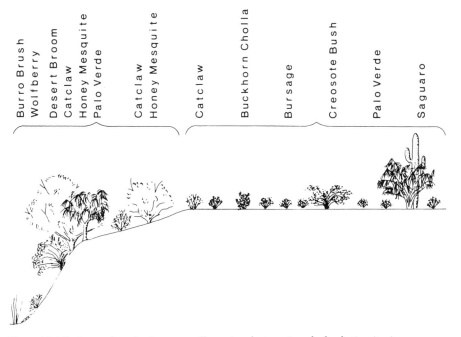

Figure 10-1B. A secondary riparian system illustrating the intrusion of a facultative riparian species, blue palo verde, into the riparian belt.

Figure 10-1C. A wide drainage system where an island of upland desert has either been cut away during periods of high water flow or alluvial soil deposits have built the island to an elevation where xerophytes have successfully invaded it.

Alluvial soils are commonly stratified, have textures that run the gamut from gravel to clay, and often are underlain by sand or gravel. Usually they are calcareous to the surface and contain moderate to low amounts of salt. Little or no development is apparent in them although they may contain moderate amounts of organic matter in the surface.

The high but fluctuating water table in the drainages generally contains enough soil moisture to allow decomposers to reduce the accumulated and transported organic materials to humus, which in turn provides soil nutrients and increased soil moisture holding capacity. The heterogeneous soil texture provides relatively good aeration, which promotes aerobic decomposition.

Flooding

In small riparian systems that drain desert uplands, flooding is highly erratic since the only input is from rainfall. Water enters the drainage as percolated or as surface flow. The magnitude of the riparian flow depends on the intensity and duration of the rainfall, the size of the watershed, and the width of the drainage system.

In permanent river systems, flooding is usually an annual occurrence, except during years of low snowpack on the watersheds. Depending on the depth of the snowpack and the duration of melt, the summer floods in larger systems can be of a relatively long duration, with peak flows usually occurring in June or July. Heavy flooding of long duration results in dramatic physical and biological changes. New channels are cut, and sharp river bends are frequently cut off to become ephemeral oxbow lakes. Previously formed oxbow lakes may be filled with transported sediment, boulders, and vegetation. Entire plant communities may be dislodged and washed away. As the flooding waters recede, new sediments rich in organic material are deposited to form the seedbeds for future plant communities.

The water in some of these streams may carry heavy loads of totally dissolved solids. Salt levels may be high enough to render the water unpotable. Input may come from the substrate over which the stream flows (Cole 1963; Sommerfield et al. 1974), from springs that feed into the river, and from agricultural drains with return flow into the river. The amount of salt input from each of these sources varies depending on the geological formations, the nature of springs that feed the system, and the amount of cultivated land along the stream course. For example, Feth and Hem (1963) estimated as high as 45,454 metric tons (50,000 tons) per year of dissolved solids being carried by the Salt River in central Arizona. Physical components such as timing of flooding, levels of salt, soil moisture availability, extreme winter temperatures, and duration of winter extreme temperatures appear to be some of the major factors that dictate plant species composition and vertical structure of the vegetation along riparian systems.

Drainage Types

North American desert drainage systems are a continuum ranging from mesic (with a permanently flowing river system) to xeric (where the stream channel seldom, if ever, carries surface flow).

PRIMARY DRAINAGES

For ease of discussion, we will term those drainages having headwaters located in mountain areas as primary drainages in which surface flow may be permanent or intermittent. Two or more primary drainages may converge to form a larger primary drainage; for example, of the Salt and Gila rivers in Arizona, the latter ultimately converges with the Colorado River.

SECONDARY AND LESS DRAINAGES

These are intermittent or nonsurface flowing systems that drain upland desert or semiarid areas and converge with primary drainage systems. Tertiary and quaternary systems drain smaller desert or semiarid watersheds and converge to form secondary drainage systems.

This very general classification allows us to visualize how major and minor drainages converge and transport water from desert areas. These drainage ways house the water- and wind-transported soil, which supports riparian-adapted plant species.

FLORAL CONSIDERATIONS

Riparian systems support a distinct flora, described by Lowe (1964:62) as "an evolutionary entity with an enduring stability equivalent to that of the landscape drainageways which form its physical habitat." Axelrod (1958) and Darrow (1961) have examined the origin and development of the vegetational communities of the Southwest, and Darrow has traced the phylogenetic relationships of the riparian vegetational complex.

Historical Development

The native tree and shrub genera that currently comprise our desert riparian ecosystems were derived from two different floras, the Arcto-Tertiary and Madro-Tertiary. The Arcto-Tertiary flora was a temperate mesophytic forest and was widespread over Canada, extending north to the Arctic Sea in the Early Cenozoic era. A number of genera were representative of this flora, but *Populus, Salix,* and *Atriplex* are some of the most important in desert riparian ecosystems. *Atriplex* became established in the central and northern Great Basin area as the forest and woodlands retreated.

The Madro-Tertiary flora appear to have evolved from subtropical species that persisted along dry margins in the tropics (Darrow 1961). Increasing aridity during the Oligocene epoch of the Cenozoic allowed strong differentiation of the Madro-Tertiary flora (Axlerod 1958), known first from fossil evidence in the Green River formation of the Eocene. Aridity continued to increase, and by the Late Cenozoic a number of important genera such as *Acacia, Baccharis, Cercidium, Condalia, Lycium,* and *Prosopis* were established in the Sonoran Desert.

By the end of the Cenozoic era vegetational patterns were well established in the desert regions of North America. Increasing aridity promoted the separation of the sagebrush of the Arcto-Tertiary flora in the Great Basin Desert and the development of the Mojave, Sonoran, and Chihuahuan deserts from the subtropical Madro-Tertiary flora. Some riparian-adapted species were not restricted from southward movement, and genera such as *Populus* and *Salix* penetrated southward into the Sonoran and Chihuahuan deserts. Why more of the Madro-Tertiary flora is not found in the Mojave and Great Basin deserts appears to be related to soil moisture availability and extreme winter temperatures. However, our knowledge of the requirements of these species is so imperfect that this is only speculation.

Fossil plants from the Pleistocene indicate an extension and reduction of floral communities rather than major shifts or extirpation of various plant species (Van Devender 1977). Riparian species from higher elevations may have descended to lower elevations along drainages, but as conditions became hotter and drier, these individuals died out, reducing the species' distribution to higher elevations.

Through evolutionary time, a number of plant species have become highly adapted to the rigors of living in riparian areas. Each species has its own limits relative to soil moisture availability, temperature, substrate type, and other physical characteristics. Systems with permanently high water tables and low to moderate salinities support such species as Fremont cottonwood (*Populus fremontii*), Goodding willow (*Salix gooddingii*), quail bush (*Atriplex lentiformis*), arrowweed (*Tessaria sericea*), screwbean mesquite (*Prosopis pubescens*), honey mesquite (*P. glandulosa*), and velvet mesquite (*P. velutina*).

Terracing

Shown diagrammatically in Figure 10-1a, the broad floodplain along permanent rivers (our example is drawn from the Colorado River) tends to contain two terraces. The first terrace may be 1 or more meters (m) below the level of the second terrace and is the primary area of river meanderings and water transport in normal flood years. Consequently, the first terrace continually undergoes change through natural bank cutting and soil deposition and from floodwaters cutting off oxbows to form new oxbow lakes. Older lakes with well-developed riparian communities slow the silt-laden floodwater, allowing the deposition of new soils to expedite hydric to xeric vegetational succession. Annual floods constantly changed the substrate along the first terrace, resulting in a heterogeneous assortment of first-terrace trees and shrubs.

FIRST TERRACE

At first the plant species occupying the first terrace in broad floodplains seem disorderly, but on closer examination, a relatively discrete separation of the available substrate by first-terrace plant species is seen (figure 10-1a). Willows and cottonwoods occupy the saturated soils lining the river's edge. The distribution of these trees depends on a number of factors, but an important one is the soil elevation above the water table. If soil elevation is too high, the air- and water-transported seeds of cottonwoods and willows cannot germinate. Sandy, water-saturated soils around newly formed oxbow lakes provide an ideal habitat for these rapidly invading species. Also, as the meandering river cuts into one tree-lined bank, the opposite shore is being formed as a sandy soil deposit. These new soils with a high water table are quickly invaded by cottonwoods and willows. Some bank areas may support either cottonwoods or willows as a monoculture, while in other areas the two species may occur as a mixed community.

Arrowweed forms extensive tracts parallel to the cottonwood and willow communities along higher portions of the first terrace and in slightly better drained soils. The roots are located only slightly below the soil surface and sprout to form dense stands (Gary 1963). These underground roots or rhizomes aid in rapid invasion of new habitats, along with the air- and water-transported seeds. The stems of this shrub may be from 2 to 3 m tall and straight as an arrow (they were used by early man for that purpose). This species usually forms a monoculture, and mature plants in a stand show little variation in height or stem size.

Farther from the river, the slower growing and maturing screwbean mesquite and quail bush occupy the remainder of the first terrace. Narrow or wide belts of screwbean mesquite flank either the arrowweed communities or the wider belts of cottonwood and willows. This slow-growing mesquite species appears to have been highly restricted as a small band in the more stable soils along the lateral edge of the first terrace. The distance to the water table is less than on the second terrace, and the more stable soils provide adequate time to allow the tree species to mature and produce fruit. The fruit of screwbean mesquite is corkscrew shaped and must be water or animal transported. Fruits are dropped year round, with a peak in early July. Rodents and coyotes (*Canis latrans*) are active dispersal agents. Well-developed spines on the trees appear to be a deterrent to browsing species such as mule deer (*Odocoileus hemionus*) and desert bighorn sheep (*Ovis canadensis*).

Quail bush is a shrub that may occur in a pure stand but usually forms a belt of scattered shrubs parallel to the river. A mature shrub seldom exceeds 2.5 m in height and may occupy an area of 10 square m. It is a dense evergreen that can tolerate and may prefer soils with a high salt content. Consequently low, moist areas with high salt levels some distance from the river's edge frequently support numerous shrubs of this species. The winged fruits are dispersed primarily by water or vertebrates.

Intermixed with the above dominants are such species as wild grape (*Vitis* spp.), wolfberry (*Lycium* spp.), inkweed (*Suaeda torreyana*), sprangletop (*Leptochloa fascicularis*), finger grass (*Chloris virgata*), careless weed (*Amaranthus palmeri*), and many others.

Marsh or wetland vegetation frequently can be found around the edges of backwaters, oxbow lakes, seeps, springs, and other slow-moving or standing water situations. Where surface water is shallow and relatively calm, the dominant emergent vegetation is usually cattail (*Typha domingensis*), but deeper slow-moving waters may support various species of bulrush (*Scirpus* spp.). Receding floodwaters annually leave cutoff oxbows replete. Those with high perennial water tables located near the river support mixtures of various species of emergent vegetation. Low areas with a high water table and saline conditions support a dense mat of saltgrass (*Distichlis* spp.).

SECOND TERRACE

Because of its increased elevation above the water table, the second terrace supports only those species with a highly developed and rapidly growing taproot that can follow the receding water table as flood levels wane. There is a more orderly appearance of the communities on this terrace, since the area is seldom cut by the river. The plant species occupying this terrace are slow-maturing and require relatively stable soils. Honey and velvet mesquite, which frequently form extensive forests, are the most highly adapted riparian species in this terrace. Plant species such as quail bush, wolfberry, and inkweed frequently become established in low areas along the secondary terrace where water may be trapped in flood periods or from adjacent upland runoff. Other more xerically adapted opportunists from the adjacent upland, such as creosote bush (*Larrea tridentata*) and xerophylic species of *Atriplex* and blue palo verde (*Cercidium floridum*), may invade these drier sites. But the primary riparian plant species along the second terrace are honey and velvet mesquite. These large-spined trees may reach a height of 6 to 8 m. The sweet pod surrounding the seed is dropped in July, and birds and mammals are the primary seed dispersers, but high floodwaters can also act as a dispersal agent.

Secondary and Lesser Drainages

Terracing, if it occurs, is usually poorly defined in secondary drainages. The lack of permanent surface flow, annual floods, and a broad alluvial floodplain does not promote terracing in secondary or lesser drainages. Along narrow secondary drainages, the riparian vegetation, if present, is forced back against the bedrock on the edge of the drainage. When surface flow does occur in narrow drainages, it is usually a high volume flow of short duration, and any vegetation that has become established in the middle of the drainage area since the end of the last erosion cycle is scoured from the wash. Broad secondary or lesser drainages usually have edges of riparian vegetation bordering them and may have islands of riparian vegetation scattered across the bed (figure 10-1c). Heavy floods frequently remove the vegetation and the islands, whereas gentle floods increase island size by depositing sediment on the downstream side.

Plant species composition is highly diversified on these islands, varying from a monoculture to a mixture of species. In many washes in the Sonoran Desert, facultative species such as palo verde and ironwood (*Olneya tesota*) may comprise a large percentage of the species composition. On older and more stable islands where soil deposition has been adequate to form higher and better drained soils, some of the adjacent desert species, such as creosote bush, may become established. Some of these islands are also the product of new channel formation, which has separated some of the adjacent desert vegetation. Riparian species then become established around the base of the island.

Lesser drainages that feed permanent riparian systems also support a distinct flora. In secondary drainages, cottonwood and willows may extend a short distance up the drainage but soon give way to other riparian plant species. In lesser drainages such as secondary, tertiary, and quaternary systems, plant species richness and vertical vegetation structure decline as compared to the primary drainage. Depending on the desert involved and many other factors, smaller riparian drainages may support honey mesquite, velvet mesquite, desert willow (*Chilopsis linearis*), catclaw (*Acacia greggii*), white thorn (*A. constricta*), seep willow (*Baccharis* spp.), brickel bush (*Brickellia californica*), Apache-plume (*Fallugia paradoxa*), burroweed (*Hymenoclea monogyra*), cheesebush (*H. salsola*), common saltbush (*Atriplex polycarpa*), rabbitbush (*Chrysothamnus paniculatus*), and others.

Some of the plant species from the adjacent desert upland frequently invade tertiary and quaternary drainage systems, as well as smaller drainages. The increased soil moisture level produces more robust individuals than conspecies on the adjacent upland. Examples in the Sonoran Desert are palo verde, ironwood, and saguaro (*Cereus giganteus*). Dick-Peddie and Hubbard (1977) have suggested differentiating between the true riparian forms as being obligate species and the opportunistic forms as being facultative species. Others have termed the latter as pseudoriparian species.

Salt Cedar

In the early 1800s an Old World plant species called salt cedar (*Tamarix chinensis*) was introduced into the United States for ornamental uses (Horton 1964). The species was preadapted to desert riparian systems, and in the intervening years it has spread throughout the deserts of North America. Graf (1978) has estimated the rate of spread in the Colorado Plateau region at about 20 kilometers (km) per year. Today salt cedar is the dominant species in many riparian plant communities. It is an aggressive species that invades rapidly on newly deposited alluvial soils, in many instances out-competing native willows and cottonwoods. Not only has it invaded primary drainages, but its soil moisture tolerance is such that it can also thrive in secondary and tertiary drainages. It matures rapidly and begins producing small wind- and water-transported seeds within a year. Robinson (1958) has estimated production at 600,000 seeds per plant per year. Because flowering and fruiting are not synchronous within a stand, seeds are produced for many months through the growing season.

Glands for excreting salt are located on the leaves, allowing the species to invade saline soils. The continual transport of salts to the leaves promotes salt accumulation on the soil surface, which deters the germination and growth of native species.

Salt cedar is deciduous and grows in dense stands. After fifteen to twenty years of establishment, the fuel level builds to such a point that fire is imminent. Heat from the fire kills the trunk portion of the mature tree, and within a few days suckers sprout from the root crowns. New growth is about 2 or 3 m the first year. Even flooding does not harm the stand, since all of the woody portions will develop adventitious roots if kept moist (Horton 1977). Consequently the species is highly adapted for both flooding and fires.

Examples of the dispersal rate of salt cedar are reported on the lower Gila River where in the early 1970s the species occupied over 50 percent of the total bottomland area (Haase 1972). Campbell and Dick-Peddie (1964) reported that salt cedar communities were dominant on the Rio Grande in southern New Mexico. Robinson (1965) reported salt cedar communities along the Pecos River in New Mexico and the lower Rio Grande in Texas. Anderson and Ohmart (1977a) show it as the dominant community type on the lower Colorado River. Christensen (1962) reported that the invasion of salt cedar around Utah Lake, the Great Salt Lake, the upper Colorado River, and Green River in Utah was rapid and occurred between 1925 and 1960.

Nomenclature

Numerous efforts have been made to develop a workable nomenclatural system to approach the difficult problem of classifying riparian communities. Many of these efforts have received little acceptance because they have lacked an evolutionary or phylogenetic approach.

Systems Level

Brown and Lowe (1974) developed a classification based on the phylogenetic relationships of riparian communities as we understand their origin (Darrow 1961). In the Brown and Lowe system, the Arcto-Tertiary-derived riparian habitat is included in their forest formation and separated at the next lower level as riparian deciduous forest, which contains the cottonwood-willow communities. The Madro-Tertiary element of the riparian community is in their woodland formation. They present a number of woodland types but the subtropical woodland contains the riparian deciduous woodland, which includes the mesquite community complex. Pase and Layser (1977) have further refined this classification system into additional subdivisions of natural communities. Additional refinement (Brown et al. 1979) has led to a digitized classification system, which includes the biotic communities of North America.

Community Level

Shreve (1942, 1951) in his classical works saw structure as an important framework in studying desert communities. Johnson (1976:137), as well as others, has commented on the importance of understanding and recognizing the structural component:

Justification for viewing desert vegetation on the basis of structure and function in addition to the traditional communities of species distribution approach is supported by the fact that species and communities of species vary structurally and functionally in both time and space. This becomes important if we accept the proposition that a principal goal of biology is to gain an understanding of the relationship between structure and function at all levels of biological organization ranging from organelles up to and including whole communities.

In an effort to understand better how vegetational structure and plant species composition in riparian communities affect animal species at the community level, Anderson and Ohmart (1977a) developed a classification system for the riparian vegetation along the lower Colorado River. Quantitative data on foliage structure and foliage volume from over one-hundred sampled communities were used in a cluster analysis to separate the various structural community types, regardless of species composition. Six structural types were recognized (figure 10-2), with type VI being the earliest successional stage. As the newly developing community matures, it becomes a type V, IV, and so forth. Since all plant species do not attain the same height at maturity and some tall-growing species may have their vertical growth arrested by abiotic and/or biotic factors, each structural group may contain mature or arrested structural types in addition to developing communities. For example, Anderson and Ohmart's (1977a) type VI structural grouping contains the mature arrowweed community and arrested honey mesquite and salt cedar communities, plus early successional stages of all other community types. Type III contains mature honey mesquite with overstory and understory. Type II contains the mature salt cedar community, characterized by a heavy overstory and no understory, which is structurally analogous to a dense cottonwood gallery forest. Type I contains the mature cottonwood forest where some cottonwoods have died and opened the canopy; if mixed with willow, there may be a well-developed midstory as well as a shrub understory.

This classification is easily understood and can be committed to memory for use in the field. Each division has quantitative limits. It is a happy medium between pooling all community and structural types and assuming there is no difference in wildlife use values, water evapotranspiration rates, or metabolic rates of young, developing communities versus mature ones. An alternative would be to develop specific profile classifications for each plant species and then make biological comparisions between all specific structural types. Eventually this may have to be done, but it will probably be too cumbersome to commit to memory for field operational purposes.

Regardless of how one approaches the problem of community structural classification, there is a definite need to recognize vegetation structure at the community level if ecological relationships in riparian communities are to be fully understood. The importance of sorting out vegetation structure and wildlife use values has been well documented (MacArthur and MacArthur 1961: Anderson and Ohmart 1977a, 1977b).

In summary, riparian systems are discrete ecological units within our North American deserts and can be defined by both physiographic and biological criteria. Physically they are drainage courses throughout deserts and transport water, which is a limited resource; flow may be surface or totally subterranean. Within these drainage systems there has evolved a number of specialized plant species derived from two distinct paleofloras.

GENERAL DISTRIBUTIONS AND STRUCTURAL DEVELOPMENT

Many factors operate either singly or in concert to limit the distribution of the riparian plant species that are found in the four North American deserts. The availability and timing of the availability of soil moisture is a very critical factor, and chemical or physical barriers that affect the limitation of soil moisture play a key role in dictating the presence or absence of riparian species. When available soil moisture and duration of its presence are coupled with extreme winter temperatures, duration of temperature extremes, levels of total dissolved solids, and other important physical constraints, it becomes difficult to understand the importance of each factor and which, if not all, are limiting distributions.

Soil moisture also affects structural development in communities. Permanent standing water prevents the establishment and growth of large woody species and supports low-growing sedges and grasses. Low soil moistures or short durations of availability disallow the establishment of large woody species as well. In instances of marginal soil moisture conditions, woody species may become established, but their growth and development is curtailed because of limited soil moisture. Other physical factors function singly or coupled to reduce the growth of woody species. The reduction in growth of these woody species prevents vertical development of the community, as well as foliage volume development on the horizontal plane.

Mojave Desert

Riparian habitats in this desert are poorly developed because of low annual rainfall. Orographic barriers produce a rain shadow effect, which reduces the amount of winter rainfall (Reitan and Green 1968). Summer rainfall from the Gulf of Mexico does not penetrate far enough northward to bring precipitation during the summer period; consequently the Mojave Desert receives most of its rainfall in the winter months (Reitan and Green 1968).

Soil moisture availability, possibly coupled with extreme winter temperatures, has limited the plant species composition and vertical complexity of riparian habitats in the Mojave Desert. Soil moisture availability appears to be the major factor since the Mojave River, a primary drainage through the Mojave Desert, supports a riparian forest of cottonwoods and willows (D. E. Brown personal communication).

In general, washes in this desert support nearly pure stands of cheesebush (Hunt 1966), or mixed communities of cheesebush, common saltbush, rabbitbush, and catclaw (Johnson 1976). On the eastern side of the Mojave Desert, in Death Valley, Hunt (1966) described mixed riparian habitats as containing honeysweet (*Tidestromia oblongifolia*), spurge (*Euphorbia* sp.), pygmy cedar (*Peucephyllum schottii*), desert trumpet (*Eriogonum inflatum*), bebbia (*Bebbia juncea*), stephanomeria ((*Stephanomeria parryi*), stingbush (*Eucnide urens*), sticky-ring (*Boerhaavia annulata*), and cheesebush.

More mesophylic riparian species were recorded near springs on the gravel fans (Hunt 1966). Species such as desert baccharis (*Baccharis sergiloides*), willow (*Salix* spp.), screw-

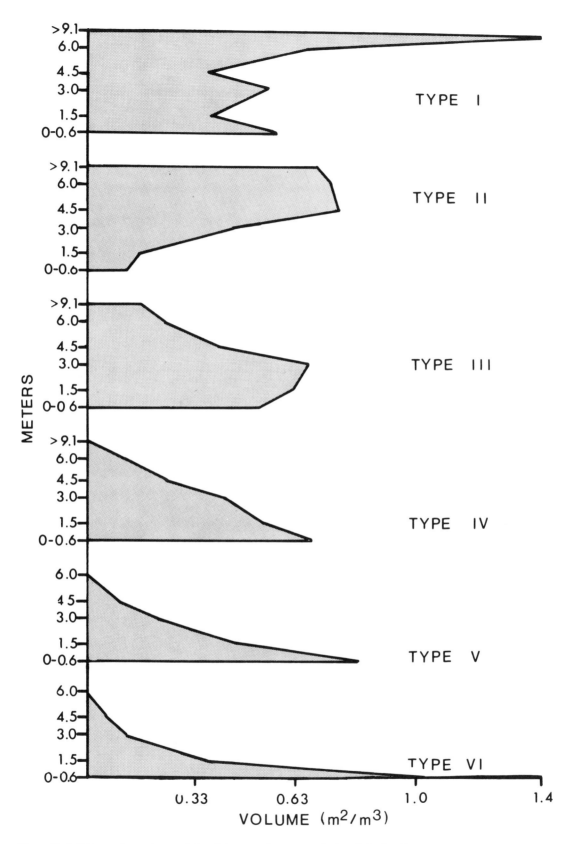

Figure 10-2. Foliage volume characteristics of six vegetation structural types of riparian plant communities found along the lower Colorado River.
Prepared by Lauren Porzer.

bean mesquite, and common reed (*Phragmites communis*) were found in these sites where soil moisture availability was higher.

Areas with high salinity and the water table only a short distance below the soil surface supported honey mesquite, arrowweed, four-winged saltbush (*Atriplex canescens*), salt cedar, inkweed, desert saltgrass (*Distichlis stricta*), rush (*Juncus cooperi*), and pickleweed (*Allenrolfea occidentalis*). The more halophytic species such as saltgrass, pickleweed, salt cedar, and inkweed were located in areas of higher salt concentrations, while honey mesquite and arrowweed were located in sites containing less salts (Hunt 1966).

The Mojave Desert is characterized by a poorly developed riparian flora. Willows and other trees or treelike species are restricted to areas around springs where soil moisture is high. Though some Mojave Desert riparian situations may support trees or treelike species, there is a general absence of riparian trees that provide foliage layers and foliage volume in the vertical profile. This absence of vegetational layering reduces the structural complexity of the riparian communities and hence lessens the importance and suitability of these habitats to many species of wildlife. This is not to imply that riparian habitats in the Mojave Desert are not important; they are, and they support a distinct and varied fauna as compared to the adjacent upland. The fact that they are so limited amplifies their importance to wildlife, and in many instances the vertebrate species that would be classified as riparian obligates primarily occur in man-made habitats (such as sewage ponds, agricultural fields, and drainage ways).

Great Basin Desert

Physiographically this desert is a large basin comprised of over a hundred smaller basins separated by fault-block mountain ranges (Billings 1951). Much of the drainage is into the many small interior basins, which are permanent lakes, playas, or saline flats. The Great Basin itself is bounded on the west by the Sierra Nevada and on the east by the Wasatch Range.

The tall and extensive north-south range of the Sierra Nevada lies in the path of Pacific cyclonic storms, which produce the rain shadow climates of the Great Basin Desert (Billings 1951). This orographic barrier is the most important factor affecting the distribution of winter and spring rainfall (Reitan and Green 1968). Summer moisture is infrequent because as cyclone tracks shift northward, the moist air brought from the Gulf of Mexico cannot extend far enough to benefit the Great Basin Desert consistently (Reitan and Green 1968).

Riparian vegetation in the Great Basin Desert is comprised of two types: emergent or wetland communities growing around the permanent lakes and playas, and the forests growing along the primary drainages that transverse this desert. Examples of the latter drainages are the Truckee River and the upper Colorado River.

The primary drainages transversing the Great Basin Desert support a flora with a well-developed structural profile. The presence of both the Fremont and narrowleaf cottonwood (*P. angustifolia*) provide a tall vertical component, while a number of species of willow (*S. lascolepis, S. nigra, S. exigua,* and *S. gooddingii*) form the midstory and understory. Daubenmire (1942:642) studied the vegetation in the northern portion of the Great Basin Desert and though the riparian descriptions are minimal, he reported "species of *Populus* and *Salix* form a thin and very discontinuous fringe along the banks of permanent streams, or other habitats where a relatively non saline water is near the surface at all times." He also lists *Cornus, Betula, Ribes,* and other ligneous genera forming tangled thickets along some stream sides. Principal species along other streams were *Crataegus douglasii, Betula microphylla, Alnus tenuifolia,* and *Amelanchier florida*.

More extensive descriptions of the riparian flora along the upper Colorado River are given by Hayward et al. (1958). They reported three lakes on Kanab Creek in Utah as supporting Fremont cottonwood, gambel oak (*Quercus gambelii*), willow clumps (*Salix* spp.), cattails, rushes (*Juncus* spp.), sedges (*Carex* spp.), and salt cedar. Other studies provide riparian community descriptions in the Glen Canyon Reservoir Basin on the Colorado River (Woodbury et al. 1959), the Flaming Gorge Reservoir basin in Utah and Wyoming (Woodbury et al. 1960; Flowers 1960), the Navajo Reservoir basin on the San Juan River in Colorado and New Mexico (Flowers 1961), and the Curecanti Reservoir basins on the Gunnison River in western Colorado (Woodbury et al. 1962). All of these papers are replete with descriptions of the riparian cottonwood-willow complex, with its rich understory of shrubs that line these primary drainages and form broad forests where the floodplain widens.

The smaller basins within the Great Basin contain few, if any, halophilic mesophytes (Shreve 1942). Kearney et al. (1914) studied the zonation of the vegetation as it related to soil salt content along the shore of the Great Salt Lake. Soils along the shore, where salt concentrations were highest, supported low-growing and open stands of *Salicornia rubra, S. utahensis,* and *Allenrolfea occidentalis*. Soils with lower salinities supported colonies of saltgrass (*Distichlis spicata*) and alkali sacaton (*Sporobolus airoides*). Greasewood (*Sarcobatus vermiculatus*) and shadscale (*Atriplex confertifolia*) occurred as dense, low-growing shrubs in less saline soils.

Sonoran Desert

The added possibility of both winter and summer rainfall in the Sonoran Desert has allowed a more diverse species composition and vertically developed riparian component in secondary and lesser drainages. Along the western edge of the Sonoran Desert, the vertical component of the riparian vegetation is increased by the invasion of facultative species such as ironwood and blue palo verde. Shreve and Wiggins (1964) pointed out that the height of this vegetation varies from 1 to 5 m and that height is proportional to the width of the drainage.

Winter rainfall is the major source of precipitation along the western edge of the Sonoran Desert (Reitan and Green 1968), and not many riparian obligates can tolerate such low levels of soil moisture. The major shrub or treelike species is the smoke tree (*Dalea spinosa*), which grows in the middle of the gravel and sandy bottoms of large drainages (figure 10-3). It appears to be able to invade and survive in the wash bottom for one of two reasons. First, it grows primarily in broader wash bottoms, which are less prone to highly destructive floods. Second, the rainfall occurs primarily in the winter, and unlike summer rains, it is widespread and gentle. Smoke tree must be able to survive periods of very low soil moisture availability because the other dominant species in these riparian ecosystems are ironwood and blue palo verde, which grow mostly along the edge of the drainage.

In drainages with a permanent water supply in southwestern California and possibly as far east as central Arizona, the native Washington palm (*Washingtonia filifera*) has been able to persist (figure 10-4). Like giants marching down a drainage, this relict species (Axelrod 1950) can still be found in a few localities. Fire may be beneficial or even essential to this species. Following fire, seedlings become established after the dense understory has been removed (Vogl and McHargue

1966). Possible competition for water may also be reduced following fire.

To the east toward central Arizona, as the probability of summer rainfall increases, other shrub and tree species such as desert willow and mesquite make up a greater percentage of the riparian flora. Primary drainages in the Sonoran Desert support a complex and species-rich riparian flora. Gallery forests of cottonwood intermixed with willow line the primary drainages of the Colorado River in western Arizona, as well as other primary drainages throughout the Sonoran Desert.

Chihuahuan Desert

Primary drainages in this desert also support a species-rich riparian flora. Forests of cottonwood and willow line the river banks. Screwbean mesquite, honey mesquite, wolfberry, and species in the genus *Atriplex* may occur along these drainages as well.

Secondary and lesser drainages in the Chihuahuan Desert are similar to those in the Sonoran Desert. Many of the same genera and species are shared, and mesquites and desert willow provide the tallest vertical height in these drainage systems. Palo verde and ironwood are absent in the Chihuahuan upland desert flora, and the absence of these facultative riparian species reduces the complexity of secondary drainages in the Chihuahuan Desert. Shrubs in the genera *Acacia*, *Mimosa*, *Baccharis*, *Condalia*, and *Hymenoclea* frequently mix with

Figure 10-3. Broad riparian ecosystem dominated by smoke tree, with blue palo verde and iron wood bordering the floodplain.
Photograph taken near the Arizona-California border by R. Ohmart.

Figure 10-4. Washington palms at Alkali Springs, Yavapai County, Arizona.
Note the typical Sonoran vegetation on the dry hillsides. Photograph by David E. Brown.

desert willow and/or mesquite to provide an overstory and understory in the secondary drainages of the Chihuahuan Desert.

GENERAL CONSIDERATIONS

Some riparian species decline or increase in commonness and community extent from west to east, or east to west, across the Sonoran and Chihuahuan deserts. Arrowweed, for example, declines in abundance from the Colorado River to the Rio Grande in west Texas, although farther south and east along the Rio Grande at lower elevations it becomes more common (D. E. Brown personal communication). This type of distribution indicates that extreme winter temperatures may be limiting the species abundance in the Chihuahuan Desert. Desert willow is a common and robust tree in the Chihuahuan Desert, but westward into the Sonoran Desert it becomes less common. In this case, soil moisture availability at critical times of the annual cycle appears to limit its distribution. These intriguing distributional questions pose many interesting ecological problems for future research in riparian ecology.

FAUNAL CONSIDERATIONS

It is significant that riparian ecosystems are depicted as lines on maps of our deserts, for they are truly the lifelines for many vertebrate species, including man. The dependence of animals on riparian habitats ranges from none to total. In this range of animal use, riparian habitats may serve as travel corridors through the deserts, as a temporary refugium during extreme desert conditions, and as vital habitat providing all of its needs for its existence.

Interestingly enough, while man tends to avoid the deserts during the hot and dry summer, many vertebrates have evolved behavioral adaptations in order to avoid the winter period, which man considers the most hospitable. Many of the poikilotherms overwinter in a dormant state, either singly or in hibernacula, and a number of birds and bats move south to more favorable environments.

A primary reason for desert-dwelling animals not leaving the desert during the hot, dry summer months is that most never experience the extreme summer temperatures of the desert. Activity patterns are adjusted so as to avoid high ambient temperatures, and refugia such as cool, moist burrows or heavily shaded, moist areas are used. Many species live exclusively in the moist and heavily shaded habitats where primary and secondary production is high in a warm, arid environment. These habitats are riparian ecosystems. The amount of dependence on the riparian community by desert animals is highly variable. In attempting to classify the species in each major group into obligate, facultative, or nonriparian, the extremes are easy, but difficulty arises with species that are intermediate.

Fishes

The distribution of native desert fishes is limited by geographic barriers or by need of tolerance to some environmental feature (Deacon and Minckley 1974). Of the approximately two hundred native species in western North America, only the speckled dace (*Rhinichthys osculus*) can be considered widespread (Miller 1959). Desert fishes are obligate to riparian habitats, being found in permanent streams or springs. During ephemeral wet periods, they may expand into intermittent drainages, but as drier periods return, their distribution is reduced.

Hubbard (1977) examined native fish numbers in Arizona and New Mexico and pointed out that the nearly thirty native fish species in Arizona (Minckley 1973) represent 5 percent of the fish fauna of the United States. New Mexico, with fifty-nine recorded species, represents 9 percent of the total United States fish fauna. These totals are impressive for two arid states: about half of Arizona is in the Sonoran Desert and about half of New Mexico is in the Chihuahuan Desert.

Many of the fish species found in the North American deserts are in the minnow family (Cyprinidae) and range in size from tiny minnows to the Colorado River squawfish (*Ptychocheilus lucius*), which has attained weights greater than 25 kg (Minckley 1973).

Desert fishes show a broad set of morphological characters for existing in desert riparian ecosystems. Fishes in torrential streams may have sucking discs or hooks to prevent them from being washed away, while in slower-moving or quiet waters, the fish may be deep-bodied and lethargic (Deacon and Minckley 1974). In slow-moving, turbid waters, barbels or other types of sense organs may be developed for nonvisual sensory input. In fast-flowing, turbid streams, fish may possess a leathery skin with minute, embedded scales, presumably to reduce the abrasive nature of the sand-laden waters (Hubbs 1940, 1941).

In desert streams with small surface flows in summer months, the processes of evapotranspiration may demand all the surface flow during the daylight hours (Campbell and Green 1968). This would essentially eliminate the fish fauna, but some species, such as the longfin dace (*Agosia chrysogaster*), can survive these dry periods under saturated mats of algae and debris (Deacon and Minckley 1974). Minckley and Barber (1971) observed an alternating wet and dry fourteen-day cycle in Sycamore Creek in Arizona; a small rain in August rejuvenated the flow, allowing the longfin dace that had survived in moist algal mats to repopulate the stream.

Because evapotranspiration from desert streams is so high and salinities can range from low to very high, it is not surprising that desert fish have evolved the capacity to tolerate high salt levels. La Bounty and Deacon (in Deacon and Minckley 1974) reported that a small *Cyprinodon* in Death Valley can survive in a salinity of 78 gm per liter, and Hunt et al. (1966) implied that these fish may experience salinities of

160 gm per liter at times. Deacon and Minckley (1974) reported *Cyprinodon atrorus* surviving concentrations up to 95 gm per liter in the Cuatro Cienegas basin. It appears that salinity levels in some closed basins limit the distribution of some fish. Freshwater fish are found in the tributary streams and deltas feeding the Great Salt Lake, but there are no fish in the open waters of the lake (Sigler and Miller 1963).

Amphibians and Reptiles

Species in these two groups have been combined because there are so few species of desert amphibians. The reproductive habits of these two classes differ greatly. Amphibians are closely tied to aquatic environments, whereas reptiles, with a cleidoic egg, are less dependent on surface aquifers. There are few reptiles that can be considered obligate riparian species.

The species list in appendix 10-A provides a compilation of the amphibian and reptilian species characteristic of each desert. Species were excluded whose distribution penetrated only a small portion of a particular desert or included only an isolated mountain range or two within a desert. Species found in the Anza Borrego region in southern California were also excluded. This region has a distinct herpetofauna and, if included, should possibly be considered a fifth desert. Only binomials were used, since each desert usually has its own subspecies. *Rana pipiens* sp. was used since the taxonomic status of this sibling species group is not clear. Subspecies were used only in the case of *Trimorphodon* because the specific epithet has recently been changed to *biscutatus* and previously held species are all considered subspecies currently.

The Great Basin and Mojave deserts, with a poorly represented riparian fauna, contain two and one species of amphibians, respectively. The Sonoran and Chihuahuan deserts, with more highly developed riparian ecosystems, more subtropical flora, and less extreme winter temperatures, contain eleven and ten amphibian species, respectively. In the Sonoran Desert about 80 percent of the total amphibians (eleven) are obligate or facultative, while in the Chihuahuan Desert these two groupings constitute 70 percent of the total (ten). Treefrogs (*Hyla* spp.) and species of frogs in the genus *Rana* contain those amphibians considered to be riparian obligates.

Of the reptiles, only two species of turtles (Sonoran mud turtle, *Kinosternon sonoriense,* and the spiny softshell, *Trionyx spinifer*), one species of lizard (*Holbrookia maculata*), and three snakes in the genus *Thamnophis* adhere to the criteria established to denote riparian obligates.

An example of productivity in riparian habitats is seen with respect to the Sonoran mud turtle. A life history study (Hulse 1974) in central and southern Arizona revealed that the species was opportunistic with respect to diet but tended to be carnivorous. Densities were 330 animals per hectare (ha-0.4 acre) at the Tule Stream study site. The densities attained by this population are an example of how much food must be produced in the stream to support such a population.

Mammals

A number of mammalian species are highly dependent on riparian habitats for food, water, shade, and cover. Many species can be considered facultative or even obligate in riparian habitats. Most mammals that use desert riparian habitats are wide ranging in the continental United States or are common at higher and moister elevations in desert mountain ranges. They occur primarily in the desert along moist riparian corridors. They may be found in parts of the upland deserts, but highest densities are attained in and around riparian ecosystems. Some of these are the raccoon (*Procyon lotor*), striped skunk (*Mephitis mephitis*), mule deer, cotton rat (*Sigmodon hispidus*), rock squirrel (*Spermophilus variegatus*), hog-nosed skunk (*Conepatus mesoleucus*), desert pocket mouse (*Perognathus penicillatus*), beaver (*Castor canadensis*), cactus mouse (*Peromyscus eremicus*), deer mouse (*Peromyscus maniculatus*), and the white-footed mouse (*Peromyscus leucopus*). Others, such as the desert cottontail (*Sylvilagus audubonii*), coyote, gray fox (*Urocyon cinereoargenteus*), bobcat (*Felis rufus*), and ring-tailed cat (*Bassariscus astutus*), may occur throughout the desert but are usually found in higher densities along riparian ecosystems. Many carnivores concentrate their predatory activities in and adjacent to riparian habitats where densities of many prey species are highest. Other small mammals, such as bats, feed extensively over and in riparian habitats. A number of bats also utilize the tall, dense riparian vegetation for daytime roosts. Riparian corridors may also be important in providing food and/or water during bat migration.

Birds

A compilation of the avian species found in the four North American deserts provides a comparison of the relative value of the riparian habitats within each of the deserts and their adjacent uplands (appendix 10-B). Other interesting comparisons emerge, such as the total number of breeding species in riparian habitats in each desert and total species in each desert. Some species vary their dependence classification throughout an annual cycle. Therefore, to simplify the data presentation, the category of highest dependency in the annual cycle was selected.

MOJAVE DESERT

A total of 215 avian species regularly occur within the Mojave Desert (Grinnell and Miller 1944; Small 1974). Of this total 68 species (32 percent) breed. Fourteen species (6 percent) are riparian obligates, and of these 12 are breeding species, 164 are considered facultative (76 percent) with 46 breeding, and 37 species (17 percent) are nonriparian forms with 15 as breeding species.

A number of the species that are considered riparian

obligates in other deserts are considered facultative species in the Mojave Desert. Their riparian habitats are either poorly developed or absent in the Mojave Desert, and when the species do occur they are present in small numbers in limited habitats such as sewage or domestic livestock ponds.

GREAT BASIN DESERT

A total of 262 avian species regularly occur within the Great Basin Desert (Burleigh 1972; Hayward et al. 1976; Johnson 1978). Of the 262 species, approximately 148 (57 percent) are breeding species. Fifty-five species (21 percent) are obligate riparian species with 50 breeding, 159 (61 percent) are facultative riparian species with 81 breeding, and 48 (18 percent) are nonriparian species with 17 breeding. Of the total avian species regularly occurring in the Great Basin Desert, 82 percent are either totally or partially dependent on riparian habitats.

SONORAN DESERT

A total of 308 avian species regularly occur within the Sonoran Desert (Phillips et al. 1964 and updated records). Of this total, 56 (18 percent) are obligate riparian, 197 (65 percent) are facultative riparian species, and 55 (18 percent) are nonriparian species. In the Sonoran Desert, approximately 82 percent of the bird species that regularly occur are dependent on riparian habitats to a great extent.

Coarser grained comparisons of the relative value of the riparian habitats have been made by determining what percentage of the total number of nesting species in an arid environment is dependent on water-related habitats. Johnson et al. (1977) did this for the Sonoran Desert, and of the 166 total breeding species, 127 (77 percent) in some manner were dependent on water-related habitats. Slightly over 50 percent (84 of 166 species) were totally dependent on water-related

habitats. The percentages reported by Johnson et al. (1977) for species that are dependent in some manner on water-related habitats (77 percent) are close to the 87 percent total, which represents those species pooled as obligate and facultative riparian species. Wauer (1977) presented similar results but confined his totals to those species known to breed in southwestern riparian habitats below 5500 feet. He listed 94 avian species breeding in southwestern riparian habitats, and of this total, 39 species (40 percent) nested in riparian habitats along the Rio Grande in west Texas. These studies indicate the general significance of riparian habitats to nesting birds.

CHIHUAHUAN DESERT

A total of 322 species regularly occur within this desert, with approximately 127 (39 percent) listed as breeding (Ligon 1961; Wauer 1973; Oberholser 1974; Hubbard 1978). Of the 322 species, 45 (14 percent) are riparian obligates with 38 breeding, 228 (70 percent) are facultative with 75 breeding, and 49 (15 percent) are nonriparian with 14 breeding. A total of 273 (85 percent) avian species in the Chihuahuan Desert are obligate or facultative species to riparian habitats.

Table 10-1 summarizes the totals for each desert. The largest percentage of riparian obligate species occurs in the Sonoran Desert and is closely followed by the Great Basin Desert. The Mojave Desert with a poorly developed riparian ecosystem has the fewest riparian obligates.

GENERAL AVIAN CONSIDERATIONS

The number of species and densities of birds are usually much higher in riparian habitats when compared to adjacent desert uplands. There is an observable disparity in primary productivity (gross biomass) between the riparian habitat and the adjacent desert upland.

Table 10-1
Summary of the dependency of the avifauna on riparian ecosystems in each North American desert

| DESERT | RIPARIAN DEPENDENCY | | | |
	OBLIGATE	FACULTATIVE	NONRIPARIAN	TOTAL
Mojave	14 (6)	164 (76)	37 (17)	215
Breeding	12 (18)	46 (68)	10 (15)	68
Great Basin	55 (21)	159 (61)	48 (18)	262
Breeding	50 (34)	81 (55)	17 (11)	148
Sonoran	56 (18)	197 (64)	55 (18)	308
Breeding	50 (37)	68 (51)	16 (12)	134
Chihuahuan	45 (14)	228 (71)	49 (15)	322
Breeding	38 (30)	75 (60)	14(11)	127

NOTE: Numbers in parentheses are percentages.

Density

Although no comparative data are available on insect biomass in riparian and desert ecosystems, an indirect measure of high insect abundance in riparian ecosystems is indicated by the density of breeding birds. Carothers and Johnson (1975) reported a breeding density of 1059 pairs per 40 ha on the Verde River in central Arizona, representing the highest reported breeding density of birds in the continental United States. Not only do these figures provide insight into insect productivity in riparian habitats, they also clearly demonstrate the value of these habitats to the breeding avifauna in our deserts.

Raptors

Another indirect indication of gross productivity in riparian habitats is the number of avian predators that are obligates to riparian habitats in desert ecosystems. The black hawk (*Buteogallus anthracinus*), gray hawk (*Buteo nitidus*), bald eagle (*Haliaeetus leucocephalus*), cooper hawk (*Accipiter cooperii*), zone-tailed hawk (*Buteo albonotatus*), and Mississippi kite (*Ictinia mississippiensis*) feed and nest in riparian habitats in desert communities. These raptors may require free drinking water and large trees or cliffs for nesting in most desert situations; however, in the Sonoran Desert where these species reach their highest densities in desert habitats, there are alternate nest sites available in the tall saguaros and palo verde or ironwood trees in the adjacent upland.

Preliminary food habit studies of these large raptors indicate that the major portion of their food resources comes from the biotically rich riparian habitats. The small breeding population of bald eagles (eight to ten active nests) in the Sonoran Desert along the Salt and Verde rivers in Arizona feed primarily (about 75 percent) on channel catfish (*Ictalurus punctatus*) and carp (*Cyprinus carpio*) (Hildebrandt and Ohmart ms.). The nesting black hawk feeds primarily on a mixture of aquatic and terrestrial riparian vertebrates (Glinski and Ohmart ms). The gray hawk specializes in terrestrial vertebrates from the riparian habitat (Glinski ms.), while the recently invading Mississippi kite primarily eats cicadas (*Diceroprocta apache*) that emerge from the floodplain soils (Glinski and Ohmart ms.).

Migration

It has been postulated that primary drainages may play an important and possibly a vital role for species that migrate across North American deserts. The lush riparian habitats provide needed water, food, or resting areas for migrants. They may also serve as north-south navigational guides through arid habitats. Little quantitative data are available, but there are few continuous mountain chains in desert areas, and the lasting physiographic feature of a primary stream course might well be used as an orientation cure for migrating species. Johnson et al. (1977) examined densities of migrant birds in the adjacent desert upland versus riparian areas in the Sonoran Desert and reported 10.6 times as many migrants per hectare in riparian habitats as in adjacent desert habitats. These data are convincing as to the value of riparian habitats to migrants, but more research is needed regarding the importance of the riparian habitat to each migratory species. Western desert riparian habitats may be more valuable to some species, while eastern desert riparian corridors may be more important to others. We also need to know the total number of migrants that use these corridors.

While many species migrate up and down riparian corridors, a number of species overwinter in these habitats. In a study (Anderson and Ohmart ms.) spanning more than seven years on the Colorado River in Arizona, researchers have examined habitat and niche breadth of visitors, including both winter and breeding, versus permanent residents. It was found that avian visitors in terrestrial habitats are generally much more specialized in their habitat requirements than are permanent residents. Consequently any habitat modifications of the riparian vegetation tend to have a more direct and dramatic effect on breeding and wintering visitors than on permanent residents. Species such as the bell vireo (*Vireo bellii*), yellow-billed cuckoo (*Coccyzus americanus*), clapper rail (*Rallus longirostris*), black rail (*Laterallus jamaicensis*), black hawk, and gray hawk are much more sensitive to habitat change than are the Abert towhee (*Pipilo aberti*), Crissal thrasher (*Toxostoma dorsale*), and black-tailed gnatcatcher (*Polioptila melanura*).

Wetlands

Waterfowl and shorebirds migrating and wintering in desert environments are highly dependent on wetlands and riparian habitats. The arid Southwest is generally not thought of as an important flyway or wintering area, but a number of species with relatively high densities rely on rivers, backwaters, and impounded bodies of water, either during migration or for overwintering sites. Little quantitative data exist considering the importance of such habitats, but a few studies do focus on this point.

Quantitative data of timing in use of various wetland habitats, number of species using them, and total number of birds have been reported by Ohmart and Anderson (1980) for the lower Colorado River. The eight riparian types of habitats examined ranged from natural wetland communities to those highly modified by man and included such habitats as reservoirs, riprapped and dredged areas, rivers in canyons, undisturbed river with adjacent riparian vegetation, old river channel, river immediately below dams. *Phragmites* marshes, bulrush (*Scirpus* spp.) marshes, dense cattail (*Typha* spp.) marshes, and moderately dense cattail marshes.

A summary of these data in table 10-2 demonstrates the value of these wetland habitats to birds. There is a general tendency for more avian species and higher densities to be present in aquatic habitats, where there is dense emergent vegetation, than in an open water situation.

Both species richness (number of species) and density must be used in evaluating these habitats, since high avian densities in different kinds of marsh habitats are a reflection of feeding and roosting densities of species such as red-winged blackbirds (*Agelaius phoeniceus*) and yellow-headed blackbirds (*Xanthocephalus xanthocephalus*). For example, a moderately dense cattail marsh will support a mean of 35.9 species and a mean density of 345 birds per 40 ha, whereas a bulrush marsh may contain 469 birds per 40 ha, with only 13 species.

The lower Colorado River and its associated habitats support between 75,000 and 100,000 water-related birds during peak months of use, with highest densities found in undisturbed wetlands. Although some of the man-made habitats studied did not support species numbers or densities equaling undisturbed habitats, they did provide a habitat that, before modern technology, was absent. Deep-water habitats (reservoirs and areas below dams) support species such as loons, grebes, and diving ducks, which do not normally inhabit shallow-water habitats.

VEGETATION LAYERING

The tall and multilayered habitats that grow along the alluvial floodplain are extremely important to the avian species. Tall riparian forests provide a number of layers of foliage that attract and support a high density of breeding and wintering avifauna. Anderson and Ohmart (ms) recently found that riparian birds responded to four layers of vegetation as opposed to the usually considered three layers (MacArthur and MacArthur 1961). Using Pearson product moment correlations, they correlated the density of each avian species, assuming no layers, three layers, and four layers. They reported nineteen species associated with layers 7.6 m or more, ten species associated with the 4.6 to 7.6 m layer, thirteen species associated with the 1.5 to 4.6 m layer, and eleven species with the 0.15 to 1.5 m layer. The overstory group, comprised of nineteen species, was composed of specialists that were generally missing when this layer was absent or poorly represented. The twenty-three species in the two middle layers and some of the eleven in the understory group were generalists. These species might be present even if their layers were absent or poorly represented. Therefore vertical structure of the vegetation in riparian communities, as well as the volume of vegetation present in these layers, is important for high avian densities, Using principal components analysis, Anderson, Rice, and Ohmart (ms) reported that vegetation volume was generally more important as a predictor of both bird density and diversity than was foliage height diversity.

PATCHINESS

Another important vegetational component in these communities that has recently been quantified is intracommunity patchiness (Anderson and Ohmart ms). The data spanned fourteen seasons and the patchiness index correlated with avian species richness in twelve seasons and bird species diversity in

TABLE 10-2
Summary of the mean number of species and densities of birds in ten types of wetland habitats along the lower Colorado River

HABITAT TYPE	NUMBER OF SPECIES	MEAN DENSITY (NUMBER PER 40 HA)
Reservoirs	4.9	20
Riprapped and dredged	6.1	36
River in canyons	6.6	25
Undisturbed river with adjacent riparian vegetation	10.5	101
Old river channel	11.4	149
River immediately downstream from dams	11.8	164
Phragmites marsh	13.0	237
Bulrush marsh	14.0	469
Dense cattail marsh	30.5	204
Moderately dense cattail marsh	35.9	354

SOURCE: Wildlife use values of wetlands in the arid southwestern United States, Robert Ohmart and Bertin Anderson, *Proceedings of the National Symposium on Wetlands 1978*, 1980.

eleven seasons. They reported that the patchiness index was strongly correlated with total insect biomass and was significantly correlated with the number of insect families present. Thus relatively homogeneous communities with a lot of patchiness in the form of canopy or understory discontinuity or a few dead trees scattered throughout would have higher insect biomass and hence support more bird species than would a homogeneous monoculture with little patchiness. Monocultural communities tend to have little patchiness, while communities mixed with trees and shrubs have high patchiness values. -

Dead trees or snags within a community promote patchiness. Anderson and Ohmart (ms) counted dead trees with a diameter breast high 15 or more cm but did not find a significant correlation between numbers of dead trees and patchiness. By grouping communities into high, medium, and low snag numbers, there was a significant correlation between patchiness and those communities having a high number of snags. It has been demonstrated that the presence of dead trees in a community attracts a number of cavity-nesting and insect-probing species (Yeager 1955; Flack 1976).

Another finding by Anderson and Ohmart (ms) was the predictive capabilities of the patchiness index with respect to avian densities, species diversity, and species richness. Predictions of avian densities had an average error of 29 percent. Climate was an important variable with respect to predicting avian densities in desert riparian communities, since harsh winters dramatically reduced avian numbers. Harsh winters also affected permanent residents; there were lowered population numbers carried into the next breeding season. Socially regulated populations were less severely affected than species that did not appear to have evolved social mechanisms for population regulation (Anderson, Fretwell, and Ohmart ms). It is not surprising that climate has such an important effect on avian densities in desert riparian habitats, since insect availability was found to be very low during harsh winters (Anderson and Ohmart unpubl.) and many of the wintering species are insectivores.

Anderson and Ohmart (ms), using the patchiness index as a predictor of bird species diversity, reported an average error of 5 percent. The index overpredicted in nineteen of twenty-three plant community and structural types and was exceedingly high in the introduced communities of salt cedar. This exotic species does not support the avian species diversity predicted from high foliage height diversity and patchiness index values.

In predictions of species richness (number of species) using the patchiness index, there was an average error of 15 percent. The predictions were too high in salt cedar communities and too low in cottonwood-willow communities.

In general, riparian habitats support the highest number of avian species, the highest densities, and bird species diversities of any other types of desert habitats. The values in table 10-3

provide some insight into the densities, number of species, and bird species diversities in these riparian communities.

FLORAL AND FAUNAL SPECIALISTS IN RIPARIAN HABITATS

Not only are there plant community attributes that are valuable to wildlife, but there are also intrinsic values of a particular plant species to specific wildlife species. It is widely accepted that an animal species, needs food, water, cover, and space to exist. Many individual plant species may provide one or more of these components, but if a particular plant species or community provides all of these components for an animal species, the animal may then be thought of as a habitat or plant species specialist. Because of the evolutionary history of riparian communities and the specialization that has occurred in riparian plant species, it is not surprising that some animal specialists have evolved with this flora.

Honey Mesquite

This important riparian species occurs in southwestern Utah, southern Nevada, extreme western Arizona, and southeastern California into Baja, California. It also occurs in southern, central, and eastern New Mexico, as well as in all of west and central Texas and extends throughout Chihuahua, Mexico in the Chihuahuan Desert (Simpson and Solbrig 1977). Velvet mesquite is found in western and southern Arizona south into Sonora, Mexico.

The rapidly growing and highly specialized mesquite taproot has been demonstrated to grow up to 50 m (Phillips 1963). Lateral roots measured at over 18 m long (Fisher et al. 1959) grow just under the soil surface to utilize subsurface moisture. The highly developed root system allows this slow-maturing species to invade successfully second terraces along desert rivers where the water table may be 6 m or more below the soil surface or to grow along the edge at secondary and tertiary drainage systems that may have deep and highly fluctuating water tables.

Leaf bud burst and leaf drop show genetically based variation in timing (McMillan and Peacock 1964; Peacock and McMillan 1965). Individuals from widely separated localities reared under uniform conditions retained leaf initiation and leaf drop characteristics similar to individuals at the natural sites. Fall dormancy in southern mesquite populations appears to be induced by temperature rather than photoperiod. Soil moisture, prior drought, and temperature appear to act in concert to initiate leaf drop in other populations.

The leaves of velvet and honey mesquite are grazed by insect herbivores. One specialist, reported by Cates and Rhoades (1977) in two study sites in Arizona, was a meloid beetle (*Epicauta arizonica*). A variety of sprays from all of the woody legumes were presented to the meloid beetle, it consumed only the leaves of velvet mesquite. The insect herbivores grazing on mesquite leaves provide a food source for higher trophic levels such as predaceous insects, birds, and small mammals.

Both honey and velvet mesquite produce tiny, greenish-white flowers that are clustered along a spike. Each spike is about 7 cm long and supports hundreds of small flowers. The spikes may occur singly, or there may be up to eighteen spikes per leaf axil. It has been estimated that a single highly branched velvet mesquite tree will produce about 6000 inflorescences or spikes during an annual cycle (Simpson et al. 1977). Time of flowering for velvet mesquite is highly variable in Arizona. Flowering was reported in late April one year and in about the first week in June the previous year (Simpson et al. 1977). These abundant and energy-rich flowers provide an important resource for insects, and the resultant fruits are an important food source for vertebrates.

On a per-tree basis, velvet mesquite provides one of the richest pollen and nectar sources in the Arizona desert (Neff, et al. 1977). Simpson et al. (1977) reported that velvet mesquite produces more pollen per floral unit than most other insect-pollinated desert shrubs in North America. During peak production each inflorescence produces 2.4 mg of nectar per day. The pollen is high in protein, which is important for insect reproduction and larval growth.

A large number of insects utilize this rich food resource while it is available. Simpson et al. (1977) found that more animals were associated with *Prosopis* flowers than with any other plant species in Sonoran Desert habitats. They reported that a single inflorescence may support hundreds of thrips and a host of other species. Solitary bees are a principal group, and over 160 different species have been recorded visiting the flowers of *Prosopis* (Moldenke and Neff 1974). At least five species of small bees have been classified as *Prosopis* flower specialists (Simpson et al. 1977), and it is common to see thousands of individuals in the subgenus *Perdita* hovering around and alighting on a flowering *Prosopis* (Cockerell 1900). The drab-colored flowers may not attract human eyes, but the pollen of *Prosopis* reflects ultraviolet light and appears as bright dots to the compound eye of an insect that can detect short wavelength light (Simpson et al. 1977).

Not all mesquite flowers produce fruit. It has been estimated that fewer than 3 percent of the millions of flowers produced initiate fruit development and from one-half to one-third of this group produce fruit (Solbrig and Cantino 1975). Apparently the level of soil moisture and rainfall during the

Table 10-3
Breeding bird densities, species richness, and bird species diversity in various riparian communities in North American deserts

COMMUNITY	DENSITY (BIRDS PER 40 HA)	SPECIES RICHNESS	BIRD SPECIES DIVERSITY	DESERT	SOURCE
Cottonwood	847	26	2.98	So	Carothers et al. 1974
	684	28	3.15	So	Stamp 1978
	425	20	2.68	So	Carothers et al. 1974
Cottonwood-willow	354	27	2.67	Ch	Engel-Wilson and Ohmart 1978
	197	10	2.76	So	Ingles 1950
Cottonwood-willow-mesquite	197	37	3.15	So	Goldwasser 1978
	435	48	2.77	So	Anderson and Ohmart 1977a
Willow-mesquite	520	8	1.62	So	McKernan 1978
Mesquite	11	13	2.35	Mv	Austin 1970
	49	12	2.33	Mv	Austin 1970
	244	19	2.60	So	Stamp 1978
	190	31	2.74	So	Anderson and Ohmart 1977b
Salt/Cedar	243	28	2.16	Ch	Engel-Wilson and Ohmart 1978
	199	22	2.03	So	Anderson and Ohmart 1977a
Screwbean mesquite-salt cedar	1176	26	1.40	So	Anderson and Ohmart 1977a
	427	32	1.98	So	Anderson and Ohmart 1977a
	278	34	2.43	So	Anderson and Ohmart 1977a
Arrowweed	87	25	2.67	So	Anderson and Ohmart 1977a

SOURCE: Wildlife use values of wetlands in the arid southwestern United States. Robert Ohmart and Bertin Anderson, *Proceedings of the National Symposium on Wetlands 1978*, 1980.

NOTE: So = Sonoran Desert; Mv = Mojave Desert; Ch = Chihuahuan Desert.

flowering period affects flower development and fruit production (Mooney et al. 1977). Low soil moisture promotes heavy flowering and good fruit production, while the opposite condition produces low fruit production. Rainfall during the flowering period also reduces flowering and fruit production.

The fruits of both velvet and honey mesquite are indehiscent pods with a thin outer layer (exocarp), a thick and spongy middle layer (mesocarp), and a thin inner layer (endocarp) surrounding the hard and stony seed. Pods vary in number and length, but when mature they constitute a large biomass rich in stored nutrients. Glendening and Paulsen (1955), over a four-year period, sampled thirty young velvet mesquite trees with a crown diameter of 4 m, and these trees produced a mean of 0.7 kg dry weight of pods per tree per year. The pods contained an estimated 5000 seeds. They estimated that more mature trees with crown diameters around 6 m would produce more than 16 kg of pods per tree per year, or about 140,000 seeds.

Chemical analysis data amplify the importance of this food resource to desert animals. *P. velutina* contains 16 percent sugar and 12 percent protein in total pod analysis, and *P. glandulosa* shows values ranging from 19 percent (Texas) to 31 percent (California) in sugars and 13 percent (Texas) to 9.5 percent (California) in protein (Walton 1923). The mesocarp contains the major portion of the sugars and starches. The pods are highly sought after by a number of vertebrate species, and once the pod is consumed, it is transported from the parent tree; eventually the seeds are deposited in a fecal pile. This process selectively benefits the mesquite in two ways: the seeds are dispersed in a nutrient-rich source, and seed density around the tree is reduced, thereby avoiding mass insect destruction, especially by bruchid beetles (Coleoptera: Bruchidae).

Bruchid beetles, both in number of species and individuals, are by far the most specialized insects that feed on the seeds of *Prosopis* (Kingsolver 1964, 1967, 1972). Of the estimated 1000 species in the Bruchidae, most are known to develop in the pods of the Leguminosae (Fabaceae) (Cushman 1911; Zacher 1952). Three genera of bruchids are obligates to *Prosopis* (Kingsolver et al. 1977). Kingsolver et al. (1977) reported that twenty-nine species have been found in *Prosopis* fruits and that twenty-seven species (93 percent) are obligates. The larvae of the bruchid (*Algarobius prosopis*) inflicted up to 43 percent seed mortality for a given tree (Kingsolver et al. 1977). A pod predator, the leaf-footed bug (*Mozena obtusa*), occurs in both the Chihuahuan and Sonoran deserts; damage caused by this insect to honey mesquite was estimated at 50 percent of the potential fruit production (Swenson 1969). The adult leaf-footed bug pierces the buds and sucks out the fluids, causing a premature fruit drop.

The value of *Prosopis* as a niche component has been examined by Mares et al. (1977:149), who reported, "We doubt that the other desert trees have as persuasive an influence upon other organisms as does mesquite." They listed the small mammals associated with mesquite communities in the Mojave, Sonoran, and Chihuahuan deserts and discussed the species diversity of small mammals associated with this community type. No differences in the three parameters of food, body size, and adaptation were found between the arid creosote bush and the velvet mesquite habitats (Mares et al. 1977). They did not report on the importance of the mesquite pod resource to large mammals.

Both mule deer and desert bighorn sheep consume the pods of mesquites, and recent data (Ohmart and Anderson ms.) from scat analysis of coyotes along the Colorado River in California and Arizona show that honey mesquite pods may comprise as much as 90 percent of the total volume in the coyote diet during the fall months. Woodward and Ohmart (1976) reported honey mesquite comprising over 20 percent of the July and August diet of wild burros (*Equus asinus*) adjacent to the Colorado River.

The availability of pod resource to these large vertebrates might or might not be significant to each animal species, but it does represent a mechanism by which large mammals transport the seeds away from the parent tree, thus enhancing the possibility of seedling establishment. The percentage of seeds that are destroyed by large mammals in the mastication and digestive process is unknown but may not be too great. Feeding trials with horses, cows, and ewes revealed 91, 79, and 16 percent, respectively, of the seeds consumed passed through the digestive system unharmed (Fisher et al. 1959). Of the seeds that passed through unharmed, 82 percent in the horse, 69 percent in the cow, and 25 percent in the ewe germinated (Haase et al. 1973). Germination success in shelled, undigested seeds was only 26 percent. If the endocarp is not broken during the digestive process, the seeds may remain viable for up to forty-four years (Martin 1948).

The action of mastication and digestion in vertebrates simulates the natural scarification process that occurs when mesquite pods are water transported during floods. The wetted mesocarp is softened, and as the pod is washed along the stream, the stony endocarp is scaled by rocks and gravel. Scarification dramatically improves seed germination, and pods used to grow mesquites in nurseries are either given a light acid treatment or notched with a knife or file to allow water penetration of the endocarp.

Phainopepla

Along the lower Colorado River between southern California and western Arizona, individuals or whole communities of honey mesquite may become parasitized by mistletoe (*Phoradendron californicum*). Although mistletoe parasitizes other species of legumes, it grows into larger clumps when growing on honey mesquite trees. Mistletoe produces a berry that ripens in the winter. Species of fruit-eating birds consume the berries, but one in particular, the Phainopepla (*Phainopepla nitens*), has a specialized digestive tract for processing mistletoe berries (Walserg 1975).

The Phainopepla has been recognized as being capricious, appearing in an area to breed and then disappearing. Walsberg (1977) studied the species in the Mojave Desert in southeastern California and reported a buildup of the population in April. Along the Colorado River where the Mojave and Sonoran deserts merge, Anderson and Ohmart (1978) reported that for three years Phainopeplas began arriving along the river in September, and after breeding began to leave the Colorado River valley in April and May of the following year. Walsberg (1977) has developed a working hypothesis relative to these movements. He suggests that the climatological shift to increased aridity over the past few thousand years caused changes in plant community patterns. During this period, breeding Phainopeplas might have been very local or altitudinal migrants, but with geographical alterations in plant communities, the birds adapted by making longer and seemingly strange migratory movements. Present migratory patterns may seem nonsensical but are more clearly understood when viewed in the context of shifting vegetational patterns through time.

During the period of September through May when the Phainopeplas live and breed in the Colorado River valley, they show distinct habitat preferences in the mesquite communities (Anderson and Ohmart 1978). On their arrival, the birds are loosely associated with mesquite trees, with and without mistletoe. Through the winter, there is a significant relationship among honey mesquite trees, mistletoe, and Phainopeplas. During March and April, Phainopeplas are associated with mistletoe and dense vegetation for nesting. Birds without young begin leaving the area in April, while those feeding young are significantly associated with wolfberry (*Lycium* spp.), whose spring-produced fruits are ripening.

During a three-year study of the Phainopepla (Anderson and Ohmart 1978), a severe winter occurred in which the major portion of the mistletoe berry crop froze and was unavailable as a food resource. The Phainopepla population declined dramatically during that winter, and there was little successful reproduction the following spring. The subsequent winter the returning population was lower, indicating that some or all of the population decline may have been due to mortality.

The relationship between mistletoe and mesquite is a parasitic one. Older mesquite trees appear to be most vulnerable to being parasitized by mistletoe. The mistletoe, in turn, has a symbiotic relationship with the Phainopepla. The densely branched mistletoe occurs in varying sized clumps, with mature plants occupying about 0.3 m³ of space. The numerous berries produced by each plant are eaten by Phainopeplas, and then as the birds perch on the upper branches of the host tree or adjacent trees, the seeds may be defecated onto suitable sites for seedling development. The mistletoe provides a winter and early spring food source for the Phainopepla, as well as a dense sphere of vegetation for Phainopepla nest placement.

Cottonwood-Willow

Much less is known about other riparian plant species and exactly how animals are dependent on these plants, but one very important community to animals is the cottonwood-willow habitat. Both plant species at maturity provide a tall vertical community profile combined with a high canopy volume. The two tree species also support a high insect biomass (Anderson and Ohmart unpubl.). Vertical profile, high foliage volume in the canopy layer, and high insect productivity in communities dominated by either cottonwoods, willows, or a mixture of the two species promotes higher annual bird species numbers and densities than do other riparian communities. Patchiness index predictions of bird species richness were consistently too low, and deviations from the predictions showed cottonwood and/or willow having an additive effect to species richness in early summer and a negative effect in late summer.

The Summer tanager (*Piranga rubra*) and the yellow-billed cuckoo (*Coccyzus americanus*) are habitat specialists within the mature or nearly mature cottonwood-willow communities. The two bird species are closely and consistently associated with mature or maturing cottonwood-willow communities. Exactly why this bird-community relationship exists remains obscure, but the vertical height of the communities, a large foliage volume in the canopy layer, and the high insect production (Anderson and Ohmart unpubl.) appear to be essential factors.

Salt Cedar

The introduced salt cedar, though not supporting any animal specialists, does have an interesting relationship with wildlife. Cohan et al. (1978) in a three-year study concluded that native birds responded to salt cedar communities in a number of ways. Some species showed a preference for or did not avoid salt cedar; these were primarily ground feeders, granivores, or species such as doves. A large number of those species utilizing salt cedar belong to genera of the Old World where salt cedar evolved. Frugivores did not utilize salt cedar at all, which is not surprising since no fruits or berries are produced by this plant. Salt cedar also frequently occurs as a monoculture. Very few insectivores used salt cedar communities, although a number of palatable insects (cicadellids) may be present temporarily in large numbers. Cohan et al. (1978) concluded that the New World avifauna, especially insectivores, may be avoiding salt cedar monocultures because of the sticky exudate produced by the salt glands on the leaves. Possibly native bird species can tolerate intermittent foraging in salt cedar when it is mixed with native vegetation; mixed communities containing salt cedar supported more bird species and had higher densities than did salt cedar monocultures.

Salt cedar is most valuable to wildlife in those areas that are in the geographical distribution of the white-winged dove (*Zenaida asiatica*). This dove is a plant structural specialist and selects nest sites in communities that have maximum foliage volumes between 3 and 6 m high with a lack of vegetation above 9 m. At maturity salt cedar, honey mesquite, and screwbean mesquite provide the vegetative profile required by this dove for nest placement. Butler (1977) examined nest placement and nest density of white-winged doves in salt cedar communities along the Colorado River and found a strong correlation between nest placement and the preferred maximum foliage volume between 3 and 6 m. Nests located above or below maximum foliage volume suffered much higher rates of predation or had reduced fledging success. Similar results have been reported on the Rio Grande in the Chihuahuan Desert in west Texas (Engel-Wilson and Ohmart 1978).

Doves comprise such a large portion of the avifauna inhabiting mature salt cedar communities that this disproportionate number depresses bird species diversity values computed for these communities. A correction factor has been developed by Anderson, Higgins, and Ohmart (1977), based on random counts and behavior of doves in salt cedar communities. The correction factor was derived by estimating the number of doves that were actually feeding in salt cedar communities versus the percentage that were nesting or involved in some behavior other than feeding. The correction factor of removing 90 percent of the doves when calculating diversity values provides a more realistic bird species diversity value for salt cedar communities as compared with native communities.

Along the lower Colorado River (Anderson, Higgins, and Ohmart 1977) and the Rio Grande in west Texas (Engel-Wilson and Ohmart 1978), salt cedar communities do not support as high a density of native bird species, do not provide significant habitat for rare or unique species, and, even with a correction factor of using only 10 percent of the doves, do not have as high a bird species diversity value as most native communities. During the winter period, the salt cedar is leafless, and very few birds forage in it. Other than its value for nesting doves, it is more comparable in wildlife use values to the native arrowweed, which generally forms a monocultural stand and seldom grows taller than about 3 m.

Mature salt cedar stands generally attain a height of about 8 m and are a monoculture with a heavy overstory and very little, if any, understory. The wildlife use values of the mature communities are higher than those of the younger or less-developed stands. But because of the deciduous nature of the species and in the absence of flooding to remove the litter during the fifteen to twenty years it takes the community to mature, there is a tremendous accumulation of litter. This litter buildup renders these communities highly susceptible to accidental or intentional fires. The heat from these fires is intense enough to kill the above-ground portion of the tree, resulting in heavy sucker growth from the root crown. The higher wildlife use value of the mature community is then lost for another ten to fifteen years until the community again reaches maturity. This fire cycle and loss of wildlife values has prompted Ohmart and Anderson (ms) to suggest periodic burning of this community type at five- or seven-year intervals. This would eliminate the litter accumulation prior to the point where the heat produced by the fire kills the above-ground portions of the tree, thus extending the potential productivity for dove nesting and other wildlife use values.

Quail Bush

Anderson et al. (1978) reported quail bush as another example of the value of a plant species to birds. Similar plant community types were studied with comparable foliage structures that differed only in the presence or absence of quail bush. Communities containing quail bush contained significantly ($p < 0.001$) higher densities of Abert towhees, blue grosbeaks (*Guiraca caerulea*), blue-gray gnatcatchers (*Polioptila caerulea*), brown-headed cowbirds (*Molothrus ater*), crissal thrashers, gambel quail, orange-crowned warblers (*Vermivora celata*), ruby-crowned kinglets (*Regulus calendula*), verdins (*Auriparus flaviceps*), and white-crowned sparrows (*Zonotrichia leucophrys*). This dense evergreen shrub provides cover and shade, and it may also harbor high densities of foliage arthropods, as well as numerous arthropods in the moist litter under the dense foliage. Though the plant-animal relationship is not clear, the presence of quail bush in a community enhances the habitat for a number of avian species.

Inkweed

Anderson et al. (1978) also reported another interesting bird-plant association: the relationship of inkweed (*Suaeda torreyana*) and the sage sparrow (*Amphispiza belli*). Why this relationship exists is not known, but wintering sage sparrows are consistently present in riparian areas that support inkweed and are largely absent in areas without inkweed along the lower Colorado River.

AERIAL ESTIMATES OF RIPARIAN COMMUNITIES

To determine the amount of riparian habitat in each desert would be a monumental task, and we do not know of published works that contain this information. Lowe (1968) has estimated that North American deserts cover about 176,000 (440,000 ha) square miles. Szaro (in press) reported that less than 0.005 percent of the land area in Arizona is represented by riparian ecosystems. We have taken aerial photographs at random in the Sonoran Desert and planimetered them to determine riparian zones versus desert vegetation. Preliminary estimates indicate that about 1 percent of the Sonoran Desert is comprised of riparian vegetation. Numerous random blocks should be planimetered in this manner for more definitive estimates, but preliminary figures indicate the paucity of riparian habitats in the Sonoran Desert. This figure is probably an overestimate for the Mojave and Great Basin deserts and may be slightly conservative for the Chihuahuan Desert. It has been estimated that riparian vegetation constitutes about 138,800 ha (347,000 acres) or less than 0.005 percent of the total land area in Arizona (Smith in Sands and Howe 1977). An estimate of the amount of riparian habitat in 117,554 ha (293,866 acres) of Lower Sonoran Desert in the Aquarius Mountain Planning Unit in Arizona shows that less than 7 percent is riparian and only 0.5 percent is cottonwood-willow (Bureau of Land Management 1980). Numerous drainages arise in the Aquarius Mountains, making the desert area rich in riparian habitats and inflating the riparian habitat-upland desert ratio. Although definitive values on areal extent of riparian habitats are lacking, they are highly restricted in desert environments.

IMPORTANCE AND MODIFICATION OF RIPARIAN ECOSYSTEMS BY MAN

Riparian ecosystems are vital to man. Without riparian communities we could not survive in desert environments. Like many other animals, we are dependent on riparian ecosystems for water, food, and living space. These ecosystems provide drinking water, rich lands for agricultural purposes, forage for livestock production, recreational opportunities, and a host of other uses. Indians, Spanish, and white settlers congregated along these riparian habitats and could not have survived in the desert environment without the water and high plant and animal productivity provided by these ecosystems.

Indians who lived in the southwestern riparian communities modified riparian habitats in varying degrees. Some tribes farmed and cleared parcels of land (Kroeber 1939; Forbes 1965), depending on their industriousness at farming. Because they did not practice flood control or river management, there were no significant modifications of the riparian floodplain.

Spanish influence on riparian habitats was of greater significance than that of the Indians. Permanent settlements were established that concentrated the Spanish and Indians around forts and churches. Prior to the Spanish influence, the Indian was more nomadic, except in tribes with sophisticated farming technology, moving along the riparian corridor as resource needs dictated. Once permanently settled by the Spaniards, the local area was used more intensively to produce natural foods such as mesquite pods, game, and fish. Agricultural practices

were also heavily relied on for food production. The domestic livestock brought by the Spanish were dependent on the forage along the rivers. Riparian trees were used for fuel, building materials, and timbers for mining operations.

Though localized, these permanent Spanish settlements began to have a significant impact on the riparian resources. Large numbers of domestic livestock depleted forage reserves. Mesquite pods were highly prized by both livestock and Indians, thus causing constant turmoil between the two groups along the lower Colorado River. Spanish expeditions herding large numbers of domestic livestock continually arrived on the Colorado River with supplies depleted. The animals were grazed on the best lands and consumed the mesquite crops. In July 1871, the Indians revolted and killed the Spanish settlers; a major reason was the constant conflict caused by the Spanish settlers and their domestic livestock competing with the Indians for food (Forbes 1965). Farming operations do not mix well with domestic livestock without good fencing, and this interaction must have been a continuing source of conflict since agricultural practices were relegated to the Indians.

White settlement began in about 1860, and by 1900 ranching and farming were major activities in the Southwest. For example, in 1886 the governor of the Arizona Territory reported, "In Arizona by 1883-84 every running stream and spring was settled upon, ranch houses built, and adjacent ranges stocked." (Report of Governor 1896:21). Mining flourished during the 1870s, and goods had to be transported overland or, where available, along navigable streams on wood-powered steamboats. Any type of activity in the riparian floodplain was constantly hampered by the annual floods in primary systems. Consequently river management was deemed essential, and attempts began in the early 1900s, but it was only with the construction of major dams such as Roosevelt Dam on the Salt River, Hoover Dam on the Colorado River, and Elephant Butte Dam on the Rio Grande that true river management was achieved. Other storage reservoirs promptly followed, and new river channels were developed to expedite water transport. The banks of the new channels were riprapped or lined with rocks to prevent bank cutting and river meandering.

Livestock Grazing

Massive domestic livestock grazing by white settlers began about 1880. Examples from the Sonoran Desert illustrate the growth of this industry and provide an indication of the density of domestic livestock that were introduced to desert rangelands. In 1870 there were only an estimated 5000 cattle in the entire Arizona Territory (U.S. Bureau Census 1872:III, 75). New settlers and army posts increased the demand for red meat over the supply, and in 1872 four herds totaling about 15,500 head (Wagoner 1952:36) were brought into Arizona. By 1885 there were about 652,500 cattle in the territory (Report of

Governor 1896:21). A census report in 1890 showed 1,095,000 cattle in the Arizona Territory (U.S. Bureau of Census 1895:I, 29). Overstocking of the range was obvious, and between depleted forage reserves and reduced summer rains in 1883, the herds began dying. "Dead cattle lay everywhere. You could actually throw a rock from one carcass to another" (Land 1934). This did not deter the cattlemen, and Ames (1977) reported that in 1900 there was a peak number of 173,000 head grazing Pima and Santa Cruz counties in southern Arizona.

Overstocking of desert range lands continued even after national forests were established (circa 1895) and state lands were assigned by the federal government and turned over to the respective states. The Bureau of Land Management (BLM) was established as a livestock managing agency under the Taylor Grazing Act of 1934. State and federal agencies were charged with the wise use and management of the natural resources on these public lands. But the cattle industry was and still is very powerful, and in most instances management agencies have either made no effort to regulate and control domestic livestock numbers on their lands or have failed.

Overgrazing of riparian habitats and their watersheds results in a number of degradative modifications of the ecosystem. When one or more portions of an ecological system are disrupted, other portions are affected as well. When a number of interactive parts are modified, the system may be severely interrupted, and the health of the ecosystem is impaired.

The action of overgrazing reduces the roughness coefficient in the watershed and results in more surface runoff, greater amounts of soil erosion, and ultimately massive floods in the riparian habitats. Davis (1977:61) stated that "overstocking and the consequent loss of vegetative ground cover on the adjacent watersheds is probably the main reason for the frequency of high intensity floods resulting in drastic changes in the density and composition of riparian bottoms." The watersheds degenerate from overgrazing as plant health is lost, and ultimately native perennial grasses are replaced by annual grasses, forbs, and shrubs. Reduction of the perennial grass cover, which serves as a soil stabilizer, allows more soil erosion through greater amounts of surface runoff from the watershed. Conversion from a perennial grass type to annuals and shrubs also results in a reduced carrying capacity for domestic livestock and wildlife.

Perennial streams with overgrazed watersheds are continually charged with heavy sediment loads that reduce the productivity in the stream. Suspended sediments reduce light penetration to plants needing solar energy for photosynthesis and reduce the oxygen carrying capacity of the water. This results in selection for fish and invertebrates species that can tolerate lower dissolved oxygen levels (see Kennedy 1977 for a review). All of these impacts reduce the productivity of desert streams.

Sediment loads from abused watersheds in dry or intermit-

tent desert streams remain in the stream channel until another flood pulse carries them farther downstream and closer to a reservoir, they displace space that could be used for water storage, thereby reducing the life and capacity of the impoundment; eventually these sediments must be mechanically removed, at high cost.

Cattle not only overgraze the watershed but also consume the succulent and verdant foliage of the riparian vegatation. Most desert habitats are grazed all year, and during the summer months the riparian ecosystems provide forage, shade, and water to domestic livestock. Intensive, concentrated use by livestock in the floodplain prevents regeneration of riparian vegetation. Davis (1977:60) wrote, "Continued over use of riparian bottoms eliminates essentially all reproduction as soon as it becomes established." Carothers (1977:3) wrote that "the most insidious threat to the riparian habitat type today is domestic livestock grazing." And Ames (1977:49) concluded that "protection of the riparian type where grazing is an established use only can be effectively achieved through fencing."

As the watershed deteriorates from overgrazing and the tree density in riparian systems is reduced by grazing and through natural loss of old trees, the magnitude and destructive forces of floods are increased. A healthy watershed absorbs most of the falling rain, and a dense riparian habitat dissipates the energy of flood waters. But the poor watersheds and thinned riparian vegetation are more vulnerable to floods; large trees and whole riparian communities may be ripped out and washed downstream during a flood. This expedites riparian degradation since cattle are concentrated in the riparian areas each spring to consume newly established seedlings. Riparian communities are quick to recover following livestock removal. Unfortunately perennial grasses are slow to reinvade the dry and degradated soils on the eroded watershed.

Overgrazing also aids the spread and establishment of exotic species such as salt cedar. Domestic livestock will browse the new growth of salt cedar, but they do not take enough new material to reduce its vigor (Gary 1960). Consequently overgrazing by domestic livestock promotes the elimination of native riparian species and aids the establishment of the exotic salt cedar. Ultimately salt cedar attains such tall, dense growth that domestic livestock are reluctant to enter it (Gary 1960).

Arroyo Formation

Domestic livestock grazing has been interpreted as being the primary cause of entrenching or arroyo formation in the West (Dodge 1910; Thornber 1910; Rich 1911; Duce 1918; Leopold 1921; Cooperrider and Hendricks 1937; Cottam and Stewart 1940). Many primary and secondary streams in desert areas are entrenched or are beginning to show signs of entrenchment. This is not to be confused with erosion patterns associated with climatic changes that have been traced through alluvial stratigraphy since the last glaciation (Haynes 1968). These entrenchments are the vertical walled channels that have been cut since the beginning of white settlements (Cooke and Warren 1973). The increased cutting is a result of increased flow velocity, an increase in erodibility of soils, or a combination of both (Cooke and Warren 1973). Flow velocity could be increased by increasing discharge or slope, reducing surface roughness, or increasing depth of flow. "Increased erodibility of valley-floor materials might have arisen through the reduction of riparian vegetation" (Cooke and Warren 1973:170). Altering surface roughness by removal of perennial grasses on the watershed would result in increased surface flow, which would increase erosion and ultimately increase slope. As Cooke and Reeves (1976) point out changes in depth of flow were probably influenced by man-made modifications in the drainage system. Increased rainfall (Huntington 1914) would produce similar, but possibly not as drastic, surface flows. But some argue that the climate may have become drier (Bryan 1925). If the climate had become wetter or drier, there should have been compensating changes in the vegetation, but to our knowledge only degradational changes have been reported (Hastings and Turner 1965).

There are probably many causes of entrenching, and they appear to vary from drainage to drainage (Cooke and Reeves 1976), but there is little doubt that most entrenchment is a result of past and present land use practices by white settlers. One strong influence of entrenching in many desert drainages appears to have been the destruction of western watersheds by overgrazing of domestic livestock.

An example of entrenchment is seen on Sonoita Creek (figures 10-5 and 10-6), which flows into the Santa Cruz River a few kilometers north of the U.S.-Mexican boundary. The history of the land use of this stream has not been synthesized, but Glinski (1977:121) commented that "the overgrazed hillsides that border Sonoita Creek no doubt assist in increasing the floodwater velocity since they support relatively less vegetation to intercept precipitation."

Entrenchment not only results in habitat removal for riparian vegetation, but it results in a lowered water table, causing vegetational species shifts and collapse of mature trees adjacent to the bank as the soil erodes away from their roots. Following floods, seedlings of riparian species attempt reestablishment but are quickly eaten back by domestic livestock. Glinski (1977:121) observed numerous instances of seedlings

Figure 10-5. Entrenchment along Sonoita Creek.
1976 photograph by R. L. Glinski.

being eaten by cattle and stated that "grazing of small seedlings by cattle was the most obvious factor preventing regeneration of cottonwood."

The San Simon Basin lies in the Chihuahuan Desert and extends from extreme southwest New Mexico into eastern Arizona. The adjacent uplands currently support such woody species as creosote bush, mesquite, joint-fir (*Ephedra* spp.), and tarbush (*Flourensia cernua*). The valley floor sediments have been entrenched over most of the area from San Simon to Safford, Arizona, a distance of over 100 km. The entrenched drainage supports a few scattered cottonwood and salt cedars. Early descriptions of the valley indicate that the area was heavily vegetated with grasses, with water readily available in the stream for livestock. Hinton (1878), an early topographer, reported good grazing lands consisting of sacaton (*Sporobolus* spp.) and gramas (*Bouteloua* spp.) and some agricultural lands present along the valley floor. Barnes (1936) interpreted early records and described the valley in 1882 as being composed of meadows covered with knee-high grass, open areas dominated by tall gramas, and sacaton stirrup-high along the washes.

By the early 1900s the picture was changed from "pristine beauty" (Peterson 1950) to an upland habitat dominated by woody vegetation and a riparian system that was heavily entrenched and contained only scattered cottonwood trees. Olmstead (1919:79) was so disheartened that he stated, "Oh, Liberty, how many crimes are committed in thy name!" Peterson (1950:410) wrote, "Today's picture of the valley, from both the conservation and the range-use viewpoint, is one of devastation." Figures 10-7 and 10-8 show some of the entrenchment that has occurred along this drainage.

Entrenchment has occurred along some portions of the San Pedro River in southeastern Arizona. Evidence presented by Cooke and Reeves (1976) argues that some entrenching occurred on the San Pedro River prior to white settlement. The possible causes of further entrenchment after about 1880 are numerous, but the most plausible ones are man-made structures in the floodplain, which caused drainage concentration, climatic change, which caused severe floods during this period, and vegetation modifications from overgrazing by domestic livestock.

Figure 10-6. Vertical entrenchment on Sonoita Creek.
1976 photograph by R. L. Glinski.

Figure 10-7. Entrenchment in the San Simon Valley in eastern Arizona.
The vertical walls are 3 to 4 m. The photograph was taken about 12 km upstream from the
mouth in the mid-1960s by Bureau of Land Management personnel.

Figure 10-8. Aerial oblique of entrenchment conditions, San Simon Valley, eastern Arizona.
Note the riparian vegetation that has developed in the entrenched channel. The photograph was
taken about 12 km upstream from the mouth in the mid-1960s by Bureau of Land Management
personnel.

Hastings and Turner (1965) provide numerous matched photographs that illustrate extensive changes in the vegetative ground cover. Rodgers (1965) argues against climatic change before or after 1885, but the records are inadequate for firm conclusions. Structures were built in the floodplain, and these probably acted in concert with the other factors to cause entrenchment. Cooke and Reeves (1976:46) stated:

There is a strong possibility that vegetation changes resulting from overgrazing within the watershed (especially south of Benson), cattle damage along trails and the river, and deforestation of some catchment areas for mining timber may have promoted entrenchment.

Caution must be used in interpreting the causes for entrenchment, and each case must be viewed with as much historical evidence as possible before coming to a conclusion. Even when this is done, the data are less than satisfactory.

River Management

The significance of the effect of agricultural practices on riparian habitats was relatively small until river management was utilized in the early 1900s. Until river management became a reality, farmers cleared and worked portions of the alluvial floodplain, but because of the expense and constant threat of flooding, only small tracts were worked, and these were confined to the second terrace. The primary terrace was flooded annually, and in extreme flood years, the second terrace was inundated.

Many methods were attempted to avoid the natural occurrence of floods. One strategy was to use water transport systems to conduct water away from the floodplain to other fertile areas. With this in mind, the All American Canal was built, stretching from Yuma on the Colorado River to the Mojave Desert in Imperial Valley, California, near the Salton Sink. But even these water transport efforts were not foolproof without flood control structures. The massive floods along the Colorado River in 1905 broke through the floodgates and became diverted into the All American Canal. The flooding waters filled the dry Salton Sink, converting it into the Salton Sea (Sykes 1937).

Major dams were built that converted riparian habitats into desert reservoirs and virtually halted natural flooding. This action slowed or stopped the reproductive cycle of some native riparian plant species. Broad, meandering rivers were channelized, banks were riprapped with stones, and the riparian communities along the second terrace were cleared for agricultural production. Levees were placed along the river at the natural division of the first and second terrace. Anything inside the levees was considered a potential flood zone, and this preserved some riparian habitats in the first terrace. Those communities on the second terrace were ripped out, raked into

piles, and burned. One example of the magnitude of removal occurred in 1961 when 32,000 ha (80,000 acres) were cleared for farming along the lower Colorado River (Fox 1977). Anderson and Ohmart (1978) estimated that the mean removal rate of riparian vegetation along the lower Colorado River has been about 1200 ha per year. Ohmart et al. (1977) conservatively estimated that there were over 2000 ha (5000 acres) of cottonwood habitat along the lower Colorado River prior to white settlement. Anderson and Ohmart (1977a) estimated that fewer than 1100 ha (2800 acres) of cottonwood-willow communities remain, and that fewer than 200 ha (500 acres) are pure cottonwood habitats.

Similar examples exist along the Rio Grande slightly below El Paso, south and east to Presidio, Texas, a distance of about 450 km (275 mi) (Engel-Wilson and Ohmart 1978). Using written and photographic documentation, the cottonwood and willow communities were traced from about 1850 to the present; in slightly over one-hundred years these communities have been virtually extirpated. Water use upstream has left this reach of the river dry, and now the abandoned farmland in the alluvial floodplain supports only a monoculture of salt cedar. Cottonwoods are found only as scattered mature trees. There is no reproduction of native cottonwoods along this reach of river, and when the mature trees die, the area will be devoid of cottonwood.

A secondary impact of agricultural practices on riparian systems is the return flow to the river caused by flood irrigation. This irrigation practice leaches salts, fertilizers, and insecticides out of the land and deposits them in the river. The phosphates in fertilizers may help enrich some rivers, but the salts and pesticides pose future problems for water managers. As levels build, the aquatic biota will be modified and possibly eliminated.

In many instances, there is a general ecological succession in primary drainages where there is a high degree of river management. Natural plant communities are converted to agriculture and then to urban developments. If the Rio Grande in west Texas is any indication, urban development may not be the final stage. As water supplies are exhausted upstream, the lower river reaches become cesspools laden with industrial, municipal, and agricultural wastewater polluted with various levels of salts, heavy metals, pesticides, and other contaminants.

As river management became a reality and water storage became possible, many hydrologists began to cast about for further ways to ensure that every drop of water was conserved for man's use. Riparian habitats were identified as useless and wasteful because these plant communities transpired large quantites of water from the groundwater table to the dry atmosphere. Massive programs were designed, and many undertaken, to remove phreatophytes for water salvage. Along the Pecos River in New Mexico, over 20,000 ha (50,000 acres) were cleared and have been maintained solely for the purpose of water salvage.

The actual amount of water saved through clearing phreatophytes to increase water flow is still unresolved. Lysimeters have been used to attempt to determine if water has been saved through riparian vegetation removal, but the data are confusing and conflicting. Many argue the merits of phreatophyte removal for water savings, while others question the validity of these claims.

Varied Effects

Mining activities were widespread throughout the desert Southwest, and the scars of these old mines are common along many of the desert drainages. The primary modification that mining causes in riparian habitats is stream pollution, but large operations have extensive spoil materials that, in some instances, have been dumped along streams covering existing riparian communities. In these instances and when stream pollution is high, the entire stream biota is killed, and the salts, heavy metals, and other toxins leach into the soil. This activity results in changes or elimination of the submerged and terrestrial plant communities along the stream.

Ultimately these polluted waters empty into reservoirs and lakes to become the water supply of metropolitan areas. In time, these pollution sources may become numerous enough or concentrated enough to affect the quality of the water supplies of large cities.

Avenues of transportation such as roads, highways, and railroads generally are placed adjacent to streams in the riparian ecosystem. These drainages provide the easiest and cheapest routes to cross areas with great topographic relief. The grade is usually not too steep, and the roadbed is built along the contour of the drainage.

Spoil materials from road construction activities are pushed downhill and eventually erode into the stream, causing turbidity problems. Unpaved areas cause the greatest problems because soil from these denuded areas is continually eroded into the stream.

Riparian areas have always been the focal points for recreation in arid environments, but it has become a serious problem only in the past two decades as human populations have increased in desert environments. Humans, like many other animals, concentrate in the primary drainages during long, hot summers. Areas receiving large concentrations of people are often denuded as wood is gathered for camp fires, and riparian areas became a place where all-wheel drive vehicles and egos are tested. Children and pets catch and usually kill any form of animal life discovered.

The exact faunal changes that take place when an area is converted to a campground and receives heavy human use are poorly understood. Aitchison (1977) examined a site in Arizona slightly higher in elevation than desert riparian habitats and found little difference between bird species composition in a control site as compared with a constructed campground that had not been opened. After the campground was opened, bird species composition and diversity decreased. Larger species of birds tended to persist after the area was opened. Information of this nature concerning desert riparian habitats is lacking.

Waterfowl use on the lower Colorado River appears to be strongly influenced by human use and river channel modification (Anderson and Ohmart unpubl.). As each factor increases, there is a direct and predictable reduction in the number of waterfowl, primarily ducks.

Current Status and Future

All primary and most secondary drainages in desert environments have been modified by man. The range of modification varies from moderate to severe destruction. All primary drainages are impounded, regulated, and managed to the point where any hope of natural regeneration of the most productive riparian habitats is unrealistic. Extensive mesquite bosques, which once covered the entire second terrace, have been or are being cleared for agricultural production. Even more alarming, river management has been so effective and enforcement so lax that agriculture and urban development have begun to penetrate into the areas between the levees. These areas are now being destroyed, and if water releases become necessary, homes, equipment, and crops will be washed away. Further, many planning and zoning commissions have allowed land developers to parcel and sell lots and homes to unsuspecting buyers. After a flood, the home owner is financially ruined and has little, if any, hope of restitution.

The once-extensive gallery forests of cottonwood and willow that lined our desert streams are either completely gone or persist as small relics. The dense and varied fauna they once supported have dwindled and now are state or federally listed as sensitive, threatened, or endangered species.

Barren and eroded watersheds supporting only annual plants and shrubs attest to the hoards of livestock that once trampled their slopes. Silt-laden reservoirs catch the eroded soils at lower elevations, and the newly deposited silt displaces the water-holding capacity of the reservoirs. Ultimately the life of the reservoir is shortened or the materials have to be removed; either alternative is very costly.

Heavily managed river systems that once supported extensive communities of cottonwood and willow now support dense communities of salt cedar. Cessation of natural floodings, higher salinities from return irrigation flow practices, domestic livestock grazing, and a reduction of in-stream flow amounts have favored the spread of the exotic salt cedar.

The wildlife these native communities once supported is now gone, but the total loss is unknown since there were no quantitative records of its existence. Nevertheless the existing riparian communities are still the most productive and impor-

Figure 10-9. Year-old cottonwood and willow trees revegetated with drip irrigation along the lower Colorado River in California.
Photograph taken in 1980 by R. Ohmart.

Figure 10-10. Two-year-old cottonwood trees revegetated with drip irrigation along the lower Colorado River in California.
Photograph taken in 1980 by R. Ohmart.

tant to desert wildlife in spite of their degraded state. Further degradation can be expected; Johnson et al. (1977) predicted the extirpation of 47 percent of the 166 species of nesting birds in the area studied. Many mammalian species would be reduced in density and others eliminated from our deserts. The same holds true for species of both reptiles and amphibians in desert environments.

The future of riparian communities is mixed. Lands under federal management are gaining more attention with respect to proper livestock use levels, restriction of recreational use and woodcutting, reduced use of off-road vehicles, and the elimination of the myriad of other destructive uses imposed on riparian systems.

Not only have there been federal laws passed that have reduced the destructive processes on federal lands (among them, the National Environmental Policy Act, the Endangered Species Act, and the National Wetlands Act), but there is a growing recognition among federal land managers of the value of the riparian vegetation to man and wildlife. Numerous federal agencies have begun to take steps to preserve existing remnants of riparian habitat, while others have begun the costly and slow process of attempting to revegetate bare areas with riparian plant communities (figures 10-9 and 10-10) valuable to wildlife (Anderson et al. 1978).

The greatest threat of continued destruction of riparian habitat is at the state, local, and private ownership level. When water rights were allocated, no one thought of or represented riparian communities or their dependent wildlife. Consequently no water was allocated to this valuable resource. Water managers and users view these riparian communities as stealing their water and believe that removal of this vegetation saves water. This type of attitude generally prevails at the state and local levels, and few desert states have laws that provide any protection to riparian ecosystems. Therefore, riparian ecosystems on state, county, and private lands are and will continue to be in jeopardy until an educated public and legislators act on this problem.

A classic case of finally recognizing the value of the riparian system and the constant problem of recurrent floods at the local level occurred in the mid-1970s in Scottsdale, Arizona. Virtually every method to control the flooding along the Indian Bend Wash had been exhausted when someone proposed making a green belt through the city and using it as open space instead of periodically inundated home sites. When the wash is dry, people picnic, ride horses and bicycles, and engage in other recreational activities along the wash. In three recent years of heavy flooding, it has acted as a conduit to drain the upland desert of its overburden of rainfall.

If the riparian resource is to be saved, it will only be through informative programs that make the public aware of the value of this habitat and the need to maintain as much of it is as possible. We can only hope that this awareness comes before it is too late.

REFERENCES

Aitchison, S. W. 1977. Some effects of a campground on breeding birds in Arizona. *USDA Forest Service, Gen. Tech. Rept.* RM-43:175-182.

Ames, C. R. 1977. Wildlife conflicts in riparian management: grazing. *USDA Forest Service, Gen. Tech. Rept.* RM-43:49-51.

Anderson, B. W.; Engel-Wilson, R. W.; Wells, D.; and Ohmart, R. D. 1977. Ecological study of southwestern riparian habitat: Techniques and data applicability. *USDA Forest Service, Gen. Tech. Rept.* RM-43:146-155.

Anderson, B. W.; Higgins, A.; and Ohmart, R. D. 1977. Avian use of salt cedar communities in the lower Colorado River valley. *USDA Forest Service, Gen. Tech. Rept.* RM-43:128-136.

Anderson, B. W., and Ohmart, R. D. 1977a. *Wildlife use and densities report of birds and mammals in the lower Colorado River valley.* USDI Bur. Rec., Lower Colo. Region.

———. 1977b. Vegetation structure and bird use in the lower Colorado River valley. *USDA Forest Service Gen. Tech. Rept.* RM-43:23-34.

———. 1978. Phainopepla utilization of honey mesquite forests in the Colorado River valley. *Condor* 80:334-338.

Anderson, B. W.; Ohmart, R. D.; and Disano, J. 1978. Revegetating the riparian floodplain for wildlife. *USDA Forest Service Gen. Tech. Rept.* WO-12:318-331.

Austin, G. T. 1970. Breeding birds of desert riparian habitat in southern Nevada. *Condor* 72(4):431-436.

Axelrod, D. I. 1950. The Piru Gorge flora of southern California. *Carnegie Inst. Wash. Publ.* 590:159-224.

———. 1958. Evolution of the Madro-Tertiary geoflora. *Bot. Rev.* 24:433-509.

Barnes, W. C. 1936. Herds in San Simon Valley. *Am. Forests* 42:456-457, 481.

Billings, W. D. 1951. Vegetational zonation in the Great Basin of western North America. *International Union Biol. Sci. Ser. B. Colloquia* 9:101-122.

Brown, D. E., and Lowe, C. H. 1974. The Arizona system for natural and potential vegetation—illustrated summary through the fifth digit for the North American Southwest. *J. Arizona Acad. Sci. 9 Suppl.* 3:1-7.

Brown, D. E.; Lowe, C. H.; and Pase, C. P. 1979. A digitized classification system for the biotic communities of North America, with community (series) and association examples for the Southwest. *Arizona-Nevada Acad. Sci. 14, Suppl.* 1:1-16.

Bryan, K. 1925. Date of channel trenching (arroyo cutting) in the arid Southwest. *Science* 62(1607):338-344.

Bureau of Land Management. 1980. *Aquarius unit resource analysis.* Phoenix, Arizona.

Burleigh, T. D. 1972. *Birds of Idaho.* Caldwell, Idaho: Caxton Printers.

Butler, W. I., Jr. 1977. A white-winged dove nesting study in three riparian communities on the lower Colorado River. Master's thesis, Arizona State University.

Campbell, C. J., and Dick-Peddie, W. A. 1964. Comparison of phreatophyte communities on the Rio Grande in New Mexico. *Ecology* 45(3):492-502.

Campbell, C. J., and Green, W. 1968. Perpetual succession of stream-channel vegetation in a semi-arid region. *J. Ariz. Acad. Sci.* 5(2):86-89.

Carothers, S. W. 1977. Importance, preservation and management of riparian habitat: An overview. *USDA Forest Service, Gen. Tech. Rept.* RM-43:2-4.

Carothers, S. W., and Johnson, R. R. 1975. Water management practices and their effects on nongame birds in range habitats. In D. Smith, *Proc. of Symp. on Management of Forests and Range Habitats for Nongame Brids*, pp. 210-222. *USDA Forest Service, Gen. Tech. Rept.* 1.

Cates, R. G., and Rhoades, D. F. 1977. *Prosopis* leaves as a resource for insects. In *Mesquite* ed. B. B. Simpson, pp. 61-83. *US/IBP Synthesis series*, 4. Stroudsburg, Pa: Dowden, Hutchinson & Ross.

Christensen, E. M. 1962. The rate of naturalization of *Tamarix* in Utah. *Am. Midl. Nat* 68(1):51-57.

Cockerell, T. D. A. 1900. Some bees visiting the flowers of mesquite. *Entomologist* 33:243-245.

Cohan, D. R.; Anderson, B. W.; and Ohmart, R. D. 1978. Avian population responses to salt cedar along the lower Colorado River. *USDA Forest Service, Gen. Tech. Rept.* WO-12:371-382.

Cole, G. L. 1963. The American Southwest and middle America. In D. G. Frey, ed., *Limnology in North America*, pp. 393-434, Madison: University of Wisconsin Press.

Cooke, R. U., and Reeves, R. W. 1976. *Arroyos and environmental change in the American Southwest*. Oxford Res. Studies in Geog. Oxford: Clarendon Press.

Cooke, R. Un., and Warren, A. 1973. *Geomorphology in deserts.* San Francisco: University of California Press.

Cooperrider, C. K., and Hendricks, B. A. 1937. Soil erosion and stream flow on range and forest lands of the upper Rio Grande watershed region in relation to land resources and human welfare. *USDA Tech. Bull.* 567.

Cottam, W. P., and Stewart, G. 1940. Succession as a result of grazing and of meadow desiccation by erosion since settlement in 1862. *J. Forestry.* 38:613-626.

Cushman, R. A. 1911. Notes of the host plants and parasites of some North American Bruchidae. *J. Econ. Entomol.* 4:489-510.

Darrow, R. A. 1961. Origin and development of the vegetational communities of the Southwest. Bioecology of the arid and semi-iarid lands of the Southwest. *New Mexico Highlands Univ. Bull.* 212:30-47.

Daubenmire, R. F. 1942. An ecological study of the vegetation of southeastern Washington and adjacent Idaho. *Ecol. Monogr.* (12)1:53-79.

Davis, G. A. 1977. Management alternatives for the riparian habitat in the Southwest. *USDA Forest Service, Gen. Tech. Rept.* RM-43:59-67.

Deacon, J. E., and Minckley, W. L. 1974. Desert fishes. In *Desert biology*, G. W. Brown, Jr., vol. 2, pp. 385-488. New York: Academic Press.

Denny, C. S. 1965. Alluvial fans in the Death Valley region, California and Nevada. *USGS Prof. Paper* 466.

Dick-Peddie, W. A., and Hubbard, J. P. 1977. Classification of riparian vegetation. *USDA Forest Service, Gen. Tech. Rept.* RM-43:85-90.

Dodge, R. F. 1910. The formation of arroyos in adobe filled valleys

in the southwestern United States (abs.). *Brit. Assoc. Advanc. Sci. Rept.* 79:531-532.

Dregne, H. E. 1968. Appraisal of research on surface materials of desert environments. In *Deserts of the world*, ed. W. G. Mc-Ginnis, B. J. Goldman, and P. Paylore, pp. 287-316. Tucson: University of Arizona Press.

Duce, J. T. 1918. The effect of cattle on the erosion of canyon bottoms. *Science* 47:450-452.

Engel-Wilson, R. W., and Ohmart, R. D. 1978. Assessment of vegetation and terrestrial vertebrates along the Rio Grande between Fort Quitman, Texas and Haciendita, Texas. Submitted to Int. Boundary and Water Comm., El Paso, Texas. In fulfillment of Contract No. IBM 77-17.

Feth, J. H., and Hem, J. D. 1963. Reconnaissance of headwater springs in the Gila River drainage basin, Arizona. *USGS Water-Supply Paper* 1619-H.

Fisher, C. E.; Meadors, C. H.; Behrens, R.; Robison, E. D.; Marion, P. T.; and Morton, H. L. 1959. Control of mesquite on grazing lands. *Texas Agric. Exp. Sta. Bull.* 935:1-23.

Flack, J. A. D. 1976. Bird populations of aspen forests in western North America. *Ornith. Monogr.* 19:1-97.

Flowers, S. 1960. Vegetation of Flaming Gorge reservoir basin. In *Ecological studies of the flora and fauna of Flaming Gorge Reservoir basin, Utah and Wyoming*, ed. C. E. Dibble. *Univ. Utah Anthropol. Pap.* 48:1-48 (Upper Colorado Series 3).

————. 1961. Vegetation of the Navajo reservoir basin in Colorado and New Mexico. In *Ecological studies of the flora and fauna of Navajo Reservoir basin, Colorado and New Mexico*, ed. D. M. Pendergast. *Univ. Utah Anthropol. Pap.* 56:1-98 (Upper Colorado Series 5).

Forbes, J. 1965. *Warriors of the Colorado.* Norman: University of Oklahoma Press.

Fox, K. 1977. Importance of riparian ecosystems: Economic considerations. *USDA Forest Service, Gen. Tech. Rept.* RM-43:19-22.

Gary, H. L. 1960. Utilization of five-stamen tamarisk by cattle. *USDA Forest Service, Rocky Mt. Forestry and Range Exp. Sta., Res.* Note 51.

————. 1963. Root distribution of five-stamen tamarisk, seepwillow, and arrowweed. *For Sci.* 9:311-314.

Glendening, C. E., and Paulsen, Jr., H. A. 1955. Reproduction and establishment of velvet mesquite as related to invasion of semidesert grasslands. *USDA Tech. Bull.* 1127.

Glinski, R. L. 1977. Regeneration and distribution of sycamore and cottonwood trees along Sonoita Creek, Santa Cruz County, Arizona. *USDA Forest Service, Gen. Tech. Rept.* RM-43:116-123.

Goldwasser, S. 1978 Desert riparian-freshwater marsh and ponds. *Amer. Birds* 32(1):113-114.

Graf, W. L. 1978. Fluvial adjustments to the spread of tamarisk in the Colorado Plateau region. *Geol. Soc. of Amer. Bull., Doc.* 81005, 89:1491-1501.

Grinnell, J., and Miller, A. H. 1944. The distribution of the birds of California. *Pacific Coast Avif.* 27.

Haas, R. H.; Meyer, R. E.; Scifres, C. J.; and Brock, J. H. 1973. Growth and development of mesquite. *Texas Agric. Exp. Sta. Res. Monogr.* 1:10-19.

Haase, E. F. 1972. Survey of floodplain vegetation along the lower Gila River in southwestern Arizona. *J. Ariz. Acad. Sci.* 7(2):66-81.

Hastings, J. R., and Turner, R. M. 1965. *The changing mile.* Tucson: University of Arizona Press.

Haynes, C. V., Jr. 1968. Geochronology of lake-Quaternary alluvium. In *Means of correlation of Quaternary successions,* ed. R. B. Morrison and H. E. Wright, Jr., pp. 591-615. Salt Lake City: University of Utah Press.

Hayward, L.; Beck, D. E.; and Tanner, W. W. 1958. Zoology of the Upper Colorado River Basin. *Brigham Young University Publ. Biol. Series* 1(3).

Hayward, C. L., Cottam, C.; Woodbury, A. M.; and Frost, H. H. 1976. Birds of Utah. *Great Basin Nat. Memoirs* 1:1-229.

Hinton, R. J. 1878. *The hand-book of Arizona.* Reprinted 1954, Tucson: Arizona Silhouettes.

Horton, J. S. 1964. Notes on the introduction of deciduous-tamarisk. *USDA Forest Service Res. Note* RM-16, Fort Collins, Colo. Rocky Mt. For. and Range Exp. Stn.

———. 1977. The development and perpetuation of the permanent tamarisk type in the phreatophyte zone in the Southwest. *USDA Forest Service, Gen. Tech. Rept.* RM-43:124-127.

Hubbard, J. P. 1977. Importance of riparian ecosystems: Biotic considerations. *USDA Forest Service, Gen. Tech. Rept.* RM-43:14-18.

———. 1978. Revised check-list of the birds of New Mexico. *New Mexico Ornith. Soc. Publ.* 6.

Hubbs, C. L. 1940. Speciation of fishes. *Amer. Nat.* 74:198-211.

———. 1941. The relation of hydrological conditions to speciation in fishes. In *A symposium on hydrobiology,* ed. J. G. Needham, pp. 182-195. Madison, Wisc.: University of Wisconsin Press.

Hulse, A. H. 1974. An autoecological study of Kinosteron sonoriense Le Conte (Chelonia: Kinosternidae). Ph.D. dissertation, Arizona State University.

Hunt, C. B. 1966. Plant ecology of Death Valley, California. *USGS Prof. Paper* 509.

Hunt, C. B.; Robinson, T. W.; Bowles, W. A.; and Washburn, A. L. 1966. Hydrologic basin, Death Valley, California. *USGS, Prof. Paper* 494-A:1-162.

Huntington, E. 1914. The climatic factor as illustrated in arid American. *Carnegie Inst. Wash. Pub.* 192.

Ingles, L. G. 1950. Nesting birds of the willow-cottonwood community in California. *Auk* 67(3):325-332.

Johnson, H. B. 1976. Vegetation and plant communities of southern California deserts. In *Plant communities of southern California. Spec. Publ.* 2. Berkeley, Calif.: Calif. Native Plant Soc.

Johnson, N. K. 1978. Patterns of avian geography and speciation in the intermountain region. *Great Basin Nat. Memoirs* 2:137-160.

Johnson, R. R.; Haight, L. T.; and Simpson, J. M. 1977. Endangered species versus endangered habitats: A concept. *USDA Forest Service, Gen. Tech. Rept.* RM-43:68-79.

Kearney, T. H.; Briggs, L. J.; Shantz, H. L.; McLane, J. W.; and Piemeisel, R. L. 1914. Indicator significance of vegetation in Tooele Valley, Utah. *J. Agric. Res.* 1(5):365-417.

Kennedy, C. E. 1977. Wildlife conflicts in riparian management: water. *U.S. Forest Service, Gen. Tech. Rept.* RM-43:52-58.

Kingsolver, J. M. 1964. The genus *Neltumius* (Coleoptera: Bruchidae). *Coleop. Bull.* 18:105-111.

———. 1967. On the genus *Ripibruchus* Bridwell, with descriptions of a new species and a closely related new genus. *Proc. Entomol. Soc. Wash.* (D.C.) 69:318-327.

———. 1972. Description of a new species of *Algarobius* Bridwell (Coleoptera: Bruchidae). *Coleop. Bull.* 26:116-120.

Kingsolver, J. M.; Johnson, C. D.; Swier, S. R.; and Teran, A. L. 1977. *Prosopis* fruits as a resource for invertebrates. In *Mesquite,* ed. B. B. Simpson, pp. 108-122. *US/IBP Synthesis Series* 4. Stroudsburg, Pa.; Dowden, Hutchinson & Ross.

Kroeber, A. L. 1939. Cultural and natural areas of native North America, lower Colorado River. *Univ. Calif. Publs. in American Archaelogy and Ethnology,* 38:42-45.

Land, E. 1934. Reminiscences. Ms. in Arizona Pioneers' Historical Society, Tucson.

Leopold, A. 1921. A plea for recognition of artificial works in forest erosion and control policy. *J. Forestry* 19:267-273.

Ligon, J. S. 1961. *New Mexico birds and where to find them.* Albuquerque, N.M.: University of New Mexico Press.

Logan, R. F. 1968. Causes, climates and distribution of deserts. In *Desert biology,* ed. G. W. Brown, Jr., Vol. 1, pp. 21-51. New York: Academic Press.

Lowe, C. H. 1964. *The vertebrates of Arizona.* Tucson: University of Arizona Press.

———. 1968. Appraisal of research on fauna of desert environments. In *Deserts of the world,* ed. W. G. McGinnis, B. J. Goldman, and P. Paylore, pp. 569-604. Tucson: University of Arizona Press.

MacArthur, R. H., and MacArthur, J. 1961. On bird species diversity. *Ecology* 42:594-598.

McKernan, R. L. 1978. Willow riparian. *American Birds* 32(1):85.

McMillan, C., and Peacock, J. T. 1964. Bud-bursting in diverse populations of mesquite (Prosopis: Leguminosae) under uniform conditions. Southwest Nat. 9:181-188.

Mares, M. A.; Enders, F. A.; Kingsolver, J. M.; Neff, J. L.; and Simpson, B. B. 1977. *Prosopis* as a niche component. In *Mesquite,* ed. B. B. Simpson, pp. 123-149. *US/IBP Synthesis Series* 4. Stroudsburg, Pa.: Dowden, Hutchinson & Ross.

Martin, S. C. 1948. Mesquite seeds remain viable after 44 years. *Ecology* 29:393.

Melton, M. A. 1958. Geometric properties of mature drainage systems and their representation in an E_4 phase space. *J. Geol.* 66:35-54.

Miller, R. R. 1959. Origin and affinities of the freshwater fish fauna of western North America. In *Zoogeography,* ed. C. L. Hubbs, pp. 187-222. Washington, D. C.: American Association for Advanced Science.

Minckley, W. L. 1973. *Fishes of Arizona.* Phoenix: Arizona Game and Fish Department.

Minckley, W. L., and Barber, W. E. 1971. Some aspects of the biology of the longfin dace, a cyprinid fish characteristic of streams in the Sonoran Desert. *Southwest Nat.* 15.

Moldenke, A. R., and Neff, J. L. 1974. The bees of California: A catalogue with special reference to pollination and ecological research. *Origin and Structure of Ecosystems Tech. Repts.* 74-1, 74-2, 74-3, 74-4, 74-5, 74-6.

Mooney, H. A.; Simpson, B. B.; and Solbrig, O. T. 1977. Phenology, morphology, physiology. In *Mesquite,* ed. B. B. Simpson, pp. 26-43. *US/IBP Synthesis Series* 4. Stroudsburg, Pa.: Dowden, Hutchinson & Ross.

Neff, J. L., Simpson, B. B.; and Moldenke, A. R. 1977. Flowers-flower visitor system. In *Convergent evolution in warm deserts,* ed. G. H. Orians and O. T. Solbrig, pp. 204-224. *US/IBP*

Synthesis Series 4. Stroudsburg, Pa.; Dowden, Hutchinson & Ross.

Oberholser, H. C. 1974. *The birdlife of Texas*. Edited by E. B. Kincaid. Jr. Austin: University of Texas Press.

Ohmart, Robert D., and Anderson, Bertin W. 1980. Wildlife use values of wetlands in the arid southwestern United States. *Proc. Nat. Symp. Wetlands, 1978*.

Olmstead, F. H. 1919. Gila River flood control. 65th Cong. 3d sess., Senate Document 436.

Pase, C. P., and Layser, E. F. 1977. Classification of riparian habitat in the Southwest. *USDA Forest Service, Gen. Tech. Rept.* RM-43:5-9.

Peacock, J. T., and McMillan, C. 1965. Ecotypic differentiation in *Prosopis* (mesquite). *Ecology* 46:35-51.

Peterson, N. V. 1950. The problem of gullying in western valleys. In *Applied sedimentation*, P. D. Trask, pp. 407-434. John Wiley.

Phillips, A.; Marshall, J.; and Monson, G. 1964. *The birds of Arizona*. Tucson: University of Arizona Press.

Phillips, W. S. 1963. Depths of roots in soil. *Ecology* 44:424.

Reitan, C. H., and Green, C. R. 1968. Appraisal of research on weather and climate of desert environments. In *Deserts of the world*, ed. W. G. McGinnis, B. J. Goldman, and P. Paylore, pp. 3-17. Tucson: University of Arizona Press.

Report of the Governor of Arizona to the Secretary of the Interior. 1896. Washington, D.C.: Government Printing Office.

Rich, J. L. 1911. Recent stream trenching in the semi-arid portion of southwestern New Mexico, a result of removal of vegetation cover. *Am. J. Sci.* 32:237-245.

Robinson, T. W. 1958. Phreatophytes. *USGS Water Supply Paper* 1423.

————. 1965. Introduction, spread and areal extent of salt cedar (*Tamarix*) in the western states. *USGS Prof. Paper* 491-A. Washington, D.C.: Government Printing Office.

Rodgers, W. M. 1965. Historical land occupance of the upper San Pedro Valley since 1870. Master's thesis, University of Arizona.

Sands, A., and Howe, G. 1977. An overview of riparian forests in California: Their ecology and conservation. *USDA Forest Service, Gen. Tech. Rept.* RM-43:98-115.

Schumm, S. A., and Hadley, R. F. 1961. Progress in the application of land form analysis in studies of semiarid erosion. *U.S. Geol. Surv. Circular* 437.

Schumm, S. A., and R. W. Lichty. 1965. Time, space and causality in geomorphology. *Am. J. Sci.* 263:110-119.

Shreve, F. 1942. The desert vegetation of North America. *Bot. Rev.* 8:195-246.

————. 1951. Vegetation of the Sonoran Desert *Carnegie Inst. Publ.* 591:1-192.

Shreve, F., and Wiggins, I. L. 1964. *Vegetation and flora of the Sonoran Desert*. 2 vols. Stanford, Calif.: Stanford University Press.

Sigler, W. F., and Miller, R. R. 1963. *Fishes of Utah*. Salt Lake City: Utah State Department of Fish and Game.

Simpson, B. B.; Neff, J. L.; and Moldenke, A. R. 1977. *Prosopis* flowers as a resource. In *Mesquite*, ed. B. B. Simpson, pp. 84-107. *US/IBP Synthesis Series* 4. Stroudsburg, Pa.: Dowden, Hutchinson & Ross.

Simpson, B. B., and Solbrig, O. T. 1977. Introduction to *Mesquite*, ed. B. B. Simpson. *US/IBP Synthesis Series* 4. Stroudsburg, Pa.: Dowden, Hutchinson & Ross.

Small, A. 1974. *The birds of California*. New York: Colliers Books.

Solbrig, O. T., and Cantino, P. D. 1975. Reproductive adaptations in *Prosopis* (Leguminosae, Mimosoideae). *J. Arnold Arb.* 56:185-210.

Sommerfeld, M. R.; Olsen, R. D.; Olsen, and Love, T. D. 1974. Some chemical observations on the upper Salt River and its tributaries. *Ariz. Acad. Sci.* 3:78-81.

Stamp, N. E. 1978. Breeding birds of riparian woodland in south-central Arizona. *Condor* 80(1):64-71.

Swenson, W. H. 1969. Comparisons of insects on mesquite in burned and unburned areas. Master's thesis, Texas Technical College, Lubbock, Texas.

Sykes, G. 1937. The Colorado Delta. *Amer. Geographical Soc. Spec. Pub.* 19. Published jointly by Carnegie Inst. Wash. and Amer. Geographical Soc. of New York.

Szaro, R. C. 1980. Factors influencing bird populations in Southwestern riparian forests. In *Proc. of the workshop on management of western forests and grasslands for nongame birds. USDA Forest Serv. Tech. Rept.*, Intermountain Forest and Range Exp. Sta., Ogden, Utah. 86:403-418.

Thornber, J. J. 1910. The grazing ranges of Arizona. *Ariz. Agric. Expt. Sta. Bull.* 65.

U.S. Bureau of the Census. 1872. *Ninth census of the United States: 1870*.

————. 1895. *Eleventh census of the United States: 1890*.

Van Devender, T. R. 1977. Holocene woodlands in the southwestern deserts. *Science* 198(4313):189-192.

Vogl, R. J., and McHargue, L. T. 1966. Vegetation of California fan palm oases on the San Andreas fault. *Ecology* 47:532-540.

Wagoner, J. J. 1952. History of the cattle industry in southern Arizona, 1540-1940. *Univ. Ariz. Soc. Sci. Bull.* 20.

Walsberg, G. E. 1975. Digestive adaptations of *Phainopepla nitens* associated with the eating of mistletoe berries. *Condor* 77:169-174.

————. 1977. Ecology and energetics of contrasting social systems in *Phainopepla nitens* (Aves: Ptilogonatidae), *Univ. Calif. Publ. Zool.* 108:1-63.

Walton, G. P. 1923. A chemical and structural study of mesquite, carob, and honey locust beans. *USDA Bull.* 1194.

Wauer, R. H. 1973. *Birds of Big Bend National Park and vicinity*. Austin: University of Texas Press.

————. 1977. Significance of Rio Grande riparian systems upon the avifauna. *USDA Forest Service, Gen. Tech. Rept.* RM-43:165-174.

Woodbury, A. M.; Durrant, S. D.; and Flowers, S. 1959. Survey of vegetation in the Glen Canyon reservoir basin. *Univ. Utah Anthropol.* 36 (Glen Canyon Series 5).

————. 1960. A survey of vegetation in the Flaming Gorge basin. *Univ. Utah Anthropol. Pap.* 45. (Upper Colorado Series 21).

————. 1962. A survey of the vegetation in the Curecanti Reservoir Basins. *Univ. Utah Anthropol. Pap.* 56:1-106 (Upper Colorado Series 6).

Woodward, S. L., and Ohmart, R. D. 1976. Habitat use and fecal analysis of feral burros (*Equus asinus*), Chemehuevi Mountains, California, 1974. *J. Range Mgmt.* 29:482-485.

Yeager, L. E. 1955. Two woodpecker populations in relation to environmental change. *Condor* 57:148-153.

Zacher, F. 1952. Die Nahrpflanzen der Samenkafer. *Zeitschr. Angew. Entomol.* 33:460-479.

HABITAT RELATIONSHIPS OF NATIVE REPTILES AND AMPHIBIANS OF NORTH AMERICAN DESERTS

Obligate riparian species are those found exclusively along desert water courses. Most often, these are amphibians that depend on riparian situations for breeding. Facultative riparian species are those that are often found in riparian habitats but occur elsewhere and do not totally rely on riparian habitats. Nonriparian species are those that seldom, if ever, use riparian habitats. The four deserts are indicated as follows: GB, Great Basin Desert; Ch, Chihuahuan Desert; Mv, Mojave Desert; and So, Sonoran Desert.

SPECIES	RIPARIAN DEPENDENCY			
	DESERT	OBLIGATE	FACULTATIVE	NONE
Amphibians				
Scaphiopus couchi	So, Ch			x
S. bombifrons	Ch			x
S. hammondi	So, Ch			x
S. intermontanus	GB			x
Rana pipiens (sp.)	GB, So, Ch	x		
R. catesbeiana	So, Ch	x		
R. arenicolor	Ch	x		
Bufo punctatus	Mv, So		x	
B. debilis	Ch		x	
B. woodhousei	So, Ch		x	
B. cognatus	So, Ch		x	
B. retiformis	So		x	
B. alvarius	So		x	
Pternohyla fodiens	So			x
Reptiles				
Terrapene ornata	Ch			x
Kinosternon flavescens	Ch		x	
K. sonoriense	Ch, So	x		
Trionyx spinifer	Ch	x		
Coleonyx brevis	Ch			
C. variegatus	Mv, So			x
Holbrookia maculata	Ch	x		
Gopherus agassizi	Mv, So			x

SPECIES	RIPARIAN DEPENDENCY			
	DESERT	OBLIGATE	FACULTATIVE	NONE
Xantusia vigilis	Mv, So			x
Sauromalus obesus	Mv, So			x
Dipsosaurus dorsalis	Mv, So			x
Callisaurus draconoides	Mv, So		x	
Uma scoparia	Mv			x
U. inornata	Mv, So			x
Phrynosoma m'calli	So			x
P. douglassii	GB, Ch			x
P. solare	So			x
P. coronatum	Ch			x
P. platyrhinos	GB, So, Mv			x
P. modestum	Ch			x
Cophosaurus texanus	Ch		x	
Cambelia wislizeni	GB, Mv, So, Ch		x	
Crotaphytus collaris	Ch		x	
C. insularis	GB, Mv, So			x
Sceloporus clarki	Ch		x	
S. occidentalis	GB		x	
S. graciosus	GB		x	
S. poinsetti	Ch		x	
S. undulatus	Ch		x	
Urosaurus ornatus	So, Ch		x	
U. graciosus	Mv, So		x	
Uta stansburiana	GB, Mv, Ch, So		x	
Eumeces obsoletus	Ch		x	
E. skiltonianus	GB		x	
Cnemidophorus tesselatus	Ch		x	
C. tigris	GB, Mv, So, Ch		x	
C. uniparens	Ch		x	
C. neomexicanus	Ch			x
C. inornatus	Ch			x
C. gularis	Ch			x
C. exsanguis	Ch		x	
C. sonorae	So, Ch		x	
C. flagellicaudus	So, Ch		x	
Charina bottae	GB		x	
Leptotyphlops humilis	Mv, So, Ch		x	
L. dulcius	Ch		x	
Lichanura trivirgata	Mv, So			x
Heterodon nascius	Ch		x	
Phyllorhynchus decurtatus	Mv, So			x
P. browni	So			x
Heloderma suspectum	So			x
Diadophis punctatus			x	
Masticophis flagellum	Mv, So, Ch		x	
M. taeniatus	GB, Ch		x	
Coluber constrictor	GB		x	
Salvadora grahamiae	Ch			x
S. hexalepis	GB, Mv, So, Ch			x
Pituophis melanoleucas	GB, Mv, So, Ch		x	
Arizona elegans	Ch, Mv, So			x
Elaphe subocularis	Ch		x	
Lampropeltis getulus	Mv, So, Ch		x	

SPECIES	RIPARIAN DEPENDENCY			
	DESERT	OBLIGATE	FACULTATIVE	NONE
Rhinocheilus lecontei	GB, Mv, So, Ch			x
Thamnophis marcianus	So, Ch		x	
T. sirtalis	GB	x		
T. eques	Ch	x		
T. elegans	GB		x	
T. cyrtopsis	So, Ch	x		
Sonora semiannulata	GB, Mv, So, Ch		x	
S. episcopa	Ch		x	
Ficimia cana	Ch		x	
Chionactis occipitalis	Mv, So			x
C. palarostris	So			x
Chilomeniscus cinctus	So		x	
Tantilla nigriceps	Ch		x	
T. planiceps	Mv, So, Ch		x	
Hypsiglena ochrorhyncha	GB, Mv, So, Ch			x
Trimorphodon biscutatus vilk- *insoni*	Ch			x
T. b. lambda	So			x
Micruroides euryxanthus	So		x	
Sistrurus catenatus	Ch		x	
Crotalus molossus	Ch		x	
C. lepidus	Ch		x	
C. viridis	GB, Ch		x	
C. scutulatus	Mv, Ch		x	
C. atrox	Mv, Ch		x	
C. cerastes	Mv		x	

APPENDIX 10-B

HABITAT RELATIONSHIPS OF NATIVE BIRDS IN NORTH AMERICAN DESERTS

If a species regularly occurred in the desert area, it was included in the total; species with irregular sightings were excluded. An obligate riparian species was defined as one whose major portion of the species population is dependent upon the riparian habitat in the desert being considered. A facultative species is one that utilizes riparian habitat but is not totally dependent on it; it may use urban, desert, or agricultural habitats as well. The definition of obligate and faculative riparian species results in a different classification of riparian dependence in one desert versus another. For example, a species that may be listed as having obligate riparian habitat dependence in one desert may be listed as facultative in another in which its riparian habitat is not present, or is present to such a limited extent that the major portion of the population is using another habitat type such as urban or agricultural. Nonriparian species are those which occur within the desert without utilizing riparian habitats to any great extent.

An asterisk indicates a breeding species. The four deserts are indicated as follows: GB, Great Basin Desert; Ch, Chihuahuan Desert; Mv, Mojave Desert; and So, Sonoran Desert.

AVIAN SPECIES	RIPARIAN DEPENDENCY		
	OBLIGATE	FACULTATIVE	NONE
Common loon			Ch, GB, Mv, So
Arctic loon			GB, Mv, So
Red-throated loon			GB
Horned grebe			Ch, GB, So, Mv
Eared grebe	GB*	So	Ch, Mv
Least grebe		Ch, So	
Western grebe	GB*, So*		Ch, Mv
Pied-billed grebe		GB*, Ch, Mv*, So*	
White Pelican		GB	Ch, Mv, So
Brown Pelican			So
Double-crested cormorant	Ch*, So*	GB*, Mv	
Olivaceous cormorant	So*, Ch*		
Great blue heron	Ch*, GB*, So*	Mv	
Green heron	Ch*, Mv*, So*, GB*		

470

AVIAN SPECIES	RIPARIAN DEPENDENCY		
	OBLIGATE	FACULTATIVE	NONE
Little blue heron		Ch	
Cattle egret		So, Ch, GB, Mv	
Great egret	Ch*, So*	GB, Mv	
Snowy egret	Ch*, GB*, So*	Mv	
Louisiana heron		Ch	
Black-crowned night heron	Ch*, So*, GB*	Mv	
Yellow-crowned night heron		Ch	
Least bittern	Ch*, GB*, Mv*, So*		
American bittern	Ch*, GB*, So*, Mv		
Wood stork		Ch, So	
White-faced ibis	GB*	Ch, Mu, So	
White ibis		Ch	
Whistling swan		GB, So, Ch	Mv
Trumpeter swan		GB	
Canada goose	GB*, Ch*	So	Mv
White-fronted goose		GB, So, Ch	Mv
Snow goose		So, GB, Ch	Mv
Ross goose		GB, Ch	
Black-bellied whistling-duck	So*		
Fulvous whistling-duck		Ch, So	
Mallard		Ch*, GB*, Mv*, So*	
Mexican duck	So*, Ch*,		
Gadwall	GB*, So*, Ch*	Mv	
Pintail	GB*, Ch*	Mv, So	
Green-winged teal	Ch*, GB*	Mv, So	
Blue-winged teal		Ch*, GB*, Mv, So	
Cinnamon teal	So*	Ch*, Mv*, GB*	
American wigeon	GB*, Ch*	Mv, So	
Northern shoveler	GB*, Ch*	Mv, So	
Wood duck		Ch, Mv, So, GB	
Redhead	GB*, So*	Mv, Ch	
Ring-necked duck	GB*	Mv, So, Ch	
Canvasback	GB*		So, Ch, Mv
Greater scaup			GB, So
Lesser scaup			Ch, Mv, So, GB
Common goldeneye			Ch, GB, Mv, So,
Barrow goldeneye			GB, So
Bufflehead			Ch, GB, Mv, So
Oldsquaw			GB
White-winged scoter			GB, Ch
Surf scoter			So, Ch
Ruddy duck	GB*, So*, Ch*	Mv*	
Hooded merganser		GB, So, Ch	
Common merganser			Ch, GB, So
Red-breasted merganser			Ch, GB, Mv, So
Turkey vulture		Ch*, Mv*, So*, GB*	
Black vulture		So, Ch*	
Mississippi kite	Ch*, So*		
Goshawk		GB, So, Ch	

AVIAN SPECIES	RIPARIAN DEPENDENCY		
	OBLIGATE	FACULTATIVE	NONE
Sharp-shinned hawk		Ch, GB, Mv, So	
Cooper hawk	GB*, So*, Mv*, Ch*		
Red-tailed hawk		Ch*, GB*, Mv*, So*	
Broad-winged hawk		Ch	
Swainson hawk		Ch*, GB*, Mv, So*	
Zone-tailed hawk	Ch*, So*		
Rough-legged hawk			Ch, GB, Mv, So
Ferruginous hawk			Ch, GB*, Mv, So
Gray hawk	Ch, So*		
Harris hawk		Ch*, So*	
Black hawk	Ch*, So*		
Golden eagle			Ch*, GB*, Mv*, So*
Bald eagle	GB*, So*	Ch, Mv	
Marsh hawk		GB*, Ch*, Mv, So	
Osprey	GB*	Ch, Mv, So	
Caracara		Ch*, So*	
Prairie falcon		Ch*, GB*, Mv,* So*	
Peregrine falcon		Ch*, GB*, Mv, So*	
Aplomado falcon			Ch
Merlin		Ch, GB, Mv, So	
American kestrel		Ch*, GB*, Mv*, So*	
Sharp-tailed grouse			GB*
Sage grouse			GB*
Scaled quail			So*, Ch*
Gambel quail		Ch*, Mv*, So*	
Ring-necked pheasant			Ch*, GB*, So*
Chukar			GB*, Mv*
Sandhill crane		So, Ch, GB*	
Clapper rail	So*		
Virginia rail	Ch*, GB*, Mv*, So*		
Sora	Ch*, GB*, Mv*, So*		
Black rail	So*		
Common gallinule	Ch*, GB*	So*, Mv	
American coot		Ch*, GB*, Mv*, So*	
Semipalmated plover		Ch, GB, So	
Snowy plover		Ch*, GB*, So*	Mv
Killdeer		Ch*, GB*, Mv*, So*	
Mountain plover			Ch, GB, Mv, So
Golden plover		GB, So	
Black-bellied plover		Ch, GB, So	
Ruddy turnstone			GB

AVIAN SPECIES	RIPARIAN DEPENDENCY		
	OBLIGATE	FACULTATIVE	NONE
Common snipe		Ch, GB*, Mv, So	
Long-billed curlew		Ch, GB*, Mv, So	
Whimbrel		Ch, GB, So	
Upland sandpiper			Ch
Spotted sandpiper		Ch, GB*, Mv, So	
Solitary sandpiper		Ch, GB, Mv, So	
Willet	GB*	Ch, Mv, So	
Greater yellowlegs		Ch, GB, Mv, So	
Lesser yellowlegs		Ch, GB, Mv, So	
Pectoral sandpiper		Ch, GB, Mv, So	
Baird sandpiper		Ch, GB, Mv, So	
Least sandpiper		Ch, GB, Mv, So	
Dunlin		Ch, GB, So	
Short-billed dowitcher		GB, So	
Long-billed dowitcher		Ch, GB, Mv, So	
Stilt sandpiper		Ch, GB, So	
Semipalmated sandpiper		Ch, GB	
Western sandpiper		Ch, GB, Mv, So	
Marbled godwit		Ch, GB, Mv, So	
Sanderling American avocet		Ch*, Mv, So, GB*	Ch, So, GB
Black-necked stilt		Ch*, Mv*, So*, GB*	
Wilson phalarope		Ch, Mv, So, GB*	
Red phalarope			GB, So
Northern phalarope			GB, Mv, So, Ch
Parasitic jaeger			GB, So
Herring gull			GB, So
California gull			Mv, So
Ring-billed gull		GB*	Ch, Mv, So
Franklin gull	GB*		Ch, So
Bonaparte gull			Mv, So, Ch, GB
Sabine gull			GB, So, Ch
Forster tern	GB*	Ch, Mv, So	
Common tern			So, Ch, GB
Least tern		Ch*	
Caspian tern	GB*	So, Ch	
Black tern	GB*	Ch, So,	
Band-tailed pigeon		GB	
Rock dove			Ch, GB, Mv, So
White-winged dove		Mv*, So*, Ch*	
Mourning dove		Ch*, GB*, Mv*, So*	
Ground dove		Ch*, So*	
Inca dove			Ch*, So*
Yellow-billed cuckoo	Ch*, GB*, So*		
Black-billed cuckoo		GB*	
Roadrunner		Ch*, GB*, Mv*, So*	
Groove-billed ani		Ch	

AVIAN SPECIES	RIPARIAN DEPENDENCY		
	OBLIGATE	FACULTATIVE	NONE
Barn owl		Ch*, GB*, Mv*, So*	
Screech owl		Ch*, GB*, Mv*, So*	
Great horned owl		Ch*, GB*, Mv*, So*	
Pygmy owl		GB	
Ferruginous owl		So*	
Elf owl		Ch*, So*	
Burrowing owl			Ch, GB, Mv, So*
Long-eared owl		Ch*, GB*, Mv*, So*	
Short-eared owl		Ch, GB*, Mv, So	
Buff-collared nightjar			So
Poor-will			So*, Mv*, GB*, Ch*
Common nighthawk		Ch*, GB*	
Lesser nighthawk		Ch*, Mv*, So*	
Chimney swift			Ch*, So
Vaux swift		GB, Mv, So	
White-throated swift			Ch*, GB*, Mv*, So*
Lucifer hummingbird		Ch*	
Black-chinned hummingbird		Ch*, GB*, Mv*, So*	
Costa hummingbird			Ch*, Mv*, So*
Anna hummingbird		Ch*, Mv*, So*	
Broad-tailed hummingbird		GB, Ch, So	
Rufous hummingbird		Ch, GB, Mv, So	
Calliope hummingbird		GB, So	
Broad-billed hummingbird		Ch, So*	
Belted kingfisher		Ch*, GB*, Mv, So	
Green kingfisher	Ch*, So*		
Common flicker		Ch*, GB*, Mv*, So*	
Gila woodpecker		So*	
Red-headed woodpecker		Ch*	
Acorn woodpecker		So	
Lewis woodpecker		GB, Mv, So, Ch	
Yellow-bellied sapsucker		Ch, GB*, Mv, So	
Hairy woodpecker		GB*, Ch	
Downy woodpecker		GB*	
Ladder-backed woodpecker		Ch*, Mv*, So*	
Eastern kingbird	GB*	Ch	
Tropical kingbird	So*	Ch*	
Western kingbird		Ch*, GB*, Mv*, So*	
Cassin kingbird		Ch, GB, Mv, So	
Thick-billed kingbird	So*		
Scissor-tailed flycatcher		Ch*	

AVIAN SPECIES	RIARIAN DEPENDENCY		
	OBLIGATE	FACULTATIVE	NONE
Wied crested flycatcher	Ch	So*	
Ash-throated flycatcher		Ch*, GB*, Mv*, So*	
Eastern phoebe		Ch, So	
Black phoebe	Ch*, GB*, So*	Mv*	
Say phoebe		Ch*, GB*, Mv*, So*	
Willow flycatcher	GB*, So	Ch, Mv	
Least flycatcher		Ch	
Hammond flycatcher		Ch, GB, Mv, So	
Dusky flycatcher		Ch, GB*, Mv, So	
Gray flycatcher		Ch, GB*, Mv, So	
Western flycatcher		Ch, GB, Mv, So	
Coues flycatcher		So	
Western wood pewee		Ch*, GB*, Mv, So*	
Olive-sided flycatcher		Ch, GB, Mv, So	
Vermilion flycatcher	So*	Ch*, Mv	
Beardless flycatcher	So*		
Horned lark			Ch*, GB*, Mv*, So*
Violet-green swallow		Ch, GB*, Mv, So*	
Tree swallow		Ch*, GB*, Mv, So	
Bank swallow		Ch*, GB*, Mv, So	
Rough-winged swallow		Ch*, GB*, Mv*, So*	
Barn swallow		Ch*, GB*, Mv, So*	
Cliff swallow		Ch*, GB*, Mv*, So*	
Cave swallow		Ch*	
Purple martin		Ch, GB, So*	
Blue jay		GB*, Ch	
Steller jay		GB	
Scrub jay		Ch, GB, Mv, So	
Black-billed magpie		GB	
Common raven		Ch*, GB*, Mv*, So*	
White-necked raven		Ch*, So*	
Common crow		GB*, So	
Black-capped chickadee		GB*	
Mountain chickadee		GB	
Black-crested titmouse		Ch*	
Plain titmouse			GB*
Bridled titmouse	So*		
Verdin		Ch*, Mv*, So*	
Bushtit		So, GB*, Ch	
White-breasted nuthatch	So*	Ch, GB	
Red-breasted nuthatch		Ch, GB, Mv, So	

AVIAN SPECIES	RIPARIAN DEPENDENCY		
	OBLIGATE	FACULTATIVE	NONE
Brown creeper		Ch, GB, Mv, So	
Dipper		GB	
House wren		Ch, GB*, Mv, So	
Winter wren	So	Mv, GB, Ch	
Bewick wren	So*	Ch, GB*, Mv*	
Cactus wren		Ch*, Mv*, So*	
Long-billed marsh wren	Ch, GB*, Mv*, So*		
Canyon wren		Ch*, GB*, Mv*, So*	
Rock wren			Ch*, GB*, Mv*, So*
Mockingbird		Ch*, GB*, Mv*, So*	
Gray catbird	GB*		
Brown thrasher		Ch	
Bendire thrasher			Mv*, So*
Curve-billed thrasher		Ch*, So*	
Le Conte thrasher			Mv*, So*
Crissal thrasher	So*	Ch*, Mv*	
Sage thrasher		Ch, GB*, Mv, So	
Rufous-backed robin		So	
American robin		Ch*, GB*, Mv, So*	
Hermit thrush		Ch, GB, Mv, So	
Swainson thrush		Ch, GB, Mv, So	
Veery	GB*		
Eastern bluebird		Ch*	
Western bluebird		Ch, GB, Mv, So	
Mountain bluebird		Ch, GB, Mv, So	
Townsend solitaire		Ch, GB, Mv, So	
Blue-gray gnatcatcher		Ch, GB*, Mv, So	
Black-tailed gnatcatcher		Ch*, Mv*, So*	
Golden-crowned kinglet		Ch, GB, Mv, So	
Ruby-crowned kinglet		Ch, GB, Mv, So	
Water pipit		Ch, GB, Mv, So	
Sprague pipit			Ch, So
Bohemian waxwing			GB
Cedar waxwing		Ch, GB*, Mv, So	
Phainopepla		Ch*, Mv*, So*	
Northern shrike			GB
Loggerhead shrike		Ch*, GB*, Mv*, So*	
Starling		Ch*, GB*, Mv*, So*	
Black-capped vireo		Ch	
Hutton vireo	So	Ch	
Bell vireo	Ch*, Mv*, So*		
Gray vireo			GB*, So
Solitary vireo		Ch, GB*, So	

AVIAN SPECIES	RIPARIAN DEPENDENCY		
	OBLIGATE	FACULTATIVE	NONE
Red-eyed vireo	GB	Mv, So, Ch	
Warbling vireo	GB	Ch, Mv, So	
Black-and-white warbler		Ch, Mv, So, GB	
Worm-eating warbler		Ch	
Tennessee warbler		Mv, Ch, So	
Orange-crowned warbler		Ch, GB*, Mv, So	
Nashville warbler		Ch, GB, Mv, So	
Virginia warbler		Ch, GB*, Mv, So	
Lucy warbler	Ch, GB, So*	Mv	
Northern parula		Ch, So, Mv	
Yellow warbler	Ch*, GB*, So*	Mv	
Black-throated blue warbler		Ch	
Yellow-rumped warbler		Ch, GB, Mv, So	
Black-throated gray warbler		Ch, GB*, Mv, So	
Townsend warbler		Ch, GB, Mv, So	
Black-throated green warbler		Ch, So	
Hermit warbler		Ch, GB, Mv, So	
Chestnut-sided warbler		Ch	
Blackpoll warbler		Ch	
Palm warbler		Ch	
Ovenbird		Ch	
Northern waterthrush	Ch, GB, So, Mv		
Kentucky warbler		Ch	
MacGillivray warbler		Ch, GB*, Mv, So	
Common yellowthroat	Ch*, GB*, Mv*, So*		
Yellow-breasted chat	Ch*, GB*, Mv*, So*		
Hooded warbler		Ch	
Wilson warbler		Ch, GB, Mv, So	
American redstart	GB*	Mv, So, Ch	
House sparrow			Ch, GB, Mv, So
Bobolink		GB*, Mv, Ch	
Eastern meadowlark		Ch	So
Western meadowlark		Ch, Mv, So	GB
Yellow-headed blackbird	GB*, So*	Ch, Mv	
Red-winged blackbird		Ch*, Mv*, So*, GB*	
Orchard oriole		Ch*	
Hooded oriole	Ch*	So*, Mv*	
Scarlet-headed oriole		So	
Scott oriole			Ch*, GB*, Mv, So*
Northern oriole	GB, So	Ch, Mv	
Brewer blackbird		GB*	So, Ch, Mv
Great-tailed grackle		Ch*, So*	
Brown-headed cowbird		Ch*, GB*, Mv*, So*	
Bronzed cowbird	Ch*	So*	
Western tanager		Ch, GB, Mv, So	

AVIAN SPECIES	RIPARIAN DEPENDENCY		
	OBLIGATE	FACULTATIVE	NONE
Summer tanager	Ch*, So*		
Cardinal	Ch*	So*	
Pyrrhuloxia		Ch*, So*	
Rose-breasted grosbeak		Ch, GB, Mv, So	
Black-headed grosbeak		Ch, GB*, Mv, So	
Blue grosbeak	Ch*, GB*, Mv*, So*		
Indigo bunting	GB*, So*, Ch*		
Lazuli bunting		Ch, GB*, Mv, So*	
Varied bunting		Ch*, So*	
Painted bunting	Ch*		
Dickcissel		Ch	
Evening grosbeak		GB, So	
Purple finch		So	
Cassin finch		GB, Ch, So	
House finch		Ch*, GB*, Mv*, So*	
Common redpoll		GB	
Pine siskin		Ch, GB, Mv, So	
American goldfinch	GB*	Ch, Mv, So	
Lesser goldfinch	GB*, So*	Ch*, Mv*	
Lawrence goldfinch	So*	Ch, Mv	
Green-tailed towhee		Ch, GB, Mv, So	
Rufous-sided towhee		Ch, GB, Mv, So	
Brown towhee		Ch*, So*	
Abert towhee	Mv*, So*		
Lark bunting			Ch, GB*, So
Savannah sparrow		Ch, GB*, Mv, So	
Grasshopper sparrow			Ch, GB*, So
Baird sparrow			Ch
Le Conte sparrow		Ch	
Vesper sparrow			Ch, GB*, Mv, So
Five-striped sparrow			So*
Lark sparrow		Ch*, GB*, Mv, So*	
Rufous-winged sparrow		So*	
Rufous-crowned sparrow			Ch*, So*
Botteri sparrow			So, Ch
Cassin sparrow			Ch*, So*
Black-throated sparrow			Ch*, GB*, Mv*, So*
Sage sparrow		Ch, GB*, Mv, So	
Dark-eyed junco		Ch, GB, Mv, So	
Gray-headed junco		Ch, GB, Mv, So	
Tree sparrow		GB	
Chipping sparrow		Ch, GB*, Mv, So	
Clay-colored sparrow		Ch	
Brewer sparrow		Ch, GB*, Mv,	

AVIAN SPECIES	RIPARIAN DEPENDENCY		
	OBLIGATE	FACULTATIVE	NONE
Field sparrow		Ch	
Worthen sparrow			Ch*
Black-chinned sparrow			Ch, So
Harris sparrow		Ch, GB, So	
White-crowned sparrow		Ch, GB, Mv, So	
Golden-crowned sparrow		GB, Mv, So	
White-throated sparrow		Ch, GB, Mv, So	
Fox sparrow	GB*	Ch, So	
Lincoln sparrow		Ch, GB, Mv, So	
Swamp sparrow	Ch, So, GB		
Song sparrow	Ch, GB*, Mv*, So*		
Lapland longspur			GB
Chestnut-collared longspur			Ch, GB, So
Snow bunting			GB

11

SAND DUNES IN THE NORTH AMERICAN DESERTS

ROGER S. U. SMITH

INTRODUCTION

SAND DUNES ARE sparse in the North American deserts, as is the geomorphological literature describing them, particularly outside California. My purpose is to summarize the available descriptive and interpretive literature on sand-dune localities in the desert areas of the United States and Mexico. The term *desert* is applied loosely here to include semiarid areas with notable dune accumulations, but dunes on the Great Plains are excluded, as are coastal dunes other than those in the true deserts of Baja California and Sonora, Mexico. In general, the active dunes described here are confined to areas that now receive less than 25 centimeters (cm) of precipitation per year. Dunes in the semiarid and polar deserts of Alaska and Canada have been omitted because I have had no experience with them.

The most detailed available summary of dune areas in North America is Thorp and Smith's (1952) map of eolian deposits in the United States. This map was compiled state by state, with a different group of geologists and soil scientists responsible for each state. The text that was planned to accompany the map was never completed, and the map bears no literature citations to document the delineation of its various eolian units. Much earlier, Free (1911) summarized existing literature on American dunes and compiled an exhaustive bibliography on eolian phenomena. More recently, Warren (1969) published a bibliography on desert dunes organized by decades, but this list was not limited to North America. Lustig's (1968) inventory of research on arid geomorphology is excellent but devotes scant attention to North American dunes. Paylore's (1967) bibliography of arid-land bibliographies lists none devoted to North American dunes. Similarly

*This study has benefited from discussions with colleagues too numerous to mention and from relevant literature and materials assembled by the late H. T. U. Smith, but I take sole responsibility for any errors or omissions presented here. Some of my studies described in the text were supported in part by NASA grants NSG-7410 and NSG-7551, whose financial assistance I gratefully acknowledge.

dunes are nearly ignored in textbooks on regional geomorphology of the United States (such as Hunt 1974; Thornbury 1965; Fenneman 1931) and in Kolb and Van Lopik's (1963) terrain-analysis atlas of the southwestern United States. Except for Eymann's (1953) catalog of dune areas in the California desert, most literature deals with individual dune fields or describes dunes in passing as isolated features of a larger geographic region. My purpose is to summarize this scattered literature on the geomorphology of North American sand dunes. The chapter is organized by states within the United States and by provinces in Mexico to facilitate access to information about dunes in specific areas, and the reference list is similarly organized. The major dune fields described in the text are keyed to the index map (figure 11-1) as a state abbreviation followed by a number; (for example, NM-1 is the White Sands of New Mexico). Some fields shown on Thorp and Smith's (1952) map have been omitted from figure 11-1 because they are small or little is known about them; others have been omitted because I do not believe that their morphology justifies the term *dune*, based on study in the field or on aerial photographs; this is true for most of the areas of shrub coppice dunes shown on their map. The scale of the map presented here is too small to show the geometry of each field or describe its precise location, but readers should be able to use the information supplied to locate any field on a larger-scale map on which dunes are commonly depicted by stippling. Best suited for this purpose are topographic maps at a scale of 1:250,000, such as the Army Map Service series distributed by the U.S. Geological Survey and its equivalent in Mexico, Joint Operations Graphic (Air), distributed by DETENAL (appendix 11-A); the sheets covering each dune are listed in table 11-1).

Dune Processes

Sand dunes are built principally of well-sorted grains of fine to medium sand (0.125-0.5 millimeters), whose dominant

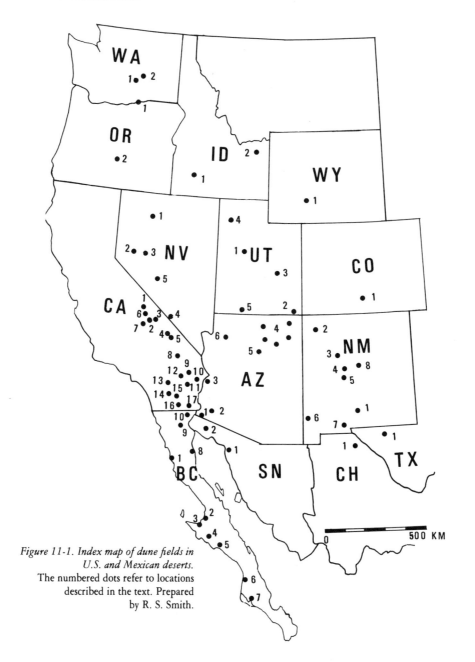

*Figure 11-1. Index map of dune fields in
U.S. and Mexican deserts.
The numbered dots refer to locations
described in the text. Prepared
by R. S. Smith.*

form of eolian transport is by saltation or bouncing along the surface. Finer material (mostly silt and clay) is carried by suspension in the wind; coarser material up to several millimeters in diameter can move slowly by surface creep, driven directly by strong winds or indirectly by the impact of finer saltating grains entrained by strong winds. Except in unusual localities or occurences, grains larger than several millimeters in diameter are immobile in the wind. Appendix 11-B gives sources of wind and climatic data.

Within a fully developed turbulent boundary layer, the velocity of the wind increases upward with the logarithm of height from the surface. Below a finite distance k from the surface, wind velocity decays to zero because of drag against the surface; this distance k is a linear function of the average size of surface roughness elements and can be taken as one-thirtieth of the average diameter of grains on the surface (Bagnold 1941:52-57, 172-174). Drag on the surface is most simply represented by the shear velocity U_* (also called drag velocity or friction velocity), which is the inverse slope of the line representing log height z as a function of velocity u (figure

TABLE 11-1
Dune Fields in U.S. and Mexican deserts

		1:250,000 map sheets			
AZ-1	Fortuna	El Centro	ID-1	Bruneau	Twin Falls
2	Mohawk	Ajo	2	Juniper Buttes	Dubois, Ashton, Idaho Falls, Driggs
3	Parker	Needles			
4	NE AZ	Marble Cnyn, Shiprock, Flagstaff, Gallup	NV-1	Winnemucca	Vya, McDermitt
			2	Carson Desert	Reno
			3	Sand Mountain	Reno
5	SF Pks	Flagstaff	4	Big Dune	Death Valley
6	Grand Canyon	Williams	5	Tonopah (Crescent Dunes)	Tonopah
			NM-1	White Sands	Tularosa, Las Cruces
BC-1	Bahia San Quintin	H 11-5, H 11-6			
2	Bahia Sebastian Vizcaino	G 11-12*, G11-3*	2	NW-Chaco-San Juan Rivers	Shiprock, Gallup
3	Sierra Vizcaino-Laguna Ojo Libre	G 11-3*	3	Albuquerque-Grants	Albuquerque, Socorro
4	Laguna Ojo Libre-Laguna San Ignacio	G 12-1*, G 12-4*			
5	Laguna Ojo Libre-Laguna San Ignacio	G 12-1*, G 12-4*	4	Rio Puerco-Rio Grande	Socorro
6	Laguna Santo Domingo	G 12-7*, G 12-8*	5	Rio Salado-Rio Grande	Socorro
7	Bahia Magdalena	G 12-11*	6	Animas Valley	Silver City
8	San Felipe	H 11-6	7	W of Las Cruces	Las Cruces, El Paso
9	Sierra de las Pintas	H 11-3	8	Estancia Valley	Socorro, Fort Sumner
10	Colo R delta	I 11-12			
CA-1	Eureka Valley	Goldfield	OR-1	Columbia River	The Dalles, Pendleton
2	Panamint Valley	Death Valley	2	Christmas & Fossil Lakes	Crescent
3	Stovepipe Wells	Death Valley			
4	Saratoga Springs	Trona	SN-1	SE of Puerto Penasco	H 12-1, H 12-4
5	Dumont	Trona	2	NW of Puerto Penasco	H 11-3, H12-1, I 11-12, I 12-10
6	Saline Valley	Death Valley			
7	Owens Valley	Death Valley			
8	Devil's Playground/Kelso	Trona, Kingman, San Bernardino, Needles	TX-1	Salt Basin	El Paso
9	Cadiz Valley	Needles	UT-1	Delta	Delta
10	Rice	Needles	2	Aneth-Mexican Hat	Escalante, Cortez
11	Desert Center	Salton Sea	3	Green River Desert	Salina
12	Dale Dry Lake	Needles	4	Great Salt Lake Desert	Cedar City
13	Coachella Valley	Santa Ana	5	Coral Pink Sand Dunes	Brigham City, Tooele
14	Clark Valley	Santa Ana			
15	Salton Sea Barchans	Salton Sea			
16	Superstition Mountain	El Centro	WA-1	Columbia River	Walla Walla
17	Algodones Dunes	El Centro, Salton Sea, I 11-12	2	Moses Lake	Ritzville, Walla Walla
CH-1	Samalyuca	H 13-1	WY-1	Kilpecker	Lander, Rock Springs
CO-1	Great Sand dunes NM	Durango			

NOTE: Asterisked sheets have not yet been published. Includes all fields listed on figure 11-1.

11-2). Under most conditions, shear velocity is about an order of magnitude smaller than wind velocity measured at typical weather-station heights, so the two must not be confused. As wind velocity above the boundary layer increases, shear velocity beneath it and drag on the surface also increase, but all profiles of velocity u versus log height z still converge at the focus $z = k$, $u = 0$. As shear velocity increases, drag on the surface increases, and at the threshold shear velocity U_{*t}, this surface drag is sufficient to overturn sand grains and initiate transport by saltation (Bagnold 1941; see Chepil and Wood-

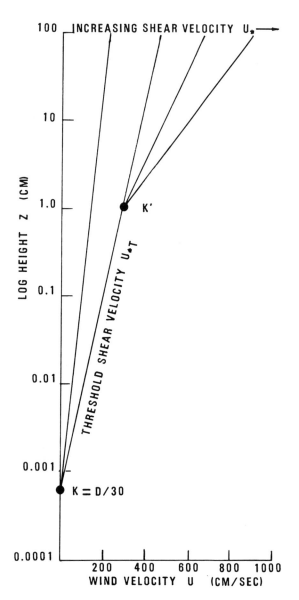

Figure 11-2. Relations between wind velocity, height, and shear velocity.
No sand transport occurs at shear velocities below U_{*t} This diagram is for 0.20 mm sand. Taken from *The physics of blown sand and desert dunes*, R. A. Bagnold, Methuen, London, 1941.

ruff 1963, for summary). At higher shear velocities, the rate of sand transport is a function of the cube of shear velocity and of the square root of grain diameter (Bagnold 1941; see Hsu 1973, for summary). Conservation of momentum between the wind and the much-denser sand grains exerts enormous drag on the wind, drag that can be represented as that of obstacles much larger than the grains on the surface, but these "obstacles" move downwind. Bagnold (1941:57-61) found experimentally that the new focus for velocity rays at height k' was much taller ($k' = 0.8 - 1.0$ cm) than k, at a velocity U_t corresponding to the intersection of the height k' and the threshold shear velocity U_{*t}, and that all stronger shear velocities were focused at that point ($z = k'$, $u = U_t$; figure 11-2).

Bagnold (1941:172) assumed that the drag of sand movement across a pebble surface was negligible compared with that across a sand surface, apparently because much of the drag associated with saltation across a sand surface was associated with energy lost in splashing on the surface upon the impact of saltating grains, a phenomenon absent in impact onto a pebble surface. This occurs despite the greater height of individual grain trajectories above a pebble surface and transport rates as much as twice those over a sand surface at the same shear velocity. If so, drag on the wind exerted by a pebble surface is controlled solely by its inherent roughness, with no drag contributed by the curtain of sand saltating above it. Consequently the shear velocity U_* above a pebble surface is controlled solely by roughness height and is a linear function of velocity u within a fully developed turbulent boundary layer by the relation $u = 5.75$ u_* $\log_{10}(z/k)$, where k is the roughness height, taken here to be one-thirtieth of the average grain diameter on the surface (Bagnold 1941:52-57, 172-174). Because any given wind condition can be characterized by a single value of shear velocity U_* regardless of the nature of the surface it blows across, the shear velocity measured within the fully developed turbulent boundary layer over a pebble layer does not change as it passes onto a loose sand surface, despite large differences in roughness height between the two surfaces. This means that sand carried efficiently across a pebble surface may be deposited in part on any existing sand surface, where inefficient transport may be unable to cope with the rate of sand influx from the pebble surface. This occurs with no change in shear velocity and is a mechanism by which dunes can grow from sand patches (Bagnold 1941:169-174). Dunes also can form behind permeable obstacles like bushes that slow down the sand-laden wind until it is oversaturated with sand and deposits sand.

Dunes must reach a minimum critical height before they can trap sand under all conditions of wind. A sand patch downwind of a pebble surface grows only when winds are strong enough to blow sand across the pebbles, and the threshold velocity for this is higher than for the much-smoother sand surface. Bagnold (1941:170-174) defined "strong" winds to be capable of moving sand only across a sand surface. Thus a sand patch can accrete sand only during strong winds and will lose sand during gentle winds unless it is tall enough to trap sand aerodynamically. This occurs once the dune is tall and steep enough that wind separates from its leeward margin, dumping its entrained sand, rather than hugging the dune surface and carrying sand off the dune. The zone of sand deposition becomes oversteepened until it collapses to a gentler, stable slope, 30-32°. This slope is the slip face of the dune, and it traps sand to maintain the dune.

Dune Classification

The most useful scheme for classifying dunes throughout most of the North American deserts is Hack's (1941) scheme devised for northeastern Arizona (figure 11-3). Assuming a unidirectional wind, he plotted dune form as a function of three variables: wind, supply of sand, and vegetation. These factors could work with or against each other to produce three basic dune forms: longitudinal (much wind, little to moderate sand), transverse (little vegetation, little to moderate wind, moderate to copious supply of sand), and parabolic (moderate wind, vegetation and sand supply). The term *U-shaped* (Smith 1946, 1949) is used here in preference to *parabolic* because

the latter implies a mathematical perfection of form that is absent in most dunes. Although the wind regime over most of this area is not unidirectional but exhibits some seasonal variation, Hack's scheme describes most dunes adequately, perhaps with respect to resultant wind.

Longitudinal dunes are long in proportion to width and roughly symmetrical in cross-section (figures 11-4, 11-9, 11-14, 11-16, 11-17). They may be anchored to a hill (figures 11-9, 11-14) or an escarpment (figure 11-4) or may stand free (figure 11-16). Although many are vegetated and inactive, active longitudinal dunes may extend for tens of kilometers and commonly have sinuous crests, along which slip faces reverse seasonally from the right to the left side (figure 11-17). In some cases, stabilized, vegetated longitudinal dune ridges may be the abandoned trailing arms of U-shaped dunes, also symmetrical in cross-section (figure 11-24b).

Transverse dunes tend to be much shorter, more asymmetrical in cross-section, and more closely spaced than longitudinal dunes, as well as more irregular in their long dimension (figure 11-8). The brinks of many transverse dunes reverse seasonally, but complete reversal of asymmetry is uncommon (figure 11-29). Barchan dunes represent a subset of transverse dunes, apparently restricted to sites where the substrate is hard and the sand supply limited. These dunes are shaped like a crescent

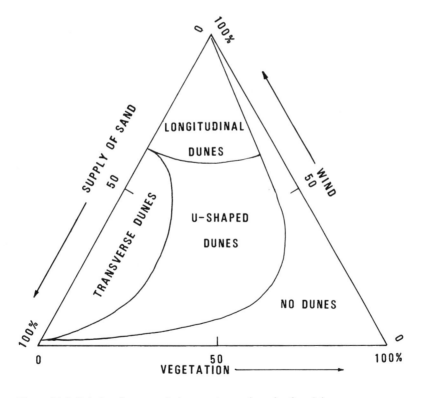

Figure 11-3. Relations between wind, vegetation, and supply of sand that generate different dune forms in a unidirectional wind.
Taken from Dunes of the western Navajo Country, J. T. Hack, *Geographic Review* 31:240-263, 1941.

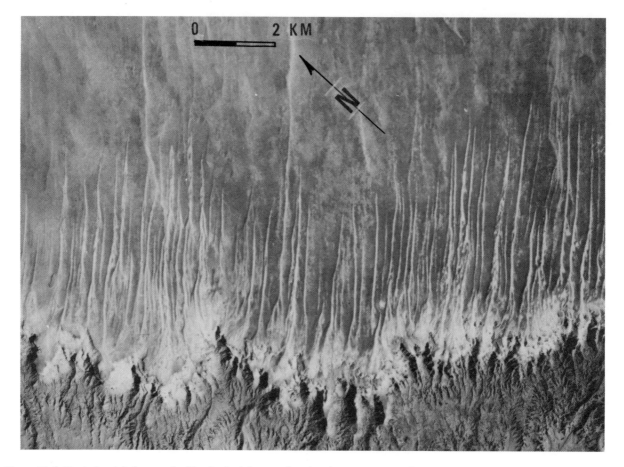

Figure 11-4. Vertical aerial photograph of longitudinal dunes anchored to the escarpment on the southwest margin of the Moenkopi Plateau at Ward Terrace east of the Little Colorado River. AZ-4, HU M14 AMS 134, frame 1606, February 21, 1954.

whose horns point downwind and enclose a slip face between them (figures 11-19, 11-20, 11-32). These typically retain their size and shape as they migrate downwind because the sand shed from their horns balances the sand that reaches the dune from upwind over its entire windward surface.

U-shaped dunes look like backward barchans in that arms on each side of the dune trail behind the advancing dune rather than preceding it (figure 11-12). Vegetation seems essential for these to develop, encroaching onto the dune mass from either side and stabilizing the flanks, but the advancing slip face smothers and kills vegetation in its path and remains free. The area between the trailing arms is also bare because it reexposes the vegetation killed by dune advance; this zone commonly is one of active deflation until it is revegetated (figure 11-24b). U-shaped dunes seem sensitive to vegetation changes, and in some cases their headward parts are transitional to barchans; these hybrid forms have both tails and horns (figure 11-24b).

The tallest dunes in the North American deserts are peaked dunes, reaching more than 200 meters (m) in height. In the

United States, most of these are short, steep-sided, slightly sinuous ridges elongate parallel to a mountain front several kilometers distant (figures 11-5a, 11-6, 11-7, 11-10, 11-21, 11-22). Rare in the United States but prevalent in northwestern Sonora are star-shaped dunes, from whose central pyramid radiate three or more arms, each roughly symmetrical in cross-section (figure 11-27). These are found both as isolated features and arranged into rows (figures 11-26, 11-28).

Climbing dunes climb the windward side of hills as sand sheets, and falling dunes cascade down the leeward side. Such dunes once enveloped many of the low ranges in the central Mojave Desert, but most are inactive now, veneered by gravel and dissected by ephemeral streams (figures 11-11, 11-30, 11-31).

Other dunes are best described as complex, meaning that their form combines elements of several simpler forms. Some appear to represent intersecting longitudinal and transverse elements; either element may appear active or subdued. The large dunes of the Algodones chain in southern California were called "megabarchans" by Norris and Norris (1961)

Figure 11-5a. Oblique aerial view toward the northeast across the Eureka Valley dune toward the last Chance Range, CA-1, February 1974).

Figure 11-5b. Ground view northward along the crest of the Eureka Valley dune. Note the approximate symmetry of this ridge, August 1967.

Figure 11-6. Telephoto view of star-shaped dunes in northern Panamint Valley from the southeast.
Note that the dunes lie several kilometers distant from the abrupt southern escarpment of Hunter Mountain. CA-2, November 1972.

Figure 11-7. Saratoga Spring's dunes from the west.
CA-4, March 1970.

Figure 11-8. Oblique aerial photograph toward the west across transverse dunes in the Saline Valley at the foot of the Inyo Mountains.
CA-6, February 1974.

because of morphologic similarity to much smaller barchans, but these occur only at the chain's southern end, and most of the large dunes are transverse ridges of two or more coalesced domal dunes (figure 11-15). These large dunes are extensively covered with "peak and hollow" topography of complex character (figure 11-18).

ARIZONA

Breed et al. (in press) have summarized the distribution of dunes in Arizona, where they occupy the southwest and northeast corners of the state. In the southwestern corner, much of the desert southeast of Yuma is sandy, but the only active dunes are near the Mexican border, where they trend N 60° E (Bryan 1925:107; Olmstead, Loeltz, and Irelan 1973:288; AZ-1). Northeast of these, a single ridge of dunes about 5 km wide extends about 25 km south-southeast up the floor of lower Mohawk Valley (Bryan 1925:107; Shreve 1938; AZ-2). About 30 m tall, this dune body is steepest on its east margin, and its level, undulating upper surface bears a network of subdued, intersecting, closely-spaced individual

dune ridges. To the south, Bryan (1925:105) and Shreve (1938) also noted wind-blown sand on the eastern margins of the Lechuguilla and Tule deserts.

East of Parker, dunes cover much of the Cactus Plain, forming a band about 10 km wide and extending more than 30 km eastward (Shreve 1938; AZ-3). Their surficial morphology is mostly parallel ridges, probably longitudinal, about 100 m apart and trending N 30°-65° E. Some ridges are oblique ("feathered") to subdued, east-trending ridges beneath them, and the field grades eastward into a complex topography of low intersecting ridges (Breed et al. in press; figures 11-14, 11-36). The eastern tip of the field is made of climbing dunes stripped away from the mountain front, and stream dissection is conspicuous along the northern and southeastern margins of the field.

The dune fields in northeastern Arizona (AZ-4) cover about 65,000 km² (Breed and Breed 1978). As mapped by Hack (1941), these are mostly northeast-trending longitudinal and U-shaped dunes with isolated fields of barchans, transverse dunes, climbing dunes, and falling dunes. The longitudinal dunes are 2 to 9 m tall, about 90 m apart, and up to about

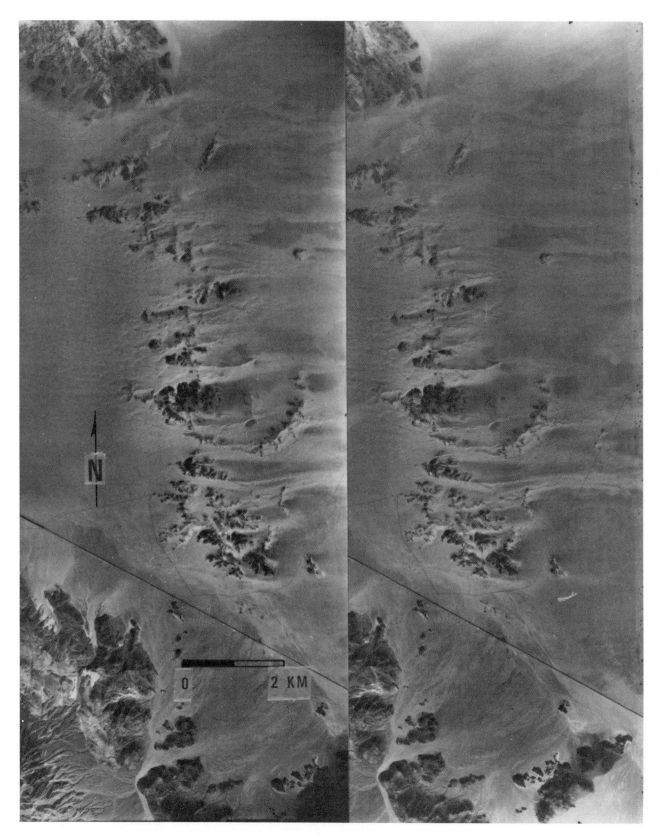

Figure 11-9. Stereogram of longitudinal dune ridges associated with bedrock hills in the Devil's Playground,
about 30 km northwest of the Kelso Dunes. CA-8 Photos FS M50 AMS 145, frames 3974-5, September 19, 1954, Old Dad Mountain 15′ quadrangle.

Figure 11-10a. Oblique air photograph of Kelso Dunes from the north.
CA-8, January 1971, Kerens and Flynn 15' quadrangle.

Figure 11-10b. Ground view toward the west along the crest of the main ridge of the Kelso Dunes.
Note that most of the sand driven from left to right hugs the ground until it streams off the brink of the slip face beyond the shovel. L. Dean (personal communication) recorded wind velocities up to about 25 m/sec at about the time this photograph was taken by the author, January 1978.

Figure 11-11a. Ground view of Cat Mountain from the south.
Note that the cat's form is delineated by gullies between the inactive dune and the mountain. This
falling dune is about 275 m tall. CA-8, July 1967.

*Figure 11-11b. View down Cat Mountain from the position of the arrow on the summit ridge in
figure 11-11a.*
The dune is veneered by coarse talus like that in the foreground. July 1967.

Figure 11-12. *Oblique aerial photograph toward the northwest across stabilized U-shaped dunes along the shoreline of Rosamond playa in the western Mojave Desert, California.*
The standing water between the dunes came from unusually heavy winter rainfall, February 1974.

Figure 11-13. Oblique aerial photograph toward the northwest across low dunes in Coachella Valley between Indio and Palm Springs.
Much of the undulating surface adjacent to the dunes is veneered by granules too coarse to move by saltation. CA-13, February 1972.

Figure 11-14. Vertical aerial photograph of longitudinal dunes anchored to Superstition Mountain.
Note the poorly developed barchans spawned from the longitudinal dunes. CA-16, Spence Air Photos AA673–133, January 18, 1934.

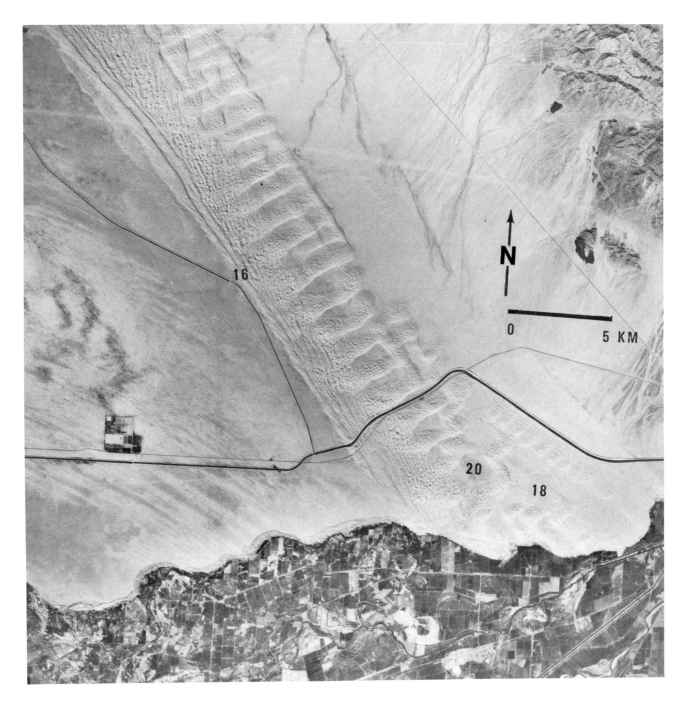

Figure 11-15. Vertical U-2 aerial photographs of the Algodones dune chain.
Numbered points represent the locations of some of the figures that follow. CA-17, USAF 665V
049, November 8, 1967.

Figure 11-16. Oblique aerial photograph toward the northwest along longitudinal dunes that form the southwest margin of the Algodones dune chain.
The ridge on the left is more than 20 km long; the subdued surface to either side of the dune is armored by granules, February 1972.

Figure 11-17. Ground view toward the northwest along the crest of a longitudinal dune on the southwest margin of the Algodones dune chain north of county road 78.
The slip face along this ridge reverses its position seasonally to give the ridge its rough symmetry, January 1976.

Figure 11-18. Oblique aerial photograph toward the south across three "megabarchans" at the south end of the Algodones dune chain.
Note that the surface of these dunes is pocked by peak and hollow topography, unlike the smooth surface of the small barchans in the foreground and in figure 11–19, January 1976.

Figure 11-19. View toward the southwest across a barchan about 2 m tall in the southern Algodones dune chain.
Note the dark gravel pavement beneath the dune, whose right horn has been deformed by recent north winds; location is in upper right corner of figure 11-20, January 1976.

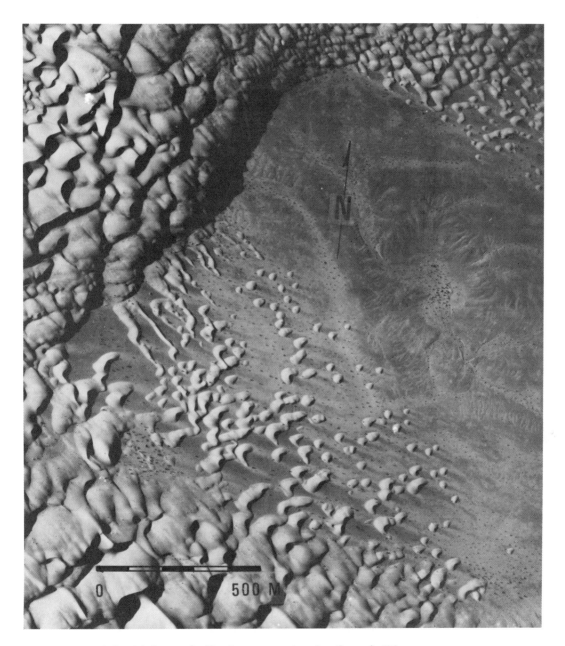

Figure 11-20. Vertical aerial photograph of barchan swarm on intradune flat on the U.S.-Mexican border, southern Algodones dune chain.
The longitudinal dune streamers spawn these barchans. Project Arizona, roll 1011M, frame 1143, May 10, 1940.

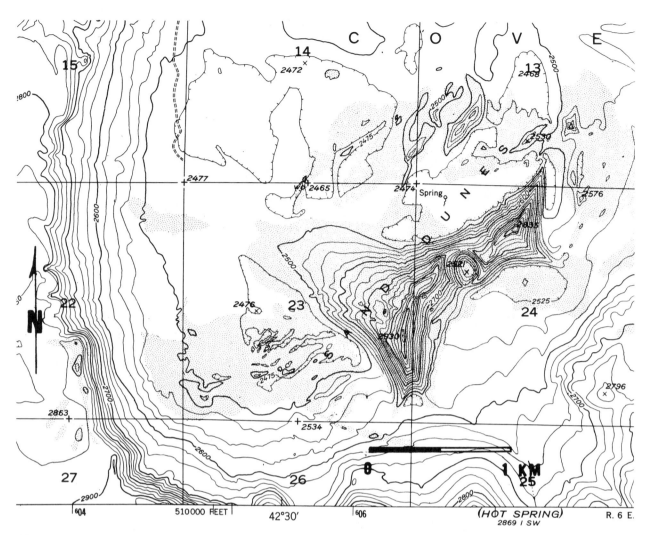

Figure 11-21. Topographic map of Bruneau Dunes.
ID-1, contour interval 25' (7.5 m); sand dunes 7½' quadrangle.

Figure 11-22. Bruneau Dunes from the northeast,
September, 1963.

Figure 11-23a. Oblique aerial photograph toward the north across closely-spaced barchans to transverse dunes in Silver State Valley.
NV-1, August 1967.

Figure 11-23b. Ground view toward the northwest across dunes just west of those in figure 23a.
The dune in the foreground is about 18 m tall; the jeep (arrow) gives scale, July 1967.

Figure 11-24a. Vertical aerial photograph of U-shaped and barchan dunes on the west side of the valley of the Little Humbolt River about 13 km north of Winnemucca.
Although these dunes generally trend and advance toward N 80° E, note how this trend is distorted at the passes on the right that sand traverses from Silver State Valley. NV-1, HU M119 AMS 109 frame 15019, June 2, 1954.

Figure 11-24b. Oblique aerial photograph toward the south across the U-shaped dunes in figures 11-23a.
Note the sand encroachment from right to left across U.S. Highway 95; the abandoned road to its right is the road on figure 11-23a, August 1967.

*Figure 11-25. Oblique aerial photograph toward the north across depressions and fixed dunes,
Laguna del Parro, Estancia Valley.
NM-8, November 1973.*

5 km long; these are most abundant on the plateaus and are mostly stabilized by vegetation, particularly at higher elevations, where annual precipitation exceeds 23 to 25 cm (Hack 1941; figure 11-4). Breed et al. (in press) noted that the slip faces on the active longitudinal dunes reversed from north facing during summer to south facing during winter. Hack (1942:39-44) inferred that the major episode of Holocene dune formation occurred between 2000 and 5000 B.C., when the climate was much drier than now and that renewed wind action has occurred since 1000 A.D. Cooley et al. (1969) noted that dissected Pleistocene longitudinal dunes were aligned along the same northeasterly trend as the Holocene dunes. Hack (1941) described the vegetation on the dunes and based his classification of dunes on the interdependence of wind intensity, sand supply, and vegetative cover (figure 11-3).

Barchans and transverse dunes are most common along the southwestern margin of the region, where sand flux is greatest (Hack 1941). McKee (1957:1721-1725) studied the internal structure of a 3 m tall barchan near Leupp and determined that most beds dipped 31 to 32 degrees, parallel to the slip face, except very near the windward surface. Cotera and McCauley (1977) described the grain-size distribution in barchans near Cameron, Moenkopi, and Tuba City and noted that sand sorting improves uphill on a climbing dune below Paiute Trail Point on the Moenkopi Plateau. Breed et al. (in press) noted that a falling dune on Bad Medicine Butte was covered with gravel.

Elsewhere in northern Arizona, Cooley et al. (1969:30-31) noted dunes of volcanic cinders on northeast-facing slopes within the San Francisco volcanic field but did not describe their morphology (AZ-5). Within the Grand Canyon at the mouth of Fall Canyon, Breed and Breed (1978) described 1973-1977 variations in a small star dune 5 m tall, apparently built by winds following intersecting canyons (AZ-6).

BAJA CALIFORNIA

The northernmost active dunes on the Pacific coast of Baja California are those near Bahía San Quintin, about 250 km south of the border between the United States and Mexico (BC-1). Here, a body of dunes about 1 km wide extends inland about 6 km across Highway 1 to spill into the valley of Arroyo Socorro (Cooper 1967:103-106). Vegetated ridges parallel to the S 60° E trend of the body enclose a central area of transverse dunes up to 30 m tall. This main dune body is about 15 km east of Cabo San Quintin.

About 300 km southeast of Cabo San Quintin, dunes occupy much of the shoreline of Bahía Sebastian Vizcaíno, which opens to the northwest (BC-2). Gastil, Phillips, and Allison (1975, pl. 1C) mapped, but did not describe, coastal dunes extending about 20 km northeastward from Punta Santo Domingo. Just south of this point, Inman, Ewing, and Corliss (1966) described a large dune field that occupies a 50 km stretch of coast and encloses (from north to south) lagunas

Manuela, Guerrero Negro, and Ojo de Libre (Scammon's Lagoon). Barchan and transverse dunes form a band 5 to 10 km deep along the coast but yield inland to southeast-trending, parallel "sand stripes" about 100 m apart; these longitudinal features extend 30 to 40 km inland. More detailed studies of a 40 square km field of barchans north of Laguna Guerrero Negro showed that these dunes, uniformly 6 m tall, migrated southeastward at a rate of eighteen m per year from October 1956 until February 1960. These barchan dunes are unique among desert barchans because periodically their toes are flooded by high tides, but their form seems unaffected by this flooding.

Beal (1948:83) noted that dunes on the south edge of Bahía Sebastian Vizcaíno extend up to 100 feet (300 m) elevation on the northwest edge of Sierra Vizcaíno, and his map (plate 1) shows them to cover much of the area between that range and Laguna Ojo de Libre (BC-3). Beal (1948:83, pl. 1) also noted dune areas on both the northwest and southeast sides of an unnamed mountain range between these lagoons and Laguna San Ignacio, about 150 km to the southeast (BC-4; BC-5). Two hundred km farther southeast, Phleger and Ewing (1962:169) noted barchan dunes up to 12 to 15 m tall on the lagoonal barrier seaward of Laguna Santo Domingo (BC-6). Fifty to 100 km farther southeast, Beal (1948:83) noted that dunes cover much of the region east of Bahía de Magdalena (BC-7).

Dunes have not been described on the Gulf of California side of Baja California except at the north end adjacent to the delta of the Colorado River. Gastil et al. (1975:pl. 1-A) mapped but did not describe an area of coastal dunes 30 to 40 km south of San Felipe, and Kniffen (1932:map 1) and Sykes (1937:pl. 1) mapped a thin band of coastal dunes extending about 30 km north of San Felipe (BC-8). In some places the latter is a single, north-trending symmetrical ridge directly adjacent to the shore (Kniffen 1932:201, pl. 27). Climbing and falling dunes occur on the north end of Sierra de las Pintas, and a nearly continuous band of low, east- to northeast-trending dune ridges less than 10 m tall extends about 100 km northward along the western margin of the Laguna Salada basin almost to the U.S. border (Kniffen 1932:201-202, map 1; Sykes 1937:pl. 1; BC-9). Small patches of dunes, mostly stabilized by vegetation, occur on the northern part of the delta (Kniffen 1932:201-203, map 1; Sykes 1937:pl. 1; BC-10). Many of these have since been leveled for fields, but some up to 10 m tall remain as vegetation-choked, symmetrical mounds of dirty sand. The southern tip of the Algodones dune chain extends into Baja California but is described in the next section.

CALIFORNIA

Sand dunes are locally significant land forms in each of the three physiographic provinces of the California desert even though they cover only about 1 percent of its surface (Clements et al. 1957:106). Eymann (1953) and Dean (1978) have cataloged many of these dune fields and their vegetation. The typical character and distribution of dunes vary by province. Within the northernmost province, the basin and range province, small, isolated fields of tall, peaked dunes are nestled between the large, steep, continuous ranges that rise 1000 to 3000 m above north-trending valleys. The middle province, the Mojave Desert province, is characterized by discontinuous ranges that typically rise less than 1000 m above extensive pediments or broad valleys; much of the eastern Mojave is blanketed by inactive climbing and falling dunes, now veneered by gravel and trenched by ephemeral streams. The southernmost province, the Colorado Desert province, occupies the Salton trough, whose broad, flat floor bears the largest field of active dunes in the United States, the complex Algodones dunes, and scattered fields of barchans and longitudinal dunes.

Basin and Range Province

Tall, peaked dunes are found at the south end of Eureka Valley (CA-1), at the north end of Panamint Valley (CA-2), near Stovepipe Wells in central Death Valley (CA-3), south of Saratoga Springs in southern Death Valley (CA-4), and in the Dumont Dunes at the north end of Silurian Valley (CA-5). The tallest of these, also the tallest in California, rises 208 m above the floor of Eureka Valley (Berkstresser 1974; CA-1). The tallest dune is a T-shaped ridge whose tail descends south-southwestward into several small star dunes (figure 11-5a). This main ridge, about 2 km long, is steepest on its west side. The Panamint Valley dunes are two stars, 50 to 60 m tall, surrounded by low, northeast-trending ridges, symmetrical in cross-section (figure 11-6; CA-2). Unlike the other peaked dunes within the province, these lie on dissected alluvium more than 300 m above the valley floor.

The Stovepipe Wells dunes in Death Valley were first noted by Ball (1907:196; CA-3). The tallest of these, a well-formed star dune, rises about 40 m above a field of smaller star dunes and long, symmetrical, northeast-trending transverse dunes. On the eastern edge of this field, Clements et al. (1963: 20-24) documented seasonal reversal of form and migration azimuth of a small barchan, 1 m tall. During April through September 1953, this dune moved 13 m northwest but reversed its form during October and moved about 9 m southeast between then and mid-December; by late February 1954, it had again reversed toward the northwest. This balanced, seasonal reversal of wind may explain in part the unusual length and symmetrical form of northeast-trending transverse dunes in a field that extends about 10 km northwestward from here along the floor of Death Valley. Five km west of the main field lies a field of east-trending longitudinal dunes, each about 1 km long.

At the south end of Death Valley, a small field of star dunes was called the Saratoga Springs Dunes by Wilcoxon. (1962)

and Garrett (1966), who studied them, but the Ibex Dunes by MacDonald (1966), who studied the nearby Dumont Dunes. These dunes, up to 50 m tall, lie in a north-trending band 3 km long on the southwestern slopes of the Saddle Peak Hills (figure 11-7; CA-4). Both of these dunes and the Dumont Dunes were briefly noted by Thompson (1929: 580).

The Dumont Dunes (CA-5) contain a northeast-trending core of star and complex dunes up to 120 m tall flanked on the south by east-trending transverse dunes and on the east by north-facing barchans (MacDonald 1966:37-46, 1970). Some north- to northeast-trending longitudinal dunes head at rock knobs in the western part of the field. Based on MacDonald's (1966:70-79) measurements during May through December 1964, the orientation of dune slip faces reverses seasonally from south facing during fall through spring to north facing during summer, but little net movement results. However, MacDonald inferred that transverse dunes along the south margin move north, barchans face north, and longitudinal dunes point north to northeast from rock knobs, all implying net northward sand transport. Nevertheless, seasonal reversal and little net transport are indicated by MacDonald's (1966:29-35) crude analysis of potential sand transport based on 1948-1950 wind data from Silver Lake, 40 km south of the dunes. These data suggest that north winter winds dominate south summer winds but are equaled by spring west winds. The topographic setting of the Dumont Dunes is less open than that of Silver Lake, and the dunes' local wind regime may well be controlled by topography.

Small fields of reversing, east-trending transverse dunes occur on the west side of the Salt Springs Hills west of the Dumont Dunes and on the west side of the Valjean Hills east of the Dumont Dunes. At the north end of the first field, a small longitudinal dune streaming northward from a rock knob has a slip face that reverses from west facing during summer to east facing during winter (MacDonald 1966:77). Moving east-northeastward from here across the main Dumont dune field, the azimuth of seasonal reversal shifts clockwise to N 60° W- S 60° E at a point north of the field's tallest dune and to N 20° W- S 20° E at the eastern edge of the field (MacDonald 1966:70-79).

Elsewhere in the province are fields of predominantly transverse dunes on the floor of Saline Valley (CA-6; figure 11-8) and south of Owens Lake (CA-7) in southern Owens Valley. The latter, which Eymann (1953:31) took as his type example of transverse dunes in the California desert, are taller and more transitional to peaked dunes than are the former. Dunes in both fields tend roughly east-west.

Mojave Desert Province

The most active dunes in the Mojave Desert are in its eastern part rather than in its western part. The main areas of dunes are the Devil's Playground-Kelso Dunes complex (CA-

8), Cadiz Valley (CA-9), Rice (CA-10), and Desert Center (CA-11), all first noted by Thompson (1929:113-114) and later cataloged by Eymann (1953). The largest and most studied of these is the Devil's Playground-Kelso Dunes complex, across which sand from the Mojave River drainage streams about 100 km discontinuously eastward from Pleistocene Lake Manix across the Cady Mountains and the Mojave River wash to the Kelso Dunes. Fresh river sand from Afton Canyon becomes well sorted within 20 km of eolian transport and becomes noticeably better rounded within 10 km; roundness improves progressively over the 50 km distance to the Kelso Dunes (Sharp 1966:1067-1070). Anders (1974) mapped the distribution of dunes throughout this area and described the different dune forms associated with varying sand flux across and around topographic obstacles of different sizes, shapes, and orientations. Driven by predominant west wind, sand envelops low hills as climbing and falling dunes with longitudinal dunes streaming off the ends of hills and from gaps between hills (figure 11-9). Dunes seem unable to climb the tallest mountains, however, but accumulate as discrete masses on the windward piedmonts of mountains, probably because of the flow pattern of wind around and over the mountains.

The Kelso Dunes at the east end of the area are the best example of this phenomenon, sited 3 km northwest of the steep northwestern escarpment of the Granite Mountains (figure 11-10a; Thompson 1929:551-552; Sharp 1966; Anders 1974: 51-56). These dunes contain a core of three parallel, complex, northeast-trending dune ridges surrounded by low, subdued, vegetated dune ridges now traversed by incised ephemeral streams. The southernmost dune ridge is the tallest, reaching about 170 m above the foot at the highest point, where it has slip faces on both sides above 20° to 25° slopes (Sharp 1966; figure 10b). Based on twelve years' measurement of four transverse ridges trending north to N75° E, Sharp noted little net movement despite seasonal form reversal and oscillatory migration within a 10 m zone. These data, plus the orientation of transverse dunes and slip faces throughout the field, led Sharp to conclude that the west winds that clearly prevail west of the dunes were opposed effectively within the dune field by winds from other directions, probably controlled by the adjacent mountains, rising about 1000 m above the dunes. This relation to topography was earlier postulated by Thompson (1929:551-552) and later elaborated on by Anders (1974:51-56).

Winds that affect the Kelso Dunes now seem to differ in azimuth from those that formed the subdued, stabilized dunes that surround the active dunes. Sharp (1966:1049-1050) inferred that typical modern winds are northwesterly, but typical ancient winds were southwesterly and southeasterly. He further inferred that the dunes are now being reactivated after an indeterminant period when more of the dune field was stabilized. Smith (1967:10-12) concurred that the dunes were stabilized during a climatic episode wetter than modern and

suggested that ephemeral stream channels became entrenched across the dunes during that wetter period.

Anders (1974) mapped the distribution of the climbing and falling dunes that mantle many of the low mountains north of the Devil's Playground and the Mojave River wash. Cat Mountain, the best known of these, is a dune falling 275 m off the steep southeast face of the Cronese Mountains (Evans 1962; figure 11-11a). This dune is inactive, eroded, and veneered by talus, as is the climbing dune that fed it (Smith 1967: 14-17; figure 11-11b). Slopes on the climbing dune are much gentler (8°) than those on the falling dune (24°; Evans 1962). Smith inferred the following sequence of events leading to the present configuration of Cat Mountain: (1) active sand drifting and dune development; (2) little or no sand drifting, accompanied by downslope movement of debris to veneer dune surfaces; (3) fluvial erosion of veneered dunes; and (4) modest renewal of sand drifting. He believed that this general sequence was valid throughout the Mojave Desert and speculated that the interval of intense wind action corresponded to a period during the Holocene when the climate was more arid than it is now (Smith 1967:17-22).

Bassett and Kupfer (1964; Kupfer and Bassett 1962) mapped the geology and dune distribution over a large area to the south and east of the Devil's Playground. As mapped, these dunes and sand sheets are best developed in and around Cadiz Valley (CA-9) and in a 10-km wide band extending 40 km eastward from Twentynine Palms across Dale Dry Lake to Clarks Pass (CA-12). Climbing and falling dunes envelop the Calumet Mountains west of Cadiz Valley and spill over the Kilbeck Hills east of the valley to form the south margin of Danby Dry Lake. This northwest-trending band of dunes and sand sheets is about 50 km long and 10 to 20 km wide. Its most active dunes are on the floor of Cadiz Valley, where they are mostly northeast-trending transverse dunes with some incipient star dunes near the north end. Some stabilized U-shaped dunes encroach southeastward onto the playa's north end. Dunes in the Twentynine-Palms/Dale field are best developed south and east of Dale Dry Lake, where Smith (1967:7-9) described climbing dunes whose windward faces slope 4° to 6°. Their upper surface is stable and veneered by detritus, apparently washed from the adjoining mountains, but streams have dissected the dunes to depths approaching 40 m to expose a great thickness of fine dune sand with little interbedded gravel beneath the gravel veneer. Dissection is particularly intense along the margin between the dune and the exhumed range front. The dunes west of Dale Dry Lake are mostly low, vegetated dunes of uncertain morphology on the valley floor. Elsewhere within the region, small sand sheets climb the mountains east of Lavic Lake and east of Broadwell Lake (Kupfer and Bassett 1962).

The Rice dunes (CA-10) extend 50 km southeastward along the floor of Rice Valley from Danby Dry Lake to the Big Maria Mountains and the Colorado River. These are mostly southeast-trending dune ridges atop a fluvially dissected sand sheet of climbing and falling dunes. The Desert Center dunes (CA-11) surround Palen Dry Lake, 20 km east of Desert Center. East of the Desert Center Airfield are southeast-trending longitudinal dunes and barchans. Eymann's (1953) map records climbing, falling, and fixed dunes here as well.

Elsewhere in the Mojave Desert, most dunes are either relict, eroded climbing and falling dunes, or low dunes mostly fixed by vegetation, typically lying east of playas. Eymann (1953) reported these at Coyote, Harper, and Koehn dry lakes in the northern Mojave Desert; at Rosamond and Buckhorn dry lakes and near Lenwood in the western Mojave Desert; and at Coyote, Deadman, Emerson, Means, and Mesquite dry lakes in the central Mojave Desert (figure 11-12).

Colorado Desert Province

Sand dunes in the Colorado Desert occur in the Coachella Valley (CA-13), in Clark Valley (CA-14), west of the Salton Sea (CA-15), at Superstition Mountain (CA-16), and in the Algodones dune chain (CA-17). Coachella Valley (CA-13) contains few dunes despite its notoriety for frequent, very strong winds sweeping eastward from San Gorgonio Pass and intense sandblasting of both natural and artificial objects (Mendenhall 1909a; Russell 1932; Sharp 1963, 1964; Beheiry 1967; Dokka 1978; Sharp and Saunders 1978). Russell (1932: 87-100) described climbing and falling dunes at Whitewater Point and Windy Point on the steep south wall of the valley northwest of Palm Springs. Here the median size of sand grains decreases uphill and sorting improves uphill on the climbing dune; grain size is smallest and sorting best halfway down the falling dune (Eymann 1953; 83-86). The valley's center from here 40 km to Indio is mostly covered by an undulating sand sheet with drifts behind obstacles and bushes (Russell 1932; 106-114; Behiery 1967). Much of this surface is veneered by granules and protected from wind erosion, but wind has excavated long, deep furrows behind houses built on this surface. Farther down valley, past Thousand Palms, several small fields of transverse to barchanoid dunes lie on either side of the valley (Beheiry 1967; figure 11-13).

Clark Valley (CA-14), north of Borrego Valley, contains some south-southeast-facing barchans of ordinary sand along its eastern edge, but it also contains drifts of slit-clay aggregates (Eymann 1953: 18, 80-83; Clements et al. 1957: 80-1; Roth 1960; Theilig et al. 1978: 125-132). The latter are made of sand-sized fragments of mud curls from Clark Dry Lake that have drifted behind bushes. Once deposited and wetted by subsequent rain, the deposits become indurated and immobile bodies of silt and clay whose median diameter is about 0.002 mm.

The Salton Sea barchans (CA-15) are the most studied dune field in the North American desert. These constitute a main field on the west side of the Salton Sea, 15 km southeast of Salton City, and several isolated dunes, notably a large barchan on the edge of Tule Wash, 10 km west-northwest of the main field's center. Between November 1955 and March 1964, Norris (1966) completed fifteen plane-table surveys of this dune to determine its migration and changes in height and volume. During this interval, its height varied between 9.4 and 11.2 m, and its volume varied between 33,000 and 46,000 cubic m, but its final height of 9.9 m and its final volume of 33,400 cubic m were near the starting values (9.4 m and 33,700 cubic m, respectively). Norris suggested that the dune shrank after 1958 because sand flux from upwind decreased after floods in 1959, 1960, and 1962 but did not speculate on why the dune grew between 1955 and 1958. Norris's data show that the dune grew taller during the winter and spring windy seasons, regardless of whether its volume increased and decreased, and got shorter during summer and fall. Between 1955 and 1964, the dune moved 152 m due east (18 m per year), the same azimuth and about the same rate (15 m per year) that Shelton (1966: 198-199) determined from aerial photographs for the period 1932-1962. Based on descriptions by Mendenhall (1909b: 84) and Brown (1923: 30), Norris (1966: 298) speculated that the present dune was the sole survivor of three that were present in 1904, the first of which vanished before 1917-1918 and the second before 1932. He reasoned that they migrated into arroyos and were washed away and that the remaining dune should suffer a similar fate in 1973 or 1974, but it was still there in May 1977 (Thelig et al. 1978: 119-121).

Barchans within the main field of the Salton Sea barchans have also migrated eastward. Long and Sharp's (1964) resurvey indicated net migration toward N 90° E during 1941-1956 but toward N 80° E during 1956-1963. Nine of the best formed of these moved toward S 87° E during 1949-1973, as determined from aerial photographs (Smith, unpub. data), and Rempel (1936) observed migration toward S 70° E during 1933-1934. Within uncertainty limits, this easterly azimuth accords with the S 75° E azimuth of potential sand transport computed from 1942-1945, sixteen-point wind data from the Salton Sea Naval Base by Smith (1979) using the methods of Bagnold (1941). Most dune movement occurs during spring, when strong west winds predominate over all other winds, based on Rempel's 1933-1934 wind measurements and Long and Sharp's (1964) observations.

Rempel (1936) noted that small dunes within the field moved faster than did large ones, a relation that Long and Sharp (1964) expressed mathematically in two forms:

$$1/D = 0.00048 + 0.0000595\,H$$

and $\quad D = 1245\,e^{-.0317\,H}$

where H is dune height in feet and D is distance moved (feet) during 1956-1963. The measured dunes ranged from about 3 to 12 m tall (the median was 6 m).

These barchans probably formed from sand blown off beaches of Holocene Lake Cahuilla (Brown 1923: 30) combined with material eroded from sediments exposed upwind (Long and Sharp 1964; Norris 1966; Christensen 1973). The dunes could have reached their present positions in fewer than three hundred years if they started at the beaches and moved at their modern rate (Long and Sharp 1964). Eventually they will migrate into the Salton Sea and be destroyed, but some barchans submerged and deformed by the 1907 rise of the Salton Sea had regained their form by 1915, after the sea had receded (Skyes 1937: 75).

Christensen (1973) and Eymann (1953) studied the sand within these barchans. The sand is 66 to 68 percent quartz, 16 to 17 percent feldspar, 10 percent rock fragments, and 3 to 4 percent heavy minerals and shell fragments, proportions that closely resemble those of sediments now exposed upwind (Christensen 1973: 42-45, 51). The median size of sand grains on the dune surface is finest at the brink (0.16-0.32 mm) and coarsest at the windward toe and horns (0.42-0.53+ mm); the sand is best sorted near the top and less well sorted along the windward edge and horns (Eymann 1953: 68-78; Christensen 1973: 63-77). Sand within a dune is finer (0.14-0.20 mm median) and tends to be better sorted than surface sand (Christensen 1973: 78-92). The bulk specific gravity of sand is greatest along the windward surface (1.61), less in the dune interior (1.51), and lowest on the slip face (1.45) (Eymann 1953: 60-63).

Other studies of these dunes have either been specialized or cursory. Howard (1977; Howard et al. 1977, 1978) used detailed field studies of wind and sand movement over a 3 m barchan to calibrate simulation studies of barchan dynamics and equilibrium and the orientation of ripples on sloping surface with respect to wind. Rempel (1936) described the vegetation, its rapid stem growth to thwart burial beneath encroaching dunes, and evidence that some barchans started as bush-trapped sand bodies. Russell (1932: 115-120) noted little deformation by north winds, but Norris (1956: 1684) noted a south-facing slip face on the south side of a 9 m barchan after twenty-four hours of north wind. Other authors who have noted the dunes include MacDougal (1908: 38-39, pl. 39), who inferred northeasterly migration; Free (1911: 61-62, pl. 2, 1914: 24, pl. 6b), who measured slip-face slopes at 31.5° to 32°; and Theilig et al. (1978: 119-125), who prepared a guidebook.

Spectacular longitudinal dune ridges cover much of the southern part of Superstition Mountain, which lies on the valley's west side about 25 km meters west of Brawley (CA-16; figure 11-14). This hill, 70 to 150 m tall, lies downwind of a field of eastward-moving barchans and longitudinal dunes, whose encroachment necessitates continual sand removal from the narrow-gauge railroad between Plaster City and a gypsum mine in the Fish Creek Mountains (Clements et al. 1963: 25-27). The irregular, knobby surface of Superstition Mountain has been overrun by dunes climbing its southwest side to form sinuous longitudinal dunes along its crest, and falling dunes

feed straight longitudinal dunes on its east end (Brown 1923: 31-32). The latter extend 0.5 to 1.5 km due east from the mountain front, and their individual forms and distal positions seem to have changed little during the last forty years, based on aerial photographs taken in 1934, 1953, and 1973 (Smith 1978c). Although the dunes' lower flanks stay put and slope 10° to 14° on both sides, the slip face along the crest reverses from south facing during winter to north facing during summer (Eymann 1953: 30-36). When the slip faces faced south, Eymann (1953: 70-75) found that sand was coarser (0.30 mm) and less well sorted on the north side than on the south side (0.19 mm), and his data show that sand was coarsest along the crest line (median 0.29-0.36 mm). Norris (1956) described development of crescentic ripples on the dunes after heavy rain wetted the surface.

Barchans spawned from the longitudinal dunes have migrated consistently toward N 80° E at rates of ten to fifteen m per year for dunes 3 to 4 m tall, based on aerial photographs taken in 1937, 1949, 1953, 1959, and 1965 (Smith 1979). The 10° disparity between the trend of longitudinal dunes and barchan migration azimuth may be explained in part by the position of Superstition Mountain very near the southern limit of winds fanning east-southeast from Borrego Valley and near the northern limit of winds fanning east-northeastward from the drainages of Carrizo and Coyote washes.

The Algodones dune chain (CA-17; figure 11-15) represents one of the largest and most accessible fields of unstabilized dunes in the United States. Its variety of forms and ease of access from two major highways render it well suited to both casual and serious study of the eolian phenomena of sand transport and the development, migration, and destruction of sand dunes. The dune chain extends 70 km southeastward from near the southeast corner of the Salton Sea to the modern floodplain of the Colorado River in Mexico. It contains roughly 6 to 11 billion cubic m of sand (Eymann 1953: 20; McCoy, Nockelberg and Norris 1967). Norris and Norris (1961) first described the dunes in detail.

The main dune mass, about 5 km wide and 30 to 90 m tall, is distinct from lower dunes, which form a fringe along the chain's northeastern flank and occupy much of East Mesa, a triangular area south and west of the chain. The dune chain rises sharply northeastward from East Mesa as a granule-covered ramp surmounted by sinuous longitudinal dune ridges, which extend up to 20 km in southeastward direction (Smith 1978d; figures 11-16, 11-17). The chain's northeastern boundary is less distinct because it is embayed, and fringing dunes encroach on the main dune mass. The main dune mass constitutes a series of large, complex, coalesced domal and barchanoid dunes whose surface displays extensive development of peak-and-hollow topography. Gravel-floored intra-dune flats separate the larger dune forms, which have moved southeastward at 35 to 40 cm per year between 1964 and 1977 (Sharp 1979). This gravel surface becomes more continuous to the southeast, where it is overlain by large isolated barchanoid dunes ("megabarchans" of Norris and Norris 1961; figure 11-18).

Fields of barchans occupy most of the intradune flats in the southern end of the dune chain (Norris 1966). Within a representative field, these barchans are 0.5 to 6 m tall and have migrated toward S 60-65° E at average annual rates of about 5 m per year for large dunes and 20 m per year for small dunes (Smith 1970, 1972; figures 11-19, 11-20). These barchans are severely deformed by summer wind reversal, but their heights are steady over the long term and fluctuate over seasonal or short-term time scales. Field measurement of more than one hundred dunes indicates that most neither grew nor shrank from June 1968 through May 1978 and from January 1976 to January 1978, but nearly all shrank during January through May 1978 and during May 1978 to January 1979 (Smith 1977, 1979).

Dunes on East Mesa and along the northeast margin of the dune chain are mostly stabilized by vegetation. The northeast margin is a zone about 2 to 3 km wide of choked transverse dunes and vegetation-shadow dunes less than 6 m tall and some longitudinal ridges trending southeast. Dunes on East Mesa are mostly low, broad ridges that trend east to east-southeast on the west edge of the mesa but swing clockwise toward southeast as followed eastward. The pattern of sand encroachment off the mesa's south edge onto the Colorado River delta indicates that net sand transport has been longitudinal along these features (Smith 1978a). However, 1978 data from sand traps and from the paucity of sand encroachment onto cultural features suggest that little sand now traverses East Mesa (Smith 1979).

The dune chain is made of sand ultimately derived from the Colorado River, based on similarity of sand (Merriam 1969; van de Kamp 1973: 840-841). This sand was probably reworked by wind from the beaches of a series of large lakes that formed by intermittent diversion of the Colorado River into the Salton Basin during Quaternary time. This sand was blown either eastward onto the dunes from beaches along the west (Brown 1923: 28-29) or northeast side (Loeltz et al. 1975: 11) of East Mesa, or southeastward from a point source near the mesa's north end to form an extending dune mass (Norris and Norris 1961; McCoy, Nockelberg, and Norris 1967).

The dunes are driven mostly by strong westerly downslope winds that follow the major stream drainages heading in the peninsular ranges; these winds are strongest and most frequent during spring. They diminish in intensity as they cross the flat floor of Imperial Valley and shift gradually clockwise to northwesterly as they cross East Mesa. At Yuma, Arizona, 20 km past the dunes, these winds are effectively opposed by persistent strong summer winds from the south-southeast, so Yuma now seems to occupy the regional sump for potential sand transport. These conclusions are based on a nearly continuous hourly record of wind speed and direction at six stations over the last fifteen years and at five stations during World War II (Smith 1970, 1978a, 1978b, 1979).

The dunes were also described briefly by Olmstead, Loeltz, and Irelan (1973: 28) and illustrated in a guidebook by Theilig et al. (1978: 134-142). Their appearance on side-looking radar from aircraft has been described by Brown and Saunders (1978) and on SEASAT radar from orbit by Blom et al. (1979). Cutts and Smith (1973) and Breed (1977) compared features of a dune field in the Hellaspontus region of Mars with the Algodones dunes.

CHIHUAHUA

The only major dune field in Chihuahua is the Samalyuca field about 40 km south of Ciudad Juarez (CH-1). Shreve's (1938) map shows this as a field nearly 150 km long, but more recent work (Webb 1969) indicates that prominent dune features are confined to an area of about 250 square km but are fed sand from extensive Pleistocene lake beds to the west. The main dune mass consists of tall, peaked dunes up to 165 m tall, and some sand blows east through a pass in the mountains to the east into the next valley. MacDougal (1908: 10-11) described the vegetation in the dunes' western part south of Samalyuca, where they reach 12 m tall.

COLORADO

The Great Sand Dunes of Colorado cover about 400 square km of the San Luis Valley of south-central Colorado (CO-1) and reach a maximum height of more than 200 m (Johnson 1967; Andrews 1978). The southwestern half of the dune field is northeast-trending U-shaped dunes, mostly fixed by vegetation, along which sand has passed from its source, probably abandoned meanders of the Rio Grande (Johnson 1967), or possibly sand reworked from the Plio-Pleistocene Santa Fe Group (Siebenthal 1910; Hayford 1947).

The main mass of active dunes covers about 100 square km and rises abruptly from the northeast toe of the U-shaped dunes as a series of tall, steep, northwest-trending transverse ridges (Johnson 1967). The tallest dunes of the main dune mass are generally described as transverse (Merk 1960; Johnson 1967), but star dunes may develop on the highest crests (Andrews 1978). Northeast-trending longitudinal dunes form the southeast margin of the dune mass along with isolated barchans (Merk 1960; Johnson 1967). Although the dunes are encroaching northeastward onto forest, movement of the dunes has not been detected from sequential aerial photography (Merk 1960). The prevailing southwest winds are opposed by occasional northeast storm winds, so the position of the slip face on the crest of transverse dunes alternates between the northeast and southwest side (Merk 1960, 1973). Piling up of sand by opposing winds may be the mechanism that has allowed the dunes to reach their unusual height (McKee 1966: 61).

IDAHO

Large active dunes in Idaho occur on the western Snake River Plains at Eagle Cove about 10 km east of Bruneau (ID-1) and on the eastern Snake River Plain at Juniper Buttes west of St. Anthony (ID-2). Eagle Cove, a large embayment on the south side of the Snake River valley, largely encloses the dune field now protected as Bruneau Dunes State Park (Murphy 1975). The central feature of this dune field is a large peaked dune about 137 m tall (figures 11-21, 11-22). This dune continues northeastward to a lower summit, about 95 m tall. Murphy (1973: 13) claimed respective heights of 160 and 110 m; these figures seem to represent imprecise conversion of respective 450 and 310 foot heights to meters). Although the dune's crest reverses seasonally, Murphy (1973: 62) detected no net migration in twenty-six years, based on aerial photographs taken in 1947 and 1973. To the north and southwest of the main dune, smaller transverse and barchanoid dunes migrate southeast during summer but westward during winter (Murphy 1973: 17). These observations accord with 1951-1965 wind data at Mountain Home Air Force Base, 25 to 30 km to the north, where west-to-northwest spring and summer winds oppose east-to-southeast fall and winter winds, but local topography probably affects wind flow near the dunes (Murphy 1973: 20, 39-45). The sand within the dunes, probably derived from local Pleistocene sediments, is 65 percent quartz, 25 percent feldspar, 6 to 8 percent basalt fragments and heavy minerals, and 2 to 4 percent magnetite (Murphy 1973: 45).

Koscielniak (1973) has described and mapped the dunes at Juniper Buttes (ID-2), earlier noted in passing by Bradley (1873), Russell (1902), Stearns et al. (1939), and Dort (1959) and later photographed from the air by Greeley (1977: 101-108). The form of the dune field is that of a tuning fork, 20 to 30 km long by 10 to 15 km wide, splitting around Juniper Buttes and opening to the northeast. The southern arm of the fork is about twice as long as the northern arm and contains much larger fields of active dunes than does the northern arm. Dune form within the southern arm is mostly transverse, 5 to 20 m tall, but dune form within the northern arm is mostly U-shaped and lower in height. A dune approaching 100 m in height is the tallest of three east-trending ridges along the south flank of the Juniper Buttes. Transverse and U-shaped dunes climb the northern Juniper Buttes to spill through saddles into nested, active U-shaped dunes called lobate dunes by Koscielniak (1973: 11, 14, 21-22); between 1941 and 1972, slip faces on lobate dunes advanced 13 to 28 m, while those on climbing dunes advanced 5 to 50 m (he did not specify heights of individual slip faces). These rates of advance accord with the rate of 3 m per year determined by

Chadwick and Dalke (1965). The dune sand is mostly quartz, but some samples contain significant proportions of fragments of basalt, rhyolite and obsidian (Koscielniak 1973: table 11-2). During Pleistocene time, this sand was probably washed into Mud Lake from the mountains to the north and then blown northeastward as dunes (Stearns et al. 1939: 41; Dort 1959).

The active dunes lie atop stabilized longitudinal and U-shaped dunes that extend about 100 km northeastward from near Arco to Juniper Buttes and form a band about 30 km wide (Dort 1959). Similar, mostly stabilized dunes along the south side of the Snake River form a band 2 to 5 km wide extending northeastward from Blackfoot to Idaho Falls. Chadwick and Dalke (1965) described five successive stages of plant communities that develop on dunes at Juniper Buttes to stabilize them and estimated that it takes up to 1000 years for the climax community to take over.

NEVADA

Nevada's largest dune field, 10 to 15 km wide, extends 60 km eastward from Desert Valley, across Silver State Valley into the valley of the Little Humbolt River north of Winnemucca (NV-1; Russell 1885: 154). The floor of Desert Valley is covered by stabilized U-shaped dunes trending N 70-75° E, but on the valley's eastern slopes these give way to active U-shaped dunes. Mostly north of these U-shaped dunes, large and small barchans and some transverse dunes cover an area of about 50 km. The U-shaped dunes spill across a broad saddle at the south end of the Slumbering Hills into Silver State Valley, floored mostly by barchans and transverse dunes up to 20 m tall (figure 11-23); Smith (1975) has described eolian processes on one of these dunes. On the east slope of the valley, dune spacing gets tighter, and the dunes are mostly transverse with some peaked forms. These break up into the active U-shaped dunes, which cross a pass between the Krum Hills and the Bloody Run Hills and descend into and across the valley of the Little Humbolt River, where some are transitional in form to barchans (figure 11-24).

The other dune fields in Nevada are small and isolated. A small field of active dunes lies between Pyramid Lake and Winnemucca dry lake (Russell 1885: 155). Morrison (1964: 81-85) described dunes in the Carson Desert around Fallon, the sump of the Carson River (NV-2). The desert floor is partly covered by stabilized U-shaped dunes trending N70-85° E, which are several thousand years old. The few active dunes trend N 80° E. These dunes are mostly 6 to 9 m tall but reach 15 m. South of the Carson Desert, climbing and falling dunes cover much of the Blowsand Mountains. Southeast of the Carson Desert is a large peaked dune, 120 m tall called Sand Mountain (NV-3). First noted by Russell (1885: 155), this is a "booming" dune, whose sound has been studied by Criswell, Lindsay and Reasoner (1975) and Lindsay et al.

(1976). Sand Mountain is actually two northeast-trending ridges arranged in echelon and steep on both sides. North of these ridges, dune topography is peaked and complex, giving forth uphill to U-shaped dunes that sweep sinuously eastward through a saddle. The western part of the main dune mass has been cut by a stream.

In southern Nevada, Big Dune south of Beatty is a large, isolated feature in the Amargosa Desert (NV-4). First noted by Ball (1907: 36), Big Dune is a series of joined star dunes, of which the largest two are about 75 m tall and elongate in an east-west direction. The lower dunes are mostly sinuous east-west ridges, symmetrical in cross-section and transitional to three-pointed star dunes at their bends. The position of the active slip face on these ridges seems to alternate seasonally between their north and south sides, much as it does in the Dumont and Stovepipe Wells dunes to the south and west in Death Valley.

In west-central Nevada, Meinzer (1917: 48-49, 61, pl. 1, 13) noted and mapped small, scattered dune fields in Big Smoky Valley, mostly on its east side, and in Clayton Valley. The tallest of these, about 70 m tall, is a peaked dune in a small field labeled Crescent Dunes on the Lone Mountain 15 foot quadrangle (NV-5). This dune is about 20 km north-northwest of Tonopah.

NEW MEXICO

The white sands of south-central New Mexico are the best known and most studied dunes in the state. Unlike most other dunes, their constituent grains are gypsum rather than quartz, although fields of quartz-sand dunes adjoin the gypsum dunes on the north (Meinzer and Hare 1915; Hunt 1978b). The gypsum dunes (NM-1), as mapped by McKee (1966: figure 1), occupy 700 square km, and their southern part can be divided downwind (east-northeast) into four types, each in a different region: (1) low, broad, gentle dome-shaped dunes at the southwestern margin against Lake Lucero, the playa that supplies the gypsum sand; (2) transverse dunes up to 12 m tall in the middle of the field; (3) barchan dunes up to 8 m tall northeast of the transverse dunes; and (4) U-shaped dunes at the northeastern and eastern margins of the field. These dunes are moving east-northeastward, driven by strong winds, which blow mostly out of the west to southwest (McKee 1966: figures 2-5). Between 1962 and 1967, McKee and Douglass (1971) reported that transverse dunes moved 1.3 to 3.7 m, barchans 1.8 to 3.9 m, and U-shaped dunes 0.7 to 2.4 m per year; these rates cannot be compared directly because dune heights were not given except for two barchans, 6.3 and 5.6 m tall, whose annual rates were 3.0 and 2.7 m, respectively. The dunes seem to overlie older dune deposits 6 to 9 m thick (McKee and Moiola 1975). The internal structure of the dunes has been intensively studied by trenching the cohesive sand (McKee 1966), and the unique gypsum sand within the

dunes has been studied by several authors (MacDougal 1908: 14-16; Jones 1959; McKee 1966).

Longitudinal dunes form a large field, 50 km on a side in northwestern New Mexico southwest of Farmington between the Chaco and San Juan rivers and continuing into Arizona (NM-2). These dunes trend N 65° - 75° E and are typically 1.8 to 9.0 m tall, 60 to 3000 m long, and about 90 m apart, with vegetation covering all but their crests (Hack 1941: 243-244). Hunt (1978a) shows the dune field extending well east of the limits of Hack's map, as well as smaller fields of longitudinal dunes between Albuquerque and Grants (NM-3) and a narrow field between Rio Puerco and Rio Grande, 20 to 40 km north of their confluence (NM-4). Within the former area just south of Laguna lies a small field of east-facing barchans and climbing and falling dunes. Near Grants, Bryan and McCann (1943) described modern blowout dunes and interbedded Holocene eolian and alluvial sediments. In all these areas the predominant dune trend is N 70° E. North of Socorro at the confluence of Rio Salado and Rio Grande, Denny (1941: 255-256) and Evans (1963) mapped a small field of northeast-trending dunes less than 12 m tall (NM-5). Evans (1963) noted that barchans here seemed to move northeast, although the strongest winds during 1962 were winter winds from the northwest; perhaps the dunes were sheltered from these winds.

In southwestern New Mexico, vegetated dunes extend northeastward off the shorelines of the pluvial lake that once occupied Animas Valley west of Lordsburg (NM-6; Schwennesen 1918). The "dunes" west of Las Cruces (NM-7; Gile 1966) cover a large area but lack distinctive dune morphology, being mostly small sand accumulations around short bushes.

A unique complex of stabilized gypsum and clay dunes and closed hollows occupies much of the east margin of Laguna del Parro in Estancia Valley, 60 km southeast of Albuquerque (NM-8; figure 11-25). The larger dunes, centered on the hollows, are U-shaped forms up to 2.5 km wide and 24 to 39 m tall (Everard 1964; Titus 1969: 101, 106-110; Bachhuber 1971: 80). The sand is about one-third gypsum, one-third clay, and one-sixth calcium carbonate, and the dune surface is crusted and inactive except for minor longitudinal ridges that head at the centers of the old slip faces (Titus 1969: 106-110; Bachhuber 1971: 80). The playa-floored depressions, 6 to 12 m deep, probably represent basins deflated by wind to provide the dune sediment (Keyes 1908; Meinzer 1911: 26; Everard 1964), but the initial basins may be sinkholes into cavities dissolved into gypsum in the substrate (Smith 1957; see Titus 1969: 110-118, and Bachhuber 1971: 87-92, for discussion). The development of both dunes and hollows postdates the last pluvial-lake stage in the basin ("Lake Meinzer," 4,000-5,000 B.P., of Bachhuber 1971: 78-80, 118). Earlier dunes, probably dating from about 4000 to 6000 B.P., are made of sand that is about two-thirds gypsum and one-third clay and calcium carbonate and have a hard, crusted surface; these are mostly less than 3 m tall and are most commonly transverse and elliptical to circular forms with rare barchans (Titus 1969: 101-105; Bachhuber 1971: 77, 198). The older dunes seem aligned with wind from S 62°-67° W and the younger with wind from S 68°-81° W (Titus 1969: 101-110). At the southeast edge of the valley, a stabilized tongue of quartz sand 2 to 4 m thick by 3 km wide, extends northeastward, spawning "poorly developed barchan and longitudinal dunes" in Pinos Wells Valley (Bachhuber 1971: 62,83). Encino Valley east of Estancia Valley has gypsum-clay dunes and hollows like those of Estancia Valley (Meinzer 1911: 79).

OREGON

Sand dunes occur in two general areas of Oregon east of the Cascade Mountains: in north-central Oregon, where they form a narrow band along the Columbia River (OR-1) and in many of the closed desert basins of south-central to southeastern Oregon. Dunes along the Columbia River are mostly vegetated ridges trending N 45°-65° E; the easternmost of those were mapped by Lewis (1960). About 1 percent of these dunes are active U-shaped dunes, and a few active barchans are found near Biggs, Eightmile Canyon, and Blalock Island.

Dunes in southern Oregon tend to occupy the eastern margins of closed basins like Alvord Valley, Catlow Valley, and Silver Lake Valley (Greenup 1941: 23). Topographic maps also show dunes east of Alkalai Lake, Coleman Lake, Coyote Lake, Summer Lake, and Warner Lakes. Allison (1966) has described the dune field that extends more than 10 km eastward from Christmas Lake and Fossil Lake (OR-2). These dunes, up to 13 m tall, are mostly choked barchans, transverse dunes, and active U-shaped dunes overriding older, stabilized U-shaped dunes on the same easterly trend. The sand, apparently deflated from the lake basins, in roughly half feldspar and half volcanic glass, with minor amounts of other materials but no quartz (Dole 1942, cited by Allison 1966: 15). Allison also described stable, east-west ridges, 1.5 to 6 m tall, made of pumice sand from Mount Mazama (ancestral Crater Lake); he interpreted these to be old transverse dunes aligned with wind perpendicular to modern, but as seen on aerial photographs, these appear to be longitudinal or U-shaped dunes aligned with the modern wind. Allison described the modern wind as blowing from the southwest during winter and from the northeast during summer.

SONORA

Although North America's largest field of active sand dunes lies at the head of the Gulf of California, literature describing them is sparse; the ensuing description is compiled mostly from topographic maps, aerial and orbital photographs, and sparse personal observations. The general area of dunes as noted by

Shreve (1938) can be divided into two parts, separated by Puerto Peñasco on the coast and by the Pinacate volcanic field inland. The southeastern part (SN-1) extends about 150 km along the coast from Puerto Peñasco to Cabo Tepoca and 20 to 50 km inland. It consists mostly of large but discontinuous fields of long, northeast-trending, probably longitudinal, dune ridges spaced several hundred meters apart and mostly less than 10 to 20 m tall. Some mountain ridges poke through the sand cover, but others seem to have been enveloped by dunes. The character of the dunes and the geometry of dune-free areas suggest that these dunes have blown northeastward, inland from the coast, to their present maximum elevation of 200 to 300 m.

The northwestern half of the dune field has more varied morphology and much taller dunes than does the southeastern part. It extends about 150 km northwestward from Puerto Peñasco into the Gran Desierto and is 20 to 50 km wide (SN-2). Although this region is better studied than the southeastern part, the literature on it is sparse and consists mostly of brief explorers' notes (Hornaday 1908; Lumholtz 1912) supplemented by brief reconnaissance descriptions (Shreve 1938; Ives 1959; Merriam 1969; May 1973; Arvidson and Mutch 1974; McKee and Breed 1976; Breed and Breed 1978). The predominant dune forms in the western half of this field are spectacular star-shaped dunes arranged into rows trending N 60° W (figures 11-26, 11-27, 11-28). Individual dunes are mostly 80 to 120 m tall and the rows about 2 km apart, separated by low dunes or by the gravel floor of the underlying mesa. In some places, individual star dunes seem to have formed on a sand surface of low relief by excavating sand around the perimeter of the star and piling it up in the center, but this mechanism remains unverified. Smaller star dunes and low, northeast-trending ridges surround the core of large star dunes on the north and west, giving way outward to low, stabilized dunes of complex pattern and northwest-trending streaks (figures 11-29, 11-30, 11-31).

The south margin is similar except that streaks and low ridges trend north and give way eastward to the large, north-facing barchanoid dunes, which predominate in the eastern half of the dune field. These dunes, 1 to 2 km wide and about 50 m tall, have been described by Ives (1959), who found that some are made mostly of shell fragments blown north from the gulf. Most dunes, however, are made of quartz sand with minor amounts of other materials, indicating that this sand originated in the tributary drainage of the Colorado River (Merriam 1969). North of the barchanoid dunes, a tongue of parallel, northeast-trending longitudinal ridges extends along the northwest margin of the Pinacate volcanic field.

The orientation of the dune ridges and barchanoid dunes, as well as the shell fragments within some of them, suggest that sand has been fed northward from the coast into the main dune mass. However, the pattern of streaks and sand encroachment around hills on the north and west margins suggests that

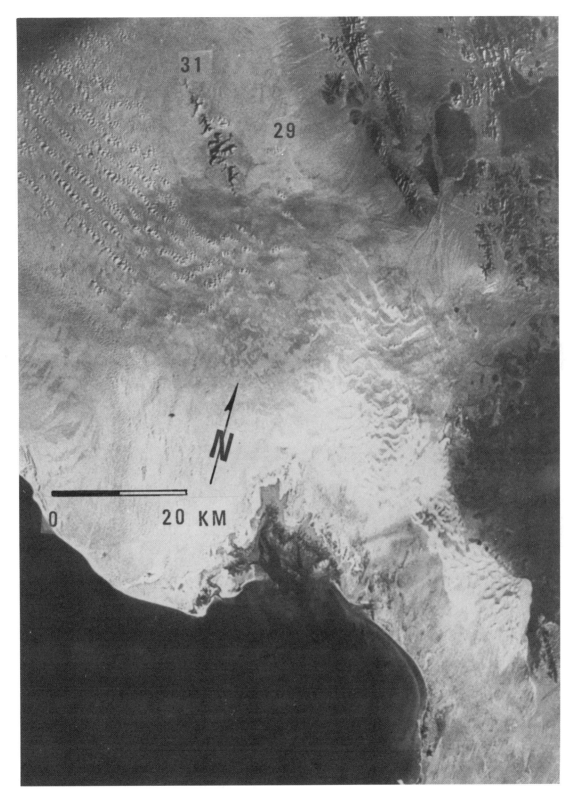

Figure 11-26. Orbital view of dunes in northwestern Sonora, Mexico.
Star-shaped dunes predominate west of the north arrow; north-facing transverse to barchanoid
dunes, unusually large, predominate to the right. Numbered points represent the locations of
some of the figures that follow. SN-1, Apollo 9 AS-923 3559.

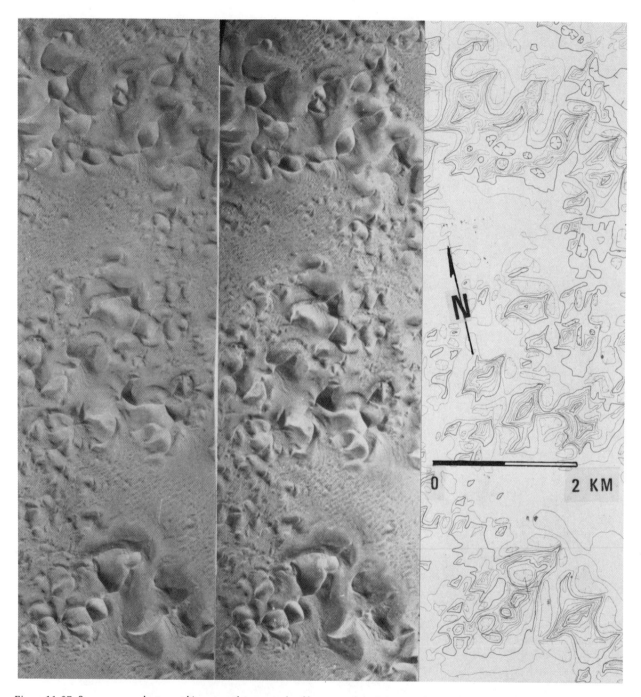

Figure 11-27. Stereogram and topographic map at the same scale of large star-shaped dune south
in the western Gran Desierto,
centered at 32° 07′N, 114° 30′W. Photographs from DETENAL Zona 69 F4 R1566, frames
22-23, May 1973. Map I 11 D 88, contour interval 10 m.

Figure 11-28. Ground view toward the north across star-shaped dunes in the western Gran
Desierto
(about 31° 55′N, 114° 30′W), April 1968.

Figure 11-29. Ground view toward the southwest across transverse dunes and low star-shaped
dunes on the east side of Sierra del Rosario
(32° 05′N, 114° 07′W), March 1968.

Figure 11-30. Ground view toward the southwest across sand-mantled hills of El Capitan (32° 17′N, 114° 25′W). Photograph by the author, March 1969.

Figure 11-31. Ground view toward the southeast across sand-enveloped hills of northern Sierra del Rosario (32°08′N, 114° 16′W), March 1969.

sand has been fed into the main dune mass from the northwest. If so, the dune field at some time occupied the regional sump for wind-blown sand. This relation of past (and perhaps present) sand flow from opposing directions into the dune field accords with the premise that star dunes develop where winds of equal intensity blow from several opposing directions (see Cooke and Warren 1973: 304-306, for critical discussion of this mechanism and for literature references).

TEXAS

No significant areas of dunes occur in the desert part of Texas west of the Guadalupe Mountains. King (1948: 138) described dunes on the east margin of Salt Basin just west of Guadalupe Peak (TX-1). Here, dunes of quartz sand reach 9 m in height and are moving eastward up the piedmont, but dunes of gypsum sand are confined to the valley floor. Between here and El Paso occur very sparse, north-trending transverse dune ridges of reddish sand.

UTAH

The largest field of active dunes in Utah is "Little Sahara," about 10 km north of Lynndyl and about 40 km northeast of Delta (UT-1). Northwest-trending transverse dunes mostly, 5 to 15 m tall climb the southwest flank of the 100 to 200 m tall Sand Hills and sweep through low passes as transverse dunes and active U-shaped dunes transitional to choked barchans and some peaked dunes. These dunes, mostly less than 10 m tall, cover a swath about 5 km wide extending about 15 km N30-40° E from the Sand Hills. The active dunes are fringed by stabilized U-shaped dunes, which also cover most of a corridor about 20 km wide extending southward from the Sand Hills, between Delta and Oak City to a point about 25 km south of these towns. These dunes trend N 30°-40° E, and contain isolated patches of U-shaped dunes and barchans. Beckwith (1951) described problems of sand encroachment onto roads and noted that some isolated barchans of unspecified height migrated northeastward at minimum rates of 1 to 2 m per year.

The other area of locally extensive dunes is in southeastern Utah. Hack (1941) mapped dune areas south of the San Juan and Colorado rivers, where northeast-trending longitudinal dunes cover much of the terrain from west of Mexican Hat east to Aneth (UT-2). These dunes, including some U-shaped dunes here and northeast of Navajo Mountain and patches of longitudinal dunes between there and Mexican Hat, represent continuations of the large dune fields of northeastern Arizona (Hack 1941; Cooley et al. 1969: 31, pl. 1). Farther north, Stokes (1964) described northeast-trending longitudinal dunes in the Green River Desert (UT-3) and inferred that the trellis drainage pattern in the area was controlled by dune orientation. Some small fields of transverse and barcan dunes also occur in the Green River Desert.

In northwestern Utah, thirty-one fields of transverse and U-shaped dunes cover about 11 percent of the Great Salt Lake Desert (Dean 1978; UT-4). In the central part of that desert, Jones (1953) and Eardley (1962) described north-trending transverse dunes, 3 to 9 m tall, whose sand is about 60 percent gypsum and 30 percent oolites of calcium carbonate. Along the desert's margin, however, dunes form smaller fields, are made mostly of quartz sand, and are 1.8 to 12.2 m tall (Dean 1978).

Small dune fields occur elsewhere. Near Kanab, the Coral Pink Sand Dunes State Park (UT-5) contains a small field of partly vegetated transverse and U-shaped dunes made of unusually colored quartz sand reworked from underlying sandstone. Northwest of here, near Cedar City, Thomas and Taylor (1946: 36) noted small dune fields north and west of the town, mostly associated with ephemeral lakes. In south-central Utah, Gregory and Moore (1931: 145) noted a few small fields of dunes on the Glen Canyon platform and in Escalante Valley.

WASHINGTON

Dunes in eastern Washington occur along the Columbia River (WA-1) and to the south and west of Moses Lake (WA-2). Isolated dune fields extending 5 to 20 km eastward from the river are mostly U-shaped dunes enclosing some barchans up to 6 m tall (Petrone 1970: 36; figure 11-32). These dunes are made almost entirely of quartz grains, presumably derived from the river (Petrone 1970: 38). Where the river crosses the Saddle Mountains, the dunes trend about S 70° E, but farther downstream, near Pasco, they trend northeast and are longer and more stabilized by vegetation than those upstream. The latter have been mapped by Lewis (1960). Still farther downstream, dune trends along the river are parallel to the enclosing valley (roughly east-west).

The dunes near Moses Lake cover more area than those along the river but are mostly stabilized U-shaped dunes except in their eastern margin east of Potholes Reservoir (WA-2). East of the reservoir, active U-shaped dunes yield eastward to barchans 12 to 18 m tall; these moved 3.3 m per year toward N 64° E between 1941 and 1961 (Petrone 1970: 54). West of the reservoir, an isolated barchan 2 to 3 m tall moved 7 m per year during the same interval (Petrone 1970: 54). Unlike the dunes along the river, dunes along the leading edge contain only 20 to 40 percent quartz, the rest being mostly basalt fragments that give these dunes their unusually dark appearance (figure 11-33). The proportion of quartz sand increases westward toward the river (Petrone 1970: 41-46).

WYOMING

The largest and most active dune field in Wyoming is the Kilpecker field (WY-1) north of Rock Springs, which covers about 274 square km and extends 90 to 100 km from Eden

Figure 11-32. Oblique aerial photograph toward the southeast across barchans moving eastward from the Columbia River near Wanapum.
These dunes are made mostly of quartz sand. WA-1, June 1978.

Figure 11-33. Ground view toward the south along the crest of a U-shaped dune, 15–20 m tall, southeast of Moses Lake.
The dune is unusually dark because most of its sand is composed of fragments of basalt. The large haystacks were probably placed on the dune by a farmer to dissipate the dune's sand. WA-2, June 1978.

Valley eastward across the continental divide as mapped by Hack (1943) and Ahlbrandt (1974, 1975). Hack (1943) described the dunes as being mostly transverse forms with a band of active and stabilized U-shaped dunes at the windward (western) edge and vegetated, stabilized longitudinal dunes at the leeward edge in the Red Desert. Ahlbrandt (1973: 61-64) believed that active and stabilized U-shaped dunes were the most common, but within the western, most active part of the field, he noted the following downwind sequence of forms: domical dunes up to 15 m tall, transverse dunes up to 45 m tall, barchans up to 27 m tall, and U-shaped dunes up to 15 m tall. Sedimentological studies indicate that the dune sand was derived from the Laney Member of the Eocene Green River formation rather than from glacio-fluvial deposits (Ahlbrandt 1974). These dunes occur in a cold dry climate above 2000 m elevation and sometimes can preserve buried snow through the summer (Steidtmann 1973).

SUMMARY AND CONCLUSIONS

The predominant dune form in the North American deserts is longitudinal, although many of these dunes are now vegetated and inactive and some probably represent the tails of U-shaped dunes. U-shaped dunes are next most common, followed by climbing and falling dunes. Complex peaked and star dunes, although uncommon outside of Sonora and southern California, probably contain more sand than the more common forms because they are much taller. Transverse dunes are locally significant and barchans are rare but widely distributed, suggesting that they are more sensitive to local conditions of their upwind surface and of sand supply than to a restricted range of wind direction.

In most places, dune orientation suggests an easterly component of sand transport. Barchans migrate N 64° E in Washington, east-northeast in southern New Mexico, southeast in Baja California, and N 80° E to S 65° E in the Colorado Desert of southern California at rates of about 3 to 40 m per year, depending on location and size. The fastest barchans are west of the Salton Sea (CA-15), only 120 km from the slowest barchans (CA-17), so much work remains to be done on regional patterns of barchan migration before these data can be used to determine regional patterns of sand flow. The advance of U-shaped dunes can also be used for this, but sand-flow patterns from longitudinal dunes are usually ambiguous. Their orientation, however, seems generally consistent with directions of barchan and U-shaped dune migration, toward the northeast over most of the western United States.

Peaked and star dunes invariably seem to be associated with sites of oscillatory movement of lower dunes and no net dune movement; the peaked dunes may grow because they are local sumps for sand. Except in Sonora, all of these dunes occur in areas where wind is probably controlled by topography. This control may represent channeling of wind by valleys; (for example, the Stovepipe Wells dunes (CA-5) seem to occupy a site where winds blowing eastward from the Cottonwood Mountains meet winds channeled north and south along Death Valley. Star dunes like these and Big Dune, Nevada (NV-4), occur mostly on valley floors away from mountains and may require the interplay of wind from three or more directions. Peaked dunes, on the other hand, invariably occur near mountain fronts and are ridges elongate parallel to the mountain front; (examples are, Eureka Valley (CA-1), Kelso (CA-8), Great Sand Dunes (CO-1), Bruneau (ID-1) and Juniper Buttes (ID-2). These relations suggest that the dune form is controlled by winds from two opposite directions and the presence of the mountain front may generate reverse flow patterns in a predominant wind to maintain dune position "windward" of the mountain front, as proposed by Anders (1974). Although existing studies indicate that these dunes remain fixed in position, no studies have been made of whether they grow taller, a subject for further study. The development and maintenance of star dunes also need much further study.

The wind regime of many dune fields is poorly known, partly because most are distant from weather stations and partly because of local variations in wind direction and intensity that are partly controlled by topography. In general, dunes seem to be aligned with the resultant azimuth of potential sand transport predicted from wind data, but most dunes seem quite tolerant of winds blowing from "improper" directions. Seasonal wind reversal has been documented throughout the western United States, and this reversal is sufficient in some places during some years to reorient small barchans. Because dunes generally seem to be aligned with the resultant wind, it is impossible to be sure of the relative strengths and directions of the component winds from dune form alone with no wind records. There seems to be little evidence to suggest that the resultant transport azimuth has changed significantly during the Holocene, and it may now be the same as during much of the Pleistocene. These inferences can only represent periods of active sand movement, however, not periods of dune stabilization, which may well represent climatic changes and accompanying changes in wind regime.

In many parts of the North American deserts, active dunes seem to override stabilized dunes, suggesting that wind action has been much more effective during part of the Holocene than now. Large areas of Washington, Oregon, Idaho, Nevada, Utah, Arizona, and New Mexico are covered by dune ridges, mostly longitudinal or U-shaped, now stabilized by vegetation. Much of the Mojave Desert is covered by climbing and falling dunes that have been stabilized by a veneer of gravel and deeply incised and stripped from mountain fronts by running water. In all areas these stabilized dunes seem to have been reactivated in the recent past, although not to their former high level of activity. The earlier episode of intense wind action was probably a drier, mid-Holocene event, but it need not have been synchronous throughout western North America.

Dunes are made of whatever sand-sized material is available locally. In most places, this sand is mostly quartz that has been presorted and concentrated by running water or wave action. Volcanic terranes bear little crystalline quartz, however, so their dunes are made of other minerals and rock fragments. The Moses Lake dunes (WA-2) are 60 to 80 percent fragments of fluvially reworked basalt, and the Fossil Lake dunes (OR-2) are half feldspar and half volcanic glass, both deflated from the playa upwind. Dunes downwind of playas tend to be made mostly of sand-sized fragments of gypsum and mud blown off the playas, and some dunes downwind of ancient or modern shorelines bear a significant proportion of sand-sized shell fragments. Quartz sand is the predominant constituent of most dunes only because its durability allows it to outlast grains of most other compositions.

REFERENCES

Introduction

Bagnold, R. A. 1941. *The physics of blown sand and desert dunes.* London: Methuen.

Changery, M. J.; Hodge, W. T.; and Ramsdell, J. V. 1977. Index-summarized wind data: *Battelle Pacific Northwest Laboratories, pub.* BNWL-2220 WIND-11 UC-60. Richland, Washington.

Chepil. W. S., and Woodruff, N. P. 1963. The physics of wind erosion and its control. *Advances in Agronomy* 15: 211-302.

Eymann, J. L. 1953. A study of sand dunes in the Colorado and Mojave deserts. Master's thesis, University of Southern California.

Fenneman, N. M. 1931. *Physiography of the western United States.* New York: McGraw-Hill.

Free, E. E. 1911. The movement of soil material by the wind (with a bibliography of eolian geology by S. C. Stuntz and E. E. Free). *U.S. Dept. Agriculture, Bur. Soils. Bull.* 68.

Hack, J. T. 1941, Dunes of the western Navajo Country. *Geog. Rev.* 31: 240-263.

Hastings, J. R., and Turner, R. M. 1965. Seasonal precipitation regimes in Baja California, Mexico. *Geografiska Annaler* 47, ser. A: 204-223.

Hsu, S. A. 1973. Computing eolian sand transport from shear velocity measurements. *J. Geol.* 81: 739-743.

Hunt, C. B. 1974. *Natural regions of the United States and Canada.* San Francisco: W. H. Freeman.

Kolb, C.R., and Van Lopik, J. R. 1963, Analogs of Yuma terrain in the southwest United States. *U.S. Army Engineer Waterways Experiment Station, Vicksburg, Miss., Tech. Report* 3-630, Report 5.

Lustig, L. K. 1968. Inventory of research on geomorphology and surface hydrology of desert environments. In eds. W. G. McGinnies, B. G. Goldman, and Patricia Paylore, *Deserts of the world: An appraisal of research on their physical and biological environments;* pp. 93-283. Tucson: University of Arizona Press.

Norris, R. M. and Norris, K. S. 1961. Algodones dunes of southeastern *Calif. Geol. Soc. America Bull.* 72:605-620.

Paylore, Patricia. 1967. A bibliography of arid-lands bibliographies.

U.S. Army Natick Labs. Earth Sci. Lab. Tech. Rep. 68-27-ES. Natick, Mass.

Schmidt, R. H., Jr. 1975. The climate of Chihuahua, Mexico. *Univ. Arizona Inst. Atm. Phys., Tech. Rept. Meteorol. Climatol. Arid Regions,* no. 23.

Smith, H. T. U. 1946. Sand dunes (abs.). *New York Acad. Sci. Trans.,* ser. 2, 8: p. 197-199.

————. 1949. Physical effects of Pleistocene climatic change in nonglacial area. *Geol. Soc. America Bull.* 60: 1485-1515.

Thornbury, W. D., 1965. *Regional geomorphology of the United States.* New York: John Wiley.

Thorp, James, and Smith, H. T. U., eds. 1952. *Pleistocene eolian deposits of the United States, Alaska and parts of Canada.* New York: Geological Society of America.

Warren, Andrew. 1969. A bibliography of desert dunes and associated phenomena. In eds. W. G. McGinnies, and B. J. Goldman, *Arid lands in perspective,* pp. 75-99. Tucson: University of Arizona.

Wilson, I. G. 1972. Aeolian bedforms—their development and origins. *Sedimentology* 19: 173-210.

Arizona

Breed, C. S., and Breed, W. J. 1978. Field studies of sand-ridge dunes in central Australia and northern Arizona (abs.). *NASA Tech. Memo.* 79729: 216-218.

Breed, C. S.; McCauley, J.F.; Breed, W. J.; Cotera, A. S., Jr; and McCauley, C. K. in press. Eolian (wind-formed) landscapes. Chapter in ———— ed, *Arizona landscapes: the geologic story.* Tucson: Univ. Arizona Press.

Breed, W. J., and Breed, C. S. 1978. Star dunes as a solitary feature in Grand Canyon, Arizona, and in a sand sea in Sonora, Mexico (abs.). *NASA Tech. Memo.* TM-78, 455, pp. 11-12.

Bryan, Kirk. 1925. The Papago country, Arizona. *U.S. Geol. Survey Water Supply Paper* 499.

Cooley, M. E.; Harshbarger, J. W.; Akers, J. P.; and Hardt, W. F. 1969. Regional hydrogeology of the Navajo and Hopi Indian Reservations, Arizona, New Mexico, and Utah. *U.S. Geol. Survey Prof. Paper* 521-A.

Cotera, A. S., and McCauley, C. K. 1977. Comparative analysis of fluvial versus aeolian sources for wind deposits (abs.) *NASA Tech Memo.* TM X-3511: 153-154.

Gregory, H. E. 1917. Geology of the Navajo country: A reconnaissance of parts of Arizona, New Mexico, and Utah. *U.S. Geol. Survey Prof. Paper* 93.

Hack, J. T. 1941. Dunes of the western Navajo Country. *Geog. J.* 31: 240-263.

————. 1942. The changing physical environment of the Hopi Indians of Arizona. *Peabody Museum American Archaeology & Ethnology* 35(1), Harvard University.

McKee, E. D. 1957. Primary structures in some recent sediments. *Amer. Assoc. Petroleum Geologists Bull.* 41: 1704-1747.

Olmstead, F. H.; Loeltz, O. J.; and Irelan, Burdge. 1973. Geohydrology of the Yuma area, Arizona and California. *U.S. Geol. Survey Prof. Paper* 486-H.

Shreve, Forrest. 1938. The sandy areas of the North American desert: *Assoc. Pacific Coast Geographers Yrbk.* 4:11-14.

Baja California

Beal, C. H. 1948. Reconnaissance of the geology and oil possibilities of Baja California, Mexico. *Geol. Soc. America Memoir* 31.

Cooper, W. C. 1967. Coastal dunes of California. *Geol. Soc. America Memoir* 104.

Gastil, R. G.; Phillips, R. P.; and Allison, E. C. 1975. Reconnaissance geology of the state of Baja California. *Geol. Soc. America Memoir* 140.

Inman, D. L.; Ewing, G. C.; and Corliss, J. B. 1966. Coastal sand dunes of Guerrero Negro, Baja California, Mexico. *Geol. Soc. America Bull.* 77: 787-802.

Kniffen, F. B. 1932. Lower California studies IV. The natural landscape of the Colorado Delta. *Univ. Calif. Pubs. Geog.* 5: 149-244.

Phleger, F. B., and Ewing, G. C. 1962. Sedimentology and oceanography of coastal lagoons in Baja California, Mexico. *Geol. Soc. America Bull.* 73 145-182.

Skyes, Godfrey 1937. The Colorado delta: *Carnegie Inst. Washington Pub.* 460.

California

Anders, F. J. 1974. Sand deposits are related to interactions of wind topography in the Mojave desert, near Barstow, California. Master's thesis, University of Virginia.

Bagnold, R. A. 1941. *The physics of blown sand and desert dunes.* London: Methuen.

Ball, S. H. 1907. A geologic reconnaissance in southwestern Nevada and eastern California. *U.S. Geol. Survey Bull.* 308.

Bassett, A. M., and Kupfer, D. H. 1964. A geologic reconnaissance in the southeastern Mojave Desert, California. *Calif. Div. Mines & Geol. Spec. Rep.* 83.

Beheiry, S. A. 1967. Sand forms in the Coachella Valley, southern California. *Amer. Assoc. Geog. Ann.* 57: 25-48.

Berkstresser, C. F. 1974. Tallest (?) sand dune in California *California Geol.* 27: 187.

Blom, R. G.; Daily, M. I.; Elachi, C.; and Saunders, R. S. 1979. Analysis of SEASAT SAR images of sand dunes (abs.). *NASA Tech. Memo.* 80339. pp. 359-361.

135Breed, C. S. 1977. Terrestrial analogs of the Hellaspontus dunes, Breed, C. S. 1977. Terrestrial analogs of the Hellaspontus dunes, Mar. *Icarus* 30: 326-340.

Brown, J. S, 1923. The Salton Sea region, Calif., a geographic, geologic, and hydrologic reconnaissance, with a guide to desert watering places. *U.S. Geol. Survey Water Supply Paper* 497.

Brown, W. E., and Saunders, R. S. 1978. Radar backscatter from sand dunes (abs.). *NASA Tech, Memo.* 79729: 137-139.

Christensen, R. J. 1973. Petrographic and textural analysis of a barchan dune southwest of the Salton Sea, Imperial County, California. Master's thesis, California State University. San Diego.

Clements, Thomas; Merriam, R. H.; Stone, R. O.; Eymann, J. L.; and Reade, A. B. 1957. A study of desert surface conditions. *U.S. Army Quartermaster Research & Development Center* (Natick, Mass.), *Environmental Protection Res. Div. Tech. Rept.* EP-53.

Clements, Thomas; Mann, J. F., Jr.; Stone, R. O.; and Eymann, J. L. 1963. A study of windborne sand and dust in desert areas. *U.S. Army Natick Labs., Earth Sci. Div. Tech. Rept.* ES-8.

Cooper, W. S. 1967. Coastal dunes of California. *Geol. Soc. America. Men.* 104.

Cutts, J. A., and Smith, R. S. U. 1973. Eolian deposits and dunes on Mars. *J. Geophys. Research* 78: 4139-4154.

Dean, Les. 1978. California desert sand dunes. Univ. Calif., Riverside, Earth Sci. Dept. Unpub. report.

Dokka, R. K. 1978. A method for determining the direction of strong winds: northwestern Cocahella Valley, California. *Assoc. Eng. Geologists Bull.* 15: 375-381.

Evans, J. R. 1962. Falling and climbing sand dunes in the Cronese ("Cat") Mountain area, San Bernardino County, California. *J. Geol.* 7: 107-113.

Eymann, J. L. 1953. A study of sand dunes in the Colorado and Mojave deserts. Master's thesis, University of Southern California.

Free, E. E. 1911. The movement of soil material by the wind (with a bibliography of eolian geology by S. C Stuntz and E. E. Free). *U.S. Dept. Agriculture, Bur. Soils Bull.* 68.

———. 1914. Sketch of the geology and soils of the Cahuilla Basin. In ed. D. T. MacDougal, *The Salton Sea: a study of the geography, the geology, the floristics and the ecology of a desert basin;* pp. 21-34. Carnegie Inst. Washington Pub. 193.

Garrett, D. M. 1966. Geology of the Saratoga Springs sand dunes, Death Valley National Monument, California. Master's thesis, University of Southern California.

Howard, Alan D. 1977. Effect of slope on threshold of motion and its application to orientation of wind ripples *Geol. Soc. America Bull.* 88: 853-856.

Howard, Alan D.; Morton, J. B.; Gad-El-Hak, Mohamed; and Pierce, D. B. 1977. Simulation model of erosion and deposition on a barchan dune. *NASA Contractor Rep.* CR-2838.

———. 1978. Sand transport model of barchan dune equilibrium. *Sedimentology* 25: p. 207-338.

van de Kamp, P. C. 1973. Holocene continental sedimentation in the Salton basin, California: A reconnaissance. *Geol. Soc. America Bull.* 84: 827-848.

Kupfer, D. H., and Basset A. M. 1962. Geologic reconnaissance map of part of the southeastern Mojave Desert, California. *U.S. Geol. Survey Min. Inv. Field Stud.* Map MF-205.

Loeltz, O. J.; Irelan, Burdge; Robinson, J. H. and Olmsted, F. H. 1975. Geohydrologic reconnaissance of the Imperial Valley, California. *U.S. Geol. Survey Prof. Paper* (468-K.

Long, J. T., and Sharp, R. P. 1964. Barchan dune movement in Imperial Valley, California. *Geol. Soc. America Bull.* 75: 149-156.

McCoy, F. W., Jr.; Nokleberg, W. J.; and Norris, R. M. 1967. Speculations on the origin of the Algodones dunes, California. *Geol. Soc. America Bull.* 78: 1039-1044.

MacDonald, A. A., 1966. The Dumont dune system of the northern Mojave Desert, California. Master's thesis, California State University, Northridge.

———. 1970. The northern Mojave Desert's little Sahara. *Min Info. Service* 23:3-6.

MacDougal. D. T. 1908. Botanical features of North American deserts. *Carnegie Inst. Washington Pub.* 99.

Mendenhall, W. C. 1909a. Ground waters of the Indio region, California, with a sketch of the Colorado Desert. *U.S. Geol. Survey Water Supply Paper* 225.

————. 1909b. Some desert watering places in southeastern California. *U.S. Geol. Survey Water Supply Paper* 224.

Merriam, Richard. 1969. Source of sand dunes of southern California and northwestern Sonora, Mexico. *Geol. Soc. America Bull.* 80: p. 531-4.

Norris. R. M. 1956. Crescentic beach cusps and barchan dunes. *Amer. Assoc. Petroleum Geologists Bull.* 40: 1681-1686.

————. 1966. Barchan dunes of Imperial County, California. *J. Geol.* 74: 292-306.

Norris, R. M., and Norris, K. S. 1961. Algodones dunes of southeastern California. *Geol. Soc. America Bull.* 72: 605-620.

Olmstead, F. H.; Loeltz, O. J.; and Irelan, Burdge. 1973. Geohydrology of the Yuma area, Arizona and California. *U.S. Geol. Survey Prof. Paper* 486-H.

Rempel, P. J. 1936. The crescentic dunes of the Salton Sea and their relation to vegetation. *Ecology* 17: 347-358.

Roth, E. S. 1960. The slit-clay dunes at Clark Dry Lake, California. *Compass* 38: 18-27.

Russell, R. J. 1932. Land forms of San Gorgonio Pass, southern California. *Univ. Calif. Pubs. Geography* 6: 23-121.

Sharp, R. P. 1963. Wind ripples. *J. Geol.* 71: 617-636.

————. 1964. Wind-driven sand in Coachella Valley, California. *Geol. Soc. America Bull.* 75: 785-804.

————. 1966. Kelso Dunes, Mojave Desert, California. *Geol. Soc. America Bull.* 77: 1045-1074.

————. 1979. Intradune flats of the Algodones chain, Imperial Valley, California. *Geol. Soc. America Bull.* 90:908-916.

Sharp, R. P., and Saunders, R. S. 1978. Aeolian activity in westernmost Coachella Valley and at Garnet Hill. In eds. Ronald Greeley and M. B. Womer, *Aeolian features of southern California: A comparative planetary geology guidebook,* pp. 9-27. Washington, D.C.: NASA Office Planetary Geol.

Shelton, J. S. 1966. *Geology illustrated.* San Francisco: Freeman.

Smith, H. T. U. 1967. Past versus present wind action in the Mojave Desert region, California. *U.S. Air Force Cambridge Res. Labs.* (Bedford, Mass.) *pub.* AFCRL-67-0683.

Smith, R. S. U. 1970. Migration and wind regime of small barchan dunes within the Algodones dune chain, southeastern Imperial County, California. Master's thesis, University of Arizona.

————. 1972. Barchan dunes in a seasonally-reversing wind regime, southeastern Imperial County, California: *Geol. Soc. America, Abs. Programs* 4: 240-241.

————. 1977. Barchan dunes: development, persistence and growth in a multi-directional wind regime, southeastern Imperial County, California. *Geol. Soc. America, Abs. Programs* 9: 502.

————. 1978a. Actual versus inferred wind regime of dunes: a test in the Algodones dune chain, southeastern California (abs.) *Geol. Soc. America Abs. Programs* 10: 494-495.

————. 1978b. The Algodones dune chain, Imperial County, California (abs.). *NASA Tech. Memo.* TM-78,455: pp. 43-44.

————. 1978c. Field trip to dunes at Superstition Mountain. In eds. Ronald Greely, M. B. Womar, R. P. Papson, and P. D. Spudis, *Aeolian features of southern California: A comparative planetary geology guidebook* pp. 66-71. Washington, D.C.: NASA Office Planetary Geol.

————. 1978d. Guide to selected features of aeolian geomorphology in the Algodones dune chain, Imperial County, California. In

eds. Ronald Greeley, M. B. Womar, R. P. Papson, and P. D. Spudis, *Aeolian features of southern California. A comparative planetary geology guidebook,* pp. 74-98. Washington, D.C.: NASA Office Planetary Geol.

————. 1979. Wind regime of sand dunes in Imperial Valley, California (abs.). *NASA Tech. Memo.* 80339: 275-276.

Sykes, Godfrey. 1937. The Colorado delta. *Carnegie Inst. Washington pub.* 460.

Theilig, Eilene; Womer, M.B.; and Papson, R. P.; 1978. Geological field guide to the Salton Trough. In eds. Ronald Greeley, M. B. Womer, R. P. Papson, and P. D. Spudis, *Aeolian features of southern California: A comparative planetology guidebook,* pp. 100-159. Washington, D.C.: NASA Office of Planetary Geology.

Thompson, D. G. 1929. The Mohave Desert region, California, a geographic, geologic and hydrologic reconnaissance. *U.S. Geol. Survey Prof. Paper* 578.

Wilcoxon, J. A. 1962. The relationship between sand ripples and wind velocity in a dune area. *Compass* 39: 65-76.

Chihuahua

MacDougal, D. T. 1908. Botanical features of North American deserts. *Carnegie Inst. Washington Pub.* 99.

Shreve, Forrest. 1938. The sandy areas of the North American desert. *Assoc. Pacific Coast Geographers Yrbk.* 4: 11-14.

Webb, E. L. 1969. Geology of Sierra de Samalyuca, Chihuahua, Mexico. In eds. D. A. Cordoba, S. A. Wengard, and John Shomaker, *Guidebook of the border region: Albuquerque,* pp. 176-181. New Mexico Geol. Soc., 20th field conf.

Colorado

Andrews, Sarah. 1978. Geometry and dynamics of Great Sand Dunes, San Luis Valley, Colorado (abs.). *Geol. Soc. America Abs. Programs,* 10: 209.

Hayford, F. S. 1947. A petrographic analysis of the Great Sand Dunes of Colorado with conclusions as to source. Master's thesis, Colorado College.

Johnson, R. B. 1967. The Great Sand Dune of southern Colorado. *U.S. Geol. Survey Prof. Paper* 575-C. 177-183.

McKee, E. D. 1966. Structures of dunes at White Sands National Monument, New Mexico (and a comparison with structures of dunes from other selected areas). *Sedimentology* 7: 1-69.

Merk, G. P. 1960. Great Sand Dunes of Colorado. In R. J. Weimer, and J. D. Haun, *1960, guide to the geology of Colorado,* pp. 127-129. Denver: Rocky Mtn. Assoc. Geologists.

————. 1973. The reversing transverse dunes in the San Luis Valley of Colorado (abs.). *Geol. Soc. America Abs. Programs,* 5: 737.

Siebenthal, C. E. 1910. *Geology and water resources of San Luis Valley, Colorado.* U.S. Geol. Survey Water Supply Paper 240.

Idaho

Bradley, F. H. 1873. Report. *U.S. Geol. Geog. Surv Terr., 6th Ann. Rept.,* pp. 190-274.

Chadwick, H. W., and Dalke, P. D. 1965. Plant succession on dune sands in Fremont County, Idaho. *Ecology* 46: 765-780.

Dort, Wakefield, Jr. 1959. Sand dunes of northeastern Snake River Plain, Idaho (abs.). *Geol. Soc. America Bull.* 69: 1555.

Greeley, Ronald. 1977. Aerial guide to the geology of the central and eastern Snake River Plain. In eds. Ronald Greeley, and J. S. King, *Volcanism of the eastern Snake River Plain, Idaho: A comparative planetary geology guidebook, pp. 60-111. NASA Conf. Report* CR-154621.

Koscielniak, D. E. 1973. Eolian deposits on a volcanic terrain near Saint Anthony, Idaho. Master's thesis, State University of New York at Buffalo.

Murphy, J. D. 1973. The geology of Eagle Cove at Bruneau, Idaho. Master's thesis, State University of New York at Buffalo.

———. 1975. The geology of Bruneau Dunes State Natural Park, Idaho (abs.). *Geol. Soc. America, Abs. Programs* 7: 633.

Russell, I. C. 1902. Geology and water resources of the Snake River Plains of Idaho. *U.S. Geol. Survey Bull.* 199.

Stearns, H.T.; Bryan, L. L.; and Crandall, Lynn. 1939. Geology and water resources of the Mud Lake region, Idaho, including the Island Park area: *U.S. Geol. Survey Water Supply Paper* 818.

Nevada

Ball, S. H. 1907. A geologic reconnaissance in southwestern Nevada and eastern California. *U.S. Geol. Survey Bull.* 308.

Criswell, D. R.; Lindsay, J. F.; and Reasoner, D. L. 1975. Seismic and acoustic emissions of a booming dune. *Geophys. Res.* 80: 4963-4974.

Lindsay, J. F.; Criswell, D. R.; Criswell, T. L.; and Criswell, B. S. 1976. Sound-producing dune and beach sands. *Geol. Soc. America Bull* 87: 463-473.

Meinzer, O. E. 1917. Geology and water resources of Big Smoky, Clayton, and Alkali Spring valleys, Nevada: *U.S. Geol. Survey Water Supply Paper* 423.

Morrison, R. B. 1964. Lake Lahontan: Geology of southern Carson Desert, Nevada. *U.S. Geol. Survey Prof. Paper* 401.

Russell, I. C. 1885. Geological history of Lake Lahontan, a Quaternary lake of northwestern Nevada. *U.S. Geol. Survey Mon.* 11.

Smith, R. S. U. 1975. Eolian transport of sand on the actively-accreting slip face of a sand dune northwest of Winnemucca, Nevada (abs.). *Geol. Soc. America Abstr. Programs* 7: 377.

New Mexico

Bachhuber, F. W. 1971. Paleolimnology of Lake Estancia and the Quaternary history of the Estancia Valley, central New Mexico. Ph.D. dissertation, University of New Mexico.

Bryan, Kirk, and McCann, F. T. 1943. Sand dunes and alluvium near Grants, New Mexico. *Amer. Antiquity* 8: 281-290.

Denny, C. S. 1941. Quaternary geology of the Santa Acacia area, New Mexico. *J. Geol.* 49: 225-260.

Evans, G. C. 1963. Geology and sedimentation along the lower Rio Salado in New Mexico. In: eds. F. J. Kuellmer, *Guidebook of the Socorro region, New Mexico.* pp. 209-216. New Mexico Geol. Soc. 14th field conf.

Everard, C. E. 1964. Playas and dunes in the Estancia Basin, New Mexico (abs.). *Int. Geog. Cong. 20th, Abtracts* 89-90.

Gile L. H. 1966. Coppice dunes and the Rotura soil. *Soil Sci. Soc. America* 30: 657-660.

Hack, J. T. 1941. Dunes of the western Navajo country. *Geog. Rev.* 31: 240-263.

Hunt, C. B. 1978a. Surficial geology of northwest New Mexico. *New Mexico Bur. Mines Min. Resources Geol.* Map 43.

———. 1978. Surficial geology of southwest New Mexico. *New Mexico Bur. Mines Min. Resources, Geol.* Map 42.

Jones, B. R. 1959. A sedimentary study of dune sands, Lamb and Bailey counties, Texas and White Sands National Monument, New Mexico. Master's thesis Texas Technical University.

MacDougal, D. T. 1908. Botanical features of North American deserts. *Carnegie Inst. Washington Pub.* 99.

McKee, E. D. 1966. Structures of dunes at White Sands National Monument, New Mexico (and a comparison with structures of dunes from other selected areas). *Sedimentology,* 7: 1-69.

McKee, E. D., and Douglass, J. R. 1971. Growth and movement of dunes at White Sands National Monument. *U.S. Geol. Survey Prof. Paper* 750-D. 108-114.

McKee, E. D., and Moiola, R. J. 1975. Geometry and growth of the White Sands dune field, New Mexico. *U.S. Geol. Survey J. Research,* 3: 59-66.

Meinzer, O. E. 1911. Geology and water resources of Estancia Valley, New Mexico. *U.S. Geol. Survey Water Supply Paper* 275.

Meinzer, O. E., and Hare, R. F. 1915, Geology and water resources of Tularosa Basin, New Mexico. *U.S. Geol. Survey Water Supply Paper* 343.

Schwennesen, A. T. 1918. Ground water in the Animas, Playas Hachita and San Luis basins, New Mexico. *U.S. Geol. Survey Water Supply Paper* 422.

Smith, R. E. 1957. Geology and ground-water resources of Torrance County, New Mexico. *New Mexico Bur. Mines Min. Resources, Ground-Water Rept.* 5.

Titus, F. B. 1969. Late Tertiary and Quaternary hydrogeology of Estancia Basin, central New Mexico. Ph.D. dissertation, University of New Mexico.

Oregon

Allison, I. S. 1966. *Fossil Lake Oregon: Its geology and fossil faunas. Oregon State Monographs, Studies in Geology,* No. 9 Corvallis: Oregon State University Press.

Dole, H. M. 1942. Petrography of Quaternary lake sediments of northern lake County, Oregon. Master's thesis, Oregon State University.

Greenup, W. E. 1941. Physiography and climate (past and present) in the Oregon portion of the Great Basin. Master's thesis, University of Oregon.

Lewis, P. F. 1960. Linear topography in the southwestern Palouse, Washington-Oregon. *Assoc. Amer. Geographers. Ann.* 50: 98-111.

Sonora

Arvidson, R. E., and Mutch, T. A. 1974. Sedimentry patterns in and around craters from the Pinacate volcanic field, Sonora, Mexico; some comparisons with Mars. *Geol. Soc. America Bull.* 85: 99-104.

Breed, W. J., and Breed, C. S. 1978. Star dunes as a solitary feature in Grand Canyon, Arizona, and in a sand sea in Sonora, Mexico: (abs.) *NASA Tech. Memo* TM-78, 455: 11-12.

Cooke, R. U., and Warren, Andrew. 1973. *Geomorphology in deserts.* Berkeley: University of California Press.

Hornaday, W. T. 1908. *Camp-fires on desert and lava.* New York: Scribners.

Ives, R. L. 1959. Shell dunes of the Sonoran shore. *Amer. J. Sci.* 257: 449-457.

Lumholtz, Carl. 1912. *New trails in Mexico.* New York: Scribners.

McKee, E. D., and Breed, C. S. 1976. Sand seas of the world. *U.S. Geol. Survey Prof. Paper* 929. 81-88.

May, L. A. 1973. Geological reconnaissance of the Gran Desierto region, northwestern Sonora, Mexico. *Arizona Acad. Sci. J.* 8: 158-169.

Merriam, Richard. 1969. Source of sand dunes of southeastern California and northwestern Sonora, Mexico. *Geol. Soc. America Bull.* 80: 531-534.

Shreve, Forrest. 1938. The sandy areas of the North American desert. *Assoc. Pacific Coast Geographers Yrbk.* 4: 11-14.

Texas

King, P. B. 1948. Geology of the southern Guadalupe Mountains, Texas. *U. S. Geol. Survey Prof. Paper 215.*

Utah

Beckwith, Frank. 1951. Fighting moving sand dunes in Utah. *Rocks and minerals* 26:592-594.

Cooley, M. E.; Harshbarger, J. W.; Akers, J. P.; and Hardt, W. F. 1969. Regional hydrogeology of the Navajo and Hopi Indian Reservations, Arizona, New Mexico and Utah. *U.S. Geol. Survey Prof. Paper* 521-A.

Dean, L. E., 1978. Eolian sand dunes of the Great Salt Lake basin: *Utah Geol.* 5:103-11.

Eardley, A. J. 1962. Gypsum dunes and evaporite history of the Great Salt Lake Desert. *Utah Geol. Minerol. Surv. Spec. Stud.,* no. 2.

Gregory, H. E., and Moore, R. D. 1931. The Kaipairowits region: A geographic and geologic reconnaissance of parts of Utah and Arizona. *U.S. Geol. Survey Prof. Paper* 164.

Hack, J. T. 1941, Dunes of the western Navajo country. *Geog. Rev.* 31:240-63.

Jones, D. J. 1953. Gypsum-oolite dunes, Utah. *Amer. Assoc. Petroleum Geologists Bull.*37:2530-2538.

Stokes, W. L. 1964. Incised, wind-aligned stream patterns of the Colorado Plateau: *Amer. J. Sci.* 262:808-16.

Thomas, H. E., and Taylor, G. H. 1946. Geology and ground-water resources of Cedar City and Parowan valleys, Iron County, Utah. *U.S. Geol. Survey Water Supply Paper* 993.

Washington

Lewis, P. F. 1960. Linear topography in the southwestern Palouse, Washington-Oregon. *Assoc. Amer. Geographers Ann.* 50:98-111

Patrone, Anthony, 1970. The Moses Lake sand dunes: Master's thesis, Washington State University

Wyoming

Ahlbrandt, T. S. 1974. The source and sand for the Killpecker sand-dune field, southwestern Wyoming. *Sedimentary Geol.* 11:39-57.
———. 1975. Comparison of textures and structures to distinguish eolian environments, Killpecker dune field, Wyoming. *Mountain Geologist,* 12:61-73.

Hack, J. T. 1943. Antiquity of the Finley site. *Amer. Antiquity,* 8:235-241.

Steidtmann, J. R. 1973. Ice and snow in eolian sand dunes of southwestern Wyoming . *Science* 179:796-798.

SOURCES OF MAPS AND AERIAL PHOTOGRAPHS

In compiling this report, the available literature was supplemented by first-hand field observations of many of the dune fields described. Other dune fields were studied on aerial photographs, photomosaics, and topographic maps. These materials, essential for any serious study of dunes, are available for most dune areas from governmental sources. Within the United States, state index maps of topographic mapping are available free from:

Branch of Distribution, U.S. Geological Survey, Federal Center, Denver, Colorado 80225.

Where available, the 7.5 foot, 1:24,000 topographic quadrangles are most useful, but contours commonly fail to present a clear sense of dune form, particularly where form is complex. Aerial photographs usually provide a good sense of dune form, particularly if viewed stereoscopically and/or if part of the dune lies in shadow to accentuate form; this requires that the elevation angle of the sun be lower than about 40° above the horizon. Because most photography is flown near midday to minimize shadows, photography taken within a few months of the winter solstice usually shows dune form better than that taken at any other time of year. In the western United States, where the steeper side of most dunes faces roughly east, afternoon photography shows form better than morning photography. A microfiche summary of photography by most government agencies is available at nominal cost as the Aerial Photography Summary Record System (APSRP) from:

National Cartographic Information Center, U.S. Geological Survey, 507 National Center, Reston, Virginia 22092.

If the latitude and longitude of an area of interest are known, a printout (similar to APSRP) of a geographic computer search can be obtained from:

EROS Data Center, U.S. Geological Survey, Sioux Falls, South Dakota 57198.

Landsat and other satellite imagery can also be obtained from the EROS Data Center, but its scale is too small to be useful for most dune studies in the United States. Easier to use are the free U.S. Department of Agriculture listings by state and county. "Aerial Photography Status Maps" and "Comprehensive Listing of Aerial Photography" are published by:

Aerial Photography Field Office, Agricultural Stabilization and Conservation Service, U.S. Department of Agriculture, 2222 West 2300 South, P.O. Box 30010, Salt Lake City, Utah 84125

and "Status of Aerial Photography" is published by:

Cartographic Division, Soil Conservation Service, U.S. Department of Agriculture, Hyattsville, Maryland 20782.

Most Agriculture Department photography is at the fairly large scale of 1:20,000 and sequential coverage is available for many areas. The department's older photography, mostly dating from before World War II, is listed in "Aerial Photographs in the National Archives," Special List No. 25, available free from:

Cartographic Division, National Archives and Records Service, Washington, D.C. 20408.

The National Archives also distributes military photography of the United States except for that within fifty to one hundred miles of the Mexican border, at this writing (8/79) available only by invoking the Freedom of Information Act to:

Freedom of Information Act Officer, Defense Intelligence Agency, Washington, D.C. 20301.

Most of the photography from all the above sources can be identified to order only after the index mosaics listed in the various cited publications have been ordered and received, which means that mosaics must be ordered about four months before the individual prints are needed. This lead time can be shortened by inspecting the mosaics directly to determine individual frame number. Government facilities in each state where this can be done are listed in "Availability of Earth Resources Data," available free from the EROS Data Center. Commercial photography is available for some dune areas and usually can be obtained much more quickly than can governmental photography.

Aerial photography of the entire Mexican desert is now available from the Mexican government agency DETENAL (Dirección General de Estudios del Territorio Nacional). Most of this area, except Baja California Sur, has also been mapped on modern, photogrammetric topographic quadrangles at a scale of 1:50,000 (figure 11-27). Each topographic sheet contains an index map of the individual photographs covering its area; index mosaics of larger areas are also available. Maps and photographs can be ordered from:

Agencia DETENAL, San Antonio Abad No. 124, Col. Tránsito México 8, D.F., Mexico.

Maps of Baja California can be purchased over the counter and photographs ordered from:

Agencia DETENAL, Av. de los Pioneros y Andador Cholula No. 2, Centro Cívico Comercial Mexicali, Mexicali, B.C.N., Mexico.

An index map of the country is available free from either office.

SOURCES OF WIND AND CLIMATIC DATA

The National Climatic Center (Federal Building, Asheville, North Carolina 28801) is the repository for most climatic and wind data collected by governmental and military agencies in the United States. Daily temperature, precipitation, and some evaporation data are published monthly (with an annual summary) as "Climatological Data" for each state, available individually or by subscription. Ten to twenty-year tabulations of these monthly data are published as "Climatic Summary of the United States" for each state, listed as supplements to the 1930 (Bulletin W) summary; these lag many years behind current observations. Some stations not included in these summaries have their precipitation listed in "Storage-Gage Precipitation Data for the Western United States," published annually. For first-order weather service stations, "Local Climatological Data" are published monthly and list wind speed and direction recorded eight times daily, along with more detailed observations of other parameters than elsewhere available. Changery, Hodge, and Ramsdell (1977) compiled a list by states of tabulated summaries of wind speed and direction, all of them available from the National Climatic Center for the cost of reproduction.

If original records are needed, their character and completeness can be ascertained from "Index of Original Surface Weather Records," published for each state. The original records, typically daily log sheets, can be obtained for the cost of reproduction, and many are available more cheaply and conveniently on microfilm or microfiche.

These data can be used most easily if their original conventions and character are known, as listed in "History of Weather Bureau Climatological Record Forms for Surface Synoptic and Airway Observations." Complete hourly data for many major stations and military installations are available on magnetic tape, whose recording conventions are listed in "TDF-14: Surface Observations." Also useful is the "Substation History," published for each state.

A few very detailed summaries are available for specific regions; an example is the "Climatological Handbook, Columbia Basin States, Hourly Data, Volume 3, Part A" published in 1968 by the Pacific Northwest River Basins Commission. Climatic summaries published by state climatologists are usually less detailed, but the "Climatic Atlas of the United States" prepared by the National Weather Service provides a good regional overview. Hastings and Humphrey (1964a, 1964b; 1969a, 1969b) have prepared monthly tables of temperature and precipitation for most stations in Sonora and Baja California, and Schmidt (1975) prepared a summary of the climate of Chihuahua. Some limited-distribution summaries of data are available from military installations, water and flood control districts, and industrial sites. Daily records of runoff and some precipitation and evaporation data are published annually for each state by the U.S. Geological Survey, Water Resources Division, 855 Oak Grove Ave., Menlo Park, California, as "Water Resources Data."

12

DESERT VARNISH

CARLETON B. MOORE AND CHRISTOPHER ELVIDGE

DESERT VARNISH IS the brown to black coat, usually 10 to 200 μm thick that forms on rock surfaces in certain areas of arid regions throughout the world. Reports of desert varnish come from the Sahara desert (Lucas 1905), Israel (Krumbein 1969), the cold dry deserts of Antartica (Tedrow and Ugolini 1966), the Chilean deserts (Grolier, personal communication), and Australia. We have observed desert varnish in many areas of the southwestern United States. Desert varnish is also found in the arid high-altitude areas of mountains in Germany (Krumbein 1969), Alaska (Smith personal communication), and in Rocky Mountain National Park. Whether all of these coatings are identical in composition and origin has not been ascertained, but they all have similar outward appearances.

Clay makes up at least 70 percent of typical desert varnish (Potter and Rossman 1977), but the striking visual characteristics of the varnish are due to the presence of iron and manganese oxides. Similar ferromanganese deposits occur as nodules and stains in soils, stream-bed deposits, and deep sea nodules.

The desert varnish forms on stable, exposed rock surfaces from the tops and ridges of mountains, to alluvial fan materials. The gross composition of the varnish does not vary significantly from rock type to rock type or from one region to another. These characteristics suggest that the varnish forms from the reaction of a few ubiquitously distributed components of airborne origin. The unexposed surfaces of varnished rocks are either bare or have a red-orange coat. The orange coat on the cracks and undersides of rocks is similar to the black desert varnish except that it is low in manganese content (Potter and Rossman 1978). Varnished rocks on alluvial fans often have a dark ground line band at the soil-air interface.

The summits of the pyramids of Cheops in Egypt show some incipient varnish. Chisel marks in the pyramid quarries have also started to revarnish (Lucas 1905:9). Thus in Egypt it has taken 5000 years for a perceptible coat of varnish to form. A fully developed coat might take 20,000 years at this rate. Petroglyphs chipped through varnish in the southwestern United States show varying degrees of revarnishing. Engel and

Sharp (1958:515) report varnish forming in twenty-five years on an abandoned road across an alluvial fan in the Mojave Desert. This short time for varnish formation has been suggested as resulting from wind deflation of soils exposing buried alluvial rocks that were at the surface at some time and already had varnish coats.

There is considerable interest in using varnish for age dating. As yet no absolute age dating techniques have been developed. Bard et al. (1976) and Hayden (1976) have been working on a relative age-dating technique based on degree of patination. Relative age dating is possible but must be used carefully due to microenvironmental effects on patination rates.

Considering the long time required for varnish to form, the importance of surface stability becomes apparent. Soluble surfaces of carbonates will not support varnish. Friable rocks will also have less varnish on them than will adjacent stable surfaces. Varnish forms an impermeable layer on rocks that hold surface grains together and tends to protect the rock surface from erosion.

Large quartz masses resist varnish formation. This is due to a lack of surface porosity and a lack of cation exchange capacity. Sandstones, however, are readily varnished. Varnish forms most readily on volcanic rocks like basalt. These rock surfaces provide abundant soil iron and manganese and are found varnished in areas where other rocks are relatively devoid of varnish.

CHEMICAL ANALYSES

Difficulty in analyzing desert varnish arises because varnish is so thin it is not easy to get uncontaminated samples and because x-ray photographs of varnish show no crystalline pattern (Engel and Sharp 1958:489). There are several incomplete analyses in the literature (Loew 1876; Lucas 1905; Merrill 1889; White 1924; Laudermilk 1931). The first complete analyses are reported by Engel and Sharp (1958). The varnish was mechanically abraded off the rock, resulting in an approximate 50 percent contamination of the varnish samples. The contamination was corrected for by subtracting a percentage of the rock composition. Wet chemical methods were used for the major elements, and semiquantitative spectrographic methods were used for trace elements. Lakin et al. (1963) and Hunt and Mabey (1966:91) made semiquantitative spectrographic analyses. In situ microprobe analyses have been made by Hooke et al. (1969), Perry and Adams (1978), and Allen (1978). Fused bead microprobe analyses are reported by Potter and Rossman (1978). Neutron activation analyses of varnish scraped off andesite are presented by Bard et al. (1976). Potter and Rossman (1977, 1978) have used infared spectroscopy to characterize the minerals of desert varnish.

MINERALOGY

Potter and Rossman (1977) determined that 70 percent or more of the desert varnish is clay. The mixed layer illite-montmorillonite minerals act as a substrate for the iron and manganese oxides. The iron mineral of the varnish is largely hematite, and the common manganese mineral is birnessite (Potter and Rossman 1978). In the varnish the birnessite is characterized by a high degree of order and a small particle size.

TRACE ELEMENTS

Desert varnish is enriched in numerous trace elements. There are several ways that trace elements can be incorporated into varnish. Trace elements that are not normally precipitated by Ep-pH changes can be coprecipitated with iron and manganese at the time of varnish formation. Hem and Skougstad (1960) report on the coprecipitation of iron and copper. Absorption onto the oxide minerals also occurs. The clay in varnish brings trace elements with it into the varnish and through cation exchange may accumulate more trace elements after deposition. The iron-manganese source will also bring other elements to the varnish source that can become incorporated.

Engel and Sharp (1958) did analyses of varnish from the Mojave Desert and determined the following trace elements in order of decreasing abundance: titanium, barium, strontium, copper, nickel, zirconium, lead, vanadium, cobalt, lathanum, yttrium, boron, chromium, scandium, and ytterbium. Some of the varnishes also contained cadmium, wolfram, silver, neodymium, tin, gallium, molybdenum, beryllium, and zinc. Most of the trace elements were enriched compared to the associated rocks and soils. Particularly enriched are copper, cobalt, nitrogen, lead, barium, chromium, ytterbium, boron, yttrium, strontium, and vanadium.

Marshall (1962) compared the lead isotope ratios of a varnish, its fresh rock, an acid leach of the crushed fresh rock, and average surface lead. The study found that the ratios of the acid leach of fresh rock were most similar to the varnish.

Lakin et al. (1963) determined trace elements in varnish from Death Valley and northeastern Nevada as a possible geochemical prospecting tool. Cobalt content showed a close correlation to manganese content, and barium, lanthanum, molybdenum, nitrogen, lead, and yttrium showed a general correlation. Death Valley samples had a high barium content. Arsenic, antimony, and copper had high values in mineralized areas containing these metals.

Hunt and Mabey (1966) also studied varnish from Death Valley. Besides iron and manganese, the other elements reported included calcium, magnesium, sodium, potassium, titanium, barium, copper, lithium, boron, silver, cobalt, chromium, gallium, molybdenum, niobium, lead, scandium, strontium, tin, vanadium, yttrium, zinc, zirconium, beryllium, cerium, lanthanum, neodymium, and ytterbium.

Bard et al. (1976) determined trace elements from successive scrapings off a varnished andesite from western Nevada. The greatest enrichments in the varnish compared to the andesite were for the elements manganese, uranium, thorium, and cerium, though all of the trace elements detected were enriched. Other elements detected included aluminum, manganese, calcium, sodium, titanium, lanthanum, neodymium, samarium, europium, terbium, dysprosium, ytterbium, lutecium, barium, chromium, copper, cesium, hafnium, scandium, tantalum, vanadium, zinc, and arsenic.

Broad similarities in trace element content exist between desert varnish and other ferromanganese deposits. Bauman (1976) compares the trace elements in varnish and marine ferromanganese nodules. Soil and stream-bed ferromanganese deposits are enriched in a similar manner (Levinson 1974).

The studies of Engel and Sharp, Marshall, and Lakin et al. indicated that some of the varnish trace elements came from the local environment, probably from soils.

LAYERING IN VARNISH

Varnishes often appear to be weathered or deteriorated. As the rock underneath begins to show through, one may see the gross layering that occurs in varnish. The outer coat is enriched in manganese and is black to brown. The varnish closest to the rock often has less manganese and looks red. The outer layer of the rock is often impregnated with iron and/or manganese (figure 12-1).

Hooke et al. (1969) electron microprobed cross-sections of varnish from alluvial fans in the Mojave Desert. The concentrations of iron and manganese were found to increase away from the rock. Manganese content, however, increased more rapidly than did the iron content. The concentrations of silicon, aluminum, and potassium decreased away from the rock. These relations form the basis for defining an outer main layer rich in iron and manganese, and an inner subordinate layer rich in silicon and aluminum. The subordinate layer is the weathered rind of the rock that has become impregnated with iron and manganese (figure 12-2).

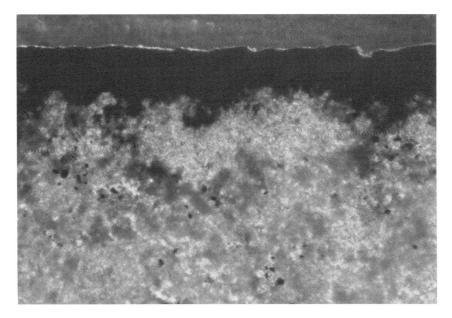

Figure 12-1. Transmitted light photomicrograph of desert varnish on felsic dike rock from South Mountain Park, Phoenix, Arizona.
The varnish forms a layer that is distinct from the rock substrata, but the weathered surface below the varnish is also impregnated with iron and manganese minerals. Prepared by the authors.

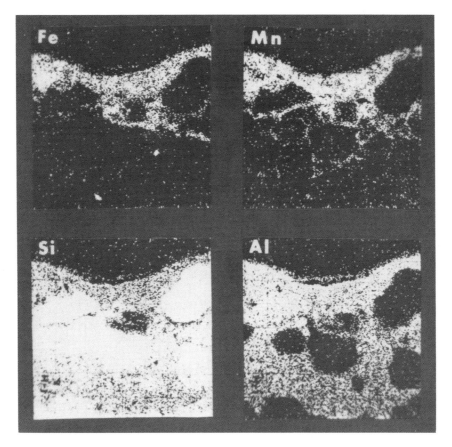

Figure 12-2. Electron microprobe x-ray maps of desert varnish on schist from Mummy Mountain, Paradise Valley, Arizona.
The length and width of the maps are 300 μm. Prepared by the authors.

Perry and Adams (1978) describe alternating red and black layers seen in ultrathin (less than 20 μm) sections. The black layers are rich in manganese while the red layers are relatively enriched in iron, silicon, aluminum, and potassium. The sequential layering is interpreted as the primary depositional feature resulting from changing climatic conditions. Allen (1978) emphasized that lateral changes in the external metal rich layer are common.

ORIGIN OF DESERT VARNISH

As of yet, no one has observed desert varnish in the process of forming. Proposed mechanisms of varnish formation may be classed as endogenetic (from rock weathering), exogenetic (from soil and airborne material), or biogenetic (organic deposition).

Loew (1876:179) examined desert varnish from the Mojave Desert and concluded that the manganese was derived from the disintegration of the surface rocks while they were covered by a shallow sea. The dissolved manganese was deposited on the rocks as insoluble $MnCO_3$ as the sea receded. The $MnCO_3$ was converted to MnO_2 by the influence of the air and sun.

Thoughts on varnish soon turned to evaporation. Merrill (1889) attributed varnish formation to the deposition of material from evaporation groundwater brought to the surface by capillary action. The iron and manganese are originally deposited as carbonates and are later converted to hydrated oxides. Blake (1904), Lucas (1905:24), and Surr (1909) call on rock and soil transpiration to bring iron and manganese to the surface where it is oxidized by the heat of the sun. The idea of weathering solutions bearing iron and manganese from the soil onto rock surfaces has been used recently by Engel and Sharp (1958) and Hooke et al. (1969).

Linck (1900) proposed that the iron and manganese are derived from the decomposition of minerals at the surface of the rock. Dew covering the rock dissolves the released ions. Oxidation of the iron and manganese is aided by electrolyte salts from airborne dust.

Hume (1925) studied varnish in Egypt and proposed that the intense heat and sunlight of the desert is enough to darken rocks. Experiments are sited where rocks are heated and darkening occurs.

The ideas of the early workers were often speculation and did not solve the question of varnish formation. The next wave of ideas evoked the action of organisms.

Although finding almost no organics in varnish and little iron or manganese in pollen, White (1924) proposed that varnish forms from pollen. Settling and sticking to rock surfaces, the pollen is burned in the desert heat, releasing CO_2 and volatile organics. The pollen ash is washed away with the next rain, leaving iron and manganese as insoluble oxides. Laudermilk (1931), noticing patches of varnished rocks on alluvial fans, suggested that lichens are the varnish-forming

agents. Acids secreted by the lichens corrode the rock, releasing iron and manganese, which are precipitated on the rock surface. As the desert varnish accumulates, the lichens are choked out.

Scheffer et al. (1963) advocate the concentration of iron and manganese by blue-green algae. Decomposition of the algae leaves the desert varnish coat. Hunt and Mabey (1966: 91) site iron and manganese stains in wet areas and suggest that varnish is a relict formed in some past pluvial environment, with selective precipitation of iron and manganese by microorganisms.

Recent advocates of a biologic origin for desert varnish are Krumbein (1969) and Bauman (1976). Krumbein cultured organisms associated with varnish in Israel and in the mountains of Germany and found that some of the cultured organisms aid in the precipitation of dissolved iron and manganese. Bauman notes the similarity of desert varnish to marine ferromanganese nodules. Chemolithotropic bacteria aid in the formation of marine ferromanganese nodules. Therefore Bauman suggests that similar organisms may aid varnish formation.

The chemistry and distribution of rock surface organisms in the desert indicate that they are not involved in varnish formation. The rock organisms grow best on north-facing surfaces, under bushes, and in mesic microenvironments. Their distribution is restricted in a way unlike that of varnish. Tests of surface pH for rocks with organisms and without organisms from sites in Arizona show that the organisms make the surfaces acidic. Lichens are visibly corrosive to varnish because of their acid production.

Recent geologic researches on desert varnish point to an exogenetic formation mechanism. Engel and Sharp (1958), finding no leaching gradient in the rock, suggest that weathering solutions from nearby debris may be the source of varnish material. The airborne material collected by Engel and Sharp was similar to the rock in iron and manganese content, while the soils tested were enriched in iron and manganese compared to the rock. Surface soil, where available, is concluded to be the best source of the iron and manganese. Hooke et al. (1969) also favor a soil source for the iron- and manganese-bearing solutions that impregnate the outer rock weathered rind and form an iron-manganese rich coat over the surface.

Potter and Rossman (1977) determined varnish to be 70 percent or more clay. A wind-blown source is advocated for the clay, while weathering solutions bring iron and manganese to the rock surface, aided by capillary movement in the clay. The clay and oxides are envisioned as being mutually dependent. The oxides cement the clay to the rock, and the clay aids oxide transportation and deposition.

As Potter and Rossman (1978) noted, the varnish differs from the rock it is on in its chemistry, mineralogy, and morphology, suggesting an external source for the varnish material. Allen (1978) includes some interesting discussion of the unreasonable thickness of weathered substrate rock required to provide the manganese and iron in the varnish layer. In summary, varnish is not a weathering phenomenon but a deposition of weathering products.

It has been suggested that weathering solutions enriched in iron and manganese migrate from the soil out over exposed rock surfaces (Engel and Sharp 1958; Hooke et al. 1969). Soil weathering solutions might migrate short distances along the bare rock-soil interface and account for the ground line band. In the arid desert it is unlikely that soil solutions migrate and form a uniform coat on even desert pavement rocks. To varnish bedrock outcrops the model of migrating solutions breaks up. This indicates that the iron and manganese of desert varnish is from wind-blown sources.

VARNISH REACTION

The materials for varnish formation are largely wind blown to the varnish site. Figure 12-3 shows the Eh-pH relations of the iron and manganese at low temperatures. Both are soluble under acidic conditions but precipitate under alkaline conditions. This means that desert varnish forms in areas with windblown alkaline material.

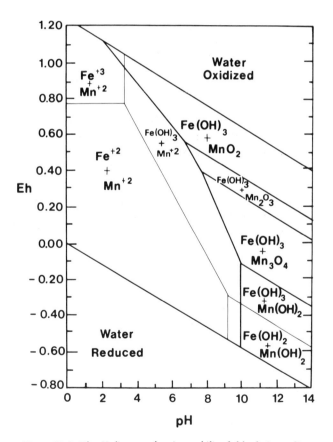

Figure 12-3. Eh-pH diagram showing stability field relations of iron and manganese at 25°C and atmospheric pressure.
The concentration of Fe^{+2} and Mn^{+2} are 10^{-5} molar.
Prepared by the authors.

One of the distinguishing characteristics of deserts is alkaline surface material. Soils are the major source for the alkalinity for varnish formation, though playa lake deposits may be locally important. Desert varnish is associated with alkaline soils in the Sonoran and Mojave deserts. Varnish in Antartica is also associated with alkaline soils (Tedrow and Ugolini 1966).

Alkalinity explains why varnish forms in the desert and not in acidic areas. If soils become less alkaline, the varnish would have a slower growth rate and begin to deteriorate.

The principal reactions of varnish formation involve the oxidation and precipitation of soluble Fe^{+2} and Mn^{+2} onto mineral surfaces. Simplified reactions are:

$$Mn^{+2} + 2H_2O \rightarrow 4H^+ + 2e^- + MnO_2$$
$$2Mn^{+2} + 3H_2O \rightarrow 6H^+ + 2e^- + Mn_2O_3$$
$$3Mn^{+2} + 4H_2O \rightarrow 8H^+ + 2e^- + Mn_3O_4$$
$$Fe^{+2} + 3H_2O \rightarrow 3H^+ + e^- + Fe(OH)_3$$

At low pH's, MnO_2 is the favored manganese precipitate, while at high pH's, more Mn_2O_3 and Mn_3O_4 are formed (Hem 1963:18). Manganese precipitaton on mineral surfaces also favors MnO_2 formation (Hem 1963b).

FRACTIONATION OF IRON AND MANGANESE

The iron and manganese ratio of crustal rocks is about fifty to one (Krauskopf 1957), while the ratio of desert varnish is about one to one. The chemistry of the two elements is very similar, but the iron compounds are less soluble than their manganese counterparts (Krauskopf 1957).

Krauskopf (1957) artificially leached lava with various solutions. The resulting weathering solutions had iron-to-manganese ratios similar to the lava. Both elements in this experiment were released from the breakdown of minerals. The results reflect the ratio of iron and manganese of the decomposing minerals. This confirms the idea that the iron and manganese in desert varnish do not come directly from the rock they are on.

The problem is how to get a deposit with equal iron and manganese concentrations. Krauskopf (1957) examined this problem and concluded that early precipitation of iron will yield solutions enriched in Mn. Figure 12-3 shows that Mn^{+2} is stable at higher pH's than is Fe^{+2}. In nature Mn^{+2} is often in the form of a soluble sulfate complex ($MnSO_4$) that shifts the Mn^{+2} boundary to even higher pH's (Hem 1963a). Thus as a solution containing Fe^{+2} and Mn^{+2} is slowly made more alkaline, $Fe(OH)_3$ begins to form and the solution becomes enriched in Mn^{+2}.

Engel and Sharp (1958) first applied this idea to desert varnish formation by suggesting that the weathering solutions involved in varnish formation deposit lose most of their iron before arriving at the varnish site. They noted that such a mechanism is not suitable for varnish on bedrock outcrops. Hooke et al. (1969) applied this fractionation concept to explain the orange undercoat, groundline band, and the more rapid outward increase of manganese content than iron content in varnish.

Another fractionation mechanism is described by Hem (1963b). The iron of many surface waters is largely in the unreactive form of suspended $Fe(OH)_3$ colloid, while the manganese is in the reactive 2 state. Increasing the Eh or pH of such a solution precipitates manganese with little of the solutions total iron.

Collins and Buol (1970) investigated the solubility of ferromanganese precipitates and found that acidification or lowering the Eh brings some of the manganese into solution, but none of the iron dissolves. The reactions Collins and Buol attribute this fractionation process to are:

$$Mn_3O_4 + 3H^+ + Mn_2O_3 + Mn^{+2} + H_2O$$
$$Mn_2O_3 + 2H^+ + MnO_2 + Mn^{+2} + H_2O$$

Thus manganese enrichment can be associated with fluctuations in the pH and Eh of systems with ferromanganese deposits.

MANGANESE PRECIPITATION

Hemstock and Low (1953) describe the precipitation of manganese oxides onto clay surfaces from Mn^{+2} solutions. The mechanism of precipitation of manganese at low concentrations onto mineral surfaces has been studied by Hem (1963b). Exposing Mn^{+2} solutions to feldspar grains at various pH's, Hem discovered a number of features pertinent to varnish formation. Solutions in contact with the feldspar precipitated manganese at lower pH's than solutions not in contact with mineral surfaces. The rate of loss for the Mn^{+2} is rapid at first for all pH's as the cation exchange sites of the feldspar surfaces come into equilibrium with the solution. At high pH's, the loss of Mn^{+2} from the solutions continues as the Mn^{+2} is oxidized and precipitated onto the feldspar surfaces. The oxidation of the Mn^{+2} is catalyzed by the mineral surface. The manganese oxide coat that forms on the mineral surface has an autocatalytic effect on manganese precipitation. The autocatalytic effect on the manganese oxides is greatly increased by thorough drying before the surface becomes wet with Mn^{+2} bearing solutions.

SOURCES OF Fe^{+2} AND Mn^{+2}

Atmospheric sea salt forms from bursting white water bubbles. The salt particles are carried to the interiors of all continents and reach the surface as dryfall or are dissolved in rainfall, at a rate of about 0.5 grams per square meter a year (Duce 1967). The origin of much of the salt that accumulates in the desert is atmospheric sea salt (Eriksson 1960, Dregne 1968). Atmospheric sea salt provides bare rock surfaces with Na^+, Ca^{+2}, Mg^{+2}, K^+, Cl^-, SO_4^{-2}, and H_3BO_3 for incorporation into varnish. All desert varnish is enriched in boron. Atmospheric sea salt may be providing the boron to the varnish. Fe^{+2} and Mn^{+2} in the sea salt will wash off acidic rock surfaces but will precipitate onto rocks with alkaline surface conditions.

The sea is alkaline, having a pH of about 8.2. The iron in the ocean is predominately in the form of $Fe(OH)_3$ and is suspended as a hydrated colloid (Horne 1969:183). The manganese in the ocean is in the form of Mn^{+2} and the soluble $MnSO_4$ complex (Mason 1966:175). Once airborne, the sea salt reacts with the atmosphere to become acidic (Eriksson 1960). Thus sea salt does not provide the alkalinity for varnish formation.

The composition of the sea surface microlayer influences the composition of the atmospheric sea salt more directly than the composition of bulk seawater (MacIntyre 1970). Biologic activity or organic surfacents may be responsible for the enrichment of certain elements in this layer. The transition elements iron, manganese, and vanadium, are enriched in the top 300 μm of the ocean (Hoffman et al. 1974) as is Cu (Duce et al. 1972). Cattell and Scott (1978) report that copper is enriched in atmospheric sea salt by a factor of 20,000 relative to Mg^{+2} or SO_4^{-2}.

The manganese content of atmospheric sea salt is not known, but evidence presented by Duce et al. (1974) and Zoller et al. (1974) does not suggest the sort of dynamic enrichment found for copper. Using an oceanic Mn^{+2} content of 2×10^{-6} grams (Riley and Chester 1971:65) per 35 gm of salt and a rate of 0.5 gm per square m per year, the amount of Mn^{+2} arriving at a rock surface in 20,000 years is only 5.7×10^{-4} gm per m^2. An enrichment factor of 1000 still only gives 0.57 gm per square m. Atmospheric sea salt is only a minor contribution to varnish manganese.

Volcanic activity also contributes to the composition of the world's atmospheric particles. Bromine, selenium, antimony, and zinc are enriched in volcanic emissions (Mroz and Zoller 1975). Volcanic particulates also contain iron and manganese, but at near crustal levels (Mroz and Zoller 1975).

Windblown desert soils provide clay and alkalinity to varnish formation. The small amount of iron and manganese available from atmospheric sea salt and volcanic particulates indicates that soils are the major source of these elements for desert varnish formation. Ferromanganese deposits occur in soils as coatings on mineral surfaces and as silt-sized nodules (Jenne 1968).

Alkaline soils are frequently deficient in Fe^{+2} and Mn^{+2} for plant growth (Ryan et al. 1974). This deficiency is because the soils' iron and manganese are in the oxidized forms. The amount of Fe^{+2} and Mn^{+2} from the small quantities of windblown soils is insufficient to form varnish in a reasonable period of time. The Fe^{+2} and Mn^{+2} precipitate as $Fe(OH)_3$ and manganese oxides, cementing the clay to the rock, but they provide little of the total iron and manganese, most of which come to the varnish as soil-formed ferromanganese coats on clay minerals. The concentration and fractionation of iron and manganese for desert varnish formation may occur largely in the soil before it is windblown. Soil organisms are likely involved in the Eh and pH fluctuations that result in ferromanganese coats on clay mineral surfaces.

VARNISH SCENARIOS

There are three plausible ways to cement the ferromanganese-coated clay to the rock surface with small quantities of Fe^{+2} and Mn^{+2}:

1. Making a soil more alkaline with playa salt, precipitating some of the soil's soluble Fe^{+2} and Mn^{+2} to cement the clay.
2. Mixing soils with different alkalinities on a rock surface. Some of the soluble Fe^{+2} and Mn^{+2} in the soil precipitates to cement the clay
3. In situ pH fluctuation.

Alkaline soils are buffered systems that respond chemically to acidification to regain their high pH (Ryan et al. 1974). During this chemical response small quantities of the soil's manganese come into solution (Ryan et al. 1974) and are redeposited as the high pH is recovered. The soil's ferromanganese deposits do not release Fe^{+2} upon acidification.

When a small amount of windblown alkaline soil mixes with rainwater with a pH of 5.5 on a rock surface, the soil will respond chemically. During this chemical response, a small quantity of Mn^{+2} comes into solution and is redeposited after the soil has responded to the rainwater. Sulfate ions in the rainwater may aid in this process by forming the $MnSO_4$ complex as an intermediary compound. This is an in situ fractionation process that produces Mn^{+2} to impregnate the rock surface and cement clay onto the rock.

ORANGE COATS

The surfaces of unexposed cracks of varnished rocks often collect windblown soil. An orange coat forms on these surfaces as it does on the undersides of some varnished rocks imbedded in soil on desert pavement. Similar orange coats can be found that form underground. The orange undercoats and crackcoats contain clay and iron but little manganese (Potter and Rossman

1978). These coats form in prolonged contact with soils. Being unfractionated, the formation of these coats is associated with lower Eh or pH conditions than the exposed rock surfaces.

FORMATION OF HEMATITE AND BIRNESSITE

Birnessite is a common sedimentary manganese mineral. It is found in soil ferromanganese stains and nodules where there are alternating wet and dry conditions (Levinson 1974:135). Marine ferromanganese nodules also contain birnessite (Bauman 1976). Potter and Rossman (1978) suggest that the tetrahedral layer of the clay acts as a template for the birnessite structure. The varnish birnessite has the general formula (Na, Ca, K) $Mn_7O_{14} \cdot 3H_2O$ (Potter and Rossman 1978).

A number of factors favor the formation of protobirnessite (MnO_2) instead of Mn_2O_3 or Mn_3O_4 in desert varnish. MnO_2 is favored when the manganese precipitates onto a mineral surface (Hem 1963b). MnO_2 is favored when precipitation occurs at slightly alkaline conditions (Hem 1963a). The pH of desert soils is buffered and the manganese oxides are a mixture of MnO_2, Mn_2O_3, and Mn_3O_4. The pH of varnish is not buffered, and rainfall without duststorms leaches Mn^{+2} from any Mn_2O_3 or Mn_3O_4, leaving a residue of MnO_2.

The $Fe(OH)_3$ could crystallize to goethite or hematite. Schellmann (1959) investigated the crystallization of iron hydroxide gel ($Fe(OH)_3$) and found goethite forms under very alkaline conditions, hematite forms under acidic conditions, and a mixture of goethite and hematite forms under neutral to slightly alkaline conditions. Goethite dehydrates to hematite only at temperatures above 180°C, high above the temperature of exposed desert rocks. The presence of hematite or goethite crystals hastens the crystallization but does not influence the direction of crystallization. What factors, then, favor the crystallization of hematite even under alkaline conditions?

At pH's above 5.5 cations are coprecipitated with iron hydroxide (Hem and Skougstad 1960). Schellmann (1959) found the presence of the divalent cations Mg^{+2} and Ca^{+2} to result in the crystallization of hematite up to high pH's where the effect of OH^- ions again predominate. The other divalent cation tested, Ba^{+2}, did not have this effect, nor did Na^+ or K^+. The action of the ions is to be adsorbed to the surface of the $Fe(OH)_3$ and give it a charge. The adsorption of OH^- gives the gel a negative charge and results in goethite. The adsorption of Mg^{+2} or Ca^{+2} is more strongly adsorbed than the monovalent ions Na^+, K^+, or OH^-. Ba^{+2} is too large to be strongly adsorbed. Both manganese and calcium are found in desert varnish and may influence the crystallization of $Fe(OH)_3$ to hematite. Atmospheric sea salt supplies Mg^{+2} and Ca^{+2} ions in a readily available form to surfaces where varnish is forming, and these ions could be adsorbed to the $Fe(OH)_3$ at the time of the varnish reaction.

The time required for crystallization of the iron hydroxide

varies with pH and temperature. Schellmann (1959) found that at the pH extremes, crystallization is complete in a few weeks. At neutral to moderate pH's, crystallization has begun only after eight months. Higher temperatures not only speed up the crystallization but also result in more hematite formation. Crystallization of the Fe(OH)$_3$ is slowed by drying of the gel.

The goethite-favoring effect of the OH$^-$ ion extends throughout the crystallization process (Schellmann 1959). At the alkaline pH and infrequently wet conditions of desert varnish formation, crystallization of the iron hydroxide gel may take months. In a case where divalent cations are initially outnumbered by OH$^-$ ions, if enough of the OH$^-$ ions are remobilized by solutions of lower pH before crystallization occurs, hematite will form.

Factors that favor the crystallization of hematite from iron hydroxide gel under the alkaline conditions of varnish formation are adsorption of divalent cations to the Fe(OH)$_3$ surface, high desert temperatures, and loss of adsorbed OH$^-$ by fluctuations of the rock surface pH.

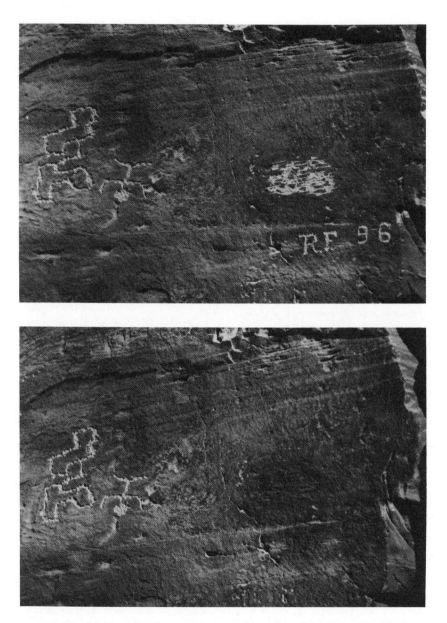

Figure 12-4. Defaced petroglyph in a manganese stained cliff of the Triassic Supai Formation near Montezuma Castle National Monument.
The defacements were covered with artificial desert varnish. Prepared by the authors.

ARTIFICIAL DESERT VARNISH

Man-made disturbances in areas with desert varnish leave rock scars. The bare rock exposed by construction, mining, and roads might take 10,000 years to revarnish. If the varnish is a relict from a previous, more alkaline environment, no varnish will regenerate. Artificial desert varnish can be used to cover up unsightly scars.

Figure 12-4 shows before and after pictures of a varnish restoration covering up recent graffitti on ancient Indian rock art near Montezuma's Castle National Monument, Arizona.

The process of making artificial desert varnish resembles the formation of real desert varnish. First, the scar surface is made alkaline with a NaOH solution. Then a solution of Fe^{+2} and Mn^{+2} is applied. The Fe^{+2} and Mn^{+2} may be in the form of a chloride or a sulfate. By varying the concentration of Fe^{+2} and Mn^{+2} the resultant color can be controlled. A solution 0.5 M in Fe and 1.0 M in Mn gives a good dark brown color.

The iron and manganese precipitate as $Fe(OH)_3$ and MnO_2, impregnating the outer surface of the rock. This stain covers up the scar. It also provides an autocatalytic surface for a head start on the regeneration of real varnish. Thus the artificial desert varnish will be a seed coat for a real varnish in areas where varnish is actively forming.

CONCLUSION

Desert varnish forms in alkaline areas of the world. The main source of the varnish material is windblown soils. Most of the ferromanganese oxides of the varnish are transported to the rock surface as soil-formed coatings on clay minerals. The ferromanganese-coated clay is cemented to the rock surface with $Fe(OH)_3$ and MnO_2 formed from the oxidation and precipitation of Fe^{+2} and Mn^{+2}.

Optimal conditions for desert varnish formation are met in the typical desert monsoon. The cycle begins with a dry dust storm that coats rocks with fine material. The subsequent shower sweeps atmospheric particles from the air. Such dust storms are common in southern Arizona (Idso 1974). The varnish reaction occurs as the rainwater mixes with the windblown dust on the rock surface.

REFERENCES

Allen, C. C. 1978. Desert varnish of the Sonoran Desert—Optical and electron probe microanalysis, *J. Geol.* 86:743-752.

Bard, J. C.; Asaro, F.; and Heizer, R. F. 1976. Perspectives on dating of Great Basin petroglyphs by neutron activation analysis of the patinated surfaces, *Lawrence Berkeley Laboratory Report* LBL-4475.

Bauman, A. J. 1976. Desert varnish and marine ferromanganese nodules: congegeric phenomena, *Nature* 5:387-388.

Blake, W. P. 1904. Superficial blackening and discoloration of rocks especially in desert regions. *Amer. Inst. of Mining Engineers—Transactions* 35:371-375.

Cattell, F. C. R.; and Scott, W. D. 1978. Copper in aerosol particles produced by the ocean, *Science* 202:429-430.

Collins, J. F., and Buol, S. W. 1970. Effects of fluctuation in the Eh-pH environment of iron and/or manganese equilibria. *Soil Science* 110:111-118.

Dregne, H. E. 1968. Appraisal of research on surface materials of desert environments. In *Deserts of the world,* ed. W. G. McGinnies, B. J. Goldman, and P. Paylore. pp. 287-316. Tucson: University of Arizona Press.

Duce, R. A. 1967. Salt nuclei. In *The encyclopedia of atmospheric sciences and astrogeology,* ed. R. W. Fairbridge. New York: Reinhold.

Duce, R. A.; Hoffman, G. L.; and Zoller, W. H. 1974. Atmospheric trace metals at remote northern and southern hemisphere sites: Pollution or natural. *Science* 187:59-61.

Duce, R. A.; Quinn, J. G.; Olney, C. E.; Piotrowicz, S.; Ray, B. J.; and Wade, T. L. 1972. Enrichment of heavy metals and organic compounds in the surface microlayer of Narragansett Bay, Rhode Island. *Science* 176:161-163.

Elvidge, C. D. 1979. Distribution and formation of desert varnish in Arizona. Master's thesis, Arizona State University.

Engel, Celleste G.; and Sharp, R. P. 1958. Chemical data on desert varnish. *GSA Bulletin,* 69:487-518.

Eriksson, E. 1960. The yearly circulation of chloride and sulfur in nature; Meteorological, geochemical, and pedological implications. *Tellus* 12:63-109.

Hayden, J. D. 1976. Pre-Altithermal archaeology in the Sierra Picante Sonora, Mexico. *American Antiquity* 41:274-289.

Hem, J. D. 1963a. Chemistry of manganese in natural water: Chemical equilibria and rates of manganese oxidation. *USGS Water-Supply Paper* 1667-A:1-63.

————. 1963b. Chemistry of manganese in natural waters: Deposition and solution of manganese oxides. *USGS Water-Supply Paper* 1667-B:1-42.

Hem, J. D., and Skougstad, M. W. 1960. Coprecipitation effects in solutions containing ferrous, ferricm and cupric ions. *USGS Water-Supply Paper* 1459-E.

Hemstock, G. A., and Low, P. F. 1953. Mechanisms responsible for the retention of manganese in the colloidal fraction of soil. *Soil Science* 76:331-343.

Hoffman, G. L.; Duce, R. A.; Walsh, P. R.; Hoffman, Eva J.; Ray, Barbara J.; and Fasching, J. L. 1974. Residence time of some particulate trace metals in the oceanic surface microlayer: Significance of atmospheric deposition. *J. Recherches Atmospheriques* 8:745-759.

Hooke, R. L.; Yang, H.; and Weiblen, P. W. 1969. Desert varnish: An electron probe study. *J. Geology* 77:275-288.

Horne, R. A. 1969. *Marine chemistry.* New York: Wiley.

Hume, W. F. 1925. *Geology of Egypt,* vol. 1. Cairo: Government Press.

Hunt, C. B., and Mabey, D. R. 1966. Stratigraphy and structure: Death Valley, California. *USGS Prof. Paper* 494-A:90-92.

Idso, S. B. 1974. Summer winds drive dust storms across the desert, *Smithsonian* 5(9)68-73.

Jenne, E. A. 1968. Controls on Mn, Fe, Co, Ni, Cu, and Zn concentrations in soils and water: The significant role of hydrous

Mn and Fe oxides. In *Trace inorganics in water,* ed. R. F. Gould, pp. 337-387. Advanced Chemistry Series. Washington, D.C.: American Chemical Society.

Krauskopf, K. B. 1957. Separation of manganese from iron in sedimentary processes. *Geochimica et Cosmochimica Acta* 12:61-84.

Krumbein, W. E. 1969. Uber den Einfluss der Mikroflora auf die exogene Dynamik. *Geologische Rundschau* 58(2):333-363.

Lakin, H. W.; Hunt, D. B.; Davidson, D. F.; and Oda, Uteana. 1963. Variation in minor-element content of desert varnish. *USGS Prof. Paper* 475-B:28-31.

Laudermilk, J. D. 1931. On the origin of desert varnish. *American J. Science* 5:51-66.

Levinson, A. A. 1974. *Introduction to exploration geochemistry.* Calgary: Applied Publishing.

Linck, G. 1900. Ueber die dunkel Rinden der Gesteine der Wusten, *Jenaische Zeitsch. Natwuv.* 35:1-8.

Loew, G. 1876. Report on the geological and mineralogical character of southeastern California and adjacent regions. In *Annual report upon the geographical surveys west of the 100th meridian, Appendix JJ of the Annual report of the Chief of Engineers.*

Lucas, A. 1905. The blackened rocks of the Nile cataracts and of the Egyptian deserts. *Egyptian Geologic Survey Report.*

MacIntyre, F. 1970. Geochemical fractionation during mass transfer from sea to air by breaking bubbles *Tellus* 22:451-462.

Marshall, R. R. 1962. Natural radioactivity and the origin of desert varnish. *American Geophysical Union Transactions* 43:446-447.

Mason, B. 1966. Principles of geochemistry. New York: Wiley.

Merrill, G. P. 1889. Desert varnish. *USGS Bulletin* 150:389-391.

Mroz, E. J., and Zoller, W. H. 1975. Composition of atmospheric particulate matter from the eruption of Heimaey, Iceland. *Science* 190:461-463.

Perry, R. S., and Adams, J. B. 1978. Desert varnish: Evidence for cyclic deposition of manganese. *Nature* 276:489-91.

Potter, R. M., and Rossman, G. R. 1977. Desert varnish: The importance of clay minerals. *Science* 196:1446-1448.

————. 1979. The manganese and iron oxide mineralogy of desert varnish. *Chemical Geology.* 25:79-94.

Riley, J. P., and Chester, R. 1971. *Introduction to marine chemistry.* New York: Academic Press.

Ryan, J.; Miyamoto, S.; and Stroehlein, J. L. 1974. Solubility of manganese, iron, and zinc as affected by application of sulfuric acid to calcareous soils. *Plant and Soil* 40:421-427.

Scheffer, F.; Meyer, B.; and Kalk, E. 1963. Biologische Ursachen der Wusternlackbildung, Z.F. *Geomorphologie* 7:112-119.

Schellmann, W. 1959. Experimentelle Untersuchengen uber die sedimentare Bildung von Goethit und Hamatit. *Chemie der Erde* 20:104-135.

Surr, G. 1909. Granites. *Mining and Scientific Press* 99:712-714.

Tedrow, J. C. F., and Ugolini, F. C. 1966. Antarctic soils. In *Antarctic soils and soil forming processes,* ed. J. C. F. Tedrow, pp. 161-177. Washington, D.C.: American Geophysical Union.

White, C. H. 1924. Desert varnish. *American J. Science* 207:413-420.

Zoller, W. H.; Gladney, E. S.; and Duce, R. A. 1974. Atmospheric concentrations and sources of trace metals at the South Pole. *Science* 183:198.

13

RESEARCH AREAS AND FACILITIES

GORDON L. BENDER

ALTHOUGH MAN HAS lived in deserts since earliest times, only relatively recently have we made a concerted effort to understand them. In 1950 the United Nations Advisory Committee on Arid Zone Research was established to stimulate interest in the arid lands of the world. Symposia were held in various countries on a variety of arid land topics, and the papers were published in a series of symposium volumes. The results were so successful that in 1957 the Major Project on Scientific Research on Arid Lands was launched by UNESCO. The project was empowered to collect and disseminate information concerning arid zone research, encourage new research, contribute toward the development of research institutes, and stimulate interest in arid zone projects.

UNITED STATES COMMITTEES

In the United States the Southwestern and Rocky Mountain Division of the American Association for the Advancement of Science established the Committee on Desert and Arid Zone Research (CODAZR) in 1951. The committee's major goal was the dissemination of information on arid lands and the stimulation of interest in arid land problems. An international meeting, Future of Arid Lands, was held in Socorro, New Mexico, in 1955, sponsored by UNESCO, the National Science Foundation, and the American Association for the Advancement of Science. In 1957 CODAZR began a series of symposia on arid land topics as part of the annual meetings of the Southwestern and Rocky Mountain Division of the AAAS. These symposia have continued to the present time. The papers presented have been published in symposium volumes available from the executive officer of the division. The name of the executive officer can be obtained from a current copy of *Science* magazine. A list of the symposium volumes published to date is included in appendix 13-A.

A national Committee on Arid Lands (COAL) was established by the AAAS in 1965. Its charge was to increase public understanding of arid lands problems, contribute to the solution of these problems, promote communication among scientists engaged in arid lands research, and serve as a catalyst for international cooperation on arid lands problems. The committee has organized symposia at national AAAS meetings, has organized and participated in international meetings on arid lands, and has published a number of books on arid land topics. A list of these publications is included in appendix 13-B.

RESEARCH FACILITIES

Although much of southwestern United States is in federal or state ownership and thus, theoretically, available for study and research, increasing restrictions on the use of public lands frequently have made it difficult to obtain access for long-term research projects. Over the years a variety of facilities and areas have been established to facilitate access for research purposes. They can be grouped into several categories: desert research institutes, federal research stations, state research stations, international biology program research sites, scientific natural areas, biosphere reserves, national parks and monuments, and university research reserves.

Although a number of colleges and universities are found in desert areas, they are included in the discussion only if there are specialized units devoted specifically to arid land topics. Only those facilities within or at the borders of deserts are included.

Desert Research Institutes

One of the first facilities established specifically for desert studies was the Carnegie Institute Desert Laboratory in Tucson, Arizona (1903). Since that time a number of desert research facilities have been established by state or federal agencies or private foundations. They are listed in appendix 13-C. A complete listing of such facilities worldwide is found in *The World Directory of Arid Lands Research Institutions* (Paylore 1977).

Federal Research Stations

Various departments of the U.S. government maintain research programs concerned with desert problems. Included are the Department of Agriculture, within which the Forest Service, Soil Conservation Service and the Agricultural Research Service maintain research facilities; the Department of Health and Human Services, with its Indian Health Offices; the Department of Interior, which includes the Fish and Wildlife Service, Geological Survey, Bureau of Land Management, Bureau of Mines, National Park Service, and the Bureau of Reclamation; and the Department of Commerce, which oversees the National Oceanic and Atmospheric Administration, which has the National Weather Service as one of its component parts.

Each of these agencies maintains offices in many communities throughout the Southwest. It is impractical to attempt to list each of them. For a given community they may be determined by looking under the appropriate listings in the local telephone directory.

A number of the research facilities that seem particularly appropriate because of their location, longevity of records, or type of research have been selected for a brief description. Each is located within or at the boundary of one of the deserts. They are listed in appendix 13-D.

State Research Facilities

Each state has a number of agencies concerned with environmental matters. Among them are departments of agriculture, energy, water resources, conservation, game and fish, land, natural resources, highways, environmental quality, and health. All maintain extensive records and many have active research programs and facilities.

Scientific Natural Areas

When the United States achieved its independence from England, the new government instituted a policy of land acquisition by treaty, purchase, and cessions. The public domain grew rapidly.

Other programs were designed to remove lands from federal ownership to state and private ownership. Land grants were made to railroads as incentives to expansion and to states to finance education and the building of roads. Private individuals were also able to purchase public lands for development. The Homestead Act of 1862, the Mining Law of 1872, and the Desert Land Act of 1877 resulted in massive transfers of federal lands to private ownership, with the resulting destruction of the natural resources. Waste and destruction of these resources became a national scandal.

In 1894 the Department of Interior was created. In 1872 Yellowstone Park was set aside, and in 1897 a bill to set aside timberlands as forest reserves was passed by the U.S. Congress.

The reversal of the policy of disposing of federal lands was underway.

Since that time there has been a steady increase in the number of federal agencies concerned with the preservation and wise use of natural resources. Examples of such agencies are the National Park Service, Forest Service, Bureau of Land Management, Fish and Wildlife Service, and the Soil Conservation Service. In addition, a number of federal programs have been developed, including the Wilderness Preservation System, Wild and Scenic Rivers System, National Trails System, and the National Landmarks Program. All were designed to reduce destruction and to promote conservation and better use of natural resources

Despite these efforts, there has been a steady erosion and destruction of natural resources. Increasing population, more leisure time, greater materialistic demands, greater energy demands, more destructive technology, and human greed have all taken a toll. Habitats have been greatly altered or destroyed, and species of plants and animals have been totally eliminated. Thoughtful persons recognized that additional steps needed to be taken.

The concept of ecosystems as highly organized, relatively stable entities has been growing despite the fact that little is known about the details of how they function. The Analysis of Ecosystems Program of the International Biological Program focused on this problem.

One of the questions still to be answered is the importance of diversity to an ecosystem. Whatever their exact function, naturally diverse ecosystems function reliably over long periods of time. It seems important to maintain examples of this diversity as long as possible so they can be studied more rigorously. One way to accomplish this is to set aside examples of various kinds of ecosystems in their natural state to provide a system of natural areas for study and research.

Well before the current interest in maintaining diversity, the U.S. Forest Service had initiated the designating of Research Natural Areas on Forest Service lands. The first was established in 1927. Today the Forest Service has 117 such areas. Its definition of a Research Natural Area is:

An area where natural processes are allowed to predominate and which is preserved for the primary purposes of research and education. These areas may include: 1. typical or unusual faunistic and/or floristic types, associations or other biotic phenomena; 2. characteristic or outstanding geologic, pedologic or aquatic features and processes.

Other federal agencies established similar areas, and in 1966 the Federal Committee on Research Natural Areas was established. Its name was later changed to the Federal Committee on Ecological Reserves to allow scientific manipulation and experimentation on certain areas. These areas were designated as experimental ecological areas.

The purposes served by these programs include providing baseline areas against which the effects of human activities in similar environments can be measured; providing sites for study of natural processes in undisturbed ecosystems; providing gene pool preserves for plant and animal species, particularly of rare and endangered types; providing educational and research areas for the study of ecology and successional trends; and providing areas where experiments or management practices can be carried out on natural ecosystems.

Natural area designations are somewhat transitory, and many areas have been declassified under pressure for other uses. There is no effective way of protecting these areas under current regulations. The natural areas set aside in North American deserts are listed in appendix 13-E.

STATE NATURAL AREAS

Following the lead of federal agencies and private organizations, various states have developed natural area programs similar in purpose to those of the federal agencies. A brief review of the status of programs in each of the desert area states is included in appendix 13-F.

PRIVATE ORGANIZATIONS

Certainly the most important single factor in the recent stimulation of interest in natural areas has been the work of the Nature Conservancy. Much of the advance in federal and state programs came as a result of efforts of the conservancy. Since 1951 it has acquired more than 1200 parcels of land to protect them from destruction by development. Many of these were later transferred to governmental agencies for management. The conservancy owns and manages more than 500 areas.

The National Audubon Society has been active in natural area programs since 1905. It has acquired more than 150 areas, some of which have been transferred to governmental ownership. It retains and manages more than 50 natural areas.

The Society of American Foresters presented its first listing of natural areas in 1949. Today it lists more than 280 natural areas representing a wide range of forest cover types.

NATIONAL NATURAL LANDMARKS PROGRAM

The objective of this program is to assist in the preservation of natural diversity by setting aside a variety of natural areas as landmarks. The program began in the 1960s in the National Park Service. An area selected as a natural landmark is listed on the National Registry of Natural Landmarks. Several hundred areas have been listed throughout the country.

A major problem with the program is that there is no legislation with specific regulations to prevent harmful uses of the areas. The administration of an area is the responsibility of the landowner, who may terminate the status of the area simply by notifying the National Park Service of this intention. There is no way to protect the area from destruction.

NATIONAL HERITAGE POLICY ACT

This act directs the secretary of interior to establish a Natural Heritage Program, a Historic Preservation Program, and a National Register of Natural Areas, and to expand and administer the National Register of Historic Places. It provides for state participation by way of state natural heritage programs.

The act attempts to consolidate a variety of closely related federal, state, and private programs under a single administrative unit and to increase the protection afforded these areas. Areas currently on the National Registry of Natural Landmarks will be incorporated into the new register. The act emphasizes the national importance of identifying and protecting natural and historic resources of significance to the nation's heritage and continuity.

DESERT BIOME VALIDATION SITES

An important component of the International Biological Program's Analysis of Ecosystems was the Desert Biome project, designed to promote understanding of the functioning of desert ecosystems. To this end, research sites were selected within each of the major deserts of the Southwest. Detailed studies were undertaken at each site, and an enormous amount of data was accumulated. These data are stored on computer tape at Utah State University and are available to scientists upon request. The Desert Biome Validation sites are listed in appendix 13-G.

BIOSPHERE RESERVES

The Man and the Biosphere Program (MAB) was initiated in 1970 by the General Conference of the United Nations Educational, Scientific and Cultural Organization. It is an intergovernmental, interdisciplinary, problem-focused research approach to management problems arising from interactions between human activities and natural systems. The U.S. role in the program began in 1972 when the State Department established a national committee of forty persons representing state and federal departments and agencies, nongovernmental organizations, and research and educational institutions.

One of the MAB projects is the establishment of biosphere reserves. The objectives of this project are (Risser and Cornelison 1979):

To conserve, for present and future human use, the diversity and integrity of biotic communities of plants and animals within natural ecosystems, and to safeguard the genetic diversity of species on which their continuing evolution depends.

To provide areas for ecological and environmental research including, particularly, baseline studies both within and adjacent to these reserves, such research to be consistent with the first objective. To provide facilities for education and training.

Criteria for the selection of areas for biosphere reserve status include representativeness, diversity, naturalness, and effectiveness as a conservation unit. To date twenty-eight reserves have been established in the United States, including seven in desert areas. A general description of these desert biosphere reserves is included in appendix 13-H.

NATIONAL PARKS AND MONUMENTS

The management policies of the National Park Service permit the use of parks by qualified investigators for scientific studies when such use is consistent with service policies and contributes to the attainment of park objectives. Research permits are required. Requests for research permits should be submitted in the form of a research proposal to the superintendent of the park in which the work will be done. A list of the national parks and monuments in desert areas is included in appendix 13-I.

REFERENCES

Bostick, V. B., and Niles, W. E. 1975. *Inventory of natural landmarks of the Great Basin.* 2 vols. University of Nevada, Las Vegas, and the National Park Service. Washington, D.C.: U.S. Dept. of Interior.

Buckman, R. E. and Quintus, B. L. 1972. *Natural areas of the Society of American Foresters.* Washington, D.C.: Society of American Foresters.

MAB Information System. 1979. *Biosphere reserves.* UNESCO. Paris: Man and the Biosphere Secretariat.

Nature Conservancy. 1975a. *The preservation of natural diversity: A survey and recommendations.* Washington, D.C.: Nature Conservancy.

————. 1975b. *Preserving our natural heritage: Federal activities.* Washington, D.C.: U.S. Dept. of Interior, National Park Service.

————. 1977. *Preserving our natural heritage, State activities.* Washington, D.C.: Government Printing Office.

Rabe, F. W., and Savage, N. L. 1978. *Aquatic natural areas in Idaho.* Moscow: Idaho Water Resources Institute, University of Idaho.

Risser, P. G., and Cornelison, K. D. 1979. *Man and the biosphere.* Norman: Oklahoma Biological Survey.

Smith, E. L. 1974. *Established scientific natural areas in Arizona.* Phoenix: Arizona Academy of Science and the Office of Economic Planning and Development, State of Arizona.

PUBLICATIONS OF THE COMMITTEE ON DESERT AND ARID ZONE RESEARCH

Climate and man in the southwest. University of Arizona, Tucson, Arizona. Terah L. Smiley, editor. 1957.

Bioecology of the arid and semiarid lands of the Southwest. New Mexico Highlands University, Las Vegas, New Mexico. Lora M. Shields and J. Linton Gardner, editors. 1958.

Agricultural problems in arid and semiarid environments. University of Wyoming, Laramie, Wyoming. Alan A. Bettle, editor. 1959.

Water yield in relation to environment in the southwestern United States. Sul Ross College, Alpine, Texas. Barton H. Warnock and J. Linton Gardner, editors. 1960.

Ecology of groundwater in the southwestern United States. Arizona State University, Tempe, Arizona. Joel E. Fletcher, editor. 1961.

Water improvement. American Association for the Advancement of Science, Denver, Colorado. J. A. Schufle and Joel E. Fletcher, editors. 1961.

Indian and Spanish-American adjustments to arid and semiarid environments. Texas Technological College, Lubbock, Texas. Clark S. Knowlton, editor. 1964.

Native plants and animals as resources in arid lands of the southwestern United States. Arizona State College, Flagstaff, Arizona. Gordon L. Bender, editor. 1965.

Social research in North American moisture-deficient regions. New Mexico State University, Las Cruces, New Mexico. John W. Bennett, editor. 1966.

Water supplies for arid regions. University of Arizon, Tucson, Arizona. J. Linton Gardner and Lloyd E. Meyers, editors. 1967.

International water law along the Mexican-American border. University of Texas at El Paso, El Paso, Texas. Clark S. Knowlton, editor. 1968.

Future environments of arid regions of the Southwest. Colorado College, Colorado Springs, Colorado. Gordon L. Bender, editor. 1969.

Saline water. American Association for the Advancement of Science, New Mexico Highlands University, Las Vegas, New Mexico. Richard B. Mattox, editor. 1970.

Health related problems in arid lands. Arizona State University, Tempe, Arizona. M. L. Riedesel, editor. 1961.

The high plains: Problems in a semiarid environment. Colorado State University, Fort Collins, Colorado. Donald D. McPhail, editor. 1972.

Responses to the dilemma: Environmental quality vs. economic development. Texas Tech University, Lubbock, Texas. William A. Dick-Peddie, editor. 1973.

The Reclamation of disturbed arid lands. University of New Mexico, Albuquerque, New Mexico. Robert A. Wright, editor. 1978.

Energy resource recovery in arid lands. University of New Mexico, Albuquerque, New Mexico. Klaus Timmerhaus, editor. 1980.

PUBLICATIONS OF THE COMMITTEE ON ARID LANDS (COAL)

Aridity and man. American Association for the Advancement of Science, Washington, D.C. Carle Hodge, editor. 1963.

Water importation into arid lands. University of Arizona Press, Tucson. Jay M. Bagley and Terah L. Smiley, editors. 1969. In *Arid lands in perspective.*

Urbanization in the arid lands. Texas Tech Press, Lubbock. Carle O. Hodge and C. N. Hodges, editors. 1974.

Polar deserts and modern man. University of Arizona Press, Tucson. T. L. Smiley and J. H. Zumberge, editors. 1974.

Arid lands in a changing world. American Association for the Advancement of Science, Washington, D.C. Harold Dregne, editor.

Directory of North American arid land scientists. American Association for the Advancement of Science, Washington, D.C. Gordon L. Bender, Patricia Paylore, and Fernando Medellin-Leal, editors. 1977.

DESERT RESEARCH INSTITUTES

CHIHUAHUAN DESERT

Centro Nacional de Investigacion Para el Desarrollo de Zonas Aridas.
Buenavista, Saltillo, Coahuila, Mexico

Interests include dry land farming, utilization of native arid and semiarid zone plants, range management, and development of water supplies. Experimental stations are maintained at Ocampo, Durango, San Luis Potosí, Zacatecas, and Nuevo Leon. Library and laboratory facilities are maintained.

Chihuahuan Desert Research Institute
Alpine, Texas 79830

The institute is supported by private donations and grants. It was founded in 1973 to promote public awareness and research in the Chihuahuan Desert. Laboratory space, bunkhouse facilities, and limited library holdings are available to qualified scientists. Field seminars are held seasonally.

Institute de Investigacion de Los Zonas Deserticas. San Luis Potosí, S.L.P., Mexico.

Established in 1954, it is the oldest arid land institute in Mexico. Research interests include vegetation, water resources, soils, geology, and ecology of Mexican arid lands. Library and laboratory facilities are maintained.

International Center for Arid and Semi Arid Lands Studies,
Texas Tech University
Lubbock, Texas 79409.

The center was established in 1966 to develop and disseminate information about arid and semiarid lands. It is located on the Texas Tech campus with off-campus centers at Amarillo and Junction, Texas. The ICASALS's newsletter is published regularly.

Rancho Experimental La Campana
Calle Victoria 310
Chihuahua, Chihuahua, Mexico.

The station is 82 km north of Chihuahua City on the Pan American Highway. It is devoted to ranch and livestock management, including grazing systems, revegetation of range lands, control of undesirable plant species, desertification, and selection of range grasses. Library facilities and laboratories are maintained. Living accommodations for visiting scientists are available.

Southwestern Research Station
Portal, Arizona 85632.

The station was established in 1955 by the American Museum of Natural History. It is located in Cave Creek Canyon on the eastern slope of the Chiricahua Mountains in one of the most biologically diverse areas of the United States. The station is located at 1646 m between the Chihuahuan Desert to the east and the Sonoran Desert to the west. Laboratory facilities and living accommodations are available for visiting scientists.

GREAT BASIN DESERT

Desert Research Institute
University of Nevada
Reno, Nevada 78507.

Established in 1959, the institute research programs include water resources, energy atmospheric environment, land use, resource management, energy conservation, and aquatic biology. Laboratory and library facilities are maintained.

Hanford Reservation Ecology Reserve
Hanford, Washington.

A facility of the Atomic Energy Commission. Research on the ecology reserve has been carried on under the auspices of the Battelle Memorial Institute.

Malheur Environmental Field Station
Burns, Oregon 97720.

The station is operated by a consortium of Oregon state universities and colleges, private colleges, and community colleges. It is located on the western edge of the Malheur National Wildlife Refuge near the northern border of the Great Basin Desert. Courses are offered during the summer. Laboratory space and housing are available to qualified scientists.

MOJAVE DESERT

Desert Research Institute
Boulder City, Nevada.

Part of the Desert Research Institute of the University of Nevada. The Boulder City laboratory specializes in research in physiology and desert ecology.

SONORAN DESERT

Arizona-Sonora Desert Museum
Tucson, Arizona 85703.

A private, living museum founded in 1952. Native plants and animals are displayed in natural surroundings. An educational and research program is maintained.

Boyce Thompson Southwestern Arboretum
Superior, Arizona.

Originally established by the Boyce Thompson Institute in 1929 to study semiarid plants, it is now managed cooperatively by the Boyce Thompson Institute, the University of Arizona, and the Arizona State Parks Board. An active research and education program is in operation. In 1979 a publication on desert plants was initiated.

Carnegie Institution Desert Laboratory
Tucson, Arizona.

Established in 1903 on Tumamoc Hill in Tucson, Arizona, it was the first important laboratory in the United States established for desert studies. It was here that some of Forrest Shreve's pioneering work on deserts was done. Although the desert laboratory is no longer in existence, the buildings on Tumamoc Hill are still there and are being utilized by the Geochronology Laboratory of the University of Arizona. The site has been declared a national historic site, and the grounds where much of the early work was carried out have been set aside for preservation.

Centro de Investigaciones Cientificas y Technologicas
Universidad de Sonora
Hermosillo, Sonora, Mexico.

Research interests include marine sciences, problems of arid lands, and utilization of native desert plants. Experiment stations are maintained at Tubutama, Puerto Penasco, and Bahia Kino, Sonora. Laborarory space and limited library facilities are available to visiting scientists.

Desert Botanical Garden
Phoenix, Arizona 85010

The garden was established in 1939 to promote interest in desert plants. Plants from desert areas of the world are on display in a natural setting. Active research and educational programs are in operation. A cactus show is an annual event.

Environmental Research Laboratory
University of Arizona
Tucson, Arizona 85706.

The laboratory is devoted to the development of controlled environment systems for food, water, and energy production in deserts of the world. A specialty is controlled environment greenhouse agriculture.

Geochronology Laboratory
University of Arizona
Tucson, Arizona 85719.

Located on Tumamoc Hill at the site of the former Carnegie Desert

Laboratory, the laboratory carries on work in pollen analysis and radioactive carbon dating.

Institute of Arid Lands Research
University of Arizona
Tucson, Arizona 85719.

Established in 1964 at the University of Arizona, its activities have included bibliographic inventories of world deserts, directories of arid lands research institutions and arid lands research scientists, and studies on economic development of native desert plants, including jojoba and guayule.

Laboratory of Tree Ring Research
University of Arizona
Tucson, Arizona 85721.

Established in 1937 as a result of the pioneering work of Dr. Andrew E. Douglas in dendrochronology, the research interests include archeological tree-ring dating and the relationships of climate and tree rings.

Lower Colorado River Basin Research Laboratory
Arizona State University
Tempe, Arizona 85281.

Established to study abiotic and biotic factors in riparian situations in the Colorado River Basin. Research includes methods of reestablishing native plant cover in riparian areas.

Philip L. Boyd Desert Research Center
University of California
Riverside, CA 92502

A 3048 ha preserve at the base of the Santa Rose-San Jacinto mountains south and west of the Salton Sea. Elevations range from 213 m to 1402 m. Rainfall averages 83 mm per year. Facilities include laboratories and limited living accommodations for visiting scientists.

Puerto Penasco Research Station
Puerto Penasco
Sonora, Mexico
31°20'N 113°35'W.

A cooperative venture between the University of Sonora and the University of Arizona, located on the Gulf of California. Research includes marine biology, water desalination, and controlled environment aquaculture. Laboratory space and dormitory facilities are available to educational groups and visiting scientists.

ARCTIC DESERT

Arctic Environmental Information and Data Center
Anchorage, Alaska 99501.

The center is a clearinghouse for environmental information and data on physical, biological, meteorological, and related research in Alaska. A yearly publication, "Current Research Profile for Alaska," lists ongoing research for the current year.

Arctic Research Institute
Point Barrow, Alaska
71°20'N 156°00'W.
University of Alaska
Fairbanks, Alaska 99701.
Geophysical Institute
Institute of Arctic Biology

USA Cold Regions Research and Engineering Laboratory
Fort Wainwright, Alaska 99703.

FEDERAL EXPERIMENT STATIONS

Desert Experimental Range
Intermountain Forest and Range Experiment Station
Milford, Utah.
   Listed under Biosphere Reserves, appendix 13-H.

Great Basin Experimental Range
Intermountain Forest and Range Experiment Station
Ephraim, Utah.

Jornada Experimental Range
Las Cruces, New Mexico
   Listed under Biosphere Reserves, appendix 13-H.

Santa Rita Experimental Range
Southwestern Forest and Range Experiment Station
Tucson, Arizona.
   Area: 2000 ha
   Elevation: 884-1372 m
   The oldest experimental range in the United States, having been established in 1903. Plant cover is classified as desert grassland. Housing accommodations available to qualified scientists.

Sierra Ancha Experimental Forest
Rocky Mountain Forest and Range Experiment Station
Tempe, Arizona.
   Area: 5600 ha
   Elevation: 1067-2134 m
   Three experimental watersheds were established in 1932. Climatic and water yield records are continuous since that time. Vegetation ranges from desert grassland to pine-fir. Housing accommodations are available to qualified scientists.

United States Water Conservation Laboratory
Agricultural Research Service
Phoenix, Arizona.

Walnut Gulch Experimental Watershed
Southwest Watershed Research Center
Tombstone, Arizona.
   Area: 35 sq. kilometers
   Elevation: 1300-1850 m
   Experiments on water yield, flash flood, and sediment damage.

NATURAL AREAS AND LANDMARKS

GREAT BASIN DESERT

Bear Natural Area
Boise National Forest
Boise, Idaho.
   Sagebrush, grass.

Berlin-Ichthyosaur State Park
Nevada State Parks
Nye County, Nevada.
T12NR39E, Ione 15' quadrangle.
Area: 200 ha
Elevation: 2134 m
   Outstanding fossils. A registered natural landmark.

Big Dune Natural Area
Bureau of Land Management
Nye County, Nevada.
T155R48E, Big Dune 15' quadrangle
Area: 2300 ha
Elevation: 832 m
   An active dune.

Budsage Natural Area
Bureau of Land Management
Lander County, Nevada.
Area: 16 ha
   Stands of budsage.

Cottonwood Canyon Natural Area
Bureau of Land Management
Washington County, Utah.
Area: 960 ha
   Desert riparian ecosystem.

Dautrich Memorial Natural Area
Nature Conservancy
Canyon County, Idaho.
   Dunes, horsebrush, shadscale, big sage.

Desert Shrub-Blackbrush Natural Area
Bureau of Land Management
Lincoln County, Nevada.
Area: 400 ha
   Desert shrub.

Big Sage Research Natural Area
Bureau of Land Management
Coconino County, Arizona.
Area: 64 ha
Elevation: 1768-1829 m
   Great Basin desert scrub.

Goodlow Mountain Natural Area
Fremont National Forest
Klamath County, Oregon.
T39SR13E, Gerber Reservoir 15' quadrangle
Area: 500 ha
Elevation: 1372 m
   Sagebrush to ponderosa pine-white fir.

Greasewood Natural Area
Bureau of Land Management
Lander County, Nevada.
Area: 16 ha
   Greasewood (*Sarcobatus vermiculatus*).

Independence Flat Natural Area
Bureau of Land Management
Elko County, Nevada.
Area: 256 ha
   Atriplex and *Sarcobatus.*

Kipuka Natural Area
National Park Service
Blaine County, Idaho.
   Sagebrush, fescue.

Lost Forest Natural Area
Bureau of Land Management
Lake County, Oregon.
Area: 11,600 ha
  Stands of pine mingled with sand dunes.

Lunar Crater Natural Area
Bureau of Land Management
Nye County, Nevada.
T5,6NR52,53E. Lunar Crate 15' quadrangle
Area: 1020 ha
  Craters, lava flows, desert mallow (*Sphaeralcea spp.*).
  A registered natural landmark.

Ripple Arch Natural Area
Bureau of Land Management
Washington County, Utah.
T40S R18,19W. Cedar City quadrangle.
Area: 300 ha
Elevation: 1311 m
  Natural arch, blackbrush (*Coleogyne ramosissima*).

Ruby Lake National Wildlife Refuge
United States Fish and Wildlife Service
Elko and White Pine counties, Nevada.
T26-28N R57, 58E. Elko quadrangle.
Area: 15,050 ha
Elevation: 1798 m
  Nesting area for sandhill cranes and trumpeter swans.
  A registered natural landmark.

Shoshone Pygmy Sage Natural Area
Bureau of Land Management
White Pine County, Nevada.
SW 1'4 S33T14NR67E. Sacramento pass 15' quadrangle.
Area: 8 ha
Elevation: 1762 m
  Small area of pygmy sage (*artemisia pygmaea*).

Sweetwater Natural Area
Yoiyabe National Forest
Lyon County, Nevada.
Area: 894 ha
  Pinyon-juniper and sagebrush stands.

Tobar Siding Natural Area
Bureau of Land Management
Elko County, Nevada.
  Greasewood (*Sarcobatus vermiculatus*).

Trough Spring Raised Bog Natural Area
Bureau of Land Management
Humboldt County, Nevada.
T39NR33N Bottle Creek 15' quadrangle
Area: 0.25 ha
Elevation: 1341 m
  Unique raised bog.

Westside Sand Mountains Natural Area
Bureau of Land Management
Joab County, Utah.
Area: 512 ha
  Sand dunes.

MOJAVE DESERT

Desert National Wildlife Range
United States Fish and Wildlife Service
Clark and Lincoln counties, Nevada.
T 9-18S R55-63E. Caliente and Las Vegas 1:250,000 quadrangle.
Area: 57,720 ha
Elevation: 1280-2972 m
  Nelson's desert bighorn (*Ovis canadensis nelsoni*)

Joshua Tree Natural Area
Bureau of Land Management
Mohave County, Arizona.
Area: 64 ha
  Joshua trees (*Yucca brevifolia*)

Joshua Tree Natural Area
Bureau of Land Management
Southern Washington County, Utah.
T43S R18W. Beaver Dam Mountains SE 7 ½' quadrangle
Area: 400 ha
Elevation: 1097-1615 m
  Only Joshua tree forest in Utah.
  A registered natural landmark.

Pinyon-Joshua Tree Natural Area
Bureau of Land Management
Nye County, Nevada.
Area: 256 ha
  Pinyon and joshua tree stands.

Silver Peak Natural Area
Bureau of Land Management
Esmeralda County, Nevada.
Area: 256 ha
  Coniferous forest-joshua tree ecotone.

Sunrise Mountain Natural Area
Bureau of Land Management
Clark County, Nevada.
T20, 21S, R52, 53E. Las Vegas and Henderson 15' quadrangle
Area: 12,660 ha
  Geology, trilobite beds, desert pavement, gypsum cave.

Valley of Fire State Park
State of Nevada Division of State Parks
Clark County, Nevada
T15-18S R66-68E. Moapa, Muddy Peak and Overton Beach 15' quadrangle
Area: 12,000 ha
  Geologic formations, desert ecology.
  A registered natural landmark.

SONORAN DESERT

Antelope Flat Natural Area
Bureau of Sports Fisheries and Wildlife
Cabeza Prieta Game Range
Yuma County, Arizona.
Area: 23 ha
Elevation: 241 m
  Endangered Sonoran pronghorn (*Antilocarpa americana sonoriensis*). Sonoran Desert vegetation. Part of the area is on a U.S. Air

Force bombing and gunnery range. Permission must be obtained before entering.

Aravaipa Canyon Natural Area
Defenders of Wildlife
Pinal County, Arizona.
Area: 3328 ha
Elevation: 792 m
Desert riparian community; rare or endangered fish. Entry by permit only well in advance.

Fishtail Canyon Natural Area
Bureau of Sport Fisheries and Wildlife
Kofa Game Range
Yuma County, Arizona.
Area: 81 ha
Elevation: 640-1036 m
Native California palms (*Washingtonia filifera*).

Palm Canyon
Bureau of Sport Fisheries and Wildlife
Kofa Game Range
Yuma County, Arizona.
Native California palms (*Washingtonia filifera*).

Pinacate Natural Area
Bureau of Sport Fisheries and Wildlife
Cabeza Prieta Game Range
Yuma County, Arizona.
Area: 2073 ha
Elevation: 183-274 m

Lava flows, playas, sand dunes, Sonoran Desert vegetation. Part of the area is on a U.S. Air Force bombing and gunnery range. Permission must be obtained before entering.

Research Ranch
Elgin, Arizona.
Area: 3328 ha
Elevation: 1436-1585 m
A nonprofit private foundation established in 1969 for purposes of conservation, research, and education. Desert grassland, oak woodland. Housing available to qualified scientists.

Sierra Pinta Natural Area
Bureau of Sport Fisheries and Wildlife
Cabeza Prieta Game Range
Yuma County, Arizona.
Area: 2973 ha
Elevation: 304-608 m
Ironwood (*Olneya tesota*), elephant trees (*bursera microphylla*), Senita cactus (*Lophocereus schotii*). Part of the area is on a U.S. Air Force bombing and gunnery range. Permission must be obtained before entering.

Sonoita Creek Natural Area
Nature Conservancy
Sonoita, Arizona.
Area: 126 ha
Elevation: 1219 m
Nature sanctuary; riparian vegetation; rare or endangered birds.

STATE NATURAL AREA PROGRAMS

ARIZONA

A natural areas coordinator operates within the state Parks Department. A natural areas registry has been established. A publication, *Established Natural Areas in Arizona,* lists thirty-six private and federal areas already designated as natural areas. A second publication, *Proposed Natural Areas in Arizona,* lists seventy-five sites suitable for natural area designation.
Contact: Natural Areas Coordinator, State Parks Department, Phoenix, Arizona 85001.

CALIFORNIA

The University of California system maintains a system of land and water reserves. A Natural Areas office has been established within the Department of Fish and Game, a Natural Areas Advisory Committee has been established, and an interagency task force on significant natural areas has been established. A computerized data base management system is being developed by the Nature Conservancy. Fragmented programs in various state agencies are being centralized and coordinated.
Contact: Program Coordinator, Significant Natural Areas Program, Department of Fish and Game, Sacramento, California 95814.

IDAHO

The Idaho Natural Areas Program was founded in 1973. There is a Natural Areas Council, a Coordinating Committee, and a series of technical committees. Inventories have been carried out, and a classification system has been developed. *Aquatic Natural Areas in Idaho* was published in 1977.
Contact: Idaho Water Resources Research Institute, University of Idaho, Moscow, Idaho 83843.

NEVADA

The State Land Use Planning Agency has authority to develop a natural areas program. The Department of Conservation and Natural Resources and the Fish and Game Department are concerned with endangered species. A list of recommended areas for natural landmark designation has been compiled (Bostick and Niles 1975).
Contact: Department of Conservation and Natural Resources, Carson City, Nevada 89701.

NEW MEXICO

Various state agencies and departments such as Game and Fish, Parks and Recreation, Natural Resource Commission, Heritage Program, and the Wilderness Commission are concerned with uses of natural resources.
Contact: One of the agencies listed above, Santa Fe, New Mexico 87501.

OREGON

A 1973 statute established the Natural Area Preserves Advisory Committee, which has authority to establish natural area preserves. The Oregon Natural Heritage Program of the Nature Conservancy inventories lands and classifies them according to their need for preservation.
Contact: Chairman, Natural Area Preserves Advisory Committee, Corvallis, Oregon 97330.

UTAH

Various divisions of the Department of Natural Resources administer a variety of land holdings. A list of recommended areas for natural landmark designation in Utah has been compiled (Bostick and Niles 1975).
Contact: Department of Natural Resources, Salt Lake City, Utah 84116.

IBP VALIDATION SITES

DESERT BIOME

Curlew Valley, Utah.

Terrestrial
A drainage basin on the Utah-Idaho border. Big sage (*Artemisia tridentata*), shadscale (*Atriplex confertifolis*), greasewood (*Sarcobatus vermiculatus*).

Aquatic
Deep Creek, a cold desert stream.
Locomotive Springs, desert springs with a flow of 2.3 to 9.0 cubic feet per second.

Saratoga Springs, California.
Southern end of Death Valley.
A spring pond and marsh.
Rock Valley, Nevada.
Mojave Desert vegetation.
Pine Valley, Utah.
Desert Experimental Range

Milford, Utah.
Listed under Biosphere Reserves, appendix 13-H.
Deep Canyon Desert Research Area
Palm Desert, California.
Sonoran Desert vegetation.
Silverbell Bajada
Tucson, Arizona.
Sonoran Desert vegetation.
Jornada Experimental Range
Las Cruces, New Mexico.
Chihuahuan Desert vegetation.
Listed under Biosphere Reserves, appendix 13-H.

TUNDRA BIOME

Naval Arctic Research Laboratory
Barrow, Alaska.
Prudhoe Bay, Alaska
Vegetation dominated by sedges and mosses.

BIOSPHERE RESERVES

CHIHUAHUAN DESERT

Big Bend National Park
Big Bend, Texas 79834.
Location: N29°30′ E 102/103°
Area: 283,247 ha
Elevation: 533-2338 m
    Located in the bend of the Rio Grande River, the area contains examples of Chihuahuan Desert vegetation. Vegetation zones include floodplains, shrub desert, grasslands, and woodlands. Housing, campgrounds, and restaurants are in the park. Library and research space are available to qualified scientists.

Jornada Experimental Range
Las Cruces, New Mexico 88001.
Location: N32°37′ W 106°45′
Area: 78,397 ha
Elevation: 1260-2830 m
    Located at the extreme northern end of the Chihuahuan Desert, the range includes semidesert grassland, shrub desert, and pinon-juniper vegetation types. Climatic, vegetation, and grazing records are continuous from 1914. One of the validation sites for the IBP Desert Biome Program.

Mapimi Reserve
Instituto de Ecologia
Apartado Postal 18-845
Mexico 18, DF
Mexico.
Location: N 26°40′ W 103°56′
Area: 20,000 ha
Elevation: 1100-1350 m
    A nature reserve containing examples of Chihuahuan Desert vegetation. An active research program is carried on.

GREAT BASIN DESERT

Desert Experimental Range
Intermountain Forest and Range Experiment Station
Milford, Utah.
Location: N 38°40′ W 113°45′
Area: 22,513 ha
Elevation: 1547-2577 m
    Located in Pine and Antelope valleys in western Utah, the range contains typical examples of Great Basin plant communities, particularly shadscale, winterfat, pinon-juniper, and perennial grasses. Climatic and grazing records for more than forty years are available. A validation site in the IBP Desert Biome Program.

SONORAN DESERT

Beaver Creek Experimental Watershed
Coconino National Forest
Flagstaff, Arizona 86002.
Location: N 34°32′ W 11°65′
Size: 111,300 ha
Elevation: 900-2400 m
    Concerned with managing vegetation for water yield. A cooperative comparative study is underway with the La Machilia biosphere reserve in Durango, Mexico. Both are located in dry pine-oak forests.

Organ Pipe Cactus National Monument
Ajo, Arizona 85321.
Location: N 32°00′ E 112°50′
Area: 133,278 ha
Elevation: 305-1472 m
    Contains a very rich Sonoran Desert flora with creosote bush and bursage predominating. Organ pipe and senita cactus are near the

northern limit of their ranges. Facilities include an interpretive center, nature trails, and campground.

ARCTIC DESERT

Noatak National Arctic Range
Bureau of Land Management
Fairbanks, Alaska 99707.
Location: N68°0' W 16°0'
Area: 3,035,200 ha
Elevation: 0-1650 m
   Contains examples of arctic tundra vegetation.

Northeast Greenland National Park
The Ministry for Greenland
DK-1128
Copenhagen K,
Denmark.
Location: N 77°0' W 39°0'
Area: 70,000,000 ha
Elevation: 0-3000 m
   A national park established by Denmark. The area is designated as barren arctic desert and ice cap. An active research program and survey work are in progress.

Northeast Svalbard Nature Reserve
International Division
Ministry of Environment
Myntgaten 2
Oslo-Dep., Oslo
Norway.
Location: N 78°40' E 80°50'
Area: 1,555,000 ha
Elevation: 0-835 m
   Classified as tundra and barren arctic desert. A research program is being carried on.

NATIONAL PARKS AND MONUMENTS

CHIHUAHUAN DESERT

Amistad National Recreation Area
Del Rio, Texas.
 Primarily a reservoir project along the U.S.-Mexico border. Chihuahuan Desert vegetation is present.

Big Bend National Park
Big Bend, Texas.
 Listed under Biosphere Reserves, appendix 13-H.

White Sands National Monument
Alamagordo, New Mexico 88310.
 Gypsum sand dunes.

GREAT BASIN DESERT

Arches National Monument
Moab, Utah 84532.

Bryce Canyon National Park
Cedar City, Utah 84717.

Canyon de Chelly
Chinle, Arizona 86503.

Canyonlands National Park
Moab, Utah 84532.

Capital Reef National Park
Torrey, Utah 84775.

Chaco Canyon National Monument
Bloomfield, New Mexico 87413.

Glen Canyon National Recreation Area
Page, Arizona 86040.

Hubbell Trading Post Historic Site
Ganado, Arizona 86505.

Navajo National Monument
Tonolea, Arizona 86044
 Prehistoric cliff dwellings.

Petrified Forest National Park
Holbrook, Arizona 86025.

Pipe Springs National Monument
Moccasin, Arizona 86022.
 Early Mormon Fort. Great Basin Desert vegetation.

Zion Canyon National Park
Springdale, Utah 84767.

MOJAVE DESERT

Death Valley National Monument
Death Valley Junction, California 92328

Joshua Tree National Monument
Twentynine Palms, California 92277.

Lake Mead National Recreation Area
Boulder City, Nevada 89005.

Valley of Fire State Park
Overton, Nevada.

SONORAN DESERT

Anza Borrego Desert State Park
Borrego Springs, California.

Casa Grande Ruins
Coolidge, Arizona 85228.
 Hohokam architecture.

Chiricahua National Monument
Willcox, Arizona 85643.

Coronado National Monument
Hereford, Arizona 85615.

Grand Canyon National Park
Grand Canyon, Arizona 86023.

Organ Pipe Cactus National Monument
Ajo, Arizona 85321.
    Listed under Biosphere Reserves, appendix 13-H.

Saguaro National Monument
Tucson, Arizona 85710.

Salton Sea State Park
North Shore, California.

Tonto National Monument
Roosevelt, Arizona 85545.

Tumacacori National Monument
Tumacacori, Arizona 85640.
    Mission church.

14

DESERTIFICATION

GORDON L. BENDER

THE UNITED STATES National Plan of Action to Combat Desertification has defined desertification as

the decline or destruction of the potential and actual biological productivity of arid and semi-arid lands caused by certain natural and man-made stresses. Such stresses, if continued or unchecked, over the long term may lead to ecological degradation and ultimately to desert-like conditions.

The definition emphasizes that several factors—climatic, physical, and biotic—may be involved in the process. Any one of these factors alone may produce desertification. An extended period of drought may bring about desertification in spite of the very best land management practices; severe mismanagement of land may produce desertification even in the face of ideal climatic conditions; highly saline soils resulting from the evaporation of water from natural bolsons may make plant growth virtually impossible. More often, however, desertification is due to a combination of these factors. A land management practice that is tolerable under a given set of physical or climatic conditions becomes intolerable when these conditions change even slightly. Dregne (1976) has indicated the extent of desertification in North America (figure 14-1).

EFFECTS OF NATIVE PEOPLE

It has been widely assumed that desertification is a fairly recent phenomenon resulting from man's unwillingness to live in harmony with nature, preferring instead to conquer nature by means of technology. It is often stated that early man knew how to live with his environment and had relatively little impact on it.

It is true that early man did gear his activities to the rhythms of his environment. As he was primarily a hunter and gatherer, he followed the seasons, moving from area to area as the fruits and other edible portions of plants became available and as the animals he hunted congregated in localized areas. Because there were so few people and because they stayed in a

given location for a short period of time, it was believed that they had very little permanent effect on the environment. An occasional paper, however, raised some questions in this regard.

McGee (1898) in his paper on the relationship between saguaro cactus and a very small band of Seri Indians living at the head of the Gulf of California pointed out the impact a small group can have if it concentrates on a particular resource in a limited geographic area. This paper is notable for the detail that McGee develops to show the importance of a single species of desert plant to the welfare of these people. As positive as this relationship was for man, a study of the data indicates that there may have been a negative effect on the number of saguaro seedlings being established each year.

Martin (1963) has suggested that the extinction of the large mammals in southern Arizona in recent geologic time was due to the hunting activities of man. Even though there were very few men and their hunting weapons were crude, they may have had a profound effect on the populations of these animals. By stalking them at their feeding sites and waterholes where they congregated, man greatly increased his chances of success. Martin believes man was responsible for their extinction.

York and Dick-Peddie (1969) pointed out the correlation between invading mesquite stands and sites of Indian encampments in the Chihuahuan Desert in New Mexico. The mesquite beans collected and transported as food for man and animals served as colonizing agents when the seeds passed unharmed through the digestive tracts of humans and animal. In fact, the experience may have increased the ability of the seeds to germinate. Over the years these colonizing stands of mesquite expanded into the desert grasslands.

The Hohokam people occupied sites along rivers in central Arizona. They were agriculturists who developed extensive irrigation systems and canals along the Salt and Gila rivers. By 1000 A.D. there were several hundred miles of canals, and thousands of acres were under irrigation growing cotton, corn, beans, and squash. By 1400 A.D. the villages were abandoned, the fields unplanted, the canals neglected, the people gone.

NORTH AMERICA

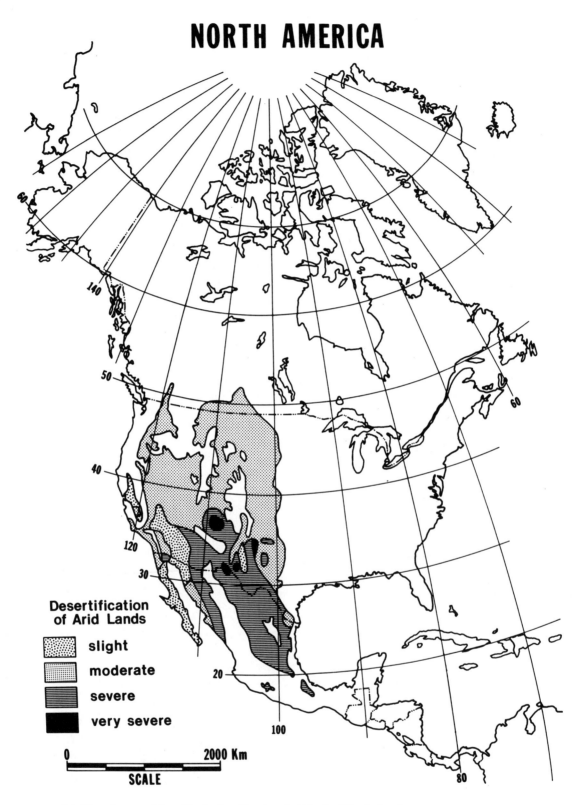

Desertification of Arid Lands

- ░ slight
- ▒ moderate
- ▤ severe
- ■ very severe

0   2000 Km

SCALE

Figure 14-1. Map of extent of desertification in arid and semiarid lands of North America.
Figure taken from Desertification of arid lands, Harold E. Dregne, *Economic Geography* 53:
322-331, Clark University, Worcester, Mass., 1977. Reprinted with permission.

Although no one knows for certain, the presumption is that increasing problems with salinity had made the fields unusable for agriculture.

EFFECTS OF SPANISH OCCUPATION

Whatever effects native peoples may have had on the environment were insignificant compared to those of later arrivals to the desert. With the fall of the Aztec Empire in 1521, the era of Spanish domination began in the Southwest. Spanish explorers and priests moved throughout the region exploring, mapping, searching for minerals, subduing the natives, and establishing missions. Of all their activities the last was perhaps the most significant in terms of final effects on the environment.

When a mission was established, a church was built, the Indians were settled in a village, and a herd of domestic cattle was provided. These actions had profound environmental effects. Settling people in a village provided opportunities for epidemic diseases, and diseases endemic to the Spanish but previously unknown to the native people took a fearsome toll in human lives. Trees were cut for firewood, for timbers for the mines, for building materials, and for corrals for cattle, and the forests gradually disappeared. Cattle destroyed the grass near the village by eating it or trampling it. Native vegetation was cleared to provide fields for agricultural crops. Man's effects on his environment had been greatly amplified. Ohmart and Andersen in chapter 10 of this book describe in some detail the environmental effects of settlers who lived along the desert streams. Cattle grazing and its relation to arroyo cutting are discussed.

EFFECTS OF IRRIGATION

Although the irrigation practices of the Hohokam produced some insignificant environmental effects, the effects of irrigation by more recent inhabitants of the desert are much more striking. The first canal company in central Arizona was formed in 1892. Many of the canals followed the same channels as the old Hohokam canals, testifying to the engineering skills of these early people. The first project approved under the Land Reclamation Act was Theodore Roosevelt Dam on the Salt River. Other dams followed on both the Salt and Verde rivers, and today the Salt River Project manages six large reservoirs for irrigation and power production. A number of other irrigation projects in other areas of the state soon followed.

However, the impounded surface waters in the reservoirs were not sufficient to meet the needs of Arizona, and pumping of underground water was necessary. In 1977 Arizona utilized 7.3 million acre-feet of water for all purposes. Of this 1.8 million acre-feet was surface water from reservoirs; 5.5 million acre-feet was underground water from the water table. Re-

charge of the underground water fell far short of the demand.

The result of this tremendous overdraft of groundwater has been a rapid lowering of the water table. Each year many acres of agricultural land go out of production because the cost of pumping water from an ever-deepening water table is greater than the value of the crops that can be grown.

Another result of this groundwater overdraft is land subsidence. As more and more water is withdrawn, spaces are left in the soil. Eventually the overlying soil settles into these spaces, and there is a settling or subsidence of the land surface. Huge cracks 20 to 25 feet across, 30 to 40 feet deep, and several hundred yards long may develop. Fields have been destroyed, houses displaced, and freeways damaged. These problems are becoming more and more frequent as underground pumping continues.

The largest river in the Southwest is the Colorado, running through or along the borders of seven states and Mexico. The Colorado River Compact of 1922 divided the river into the upper basin and lower basin with Lee's Ferry as the dividing point. It also allocated 7.5 million acre-feet of water annually to each basin. Upper basin states were Colorado, New Mexico, Wyoming, and Utah. Lower basin states were Arizona, California, and Nevada. All of the states but Arizona signed the compact. In 1944 Arizona and Mexico joined the Compact. Mexico had not been included in the original compact even though the Colorado River passes through part of Mexico on its way to the Gulf of California. The 1944 agreement allocated 1.5 million acre-feet of water to Mexico annually. It did not specify whether it was to come from the allotment of the upper basin or lower basin states, nor did it specify the quality of the water that was to be delivered.

In 1961 Mexico began to complain that the Colorado River water reaching it was so salty that it was damaging the agricultural crops. Analysis revealed a dissolved solids content of 1500 parts per million (ppm). This increased salinity had been brought about by two events: the filling of Lake Powell on the Colorado River, which reduced the volume of water reaching the lower basin, and the start of pumping saline water from the Wellton-Mohawk Irrigation Project.

The Wellton-Mohawk Project along the lower Gila River was designed to utilize water from the Colorado River. By 1960 75,000 acres were under irrigation. The water entering the project contained 850 ppm of dissolved solids. When this water was put on the land by flood irrigation, it percolated through the soil, picking up additional salts, and finally came to rest in the water table. By 1960 the water table had risen dramatically, and the soil was becoming waterlogged. Pumps were installed to lower the water table.

The water was pumped into a salinity canal and returned to the Colorado River. This water now contained 5700 ppm dissolved solids and was responsible for the increase in salinity of the water reaching Mexico.

Although there had been no commitment as to the quality

of water in the allotment to Mexico, it was deemed politically expedient to accommodate their demand for better-quality water. In 1975 an agreement was signed by the United States and Mexico specifying the quality of water to be delivered and outlining the methods to be used to reduce the salinity to tolerable levels. Ten thousand acres were to be retired from production in the Wellton-Mohawk Project to reduce the amount of saline water being returned to the Colorado River. More efficient means of irrigation such as sprinkler or drip irrigation were to be employed on the remaining lands of the Wellton-Mohawk Project to reduce the volume of saline water significantly. Some of the saline water would be bypassed into a canal and emptied into Santa Clara Slough in Mexico. A desalting plant would be built south of Yuma, Arizona, to remove much of the salt from the water. This desalted water would be combined with Colorado River water so as to ensure an acceptable level of salinity and delivered to Mexico in canals. As of 1980, these provisions are in the process of being implemented.

ANTIDESERTIFICATION PROJECTS

The Vale District Rangeland Rehabilitation program of the Bureau of Land Management is often cited as an example of successful rehabilitation of a desertified range. The Vale District lies in the southeastern corner of Oregon at an elevation of 600 to 2400 meters. It occupies 2.6 million hectares that have been subject to grazing pressure since 1824. By 1880 increasing numbers of horses, cattle, and sheep resulted in severe overgrazing, which continued into the 1930s. The original vegetation consisted of native perennial grasses, with 25 percent woody shrubs. By 1900 most of the grass had disappeared, and sagebrush had increased dramatically.

Rehabilitation efforts began in 1962 and continued until 1973. They included spraying of woody shrubs, plowing, seeding, and burning, together with more effective management of livestock. Fences were erected, wells drilled, springs developed, reservoirs constructed, and roads built. Grazing management plans were implemented. As a result of these measures, the grazing capacity for livestock is estimated to have doubled. Grass cover has increased, while sagebrush has declined. Plant succession has continued toward a grass-sagebrush climax. The project has been judged to have been highly successful.

The California Desert Plan is concerned with the California Desert Conservation Area, 25 million acres extending from Death Valley on the north to the Mexican border. The area has received tremendous pressures from special interest groups over the years, beginning with mining interests. Initially gold and silver were the lures, but over the years other minerals have been discovered and mined. The 1980 mineral production from the area is estimated at more than $300 million.

In order to stimulate development of the West, vast acreages were granted to individuals, companies, and agencies. Railroads received huge grants of land, military reservations were set aside, homesteads were granted, and settlements were established creating demands for services and utilities. Today 2100 kilometers of power lines, 7200 km of pipelines, and many microwave stations dot the landscape. Interest is increasing in the desert as a source of energy. Solar energy is an obvious possibility; geothermal sites have been identified and a number of sites for thermal and nuclear energy installations have been selected.

However, one of the greatest impacts has been the recognition of the desert as a recreation area; as many as 4 million recreationists pour into the desert each year. Dune buggies, recreational vehicles, four-wheel drives, and motorcycles crisscross the desert with no interest in or regard for the fragile environment. Off-road races involving hundreds or even thousands of vehicles have been held annually. The victim of this destruction has been the desert environment, including soil, vegetation, cultural sites, and animals. The damage already done can never be repaired. All that can be done is to develop a management plan that will minimize future destruction.

The complexities of developing such a plan are illustrated by the laws and regulations already on the books, which must be adhered to. They include the Mining Law of 1872, Antiquities Act of 1906, Mineral Leasing Act of 1920, Taylor Grazing Act of 1934, Historic Sites Act of 1935, Wilderness Act of 1964, National Environmental Policy Act of 1969, Endangered Species Act of 1973, Federal Land Policy and Management Act of 1976, Clean Air Act, Fish and Wildlife Coordination Act, Federal Water Pollution and Control Act, Safe Drinking Water Act, and various executive orders and regulations on off-road vehicles. A draft plan and environmental statement were prepared by the Desert Plan staff of the Bureau of Land Management in 1979, based on an intensive three-year study utilizing the latest technologies in inventorying, mapping, and remote sensing. The importance of this management plan and its effective implementation cannot be overemphasized. It attempts to halt the ultimate in desertification: the destruction of the desert itself.

UNITED NATIONS CONFERENCE ON DESERTIFICATION

The United Nations Conference on Desertification held in Nairobi, Kenya, in 1977 focused world attention on the desertification problem. Approximately seven hundred official delegates from one hundred countries deliberated for two weeks before approving a plan of action designed to achieve zero desertification growth by the year 2000. Among the twenty-six principal recommendations of the conference was the formation of national plans of action to combat desertification in each country. The conference also approved three maps: Status of Desertification in the Hot Arid Regions,

Climate Aridity Index Map, and Experimental World Scheme of Aridity and Drought Probability.

UNITED STATES NATIONAL PLAN OF ACTION TO COMBAT DESERTIFICATION

The U.S. plan of action began with an interdepartmental task force meeting in December 1978. The Bureau of Land Management was designated to lead the effort with the participation and cooperation of other federal agencies and nongovernmental organizations. At almost the same time, discussions were being carried on regarding bilateral cooperation between the United States and Mexico in combating desertification. Both projects were carried forward.

The objectives of the U.S. national plan were to promote understanding of the desertification process and to support the development of practices and decision-making procedures necessary to manage renewable resources in arid and semiarid lands consistent with the restoration, maintenance, and enhancement of productivity in accordance with national needs. Two specific goals were to prevent and to arrest the advance of desertification, and to reclaim, where possible, desertified land for productive use. The area considered in the national plan extends across the seventeen continental states west of the 100° longitude. Alaska and Hawaii were not included.

UNITED STATES-MEXICO BILATERAL AGREEMENT

This agreement is a cooperative effort between the United States and Mexico to combat desertification. Resources and personnel have been mobilized to meet the problem on both sides of the border.

DESERTIFICATION INDICATORS

Prior to the 1977 United Nations Conference on Desertification, the American Association for the Advancement of Science, the Indian Science Congress, the East African Academy, Interciencia, the British Academy of Science, and the Association française pour l'avancement des sciences assembled a group of scientists to discuss how to monitor the desertification process. The result was *Handbook on Desertification Indicators* (Reining 1978). The major categories of indicators are physical, biological-agricultural, and social, with many subdivisions of each. The indicators have been tested in centers in various parts of the world and have been revised or modified as needed. They form the backbone of many national plans to combat desertification. The *Directory of North American Arid Lands Research Scientists,* prepared by the Committee on Arid Lands of the American Association for the Advancement of

Science, lists scientists in Mexico, the United States, and Canada who are active in arid land research. (Bender 1977).

DESERTIFICATION AND THE FUTURE

The process has been recognized, it has been defined, its indicators have been identified, its results are only too well known, the corrective methods are available. It would seem that desertification should be on the way to extinction. However, exploitative pressures on arid lands are mounting steadily. These regions contain the major portion of U.S. energy resources for the future. Uranium, coal, natural gas, oil shales, geothermal, and solar energy sources are all common in arid regions of the west and the arctic.

A symposium on Energy Resource Recovery in Arid Lands was held in 1979 by the Committee on Desert and Arid Zones Research. This symposium identified the energy resources in arid regions of North America, discussed the processes for energy resource recovery, and the social, economic, cultural, and environmental impacts of such recovery (Timmerhaus 1980). A 1978 symposium considered the problems of reclaiming arid lands that have been disturbed. With strip mining becoming increasingly important in the West, papers presented in the symposium considered the effects of oil shale exploitation and strip mining of coal on the natural environment. Social, economic, and water quality impacts were also considered (Wright 1978).

Enough is known to approach the problems of desertification and land reclamation in an intelligent and rational manner. However, given the history of man's relationships with the environment, his insatiable appetite for material things, and the often demonstrated greed of individuals and corporations, the prospects for success are not good.

REFERENCES

Bender, G. L.; Paylore P.; and Medellín-Leal, F. 1977. *A directory of North American arid lands scientists.* Washington, D.C.: American Association for the Advancement of Science.

Bureau of Land Management. 1976. *The California desert.* Riverside, Calif.: Bureau of Land Management.

Dregne, Harold E. 1976. Desertification: Symptom of a crisis. In *Desertification: Process, problems, perspectives.* Seminar Series. Tucson: University of Arizona Office of Arid Lands Studies.

Heady, H. F., and Bartolomé, J. W. 1976. Desert repaired in southeastern Oregon: A case study in range management. In *Desertification: Process, problems, perspectives.* Seminar Series. Tucson: University of Arizona Office of Arid Lands Studies.

Johnson, J. D. 1977. *Desertification in the United States.* Paris: UNESCO.

McGee, W. J. 1898. *The Seri Indians: Annual report of the Bureau of Ethnology* 17:1895-1896. Washington, D.C.: Government Printing Office.

Martin, P. S. 1963. *The last 10,000 years.* Tucson: University of Arizona Press.

Medellín-Leal, F. 1978. *La Désertification en Mexico.* San Luis Potosí, Mexico: Instituto de Investigaciones de Zonas Deserticas, Universidad Autonoma de San Luis Potosi.

Paylore, P. 1977. *Arid lands research institutions: A world directory.* Tucson: Office of Arid Lands Studies, University of Arizona.

Reining, P. 1978. *Handbook on desertification indicators.* Washington, D.C.: American Association for the Advancement of Science.

Report of the United Nations Conference on Desertification, 1977. Paris: UNESCO.

Timmerhaus, K. 1980. *Energy resource recovery in arid lands.* Albuquerque: University of New Mexico.

U.S. National Plan of Action to Combat Desertification, 1979. Proposed outline. Unpublished.

Wright, R. A. 1978. *The reclamation of distubed arid lands.* Albuquerque: University of New Mexico.

York, J. C., and Dick-Peddie, W. A. 1969. Vegetation changes in southern New Mexico during the past hundred years. In *Arid lands in perspective,* ed. W. G. McGinnies and B. J. Goldman. Tucson: University of Arizona Press.

AUTHOR INDEX

This index contains names of authors cited in the chapter texts.

Adam, K. R., 248
Adams, J. B., 528, 530
Adams, M. S., 192
Adams, R. H., 131
Ahearn, G. A., 247, 412
Ahlbrandt, T. S., 519
Aiken, C.L.V., 181
Aikens, C. M., 9, 27, 45, 46, 48, 130
Aitchison, S. W., 460
Al-Ani, H. A., 194, 202
Albers, J. P., 19
Alcorn, J. R., 28
Alcorn, S. M., 194, 209, 211
Alden, P., 237
Alderfer, R., 191
Aldous, A. E., 217
Alexander, C. E., 241
Allenbaugh, G. L., 32, 36
Allison, E. C., 504
Allison, I. S., 512
Allred, D. M., 41, 125, 212
Alpert, P., 210
Alsoszatai-Petheo, J. A., 130
Alvarez, J. M., 354
Alvarez, T., 231
Amant, J. A., 245
Amen, R. D., 209
Amin, O., 245
Amsden, C. A., 130
Anders, F. J., 506
Anderson, A., 237, 238
Anderson, A. H., 237, 238
Anderson, B. W., 439, 440, 448, 450, 453, 460, 463
Anderson, D., 206
Andreev, S. F., 239
Andrews, S., 509
Antevs, E., 9, 29, 30, 48, 130
Aranda, J. J., 321

Armstrong, D., 236
Armstrong, D. M., 29
Armstrong, R. L., 18, 19
Arnal, R. E., 187
Arnett, R. H., Jr., 247
Arnold, J. F., 205
Arnold, R. E., 242
Arvidson, R. E., 512
Aschmann, H., 133, 193, 232, 240, 242, 246
Asplund, K. K., 240, 241
Atstatt, S. R., 239
Atwater, T., 18, 19
Austin, G. T., 26, 238
Axelrod, D. I., 3, 29, 30, 73, 163, 185, 191, 192, 436, 437, 443
Axler, R. P., 22
Axtell, R. W., 363
Ayres, J. E., 186

Bachhuber, F. W., 511, 512
Bagnold, R. A., 482, 484, 507
Bailey, V., 28
Bailey, V. E., 234
Baird, J. W., 132
Baird, S. F., 26, 27
Baker, D., 424
Baker, E. W., 248
Baker, J. R., 22
Baker, R. H., 363
Balda, R. P., 235
Baldwin, R., 243
Balinski, J. B., 242
Ball, E. D., 247
Ball, S. H., 505, 511
Bamberg, S., 422, 423
Bancroft, G., 230
Bannister, P., 420
Banta, B. H., 26, 125

Barber, W. E., 243, 244, 445
Barbour, M. G., 116
Barbour, R. W., 209, 236
Bard, J. C., 527, 528
Barker, J. P., 131, 133, 134
Barlow, G. W., 244
Barlow, J. C., 363
Barnes, W. C., 458
Barry, R. G., 383
Bartels, P., 210
Bartholomew, G. A., 235, 237, 238, 240, 241
Barton, L., 209
Bartram, E. B., 210
Bassett, A. M., 506
Bateman, G. C., 233
Bauman, A. J., 528, 530, 533
Baumhoff, M. A., 46, 48, 133
Beal, C. H., 504
Beard, R. L., 246
Beatley, J. C., 41, 42, 45, 112, 117
Beck, B. B., 238
Beck, R. F., 194
Becker, H. T., 29
Beckwith, E. G., 21
Beckwith, F., 517
Bedwell, S. F., 46, 131
Beers, G. D., 245
Beetle, A. A., 38
Beheiry, S. A., 507
Behle, W. H., 27, 28
Beland, R. D., 243
Belk, D., 188, 248
Bell, W. H., 232
Beltrami, C., 321
Beltrán, E., 243, 321, 363
Bemis, W. P., 197
Bender, G. L., 559
Bendire, C. E., 27, 70

Benes, E. S., 241
Bennett, A. F., 410, 415
Bennett, E. I., 210
Benson, L., 193, 194
Bentley, E., 70, 73, 75
Bentley, P. J., 237, 239, 242
Benzioni, A., 422
Berger, R., 28, 45, 46
Bernstein, L., 204
Berry, E. L., 211
Berry, E. W., 29
Berry, K. H., 240, 242
Bersell, P. O., 243
Bettinger, R. L., 46, 130
Beutner, E. L., 206
Beyenbach, K. W., 248
Bidwell, G. L., 195
Bigler, W. J., 232
Bikerman, M., 178, 180
Billings, W. D., 30, 32, 33, 34, 35, 36,
    38, 39, 40, 41, 45, 442
Bingham, S. B., 207
Birkeland, P. W., 30
Birman, J. H., 30
Bjorkman, O., 427
Black, C. C., 424
Blackburn, W. H., 36
Blackwelder, E., 30
Blair, W. F., 3, 230, 243
Blake, J., 70
Blake, W. P., 175, 530
Blanchard, C. J., 39
Blanchard, W. E., 233
Blaney, H. F., 187
Blank, G. B., 211
Blaylock, L. A., 410
Blewett, M., 246
Blom, R. G., 509
Blossom, P. M., 231
Blount, R. F., 241
Blumer, J. C., 199, 208
Bogert, C. M., 230, 238, 239, 240, 242
Bogusch, E. R., 194
Bolyshev, N. N., 190
Bonner, J., 210
Booth, J. A., 209
Booth, K., 242
Borden, F. W., 131
Borror, D. J., 246
Bortz, L. C., 11
Borut, A., 409, 415
Bostic, D. L., 239, 241
Boulaine, J., 357
Boulenger, G. A., 26
Bovis, M. J., 2, 383
Bowen, B. L., 423
Bowers, J. E., 186
Boyer, D. R., 240

Boyer, J. S., 423
Boyer, K. B., 125
Bradley, E. C., 235
Bradley, F. H., 510
Bradley, W. G., 28, 33, 37, 41, 103, 117,
    119, 125, 242
Bradshaw, G., 238
Brainerd, G. W., 131
Brand, D. B., 193
Brandt, H., 237
Branson, B. A., 243, 245
Branson, F. A., 204
Brattstrom, B. H., 26, 239, 242
Braun, E. J., 413
Bray, W. L., 192
Breazeale, J. F., 189
Breckle, S. W., 422
Breed, C. S., 489, 504, 509, 512
Breed, W. J., 489, 512
Breese, C. R., 22
Breitung, A. J., 195
Brenchley, J., 27
Britton, N. L., 194
Brock, K., 35
Brodkorb, W. P., 27
Broecker, W. S., 21, 30
Brooks, R. H., 133, 134, 136
Brott, C. W., 130
Brown, D. E., 117, 192, 193, 195, 208,
    211, 229, 232, 439, 440, 445
Brown, G., 405
Brown, H. A., 243
Brown, J. E., 29
Brown, J. H., 233, 234
Brown, J. R., 264
Brown, J. S., 507, 509
Brown, K. H., 130
Brown, R. W., 29, 30
Brown, V., 237, 239
Brown, W. E., 509
Brown, W. H., 205
Bruder, J. S., 259
Brugge, D. M., 258
Bruhn, J. G., 194
Brum, G. D., 210
Brusca, R. C., 243, 248
Bryan, A. L., 45
Bryan, K., 178, 192, 205, 216, 220, 255,
    457, 489, 511
Bryant, H. C., 237, 241
Budyko, M. I., 1
Bull, C. S., 134
Bullard, R. W., 405
Bunch, T. D., 231
Buol, S. W., 532
Burden, J. D., 194, 202
Burge, B. L., 242
Burgess, R. L., 194

Burgess, T. L., 360, 361
Burk, J. H., 423
Burkart, A., 194
Burleigh, T. D., 447
Burleson, B., 359
Burleson, G. L., 241
Burt, C. E., 26, 241
Burt, W. H., 27, 28, 231, 236
Butler, W. I., Jr., 454
Butterworth, B. B., 235

Cable, D. R., 190, 207, 210
Cadle, T. J., 237, 238
Caire, W., 363
Calder, W. A., 206, 237
Caldwell, M. M., 424
Callahan, J. R., 235
Cameron, R. E., 190, 207, 211
Camp, C. L., 239
Campbell, C. J., 439, 445
Campbell, E.W.C., 130, 131
Campbell, W. H., 130
Canfield, E. R., 211
Cannon, W. A., 195, 199, 204, 205, 207,
    217
Cantino, P. D., 451
Capocaccia, L., 241
Capon, B., 206
Carey, C., 238
Carlisle, D. B., 248
Carothers, S. W., 448
Carpelan, L. H., 187
Carpenter, C. C., 235, 240, 241
Carpenter, R. E., 236
Carr, A., 242
Carr, J. N., 232
Carr, W. H., 233
Carroll, C. R., 246
Carroll, P. H., 189
Carter, G. F., 131
Casebier, D. G., 134
Castetter, E. F., 232, 250, 252, 255, 256,
    257
Cates, R. G., 427, 450
Cattell, F.C.R., 532
Cavender, T. M., 25
Cavill, G.W.K., 246
Chadwick, H. W., 510
Chalif, E. L., 237
Chan, K.K.O., 240
Chaney, R. W., 29, 30
Charley, J. L., 426
Chase, V. C., 206, 422
Chesky, J., 259
Chester, R., 532
Chew, A. E., 194, 231
Chew, R., 231, 235, 248, 415
Chilcote, W. W., 38

Childs, T., 209, 252, 260, 261, 262, 263
Christensen, E. M., 38, 439
Christensen, J. L., 241
Christensen, R. J., 508
Clark, E. D., 232
Clark, J. B., 28
Clark, J. E., 41
Clark, W. J., 26
Clarke, R. S., 363
Clements, F. E., 37
Clements, L., 130
Clements, T., 130, 504, 507, 508
Cliffton, K., 242
Cline, G. G., 7, 21
Clokey, I. W., 32, 33, 35, 41, 117
Cloudsley-Thompson, J. L., 239, 246, 247, 405, 409
Clyna, W., 186
Cochran, D. M., 243
Cockerell, T.D.A., 230, 451
Cockrum, E. L., 231, 236
Cohan, D. R., 453
Colbert, E. H., 26
Cole, G. A., 187, 188, 246, 248, 436
Cole, L. C., 239, 243
Coleman, G. A., 187
Collins, B. J., 193
Collins, J. F., 532
Collins, J. P., 243
Collins, M. S., 247
Compton, R. R., 74
Conant, R., 363
Condie, K. C., 28
Coney, P. J., 179
Congdon, J. D., 409, 415
Conley, W. J., 363
Contreras-Balderas, S., 338, 363
Coogan, A. H., 11
Cooke, R. U., 434, 457, 458, 460, 517
Cooley, D. B., 189
Cooley, M. E., 504, 517
Coombs, G. B., 133, 134, 136
Cooper, 194
Cooper, W. S., 504
Cooperrider, C. K., 456
Cope, E. D., 25, 26, 239
Copple, R. F., 195
Corliss, J. B., 504
Cotera, A. S., 504
Cottam, C., 27
Cottam, W. P., 35, 37, 38, 456
Coues, E., 239
Coutchie, P. A., 414
Cowles, R. B., 239, 240, 241, 244
Cox, T. J., 244
Coyle, J., 193, 228
Crabtree, R., 130
Craig, R. T., 194

Crawford, C. S., 246, 409
Crawford, E. C., 406
Crawford, J. P., 11
Crider, F. J., 191
Criswell, D. R., 510
Critchfield, W. B., 32, 36
Cronquist, A., 32, 33, 34, 36, 38, 39, 41, 45, 117
Cross, S. P., 236
Crossin, R. S., 237
Crosswhite, C. D., 238
Crosswhite, F. S., 188, 238, 252, 254, 256, 258
Crowe, J. H., 414
Cruden, R. W., 192
Cuellar, O., 241
Cumming, K. J., 208
Cunningham, G. L., 201, 202, 423
Curry, H. D., 28
Cushing, F., 250
Cushman, R. A., 452
Cutler, H. C., 195
Cutter, W. L., 241
Cutts, J. A., 509

Dale, B. A., 236
Dalke, P. D., 510
Dalley, G. F., 46
Dalton, P. D., 194
Damon, P. E., 176, 178, 180, 181, 182, 183, 187
Dantzler, W. H., 242, 413
Darrow, R. A., 192, 205, 206, 436
Darrow, R. H., 437, 439
Dasmann, R., 192
Daubenmire, R. F., 230, 442
David, L. R., 25
Davidson, P., 25
Davidson, R. W., 211
Davis, E. H., 259
Davis, E. L., 11, 45, 130, 131
Davis, G. A., 456
Davis, G. H., 180
Davis, J. T., 133, 134
Davis, W. H., 236
Davis, W. S., 195
Dawson, M., 28
Dawson, W. R., 234, 236, 237, 238, 239, 240, 413, 414
Day, G. I., 232
Deacon, J. E., 29, 32, 33, 37, 41, 103, 117, 119, 125, 243, 445, 446
Dean, L., 505, 517
Dechel, W. C., 202, 210
DeCosta, J., 131
de Cuéller, J. T., 321
Deevey, E., 230
Deffeyes, K. S., 19

Degenhardt, W. G., 363
del Campo, R. M., 238, 240
DeMartone, E., 1
Denny, C. S., 434, 511
Denny, W. F., 248
Densmore, F., 259
Desautels, R. J., 133
Despain, D. G., 209
Devine, J. S., 67
DeWald, T., 258
DeWitt, C. B., 240
Díaz-de-León, E., 354
Dice, L. R., 192, 212, 231, 322
Dickerman, R. W., 230
Dick-Peddie, W. A., 439, 555
Dickson, L. L., 240
DiFeo, D. R., 422, 429
Dilks, T.J.K., 210
Dill, D. B., 240, 243
Dillaha, C. J., 248
Dina, S. J., 210
Dinger, B. E., 194
Dirst, V., 252
Dittert, A. E., 259
Dobie, J. F., 233
Dodge, N. N., 193
Dodge, R. E., 456
Dokka, R. K., 507
Dole, H. M., 512
Donnan, C. B., 131, 133
Dort, W., Jr., 510
Douglas, J. R., 511
Douglas, P. A., 245
Drabek, C. M., 231, 235
Dregne, H. E., 345, 363, 435, 532, 555
Drewes, R. C., 410, 412
Driscoll, R. S., 38
Drover, C. E., 133, 134
DuBois, S. M., 181, 183
Duce, J. T., 456
Duce, R. A., 532
Duewer, E.A.H., 211
Dunbar, C. O., 323, 350
Dunbier, R., 164, 186, 192, 193, 221, 222, 223, 224
Durrant, S. D., 28, 29
Dutton, C., 7

Eardley, A. J., 11, 517
Easterla, D. A., 248
Eastman, C. R., 25
Eastwood, R. L., 180
Eaton, G. P., 183
Ebeling, W., 410
Eberly, L. D., 181
Echternacht, A. C., 241
Eddy, T. A., 232
Edmondson, W. T., 19

Edney, E. B., 245, 246, 247, 405, 408, 410, 412, 413, 414
Edwards, E. P., 237
Edwards, H. T., 240
Ehrelinger, J. R., 427
Ehrler, W. L., 191, 194
Eisen, G., 133
Eisner, T., 245
Elbein, A. D., 211
Elder, H. B., 232
Elliot, R. R., 21
Ellis, D., 28, 45
Emberger, L., 1
Emery, J. A., 237, 242
Emery, L. A., 30, 41
Engel, C. G., 527, 528, 530, 531
Engelmann, H., 21
Engel-Wilson, R. W., 454, 460
Engen, C. W., 238
England, A. S., 125
Ericson, J. E., 132
Eriksson, E., 532
Essig, E. O., 246
Evans, F. R., 35
Evans, G. C., 511
Evans, J. R., 506
Evans, K. J., 240
Evans, R. A., 38
Evenari, M., 209, 419
Everard, C. E., 511
Everndan, J. F., 29, 30
Ewing, G. C., 504
Eymann, J. L., 481, 505, 506, 507, 508
Ezell, P. H., 260, 265

Fagan, J. L., 48
Fahn, A., 209
Farley, R. D., 247
Farmer, M. F., 134
Felger, R. S., 193, 194, 195, 197, 211, 223, 225, 226, 241, 242, 261, 262
Fenneman, N. F., 176
Fenneman, N. M., 481
Ferdon, E. N., Jr., 261
Ferguson, C. W., 194
Ferguson, G. W., 241
Feth, J. M., 21, 22, 436
Findley, J. S., 229, 231, 363
Findley, R., 322
Fink, B., 211
Finley, R. B., 234
Fish, E. B., 192
Fisher, A. K., 27
Fisher, C. E., 450
Fisher, H. I., 236
Fisher, S. G., 187
Fisk, L. O., 243
Fitch, H. S., 240

Flack, J.A.D., 450
Flake, L. D., 234
Fletcher, J. E., 189, 211
Flint, R. F., 19, 30
Flohn, H., 350
Flores-Mata, G., 193
Flowers, S., 39, 40, 42, 43, 44, 442
Follett, W. I., 243
Fontana, B. L., 194, 255, 256, 259, 260, 261, 262, 263
Forbes, J., 455
Fowler, C. S., 46, 52
Fowler, D. D., 46, 48, 50, 52, 131, 133, 134
Fowlie, J. A., 241
Fox, K., 460
Fraenkel, G., 246
Free, E. E., 481, 508
Fremont, J. C., 7, 21
French N. R., 233
French, P., 67
Friauf, J. J., 247
Friedmann, E. I., 211
Frost, H. H., 27
Fuller, R. E., 72, 73
Fuller, W. H., 188, 189, 190, 191, 211
Funk, R. S., 240
Furlong, E. L., 28

Gaines, J. F., 117
Gallegos, D. C., 133, 134, 136
Gallizioli, S., 231, 233
Galum, M., 211
García, E., 345, 347
García-Castañeda, F., 193, 333
Garman, S., 26
Garrett, D. M., 505
Gary, H. L., 204, 437, 456
Gates, D. M., 191, 424, 425
Gates, G. O., 239
Gatewood, J. S., 204
Gavin, T. A., 194, 220
Gearhart, P. L., 133
Gelfand, P. W., 238
Geluso, K. N., 236
Gentry, H. S., 193, 194, 195, 202, 227
Gerard, J. B., 194
Gersberg, R. M., 22
Gersh, I., 235
Gertsch, W. J., 248
Ghiselin, J., 27
Ghobrial, L. I., 234
Ghosh, P. K., 233
Gibble, W. P., 194
Gibbs, J. G., 191, 206
Gidley, J. W., 28

Gilbert, B., 233
Gilbert, G. K., 18, 19, 30
Gilbertson, R. L., 211
Gile, L. H., 189, 511
Giluly, 204
Gindel, I., 192
Girard, C., 26
Givnish, T., 425
Gladwin, H. S., 249
Glantz, M., 321
Glendening, G. E., 209, 215, 452
Glenn, W. G., 248
Glennan, W. S., 130
Glinski, R. L., 237, 457
Goldberg, S. R., 241
Goldman, B. J., 321, 322, 342, 345, 347, 357
Goldman, E. A., 193, 230, 233
Gómez-Gonzáles, A., 322, 347, 357, 360, 361
Gomez-Pompa, A., 193
Gonzáles-Cortés, A., 321
Goodall, D. W., 209, 405
Goodpaster, W. W., 236
Goodspeed, T. H., 195
Goodyear, A. C., 259
Gorbman, A., 243
Gordon, S., 237
Gorsuch, D. M., 238
Goss, J. A., 133
Gould, R. A., 52
Graf, W. L., 439
Graham, A. E., 29
Grant, C., 132, 242
Grant, K. A., 238
Grant, R. L., 239
Grant, T. A., 11
Grant, V., 238
Gray, A. B., 261
Gray, R., 210
Greaves, J. E., 190
Greeley, R., 510
Green, C. R., 440, 442, 443
Green, R. A., 207
Green, W., 445
Greenberg, B., 240
Greene, J. L., 232, 234, 236, 262
Greenup, W. E., 512
Greenwald, L., 237
Greenwood, R. S., 134
Gregory, H. E., 517
Griffith, J. S., 131
Grinnell, J., 230, 237, 239, 446
Groepper, K. L., 234
Grubits, G. III, 241
Guilliam, A. M., 321
Gulmon, S. L., 424, 426
Gunst, M. L., 259

Haase, E. F., 192, 194, 204, 208, 217, 439, 452
Hack, J. T., 489, 504, 511, 517, 519
Hadley, N. F., 230, 232, 246, 248, 405, 408, 409, 410, 412, 413, 415
Hadley, R. F., 434
Haefer, J. R., 259
Hagmeier, E. M., 230
Hague, A., 21
Hahn, W. E., 241
Haight, L. T., 237
Hailey, T. L., 363
Haines, H. B., 234
Hainsworth, F. R., 408
Hales, J.R.S., 407
Hall, E. R., 28, 29, 133, 231, 234, 235, 236
Hall, H. M., 195, 230
Hall, M. G., 131, 134
Hall, S. M., 250
Halloran, A. F., 232, 233, 234
Halvorson, W. L., 194, 212
Halvorson, W. T., 423
Hamilton, L. B., 193
Hamilton, W. J. III, 246, 408, 414
Hanna, W. C., 27
Hanford, F. S., 27
Hansen, C. G., 70
Hansen, R. M., 240, 242
Hardman, G., 21, 22
Hardy, R., 242
Hardy, W. H., 321
Hare, R. F., 511
Haring, I. M., 210
Harner, R. F., 36
Harper, K. T., 36
Harrington, M. R., 27, 28, 45, 46, 130
Harris, A. H., 363
Harris, D., 244
Harris, J. A., 205
Harrison, A. T., 198
Harshberger, J. W., 192, 322
Hart, J. S., 415
Haskell, W. L., 245
Haskill, 43
Hastings, J. R., 191, 192, 205, 206, 208, 212, 217, 218, 247, 457, 460
Hatch, D. E., 408
Hatch, M. D., 424
Hatfield, D. M., 231
Hattingh, J., 410
Hattori, E. M., 26, 46, 48
Hauck, F. R., 134
Haury, E. W., 186, 231, 232, 239, 249, 259, 263
Haverty, M. I., 247
Hawke, D. D., 247
Hay, O. P., 25, 28, 241

Hayden, J. D., 130, 259, 527
Hayford, F. S., 509
Haynes, C. V., 45, 130, 457
Hayward, C. L., 27, 29, 447
Hayward, H. E., 204
Hayward, L., 442
Heath, J. E., 241, 243, 247, 406, 407
Heath, W. G., 244, 245
Hecht, M. K., 26
Heid, J., 131
Heifetz, W., 240
Heimlich, E. M., 241
Heimlich, M. G., 241
Heinrich, B., 246, 407
Heizer, R. F., 26, 28, 45, 46, 48, 131, 13
Heller, A. A., 33, 35, 239
Helmberger, D. V., 177
Hem, J. D., 436, 528, 531, 532, 533
Hemstock, G. A., 532
Henckel, P. A., 191
Hendricks, B. A., 191, 456
Hendrickson, J. R., 243, 244
Henning, D., 350
Henrickson, J., 195, 322, 323, 333, 360, 361
Henshaw, H. W., 26, 27
Henshaw, P. C., 28
Hensley, M. M., 237, 240
Henwood, K., 408
Heppner, F., 237, 408
Herre, A. W., 44
Hesketh, J., 424
Hester, T. R., 130, 131
Hevly, R. H., 210
Hibbard, C. W., 28
Hickman, P. P., 136
Hill, J. E., 234
Hill, R. D., 248
Hillebrand, T. S., 131, 133
Hinckley, A. L., 206
Hinde, R. B., 322
Hinds, D. S., 230, 236, 237, 409
Hinton, R. J., 458
Hitchcock, C. L., 195
Hodge, R. P., 399
Hoffman, G. L., 532
Hoffman, W. J., 27
Hoffmeister, D. F., 236
Hooke, R. L., 528, 530, 531, 532
Hoover, F. G., 245
Hornaday, W. T., 232, 512
Horne, F., 248
Horne, R. A., 532
Horton, J. S., 204, 205, 439
Horvath, S. M., 405
Hotton, N., 239
Hotz, P. E., 9
Houghton, J. G., 22, 23, 24

Houghton, S. G., 19, 21, 30
Householder, V. H., 242
Howard, A. D., 508
Howard, B., 238
Howard, H., 27, 204, 237
Howe, G., 455
Howell, A. B., 235
Howell, A. H., 235
Howell, D. J., 236
Howell, T., 70
Howes, P. G., 194
Hsiao, T. C., 420, 421
Hubbard, J. P., 229, 237, 445, 447
Hubbert, M. K., 11
Hubbs, C. L., 19, 21, 22, 25, 26, 243, 245, 445
Hudson, J. W., 235
Huey, L. M., 230, 231, 232, 235, 236
Huibregtse, W., 236
Hull, H. M., 202, 206
Hulse, A. H., 446
Hume, W. F., 530
Humphrey, R. R., 193, 194, 195, 204, 210, 220, 221, 228
Hungerford, C. R., 235
Hunt, A. P., 131, 133
Hunt, C. B., 117, 133, 440, 442, 445, 481, 511, 528, 530
Hunt, J. H., 246
Hunt, W. G., 363
Huntington, E., 457
Hunziker, J. H., 125, 192, 422, 429
Hurd, P. D., 246, 247
Hutchinson, G. E., 21, 22

Idso, S. B., 535
Ingram, D. L., 405, 406, 504
Irelan, B., 489, 509
Irwin, T. D., 181
Isely, D., 195
Itai, C., 422
Ives, R. L., 181, 187, 512

Jackson, D. G., Jr., 242
Jackson, H.H.T., 233
Jackson, I. L., 410
Jaeger, E. C., 38, 41, 116, 192, 193, 234, 321, 322
James, A. E., 243
James, G. T., 29, 30
Jameson, D. L., 243
Janzen, D. H., 246
Jasinski, A., 243
Jáuregui, E., 342, 350
Jehl, J. R., Jr., 27
Jenne, E. A., 532
Jennings, J. D., 27, 46, 131
Jenny, H., 355
Joern, A., 247

John, K. R., 243, 244
Johnson, A. W., 3, 192
Johnson, C. R., 240
Johnson, D. E., 194, 208, 211
Johnson, D. R., 234
Johnson, D. W., 243
Johnson, H. B., 38, 124, 193, 439
Johnson, H. R., 234
Johnson, J. E., 245
Johnson, N. K., 27, 28, 447
Johnson, O. W., 413
Johnson, R. B., 509
Johnson, R. R., 237, 448, 463
Johnston, I. M., 192, 194
Johnston, M. C., 195, 322, 333, 360, 362
Jollie, M. T., 27
Jones, D. J., 517
Jones, G. N., 210
Jones, J., 236
Jones, J. C., 21
Jones, L. W., 190
Jones, R. M., 414
Jones, W., 186, 206
Jordan, D. S., 25
Jorgensen, C. B., 240
Jorgensen, C. D., 26, 29
Jorgensen, C. J., 233
Joseph, A., 259
Judd, B. I., 205

Kappen, L., 420
Kasheer, T., 193
Kauffield, C. F., 239
Kaufmann, J. H., 233
Kavanau, J. L., 241
Kay, M., 11
Kearney, T. H., 117, 193, 195, 419, 442
Keck, D. D., 116, 193
Keegan, H. L., 248
Keil, D. J., 194
Kelleher, J., 28
Kellogg, L., 28
Kelson, K., 29, 231, 234, 235, 236
Kendeigh, S. C., 237
Kennedy, C. E., 456
Kennedy, P. B., 32
Kessler, S. J., 187
Khudairi, A. K., 190
Kidd, D. E., 211
Kilcrease, A. T., 250, 261
Kilgore, D. L., Jr., 406
Kilpack, M. L., 27
Kim, J., 187
Kimmel, B. L., 22
Kimsey, J. B., 243
King, C., 9, 18, 21, 131, 133, 134
King, J. E., 192
King, P. B., 11, 30, 517
King, W. W., 415

Kingsolver, J. M., 452
Kinne, O., 244
Kissell, M. L., 258, 259
Klauber, L. M., 26, 239, 241
Klebenow, D. A., 28
Kleiner, E., 28
Klemmendson, J. U., 426
Klikoff, L. G., 205, 210
Klots, A. B., 246
Kluge, M., 424, 425
Kluger, M. J., 407
Knapp, R., 116
Kniffen, F. B., 504
Knipe, O. D., 209, 232
Koch, D. L., 21
Koehn, R. K., 245
Koenig, E. R., 22
Kolb, C. R., 481
Koller, D., 191
Kornalik, F., 240
Kornoeldge, T. A., 117
Koscielniak, D. E., 510
Kowta, M., 131
Kramer, R. H., 245
Kramer, R. J., 207
Krauskopf, K. B., 531
Kreiger, M. H., 180
Krieger, A. D., 130
Krochmal, A., 34, 38, 41
Krochmal, C., 38, 41
Kroeber, A. L., 455
Krumbein, W. E., 527, 530
Krumholz, L. A., 245
Kuchler, A. W., 116, 193
Kupfer, D. H., 506
Kurtz, E. B., 209

Lacey, J. R., 192, 204
Lakin, H. W., 528
Lamb, B., 194
Lamb, E., 194, 211
LaMotte, R. A., 29
Lange, O. L., 210, 211, 420
· Langebartel, D. A., 239
Langenheim, R. L., Jr., 10
Langman, I. K., 193
Langston, G. A., 177
Lanning, E. P., 131, 133
Larcher, W., 424
LaRivers, I., 21, 22, 25, 26
Larrison, E. J., 27, 29
Larson, E. R., 10
Lasiewski, R. C., 237, 407, 408
Lathrap, D. W., 133
Lauber, P., 194
Laudenslayer, W. F., Jr., 25
Laudermilk, J. D., 527, 530
Lawlor, T. E., 234
Lawson, C. L., 28

Layser, E. F., 192, 439
Layton, T. N., 46
Leakey, L.S.B., 130
Leary, P. J., 117
Lechleitner, R. R., 236
Leding, A. R., 256
Lee, A. K., 234, 410, 413
Lee, H. H., 242
Lee, V. F., 248
Lehner, R. E., 11
Lehr, J. H., 193
Leidy, J., 26
Lemen, C. A., 233
Leopold, A. S., 193, 230, 234, 236, 456
Levi, H. W., 248
Levine, M., 211
Levinson, A. A., 528, 533
Lew, L., 243
Lewis, M. A., 243
Lewis, P. F., 517
Leydet, F., 233
Licht, P., 239, 410
Lieberman, G. L., 233
Ligon, J. D., 238, 363, 447
Linck, G., 530
Lindsay, G., 193, 194
Lindsay, J. F., 510
Lindsey, J. P., 211
Lindsey, L., 48
Lindstedt, S. L., 236
Linsdale, J. M., 26, 27, 235, 239
Linsdale, M. A., 33, 35
Linsley, E. G., 246, 247
Lintz, J., Jr., 10
Lisk, R. D., 240
Lister, P. B., 235
Little, E. L., 35, 193, 233
Livingston, B. E., 190, 205
Lloyd, R. M., 35
Lockard, J. S., 235
Lockard, R. B., 235
Locke, M., 410
Loeb, L., 240
Loeltz, P. J., 489, 509
Loew, G., 527, 530
Logan, R. F., 117, 435
Lommen, P. W., 424
Long, J. T., 507, 508
Louderback, G. D., 30
Love, L. D., 422
Lovejoy, E. M., 30
Loveridge, J. P., 412
Low, P. F., 532
Low, W. A., 232
Lowe, B. S., 243
Lowe, C. H., 117, 192, 193, 194, 205, 206, 208, 209, 210, 211, 212, 215, 220, 223, 226, 229, 230, 231, 237, 238, 239, 241, 243, 244, 362, 408,

431, 433, 436, 439, 454
Lucas, A., 527, 530
Lucas, F. A., 25
Lugaski, T., 21, 25, 26, 28, 30
Lumholtz, C., 231, 232, 250, 254, 258, 259, 260, 261, 262, 263, 512
Lund, E. H., 70, 73, 75
Lustig, L. K., 481
Lyford, F. P., 191
Lynch, D. J., 181
Lyneis, M. M., 137
Lynn, R. T., 241

Mabey, D. R., 528, 530
Mabry, T. J., 194, 422, 429
MacArthur, J., 440, 449
MacArthur, R. H., 440, 449
MacDonald, A. A., 505
MacDonald, J. R., 28
MacDougal, D. T., 187, 191, 195, 205, 216, 225, 508, 509
MacFarland, W. N., 406, 413
MacGregor, A. N., 211
MacIntyre, F., 532
Mackay, W. P., 240
MacMahon, J. A., 229, 230, 231, 245
MacMillan, R. E., 234, 238, 413
MacNeish, R. S., 45
Madsen, D. B., 46, 48
Madsen, R. L., 235
Mahmoud, S. A., 190
Major, J., 116, 192
Mallery, T. D., 207
Maloiy, G.M.O., 405
Malouf, A. A., 52
Manje, J. M., 231, 260
Manville, R. H., 233
Marder, J., 408
Mares, M. A., 452
Mark, A. F., 35
Marks, J. B., 194, 206
Marroquin, J. S., 321
Marshall, R. R., 528
Martin, P. S., 28, 209, 230, 555
Martin, R. C., 205
Martin, S. C., 195, 209, 452
Martin, W. P., 211
Marvin, R. F., 180
Maslin, T. P., 241
Mason, B., 532
Mason, H. L., 36
Mason, J. A., 259
Master, R. W., 248
Mathews, M. M., 255
Mathews, T. W., 232, 234, 236
Mathiot, M., 259
Mauer, R. A., 235
Mawby, J. E., 28
Maximov, N. A., 419, 422

Maxwell, J. C., 9, 11
May, C., 211
Mayhew, W. W., 239, 240, 241, 243
Maze, R. L., 237
McCann, F. T., 511
McCauley, C. K., 504
McClanahan, L. L., 242, 243, 410, 412, 414
McCleary, J. A., 209, 210
McConnell, W. J., 26, 245
McCoy, F. W., Jr., 508, 509
McCrone, J. D., 248
McDonough, W. T., 206, 210
McGee, W. J., 555
McGeorge, W. T., 189
McGinitie, D., 29
McGinnies, W. G., 192, 194, 205, 321, 322, 342, 345, 347, 357
McGinnis, S. M., 240
McHargue, J. T., 443
McHenry, J., 211
McIntyre, M. J., 134
McKee, E. D., 504, 510, 511, 512
McKell, C. M., 38
McKelvey, S. D., 195
McKibbin, N., 244
McKinney, A., 133
McKusick, C., 232, 237, 239
McLaughlin, A. M., 211
McMillan, C., 450
McNab, B. K., 230, 234
Mearns, E. A., 231, 232
Medellín-Leal, F., 322, 347, 357, 360, 363
Medin, D. E., 35
Meek, S. E., 243
Mehringer, P. J., Jr., 9, 48, 130
Meighan, C. W., 130, 131, 133
Meigs, P., 2, 345, 350
Meinwald, J., 245
Meinzer, O. E., 511, 512
Meléndez, J., 363
Melton, M. A., 178, 183, 434
Mendenhall, W. C., 507
Mercer, E. M., 410
Merk, G. P., 509
Merriam, C. H., 26, 27, 193, 230, 235
Merriam, J. C., 26, 27, 28
Merriam, R., 509, 512
Merrill, G. P., 527, 530
Metcalf, A. L., 363
Meyer, S. E., 41
Mielke, J. L., 211
Mifflin, M. D., 9, 48, 130
Miller, A. H., 27, 237, 241, 446
Miller, L., 242
Miller, M. R., 241
Miller, R. R., 19, 21, 22, 25, 26, 243, 244, 245, 338, 363, 445, 446

Milstead, W. W., 363
Minckley, W. L., 187, 188, 243, 244, 245, 338, 363, 445, 446
Minnich, J. E., 240
Minton, S. A., 242
Miranda, F., 322, 323
Mitchell, G. C., 230
Mitchell, R. S., 35
Mitichi, L. W., 194
Moberly, W. R., 240
Moiola, R. J., 511
Moldenke, A. R., 451
Monger, J.W.H., 10
Monson, G., 232, 234, 237
Mooney, H. A., 191, 192, 198, 424, 425, 426, 427, 429
Moore, R. D., 517
Moore, R. T., 230
Moore, W. S., 245
Morafka, D. J., 321, 322, 323, 342, 345, 350, 354, 360, 361, 363
Moran, R., 193, 194, 195
Morello, J., 422
Morgan, S., 209
Morrison, P., 234
Morrison, R. B., 7, 19, 21, 30, 510
Morse, N., 48
Morton, M. L., 238
Mosauer, W., 239
Mosiño, P. A., 350
Mount, L. E., 405, 406
Mozingo, H., 44
Mroz, E. J., 532
Muhktar, H.A.M., 209
Mullen, R. K., 415
Muller, C. H., 210
Muller, S. W., 11, 18
Mulroy, T. W., 192, 194
Muma, M. H., 248
Munsey, L. D., 424
Munz, P. A., 116, 193
Murdock, J., 133
Murie, A., 233
Murie, M., 234
Murphy, J. D., 510
Murphy, R. W., 239, 240, 241
Murray, K. F., 239
Musgrave, M. E., 243
Musick, H. B., 206
Mutch, T. A., 512

Nabhan, G. P., 204
Nagy, K. A., 240, 412, 413, 414, 415
Nappe, L., 28
Napton, L. K., 46
Nash, T. H., 194, 211
Neal, B. J., 232, 235
Neales, T. F., 191
Nebeker, G. T., 211

Neff, J. L., 451
Nelson, E. W., 180, 231
Neuenschwander, L. F., 195
Nichol, A. A., 193, 220
Nichols, D., 363
Nichols, J., 130
Nichols, U. G., 242
Nicot, J., 190
Niering, W. A., 208
Nisbet, R. A., 194
Nobel, P. S., 194
Nockleberg, W. J., 508, 509
Nolan, T. B., 9, 10, 18, 30
Nolf, B., 70
Norbeck, E., 131
Norr, M., 210
Norris, K. S., 236, 239, 240, 241, 486
Norris, R. M., 486, 507, 508, 509
Northey, W. T., 248
Northington, D. K., 360, 361
Norwood, R. H., 134
Nour, T. A., 234
Novick, A., 236
Noy-Meir, I., 405
Nutting, W. L., 247

Oberholser, H. C., 447
Oberstein, D., 235
Ochoterana, I., 193
O'Connell, J. E., 46
Odening, W. R., 205, 423
O'Donnell, M. J., 414
Odum, E. P., 36
Oechel, W. C., 194, 423
O'Farrell, T. P., 30, 41
Ohmart, R. D., 234, 237, 408, 413, 439,
    440, 448, 450, 452, 453, 460
Okasha, A.Y.K., 413
Olin, G., 231
Oliver, J. A., 230, 239
Olmstead, F. H., 458, 489, 509
Olsen, A. R., 246
Olsen, S. J., 234, 245
Oppenheimer, H. R., 422
Ore, H. T., 131
Orians, G. H., 192, 405, 424, 425, 426,
    428
Ornduff, R., 116
Orr, P. C., 21, 28, 30
Ortenburger, A. L., 239
Ortenburger, R. D., 239
Ortiz-Villanueva, B., 355
Osgood, W. H., 235
Osmond, C. B., 424
Otte, D., 247
Otto, R. G., 245

Pack, H. J., 26

Packard, R. L., 363
Pakiser, L. C., 177
Panlaqui, C., 131
Parish, S. B., 193, 212
Parker, K. F., 193
Parker, L. R., 29
Parker, R. H., 243
Parker, W. S., 240, 241
Pase, C. P., 192, 193, 195, 229, 439
Patten, D. T., 191, 194, 206, 212, 423
Patterson, R. A., 242, 248
Patton, D. R., 192
Patton, J. L., 235
Paulsen, H. A., 190, 209, 452
Payen, L. A., 130
Paylore, P., 164, 321, 322, 342, 345, 347,
    357, 481, 537
Peacock, J. T., 450
Pearson, O. P., 234
Peck, S. L., 133
Peebles, R. H., 117, 193
Peirce, H., 183
Perry, R. S., 528, 530
Peterson, H. B., 189
Peterson, N. V., 458
Peterson, R. T., 237
Petrone, A., 517
Pettus, D., 243
Pewe, T. L., 383
Phelps, J. S., 232
Phillips, A. R., 230, 236, 237, 239, 363,
    447
Phillips, J. E., 412
Phillips, O. L., 207
Phillips, R. P., 504
Phillips, W. S., 206, 450
Philpott, C. W., 240
Pianka, E. R., 240, 241
Piemeisel, L. N., 218, 424
Pimm, S. L., 363
Pinckney, F. C., 195
Pinkava, D. J., 360
Pinkham, C.F.A., 235
Pippin, L. C., 46
Pitelka, F. A., 398
Pocock, R. I., 233
Porter, L. R., 187, 239
Potbury, S. S., 29
Potter, J. M., 248
Potter, R. M., 527, 528, 530, 533
Pough, F. H., 406
Powell, M. A., 360, 361
Prakash, I., 233
Price, H. M., 233
Price, M. V., 234
Prigge, B. A., 117
Pringle, J. K., 132

Probstein, R. F., 354
Proctor, M.C.F., 210
Pugach, S., 409
Putnam, W. C., 30

Quay, W. B., 234

Raisz, E., 322, 323, 333
Raitt, R. J., 237, 363
Ramós-Arizpe, M., 321
Randall, D. C., 117
Rapoport, A., 419
Rath, L., 240
Raven, P. H., 192
Ray, G. C., 192
Ray, H., 189
Rea, A., 28, 192, 220, 234, 235, 264, 265
Reasoner, D. L., 510
Rector, C. H., 131, 133, 134
Redfield, A. C., 241
Reed, C. A, 28, 233
Reeder, W. G., 236
Reeve, W. L., 241
Reeves, R. W., 457, 458, 460
Rehn, J.A.G., 247
Reichman, O. J., 231, 233, 234, 235, 237,
    245, 246, 248
Reifschneider, O., 32
Reining, P., 559
Reitan, C. H., 440, 442, 443
Rempel, P. J., 507, 508
Remy, J., 27
Rendell, D., 46
Repenning, C. A., 72, 73
Reveal, J. L., 195
Reynard, C., 207
Reynolds, H. G., 234, 236
Reynolds, S. J., 179
Rhoads, D. F., 427, 450
Riazance, J., 246, 247
Rice, A., 188
Rich, J. L., 456
Richards, A. G., 414
Richards, D., 210
Richards, G. L., 27, 28
Richardson, C. H., 26
Richardson, W. B., 233
Richmond, G. M., 30
Riddell, H. S., Jr., 133
Riddle, W. A., 409, 414
Ridgeway, R., 27
Riley, J. P., 532
Rinne, J. N., 244
Riskind, D. H., 321
Ritter, E. W., 133, 134
Roberts, F. H. H., Jr., 131
Roberts, N. C., 193, 228
Roberts, R. J., 9, 11

Robertson, J. H., 38
Robertson, P., 246
Robinson, R. W., 130, 132
Robinson, T. W., 204, 439
Robison, W. G., 240
Rocchio, P., 131
Roden, G. I., 243
Rodgers, R. W., 211
Rodgers, W. M., 460
Roest, A., 233
Rogers, C. L., 354
Rogers, M., 130, 131, 133
Romney, E. M., 422
Rose, J. N., 194
Rosenzweig, M. L., 230, 233, 235
Ross, A., 236
Rossman, G. R., 527, 528, 530, 533
Roth, E. S., 507
Roth, L. M., 245
Rowlands, P. G., 108, 116, 117, 119
Royse, C. F., 179
Ruben, J., 26
Rubey, W. W., 11
Ruby, J., 133
Rue, L. L., 232
Ruibal, R., 243
Rundell, P. W., 192
Runyon, E. H., 202
Russell, F., 232
Russell, F. E., 242, 248, 250, 252, 254,
    257, 258, 261, 262, 263, 264
Russell, I. C., 21, 22, 30, 510
Russell, R. J., 507, 508
Russell, S. M., 237
Russo, J. P., 231
Ryan, J., 532, 533
Ryan, R. M., 231, 234
Ryser, F. A., 28
Rzedowski, J., 3, 193, 321, 350, 354, 360,
    361, 362

Salisbury, F. B., 34
Salt, G. W., 237, 238
Sammis, T. W., 195
Sampson, A. W., 32, 33
Sands, A., 455
Saunders, R. S., 507, 509
Saunier, R. E., 194
Savage, J. M., 239
Savory, T., 248
Saxton, D., 248, 250, 254, 259
Sayers, S. J., 193
Scantling, F. H., 259
Scarborough, R. B., 179, 183
Scarlett, P. L., 194
Schaut, G. G., 243
Scheffer, F., 530
Schellmann, W., 533, 534

Schimpfer, A.F.W., 419
Schmidly, D. J., 363
Schmidt, K. P., 239
Schmidt-Nielsen, B., 234
Schmidt-Nielsen, K., 234, 235, 236, 237,
    239, 242, 406, 412, 414
Schmutz, E. M., 193, 194
Schroder, G. D., 235
Schroeder, A. H., 131
Schuiling, W. C., 130
Schulze, E. D., 32, 420
Schumm, S. A., 434
Schuster, J. L., 194
Schwennesen, A. T., 511
Scott, I. M., 234
Scott, W. D., 532
Scott, W. F., 11
Scudday, J. F., 363
Seely, M. K., 414
Segalen, P., 357
Selander, R. K., 27
Self, J., 210
Sell, J. D., 183
Sellers, W. D., 2, 230
Serventy, D. L., 237
Seymour, R. S., 248
Shafiqullah, M., 176, 180, 181, 183
Shantz, H. L., 38, 39, 40, 41, 194, 205,
    217, 218, 419, 422, 424
Shapovalov, L., 243
Sharp, R. P., 30, 506, 507, 508, 527, 528,
    530, 531
Shaw, A. M., 265
Shaw, C. E., 240
Shaw, J., 246
Shefi, R. M., 117
Shelton, J. S., 507
Shenk, L. O., 136
Sherbrooke, W. C., 194, 230, 235
Sheridan, M. F., 179, 180
Sheridan, T. L., 204
Sherwood, S., 230, 235
Shields, L. M., 190, 202, 211
Shive, J. W., 205
Shkolnik, A., 409, 414, 415
Shoemaker, V. H., 240, 243, 410, 412,
    413, 414, 415
Short, L. L., 237
Shotwell, J. A., 28
Shreve, F., 43, 163, 164, 178, 187, 190,
    191, 192, 193, 195, 196, 197, 198,
    201, 202, 203, 206, 207, 208, 209,
    211, 215, 216, 217, 219, 220, 224,
    229, 258, 321, 322, 323, 439, 442,
    443, 489, 509, 512
Shultze, J. C., 245
Shutler, M. E., 46, 48
Shutler, R. J., Jr., 46, 131, 133

Siebenthal, C. E., 509
Siebert, B. D., 413
Sigleo, A. C., 133
Sigler, W. F., 26, 446
Silberling, N. J., 11, 18
Simmons, N. M., 194, 231
Simms, S. R., 46
Simpson, B. B., 194, 209, 450, 451
Simpson, G. G., 28
Simpson, J. H., 21, 26
Simpson, R. D., 45, 130, 131
Singer, C. A., 130, 132
Skadhauge, E., 237
Skougstad, M. W., 528, 533
Skujins, J., 426
Slayter, R. O., 420
Slemmons, D. B., 19
Slevin, J. R., 26, 239
Sloan, A. J., 239, 240
Slobodkin, L. B., 419
Small, A., 446
Smart, E. W., 26
Smith, A. C., 116
Smith, D. S., 248
Smith, E. L., 192, 193
Smith, E. M., 191, 194, 206
Smith, E. R., 46
Smith, G. A., 131, 132, 133
Smith, G. L., 32, 36
Smith, G. R., 25, 26, 245
Smith, H. D., 233
Smith, H. M., 239, 240, 241, 243
Smith, H.T.U., 481
Smith, H. V., 189
Smith, R. B., 239, 241, 243
Smith, R.S.U., 506, 507, 508, 509, 510,
    511
Smyth, M., 238
Snelling, R. R., 246
Snyder, C. T., 21, 22
Snyder, J. M., 211
Snyder, J. O., 25, 26
Sohns, E. R., 195
Solbrig, O. T., 192, 405, 419, 424, 425,
    426, 428, 450, 451
Soleglad, M., 248
Somero, G. N., 243
Sommerfeld, M. R., 211, 436
Soto-Mora, C., 342, 350
Soulè, M., 239, 240, 241
Souther, J. G., 10
Sowls, L. K., 232, 236
Spalding, E. S., 205
Spalding, V. M., 194, 205, 422
Spaulding, W. G., 130, 192
Spencer, A. L., 234
Spencer, D. A., 234
Sperry, C. C., 233, 234

Spicer, E. H., 249, 252
Spurr, J. E., 18
Stadelman, E., 204, 205
Stager, K. E., 237
Stahnke, H. L., 240, 248
Stälfelt, M. G., 39, 40
Stamp, N. E., 234
Standley, P. C., 193
Stanley, T. B., 181
Stansbury, H., 19
Stark, N., 422
Stearns, H. T., 510
Stebbins, G. L., 192
Stebbins, R. C., 26, 239, 240, 241, 243
Steen, E., 70
Steenbergh, W. F., 194, 206, 208, 209, 431
Steidtmann, J. R., 519
Steineger, L. H., 26
Sternberg, L., 124
Sterud, E. L., 133
Stewart, G., 37, 456
Stewart, J. H., 10, 46, 50
Stewart, O. C., 70
Stickel, E. G., 134
Stirton, R. A., 28
Stobbart, R. H., 246
Stock, A. D., 235
Stock, C., 28
Stocker, O., 419, 424
Stockwell, W. P., 36
Stokes, W. L., 18, 26, 28, 517
Storer, T. I., 243
Stout, G. A., 236, 243
Strain, B. R., 192, 194, 201, 202, 206, 422, 423
Straw, R. M., 322, 323, 333
Strecker, J. K., 243
Stretta, E.J.P., 350, 354
Strong, W. D., 131
Stuart, G. R., 242
Stuart, L. C., 230
Stuckless, J. S., 179
Stull, E. A., 187
Sturgal, J. R., 181
Stutz, C. D., 230
Stutz, H. C., 40
Styblova, Z., 240
Sugden, J. W., 27
Sumner, J. S., 181, 183
Suneson, N. H., 179
Surr, G., 530
Sutton, G. M., 237
Sutton, M. O., 132, 133
Swarth, H. S., 230, 237
Sweet, J. G., 244
Swenson, W. H., 452
Sykes, G., 176, 460, 504, 508

Syvertsen, J. P., 422, 423
Szarek, S. R., 192, 424, 429
Szaro, R. C., 454

Taber, F. W., 233
Tadmor, N., 419
Tanner, W. W., 26, 125, 240, 241
Taylor, C. R., 406, 407, 414
Taylor, E., 191
Taylor, E. H., 240
Taylor, E. S., 133
Taylor, G. H., 517
Taylor, R. E., 130
Taylor, S. E., 425
Taylor, W. K., 238
Taylor, W. P., 26, 27, 233, 234, 235, 236
Tavara-Alfaro, X., 321
Teague, G. A., 136
Tedrow, J.C.F., 527, 531
Teeri, J. A., 424
Templeton, J. R., 240, 405, 415
Terron, C. C., 241
Tevis, L., 209, 239, 243, 246
Thackery, F. A., 256
Theilig, E., 507, 509
Thoday, D., 202
Thomas, C. E. 23
Thomas, D. H., 46
Thomas, H. E., 517
Thompson, D. G., 505, 506
Thompson, H. J., 195
Thomson, D. A., 243, 244
Thornber, J. J., 456
Thornbury, W. D., 481
Thorne, D. W., 189
Thorne, R. F., 116
Thornthwaite, C. W., 1
Thorp, J., 481
Thorp, R. W., 248
Tidestrom, I., 117
Tidwell, W. D., 29
Tiedmann, A. R., 194, 426
Timmerhaus, K., 559
Ting, I. P., 45, 424, 425
Ting, W. S., 30
Tinker, B., 230
Tinkle, D. W., 241
Titus, F. B., 511, 512
Tomlinson, R. E., 238
Tomoff, C. S., 237, 238
Toolson, E. C., 410
Trainor, F. R., 211
Trelease, T. J., 25
Tripton, A. R., 363
Trost, C. H., 238
True, D. L., 131, 133
Truxal, F. S., 246
Tryon, A., 210

Tschirley, F. H., 194, 195, 209
Tucker, J. L., 27
Tucker, V. A., 239
Tueller, P. T., 36, 38, 41
Tuohy, D. R., 45, 46, 130
Turkowski, F. J., 236
Turnage, W. V., 190, 206
Turner, B. L., 360, 361
Turner, H. W., 9
Turner, R. M., 191, 192, 194, 204, 206, 208, 212, 217, 218, 457, 460
Tuttle, D. M., 248
Twisselman, E., 116
Tyler, A., 240

Ugolini, F. C., 527, 531
Underhill, R. M., 250, 252, 254, 255, 256, 257, 259
Ungar, I. A., 204

Valentine, K. A., 194
Valera-Adam, J., 350
Valverde, J. M., 231
Van Asdall, W., 206
Vance, V. J., 241
Van De Graaff, K. M., 231, 234, 235
Van de Kamp, P. C., 509
Van Denburgh, J., 26, 239
Van Devender, T. R., 3, 117, 130, 192, 363, 437
Vanicek, C. D., 245
Vanjonack, W. J., 234
Van Lopik, J. R., 481
Van Rossem, A. J., 27, 237
Vasek, F. C., 35, 116, 124
Venner, W. T., 131
Vieira de Silva, J., 422
Villalobos, C. I., 354
Villa-Ramírez, B., 236, 363
Vinegar, A., 248
Vogl, R. J., 443
Vorhies, C. T., 234, 235, 236, 238
Voth, A., 195
Vreeland, H., 26, 30
Vreeland, P., 26, 35

Waddell, J. O., 252
Wade, W. E., 211
Wadleigh, C. H., 204
Wagle, R. F., 194
Wagoner, J. J., 455
Waisel, Y., 204
Waitman, J., 233
Walker, B. W., 187, 243
Walker, E. P., 235
Walker, G. W., 72, 73
Wall, B. J., 413
Wallace, A., 422

Wallace, E., 133
Wallace, W. J., 130, 131, 133, 134
Wallmo, O. C., 233
Wallwork, J. A., 245
Walsberg, G. E., 408, 453
Walter, H., 201, 202, 204, 205, 210, 419
Walton, G. P., 452
Wang, L. C., 235
Warburg, M. R., 248
Warren, A., 434, 457, 481, 517
Warren, C. N., 130, 131, 133, 134
Warren, D. K., 192
Warren, J. W., 241
Warren, P. L., 194, 234
Warren, R. E., 177
Watson, B. N., 180
Watt, D., 248
Watters, D. R., 45
Wauer, R. H., 321, 363, 447
Weathers, W. W., 240
Webb, G. W., 10
Webb, P. M., 232
Webb, R. L., 509
Webb, R. W., 133
Webber, J. M., 195
Webster, J. D., 363
Wedel, W. R., 45
Weese, A. O., 241
Weide, D. L., 9, 48, 131
Weide, M. L., 9, 46, 48
Weinman-Roberts, L. J., 134
Weiss, C., 248
Wells, P. V., 3, 125, 192, 354, 361, 362
Welsh, R. G., 194
Wemple, E. M., 25
Went, F. W., 33, 194, 209
Werker, E., 209
Werner, F. G., 246, 247
West, N. E., 36, 426
Whaley, J. W., 238
Whaley, W. H., 211
Wharton, G. W., 414

Wheat, M. M., 9, 48, 130
Wheeler, G. C., 246
Wheeler, H. E., 11, 25
Wheeler, J., 246
Wheeler, S. M., 46
Wheeler, S. S., 32, 43
Whitaker, T. W., 197
White, C. H., 527, 530
White, F. N., 406
White, L. D., 194
White, R. E., 246
White, S. S., 193
Whiteside, M. C., 188
Whitford, W. B., 241
Whitford, W. G., 233, 241, 246, 247
Whittaker, R. H., 37, 38, 119, 208
Whittemore, F. W., Jr., 248
Wien, J., 211
Wiggins, I. L., 164, 192, 193, 206, 224, 443
Wilcoxon, J. S., 505
Wilkerson, W. L., 74
Wilkin, P. J., 247
Willden, R., 9
Willet, G., 237
Williams, H., 74
Williams, K. B., 195
Williams, S. C., 248
Williams, S. E., 195
Willoughby, E., 238
Wilson, E. D., 164
Wilson, R. W., 28
Winakur, J., 230, 235
Winkler, P., 245
Withers, A. M., 259
Wolfe, J. A., 29, 30
Wondallek, J. T., 235
Wood, A. E., 28
Wood, C. W., 194
Woodbury, A. M., 26, 27, 36, 242, 442
Woodell, S.J.R., 194

Woodhouse, R. M., 192, 429
Woodin, W. H., 241
Woods, F. W., 210
Woods, W. F., 35
Woodson, W., 248
Woodward, A. F., 131
Woodward, S. L., 452
Wooton, E. O., 195
Work, J., 70
Worley, D., 28
Wormington, H. M., 28, 45
Worthington, R. D., 3, 363
Wright, A. A., 243
Wright, J. W., 241
Wright, R. A., 194, 559
Wullstein, L. H., 211

Yang, T. W., 124, 194, 206, 210, 215, 220
Yarrow, H. C., 26
Yeager, L. E., 450
Yeates, N.T.M., 237
Yeaton, R. I., 210
Yermanos, D. M., 204
Yesilsoy, M., 190
Yocum, C. S., 424
Yonkers, T., 124
York, J. C., 555
Young, J. A., 30, 38
Young, R. A., 186
Young, S. P., 233
Yousef, M. K., 234, 235, 405

Zacher, F., 452
Zdendek, F. F., 21, 22
Zelitch, I., 424
Zeller, E. A., 242
Zervanos, S. M., 232, 409
Zeuthen, E., 238
Zoller, W. H., 532
Zweifel, R. G., 26, 239, 241

INDEX TO COMMON NAMES

This index contains the names of organisms appearing in some significant manner in the text of the chapters. Names appearing in the various appendixes have not been considered.

ANIMALS

Abert towhee, 448, 454
alligator, 397
Alvord chub, 79
Amargosa toad, 26
American kestrel, 80
American peregrine falcon, 28
American widgeon, 80
antelope jackrabbit, 235, 409

badger, 81
bald eagle, 448
banded gecko, 240
bark scorpion, 248
bass, 245
beaver, 81, 233, 446
bedouin goat, 414
Bell vireo, 238, 448
bighorn sheep, 80, 81, 231
blackfish, 233
black hawk, 238, 448
black rail, 448
black-tailed gnatcatcher, 448
black-tailed jackrabbit, 80, 235, 409
black toad, 26
bluegill, 245
blue-gray gnatcatcher, 454
blue grosbeak, 454
bobcat, 81, 233, 446
bonytail, 244
Brewer's blackbird, 80
brown bat, 407
brown-headed cowbird, 454
brown pelican, 236
bruchid beetle, 452
buffalo (fish), 245
bullfrog, 27, 79

burro, 125, 452
bushy-tailed woodrat, 81

cactus fly, 246
cactus mouse, 446
cactus wren, 238
California condor, 27, 239
California quail, 238
camel, 28, 407, 413, 414
Canada goose, 27, 80
canvasback, 80
carp, 245, 448
catfish, 79, 245
centipede, 248
channel catfish, 448
chipmunk, 233
chisel-toothed kangaroo rat, 81
chuckwalla, 240
cicada, 448
clapper rail, 448
clingfish, 244
clinid, 243
coatimundi, 233
collared lizard, 79, 240
collared peccary, 409
Colorado river squawfish, 244, 445
columbian ground squirrel, 81
convict cichlid, 245
Cooper's hawk, 448
cotton rat, 233, 234, 265, 446
cottontail rabbit, 265
cowbird, 238
coyote, 80, 81, 233, 438, 446
crappie, 245
Crissal thrasher, 448, 454

damselfish, 243

dark kangaroo mouse, 81
deer, 231, 232
deer mouse, 81, 446
desert bighorn sheep, 125
desert cicada, 247
desert cockroach, 414
desert cottontail, 446
desert ground squirrel, 235
desert iguana, 240
desert locust, 409
desert pocket mouse, 446
desert pupfish, 244
desert scorpion, 410
desert shrew, 236
desert toad, 415
desert tortoise, 125, 242
desert wood rat, 81
Devil's hole pupfish, 105
dolphin, 233
donkey, 407, 413
double crested cormorant, 27
Douglas' squirrel, 81

eagle, 237
eared grebe, 237
edestid shark, 25
elephant seal, 232
elf owl, 238
elk, 81

fairy shrimp, 248
ferruginous hawk, 28
fin-backed whale, 233
fish-eating bat, 236
flathead minnow, 245
free-tailed bat, 236
frigatebird, 236

Gambel's quail, 238, 265, 454
Gapper's red-backed mouse, 81
gazelle, 413
giant condor, 27
Gila chub, 244
gila monster, 240
Gila topminnow, 245
golden eagle, 80, 125
golden shiner, 245
gopher snake, 79
goshawk, 28
grasshopper mouse, 233
gray fox, 446
gray hawk, 448
gray whale, 233
gray wolf, 81
Great Basin pocket mouse, 81
Great Basin rattlesnake, 26
Great Basin skink, 26
Great Basin spadefoot, 79
greater sandhill crane, 28
great horned owl, 80
green swordtail, 245
ground sloth, 28
ground squirrel, 233
grunt, 243
guppy, 245

harbor seal, 232
Harris' hawk, 238
harvester ant, 246
harvest mouse, 233
heather mouse, 81
Hepburn's finch, 70
hog-nosed skunk, 446
hooded skunk, 233
horned lizard, 241, 406
horse, 28
house finch, 408
hualapai tiger, 247
hummingbird, 238

isopod, 248

jackrabbit, 265
jaguarundi, 233
javelina, 201, 231, 232

kangaroo rat, 207, 233
killer whale, 233
kissing bug, 247
kit fox, 81, 233

Lahontan leopard lizard, 26
leaf-cutter ant, 246
leaf-footed bug, 452
leaf-nosed bat, 236
lemming, 398

leopard lizard, 79
lesser sandhill crane, 80
little pocket mouse, 81
loach minnow, 244
longfin dace, 244, 445
long-nosed bat, 236
long-toed salamander, 79
lyre snake, 242

mammoth, 28
marsh hawk, 80
masked bobwhite quail, 238
meloid beetle, 450
Mexican stoneroller, 244
millipede, 414
Mississippi kite, 448
mite, 248
mole, 80
montane scorpion, 410
moray, 243
mosquito fish, 245
mountain lion, 233
mourning dove, 238
mule deer, 81, 232, 438, 446
musk ox, 398
muskrat, 233, 234

Nevada amazon sucker, 25
Nevada killifish, 25
Nevada pirate perch, 25
Nevada stickleback, 25
nighthawk, 90
night snake, 79, 242
northern flying squirrel, 81
northern grasshopper mouse, 81
northern shoveler, 80

ocelot, 233
olive ridley turtle, 242
opossum, 236
orange-crowned warbler, 454
Ord's kangaroo rat, 81
ornate box turtle, 407
osprey, 28

Pacific green turtle, 242
Pacific hawksbill turtle, 242
Pacific leatherback turtle, 242
Pacific loggerhead turtle, 242
Pacific treefrog, 79
pack rat, 207, 265
Panamint alligator lizard, 26
Panamint kangaroo rat, 125
parasitic jaeger, 236
peccary, 413
peninsula chipmunk, 235
perch, 79
peregrine falcon, 80

phainopepla, 238, 452
pigeon hawk, 28
pigmy mouse, 233
pintail, 79
pocket gopher, 233
pocket mouse, 233
polar bear, 398
pomarine jaeger, 236
porcupine, 81, 233, 234
porpoise, 233
prairie falcon, 27, 28, 80
pronghorn, 81, 231
pseudoscorpion, 248
pygmy goose, 27
pygmy sperm whale, 233

quail, 239

raccoon, 265, 446
razorback sucker, 245
red fox, 81
red-legged frog, 27
red shiner, 245
red squirrel, 81
red-tailed hawk, 80
red-winged blackbird, 80, 449
rice rat, 233
Richardson's water vole, 81
right whale, 233
ringtail, 233, 446
roadrunner, 237
rock squirrel, 446
rough-legged hawk, 80
ruby-crowned kinglet, 454

sage grouse, 27, 80
sage sparrow, 454
sailfin molly, 245
scorpion, 248, 409, 414
sea bass, 243
sea lion, 232
sea otter, 233
shark, 244
sharp-tailed grouse, 28
shrew, 80
skunk, 233
snow goose, 79
snowshoe hare, 80
Sonoran chub, 244
Sonoran mud turtle, 242, 446
Sonoran pronghorn, 232
Sonoran shiner, 244
southern bald eagle, 28
southwestern earless lizard, 240
spadefoot toad, 414
speckled dace, 244, 445
sperm whale, 233
sphinx moth, 407

spider, 248
spikedace, 244
spiny pocket mouse, 233
spiny softshelled turtle, 446
spotted frog, 79
spotted owl, 27
striped bass, 245
striped skunk, 446
striped whip snake, 79
summer tanager, 453
sunfish, 79
Switak's barefoot gecko, 240

tadpole shrimp, 248
tarantula, 248
tenebrionid beetle, 408, 414
termite, 247
Texas softshell turtle, 242
thick-billed parrot, 239
threadfish shad, 245
thysanura, 414
totoaba, 244
treefrog, 446
trumpeter swan, 80
tube blenny, 244
turkey vulture, 27, 80, 408
turtle, 397

Utah mountain kingsnake, 26

variable platyfish, 245
Vegas valley leopard frog, 26
verde trout, 244
verdin, 238, 454
vespertilionid bat, 236
vine snake, 242

walleye, 245
western fence lizard, 79
western grebe, 80
western rattlesnake, 79
western red-tailed skink, 26
western toad, 79
western whiptail, 79
western yellow-billed cuckoo, 28
whistling swan, 79, 80
white bass, 245
white-crowned sparrow, 454
white-faced ibis, 28, 80
white-footed mouse, 233, 446
white-fronted goose, 79
white pelican, 27, 28, 80
white-winged dove, 238, 454
wolverine, 81
wood frog, 399
wood rat, 233
worm lizard, 241
woundfin, 244
wrasse, 243

Yaqui chub, 244
Yaqui topminnow, 245
yellow bass, 245
yellow-billed cuckoo, 448, 453
yellow box turtle, 242
yellow-eared pocket mouse, 125
yellow-headed blackbird, 80, 449
yellow mud turtle, 242

zebra-tailed lizard, 240
zone-tailed hawk, 448

PLANTS

adelia, 220, 223
alder, 76
alkali sacaton, 442
alkali sacaton grass, 209
allscale, 122
alpine bluegrass, 400
alpine prickly currant, 78
alpine sorrel, 78
antelope bitterbrush, 78
Apache-plume, 438
Arizona ash, 218
Arizona grape, 220
Arizona rosewood, 208, 221
Arizona sycamore, 219
Arizona walnut, 218
arroweed, 204, 220, 263, 265, 437, 440,
    442, 445
ash, 203
awpon, 265

baccharis, 440
Baja palo verde, 229
Baltic rush, 76
banana yucca, 220
barrel cactus, 199, 205, 220, 223, 224
batamote, 204
beargrass, 198, 221
beavertail cactus, 220
bebbia, 440
beehive cactus, 216
big galleta grass, 124, 197, 215, 216
big sage, 38, 76, 78
big-tooth maple, 34
blackbrush, 41
black cottonwood, 76, 78
bladder sage, 220
blood leaf, 223
blue bonnet, 196
blue-green algae, 190
blue palo verde, 212, 215, 218, 221, 223,
    226, 438, 443
blue sand lily, 197
blue spruce, 34
bluestem grass, 209
boojum tree, 193, 202, 225

borage, 204, 216
branching cholla, 215
brickell bush, 216, 438
bristlecone pine, 32, 33, 122
brittlebush, 201, 205, 208, 210, 216, 217,
    220, 222, 223, 224, 226, 228, 229
broad-leaved arrowhead, 76
broad-leaved bursage, 206
brome grass, 197, 220
broom rape, 204
broom rush, 217
buckhorn cholla, 216, 217
buckwheat bush, 207, 215, 221, 228
bulrush, 220, 438, 449
burreed, 76
burrobrush, 215, 219, 226
burroweed, 195, 438
bur-sage, 41, 201, 204, 208, 212
bush muhly, 197, 218, 220
bush penstemon, 208, 210, 221
buttercup, 399

California juniper, 123
canaigre dock, 197, 219
candelilla, 200, 226, 228
cane cholla, 218
canescent borage, 195, 228
canescent rhatany, 218
cañutillo, 201, 216, 220
canyon hackberry, 219
cardón, 194, 206
cardón hecho, 199, 223, 225
cardón pelón, 199, 226, 228, 229
careless weed, 196, 262, 438
carrizo, 220, 265
catclaw, 117, 203, 209, 438, 440
catclaw acacia, 215, 218, 219, 223, 224
caterpillar cactus, 229
cattail, 76, 438, 442, 449
cedar, 208
century plant, 195, 198
chaparral, 208
chain fruit cholla, 194, 218, 220, 223
cheat grass, 76, 77, 78
cheesebush, 122, 438, 440
chokecherry, 76
cholla, 200, 217, 229
Christmas cholla, 220, 223
chuhugia, 265
chuparosa, 203, 215, 226
cirio, 202, 210, 227
ciruelo, 202
cliff fern, 221
climbing milkweed, 220
cloak fern, 221
cocklebur, 196
cocotera, 199
common saltbush, 438
compass barrel cactus, 194, 208, 215

copalquin, 202, 226, 228
copperleaf, 215
corn, 265
cottongrass, 399
cottonwood, 203, 218, 220, 437, 440,
    443, 458, 460, 461
covena, 197, 262
coyotillo, 224
creeping sibbaldia, 78
creeping spike-rush, 76
creosote bush, 41, 42, 112, 122, 124, 191,
    194, 202, 205, 206, 207, 208, 210,
    212, 215, 217, 218, 220, 223, 226,
    228, 429, 438, 458
crownbeard, 196
crowngall, 211
crucifixion thorn, 218, 220
crucillo, 202, 220, 223, 224
currant, 76
Cusick's draba, 78
Cusick's horsemint, 78

datil, 199
datilillo, 227, 228
Davidson's penstemon, 78
desert anemone, 197
desert broom, 203, 215, 219, 223
desert buckwheat, 216
desert hackberry, 195, 202, 215, 217, 218,
    223, 224, 225
desert hopbush, 224
desert larkspur, 197
desert lavender, 216, 217, 223, 226
desert lily, 197
desert marigold, 196
desert mescal, 216
desert mistletoe, 201, 204, 238
desert needle grass, 123
desert peach, 45
desert penstemon, 196
desert saltbush, 204, 207, 217, 218, 228,
    229
desert saltgrass, 78, 442
desert trumpet, 440
desert willow, 117, 215, 218, 219, 438,
    443, 445
desert zinnia, 218, 220
devil's claw, 196, 265
ditch grass, 226
dodder, 204, 228
dogwood, 76
Douglas fir, 34
downy oat grass, 78
dwarf desert knotweed, 78
dwarf fireweed, 400
dwarf quelite salado, 207
dye bush, 216

eelgrass, 226

elderberry, 76, 219
elephant tree, 202, 208, 216, 220, 223,
    224, 225, 226, 229
Engelmann prickly pear cactus, 221, 223
Engelmann spruce, 34
evening primrose, 196
espino, 223, 224, 225
estafiate, 228

feather acacia, 224
feather duster, 220
fern-leaf elephant tree, 202
fiddleneck, 196
finger grass, 438
fishhook barrel cactus, 217
flor de sol, 227
foothills palo verde, 205, 208, 212, 216,
    217, 218, 220, 221, 222, 223, 228
four-o-clock, 196
four-winged saltbush, 207, 215, 216, 224,
    226, 442
Fremont cottonwood, 437, 442

galleta grass, 124
gambel oak, 34, 442
garambulla cactus, 236
giant wild rye, 76, 78
globe mallow, 196
goldenhead, 220
Goodding willow, 437
goosefoot, 265
goosefoot bursage, 227, 228
grama, 458
grape ivy, 197
gray molly, 37
gray rabbitbrush, 77, 78
graythorn, 203, 215, 218, 223, 224, 228
greasewood, 76, 78, 442
green rabbitbrush, 78
ground cholla, 215
guapilla, 226
guayacan, 228
gum bush, 218

hardstem bulrush, 76
heart leaf bursage, 215, 220, 224, 225
heather, 399
hedgehog cactus, 200, 215, 217, 223
hen and chicks, 198
heno pequeno, 226, 227
hinds nightshade, 215
honey bush, 223
honey mesquite, 437, 438, 440, 442, 443,
    450
honeysweet, 440
hopbush, 208
hopsage, 37
horned pondweed, 76
horsebrush, 76, 78

horsetail, 400
hummingbird bush, 215, 223

Indian rice grass, 76, 123
Indian wheat, 196
inkweed, 438, 442, 454
iodine bush, 78, 122
ironwood, 438, 439, 443. *See also* palo
    fierro

jeffrey pine, 36
jimmyweed, 201, 218
jimson weed, 196
jito, 202, 223
jointfir, 201, 458
jojoba, 194, 202, 204, 205, 206, 207,
    208, 210, 215, 216, 220, 221, 226,
    228
Joshua tree, 104, 220
juniper, 37

kidneywood, 223, 224
ku'ukpark, 265

lambsquarters, 262
lantana, 223, 225
large-leaf bursage, 219, 223, 225, 226
large-leaf ragweed, 215, 218
limber pine, 32, 33, 122
lip fern, 221
little-leaf palo verde, 191
little rabbitbrush, 38
lodgepole pine, 34
lomboy, 203, 225, 226, 228, 229
lousewort, 399
lovegrass, 209
low sagebrush, 76

magdalena bursage, 226, 228
maguey, 227, 228
mallow, 216
mangle dulce, 204, 224, 226
mangrove, 224
maravilla, 197
mariposa lily, 197
mat saltbrush, 37
mauto, 224, 226
meadow barley, 76
mescal, 208
mesquite, 123, 125, 190, 194, 203, 204,
    206, 207, 208, 209, 210, 212, 215,
    218, 219, 220, 222, 223, 224, 226,
    229, 443, 445, 458
Mexican elderberry, 210
mistletoe, 452
Mojave saltbush, 122
Mojave yucca, 125
Mormon tea, 78
mountain hemlock, 36

mountain mahogany, 78

narrowleaf cottonwood, 442

ocotillo, 202, 205, 206, 208, 210, 215,
    216, 217, 222, 223, 226, 228, 238
octopus cactus, 200, 223, 224, 225
onk livak, 265
orange sneezeweed, 78
organ-pipe cactus, 194, 200, 210, 216,
    220, 222, 224, 229

palma, 199
palma ceniza, 199, 228
palma de taco, 199
palo adan, 222, 224, 228, 229
palo blanco, 226, 229
palo brea, 222
palo de asta, 223, 225
palo estribo, 222, 224
palo fierro, 206, 208, 209, 212, 215, 216,
    217, 218, 220, 222, 223, 226. *See also*
    ironwood
palo hierro, 229
palo santo, 203, 223, 224, 225
palo verde, 188, 190, 199, 201, 203, 206,
    207, 209, 230
panic grass, 209
paperflower, 220
paperfruit, 223, 226
passion flower, 223
patata, 262, 265
Patterson's bluegrass, 78
pencil cholla, 218, 223
peppergrass, 76
pickleweed, 43, 46, 442
pincushion cactus, 194, 200, 220, 224
pitahaya, 206, 228, 229, 236
plumeria, 203
pochote, 203, 223
ponderosa pine, 34
poppy, 196
prickly pear cactus, 194, 200, 208, 217,
    226
purple mountain saxifrage, 399
purslane, 199
pygmy cedar, 216, 440

quail bush, 437, 438, 454
quaking aspen, 34, 77
queen's wreath, 223
quelite salado, 204, 207, 224, 226

rabbitbrush, 37, 76, 438, 440
resurrection plant, 188, 221
rhatany, 203, 216, 217, 220, 223, 226

rhyolite bush, 221
rock purslane, 199
rush broom, 203, 215, 226
rushes, 442
Russian thistle, 78

sacaton grass, 197, 220, 458
sagebrush, 35, 37, 67
sago pondweed, 76, 80
saguaro, 194, 199, 205, 206, 208, 209,
    215, 217, 220, 222, 223, 431, 439
saltbush, 76, 78, 195, 201, 207, 440
salt cedar, 204, 219, 439, 440, 442, 453,
    461
saltgrass, 124, 197, 220, 442
sandpaper plant, 216
sand root, 204, 262
sangre de cristo, 203, 205, 220, 223, 224
screwbean mesquite, 437, 442, 443
scrub oak, 208, 219, 220, 221
sedge, 76, 442
seepwillow, 219, 438
senita cactus, 194, 195, 200, 211, 220,
    222, 228, 229
shadscale, 37, 112, 442
sheep fescue, 78
short-leaved cinquefoil, 78
short sagebrush, 78
shrubby bedstraw, 216, 221
shrubby cinquefoil, 78
shrubby spurge, 226
shrub mallow, 215
siempreviva, 198, 227
smoke brush, 45
smoke tree, 201, 203, 215, 443
smooth mimosa, 220, 224
snail seed, 220, 223
snow buttercup, 78
soapberry, 203, 225, 226
Sonora caper, 202, 223
sotol, 198, 208
Spanish dagger, 195
spiny gourd, 220
spiny hopsage, 76, 78
sprangletop, 438
spurge, 196, 203, 440
squirrel-tail barley, 78
staghorn cholla, 23, 218
Steen's mountain paintbrush, 78
Steen's mountain thistle, 78
stephanomeria, 440
sticky ring, 440
stingbush, 440
subalpine fir, 34
sulfur buckwheat, 78
sweet acacia, 203, 224

tanglehead grass, 197, 209, 216, 217
tarbush, 458
teddy bear cholla, 221
tepary bean, 255
thorny buffalo berry, 78
tobosa grass, 197, 216, 218
tomatillo, 207, 215, 217, 223, 226, 228
torote prieto, 203, 220, 226, 229, 261
tree beargrass, 216
tree ocotillo, 223
tree tobacco, 238
triangle-leaf bursage, 205, 215, 218, 220,
    223
trumpet tree, 203
tufted hairgrass, 78
turpentine broom, 220
turpentine bush, 216, 221

Utah juniper, 123, 220

velvet mesquite, 437, 438, 451
virgin's bower, 220, 223

walnut, 203
Washington palm, 443
Washoe pine, 35
water leaf, 196
water milfoil, 76
water plantain, 76
western chokecherry, 77
western false hellebore, 78
western honey mesquite, 117
western juniper, 77, 78
western white pine, 36
whitebark pine, 36
white bursage, 216, 217, 218, 223, 226
white fir, 32, 34, 78, 123
white rhatany, 215
whitethorn acacia, 217, 218, 223, 438
wild bean, 223
wild buckwheat, 201, 204
wild cucumber, 197
wild grape, 438
wild onion, 197
willow, 76, 77, 203, 265, 400, 437, 440,
    442, 460, 461
winter fat, 37
wolfberry, 438, 443, 453

yellow bells, 223
yerba de la flecha, 202, 215, 216
yerba del venado, 203, 217, 220
yerba reuma, 204, 226, 227

zizyphus, 217

INDEX TO SCIENTIFIC NAMES

This index contains the scientific names appearing in the text of the various chapters. The appendexes have not been considered.

Abies concolor, 32, 34, 35, 36, 37, 78, 123
Abies lasiocarpa, 34
Abies magnifica, 36
Abronia, 45
Acacia, 195, 204, 206, 207, 357, 359, 360, 361, 362, 429, 437, 443
Acacia constricta, 217, 218, 223, 226, 438
Acacia cymbispina, 223, 224
Acacia farnesiana, 203, 224
Acacia greggii, 117, 202, 209, 215, 218, 223, 224, 226, 438
Acacia occidentalis, 223
Acacia pennatula, 224
Acacia willardiana, 223, 226
Acalypha californica, 215
Acamtopappus sphaerocephalus, 220
Acanthodactylus boskianus, 415
Acarospora, 211
Accipiter cooperii, 448
Accipiter gentilis atricapillus, 28
Acer, 30
Acer glabrum, 122, 123
Achillea millefolium, 34
Acromyrmex versicolor, 246
Adolphia, 360, 361
Adolphia infesta, 362
Aechmophorus occidentalis, 80
Agastache cusickii var. cusickii, 78
Agave, 122, 195, 198, 227, 345, 359, 360, 361, 362
Agave chrysantha, 208
Agave dentiens, 226
Agave deserti, 216, 228, 431
Agave lechuguilla, 359
Agave margaritae, 29
Agave shawii, 227
Agave sobria, 226
Agave striata, 359

Agave stricta, 359
Agave utahensis, 37, 41, 134
Agelaius phoeniceus, 80, 449
Agoseris, 34
Agosia chrysogaster, 445
Agropyron, 34, 38
Agropyron scribneri, 33
Agropyron spicatum, 34
Algarobius prosopis, 452
Alisma, 76
Allenrolfea, 192, 217, 226, 227
Allenrolfea occidentalis, 40, 43, 46, 78, 122, 124, 442
Allium, 197
Alnus, 76
Alnus tenuifolia, 442
Aloina pilfera, 44
Aloysia, 360
Aloysia lycioides, 223
Amaranthus, 196
Amaranthus fimbriatus, 41
Amaranthus palmeri, 262, 265, 438
Ambrosia, 201, 204, 206, 216
Ambrosia ambrosioides, 215, 218, 223, 225, 226
Ambrosia camphorata, 228
Ambrosia chenopodifolia, 227
Ambrosia cordata, 215
Ambrosia cordifolia, 220, 224
Ambrosia deltoidea, 205, 208, 215, 218, 220, 223
Ambrosia dumosa, 42, 119, 122, 123, 212, 218, 223, 226, 228
Ambrosia eriocentra, 122
Ambrosia magdalenae, 226, 228
Ambystoma macrodactylum, 79
Amelanchier, 361
Amelanchier alnifolia, 34

Amelanchier florida, 442
Amelanchier pallida, 36
Amelanchier utahensis, 37, 42
Amitermes, 247
Ammobroma, 204
Ammobroma sonorae, 262
Ammospermophilus, 233
Amphibolurus maculosus, 412
Amphispiza belli, 454
Amsinckia, 196
Amyzon mentalis, 25
Ana bernicula oregonensis, 27
Anarbylus, 240
Anarbylus switaki, 240
Anas acuta, 79
Andropogon, 209, 362
Anemone tuberosa, 197
Anemopsis californica, 124
Anisacanthus thurberi, 215
Anisacanthus wrightii, 223
Anniella, 240
Anostraca, 248
Anser albifrons, 79
Antennaria, 34
Antennaria alpina, 32
Antigonon leptopus, 223
Antilocapra, 231
Antilocapra americana, 81
Antrozous, 236
Aplopappus, 195, 196, 201, 207, 210, 218, 226, 361
Aplopappus laricifolius, 216
Aquila chrysaetos, 80
Aquilegia scopulorum, 33
Araucaria, 29
Arbutus, 361
Arctostaphylos, 30, 35, 361
Arctostaphylos patula, 32

Arctostaphylos pungens, 37, 42
Arenaria, 33
Arenaria kingii, 33
Arenivaga, 247, 414
Aristida, 360, 362
Arizona, 241
Artemisia, 67, 362
Artemisia arbuscula, 34, 76
Artemisia frigida, 34
Artemisia nova, 42, 119
Artemisia spinescens, 40, 41, 42, 119
Artemisia tridentata, 34, 35, 37, 38, 40,
    42, 45, 76, 119, 123
Artemisia tridentata vaseyana, 35
Arvicola richardsoni, 81
Asclepias, 361
Asclepias subverticillata, 361
Aster foliaceus, 34
Asterophyllites, 29
Astragalus, 42
Astragalus funereus, 124
Astragalus panamintensis, 124
Astragalus platytropis, 33
Atamisquea emarginata, 202, 223
Athya valisineria, 80
Atriplex, 41, 43, 44, 195, 201, 217, 226,
    361, 362, 436, 438, 443
Atriplex canescens, 42, 45, 119, 122, 195,
    207, 215, 224, 226, 442
Atriplex confertifolia, 38, 39, 40, 41, 76,
    112, 119, 122, 422, 424, 425, 442
Atriplex corrugata, 39
Atriplex elegans, 207
Atriplex hastata, 44
Atriplex hymenelytra, 122
Atriplex lentiformis, 437
Atriplex nuttallii, 207
Atriplex parryi, 122
Atriplex polycarpa, 40, 122, 123, 204,
    207, 217, 218, 228, 229, 438
Atriplex spinescens, 38
Atriplex spinifera, 122
Atriplex spinosa, 76
Atriplex wrightii, 265
Auriparus flaviceps, 454
Avicennia germinans, 224, 226

Baccharis, 266, 360, 362, 437, 438, 443
Baccharis glutinosa, 204
Baccharis sarothroides, 203, 215, 219, 223
Baccharis sergiloides, 440
Baileya multiradiata, 196, 361
Baiomys, 233
Balaenoptera, 233
Barbula, 44
Bartramia umatilla, 27
Bassariscus, 233
Bassariscus astutus, 446

Batis, 226
Bebbia juncea, 203, 215, 226, 440
Berberis, 361
Berberis repens, 34
Berginia virgata, 215
Bersera laxiflora, 229
Betula, 442
Betula microphylla, 442
Betula occidentalis, 34, 37
Bipes, 240
Boerhaavia, 196
Boerhaavia annulata, 440
Bouteloua, 360, 362, 458
Bouteloua breviseta, 361
Bouteloua chasei, 361
Bouteloua erippoda, 124
Bouteloua gracilis, 124
Brahea armata, 199, 228
Brahea brandegeei, 199
Brahea roezlii, 199
Branta, 27
Branta canadensis, 27, 80
Branta esmeralda, 27
Breagyps clarki, 27
Brickellia, 362
Brickellia atractyloides, 216
Brickellia californica, 438
Brickellia desertorum, 124
Brickellia incana, 122
Brodiaea, 197
Brodiaea capitata, 262
Bromus marginatus, 34
Bromus rubens, 197, 220, 221
Bromus tectorum, 38, 40, 44, 76
Bruchus prosopis, 247
Bryum, 43, 44
Bubo virginianus, 69, 80
Buchloe, 362
Buchloe dactyloides, 361
Bufo, 242
Bufo alvarius, 243
Bufo boreas, 79
Bufo boreas exsul, 26
Bufo boreas nelson, 26
Bufo punctatus, 26, 243
Bulnesia retama, 429
Bumelia occidentalis, 218
Bursera, 191, 359
Bursera filicifolia, 202
Bursera hindsiana, 225, 226
Bursera laxiflora, 223, 224
Bursera microphylla, 202, 208, 216, 220,
    223, 224, 225, 229
Bursera odorata, 229
Buteo albonotatus, 448
Buteo jamaicensis, 80
Buteo lagopus, 80
Buteo nitidus, 448

Buteo regalis, 28
Buteogallus anthracinus, 448

Caesalpinia, 207
Caesalpinia pumila, 220, 223
Calamogrostis canadensis, 34
Calamites, 29
Calandrinia ambigua, 199
Calandrinia ciliata, 199
Calliandra, 207
Calliandra eriophylla, 220
Callipepla gambellii, 265
Callisaurus, 240
Calocedrus decurrens, 35
Calochortus kennedyi, 197
Calyptridium monandrum, 199
Camelops, 28
Camissonia, 42
Camissonia claviformis, 427, 429
Canis, 223
Canis latrans, 81, 438
Canis lupus, 81
Canotia holacantha, 123, 201, 220
Caretta, 242
Carex, 33, 76, 78, 123, 442
Carex belleri, 32
Carnegiea gigantea, 194, 199, 205, 206,
    208, 209, 215, 217, 220, 222, 223,
    431
Cassia, 359, 362
Cassia aphylla, 429
Cassia armata, 122
Cassiope mertensiana, 32
Castela, 362
Castilleja, 34
Castilleja applegatei, 33
Castilleja lapidicola, 33
Castilleja miniata, 34
Castilleja steenensis, 78
Castor, 233
Castor canadensis, 446
Castor fiber, 81
Cathartes aura, 27, 80
Catostomus, 245
Catostomus ardens, 26
Caulanthus, 42
Ceanothus, 30, 35, 37, 42
Ceanothus cordulatus, 36
Ceanothus fendleri, 34
Ceanothus velutinus, 32, 34
Ceiba acuminata, 203, 223
Celerio lineata, 247
Celtis, 360
Celtis pallida, 195, 202, 215, 218, 223,
    224
Celtis reticulata, 219
Centrocercus urophasianus, 27, 80
Centruroides sculpturatus, 248, 410

Ceratoides lanata, 38, 39, 40, 41, 42, 45, 122, 422, 424, 425
Cercidium, 188, 190, 192, 199, 201, 206, 230, 437
Cercidium floridum, 203, 212, 218, 223, 226, 429, 438
Cercidium microphyllum, 191, 203, 205, 207, 208, 209, 212, 217, 218, 220, 222, 228, 429
Cercidium peninsulare, 229
Cercidium praecox, 222
Cercidium sonorae, 203, 222, 224
Cercocarpus, 30
Cercocarpus intricatus, 124
Cercocarpus ledifolius, 35, 37, 42, 78
Cereus giganteus, 439
Cervus elaphis, 81
Chasmistes, 25
Cheilanthes, 124, 221
Chelonia, 242
Chen hyperborea, 79
Chenopodium, 42, 265
Chenopodium murale, 262
Chilomeniscus, 241
Chilopsis linearis, 117, 215, 218, 219, 362, 438
Chionactis, 241
Chiromantis, 410, 412
Chloris virgata, 438
Choeronycteris, 236
Choeronycteris mexicana, 236
Chordeiles minor, 80
Chrysolepis sempervirens, 32
Chrysothamnus, 37, 38, 44, 45, 76
Chrysothamnus graveolens, 40
Chrysothamnus nauseosus, 42, 77
Chrysothamnus paniculatus, 438
Chrysothamnus puberulus, 38
Chrysothamnus viscidiflorus, 33, 34, 35, 38, 42, 78, 119, 124
Cinnamomum, 29
Circus cyaneus, 80
Circium, 33, 34
Circium peckii, 78
Cissus trifoliata, 197
Clematis drummondii, 220, 223
Clematis pseudoalpina, 33
Clemmys marmorata, 26
Cleome sparsifolia, 41
Clethrionomys gapperi, 81
Cnemidophorus, 241
Cnemidophorus tigris, 26, 79, 241
Cocculus diversifolius, 220, 223
Coccyzus americanus, 448, 453
Coccyzus americanus occidentalis, 28
Cochemiea pondii, 228
Cocos nucifera, 199
Coldenia, 40, 362

Coldenia hispidissima, 361
Coleogyne, 37
Coleogyne ramosissima, 39, 41, 119, 123
Coleonyx, 240
Colinus virginianus ridgwayi, 238
Coluber constrictor, 26, 79
Condalia, 195, 217, 362, 437, 443
Condalia lycioides, 203, 215, 218, 223, 224, 228
Condalia warnockii, 202, 220, 223, 224
Conepatus, 233
Conepatus mesoleucus, 446
Constrictor, 241
Coragyps occidentalis, 27
Cordia parvifolia, 223
Cordia sonorae, 225
Cornus, 76, 442
Cottus beldingi, 25
Cottus extensus, 26
Cottus hairdii, 26
Cowania mexicana, 37, 42
Crataegus douglasii, 442
Crossidium, 44
Crossopteris, 29
Crossosoma bigelovii, 221
Crotalus atrox, 26
Crotalus mitchelli, 26
Crotalus viridis, 26, 79
Crotalus viridis lutosus, 26
Crotaphytus, 240
Crotaphytus collaris, 26, 79, 415
Crotaphytus wislizenii, 79
Crotaphytus wislizenii maculosus, 26
Croton sonorae, 220, 223
Cryptantha, 40
Cryptantha circumscissa, 45
Cryptoglossa verrucosa, 408
Ctenolepisma terebrans, 414
Cucurbita, 197
Cucurbita moschata, 197
Cuscuta, 204
Cuscuta veatchii, 228
Cylindropuntia, 200
Cynoscion macdonaldi, 244
Cyprinodon, 445
Cyprinodon atrorus, 446
Cyprinodon macularis, 244
Cyprinus carpo, 448
Cyrtocarpa edulis, 202

Dalea, 41, 204, 360, 361, 362
Dalea emoryi, 216
Dalea filiciformis, 361
Dalea spinosa, 201, 203, 215, 443
Dasylirion, 198, 359
Dasylirion wheeleri, 208, 258
Dasypterus, 236
Datura, 196

Delphinium nelsonii, 34
Delphinium occidentale, 34
Delphinium scaposum, 197
Dermochelys, 242
Deschampsia cespitosa, 33, 34, 78
Diadophis, 241
Diceroprocta apache, 247, 448
Didelphis, 236
Dipodomys, 233, 234, 235
Dipodomys agilis, 235
Dipodomys deserti, 235
Dipodomys heermanni, 81
Dipodomys merriami, 235
Dipodomys ordi, 81
Dipodomys spectabilis, 207, 235
Dipsosaurus dorsalis, 240, 415
Distichlis, 362
Distichlis spicata, 40, 43, 44, 124, 197, 220, 361, 442
Distichlis stricta, 38, 78, 442
Dodecatheon alpinum, 33
Dodonaea, 362
Dodonaea viscosa, 208, 224
Draba densifolia, 32
Draba oligosperma, 33
Draba sphaeroides var. cusickii, 78
Drosophila nigrospiracula, 408
Dryophyllum, 29
Dudleya, 198, 227
Dyssodia porophylloides, 217

Echeveria, 359, 362
Echinocactus visnaga, 359
Echinocarpus engelmannii, 123
Echinocereus, 122, 194, 200
Echinocereus barthelowanus, 229
Echinocereus engelmannii, 215, 223
Echinocereus grandis, 226
Echinocereus websterianus, 226
Echinomastus johnsonii, 216
Echinopepon wrightii, 220
Echinopsilon hyssopifolius, 38
Elaphe, 241
Eleocharis palustris, 76
Eleodes armata, 410
Eleodes hispilabris, 414
Elephantella attolens, 32
Elymus cinereus, 34, 41, 76, 78
Encelia, 201
Encelia californica, 427
Encelia farinosa, 122, 201, 205, 208, 210, 216, 220, 222, 224, 228, 229, 427
Encelia frutescens, 215, 217
Enhydra, 233
Eorupeta nevadensis, 26
Ephedra, 37, 38, 41, 195, 201, 217, 357, 359, 362, 458
Ephedra nevadensis, 38, 42, 119

Ephedra trifurca, 216, 220
Ephedra viridis, 78, 216
Epicauta arizonica, 450
Epilobium, 36
Eptesicus, 236
Eptesicus fascus, 407
Equus, 28
Equus asinus, 452
Eragrostis, 209, 362
Eragrostis obtusiflora, 361
Erethizon, 233
Erethizon dorsatum, 81
Eretmochelys, 242
Eridiphas, 241
Erigeron, 33, 34, 42
Erigeron clokeyi, 33
Erigeron leiomerum, 33
Erigeron pygmaeus, 33
Erigeron trifidus, 32
Eriogonum, 40, 42, 124, 195, 201, 204
Eriogonum deserticola, 216
Eriogonum fasciculatum, 207, 215, 221, 228
Eriogonum gracilipes, 33
Eriogonum inflatum, 440
Eriogonum neglectum, 33
Eriogonum ovalifolium, 33, 45
Eriogonum umbellatum var. glaberrimum, 78
Erioneuron, 362
Errazurizia megacarpa, 226
Erwinia carnegieana, 211, 254
Eschrichtius, 233
Eschscholtzia, 196
Eubalaena, 233
Eucnide urens, 440
Euderma, 236
Eumeces gilberti rubricaudatus, 26
Eumeces skiltonianus utahensis, 26
Eumops, 236
Euphagus cyanocephalus, 80
Euphorbia, 41, 196, 359, 440
Euphorbia antisyphilitica, 359
Euphorbia californica, 229
Euphorbia colletiodes, 203
Euphorbia magdalenae, 229
Euphorbia misera, 226
Euphorbia xanti, 203
Eutamias, 233
Eysenhardtia, 207, 360
Eysenhardtia orthocarpa, 223, 224

Fagonia californica, 216
Falco columbarius, 28
Falco mexicanus, 27, 28, 80
Falco peregrinus, 80
Falco peregrinus anatum, 28
Falco tinnunculus, 80

Fallugia paradoxa, 37, 438
Felis, 232
Felis rufus, 446
Ferocactus, 122, 224, 359
Ferocactus acanthodes, 41, 194, 199, 208, 215
Ferocactus chrysacanthus, 199, 228
Ferocactus covillei, 199, 223
Ferocactus diguetii, 226
Ferocactus gatesii, 226
Ferocactus gracilis, 199
Ferocactus johnstonianus, 226
Ferocactus wislizenii, 199, 205, 217
Festuca, 34
Festuca ovina, 33, 78
Ficimia, 241
Ficus, 29
Flaveria, 362
Flaveria oppositifolia, 361
Flourensia, 362
Flourensia cernua, 357, 458
Fomes robustus, 211
Forchammeria watsoni, 202, 223
Forestiera neomexicana, 220, 223
Fouquieria, 195, 202, 210, 359, 362
Fouquieria diguetii, 224, 226, 228, 229
Fouquieria macdougalii, 222
Fouquieria splendens, 205, 206, 208, 215, 217, 222, 224, 226, 228, 238
Frankenia, 362
Frankenia palmeri, 226
Franseria, 362. *See also* Ambrosia
Franseria dumosa, 41
Franseria speciosa, 33, 34
Frankenia palmeri, 204
Fraxinus, 203
Fraxinus velutina, 218
Funaria, 43, 44
Funastrum, 220
Fundulus lariversi, 25
Fundulus nevadensis, 25
Fusarium oxysporum, 211

Galium stellatum, 216, 221
Gambusia affinis, 245
Garotia lanata, 119
Gasterosteus doryssus, 25
Gastrophryne, 242
Geonomites, 29
Geranium fremontii, 34
Gerrhonotus, 240
Gerrhonotus panamintinus, 26
Gila atrarium, 26
Gila bicolor, 25
Gila ditaenia, 244
Gila elegans, 244
Gila esmeralda, 25
Gila intermedia, 244

Gila robusta, 244
Gila purpurea, 244
Gila traini, 25
Gilia, 40, 41, 42
Gilia aggregata, 34
Gilia alvordensis, 79
Gilia leptomeria, 45
Ginkgo, 29
Glaucomys sabrinus, 81
Globicephala, 233
Glyptopleura, 40
Gnathamitermes perplexus, 247
Gopherus, 242
Gopherus agassiz, 125
Graptopelatum, 198
Grayia spinosa, 39, 40, 41, 42, 45, 119
Grimmia, 44
Grus canadensis, 80
Grus canadensis tabida, 28
Guaiacum coulteri, 224
Guiraca caerulea, 454
Gulo gulo, 81
Gutierrezia, 262
Gutierrezia microcephala, 44
Gutierrezia sarothrae, 39, 119
Gymnogyps californianus, 27, 239

Hadrurus arizonensis, 248, 410, 412
Haliaeetus leucocephalus, 448
Haliaeetus leucocephalus leucocephalus, 28
Haliotus, 132
Halogeton glomeratus, 39
Haplopappus, 41
Haplopappus acaulis, 33
Haplopappus suffruticosus, 33
Hechtia, 359, 362
Hechtia glomerata, 359
Hechtia montana, 226
Helenium hoopesii, 78
Helicoprion nevadensis, 25
Heliotropium curassivacum, 38
Heloderma, 240
Heloderma suspectum, 26
Hermidium alipes, 40
Hesperaloe, 359
Hesperaloe sonorensis, 198
Hesperocaulis undulata, 197
Heteropogon, 360
Heteropogon contortus, 197, 209, 216
Heterotermes aureus, 247
Hibiscus denudatus, 216
Hilaria, 195, 362
Hilaria jamesii, 124
Hilaria mutica, 197, 216, 218
Hilaria rigida, 124, 197, 215
Hoffmanseggia, 192
Holacantha, 192
Holacantha emoryi, 201, 218, 357, 362

Holbrookia, 240
Holbrookia maculata, 446
Holodiscus, 35
Holodiscus dumosus, 34, 35
Hordeum brachyantherum, 76
Hordeum jubatum, 78
Horsfordia alata, 215
Hulsea algida, 32
Hulsea caespitosa, 32
Hulsea nana, 33
Hyla, 446
Hyla regilla, 79
Hymenoclea, 443
Hymenoclea monogyra, 215, 219, 438
Hymenoclea pentalepis, 215, 226
Hymenoclea salsola, 122, 438
Hypsiglena, 241
Hypsiglena torquata, 79
Hyptis emoryi, 216, 223, 226

Ictalurus punctatus, 448
Ictinia mississippiensis, 448
Idria, 210
Idria columnaris, 193, 202, 225, 227
Ipomaea arborescens, 203, 223, 224
Iresine interrupta, 223
Ivesia gordonii, 33

Jacobinia californica, 215, 226
Jacquinia pungens, 202
Janusia gracilis, 224
Jatropha, 191, 203, 362
Jatropha cardiophylla, 205, 220, 223, 224
Jatropha cinerea, 203, 225, 228, 229
Jatropha cordata, 203, 220
Jatropha cuneata, 203, 216, 220, 226, 229
Jouvea, 226
Juglans, 29, 30, 203
Juglans major, 218
Juncus, 123, 442
Juncus balticus, 76
Juncus cooperi, 122, 124, 442
Juncus drummondii, 33
Juncus parryi, 32
Juniperus, 208, 361, 362
Juniperus californica, 123, 124
Juniperus communis, 32, 122
Juniperus occidentalis, 77
Juniperus osteosperma, 42, 119, 123, 124
Juniperus utahensis, 44, 220
Justicia californica, 203

Karwinskia, 359
Karwinskia humboldtiana, 224
Kinosternon, 242
Kinosternon sonoriense, 446
Kochia, 39
Kochia americana, 39

Koeberlinia spinosa, 201, 218, 357
Kogia, 233
Krameria, 192, 203, 207, 216, 217, 220, 359
Krameria grayi, 215, 218
Krameria parvifolia, 42, 223, 226

Lagascea, 226
Lagenorhynchus, 233
Laguncularis racemosa, 224, 226
Lampropeltis, 241
Lampropeltis getulus, 26
Lampropeltis pyromelana, 26
Lampropeltis pyromelana infralabialis, 26
Lantana horrida, 223
Larrea, 42, 192, 217, 359, 361, 362
Larrea cuneifolia, 422, 425
Larrea divaricata, 41, 357, 422
Larrea tridentata, 112, 119, 122, 123, 124, 191, 202, 205, 206, 207, 208, 212, 217, 223, 228, 422, 423, 429, 438
Lasiurus, 236
Laterallus jamaicensis, 448
Lemaireocereus, 360
Lepidium, 76
Lepidochelys, 242
Lepidodendron, 29
Leptochloa fascicularis, 438
Leptodactylon pungens, 42
Leptolepis nevadensis, 25
Leptotyphlops, 241
Leptonycteris, 236
Lepus, 235
Lepus alleni, 265, 409
Lepus americanus, 80
Lepus californicus, 80, 265, 409
Leuciscus turneri, 25
Leucophyllum, 359
Leuchtenbergia, 359
Leucopoa kingii, 33
Leucosticte tephrocotis, 70
Lewisia pygmaea, 33
Lichanura, 241
Lindleyella mespiloides, 362
Liomys, 233, 235
Lithophyagma parviflora, 34
Lophocereus, 194
Lophocereus schottii, 195, 200, 220, 222, 228
Lophortyx shotwelli, 27
Lupinus, 34, 42, 196
Lupinus argenteus, 33
Lycium, 42, 195, 207, 219, 228, 360, 362, 437, 438, 453
Lycium andersonii, 42, 119, 122, 124, 215
Lycium brevipes, 223
Lycium cooperi, 40

Lycium fremontii, 226
Lycium pallidum, 42, 119
Lycium shockleyi, 42, 119
Lycurus, 362
Lynx, 232
Lynx rufus, 81
Lyrocarpa linearifolia, 226
Lysiloma, 203, 207
Lysiloma candida, 226, 229
Lysiloma divaricata, 224, 226

Machaeranthera gracilis, 429
Machaerocereus eruca, 229
Machaerocereus gummosus, 200, 228, 229
Macrotis, 236
Magnolia, 29
Mahonia, 30
Mammilaria, 194, 200, 224
Mammilaria cerralboa, 226
Mammilaria estebanensis, 226
Mammilaria goodridgei, 228
Mammilaria insularis, 226
Mammilaria microcarpa, 220
Mammilaria multidigitata, 226
Mammilaria tayloriorum, 226
Mammuthus columbi, 28
Manducta sexta, 407
Marah gilensis, 197
Mareca americana, 80
Mariopteris, 29
Mascagnia macroptera, 223, 226
Masticophis, 241
Masticophis flagellum, 26
Masticophis taeniatus, 26, 79
Maytenus phyllanthoides, 204, 224, 226
Megaptera, 233
Megathura, 132
Menispermum, 29
Menodora, 362
Mentzelia, 42
Mephitis, 233
Mephitis mephitis, 446
Mesembryanthemum, 227
Microdipodops megacephalus, 81
Micruroides, 241
Mimosa, 359, 362, 443
Mimosa laxiflora, 220, 224
Mimulus implexus, 32
Miopelodytes gilmorei, 26
Mirabilis bigelovii var. retrorsa, 78
Mirabilis multiflora, 197
Mirabilis pudica, 42
Molothrus ater, 454
Monanthochloe, 226
Monardella odoratissima, 34
Monolepis nuttallii, 262, 265
Mormoops, 236
Mortonia, 359, 362

Mozena obtusa, 452
Muhlenbergia, 34, 360, 362
Muhlenbergia porteri, 124, 197, 218, 220
Muhlenbergia purpussi, 361
Muhlenbergia richardsoni, 32
Myotis, 236
Myriophyllus spicatum, 76
Myrtillocactus, 236, 357, 360
Myrtophyllum, 29
Mytilus, 132

Nasua, 233
Neotoma, 233, 234
Neotoma albigula, 207, 265
Neotoma cinerea, 81
Neotoma lepida, 81
Nettion bunkeri, 27
Neuropteris, 29
Nicotiana, 195
Nicotiana glauca, 238
Nitrophila occidentalis, 199
Nolina, 122, 198, 221, 359
Nolina bigelovii, 216
Nortbrotherium shastense, 28
Notholaena, 221, 362
Notholaena sinuata, 210
Notholaena sinuata var. cochisensis, 361
Notiosorex, 236
Notostraca, 248
Nymphaeites, 29

Odocoileus, 231
Odocoileus hemionus, 81
Odontoloxozus longicornis, 246
Oenothera, 41, 196
Oligomeris linifolia, 41
Olivella, 132
Olneya, 191, 207
Olneya tesota, 206, 208, 209, 212, 217,
    218, 222, 226, 438
Olor buccinator, 80
Olor columbianus, 79
Ondatra, 233
Onychomys, 233
Onychomys leucogaster, 81
Onymacris plana, 408
Opuntia, 40, 41, 42, 122, 217, 345, 357,
    360, 361, 362
Opuntia acanthocarpa, 216
Opuntia arbuscula, 218, 223
Opuntia basilaris, 220
Opuntia bigelovii, 221, 226
Opuntia brevispina, 226
Opuntia cholla, 226, 229
Opuntia clavellina, 226, 228
Opuntia echinocarpa, 215
Opuntia fulgida, 194, 218, 223, 224, 262
Opuntia leptocaulis, 220, 223

Opuntia phaeacantha, 208, 223
Opuntia pycnantha, 229
Opuntia ramosissima, 215, 226
Opuntia santamaria, 229
Opuntia spinosior, 218
Opuntia tesajo, 226
Opuntia thurberi, 223, 224, 225
Opuntia versicolor, 218, 223
Opuntia wrightiana, 215
Orcinus, 233
Orobanche, 204
Orthoporus ornatus, 414
Orthosphenia mexicana, 362
Oryctes nevadensis, 45
Oryzomys, 233
Oryzopsis, 38
Oryzopsis hymenoides, 40, 42, 76, 119,
    123, 124, 134
Ovis, 231
Ovis canadensis, 80, 81, 438
Ovis canadensis nelsoni, 125
Oxybelis, 241
Oxyria digyna, 32, 78
Oxytheca dendroidea, 45

Pachystima myrsinites, 34
Pachycereus, 194, 206
Pachycereus pecten-aboriginum, 199, 223
Pachycereus pringlei, 199, 226, 228, 229
Pachycormus, 191
Pachycormus discolor, 202, 226, 228
Pandion haliaetus carolinensis, 28
Panicum, 209, 362
Pantosteus, 245
Paraneotermes simpliciocornis, 247
Parmelia, 211
Parosela polyadenia, 45
Parthenium, 361
Parthenium argentatum, 359
Paruroctonus aquilonalis, 409, 414
Paspalum, 362
Passiflora mexicana, 223
Pecari tajacu, 201
Pedilanthus macrocarpus, 200, 226, 228
Pedioecetes phasianellus columbianus, 28
Peganum, 362
Peganum harmala, 361
Pelarais, 241
Pelecanus erythrorhynchos, 27, 28, 80
Pelecyphora, 359
Pellaea, 221
Penstemon, 33, 36, 124
Penstemon cedrosensis, 228, 238
Penstemon davidacnii, 32
Penstemon davidsonii var. praeteritus, 78
Penstemon microphyllus, 208, 210, 221
Penstemon parryi, 196
Penstemon rydbergii, 34

Pepsis, 246
Perityle emoryi, 124
Perognathus, 233, 235
Perognathus longimembris, 81
Perognathus parvus, 81
Perognathus penicillatus, 446
Peromyscus, 233, 234
Peromyscus eremicus, 234, 446
Peromyscus leucopus, 446
Peromyscus maniculatus, 81, 446
Petalonyx, 195
Petalonyx thurberi, 216
Petriwollelo, 440
Peucephyllum schottii, 124, 216, 440
Phacelia, 196
Phacelia alpina, 33
Phacelia bicolor, 45
Phacelia fremontii, 41
Phacelia frigida, 32
Phainopepla nitens, 453
Phalacrocorax auritus, 27
Phalacrocorax macropus, 27
Phaseolus acutifolius var. latifolius, 255
Phaseolus atropurpureus, 223
Phaulothamnus, 224
Phaulothamnus spinescens, 215
Phenacomys intermedius, 81
Phlox, 33
Phlox covillei, 33
Phlox dejecta, 32
Phoca, 232
Phocoena, 233
Phoenicopterus copei, 27
Phoenix dactylifera, 199
Phoradendron californicum, 201, 204, 238
Phoradendron villosum, 204
Phragmites, 448
Phragmites australis, 123
Phragmites communis, 220, 265, 442
Phrynosoma, 241
Phrynosoma coronatum, 406
Phrynosoma platyrhinos, 26
Phyllodactylis, 240
Phyllomedusa, 410, 412
Phyllomedusa sauvagei, 410
Phyllorhynchus, 241
Phymototrichum omnivorum, 211
Physeter, 233
Physocarpus malvaceus, 34
Picea engelmannii, 34
Picea pungens, 34
Pilio aberti, 448
Pinus, 359, 361, 362
Pinus albicaulis, 36
Pinus contorta, 35, 36
Pinus contorta var. latifolia, 34
Pinus flexilis, 32, 34, 122, 123
Pinus jeffreyi, 35

Pinus lambertiana, 35
Pinus longaeva, 3, 4, 122
Pinus longaeva ar istata, 32
Pinus monophylla, 35, 36, 42, 119, 122, 123, 134
Pinus monticola, 35, 36
Pinus ponderosa, 34, 35, 37, 69
Pinus washoensis, 35
Pipistrellus, 236
Piranga rubra, 453
Piscidia mollis, 224
Pithecellobium, 207
Pithecellobium sonorae, 224, 225, 226
Pituophis catenifer, 26
Pituophis melanoleucus, 79
Pituophis melanoleucus catenifer, 410
Pizonyx, 236
Pizonyx vivesi, 236
Plantago, 196
Plantago insularis, 210
Platanus wrightii, 219
Platyopuntia, 201
Plecotis, 236
Plegadis chihi, 28, 80
Pluchea, 204, 217
Pluchea sericea, 123, 204, 220, 263, 265
Plumbago scandens, 225
Plumeria, 203
Plumeria acutifolia, 203
Poa, 34, 38, 44
Poa pattersonii, 78
Poa pratensis, 34
Poa rupicola, 33
Poeciliopsis, 245
Polemonium montrosense, 32
Polemonium pulcherrimum, 32
Polioptila caerulea, 454
Polioptila melanura, 448
Polygonum heterosepalum, 78
Polygonum histortoides, 33
Populus, 30, 37, 203, 220, 436
Populus angustifolia, 442
Populus fremontii, 123, 204, 437
Populus grandifolia, 429
Populus macdougallii, 123
Populus tremuloides, 24, 36, 77, 123, 218
Populus trichocarpa, 76
Poria carnegiea, 211
Porophyllum gracile, 203, 217, 220
Portulaca suffrutescens, 199
Portulaca lanceolata, 199
Potamogeton pectinatus, 76, 80
Potentilla, 34
Potentilla brevifolia, 78
Potentilla crinita, 33
Potentilla fruticosa, 78
Pottia, 43
Pottia nevadensis, 44

Primula nevadensis, 33
Primula parryi, 33
Primula suffrutescens, 32
Proboscidea, 196
Proboscidea parviflora, 265
Procyon, 233
Procyon lotor, 265, 446
Prosopis, 190, 192, 194, 203, 204, 217, 226, 357, 359, 360, 362, 437, 451
Prosopis glandulosa, 123, 429, 431, 437, 452
Prosopis juliflora, 41, 117, 125, 133, 134
Prosopis palmeri, 229
Prosopis pubescens, 41, 123, 124, 437
Prosopis torreyana, 229
Prosopis velutina, 206, 208, 209, 212, 217, 218, 222, 224, 437, 452
Prosopium gemmiforum, 26
Prosopium spilonotus, 26
Protophyllocladus, 29
Prunus, 30, 36, 76
Prunus andersonii, 42, 45
Prunus fasciculata, 37
Prunus virginiana, 34, 77
Pseudemys, 242
Pseudorca, 233
Pseudotsuga menziesii, 34
Pseudotsuga menziesii var. glauca, 37
Psilostrophe cooperi, 220
Psilostrophe gnaphalodes, 361
Psorothamnus fremontii, 42
Psorothamnus kingii, 45
Pternohyla, 242
Pterygoneurum, 43, 44
Ptychocheilus lucius, 445
Purshia, 30, 37, 38
Purshia glandulosa, 42
Purshia tridentata, 34, 78

Quercus, 35, 37, 361, 362
Quercus arizonica, 219
Quercus emoryi, 219
Quercus gambelii, 34, 37, 42, 442
Quercus grandidentatum, 34
Quercus hypoleucoides, 219
Quercus oblongifolia, 219
Quercus turbinella, 37, 123, 208, 219, 220
Quercus vaccinifolia, 32

Rallus longirostris, 448
Rana, 242, 446
Rana aurora, 27
Rana catesbiana, 79
Rana johnsoni, 26
Rana pipiens, 26, 243, 446
Rana pipiens fisheri, 26
Rana plax, 26
Rana pretiosa, 79

Rana sylvatica, 399
Randia echinocarpa, 225
Randia obcordata, 224
Randia thurberi, 223
Ranunculus escholtzii, 78
Ranunculus oxynotus, 32
Rathbunia, 200
Rathbunia alamosensis, 223, 225
Regulus calendula, 454
Reithrodontomys, 233
Rhea americana, 406
Rhinichthys osculus, 445
Rhinocheilus, 241
Rhizoctonia solani, 211
Rhizophora mangle, 224, 226
Rhodea, 29
Rhodiola integrifolia, 32
Rhonobatos productus, 244
Rhus, 30, 360, 361
Rhynopsitta pachyrhyncha, 239
Ribes, 36, 37, 38, 76, 442
Ribes cereum, 33, 34, 122
Ribes inebrians, 32
Ribes montigenum, 34, 78
Rosa woodsii, 34
Roseocactus, 359
Rubus parviflorus, 34
Rudbeckia occidentalis, 34
Rumex hymenosepalus, 197, 219
Rumex venosus, 45
Ruppia maritima, 226

Sabalites, 29
Sagittaria cuneata, 76
Salazaria mexicana, 220
Salicornia, 43, 226
Salicornia rubra, 40, 442
Salicornia utahensis, 40, 122, 442
Salix, 29, 30, 34, 36, 37, 76, 77, 78, 203, 436, 440
Salix caespitosa, 32
Salix exigua, 123, 442
Salix gooddingii, 123, 204, 219, 265, 437, 442
Salix lasiolepis, 123, 442
Salix nigra, 442
Salmo clarki, 25, 26
Salmo cyniclope, 25
Salmo esmeralda, 25
Salsola kali, 38, 78
Salsola pestifer, 44
Salvadora, 241
Salvelinus, 25
Salvia, 41, 134
Sambucus, 37, 76, 210
Sambucus cerulea, 34
Sambucus mexicana, 219
Sambucus racemosa, 34

Sanvitalia, 362
Sapindus, 203
Sapindus saponaria, 225, 226
Sapium biloculare, 202, 215
Sarcobatus, 43, 44
Sarcobatus baileyi, 39
Sarcobatus vermiculatus, 39, 40, 42, 76, 442
Sarcodes sanguinea, 35
Sator, 240
Sauromalus obesus, 26, 240, 414
Saxifraga caespitosa, 33
Scaphieri alexanderi, 26
Scaphiopus, 242
Scaphiopus couchii, 243, 414
Scaphiopus hammondi, 243
Scaphiopus intermontanus, 79
Sceloporus, 240
Sceloporus jarrovi, 409
Sceloporus occidentalis, 69, 79
Scirpus, 220, 438, 449
Scirpus acutus, 76
Schinus molle, 360
Scolopendra heros, 248
Scrophularia lanceolata, 34
Scytonema, 190
Selaginella, 221, 359
Selaginella arizonica, 188
Selaginella densa, 33
Sesuvium, 361
Sequoia, 29
Shepherdia argentee, 78
Sibbaldia, 233
Sibbaldia procumbens, 78
Sigmodon, 233
Sigmodon hispidus, 265, 446
Simmondsia, 191
Simmondsia chinensis, 194, 202, 205, 206, 207, 208, 215, 220, 228
Sitanion, 38
Sitanion hystrix, 42, 124
Smilacina, 34
Solanum, 362
Solanum Hindsianum, 215
Solidago spectabilis, 41
Sonora, 241
Sophora, 359, 361
Sorbus scopulina, 34
Sorex, 80
Sparganium eurycarpus, 76
Spartina spartinae, 361
Spatula clypeata, 80
Spermophilus, 233, 234
Spermophilus columbianus, 81
Spermophilus harrisi, 235
Spermophilus tereticaudus, 235
Spermophilus variegatus, 446
Sphaeralcea, 42, 195, 196

Sphaeralcea ambigua, 40, 42
Sphenopteris, 29
Spilogale, 232
Spiraea densiflora, 36
Spizaetus willetti, 27
Sporobolus, 209, 220, 226, 362, 458
Sporobolus airoides, 40, 44, 122, 123, 124, 442
Sporobolus nealleyi, 361
Sporobolus pyramidatus, 361
Sporobolus wrightii, 197, 220, 361
Stanleya, 42
Stanleya elata, 45
Stegnosperma, 224
Stegnosperma halimifolium, 215, 226
Stenocereus, 206, 236
Stenocereus thurberi, 194, 200, 210, 216, 220, 222, 229
Stephanomeria parryi, 42, 440
Stipa, 38, 360
Stipa columbiana, 34
Stipa comata, 42
Stipa speciosa, 123, 124
Strix occidentalis, 27
Stylophyllum attenuatum, 199
Suaeda, 43, 123, 204, 207, 217, 224, 226, 361, 362
Suaeda fruticosa, 122
Suaeda ramosissima, 226
Suaeda torreyana, 41, 122, 438, 454
Sylvilagus, 235
Sylvilagus audubonii, 265, 446
Symphoricarpos, 35, 37, 38
Symphoricarpos longiflorus, 42
Symphoricarpos oreophilus, 34

Tabebuia palmeri, 203
Tadarida, 236
Talinum paniculatum, 199
Tamarix, 123, 204, 217, 219
Tamarix chinensis, 439
Tamarix pentandra, 217
Tamiasciurus douglasii, 81
Tamiasciurus hudsonicus, 81
Tantilla, 241
Tayassu, 231
Tayassu tajacu, 409
Taxidea, 232
Taxidea taxus, 81
Tecoma stans, 223
Teratornis, 27
Teratornis incredibilis, 27
Terrapene ornata, 408
Tessaria sericea, 437
Tetradymia, 37, 38, 41, 76
Tetradymia axillaris, 40, 42
Tetradymia comosa, 45
Tetradymia glabrata, 39, 42, 122

Thamnophis, 26, 241, 446
Thamnosma montana, 210, 220
Thomomys, 233, 234
Tidestromia oblongifolia, 122, 440
Tilapia, 245
Tillandsia recurvata, 226, 227, 359
Tiquilia, 204, 216
Tiquilia canescens, 195, 228
Tiquilia nuttallii, 45
Tiquilia palmeri, 216
Tivella, 132
Tortula, 43
Tortula ruralis, 44
Toxostoma dorsale, 448
Trapa, 29
Trebouxia, 211
Trianthema portulacastrum, 265
Triatoma, 247
Tricerma, 226
Trichophanes hians, 25
Trifolium, 34
Trifolium monoense, 33
Trimorphodon, 241
Trionyx, 242
Trionyx spinifer, 446
Trisetum spicatum, 78
Triteliopsis palmeri, 197
Trixis californica, 208, 216, 217
Tsuga mertensiana, 36
Tursiops, 233
Typha, 123, 217, 449
Typha domingensis, 438
Typha latifolia, 76

Uma, 240
Uroctonus apacheanus, 410
Urocyon, 233
Urocyon cinereoargenteus, 446
Urosaurus, 240
Urosaurus ornatus, 241
Uta, 240, 241
Uta stansburiana, 241

Vauquelinia, 195
Vauquelinia californica, 208, 221
Vejovis, 248
Veratrum californicum, 78
Verbesina, 360
Verbesina encelioides, 196
Veromessor pergandei, 246
Verticillium albo-atrum, 211
Vermivora celata, 454
Viguieria deltoidea, 228
Viola, 34
Viola nephrophylla, 33
Vireo bellii, 448
Vitis, 438
Vitis arizonica, 220

Vizcainoa geniculata, 228
Vulpes, 233
Vulpes velox, 81
Vulpes vulpes, 81

Washingtonia, 195
Washingtonia filifera, 123, 199, 443
Washingtonia robusta, 199
Wyethia amplexicaulis, 34

Xanthium, 196, 361

Xanthocephalus xanthocephalus, 80, 449
Xantusia, 241
Xantusia vigilis, 241

Yucca, 40, 122, 195, 198, 209, 357, 359,
    360, 361, 362
Yucca baccata, 37, 123, 220
Yucca brevifolia, 41, 42, 119, 122, 123,
    124, 220
Yucca carnerosana, 359
Yucca schidigera, 41, 125

Yucca valida, 227

Zalophus, 232
Zannichellia palustris, 76
Zea mays, 265
Zenaida asiatica, 454
Zinnia, 357, 359, 362
Zinnia grandiflora, 218
Zizyphus sonorensis, 225
Zonotrichia leucophrys, 454
Zostera marina, 226

SUBJECT INDEX

This index includes materials from the texts and appendixes of the various chapters.

Adaptations, animals, 3;
—to cold, 397, 408; supercooling, 408
—to heat: cardiovascular adjustments, 406; countercurrent heat exchange, 406; differential blood shunting, 406; evaporative heat loss, 407; hyperthermia, 406; surface coloration, 408; temperature balance, 405; thermal tolerance, 405; thermo-regulatory control, 406; torpor, 409; vasodilation, 407
—water balance, 405: dehydration, 406, 409; dehydration tolerance, 413, 414; epicuticular lipids, 410; excretory water loss, 412, 413; integumentary transpiration, 410; respiratory transpiration, 412; water relations, 409; water uptake, 414
Adaptations, plants: 419; carbon capture, 423; carbon dioxide flux, 424; competition, 210; dormancy, 210; drought avoidance, 422; drought enduring, 429; drought evasion, 428; drought resistance, 422; drought tolerance, 422; ecotypic variation, 210; elevation influences, 208; energy capture, 423; ephemerals, 427; form and structure, 427; germination, 209, 451; growth regulators, 210; leaf angle, 425; leaf shape, 425; light intensity, 424; mesophyll resistance, 424; mineral availability, 420; mineral nutrition, 426; natural selection, 419; pH, 207; photosynthesis, 423; CAM photosynthesis, 425, 431; C_3 photosynthesis, 424; C_4 photosynthesis, 424; regeneration, 419; resource availability, 419; site orientation, 208; soil relations, 206,

207; succulents, 425, 431; survival, 209; transpiration, 112, 420; turgor pressure, 421; water economy, 420; water relations, 205; water stress, 422
Aerial Photography Field Office, 525
Age of deserts, 2; Chihuahuan, 354; desert islands, 173, 174; Sonoran, 192
Aguascaliente, 322
All American canal, 460
Admundsen, Roald, 402
Amphibians: fossil, 26; in Arctic desert, 399; in Big Bend National Park, 373; in Chihuahuan desert, 467; in Death Valley National Monument, 152; in Great Basin desert, 26, 79, 96, 97, 467; in Joshua Tree National Monument, 155; in Mojave desert, 125, 467; in riparian communities, 446; in Sonoran desert, 317, 318, 467; water loss in, 410
Anasazi, 48, 361
Animals: in Arctic desert, 398; in Chihuahuan desert, 362, 363; in Great Basin desert, 25–29, 79, 96, 97, 467; in Mojave desert, 125, 129, 148–62; Northern Great Basin desert, 79–81, 96–102; refuges, 67, 69, 79, 404; in Sonoran desert, 229–48, 317, 467. *See also* Adaptations, animals; Mammals
Arachnids, 248
Archeological sites, 136, 397: Amy's Shelter, 46; Black Rock Caves, 46; Calico, 130; Conaway, 48; Council Hall, 46; Crypt Cave, 27; Danger Cave, 27, 46; Deadman Cave, 46; Deer Creek Cave, 46; Dirty Shame Rock Shelter, 46; Eastgate Shelter, 46; Etna Cave, 46; Fort Rock Cave, 46, 67; Gatecliff Shelter, 46; Gypsum Cave, 26, 27, 28, 45; Hidden Cave, 46; Hogup Cave, 27, 46; Kachina

Cave, 46; Last Supper Cave, 46; Leonard Rock Shelter, 46; Newark Cave, 46; Newberry Cave, 132; O'Malley Shelter, 46, 48; Promontory Cave, 46; Rancho La Brea pits, 27; Smith Creek Cave, 26, 27, 45, 46; South Fork Shelter, 46; Swallow Shelter, 46; Tule Springs, 28, 45; Ventana Cave, 232, 249; Wagon Jack Shelter, 46
Archeology, in Mojave desert, 130–37
Arctic desert: animals, 398; climate, 394–407; definition of, 383; glacial history, 387–88; land forms and frost processes, 388–91; location, 383; man, 400–402; permafrost, 388; plants, 399–400; regional geology, 391–94; regional physiography, 383–87; soil, 397
Arctic National Wildlife Refuge, 404
Arid lands research, 383, 525, 526, 537, 541, 543, 544, 545, 559
Aridity index. *See* Indexes of dryness
Arroyo formation, 456
Arroyo Socorro, 504
Ash Meadows, 22
Aztec Empire, 321

Bad Water, 22
Basins: Alvord, 67, 69; Big Bend, 349; Bridger, 48; Cuatro Cienegas, 446; Great, 7; Harney, 67; Panuco, 338; Pecos, 334; Pennsylvanian Ely, 11; San Simon, 458; Toyah, 334; Uintah, 50
Bering Strait, 400, 402
Bering, Vitus, 401
Big Bend National Park, 349; checklists of vertebrates, 373–82
Biosphere reserves, 551–52
Birds: breeding density, 448, 451; in Arctic desert, 398; in Big Bend National Park,

376–78; in Chihuahuan desert, 363, 447, 470–79; in Death Valley National Monument, 148–51; fossil, 27; in Great Basin desert, 27, 28, 79, 97–101, 447, 470–79; in Joshua Tree National Monument, 159–62; in Mojave desert, 125, 446, 470–79; migration of, 448; physiology of, 237; rare and endangered species, 28; in riparian communities, 446, 447; in Sonoran desert, 303–11, 447, 470–79; water loss in, 410

Bolsons, 339; de Cedros, 339; de Coahuila, 339; del Diablo, 339; de los Lipanes, 339; de Mapimi, 339; de Monclova, 339; de Parras, 339; del Salado, 339; de Viesca, 339; San Carlos, 339; White Sands, 339

Boyce Thompson Southwestern Arboretum, 190, 192, 207

Briggs and Shantz wilting coefficient, 205

Budyko radiational index of dryness, 1, 350, 383

Burns, Oregon, 67, 69

Cabeza Prieta Game Refuge, 194, 231

Cactus Plain, 489

California desert plan, 558

Canyons: Big Indian Creek, 74, 75; Kiger Gorge, 70, 74, 75; Little Blitzen, 74, 75; Little Indian Creek, 74, 75; Little Wildhorse Creek, 74; Pike Creek, 73; Wildhorse Creek, 74

Cape Farewell, 393

Carbon capture, 423

Carbon dioxide flux, 424

Carnegie Institute Desert Laboratory, 537

Casa Grande National Monument, 205

Ceramics: Paiute, 48; Sevier Branch, 48; Virgin Branch, 48

Chihuahuan desert: age, 354; amphibians, 467; causes, 323; climate, 339, 342, 345, 347, 349; communications, 366; desertification, 366; early man, 363; fauna, 362, 363; geology, 350, 354; hydrography and drainage, 334, 338, 339; location, 322; name, 322; physiography, 323, 334; size, 323; soils, 355, 357; special phenomena, 366; tourism, 366; vegetation, 357, 359–62

Chubascos, 108

Cirques, 75

Climate: air-sea interaction, 394; of Arctic desert, 394; changes in, 29; of Chihuahuan desert, 339; continental, 23; desert, 420; icelandic low, 395; comparative analysis of North American deserts, 350; geographic factors, 22; of Great Basin desert, 22, 69; of Mojave

desert, 107; of Sonoran desert, 183; post-pleistocene, 48; summer solstice, 397; types of, 347, 349; winter solstice, 397

Climatic and wind data sources, 526

Coahuila, 322

Cochise, 363

Committee on desert and arid zone research, 537, 541

Communications, 366

Continental drift, 164

Craters, 69, 79

Creeks: Alvord, 72; Big Alvord, 73; Big Fir, 78; Cottonwood, 73, 75; Indian, 72; Kanab, 442; Little Fir, 78; Muddy, 78; Pike, 72; Sonoita, 457; Steamboat, 22; Sycamore, 445; Toughey, 73

Cremation, 133

Cultures: developmental history of, 130; Anasazi, 363; Anasazi, Virgin Branch, 48; Clovis lithic tradition, 45; Cochise, 363; Dorset, 401; early and Paleo-Indian, 45; Eskimo, 400, 401, 403; Folsom, 363; Fremont, 48; Great Basin, 48; Hohokam, 232, 234, 237, 555; Inuit, 401; Inupiat, 401; Kitanemuk, 133; Lovelock, 46; lowland Patayan, 133; Numic, 50; O'Odham, 248–65; Prehorse, 50; Puebloid, 48; Real People, 401; Sandia, 363; San Juan, 363; Seri, 555; Sevier, 48; Southwest Basketmaker, 132; Thule, 401; Washo, 50; Western Archaic tradition, 46

Dale dry lake, 506

Dams: Butte, 338; Caballo, 338; Elephant Butte, 455; Hoover, 455; Lahontan, 22; Roosevelt, 455

Death Valley, 7, 22, 105, 106, 445

Death Valley National Monument, 125; checklists of vertebrates, 148–54

Dehydration, 406, 409, 413, 414

DeMartone aridity index, 1

Desert biome validation sites, 539, 550

Desert definition, 1; arctic desert, 383

Desertification, 3, 339, 360, 363, 366, 457, 458; definition of, 555; future of, 559; indicators of, 559; projects combating, 558; United Nations conference on, 558; United States national plan of action on, 555, 559; United States-Mexico bilateral agreement on, 559; Vale district rangeland rehabilitation, 558

Desert pavement, 191

Desert Research Institutes, 537, 543, 544

Desert tortoise preserve, 106

Desert varnish, 3: age dating, 527;

chemical composition, 527; definition of, 527; formula for artificial, 535; layering in, 528; manganese and iron importance, 532; mineralogy, 528; occurrence, 527; orange coats, 533; origin of, 530; reactions of, 531; relation to petroglyphs, 527; trace elements in, 528; varnish scenarios, 533

Deserts: age, 2, 173, 174, 192, 354; Alvord, 72, 75, 78, 81; Amargosa, 105, 511; Arctic, 6, 383–404; Atacama, 425, 428; Black Rock, 25, 43; Carson, 45; Chihuahuan, 6, 321–82; Forty Mile, 43; Gran Desierto, 512; Great Basin, 6, 7–102; Green River, 517; islands, 173, 179; Lechuguilla, 489; Mojave, 6, 103–62; Monte, 422, 429; Red, 38, 519; Smoke Creek, 43; Sonoran, 6, 163–320; Tule, 489

DETENAL, 525

Devil's Hole, 105

Dew, 420

Directory of North American Arid Lands Research Scientists, 559

Dormancy, of plants, 210

Dorset culture, 401

Drainage: Chihuahuan desert, 334–39, 354; Mojave desert, 106; Sonoran desert, 181

Drought, plant adaptations to, 422, 428, 429

Dryness index. *See* Indexes of dryness

Durango, 322

Earthquakes, 181

Edwards Air Force Base, 106

Emberger quotient of dryness, 350

Endangered species: Act, 463; birds, 28

Epicuticular lipids, 410

Eric the Red, 401

Eskimo culture, 400, 401, 403

Ethnography, 48. *See also* Cultures

Evaporation. *See* Indexes of dryness

Federal research stations, 538, 545

Fiaerol Islands, 401

Field offices, 68, 70, 525, 538, 545

Fish Creek Station, 70

Fishes: fossil, 25; in Chihuahuan desert, 363; in Death Valley National Monument, 152; in Great Basin desert, 25, 64, 65, 66, 79, 96; in riparian ecosystems, 445; in Sororan desert, 243–45, 319, 320

Fluvial systems, 434. *See also* Riparian ecosystems

Fog, 350

Folsom culture, 363

Fossils, 25, 26, 27
Foster Flat, 80
Franklin, Sir John, 402
Fremont culture, 48
Frenchglen, 67
Frost, 342, 388-91, 420

Geology
—Arctic desert: Barrow arch, 392;
beaded streams, 390; Caledonian
orogeny, 393; Canadian shield, 392;
cretaceous rocks, 392; frost mounds,
391; glaciation, 387, 388; ice wedge
polygons, 388; igneous rocks, 392;
Innutian fold, 393; isostatic rebound,
388; little ice age, 401; mesozoic
rocks, 392; metamorphic rocks, 392;
paleozoic rocks, 392; permafrost, 388;
pingos, 385, 390; pleistocene rocks,
392; string bogs, 391; thaw lakes,
385, 389; thermokarst topography,
389
—Chihuahuan, 350, 354
—Great Basin: Alvord creek beds, 72;
Antler orogeny, 11; ash flow tuffs, 74;
cordilleran geosyncline, 9; Cortez
window, 10; Danforth formation, 74;
diamond craters, 69, 79; Dunlap
formation, 18; geothermal activity,
69; Golconda thrust, 11; Havallah
formation, 11; history, 9-19; hot
springs, 69; Luning embayment, 18;
Pike creek volcanic series, 73; Pike
creek formation, 73; plate tectonics,
18; San Andreas fault, 19; Sevier
orogeny, 18; Sonoma orogeny, 11;
Steens Mountain uplift, 70; Valmy
formation, 10, 11; Vinini formation,
10
—Sonoran desert: Arizonan Revolution,
164; Baja California formation, 173;
basin and range disturbance, 164,
181; block rotation, 180; continental
drift, 164; earth fissures, 183;
earthquakes, 181; erosion and
deposition, 164, 176, 183; faults,
181; Grand Canyon disturbance, 164;
history, 164-83; ignimbrite flare-up,
178; Laramide orogeny, 164, 178;
Mazatzal Revolution, 164; Nevadan
Revolution, 164; Salton trough, 174;
San Andreas fault, 173, 181; sea level
changes, 173; sedimentary rocks, 164,
173; volcanism, 180
Geothermal activity, 69
Germination, 209, 451
Glaciation, 30, 74
Glaciers, 75, 386, 387, 398; Blitzen

advance, 75; cirques, 75; Fish Lake
advance, 75; post-Blitzen advance, 75;
Kiger glacier, 75; Little Wildhorse
glacier, 75
Grazing, 456
Great Basin desert: amphibians, 26, 79, 96,
97, 467; climate, 22-25; ethnography,
48-52; fauna, 25-29; location, 7; name,
7; physiography, 7; pleistocene lakes and
drainage systems, 19, 21, 22; prehistory,
45-48; size, 7; structural geologic
history, 9-19; vegetation, 29-45
Greenland ice cap, 386
Guanajuato, 322
Gulf of California, 187

Hail, 345
Halophytes, 38, 39, 40, 42, 119, 122, 204,
361, 440, 442, 443
Hart Mountain National Antelope Refuge,
69
Hearne, Samuel, 401
Hidalgo, 322
Hot springs, 69
Hudson, 401
Hudson Bay, 402
Hudson's Bay Company, 401
Humidity: in Chihuahuan desert, 345; in
Mojave desert, 112
Hydrography and drainage, 334, 338, 339
Hyperthermia, 406

Ice age, 401
Ice cap, 386
Icelandic low, 395
Indexes of dryness: Budyko radiational
index of dryness, 1, 350, 383;
DeMartone aridity index, 1; Emberger
quotient of dryness, 350; Köppen
climatic system, 347; Meigs climatic
system, 347; Thornthwaite-Mather
moisture index, 1, 383; Turc evaporation
formula, 383
Indian and early cultures: Anasazi, 363;
Anasazi, Virgin Branch, 45; Clovis lithic
tradition, 45; Cochise, 363; Dorset, 401;
early and Paleo-Indian, 45; Eskimo,
400, 401, 403; Folsom, 363; Fremont,
48; Great Basin, 48; Hohokan, 232,
234, 237, 555; Inuit, 401; Inupiat,
401; Kitanemuk, 133; Lovelock, 46;
lowland Patayan, 133; Numic, 50;
O'Odham, 248-65; Prehorse, 50;
Puebloid, 48; Real People, 401; Sandia,
363; San Juan, 363; Seri, 555; Sevier,
48; Southwest Basketmaker, 132; Thule,
401; Washo, 50; Western Archaic
tradition, 46

Indian languages: Hokan, 50; Numic, 50;
Uto-Aztekan, 50
Insects, 245
Inuit culture, 401
Inupiat culture, 401
Irrigation, 557

Joshua Tree National Monument, 106,
125; checklists of vertebrates, 155-62

Kangerdlugsuak, 393
Kitanemuk culture, 133
Kola peninsula, 401
Köppen climatic system, 347

Laboratories, 386, 557
Labrador peninsula, 386
Lagunas, 339: de Alamos, 339; de
Guzman, 339; de la Media Luna, 334;
de Mayran, 339; de Patos, 339; de Santa
Maria, 339
Lakes, pluvial, 130, 131: Ayer, 245; Big
Soda, 22; Bonneville, 30, 40; Buffalo,
21; Carson, 21; China, 131; Clover, 21;
Columbus, 22; Coyote, 131, 512;
Diamond, 21; Dixie, 21; Eagle, 46;
Fossil, 27, 512; Franklin, 21; Gale, 21;
Gilbert, 21; Honey Lake, 46; Hot
Creek, 22; Hubbs, 21; Ivanpah, 131;
Lahontan, 21, 30, 40; Lucero, 510;
Manly, 22, 107; Manix, 106, 131;
Newark, 21; Owens, 22, 106, 505;
Panamint, 106, 131; Railroad, 22;
Reveille, 22; Rogers, 107; Searles, 106;
Silver, 106, 131; Soda, 106, 131;
Steptoe, 21; Toiyabe, 22; Tonopah, 22;
Troy, 131; Waring, 21; Washoe, 22,
27
—present: Alkalai, 512; Amistad, 338;
Bear, 21; Borax, 79; Cahuilla, 124,
174, 175, 507; China, 45, 106;
Christmas, 512; Coleman, 512;
Curecanti, 442; Elephant Butte, 338;
Falcon Reservoir, 338; Fish, 67, 75;
Flaming Gorge, 442; Glen Canyon,
442; Great Salt, 439, 442, 446;
Harney, 67, 69, 80; Malheur, 67, 76,
80; Mead, 103; Navajo, 442;
Pyramid, 21, 39, 46, 50; Sevier, 21;
Silver, 505; Soda, 21; Soap, 79;
Summer, 512; Tahoe, 21, 50;
Teshekput, 390; Utah, 21, 439;
Walker, 21, 50; Warner, 512;
Wildhorse, 75; Winnemucca, 21, 46
Land bridge, 400
Lichen crusts, 190
Limnology, 187; Quitobaquito springs,
188; Sonoita creek, 188

Lithic industry: Malpais, 130; Lake Manly, 130
Lovelock culture, 46
Lowland Patayan culture, 133

Mackenzie, Alexander, 401
Mackenzie: delta, 385; lowland, 385
Malheur Environmental Field Station, 68
Malheur National Wildlife Refuge, Oregon, 67, 79
Mammals: extinct, 28, 45; in Arctic desert, 398; in Arctic National Wildlife Refuge, 404; in Big Bend National Park, 379–82; in Chihuahuan desert, 363; in Death Valley National Monument, 153–54; extinction of, 555; fossil, 28; in Great Basin desert, 28, 101, 102; in Joshua Tree National Monument, 157–58; in Mojave desert, 125; physiology of, 406–14; in riparian communities, 446; in Sonoran desert, 231–36, 296–302
Man: early, 130; in arctic, 400; in Chihuahuan desert, 363; migration of, 400
Man and the Biosphere Program, 539
Map sources, 525
Marine shell beads, 132
McClure Strait, 402
Meigs climatic system, 347
Mesophyll resistance in plants, 424
Meteorites, 363
Mexico-United States bilateral agreement on desertification, 559
Milling tools, 133
Minerals: availability in plants, 420, 426; in desert varnish, 528, 532
Mining, 460
Mitchell Caverns, 104
Mojave desert: amphibian, 125, 467; animals, 125, 129, 148–62; archaeology, 130–37; climate, 107–12; drainage, 106, 107; land forms, 112; location, 103; physical setting, 103–6; soils, 114; vegetation, 116–25, 146–47
Mountains, 3: Amargosa Range, 105; Antler Peak, 9; Argus, 125; Black, 103, 105; Blue, 69, 81; Brooks Range, 396, 397; Calico, 106; Carson Range, 35, 36; Cascade, 22, 69; Catlow Rim, 72; Cerbat, 103; Chalk, 349; Charleston, 32, 35, 103; Christmas, 349; Chuckawalla, 212; Colorado Plateau, 7; Columbia Plateau, 7; Coso Range, 132; Cottonwood, 105; Cross Range, 333; David, 349; Eagle, 106; Edna, 9, 11; Funeral, 105; Granite, 232; Grapevine, 105; Guadalupe, 333, 349; Hart, 80;

Havingdon Peak, 11; Humboldt Range, 11; Little Cow, 104; Little San Bernardino, 106; Mazama, 512; New York, 104, 125; Moenkopi Plateau, 504; Osgood, 9, 11; Panamint, 105; Providence, 104; Pueblo, 70, 72; Rattlesnake Hills, 27; Richardson, 385; Roberts, 10, 11; Rose, 36; Ruby, 21, 35; San Gabriel, 106; Santiago, 349; Sheep, 103; Sierra de los Alamitos, 333, 350; Sierra de Alvarez, 334; Sierra de Parras, 349; Sierra de San Miguelito, 334; Sierra Madre Occidental, 338, 339, 354; Sierra Madre Oriental, 349, 354; Sierra Nevada, 7, 9, 22, 35, 36; Sierra San Pedro Martir, 198; Slide, 36; Snow, 70; Sonoma Range, 11; Spring, 27, 103; Steens, 67, 69; Tehachapi, 106; Timber, 22; Toiyabe Range, 9, 21, 33; Tule, 232; Virginia, 36; Warner Peak, 69; Wasatch, 7, 22; White, 35; White Tank, 194

NASA Goldstone Tracking Station, 106
National Audubon Society, 539
National Climatic Center, 526
National Committee on Arid Lands, 537, 542
National Environmental Policy Act, 463
National Heritage Policy Act, 539
National Natural Landmarks Program, 539
National parks and monuments, 106, 125, 194, 205, 349, 539, 553, 554
National Wetlands Act, 463
National areas, 538, 539, 546–48
Natural gas, 402
National selection, in plants, 419
Nature Conservancy, 539
Nevada test site, 27, 29, 30, 41, 105, 112, 125
Nordenskiold, A. E., 401
Northern Great Basin: geology, 70–74; glaciation, 74, 75; physiography, 70, 72; size, 67; unique features, 67–69; vegetation, 76–79, 83–95; vertebrate animals, 79–81, 96–102; weather and climate, 69, 70
Nortes, 342
Northwest Company, 401
Nuevo Leon, 322
Nugssuak Peninsula, 393
Numic peoples, 48, 133; Bannock, 50; Comanche, 50; Kawaiisu, 50; Mono, 50; Paiute, 50, 133; northern Paiute, 70; Panamint Shoshoni, 50, 133; Shoshone, 50; Ute, 50

Ohm's law, 424

O'Odham culture, 245–68
Oil, 402
Organ Pipe Cactus National Monument, 194
Overgrazing, 456

Palma de la Cruz, 357
Parry, William, 402
Pearson product moment correlations, 449
Peary, Robert, 402
Peary Land, 393
Permafrost, 388
Petroglyphs, 134, 527
Phreatophytes, 41, 123, 204, 429, 437, 438, 439, 440, 442, 443, 445, 450, 453, 454, 461
Physiographic regions, Arctic Desert: Alaska's arctic slope, 383; Axel Heiberg island, 386; Baffin island, 393; Banks island, 386, 393; Canadian Arctic archipelago, 393; Canadian continental arctic region, 385; Devon island, 386; Disko island, 393; Ellesmere island, 386; Greenland, 386; Meighan island, 393; Melville island, 386; Parry islands, 393; Prince Patrick island, 393; Victoria island, 386
—Chihuahuan desert: basin and ranges, 333; great plains, 334; central highlands, 333; cross range, 333; Gulf coast lowland, 333; lower ranges of the Sierra Madre Oriental, 333; northern Sierra Madre Oriental, 333; upland with basins of the Sierra Madre Occidental, 333
—Great Basin: basin and range province, 7, 70, 72
—Mojave desert: central Mojave, 106; eastern Mojave, 103, California section, 104, northwestern Arizona section, 103, 104, southern Nevada section, 103; northern Mojave, 105, California section, 105, Nevada section, 105; southcentral Mojave, 106; southwestern Mojave, 106
—Sonoran desert: Arizona upland, 217–21; central coast, 225–26; lower-Colorado-Gila, 211–17; Magdalena, 228, 229; Sonoran foothills, 223–24; Sonoran plains, 221–23; Vizcaino, 227–28
Photosynthesis, 423; CAM photosynthesis, 425, 431; C_3 photosynthesis, 424; C_4 photosynthesis, 424
Pinacate region, 181
Pingos, 305
Plants, 399–400. *See also* Adaptations, plants; Vegetation

Pollination: anemophilous, 360, 361, 362; entomophilous, 359, 361; zoophilous, 360, 362

Pottery: Anasazi gray ware, 133; Colorado buff, 133; Tizon brown, 133

Prehorse culture, 50

Projectile points, 130: Cottonwood, 133; Eastgate, 133; Gypsum Cave, 131; Little Lake, 131; Rose Spring, 133

Prudhoe Bay, 388

Puebloid culture, 48

Queretaro, 322

Rainfall, 405, 420, 428; in Arctic desert, 394; in Chihuahuan desert, 342; in Great Basin desert, 23, 24, 69, 70, 442; in Mojave desert, 107, 108, 440; rain shadow, 22; in Sonoran desert, 185, 212, 215, 217, 222, 224, 225, 226, 227, 229, 442, 443

Rangeland rehabilitation, 558

Real People, 401

Recreation, 460

Reptiles: in Big Bend National Park, 373–75; in Chihuahuan desert, 363, 467–69; in Death Valley National Monument, 152–53; in Great Basin desert, 79, 97, 467–69; in Joshua Tree National Monument, 155–56; in Mojave desert, 125, 467–69; physiology of, 239, 405, 407–11, 414, 415; in Sonoran desert, 240–42, 312–17, 467–69

Research: climatic and wind data sources, 526; committee, 541; directory, 559; institutes, 537, 538, 543, 544; scientific natural areas, 538, 539, 546–48; state facilities, 538

Resource availability, to animals, 419

Riparian communities, amphibians in, 446

Riparian ecosystems, 3, 435; classification of, 439; extent of, 454; flooding of, 436; fluvial systems, 434; modification of, 455; patchiness in, 449; phreatophytes, 429, 433, 460; primary drainages, 436; river management, 460; runoff, 434; salinity, 445; secondary drainages, 436, 438; terracing, 437; total dissolved solids, 436; vegetation layering, 449; wetlands, 448

Rivers: Amargosa, 22, 106; Bear, 48; Bill Williams, 186; Carson, 21, 22; Colorado, 106, 186, 439, 442, 445, 448, 453; Colville, 385; Donner and Blitzen, 68; Gila, 186, 204, 439; Green, 439; Gunnison, 442; Humboldt, 21; Kuk, 385; Mackenzie, 386, 402; Moapa, 22, 103; Mojave, 106, 132; Muddy, 133; Owens, 106; Pecos, 338, 349, 439, 460; Reese, 21; Rio Aguanaval, 339; Rio Bavispe, 224; Rio Bravo, 338; Rio Concepcion, 225; Rio Conchos, 338; Rio del Carmen, 339; Rio Extoraz, 334; Rio Grande, 338, 349, 354, 439, 445, 447, 454; Rio Magdalena, 187; Rio Moctezuma, 224, 334; Rio Nazas, 339; Rio Panuco, 334; Rio Pecos, 354; Rio Sabinas, 338; Rio Salado, 338; Rio San Carlos, 338; Rio San Miguel, 221; Rio Santa Catarina, 334; Rio Santa Maria, 334; Rio Sonora, 187, 221; Rio Sonoyta, 187; Rio Tamuin, 334; Rio Tula, 334; Rio Verde, 334; Rio Yaqui, 187, 224, 225; Sagavanirktok, 387; Salt, 186, 436; San Juan, 442; San Pedro, 458; Santa Cruz, 186; Silvies, 68; Truckee, 21, 442; Verde, 186, 448; Virgin, 103; Walker, 21, 22; White, 22, 186

Rocks. See Geology

RUNWILD, 192

Salinity, 354; anions, 354; cations, 354; Colorado River, 557; total dissolved solids, 354

Salton Sea, 187

San Juan culture, 363

San Luis Potosi, 322

Sand dunes, 3, 481: Arctic, 385; Algodones, 486, 505, 507, 508; Arizona, 489; Bad Medicine, 504; Bahia Sebastian Vizcaino, 504; Baja California, 504; Big, 511; "booming," 510; Bruneau, 510; Cactus plain, 489; California, 504; Carson Desert, 510; Chihuahua, 509; classification of, 485, 486; Colorado, 509; Coral pink, 517; Desert Center, 507; Desert valley, 510; Devil's Playground, 505; Dumont, 505; Great, 509; gypsum, 511; Ibex, 505; Idaho, 510; Juniper Buttes, 510; Kelso, 506; Kilpecker, 517; Little Sahara, 517; Mohawk valley, 489; Nevada, 510; New Mexico, 511; Oregon, 512; processes of, 481; Rice, 506; Salton Sea, 507; Sand Mountain, 511; Saratoga Springs, 505; Samalyuca, 509; Sonora, 512; Stansbury Island, 44; Stovepipe Wells, 505; summary, 519; Texas, 517; Utah, 517; vegetation on, 44, 45, 126, 216; Washington, 517; White Sands, 511; Wyoming, 517

Sandia culture, 363

Santa Rita Experimental Range, 190

Scammon's Lagoon, 233

Scientific natural areas, 538, 539, 546–48

Serranias del Burro, 349

Serriculture, 555

Sevier culture, 48

Sinks: Carson, 22; Humboldt, 46; Mohave, 106; Salton, 460

Smith Flat, 72, 74

Snow: in Arctic desert, 394; in Chihuahuan desert, 345; in Great Basin desert, 24, 69, 70; in Mojave desert, 108

Society of American Foresters, 539

Soil: in Arctic desert, 397; caliche, 189; in Chihuahuan desert, 355; lichen crusts, 190; in Mojave desert, 114; moisture, 190, 428, 440; organic content, 190; relations in plants, 206, 207; riparian, 435; in Sonoran desert, 188, 189; temperature, 190

Solar radiation, 345, 396, 420

Solstice, 397

Sonora, 322

Sonoran desert: age, 163; amphibians, 317, 318, 467; birds, 303–11, 447, 470–79; drainage, 181; environmental factors, 205–11; fauna, 229–48, 296–320; fishes, 243–45, 319, 320; formation of, 143; geographic divisions of vegetation, 211–29; historical geology, 164–83; O'Odham People, 248–66; soils, 188–91; vegetation, 191–205

Southhampton Island, 401

Southwest Basketmaker culture, 132

Southwest trade network, 133; marine shell beads, 132; turquoise, 133

Spanish occupation of the Southwest, 455, 557

Springs: Quitobaquito, 211; Roaring, 70; Tzindeje, 334; thermal, 354

Squaw Butte Experimental Station, 69

State natural areas, 539, 549

State research facilities, 538

Succulents, 425, 431

Sun Dance, 50

Tamaulipas, 322

Tarascan Kingdom, 321

Taylor Grazing Act, 456

Temperature: in Arctic desert, 383, 394; in Chihuahuan desert, 342; cold air inversion, 112; effects of polar sea ice, 394; effects of North Atlantic Gulf Stream, 394; in Great Basin desert, 24, 69, 70; in Mojave desert, 112; in Sonoran desert, 186

Thornthwaite-Mather moisture index, 1, 383

Thule culture, 40

Tonto National Monument, 194

Tortoise preserve, 106

Tourism, 363, 366

Transpiration, of plants, 112, 420

Tropic of Cancer, 321

Truckee Meadows, 50
Turc evaporation formula, 383
Turquoise, 133
Twenty-nine Palms, 106

Ungava Peninsula, 386
United Nations Conference on
    Desertification, 558
United States: bilateral agreement with
    Mexico on desertification, 559; federal
    research stations, 538, 545; Geological
    Survey, 526; national parks and
    monuments, 106, 125, 196, 205, 349,
    539, 553, 554; national plan of action
    on desertification, 555

Valley of Fire State Park, Nevada, 103
Valleys: Antelope, 106; Apple, 106;
    Catlow, 67, 70, 72, 80; Cottonwood
    Creek, 73; Death, 7, 22, 105, 106, 445;
    Dixie, 19; Fishlake, 26, 27; Fenner-
    Chemehuevi, 125; Fremont-Stoddard,
    125; Imperial, 460; Ivanpah, 125;
    Jersey, 25; Lahontan, 28; Lucerne, 106;
    Owens, 46, 50; Pahranagat, 22; Reese
    River, 46, 50; Rock, 423; Snake, 46,
    48; Surprise, 46; Upper Blitzen, 72;
    Valle de Mezquital, 349; Virgin River,
    48; Wagner, 46; Warm Springs, 22;
    Warner, 69
Vegetation, Arctic desert: boreal forest,
    398; herbs, 399; grasses, 399; lichens,
    399; mosses, 399; shrubs, 399; tundra,
    388, 399
    —Chihuahuan desert: aciculifolious-
    squamifolious low forest, 361;
    crassicauleous brushwood, 360; desert
    grassland, 360; gypsophilous
    grassland, 361; microphyllous desertic
    brushwood, 357; rosettophyllous
    desertic brushwood, 359;
    sclerophyllous brushwood, 361
    —geofloras: Aldrich Station flora, 30;
    Alturus flora, 30; Alvord Creek flora,
    29; arcto-tertiary geoflora, 30, 436,
    437, 439; Cache Valley flora, 30;
    Chlorophagus flora, 29; Eocene
    Copper Basin flora, 29; Esmeralda
    flora, 29; Fallon flora, 30; Fingerrock
    Wash flora, 29; madro-tertiary
    geoflora, 29, 436, 437, 439; Middle-
    gate flora, 29; Pyramid flora, 29;
    Succor Creek flora, 29; Sutro flora,
    28; Trapper Creek flora, 29; Trout
    Creek flora, 29; Upper Cedarville
    flora, 29; Upper Eocene Bull Run
    flora, 29; Verdi flora, 30
    —Great Basin: alpine bunchgrass
    community, 78; boxthorn-shadscale

community, 42; chaparral, 30, 79;
    cottonwood community, 78; creosote
    bush-boxthorn community, 42; desert
    saltbrush community, 40; Engelmann
    spruce-alpine fir community, 34;
    fungi, 45; grasslands, 79; gray molly
    community, 39; greasewood
    community, 39; greasewood-shadscale
    community, 39; hopsage-boxthorn
    community, 42; hopsage-coleogyne
    community, 39; joshua tree
    community, 41; lichens, 44; live oaks,
    79; lodgepole pine-mountain hemlock
    community, 36; mat salt bush
    community, 39; mesquite community,
    41; mosses, 43; mountain mahogany
    community, 35; northern Great Basin,
    83–95; pine-deciduous forest
    community, 34; pinyon-juniper
    community, 36; rabbitbrush
    community, 38, 40; red fir
    community, 36; sagebrush community,
    38, 42, 76; saline-alkaline community,
    40; savanna, 79; sedges, 78; shadscale
    community, 38; upper sagebrush
    community, 35; whitebark pine
    community, 36; white fir-douglas fir-
    blue spruce community, 34; winter fat
    community, 39; yellow pine-white fir
    community, 35
    —Mojave desert: allscale-alkali scrub,
    122; big galleta scrub-steppe, 124;
    blackbrush scrub, 119; calciphyte
    saxicole subscrub, 124; california
    juniper-one leaf pinyon woodland,
    123; cheesebush scrub, 122;
    classification of, 117; cottonwood-
    willow-mesquite bottomland, 123;
    cottonwood-willow streamside
    woodland, 123; creosote bush clones,
    124; desert holly scrub, 122; desert
    needle grass scrub steppe, 123; desert
    oasis woodland, 123; desert
    psammophyte complex, 124; flora,
    146–47; galleta scrub steppe, 124;
    great basin bristlecone pine woodland,
    122; hopsage scrub, 119; important
    families, 116; indian rice grass scrub
    steppe, 123; iodine bush alkali scrub,
    122; limber pine woodland, 122;
    mojave-colorado desert scrub, 122;
    mojave saltbush-allscale scrub, 122;
    noncalciphyte saxicole subscrub, 124;
    sagebrush scrub, 119; saltgrass
    meadow, 124; shadscale scrub, 119;
    streamside-woodland, 123; succulent
    scrub, 122; Utah juniper-one leaf
    pinyon woodland, 123; vascular plant
    species, 116; white fir-pine forest, 123

—on sand dunes, 44, 45, 126, 216
—Sonoran desert: age of, 192; algae,
    211; antibiotics, 211; arborescent
    oligopodial stem succulents, 199; bulb
    perennials, 197; clustering stem
    succulents, 200; cryptogamic plants,
    210, 211; cylindrocaulescent shrubby
    stem succulents, 200; digitized
    classification of, 229; drought
    deciduous indurate base perennials,
    202; drought deciduous large leaved
    perennials, 203; drought deciduous
    sarcocaulescent perennials, 203;
    endemism, 191; evergreen hardwood
    bushes, 202; facultatively perennial
    ephemerals, 196; ferns, 205, 210;
    fungi, 211; geographic divisions of,
    211–29; green stemmed microphyllous
    perennials, 203; grasses, 195, 210;
    halophytes, 204; leafless green
    stemmed plants, 201; leaf
    nonsucculent caudiciforms, 198; leaf
    succulent caudiciforms, 198; leaf
    succulent non-candiciforms, 199;
    lichens, 211; low leafy softwood
    bushes, 201; mangrove scrub, 226;
    monopodial stem succulents, 199;
    mosses, 210; palms, 198; parasites,
    204; perennial grasses, 197;
    phreatophytes, 204; platycaulescent
    shrubby stem succulents, 201;
    riparian, 192; root perennials, 196;
    sinaloan thornscrub, 209; summer
    ephemerals, 196; winter deciduous
    broad leaved perennials, 203; winter
    deciduous microphyllous perennials,
    203; winter ephemerals, 195
Vertebrates, 79–81, 96–102, 148–54,
    155–62, 373–82
Vikings, 401

Water relations: animal adaptations, 405;
    plant adaptations, 205. *See also*
    Adaptations, animal and plant
Wetlands, 448
Washo culture, 50
Western Archaic tradition, 46
White Horse Rand, 69
White settlement, 455
Whiteout, 396
Wildlife regues, 67, 69, 106, 194, 231,
    404
Wind: channeling, 396; chubascos, 108; in
    Arctic desert, 395; in Chihuahuan desert,
    345; data sources, 526; in Mojave
    desert, 112; nortes, 342; whiteout, 396

Zacatecas, 322
Zone of silence, 363

ABOUT THE CONTRIBUTORS

Editor:

GORDON L. BENDER, Professor Emeritus of Zoology, Arizona State University. An entomologist by training, Dr. Bender has studied the deserts of the Southwest for the past twenty-five years. He has served as president of the Arizona Academy of Science, president of the Southwestern and Rocky Mountain Division of the American Association for the Advancement of Science, director of the Summer Institute in Desert Biology supported by the National Science Foundation, member and chairman of the Committee on Desert and Arid Zone Research of the Southwestern and Rocky Mountain Division of the AAAS, and a member and chairman of the national Committee on Arid Lands of the AAAS. His research interests are the ecology of desert insects.

Great Basin Desert

DON D. FOWLER
DAVID KOCH

Both authors have worked with the Desert Research Institute of the University of Nevada, Reno, one of the first of the desert institutes to be established in the United States. It has active research programs in hydrology, human systems, atmospheric science, ecology, and physiology. Because of its location in the Great Basin, the institute is an important source of information on this area. Dr. Fowler is a Professor in the Historic Preservation Program of the University of Nevada, Reno. He is an anthropologist and has published widely on early man in the Great Basin. Dr. Koch is an ecologist with the Wright Energy Nevada Corporation.

Northern Great Basin

DON McKENZIE is Associate Professor of Biology at Lewis and Clark College in Portland, Oregon. He has been interested in deserts for a number of years and has taught classes and carried on research at the Malheur Environmental Center in the Great Basin since its establishment. His detailed study of

the Steen's Mountain area gives an excellent insight into the character of the northern Great Basin.

Mojave Desert

HYRUM JOHNSON
PETER ROWLANDS
ERIC RITTER
ALBERT ENDO

These authors were members of the team of scientists gathering data for the California Desert Conservation Area Use Plan prepared by the Bureau of Land Management. Hyrum Johnson and Peter Rowlands are botanists; Eric Ritter is an anthropologist; and Albert Endo is a soil scientist.

Sonoran Desert

FRANK S. CROSSWHITE
CAROL D. CROSSWHITE

Frank Crosswhite is Curator and Research Scientist with the Boyce Thompson Arboretum at Superior, Arizona. He is the editor of *Desert Plants,* a quarterly journal published by the Boyce Thompson Arboretum. He is a botanist who is intensely interested in man's utilization of native desert plants. Carol Crosswhite is Curator and Educational Coordinator for the Boyce Thompson Arboretum. An entomologist by training, she is concerned with the native desert animals.

Chihuahuan Desert

FERNANDO MEDELLIN-LEAL is Director of the Instituto de Investigacion de Las Zonas Deserticas of the Universidad Autonoma de San Luis Potosí, San Luis Potosí, Mexico. It is the oldest institute in Mexico devoted exclusively to the study of arid zones. Its research interests include the vegetation, soils, water, fauna, ecology, and use of renewable natural resources. Founded in 1954, the institute has developed an enormous amount of data on the Chihuahuan Desert. Dr. Medellín-Leal is a member of the Committee on Arid Lands of the AAAS and has been active in international organizations interested in the study of deserts. His research interests are in desert plants.

Animal Adaptations

NEIL F. HADLEY is Professor of Zoology at Arizona State University. He is an environmental physiologist interested in the adaptations of desert animals to their environment. He is the author of several research articles and has served as editor for an AIBS symposium volume on animal adaptation to desert conditions. He is an editor for *Ecology*, a journal of the Ecological Society.

Plant Adaptations

OTTO T. SOLBRIG is Director of the Gray Herbarium of Harvard University. He has been active in the Desert Biome Project of the International Biology Program. He has published extensively on the adaptations and evolution of desert plants.

Arctic Desert
DAVID M. HICKOK
JOSEPH C.LaBELLE
ROBERT G. ADLER

Dr. Hickok is Director of the Arctic Environmental Information and Data Center of the University of Alaska. He has served as president of the Alaska Division of the American Association for the Advancement of Science. The Data Center serves as a clearinghouse for environmental information on the Arctic and Alaska. There is an annual publication listing current research in these regions. Joseph LaBelle is geo-

morphologist with the Arctic Environmental Information and Data Center. Robert Adler is a geographer with the center.

Riparian Ecosystems

ROBERT D. OHMART
BERTIN W. ANDERSON
Dr. Ohmart is Associate Professor of Zoology and Associate Director of the Institute of Ecology at Arizona State University. He has been the recipient of a number of grants to study riparian habitats in the Southwest. He has active research projects on the Colorado and Rio Grande rivers and at other smaller riparian habitats. Bertin Anderson is a waterfowl specialist who has done much of the fieldwork on bird habitats in riparian situations.

Sand Dunes

ROGER S. U. SMITH is Assistant Professor of Geology at the University of Houston. He has studied the dune areas of the southwestern United States in detail.

Desert Varnish

CARLETON B. MOORE
CHRISTOPHER ELVIDGE
Carleton Moore is Professor of Chemistry and Geology and Director of the Center for Meteorite Studies at Arizona State University. He has been interested in natural and synthetic desert varnishes for a number of years. Christopher Elvidge is a graduate student working with Dr. Moore on the desert varnish problem. His master's thesis was concerned with this topic.